PHYSICS OF SEMICONDUCTORS

To learn more about the AIP Conference Proceedings, including the
Conference Proceedings Series, please visit the webpage
http://proceedings.aip.org/proceedings

PHYSICS OF SEMICONDUCTORS

27th International Conference on the Physics of Semiconductors

ICPS-27

Flagstaff, Arizona   26 – 30 July 2004

PART B

EDITORS
José Menéndez
Arizona State University
Tempe, Arizona

Chris G. Van de Walle
University of California, Santa Barbara
Santa Barbara, California

All papers have been peer-reviewed

SPONSORING ORGANIZATIONS
International Union of Pure and Applied Physics (IUPAP)
National Science Foundation (NSF)
Air Force Office of Scientific Research (AFOSR)
Office of Naval Research (ONR)
Army Research Office (ARO) (sponsor of the "Special Session on Cyclotron Resonance")
Defense Advanced Research Projects Agency (DARPA)
European Office of Aerospace Research and Development (EOARD)
Asian Office of Aerospace Research and Development (AOARD)
Lawrence Berkeley National Laboratory (LBNL)
National Institute of Standards and Technology (NIST)
Forum on Industrial and Applied Physics (FIAP) of the American Physical Society (APS)

Melville, New York, 2005
AIP CONFERENCE PROCEEDINGS ■ VOLUME 772

Editors:

José Menéndez
Arizona State University
Department of Physics and Astronomy
Tempe, AZ 85287-1504
USA

E-mail: jose.menendez@asu.edu

Chris G. Van de Walle
University of California, Santa Barbara
Materials Department
Santa Barbara, CA 93106-5050
USA

E-mail: vandewalle@mrl.ucsb.edu

Cover: Photograph by David G. Seiler

The articles on pp. 243–244 and 1451–1452 were authored by U. S. Government employees and are not covered by the below mentioned copyright.

Authorization to photocopy items for internal or personal use, beyond the free copying permitted under the 1978 U.S. Copyright Law (see statement below), is granted by the American Institute of Physics for users registered with the Copyright Clearance Center (CCC) Transactional Reporting Service, provided that the base fee of $22.50 per copy is paid directly to CCC, 222 Rosewood Drive, Danvers, MA 01923, USA. For those organizations that have been granted a photocopy license by CCC, a separate system of payment has been arranged. The fee code for users of the Transactional Reporting Services is: ISBN/0-7354-0257-4/05/$22.50.

© 2005 American Institute of Physics

Permission is granted to quote from the AIP Conference Proceedings with the customary acknowledgment of the source. Republication of an article or portions thereof (e.g., extensive excerpts, figures, tables, etc.) in original form or in translation, as well as other types of reuse (e.g., in course packs) require formal permission from AIP and may be subject to fees. As a courtesy, the author of the original proceedings article should be informed of any request for republication/reuse. Permission may be obtained online using Rightslink. Locate the article online at http://proceedings.aip.org, then simply click on the Rightslink icon/"Permission for Reuse" link found in the article abstract. You may also address requests to: AIP Office of Rights and Permissions, Suite 1NO1, 2 Huntington Quadrangle, Melville, NY 11747-4502, USA; Fax: 516-576-2450; Tel.: 516-576-2268; E-mail: rights@aip.org.

L.C. Catalog Card No. 2005927643
ISBN 0-7354-0257-4
ISSN 0094-243X

Printed in the United States of America

CONTENTS

Preface xli
Program xliv
Special Session on the 50th Anniversary of Cyclotron Resonance lii
IUPAP Young Author Best Paper Awards liii
Proceedings, Social Events, Related Conferences liv
Acknowledgments lv
Sponsors lvii
Committees lix
Opening Address lxiii
    F. A. Ponce
Opening Ceremony lxv
    M. Cardona
Opening Ceremony, ICPS-27 lxvii
    J. Franz

SPECIAL SYMPOSIUM ON THE 50th ANNIVERSARY OF CYCLOTRON RESONANCE

Cyclotron Resonance and Quasiparticles 3
    M. L. Cohen
Personal Reminiscences of Cyclotron Resonance 7
    G. Dresselhaus
Cyclotron Resonance Spectroscopy of Semiconductors 11
    B. Lax
Investigations of Cyclotron Resonance in InSb and PbTe: Intraband Transitions between Landau Levels 17
    E. Burstein

PLENARY TALKS

Semiconducting Carbon Nanotubes 25
    M. S. Dresselhaus, R. Saito, and A. Jorio
Semiconductor Quantum Dots for Quantum Information Processing: An Optical Approach 32
    M. V. Gurudev Dutt, Y. Wu, X. Li, D. G. Steel, D. Gammon, and L. J. Sham
Silicon Quantum Computer 38
    K. M. Itoh
Experiments and Simulations on a Few-Electron Quantum Dot Circuit with Integrated Charge Read-Out 44
    R. Hanson, J. M. Elzerman, L. H. Willems van Beveren, L. M. K. Vandersypen, L.-X. Zhang,
    P. Matagne, J. P. Leburton, and L. P. Kouwenhoven
Physics and Applications of Self-Assembled Semiconductor Quantum Dots ... 50
    M. S. Skolnick and D. J. Mowbray
Spintronics and Ferromagnetism in Wide-Band-Gap Semiconductors 56
    T. Dietl

1. BULK SEMICONDUCTORS

1.1. Group-IV and Elemental Semiconductors

Compositional Dependence of Critical Point Transitions in $Ge_{1-x}Sn_x$ Alloys 65
    C. S. Cook, V. D'Costa, J. Kouvetakis, S. Zollner, and J. Menéndez

Progress in Semiconductor Spectroscopy Using Isotopically Enriched Si 67
    M. L. W. Thewalt, T. A. Meyer, D. Karaiskaj, M. Cardona, E. E. Haller, J. W. Ager III, and H. Riemann

High Precision Measurement of the Avogadro Constant Based on Silicon 69
    P. Becker

Isotopic Effects in the Indirect Excitonic Transitions of Isotopically Enriched Silicon 73
    S. Tsoi, H. Alawadhi, X. Lu, J. W. Ager III, C. Y. Liao, H. Riemann, E. E. Haller, S. Rodriguez, and
    A. K. Ramdas

Pr-Silicate Formation on SiO_2 Covered 3C-SiC(111) 75
    D. Schmeißer and H.-J. Muessig

Defect Kinetics of O-V-N Complexes in Si Ingot Growth Based on First-Principles Calculations and Thermodynamics 77
    H. Kageshima, A. Taguchi, and K. Wada

Piezoresistance Effect in p-Type Silicon 79
    Y. Kanda and K. Matsuda

Antiferromagnetic Spin Glass Ordering in the Vicinity of Insulator–Metal Transition in n-Ge:As 81
    A. I. Veinger, A. G. Zabrodskii, T. V. Tisnek, and S. I. Goloshchapov

Electronic Transport Properties of Si Thin Film from Bulk to Sub-nm Thickness: A First-Principles Study 83
    J. Yamauchi

Structural Instability Depending on a Structural Configuration at the Surroundings of the Si-H Bond in a-Si:H 85
    H. Oheda

Crystal Stability of α- and β-Boron 87
    A. Masago, K. Shirai, and H. Katayama-Yoshida

Deep States of Pt, Ir, and Os in Silicon Carbide 89
    J. Grillenberger, U. Grossner, B. G. Svensson, F. Albrecht, W. Witthuhn, and R. Sielemann

Quantum Mechanical Modeling of the Structure and Doping Properties of Defects in Diamond 91
    J. P. Goss, P. R. Briddon, R. Sachdeva, R. Jones, and S. J. Sque

Energetics of Various Electrically Deactivating Defects in Heavily n-type Si 95
    C.-Y. Moon, Y.-S. Kim, and K. J. Chang

Diffusion of Silicon in Germanium 97
    H. H. Silvestri, H. Bracht, J. Lundsgaard Hansen, A. Nylandsted Larsen, and E. E. Haller

Deep Level Defects E_1/E_2 in n-Type 6H Silicon Carbide Induced by Electron Radiation and He-Implantation 99
    C. C. Ling, X. D. Chen, S. Fung, C. D. Beling, G. Brauer, W. Anwand, W. Skorupa, and M. Gong

Piezospectroscopy of the I Lines of Gallium in Germanium 101
    P. Fisher, R. E. M. Vickers, and P. S. Bowdler

Aspects of Iron Contamination Studies in Silicon by Photoluminescence Correlation to Other Techniques 103
    I. Rapoport, P. Taylor, B. Orschel, J. Kearns, F. Kirscht, A. Buczkowski, and S. Hummel

A Theoretical Study of Cu Gettering in Si 105
    K. Shirai, T. Michikita, and H. Katayama-Yoshida

Self-Interstitials and Nanoclusters in Crystalline Silicon 107
    B. N. Mukashev, Y. V. Gorelkinskii, and K. A. Abdullin

Optical Investigations of Semiconducting $FeSi_2$ Using Photoreflectance Spectroscopy 109
    A. G. Birdwell, A. Davydov, C. L. Littler, R. Glosser, F. H. Pollak, and L. Malikova

Birefringence and VIS/VUV Optical Absorption of Graphite-Like Amorphous Carbon 111
    S. Zollner, W. Qin, R. B. Gregory, J. Kulik, N. V. Edwards, K. Junker, and T. E. Tiwald

Infrared Response of Heavily Doped p-Type Si and SiGe Alloys from Ellipsometric Measurements 113
    J. Humlíček and V. Křápek

A Study of the Effect of the Hydrogen Gas Flow on the Quality of Thin CVD Diamond Films Deposited on Silicon Substrates in CH_4/H_2 Gas Mixture 115
    M. Bataineh, S. Khatami, and J. Asmussen, Jr.

1.2. Arsenides, Antimonides, and Phosphides

Modulation of GaAs(100) Surface Morphology by Ultra-Thin MOVPE Grown InP Layers 117
    M. Mattila, M. Sopanen, and H. Lipsanen

Determination of Piezo-Electric Fields in Spontaneously Ordered InGaAsP 119
    S. Krämer, S. Neumann, W. Prost, F. J. Tegude, S. Malzer, and G. H. Döhler

ESR Study of Zn-Codoping Effect on the Luminescence Efficiency of the Er-2O Center in GaAs:Er,O 121
    M. Yoshida, K. Hiraka, H. Ohta, Y. Fujiwara, A. Koizumi, and Y. Takeda

Evolution of the Band Gaps and Band Edges of Quaternary $Ga_{1-y}In_yAs_xSb_{1-x}$/InAs Alloys as a Function of Composition 123
    R. Magri, A. Zunger, and H. Kroemer

Optical Properties of AlGaSb Alloys Grown by MBE 125
    S. G. Choi, S. K. Srivastava, and C. J. Palmstrøm

Unusual Behavior of Far-Infrared Magneto-Photoconduction in Electric Field for n-GaAs Near Metal-Insulator-Transition 127
    H. Kobori, M. Inoue, K. Fujii, and T. Ohyama

Photoluminescence Brightening Due to Partial Auger Recombination Process of the (D^0,X) Complex by the Impact Ionization Avalanche in n-GaAs 129
    K. Aoki and D. Hosokawa

Terahertz Radiation from Er,O-Codoped GaAs Surface 131
    M. Suzuki, K. Nakamura, Y. Fujiwara, A. Koizumi, Y. Takeda, and M. Tonouchi

Exciton Relaxation and Spin Dynamics in $Al_xGa_{1-x}As$ Films 133
    A. Amo, M. D. Martin, Ł. Kłopotowski, L. Viña, A. I. Toropov, and K. S. Zhuravlev

Anti-Stokes and Stokes Hot Luminescence from Bulk InAs and InSb 135
    N. Wada, A. C. H. Rowe, and S. A. Solin

Spin Orientation of Geminate Excitons in High Purity AlGaAs Layers 137
    A. V. Efanov, T. S. Shamirzaev, K. S. Zhuravlev, J. Fuerst, and H. Pascher

Pump and Probe Reflection Study on Photoexcited Carrier Dynamics in Er,O-Codoped GaAs 139
    Y. Fujiwara, K. Nakamura, A. Koizumi, Y. Takeda, M. Suzuki, and M. Tonouchi

Acceptor Photoluminescence Kinetics of GaAs under Resonant Photoexcitation of Shallow Donors 141
    A. E. Nickolaenko, A. M. Gilinsky, O. A. Shegai, T. S. Shamirzaev, and K. S. Zhuravlev

Deep Levels in Ruthenium Doped p-Type MOCVD GaAs 143
    A. Majid, M. Zafar Iqbal, A. Dadgar, and D. Bimberg

The Structure of Intrinsic Stacking Faults in GaAs 145
    S. P. Beckman and D. C. Chrzan

Deep Levels in Osmium Doped p-Type GaAs Grown by Metal Organic Chemical Vapor Deposition 147
    M. Zafar Iqbal, A. Majid, A. Dadgar, and D. Bimberg

Determination of Optical Cross Sections for Electron and Hole Emission from Deep Defects in Low-Temperature-Grown GaAs 149
    S. Malzer, C. Steen, P. Kiesel, and G. H. Döhler

Structural Properties and Electronic States of Carbon Delta-Layers in GaAs 151
    L. Winking, M. Wenderoth, T. C. G. Reusch, R. G. Ulbrich, P.-J. Wilbrandt, R. Kirchheim, S. Malzer, and G. Döhler

1.3. Tellurides, Sulfides, and Selenides

ZnS Deposition onto Bare and GaSe Terminated Silicon-(111)-Surfaces 153
    B. Jaeckel, R. Fritsche, A. Klein, and W. Jaegermann

Formation and Dynamics of Current Filaments in AC-Driven Thin-Film ZnS:Mn Structures 155
    T. Raker and T. Kuhn

Nanoscale Cubic Growth of SrS Codoped with Eu and Sm on MgO (001) for Optical Memory Materials 157
    C. Chen, K. L. Teo, and T. C. Chong

Crystal Growth, Characterization and Anisotropic Electrical Properties of GaSe Single Crystals for THz Source and Radiation Detector Applications.................159
    K. C. Mandal, C. Noblitt, M. Choi, A. Smirnov, and R. D. Rauh

Effect of Cu Deficiency on the Optical Properties and Electronic Structure of $CuIn_{1-x}Ga_xSe_2$ ($x=0, 0.2$) Determined by Spectroscopic Ellipsometry.................161
    S.-H. Han, F. S. Hasoon, H. A. Al-Thani, A. M. Hermann, and D. H. Levi

Phase Jumps in the Reflected Light of a ZnSe Layer.................163
    G. Manzke, M. Seemann, F. Kieseling, H. Stolz, R. Franz, and K. Henneberger

Combined Raman Scattering, X-Ray Fluorescence and Ellipsometry *In-Situ* Growth Monitoring of $CuInSe_2$-Based Photoabsorber Layers on Polyimide Substrates.................165
    C. Bundesmann, M. Schubert, N. Ashkenov, M. Grundmann, G. Lippold, and J. Piltz

Molecular-Beam Epitaxial Growth of α-$Zn_{0.05}Sr_{0.95}S$ Codoped with Eu and Sm on MgO(001) Substrate via α-MnS Buffer Layer.................167
    C. Chen, K. L. Teo, and T. C. Chong

Muonium in ZnTe as a Model for Isolated Hydrogen.................169
    R. C. Vilão, J. M. Gil, H. V. Alberto, J. Piroto Duarte, N. Ayres de Campos, A. Weidinger, R. L. Lichti, K. H. Chow, and S. F. J. Cox

Electrical and Infrared-Optical Investigations of N-Doped (Zn,Mn)Se Epilayers.................171
    B. Daniel, K. C. Agarwal, C. Klingshirn, and M. Hetterich

Chirality and Electronic/Optical Properties of Screw-Vacancy In_2Se_3.................173
    T. Nakayama and K. Fujita

Band Bowing in $BeSe_xTe_{1-x}$.................175
    T. Sandu and W. P. Kirk

1.4. Oxide Semiconductors

Electronic and Optical Properties of TiO_2.................177
    A. Ferreira da Silva, N. Souza Dantas, E. F. da Silva, Jr., I. Pepe, M. O. Torres, C. Persson, T. Lindgren, J. Souza de Almeida, and R. Ahuja

Photoconductivity Spectra of Bulk ZnO.................179
    U. Röder, D. Klarer, R. Sauer, and K. Thonke

Palladium Schottky Barrier Contacts to the $(000\bar{1})$- and $(10\bar{1}0)$-Face of Hydrothermally Grown n-ZnO.................181
    U. Grossner, S. Gabrielsen, T. M. Borseth, J. Grillenberger, A. Y. Kuznetsov, and B. G. Svensson

Incorporation and Electrical Activity of Group V Acceptors in ZnO Thin Films.................183
    H. von Wenckstern, S. Heitsch, G. Benndorf, D. Spemann, E. M. Kaidashev, M. Lorenz, and M. Grundmann

Transparent p-ZnO by Oxidation of Zn-Based Compounds.................185
    E. Kaminska, J. Kossut, A. Piotrowska, E. Przezdziecka, W. Dobrowolski, E. Dynowska, R. Butkute, A. Barcz, R. Jakiela, M. Aleszkiewicz, E. Janik, and E. Kowalczyk

Properties of Transparent Zinc-Tin Oxide Conducting Films Prepared by Chemical Spray Pyrolysis.................187
    A. I. Martinez, B. A. Garcia, and D. R. Acosta

Band Structure Derived Properties of HfO_2 from First Principles Calculations.................189
    J. C. Garcia, A. T. Lino, L. M. R. Scolfaro, J. R. Leite, V. N. Freire, G. A. Farias, and E. F. da Silva, Jr.

Ab Initio Study of Native Point Defects in Delafossite $CuAlO_2$.................191
    I. Hamada and H. Katayama-Yoshida

Hydrogen in Oxides, Modelled by Muonium.................193
    S. F. J. Cox, S. P. Cottrell, J. S. Lord, R. C. Vilão, H. V. Alberto, J. Piroto Duarte, J. M. Gil, N. Ayres de Campos, E. A. Davis, M. Charlton, D. P. Van der Werf, D. J. Keeble, R. L. Lichti, and A. Weidinger

Role of Intentionally Incorporated Hydrogen in Wide-Band-Gap ZnO Thin Film Prepared by Photo-MOCVD Technique.................195
    S. Y. Myong and K. S. Lim

Static and Transient Capacitance Spectroscopy on ZnO.................197
    H. von Wenckstern, S. Weinhold, G. Biehne, R. Pickenhain, E. M. Kaidashev, M. Lorenz, and M. Grundmann

Role of Oxygen Vacancies in the O_2 Photoreduction at the (100) TiO_2 (Anatase) Surface.................199
    A. Amore Bonapasta and F. Filippone

Band-to-Band Transitions and Optical Properties of $Mg_xZn_{1-x}O(0 \leq x \leq 1)$ Films 201
    R. Schmidt-Grund, D. Fritsch, M. Schubert, B. Rheinländer, H. Schmidt, H. Hochmut, M. Lorenz,
    D. Spemann, C. M. Herzinger, and M. Grundmann

Optical Properties of p-Type ZnO Doped by As Ion Implantation.................... 203
    T. S. Jeong, C. J. Youn, M. S. Han, Y. S. Park, and W. S. Lee

Photoluminescence Properties of Ultra-High-Quality ZnO Thin Films Grown by
Sputtering Deposition 205
    D. Kim, T. Shimomura, and M. Nakayama

A New Radiation Hard Semiconductor-Semi-Insulating GaN: Photoelectric Properties.................... 207
    J. Vaitkus, E. Gaubas, V. Kazukauskas, A. Blue, W. Cunningham, M. Rahman, K. Smith, and S. Sakai

1.5. Nitride Semiconductors

1.5.1. Growth and Structural Properties of Nitride Semiconductors

Relation between Structural and Optical Properties of InN and $In_xGa_{1-x}N$ Thin Films.................... 209
    Z. Liliental-Weber, D. N. Zakharov, J. Jasinski, K. M. Yu, J. Wu, J. W. Ager III, W. Walukiewicz,
    E. E. Haller, H. Lu, S. X. Li, and W. J. Schaff

Growth of High Quality AlN Single Crystals and Their Optical Properties.................... 211
    M. Strassburg, J. Senawiratne, N. Dietz, U. Haboeck, A. Hoffman, V. Noveski, R. Dalmau,
    R. Schlesser, and Z. Sitar

Electrostatic Fields and Compositional Fluctuations in InGaN Quantum Wells 213
    M. Stevens, A. Bell, H. Marui, S. Tanaka, and F. A. Ponce

Strain Relaxation Mechanisms in InGaN Epilayers.................... 215
    R. Liu, J. Mei, F. A. Ponce, Y. Narukawa, H. Omiya, and T. Mukai

The Nature of Crystalline Defects in *a*-plane GaN Films 217
    R. Liu, A. Bell, V. R. D'Costa, F. A. Ponce, C. Q. Chen, J. W. Yang, and M. A. Khan

In-situ Multiple Wavelength Ellipsometry for Real Time Process Characterization of
Nitride MOCVD.................... 219
    A. Bonanni, D. Stifter, K. Hingerl, H. Sitter, and K. Schmidegg

Resonant Raman Study of Strain and Composition in InGaN Multiquantum Wells 221
    S. Lazić, J. M. Calleja, F. B. Naranjo, S. Fernández, and E. Calleja

Fabrication of GaN_xAs_{1-x} Quantum Structures by Focused Ion Beam Patterning.................... 223
    K. Alberi, A. Minor, M. A. Scarpulla, S. J. Chung, D. E. Mars, K. M. Yu, W. Walukiewicz,
    and O. D. Dubon

Stabilization of Zinc-Blende Semiconductors through 3d Impurities and Holes 225
    G. M. Dalpian and S.-H. Wei

1.5.2. Vibrational Properties and Transport in Nitride Semiconductors

Electron-Phonon Interaction of Wurtzite GaN and Its Effect on High Field Transport.................... 227
    S. Yamakawa, J. M. Barker, S. M. Goodnick, D. K. Ferry, and M. Saraniti

Noise Measurements in Mg-doped GaN.................... 229
    D. Seghier and H. P. Gislason

Harmonic Enhancement of Gunn Oscillations in GaN.................... 231
    C. Sevik, D. E. Yilmaz, and C. Bulutay

Intermediate-Coupling Polarons in GaN, AlN, and InN 233
    Z. W. Yan, X. X. Liang, and S. L. Ban

Ultrafast Carrier Dynamics in Nitrogen Containing GaAs 235
    T. Dekorsy, S. Sinning, M. Helm, G. Mussler, L. Däweritz, and K. H. Ploog

Influence of Indium-Nitrogen Interactions on the Local Mode Frequency of Nitrogen in
GaAs-Based Dilute Nitrides 237
    H. C. Alt, Y. V. Gomeniuk, and G. Mussler

Electron-Hole Dynamics in GaN under Short-Pulse Laser Excitation.................... 239
    Y. Wang, E. Bellotti, M. Wraback, and S. Rudin

Dislocation-Limited Lifetime of Nonequilibrium Carriers in AlGaN Epilayers 241
    G. Tamulaitis, J. Mickevicius, R. Aleksiejunas, M. S. Shur, J. P. Zhang, Q. Fareed, and R. Gaska

Hot Carrier Dynamics in GaN Studied by Subpicosecond Luminescence Spectroscopy 243
G. A. Garrett, S. Rudin, A. V. Sampath, C. J. Collins, H. Shen, and M. Wraback

1.5.3. Defects and Impurities in Nitride Semiconductors

Intrinsic Defects in GaN and ZnO: A Study by Optical Detection of Electron Paramagnetic Resonance 245
G. D. Watkins

Radiation Damage Formation and Annealing in Mg-Implanted GaN 249
S. Whelan, M. J. Kelly, J. Yan, and R. Gwilliam

Neutral Mn Acceptor in GaN Studied in High Magnetic Fields 251
A. Wolos, A. Wysmolek, M. Kaminska, A. Twardowski, M. Bockowski, I. Grzegory, S. Porowski, and M. Potemski

Photoconductivity Study of Mg and C Acceptors in Cubic GaN 253
H. Przybylińska, G. Kocher, W. Jantsch, D. As, and K. Lischka

Motion of Muonium (Hydrogen) in the III-V Nitrides 255
R. L. Lichti, Y. G. Celebi, S. F. J. Cox, and E. A. Davis

On the Built-in Spontaneous Electric Field in MOVPE Grown GaN Crystals on Sapphire 257
A. M. Witowski, M. L. Sadowski, and K. Pakuła

Ga-Interstitial Related Defects in Ga(Al)NP 259
N. Q. Thinh, I. Vorona, I. A. Buyanova, W. M. Chen, Y. G. Hong, C. W. Tu, S. Limpijumnong, and S. B. Zhang

Dominating Behaviour of the Donor-Acceptor Pair Emission in Mass-Transport GaN 261
T. Paskova, B. Arnaudov, P. P. Paskov, E. M. Goldys, S. Hautakangas, K. Saarinen, U. Södervall, and B. Monemar

1.5.4. Electronic and Optical Properties of Nitride Semiconductors

Mie Resonant Absorption and Infrared Emission in InN Related to Metallic Indium Clusters 263
T. V. Shubina, S. V. Ivanov, V. N. Jmerik, D. D. Solnyshkov, P. S. Kop'ev, A. Vasson, J. Leymarie, A. Kavokin, H. Amano, S. Kamiyama, M. Iwaya, I. Akasaki, H. Lu, W. J. Schaff, A. Kasic, and B. Monemar

High Energy Optical Transitions in Ga(PN): Contribution from Perturbed Valence Band 265
M. Felici, A. Polimeni, M. Capizzi, S. V. Dudiy, A. Zunger, I. A. Buyanova, W. M. Chen, H. P. Xin, and C. W. Tu

Dilute Nitride Ga(AsN) Alloys: An *Unusual* Band Structure Probed by Magneto-Tunneling 267
A. Patanè, J. Endicott, J. Ibáñez, L. Eaves, and M. Hopkinson

New Insight into the Electronic Properties of GaNP Alloys 271
I. A. Buyanova, W. M. Chen, and C. W. Tu

Urbach Tails of Valence and Conductivity Bands and Optical Spectra of Hexagonal InN near the Fundamental Band Gap 275
A. Klochikhin, V. Y. Davydov, V. V. Emtsev, and A. V. Mudryi

Theory of Electron Effective Mass and Mobility in Dilute Nitride Alloys 277
A. Lindsay, S. Fahy, and E. P. O'Reilly

The Electronic Nature of Metal/p-GaN Junctions 279
S. Srinivasan, H. Omiya, F. A. Ponce, S. Tanaka, H. Marui, and T. Mukai

Infrared Characterization of GaN Films Grown on Sapphire by MOCVD 281
N. Kuroda, K. Saiki, Hasanudin, J. Watanabe, and M. Cho

Photoreflectance Investigations of Temperature Dependence of the "Different" Energy Gaps in GaInNAs Compounds 283
R. Kudrawiec, J. Misiewicz, E. M. Pavelescu, J. Konttinen, and M. Pessa

On the Nature of Near Bandedge Luminescence of InN Epitaxial Layers 285
B. Arnaudov, T. Paskova, P. P. Paskov, B. Magnusson, E. Valcheva, B. Monemar, H. Lu, W. Schaff, H. Amano, and I. Akasaki

Band Structure Parameters of the Nitrides: The Origin of the Small Band Gap of InN 287
P. Carrier and S.-H. Wei

Optical Band Gap of a New Filled Tetrahedral Semiconductor Li_3AlN_2 289
    K. Kushida, Y. Kaneko, and K. Kuriyama

The Free Exciton Binding Energy in a Strained $GaN_{0.02}As_{0.98}$ Layer 291
    R. Kudrawiec, J. Misiewicz, L. H. Li, and J. C. Harmand

Raman Scattering by LO Phonon-Plasmon Coupled Modes in Heavily Doped Ga(AsN) 293
    J. Ibáñez, R. Cuscó, L. Artús, D. Fowler, A. Patanè, L. Eaves, K. Uesugi, and I. Suemune

Bandgap Bowing in $InSb_{1-x}N_x$ Investigated with a New Fourier Transform Modulated Spectroscopy Technique for the Mid-Infrared 295
    M. Merrick, T. J. C. Hosea, B. N. Murdin, T. Ashley, L. Buckle, and T. Burke

Optical Reflectance of Bulk AlN Crystals and AlN Epitaxial Films 297
    L. Chen, B. J. Skromme, C. Chen, W. Sun, J. Yang, M. A. Khan, M. L. Nakarmi, J. Y. Lin, H. X. Jiang, Z. J. Reitmeyer, R. F. Davis, R. F. Dalmau, R. Schlesser, and Z. Sitar

Sub-Picosecond Spin Relaxation in GaN 299
    T. Kuroda, T. Yabushita, T. Kosuge, A. Tackeuchi, K. Taniguchi, T. Chinone, and N. Horio

Localization Versus Carrier-Screening Effects in InGaN Quantum Wells—A Time-Resolved Cathodoluminescence Study 301
    A. Bell, J. Christen, F. Bertram, M. R. Stevens, F. A. Ponce, H. Marui, and S. Tanaka

1.6. Magnetic Semiconductors

Effects of Mn Site Location on the Magnetic Properties of $III_{1-x}Mn_xV$ Semiconductors 303
    K. M. Yu, W. Walukiewicz, T. Wojtowicz, J. Denlinger, X. Liu, and J. K. Furdyna

Influence of Light on Ferromagnetic Semiconductors: Magnetization Rotation via the Angular Momentum and the Photon Energy 307
    H. Munekata

Ab Initio Study of Transition-Metal Silicide Films on Si(001) 311
    H. Wu, P. Kratzer, and M. Scheffler

Neutron Scattering Studies of the Spin Structure of Magnetic Semiconductor Superlattices 313
    H. Kępa, P. Sankowski, P. Kacman, C. F. Majkrzak, A. Y. Sipatov, and T. M. Giebultowicz

Electronic Structure and Stability of Ferromagnetic GaN Doped with Mn 315
    J. Kang, K. J. Chang, and H. Katayama-Yoshida

New Class of high-T_C Diluted Ferromagnetic Semiconductors Based on K_2S without Transition Metal Elements 317
    M. Seike, K. Kemmochi, K. Sato, A. Yanase, and H. Katayama-Yoshida

Magnetic Doping and Characterization of n-type GaN 319
    X. M. Cai, A. B. Djurišić, M. H. Xie, H. Liu, X. X. Zhang, J. J. Zhu, H. Yang, and Y. H. Leung

Comparison of Ferromagnetic Properties between InMnP and GaMnN Quantum Wells 321
    J. W. Kim, N. Kim, S. J. Lee, Y. Shon, T. W. Kang, and T. F. George

Spin-Polarized Charge Densities in (In,Ga,Mn)As-Based Diluted Magnetic Semiconductor Ternary and Quaternary Alloy Heterostructures 323
    G. M. Sipahi, S. C. P. Rodrigues, L. M. R. Scolfaro, I. C. da Cunha Lima, and J. R. Leite

Structural Studies of Layered Magnetic Semiconductor $(La_{1-x}Ca_xO)Cu_{1-x}Ni_xS$ 325
    K. Takase, O. Shoji, K. Sato, S. Koyano, Y. Kuroiwa, S. Aoyagi, Y. Takahashi, Y. Takano, and K. Sekizawa

A New Ferromagnetic Material Excluding Transition Metals: CaAs in a Distorted Zinc-Blende Structure 327
    M. Geshi, K. Kusakabe, H. Tsukamoto, and N. Suzuki

Ferroelectric Properties of Mn-Implanted CdTe 329
    D. J. Fu, J. C. Lee, W. C. Lee, S. W. Choi, S. J. Lee, and T. W. Kang

Transition-Metal Substitution in Semiconducting $Ba_8Ga_{16}Ge_{30}$ Clathrates 331
    Y. Li, W. Gou, J. Chi, V. Goruganti, and J. H. Ross, Jr.

Magnetism in (Ga,Mn)As Thin Films with T_C up to 173K 333
    K. Y. Wang, R. P. Campion, K. W. Edmonds, M. Sawicki, T. Dietl, C. T. Foxon, and B. L. Gallagher

Magnetic Order and Fluctuations in GaMnAs 335
    O. M. Fedorych, Z. Wilamowski, and J. Sadowski

Electrical, Magnetic and Magnetooptical Properties of Bulk (Zn,Mn)Te Semimagnetic Semiconductor Doped with Phosphorus 337
    V. Khoi Le, R. R. Galazka, M. Dobrowolska, K. Yee, X. Liu, W.-L. Lin, J. K. Furdyna, and T. M. Giebultowicz

Effect of Light Irradiation on Magnetization Reversal in Ferromagnetic (Ga,Mn)As with Perpendicular Magnetic Anisotropy 339
    A. Oiwa, T. Endo, H. Takechi, and H. Munekata

Temperature Dependent Magnetization and Magnetic Phases of Conduction-Band Dilute-Magnetic-Semiconductor Quantum Wells with Non-Step-like Density of States 341
    C. Simserides

Transport Properties of Diluted Magnetic Semiconductors 343
    M. Takahashi, K. Kagami, and K. Kubo

Magnetic Properties of Undoped and N-doped $Zn_{1-x}Cr_x Te$ Grown by MBE 345
    N. Ozaki, N. Nishizawa, S. Kuroda, and K. Takita

Magnetic Anisotropy in Epitaxial InMnAs 347
    P. T. Chiu, S. J. May, A. J. Blattner, and B. W. Wessels

X-Ray Magnetic Circular Dichroism of Ferromagnetic Semiconductor $Ge_{1-x}Mn_x Te$ 349
    Y. Fukuma, K. Fujimoto, H. Sato, K. Tsuji, A. Kimura, M. Taniguchi, S. Senba, A. Tanaka, H. Asada, and T. Koyanagi

N-Conducting, Ferromagnetic Mn-Doped ZnO Thin Films on Sapphire Substrates 351
    H. Schmidt, M. Diaconu, E. Guzman, H. Hochmuth, M. Lorenz, G. Benndorf, A. Setzer, P. Esquinazi, H. von Wenckstern, D. Spemann, A. Pöppl, R. Böttcher, and M. Grundmann

ESR Study of Diluted Magnetic Semiconductor (Ga,Cr)As 353
    Y. Yamamoto, M. Nagai, S. Itaya, and H. Hori

Novel IV-Group Based Diluted Magnetic Semiconductor: $Cr_x Ge_{1-x}$ 355
    S.-i. Itaya, Y. Yamamoto, and H. Hori

HgSe and HgSe:Fe in Extreme Magnetic Fields 357
    M. Von Ortenberg, C. Puhle, and S. Hansel

Effect of Wet Etching on the Magnetic Properties of *n*-type GaMnN Layers 359
    Y. Liu, M. I. Nathan, P. P. Ruden, J. E. Van Nostrand, B. Claflin, and J. D. Albrecht

Optical Investigation of Temperature-Induced Changes in Magnetic Anisotropy in III-Mn-As Ferromagnetic Semiconductors 361
    M. Kutrowski, L. Titova, R. Chakarvorty, K. Yee, W. L. Lim, X. Liu, T. Wojtowicz, J. K. Furdyna, and M. Dobrowolska

Tuning the Ferromagnetic Properties of Hydrogenated GaMnAs 363
    A. Lemaître, L. Thevenard, M. Viret, L. Largeau, O. Mauguin, B. Theys, F. Bernardot, R. Bouanani-Rahbi, B. Clerjaud, and F. Jomard

Properties of $Ga_{1-x}Mn_x N$ Epilayers Grown by Molecular Beam Epitaxy 365
    S. Marcet, E. Bellet, X. Biquard, C. Bougerol, J. Cibert, D. Ferrand, R. Giraud, D. Halley, E. Kulatov, S. Kuroda, H. Mariette, and A. Titov

Magnetic Anisotropy of Strain-Engineered InMnAs Ferromagnetic Films and Easy-Axis Manipulation from Out-of-Plane to In-Plane Orientations 367
    X. Liu, W. L. Lim, Z. Ge, S. Shen, T. Wojtowicz, K. M. Yu, W. Walukiewicz, M. Dobrowolska, and J. K. Furdyna

2. HETEROSTRUCTURES, SURFACES, AND INTERFACES

2.1. Structural and Electronic Properties of Surfaces

Energy Barrier for Dimer Flipping at the Si(001)-2×1 Surface in External Electrostatic Fields 371
    J. Nakamura and A. Natori

First-Principles Calculations of 2×2 Reconstructions of GaN Surfaces 373
    V. T. Salinero, M. C. Gibson, S. Brand, S. J. Clark, and R. A. Abram

First-Principles Study of GaSb(001) Surface Reconstructions 375
    M. C. Righi, R. Magri, and C. M. Bertoni

The Physics of Hydrogen-Terminated Diamond Surfaces 377
    J. Ristein

Metallizing a Semiconductor Surface with Hydrogen...........381
P. G. Soukiassian and H. B. Enriquez

2.2. Structural and Electronic Properties of Interfaces

Exciton Confinement in Si/TiO$_2$ 0-2D Systems...........385
J. M. Henriques Neto, E. W. S. Caetano, J. A. P. da Costa, E. L. Albuquerque, V. N. Freire, L. Scolfaro, and J. R. Leite

Order-Parameter Dependence of Spontaneous Electron Accumulation at Ga$_{0.5}$In$_{0.5}$P/GaAs Studied by Raman-Scattering and Photoluminescence Measurements...........387
K. Yamashita, K. Oe, T. Kita, O. Wada, Y. Wang, C. Geng, F. Scholz, and H. Schweizer

First-Principles Study of Excess Si-Atom Stability around Si-Oxide/Si Interfaces...........389
H. Kageshima, M. Uematsu, K. Akagi, S. Tsuneyuki, T. Akiyama, and K. Shiraishi

Magnetic Field Effect on Bound Polarons in Semiconductor Heterojunctions under Pressure...........391
S. L. Ban and S. T. Wang

Microscopic Theory of Oxygen Reaction Mechanisms at SiO$_2$/Si(100) Interface...........393
T. Akiyama, H. Kageshima, and T. Ito

Uniform Sub-nm Nitridation on Si(100) through Strong N Condensation...........395
K. Kato, Y. Nakasaki, D. Matsushita, and K. Muraoka

Magneto-Optics in Be-δ-Doped GaAs Quantum Wells with a Back Gate...........397
M. Yamaguchi, S. Nomura, D. Sato, T. Akazaki, H. Tamura, and H. Takayanagi

Temperature Dependent Bandgap Energy and Conduction Band Offset of GaAsSb on GaAs...........399
J.-B. Wang, S. R. Johnson, S. A. Chaparro, J. A. Gupta, Y. G. Sadofyev, D. Ding, and Y.-H. Zhang

Electro-Optic Raman Observation of Low Temperature Phase Transitions in ZnO-BaTiO$_3$-ZnO Heterostructures...........401
B. N. Mbenkum, N. Ashkenov, M. Schubert, and M. Lorenz

Interface and Confined LO-Phonon Effects on 2D Strong Coupled Electron Gases...........403
M. Adames and A. S. Camacho B.

Spin-Flip Subband-Landau Level Coupling in InSb Heterostructures...........405
X. H. Zhang, R. C. Meyer, T. Kasturiarachchi, N. Goel, R. E. Doezema, S. J. Chung, M. B. Santos, and Y. J. Wang

Electronic Properties of C-Doped (100) AlGaAs Heterostructures...........407
B. Grbić, C. Ellenberger, T. Ihn, K. Ensslin, D. Reuter, and A. D. Wieck

Analysis of the $\Delta(X)$-L Intervalley Mixing in Group-IV Heterostructures...........409
A. A. Kiselev, K. W. Kim, and E. Yablonovitch

L-Valley Electrons in SiGe Heterostructures: Highly Anisotropic and Tunable Zeeman and Rashba-like Spin Splittings...........411
A. A. Kiselev, F. A. Baron, K. W. Kim, K. L. Wang, and E. Yablonovitch

Interface Band Mixing from First Principles...........413
B. A. Foreman

Interband Transitions in Modulation-Doped GaAs/AlGaAs Heterostructures under In-Plane Magnetic Field...........415
B. M. Ashkinadze, E. Linder, E. Cohen, and L. N. Pfeiffer

Photoreflectance Studies of Undoped and Si-Doped AlGaN/GaN Heterostructures with a Two Dimensional Electron Gas...........417
R. Kudrawiec, M. Syperek, J. Misiewicz, R. Paszkiewicz, B. Paszkiewicz, and M. Tlaczala

Measurement of Electric Fields in GaN/AlGaN FETs Using Photoreflectance with Different Excitation Energies...........419
D. K. Gaskill, O. J. Glembocki, D. W. Gotthold, S. P. Guo, B. E. Albert, A. V. Vertiatchikh, and L. F. Eastman

Crystal Structure of Low-Resistance Au-Ni/p-GaN Contacts...........421
H. Omiya, S. Srinivasan, F. A. Ponce, S. Tanaka, H. Marui, and T. Mukai

Atomic-Scale Modelling of the Si(100)-SiO$_2$ Interface...........423
F. Giustino, A. Bongiorno, and A. Pasquarello

Atomic Processes at and near Silicon/Silicon Dioxide Interfaces...........427
K. Shiraishi

2.3. Transport Properties of Heterostructures and Two-Dimensional Electron Gases

Enhanced Magnetoresistance of Semiconductor-Metal Hybrid Structures 431
    M. Holz, O. Kronenwerth, and D. Grundler

Low-Frequency Noise in AlGaN-Based Schottky Barriers. 433
    D. Seghier and H. P. Gislason

Magnetoconductivity of Dilute 2D Electron Systems on a Silicon Surface. 435
    S. A. Vitkalov, Y. Tsui, M. P. Sarachik, and T. M. Klapwijk

In-Plane Tunneling Spectroscopy of a Two-Dimensional Electron Gas through a One-Dimensional Barrier. 437
    J.-L. Deborde, S. F. Fischer, U. Kunze, D. Reuter, and A. D. Wieck

Subnanosecond Current Kinetics under Hot Carrier Transport in AlGaN/GaN Heterostructures. 439
    B. A. Danilchenko, S. E. Zelensky, E. Drok, S. A. Vitusevich, S. V. Danylyuk, N. Klein, H. Lüth, A. E. Belyaev, and V. A. Kochelap

Monte Carlo Study of Noise Scaling in AlGaN/GaN HFETs. 441
    M. Singh, Y.-R. Wu, T. Palacios, J. Singh, and U. Mishra

Physics and Growth of Si-Doped Two-Dimensional High Mobility Hole Gases on (110) Oriented GaAs 443
    F. Fischer, M. Grayson, D. Schuh, M. Bichler, and G. Abstreiter

Weak Localization and Interaction Effects in 2D Electron Gas in AlGaN/GaN Heterostructure in Presence of Phonons.. 445
    V. Renard, Z. D. Kvon, H.-I. Cho, J.-H. Lee, and J. C. Portal

Magnetometry of AlGaN/GaN Heterostructure Wafers 447
    K. Tsubaki, N. Maeda, T. Saitoh, and N. Kobayashi

Giant Hole-Phonon Coupling in a Strongly Interacting Two-Dimensional (2D) Hole System 449
    X. P. A. Gao, G. S. Boebinger, A. P. Mills, Jr., A. P. Ramierez, L. N. Pfeiffer, and K. W. West

Magnetotransport on Evenly Curved Hall-Bars in InGaAs/GaAs-Microtubes 451
    S. Mendach, H. Welsch, C. Heyn, and W. Hansen

Weak Localization of Dilute 2D Electrons in Undoped GaAs Heterostructures. 453
    M. P. Lilly, E. Bielejec, J. A. Seamons, and J. L. Reno

The Inertial-Mass Scale for Free-Charge-Carriers in Semiconductor Heterostructures 455
    T. Hofmann, M. Schubert, C. von Middendorff, G. Leibiger, V. Gottschalch, C. M. Herzinger, A. Lindsay, and E. O'Reilly

Aharonov-Bohm Ring Interferometer on the Basis of 2D Electron Gas in a Double Quantum Well 457
    D. V. Nomokonov, A. A. Bykov, A. K. Bakarov, A. K. Kalagin, A. I. Toropov, J. C. Portal, and A. M. Mishchenko

Low-Temperature Transport in AlGaN/GaN 2D Electron Systems 459
    S. A. Vitusevich, A. M. Kurakin, S. V. Danylyuk, N. Klein, H. Lüth, and A. E. Belyaev

The Effect of Controllable Optically-Induced Random Anisotropic Disorder on the Magnetotransport in a Two-Dimensional Electron System 461
    G. P. Melhuish, A. S. Plaut, S. H. Simon, N. Rocher, V. Robbe, M. C. Holland, and C. R. Stanley

Non-Markovian Memory Effects in Ballistic Two-Dimensional Electron System with Non-Planar Topology 463
    N. M. Sotomayor, G. M. Gusev, J. R. Leite, A. A. Bykov, A. K. Bakarov, V. M. Kudryashev, and A. I. Toropov

Contributions to the Resistivity of a 2DEG from Magnetically Ordered Array of Sub-Micron Cobalt Elements. 465
    T. McMullen, E. Skuras, K. J. Kirk, J. A. Wilson, J. Davies, and A. R. Long

Magnetization of Modulation Doped Si/SiGe Quantum Wells in High Magnetic Fields 467
    M. A. Wilde, M. Rhode, C. Heyn, F. Schäffler, U. Zeitler, R. J. Haug, D. Heitmann, and D. Grundler

Magnetotransport Measurements on the Defect-Induced Conduction Electrons near InAs(111) Surface. 469
    Y. Tsuji and T. Okamoto

High-Resolution Ballistic Electron Spectroscopy in Parallel Magnetic Fields 471
    M. Kast, W. Boxleitner, G. Strasser, and E. Gornik

Shot Noise Behavior of Cascaded Mesoscopic Structures 473
    M. Macucci, G. Iannaccone, and P. Marconcini

Chaotic Behavior in Quantum Transport Devices 475
    S. Harada, N. Kida, T. Morimoto, M. Hemmi, R. Naito, T. Sasaki, N. Aoki, T. Harayama, J. P. Bird, and Y. Ochiai

Importance of Interface Sampling for Extraordinary Effects in Metal Semiconductor Hybrids 477
    A. C. H. Rowe and S. A. Solin

Electron Transport on Hydrogen-Passivated Silicon Surfaces 479
    K. Eng, R. McFarland, and B. E. Kane

The 2d MIT: The Pseudogap and Fermi Liquid Theory 481
    T. G. Castner

Trion Assisted Tunneling in Double Barrier Diodes 483
    A. Vercik, I. Camps, Y. Galvão Gobato, M. J. S. P. Brasil, G. E. Marques, and S. S. Makler

Optimization of Electron Velocity in p-HEMT 485
    S. Mil'shtein and S. Somisetty

Temperature Dependent Sign of the Interaction-Induced Magnetoresistance in a n-Si/SiGe Heterostructure 487
    V. Renard, E. B. Olshanetsky, Z. D. Kvon, J.-C. Portal, N. J. Woods, J. Zhang, and J. J. Harris

Anisotropic Electron Transport in Pseudomorphic InGaAs Channels and Its Structural Origin 489
    J. A. Wilson, M. MacKenzie, S. MacFadzean, M. C. Holland, C. R. Stanley, T. McMullen, S. Hamill, P. Stopford, and A. R. Long

Interaction Effects in High-Mobility Two-Dimensional Systems 491
    A. K. Savchenko, E. A. Galaktionov, S. S. Safonov, Y. Y. Proskuryakov, L. Li, M. Pepper, M. Y. Simmons, D. A. Ritchie, E. H. Linfield, and Z. D. Kvon

Magnetotransport in Nonplanar Two-Dimensional Electron Gases in InAs Heterostructures 493
    S. Löhr, C. Heyn, and W. Hansen

Electron Effective Mass Enhancement in Ultrathin Gate-Oxide Si-MOSFETs 495
    M. Dragosavac, D. J. Paul, M. Pepper, A. B. Fowler, and D. A. Buchanan

Electrical Conduction Properties of Ga(AsN) Layers 497
    D. Fowler, O. Makarovsky, A. Patanè, L. Eaves, L. Geelhaar, H. Riechert, and K. Uesugi

Transport in a Two-Dimensional Electron Gas Narrow Channel with a Magnetic Field Gradient 499
    M. Hara, A. Endo, S. Katsumoto, and Y. Iye

2.4. Physics in the Quantum Hall Regime

The Absorption Spectrum around $\nu=1$: Evidence for a Small Size Skyrmion 501
    J. G. Groshaus, V. Umansky, H. Shtrikman, Y. Levinson, and I. Bar-Joseph

Relative Specific Heat at $\nu=1/2$ Measured in a Phonon Absorption Experiment 503
    F. Schulze-Wischeler, F. Hohls, U. Zeitler, D. Reuter, A. D. Wieck, and R. J. Haug

Manipulating Nuclear Spins via Quantum Hall Edge Channels 505
    S. Komiyama, T. Machida, and K. Ikushima

Local Investigation of the Classical and Quantum Hall Effect 509
    A. Baumgartner, T. Ihn, K. Ensslin, K. Maranowski, and A. C. Gossard

Microwave-Modulated Photoluminescence of a 2D-Electron Gas in a Magnetic Field 511
    B. M. Ashkinadze, E. Linder, E. Cohen, and L. N. Pfeiffer

Illuminating Electron Liquids in Two Dimensional Quantum Structures 513
    A. Pinczuk

Possibility of a Magnetic-Field Sweep-Direction-Dependent Hysteretic-Effect in the Microwave-Excited 2DES 517
    R. G. Mani

MD-ENDOR Spectroscopy in a Two-Dimensional Electron System in the Regime of the Quantum Hall Effect 519
    J. D. Caldwell, A. E. Kovalev, E. Olshanetsky, C. R. Bowers, J. L. Reno, and J. A. Simmons

Direct Measurement of the *g*-Factor of Composite Fermions 521
    F. Schulze-Wischeler, E. Mariani, F. Hohls, and R. J. Haug

Origin of Sub-Linear Width Dependence of Quantum Hall Breakdown Current 523
    K. Oto, T. Katagiri, and T. Tsubota

Interaction of Spin Excitations in Quantum Hall Systems 525
    J. J. Quinn and A. Wójs

Imaging of Intra and Inter Landau Level Scattering in Quantum Hall Devices 527
    Y. Kawano and T. Okamoto

Exact Solutions for Few-Particle Anyon Excitons 529
    D. G. W. Parfitt and M. E. Portnoi

Magneto-PL of Free and Localized Holes in Semiconductor Heterojunctions near Integer Filling Factors 531
    H. P. van der Meulen, D. Sarkar, J. M. Calleja, R. Hey, K. J. Friedland, and K. Ploog

Quantum Hall Ferromagnetism in InSb Heterostructures 533
    J. C. Chokomakoua, N. Goel, J. L. Hicks, S. J. Chung, M. B. Santos, M. B. Johnson, and S. Q. Murphy

Spin Structures in Inhomogeneous Fractional Quantum Hall Systems 535
    K. Výborný and D. Pfannkuche

The Localization and the Quantum Hall Effect on the Hofstadter Butterfly 537
    M. Koshino and T. Ando

A Unified Picture of the Scaling and Non-Scaling Behavior in Quantum Hall Plateau Transitions 539
    X. R. Wang, G. Xiong, and Q. Niu

Moore-Read States on a Sphere: Three-body Correlations and Finite-Size Effects 541
    A. Wójs and J. J. Quinn

Single Mode Approximation Spectrum for a Quantum Hall Liquid Crystal 543
    C. M. Lapilli and C. Wexler

SU(4) Skyrmions and Activation Energy Anomaly in Bilayer Quantum Hall Systems 545
    Z. F. Ezawa

Anomalous Hall Effect in a Two-Dimensional Hole System 547
    G. M. Gusev, A. A. Quivy, T. E. Lamas, J. R. Leite, and J. C. Portal

Unusual Tunneling Characteristics of Double-Quantum-Well Heterostructures 549
    Y. Lin, J. Nitta, A. K. M. Newaz, W. Song, and E. E. Mendez

Ettingshausen Effect and Electron Temperature Distribution in Quantum Hall Systems with Slowly-Varying Confining Potential 551
    T. Maeda, S. Kanamaru, H. Akera, and H. Suzuura

Peltier Effect and Electron Temperature Distribution in Quantum Hall Systems with Potential Discontinuity 553
    T. Nakagawa, H. Akera, and H. Suzuura

Ettingshausen Effect in Integer Quantum Hall Systems 555
    Y. Komori and T. Okamoto

High-Current Breakdown of the Quantum Hall Effect 557
    A. J. Matthews, K. V. Kavokin, A. Usher, M. E. Portnoi, J. D. Gething, M. Zhu, and D. A. Ritchie

Enhanced Localization in Landau-Quantized Systems Induced by Very Low Frequencies 559
    A. Buß, F. Hohls, R. J. Haug, C. Stellmach, G. Hein, and G. Nachtwei

Breakdown of the Quantum Hall Effects in Hole Systems at High Induced Currents 561
    J. D. Gething, A. J. Matthews, A. Usher, M. E. Portnoi, K. V. Kavokin, and M. Henini

Double Magnetoresistance Minima for $\nu=1$ Quantum Hall State Near Commensurate-Incommensurate Transition 563
    D. Terasawa, K. Nakada, S. Kozumi, A. Fukuda, A. Sawada, Z. F. Ezawa, N. Kumada, K. Muraki, T. Saku, and Y. Hirayama

In-plane Field Induced Anisotropy of the Longitudinal Resistance in a Bilayer Quantum Hall System 565
    M. Morino, K. Iwata, M. Suzuki, A. Fukuda, A. Sawada, Z. F. Ezawa, N. Kumada, K. Muraki, T. Saku, and Y. Hirayama

Strong to Weak Correlation Phase Transition at $\nu=1$ in a Bilayer System Detected by Nuclear-Spin Relaxation 567
    N. Kumada, K. Muraki, K. Hashimoto, T. Saku, and Y. Hirayama

Imaging Non-equilibrium Edge States in Quantum Hall Conductors 569
    K. Ikushima, H. Sakuma, S. Komiyama, and K. Hirakawa
Optical Detection of Spin Polarization of Electrons in Quantum Dot Edge Channels 571
    S. Nomura and Y. Aoyagi
Observation of Two Modes of Edge Magnetoplasmons by Selective Edge Channel Detection 573
    G. Sukhodub, F. Hohls, and R. J. Haug
The Insulator-Quantum Hall-Insulator Transitions in a Two-Dimensional GaAs System Containing Self-Assembled InAs Quantum Dots 575
    G.-H. Kim, C.-T. Liang, C. F. Huang, and D. A. Ritchie
Theory of Excitonic Filling Factor $\nu=2$ Quantum Hall Droplet in Self-Assembled Quantum Dots 577
    S.-J. Cheng, W. Sheng, and P. Hawrylak
Coulomb Oscillations of Conductance in an Open Ring Interferometer in the Quantum Hall Regime 579
    A. A. Bykov, D. V. Nomokonov, A. K. Bakarov, and J. C. Portal

3. NANOSTRUCTURES

3.1. Quantum Dots and Nanocrystals

3.1.1. Growth and Structural Properties of Quantum Dots

Determination of In(Ga)As/GaAs Quantum Dot Composition Profile 583
    A. Lemaître, G. Patriarche, and F. Glas
Determination of Quantum Dots Structural Parameters by XAFS Spectroscopy 585
    S. Erenburg, N. Bausk, A. Nikiforov, A. Yakimov, A. Dvurechenskii, G. Kulipanov, and S. Nikitenko
Silicon Quantum Dot Formation Using Photo-CVD 587
    S. S. Kim, K. I. Bang, and K. S. Lim
Structural and Electronic Properties of HgTe Quantum Dots 589
    S. J. Clark, X.-Q. Wang, and R. A. Abram
Structural Analysis of InAs Quantum Dashes Grown on InP Substrate by Scanning Transmission Electron Microscopy 591
    A. Sauerwald, T. Kümmell, G. Bacher, A. Somers, R. Schwertberger, J. P. Reithmaier, and A. Forchel
Biexciton Formation Induced by Bright-Dark Exciton Transitions in (Cd,Mg)Te/CdTe/(Cd,Mn)Te Asymmetric Quantum Wells 593
    R. Shen, H. Mino, G. Karczewski, T. Wojtowicz, J. Kossut, and S. Takeyama
Fabrication of Coupled Quantum Disks on a Patterned Substrate by MBE 595
    M. Yamaguchi, Y. Higuchi, Y. Nishimoto, and N. Sawaki
Fabrication and Characterization of PbS Quantum Dots 597
    L. Cademartiri, G. von Freymann, V. Kitaev, and G. A. Ozin
Investigating the Evolution of Dislocated SiGe Islands by Selective Wet-Chemical Etching 599
    U. Denker, D. E. Jesson, M. Stoffel, A. Rastelli, and O. G. Schmidt
Role of Strain Relaxation during Different Stages of InAs Quantum Dot Growth 601
    T. Hammerschmidt and P. Kratzer
Effect of InGaP Strain-Compensation Layers in Stacked 1.3 μm InAs/GaAs Quantum Dot Active Regions Grown by MOCVD 603
    N. Nuntawong, S. Birudavolu, C. P. Hains, S. Huang, Y. C. Xin, and D. L. Huffaker
Reverse Micelles Synthesis and Optical Characterization of Manganese Doped CdSe Quantum Dots 605
    B. C. Guo, Q. Pang, C. L. Yang, W. K. Ge, S. H. Yang, and J. N. Wang
Quantum Dot Size Tuning and Self-Organization with Pulsed Ion Beam Nucleation and Growth of Ge/Si Nanostructures 607
    A. V. Dvurechenskii, Z. V. Smagina, R. Groetzschel, V. A. Zinovyev, and P. L. Novikov
Directed Assembly of Ge Islands Grown on Au-Patterned Si(100) 609
    J. T. Robinson, J. A. Liddle, A. Minor, V. Radmilovic, and O. D. Dubon

In Situ Characterization of Ge Nanocrystals near the Growth Temperature 611
  I. D. Sharp, Q. Xu, D. O. Yi, C. Y. Liao, J. W. Beeman, Z. Liliental-Weber, K. M. Yu, D. N. Zakharov,
  J. W. Ager III, D. C. Chrzan, and E. E. Haller

Self-Organized Lattice of Ordered Quantum Dot Molecules 613
  T. van Lippen, R. Nötzel, G. J. Hamhuis, and J. H. Wolter

Engineering the Wetting Layer States to Reach Room Temperature Emission for CdTe
Quantum Dot Structures. 615
  S. Moehl, L. Maingault, K. Kheng, and H. Mariette

Raman Study of Strain Relaxation in GaN/AlN Quantum Dots. 617
  N. Garro, A. Cros, J. M. Llorens, A. García-Cristóbal, A. Cantarero, N. Gogneau, E. Monroy, and
  B. Daudin

Composition, Photoelectric Properties, and Electroluminescence of the SiGe/Si
Heterostructures with Self-Assembled Nanoclusters Grown by Molecular Beam Epitaxy with
Vapor Ge Source 619
  M. V. Kruglova, G. A. Maximov, D. O. Filatov, D. E. Nikolitchev, V. G. Shengurov, and S. V. Morozov

Hydrogenation of Stacked Self-Assembled InAs/GaAs Quantum Dots 621
  S. Mazzucato, D. Nardin, A. Polimeni, M. Capizzi, D. Granados, and J. M. García

Phase Separated Quantum Dots in GaNP and InGaN Layers 623
  R. K. Soni, R. S. Katiyar, and H. Asahi

Icosahedral Quantum Dots and 2D Quasicrystals for Group IV Semiconductors 625
  Y. Zhao, Y.-H. Kim, M.-H. Du, and S. B. Zhang

3.1.2. Optical, Vibrational, and Electronic Properties of Quantum Dots

Single Dot Spectroscopy of GaN/AlN Self-Assembled Quantum Dots 627
  S. Kako, K. Hoshino, S. Iwamoto, S. Ishida, and Y. Arakawa

Mechanism of Recombination in InAs Quantum Dots in Indirect Bandgap AlGaAs Matrices 629
  T. S. Shamirzaev, A. M. Gilinsky, A. I. Toropov, A. K. Bakarov, D. A. Tenne, K. S. Zhuravlev,
  S. Schulze, C. von Borczyskowski, and D. R. T. Zahn

Time-Resolved Optical-Phonon Emission and Electron-Hole Relaxation Dynamics in ZnCdTe
Quantum Wells and Quantum Dots. 631
  P. Gilliot, S. Cronenberger, Y. Viale, M. Gallart, B. Hönerlage, K. Kheng, and H. Mariette

Dephasing and Energy Relaxation Processes in Self-Assembled In(Ga)As/GaAs Quantum Dots 633
  M. Dworzak, P. Zimmer, H. Born, and A. Hoffmann

Shell Structures in Self-Assembled InAs Quantum Dots Observed by Lateral Single Electron
Tunneling Structures. 635
  M. Jung, K. Hirakawa, S. Ishida, Y. Arakawa, Y. Kawaguchi, and S. Komiyama

Spin-Hybridization Effects in Quantum Dots 637
  C. Trallero-Giner, A. M. Alcalde, V. López-Richard, S. J. Prado, and G. E. Marques

Room-Temperature 1.35 μm Emission from Wire-like Anisotropic InGaAs Quantum Dots
Grown on Just-(001) GaAs Substrate by Atomic Layer Epitaxy Technique. 639
  S.-J. Lee, S. K. Noh, and E. K. Kim

Magneto-Spectroscopy of Hole Levels in InAs Quantum Dots 641
  D. Reuter, U. Zeitler, J. C. Maan, and A. D. Wieck

Spin States in a Single InAs Quantum Dot Molecule Probed by Single-Electron
Tunneling Spectroscopy. 643
  T. Ota, M. Stopa, M. Rontani, T. Hatano, K. Yamada, S. Tarucha, Y. Nakata, H. Z. Song, T. Miyazawa,
  T. Usuki, M. Takatsu, and N. Yokoyama

Dynamics of Bright and Dark Excitons in a Self-Assembled Quantum Dot. 645
  P. A. Dalgarno, M. Ediger, J. M. Smith, R. J. Warburton, K. Karrai, A. O. Govorov, B. D. Gerardot, and
  P. M. Petroff

The Thermal Model for Carrier Hopping and Retrapping in Self-Organized InAs Quantum
Dot Heterostructure. 647
  Y.-F. Wu, H.-T. Shen, R.-M. Lin, T.-E. Nee, N.-T. Yeh, T.-P. Hsieh, and J.-C. Lee

Configuration Transitions between MDD and Spin-Polarized Pentagonal Molecules via Lower
Spin States in a Five-Electron Quantum Dot. 649
  Y. Nishi, P. Maksym, D. G. Austing, K. Yamada, H. Aoki, and S. Tarucha

Magneto-Capacitance Imaging of Quasi-Particle Wave Functions in Quantum Dots 651
    O. S. Wibbelhoff, A. Lorke, D. Reuter, and A. D. Wieck

Individual Quantum Dot Spectroscopy Probed by a Cryogenic Scanning Probe 653
    M. Liu, C. Li, A. Mampazhy, C.-H. Yang, and M.-J. Yang

A Very Narrow Photoluminescence Broadening (<16 meV) from ~1.5 μm Self-Assembled
Quantum Dots at Room Temperature 655
    T. Fukuda, J. Tatebayashi, M. Nishioka, and Y. Arakawa

Spatial Diffusion of Carriers in a Quantum Dot System Grown by Shadow Mask
Controlled Epitaxy 657
    S. Mackowski, T. Gurung, F. Haque, H. E. Jackson, L. M. Smith, T. Schallenberg, K. Brunner,
    and L. W. Molenkamp

Interface Optical Phonons in Triaxial Ellipsoidal Quantum Dots 659
    O. Reese, L. C. Lew Yan Voon, and M. Willatzen

Single-Photon and Photon Pair Emission from Individual (In,Ga)As Quantum Dots 661
    S. M. Ulrich, M. Benyoucef, P. Michler, J. Wiersig, N. Baer, P. Gartner, F. Jahnke, M. Schwab, H. Kurtze,
    R. Oulton, M. Bayer, S. Fafard, Z. Wasilewski, and A. Forchel

Temperature Control of the Polarisation of a Single Photon Source 665
    D. C. Unitt, A. J. Bennett, P. Atkinson, K. Cooper, D. A. Ritchie, and A. J. Shields

Controlled Generation of Neutral, Negatively Charged and Positively Charged Excitons in the
Same Single Quantum Dot 667
    M. Ediger, P. A. Dalgarno, J. M. Smith, R. J. Warburton, K. Karrai, B. D. Gerardot, and P. M. Petroff

Polarization Correlations between Single Photons Emitted by Quantum Dots in
Planar Microcavities 669
    N. Akopian, S. Vilan, U. Mizrahi, D. V. Regelman, D. Gershoni, E. Ehrenfreund, A. Shabaev, A. L. Efros,
    B. Gerardot, and P. M. Petroff

Near-Field Light Emission from Mesoscopic Complex Structures 671
    M. Pieruccini, S. Savasta, R. Girlanda, R. C. Iotti, and F. Rossi

Optical Properties of Tetrahedral Quantum Dot Quantum Wells 673
    V. A. Fonoberov, E. P. Pokatilov, V. M. Fomin, and J. T. Devreese

Polaron Effect in Semiconductor Quantum Dots: Impact on the Optical Absorption,
Up-Converted Photoluminescence and Raman Scattering 675
    M. I. Vasilevskiy, R. P. Miranda, S. S. Makler, E. V. Anda, and S. A. Filonovich

Morphology of CdTe/ZnTe Self-Assembled Quantum Dots Studied by Excitation Spectroscopy 677
    T. Nguyen, T. B. Hoang, S. Mackowski, H. E. Jackson, L. M. Smith, J. Wrobel, K. Fronc, J. Kossut,
    and G. Karczewski

Electronic and Multi-Electronic Structures in GaAs Quantum Dots 679
    M. Yamagiwa, F. Minami, S. Sanguinetti, T. Kuroda, and N. Koguchi

The Influence of Auger Processes on Recombination in Long-Wavelength InAs/GaAs
Quantum Dots 681
    I. P. Marko, A. D. Andreev, S. J. Sweeney, A. R. Adams, R. Krebs, S. Deubert, J. P. Reithmaier,
    and A. Forchel

Spin Polarized Photocurrent from a Single Quantum Dot 683
    J. M. Villas-Bôas, S. E. Ulloa, and A. O. Govorov

Electron and Hole States in Vertically Coupled Self-Assembled InGaAs Quantum Dots 685
    M. Korkusinski, W. Sheng, P. Hawrylak, Z. Wasilewski, G. Ortner, M. Bayer, A. Babinski,
    and M. Potemski

Electronic Coupling between Self-Assembled Quantum Dots Tuned by High Pressure 687
    S. I. Rybchenko, I. E. Itskevich, J. Cahill, A. I. Tartakovskii, M. S. Skolnick, G. Hill, and M. Hopkinson

Local Phonon Modes in InAs/GaAs Quantum Dots 689
    A. Paarmann, F. Guffarth, T. Warming, A. Hoffmann, and D. Bimberg

Correlation between Electronic Structure and Chemical Bond on the Surface of Hydrogenated
Silicon Nanocrystallites 691
    I. Umezu, T. Makino, M. Takata, M. Inada, and A. Sugimura

Interface Phonons of Quantum Dots in InAs/(Al,Ga)As Heteroepitaxial System:
A Raman Study 693
    D. A. Tenne, A. G. Milekhin, A. K. Bakarov, A. I. Toropov, G. Zanelatto, J. C. Galzerani, S. Schulze,
    and D. R. T. Zahn

Theoretical and Experimental Investigation of Biexcitons and Charged Excitons in InGaN Single Quantum Dots..695
    D. P. Williams, A. D. Andreev, E. P. O'Reilly, J. H. Rice, J. W. Robinson, A. Jarjour, J. D. Smith, R. A. Taylor, G. A. D. Briggs, Y. Arakawa, and S. Yasin

Path Integral Simulations of Charged Multiexcitons in InGaAs/GaAs Quantum Dots.................697
    M. Harowitz and J. Shumway

Single and Two Photon Emission from a Semiconductor Quantum Dot in an Optical Microcavity..699
    J. I. Perea, C. Tejedor, and D. Porras

Redistribution of Excitons Localized in InGaN Quantum Dot Structures............................701
    M. Dworzak, T. Bartel, M. Strassburg, A. Hoffmann, A. Strittmatter, and D. Bimberg

Enhancement of Spin Polarization in Asymmetric Double Quantum Dot Configurations Involving Diluted Magnetic Semiconductors...703
    S. Lee, M. Dobrowolska, and J. K. Furdyna

Exploitation of an Additional Infrared Laser to Modulate the Luminescence Intensity from InAs Quantum Dots..705
    E. S. Moskalenko, K. F. Karlsson, V. Donchev, P. O. Holtz, B. Monemar, W. V. Schoenfeld, and P. M. Petroff

Spectral Feature of Short Radiative Lifetime Quantum Dot..707
    E. Peter, P. Senellart, J. Hours, A. Cavanna, and J. Bloch

Spin-Polarized PL and Raman Spectroscopy of Nanocrystal Quantum Dots in High Magnetic Fields..709
    M. Furis, P. D. Robbins, T. Barrick, M. Petruska, V. I. Klimov, and S. A. Crooker

Photoluminescence Imaging of CdTe/ZnTe Self-Assembled Quantum Dots...........................711
    K. Hewaparakrama, A. J. Wilson, S. Mackowski, H. E. Jackson, L. M. Smith, G. Karczewski, and J. Kossut

Asymmetric Band Alignment of Si/Ge Quantum Dots Studied by Luminescence of p-i-n and n-i-p Structures...713
    M. Larsson, P. O. Holtz, A. Elfving, G. V. Hansson, and W. X. Ni

Polaronic Effects in Optical Transitions of Single InAs/AlAs Quantum Dots.......................715
    D. Sarkar, H. P. van der Meulen, J. M. Calleja, J. M. Becker, R. J. Haug, and K. Pierz

Control of the Anisotropic Exchange Splitting of Individual InAs/GaAs Quantum Dots with an In-Plane Electric Field..717
    K. Kowalik, O. Krebs, A. Lemaître, P. Senellart, J. Gaj, and P. Voisin

Origin of below Band-Gap Photoluminescence from GaN Quantum Dots in AlN Matrix.............719
    K. S. Zhuravlev, D. D. Ree, V. G. Mansurov, A. Y. Nikitin, M. Teisseire, N. Grandjean, G. Neu, and P. Tronc

Direct Comparison of Biexciton Binding Energy in a Quantum Well and Quantum Dots............721
    M. Ikezawa, Y. Masumoto, and H.-W. Ren

Observation of the Light-Hole Quantum Dots in a Strained GaAs Quantum Well...................723
    W. Maruyama, Y. Masumoto, and H.-W. Ren

Optical Properties of Semiconductor Quantum Dots and Pillar Microcavities.......................725
    N. Baer, J. Wiersig, P. Gartner, F. Jahnke, M. Benyoucef, S. M. Ulrich, P. Michler, and A. Forchel

Optical Study of Spatially Ordered InAs Quantum Dots in Disk-like Structures...................727
    Z. Xie, F. Wei, H. Cao, and G. S. Solomon

Tunneling Induced Dephasing and Pauli Blocking in InP Quantum Dots..............................729
    Y. Masumoto, F. Suto, M. Ikezawa, C. Uchiyama, and M. Aihara

Bleaching Dynamics in InAs Self-Assembled Quantum Dots...731
    E. W. Bogaart, J. E. M. Haverkort, T. Mano, R. Nötzel, and J. H. Wolter

Emission from Neutral and Charged Excitons in Self-Organized InAs Quantum Dots: Band Bending vs Pauli Blocking..733
    S. Lüttjohann, C. Meier, A. Lorke, D. Reuter, and A. Wieck

Molecular Like States in Coupled Magnetic Quantum Dots...735
    S. Souma, S. J. Lee, T. W. Kang, G. Ihm, and K. J. Chang

Electron-Hole Coulomb Interaction in the Stacks of ZnSe/CdSe Quantum Dots..................737
    T. Tchelidze and T. Kereselidze

Forbidden Transitions in the Emission Spectrum of Charged Excitons in a Single
Semiconductor Quantum Dot..................739
    M. Ediger, R. J. Warburton, K. Karrai, B. D. Gerardot, and P. M. Petroff

Characterization of g-Factors in Various In(Ga)As Quantum Dots..................741
    T. Nakaoka, T. Saito, H.-Z. Song, T. Usuki, and Y. Arakawa

Ultranarrow Photoluminescence Line in 1.3-1.55 μm of Single InAs/InP Quantum Dots..................743
    K. Takemoto, Y. Sakuma, S. Hirose, T. Usuki, N. Yokoyama, T. Miyazawa, M. Takatsu, and Y. Arakawa

Strain Effects on the Electronic and Optical Properties of InAs/GaAs Quantum Dots:
Tight-Binding Study..................745
    R. Santoprete, B. Koiller, R. B. Capaz, P. Kratzer, and M. Scheffler

Resonant Raman Scattering in Spherical InP QDs: The Role of the Optical
Deformation Potential Interaction..................747
    A. G. Rolo, M. I. Vasilevskiy, D. V. Talapin, and A. L. Rogach

Intraband Absorption in InAs/GaAs Self-Assembled Quantum Dots..................749
    J.-Z. Zhang and I. Galbraith

Spin-Dependent Coupling of Charged Excitons in Quantum Dots with Continuum States..................751
    B. Urbaszek, R. J. Warburton, E. J. McGhee, M. Ediger, K. Karrai, C. Schulhauser, A. Högele, X. Marie,
    A. O. Govorov, B. D. Gerardot, P. M. Petroff, and T. Amand

Anisotropy of Absorption and Luminescence of Multilayer InAs/GaAs Quantum Dots..................753
    J. Humlíček, V. Křápek, and J. Fikar

Features of Multilayer CdTe/ZnTe Quantum Dot Structures: Optical Studies..................755
    V. S. Bagaev and E. E. Onishchenko

Coherent Optics of Spherical Photonic Dots: Transition between Weak and Strong Coupling..................757
    A. Smith, N. I. Nikolaev, and A. L. Ivanov

Stark Shift and Permanent Dipole Moment of Vertically Confined Excitons in InAs/GaAs
Ring-like Quantum Dots..................759
    M. H. Degani, J. A. K. Freire, V. N. Freire, and G. A. Farias

Dynamics of Quantum Dot Clusters and Quantum State Monitoring..................761
    A. N. Al-Ahmadi and S. E. Ulloa

Anisotropic Exchange Interaction in Coupled Semiconductor Quantum Dots..................763
    Ş. C. Bădescu, Y. Lyanda-Geller, and T. L. Reinecke

Thermally Activated Carriers Transfer Process for the Success of InGaN/GaN Multi-Quantum
Well Light Emitting Devices..................765
    C. L. Yang, H. Liang, L. S. Yu, Y. D. Qi, D. L. Wang, Z. D. Lu, K. M. Lau, L. Ding, J. N. Wang,
    K. K. Fung, and W. K. Ge

Many-Particle States in Single InGaN/GaN Quantum Dots Grown on Si-Substrates..................767
    R. Seguin, S. Rodt, M. Winkelnkemper, A. Schliwa, A. Strittmatter, L. Reißmann, D. Bimberg, E. Hahn,
    and D. Gerthsen

Onion-like Growth of and Inverted Many-Particle Energies in Quantum Dots..................769
    A. Schliwa, S. Rodt, K. Pötschke, F. Guffarth, and D. Bimberg

Short Radiative Lifetime of Single GaAs Quantum Dots..................771
    J. Hours, P. Senellart, A. Cavanna, E. Peter, J. M. Gérard, and J. Bloch

3.1.3. Transport in Quantum Dots

Interference through a Single Quantum Dot..................773
    H. Aikawa, K. Kobayashi, A. Sano, S. Katsumoto, and Y. Iye

Time Resolved Single Electron Detection in a Quantum Dot..................775
    R. Schleser, E. Ruh, T. Ihn, K. Ensslin, D. D. Driscoll, and A. C. Gossard

Shot Noise in Tunneling through a Single InAs Quantum Dot..................777
    F. Hohls, A. Nauen, N. Maire, K. Pierz, and R. J. Haug

Imaging Electrons in a Single-Electron Quantum Dot..................779
    P. Fallahi, A. C. Bleszynski, R. M. Westervelt, J. Huang, J. D. Walls, E. J. Heller, M. Hanson,
    and A. C. Gossard

Spatially Resolved Manipulation of Single Electrons in Quantum Dots Using a Scanned Probe 781
    A. Pioda, S. Kičin, T. Ihn, M. Sigrist, A. Fuhrer, K. Ensslin, A. Weichselbaum, S. E. Ulloa, M. Reinwald, and W. Wegscheider

Negative Coulomb Drag in Parallel Quantum Wires Having Coupled Two Dots 783
    M. Honda, M. Yamamoto, M. Stopa, and S. Tarucha

Tunnel-Coupling Blockade in Vertical/Lateral Hybrid Dot to Study Many-Body States for Electron Number N=1, 2 and 3 785
    K. Yamada, M. Stopa, Y. Tokura, T. Hatano, T. Ota, T. Yamaguchi, and S. Tarucha

Theoretical Description of the Electronic Coupling between a Wetting Layer and a QD Superlattice Plane 787
    C. Cornet, C. Platz, J. Even, P. Miska, C. Labbé, H. Folliot, and S. Loualiche

Capacitance Spectroscopy Study of InAs Quantum Dots and Dislocations in p-GaAs Matrix 789
    P. N. Brunkov, E. V. Monakhov, A. Y. Kuznetsov, A. A. Gutkin, A. V. Bobyl, Y. G. Musikhin, A. E. Zhukov, V. M. Ustinov, and S. G. Konnikov

Resonant Scattering of a Two-Dimensional Electron Gas by Quantum Dot Levels 791
    N. M. Sotomayor, G. M. Gusev, A. C. Seabra, A. A. Quivy, T. E. Lamas, and J. R. Leite

Transitions of Dimensional States in Quantum Dots Formed by Spherical Barrier 793
    S. J. Lee, T. W. Kang, S. Souma, S. K. Noh, and J. C. Woo

Magnetotransport in HgSe:Fe Quantum-Dots 795
    M. Bayir, A. Kirste, T. Tran-Anh, S. Hansel, and M. von Ortenberg

Electron-Phonon Interaction in Si Quantum Dots Interconnected with Thin Oxide Layers 797
    S. Uno, N. Mori, K. Nakazato, N. Koshida, and H. Mizuta

Calculation of Carrier Transport through Quantum Dot Molecules 799
    T. Zibold, M. Sabathil, D. Mamaluy, and P. Vogl

Magnetoresistance Experiments and Quasi-Classical Calculations Regarding Backscattering in Open Quantum Dots 801
    R. Brunner, R. Meisels, F. Kuchar, M. Elhassan, J. P. Bird, D. K. Ferry, and K. Ishibashi

Modeling of the Magnetization Behavior of Realistic Self-Organized InAs/GaAs Quantum Craters as Observed with Cross-Sectional STM 803
    V. M. Fomin, V. N. Gladilin, J. T. Devreese, P. Offermans, P. M. Koenraad, J. H. Wolter, J. M. García, and D. Granados

Dynamical Control of Electronic States in AC-Driven Quantum Dots 805
    C. Creffield and G. Platero

Field Effect Enhanced Carrier-Emission from InAs Quantum Dots 807
    S. Schulz, T. Zander, A. Schramm, C. Heyn, and W. Hansen

Level Anticrossings in Quantum Dots 809
    J. Könemann, E. Räsänen, M. J. Puska, R. M. Nieminen, and R. J. Haug

Multi-Terminal Transport through Semi-Conductor Quantum Dots 811
    R. Leturcq, D. Graf, T. Ihn, K. Ensslin, D. D. Driscoll, and A. C. Gossard

Population Inversion and Coherent Phonon Emission in a Biased Quantum Dot System Placed in an Acoustic Cavity 813
    L. Mourokh, A. Smirnov, and A. Govorov

Anisotropy of Zeeman-Splitting in Quantum Dots 815
    J. Könemann, V. I. Fal'ko, D. K. Maude, and R. J. Haug

Propagation of Electron Waves in InAs/AlGaSb 817
    T. Maemoto, M. Koyama, A. Nakashima, M. Nakai, S. Sasa, J. P. Bird, and M. Inoue

HEMT Amplified SET Measurements of Individual InGaAs Quantum Dots 819
    K. D. Osborn, M. W. Keller, and R. P. Mirin

SU(4) Kondo Effect in Quantum Dots with Two Orbitals and Spin 1/2 821
    M. Eto

Fractal Study of Coupling Transitions in Ballistic Quantum-Dot Arrays 823
    T. P. Martin, R. P. Taylor, H. Linke, B. Murray, C. Arndt, N. Aoki, D. Oonishi, Y. Iwase, and Y. Ochiai

Imaging of Local Tunneling Barrier Height of InAs Nanostructures Using Low-Temperature Scanning Tunneling Microscopy 825
    K. Kanisawa, H. Yamaguchi, and Y. Hirayama

Resonance Backscattering in Triangular Quantum Dots Inside a Small Ring Interferometer 827
    D. V. Nomokonov, A. K. Bakarov, A. A. Bykov, and A. M. Mishchenko

Experimental Study on Far-Infrared Photoconductivity and Magnetoresistance in InAs Single Quantum Wire 829
    C. Zehnder, S. Löhr, A. Wirthmann, Y. S. Gui, C. Heyn, W. Hansen, D. Heitmann, and C.-M. Hu

3.1.4. Nanocrystals

Nanostructures from ZnO and other Semiconductors Generated via Self-Organizing Polymers 831
    K. Thonke

Amorphous Phase Emergence of Polar Nano-Crystalline Semiconductors 835
    S.-L. Zhang, S.-N. Wu, J. Shao, and Y. Song

Vapor-Phase Growth and Characterization of Luminescent Silicon Layers 837
    E. A. de Vasconcelos, J. B. da Silva, Jr., B. E. C. A. dos Santos, E. F. da Silva, Jr., W. M. de Azevedo, J. A. K. Freire, J. A. N. T. Soares, and J. R. Leite

Synthesis and Characterization of Transition Metal Ion Doped ZnS and ZnO Nanostructures 839
    S. P. Singh, O. Perales, M. S. Tomar, and O. V. Mata

The Crystalline Volume Fraction Dependence of Anisotropic Electrical Transport in nc-Si Thin Films—Theoretical and Experimental Studies 841
    F. Liu, M. Zhu, Q. Wang, and Y. Han

Crystalline and Amorphous Dualism of Polar Nano-Semiconductors 843
    S.-L. Zhang, L. S. Hoi, Y. Yan, J. Shao, X. Lu, and H.-D. Li

Strong Sub-Bandgap Absorption in GaSb/ErSb Nanocomposites Attributed to Plasma Resonances of Semimetallic ErSb Nanoparticles 845
    M. P. Hanson, D. C. Driscoll, E. R. Brown, and A. C. Gossard

Optical Phonon Modes of InP/II-VI Core-Shell Nanoparticles: A Raman and Infrared Study 847
    F. S. Manciu, R. E. Tallman, B. A. Weinstein, B. D. McCombe, D. W. Lucey, Y. Sahoo, and P. N. Prasad

Optical Resonances of Single Zinc Oxide Microcrystals 849
    T. Nobis, E. M. Kaidashev, A. Rahm, M. Lorenz, J. Lenzner, and M. Grundmann

A New Microscopic Theory of Low Frequency Raman Modes in Ge Nanocrystals 851
    W. Cheng, S.-F. Ren, and P. Y. Yu

Anomalous Stark Effect in Intraband Absorption of Silicon Nanocrystals 853
    J. S. de Sousa, J.-P. Leburton, V. N. Freire, and E. F. da Silva, Jr.

Correlation between PL Emission Band and Growth of Oxide Layer on Surface of Silicon Nanocrystallites 855
    I. Umezu, M. Koyama, T. Hasegawa, K. Matsumoto, M. Inada, and A. Sugimura

Size Dependence of the Optical Gap in Silicon Nanocrystals Embedded into a-Si:H Matrix 857
    V. A. Burdov, M. F. Cerqueira, M. I. Vasilevskiy, and A. M. Satanin

Ab-Initio Calculations of the Electronic Properties of Silicon Nanocrystals: Absorption, Emission, Stokes Shift 859
    E. Degoli, G. Cantele, E. Luppi, R. Magri, S. Ossicini, D. Ninno, O. Bisi, G. Onida, M. Gatti, A. Incze, O. Pulci, and R. Del Sole

Preparation of Surface Controlled Silicon Nanocrystallites by Pulsed Laser Ablation 861
    I. Umezu, M. Inada, T. Makino, and A. Sugimura

Light Emission from GaN Microcrystals 863
    R. Garcia, A. Bell, A. C. Thomas, and F. A. Ponce

Quantum Dot Emission from Selectively-Grown InGaN/GaN Micropyramid Arrays 865
    R. A. Taylor, J. H. Rice, J. H. Na, and J. W. Robinson

Considerable Enhancement of PL Intensity for Free Exciton in ZnO Ultra-Fine-Particles and Oscillatory Green Band Emergence in PL Spectrum for Cu-Doped ZnO Ultra-Fine-Particles with Heat Treatment 867
    H. Kobori, T. Nanao, S. Kawaguchi, I. Umezu, and A. Sugimura

3.2. Nanowires and Quantum Wires

3.2.1. Structural, Electronic, Optical, and Vibrational Properties of Quantum Wires

Charged Excitons in Modulation-Doped Quantum Wires 869
    T. Otterburg, D. Y. Oberli, M. Dupertuis, N. Moret, A. Malko, E. Pelucchi, B. Dwir, and E. Kapon

Influence of Aspect Ratio on the Lowest States of Quantum Rods 871
    M. Willatzen, B. Lassen, R. Melnik, and L. C. Lew Yan Voon

Barrier Localization in the Valence Band of Modulated Nanowires 873
    L. C. Lew Yan Voon, R. Melnik, B. Lassen, and M. Willatzen

High-Pressure Pulsed Laser Deposition and Structural Characterization of Zinc Oxide Nanowires 875
    A. Rahm, T. Nobis, E. M. Kaidashev, M. Lorenz, G. Wagner, J. Lenzner, and M. Grundmann

Catalyst-Free Growth of Semiconductor Nanowires by Selective Area MOVPE 877
    J. Motohisa, J. Noborisaka, S. Hara, M. Inari, and T. Fukui

Synthesis and Properties of ZnO Nano-Ribbon and Comb Structures 879
    Y. H. Leung, A. B. Djurišić, and M. H. Xie

The Fractional-Dimensional Excitonic Absorption Theory Applied to Real V-Groove Quantum Wires 881
    K. F. Karlsson, M.-A. Dupertuis, H. Weman, and E. Kapon

Mixing of Discrete and Continuum Excitations Induced by Nonperturbative Coulomb-Correlations 883
    V. M. Axt, J. Wühr, and T. Kuhn

Crossover from Excitons to an Electron-Hole Plasma in a High-Quality Single T-Shaped Quantum Wire 885
    M. Yoshita, Y. Hayamizu, H. Akiyama, L. N. Pfeiffer, and K. W. West

Photoluminescence Excitation Spectra of One-Dimensional Electron Systems in an N-type Doped Quantum Wire 887
    T. Ihara, Y. Hayamizu, M. Yoshita, H. Akiyama, L. N. Pfeiffer, and K. W. West

Green Photoluminescence in ZnO Nanostructures 889
    A. B. Djurišić, Y. H. Leung, Z. T. Liu, D. Li, M. H. Xie, W. C. H. Choy, and K. W. Cheah

Simple Ideas on Excitons in Quantum Wires 891
    M. Combescot and T. Guillet

Field-Effect Induced Mid-Infrared Intersubband Electroluminescence of Quantum Wire Cascade Structures 892
    S. Schmult, T. Herrle, H.-P. Tranitz, M. Reinwald, W. Wegscheider, M. Bichler, D. Schuh, and G. Abstreiter

Direct Observation of Excitonic Lasing from Single ZnO Nanobelts at Room Temperature 894
    K. Bando, T. Sawabe, K. Asaka, and Y. Masumoto

Temperature Dependent Dynamics of the Excitonic Photoluminescence Zinc in Oxide Nanorods 896
    H. Priller, R. Hauschild, J. Zeller, C. Klingshirn, H. Kalt, F. Reuss, R. Kling, C. Kirchner, and A. Waag

Purely Strain Induced GaAs/InAlAs Single Quantum Wires Exhibiting Strong Charge Carrier Confinement 898
    R. Schuster, H. Hajak, M. Reinwald, W. Wegscheider, D. Schuh, M. Bichler, S. Birner, P. Vogl, and G. Abstreiter

3.2.2. Transport in Quantum Wires

Negative Differential Conductance in Cleaved Edge Overgrown Surface Superlattices 900
    T. Feil, H.-P. Tranitz, M. Reinwald, W. Wegscheider, M. Bichler, D. Schuh, G. Abstreiter, and S. J. Allen

Electronic and Geometric Structure of One Dimensional Wires: An STM Study 902
    J. Lee, H. Kim, Y. J. Song, and Y. Kuk

Coupling of Two Localized Magnetic Moments and Its Detection 905
    V. I. Puller, L. G. Mourokh, A. Shailos, and J. P. Bird

Electron-Phonon Quantum Kinetics beyond the Second-Order Born Approximation 907
    H. Lohmeyer, V. M. Axt, and T. Kuhn

Real Space Observation of Anisotropic Scattering in the π-Chains on Si(111)-2×1 as a Quasi 1D System ... 909
    J. K. Garleff, M. Wenderoth, R. G. Ulbrich, C. Sürgers, and H. v. Löhneysen

Conductance of a Multiterminal Ballistic Wire ... 911
    Z. D. Kvon, V. A. Tkachenko, A. E. Plotnikov, V. A. Sablikov, V. Renard, and J. C. Portal

Dynamical Response in Resonance Interaction in Coupled Quantum Wires ... 913
    T. Morimoto, T. Sasaki, N. Aoki, Y. Ochiai, A. Shailos, J. P. Bird, M. P. Lilly, J. R. Reno, and J. A. Simmons

Towards a New Quantum Wire Structure Realizable by Double Cleaved-Edge Overgrowth: Characterizing the Transfer Potential ... 915
    S. F. Roth, M. Grayson, M. Bichler, D. Schuh, and G. Abstreiter

Ballistic to Diffuse Crossover in Long Quantum Wires ... 917
    J. A. Seamons, E. Bielejec, M. P. Lilly, J. L. Reno, and R. R. Du

Carrier Tunneling between Parallel GaAs/AlGaAs V-Groove Quantum Wires ... 919
    K. F. Karlsson, H. Weman, K. Leifer, A. Rudra, and E. Kapon

1D-1D Tunneling between Vertically Coupled GaAs/AlGaAs Quantum Wires ... 921
    E. S. Bielejec, J. A. Seamons, M. P. Lilly, and J. L. Reno

Magnetotransport Spectroscopy of Mode Coupling in Electron Wave Guides ... 923
    G. Apetrii, S. F. Fischer, U. Kunze, D. Schuh, and G. Abstreiter

Tunneling between Parallel Quantum Wires ... 925
    M. Yamamoto, Y. Tokura, Y. Hirayama, M. Stopa, K. Ono, and S. Tarucha

3.3. Quantum Wells and Superlattices

3.3.1. Structural, Electronic, Optical, and Vibrational Properties of Quantum Wells and Superlattices

Optical Spectroscopy on Single Localized States in an InGaN/GaN Structure ... 927
    S. Halm, G. Bacher, H. Schömig, A. Forchel, J. Off, and F. Scholz

The Effect of Inter-Well Correlations and Electric Field on Inhomogeneous Broadening of Excitons in Quantum Wells ... 929
    I. V. Ponomarev, L. I. Deych, and A. A. Lisyansky

Low Temperature Photoluminescence of GaAs/GaInP Heterostructures Measured under Hydrostatic Pressure ... 931
    T. Kobayashi, A. Nagata, A. D. Prins, Y. Homma, K. Uchida, and J.-i. Nakahara

Red to Blue Excitonic Emission with Ultra-Thin Quantum Wells and Fractional Monolayer Quantum Dots of II-VI Semiconductors ... 933
    I. Hernández-Calderón, M. García-Rocha, and P. Díaz-Arencibia

Characterization of the Emitting States in Quantum Wells with Planar Nano-Islands by Polarization Spectroscopy ... 935
    A. Reznitsky, A. Klochikhin, S. Permogorov, L. Tenishev, S. Y. Verbin, H. Kalt, and C. Klingshirn

Imaging of the Electric Fields and Charge Associated with Modulation-Doped 4H/3C/4H Polytypic Quantum Wells in SiC ... 937
    M. K. Mikhov, G. Samson, B. J. Skromme, R. Wang, C. Li, and I. Bhat

Study of InGaAs Band Structure for Nonparabolic Conduction Band and Parabolic Heavy-Hole Band Using InGaAs/InAlAs Multi-Quantum Well Structure ... 939
    N. Kotera and K. Tanaka

Wavefunction Engineering for GaN-Based Quantum Wells and Superlattices ... 941
    L. R. Ram-Mohan, A. M. Girgis, J. D. Albrecht, C. W. Litton, and T. D. Steiner

Blue-Light Emission from GaN/AlGaN Multiple Quantum Wells with an $Al_{0.5}Ga_{0.5}N$ Perturbation Monolayers Grown by Molecular Beam Epitaxy ... 943
    Y. S. Park, S. H. Lee, J. E. Oh, C. M. Park, and T. W. Kang

Theory of Raman Lasing due to Coupled Intersubband Plasmon-Phonon Modes in Asymmetric Coupled Double Quantum Wells ... 945
    S. M. Maung and S. Katayama

A^+-Centers and "Barrier-Spaced" A^0-Centers in Ge/GeSi MQW Heterostructures ... 947
    A. Ikonnikov, I. Erofeeva, D. Kozlov, O. Kuznetsov, V. Aleshkin, V. Gavrilenko, D. Veksler, and M. S. Shur

Intersubband Hole Cyclotron Resonance in Strained Ge/GeSi MQW Heterostructures................949
    A. Ikonnikov, I. Erofeeva, D. Kozlov, O. Kuznetsov, V. Aleshkin, V. Gavrilenko, D. Veksler,
    and M. S. Shur

Dielectric Properties of Ultra-Thin Films....................951
    J. Nakamura, S. Ishihara, H. Ozawa, and A. Natori

Localized and Extended States in Semiconductor Quantum Wells with Wire-like
Interface Islands....................953
    N. Shtinkov, P. Desjardins, and R. A. Masut

Photoluminescent and Kinetic Properties of A(+) Centers in Quantum Well....................955
    K. S. Romanov, N. V. Agrinskaja, N. S. Averkiev, Y. L. Ivanov, P. V. Petrov, and V. M. Ustinov

How do Electrons, Excitons, and Trions Share the Reciprocal Space in a Quantum Well?....................957
    M. T. Portella-Oberli, J. H. Berney, V. Ciulin, M. Kutrowski, T. Wojtowicz, and B. Deveaud

Effect of the Electron and Hole Scattering Potentials Compensation on Optical Band Edge of
Heavily Doped GaAs/AlGaAs Superlattices....................959
    Y. A. Pusep, F. E. G. Guimarães, M. B. Ribeiro, H. Arakaki, C. A. de Souza, S. Malzer, and G. H. Döhler

Polarization Dependence of Photosensitivity of the Schottky Barrier Diodes Based on the
InGaAs/GaAs Quantum Well and Quantum Dot Structures....................961
    D. O. Filatov, I. A. Karpovich, V. Y. Demikhovskii, D. V. Khomitskiy, and V. V. Levichev

Temperature-Dependent Conduction Band Structure of GaNAs and GaInNAs/GaAs
Quantum Wells....................963
    M. Hetterich, A. Grau, T. Passow, A. Y. Egorov, and H. Riechert

Photoluminescence from Ion Implanted and Low-Power-Laser Annealed GaAs/AlGaAs
Quantum Wells....................965
    A. Sonkusare, D. Sands, S. I. Rybchenko, and I. E. Itskevich

Vertical Correlation of Highly Organized Exciton Complex at ZnSe/BeTe Type-II
Asymmetric Superlattices....................967
    A. Fujikawa, H. Mino, K. Oto, R. Akimoto, and S. Takeyama

New Efficient Approach to Calculation Exciton Resonance Position and Width for
Quantum-Confined Stark Effect in Shallow Quantum Wells....................969
    I. V. Ponomarev, L. I. Deych, and A. A. Lisyansky

Exciton Complexes in ZnSe/BeTe Type-II Single Quantum Wells....................971
    H. Yamamoto, Z. Ji, H. Mino, R. Akimoto, and S. Takeyama

Singlet and Triplet Trion States in QW Structures....................973
    D. Andronikov, V. Kochereshko, A. Platonov, S. A. Crooker, T. Barrick, and G. Karczewski

Quantum Theory of Spatially Resolved Photoluminescence in Semiconductor Quantum Wells....................975
    G. Pistone, S. Savasta, O. Di Stefano, and R. Girlanda

Determination of InAsP/InP and InGaAs/InP Band Offsets Using Blue Shifting Type II
Asymmetric Multiple Quantum Wells....................977
    A. C. H. Lim, R. Gupta, S. K. Haywood, P. N. Stavrinou, M. Hopkinson, and G. Hill

Van Hove Singularities Detected by Photoluminescence in Doped AlGaAs/GaAs Superlattices....................979
    A. B. Henriques, R. F. Oliveira, T. E. Lamas, and A. A. Quivy

Superradiance of Excitons in Multi-Quantum-Well Structures Based on InAs-GaSb Coupled
Quantum Wells....................981
    B. Laikhtman and L. D. Shvartsman

Excitons and Trions in Heavily Doped QWs at High Magnetic Fields....................983
    V. Kochereshko, D. Andronikov, G. Karczewski, and S. A. Crooker

Investigation of Carrier Recombination Processes and Transport Properties in GaInAsN/GaAs
Quantum Wells....................985
    R. Fehse, S. J. Sweeney, A. R. Adams, E. P. O'Reilly, D. McConville, H. Riechert, and L. Geelhaar

Inhibition of Exciton Formation in Iron Doped InGaAs/InP Multiple Quantum Wells....................987
    M. Guezo, S. Loualiche, J. Even, A. LeCorre, O. Dehaese, and C. Labbe

Optical Spectroscopy of Polytypic Quantum Wells in SiC....................989
    G. Samson, L. Chen, B. J. Skromme, R. Wang, C. Li, and I. Bhat

Excitonic Properties of ZnO Films and Nanorods....................991
    A. A. Toropov, O. Nekrutkina, T. V. Shubina, S. V. Ivanov, T. Gruber, R. Kling, F. Reuss, C. Kirchner,
    A. Waag, K. F. Karlsson, J. P. Bergman, and B. Monemar

3.3.2. Transport in Quantum Wells and Superlattices

Observation of Current Resonances due to Enhanced Electron Transport through Stochastic
Webs in Superlattices 993
    S. Bujkiewicz, D. Fowler, T. M. Fromhold, A. Patanè, L. Eaves, A. A. Krokhin, P. B. Wilkinson,
    S. P. Stapleton, D. Hardwick, M. Henini, and F. W. Sheard

Mobile Potential Dots in GaAs Quantum Wells 995
    J. A. H. Stotz, T. Sogawa, F. Alsina, H. Hey, and P. V. Santos

Tuning of Transmission Function and Tunneling Time in Superlattices 997
    C. Pacher, U. Merc, and E. Gornik

Pekar Mechanism of Electron-Phonon Interaction in Nanostructures................................... 999
    B. A. Glavin, V. A. Kochelap, T. L. Linnik, K. W. Kim, and V. N. Sokolov

Electronic Band Structure and New Magneto-Transport Properties in p-Type Semiconductor
Medium-Infrared HgTe/CdTe Superlattice.. 1001
    Ab. Nafidi, A. El Abidi, A. El Kaaouachi, and Ah. Nafidi

Spin Effects in Magnetotransport of an n-$In_xGa_{1-x}As$/GaAs Double Quantum Well under
Parallel Magnetic Fields 1003
    M. V. Yakunin, G. A. Alshanskii, Y. G. Arapov, G. I. Harus, V. N. Neverov, N. G. Shelushinina,
    B. N. Zvonkov, E. A. Uskova, A. de Visser, and L. Ponomarenko

Z-Shaped Current-Voltage Characteristics of Disordered Semiconductor Superlattices....... 1005
    O. Pupysheva, A. Dmitriev, and A. Kozhanov

Quantum Interference and Localization in Disordered GaAs/AlGaAs Superlattices............. 1007
    Y. A. Pusep

Individual Band Mobilities in a Double Quantum Well... 1009
    R. Fletcher, M. Tsaousidou, T. Smith, P. T. Coleridge, Z. R. Wasilewski, and Y. Feng

An External ac Bias Induced Expansion of Dynamic Voltage Bands in a Weakly Coupled
GaAs/AlAs Superlattice... 1011
    H. T. He, Z. Z. Sun, X. R. Wang, Y. Q. Wang, W. K. Ge, and J. N. Wang

A New Class of Small Low-Field Magnetoresistance Oscillation in Unidirectional
Lateral Superlattice.. 1013
    A. Endo and Y. Iye

Interlevel Crossings in Double Period Superlattices.. 1015
    M. Coquelin, C. Pacher, M. Kast, G. Strasser, and E. Gornik

Spatiotemporal Dynamics of Carriers in Quantum Wells Modulated by Surface
Acoustic Waves.. 1017
    A. García-Cristóbal, A. Cantarero, and P. V. Santos

High Field Magnetoresistance of Strongly Coupled InAs/GaSb Superlattices 1019
    R. S. Deacon, A. B. Henriques, R. J. Nicholas, and P. Shields

3.4. Nanotubes

Magnetoresistance over 1000 % in CoFe/Carbon Nanotube/CoFe Junctions........................ 1021
    Y. Ishiwata, H. Maki, D. Tsuya, M. Suzuki, and K. Ishibashi

Laser-Resonance Chirality Selection in Single-Walled Carbon Nanotubes.......................... 1023
    K. Maehashi, Y. Ohno, K. Inoue, and K. Matsumoto

A Symmetrized-Basis Approach to Excitons in Carbon Nanotubes 1025
    G. Bussi, E. Chang, A. Ruini, and E. Molinari

Chirality Assignment of Single-Walled Carbon Nanotubes with Strain................................ 1027
    L.-J. Li, R. J. Nicholas, R. S. Deacon, P. A. Shields, C.-Y. Chen, R. C. Darton, and S. C. Baker

Inter-Tube Transfer of Electrons in Various Double-Wall Carbon Nanotubes...................... 1029
    S. Uryu and T. Ando

First-Principles Band Offsets of Carbon Nanotubes with III-V Semiconductors................... 1031
    Y.-H. Kim, M. J. Heben, and S. B. Zhang

Electron Transport in Carbon Nanotubes Using Superconducting Electrodes...................... 1033
    S. Ishii, J.-F. Lin, E. S. Sadki, K. Kida, T. Sasaki, N. Aoki, S. Ooi, J. P. Bird, K. Hirata, and Y. Ochiai

The Transport Property at Cross-Junction of Multiwall Carbon Nanotubes......................... 1035
    N. Aoki, T. Mihara, M. Kida, K. Miyamoto, T. Sasaki, and Y. Ochiai

Quasiparticle Band Structure of Carbon Nanotubes 1037
    T. Miyake and S. Saito

Magneto-Optical Properties of Aligned Single-Walled Carbon Nanotubes 1039
    M. Ichida, H. Wakida, H. Kataura, Y. Achiba, and H. Ando

Transport Properties of Charge Carriers in Single-Walled Carbon Nanotubes by Flash-Photolysis Time-Resolved Microwave Conductivity Technique 1041
    Y. Ohno, K. Maehashi, K. Inoue, K. Matsumoto, A. Saeki, S. Seki, and S. Tagawa

Modification of the Band Gaps and Optical Properties of Single-Walled Carbon Nanotubes 1043
    J. G. Wiltshire, L. J. Li, A. N. Khlobystov, M. Glerup, P. Bernier, and R. J. Nicholas

A New Type of Superlattice Based on Carbon Nanotubes 1045
    O. V. Kibis, D. G. W. Parfitt, and M. E. Portnoi

Temperature and Hydrostatic Pressure Effects on the Band Gap of Semiconducting Carbon Nanotubes 1047
    R. B. Capaz, C. D. Spataru, P. Tangney, M. L. Cohen, and S. G. Louie

Ensemble Monte Carlo Transport Simulations for Semiconducting Carbon Nanotubes 1049
    A. Verma, M. Z. Kauser, B. W. Lee, K. F. Brennan, and P. P. Ruden

Growth and Characterization of SnO_2 Micro- and Nanotubes 1051
    D. Maestre, A. Cremades, and J. Piqueras

Quantum Interference in Cross-Linked Carbon Nanotube 1053
    M. Kida, S. Harada, T. Mihara, T. Morimoto, T. Sasaki, N. Aoki, and Y. Ochiai

Effect of Interwall Interaction on the Transport Properties of Multi-Wall Carbon Nanotubes 1055
    T. Matsumoto and S. Saito

Systematic First-Principles Electronic-Structure Study of Carbon Nanotubes 1057
    Y. Akai and S. Saito

Effects of Symmetry Breaking on Perfect Channel in Metallic Carbon Nanotubes 1059
    T. Ando and K. Akimoto

Excitonic Effects and Optical Spectra of Single-Walled Carbon Nanotubes 1061
    C. D. Spataru, S. Ismail-Beigi, L. X. Benedict, and S. G. Louie

3.5. Molecular Systems and Organic Semiconductors

Magneto-Transport Studies of Fe/Alq_3/Co Organic Spin-Valves 1063
    F. J. Wang, Z. H. Xiong, D. Wu, J. Shi, and Z. V. Vardeny

Saturated Carboxylic Acids on Silicon: A First-Principles Study 1067
    C. Cucinotta, A. Ruini, M. J. Caldas, and E. Molinari

Interchain Effects on the Vibrational Properties of PPP and PPV 1069
    R. L. de Sousa and H. W. Leite Alves

Fabrication of Atomic-Scale Gold Junctions by Electrochemical Plating Technique Using a Common Medical Disinfectant 1071
    A. Umeno and K. Hirakawa

Room Temperature Polariton Photoluminescence in a Two-Dimensional Array of Inorganic-Organic Hybrid-Type Quantum-Wells 1073
    J. Ishi-Hayase and T. Ishihara

Optical Absorption, Photoluminescence, and Photoconductivity of Organic-Inorganic One-Dimensional Semiconductors $C_5H_{10}NH_2PbI_3$ and $[NH_2C(I)=NH_2]_3PbI_5$ 1075
    K. Matsuishi, Y. Kubo, T. Ichikawa, and S. Onari

Linear and Nonlinear Optical Spectroscopies of PPE/PPV Copolymer Semiconductors 1077
    M. H. Tong, Z. V. Vardeny, and Y. Pang

Raman Scattering from Organic Light Emitting Diodes 1079
    S. Guha, M. Arif, J. G. Keeth, T. W. Kehl, K. Ghosh, and R. E. Giedd

Infrared Ultrafast Optical Probes of Photoexcitations in π-Conjugated Organic Semiconductors 1081
    C. X. Sheng and Z. V. Vardeny

Nano-Scale Organic FET Fabricated with Carbon Nanotubes 1083
    K. Horiuchi, T. Kato, M. Mochizuki, S. Hashii, A. Hashimoto, T. Sasaki, N. Aoki, and Y. Ochiai

Light-Emitting Polymers: A First-Principles Analysis of Singlet-Exciton Harvesting in PPV 1085
    M. J. Caldas, G. Bussi, A. Ruini, and E. Molinari

Probing Nanoscale Pentacene Films by Resonant Raman Scattering 1087
R. He, I. Dujovne, L. Chen, Q. Miao, C. F. Hirjibehedin, A. Pinczuk, C. Nuckolls, C. Kloc, and G. B. Blanchet

Control of Amino-Acid Electronic Structures on Semiconductor Surfaces 1089
M. Oda and T. Nakayama

Transport Measurements of DNA Molecules by Using Carbon Nanotube Nanoelectrodes 1091
T. K. Sasaki, A. Ikegami, M. Mochizuki, N. Aoki, and Y. Ochiai

Electronic Transport in DNA—the Disorder Perspective 1093
D. K. Klotsa, R. A. Römer, and M. S. Turner

Molecular Signature in the Photoluminescence of α-Glycine, L-Alanine, and L-Asparagine Crystals: Detection, *ab initio* Calculations, and Bio-Sensor Applications 1095
E. W. S. Caetano, J. R. Pinheiro, M. Zimmer, V. N. Freire, G. D. Farias, G. A. Bezerra, B. S. Cavada, J. R. L. Fernandez, J. R. Leite, M. C. F. de Oliveira, J. A. Pinheiro, J. L. de Lima Filho, and H. W. Leite Alves

Absorption and Emission of Excitons in Thin PTCDA Films and PTCDA/Alq$_3$ Multilayers 1097
H. P. Wagner, A. DeSilva, V. R. Gangilenka, and T. U. Kampen

Optically Controlled Rotation of PTCDA Crystals in Optical Tweezers 1099
C. Starr, W. Dultz, H. P. Wagner, K. Dholakia, and H. Schmitzer

Ab-initio Theory of Charge Transport in Organic Crystals 1101
K. Hannewald and P. A. Bobbert

4. MICROCAVITIES

Acoustic Cavities 1105
A. Fainstein, P. Lacharmoise, and B. Jusserand

Manipulation of Photons and Electrons in Photonic Structures Using Surface Acoustic Waves 1109
P. V. Santos, M. M. de Lima, Jr., and R. Hey

Polarization of Light Emission in Semiconductor Microcavities: Dispersion Mapping 1113
Ł. Kłopotowski, A. Amo, M. D. Martín, L. Viña, and R. André

Optically Pumped Lasing from Localized States in Quantum-Well and Quantum-Dot Microdisks 1115
T. Kipp, K. Petter, C. Heyn, D. Heitmann, and C. Schüller

Dephasing in Cavity-Polariton Mediated Resonant Raman Scattering 1117
A. Bruchhausen, A. Fainstein, and B. Jusserand

5. ELECTRONIC AND INTERBAND TRANSITIONS

5.1. Band Structures

The Inverse Band Structure Approach: Find the Atomic Configuration that Has Desired Electronic Properties 1121
A. Zunger, S. V. Dudiy, K. Kim, and W. B. Jones

Energy-Band Structure of Si, Ge, and GaAs over the Whole Brillouin Zone via the k.p Method 1123
S. Richard, F. Aniel, and G. Fishman

Screened Exchange Calculations of Semiconductor Band Structures 1125
M. C. Gibson, S. J. Clark, S. Brand, and R. A. Abram

Material Design via Genetic Algorithms for Semiconductor Alloys and Superlattices 1127
K. Kim, P. A. Graf, and W. B. Jones

5.2. Intersubband and Interband Transitions

Intra-Impurity Transitions in Uniformly Iodine Doped MBE CdTe/CdMgTe Quantum Well—with No Energetic Scaling 1129
M. Szot, K. Karpierz, J. Kossut, and M. Grynberg

Radiative Processes in Layered Transition Metal Dichalcogenides........1131
L. Kulyuk, E. Bucher, L. Charron, D. Dumchenko, E. Fortin, and C. Gherman

How to Observe Charged Bosons in Quantum Wells?........1133
L. D. Shvartsman and D. A. Romanov

Intersubband Absorption from 2-7 μm Using Strain-Compensated Double-Barrier InGaAs Multiquantum Wells........1135
K. T. Lai, R. Gupta, M. Missous, and S. K. Haywood

Effects of ZnSe Interlayer on Properties of (CdS/ZnSe)/BeTe Type-II Super-Lattices Grown by Molecular Beam Epitaxy........1137
B. S. Li, R. Akimoto, K. Akita, and H. Hasama

Transparency Induced by Coupling of Intersubband Plasmons in a Quantum Well........1139
J. Li and C. Z. Ning

Nonlinear Intersubband Photoabsorption in Asymmetric Single Quantum Wells........1141
H. O. Wijewardane and C. A. Ullrich

Intersubband Tunneling without Intrasubband Relaxation in Multi-Quantum Wells........1143
G. S. Vieira, J. M. Villas-Bôas, P. S. S. Guimarães, N. Studart, J. Kono, S. J. Allen, K. L. Campman, and A. C. Gossard

Intersubband Lifetime Magnetophonon Oscillations in GaAs Quantum Cascade Lasers........1145
O. Drachenko, D. Smirnov, J. Léotin, A. Vasanelli, and C. Sirtori

Molecular Model for the Radiative Dipole Strengths and Lifetimes of the Fluorescent Levels of Mn^{2+} and Fe^{3+} in II-VI and III-V Compounds........1147
R. Parrot and D. Boulanger

5.3. Excitons and Condensates

On the Origin of Excitonic Luminescence in Quantum Wells: Direct Measure of the Exciton Formation in Quantum Wells from Time Resolved Interband Luminescence........1149
J. Szczytko, L. Kappei, J. Berney, F. Morier-Genoud, M. T. Portella-Oberli, and B. Deveaud

Why Interacting Excitons Cannot be Bosonized........1151
M. Combescot and O. Betbeder-Matibet

The Trion as an Exciton Interacting with a Carrier........1153
M. Combescot and O. Betbeder-Matibet

Resonant Coupling of an Excitonic State in a Quantum Disk with an Exciton-Polariton Mode Traveling through a Nearby Wave-Guide........1154
H. Takagi, H. Tanaka, M. Yamaguchi, and N. Sawaki

Excitonic Mott Transition in Spatially-Separated Electron-Hole Systems........1156
V. V. Nikolaev, M. E. Portnoi, and A. V. Kavokin

Investigations of Interface Excitons at p-Type GaAlAs/GaAs Single Heterojunctions in Continuous Wave and Time-Resolved Magneto Photoluminescence Experiments........1158
L. Bryja, M. Kubisa, K. Ryczko, J. Misiewicz, M. Kneip, M. Bayer, R. Stępniewski, M. Byszewski, M. Potemski, D. Reuter, and A. Wieck

Exciton Line Broadening due to Plasma Potential Fluctuations........1160
W. Bardyszewski, P. Kossacki, J. Cibert, and S. Tatarenko

Negatively Charged Excitons in a Back-Gated Undoped Heterostructure........1162
S. Nomura, M. Yamaguchi, D. Sato, T. Akazaki, H. Tamura, H. Takayanagi, T. Saku, and Y. Hirayama

Exciton Oscillator Strengths in Quantum Wells Containing a 2-D Electron Gas........1164
R. T. Cox, K. Kheng, R. B. Miller, V. Huard, C. Bourgognon, K. Saminadayar, and S. Tatarenko

The Interaction of Quantum Well Excitons with Evanescent EM Waves and the Spectroscopy of Waveguide Polaritons........1166
D. M. Beggs, M. A. Kaliteevski, S. Brand, R. A. Abram, and A. V. Kavokin

Interplay of Excitons, Biexcitons, and Charged Excitons in Pump-Probe Absorption Experiments on a (Cd,Mn)Te Quantum Well........1168
P. Płochocka, P. Kossacki, W. Maślana, C. Radzewicz, J. Cibert, S. Tatarenko, and J. A. Gaj

Free and Bound Exciton Dynamics in Bulk II-VI Semiconductors........1170
M. D. Martin, A. Amo, L. Viña, G. Karczewski, and J. Kossut

Ultrafast Dynamics of Neutral and Charged Triplet Magnetoexcitons in a High-Mobility Density-Tunable GaAs-AlGaAs Single Quantum Well........1172
P. Schröter, B. Su, D. Heitmann, C. Schüller, M. Reinwald, H.-P. Tranitz, and W. Wegscheider

Radiative Lifetime and Dephasing of Excitons Studied by Femtosecond Time Resolved
Intersubband Spectroscopy 1174
    I. Marderfeld, D. Gershoni, E. Ehrenfreund, and A. C. Gossard
Coherent Control of Bloch Oscillations by Means of Optical Pulse Shaping 1176
    R. Fanciulli, A. M. Weiner, M. M. Dignam, D. Meinhold, and K. Leo

5.4. High Frequency and Microwaves

Optical and Transport Properties of Devices Utilizing Nanoscale Deep-Centers 1178
    J. L. Pan
Intraband Carrier Dynamics in Semiconductor Optical Amplifier-Based Ultrafast Switch 1182
    A. Gomez-Iglesias, J. G. Fenn, M. Mazilu, A. Miller, and R. J. Manning
Terahertz Induced Photoconductivity of 2D Electron System in HEMT at Low Magnetic Field 1184
    A. Chebotarev and G. Chebotareva
Observation of the Quantum Correction Term for the Microwave Magnetoresistance in
Weakly Doped Ge 1186
    A. I. Veinger, A. G. Zabrodskii, T. V. Tisnek, and S. I. Goloshchapov
Microwave Time Resolved Cyclotron Resonance with Nanosecond Resolution 1188
    H. E. Porţeanu, O. Loginenko, and F. Koch
Generation and Remote Detection of Coherent Folded Acoustic Phonons 1190
    M. Trigo, T. A. Eckhause, J. K. Wahlstrand, R. Merlin, M. Reason, and R. S. Goldman
Direct Experimental Evidence of the Hole Capture by Resonant Levels in Boron
Doped Silicon 1192
    S. T. Yen, V. Tulupenko, E. S. Cheng, A. Dalakyan, C. P. Lee, K. A. Chao, V. Belykh, A. Abramov,
    and V. Ryzhkov
Two-Phonon Infrared Processes in Semiconductors 1194
    H. M. Lawler and E. L. Shirley
Time Resolved Cyclotron Resonance Studies in Semiconductors with Very Low and Very
High Mobility 1196
    H. E. Porţeanu, O. Loginenko, and F. Koch
THz Emitter based on InAs-GaSb Coupled Quantum Wells: New Prospects for THz Photonics 1198
    B. Laikhtman and L. D. Shvartsman
Theory of Heterostructural Tunnel Emitters for Ballistic Transit-Time
Terahertz-range Oscillators 1200
    Z. S. Gribnikov and G. I. Haddad
Submillimeter Radiation-Induced Persistent Photoconductivity in $Pb_{1-x}Sn_xTe(In)$ 1202
    A. Kozhanov, D. Dolzhenko, I. Ivanchik, D. Watson, and D. Khokhlov
THz/subTHz Detection by Asymmetrically-Shaped Bow-Tie Diodes Containing 2DEG Layer 1204
    D. Seliuta, V. Tamošiūnas, E. Širmulis, S. Ašmontas, A. Sužiedėlis, J. Gradauskas, G. Valušis, P. Steenson,
    W. Chow, P. Harrison, A. Lisauskas, H. G. Roskos, and K. Köhler
Terahertz Emitters based on Ion-implanted $In_{0.53}Ga_{0.47}As$ 1206
    M. Suzuki and M. Tonouchi
Low Frequency Noise Performance of Quantum Tunneling Sb-Heterostructure Millimeter
Wave Diodes 1208
    A. Luukanen, E. N. Grossman, H. P. Moyer, and J. N. Schulman
DC Field Response of Hot Carriers under Circularly Polarized Intense Microwave Fields
in Semiconductors 1210
    N. Ishida
Acoustoelectric Effects in Ge/Si Nanosystems with Ge Quantum Dots 1212
    I. L. Drichko, A. M. Diakonov, I. Y. Smirnov, Y. M. Galperin, A. I. Yakimov, and A. I. Nikiforov
Cyclotron Resonance Study of Doped and Undoped InAs/AlSb QW Heterostructures 1214
    A. V. Ikonnikov, V. Y. Aleshkin, V. I. Gavrilenko, Y. G. Sadofyev, J. P. Bird, S. R. Johnson, and Y. Zhang
Time Domain Terahertz Spectroscopy of the Magnetic Field Induced Metal-Insulator
Transition in n:InSb 1216
    J. Y. Sohn, X. P. A. Gao, and S. A. Crooker
Cyclotron Resonance Revisited: The Effect of Carrier Heating 1218
    H. Malissa, Z. Wilamowski, and W. Jantsch

The Nipnip-THz-Emitter: Photomixing based on Ballistic Transport 1220
    F. H. Renner, O. Klar, S. Malzer, D. Driscoll, M. Hanson, A. C. Gossard, G. Loata, T. Löffler, H. Roskos, and G. H. Döhler

Short Decay Times of the THz Photoresponse in Quantum Hall Corbino Detectors with
Spectral Tunability. 1222
    A. Hirsch, C. Stellmach, N. G. Kalugin, G. Hein, Y. Vasilyev, and G. Nachtwei

Acceleration Dynamics of Bloch Oscillating Electrons in Semiconductor Superlattices
Investigated by Terahertz Electro-optic Sampling Method 1224
    N. Sekine and K. Hirakawa

5.5. Light Scattering and Nonlinear Optics

Carrier and Coherent Lattice Dynamics of Si Probed with Ultrafast Spectroscopy 1226
    D. M. Riffe and A. J. Sabbah

Anharmonic Interactions in ZnO Probed with Impulsive Stimulated Raman Scattering. 1228
    C. Aku-Leh, J. Zhao, R. Merlin, J. Menéndez, and M. Cardona

Non-resonant Raman Efficiency in Semiconductors under High Pressure 1230
    C. Trallero-Giner, K. Kunc, and K. Syassen

Transient Four-Wave Mixing of Single Exciton States: Exciton-Exciton Interaction and
Rabi Oscillations 1232
    B. Patton, W. Langbein, and U. Woggon

Exchange Effects on Electronic States in QWs with e-h Plasma in an Electric Field 1234
    I. A. Fedorov, K. W. Kim, V. N. Sokolov, and J. M. Zavada

Rabi Oscillations of Ultrashort Pulses in 1.55-μm InGaAs/InGaAsP Quantum-Well
Optical Amplifiers 1236
    J.-Z. Zhang and I. Galbraith

Temperature Dependence of the Dephasing of Excitonic and Biexcitonic Polarization in a
ZnSe Single Quantum Well 1238
    T. Voss, L. Wischmeier, H. G. Breunig, I. Rückmann, and J. Gutowski

Microscopic Theory for Nonlinear Polariton Propagation 1240
    S. Schumacher, G. Czycholl, and F. Jahnke

Coherent Control of the Exciton-Biexciton System Demonstrated in
Four-Wave-Mixing Experiments 1242
    T. Voss, H. G. Breunig, I. Rückmann, J. Gutowski, V. M. Axt, and T. Kuhn

Exciton Associated Photorefractive Effect in ZnMgSe/ZnSe Quantum Wells. 1244
    H. P. Wagner, H.-P. Tranitz, and S. Tripathy

Coherent Photon Trapping in Doped Photonic Crystals and Dispersive
Semiconductor Materials 1246
    I. Haque and M. R. Singh

6. TRANSPORT

6.1. Carrier Dynamics and Magnetotransport

InAs-based Micromechanical Two-Dimensional Electron Systems 1251
    H. Yamaguchi, S. Miyashita, Y. Tokura, and Y. Hirayama

Shot Noise as a Probe of Electron Dynamics in Hopping and Resonant Tunnelling 1255
    S. S. Safonov, A. K. Savchenko, S. H. Roshko, D. A. Bagrets, O. N. Jouravlev, Y. V. Nazarov, E. H. Linfield, and D. A. Ritchie

Preserved Symmetries in Nonlinear Electric Conduction 1257
    C. A. Marlow, A. Löfgren, I. Shorubalko, R. P. Taylor, P. Omling, L. Samuelson, and H. Linke

Effects of Scale-Free Disorder on the Metal-Insulator Transition 1259
    M. L. Ndawana, R. A. Römer, and M. Schreiber

Non-perturbative Scattering of Electrons by Charged Dislocations. 1261
    L. B. Hovakimian

Picosecond Raman Studies of Electron and Hole Velocity Overshoots in a GaAs-based p-i-n Semiconductor Nanostructure 1263
    W. Liang, K. T. Tsen, C. Poweleit, J. M. Barker, D. K. Ferry, and H. Morkoc

Disorder-induced Non-Ohmic Steady-State Flow of Hopping Carriers 1265
    S. A. Baily and D. Emin

Alkali Metals Transport at High Temperatures in the Presence of an Electric Field 1267
    I. Rapoport, P. Taylor, V. Mart, J. Kearns, and F. Kirscht

A Theory of Low-Field, High-Carrier-Density Breakdown in Semiconductors 1269
    K. Kambour, H. P. Hjalmarson, and C. W. Myles

Quantum Hall Ferromagnetism in Magnetic Heterostuctures and Wires 1271
    J. Jaroszyński, T. Andrearczyk, E. A. Stringer, G. Karczewski, T. Wojtowicz, J. Wróbel, D. Popović, and T. Dietl

Imaging Transport: Monitoring the Motion of Charge through the Detection of Light 1273
    N. M. Haegel, V. D. Hoang, and W. Freeman

Calibrated Scanning Capacitance Microscopy for Two-Dimensional Carrier Mapping of n-Type Implants in p-Doped Si-Wafers 1275
    W. Brezna, B. Basnar, S. Golka, H. Enichlmair, and J. Smoliner

Evaporative Cooling of Electrons in Semiconductor Devices 1279
    T. Jayasekera, K. Mullen, and M. A. Morrison

Mesoscopic Phonon-Electric Effect 1281
    D. W. Horsell, A. K. Savchenko, Y. M. Galperin, V. I. Kozub, and V. M. Vinokur

Theory of Electric-Field-Induced Quantum Diffusion in Semiconductors 1283
    P. Kleinert and V. V. Bryksin

Quantum Electron Transport in Finite-Size Flat-Band Kagome Lattice Systems 1285
    H. Ishii and T. Nakayama

Spin-Orbit Interaction in InSb Thin Films Grown on GaAs(100) Substrates by MBE: Effect of Hetero-Interface 1287
    S. Ishida, K. Takeda, A. Okamoto, and I. Shibasaki

6.2. Spin Dynamics for Spintronics

Spin-Dependent Non-Equilibrium Transport in Mesoscopic 2D Electron Systems 1289
    A. Ghosh, M. H. Wright, K. DasGupta, M. Pepper, H. E. Beere, and D. A. Ritchie

Ferromagnetic/DMS Hybrid Structures: One- and Zero-Dimensional Magnetic Traps for Quasiparticles 1291
    P. Redliński, T. Wojtowicz, T. G. Rappoport, A. Libal, J. K. Furdyna, and B. Jankó

Controlling Hole Spin Relaxation in Charge Tunable InAs/GaAs Quantum Dots 1293
    S. Laurent, B. Eble, O. Krebs, A. Lemaître, B. Urbaszek, X. Marie, T. Amand, and P. Voisin

Whole Spectrum of the Spin Polarized Two Dimensional Electron Gas 1295
    F. Perez, B. Jusserand, and G. Karczewski

Side-Gate Control of Rashba Spin Splitting in a $In_{0.75}Ga_{0.25}As/In_{0.75}Al_{0.25}As$ Heterojunction Narrow Channel: Toward Spin-Transistor based Qubits 1297
    T. Kakegawa, M. Akabori, and S. Yamada

Microphotoluminescence Study of Disorder in a Ferromagnetic (Cd,Mn)Te Quantum Well 1299
    W. Maślana, P. Kossacki, P. Płochocka, A. Golnik, J. A. Gaj, D. Ferrand, M. Bertolini, S. Tatarenko, and J. Cibert

Spin Dynamics of Mn-ion System in Diluted-Magnetic-Semiconductor Heterostructures based on ZnMnSe 1301
    D. R. Yakovlev, M. Kneip, M. Bayer, A. A. Maksimov, I. I. Tartakovskii, A. V. Scherbakov, A. V. Akimov, D. Keller, W. Ossau, L. W. Molenkamp, and A. Waag

Charge Redistribution Spectroscopy as a Probe of Spin Phenomena in Quantum Dots 1303
    A. S. Sachrajda, M. Pioro-Ladrière, M. Ciorga, S. Studenikin, P. Zawadzki, P. Hawrylak, J. Lapointe, Z. R. Wasilewski, and J. A. Gupta

Bulk Inversion Asymmetry Spin-Splitting in L-Valley GaSb Quantum Wells 1307
    J.-M. Jancu, R. Scholz, G. C. La Rocca, E. A. de Andrada e Silva, and P. Voisin

Experimental Separation of Rashba and Dresselhaus Spin-Splittings 1309
    S. D. Ganichev, P. Schneider, S. Giglberger, W. Wegscheider, D. Weiss, W. Prettl, V. V. Bel'kov, L. E. Golub, and E. L. Ivchenko

Spectroscopy and Characteristic Energies of a Spin-Polarized Hole Gas 1311
    H. Boukari, P. Kossacki, D. Ferrand, J. Cibert, M. Bertolini, S. Tatarenko and J. A. Gaj

Observation of Coherent Hybrid Intersubband-Cyclotron Modes in a Quantum Well. 1313
    J. K. Wahlstrand, J. M. Bao, D. M. Wang, P. Jacobs, R. Merlin, K. W. West, and L. N. Pfeiffer

Room Temperature Spin Dependent Current Modulation in an InGaAs-based Spin Transistor
with Ferromagnetic Contact. 1315
    K. Yoh, M. Ferhat, A. Riposan, and J. M. Millunchick

Dynamic Nuclear Polarization In GaAs: Hot Electron and Spin Injection Mechanisms 1317
    M. J. R. Hoch, J. Lu, P. L. Kuhns, and W. G. Moulton

Investigation of a GaMnN/GaN/InGaN Structure for Spin LED 1319
    F. V. Kyrychenko, C. J. Stanton, C. R. Abernathy, S. J. Pearton, F. Ren, G. Thaler, R. Frazier,
    I. Buyanova, J. P. Bergman, and W. M. Chen

Ultrafast Exciton Spin Dynamics in $Cd_{1-x}Mn_xTe$ Quantum Wells. 1321
    K. Seo, K. Nishibayashi, Z. H. Chen, K. Kayanuma, A. Murayama, and Y. Oka

Control of Excitonic Motion in Modulation-Doped (Cd,Mn)Te QW by Magnetic and
Electric Fields. 1323
    F. Takano, T. Tokizaki, H. Akinaga, S. Kuroda, and K. Takita

Intrinsic & Phonon-Induced Spin Relaxation in Quantum Dots 1325
    C. F. Destefani, S. E. Ulloa, and G. E. Marques

Optically Induced Zero-Field Magnetization of CdMnTe Quantum Dots. 1327
    S. Mackowski, T. Gurung, H. E. Jackson, L. M. Smith, G. Karczewski, and J. Kossut

Control of Electron-Spin Precession in Quantum Well through the E Field Influence on the
Interface Asymmetry. 1329
    Y. G. Semenov and S. M. Ryabchenko

Theory of Electronic Structure and Magnetic Interactions in (Ga,Mn)N and (Ga,Mn)As. 1331
    P. Bogusławski and J. Bernholc

Electrical Properties of Ni/GaAs and Au/GaAs Schottky Contacts in High Magnetic Fields 1333
    H. von Wenckstern, R. Pickenhain, S. Weinhold, M. Ziese, P. Esquinazi, and M. Grundmann

Square-Wave Conductance through a Chain of Rings due to Spin-Orbit Interaction. 1335
    B. Molnár, P. Vasilopoulos, and F. Peeters

Magnetic Circular Dichroism in $ZnSe/Ga_{1-x}Mn_xAs$ Hybrid Structures with Be and
Si Co-Doping. 1337
    R. Chakarvorty, K. J. Yee, X. Liu, P. Redlinski, M. Kutrowski, L. V. Titova, T. Wojtowicz, J. K. Furdyna,
    B. Janko, and M. Dobrowolska

Extension of the Semiconductor Bloch Equations for Spin Dynamics 1339
    C. Lechner and U. Rössler

Self-Sustaining Resistance Oscillations by Electron-Nuclear Spin Coupling in Mesoscopic
Quantum Hall Systems 1341
    G. Yusa, K. Hashimoto, K. Muraki, T. Saku, and Y. Hirayama

Electron Spin Decoherence due to Hyperfine Coupling in Quantum Dots 1343
    L. M. Woods and T. L. Reinecke

Spin-Dependent Tunneling in III-V Semiconductors 1345
    S. Richard, H.-J. Drouhin, G. Fishman, and N. Rougemaille

Interplay between In-Plane Magnetic Fields and Spin-Orbit Coupling in InGaAs/InP 1347
    P. T. Coleridge, S. A. Studenikin, G. Yu, and P. J. Poole

Zeeman Measurement of Quantum Wire Array in High Magnetic Field Exhibiting Abrupt
Change at Elliptical Landau Orbit Formation 1349
    I. T. Jeong, T. S. Kim, S. Ahn, B. C. Lee, D. H. Kim, M. G. Sung, and J. C. Woo

Spin Relaxation in CdTe/ZnTe Quantum Dots 1351
    Y. Chen, T. Okuno, Y. Masumoto, Y. Terai, S. Kuroda, and K. Takita

Hole-Spin Reorientation in $(CdTe)_{0.5}(Cd_{0.75}Mn_{0.25}Te)_{0.5}$ Tilted Superlattices Grown on
$Cd_{0.74}Mg_{0.26}Te(001)$ Vicinal Surface 1353
    T. Kita, S. Nagahara, Y. Harada, O. Wada, L. Marsal, and H. Mariette

Towards Using Multiferroism in Optoelectronics and Spintronics: Tunnelling, Confinement,
and Optical Properties of $Si/BiMnO_3$ Systems. 1355
    L. M. Rebelo, F. F. Maia, Jr., E. W. S. Caetano, V. N. Freire, G. A. Farias, J. A. P. da Costa,
    and E. F. da Silva, Jr.

Exciton Spin Relaxation in Symmetric Self-Assembled Quantum Dots...........1357
    S. Mackowski, T. Gurung, H. E. Jackson, L. M. Smith, G. Karczewski, J. Kossut, M. Dobrowolska, and J. K. Furdyna

A Graphite-Diamond Hybrid Structure as a Half-Metallic Nano Wire...........1359
    K. Kusakabe and N. Suzuki

Exciton Spin Manipulation in InAs/GaAs Quantum Dots: Exchange Interaction and Magnetic Field Effects...........1361
    M. Sénès, B. Urbaszek, X. Marie, T. Amand, J. Tribollet, F. Bernardot, C. Testelin, M. Chamarro, and J. M. Gérard

Photo-Induced Ferromagnetism in Bulk-$Cd_{0.95}Mn_{0.05}Te$ via Exciton Magnetic Polarons...........1363
    Y. Hashimoto, H. Mino, T. Yamamuro, D. Kanbara, T. Matsusue, S. Takeyama, G. Karczewski, T. Wojtowicz, and J. Kossut

Ac Conductivity and Magneto-Optical Effects in the Metallic (III,Mn)V Ferromagnetic Semiconductors from the Infrared to Visible Range...........1365
    E. M. Hankiewicz, T. Jungwirth, T. Dietl, C. Timm, and J. Sinova

Carrier Concentration Dependencies of Magnetization & Transport in $Ga_{1-x}Mn_xAs_{1-y}Te_y$...........1367
    M. A. Scarpulla, K. M. Yu, W. Walukiewicz, and O. D. Dubon

Ferromagnetism and Carrier Polarization of Mn-doped II-IV-V_2 Chalcopyrites...........1369
    P. R. C. Kent and T. C. Schulthess

Search for Hole Mediated Ferromagnetism in Cubic (Ga,Mn)N...........1371
    M. Sawicki, T. Dietl, C. T. Foxon, S. V. Novikov, R. P. Campion, K. W. Edmonds, K. Y. Wang, A. D. Giddings, and B. L. Gallagher

Clear Spin Valve Signals in Conventional $NiFe/In_{0.75}Ga_{0.25}As$-2DEG Hybrid Two-Terminal Structures...........1373
    M. Akabori, K. Suzuki, and S. Yamada

Tunneling Magnetoresistance: The Relevance of Disorder at the Interface...........1375
    M. Wimmer and K. Richter

Comparison of Spin Injection and Transport in Organic and Inorganic Semiconductors...........1377
    P. P. Ruden, D. L. Smith, and J. D. Albrecht

Spin-Polarized and Ballistic Transport in InSb/InAlSb Heterostructures...........1379
    H. Chen, J. A. Peters, A. O. Govorov, J. J. Heremans, N. Goel, S. J. Chung, and M. B. Santos

High Spin Filtering under the Influence of In-Plane Magneto-Electric Field and Spin Orbit Coupling...........1381
    S. G. Tan, M. B. A. Jalil, K. L. Teo, T. Liew, and T. C. Chong

Spin Precession in a Model Structure for Spintronics...........1383
    M. Ghali, J. Kossut, E. Janik, F. Teppe, M. Vladimirova, and D. Scalbert

Observation of Spin Diffusion, Drift and Precession in Bulk n-GaAs Using a Spatially Resolved Steady-State Technique...........1385
    M. Beck, C. Metzner, S. Malzer, and G. H. Döhler

Optical Study of Spin Injection Dynamics in Double Quantum Wells of II-VI Diluted Magnetic Semiconductors...........1387
    K. Kayanuma, T. Tomita, A. Murayama, Y. Oka, A. A. Toropov, S. V. Ivanov, I. A. Buyanova, and W. M. Chen

Excitonic Spin Dynamics in Coupled Quantum Dots of Diluted Magnetic Semiconductors...........1389
    A. Murayama, A. Uetake, I. Souma, K. Kayanuma, T. Asahina, K. Hyomi, T. Tomita, and Y. Oka

Spin Relaxation Dynamics in Highly Uniform InAs Quantum Dots...........1391
    A. Tackeuchi, Y. Suzuki, M. Murayama, T. Kitamura, T. Kuroda, T. Takagahara, and K. Yamaguchi

Spin Transport and Spin Relaxation in Ge/Si Quantum Dots...........1393
    A. F. Zinovieva, A. V. Nenashev, and A. V. Dvurechenskii

Quantum Antidot as a Controllable Spin Injector and Spin Filter...........1395
    I. V. Zozoulenko and M. Evaldsson

Magnetic Field Induced Polarized Optical Absorption in Europium Chalcogenides...........1397
    A. B. Henriques, L. K. Hanamoto, E. Abramof, A. Y. Ueta, and P. H. O. Rappl

Spin Injection and Spin Loss in GaMnN/InGaN Light-Emitting Diodes...........1399
    I. A. Buyanova, M. Izadifard, W. M. Chen, J. Kim, F. Ren, G. Thaler, C. R. Abernathy, S. J. Pearton, C.-C. Pan, G.-T. Chen, J.-I. Chyi, and J. M. Zavada

Transient Faraday Rotation and Circular Dichroism Induced by Circularly-Polarized Light
in InGaN 1401
    T. Matsusue, D. Kanbara, H. Mino, M. Arita, and Y. Arakawa

Field Dependence of Spin Lifetimes in Nitride Heterostructures 1403
    J. A. Majewski and P. Vogl

Current-Induced Spin Polarization at a Single Heterojunction 1405
    A. Y. Silov, P. A. Blajnov, J. H. Wolter, R. Hey, K. H. Ploog, and N. S. Averkiev

Optical Studies of Spin Coherence in Organic Semiconductors 1407
    C. Yang, C. Liu, and Z. V. Vardeny

Twisted Exchange Interaction between Localized Spins in Presence of Rashba
Spin-Orbit Coupling 1409
    H. Imamura, P. Bruno, and Y. Utsumi

AC-Driven Double Quantum Dots as Spin Pumps 1411
    E. Cota, R. Aguado, and G. Platero

Spin Splitting in Open Quantum Dots 1413
    M. Evaldsson, I. V. Zozoulenko, M. Ciorga, P. Zawadzki, and A. S. Sachrajda

Square-Wave, Spin-Dependent Transmission through Periodically Stubbed
Electron Waveguides 1415
    X. F. Wang and P. Vasilopoulos

Optical Orientation of Electron and Nuclear Spins in Negatively Charged InP QDs 1417
    S. Y. Verbin, I. Y. Gerlovin, I. V. Ignatev, and Y. Masumoto

Ferromagnetism in Partially Spin-Polarized GaInAs Quantum Wells 1419
    F. Vogt, G. Nachtwei, G. Hein, and H. Künzel

Electrically Controlled Spin Device Concepts 1421
    D. Z.-Y. Ting, X. Cartoixà, and Y.-C. Chang

Strong Enhancement of Rashba Effect in Strained P-Type Quantum Wells 1423
    D. M. Gvozdić and U. Ekenberg

7. QUANTUM INFORMATION

Strongly Tunable Coupling between Quantum Dots 1427
    Y. B. Lyanda-Geller, G. Bacher, T. L. Reinecke, M. K. Welsch, A. Forchel, C. R. Becker,
    and L. Molenkamp

Optically-Generated Many Spin Entanglement in a Quantum Well 1429
    J. Bao, A. V. Bragas, J. K. Furdyna, and R. Merlin

Indirect Spin Coupling between Quantum Dots 1431
    G. Ramon, Y. Lyanda-Geller, T. L. Reinecke, and L. J. Sham

Ruderman-Kittel-Kasuya-Yosida Interaction in Quantum Dot Arrays 1433
    H. Tamura, K. Shiraishi, and H. Takayanagi

Tunable Electronic-Nuclear Dynamics and Current Instability in Double Quantum Dots 1435
    T. Inoshita, and S. Tarucha

Controlling and Measuring a Single Donor Electron in Silicon 1437
    K. R. Brown, L. Sun, B. Bryce, and B. E. Kane

Manipulation of Charge States of an Isolated Silicon Double Quantum Dot through
Microwave Irradiation 1439
    E. Emiroglu, D. Hasko, and D. Williams

Silicon-based Spin Quantum Computation and the Shallow Donor Exchange Gate 1441
    B. Koiller, R. B. Capaz, X. Hu, and S. Das Sarma

Coherent Control of Tunneling in a Quantum Dot Array 1445
    J. M. Villas-Bôas, S. E. Ulloa, and N. Studart

Fano Resonance in Quantum Dots with Electron-Phonon Interaction 1447
    A. Ueda and M. Eto

Electric Field Induced Charge Noise in Doped Silicon: Ionisation of Phosphorus Dopants 1449
    A. J. Ferguson, V. Chan, A. R. Hamilton, and R. G. Clark

Exciton Coherence Times and Linewidths in InGaAs Quantum Dots 1451
    S. Rudin and T. L. Reinecke

Anisotropic g-Factor Dependence of Dynamic Nuclear Polarization in n-GaAs/AlGaAs (110) Quantum Wells...1453
    S. Matsuzaka, H. Sanada, K. Morita, C. Y. Hu, Y. Ohno, and H. Ohno

Two-Dimensional Nuclear Magnetic Resonance in Optically Pumped Semiconductors 1455
    A. Patel and C. R. Bowers

Clean Thermal Processing at Elevated Temperatures 1457
    I. Rapoport, P. Taylor, B. Orschel, and J. Kearns

Decoherence Control of Excitons by a Sequence of Pulses 1459
    A. Hasegawa, T. Kishimoto, Y. Mitsumori, M. Sasaki, and F. Minami

Imaging Electron Interferometer...1461
    A. C. Bleszynski, K. E. Aidala, B. J. LeRoy, R. M. Westervelt, E. J. Heller, K. D. Maranowski, and A. C. Gossard

A New Concept on a Quantum Computer based on Schockley-Read-Hall Recombination Statistics in Microelectronic Devices...1463
    K. Theodoropoulos, D. Ntalaperas, I. Petras, A. Tsakalidis, and N. Konofaos

Quantum Information Processing Using Coulomb-Coupled Quantum Dots........................1465
    J. Danckwerts, J. Förstner, and A. Knorr

Decoherence of Charge Qubit Systems 1467
    A. Weichselbaum and S. E. Ulloa

Modelling of Open Quantum Devices within the Closed-System Paradigm 1469
    R. Proietti Zaccaria, R. C. Iotti, and F. Rossi

Nuclear Spin Polarizer for Solid-State NMR Quantum Computers......................................1471
    A. Goto, T. Shimizu, K. Hashi, S. Ohki, T. Iijima, S. Kato, H. Kitazawa, and G. Kido

Small Metallic Contacts in the System Metal/Barrier/Semiconductor as the Single-Electron Qubits...1473
    Z. S. Gribnikov and G. I. Haddad

Tuning Nanocrystal Properties for Quantum Information Processing Devices....................1475
    G. Medeiros-Ribeiro, E. Ribeiro, and H. Westfahl, Jr.

8. DEVICES

8.1. Electronic Devices

Remarkably Strong Image Potential Effects in SrTiO$_3$/Si and HfO$_2$/Si Tunneling Structures..........1481
    T. A. S. Pereira, J. A. K. Freire, J. S. Sousa, V. N. Freire, G. A. Farias, L. M. R. Scolfaro, J. R. Leite, and E. F. da Silva, Jr.

Electrical Properties of Modulation Doped Si/SiGe Heterostructures Grown on Silicon on Insulator Substrates...1483
    K. Alfaramawi, A. Sweyllam, L. Abulnasr, S. Abboudy, E. F. El-Wahidy, L. Di Gaspare, and F. Evangelisti

Transient Quantum Drift-Diffusion Modelling of Resonant Tunneling Heterostructure Nanodevices 1485
    N. Radulovic, M. Willatzen, and R. V. N. Melnik

The Role of Non-abrupt Interfaces in SiC MOS Devices: Quantum Mechanical Simulations and Experiments 1487
    E. L. de Oliveira, J. S. de Sousa, V. N. Freire, E. A. de Vasconcelos, and E. F. da Silva, Jr.

A Comprehensive Model for Low Frequency Noise in Poly-Si Thin-Film Transistors 1489
    I. K. Han, J. I. Lee, M. B. Lee, S. K. Chang, and A. Chovet

Hydrogen Cleaving of Silicon at the Sub-100-nm Scale...1491
    O. Moutanabbir, B. Terreault, N. Desrosiers, A. Giguère, G. G. Ross, M. Chicoine, and F. Schiettekatte

Quantised Vortex Flows and Conductance Fluctuations in High Temperature Atomistic Silicon MOSFET Devices....1493
    J. R. Barker

Raman and XRD Strain Analysis of 3D Bonded and Thinned SOI Wafers 1495
    S. Pozder, M. Canonico, S. Zollner, R. Liu, K. Yu, and J.-Q. Lu

The Impact of Soft-Optical Phonon Scattering due to High-*k* Dielectrics on the Performance of Sub-100nm Conventional and Strained Si *n*-MOSFETs 1497
    L. Yang, J. R. Watling, J. R. Barker, and A. Asenov

Quantization Conditions for Variable Operations of pHEMT 1499
    S. Mil'shtein, S. Gudimetta, and C. Gil

Design and Optimization of Vertical CEO-T-FETs with Atomically Precise Ultrashort Gates by Simulation with Quantum Transport Models 1501
    J. Höntschel, W. Klix, R. Stenzel, F. Ertl, and G. Abstreiter

Double Quantum Dot with Integrated Charge Readout Fabricated by Layered SFM-Lithography 1503
    M. Sigrist, A. Fuhrer, T. Ihn, K. Ensslin, D. D. Driscoll, and A. C. Gossard

Rad-Hard Silicon Detectors 1505
    M. Giorgi

Modeling, Fabrication and Test Results of a MOS Controlled Thyristor—MCT—with High Controllable Current Density. 1507
    E. Chernyavsky, V. Popov, and B. Vermeire

Triode with Heterostructure Filament 1509
    C. Gil and S. Mil'shtein

From Vacuum Tubes to a Semiconductor Triode 1511
    S. Mil'shtein

Steering of Quantum Waves: Demonstration of Y-Junction Transistors Using InAs Quantum Wires 1513
    G. M. Jones, J. Qin, C.-H. Yang, and M.-J. Yang

8.2. Photonic Devices

InGaN-based Nanorod Array Light Emitting Diodes 1515
    H.-M. Kim, Y. H. Cho, D. Y. Kim, T. W. Kang, and K. S. Chung

Spiral-Shaped Microcavity Laser: A New Class of Semiconductor Laser 1517
    M. Kneissl, M. Teepe, N. Miyashita, G. D. Chern, R. K. Chang, and N. M. Johnson

Room-Temperature Operation of a Green Monolithic II-VI Vertical-Cavity Surface-Emitting Laser 1521
    C. Kruse, K. Sebald, H. Lohmeyer, B. Brendemühl, R. Kröger, J. Gutowski, and D. Hommel

Terahertz Emission and Detection by Plasma Waves in Nanoscale Transistors 1523
    F. Teppe, J. Łusakowski, N. Dyakonova, Y. M. Meziani, W. Knap, T. Parenty, S. Bollaert, A. Cappy, V. Popov, F. Boeuf, T. Skotnicki, D. Maude, S. Rumyantsev, and M. S. Shur

Big Light: Optical Coherence over Very Large Areas in Photonic-Crystal Distributed-Feedback Lasers 1525
    W. W. Bewley, J. R. Lindle, C. S. Kim, I. Vurgaftman, C. L. Canedy, M. Kim and J. R. Meyer

Novel Resonant-Tunnelling Quantum Dot Photon Detectors for Quantum Information Technology 1527
    J. C. Blakesley, P. See, A. J. Shields, B. E. Kardynal, P. Atkinson, I. Farrer, and D. A. Ritchie

The Dynamical Diffraction Effect in a Two-Dimensional Photonic Crystals. 1529
    O. A. Usov and M. V. Maximov

Piezoelectric Effect on the Lasing Characteristics of (111)B InGaAs/AlGaAs Laser Diodes 1531
    G. Deligeorgis, G. E. Dialynas, N. Le Thomas, Z. Hatzopoulos, and N. T. Pelekanos

Resonant-Photon Tunneling Effect at 1.5 Micron Observed in GaAs/AlGaAs Multi-Layered Structure Containing InGaSb Quantum Dots 1533
    N. Yamamoto, K. Akahane, S.-i Gozu, and N. Ohtani

Amorphous Si/Multicrystalline Si Heterojunctions for Photovoltaic Device Applications 1535
    M. F. Baroughi and S. Sivoththaman

High-Index-Contrast Optical Waveguides on Silicon 1537
    A. Säynätjoki, S. Arpiainen, J. Ahopelto, and H. Lipsanen

Origin of Efficient Light Emission from Si pn Diodes Prepared by Ion Implantation 1539
    T. Dekorsy, J. M. Sun, W. Skorupa, A. Mücklich, B. Schmidt, and M. Helm

Temperature and Polarization Dependence of the Optical Gain and Optically Pumped Lasing in GaInNAs/GaAs MQW Structures 1541
    J. Kvietkova, M. Hetterich, A. Y. Egorov, H. Riechert, G. Leibiger, and V. Gottschalch

Enhanced Electroabsorption in MQW Structures Containing an *nipi* Delta Doping Superlattice 1543
    C. V-B. Tribuzy, M. C. L. Areiza, S. M. Landi, M. Borgström, M. P. Pires, and P. L. Souza

Carrier Recombination in InGaAs(P) Quantum Well Laser Structures: Band Gap and Temperature Dependence 1545
    S. J. Sweeney, D. A. Lock, and A. R. Adams

Growth and Characterization of 1.3 μm Multi-Layer Quantum Dots Lasers Incorporating High Growth Temperature Spacer Layers 1547
    I. R. Sellers, H. Y. Liu, D. J. Mowbray, T. J. Badcock, K. M. Groom, M. Hopkinson, M. Gutiérrez, M. S. Skolnick, R. Beanland, D. T. Childs, and D. J. Robbins

Self-consistent Calculation of Band Diagram and Carrier Distribution of Type-II Interband Cascade Lasers 1549
    P. Peng, Y.-M. Mu, and S. S. Pei

Study on Low Bias Avalanche Multiplication in Modulation Doped Quantum-Dot Infrared Photodetectors 1551
    Y. H. Kang, U. H. Lee, J. H. Oum, and S. Hong

Type-II Interband Cascade Lasers: From Concept to Devices 1553
    R. Q. Yang

Spectral Hole Burning by Storage of Electrons or Holes 1555
    T. Warming, W. Wieczorek, M. Geller, A. Zhukov, V. M. Ustinov, and D. Bimberg

Buried Long-Wavelength Infrared HgCdTe P-on-n Heterojunctions 1557
    J. Rutkowski, P. Madejczyk, W. Gawron, L. Kubiak, and A. Piotrowski

Improvement of Fabrication Method of Resonant Cavity Enhanced Photodiodes for Bi-directional Optical Interconnects 1559
    I.-S. Chung, Y. T. Lee, J.-E. Kim, and H. Y. Park

Real Time Read-Out of Single Photon Absorption by a Field Effect Transistor with a Layer of Quantum Dots 1561
    B. Kardynał, A. J. Shields, N. S. Beattie, I. Farrer, and D. A. Ritchie

The Polarization-dependence of the Gain in Quantum Well Lasers 1563
    F. Boxberg, R. Tereshonkov, and J. Tulkki

GaAs/Al$_{0.45}$Ga$_{0.55}$As Double Phonon Resonance Quantum Cascade Laser 1565
    D. Indjin, A. Mirčetić, P. Harrison, R. W. Kelsall, Z. Ikonić, V. Jovanović, V. Milanović, M. Giehler, R. Hey, and H. Grahn

Very High Temperature Operation of ~5.75 μm Quantum Cascade Lasers 1567
    A. Friedrich, G. Scarpa, G. Boehm, and M.-C. Amann

Microscopic Modeling of THz Quantum Cascade Lasers and Other Optoelectronic Quantum Devices 1569
    R. C. Iotti and F. Rossi

Terahertz Quantum Cascade Laser Emitting at 160 μm in Strong Magnetic Field 1573
    G. Scalari, S. Blaser, L. Sirigu, M. Graf, L. Ajili, J. Faist, H. Beere, E. Linfield, D. Ritchie, and G. Davies

Optical Gain Spectra due to a One-Dimensional Electron-Hole Plasma in Quantum-Wire Lasers 1575
    Y. Hayamizu, M. Yoshita, H. Itoh, H. Akiyama, L. N. Pfeiffer, and K. W. West

8.3. Frontiers in Device Physics

Chemical and Biological Sensing based on the Surface Photovoltage Measurement of the Si Surface Potential Barrier 1577
    K. Nauka, Z. Li, and T. I. Kamins

Towards a High Diffraction Efficiency of Photorefractive Multiple Quantum Wells 1579
    T. Z. Ward, P. Yu, S. Balasubramanian, M. Chandrasekhar, and H. R. Chandrasekhar

Functional Imaging Using InGaAs/GaAs Photorefractive Multiple Quantum Wells 1581
    S. Balasubramanian, S. Iwamoto, M. Chandrasekhar, H. R. Chandrasekhar, K. Kuroda, and P. Yu

Electric and Magnetic Manipulation of Biological Systems 1583
    H. Lee, T. P. Hunt, Y. Liu, D. Ham, and R. M. Westervelt

Ferroelectric Gates with Rewritable Domain Nanopatterns for Modulation of Transport Properties in GaN/AlGaN Heterostructures 1585
    I. Stolichnov, L. Malin, E. Colla, J. Baborowski, N. Setter, and J.-F. Carlin

Towards Tunneling through a Single Dopant Atom 1587
    J. Caro, G. D. J. Smit, H. Sellier, R. Loo, M. Caymax, S. Rogge, and T. M. Klapwijk

High-Speed and Non-Volatile Nano Electro-Mechanical Memory Incorporating Si Quantum Dots 1589
    Y. Tsuchiya, K. Takai, N. Momo, T. Nagami, S. Yamaguchi, T. Shimada, H. Mizuta, and S. Oda

Physics of Deep Submicron CMOS VLSI 1591
    D. D. Buss

Author Index A1

Short radiative lifetime of single GaAs quantum dots

J. Hours*, P. Senellart*, A. Cavanna*, E. Peter*, J.M. Gérard† and J. Bloch*.

*CNRS-LPN, Laboratoire de Photonique et de Nanostructures, Route de Nozay, 91460 Marcoussis, France
† CEA/DRFMC/SP2M, 17 rue des Martyrs, 38054 Grenoble, FRANCE

Abstract. We report on time resolved measurements on single monolayer fluctuation GaAs quantum dots. We measure radiative lifetimes as short as 100 ps for the exciton. Studying various single quantum dots, we demonstrate that the radiative lifetime of the exciton is controlled by the quantum dot lateral size.

INTRODUCTION

Many options to achieve optical quantum devices based on single quantum dots are investigated such as delivering Fourier transform limited single photons [1] or using the exciton-photon strong coupling regime to couple quantum bits [2]. In these matters, QDs with short radiative lifetime such as QDs formed at an interface fluctuation of a quantum well, are of great interest. Theory predicts that these QDs should present an oscillator strenght ten times larger than other III-V QDs [3]. In this paper, we report on time resolved measurements on single monolayer fluctuation QDs. We measure radiative lifetime as short as 100 *ps* for the exciton state and 60 *ps* for the biexciton state. By studying various single QDs, we evidence lifetimes ranging from 100 *ps* to 230 *ps*. We show that there is a direct correlation between the QD lateral size and its radiative lifetime.

EXPERIMENT

The sample under study is a nominally 10 monolayer GaAs quantum well embedded in $Al_{0.33}Ga_{0.67}As$ barriers. Measurements are performed at a sample temperature of 8K. Micro-photoluminescence (*μ*-PL) measurements are performed in the far Field using a microscope objective. We excite the sample with a Ti:sapphire laser delivering 3 *ps* pulses every 12 *ns*. The excitation is non-resonant with an energy around *1.73 eV*. The sample emission is spectrally dispersed using a 32 *cm* spectrometer and temporally analyzed with a streak camera system. Figure 1 presents a streak camera image obtained for an excitation power of 14 *W/cm2*. The horizontal direction corresponds to energy, the vertical direction to time. The intensity of the emission is represented using a gray scale. On top, the time integrated spectrum is plotted. The high energy broad emission corresponds to the recombination of excitons in the 2D quantum well. On the low energy side of the spectrum, we observe the emission from a spectrally well isolated single quantum dot: the exciton (X) at the energy of 1.698 *eV* and the biexciton (XX) at an energy around 1.695 *eV*. Lines are identified by their power dependence (linear for X, quadratic for XX). On

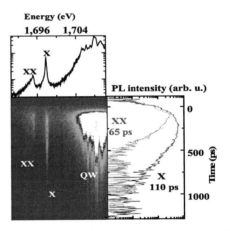

FIGURE 1. Streak camera image measured an excitation power of 14 *W/cm2*. Top: time integrated spectrum (log. scale). Right: spectrally integrated emission as a function of time for X (black) and for XX (gray). The lines are linear ⁻t to the emission decay for long time delay.

the right side of Figure 1, the temporal dependence of the X and XX emission is plotted. The XX emission rises shortly after the excitation pulse, whereas the X emission is delayed : this is a direct evidence of the radiative cascade taking place in QD emission [4]. The peculiar line shape of the XX line is addressed elsewhere[5]. For long time delays (around 500 *ps*), both X and XX emissions decay mono-exponentially with a decay time of 110 *ps* and 60 ps. As discussed in ref. [6], the ratio between the XX and X lifetime depends in general on both the spin structure of theses states and their spatial wave functions.

The 100 *ps* exciton radiative lifetime we measure is shorter than any other III-V or II-VI QD radiative lifetime reported up to now [6,7]. It corresponds to an oscillator strength $f = 75$. It means that, if one inserts such a monolayer fluctuation QD inside a micro-pillar with a radius of 0.5 μm and a quality factor of 2000, the exciton-photon system would be close to the strong coupling regime [3]. Besides, it is easy to calculate that the strong coupling regime would be reached if the monolayer fluctuation QD is inserted in a microdisc with the same radius and a standard quality factor of 8000. The Rabi splitting would be 570 *µeV*, much larger than the exciton or photon individual linewidths.

Let us now study the radiative lifetime of various QDs. It ranges from 100 *ps* to 230 *ps*. To understand these variations, we have performed photoluminescence excitation (PLE) measurements on each single QD. Figure 2a shows a typical PLE spectrum and the absorption edge from which we deduce the confinement energy. In monolayer fluctuation quantum dots, the exciton is quasi two-dimensional but its center of mass is weakly localized at the quantum well interface. The energy difference ΔE we measure between the X emission line and the 2D absorption from the quantum well is the center of mass confinement energy. In Figure 2b, we report the measured radiative lifetimes as a function of ΔE. For very small ΔE, the radiative lifetime is very short (around 100 *ps*). It then increases and reaches maximum around a confinement energy of 4-5 *meV*. For larger ΔE, the radiative lifetime decreases continuously, and reaches 100 *ps* again.

ΔE is directly related to the lateral size of the monolayer fluctuation QD. The smaller the radius of the QD is, the smaller the confinement energy of the X will be. As a result, figure 2b can also be read as follows. The radiative lifetime is very short (around 100 ps) for small and large QD radii, encountering a maximum for an intermediate radius. This observation is in agreement with the radius dependence of the oscillator strength predicted in [3]. The oscillator strength, inversely proportional to the radiative lifetime, is proportional to the area occupied by the X center of mass.

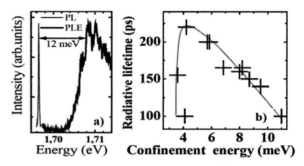

FIGURE 2. a) Single QD PL(thin line) and PLE (thick line) spectra. b) Symbols : radiative decay time as a function of the X confinement energy ΔE. The line is a guide to the eyes.

For large QD radii, the center of mass is localized in a quasi two-dimensional trap. Its wave function occupies an area increasing with the QD radius. The oscillator strength is roughly the two dimensional QW oscillator strength (per area units), multiplied by the area of the QD. It thus increases with the QD radius as we observe. For small QD radii, the center of mass is weakly localized in the QD. The electron and hole wave function expands outside the defect all the more than the QD is small. The smaller the QD is, the larger the oscillator strength gets.

CONCLUSION

To conclude, we have evidenced that the exciton radiative lifetime can be as short as 100 *ps* in single GaAs monolayer fluctuation QDs. The study of several QDs shows that the exciton radiative lifetime is directly controlled by the lateral size of the QD. The typical oscillator strength $f = 75$ should allow to achieve the strong coupling regime for a monolayer fluctuation QD inserted in typical microdisk structures.

REFERENCES

1. C. Santori et al., Nature 419, 594 (2002).
2. A. Imamoglu et al., Phys. Rev. Lett. 83, 4204 (1999).
3. L. C. Andreani, G. Panzarini and J-M. Gérard, Phys. Rev. B 60, 13276 (1999).
4. E. Moreau et al., Phys. Rev. Lett. 87, 183601 (2001).
5. P. Senellart et al. Submitted.
6. G. Bacher et al., Phys. Rev. Lett. 83, 4417 (1999).
7. C. Santori et al., Phys. Rev. B 65, 073310 (2002).

Interference through a Single Quantum Dot

H. Aikawa, K. Kobayashi, A. Sano, S. Katsumoto, and Y. Iye

Institute for Solid State Physics, University of Tokyo, 5-1-5 Kashiwanoha, Kashiwa Chiba 277-8581, Japan

Abstract. We have observed Fano-type interference in the transport through a quantum dot (QD) in a marginal regime between Coulomb-blockade and open-dot. This originates from transport through two levels in the QD, one of which has stronger coupling to the leads than the other. The sign of the Fano's asymmetriy parameter is reversed at some special points, while it is unchanged between most of successive resonances, i.e., many neighboring peaks are in-phase. We have found that the "special" points coincide with the peaks and valleys of the slow background oscillation of conductance, which can be consistently explained with a simple theoretical model.

INTRODUCTION

Phase-shift measurements of quantum dots (QDs) have been unveiling fundamental properties of the electron wave function in QDs. In the pioneering experiments [1], it was unexpectedly found that the phase shifts observed at the successive peaks are in-phase. Many theories have been proposed to solve the mystery, which are waiting for experimental verification. One idea is that a small number of levels in a QD strongly couple to the leads (strongly coupled states, SCSs) and govern the phase shifts of neighboring weakly coupled states (WCSs) [2, 3].

The basic idea is as follows [4]. Due to distortion in the dot potential, the wave function of a WCS ψ^0_j is intermixed with that of an SCS. After the mixing, the wave function of a WCS ψ_j is expressed to a first order of perturbation by the distortion potential V as

$$\psi_j \approx \psi^0_j + \psi_N \frac{\langle \psi^0_j | V | \psi_N \rangle}{E^0_j - E_N}, \quad (1)$$

where N is the index of SCS closest to the energy level E^0_j of the unperturbed state. The transport through ψ_j is dominated by the second term in the right hand side of Eq. (1) because ψ_N has much stronger coupling with the lead states. As a consequence, the phase shift through the level ψ_j becomes the same with that through ψ_N. This is applicable to all WCSs in the neighborhood of SCS in energy, leading to a range of successive in-phase Coulomb peaks.

In this work, the existence and the role of the SCSs is tested by detecting Fano-type interference effect through a QD [5], in which an SCS acts as the continuum and can be viewed as an off-resonant path. The main features of the experimental results can be well explained within the this simple model.

EXPERIMENTAL

We prepared a QD as shown in Fig. 1(a) from 2DEG at the interface of GaAs/AlGaAs heterostructure (sheet carrier density = 3.8×10^{15} m^{-2}, mobility = 80 m^2/Vs) by electron beam lithography, deposition of metallic gates and wet etching. The sample was cooled in a dilution refrigerator with base temperature of 30 mK. The conductance was measured in a two-terminal setup. To enhance the visibility of the interference effect, we chose the side gate (V_L and V_R) voltages to keep the total conductance around e^2/h so that the QD is at the border between a Coulomb blockade (CB) regime and an open-dot regime.

RESULTS AND DISCUSSIONS

The QD conductance G as a function of the gate voltage V_g shows a slow background oscillation (BO), on which a rapid oscillation (RO) is superposed (Fig.1(b)). The RO has the same period with the

Coulomb oscillation in the CB regime, while the lineshape of the peaks are strongly distorted, that can be fitted to Fano's line shape $G = (v+q)^2/(v^2+1)$ as presented in Fig. 1(d) where v and q represent the gate voltage divided by the width of the resonance and Fano's asymmetric parameter, respectively. This indicates that some interference affects the conductance at the peaks.

The sizes of the Coulomb diamonds of the RO, which are thin vertical white regions in Fig.1(c), are modified along the BO, *i.e.*, they are large at the valleys of the BO and small at peaks. This can naturally be explained if we attribute the BO to the oscillation of co-tunneling rate through the SCSs. The valleys, according to this interpretation, correspond to the switching points of the SCS closest to the Fermi level. There the WCS on resonance is farthest from the SCSs in energy, making the hybridization in Eq.(1) small, which results in a large diamond (*i.e.* wave function is more localized in the QD).

The sign of q, which stands for the relative phase of SCS and WCS, is the same for consecutive peaks except for several points indicated by arrows in Fig. 1(c), which coincide with the peaks and valleys of the BO. This again supports the interpretation because they correspond to the phase-switching points of the SCS.

Next we adjusted the side-gate voltages so as to make $|q| > 1$. The result is displayed in Fig. 2(a). The

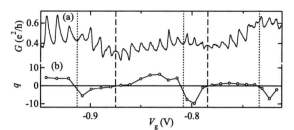

FIGURE 2. (a) Conductance and (b) the fitted value of q as a function of V_g. The conditions are slightly different from those in Fig.1.

peaks show clear Fano distortion with $|q|$'s large enough to determine their sign. In Fig. 2(b), q is plotted versus the peak position. The zero-crossing points of q again correspond to the peaks (valleys) of the BO, which confirms the presence of SCS and its influence on the phase shift of the WCSs and the legitimacy of the above interpretation.

CONCLUSION

In summary, we have observed Fano interference arising from the transmission through two energy levels in a QD. Some of the levels have stronger coupling to the leads and dominate the phase at the Coulomb peaks. This method to investigate the phase property through a QD without outer interference circuit provides a clue to solve the problem of the in-phase Coulomb peaks.

ACKNOWLEDGMENTS

This work is supported by a Grant-in-Aid for Scientific Research and by a Grant-in-Aid for COE Research from MEXT of Japan. K.K. is supported by a Grant-in-Aid for Young Scientists (B) (No. 14740186) from Japan Society for the Promotion of Science.

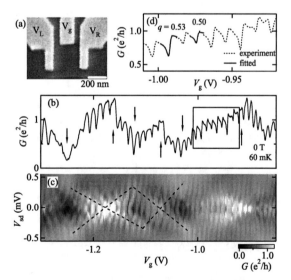

FIGURE 1. (a) SEM image of the sample. (b) Conductance shown as a function of the gate voltage V_g. The arrows indicate sign-change points of q. (c) Gray-scale plot of conductance versus bias voltage (V_{sd}) and V_g. (d) Magnified view of (b).

REFERENCES

1. Yacoby, A., Heiblum, M., Mahalu, D., and Shtrikman, H., *Phys. Rev. Lett.* **74,** 4047 (1995); Schuster, R. *et al.*, *Nature* **385,** 417 (1997).
2. Silvestrov, P., and Imry, Y., *Phys. Rev. Lett.* **85,** 2565 (2000)
3. Yeyati, A. L., and Büttiker, M., *Phys. Rev. B* **62,** 7307 (2000).
4. Nakanishi, T., Terakura, K., and Ando, T., *Phys. Rev. B* **69,** 115307 (2004).
5. Göres, J. *et al.*, *Phys. Rev. B* **62,** 2188 (2000).

Time resolved single electron detection in a quantum dot

R. Schleser*, E. Ruh*, T. Ihn*, K. Ensslin*, D. D. Driscoll[†] and A. C. Gossard[†]

*Solid State Physics Laboratory, ETH Zürich, 8093 Zürich, Switzerland
[†]Materials Department, University of California, Santa Barbara California 93106

Abstract. We have performed transport measurements on a composite nanostructure, consisting of a quantum dot and a nearby quantum point contact (QPC) used as a charge detector. For very low coupling of the dot to only one of the reservoirs and no coupling to the other one, we observe time-dependent features related to electron tunneling onto the dot and back to the reservoir. The transition frequencies visible in the QPC signal are of the order of only a few Hz. From these time-dependent data, the Fermi distribution in the lead and the temperature can be extracted, as well as the coupling to the reservoir.

Using a quantum point contact (QPC) as a detector, the charge on a quantum dot can be determined [1]. As quantum dots are possible candidates for the construction of qubits in future quantum computers [2], QPCs could become part of the readout system. High-bandwidth readout has been performed using a RF-SET as a detector [3]. However, theoretical considerations [4, 5] lead to the assumption that QPCs can obtain a higher quantum measurement efficiency. Recently, high-frequency real-time charge-readout measurements have been performed using a split-gate defined QPC [6].

We present transport measurements on a nanostructure, consisting of a quantum dot and a nearby quantum point contact used as a charge detector. The sample was created using scanning probe lithography [7, 8, 9] on a GaAs/Al$_{0.3}$Ga$_{0.7}$As heterostructure, containing a 2-dimensional electron gas (2DEG) 34 nm below the surface as well as a backgate 1400 nm below the 2DEG, isolated from it by a layer of LT-GaAs. The central part of the structure together with the designations of the gates is shown in Fig. 1(a).

We measured transport through both the QD and the QPC simultaneously. The gates G1 and G2 were used to tune the dot's chemical potential and control the coupling to the reservoirs. The gates S$_{QPC}$ and D$_{QPC}$ forming the QPC circuit were also used to tune the quantum dot. The QPC itself can be tuned to a sensitive parameter regime by applying a voltage to gate P. The time dependence of the QPC current was measured using an oscilloscope, the bandwidth of our setup (cabling, amplifiers) being currently of the order of a few kHz. All measurements were performed in a dilution refrigerator with a base temperature of 80 mK.

When the electron number on the dot is tuned using a gate voltage, the conductance of the QPC (Fig. 1(c)) dis-

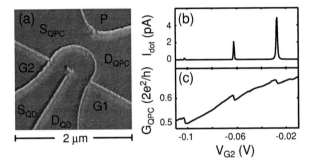

FIGURE 1. (a) AFM micrograph of the structure with designations of gates: source (S) and drain (D) of the quantum dot (QD) and the quantum point contact (QPC) used as a charge detector; lateral gates G1 and G2 to control the coupling of the dots to the reservoirs; Plunger gate (P) to tune the QPC detector. (b) Example measurement of the current through the dot. In this measurement, the QD was weakly coupled to both reservoirs. A bias voltage of 10 μV was symmetrically applied between S$_{QD}$ and D$_{QD}$. (c) Simultaneous measurement of the conductance through the QPC, where each kink corresponds to a change of the dot's charge by one electron. The width of the kink is related to the Coulomb peak's width, while its height depends only on the local slope of the QPC's I vs. V characteristics and the sensitivity to the charge on the dot.

plays distinct features at the same voltage values where the dot conductance shows Coulomb resonances (Fig. 1(b)). For the following measurements, a compensation voltage $V_P = aV_{G1} + bV_{G2}$ was applied to the gate P, with constants a and b chosen such as to avoid large changes in the QPC's conductance and the resulting changes in sensitivity.

An oscilloscope at a sampling frequency of 250 Hz was used for the time-dependent measurements. One of the QD's leads was completely closed and the other one tuned to an electron transition frequency of the order of a few Hz. Near the parameter range were the N and $N+1$ electron states are degenerate, we observe random, telegraph-noise-like switching between two values in the time dependent QPC detector current (Fig. 2(a)). The mean switching frequency can be tuned in a wide range using the gate G2 (the one used for tuning the coupling of the dot to the single reservoir). We believe that these time-dependent features are related to single electron tunneling events onto and off the dot: the amplitude of the oscillations in the QPC current is equal to the step height in time-averaged, i.e. strong coupling measurements. In addition, the position of the crossover regime in which switching occurs shows the same gate voltage dependence as the Coulomb blockade peaks in transport measurements [10].

Figure 2(b) shows a simple model in which an energy level in a QD is weakly coupled to a single lead. In this special case, the relative occupancy of the lower conductance state in the QPC is equal to the value of the energy distribution in the reservoir at the energy value of the dot's chemical potential [11, 3, 10]. Assuming a continuous density of states in the leads, the energy distribution should be the Fermi distribution. Thus, from a statistical analysis of the switching events in the QPC current, the Fermi distribution in the lead and therefore its temperature can be extracted: Figure 2(c) contains different time-dependent traces, taken for slightly different values of the gate voltage V_{G1}. From these and similar traces, the switching rate and relative occupancy have been determined for every value of V_{G1}. Using a lever arm $\alpha = d\mu_{\text{dot}}/dV_{G1}$ determined in an earlier experiment, an energy scale was estimated (see remarks in ref. [10]). Figure 2(d) shows the values of the dot's relative occupancy vs. this energy scale, together with a fit to the Fermi function, which yields a temperature value $T \approx 200$ mK. In a similar way, using the observed switching frequencies, the coupling of the dot to the reservoir can be determined quantitatively. Increasing the gate voltage V_{G2} leads to a strongly increased coupling [10], as one would expect.

In summary, we have used a quantum point contact to perform real-time readout of the electronic charge of a semiconductor quantum dot defined by scanning probe lithography.

From our measurements we conclude that after moderate improvements on amplifier and cabling bandwidth, read-out frequencies of the order of MHz should be attainable (see e.g. ref. [6]).

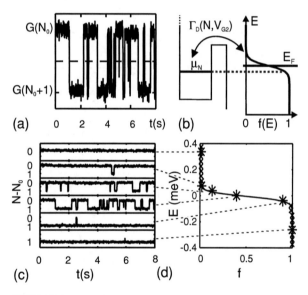

FIGURE 2. (a) Single oscilloscope trace of the QPC's conductance versus time. The dot was tuned to a point where two states differing by one electron in charge were almost degenerate. (b) Model for the transfer of individual electrons between a single revervoir and a quantum dot. The tunnel coupling is assumed to depend on the single-level state N and can be tuned via the voltage V_{G2}. (c) Changing the voltage V_{G1} changes the dot's electrochemical potential and allows a transition from the N electron state to the $N+1$ electron state. In the oscilloscope traces, this is seen as a change in the relative occupancy of the two possible states. (d) Distribution $f(E)$ extracted from oscilloscope traces (where for every point, 20 traces each of length 9 seconds where taken into account). The data points marked by large asterisks correspond directly to the traces shown in (b). For a discussion of the energy scale see text.

REFERENCES

1. Field, M., Smith, C. G., Pepper, M., Ritchie, D. A., Frost, J. E. F., Jones, G. A. C., and Hasko, D. G., *Phys. Rev. Lett.*, **70**, 1311 (1993).
2. Loss, D., and DiVincenzo, D.-P., *Phys. Rev. A*, **57**, 120 (1998).
3. Wei-Lu, Zhongqing-Ji, Pfeiffer, L., West, K. W., and Rimberg, A. J., *Nature*, **423**, 422 (2003).
4. Pilgram, S., and Büttiker, M., *Phys. Rev. Lett.*, **89**, 200401 (2002).
5. Clerk, A. A., Girvin, S. M., and Stone, A. D., *Phys. Rev. B*, **67**, 165324 (2003).
6. Elzerman, J., Vandersypen, L., Hanson, R., Vink, I., van Beveren, L. W., and Kouwenhoven, L., *in preparation* (2004).
7. Held, R., Lüscher, S., Heinzel, T., Ensslin, K., and Wegscheider, W., *Appl. Phys. Lett.*, **75**, 1134 (1999).
8. Lüscher, S., Fuhrer, A., Held, R., Heinzel, T., Ensslin, K., and Wegscheider, W., *Appl. Phys. Lett.*, **75**, 2452 (1999).
9. Nemutudi, R., Kataoka, M., Ford, C. J. B., Appleyard, N. J., Pepper, M., Ritchie, D. A., and Jones, G. A. C., *J. Appl. Phys.*, **95**, 2557 (2004).
10. Schleser, R., Ruh, E., Ihn, T., Ensslin, K., Driscoll, D. C., and Gossard, A. C., *submitted for publication* (2004).
11. Beenakker, C. W. J., *Phys. Rev. B*, **44**, 1646–56 (1991).

Shot noise in tunneling through a single InAs quantum dot

Frank Hohls*, André Nauen*, Niels Maire*, Klaus Pierz[†] and Rolf J. Haug*

*Institut für Festkörperphysik, Universität Hannover, Appelstr. 2, 30167 Hannover, Germany
[†]Physikalisch-Technische Bundesanstalt, Bundesallee 100, 38116 Braunschweig, Germany

Abstract. We examine the dynamical properties of resonant tunneling through single InAs quantum dots by measuring the shot noise of the current. We observe an approximately linear voltage dependence of both the shot noise, characterized by the Fano factor, and the tunneling current itself. We ascribe this to the three-dimensional density of states of the emitter and collector and are able to model the voltage dependence using a master equation approach.

Shot noise measurements allow us to access the dynamical properties of a resonant tunneling device which are not accessible by measuring solely the average current. This offers the possibility to study the individual tunneling rates and their dependence on e.g. the bias voltage and thus yields a complete characterization of the device. We employ this to thoroughly characterize a single electron resonant tunneling device based on InAs quantum dots and point out the influence of the density of states of the emitter and collector contacts.

Shot noise was initially discussed for vacuum tubes and later on for any kind of single tunneling barriers. For all such systems the current $I(t)$ displays a frequency independent noise power density $S = 2q\langle I \rangle$ with $q = e$ the elementary charge and $\langle I \rangle$ the mean or DC-current, further on just denoted by I. However, the shot noise in a double-barrier structure displays a reduced shot noise power below the full shot noise value $2eI$ of the single barrier system. The degree of reduction, characterized by the so called Fano factor $\alpha = S/2eI$ with S being the measured shot noise power, yields a direct measure of the ratio of the tunneling rates through both barriers.

In this paper we present noise measurements on self assembled InAs quantum dot (QD) systems. The InAs QD's are embedded into the AlAs barrier of a GaAs-AlAs-GaAs tunneling device. Due to the growth of InAs on AlAs we achieve small dot sizes of 10 – 15 nm diameter and 3 nm height. The InAs dots are grown on a 4 nm AlAs layer and overgrown by another 6 nm of AlAs, resulting into effective tunneling barrier thickness of 4 nm (bottom) and 3-4 nm (top). Graded n-doped GaAs on both sides of the barrier acts as leads. Au/Ge/Ni contacts are used as etch mask to prepare diode structures with an area of $40 \times 40\,\mu$m. Only a small fraction of the InAs dots if electrically active [1].

Throughout this paper we choose the bias polarity in

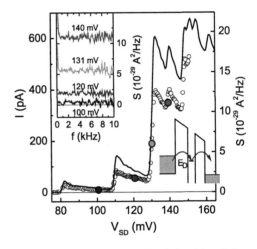

FIGURE 1. Measured current (line, left axis) and shot noise (circles, right axis) for the first few QD's ($T = 1.5$ K). Mapped onto the right axis the line corresponds to full shot noise $S = 2eI$. Inset: Spectral density measured for different voltages V_{sd} is frequency independent for $f > 0.5$ kHz.

such a way that the electrons first tunnel through the thicker (bottom) barrier and then through the thinner (top) barrier. Due to the higher collector tunneling rate $\Gamma_C > \Gamma_E$ compared to the emitter tunneling rate the dot is mostly empty which allows to examine the influence of the emitter density of states.

Without bias voltage the ground state energy of all dots lies above the Fermi energy due to the small size of the InAs dots on AlAs. The current is blocked until the energy of the largest dot with lowest ground state energy is aligned with the Fermi energy of the emitter. For the device under study this happens for a voltage $V_{sd} = 80$ mV at which a resonant tunneling current through the first dot sets in. Successive steps show up in the current with rising voltage (Fig. 1), each denoting another QD

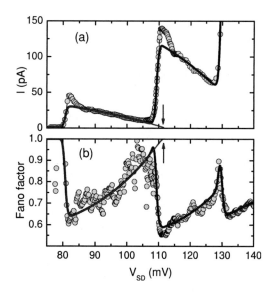

FIGURE 2. (a) Resonant tunneling current as measured (circle) and modeled (line). (b) Same for the Fano factor $\alpha = S/2eI$.

coming into resonance.

The measured noise spectra $S(f)$ are shown in the inset of Fig. 1 for different voltages. As expected for shot noise $S(f)$ is essentially frequency independent, and we determine the shot noise density S by averaging the spectra from 1 to 10 kHz. The result is displayed by the circles in Fig. 1 (right scale). Comparison to the full shot noise $2eI$ (line mapped to right scale) reveals a suppression of noise which is well understood as a consequence of Coulomb repulsion on the QD's [2].

The large spacing between the current onset through the first few dots allows us to examine the noise properties for resonant tunneling through a single InAs-QD. The resonant tunneling current and the corresponding Fano factor $\alpha = S/2eI$ characterizing the noise are shown in detail for the first two dots and the onset of the third one in Fig. 2.

One striking feature of both the current I and the Fano factor α is the nearly linear decrease of I resp. increase of α which follows the initial step. This voltage dependence is distinct from the one observed in resonant tunneling devices realized by patterning two-dimensional systems and is caused by the three-dimensional nature of the emitter and collector. It was pointed out by Liu and Aers [3] that the tunneling rates are proportional to the area $A(E_D)$ in momentum space with QD energy E_D. In three dimensions $A(E_D) \propto E_D - E_C$ with E_C the conduction band edge and thus $\Gamma(E_D) \propto E_D - E_C$ with $E_D(V_{sd}) = e\beta(V_{sd} - V_0)$ depending on the bias voltage (lever arm $\beta \approx 0.4$). Assuming $\Gamma_E \ll \Gamma_C$ the current $I \approx e\Gamma_E(V_{sd}) \propto V_{sd} - V_0$ depends approximately linearly on the voltage and vanishes at V_0, shown by the arrow in Fig. 2, where the QD energy E_D crosses the band edge of the emitter. The same behavior is also mirrored by the Fano factor which in first order of Γ_E/Γ_C follows $\alpha \approx 1 - 2\Gamma_E/\Gamma_C$. Due to the large bias we can assume $\Gamma_C \approx$ constant.

The window of transport ΔV between the onset of current when the QD energy crosses the Fermi energy, $E_D = E_F$, and the vanishing of the current through the same dot for $E_D = E_C$ allows us to estimate the Fermi energy. With $\beta \approx 0.4$ from temperature dependent measurements and $\Delta V \approx 32$ mV we find $E_F - E_C \approx 13$ meV which compares well with a Fermi energy of 14 meV estimated from growth parameters.

Closer examination of Fig. 2 reveals deviations from the simple linear dependence of current and Fano factor discussed above. To account for these we have to abandon the assumption of $\Gamma_E \ll \Gamma_C$ and calculate the full expressions using a master equation approach [4, 5], taking into account temperature and spin degeneracy. For a single quantum dot we find

$$I = \frac{2ef_E\,\Gamma_E(V_{sd})\Gamma_C}{(1+f_E)\Gamma_E(V_{sd})+\Gamma_C}, \quad (1)$$

$$\alpha = 1 - \frac{4f_E\,\Gamma_E(V_{sd})\Gamma_C}{((1+f_E)\Gamma_E(V_{sd})+\Gamma_C)^2} \quad (2)$$

with f_E being the Fermi function of the emitter. For transport through multiple dots this has to be extended to $I = \sum I_i$ and $\alpha = \sum (I_i/I)\alpha_i$ with I_i and α_i for each dot given by the above equations. The result for the best fit of the tunneling rates at current onset and the onset voltage for each dot is shown by the lines in Fig. 2. We observe a good agreement between the model and the measured data, and even the peak in the Fano factor at $V_{sd} \approx 129$ mV is nicely reproduced.

Nevertheless, we observe some deviations from our simple non-interacting model directly at the step edge of the current onset for each dot, which we discuss elsewhere and attribute to interaction effects, namely a Fermi edge singularity [5].

We thank Gerold Kiesslich for discussions and acknowledge financial support from DFG and BMBF.

REFERENCES

1. Hapke-Wurst, I., Zeitler, U., Frahm, H., Jansen, A. G. M., Haug, R. J., and Pierz, K., *Phys. Rev. B*, **62**, 12621 (2000).
2. Nauen, A., Hapke-Wurst, I., Hohls, F., Zeitler, U., Haug, R. J., and Pierz, K., *Phys. Rev. B*, **66**, R161303 (2002).
3. Liu, H. C., and Aers, G. C., *Solid State Commun.*, **67**, 1131 (1988).
4. Kiesslich, G., Wacker, A., and Schöll, E., *Phys. Rev. B*, **68**, 125320 (2003).
5. Nauen, A., Hohls, F., Maire, N., Pierz, K., and Haug, R. J., *Phys. Rev. B*, **70**, 033305 (2004).

Imaging Electrons in a Single-Electron Quantum Dot

P. Fallahi,[1] A.C. Bleszynski,[2] R.M. Westervelt,[1,2] J. Huang,[3] J.D. Walls,[3] E.J. Heller,[2,3] M. Hanson[4] and A.C. Gossard[4]

[1]*Division of Engineering and Applied Sciences,* [2]*Department of Physics,*
[3]*Department of Chemistry and Chemical Biology, Harvard University, Cambridge, MA 02138*
[4]*Materials Department, University of California, Santa Barbara, CA 93106*

Abstract. A scanning probe microscope (SPM) can be used to image a single-electron quantum dot at liquid-He temperatures by recording the Coulomb blockade conductance as a charged SPM tip is scanned above. The Coulomb blockade produces a ring in the image as the first electron is added to the dot. If the tip is sufficiently close to the surface, simulations show one may be able to extract the shape of the electron wave function inside the dot from Coulomb blockade conductance images with a resolution exceeding the size of the tip perturbation.

INTRODUCTION

A scanning probe microscope (SPM) can be used to image the flow of electron waves in a two-dimensional electron gas (2DEG) at liquid He temperatures [1] and to obtain information about the wavefunction in nanostructures [2]. A cooled SPM can image quantum dots in a carbon nanotube [3] and a single-electron dot in a GaAs/AlGaAs heterostructure [4] in the Coulomb blockade regime. This will help in the development of single-electron quantum dots and dot circuits for quantum information processing, by providing ways to image the location of electrons and to locally probe the circuit. Single-electron quantum dots have been proposed as spin qubits [5], and single dots and tunnel-coupled double dots of this type have been built and characterized [6]. Here we propose a technique that will allow SPM imaging in the Coulomb blockade regime to determine the shape of the electron wavefunction inside a quantum dot.

COULOMB-BLOCKADE IMAGING

Figure 1a illustrates the experimental set up used for Coulomb-blockade imaging of a single-electron quantum dot [4]. The dot was patterned by surface gates on a GaAs/AlGaAs heterostructure containing a 2DEG. A liquid-He cooled charged SPM tip was scanned at a fixed height above the surface of the heterostructure. The charged tip induces a perturbation, shifting the potential forming the dot as well as the energy levels inside the dot, thereby changing the dot conductance in the Coulomb blockade regime. Images are obtained by recording the dot conductance as a function of tip position.

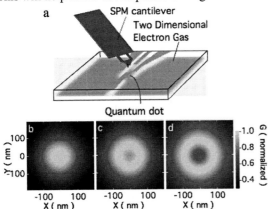

FIGURE 1. (a) Experimental set up for Coulomb blockade imaging of a quantum dot. (b)-(d) Simulated images of Coulomb blockade conductance of a single-electron quantum dot vs. tip position for tip voltages V_{tip} (b) −15mV, (c) −18mV and (d) −21mV; see Fig. 2 caption for dot and tip parameters. A less negative tip voltage causes the ring to shrink.

Figures 1b-d show simulated SPM images of a single-electron quantum dot. For these simulations, the tip was located 50nm above the 2DEG, half the spacing used in the experiment [4]. A smaller tip height enhances the imaging resolution, allowing for the extraction of information about the electron wavefunction inside the quantum dot.

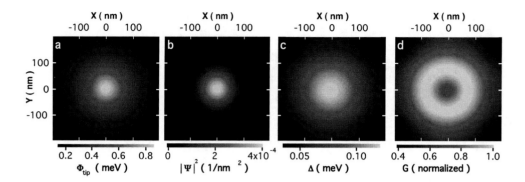

FIGURE 2. (a) Tip perturbation for a tip 50nm above the 2DEG with radius 30 nm and voltage 20 mV. (b) Calculated wavefunction of the single-electron ground state of a quantum dot with a harmonic potential with $\hbar\omega = 0.31$ meV. (c) Calculated shift of the lowest energy level in the dot vs. tip position. (d) Dot conductance vs. tip position at 0.3K.

Figure 1d contains a ring of high conductance corresponding to a Coulomb-blockade resonance through the lowest energy level in the dot. Inside the ring, there are 0 electrons in the dot; outside there is 1. Making the tip voltage V_{tip} less negative alters the size of the ring as demonstrated in Figs. 1b-d. Similar rings were observed in the experimental images, and the adjustability of the ring diameter with tip voltage was demonstrated [4].

The shift Δ in the energy level inside the quantum dot as the tip is scanned above the dot is a convolution of the tip perturbation Φ_{tip} on the potential energy that forms the dot and the square of electron wavefunction amplitude $|\Psi|^2$ from first order perturbation theory. If Φ_{tip} is known and it is comparable in width or narrower than $|\Psi|^2$, one can extract the shape of $|\Psi|^2$ by deconvoluting Δ with respect to Φ_{tip}. The energy shift Δ can be extracted from the line shape of the resonance peaks forming the rings in the conductance images [4]. In the resonant tunneling regime where only one energy level contributes to the dot conductance, the line shape of Coulomb blockade peak is given by [7]:

$$G = G_{\max}\left[Cosh(\Delta/2k_BT)\right]^{-2} \quad (1)$$

where G_{max} is the dot conductance at resonance determined by the tunneling rate through the quantum dot and the temperature T. The values of Δ can be extracted using the inverse of Eq. 1.

Figure 2 shows simulations of the Coulomb blockade conductance G (Fig. 2d) and the energy shift Δ (Fig. 2c) vs. tip position based on first order perturbation theory for a quantum dot formed by a parabolic potential with single electron ground-state wavefunction $|\Psi|^2$ (Fig. 2b). The tip perturbation Φ_{tip} in Fig. 2a is from a small conducting sphere located above the sample, using a simple model that neglects the effects of metal gates. We can deconvolute the calculated Δ with respect to Φ_{tip} to reproduce the wavefunction in Fig. 2b.

ACKNOWLEDGEMENTS

This research was supported at Harvard by DARPA grant DAAD19-01-1-0659 and by the NSEC under NSF Grant PHY-0117795, and at UC Santa Barbara by the Institute for Quantum Engineering, Science and Technology (iQUEST).

REFERENCES

1. M.A. Topinka et al., *Phys. Today*, Dec 47 (2003); M.A. Topinka *et al.*, *Science* **289**, 2323 (2000); M.A. Topinka *et al.*, *Nature* **410**, 183-186 (2001).

2. G. Salis et al., Phys. Rev. Lett. **79**, 5106 (1997); R. Crook et al., J. Phys.: Condens. Matter **12**, L735 (2000); R. Crook et al. Phys. Rev. Lett. **91**, 246803 (2003); T. Ihn, Springer Tracts in Modern Physics **192**, Ch. 17 (2004).

3. M.T. Woodside et al., Science **296**, 1098 (2002).

4. P. Fallahi, A.C. Bleszynski, et al., submitted for publication (2004).

5. D. Loss and D.P. DiVincenzo, *Phys. Rev. A* **57**, 120 (1998).

6. S. Tarucha *et al.*, *Phys. Rev. Lett.* **77**, 3613 (1996); J.M. Elzerman *et. al.*, *Phys. Rev. B* **67**, 1613081 (2003); I.H. Chan *et. al.*, *Nanotechnology* **15**, 609 (2004).

7. C.W.J. Beenakker, *Phys. Rev. B* **44**, 1646-1656 (1991).

Spatially resolved manipulation of single electrons in quantum dots using a scanned probe

A. Pioda*, S. Kičin*, T. Ihn*, M. Sigrist*, A. Fuhrer*, K. Ensslin*, A. Weichselbaum[†], S.E. Ulloa[†], M. Reinwald** and W. Wegscheider**

*Solid State Physics, ETH Zürich, 8093 Zürich, Switzerland
[†]Departement of Physics and Astronomy, Ohio University, Athens, Ohio 45701-2979
**Institut für experimentelle und angewandte Physik, Universität Regensburg

Abstract. Electron states in quantum dots have been investigated with the metallic tip of a scanning force microscope. The interaction potential between the scanning tip and the electronic states in the quantum dots has been mapped out by spatial images of the conductance resonances corresponding to single electrons moving in and out of the dot.

Semiconductor quantum dots offer the opportunity to study effects like single-electron charging and tunneling, allowing the manipulation of single electron charges, spins, and orbital quantum states [1, 2, 3, 4, 5, 6, 7]. Scanning probe techniques allow to gain local access to the electronic properties of semiconductor nanostructures. Here we show that single electrons can be made to hop on and off the dot by changing the position of the scanning tip. Spatial images of the conductance resonances allow to map out the local potential, as well as to determine quantitatively the interaction potential between the SFM tip and the individual electrons.

Samples have been fabricated on an AlGaAs-GaAs heterostructure containing a two-dimensional electron gas (2DEG) 34 nm below the surface, with density 5×10^{11} cm^2 and mobility 450'000 cm^2/Vs at 4.2 K. A quantum dot and a quantum ring have been defined by room temperature local anodic oxidation with a scanning force microscope (SFM) which allows to write oxide lines on a semiconductor surface that locally deplete the 2DEG underneath [8].

Figure 1 is a topography image of the two samples. The left one is taken at 300 mK showing the quantum dot and three adjacent quantum point contacts that are not analyzed in the present experiment. The number of electrons in the quantum dot can be controlled by the lateral plunger gate pg. The gates qpc1 and qpc2 are used for tuning the coupling of the dot to source and drain contacts. The dot has been operated in the Coulomb-blockade regime. The average charging energy is about 1.4 meV as determined from Coulomb-blockade diamonds. We estimate the spacing of single-particle quantum states from the geometry of the dot to be about

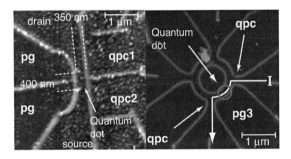

FIGURE 1. Left: Topography image of the dot taken at 300 mK. The electron gas is depleted below the bright oxide lines. Right: Room temperature image of the quantum ring, in scale with the left image. The position of the quantum dot in one of the arms of the ring is indicated in the figure. The scanning range of both images is 4×4 μm.

30 μeV.

The right part of the Fig. 1 illustrates the quantum ring sample. The quantum dot (marked by a circle) forms just before the ring segment is completely pinched off by the plunger gate pg3. The image was taken at room temperature with the AFM used for writing the nanostructure.

Experiments are performed with a SFM in a ^{3}He cryostat with a base temperature of 300 mK. The scanning sensor is a PtIr wire with an electrochemically sharpened tip (typical radius a few tens of nanometers) mounted on a piezoelectric tuning fork. Details of this low temperature SFM setup can be found in [9].

Figure 2 shows scanning gate measurements of the two samples. The tip is kept at a constant voltage and scanned over the surface, while measuring the conductance of the quantum dots. First, we discuss the left image in Fig. 2. Four different regions may be distinguished: (1) A region with weakly varying conductance, (2) a region of concentric conductance peaks, (3) a blockaded region of zero current, and (4) a peak of increased conductance above the quantum dot. The almost constant conductance in region (1) is due to the fact that the tip is too far away from the dot to change its conductance. When the tip approaches the dot [region (2)] the potential energy of the electrons is increased, and they spill over to source or drain, one by one [10]. Each of these concentric conductance peaks corresponds to one electron leaving the quantum dot, and the closed curve defined by these conductance peaks corresponds to a constant potential in the dot. Conductance peaks therefore map out the lines of constant interaction potential between the tip and the individual electrons in the quantum dot. In region (3) the tip, biased with 0V, completely pinches off the quantum point contacts which couple the dot to source and drain, and the conductance drops below the current resolution of the measurement. The tip bias of 0V can be considered negative, since the contact potential difference between the tip and the sample is compensated for a bias of about 0.56 V. This was measured by scanning the tip along a line close to the dot, and varying the tip bias. The conductance increase in region (4) may be related to screening effects.

The right part of the figure (measurement of the dot in the quantum ring) shows similar behavior. In this case, the dot is located near the right edge just outside the image. Therefore, the lines corresponding to Coulomb-blockade peaks do not form closed curves. Regions (2) and (3) can again be identified, corresponding to a region with conductance peaks and one where the dot is depleted. The structure marked 4 bears resemblance of the region 4 of enhanced conductance in the left image. However, it is not located close to the dot but rather appears around a small grain of material on the surface which takes significant influence on the electrostatics of the dot.

Our scanning gate measurements demonstrate that single electrons confined in a quantum dot can be manipulated one by one by a macroscopic scanning tip. Using the quantum dot as a detector allows us to set experimental parameters which do not inflict long term modifications to the sample, and allow to map out the interaction potentials quantitatively.

We thank T. Vančura for his contribution to the experimental setup. Financial support from the Swiss National Science Foundation (Schweizerischer Nationalfonds) and the NSF-NIRT is gratefully acknowledged.

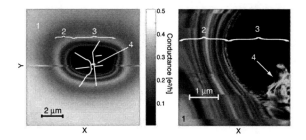

FIGURE 2. Scanning gate images measured in feedback at a distance of a few nanometers from the surface. Left: Image for the quantum dot. The oxide lines defining the structure have been drawn as white lines. Regions 1 to 4 are described in the text. Right: As a comparison: same measurement for the quantum ring: The dot is not inside the scanning range.

REFERENCES

1. S. Tarucha, et. al., *Phys. Rev. Lett.*, **84**, 2485 (2000).
2. S. Lüscher, et. al., *Phys. Rev. Lett.*, **86**, 206802 (2001).
3. L.P. Rokhinson, et. al., *Phys. Rev. B*, **63**, 035321 (2001).
4. S. Reimann, and M. Manninen, *Rev. Mod. Phys.*, **74**, 1283 (2002).
5. A. Fuhrer, et. al., *Phys. Rev. Lett.*, **91**, 206802 (2003).
6. H. Grabert, "Single-Charge Tunneling" in *Coulomb Blockade Phenomena in Nanostructures*, edited by H. Grabert and M. H. Devoret, NATO ASI Series **294**, Plenum Press, New York, 1992.
7. L. P. Kouwenhoven, et. al., "Electron Transport in Quantum Dots" in Proceedings of the Advanced Study Institute on *Mesoscopic Electron Transport*, edited by L. P. Kouwenhoven et. al., Kluwer, Dordrecht, Netherlands, 1997.
8. A. Fuhrer, et. al., *Superlattices Microst.*, **31**, 19 (2002).
9. T. Ihn, "Electronic Quantum Transport in Semiconductor Nanostructures" in *Springer Tracts in Modern Physics* **192**, edited by G. Höhler, Springer, New York, 2004.
10. M. T. Woodside, and P. L. McEuen, *Science*, **296**, 1098 (2002).

Negative Coulomb Drag in Parallel Quantum Wires Having Coupled Two Dots

Motonari Honda[1], Michihisa Yamamoto[1], Michael Stopa[2], Seigo Tarucha[1,2]

[1]*Departmen of Physics, University of Tokyo, 7-3-1 Hongo, Bunkyo-ku Tokyo 113-0033 Japan*
[2]*ERATO Mesoscopic Correlation Project, Japan Science & Technology Corporation, 3-1 Morinosato Atsugi-shi Kanagawa 243-0198 Japan*

Abstract. We have fabricated capacitively coupled two quantum wires having a series of quantum dots to study Coulomb drag between the two quantum wires. We have observed strong negative Coulomb drag when the electron transport through the dots is in the tunneling regime for both wires. This indicates that electrons in one dot of the dive wire and holes in the other wire are strongly correlated to generate exciton-like transport.

INTRODUCTION

Coulomb drag is caused by momentum transfer between two closely separated electronic systems due to the direct Coulomb interaction. Several experiments about Coulomb drag have been performed for one-dimensional (1D) [1-5] and two-dimensional (2D) [6,7] electron systems. The drag effect is generally stronger in 1D system because of the reduced Coulomb screening. Drag resistance R_D is usually defined as

$$R_D = -\frac{V_{drag}}{I_{drive}}, \quad (1)$$

where I_{drive} is an injected current into one of the electron systems and V_{drag} is an induced voltage in the other electron system. R_D is usually positive. So for parallel-coupled quantum wires, electrons in the drive wire drag those in the drag wire. However, we have recently found that R_D changes the sign when the electron densities are pretty low in the coupled wires [8]. This is understood by considering that electrons have particle-like states in the drive wire to drag correlation holes in the drag wire. The combined particle-like transport can be facilitated by replacing each wire with a series of quantum dots. In this work we prepare such a device having parallel coupled quantum wires with a series of two dots to study the negative Coulomb drag. The two wires have geometry with mirror-symmetry, and capacitively coupled to each other, particularly in and near the dot region. Note electrons have a particle nature in each wire when tunneling through the dot region.

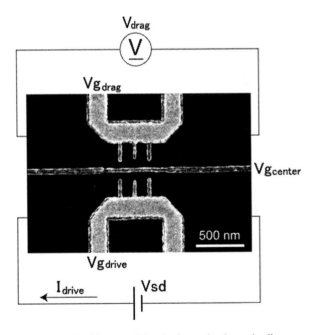

Figure 1. SEM image of the device and schematic diagram of the drag measurement.

DEVICE STRUCTURE

A device used for the experiment is prepared using a wafer containing a GaAs/n-AlGaAs heterostructure. The 2D electron gas (2DEG), which is located 90 nm below the surface, has a mobility of 6×10^5 cm^{2}V^{-1}s^{-1} and an electron density of 3×10^{11} cm^{-2} at 1.5 K. The parallel coupled wires including two dots are formed using a split Schottky gate technique (Fig. 1). The device has four ohmic contacts and three Schottky gates (two side gates and one center gate). Three narrow tips are attached to each side gate to define a series of two dots. The dimensions of the dot defined by the gate electrode pattern are 300×160 nm^2.

RESULTS AND DISCUSSION

The device is placed in a dilution refrigerator with a base temperature of 10 mK. In the Coulomb drag experiment a constant source-drain voltage Vsd is applied to one wire (drive wire), and the current I_{drive} injected into the drive wire and the open circuit voltage V_{drag} induced in the other wire (drag wire) are simultaneously measured. The measurement setup is schematically shown in Fig. 1. Two side gate voltages Vg_{drive} and Vg_{drag} are properly adjusted to define two quantum dots in each wire. Fig. 2 (a) shows the V_{drag} and R_D vs. Vg_{dirve} measured for $Vsd = 600$ µV, $Vg_{center} = -1$ V and $Vg_{drag} = -0.652$ V, respectively. Note that the tunneling current through the center barrier is negligible for a center gate voltage Vg_{center} of -1 V. The drag voltage V_{drag} has a peak at $Vg_{drive} = -0.592$ V and becomes zero away from the peak. This, to the left of the peak, is because the drive side is pinched off in the dot region and, to the right of the peak, because the dots are open in the drive wire. The conductance of each wire is smaller than e^2/h at the V_{drag} peak, so that the transport is in the tunneling regime. From the measured I_{drive} and V_{drag}, R_D is calculated to be about -2 kΩ at the V_{drag} peak. This value is about ten times larger than that in our previous report on coupled quantum wires [8]. The strong negative Coulomb drag obtained in this work can be due to the particle-like tunneling through the dot region in the two wires. An exciton formed by an electron in a dot of the drive wire and a hole in the adjacent dot in the drag wire moves through the system to generate a current flowing through the two wires in the opposite directions. From the obtained R_D value we can estimate the ratio of the exciton number to the injected electron number to be about 0.1 %.

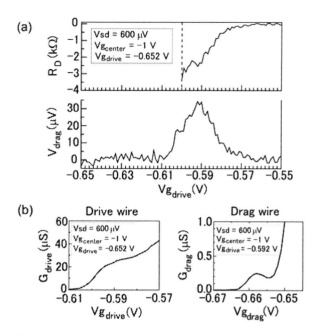

Figure 2. (a) Drag voltage V_{drag} observed for Vsd = 600 µV, $Vg_{center} = -1$ V and $Vg_{drag} = -0.652$ V at 10 mK as a function of Vg_{drive}. (b) Conductance vs. gate voltage for the drive and drag wires, respectively.

CONCLUSION

We performed Coulomb drag experiments on parallel coupled quantum wires having a series of two quantum dots. We observed strong negative Coulomb drag when the transport through the dot region is in the tunneling regime for both wires. This is assigned to exciton transport promoted by electron tunneling through the series of two dots in the drive wire.

REFERENCES

1. T. Gramila. et al., Phys. Rev. Lett. 66, 1216 (1991).
2. U. Silvan. et al., Phys. Rev. Lett. 68, 1196 (1992).
3. A. Jauho and H. Smith, Phys. Rev. B 47, 4420 (1993).
4. N. P. R. Hill. et al., Phys. Rev. Lett. 78, 2204 (1997).
5. H. Rubel. et al., Phys. Res. Lett. 78, 1763 (1997).
6. P. Debray. et al., Physica E 6, 694 (2000).
7. P. Debray. et al., J. Phys.: Condense. Matter 13, 3389 (2001).
8. M. Yamamoto. et al., Proc. of the 25th International Conference on the Physics of Semiconductors (2000).

Tunnel-coupling blockade in vertical/lateral hybrid dot to study many-body states for electron number N=1,2 and 3

K.Yamada[1*], M.Stopa[1], Y. Tokura[3], T.Hatano[1], T.Ota[1], T.Yamaguchi[1], and S.Tarucha[1,2,3]

[1]ERATO Mesoscopic Correlation Project, JST, 4S-308S, NTT Atsugi Research and Development Center, 3-1 Wakamiya, Morinosato, Atsugi, 243-0198, Japan
[2] Department of Applied Physics, University of Tokyo, Bunkyo-ku, Tokyo, 113-0033, Japan
[3]Basic research laboratories, NTT Atsugi Research and Development Center, 3-1 Wakamiya, Morinosato, Atsugi, 243-0198, Japan

Abstract. We report an asymmetric current blockade phenomenon which occurs in unique, hybrid vertical/lateral GaAs quantum dots and which modifies the nonlinear current-voltage characteristics. The data exhibit an enlargement of Coulomb blockade regions in the plane of source-drain and gate voltage, revealing suppression of elementary transitions between ground states or excited state of different electron number. This suppression indicates inaccessibility due to the eigenstate from the dot to 2DEG.

The current through the quantum dot between two leads oscillates as the function of the potential by the Coulomb blockade effect[1]. Furthermore, in the quantum Hall(QH) regime for lateral quantum dots, current blocking can occur due to inaccessibility of inner Landau levels through the tunnel barriers from the two-dimensional electron gas (2DEG)[2]. On the other hands, in vertical dots, quantum levels distribute to well-known atomic-like shell structure. We call levels for (n, l) = (0, 0), (0, 1), (0, -1), (0, 2) ,"$1s, 2p_x, 2p_y, 3d_x$" after atomic electron levels, respectively[3]. Here, n and l are the radial quantum number and the quantum number for angular momentum, respectively. We reported electron-spin and electron-orbital dependence of the tunnel in laterally coupled double vertical dots[4]. In order to investigate accessibility of atomic-like wave-function from the leads, we connected a vertical dot laterally to 2DEG. We report a new blockade due to inaccessibility for this unique hybrid-vertical lateral single dot[5].

We show the schematic drawing of the hybrid vertical-lateral quantum dot in Fig.1 and the scanning electron microscope of a typical dot in the inset figure. The white line indicates the area of the cross section. Electrons are injected laterally from the two-dimensional electron gas (2DEG) to the dot across a schottky barrier at positive source drain voltage (V_{SD}). Furthermore, electrons escape vertically to three-dimensional electron gas(3DEG) across an AlGaAs barrier. The side-gate surrounds the pillar and its voltage (V_{side}) controls the number N of electrons in the dot. The back-gate adds an extra degree of barrier and electron number control. Tunnel coupling to the dot from the 3D lead, which occurs through the AlGaAs heterostructure barrier, is effectively independent of eigenstate whereas the lateral coupling to 2DEG lead is highly state-dependent.

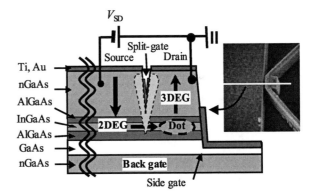

FIGURE 1. Schematic drawing of the hybrid vertical lateral dot. In the inset, the white line indicates the cross section area of the main figure. Electrons are injected from 2DEG to the dot through the schottky barrier and escape to 3DEG across the AlGaAs barrier. The inset figure shows scanning electron micrograph of a typical dot.

We measure transport properties of the hybrid vertical-lateral quantum dot at 10mK in a dilution refrigerator applying magnetic filed B perpendicularly

---

* Present address: Department of Physics, Kyushu university, 6-10-1 Hakozaki, Higashi-ku, Fukuoka, 812-8581, Japan.

to the 2DEG. We show the plot of current as functions of V_{side} and V_{SD} at the magnetic filed $B=3T$ in the Fig.2. The main structures visible are sketched in the inset of Fig.2. In the white diamond-like region (A) the current thought the dot is blocked by the Coulomb blockade effect and electrons inside the dot are fixed to the state $N=2$, $(1s^2)$. In the region (B) with $\mu_{3D}<E(1s^22p_x)-E(1s^2)<\mu_{2D}$, current flows from 2DEG thought the dot to 3DEG as the following process. Here μ_{2D}, μ_{3D} and $E(1s^22p_x)$ are the chemical potential μ_{2D} of 2DEG and 3DEG, the energy of dot with state $(1s^22p_x)$, respectively. An electron is injected to a $2p_x$-orbital of the dot from 2DEG and the state of the dot changes from $1s^2$ to $1s^22p_x$. And then a electron escapes from $2p_x$-orbital to 3DEG and the state of the dot returns from $1s^22p_x$ to $1s^2$. In the light gray region (C) with $\mu_{3D}<E(1s^22p_x)-E(1s2p_x)$, the current is suppressed. The reason of this suppression is considered as following. A $1s$-electron escapes from the dot with $N=3$ state $(1s^22p_x)$ to 3DEG and the state of the dot becomes the $N=2$ excited state $(1s2p_x)$. It is considered that $1s$-orbital is coupled weakly to 2DEG because wavefunction of $1s$ spreads over small area whereas that of $2p_x$ spreads over an large area and orbital $1s$ is further from 2DEG. The $(1s2p_x)$ state is metastable state because 2DEG cannot inject another $1s$ electron due to weak connectivity and the dot cannot go to the $N=3$ state $(1s^22p_x)$. The state $(1s2p_x)$ has parallel spins due to many body effect[6] and the relaxation to the singlet state $(1s^2)$ with anti-parallel spin takes long time (~100µs) because spin relaxation is needed. In the dot, the electrons are fixed to metastable state and the current thought the dot is suppressed. At the negative V_{SD} region (D), the state $(1s2p_x)$ is not metastable because 3DEG injects a $1s$-electron to the dot easily and electrons inside the dot can become the state $(1s^22p_x)$ from the state $(1s2p_x)$.

In conclusion, we fabricated a dot connected vertically to 3DEG and latterly to 2DEG. We observed the asymmetric tunneling due to accessibility from the dot to 2DEG and 3DEG reservoirs.

We thank T. Maruyama, T. Sato and S. Sasaki for help of fabricating device. We thank S. Teraoka, T. Inoshita, W. Izumida for various contributions. The authors acknowledge financial supports from the DARPA grant number DAAD19-01-1-0659 of the QuIST program, SORST-JST, the Grant-in-Aid for Scientific Research A (No. 40302799).

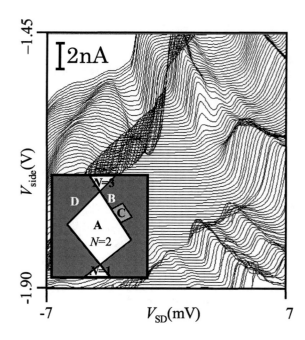

FIGURE 2. The plot of current as function of V_{side} and V_{SD} at magnetic field 3T. In the schematic drawings(the inset), the white, light gray and dark gray areas indicate the Coulomb blockade, geometric suppression and tunneling region, respectively.

REFERENCES

1. L.P. Kouenhoven, C. M. Marcus, P. L. McEuen, S. Tarucha, R. M. Westervelt, and N. S. Wingreen, in *Mesoscopic Electron Transport*, edited by L. L. Sohn, L. P. Kouwenhoven, and G. Schön, NATO ASI, Ser. E, vol. 345 (Kluwer Dordrecht, 1997), pp. 105-214.

2. M. Ciorga, M. Pioro-Ladriere, P. Zawadzki, P. Hawrylak, and A. S. Sachrajda, *Appl. Phys. Lett.* **80**, 2177 (2002).

3. S. Tarucha, D. G. Austing, T. Honda, R. J. van der Hage, and L. P. Kouenhoven, *Phys. Rev. Lett.* **77**, 3613 (1996).

4. T. Hatano, M. Stopa, T. Yamaguchi, T. Ota, K. Yamada and S. Tarucha, *Phys. Rev. Lett.* **93**, 066806 (2004).

5. K. Yamada, M. Stopa, T. Hatano, T. Ota, T. Yamaguchi, and S. Tarucha, *Superlattices and Microstructures*, Vol.34, Issue 3-6, pp. 185-189 (2003).

6. Y. Tokura, S. Sasaki, D. G. Austing, and S. Tarucha, *Physica* B **298**, 260 (2001).

Theoretical Description Of The Electronic Coupling Between A Wetting Layer And A QD Superlattice Plane

C. Cornet, C. Platz, J. Even, P. Miska[1], C. Labbé, H. Folliot, S. Loualiche

Laboratoire d'étude des nanostructures à semiconducteurs (LENS-UMR6082)
Insa de Rennes, 20 Avenue des Buttes de Coesmes, 35043 Rennes Cedex, France
[1] *Present address : IEMN-UMR8520, BP 69 F, 59652 Villeneuve d'Ascq Cedex, France*

Abstract. New results on the simulation of 2D Quantum Dots (QD) InAs/InP superlattices, emitting at 1.55 µm, telecommunication wavelength, are presented here. Using the Fourier-transformed Schrödinger equation, we describe the electronic coupling between a wetting layer (WL) and the QD superlattice. It is shown that the increase of quantum dot density, induces a fragmentation of WL density of states, and the apparition of localized states. The results shows that WL and QD have to be considered as a unique system, in strong coupling conditions.

INTRODUCTION

Quantum dots (QD) are grown on InP in our group in order to study their physical properties and propose new device applications based on the remarkable properties of these structures where emission wavelength can be tuned to the telecommunication spectrum of 1.55 µm. Growth on (113)B substrate orientation leads to good quality QD structures with high densities and low size dispersion. An original double cap method (DC) has been developed for controlling the emission energy[1]. Optical characterisation of QD ground state (GS), excited state (ES) and wetting layer states (WL) emissions has been obtained by photoluminescence (PL), Fourier Transformed Infrared (FTIR) spectroscopy[2] and time-resolved photoluminescence (TRPL) experiments[1]. In order to understand QD properties, a theoretical description of these structures is performed to establish the link between measured properties and QD organization.

EXPERIMENTAL RESULTS

Recent PL and FTIR absorption measurements, on a 12 QD stacked layers (InAs/InP) sample show the proximity between QD and WL spectra[2]. Moreover, we can clearly identify a WL structuring on FTIR and TRPL[1] spectra. Kamerrer and al.[3] have already pointed out the influence of the environment on the QD properties.

CALCULATION

An interpretation of experimental observations is done by simulating a 2D QD superlattice on a WL. In order to adequately describe the electronic coupling between QD and WL states, we resolve the Fourier-transformed (FT) Schrödinger equation for the electron or hole:

$$\left\langle \vec{k}' \left| -\frac{\hbar^2}{2} \vec{\nabla} \frac{1}{m^*(\vec{r})} \vec{\nabla} \sum_{\vec{k}} c_{\vec{k}} \right| \vec{k} \right\rangle + \left\langle \vec{k}' \left| \sum_{\vec{k}} V(\vec{r}) c_{\vec{k}} \right| \vec{k} \right\rangle = E \left\langle \vec{k}' \left| \sum_{\vec{k}} c_{\vec{k}} \right| \vec{k} \right\rangle$$

We choose anisotropic effective masses for kinetic Hamiltonian part. The FT potential is calculated analytically. Hamiltonian diagonalization is numerically performed, with a mixed basis of localized states and plane waves:

$$\Psi(\vec{r}) = \frac{1}{\sqrt{V_{crystal}}} \sum_{\vec{k}} c_{\vec{k}} e^{i\vec{k}\vec{r}} + \sum_{n} c_n \varphi_n(\vec{r})$$

RESULTS

First, we study QD/WL coupling influence on the density of states. Calculation is made for a 2D hexagonal lattice. As shown in Fig. 1 the (QD + WL) system electronic density of states is very different from sum of the two separate QD (0D) and WL (2D) densities of states, in the [-100 ; 0] meV energy range. The potential's barrier is chosen as the reference level for the energies. Energy spectrum modification is an evidence of the coupling effects. 2D WL electronic density of states is fragmented with coupling.

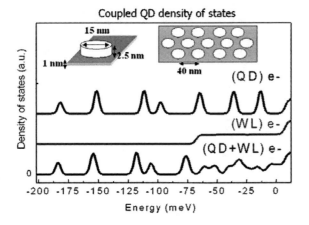

FIGURE 1. Electronic density of states for an hexagonal QD lattice on a WL. Coupling effects evidence.

By calculating the heavy hole and the electronic densities of states, we can determine the interband absorption spectrum. Calculation is made for several geometries, 2D hexagonal QD lattice, 2D square QD lattice, 2D linear Quantum wires (Qwi) lattice. Results are similar in all cases. As shown in Fig. 2, absorption spectra for high and low quantum dots density (QDD) are very different. First, WL absorption spectrum structuring, when QDD increases, is observed once more. Another consequence is a transition energy redshift.

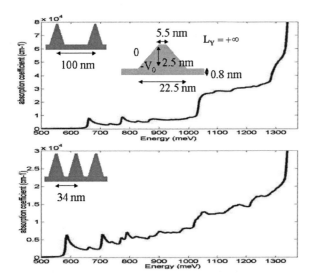

FIGURE 2. Absorption spectrum for a 2D linear Qwi lattice. WL contribution is structured for a high QDD.

In order to have a better understanding, we can plot Brillouin zone dispersion. This leads to very weak (0.1 meV) minibands formation with the studied coupling, for the QD states. The wavefunctions spatial variations show that QD ground and excited states, are still localized with the considered coupling, i.e. for a high QDD. But on the other hand, lower energy WL wavefunctions lose its 2D character. A part of WL wavefunctions, moves into the QD region. This means that WL states have no longer a purely 2D character, but they look like hybrid 2D/0D confinement states. WL pseudo-confined states appear. This explains why WL contribution in the electronic density of states spectrum is structured with coupling.

For FTIR and TRPL spectra structuring interpretation, two hypothesis can be made. As shown in the present calculation, WL structuring can be a direct consequence of QD/WL coupling. Another way is to consider that WL structuring shows different HH and LH transitions[4]. New experiments on high density specific samples and calculations in progress will help to clarify these different hypothesis.

CONCLUSIONS

The electronic coupling between a WL and a QD superlattice plane and its experimental consequences were studied using FT-Schrödinger equation on a mixed basis. We found clear evidence of coupling influence on electronic properties. We demonstrated that WL should no longer be considered as a 2D-like confining object, but as an hybrid 2D/0D object. Experimental measurements on high density samples can be analyzed using these results. Then, because of its states, that are spread laterally in QD volume, WL could help for an inter-QD carrier redistribution, i.e. for QD laser injection.

REFERENCES

1. P. Miska, C. Paranthoen, J. Even, O. Dehaese, H. Folliot, N. Bertru, S. Loualiche, M. Senes and X. Marie, *Semicond. Sci. Technol* **17**, L63-L67 (2002) and references.

2. C. Cornet, C. Labbé, N. Bertru, O. Dehaese, J. Even, H. Folliot, A. Le Corre, C. Paranthoen, C. Platz and S. Loualiche, *(submitted to Appl. Phys. Lett.)*

3. C. Kamerrer, G. Cassabois, C. Voisin, C. Delalande, Ph. Roussignol and J. M. Gérard, *Phys. Rev. Lett.* **87**, 207401 (2001).

4. S. Malzer, M. Kahl, O. Wolst, M. Schardt and G. H. Döhler, *Proceedings of the ICPS 2002*.

Capacitance spectroscopy study of InAs quantum dots and dislocations in p-GaAs matrix

P.N. Brunkov[*†], E. V. Monakhov[†], A. Yu. Kuznetsov[†], A.A. Gutkin[*], A.V. Bobyl[*], Yu.G. Musikhin[*], A.E. Zhukov[*], V. M. Ustinov[*], and S. G.Konnikov[*]

[*] *A.F.Ioffe Physico-Technical Institute RAS, 194021 St-Petersburg, Russia*
[†] *Department of Physics, Physical Electronics, University of Oslo, PO Box 1048 Blindern, N-0316 Oslo, Norway*

Abstract. Two hole traps have been identified in epitaxially grown n^+-p-GaAs diodes with embedded self organized InAs quantum dots using temperature dependent capacitance-voltage (C-V) measurement, admittance spectroscopy and deep level transient spectroscopy (DLTS). One trap is found to be located in the range of (0.18÷0.29) eV above the top of the GaAs valence band (E_V) and was assigned to the manifestation of QDs. A deeper trap located at (E_V+0.34) eV in accordance DLTS and confirmed by fitting to the *C-V* data was observed and attributed to the interaction of carriers with dislocations.

INTRODUCTION

InAs/GaAs self-organized quantum dots (QDs) have recently attracted significant interest [1]. The driving force of the self-organization process is an elastic strain due to a considerable lattice mismatch at the InAs/GaAs interface. The QDs form zero-dimensional potential in a wide band gap matrix and act as traps of charge carriers. Along with self-organization of QDs the elastic strain may result in generation of both point defects and dislocations in the vicinity of the InAs/GaAs heterointerface. In this study we have identified hole traps associated with QDs and dislocations located presumably in the p-type region of the GaAs p-n^+ structure with InAs QDs using various modes of capacitance spectroscopy.

RESULTS AND CONCLUSIONS

The p-n^+ diode structure with InAs QDs was grown by molecular beam epitaxy on p-type GaAs substrate. The InAs QDs formed after deposition of 3.3 ML of InAs on the GaAs surface at 480°C was embedded in a Be-doped GaAs matrix (p =2 × 10^{16} cm^{-3}) at a distance of ~500 nm from the p-n^+ junction. The QDs are sandwiched between two 10 nm thick undoped GaAs buffers. Ohmic contacts to the p^+ substrate and the top n^+ layer were formed by evaporation and alloying of AuZn and AuGe, respectively. Fig.1 shows temperature dependence of the *C-V* characteristics of QD structure measured at the frequency f = 1 MHz. It is seen from Fig.1 that at 150 K the *C-V* curves demonstrate a characteristic feature – a step in the range of -0.15 ÷ 0.3 V which, in accordance with Ref. 2, is related to the discharging of the QDs. Using quasi-static charging model from Ref. 2 the characteristic step at 150 K has been numerically fitted and hole ground state in the InAs QDs is found to be located at 0.23 eV above the top of the GaAs valence band (E_V). For temperatures below 120 K, the plateau in the *C-V* characteristic in Fig.1 disappears presumably because the thermionic emission rate of holes (e_p) from QDs becomes lower than the angular measurement frequency $\omega = 2\pi f$. As the temperature increases above 200 K the second feature appears in the *C-V* curves in Fig.1, which indicates the presence of one more hole trap in the vicinity of the QD layer. Further, admittance spectroscopy was used to measure the temperature dependence of the hole escape rate out of the QDs, $e_p=1/\tau_{esc}^h$, where τ_{esc}^h is a time associated with a thermal escape of holes from the ground state of the QDs. Admittance spectroscopy involves measuring the capacitance (*C*) and conductance (*G*) of the device in the *ac* mode with a tunable frequency *f*.

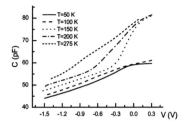

FIGURE 1. Temperature dependence of the C-V characteristics of QD structure as measured at $f = 1$ MHz.

Fig. 2 shows the Arrhenius plot for e_p and admittance spectroscopy has been measured at different reverse biases. For the bias range from +0.3 V to -0.15 V we found that the activation energy for hole emission from the ground state of the QDs (ΔE_a^{h0}) varies in the range from 0.18 to 0.29 eV above the top of the GaAs valence band. The spread in the ΔE_a^{h0} values may be attributed to the energy distribution of the QDs due to fluctuations in their size, composition and shape [2]. The average value of ΔE_a^{h0} is in agreement with the energy of the hole ground state as determined above from the quasi-static C-V simulations. To resolve the second hole trap in the vicinity of the QD layer we used the Deep Level Transient Spectroscopy (DLTS), which shows better energy resolution then the admittance spectroscopy. The DLTS spectra of the QD structure reveals two hole traps H_{QD} and H_{disl} (Fig.3). With decreasing the amplitude of the reverse bias V_b and the filling pulse V_p the activation energy of the trap labeled as H_{qd} approaches a value of 0.20 eV, which is close to the ΔE_a^{h0} value as determined both from the C-V and admittance measurements. A deeper hole trap (labeled below as H_{disl}) was found to have an activation energy of 0.34 eV above the top of the GaAs valence band E_V. The characteristics of the H_{disl} trap as determined in Fig.3 are similar to those obtained in Ref.3 for a trap labeled as H1 and associated with interaction of carriers with threading dislocations in strained InAs/GaAs heterostructures. It is worthwhile to mention that the presence of dislocations in our samples around QD layer was confirmed by microscopy studies. In addition, we have studied the DLTS peak amplitudes as a function of the duration of the DLTS filing pulse (t_p). In the range of t_p from 100 ns to 100 ms the H_{QD} peak does not show any significant dependence on the t_p duration, see Fig.3. This is an indication of the fast carrier capture in the InAs QDs, which is a characteristic feature of the self-organized QDs [1]. In contrast, the amplitude of the H_{disl} peak grows linearly with a logarithmic increase of t_p (Fig.3), which is an indication for a change of the carrier capture barrier as traps change the charge. This behavior is typical for the traps associated with dislocation [4].

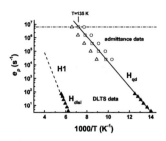

FIGURE 2. Arrhenius plots of the hole emission rates from QDs (H_{QD}) and a trap associated with dislocations (H_{disl}). Open symbols show the results of the admittance spectroscopy measured at different reverse voltages V_b: ○ – +0.2 V ; □ – 0.0 V; △ – -0.2 V. The filed triangles (▲) represent the data of the DLTS measurements with V_b = -1.0 V, V_p = +1.2 V and t_p = 1 ms. The horizontal dash-dotted line is the angular frequency for the $f = 1$ MHz. The dashed line represents the hole trap H1 from Ref.3. The solid line is a guide for eyes.

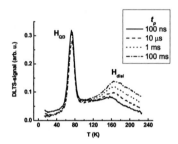

FIGURE 3. DLTS spectra of the QD structure as measured using V_b = -1.0 V and V_p = +1.2 V. Different curves represent different filling pulse durations.

In summary, two hole traps associated with QDs and dislocations has been found in our QD structure and electronic properties of the traps are determined consistently by various modes of capacitance spectroscopy. The work is supported by the INTAS project 0175-WP. The Nordic Academy for Advanced Study (NorFA) is acknowledged for the promotion of the mobility of researches.

REFERENCES

1. Bimberg, D., Grundmann, M., and Ledentsov, N. N., *Quantum Dot Heterostructures,* John Wiley & Sons Ltd: Chichester, 1998.

2. Brunkov, *et al, Phys.Rev.B* **65** 085326, (2002).

3. Du, A.Y., *et al, Appl. Phys. Lett.* **69**, 2849-2851, (1996).

4. Wosinski T., *J. Appl. Phys.* **65**, 1566-1570, (1989).

Resonant scattering of a two-dimensional electron gas by quantum dot levels

N.M.Sotomayor*, G.M.Gusev*, A.C.Seabra†, A.A.Quivy*, T.E.Lamas* and J.R.Leite*

*Departamento de Física de Materiais e Mecanica, Instituto de Fisica da Universidade de Sao Paulo, SP, Brazil
†Escola Politecnica da Universidade de Sao Paulo, SP, Brazil

Abstract. In present work we produced quantum dots separated by tunnelling $Ga_xAl_{1-x}As$ barrier with two-dimensional electron gas. The dots are formed by overlapped antidots on the front quantum well of the double-well GaAs structure. We observed the peak in the resistance which we attributed to the scattering of back well electrons by the dots level. In magnetic field we found commensurability oscillations of the electrons in the back quantum well due to the antidot lattice potential.

INTRODUCTION

The energy levels of the remote impurities in the heterostructures $GaAs/Ga_xAl_{1-x}As$ play no role in the scattering, since the charged donors has the energy levels substantially higher than the electronic Fermi level E_F. When the levels are brought close to E_F, the scattering becomes strongly enhanced. Moreover, the virtual transitions between dots and plane may results in multiple resonant scattering. Such scattering strongly affects single particle density of states and leads to a renormalization of the diffusion coefficient up to an order of magnitude. Resonant scattering is probably realized in systems with self-assembled quantum dots separated from two-dimensional electron gas (2DEG) by a tunnelling barrier[1]. It has been found that the mobility of the two-dimensional electron gas dropped by 2 order of magnitude when the thickness of the barrier between the self-assembled dots and 2DEG was reduced. Very high density of the dots ($\sim 10^{10} - 10^{11}$ and its small size ($\sim 20nm$) allow to use this system for comparison of the experimental results with theoretical models of the resonant scattering. Disadvantage of such system is the relatively wide distribution of the dots size, which leads to the broadening of the resonant scattering effect in realistic structures. In present work we produced and studied another system, when quantum dots coupled with 2D electron gas were fabricated by electron beam lithography. The ability to tune the dots size allows to fabricate, in principal, transistor, when the mobility of 2DEG can be manipulated by resonance conditions between Fermi level and levels in the dots. Disadvantage is still relatively large size of the dots and its small density.

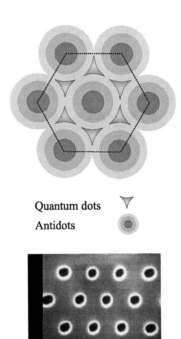

FIGURE 1. Schematic view of the hexagonal antidot lattice. The potential of the neighbor antidots is overlapped, which leads to the formations of the the array of disconnected quantum dots with triangular shape in the top quantum well. Bottom-micrography of the structure with antidot array.

EXPERIMENTAL RESULTS AND DISCUSSIONS

The structures consist of asymmetrically doped GaAs double quantum wells separated by tunnelling

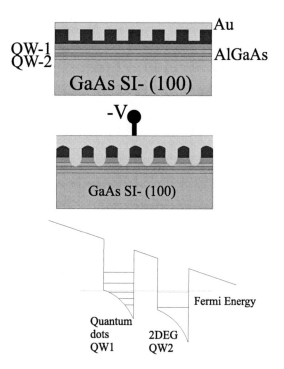

FIGURE 2. Top- profile of the device with double quantum wells. Applied negative gate voltage forms lattice with overlapped antidots and triangular disconnected dots. Bottom-schematic conduction-band profile for a double-layer 2DEG, when quantum dots are formed in front well.

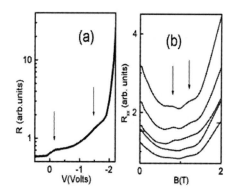

FIGURE 3. (a) Resistance as a function of the gate voltage, T=1.5 K. Arrows indicate resistance peaks associated with resonant scattering. (b) Magnetoresistance as a function of the magnetic field for different gate voltages. Arrows indicate commensurability peaks.

$Ga_xAl_{1-x}As$ barrier of 50 width. The width of the GaAs well is 140. The samples with different electronic densities in the front well has been grown. Sample A has density in back well $1 \times 10^{12} cm^{-2}$, in the front well $6 \times 10^{11} cm^{-2}$ and average mobility $\sim 240 \times 10^3 cm^2/Vs$. Sample B has density in back well $1 \times 10^{12} cm^{-2}$, in the front well $3 \times 10^{11} cm^{-2}$ and average mobility $\sim 100 \times 10^3 cm^2/Vs$. On the top of the structure the hexagonal superlattice of the antidots with periodicity 0.6 μm and diameter 0.2 μm was patterned in the PMMA resist, which was then covered by the gold gate. Fig.1 shows schematically the part of the antidot lattice and realistic micrography of this structure. When the negative gate voltage is applied, potential of the neighbor antidots is overlapped and the array of disconnected quantum dots with triangular shape has been formed in the topmost quantum well. Figure 2 shows schematically profile of the structure (top) and the conduction-band profile for a double-layer 2DEG, when quantum dots are formed in front well.

Since the concentration of the electrons in the front well is smaller than in the bottom, we believe, that electrons in the topmost well are collected into triangular dots before electrons in back well. Indeed electrons in back well are moving in antidot lattice potential [2]. Fig.3 shows the dependence of the resistance of described above structure as a function of the gate voltage.

We may see two broad peaks. We attribute these peaks to resonance scattering of the electrons in the back well by the levels of the dots formed in the front well. Density of the dots is near $10^9 cm^{-2}$ and 100 times lower than electron density. The area of the dot is approximately $0.055 \mu m^2$ which corresponds to the characteristic energy for circular dot $E_c \sim 0.5 - 1 meV$. Relatively small magnitude of the peaks results from the low quantum dot density. Further increase of the antidot periodicity is required for larger effect. Fig.3b shows magnetoresistance curves for different gate voltages. Note two oscillations, which becomes weaker for voltage decrease. High field oscillation corresponds to the cyclotron radius $R_c \approx 0.07 \mu m$, which is much smaller than the periodicity of the antidot array $0.6 \mu m$. We may assume that the peaks are due to the commensurability conditions inside of the dots in the back well, which are almost closed. Indeed in this case dots in the front well are closed, and, as we assumed, play the role of scatterers for bottom layer.

ACKNOWLEDGMENTS

The support by FAPESP and CNPq is gratefully acknowledged.

REFERENCES

1. H.Sakaki *et al*, Apll.Phys.Lett, **67**, 3444 (1995).
2. G.M.Gusev *et al*, J.Phys.Cond.Matter, **4**, L269 (1992).

Transitions of dimensional states in quantum dots formed by spherical barrier

Seung Joo Lee[1], Tae Won Kang[1], Satofumi Souma[2], S. K. Noh[3], and J. C. Woo[4]

[1]*Quantum-Functional Semiconductor Research Center, Dongguk University, Seoul 100-715, Korea*
[2]*Department of physics, University of Delaware, Delaware, USA*
[3]*Korea Research Institute of Standard and Science, Taejeon, Korea*
[4]*Department of physics, Seoul National University, Seoul, Korea*

Abstract. The control of the spherical barrier thickness or height is found to cause the dimensional transitions from the three dimensional (3D) behavior to the quasi-zero dimensional (Q0D) behavior in the density of states (DOS) and the heat capacity. When the barrier is thick enough such that the DOS shows the Q0D-like signature but is still thin enough to allow electrons to tunnel it, the temperature dependence of the heat capacity exhibits quite distinct behavior depending on the electron density.

INTRODUCTION

Recently semiconductors synthesized in colloidal solutions have attracted much interest because of their small size ranging from 1 to 10nm [1-4]. Those quantum dots grown in a colloidal solution have frequently spherical shapes and can be fabricated as multi-shell structures such as quantum dot-quantum wells and quantum dot-quantum barriers, which is very attractive for applications in novel optoelectronic devices. Motivated by such experimental techniques to fabricate multi-shell structures, we study the density of electronic states in quantum dot structures (nano-crystals) formed by a thin spherical barrier. The finite thickness of the spherical barrier layer allows electrons in the quantum dot to escape to the outside of the dot. We study the heat capacity of such systems. It is interesting to ask how are the electron density dependence and the temperature dependence of the heat capacity influenced by the thickness of the spherical barrier. Even though the fundamental importance of the heat capacity of small systems [5], the heat capacities of quantum dots have not been investigated except for the important work in which the effect of electron-electron interaction is studied in strong magnetic field[6].

THEORY AND DISCUSSIONS

To consider this problem, we extend the theoretical formulation used to treat the multi-directional double barriers [7-9] and the cylindrical quantum wire [10] to the spherical problem. To begin, we model the proposed spherical quantum dot structure by the following simplified model potential:

$V(r) = V_0 \delta(r-a)$, where $r = (x^2+y^2+z^2)^{1/2}$. In this expression, the thin $Al_xGa_{1-x}As$ spherical barrier has been replaced by δ function with strength V_0. The parameter V_0 is given by $V_0 = d\Delta E_c$, where d is the barrier width and ΔE_c is the conduction-band discontinuity between GaAs and AlGaAs.

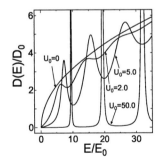

FIGURE 1. The calculated DOS $D(E)$ vs energy E for a spherical quantum dot with a = 10nm. Here $D_0 = m^*a^2/(\pi\hbar^2)$. The energy is scaled by $E_0 = \hbar^2/(2m^*a^2) = 5.68$ meV.

The use of this simplified potential form is justified if ΔE_c is very large and d is less than the de-Broglie wavelength of an electron. We now solve the time-independent Schrodinger equation with the Hamiltonian $H=-(h^2/2m^*)\nabla^2+V(r)$, where $m^*=0.067m_0$ is the effective electron mass of GaAs. We assume the vanishing boundary condition $\Psi(r = R, \theta, \varphi) = 0$ just for the normalization of the wave function along the radial direction. At the final stage of our calculation, we will take the limit $a/R \rightarrow 0$ corresponding to the bulk limit.

After getting the normalization constant and eigen energies, we get the local density of states (LDOS) and the DOS following the routine procedure [10]. Once we obtain the analytical expression for the DOS, the heat capacity can be calculated numerically as follows. Here we consider only the heat capacity contributed by electrons. Other effects such as phonons are neglected to test the dimensional crossover of the heat capacity. The heat capacity C_v is defined and calculated as Ref. [9].

Figure 1 shows the calculated DOS as a function of the energy E for several values of U_0. $U_0 = 2m^*\Delta E_c da/\hbar$. Here we consider the sphere with radius a = 10[nm]. As seen in Fig. 2, one can observe the direct transition from the square-root-type 3D-DOS to the δ-function-type Q0D-DOS by controlling only one parameter, U_0. Moreover, in between the 3D and the Q0D limiting cases, we can see that the DOS shows the step-function-like behavior (2D signature) and the saw-tooth-like behavior (1D signature) at appropriate barrier strengths. As a whole, Fig. 1 shows that the increase of U_0 results in the appearance of the low-dimensional behavior in the DOS. Here the electron tunneling through the spherical barrier can be interpreted to give the "free" degree of freedom to the DOS, while the electrons' motion along the θ and φ directions is almost "frozen" because the discreetness of m and the smallness of the radius a cause a large energy level separation between two adjacent values of m. Figures 2(a), (b) shows the electron number N dependence of the heat capacity C_v for various values of the barrier strength U_0. Detailed discussions on the temperature dependence of the heat capacity for this system will be published elsewhere [11].

This work was supported by the Korea Science and Engineering Foundation through the Quantum-Functional Semiconductor Research Center at Dongguk University and the Ministry of information and Communication (Contract No. IMT2000-B4-1).

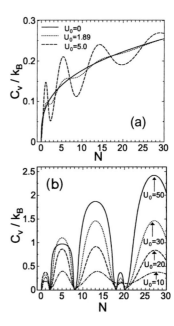

FIGURE 2. Calculated heat capacities C_v are shown as a function of the number of electrons N at the temperature T =4 K. (a): Results for smaller values of U_0. (b): Results for larger values of U_0.

REFERENCES

1. A. Mews, A. Eychmuller, M. Giersig, D. Schoos, and H.Weller, J. Phys. Chem. 98, 934 (1994).

2. A. P. Alivisatos, Science 271, 933 (1996).

3. J. Planelles, J. G. Diaz, J. Climente, and W. Jaskolski, Phys. Rev. B 65, 245302 (2002).

4. M. Achermann, M. A. Petruska, S. Kos, D. C. Smith, D. D. Koleske, and V. I. Klimov, Nature, 429, 642 (2004)

5. M. Arzberger and M.-C. Amann, Phys. Rev. B 62, 11029 (2000).

6. P. A. Maksym, and T. Chakraborty, Phys. Rev. Lett, 65 108 (1990).

7. T. B. Bahder, J. D. Bruno, R. G. Hay, and C. A. Morrison, Phys. Rev. B 37, 6256 (1988).

8. S. J. Lee, N. H. Shin, J. J. Ko, C. I. Um, and T. F. George, Phys. Rev. B 45, 9173 (1992).

9. S. J. Lee, J. H. Oh, K. J. Chang, and G. Ihm, J. Phys. Condens. Matter, 6, 4541 (1994).

10. S. Souma, S. J. Lee, N. Kim, and T. W. Kang, J. Appl. Phys. 58, 4649 (2002).

11. S. Souma, S. J. Lee, and T. W. Kang, Unpublished (2004).

Magnetotransport in HgSe:Fe quantum-dots

Mehtap Bayir, Alexander Kirste, Tuan Tran-Anh, Stefan Hansel, Michael von Ortenberg

Humboldt University at Berlin, Germany, Institute of Physics, Magnetotransport in Solids

Abstract. We report on magnetotransport measurements in intrinsically populated HgSe:Fe quantum-dot systems in magnetic fields up to 12 T. Depending on the dot density we observe positive and negative differential magnetoresistance, the later similar to hopping conductivity in impurity bands.

INTRODUCTION

In many ways modern semiconductor technology reconstructs natural systems artificially: superlattices as mesoscopic replicas of the natural crystal lattices and quantum dots similar to single atoms. To transpose the special features of interacting atomic states as manifested in *hopping conductivity* by overlapping donor or acceptor states into quantum-dot systems was the principal objective of the present investigations. Whereas most of the quantum-dot systems grown so far need artificial population with charge carriers, HgSe:Fe provides a dot system, where the bound dot states are electronically populated even at low temperature. We succeeded in growing quantum-dot systems of HgSe:Fe exhibiting different sizes of the individual quantum-dot as well as different mean distance in a statistical configuration [1]. The resulting structures resemble in many aspects two-dimensional networks giving rise to percolation manifesting in a hopping conductivity. Here we present a study of combined structural and magnetotransport investigations.

THE HgSe:Fe QUANTUM-DOT SYSTEM

We succeeded in growing HgSe:Fe quantum-dot system of different density depending on the interface strain with respect to the ZnSe(Te) buffer layer. The HgSe:Fe layer formed strained quantum dots of about 10-100 nm diameter and a height of the order of 10 nm as shown by the AFM images in fig. 1. The Fe-doping in HgSe ensures a "bulk" carrier concentration of $n_e = 5*10^{18}$ cm^{-3} in a Fermi-level pinned system [2]

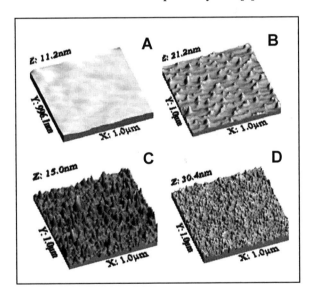

FIGURE 1. A flat HgSe:Fe-surface of sample **A** in comparison with quantum-dot systems of increasing density in samples **B** - **D** of the same material.

The Fe-doping ensures that 5-500 electronic states are occupied in each quantum dot, depending on its size,. To characterize these states electronically we applied the contact-less method of Megagaussmagnetospectroscopy using far infrared radiation [1]. These results reflect the properties within the individual quantum dot due to the magnetic field induced localization into cyclotron orbits. The principal objective of the present investigation was, however, to study the inter-quantum-dot interaction of the occupied electronic states reminding in some way to a hopping transfer in the impurity band of semiconductors. To concentrate on these features of inter-dot transfer experimentally definitely electrical contacts of the samples were necessary. To avoid the shortcomings of silver-epoxy cracking the dot layer, we managed to fix thin gold wires directly onto the surface by indium solder. All contacts of the Hall-bar were checked for ohmic behavior. Transverse magneto-resistance, longitudinal magnetoresistance, and Hall voltage were measured in magnetic fields up to 12 T at He-temperatures. Depending on the different manifestations of the dot system different features of the magneto-transport properties became predominant indicating Shubnikov-de Haas-oscillations, positive, or negative differential magnetoresistance as shown in figure 2. The sample B with low dot-density shows a strong positive magnetoresistance with small SdH-oscillations referring to a carrier concentration of about $n_e = 4*10^{18}$ cm^{-3}. For the more dense sample C the positive differential magnetoresistance is less pronounced, and for sample D with the highest dot density the positive differential magnetoresistance is limited to the magnetic field range below 1 T and followed by a pronounced negative differential magnetoresistance. This behavior is similar to the results for hopping conductivity in impurity bands of semiconductors. As a matter of fact our high-density quantum-dot sample is in many features similar to an ordinary hopping system even with the characteristic lengths parameters being of the same order of magnitude. For the interpretation of our experimental results we have therefore successfully applied a nearest-neighbor hopping approach considering spin-flip scattering as established by Movaghar and Schweitzer originally applied to ordinary impurity hopping systems [3]. For sample D we have included the corresponding theoretical result by the dashed curve.

REFERENCES

1. Tran-Anh T., Hansel S, Kirste A., Mueller H.-U., von Ortenberg M, Barner J, Rabe J.P. Journal of Alloys and Compounds **371** (2004) 198-201

2. von Ortenberg M., Journal of Alloys and Compounds **371** (2004) 42-47

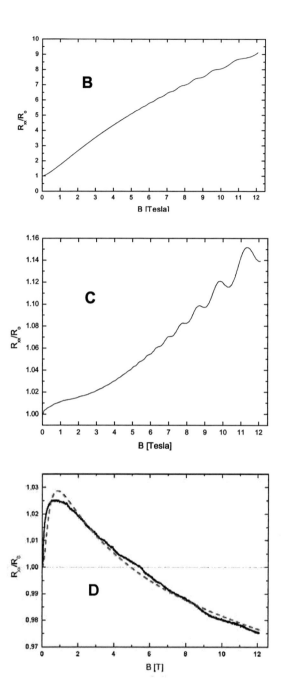

FIGURE 2. Magnetoresistance curves for the samples B, C, and D show a decrease o of the positive magnetoresistance with increasing dot density so that sample D with the highest dot density has a well pronounced negative magnetoresistance. The dashed curve for sample D is the theoretical fit to ref. [3].

3. Movaghar B. and Schweitzer L., J. Phys. **C11** (1977) 125

Electron-phonon Interaction in Si Quantum Dots Interconnected with Thin Oxide Layers

Shigeyasu Uno[*,†], Nobuya Mori[**], Kazuo Nakazato[‡,‡], Nobuyoshi Koshida[§,†] and Hiroshi Mizuta[¶,†]

[*]*Hitachi Cambridge Laboratory, Hitachi Europe Ltd., Madingley Road, Cambridge, CB3 0HE, United Kingdom*
[†]*CREST JST, Shibuya TK Bldg., 3-13-11 Shibuya, Shibuya-ku, Tokyo 150-0002, Japan*
[**]*Department of Electronic Engineering, Osaka University, 2-1 Yamada-oka, Suita, Osaka 565-0871, Japan*
[‡]*Department of Electrical Engineering and Computer Science, Graduate School of Engineering, Nagoya University, Furo-cho, Chikusa-ku, Nagoya 464-8603, Japan*
[§]*Division of Electronic and Information Engineering, Faculty of Technology, Tokyo University of Agriculture and Technology, Tokyo 184, Japan*
[¶]*Department of Physical Electronics, Tokyo Institute of Technology, 2-12-1 O-okayama, Meguro-ku, Tokyo 152-8552, Japan*

Abstract. The electron-phonon interaction in one-dimensional array of Si dots interconnected with oxide layers is theoretically examined. The folded electron/phonon energy dispersion relations and rigorous phonon normal mode calculation are incorporated. The acoustic deformation potential scattering is weakened due to the strain absorption by the oxide layers. Electron energy loss is strongly suppressed near the bottom of a miniband due to restriction of high-energy phonon emission.

INTRODUCTION

Recently, electron transport along one-dimensional array of Si quantum dots interconnected with thin oxide layers (1DSiQDA) has attracted growing attention due to the experimental observation of suppressed electron energy loss [1]. In this work, we theoretically investigate electron-phonon interaction in the 1DSiQDA.

THEORY AND RESULTS

It has been reported that, in the Si quantum wires (SiQW), six equivalent electron energy minima in bulk Si reduces to one Γ-like and two X-like valleys on the k_x axis [2] (Fig. 1 (a)). In the 1DSiQDA, however, the two X-like valleys are folded near the $k_x = 0$ due to structural periodicity, d (Fig. 1 (b)). Electronic properties might then be considered within one-valley effective mass approximation. Assuming low electric field applied along the x axes, electron wave functions are easily obtained using the Krönig-Penny model. The scattering rate due to electron-phonon interaction is calculated using the Fermi's golden rule. Because of the strong lateral electron confinement, the transition matrix has non-zero value only when the mediating phonons have lateral wave vector components smaller than the order of $\sim 1/L$,

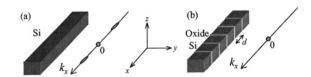

FIGURE 1. Device geometry and electron energy minima in (a) Si quantum wire (SiQW) and (b) one-dimensional array of Si dots interconnected with oxide layers (1DSiQDA). The structural periodicity is denoted by d.

where L is size of the Si dots. For $L = 4$nm, phonons satisfying the above criteria have energy of several meV in the y and z directions. As the x component of the phonon energy can range up to about 60meV in Si, majority of such phonons have wave vectors parallel to the x axis. Consequently, calculation can be reduced to one-dimensional problem at reasonable approximation. In our calculation, phonon normal modes in the x direction was numerically calculated using the linear atomic chain model, taking into account the large difference of Young's modulus for the Si dots and the oxide layers.

The calculation of electronic states gives electron wave functions and electron minibands (Fig. 2 (a)). The phonon energy dispersion shows folding of phonon

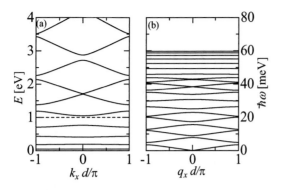

FIGURE 2. Energy dispersion relations of (a) electron and (b) phonon. Broken line: barrier height of the oxides.

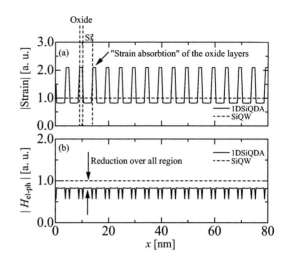

FIGURE 3. (a) Strain caused by an acoustic phonon lattice vibration. (b) Acoustic deformation potential perturbation Hamiltonian.

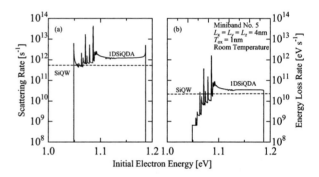

FIGURE 4. (a) Phonon scattering rate (emission) and (b) energy loss rate for electrons in the 5th lowest miniband.

branches, as well as phonon band gaps (Fig. 2 (b)). Interestingly, calculation of acoustic phonons revealed that the oxides between the Si dots "absorb" the strain caused by the lattice vibration, reducing the strain in the Si region compared to that in the SiQW (Fig. 3 (a)). As the perturbation Hamiltonian for acoustic deformaiton potential, H_{el-ph}, is proportional to the strain, the magnitude of H_{el-ph} in the Si dots decreases due to the strain absorption effect (Fig. 3 (b)). Furthermore, even with the strain absorption, H_{el-ph} in the oxide layer is also smaller than that in the SiQW, because of smaller deformation potential coefficient. As a result, H_{el-ph} is decreased over all the region. Despite the reduction of H_{el-ph} for acoustic phonon scattering, the scattering rate increases due to the increase of the electron density of states at energies within electron miniband (Fig. 4 (a)). However, the energy loss rate, defined as expectation value of electron energy loss due to phonon emission per unit time, showed significant reduction at near the bottom of a miniband (Fig. 4 (b)). This is not caused by the reduction of H_{el-ph}, but rather by the restriction of high energy phonon emission. Scattering events mediated by high energy phonon branches are not allowed because the final electron states fall in the energy gap below the miniband. While scattering rate is not strongly dependent on the initial electron energy, the energy of phonons mediating the scattering processes decreases with decreasing initial electron energy. This causes the drastic decrease in the energy loss rate.

CONCLUSION

We have calculated electronic/phononic states and the electron-phonon interaction in the 1DSiQDA. The acoustic deformation potential scattering is weakened due to the strain absorption by the oxide layers. However, increase of the electron density of states dominates the above reduction mechanism, leading to scattering rate higher than that in the SiQW. Energy loss rate has been found to decrease at near the bottom of a miniband by the restriction of high-energy phonon emission.

ACKNOWLEDGMENTS

The authors would like to thank Dr. D. Williams of Hitachi Cambridge Laboratory and Prof. S. Oda of Tokyo Institute of Technology for their support.

REFERENCES

1. Koshida *et al.*, Appl. Surf. Sci., 146, p. 371 (1999).
2. Sanders *et al.*, Phys. Rev. B, 48, p. 11067 (1993).

Calculation of carrier transport through quantum dot molecules

T. Zibold[*], M. Sabathil[*], D. Mamaluy[†], P. Vogl[*]

[*]*Walter Schottky Institute, Technische Universität München, Am Coulombwall 3, 85748 Garching, Germany*
[†]*Department of Electrical Engineering, Arizona State University, Tempe, AZ 85287-5706, USA*

Abstract. A detailed calculation is presented of the electronic structure and the ballistic current through a realistic InAs/InP open quantum dot molecule consisting of two vertically stacked InAs quantum dots. It is shown that a careful analysis of the experimental tunneling current allows one to extract a wealth of information about the energy levels, base widths, distance and lateral misalignment of the quantum dots in a quantum dot molecule.

INTRODUCTION

In this paper we present a quantitative theoretical analysis of the ballistic current through a quantum dot molecule (QDM) based on two vertically stacked quantum dots (QDs) embedded within a resonant tunneling diode (RTD). In particular, we show that measurements of the resonant tunneling current provide a wealth of unique information about the size, geometry, and the energy levels of the QDM. Our calculations are based on a fully three-dimensional model of the RTD with realistically shaped QD structures. The local strain is calculated by minimizing the total elastic energy of the entire device. We solve the Poisson equation including the piezoelectric charges and subsequently calculate the ballistic current in terms of the contact block reduction method [1]. This scheme computes the quantum states of the open system with scattering boundary conditions rigorously. The quantum states themselves are evaluated in terms of the single-band Schrödinger equation for the conduction bands, taking into account spatially varying electron masses, band offsets, deformation potentials, and the electrostatic potential due to the piezoelectric charges. All calculations are carried out within the framework of next**nano**³ [2].

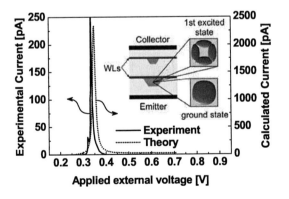

FIGURE 1. The experimental and calculated current through the InAs/InP QDM as a function of the applied bias. The inset shows the structure of the device, including the leads (black), WLs (dark gray), and the QDs (light gray), schematically. Furthermore, a top view of the 3D orbitals of the ground and first excited state of the large QD are plotted.

RESULTS AND DISCUSSION

The RTD as depicted in the inset of Fig. 1 is composed of a 33 nm thick InP barrier with two embedded InAs QDs that are grown on 0.5 nm thick wetting layers (WLs). The distance between the WLs is 15 nm. This RTD was studied experimentally by Bryllert et al. [3,4]. Upon applying a bias, pairs of bound states in the two adjacent QDs can be brought into resonance with one another, thereby increasing the electron transmission probability by several orders of magnitude.

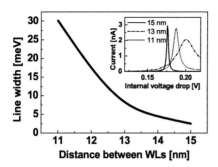

FIGURE 2. Calculated resonance line width as a function of the inter-dot distance. The inset shows the resonances for WL distances of 11, 13, and 15 nm, respectively. The peak current increases with increasing inter-dot distance, whereas the integrated current falls off exponentially.

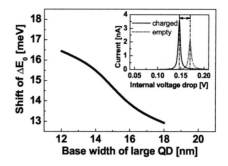

FIGURE 4. Change in the zero-bias energy difference ΔE_0 between the two QD ground states due to charging of the large QD, plotted as a function of the base width. The inset shows the shift of the current resonance for the neutral and charged situation.

Figure 1 shows the experimentally determined as well as the theoretically predicted current. The two QDs were modeled by truncated pyramids of 2.5 (5) nm height and 12 (16) nm base width, respectively. The difference ΔE_0 of the zero bias ground state energies is 87 meV. The peak height and the line width of the current resonance depend strongly on the inter-dot distance and the lateral misalignment, which may explain the differences between theory and experiment. In fact, Fig. 2 shows the line width of the resonance to decrease exponentially with increasing QD distance which reflects the exponential decrease in the QD coupling. The inset shows the shape of several current resonances explicitly.

As depicted in the inset of Fig. 3, an additional resonance may occur if the excited state **1** in the large dot lies in resonance with the ground state **0'** of the small dot. Figure 3 displays the ratio between the peak currents of the **1 → 0'** resonance and the **0 → 0'** resonance as a function of the relative lateral displacement of the two QDs.

The larger QD close to the emitter may be charged by emitter electrons. This charging decreases the energy difference ΔE_0 and shifts the resonance bias as is illustrated in the inset of Fig. 4. This shift decreases for larger base width, as shown in Fig 4, and can therefore be used to estimate the lateral dimensions of the large QD.

The peak heights and positions, and the line widths of the tunneling resonances through ground and excited states yield quantitative information about the energy levels, base widths, distance and lateral alignment of the QDs in a quantum dot molecule.

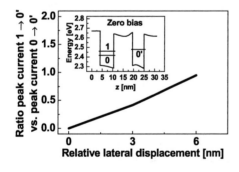

FIGURE 3. Ratio between the peak currents of the **1 → 0'** resonance and the **0 → 0'** resonance as a function of the relative lateral displacement of the two QDs. The inset shows the energies of the ground state **0** and excited state **1** in the large QD and of the ground state **0'** in the small QD.

REFERENCES

1. Hackenbuchner S., Sabathil M., Majewski J.A., Zandler G., Vogl P., Beham E., Zrenner A., Lugli P., *Physica B* **314**, 145-149 (2002).

2. Mamaluy D., Sabathil M. and Vogl P., *J. Appl. Phys.* **93**, 4628 (2003).

3. Bryllert, T., Borgstrom, M., Wernersson, L.-E., Seifert, W. and Samuelson, L., *Appl. Phys. Letters* **82**, 2655 (2003).

4. Borgstrom, M., Bryllert, T., Sass T., Gustafson B., Wernersson, L.-E., Seifert, W. and Samuelson, L., *Appl. Phys. Letters* **78**, 3232 (2001).

Magnetoresistance experiments and quasi-classical calculations regarding backscattering in open quantum dots

R. Brunner*, R. Meisels*, F. Kuchar*, M. Elhassan[†], J. P. Bird[†], D. K. Ferry[†] and K. Ishibashi**

Institute of Physics, University of Leoben, Leoben, Austria
[†]*Department of Electrical Engineering, Arizona State University, Tempe, USA*
**Semiconductor Laboratory, RIKEN, Saitama, Japan*

Abstract. Closed and open quantum dots are low-dimensional electron systems with different strength of coupling to the two-dimensional environment. In the case of a the closed dot, only transport by tunneling is possible. Our work, on the the other hand, investigates the transport of open quantum dots strongly coupled to the surrounding two-dimensional electron system. The open dot show peculiar peaks in their magnetoresistance. We compare experimental results with those of a semi-classical model using trajectories in a parabolic potential. Using a velocity distribution of the incoming electrons based on Fermi-Dirac statistics we can explain the temperature dependence of the peak heights.

INTRODUCTION

We investigate the transport of electrons through open quantum dots in the semi-classical regime at low magnetic fields. Previously, this was studied examining trajectories with specular reflections at the bondaries of the dots modelled by a hard-wall potential [1]. A soft-wall parabolic confinement was shown to better correspond to the experimental results[2]. These calculations did not include the effects of non-zero temperature. Therefore it is of interest to include $T > 0$ into the model to determine the temperature dependence.

EXPERIMENTAL METHODS

The dot systems are defined by finger gate on top of GaAs/AlGaAs heterostructures ($n_s = 2.5 \times 10^{11}$cm^{-2}, $\mu = 1.2 \times 10^6$Vs/cm^2). By applying a negative gate voltage the two-dimensional electron system (2DES) below the gates is depleted of electrons and a dot is formed between the two gates. The geometrical size of the dot is defined by the outline of the gates and is approximately 280 nm. It is connected to the surrounding 2DES by two constrictions. The experiments were performed at temperatures down to 0.3 K and at magnetic fields below 1 T.

The resistance of the dot was determined by driving a small (2 nA$_{rms}$) current through the sample and measuring the voltage between two contacts to the 2DES at opposite sides of the dot. This voltage drop is measured by a lock-in amplifier.

THEORY

To interpret these experimental findings we set up a classical model for the open quantum dot. For the confining potential we choose a parabolic relationship

$$V(r) = kr^2. \quad (1)$$

This potential is a good approximation for a smooth confinement. It has the benefit of allowing to solve the movement of an electron in this potential analytically. The differential equation of motion for an electron with charge $q = -e$ and mass m in the harmonic potential (elastic constant k) within a static magnetic field $\mathbf{B} = (0,0,B_z)$ is:

$$m\ddot{\mathbf{r}} + k\mathbf{r} - q\dot{\mathbf{r}} \times \mathbf{B} = 0 \quad (2)$$

The path of the electron can be given in a closed form (containing harmonic functions).

$$\mathbf{r} = \mathbf{r}_+ e^{i\omega_+ t} + \mathbf{r}_- e^{i\omega_- t} \quad (3)$$

Here $\mathbf{r}$ is complex. The actual motion is given by its real part $\Re(\mathbf{r})$. Applying Eq.(3) to Eq.(2) yields an eigenvalue problem with eigenvalues $\omega_\pm$:

$$\omega_+ = \sqrt{\omega_0^2 + (\omega_c/2)^2} + \omega_c/2 \quad (4)$$

$$\omega_- = \sqrt{\omega_0^2 + (\omega_c/2)^2} - \omega_c/2 \quad (5)$$

$$\omega_c = eB_z/m \quad (6)$$

Here k is replaced by the harmonic oscillator frequency $\omega_0 = \sqrt{k/m}$. $\mathbf{r}_+$ and $\mathbf{r}_-$ have to be determined according to the the initial position $\mathbf{r}(t=0)$ and velocity $\frac{d}{dt}\mathbf{r}(t=0)$ of the electron.

This model can be generalized to account for an anisotropic (*i.e.* elliptical) dot with two harmonic frequencies ω_{x0} and ω_{y0}. The x-direction is the direction between the two constrictions. In this case the two eigenfrequencies are:

$$\omega_\pm = \left(\sqrt{\omega_c^2 + (\omega_{0x} + \omega_{0y})^2} \pm \sqrt{\omega_c^2 + (\omega_{0x} - \omega_{0y})^2} \right)/2. \quad (7)$$

Given the initial conditions the trajectory of the electron can be calculated for any time in the future. To find out whether an electron will first reach the initial constriction (backscattering) or reaches first the exit constriction (transmition) the position of the electron is calculated at several times, adapting the interval between these times according to the closeness to the constrictions. Only transmitted electrons contribute to the conductance, backscattered electron yield $G = 0$. This test is performed for different directions α of the initial electron motion ($\alpha = 0$ correponds to an initial motion towards the center of the dot). Averaging over α a Lambertian weight factor $\cos \alpha$ is used. In this way, the conductance of the dot vs. the magnetic field is calculated up to a fixed factor, *i.e.* in arbitrary units. Arrays of dots can be also be simulated. When an electron reaches a constriction the calculation is restarted in the adjacent dot. Only when reaching the the initial entrance or the final exit constriction the contribution to G is determined.

For temperatures $T > 0$ the Fermi distribution is not a step function but smoothed. Classically, the absolute value of the initial velocity is not fixed at $v_F = \sqrt{2mE_F}$ but distributed around this value according to $\frac{df(E)}{dE}$. Therefore, to calculate the conductance $G(B;T)$ for $T > 0$ an addtional weighted averaging has to be performed over the absolute value of the initial velocities.

EXPERIMENTAL RESULTS

Figure 1 shows the temperature dependence of the dominant peak (at ± 0.25T) in the magnetoresistance of an open quantum dot[2] and for a triple-dot array. The height of these peaks, contrary to the SdH oscillations, is temperature independent for both systems below 5K.

DISCUSSION

The results of the calculations on the temperature dependence of the dot height for a single dot and a three dot

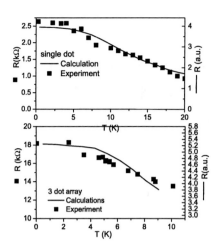

FIGURE 1. Peak height at 0.25 T vs. temperature. Experimental results are shown as symbols and results of calculation as lines (single dot: upper, array of three dots: lower diagram)

array are shown in Fig. 1. In both cases the temperature independence below about 5K and the subsequent drop of the peak resistance (corresponding to a rise in conductivity) is well reproduced. We conclude, that in addition to explaining the position of the peaks [2], our simple, semi-classical model (quantum effects are only used to determine the spread of the intial electron velocity) is able to show the temperature dependence of the peak heights.

ACKNOWLEDGMENTS

This work was supported by the "Fonds zur Förderung der wissenschaftlichen Forschung", Austria, project P15513.

REFERENCES

1. Lin, L. H., Aoki, N., Nakao, K., Ishibashi, K., Aoyagi, Y., Sugano, T., Holmberg, N., Vasileska, D., Akis, R., Bird, J., Ferry, D., and Ochiai, Y., *Physica E*, **7**, 750–1446 (2000).
2. Brunner, R., Meisels, R., Kuchar, F., ElHassan, M., Bird, J., and Ishibashi, K., *Physica E*, **21**, 491–495 (2004).

Modeling of the Magnetization Behavior of Realistic Self-Organized InAs/GaAs Quantum Craters as Observed with Cross-Sectional STM

V. M. Fomin[*,†], V. N. Gladilin[*], J. T. Devreese[*,†], P. Offermans[†], P. M. Koenraad[†], J. H. Wolter[†], J. M. García[**] and D. Granados[**]

[*]TFVS, Departement Natuurkunde, Universiteit Antwerpen, Universiteitsplein 1, B-2610 Antwerpen, Belgium
[†]eiTT/COBRA Inter-University Research Institute, Physics Department, Eindhoven University of Technology, P.O. Box 513, 5600 MB Eindhoven, The Netherlands
[**]Instituto di Microelectronica de Madrid, CNM(CSIC), C/Isaac Newton 8, 28760 Tres Cantos, Spain

Abstract. Recently, using cross-sectional scanning-tunneling microscopy (X-STM), it was shown that self-organized ring-like InAs quantum dots are much smaller in diameter than it is expected from atomic force microscopy measurements and, moreover, that they possess a depression rather than an opening in the central region. For those quantum craters, we analyze the possibility to reveal the electronic properties (like the Aharonov-Bohm oscillations) peculiar to doubly connected geometry of quantum rings.

The Aharonov-Bohm-effect-related quantum interference phenomena in quantum rings [1-4] have attracted a renewed interest due to the development of self-organizing [5, 6] and lithographic [7] techniques to fabricate nanosize semiconductor ring-like structures. The Aharonov-Bohm effect has been optically detected on a charged exciton in a single quantum ring [7]. The shape and composition profiles of the quantum ring-like structures consistent with the atomic force microscopy (AFM) data on uncapped structures have been recently proposed [8]. Using cross-sectional scanning-tunneling microscopy (X-STM), it has been shown that buried self-organized InAs quantum ring-like structures are much smaller in diameter than it is inferred from AFM measurements on the quantum rings after equilibration [9] and, moreover, that they possess a *depression* rather than an *opening* in the central region (see Fig. 1). Motivated by this fact, we address in the present paper the key question, whether or not those quantum-crater structures can effectively manifest the electronic properties (like the Aharonov-Bohm oscillations) peculiar to doubly connected geometry, which corresponded with the expected geometry of InAs quantum rings [5].

Electron states in quantum craters, subjected to a magnetic field **H** parallel to the growth axis, are found, modelling the shape of quantum craters with the function $h(\rho) = h_0 \left[1 - \left(1 - h_0^2/h_M^2\right)\rho^2/R^2\right]^{-1/2}$ (see inset of Fig. 2). Since the lateral size of quantum craters substantially exceeds their height ($R \gg h_0, h_M$), we con-

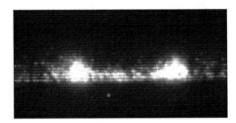

FIGURE 1. Cross-sectional scanning-tunneling microscopy (X-STM) image of a InAs/GaAs quantum crater. The shown area is 20×40 nm^2.

sider the lateral motion of electrons to be governed by the adiabatic potential related to the fast electron motion along the growth axis. This allows us to look for an electron wave function in the form of the adiabatic Ansatz: $\Psi(\mathbf{r}) = \psi(z;\rho)\chi(\rho)e^{im\varphi}$, where we have also exploited the axial symmetry of the structure under consideration; m is the magnetic quantum number of an electron.

As seen from Fig. 2, the ground-state magnetic quantum number m_{GS}, calculated as a function of the applied magnetic field for realistic parameters R, h_0, and h_M of quantum craters, forms a staircase, similar to that for the Aharonov-Bohm effect in quantum rings. An increase of the quantum-crater radius R significantly reduces the values of the applied magnetic field, which correspond to jumps of $m_{GS}(H)$, this effect becoming more pronounced with decreasing $h_M - h_0$ [see Fig. 2]. The obtained results

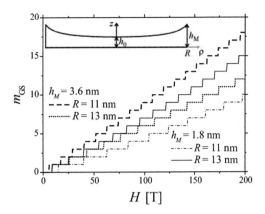

FIGURE 2. Ground-state magnetic quantum number of an electron as a function of the applied magnetic field for quantum craters with $h_0 = 1.6$ nm and different values of h_M and R. Inset: the used model for the shape of quantum craters.

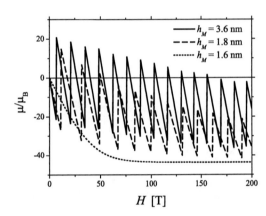

FIGURE 3. Magnetic moment as a function of the applied magnetic field for quantum craters with $R = 12$ nm, $h_0 = 1.6$ nm, and three different values of h_M (the value $h_M = 1.6$ nm corresponds to a perfectly flat disk).

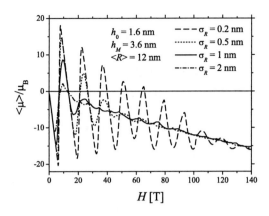

FIGURE 4. Average magnetic moment as a function of the applied magnetic field for ensembles of quantum craters with $h_0 = 1.6$ nm, $h_M = 3.6$ nm, and Gaussian distributions of radii, which are characterized by the average value $\langle R \rangle = 12$ nm and the standard deviation σ_R.

imply that the Aharonov-Bohm-effect-related electronic phenomena can appear even in quantum craters with a relatively small height of the rim ($h_M - h_0 \ll h_0$).

For single quantum craters, the calculated magnetic moment μ as a function of the applied magnetic field demonstrates an oscillating behavior typical for the Aharonov-Bohm effect (see Fig. 3). When decreasing the height h_M, the oscillation amplitude diminishes, and the results for $\mu(H)$ tend to approach the non-oscillating curve for a perfectly flat disk. Due to a strong dependence of transition magnetic fields on the lateral size of a quantum crater (see Fig. 2), size dispersion of quantum craters strongly suppresses oscillations of the ensemble average for the magnetic moment, $\langle \mu \rangle$, as a function of H, this suppression being especially pronounced at high magnetic fields. At the same time, figure 4 shows that in relatively low magnetic fields the characteristic oscillating behavior of $\langle \mu \rangle (H)$ survives for ensembles of quantum craters with radial dispersion as large as 10-20%. For a sample, which contains ~ 10 layers of stacked quantum craters with density $\sim 10^{10}$ cm^{-2} per layer, the oscillation magnitude for the total magnetic moment, related to persistent currents in quantum craters, is estimated to be $\sim 10^{10}$ μ_B/mm^2 (μ_B is the Bohr magneton), implying that the oscillations can be experimentally detected using, e.g., torsion magnetometry.

This work has been supported by the GOA BOF UA 2000, IUAP, FWO-V projects G.0274.01N, G.0435.03, the WOG WO.025.99N (Belgium) and the European Commission GROWTH Programme, NANOMAT project, contract No. G5RD-CT-2001-00545.

REFERENCES

1. N. Byers and C. N. Yang, *Phys. Rev. Lett.* **7**, 46-49 (1961).
2. M. Büttiker, Y. Imry, and R. Landauer, *Phys. Letters A* **96**, 365-367 (1983).
3. D. Mailly, C. Chapelier, and A. Benoit, *Phys. Rev. Lett.* **70**, 2020-2023 (1993).
4. L. Wendler, V. M. Fomin, and A. A. Krokhin, *Phys. Rev. B* **50**, 4642-4647 (1994).
5. A. Lorke, R. J. Luyken, A. O. Govorov, J. P. Kotthaus, J. M. Garcia, and P. M. Petroff, *Phys. Rev. Lett.* **84**, 2223-2226 (2000).
6. D. Granados and J. Garcia, *Appl. Phys. Lett.* **82**, 2401-2403 (2003)
7. M. Bayer, M. Korkusinski, P. Hawrylak, T. Gutbrod, M. Michel, and A. Forchel, *Phys. Rev. Lett.* **90**, 186801, 1-4 (2003).
8. J. A. Barker, R. J. Warburton, and E. P. O'Reilly, *Phys. Rev. B* **69**, 035327, 1-9 (2004).
9. R. Blossey and A. Lorke, *Phys. Rev. E* **65**, 021603, 1-3 (2002).

Dynamical control of electronic states in AC-driven quantum dots

Charles Creffield[*][†] and Gloria Platero[**]

[*]*Dpto. di Fisica, Università di Roma "La Sapienza", Piazzale Aldo Moro 2, I-00185 Rome, Italy.*
[†]*Dept. of Physics and Astronomy, University College London, Gower Street, London WC1E 6BT, UK*
[**]*Instituto de Ciencia de Materiales (CSIC), Cantoblanco, E-28049 Madrid, Spain.*

Abstract. We investigate the dynamics of two interacting electrons moving in a one-dimensional array of quantum dots, under the influence of an ac field. The quantum dot array is modeled as a single-band tight-binding model of Hubbard-type, and the system is analyzed by means of the Floquet approach. Our results show that the system exhibits two distinct regimes of behavior, depending on the ratio of the strength of the driving field to the inter-electron Coulomb repulsion. When the ac-field dominates an effect termed coherent destruction of tunneling occurs at certain frequencies, in which transport along the array is suppressed. In the other, weak-driving, regime we find the surprising result that the two electrons can bind together into a single composite particle — despite the strong Coulomb repulsion between them — which can then be controlled by the ac field in an analogous way. We show how calculation of the system's Floquet quasienergies explains these results, and thus how ac fields can be used to control the time evolution of entangled states in mesoscopic devices.

Recent developments of laser sources have motivated the study of coherent control of the electron dynamics in quantum systems. In parallel, the exciting field of quantum computation has encouraged the study of electron wave function coherence in different configurations of quantum dots (QDs). It is well-known that an electron initially localized in a double quantum well, or in a superlattice, can remain localized under the influence of a sinusoidal electric field, if the ratio of the field intensity to the field frequency is a root of the zeroth-order Bessel function [1]. This effect is termed coherent destruction of tunneling (CDT).

Unlike superlattice systems, electrons in QD structures are in general strongly correlated due to the Coulomb interaction between them. We therefore model the QD array as a single-band tight-binding model of Hubbard type. For the double QD case, explicit calculation has shown that such an effective model indeed reproduces the behavior of a more realistic model [2], and accordingly for an array of N QDs we use the effective lattice Hamiltonian:

$$H = -t \sum_{\langle i,j \rangle, \sigma} \left[c_{i\sigma}^\dagger c_{j\sigma} + H.c. \right] + \sum_{j=1}^{N} \left[U n_{j\uparrow} n_{j\downarrow} + eaF(t) j n_j \right]. \quad (1)$$

Here U is the Hubbard term giving the energy cost for double-occupation of a QD, and t is the hopping parameter between adjacent QDs. The operators $c_{j\sigma}/c_{j\sigma}^\dagger$ are the electronic annihilation/creation operators, $n_{j\sigma}$ is the standard number operator, and $n_j = n_{j\uparrow} + n_{j\downarrow}$. The inter-dot spacing is denoted by a, and we describe the time-dependent electric field in terms of the potential difference between neighboring sites, $E(t) = eaF(t) = E \sin \omega t$. We neglect spin-flip processes, and consider *just* the singlet sub-space, as in the triplet sub-space the Pauli principle forbids double-occupation of a QD, and the Coulomb interaction described by the Hubbard-U term is consequently irrelevant.

As the system is driven by a time-dependent field it does not possess energy eigenstates. As the driving is periodic, however, we can write its time-development as $|\Psi(t)\rangle = \sum [c_j \exp(-i\varepsilon_j t) |\phi_j(t)\rangle]$, where ε_j are the *quasienergies* and $|\phi_j(t)\rangle$ are the *Floquet states*. The Floquet states have the same periodicity as the driving field, and so it is the quasienergies that chiefly determine the long timescale behavior of the system, and in particular, quasienergy degeneracy produces CDT. In Fig.1 we compare the quasienergy spectrum for a double QD (i.e. $N = 2$) with that of a 15-site QD-array. For clarity we show only the quasienergies for states in which the electrons either occupy the same QD (solid symbols), or occupy neighboring QDs (hollow symbols). The system clearly shows two distinct regimes of behavior. In the high-field regime, $E > U$, the spectrum exhibits a set of crossings between the two classes of quasienergies. As shown in Refs.[2, 3], CDT occurs at these points, freezing the propagation of electrons along the array. For lower values of E, however, a very different behavior occurs. Here the set of doubly-occupied states form a *mini-*

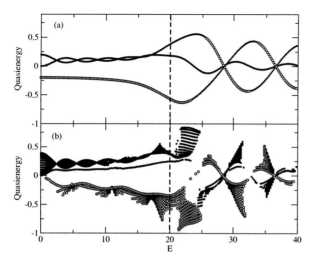

FIGURE 1. Quasienergy spectrum for (a) a two-site system, and (b) a 15-site system, for $U = 20$ and $\omega = 2$. Symbols indicate the characteristic of the corresponding Floquet state: solid for doubly-occupied states, hollow for neighbor-states. The vertical dashed line marks the boundary between the weak and strong driving regimes.

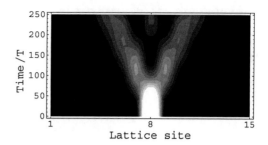

FIGURE 2. Time-dependent charge-density, $|\Psi(t)|^2$, for a 15-site lattice, with the initial condition of both particles on the central site. White / black signifies high / low values of density. Field parameters are $E = 2.35$, $\omega = 2$.

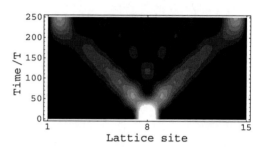

FIGURE 3. Same as Fig.2 for E=2.30

band, similar to that observed for non-interacting electrons [4], which exhibits a sequence of "band-collapses" at which the quasienergies become degenerate. In contrast the neighbor states in this regime show little structure, and remain essentially isolated from the doubly-occupied states.

In the vicinity of the band collapses the tunneling dynamics of the system are sharply quenched, and when the quasienergies become degenerate, the timescale for tunneling diverges. Thus if the system is initialized in a state in which both electrons occupy the same QD we can control the speed at which the two-electron wavepacket propagates along the array. The first band-collapse occurs when $J_0(2E/\omega) = 0$, that is, $2E/\omega = 2.4048$. In Fig.2 we show how a two-electron state initialized at the center of the array splits smoothly into pulses moving to the left and right at an almost constant velocity. In this plot $2E/\omega = 2.35$, and thus the proximity to the point of quasienergy degeneracy causes the pulse velocity to be very slow. Moving the field parameters further from the point of band-collapse causes the pulses to travel more rapidly. In Fig.3 it can be seen that for $E = 2.3$ the velocity of the pulses has almost doubled, allowing the electrons to reach the ends of the array in the same timespan. It should be noted that this arrangement generates a current of entangled, correlated electrons. It can thus be regarded as an electronic analogue of an optical beam-splitter, which divides the initial state into two spatially separated and entangled electron pairs, moving at a velocity controlled by the ac-driving field.

In summary, we have seen that the competition between the Coulomb energy and the driving field produces a rich phenomenology, which can be described by means of Floquet theory. For strong-driving, a suitably chosen ac-field is able to produce CDT by inducing crossings in the Floquet spectrum. In the weak field regime, the driving field and the Coulomb interaction combine in a counter-intuitive way to bind the two electrons together to form a composite particle resembling an exciton. The tunneling of this particle from site to site can then be finely regulated by manipulating the driving field parameters.

GP was supported by the Spanish DGES grant MAT2002-02465 and by the EU Human Potential Programme under contract HPRN-CT-2000-00144.

REFERENCES

1. F. Grossmann, T. Dittrich, P. Jung and P. Hänggi, Phys. Rev. Lett. **67**, 516 (1991).
2. C.E. Creffield and G. Platero, Phys. Rev. B **65**, 113304 (2002).
3. C.E. Creffield and G. Platero, Phys. Rev. B **69**, 165312 (2004).
4. M. Holthaus, Phys. Rev. Lett. **69**, 351 (1992).

Field effect enhanced carrier-emission from InAs Quantum Dots

S. Schulz*, T. Zander*, A. Schramm*, Ch. Heyn* and W. Hansen*

Institute for Applied Physics, University of Hamburg, Jungiusstrasse 11, 20355 Hamburg, Germany

Abstract. We probe the emission from the electron levels bound to MBE-grown self-assembled InAs/GaAs quantum dots with transient capacitance spectroscopy (DLTS). Emission from the s-state occupied with one ore two electrons is reflected by two well separated peaks. An additional broad peak observed at lower temperature is associated to emission from the p-state. We relate substructure in this peak to the p-level occupation with up to 4 electrons. The observations establish a strong dependence of the emission energies on the electric field at the dots. From an analysis of our results we conclude that the field dependence originates from a thermally assisted tunneling process through an intermediate state in a continuous band. Dot-binding energies calculated with this model are 165 ± 2 meV for the s_1-state and 151 ± 2 meV for the s_2-state in our sample and found to be nearly field independent. In addition we perform admittance spectroscopy on the same samples, which yields emission energies in good agreement with those from DLTS.

Emission energies of self-assembled quantum dots (SAQD) are determined from transient capacitance measurements [1, 2, 3, 4, 5, 6]. From the results it is concluded [2, 3, 5, 6] that the emission energies may deviate from the dot-binding energies due to indirect emission processes. Different possible scenarios are proposed for the emission process in InAs-SAQD [1, 2, 3, 5, 7, 8]. We study the electric-field dependence of the emission from electron levels in InAs SAQD [6]. From an analysis of our results with a simple model adopted from Vincent et al. [9] we infer that the escape takes place through thermal excitation to a continuous band of states and subsequent tunneling.

Here we present data of a sample in which the electric field can be varied over a wide range between 1.0 and 3.5×10^4 V/cm. In addition the dot density $N_{QD} = 4.3 \times 10^9$ cm^{-2} is about 5-times lower than in Ref. [6]. Assuming that the above thermally assisted tunneling model, in which the field at the SAQD is calculated with the one-dimensional (1d) Poisson equation, is still valid the electron occupation dependence of the electric field at the dots should be reduced. Thus more electrons can be charged into each dot and therefore higher levels are observed in DLTS. Furthermore, we present admittance measurements [7, 8] recorded from the same device.

The SAQDs studied in this work are embedded in a Si-doped Schottky diode with an effective doping of $N_D - N_A = 3.8 \times 10^{15}$ cm^{-3}. The dots were MBE-grown at 485°C depositing 2.1 ML InAs with 0.01 ML/s growth rate. They are located 750 nm below the surface. Nominally similar dots where grown on the surface. Their

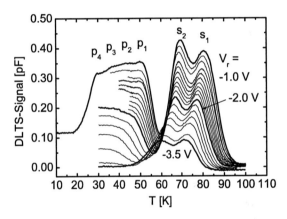

FIGURE 1. DLTS-spectra taken at different reverse voltage V_r between -3.5 V and -1.0 V with a rate window of 258 Hz and sufficiently high pulse bias to ensure the s- and p-levels completely filled. The amplitudes of the spectra varies due to decreasing DLTS-sensitivity with increasing negative V_r.

density and height are determined by AFM inspection to $N_{QD} = 4.3 \times 10^9$ cm^{-2} and 10 nm, respectively. Due to different growth conditions the dots are slightly larger than in Ref. [6]. For Schottky contacts, 50 nm chromium was evaporated with 1 mm diameter.

Typical DLTS-spectra are presented in Fig. 1. The two peaks denoted with s_1 and s_2 reflect the emission from singly and doubly occupied dots as verified by the pulse-voltage dependence of the peaks [6]. Similarly, we find that the broad peak at temperatures between 25 K and 55 K gradually evolves with increasing occupation of the dot. We thus assign its substructure to the four possible electrons in the p-level. The electric field at the dots is changed by the reverse bias. Both s-peaks decrease in temperature with increasing field.

Filled symbols in Fig. 2 depict the emission energies as determined from conventional Arrhenius analysis. They are plotted vs. the electric field at the dots during emission. The field is calculated with the 1d model using CV-curves and layer parameters of the particular device. The electrons occupying the dots are taken into account by a homogeneous sheet charge in the plane of the SAQD. We note that we expect the thus calculated fields are underestimated at low dot densities. The field in the tunnel barrier close to the dots may be better described by a 3d Coulomb potential independent of the dot density.

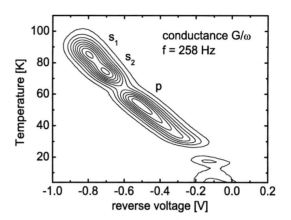

FIGURE 3. Conductance spectra taken at 258 Hz of the same device used for DLTS-measurements. The contour-scale of G/ω is 0.4 pF per step.

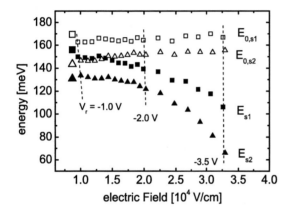

FIGURE 2. Emission energies for the s-state determined from DLTS-measurements (solid symbols) plotted vs. the electric field. The open symbols denote the binding energies E_0 calculated with the thermally assisted tunneling model. Dashed lines in Fig. 2 connect data points recorded at the same reverse voltage. The leftmost data are determined from admittance spectroscopy and denoted with bigger symbols.

The emission energies strongly decrease with increasing field. With the thermally assisted tunneling model we find binding energies $E_{0,s1} = 165 \pm 2$ meV and $E_{0,s2} = 151 \pm 2$ meV (open symbols in Fig. 2). These values are larger by about 10 % than in Ref. [6]. This we associate to the larger dot size. In good agreement with the previously found behavior, the energies are nearly field-independent, although this sample less well complies with the preconditions of the 1d model.

Emission rates can also be obtained from admittance measurements [7, 8]. Since here the Fermi level of the dots and the electron reservoir always coincide, the electric field at the dots is the lowest possible. In Fig. 3 conductance spectra are shown. A conductance peak occurs at the reverse voltage where the dot-level is equal to the Fermi-level and at the temperature, where the emission and capture rates coincide with the excitation frequency [7]. As in DLTS-spectra two distinct peaks associated to charge injection into the the s-level are observed and a broad peak for the p-level. An Arrhenius analysis of the s-peak emission rates determined at different frequencies yields energies $E_{G,s1} = 156$ meV and $E_{G,s2} = 131$ meV for the emission from the singly and doubly occupied dots, respectively. These values are in good agreement with the emission energies of the DLTS-data, as demonstrated in Fig. 2.

REFERENCES

1. Anand, S., Carlsson, N., Pistol, M.-E., Samuelson, L., and Seifert, W., *J. Appl. Phys.*, **84**, 3747 (1998).
2. Kapteyn, C. M. A., Heinrichsdorff, F., Stier, O., Heitz, R., Grundmann, M., Zakharov, N. D., Werner, P., and Bimberg, D., *Phys. Rev. B*, **60**, 14265 (1999).
3. Kapteyn, C. M. A., Lion, M., Heitz, R., Bimberg, D., Brunkov, P., Volovik, B. V., Konnikov, S. G., Kovsh, A. R., and Ustinov, V. M., *Appl. Phys. Lett.*, **76**, 1573 (2000).
4. Geller, M., Kapteyn, C., Stock, E., Müller-Kirsch, L., Heitz, R., and Bimberg, D., *Physica E*, **21**, 474 (2004).
5. Engstrom, O., Malmkvist, M., Fu, Y., Olafson, H. O., and Sveinbjornsson, E. O., *Appl. Phys. Lett.*, **83**, 3578 (2003).
6. Schulz, S., Schnuell, S., Heyn, C., and Hansen, W., *Phys. Rev. B*, **69**, 195317 (2004).
7. Chang, W.-H., Chen, W. Y., Cheng, M. C., Lai, C. Y., Hsu, T. M., Yeh, N.-T., and Chyi, J.-I., *Phys. Rev. B*, **64**, 125315 (2001).
8. Chang, W.-H., Chen, W. Y., Hsu, T. M., Yeh, N.-T., and Chyi, J.-I., *Phys. Rev. B*, **66**, 195337 (2002).
9. Vincent, G., Chantre, A., and Bois, D., *J. Appl. Phys.*, **50**, 5484 (1979).

Level anticrossings in quantum dots

J. Könemann[*], E. Räsänen[†], M. J. Puska[†], R. M. Nieminen[†] and R. J. Haug[*]

[*]Institut für Festkörperphysik, Universität Hannover, Appelstrasse 2, D-30167 Hannover, Germany
[†]Laboratory of Physics, Helsinki University of Technology, P.O. Box 1100, FIN-02015 HUT, Finland

Abstract. We have studied the single-electron transport spectrum of a localized level embedded into a double-barrier resonant tunneling structure. The experimental spectrum shows clear deviations from the Fock-Darwin model indicating a broken circular symmetry. We interprete our observations in terms of the existence of an additional multiple-charged impurity in the vicinity of the quantum dot.

Semiconductor quantum dots (QD) have attracted broad interest in experimental [1] and theoretical works [2]. The understanding of the electronic transport through such systems plays an important role in the future development of electronic devices like single-electron transistors. Due to their quantum confinement QD's can be considered as artificial atoms [3]. The modeling of QD spectra is usually based on the assumption of a perfect 2D harmonic confining potential giving the well-known Fock-Darwin (FD) single-electron spectrum [4]. Such FD spectra have been observed in several experiments [5].

Anticrossings of orbital single-electron energy levels in QD's always indicate a symmetry-breaking mechanism, which distorts the harmonic confining potential of the QD. In our experiment we have studied the single-particle transport spectrum of a single localized electronic state embedded in a double-barrier resonant tunneling structure. The experiment was performed with a highly asymmetric double barrier resonant tunneling device grown by molecular beam epitaxy on n^+-type GaAs substrate. The heterostructure consists of a 10 nm wide GaAs quantum well sandwiched between two $Al_{0.3}Ga_{0.7}As$-tunneling barriers of 5 and 8 nm. The contacts are formed by 0.5 μm thick GaAs layers highly doped with Si up to $4. \times 10^{17}$ cm^{-3} and separated from the active region by 7 nm thin spacer layers of undoped GaAs. DC measurements of the I-V-characteristics were carried out in a dilution refrigerator at 20 mK base temperature and in high magnetic fields up to 14 T.

The resulting I-V curves are presented as a function of the magnetic field oriented parallel to the tunneling current (see Fig. 1). The additional field-induced lateral confinement shifts the bias position of the first step (single-electron ground state) to higher voltages. Moreover, most of the steps appear to approach the first step

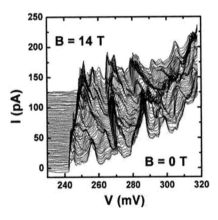

FIGURE 1. Experimental I-V curves between zero and 14 T. Curves are plotted with a vertical offset of 1 pA.

at higher magnetic fields. Additionally, we observe a fluctuation pattern reflecting the local density of states of the emitter and its highly-disordered Landau bands [6].

The grey-scale plot Fig. 2 depicts the differential conductance $G = dI/dV$ as a function of bias voltage and magnetic field (numerically obtained from the current data). The black lines $V_{n,l}$ of high differential conductance G trace the position of the states $E_{n,l}$ of the spectrum

$$V_{n,l} = V_0 + 1/(e\alpha)E_{n,l}, \quad (1)$$

with the energy-voltage conversion factor $\alpha = 0.4$ obtained from measurements of the broadening of the step edge with temperature and the onset voltage V_0 from modelling of the experimental spectrum. We observe clear deviations from an ideal single-electron FD spectrum. At zero magnetic field the lifting of the degeneracy of orbital states is observed. At intermediate magnetic fields we find strong level repulsions, whereas at high magnetic fields the traces of the electronic levels follow parallel lines with an equal spacing and a constant slope in magnetic field showing the condensation into the lowest lying 2D-Landau band.

Such avoided crossings and lifted degeneracies were studied theoretically by Halonen et al. [7] within a model of QD's distorted by repulsive Gaussian scattering centers. We have applied a more sophisticated model by expressing the realistic Coulomb-impurity as

$$V_{\text{imp}}(\mathbf{r}) = \frac{|q|}{4\pi\varepsilon_0\varepsilon_r\sqrt{(\mathbf{r}-\mathbf{R})^2+d^2}}, \quad (2)$$

with the negative charge q of the impurity, $\varepsilon_r = 13$ being the dielectric constant of GaAs, and R and d the lateral and vertical distances of the impurity from the QD center, respectively. The confining part is written as

$$V_{\text{conf}}(\mathbf{r}) = \begin{cases} \frac{1}{2}m^*\omega_0^2 r^2, & r \leq r_c \\ m^*\omega_0^2\left[s(r-r_c)^2 - r_c(\frac{r_c}{2}-r)\right], & r > r_c, \end{cases} \quad (3)$$

where the parameter s defines the strength of the rounding (softening) term. As shown below, the rounding of the confinement is crucial in obtaining a good agreement with the experimental energy spectrum. The single-electron energies are calculated numerically from the discretized eigenvalue equation on a 2D point grid using a Rayleigh quotient multigrid method [8]. The resulting calculated eigenenergies are plotted as dashed lines in Fig. 2 (a) with the model potential plotted in Fig. 2 (b). As can be seen, the agreement between the experimental traces and the calculated eigenvalues is good and the positions of the anticrossings are predicted with a good accuracy. The confinement is given by $\hbar\omega_0 = 13.8$ meV, $r_c = 15.5$ nm, and $s = -0.2$, and the impurity parameters are given by $q = -2e$, $R = 14.5$ nm, and $d = 2$ nm, whereas the voltage offset is $V_0 = 172$ mV.

The fitting parameter suggest a double-charged impurity located very close to the quantum dot plane, probably embedded inside the GaAs quantum well. The impurity might be a Si-dopant atom migrated from the highly-doped emitter through the thin spacer layer, and breaking now the spatial symmetry of the system.

We have presented an experimental single-electron spectrum with lifted level degeneracies and level repulsions indicating a broken circular symmetry. We are able to model quantitatively our experimental data with a realistic model of a harmonic QD potential containing a repulsive Coulomb-impurity term. As a result we identified the existence of a double-charged Si-dopant atom very close to the QD plane.

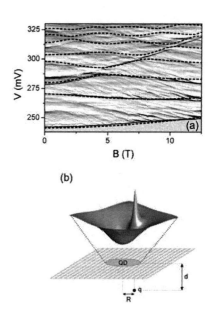

FIGURE 2. (a) Measured transport spectrum and calculated single-electron spectrum. (b) Potential profile used in the simulation.

ACKNOWLEDGMENTS

We acknowledge sample growth by V. Avrutin and A. Waag and financial support by BMBF.

REFERENCES

1. For review, see, e.g., L. P. Kouwenhoven, D. G. Austing, and S. Tarucha, Rep. Prog. Phys. **64**, 701 (2001).
2. For review, see, e.g., S. M. Reimann and M. Manninen, Rev. Mod. Phys. **74**, 1283 (2002).
3. S. Tarucha, D. G. Austing, Y. Tokura, W. G. van der Wiel, and L. P. Kouwenhoven, Phys. Rev. Lett. **84**, 2485 (2000).
4. V. Fock, Z. Phys. **47**, 446 (1928).
5. J. Weis, R. J. Haug, K. v. Klitzing, and K. Ploog, Phys. Rev. B **46**, 12837 (1992); M. Ciorga, A. S. Sachrajda, P. Hawrylak, C. Gould, P. Zawadzki, S. Jullian, Y. Feng, Z. Wasilewski, Phys. Rev. B **61**, R16315 (2000); R. H. Blick, T. Schmidt, R. H. Haug, K. von Klitzing, Semicond. Sci. Technol. **11**, 1506, (1996); J. Könemann, D. K. Maude, V. Avrutin, A. Waag, and R. J. Haug, Physica E **22**, 434 (2004).
6. P. König., T. Schmidt, and R.J. Haug, Europhys. Lett. **54**, 495 (2001).
7. V. Halonen, P. Hyvönen, P. Pietiläinen, and T. Chakraborty, Phys. Rev. B **53**, 6971 (1996).
8. E. Räsänen, J. Könemann, R. J. Haug, M. J. Puska, R. M. Nieminen, to appear in Phys. Rev. B.

Multi-terminal transport through semi-conductor quantum dots

R. Leturcq*, D. Graf*, T. Ihn*, K. Ensslin*, D. D. Driscoll[†] and A. C. Gossard[†]

*Solid State Physics Laboratory, ETH Zürich, 8093 Zürich, Switzerland
[†]Materials Department, University of California, Santa Barbara, Ca 93106, USA

Abstract. We report on a three-terminal tunneling experiment on a semiconductor quantum dot in the Coulomb blockade regime. The set-up allows us to measure directly the conductance matrix of this closed system. In the weak coupling regime, we can determine the coupling strengths of the dot to its individual leads. The independant fluctuations of these coupling strengths as a function of the electronic state are attributed to fluctuations of the shape of the wave function in the dot.

In a standard two-terminal experiment with a quantum dot in the Coulomb blockade regime, the current on a conductance resonance is determined by the average coupling of the electron wave function in the dot with the wave functions in both leads [1]. In the linear regime, such an experiment does not allow to determine the individual coupling strengths of the wave function in the dot with each lead. Here we demonstrate that it is possible to deduce the individual coupling strengths if three or more terminals are connected to the dot, giving local information on the wave function in the dot.

The sample has been fabricated on an AlGaAs-GaAs heterostructure with a two-dimensional electron gas (2DEG) 34 nm below the sample surface. The quantum dot, shown in Fig. 1(a), is defined by local oxidation using an atomic force microscope [2, 3]. The width of the four quantum point contacts connecting the dot to the four reservoirs numbered 1–4 is controlled by voltages applied to the lateral gate electrodes LG1–LG4. In this paper we focus on the dot being connected to three terminals. The point contact connecting the dot to reservoir 4 is completely closed, and gate LG4 is used for controlling the number of electrons in the dot. Measurements of Coulomb diamonds reveal a charging energy $E_C \approx 0.5$ meV and an average single-particle level spacing $\Delta \approx 35$ μeV. An electronic temperature of $T = 90$ mK is deduced from the width of the Coulomb peaks [1]. The quantum dot is tuned into the quantum Coulomb blockade regime with the level broadening $h\Gamma \ll k_B T \ll \Delta$.

Figure 1(b) shows the measurement set-up. A dc bias voltage of 10 μV is applied to one terminal of the dot (e.g. V_{bias1}), while the two other terminals are grounded (e.g. $V_{bias2} = V_{bias3} = 0$). Current-voltage converters are used to measure the currents through each terminal. By applying the bias successively to each of the three termi-

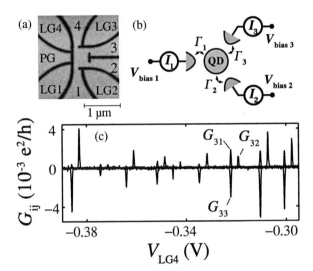

FIGURE 1. (a) Micrograph of the four terminal quantum dot, the black lines being the oxide lines below which the 2DEG is depleted. (b) Measurement set-up using three terminals. The quantum dot (QD) is connected to the three leads through tunnel barriers with tunneling rates Γ_1 to Γ_3. (c) Measurement of Coulomb blockade for three of the nine measured conductances. G_{32} is laterally offset by a constant of +3 mV to make the presentation more transparent.

nals, we obtain nine different current measurements. In linear response theory, these nine currents correspond to the nine elements of the conductance matrix of the three-terminal system:

$$\begin{pmatrix} I_1 \\ I_2 \\ I_3 \end{pmatrix} = \begin{pmatrix} G_{11} & G_{12} & G_{13} \\ G_{21} & G_{22} & G_{23} \\ G_{31} & G_{32} & G_{33} \end{pmatrix} \begin{pmatrix} V_1 \\ V_2 \\ V_3 \end{pmatrix} \quad (1)$$

In a preliminary experiment on a strongly coupled dot some of the G_{ij}'s have been measured [4]. Figure 1(c) shows three of the nine conductance signals corresponding to the G_{ij} as a function of the gate voltage V_{LG4} that controls the number of electrons on the dot. The positions of all corresponding Coulomb resonances agree within less than 1/10 of the peak width, indicating that the same energy level in the dot is probed in all configurations.

In the case of very low temperatures and weak coupling, one can calculate the conductance matrix elements using the theory for lowest order sequential tunneling [5], generalized to the case of more than two terminals:

$$G_{ij} = \frac{e^2}{4k_BT} \frac{\Gamma_i \Gamma_j}{\Gamma_1 + \Gamma_2 + \Gamma_3} \cosh^{-2}\left(\frac{\delta}{2k_BT}\right) \quad (2)$$

for $i \neq j$, and

$$G_{ii} = -\frac{e^2}{4k_BT} \frac{\Gamma_i(\Gamma_j + \Gamma_k)}{\Gamma_1 + \Gamma_2 + \Gamma_3} \cosh^{-2}\left(\frac{\delta}{2k_BT}\right) \quad (3)$$

for $i \neq j$, $i \neq k$ and $j \neq k$. Γ_k is the tunneling rate from the dot to lead k (see Fig. 1(b)), and $\delta = e\alpha_{LG4}(V_{LG4,res} - V_{LG4})$, with $\alpha_{LG4} = 0.07$ the lever arm of gate LG4, determined by the measurement of Coulomb diamonds. Each peak of the conductance curves is fitted with Eq. (2) or (3) in order to deduce its position, its maximum and its width, as shown in Fig. 2(a). Figure 2(b) shows that the width of the peaks depend weakly on V_{LG4}, as expected for a weakly coupled quantum dot. From the maxima of the peaks, we calculate the individual tunneling rates Γ_k. The tunneling rates depend on the overlap of the wave function in the dot with the wave function in lead k.

The coupling strengths $h\Gamma_k$ measured at resonance as a function of the gate voltage V_{LG4} are shown in Fig. 2(c). The values of the lead conductances fluctuate strongly and independently, also when the mean coupling strengths are similar for all leads. Peak height fluctuations attributed to fluctuations of the shape of quasi-bound states in chaotic dots have been extensively studied in two-terminal quantum dots [6, 7], and calculations based on random matrix theory are in reasonable agreement with experimental results[6, 7, 8, 9]. However, two-terminal experiments can only give information on the global conductance of the entire system. In a three-terminal setup this picture can be probed more directly by looking at the spatial distribution of the wave function. In our experiment, these spatial fluctuations are observed as independent coupling strenghts because the distance between leads (≈ 400 nm) is much larger than the Fermi wave length (≈ 40 nm) [10]. A perpendicular magnetic field is also expected to change the interference pattern in the dot. We have observed fluctuations of the coupling strengths related to fluctuations of the shape of the wave function as a function of the magnetic field [11].

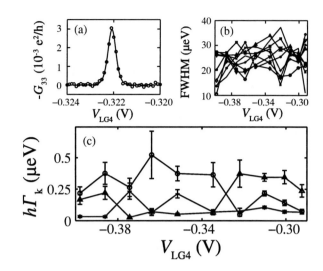

FIGURE 2. (a) Fit of a Coulomb peak in the conductance G_{33} by Eq. (3) (circles are experimental points). (b) Peak width (FWHM) for the resonance measured in the nine conductances G_{ij}. (c) Tunneling rates $h\Gamma_k$ for individual resonances for lead 1 (+), lead 2 (o) and lead 3 (Δ).

To conclude, we show that it is possible to deduce the individual coupling strengths from a weakly coupled quantum dot to its leads if three or more terminals are connected to the dot. The set-up allows then to monitor the tunneling rates from the dot onto the three terminals individually, and the changing shape of the electronic wave function can be monitored on a qualitative level.

Financial support from the Swiss Science Foundation (Schweizerischer Nationalfonds) and from the EU Human Potential Program financed via the Bundesministerium für Bildung und Wissenschaft is gratefully acknowledged. We thank C. Lewenkopf, Y. Gefen and M. Büttiker for useful discussions.

REFERENCES

1. Kouwenhoven, L. P., et al., "Electron Transport in Quantum Dots," in *Mesoscopic Electron Transport*, edited by L. P. Kouwenhoven, G. Schön, and L. L. Sohn, Kluwer, Dordrecht, 1997, pp. 105–214.
2. Held, R., et al., *Appl. Phys. Lett.*, **73**, 262 (1998).
3. Lüscher, S., et al., *Phys. Rev. Lett.*, **86**, 2118 (2001).
4. Kumar, A., et al., *Appl. Phys. Lett.*, **66**, 1379 (1995).
5. Beenakker, C. W. J., *Phys. Rev. B*, **44**, 1646 (1991).
6. Chang, A. M., et al., *Phys. Rev. Lett.*, **76**, 1695 (1996);
7. Folk, J. A., et al., *Phys. Rev. Lett.*, **76**, 1699 (1996).
8. Jalabert, R. A., Stone, A. D., and Alhassid, Y., *Phys. Rev. Lett.*, **68**, 3468 (1992).
9. Foa Torres, L. E. F., Lewenkopf, C. H., and Pastawski, H. M., *Phys. Rev. Lett.*, **91**, 116801 (2003).
10. Lewenkopf, C. H., private communication.
11. Leturcq, R., et al., *Europhys. Lett.*, **67**, 439 (2004).

Population Inversion and Coherent Phonon Emission in a Biased Quantum Dot System Placed in an Acoustic Cavity

Lev Mourokh[1], Anatoly Smirnov[2], and Alexander Govorov[3]

[1]*Department of Physics, Stevens Institute of Technology, Hoboken, New Jersey 07030, USA*

[2]*D-Wave System, Inc., 320-1985 W. Broadway, Vancouver, British Columbia, Canada V6J 4Y3*

[3]*Department of Physics and Astronomy, Ohio University, Athens, Ohio 45701, USA*

Abstract. We have analyzed electron dynamics in a semiconductor double-dot system attached to leads. The voltage bias applied to this structure gives rise to electron localization in the constituent dots. We have solved the Schrodinger equation numerically to determine the overlap integrals of the double-dot electron wave functions and the wave functions of the leads, and, accordingly, to obtain the ratios of the tunnel coupling constants in the microscopic basis. The populations of the double-dot electron levels have been found from the nonequilibrium Green's function technique. We have shown that for the appropriate bias voltage there is population inversion between the upper electron level mainly connected to the lead with higher chemical potential and the lower electron level effectively emptied to the lead with lower chemical potential. When such a system is suspended and forms an acoustic cavity having specific phonon modes, the electron-phonon interaction becomes resonant and can lead to coherent phonon generation.

INTRODUCTION

Semiconductor tunnel-coupled double-dot systems have been at the focal point of intense research activity in recent years [1], in particular due to the formation of molecular-like states. At equilibrium, symmetric double-dot electrons form coherent waves delocalized over the two dots. An applied external electric field removes the symmetry and leads to electron localization in the particular dots. In the present paper we show that population inversion between these electron levels can be achieved at the appropriate bias. If the double-dot system is placed in an acoustic microcavity, this population inversion can lead to coherent phonon generation.

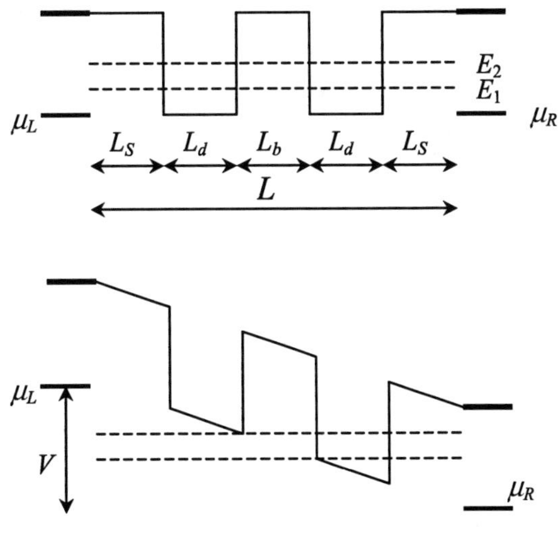

FIGURE 1. Energetic diagram of the double-dot system (a) at equilibrium; (b) with bias voltage applied to the leads.

DOUBLE-DOT ELECTRON LEVELS

CP772, *Physics of Semiconductors: 27th International Conference on the Physics of Semiconductors*, edited by José Menéndez and Chris G. Van de Walle

We model the biased double-dot system in terms of two rectangular quantum wells of widths L_d separated by a barrier of width L_b and connected to leads having equilibrium chemical potentials μ via spacing layers of widths L_S (Figure 1a). A voltage bias applied to the leads gives rise to modification of the confinement potential and to a shift of energetic levels (Figure 1b).

We solve numerically the one-dimensional Schrödinger equation to obtain the energies, E_1 and E_2, and wave functions, $\psi_1(x)$ and $\psi_2(x)$, of the tunnel-coupled dots in an electric field, corresponding to the two lowest levels. Higher levels are taken to be energetically inaccessible.

LEVEL POPULATIONS

To analyze the lead-to-lead current in such a system, we employ the nonequilibrium Green's function formalism presented and discussed in detail in [2]. The second-quantized Hamiltonian of the double-dot electrons is given by

$$H = E_1 a_1^+ a_1 + E_2 a_2^+ a_2 + U a_1^+ a_2^+ a_2 a_1 \\
+ \sum_k \left(E_{Lk} c_{Lk}^+ c_{Lk} + E_{Rk} c_{Rk}^+ c_{Rk} \right) \\
+ \sum_k \left(L_{1k} c_{Lk}^+ a_1 + R_{1k} c_{Rk}^+ a_1 + H.C. \right) \quad (1) \\
+ \sum_k \left(L_{2k} c_{Lk}^+ a_2 + R_{2k} c_{Rk}^+ a_2 + H.C. \right),$$

where a_n^+/a_n and c_{Sk}^+/c_{Sk} are the creation/annihilation operators of n-th double-dot level and S-th lead, respectively. Ratios of the coupling constant (assumed to be k-independent) can be found in the microscopic basis using the solutions of Schrödinger equation as

$$\alpha_L = \frac{L_1}{L_2} = \frac{\int_0^{L_S} \psi_1(x)dx}{\int_0^{L_S} \psi_2(x)dx}; \quad \alpha_R = \frac{R_2}{R_1} = \frac{\int_{L-L_S}^{L} \psi_2(x)dx}{\int_{L-L_S}^{L} \psi_1(x)dx}. \quad (2)$$

We have derived equations of motions for the double-dot electron Green's functions and have been able to reformulate them as a set of algebraic equations for modified double-dot level energies and level populations as ($f_{L,R}$ are Fermi distribution functions of the left and right leads)

$$E_1' = E_1 + UN_2; \quad E_2' = E_2 + UN_1 \quad (3)$$

and

$$N_1 = \frac{\alpha_L^2 f_L(E_1') + f_R(E_1')}{1+\alpha_L^2}; \quad N_2 = \frac{f_L(E_2') + \alpha_R^2 f_R(E_2')}{1+\alpha_R^2}. \quad (4)$$

Eqs. (3) and (4) have been solved self-consistently. The electric-field dependence of the population is shown in Figure 2. It is evident from this Figure that at equilibrium the population of the lower level is larger. With increasing voltage, the lower level becomes conductive (we chose the equilibrium chemical potential of the leads to be smaller than the level energies, see Figure 1a) and its population increases. With further increase of the voltage, both levels become conductive (see Figure 1b) and the first level localized in the right dot is effectively emptied by the right lead, whereas the second level localized in the left dot is effectively filled by the left lead, giving rise to the population inversion. For a double-dot structure placed in an acoustic cavity, this population inversion can lead to coherent phonon generation.

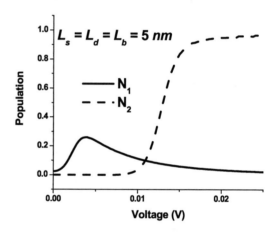

FIGURE 2. Level populations as functions of the bias voltage.

ACKNOWLEDGMENTS

L. M. gratefully acknowledges support from the Department of Defense, DAAD 19-01-1-0592. A. G. is thankful to Robert Blick for stimulated discussions.

REFERENCES

1. W.G. van der Wiel, S. De Franceschi, J.M. Elzerman, T. Fujisawa, S. Tarusha, and L.P. Kouwenhoven, *Rev. Mod. Phys.* **75**, 1 (2003).
2. L.G.Mourokh, N.J.M. Horing, and A.Yu. Smirnov, *Phys. Rev. B* **66**, 085332 (2002).

Anisotropy of Zeeman-splitting in quantum dots

J. Könemann*, V. I. Fal'ko†, D.K. Maude** and R. J. Haug*

Universität Hannover Institut für Festkörperphysik Appelstrasse 2, 30167 Hannover Germany
†*School of Physics & Chemistry, Lancaster University, LA1 4YB Lancaster, United Kingdom*
**High Magnetic Field Laboratory, CNRS, 25 Avenue des Martyrs, BP 166, 38042 Grenoble cedex 9, France*

Abstract. By single-electron tunneling spectroscopy we investigate the difference in spin splitting of single-electron resonances in a double-barrier structure subjected to a magnetic field perpendicular ($\Delta_\perp$) and parallel ($\Delta_\parallel$) to the plane of quantum well. The observed anisotropy of spin splitting is interpreted within a model of spin-orbit coupling in quantum dots.

Spin in semiconductor nanostructures like quantum dots has attracted wide interest with respect to future applications like spin transistors[1] or spin valves[2]. In quantum dots the orbital degrees of freedom and the spin degree of freedom can be tuned electrostatically and by an applied magnetic field. In our work we applied single-electron resonant tunneling spectroscopy[3] to investigate the anisotropy of spin splitting of electrons in quantum dots with respect to different configurations of an applied magnetic field and compare it to the gyromagnetic ratio, the effective Landé-factor. We explain our results by an interplay between spin-orbit coupling and quantum dot confinement of the electrons[4].

The experiment was performed with two highly asymmetric double barrier resonant tunneling devices of different pillar diameters grown by molecular beam epitaxy on n$^+$-type GaAs substrate. The heterostructures consist of a 10 nm wide GaAs quantum well sandwiched between two Al$_{0.3}$Ga$_{0.7}$As-tunneling barriers of 5 and 8 nm. The contacts are formed by 0.5 μm thick GaAs layers highly doped with Si up to 4×10^{17} cm^{-3} and separated from the active region by 7 nm thin spacer layers of undoped GaAs. We carried out DC measurements of the I-V-characteristics in a dilution refrigerator at 20 mK base temperature in high magnetic fields up to 27 T. We were able to measure the transport spectrum of single localized states with different confinement strength in both samples for $B \parallel I$ and $B \perp I$. Sample A contains a weakly confined state with a confinement energy $\hbar\omega_0$ of 13 meV[5]. Fig. 1 (a) displays the diamagnetic shift of two conductance peaks P1 and P2 found in sample A for $B \parallel I$, whereas for $B \perp I$ in Fig. 1 (b) no diamagnetic shift can be seen. In the spectrum of sample B only a single conductance peak P0 with a much weaker diamagnetic shift is analyzed attributed to a strongly bound localized state with $\hbar\omega_0 = 31$ meV. Besides the diamagnetic shift,

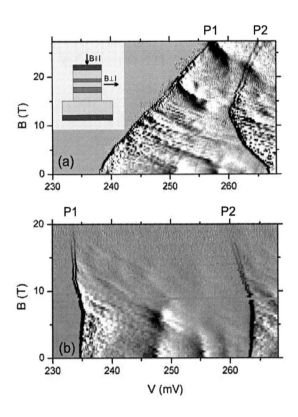

FIGURE 1. G(V,B)-plot of the transport spectrum of sample A for (a) the $\vec{B} \parallel \vec{I}$ and (b) the $\vec{B} \perp \vec{I}$-configuration. P1 and P2 indicate the first spin-split conductance-peaks.

all peaks in dI/dV resolve into two peaks at high enough field values, manifesting the spin splitting of each single localized state. These spin splitting data have been analyzed in detail for sample A, see Fig. 2, and B[6]. The graph includes splittings measured in two geometries. Full and empty symbols stand for splittings measured in the field perpendicular ($\mathbf{B} = B\hat{\mathbf{e}}_z$) and parallel to the plane of the double-barrier structure, respectively (*i.e.*, oriented in and across the tunneling directions).

The data for both samples display a distinct anisotropy of peak splitting, where the splitting caused by the out-of-plane field is systematically larger in comparison to the splitting observed with an in-plane field. In sample B, that means in the regime of strong spatial confinement ($\hbar\omega_0 > \hbar\omega_c$), we observe for the in-plane-magnetic field orientation a smaller slope of the linear spin-splitting than for the out-of-plane-magnetic field orientation. For the low-field asymptotic, that is for $\omega_c < \omega$, we assume $\Delta_\perp - \Delta_\parallel \approx \frac{-g^*}{|g^*|} B\hbar e^2 B_{\mathrm{so}}/(2\omega(m^*)^2)$. The internal magnetic field $B_{\mathrm{so}} \propto (\rho_{\mathrm{BR}}^2 - \rho_{\mathrm{D}}^2)$ reflects here the difference of the spin orbit coupling parameter ρ_{BR} for the Bychkov-Rashba[7] and ρ_{D} for the Dresselhaus-mechanism[8].

In contrary, we find in sample A, that means in the weak confinement regime ($\hbar\omega_0 < \hbar\omega_c$), the same slope of the spin splitting for both magnetic field orientations. But we find for the out-of-plane-magnetic field orientation in the spin splitting dependence a constant energy offset compared to the spin splitting in the in-plane-magnetic field orientation. We assume now for the high-field asymptotics, that is for $\omega_c < \omega$, $\Delta_\perp - \Delta_\parallel \approx \frac{-g^*}{|g^*|} \frac{e\hbar}{m^*} B_{\mathrm{so}}$ for $\omega_c \gg \omega$. That means, the anisotropy of spin splitting of few lowest quantum dot states transforms into an offset with the sign being dependent on the sign of B_{so} and on the sign of electron g^*-factor of our sample, which is known to be negative from a previous experiment[9]. So, we are able to determine the spin-orbit coupling characteristics which appeared difficult to separate in previous experiments[10].

In conclusion, we have applied single-electron resonant tunneling spectroscopy to investigate the anisotropy of spin splitting of single-electron resonances. As a result, we are able to explain the anisotropy of spin splitting with an interplay of the spin-orbit coupling characteristics and the quantum confinement of our samples.

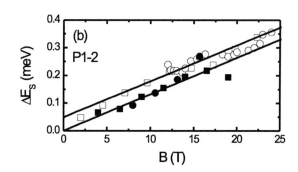

FIGURE 2. Spin splitting of the states P1 (squares) and P2 (circles) of sample A for $\vec{B} \parallel \vec{I}$ (open symbols) and (b) $\vec{B} \perp \vec{I}$-configurations (filled symbols). Solid lines are fits to the experimental data.

REFERENCES

1. S. Datta and B. Das, Appl. Phys. Lett. **56**, 665 (1990).
2. G. A. Prinz, Science **282**, 1660 (1998).
3. M. R. Deshpande, J. W. Sleight, M. A. Reed, R. G. Wheeler, and R. J. Matyi, Phys. Rev. Lett. **76**, 1328 (1996); A. S. G. Thornton, T. Ihn, P. C. Main, L. Eaves, and M. Henini, Appl. Phys. Lett. **73**, 354 (1998).
4. J.-H. Cremers, P.W. Brouwer, and V.I. Fal'ko, Phys. Rev. B 68, 125329 (2003).
5. E. Räsänen, J. Könemann, R. J. Haug, M. J. Puska, and R. M. Nieminen, cond-mat/0404581.
6. J. Könemann, P. König, and R. J. Haug, Physica E **13**, 675 (2002).
7. Yu. Bychkov and E. Rashba, JETP Lett. **39**, 78 (1984); L. Wissinger *et al*, Phys. Rev. B **58**, 15375-15377 (1998).
8. G. Dresselhaus, Phys. Rev. **100**, 580 (1955); F. Malcher, G. Lommer, U. Rossler, Superlatt. Microstruct. **2**, 273 (1986).
9. P. König, T. Schmidt, and R. J. Haug, Europhys. Lett. **54**.
10. D. M. Zumbühl *et al*, Phys. Rev. Lett. **89**, 276803 (2002); J. A. Folk *et al*, Phys. Rev. Lett. 86, 2102-2105 (2001).

Propagation of electron waves in InAs/AlGaSb

Toshihiko Maemoto[1], Masatoshi Koyama[1], Atsushi Nakashima[1], Masato Nakai[1], Shigehiko Sasa[1], J. P. Bird[2], and Masataka Inoue[1],

[1]*New Materials Research Center, Osaka Institute of Technology, 5-16-1, Omiya, Asahi-ku, Osaka, Japan*
[2]*Department of Electrical Engineering, University at Buffalo, The State University of New York, USA*

Abstract. We report on magneto-transport characteristics of InAs/AlGaSb quantum open dot and anti-dot structures. Aperiodic oscillations have been observed at low magnetic fields in addition to the Shubnikov-de-Haas oscillations measured at 4.2K. The characteristics are due to the ballistic nature of electron transport and the propagation of electron waves in InAs. Phase breaking times of electrons in InAs/AlGaSb dot and anti-dot were determined from correlation functions. From these magnetoresistance data and correlation functions, the electron waves propagation in the InAs/AlGaSb is discussed.

INTRODUCTION

Transport properties of electrons confined in low dimensional systems have been reported intensive investigation both theoretically and experimentally in recent years. Open quantum dot and anti-dot structures have been used to study quantum propagation effects, such as semi-classical orbit transport, decoherence, and wave function scarring [1]. Furthermore, the spin-orbit interaction plays a crucial role in determining persistent currents and Aharonov-Bohm (AB) effect in mesoscopic rings. InAs/AlGaSb heterostructure is a promising material for observation of these mesoscopic effects, because InAs/AlGaSb systems exhibit high mobility, long mean free path (l_e), and long coherence length (l_ϕ). Phase breaking time (τ_ϕ) is also quantity of fundamental importance in mesoscopic structures. Recently, we have reported the τ_ϕ in InAs/AlGaSb open quantum dot structures [2]. Ring geometries have been used to determine the l_ϕ and the τ_ϕ from the electron propagation such as weak localization, UCF, and AB oscillations.

In this paper, we report on the studies of the magneto-transport in comparison between InAs open quantum dot and anti-dot structures of ring geometry. The phase breaking times of electrons in InAs were determined from the magnetoresistance data and the correlation functions.

FABRICATION

Figure 1 shows an InAs/AlGaSb heterostructure grown on semi-insulating GaAs substrate by using molecular beam epitaxy. The electron mobility ranges from 2.5×10^5 to 3×10^5 cm^2/Vs at 4.2 K, which is corresponding to a mean pass length of ~ 4–5 μm. Open dot and anti-dot structures were fabricated by using e-beam lithography, wet chemical etching (H$_3$PO$_4$: H$_2$O$_2$:H$_2$O, 1:1:30), and non-alloy ohmic process. The cavity dimensions in anti-dot were varied from 0.25 to 1.00 μm^2 square. In magnetoresistance measurements, the driving sample current was kept between 3 nA and 100 nA.

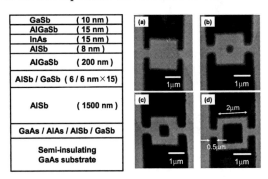

FIGURE 1. Schematic diagram of InAs/AlGaSb heterostructure grown by MBE. AFM images of open dot and three different anti-dot structures. The cavity size of (b) is 0.25 x 0.25 μm^2, that of (c) is 0.50 x 0.50 μm^2, and that of (d) is 1.00 x 1.00 μm^2.

RESULTS AND DISCUSSION

Figure 2 shows the magnetoresistance measured at 4.2 K for the open dot shown in Fig. 1 (a). From the magnetoresistance data, the amplitude of oscillations increased with decreasing the driving sample current, and reproducible fluctuations of magnetoresistance were observed at the different current level. These fluctuations are due to the propagation of electron waves in InAs open dot. The magnetic fields marked by arrows in Fig. 2 correspond to decrease magnetoresistance by the formation of skipping orbit in the cyclotron motion. The electron motion can be understood from the terms of classical ballistic trajectories.

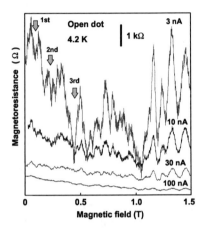

FIGURE 2. Magnetoresistance fluctuations in the open dot at the different sample current (These trace have been shifted upwards).

We analyzed the magnetoresistance oscillations by using correlation functions, and the correlation fields (B_c) were estimated from correlation function. Figure 3 shows the relationship between B_c as a function of magnetic field (B) for the different structures. The increase in the correlation fields indicates a corresponding reduction in the propagation area of electrons. When the cyclotron diameter becomes smaller than the effective dot size, the B_c increased with a linear slope at the high fields. The arrows in Fig.3 correspond to the cyclotron diameter becomes comparable to the effective conducting width of the anti-dot, which gives at around 1T. From the relationship between B_c and B, the phase breaking times in InAs open dot and anti-dot structures were estimated to be 17-42 ps as shown in Table 1. The present results indicate that the phase coherence times of electrons are influenced by the geometry of structures. As a result, the phase breaking time in InAs may be determined generally by summation of phase breaking time between dot geometry and connected wire as shown in equation (1) [3]. We consider that the ring geometry is regarded as a combination of connected wires rather than the dot structures.

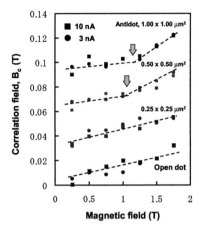

FIGURE 3. Relationship between correlation fields, B_c and magnetic fields, B for the open dot and anti-dot structures.

TABLE 1. Phase coherence times of InAs/AlGaSb open dot and anti-dot structures

Device structures	Phase breaking times τ_ϕ
Open dot : square	42 ps
Anti-dot : 0.25 x 0.25 μm²	31 ps
Anti-dot : 0.50 x 0.50 μm²	26 ps
Anti-dot : 1.00 x 1.00 μm²	17 ps

$$\frac{1}{\tau_\phi} = \frac{1}{\tau_\phi}\bigg|_{dot} + \frac{1}{\tau_\phi}\bigg|_{wire} \quad (1)$$

CONCLUSION

Magnetoresistance fluctuations due to the electron waves propagation in the open dot and anti-dot structures have been observed even at 4.2 K. Phase breaking times in InAs open dot and anti-dot structures have been estimated to be 17-42 ps from the magnetoresistance data measured at 4.2 K.

REFERENCES

1. Electron Transport in Quantum Dots, edited by J. P. Bird: Kluwer Academic Publishers, 2003.

2. T. Maemoto, T. Kobayashi, T. Karasaki, K. Kita, S. Sasa, M. Inoue, K. Ishibashi, and Y. Aoyagi, Physica B **314**, pp. 481-484 (2002).

3. D. P. Pivin, Jr. A. Andersen, J. P. Bird, and D. K. Ferry, Phys. Rev. Lett. **82**, pp. 4687-4690 (1999).

HEMT Amplified SET Measurements of Individual InGaAs Quantum Dots

K. D. Osborn, Mark W. Keller, R. P. Mirin

National Institute of Standards and Technology, 325 Broadway, Boulder, CO 80305

Abstract. A high electron mobility transistor (HEMT) is used with a single-electron transistor (SET) to measure single electrons tunnelling into individual InGaAs quantum dots. The SET detects a change in location of an electron once it tunnels from an underlying n-doped layer into a quantum dot lying in an intermediate layer. A HEMT on the He^3 stage with the SET is used to extend the measurement bandwidth to 400 kHz. We demonstrate this technique with a measurement of the Stark shift in the first electron state of the quantum dot as a function of lateral electric field.

The electron states of individual self-assembled quantum dots must be controllably occupied in order to develop a quantum dot-based electrically-triggered single-photon source. Here and in previous work [1], we measure individual electrons tunnelling into quantum dots with a single-electron transistor (SET). However, in this work we have increased the bandwidth of the measurement, which allows us to quickly obtain the energies of electron states as a function of applied fields. We demonstrate our improved technique by measuring the Stark effect in the ground state of a quantum dot as a function of the applied lateral electric field.

By themselves SETs typically have a low bandwidth since the high resistance across the SET electrodes is coupled to wiring that capacitively filters frequencies greater than approximately 1 kHz. In order to improve the bandwidth of SET measurements, high electron mobility transistor (HEMT) amplifiers [2] and RF amplifiers [3] have been used. We have selected a commercially available GaAs HEMT chip in our application. The HEMT has the advantage of being small and inexpensive compared to a RF amplifier, and therefore several SETs can be amplified with HEMTs in a single cooldown of the dewar.

A schematic of the SET and HEMT, including the capacitively coupled quantum dot and gates, is shown in Fig. 1. The SET is current-biased with a 15 MΩ chip resistor, and the SET voltage V_{SET} is sent to the gate of the HEMT chip. Along with the SET, the resistor and HEMT are on the sample stage of a He^3 dewar, which allows for a small stray capacitance to ground. Since V_{SET} is small, the source-gate bias is controlled by a voltage V_{SO} applied to the source. The HEMT is biased with a large resistor on the drain lead to keep the power dissipated by the HEMT at about 1 μW. The

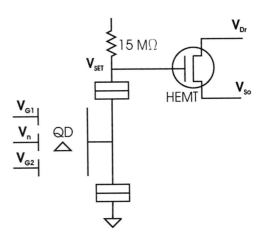

FIGURE 1. Schematic of SET coupled to the QD and amplified by a HEMT.

bandwidth in this configuration is 400 kHz, which allows us to take data significantly faster than for our previous measurements on quantum dots.

The Stark effect has been observed in excitons from InGaAs quantum dots by applying a vertical electric field to the quantum dot layer embedded within a p-n diode [4]. In addition, the vertical and lateral Stark shift have been theoretically studied in an InAs/GaAs quantum dot [5]. Here we study the lateral Stark shift in the first electron state of a quantum dot. Since the vertical Stark shift is small compared to the lateral shift in our measurement, we will avoid discussing the vertical shift below.

Measurements are taken on a heterostructure containing, from bottom to top, an n-doped layer of GaAs, a GaAs tunnelling barrier, a layer of InGaAs quantum dots, and finally GaAs and AlGaAs capping and blocking layers, where the thicknesses are given in ref. [1]. The nominal quantum dot density is 2×10^{10} cm^{-2}, as measured from an atomic force microscope image of a wafer of uncapped dots grown immediately before the wafer with the completed structure. Room-temperature photoluminescence revealed a ground-state exciton peak at 1160 nm. In this work we have fabricated an SET using e-beam lithography and Al evaporations at three angles to produce a 80 nm × 120 nm island connected to two overlapping electrodes by tunnel barriers and capacitively coupled to two neighboring gates, as shown in the inset to Fig. 2.

In the device, electrons with electrochemical potential $-eV_n$ are tunnelled from the n-doped layer into the quantum dots, near an overlying SET. The SET is amplified with the HEMT circuit described above. The voltages applied to gate 1 and 2 are respectively $V_{G1} = V_S + V_{FB}$ and $V_{G2} = -V_S + V_{FB}$, where V_{FB} is the SET feedback voltage and V_S is proportional to a lateral electric field. The chemical potential for electron N in the quantum dot is

$$\mu(N,V_S)/e = (\eta_n - 1)V_n + (\eta_{G1} - \eta_{G2})V_S + (\eta_{G1} + \eta_{G2})V_{FB} \quad (1)$$

for a lateral electric field set by V_s. For our data, $\mu(N,V_s)$ can be expressed as $\mu_0(N) - \alpha(V_S/d)^2$, where d is an effective length that creates the absolute value of the lateral field at the quantum dot $|V_S/d|$, and α describes the strength of the Stark shift. The electrochemical potential of the quantum dot is shifted by voltages applied to the n-doped layer, gate 1, and gate 2 through the dimensionless coupling coefficients, η_n, η_{G1}, and η_{G2}, respectively.

In Fig. 2, a plot of dV_{FB}/dV_S is shown as a function of V_n and V_S. The dark points fall along curves that indicate that an electron has tunnelled into a quantum dot. Approximately parallel curves correspond to electrons with a different N tunnelling into the same quantum dot. Three curves marked with pairs of solid arrows show the first 3 electrons tunnelling into a quantum dot under the body of the SET, which we analyze below. For the addition of a particular electron number, the points in V_n and V_{FB} are fit to second order polynomials in V_S. The quadratic dependence as a function of V_S in the lowest curve marked by solid arrows reflects a Stark shift in the chemical potential $\mu(1,V_S)$. To solve equation (1), we use the value $\eta_n = 1/3$, obtained in ref [2], and $(\eta_{G1} + \eta_{G2}) << 1$. This yields the chemical potential of the first electron level, $\mu_0(1) = 291$ meV, and the first two addition energies, $\mu_0(2) - \mu_0(1) = 23$ meV and $\mu_0(3) - \mu_0(2) = 38$ meV. The curvature in the N= 1 data yields $\alpha/d^2 = (1.8 \pm 0.3) \times 10^{-3}$e/V, where the uncertainty in α/d^2 is determined from the uncertainty in $(\eta_{G1} + \eta_{G2})$. If we estimate the effective distance as d=100 nm, we obtain $\alpha \approx 1.8 \times 10^{-17}$ cm^2/V, which is approximately 2.5 times smaller than a value obtained from a calculation of

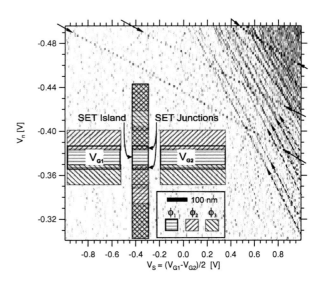

FIGURE 2. dV_{FB}/dV_S as a function of V_n and V_S. The inset shows the SET structure made with three sequential depositions at angles ϕ_1, ϕ_2, and ϕ_3.

a small cylindrical InAs/GaAs quantum dot [5]. A future experiment will use an improved SET and gate design to better quantify the lateral Stark shift.

ACKNOWLEDGMENTS

One of the authors (KDO) acknowledges fruitful discussions with S. Tessmer, A. Zunger, G. Bester, M. Hanna, and G. Austing, and technical assistance from T. Harvey, K. Ullmann and J. Beyer. Contributions of the National Institute of Standards and Technology are not subject to copyright.

REFERENCES

1. Osborn, K. D., Keller, M. W., and Mirin, R. P., *Physica E*, **21**, 501–505 (2004).
2. Visscher, E. H., Lendeman, J., Verbrugh, S. M., Hadley, P., and Mooij, J. E., *Applied Physics Letters*, **68**, 2014–2016 (1996).
3. Schoelkopf, R. J., Wahlgren, P., Kozhevnikov, A. A., Delsing, P., and Prober, D. E., *Science*, **280**, 1238–1242 (1998).
4. F. Findeis, E. B. A. Z., M. Baier, and Abstreiter, G., *Applied Physics Letters*, **78**, 2958–2960 (2001).
5. Li, S.-S., and Xia, J.-B., *Journal of Applied Physics*, **88**, 7171–7174 (2000).

SU(4) Kondo effect in quantum dots with two orbitals and spin 1/2

Mikio Eto

Faculty of Science and Technology, Keio University, 3-14-1 Hiyoshi, Kohoku-ku, Yokohama 223-8522, Japan

Abstract. We theoretically study the Kondo effect in quantum dots with two orbitals and spin 1/2. The Kondo temperature T_K is evaluated, as a function of energy difference Δ between the orbitals, using the scaling method. T_K is maximal around the degeneracy point ($\Delta = 0$) and decreases with increasing $|\Delta|$ following a power law, $T_K(\Delta) = T_K(0) \cdot (T_K(0)/|\Delta|)^\gamma$, which is understood as a crossover from SU(4) to SU(2) Kondo effect. The exponents on both sides of a level crossing, γ_L and γ_R, satisfy a relation of $\gamma_L \cdot \gamma_R = 1$. We compare this enhanced Kondo effect with that at the spin-singlet-triplet degeneracy for an even number of electrons, to explain recent experimental results.

INTRODUCTION

In quantum dots, the Kondo effect usually takes place when an electron spin 1/2 is formed due to the Coulomb blockade and couples to the electron Fermi sea in the leads through tunnel barriers. Besides this conventional Kondo effect, unconventional Kondo effects have been observed which stem from the interplay between spin and orbital degrees of freedom. An example is the enhanced Kondo effect at the energy degeneracy between spin-singlet and -triplet states for an even number of electrons (S-T Kondo effect) [1, 2]. Recently another large Kondo effect has been found around the two-orbital degeneracy for an odd number of electrons with spin 1/2 [doublet-doublet (D-D) Kondo effect] [3]. Both of S-T and D-D Kondo effects seem to indicate that a total of fourfold spin and orbital degeneracy accounts for the enhancement of the Kondo effect. We theoretically examine the enhancement mechanism of D-D Kondo effect. By comparing D-D and S-T Kondo effects, the experimental results of ref. [3] are clearly explained.

MODEL

For the D-D Kondo effect, we consider a quantum dot with an electron (spin 1/2) and two orbitals ($i = 1, 2$). We denote the energy separation between the orbitals by $\Delta = \varepsilon_2 - \varepsilon_1$, which is tunable in experiments. External leads have two conduction channels; channel i couples to orbital i by V_i via tunnel barriers, which is the case of vertical quantum dots [1, 3]. Considering the second-order tunnel processes, we obtain the effective Hamiltonian $\mathscr{H}_{\text{eff}}$ in a space of four dot-states,

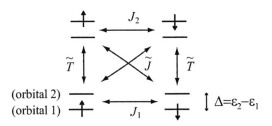

FIGURE 1. Four states in a quantum dot with an electron and two orbitals ($i = 1, 2$) and exchange processes among them. The energy difference between the orbitals is denoted by Δ.

($i = 1, 2$; $S_z = \pm 1/2$). $\mathscr{H}_{\text{eff}}$ involves spin and orbital exchange processes shown in Fig. 1. The coupling constants are given by $J_1 = V_1^2/\tilde{E}_C$, $J_2 = V_2^2/\tilde{E}_C$ and $\tilde{J} = \tilde{T} = V_1 V_2/\tilde{E}_C$, where $1/\tilde{E}_C = 1/E^+ + 1/E^-$, $E^\pm$ being addition and extraction energies. A potential scattering with $T = (V_1^2 + V_2^2)/(2\tilde{E}_C)$ is also included in $\mathscr{H}_{\text{eff}}$ since it is relevant to the Kondo effect.

SCALING CALCULATIONS

The Kondo temperature T_K is calculated as a function of Δ, using the "poor man's" scaling method [2].

We begin with the case of equivalent tunnel couplings for two orbitals ($V_1 = V_2$). Then our model is reduced to the Coqblin-Schrieffer model [4] of SU(4) symmetry, with $J_1 = J_2 = \tilde{J} = \tilde{T} = T \equiv J$, if $\Delta = 0$. We obtain the scaling equations for J in two limits.

When the energy scale D is much larger than $|\Delta|$,

$$dJ/d\ln D = -4\nu J^2, \qquad (1)$$

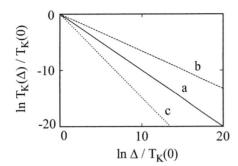

FIGURE 2. Numerical results of the Kondo temperature T_K, as a function of Δ, on a log-log scale. Both T_K and Δ are normalized by the Kondo temperature at $\Delta = 0$, $T_K(0)$. (a) $(V_2/V_1)^2 = 1$, (b) 2/3, and (c) 3/2.

where ν is the density of states of conduction electrons in the leads. The exchange coupling J develops rapidly with decreasing D, reflecting a fourfold degeneracy of spin and orbitals [SU(4) Kondo effect].

When $D \ll |\Delta|$,

$$dJ/d\ln D = -2\nu J^2. \tag{2}$$

The evolution of J is slower since the upper orbital is irrelevant. This is the conventional Kondo effect of SU(2) symmetry with single orbital and spin 1/2.

From these equations, we evaluate T_K as a function of Δ. (i) T_K is maximal around $\Delta = 0$ [$|\Delta| \ll T_K(\Delta = 0)$]. In this case, the scaling equation (1) remains valid till the scaling ends. This yields

$$T_K(\Delta = 0) = D_0 \exp[-1/4\nu J], \tag{3}$$

where D_0 is the bandwidth in the leads. (ii) T_K is minimal when $|\Delta| \gg D_0$. Then eq. (2) is valid in the whole scaling region. Solving the equation, we obtain

$$T_K(\Delta = \infty) = D_0 \exp[-1/2\nu J]. \tag{4}$$

(iii) In the intermediate region of $T_K(0) \ll |\Delta| \ll D_0$, we match the solutions of eqs. (1) and (2) at $D \approx |\Delta|$. We find that $T_K(\Delta)$ decreases with increasing $|\Delta|$, following

$$T_K(\Delta) = T_K(0) \cdot (T_K(0)/|\Delta|)^\gamma, \tag{5}$$

with exponent $\gamma = 1$. This power law was originally derived by Yamada et al. [5].

Now we examine a general case of $V_1 \neq V_2$. We set $\Delta \geq 0$ hereafter. For $D \gg \Delta$, we derive a set of scaling equations for $J_1, J_2, \tilde{J}, \tilde{T}$, and T;

$$\begin{cases} dJ_1/\nu d\ln D &= -2J_1^2 - \tilde{J}(\tilde{J} + \tilde{T}), \\ dJ_2/\nu d\ln D &= -2J_2^2 - \tilde{J}(\tilde{J} + \tilde{T}), \\ d\tilde{J}/\nu d\ln D &= -\tilde{J}(J_1 + J_2 + T) - \tilde{T}(J_1 + J_2)/2, \\ d\tilde{T}/\nu d\ln D &= -3\tilde{J}(J_1 + J_2)/2 - \tilde{T}T, \\ dT/\nu d\ln D &= -3\tilde{J}^2 - \tilde{T}^2. \end{cases} \tag{6}$$

For $D \ll \Delta$,

$$dJ_1/d\ln D = -2\nu J_1^2, \tag{7}$$

whereas the other coupling constants do not evolve.

Analyzing eqs. (6), we find that the fixed point of SU(4) Kondo effect, $J_1 = J_2 = \tilde{J} = \tilde{T} = T = \infty$, is marginal. As a result, $T_K(\Delta)$ is not a universal function although T_K is always maximal around $\Delta = 0$. To see $T_K(\Delta)$, we perform numerical calculations using eqs. (6) [eq. (7)] for $D > \Delta$ [$D < \Delta$]. Figure 2 presents the results. $T_K(\Delta)$ approximately obeys a power law with an exponent that depends on model parameters,

$$\gamma = (V_2/V_1)^2 = \Gamma_2/\Gamma_1, \tag{8}$$

as long as $\Gamma_1 \sim \Gamma_2$, where $\Gamma_i = \pi \nu V_i^2$ is the broadening of level i. For the exponents, γ_L, γ_R, on both sides of a level crossing (at which roles of the two orbitals are interchanged), we obtain a general relation of

$$\gamma_L \cdot \gamma_R = (\Gamma_2/\Gamma_1) \cdot (\Gamma_1/\Gamma_2) = 1.$$

CONCLUSIONS

We have theoretically examined the Kondo effect in quantum dot with two orbitals and spin 1/2. The Kondo temperature T_K is maximal around the degeneracy point and decreases with increasing the energy difference between the orbitals Δ by a power law. Note that, in the S-T Kondo effect, $T_K(\Delta)$ shows a similar power law with an exponent of $\gamma = 2 + \sqrt{5}$ on the triplet side, whereas T_K drops to zero suddenly on the singlet side, where Δ is the energy difference between spin-singlet and -triplet states in this case [2].

In the experiment of ref. [3], the D-D and S-T Kondo effects have been observed in the same quantum dots with $V_1 \approx V_2$. (i) T_K decreases with increasing $|\Delta|$ almost symmetrically with respect to $\Delta = 0$ in D-D case, whereas the decrease in T_K is asymmetric in S-T case. (ii) T_K decreases more slowly in D-D case ($\gamma_L \approx \gamma_R \approx 1$ is expected) than in S-T case ($\gamma = 2 + \sqrt{5}$). Both results are in good agreement with our theoretical results.

REFERENCES

1. S. Sasaki et al., Nature **405**, 764 (2000).
2. M. Eto and Yu. V. Nazarov, Phys. Rev. Lett. **85**, 1306 (2000); M. Eto and Yu. V. Nazarov, Phys. Rev. B **66**, 153319 (2002); M. Pustilnik and L. I. Glazman, Phys. Rev. Lett. **85**, 2993 (2000).
3. S. Sasaki et al., Phys. Rev. Lett. **93**, 17205 (2004).
4. B. Coqblin and J. R. Schrieffer, Phys. Rev. **185**, 847 (1969).
5. K. Yamada et al., Prog. Theor. Phys. **71**, 450 (1984).

Fractal Study of Coupling Transitions in Ballistic Quantum-Dot Arrays

T. P. Martin*, R. P. Taylor*, H. Linke*, B. Murray*, C. Arndt*, N. Aoki[†],
D. Oonishi[†], Y. Iwase[†], and Y. Ochiai[†]

*Department of Physics, University of Oregon, Eugene, OR 97403, USA
[†]Department of Materials Technology, Chiba University, 1-33, Yayoi, Inage, Chiba 263-8522, Japan

Abstract. A fractal analysis is used to characterize magnetoconductance fluctuations in an open, coupled quantum-dot array. The fractal dimension is combined with the empirical parameter Q as a probe of the energy level spectrum in the array, and a coupling transition is identified as a function of the number of modes in the coupling quantum point contact.

INTRODUCTION

A question central to the development of quantum transport concerns the ability to couple quantum-dots into an array while maintaining a "global" coherence across the entire device. As the coupling between the dots is changed, it is expected that the energy level spectrum will evolve from an array of single, "atomic" quantum systems to one combined, "molecular" quantum system. While many previous studies of quantum-dot arrays have focused on transport in the tunneling regime [1-7], a number of recent studies featured "open" quantum point contacts (QPCs) that supported fully conducting modes [8-10]. Here we continue these investigations by utilizing the fractal analysis of magnetoconductance fluctuations (MCF) in an open coupled-dot array as a non-invasive probe of the energy level spectrum as the dot coupling is varied.

EXPERIMENT AND RESULTS

A split-gate pattern (see Fig. 1 inset) was used to define a double-dot array consisting of a 1.0 x 1.0 μm^2 and 1.2 x 1.2 μm^2 dot in the two dimensional electron gas (2DEG) of an AlGaAs/GaAs heterostructure. Details of the fabrication are presented elsewhere [8], but we note a carrier density of 3.9x10^{15} m^{-2} and a mean free path of 8 μm, which ensured ballistic transport through the array. The split-gate design allowed each QPC to be individually contacted, and fixed voltages were applied such that each of the two side QPCs supported $n = 3$ conducting modes. The central gate voltage was independently tuned to allow more modes (strong coupling) or fewer modes (weak coupling) to propagate. Figure 1 shows MCF traces measured at a temperature of 100 mK for various modes of the central QPC. Low-field MCF are a direct result of electron interference arising from trajectory loops within a quantum dot [10, 11]. Since each loop will have a characteristic magnetic field

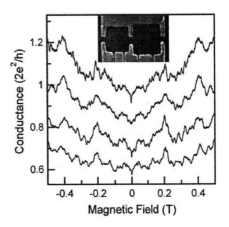

FIGURE 1. Magnetoconductance fluctuations for (bottom to top) $n = 2, 3, 5,$ and 8 modes conducting through the central QPC depicted in the inset. Inset: SEM image of the split-gate pattern used to define the coupled array.

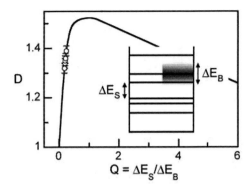

FIGURE 2. Fractal dimension D obtained from the MCF traces in the current study (white circles) vs. Q plotted on the Q curve (black line). Inset: schematic representation of the energy levels in a dot showing the average energy level spacing ΔE_S and average energy level broadening ΔE_B.

scale, the combination of all electron trajectories in the device will result in fluctuations. We employed a fractal analysis known as the box counting method to assess the fractal dimension D of the MCF, which characterizes the statistical scaling properties of the fluctuations at different magnetic field scales [11]. In quantum dots, D has been shown to depend on an empirical parameter Q that quantifies the resolution of the energy levels in the dot [11-13]. The relationship between D and Q has been found to follow a single curve, and it is plotted in Fig. 2. Q is defined as the ratio of an average energy level spacing ΔE_S to an average energy level broadening ΔE_B (see Fig. 2 inset):

$$Q = \frac{\Delta E_S}{\Delta E_B} = \frac{(2\pi\hbar^2)/(m^*A)}{\sqrt{(\hbar/\tau_q)^2 + (k_B T)^2}} \quad (1)$$

where m^* is the electron effective mass, τ_q is the phase coherence time, T is the temperature, and A is the active area of the device.

The trend of D with Q has been well charted for a number of different AlGaAs/GaAs devices [11-13]. D values of the coupled dot were mapped to a corresponding Q value based on this well-established trend. These are plotted in Fig. 2, and in Fig. 3(a) as a function of mode number n. ΔE_B was calculated from measured values of T and τ_q (calculated from a high field correlation analysis [12, 14]), and is plotted in Fig. 3(b). Finally, Q and ΔE_B can be combined to calculate ΔE_S, which is plotted in Fig. 3(b). We identify a transition occurring between $n = 2$ and 4 where ΔE_B decreases due to an increase in phase coherence. We find that ΔE_S also decreases, which is nominally consistent with a transition from isolated energy levels of artificial atoms to hybridized energy levels of an artificial molecule. We stress, however,

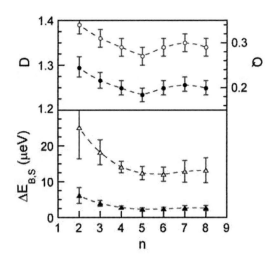

FIGURE 3. (a) D (empty circles) and Q (filled circles), and (b) ΔE_B (empty triangles) and ΔE_S (filled triangles) vs propagating mode n. All dashed lines are guides to the eye.

that this transition occurs in the "open" regime, in contrast to studies in the tunneling regime [1-7].

In conclusion, we presented a novel technique, based on fractal MCF, allowing us to probe two central characteristics of the energy spectrum of coupled quantum-dots: ΔE_S and ΔE_B.

ACKNOWLEDGMENTS

T. P. Martin is an NSF IGERT Fellow. R. P. Taylor is a Research Corporation Cottrell Scholar.

REFERENCES

1. R. H. Blick et al., *Phys. Rev. Lett.* **81**, 689 (1998).
2. R. H. Blick et al., *Phys. Rev. Lett.* **80**, 4032 (1998).
3. J. A. Folk et al., *Phys. Rev. Lett.* **87**, 206802 (2001).
4. T. Fujisawa et al., *Science* **282**, 932 (1998).
5. L. P. Kouwenhoven et al., *Phys. Rev. Lett.* **65**, 361 (1990).
6. T. H. Oosterkamp et al., *Nature* **395**, 873 (1998).
7. F. R. Waugh et al., *Phys. Rev. Lett.* **75**, 705 (1995).
8. N. Aoki et al., *Appl. Phys. Lett.* **80**, 2970 (2002).
9. M. Elhassan et al., *Phys. Rev. B* **64**, 085325 (2001).
10. J. P. Bird et al., *Rep. Prog. Phys.* **66**, 583 (2003).
11. R. P. Taylor et al., "A Review of Fractal Conductance Fluctuations in Ballistic Semiconductor Devices" in *Electron Transport in Quantum Dots*, edited by J. P. Bird, Kluwer Academic/Plenum, New York, 2003.
12. A. P. Micolich et al., *Phys. Rev. Lett.* **8703**, 036802 (2001).
13. A. P. Micolich et al., *Appl. Phys. Lett.* **80**, 4381 (2002).
14. J. P. Bird et al., *Phys. Rev. B* **51**, 18037 (1995).

Imaging of Local Tunneling Barrier Height of InAs Nanostructures Using Low-Temperature Scanning Tunneling Microscopy

Kiyoshi Kanisawa[*], Hiroshi Yamaguchi[*], and Yoshiro Hirayama[*, 1]

[*]NTT Basic Research Laboratories, NTT Corporation, 3-1 Morinosato Wakamiya, Atsugi, Kanagawa, 243-0198, Japan
[1]SORST-JST, 4-1-8 Honmachi, Kawaguchi, Saitama, 331-0012, Japan

Abstract. The local tunneling barrier height (LTBH) has been measured using low-temperature scanning tunneling microscopy to examine the local potential profile of an InAs nanostructure. The nanostructure is a faultily-stacked nanocrystal in epitaxial InAs thin film grown on GaAs(111)A substrate. It is found that averaged LTBH is consistent with the workfunction or electron affinity of InAs. The nanostructure boundary is found to have a higher LTBH than the surroundings. The resonance peak calculated using this potential wall is comparable to that measured by spectroscopy of local density of states (LDOS) in the nanostructure. A gradual LTBH decrease is additionally observed at negative sample bias voltage near the boundary indicating downward band bending, which is consistent with LDOS there.

INTRODUCTION

The characterization of the local electronic properties of semiconductor nanostructures is very important for understanding the wave feature of electrons. We have studied the local density of states (LDOS) in nanostructures using low-temperature scanning tunneling microscopy (LT-STM) [1]. The electrostatic potential profile of nanostructures reveals complimentary information and is also important for predicting the quantum mechanical nature of nanostructures. Such local potential of semiconductors has been reported to characterize doping effects at Si [2, 3] and GaAs [4] surfaces using STM. To our knowledge, however, the local potential profile of semiconductor nanostructures has not been examined.

EXPERIMENTAL

Undoped InAs thin film was grown on an n-type GaAs(111)A substrate by molecular beam epitaxy. At the surface, there are natively formed stacking fault tetrahedrons (SFTs) with surface-accumulated two-dimensional electron gas. A small SFT behaves as a zero-dimensional (0D) structure [1]. All LT-STM measurements were performed in an ultra-high-vacuum at 5 K. Variation of the tunneling current I by modulation of tip-sample separation z, i.e. dI/dz, was detected using the lock-in technique. The value of $0.952 \cdot (dI/dz)^2 / I^2$ [eV] is a good approximation of the apparent barrier height at electron tunneling, i.e. the local tunneling barrier height (LTBH) in case of STM experiments [5]. At plain SFT surface, LTBH is precisely measured due to negligible geometry dependence. According to the $I-z$ characteristics measured, tip-induced band bending effect was found to be negligible due to the surface Fermi level pinning.

RESULTS AND DISCUSSION

The local potential effective to perturb I is expected to be within a few nanometer thick subsurface region, which is shorter than the screening length of semiconductors [3, 6]. Such local potential beneath the surface is detectable almost independently of electron concentration at semiconductor surfaces [6].

When the net electron tunneling occurs from the tip to the SFT at positive sample bias voltage, averaged LTBH depends on the Fermi level or 0D level position

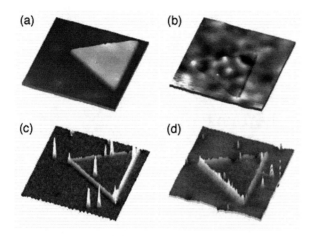

FIGURE 1. (a) Topographic LT-STM image of an SFT at InAs(111)A surface. Structure side length is 28 nm. (b) LDOS image of the SFT by dI/dV mapping at + 0.30 V. LTBH images at (c) positive bias (+0.30 V) and (d) negative bias (-1.5 V) voltages. Brighter region in (b) - (d) indicates higher LDOS or LTBH. Each image has an area of 40.2 nm × 40.2 nm. Thermal drift resulted in the relatively shifted position of the corresponding SFT in each image. Sharp LTBH peaks in (c) and (d) were due to adsorbates.

of the SFT. Measured LTBH was about 5 eV, which is consistent with the workfunction or electron affinity (4.9 eV) of InAs. Along the SFT boundary, there was a narrow region (about 0.6 nm width) where LTBH was roughly 0.4 eV higher than the surroundings (Fig. 1). This is not due to the geometry dependence, because such higher LTBH is not observed at monolayer step edges. There are donor-type localized surface states at the edge of the SFT boundary, which should lower the LTBH there [3, 6]. Since the surface states can not explain the LTBH observed, the {111} stacking-fault boundaries of the SFT beneath the top-most (111)A surface is a candidate of the local potential wall. An abrupt potential wall localized in atomically thin region is possible, if two electric double layers are mirror images of each other. Resonance peak width calculated using the observed potential wall was about 50 meV, which is comparable to the 35 meV measured by LDOS spectroscopy [7]. This suggests that the potential wall is related to the 0D confinement in SFTs.

When the net electron tunneling mainly occurs from the valence band to the tip at negative sample bias voltage, LTBH depends on the apparent valence band edge position, which is determined by both the valence band edge position on the surface and the band bending profile beneath. A gradual LTBH decrease of 0.1 eV was additionally observed near the potential wall within a distance of 1 - 3 nm (Fig. 1). Since electron affinity is an intrinsic constant, the higher LTBH near the potential wall means a lower apparent valence band edge position. The observed higher LDOS at the boundary within a distance of 2 - 3 nm also suggests a lower apparent valence band position there [1]. Therefore, the measured LTBH reflects the local potential profile of SFTs due to the polarization and the energy band bending at the boundaries.

CONCLUSION

Using LT-STM, the LTBH has been measured to examine the local potential profile of the SFT. The SFT boundary is found to show a higher LTBH than the surroundings as the confinement potential wall. The calculated resonance peak width based on the measured LTBH is comparable to that measured by LDOS spectroscopy. A gradual LTBH decrease was additionally found at the negative sample voltage near the boundary, suggesting downward band bending. This is also consistent with LDOS observed there.

ACKNOWLEDGMENTS

This work was partly supported by a Grant-in-Aid for Scientific Research from the Japan Society for the Promotion of Science.

REFERENCES

1. K. Kanisawa, M. J. Butcher, Y. Tokura, H. Yamaguchi, and Y. Hirayama, Phys. Rev. Lett. **87**, 196804 (2001).

2. S. Hosaka, K. Sagara, T. Hasegawa, K. Takata, and S. Hoseki, J. Vac. Sci. Technol. A **8**, 270-274 (1990).

3. S. Kurokawa, T. Takei, and A. Sakai, Jpn. J. Appl. Phys. **42**, 4655-4658 (2003).

4. S. Modesti, D. Furlanetto, M. Piccin, S. Rubini, and A. Franciosi, Appl. Phys. Lett. **82**, 1932-1934 (2003).

5. R. Wiesendanger, *Scanning Probe Microscopy and Spectroscopy: Methods and Applications*, Cambridge University Press, Cambridge, 1994.

6. J. F. Zheng, X. Liu, N. Newman, E. R. Weber, D. F. Ogletree, and M. Salmeron, Phys. Rev. Lett. **72**, 1490-1493 (1994).

7. K. Kanisawa, Y. Tokura, H. Yamaguchi, and Y. Hirayama, *the 29th International Symposium on Compound Semiconductors (ISCS 2002)*, Mo-P-30, (2002).

Resonance Backscattering in Triangular Quantum Dots Inside a Small Ring Interferometer

D.V.Nomokonov[1], A.K.Bakarov[1], A.A.Bykov[1], and A.M.Mishchenko[2]

[1]*Institute of Semiconductor Physics, 630090 Novosibirsk, Russia*
[2]*Novosibirsk State University, 630090 Novosibirsk, Russia*

Abstract. Quasiperiodic peaks of the resistance and quasiperiodic oscillations of e.m.f. generated by microwave radiation as the functions of gate voltage were observed in the small rings fabricated on the basis of a two-dimensional electron gas in a GaAs quantum well with the AlAs/GaAs superlattice barriers. In magnetic fields higher than 1 T, the resistance peaks and e.m.f. oscillations disappeared. The experimental results are explained by the magnetic-field-induced suppression of the resonance backscattering that appears in the triangular quantum dots situated in the branching regions of a ring interferometer.

INTRODUCTION

It is known that the dimensions of the conducting regions of a semiconductor interferometer are determined not only by the capabilities of lithography but also by the depletion regions that are formed along the boundaries of the conducting channels. The width of the depletion regions in small semiconductor rings is comparable with the mean radius of the ring fabricated by the electron beam lithography. In this situation the electron system of the interferometer is divided into two triangular virtual quantum dots at the ring input and output that connect to each other and to reservoirs through narrow channels [1].

SAMPLES

The rings were fabricated on the basis of a two-dimensional electron gas (2DEG) in a GaAs quantum well with the AlAs/GaAs superlattice barriers by electron-beam lithography and ion plasma etching. The 2DEG mobility and the concentration in the original structure at 4.2 K was $\mu = 4*10^5$ cm^2/Vs and $n_s = 1.6*10^{12}$ cm^{-2}, respectively. SEM – micrograph of the ring is shown in Fig.1a. The effective ring radius, as determined from the period of the *h/e* oscillations, was on the order of $r_{eff} \approx 0.13$ µm. The AuTi/GaAs Schottky barrier was used as a planar gate.

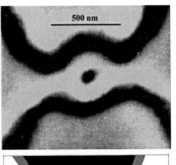

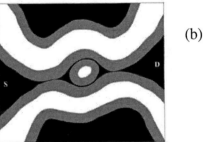

FIGURE 1. (a) SEM-micrograph of the ring structure. The dark areas are the etched regions. (b) Schematic drawing of the ring. The dark areas represent the 2DEG, the gray areas show the depletion regions, and the white areas represent the etched regions. (1-4) are the locations of the constrictions.

The schematic view of a ring with the depletion regions appearing along the etching edges is shown in Fig.1b. The interferometer has four constrictions, which divide the ring in the tunneling regime into two triangular quantum dots.

RESULTS AND DISCUSSION

The results of measuring $R_{SD}(V_g)$ for different values of a magnetic field are shown in Fig. 2a. In the zero-field dependence, one can clearly see the gate peaks separated by about 0.05–0.07 V and completely disappearing in magnetic fields higher than 2 T. The application of a relatively weak magnetic field completely destroys these resonances. The reflection resonances in the transmission regime intermediate between the open and closed rings(Fig.2b) were observed up to 30 K, and their amplitude at temperatures below 4.2 K virtually did not change.

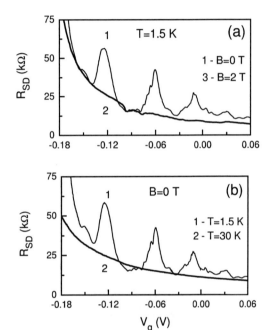

FIGURE 2. (a) The $R_{SD}(V_g)$ dependence at $T = 1.5$ K for different B values. (b) The $R_{SD}(V_g)$ dependence at $B = 0$ for different T values.

The gate peaks, on the background of the monotonically improved passage with increasing V_g, are due, in our opinion, to the resonances in the triangular quantum dots, which take place at the input and output of small rings. The negative magnetoresistance observed at the peaks is explained by the magnetic-field-induced suppression of the resonance backscattering that appears in these triangular quantum dots situated in the branching regions of a ring interferometer. In zero magnetic field, the greatest contribution to gating comes from the state with the wave function localized mainly along the triangle height and most electrons are scattered backwards from the wall of the triangular quantum well, whereas, in the presence of a magnetic field that distorts the electron wave function, the probability of passing through the triangular dot increases.

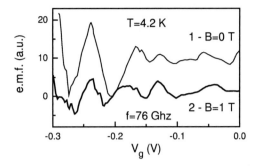

FIGURE 3. Dependences of e.m.f. at $T = 4.2$ K on the V_g: f= 76 GHz, (1) $B = 1$ T and (2) $B = 0$.

The experimental results of microwave investigations are shown in Fig. 3. It is found that e.m.f. generated in the small rings by microwave radiation exhibit quasiperiodic oscillations versus gate voltage. In magnetic fields higher than 1 T, e.m.f. oscillations disappeared. This means, the origin the observed e.m.f is the asymmetry of the confining potential of the triangular quantum dots. A possible reason for the asymmetry is the imperfections of the fabrication technology of the rings [2].

SUMMARY

This work presents the experimental results on measuring the charge-carrier transport through single submicron-sized rings based on a high-mobility 2DEG with high density. Quasiperiodic peaks have been observed as function of gate voltage. Such a behavior is explained by the resonance electron backscattering in the triangular quantum dots at the ring input and output. The experimental results of microwave investigations can be qualitatively explained by the quantum ratchet in these triangular quantum dots.

ACKNOWLEDGMENTS

This work was supported by RFBR, project # 04-02-16789 and INTAS, project # 03-51-6453.

REFERENCES

1. A. A. Bykov, *et al.*, JETP Lett. **71**, 434 (2000).
2. A. A. Bykov, *et al.*, Superlattices and Microstructures **23**, 1285 (1998).

Experimental study on far-infrared photoconductivity and magnetoresistance in InAs single quantum wire

C. Zehnder[*], S. Löhr[*], A. Wirthmann[*], Y. S. Gui[*], Ch. Heyn[*], W. Hansen[*], D. Heitmann[*] and C.-M. Hu[*]

[*]*Institute of Applied Physics, University of Hamburg, Jungiusstraße 11, 20355 Hamburg, Germany*

Abstract. By applying far-infrared photoconductivity spectroscopy, we detect for the first time spectrally resolved the dynamic excitation in a *single* InAs quantum wire, which allows us to determine its confinement energy.

INTRODUCTION

Currently there is an increasing interest in two- and one-dimensional electron systems (2DES/1DES) in InAs for spintronic applications[1]. The most frequently used tools to investigate these systems are magnetotransport measurements. The beating pattern due to the spin-orbit coupling is analyzed to extract the Rashba parameter[2].

For far-infrared (FIR) transmission experiments usally arrays of quantum wires are investigated since the response of a single quantum wire is too weak due to the small absorption area. The photoconductivity (PC) spectroscopy is a combination of transport and spectroscopy. Using the quantum wire as a detector and measuring the change of the resistivity by the FIR-radiation, excitations of such small systems can be detected.

SAMPLE PREPARATION AND EXPERIMENTAL TECHNIQUES

The sample was an inverted doped InAs step quantum well, grown by molecular beam epitaxy on a GaAs substrate. The quantum well was composed of 2.5 nm $In_{0.75}Ga_{0.25}As$, a 4 nm channel of InAs and 13.5 nm $In_{0.75}Ga_{0.25}As$. The electron gas with a density of $n_s = 3.7 \times 10^{11} cm^{-2}$ and a mobility $\mu = 180,000 \frac{cm^2}{Vs}$ was mostly confined in the InAs channel [3]. We have fabricated eight single quantum wires with different widths between 300 and 1000 nm and a length of 180 μm on the sample by electron beam lithography. For four-point measurements we prepare source/drain contacts and voltage probes, which are connected via a 2DES mesa as leads (see inset of Fig. 1).

The photoconductivity spectroscopy was performed in a variable temperature insert at 1.6 K and at magnetic fields up to 12 T. We apply a dc current of 90 nA and measure the voltage drop caused by the FIR radiation of a broadband Hg lamp. The radiation was modulated by a Michelson interferometer at a fixed magnetic field. The voltage drop was coupled into a broadband preamplifier and recorded as an interferogram. The spectrum is obtained from a Fast Fourier transformation of the interferogram.

RESULTS

In Fig. 1 the magnetoresistance measurements of the wires are shown. At magnetic fields larger than 2 T Shubnikov-de Haas oscillations are clearly observed. No deviations from the $1/B$ periodicity are found. Effects of the lateral confinement only appear in the negative magnetoresistance, which is stronger for smaller width of the wires caused by the boundary scattering.

Figure 2 shows two PC spectra of the 400 nm wide wire and, for comparison, the PC spectra of a 2DES of the same wafer. At high B fields around 8 T the influence of the confinement potential is rather small and the magnetoplasmon (MP) mode of the 1DES does not differ from the cyclotron resonance (CR) of the 2DES. At fields about 5 T the impact of the confinement potential is remarkable large and the peak is shifted to higher wave numbers. However, the lineshape is slightly asymmetric. We believe that there is a 2D like response in the PC which arises from the broad part of the mesa leads.

In Fig. 3 the experimental peak positions are shown. The amplitude of the PC spectra depends on the magnetic fields. The strongest signal can be found near the minima of the longitudinal resistance, like the PC response in a 2DES [4]. The resonances can be fitted, using

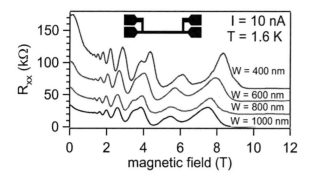

FIGURE 1. Magnetoresistance measurement of four wires with widths between 400 nm and 1 μm. The curves are shifted for clarity by 20kΩ each. Inset: schematic diagram of one wire with leads.

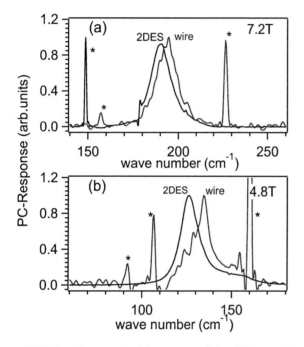

FIGURE 2. Photoconductivity spectra of the 400 nm wide wire and a 2DES at 7.2 T and 4.8 T. The spectra are magnified to a constant level for a better comparison to the PC spectra of a 2DES. (* marks artifacts from the Fourier transformation)

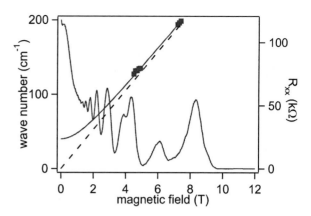

FIGURE 3. Peak position of the 400 nm wide wire versus magnetic field. The lines are fits for the CR (dashed line) of the 2DES and of the magnetoplasmon modes (solid line) of the wire. The corresponding magnetotransport measurement is shown in addition.

the effective mass $m^* = 0.035$ determined from the CR $\omega_c = \frac{eB}{m^*}$ of the 2DES in a sample of the same wafer, to the confined MP of 1DES with $\omega_{mp} = \sqrt{\omega_c^2 + \omega_0^2}$ [5], from which we determine the confinement energy $\omega_0 = 40\,\text{cm}^{-1}$ for the single quantum wire. This energy represents, due to Kohn's theorem, the external confinement potential. The origin of the PC response of the 1DES confined in the quantum wire is the bolometric effect[6].

In conclusion, we determined the confinement energy of a single quantum wire by performing spectrally resolved photoconductivity. We find that the photoresistance of the single quantum wire sensitively measures dynamic excitations of the wire.

ACKNOWLEDGMENTS

This work is financially supported by BMBF, DFG via SFB508 and the EU 6th-framework programme through project BMR-505282-1.

REFERENCES

1. Žutić, I., Fabian, J., and Sarma, S. D., *Rev. Mod. Phy.*, **76**, 323 (2004).
2. Schäpers, T., Knobbe, J., and Guzenko, V., *Phys. Rev. B*, **69**, 235323 (2004).
3. Richter, A., Koch, M., Matsuyama, T., Heyn, C., and Merkt, U., *Appl. Phys. Lett.*, **77**, 3227 (2000).
4. Zehnder, C., Wirthmann, A., Heyn, C., Heitmann, D., and Hu, C.-M., *Euro Phys. Lett.*, **63**, 576 (2003).
5. Kono, J., Nakamura, Y., Peralta, X., Černe, J., S.J. Allen, J., Akiyama, H., Sakaki, H., Sugihara, H., Sasa, S., and Inoue, M., *Superlattices and Microstructures*, **20**, 383 (1996).
6. Neppl, F., Kotthaus, J., and Koch, J., *Phys. Rev. B*, **19**, 5240 (1979).

Nanostructures From ZnO And Other Semiconductors Generated Via Self-Organizing Polymers

Klaus Thonke

Abt. Halbleiterphysik, Univ. Ulm, Albert-Einstein-Allee 45, D 89081 Ulm (Germany)

Abstract. Self-organized polymer nanostructures of various kinds are used as templates for the definition and production of nanostructures from inorganic semiconductors like ZnO, Si, and GaAs. Examples generated both in bottom-up or top-down-approaches are shown.

INTRODUCTION

For many future applications of semiconductor or metal nanostructures a good control of the size and average distance should be sufficient. In these cases, the expensive and time-consuming serial lithography techniques like electron-beam or ion-beam writing, which allow complete control over the arrangement of the patterns generated, are not really necessary. Instead, cheaper parallel techniques, which only define a good near-range order, but allow for a high throughput, are an attractive alternative. In this contribution, the feasibility of some of these pathways is demonstrated.

DIBLOCK-COPOLYMER MICELLES AS "NANOREACTORS"

Diblock copolymers are polymer chains which consist of two chemically different parts, typically a hydrophobic part and a hydrophilic part, joined end to end. In our case, we used macromolecules consisting of longer polystyrene (PS) and shorter poly(2vinyl-pyridine) (PV2P) partials (e.g. $PS_{1400}PV2P_{450}$). When such chains are brought into a non-polar solvent like toluene, the non-polar PS chains stay in contact with the solvent, whereas the polar P2VP ends stick together forming so-called "inverse micelles". The polar core of these can be loaded selectively by metal salts [1] like, e.g., $HAuCl_4$, $ZnCl_2$, and others. By simple dip-coating a monolayer of micelles can be deposited on virtually any substrate like glass, silicon, sapphire etc., where they form a pattern with hexagonal short-range order. Depending on the chain length of the diblock copolymer constituents, the diameter and distance of the micelles can be varied between 1-20 nm and 10-150 nm, respectively [2]. By treatment in an oxygen (or hydrogen) plasma, the polymer film can be removed completely, and at the same time the metal salt is reduced to pure metal as in the case of gold, or metal oxide is generated for most other metals. If the polymer length distribution is narrow, and the process parameters are optimized, metal dot arrays can be generated with narrow size distributions [3].

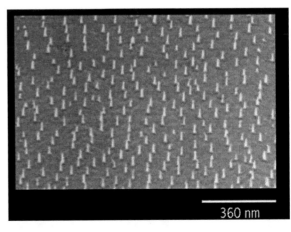

FIGURE 1. SEM micrograph (tilted view) of nano-pillars etched from bulk GaAs.

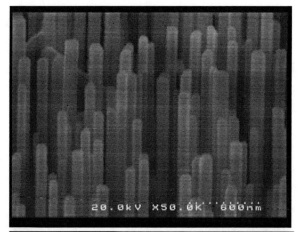

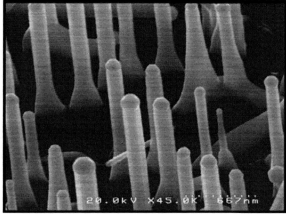

FIGURE 2. SEM micrograph (tilted view) of ZnO nano-pillars grown by the VLS method on A-plane sapphire under slightly different conditions. The Zn/Au alloy on top of the pillars is clearly visible.

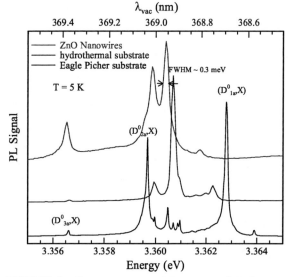

FIGURE 3. Low-temperature PL spectra of ZnO nanowires (upper trace), and bulk ZnO substrates grown by a hydrothermal method (middle trace) or chemical vapor transport (lower trace).

The Au dots generated in this way can subsequently be used in various ways. In a top-down approach, they acted as nano-masks on GaAs substrates for highly anisotropic reactive ion etching. With optimized etching parameters, a dense array of pillars with diameters of ≈ 8 nm and heights of up to 80 nm could be produced. Figure 1 shows such an example [4]. TEM cross-sections of these pillars show, that the Au nanocrystals serving as etching mask are still on top of the pillars. When instead of the GaAs substrate quantum well (QW) structures with the QW embedded within the top 70 nm are etched, a dense array of quantum dots (QDs) embedded in the pillars can be produced, which are not interconnected with a "wetting layer" like in the case of Stranski-Krastanov growth.

If these pillar arrays are evaporated afterwards by a thin metal film (e.g. 50 nm Au), the thickness of which is less than the height of the pillars, and the pillars are removed by simple wet chemical etching, a thin metal sheet is left behind with a dense network of nanoholes, the diameter and position is controlled by the position of the GaAs pillars, and thus via the intermediate dry etching step by the initially used micelle pattern [5].

The micelle-generated Au nano-dots open as well a pathway to a bottom-up approach to nano-pillars. We used them successfully on A-plane sapphire substrates as catalysts for the growth of ZnO nano-pillars with the so-called "vapor liquid solid" (VLS) process [6]. Fig. 2 shows an example of the structures we obtained. The Au dots allow a good control of the diameters of the pillars, which are around 30 nm. The height is up to ≈ 800 nm. Under close inspection the Zn/Au alloy, which is formed initially and remains on top of the pillars during growth, can still be seen with the SEM. With optimization of the growth parameters, straight, upright ZnO pillars are obtained. In low-temperature photoluminescence (PL) experiments (see Fig. 3), we find very bright emission from the ZnO pillars, which is more intense than that from state-of-the-art ZnO bulk crystals, although only a part of the total area (≈ 1 mm^2) excited by the laser is covered by ZnO. The PL spectra with very sharp dominant donor-related emissions indicate very good crystal quality and low uncontrolled defect concentration. Any low-energy "green" or donor-acceptor pair bands are extremely weak, thus the concentration of acceptors or deep defects is obviously very low. We do not see any indication of incorporation of the Au catalyst into the crystal.

NANOPOROUS MASKS

Structures with typical sizes of 100 nm to ≈ 500 nm can be obtained with the aid of so-called nanoporous masks [7]. For this purpose, a hydrophobic monomer is poured on water together with especially coated silica colloids. The silica colloids order into a hexagonal closest package with the monomer wetting the surface of the spheres and filling the space in between. The monomer is then cross-linked by UV irradiation. Subsequently, the floating film is picked up, and the silica colloids are removed by HF vapor, leaving a nanoporous membrane. The thickness of this film is controlled by the initial amount of monomer, and the hole diameters are dictated by the diameters of the silica colloids used. For our experiments, membranes with typical hole diameters of 300 nm were prepared.

FIGURE 4. Upper: SEM tilted view of the nanopourous polymer mask on Si after evaporation of ≈ 50 nm Au and removal of the top metal film. Au disks located in the holes remain. **Lower:** Si pillars generated with these Au disks as masks in a highly anisotropic etching process.

We used such masks on Silicon wafers, and evaporated a thin metal layer (e.g. 50 nm of Au) on the mask. After careful removal of the top Au net, tiny Au disks remain in the nano-pores (see Fig. 4, upper). These can be used in a specialized, highly anisotropic dry etching process to cut out pillars from the Si substrate [8] (see Fig. 4, lower). These hexagonally ordered pillars have diameters of ≈ 150 nm, and heights of ≈1 μm. Due to their high refractive index, they are good candidates for 2D photonic bandgap structures.

Starting with the same masks, it is possible to generate in a bottom-up approach ring-shaped objects from (Zn,Cd)O. For this purpose, the mask was transferred to a sapphire substrate, and then Zn (and/or Cd) acetate was filled into the pores, which due to special treatment of the initial silica colloids wets preferentially to the pore side walls. Treatment in an oxygen plasma removes the polymer mask and converts the (Zn,Cd) acetate into ZnO/CdO. The resultant "donuts" have an outer diameter of ≈ 300 nm, and a height of ≈ 150 nm (see Fig. 5). Annealing at around 900°C is required to get bandgap-related donor-bound exciton PL from these rings. It is less bright than in the case of the ZnO pillars and shows a higher incorporation of deep defects [9]. Close inspection with the SEM reveals, that the rings consist of numerous nano-crystals attached to each other. Depending on the amount of filling of the pores with the metal salt, almost any desired shape ranging from thin rings to complete disks can be produced.

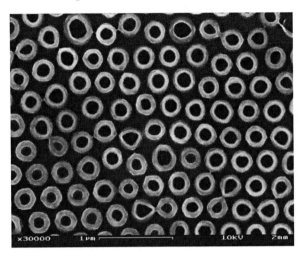

FIGURE 5. ZnO donuts of ≈ 300 nm outer diameter and ≈ 150 nm height created via nanoporous membranes.

BREATH FIGURES

A third category of polymer masks we used are so-called "breath figures". Here, water vapor condenses on a special polymer, which is dissolved in a hydrophobic solvent with high evaporation rate. The water droplets are stabilized by the polymer and finally leave a hexagonal pattern of imprints after complete evaporation of both the water and solvent [10,11]. Depending on the process parameters, structures aiming to miniaturized honeycombs with holes from 2 μm down to 200 nm can be produced (see. Fig. 6). Onto these templates a thin metal film is evaporated. Removal of the top part of this film yields a hexagonal net. In one application, we used this as an optical short-pass filter, since only light with roughly $\lambda/2 < d$ (d = hole diameter) can pass through. Transferred to a substrate, the net allows to etch regular holes (see Fig. 6, middle part). Similarly to the arrangement shown in Fig. 4, (lower part) regular arrays of Si pillars were etched. Using isotropic etching, mushroom-type structures can be obtained under the metal disks.

In summary, quite numerous ways exist to transfer the self-organized patterns of polymer nanostructures in an intermediate step first into metal structures (which might be of interest in their own right) and then into pillars, holes and rings of inorganic semiconductors.

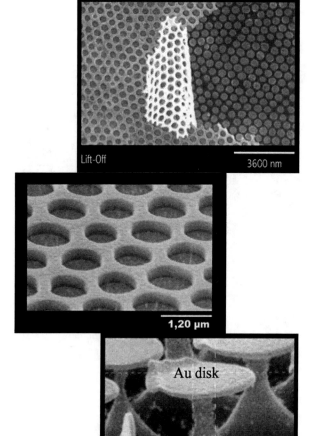

FIGURE 6. Upper: SEM micrograph (top view) of polymer breath figures, onto which a Au film was evaporated. The top part of the film can be removed, yielding a hexagonal Au net and leaving Au disks on the substrate. **Middle:** Etching results on GaAs with the Au net as mask. **Lower:** result of *isotropic* etching into a SiGe layer.

ACKNOWLEDGMENTS

The author wants to express his deep thanks to: R. Sauer, A. Waag, M. Haupt, A. Ladenburger, S. Miller, X. Cao, A. Reiser (Abt. Halbleiterphysik); M. Möller[a], J. P. Spatz[b], W. Goedel, A. Mourran[a], S. Riethmüller, R. Glass[b], H. Xu, F. Yan, C. Hartmann (Abt. Org. Chem. III); F. Banhart[c] and P. Walther (Zentrale Einrichtung Elektronenmikroskopie), and H. Rauscher, J. Behm (Abt. für Oberflächenchemie und Katalyse); P. Unger and H. Wolff (Abt. Optoelektronik) (all departments at the University of Ulm, Germany); J. Konle and H. Presting (Daimler Chrysler AG). The financial support of the SFB 569 of the Deutsche Forschungsgemeinschaft and the Graduiertenkolleg "Molecular organisation and dynamics at interfaces" is gratefully acknowledged.

REFERENCES

1. J. P. Spatz, T. Herzog, S. Mößmer, M. Möller, P. Ziemann, *ACS Symp. Ser.* **706**, 1997, p. 12.
2. J. P. Spatz, P. Eibeck, S. Mößmer, M. Möller, T. Herzog, P. Ziemann, *Adv. Mater.* **10**, 1998, p. 849.
3. G. Kästle, H.G. Boyen, F. Weigl, G. Lengl, T. Herzog, P. Ziemann, S. Riethmüller, O. Mayer, C. Hartmann, J. Spatz, M. Möller, M. Ozawa, F. Banhart, M.G. Garnier, P. Olehafen, *Adv. Funct. Mat.* **13**, 2003, p. 1.
4. M. Haupt, S. Miller, A. Ladenburger, R. Sauer, K. Thonke, J. P. Spatz, S. Riethmüller, M. Möller, F. Banhart, *J. Appl. Phys.* **91**, 2002, p.6057-9.
5. M. Haupt, S. Miller, K. Thonke, R. Sauer, M. Möller, R. Glass, M. Arnold, J.P. Spatz, *Adv. L, Mat.* **15**, 2003, p.829.
6. M. H. Huang, S. Mao, H. Feick, H. Yan, Y. Wu, H. Kind, E. Weber, R. Russo, P. Yang, *Science* **292**, 2001, p. 1897.
7. H. Xu, W. A. Goedel, *Langmuir* **19**, 2003, p. 12.
8. A. Ladenburger et al., to be published.
9. A. Ladenburger, M. Haupt, R. Sauer, K. Thonke, H. Xu, W.A. Goedel, *Physica* **E17**, 2003, p. 489.
10. A. Mourran, S.S. Sheiko, M. Krupers, M. Möller, PMSE Proceedings of the Am. Chem. Soc. **80**, 1999, p. 175; A. Mourran, S.S. Sheiko, M. Möller, PMSE Proceedings of the Am. Chem. Soc. **81**, 1999, p. 426.
11. M. Haupt, S. Miller, R. Sauer, K. Thonke, A. Mourran, M. Moeller, J. Appl. Phys., 2004 (in print).

[a] now with Inst. für Textilchemie und Makromolekulare Chemie, RWTH Aachen, Germany.
[b] now with Inst. für Phys. Chem. und Biophys. Chemie, Univ. Heidelberg, Germany.
[c] now with the Institut für Physikalische Chemie, Univ. Mainz.

Amorphous Phase Emergence of Polar Nano-crystalline Semiconductors

Shu-Lin Zhang,[*] Song-Nan Wu, Jia Shao, Yang Song

School of Physics, Peking University, Beijing 100871, China

Abstract. From the view of Raman spectroscopy, the emergence of amorphous nature was shown in nano-crystalline polar-semiconductors, e.g. ZnO nano-particulars and SiC nano-rods.

The crystallographic property of nano-semiconductors is determined usually by their X-ray measurement, which is consistent with that determined by Raman spectroscopy in bulk materials. This consistence has been confirmed to be remained in silicon nano-wires [1] and carbon nano-diamond [2]. One of the evidences in confirmation of this consistence is that there appears the resonance size-selection effect (RSE) in their Raman spectra. The Raman feature deduced from RSE is that Raman frequency shifts with excitation wavelength [2, 3].

However, in polar nano-semiconductors, e.g. ZnO nano-particles (NPs) and SiC nano-rods (NRs) we did not observed the Raman frequency shifting with excitation wavelength. In this letter we will report on this phenomenon and discussion to it.

Figure 1 shows the Raman spectra excited with different excitation wavelengths for ZnO nano-particulars (NPs) and SiC nano-rods (NRs).

The variations of Raman frequencies from the fitting of Fig. 1 with excitation wavelength are shown in Fig. 2.

From Fig. 1 and 2, we can see that there is no Raman frequency change with excitation wavelength, against the Raman feature induced by RSE. The similar results are also found in other polar nano-semiconductors, e.g., ZnO nano-tubes and GaN NPs. The above result indicates that the Raman scattering of polar nano-semiconductors is quite different from the scattering of non-polar nano-semiconductors such as Si NWs and CNT.

It has been verified that the Raman scattering of polar nano-semiconductor SiC NRs is amorphous type scattering relative to the phonon density of states (PDOS) [4]. The PDOS, $g(\omega)$, can be expressed as

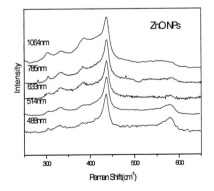 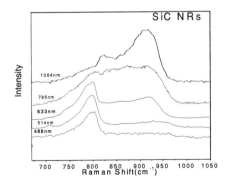

FIGURE 1. Raman spectra excited with different excitation wavelengths for ZnO NPs and SiC NRs.

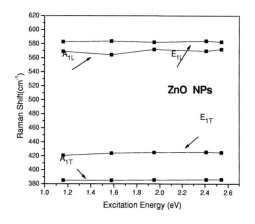 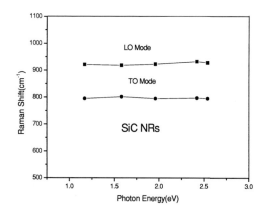

FIGURE 2. Variation of Raman frequency with excitation wavelengths for ZnO NPs and SiC NRs

$$g(\omega) = \sum_s \frac{1}{\nabla_q \omega_s(q)} \frac{ds}{(2\pi)^3} \qquad (1)$$

where $\omega_s(q)$ is phonon dispersion relation. The summation is performed on the first Brillouin zone. The formula (1) indicates that the PDOS is not directly relative to the sample size. Recently the PDOS calculation of nano-scale materials has also exhibited that there is no obvious variation with sample size for PDOS [5]. Thus it is easy to understand that the Raman feature shown in Fig 1 and 2 is due to that the scattering is relative to the PDOS rather than phonon dispersion relation. On the other hand, it is well known that the Raman spectrum of amorphous materials is relative to the PDOS [6]. Therefore from the view of Raman spectroscopy the Raman spectra shown in Fig 1 and 2 indicate clearly that an amorphous structure emerges in polar nano-semiconductors, although they are crystalline structure from the view of X-ray measurement.

We acknowledge the assistance from the nano-chemical group and bio-chip group at Peking University and the support from the National Science Foundation of China under grants NOs. 60290083, 50334040 and 50272017, RGC of Hong Kong under the grant No. 401003, as well as, from the State Key Lab for Inferred Physics and the Beijing Key Laboratory for Nano-Photonics and Nano-Structure.

REFERENCES

* Email: slzhang@pku.edu.cn.
1. S.L. Zhang, Y. Hou, H.-S. Ho, B. Qian, and S. Cai, J. Appl. Phys. **72**, 4469 (1992)
2. Y. Yan, X. T. Zhang, S. K. Hark and S. L. Zhang, in *Proceedings of The International Conference on Raman Spectroscopy, 8-13 August, 2004, Gold coast, Australia*, p.112.
3. Shu-Lin Zhang, Wei Ding, Yan Yan, Jiang Qu, Bibo Li, Le-yu Li, Kwok To Yue, and Dapeng Yu, Appl. Phys Lett, **81**, 4446(2002)
4. Shu-Lin Zhang, Bang-Fen Zhu, Fuming Huang, Yan Yan, Er-yi Shang, Shoushan Fan, and Weigiang Han, Solid State Commu. **111**, 647 (1999)
5. X. Hu, G. Wang, W. Wu, P. Jiang and J. Zi, J. Phys: Condens. Matter, **13**, L835(2001)
6. R. Shuker and R. Gammon, Phys. Rev. Lett. **25**, 222(1970).

Vapor-Phase Growth and Characterization of Luminescent Silicon Layers

E. A. de Vasconcelos[1], J. B. da Silva Jr.[1], B. E. C. A. dos Santos[1], E. F. da Silva Jr.[1], W. M. de Azevedo[2], J. A. K. Freire[3], J. A. N. T. Soares[4], J. R. Leite[4]

Departamento de Física, Universidade Federal de Pernambuco, 50670-901 Recife-PE, Brazil
Departamento de Química Fundamental, Universidade Federal de Pernambuco, 50670-901 Recife-PE, Brazil
Departamento de Física, Universidade Federal do Ceará, Campus do Pici, 60455-900 Fortaleza-CE, Brazil.
Instituto de Física, Universidade de São Paulo, Cidade Universitária, 05315-970 São Paulo-SP, Brazil.

Abstract: We discuss the vapor-phase technique for porous silicon formation and the structural, optical and electrical properties of the layers with emphasis on the statistical and fractal analysis of Atomic Force Microscopy (AFM) images. First-order and second order parameters of different AFM images of vapor-phase grown layers were calculated. We found a trend towards slightly lower roughness exponents for vapor-phase samples, possibly related with improved structural reproducibility. The current-voltage characteristics show an exponential dependence with voltage, followed by a power-law relationship, typical of space charge limited currents in high resistivity materials.

INTRODUCTION

The development of a new vapor-phase technique to produce luminescent porous silicon makes it possible to form, in less than 5 minutes, luminescent layers in p- and n- type substrates of any resistivity, with minimum apparatus and maximum simplicity [1]. Here, we report on the properties of the layers, with emphasis on statistical analysis of the AFM images, including first-order and second-order statistics.

EXPERIMENTAL DETAILS

We expose the Si surface to the vapor generated by the dissolution reaction of a metal or a semiconductor in a 4:HF/1:HNO_3 mixture. The mixture is not heated and no technical equipment (etching cell, current source, illumination, temperature controller, etc.), formation of electrical contacts, or addition of surfactants is needed. After exposure to this vapor, a porous silicon (PS) layer is formed, which looks very homogeneous to the naked eye and presents reproducible and uniform photoluminescence (PL). One can easily obtain luminescent layers in both p- and n-type substrates of high or low resistivity. We obtain red-orange photoluminescence with good lateral uniformity across the entire area of a 2-inch diameter wafer or greater. The structure of the layers was investigated by scanning electron microscopy (SEM) and atomic force microscopy (AFM). The PL spectroscopy was performed at room and low temperatures. For the sake of comparison, we also fabricated porous silicon layers using the standard electrochemical technique in a 1:HF/1:Ethanol solution.

RESULTS AND DISCUSSION

The porous silicon surface grown by the vapor-phase method contains irregularly shaped, rough regions with sizes in the order of tens of microns, bounded by cracks a few microns long and approximately one micron wide. Figs. 1(a) and 1(b) show AFM topographical images showing hillocks with columnar, rodlike aspect. We performed first-order (surface height distribution, 2nd, 3rd and 4th-order moments), as well as second-order statistics (height-height correlation function) of AFM images of vapor-phase and electrochemically etched samples. Fig. 2 shows an example of the second-order analysis performed. It shows the height-height correlation as a function of the translation for the image in Fig. 1(b).

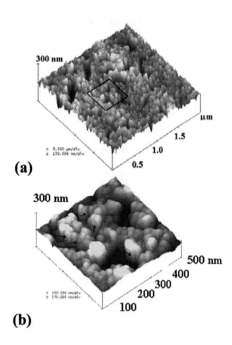

Figure 1. AFM images of vapor-phase grown layers. (a) 2.0 x 2.0 µm. (b) zoom of marked region in (a).

The characteristic shape of the plot shows that the surface is self-affine. From the height-height correlation, as shown in Fig. 2, one can obtain the seconder order statistical parameters, namely, the RMS roughness (w), the correlation length (ξ) and the roughness exponent (α) [2]. For the images in Figs. 1(a) and 1(b), both first-order and second order parameters have similar values. We also compared the parameters of vapor-phase grown layers with those from electrochemically grown layers. We found a trend towards slightly lower roughness exponents for the vapor-phase samples, possibly related with improved structural reproducibility. In fact, vapor-phase formed porous layers look very homogeneous to the naked eye and present reproducible and uniform PL. The PL spectra of the vapor-phase grown layers have peaks from 1.85 eV to 2.1 eV at room temperatures, corresponding to red-orange PL. At lower temperatures, the spectrum is slightly blue-shifted. We also fabricated Au/vapor-phase PS/Si/Al devices and investigated their electrical properties. The current-voltage characteristics are rectifying. The forward bias corresponds to the case where a negative bias is applied to the gate. There are two different regimes for the forward current. At voltages below 1 V, it follows an ideal diode dependence with ideality factor approximately between 2 and 4. At higher voltages, it follows a power law, a clear indication of space-charge limited current mechanisms.

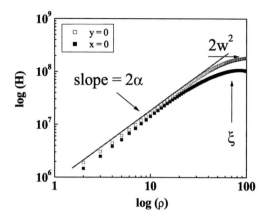

Figure 2. Height-height correlation for the AFM image in Fig. 1(b).

CONCLUSIONS

We developed a vapor-phase technique for porous silicon fabrication. This method has advantages in simplicity and cost and gives improved uniformity and repeatability of characteristics.

ACKNOWLEDGMENTS

This work was supported by CNPq under contracts NanoSemiMat/CNPq #550.015/01-9 and the Ministry of Planning (FINEP) under contract CTPETRO/FINEP # 65.00.02.80.00.

REFERENCES

1. de Vasconcelos, E. A., da Silva Jr., E. F., dos Santos, B. E. C. A., de Azevedo, W. M., and Freire, J. A. K., "A New Method for Luminescent Porous Silicon Formation: Reaction-Induced Vapor-Phase Stain Etch", in: Proceedings of the 4th International Conference on Porous Silicon Science and Technology, Cullera-Spain, 2004, pp. 236-237.

2. Zhao, Y., Wang, G-C., and Lu, T-M., "Statistical Nature of Rough Surfaces", in Characterization of Amorphous and Crystalline Rough Surface: Principles and Applications, San Diego: Academic Press, 2001, pp. 7-32.

Synthesis and Characterization of Transition Metal ion Doped ZnS and ZnO Nanostructures

S.P. Singh, O. Perales*, M.S. Tomar, and O.V. Mata

Department of Physics, University of Puerto Rico, Mayagüez, PR - 00681-9016.
*General Engineering Department-Materials Science and Engineering, University of Puerto Rico, Mayagüez, PR - 00681-9044

Abstract. Nanocrystalline films and powders of semiconducting ZnS, with and without doping with Cr, Co and Ni ions, have been synthesized using a low-temperature chemical bath deposition approach in aqueous solution. X-ray diffraction analyses evidenced that ZnS structures consisted of cubic or a mixture of a cubic/hexagonal phase. Average crystallite size was estimated to be 3-4 nm using Scherrer's equation. ZnS and transition metal-doped ZnS were thermally oxidized into corresponding oxides. Magnetic and optical behavior of the sulfide films and powders were investigated using SQUID and UV-visible spectrometry, respectively. Preliminary magnetic measurements suggest antiferromagnetic ordering in $Zn_{0.9}Cr_{0.1}S$.

INTRODUCTION

ZnS is an important II-VI wide band gap semiconductor that is used in luminescent devices, solar cells and other optoelectronics devices. The size dependent optical properties of semiconductor materials at the nanoscale opened a new thrust area in optoelectronics and biological sciences. In turn, the discovery of carrier induced ferromagnetism in III-V semiconductors opened the possible application of these so-called diluted magnetic semiconductors, DMS, in spintronics[1]. As solubility of transition metal ion in III-V semiconductors is low, their use in spintronic devices would be limited. On the contrary, II-VI semiconductors can dissolve higher levels of transition metal ions and, therefore, exhibit a higher potential to be used in magneto optics and spintronic-based devices. So far, it is controversial to find an II-VI DMS to show the ferromagnetism at room temperature.

Recently K. Sato *et. al.* [2] proposed to design transitional metal ion doped-calcogenides. Their *ab-initio* calculations suggested that V and Cr doped-ZnS, ZnSe and ZnTe should exhibit room-temperature ferromagnetic behavior independent of doping treatment. Also, Mn, Fe, Co, and Ni doping in these compounds would show spin glass behavior.

Several methods have been reported in order to make thin films and powders of nanocrystalline II-VI semiconductors. In addition to physical routes, the preparation of metal chalcogenides thin films has been attempted by wet chemistry approach like chemical bath deposition (CBD). Depending on synthesis conditions, metal sulfide nanoparticles can also be produced. The present work attempted the synthesis of transition metal-doped nanocrystalline ZnS powders and thin films by the CBD technique and to obtain information about their optical and magnetic properties. The experimental conditions were the same as reported earlier [3].

RESULTS AND DISCUSSION

Figure 1 shows the typical XRD patterns for nanocrstalline ZnS-based powders. The stoichiometry of the solids is proposed based on the ions concentrations in starting solutions. Peaks of bulk ZnS were included only for comparison. Diffraction peaks for nanocrystalline powders were very broad, suggesting the nanocrystalline nature of these materials, and matched with the corresponding Bragg angles for bulk ZnS. The average crystallite size was estimated to be 3-4 nm by using Scherrer's equation.

As shown in fig. 2, the optical absorption spectra showed blue shifted excitonic peak for ZnS.

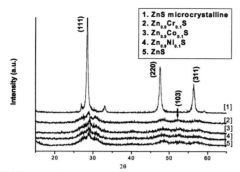

FIGURE 1. X-ray pattern for ZnS and metal ion doped ZnS nanocrystalline powders

The corresponding transition metal doped-ZnS and ZnS, absorption peaks reveals quantum confinement effect due to particle size.

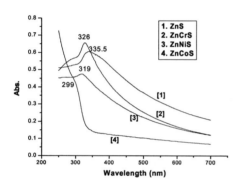

FIGURE 2. Absorption spectrum for ZnS and metal ion doped ZnS nanocrystalline thin fims.

ZnS nanocrystalline powder was annealed at different temperature in a box furnace. Successive conversion of ZnS powders in to ZnO powders is shown in fig. 3. XRD pattern suggests the growth of crystallites but still the broadening in ZnO peaks suggest nanocrystalline behavior of the ZnO powder.

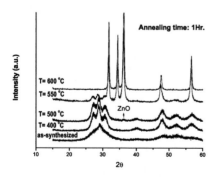

FIGURE 3. X-Ray pattern of ZnS powder at different annealing temperatures.

Magnetic measurements were performed using SQUID magnetometer (Quantum design MPMS).

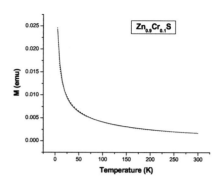

FIGURE 4. ZFC and FC magnetization versus temperature plot with a fit (doted line) to Curie Weiss behavior.

Figure 4 shows the field cooled (FC) and Zero field cooled (ZFC) curves for $Zn_{0.9}Cr_{0.1}S$ nanocrystalline powder. There is no splitting in ZFC and FC curves. Given data was fitted to the relation $\chi = C/T + C/(T+\theta)$, which corresponds to a Curie Weiss behavior. The good fitting of experimental data suggests an antiferromagnetic behavior with a Neel temperature of about 55.6 K.

In summary, we have synthesized nanocrystalline $Zn_{1-x}M_xS$ (M=Co, Ni, Cr) powders and thin films by CBD technique. Nanocrystalline nature of the produced powder was evidenced by XRD and optical measurements. The systematic development of oxide phase was evidenced by XRD. Preliminary magnetic measurements reveal that the nature of $Zn_{0.9}Cr_{0.1}S$ is antiferromagnetic in contrast to theoretical predictions [2]. Detailed studies on magnetic measurements and optical behavior will be published elsewhere.

ACKNOWLEDGMENTS

This work was supported by DoD AFSOR Grant No.: F49620-01-10454, UPRM CID Grant No.: SMDS-03 and NSF Grant No.: CTS-0320534 is gratefully acknowledged. Authors also express their thanks to Dr. Juan Lopez Gariga and Dr. Carlos Rinaldi for optical and magnetic measurements.

REFERENCES

1. H. Ohano, *Science*, **281**, 951 (1998).
2. K.Sato and H. Katayama-Yoshida, *Semicond. Sci. Technol.*, **17**, 367 (2002)
3. S.P. Singh, O.J. Perales-Perez, M.S. Tomar and O.V. Mata, *Phys. Stat. Sol (c)* **1**, 4, 811-814, 2004.

The Crystalline Volume Fraction Dependence of Anisotropic Electrical Transport in nc-Si Thin Films ---Theoretical and Experimental Studies

F. Liu, M. Zhu, Q. Wang[1], Y. Han

Department of Physics, Graduate School, Chinese Academy of Sciences, Beijing, 100039, China
[1]NREL, 1617 Cole Blvd., Golden, CO 80401, USA

Abstract: The three-dimensional cube-networks with different anisotropic distributions of crystallites and amorphous cubes were adopted to simulate the columnar growth character of nc-Si thin films with different crystalline volume fractions (Xc). Take the interface into account, the numerical calculations for the parallel and perpendicular conductivity ($\sigma_{//}$ and $\sigma_\perp$) as a function of Xc suggest that as the structural anisotropy increasing, the $\sigma_\perp$ percolation threshold of nc-Si film shifts to a lower value of Xc, but it is not in the case for $\sigma_{//}$. As a result, the transport anisotropy becomes apparent when Xc < 0.8. The simulation results agree well with the experimental data.

INTRODUCTION

Nano-crystalline Si film by chemical vapor deposition as a promising material for the large area electronics applications has been extensively studied [1]. The correlation between the electrical transport and microstructure is still poorly understood because of the structural complexity of nc-Si [2]. Overhof and Otte [3] applied the percolation model with an isotropic random distribution system to deal with the DC conductivity and Hall mobility of nc-Si films. Shimakawa calculated the conductivity of a random mixture of particles with the effective medium approximation [4]. Both present the phase transition threshold from amorphous to crystalline at Xc~0.32 [3,4]. The anisotropy of electrical transport in nc-Si films is widely observed in the experiments [2,5]. Above calculations based on an isotropic disordered distribution did not take into account of the transport anisotropy. Based on a three-dimensional network model with an anisotropic distribution of crystallite and amorphous cubes, the numerical studies of $\sigma_{//}$ and $\sigma_\perp$ of nc-Si films with different Xc were presented and compared with the experimental data.

SIMULATION NETWORK MODEL

A three-dimensional network, consisting of Nx·Ny·Nz unit cubes, was constructed to simulate the nc-Si film. The black and white cubes represent the crystalline and amorphous regions, respectively, as shown in Fig.1. For a nc-Si film with a Xc, cubes are thought as being distributed with the probabilities of Xc for crystalline and 1-Xc for amorphous. Figure1 (a) shows the schematic picture of an isotropic network with Xc=0.5 as Overhof's model [3]. Assuming the film growing in z direction, the columnar grain was constructed with sequentially connecting n (integer) black cubes in z direction. Considering the distribution of the grain size in the actual film, the corresponding probability of n was described by a Gaussian function $P(n) = \sqrt{2/\pi}/n_0 \exp[-(n-n_0)^2/n_0^2/2]$. The n is defined as the anisotropic degree and n_0 is the mean anisotropic degree. In the x- y plane, cubes are random distributed as those in the isotropic model. Figure 1 (b) draws an anisotropic network with Xc = 0.5 and n_0 = 5.

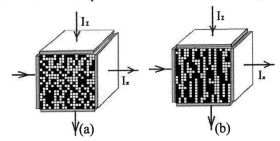

FIGURE 1. Three-dimensional networks (Nx=Ny=Nz=20). (a) Isotropic network. (b) Anisotropic network.

The conductivity of the above three-dimensional network was calculated by Kirchhoff's equations, in which the local crystalline and amorphous conductivities were assumed to be $\sigma_c = 1\times10^{-5}$ $\Omega^{-1}cm^{-1}$ and $\sigma_a = 1\times10^{-10}$ $\Omega^{-1}cm^{-1}$, respectively. During the calculation, with a given set of random parameters at same Xc, the computer will generate different network configurations. Therefore, for each Xc, the conductivity was determined by average of calculated results from twenty different network realizations.

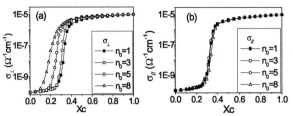

FIGURE 2. The Xc dependence of $\sigma_\perp$ (a) and $\sigma_{//}$ (b).

RESULTS AND DISCUSSIONS

The Xc dependence of $\sigma_{//}$ and $\sigma_\perp$ calculated from the isotropic network shown in Fig.1(a) exhibits similar percolation threshold of Xc as reported [3,4]. Figure 2 (a) and (b) show the $\sigma_{//}$ and $\sigma_\perp$ as the functions of Xc calculated from the anisotropic networks for n_0 =1, 3, 5, 8. With an increasing in n_0, the phase transition threshold for $\sigma_\perp$ shifts to the lower Xc value, which is due to the large amount of linked n_0 crystallite cubes in the z direction. But, the percolation threshold for $\sigma_{//}$ is almost independent of n_0. To evaluate the transport anisotropy, the ratios of $\sigma_\perp$ to $\sigma_{//}$ as a function of Xc for different n_0 were calculated. As $n_0 >1$, the difference between $\sigma_\perp$ and $\sigma_{//}$ appears which indicates anisotropic transport. The value of $\sigma_\perp/\sigma_{//}$ exhibits considerably large even for n_0=3, which does not agree with the experimental results [2,5]. It is noted that, in the real structure of nc-Si film, carriers transport will be affected by the possible defects, electron-phonon scattering and quantum reflection etc. at the interfaces [6]. An interface factor, F (F ≤ 1), was introduced into the model to improve the calculation. Figure 3 plots the $\sigma_\perp/\sigma_{//}$ as a function of Xc for F=0.95 and 1 as n_0=8. The $\sigma_\perp/\sigma_{//}$ is in the range of 5~2 for Xc of 0.35~0.8 as F=0.95. However as Xc> 0.8, the transport anisotropy vanishes.

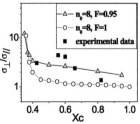

FIGURE 3. Xc dependence of $\sigma_\perp/\sigma_{//}$ for n_0=8 and F=0.95 and 1. The scatter plots were experimental data.

To exam the simulation model, a serious of nc-Si samples with different Xc (from 0.43 to 0.82) were prepared by hot wire chemical vapor deposition. The filament and substrate temperatures are kept at 2000^0C and 250^0C, respectively. To obtain different Xc, the hydrogen dilution was varied from 67% to 93%. For comparison, the thicknesses of all films were controlled at 1µm. The $\sigma_{//}$ was obtained from DC measurement in a coplanar configuration. The AC conductivity, $\sigma_{ac}(f)$, in a sandwich configuration, was measured in the frequency range of 10 to 10 MHz as plotted in Fig.4(a). The $\sigma_\perp$ equals to $\sigma_{ac}(f_0)$ that the plateau appears at f_0 in the frequency dependence of AC ε_{eff} curve as shown in Fig. 4(b). The experimental data of $\sigma_\perp/\sigma_{//}$ and the simulation curve for n_0=8 and F = 0.95 showing in Fig. 3 exhibit the similar value and tendency.

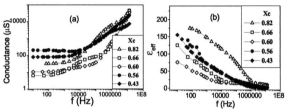

FIGURE 4. The frequency dependence of $\sigma_\perp$ (a) and ε_{eff} (b).

CONCLUSION

With a proper approximation of the complicated nc-Si film structure, the transport anisotropy in nc-Si film was explained by a geometrical columnar growth effect. Increasing n_0 mainly affects the perpendicular transport percolation threshold but not for the parallel transport. Asymmetry distribution of the interfaces in nc-Si film is an important factor for anisotropic transport.

ACKNOWLEDGEMENTS

This work is supported by National Basic Research Priorities Program G2000028208 and project yzjj 200202.

REFERENCES

1. M. Kondo and A. Matsuda, Current Opinion in Solid State & Materials Science, **6**, 445-453 (2002).
2. K. Nakahata, T. Kamiya, C. M. Fortmann, I. Shimizu, H. Stuchlikova, A. Fejfar and J. Kocka, J. Non-Cryst. Solids, **266-269**, 341-345. (2000)
3. H. Overhof and M.Otte, Proceeding of the Ninth International School on Condensed Matter Physics, Singapore, World Scintific, 1997, pp 23~31.
4. K. Shimakawa, J. Non-Cryst. Solids, **266-269**, 223-226 (2000).
5. J. Kocka, A. Fejfar, Int'l PVSEC-**11**, 207, (1999).
6. S. Lombardo, S. U. Campisano, F. Baroetto: Phys. Rev. **B47,** 13561-13565, (1993).

Crystalline and amorphous dualism of polar nano-semiconductors

Shu-Lin Zhang,[*] L.S. Hoi, Yan Yan, Jia Shao, Xin Lu, Hong-Dong Li

School of Physics, Peking University, Beijing 100871, China

Abstract. It is found that the polar semiconductor nano-materials, e.g., GaN nano-particles and SiC nano-roads, can appear as crystalline and amorphous dualism in Raman spectral measurements.

The Raman spectral features of crystalline bulk materials (CBMs) and amorphous bulk materials (ABMs) are sharply different, which originates essentially from the difference in structure and symmetry of materials. The CBMs possess of the lattice periodicity and translational symmetry, resulting in the wave-vector selection rules in the light scattering and the Raman spectrum consisting of discrete set of lines [1]. While in ABMs, the lattice periodicity and the translational symmetry lose, implying the breakdown of the wave-vector selection rule and the Raman spectrum appearing as a continuously distributed fashion relative to the phonon density of states (PDOS) [1]. Such different Raman feature has been utilized to identify the crystallography structure of materials. In retrospect, the identified result by Raman spectroscopy is consistent with that by the X-ray refraction (XRD) in bulk materials and in porous Si (PS) [2] and Si nano-wires [3]. In this letter we will report on crystallographic study by Raman spectroscopy for polar-semiconductors nano-materials.

It is well known that in the nano-crystalline materials (NCMs), the size confinement effect (SCE) play a key role. Due to the SCE, Raman intensity at frequency ω and wavevector q_0, $I(\omega, q_0)$, can be expressed by micro-crystalline model (MCM) as [4]:

$$I(\omega) = \int \frac{d^3q \cdot |C(\boldsymbol{q_0},\boldsymbol{q})|^2}{[\omega-\omega(\boldsymbol{q})]^2 + (\Gamma_0/2)^2} \quad (1)$$

where the $C(\boldsymbol{q_0}, \boldsymbol{q})$ is the Fourier coefficients of the weight function describing SCE, $w(\boldsymbol{r}, L)$, $\omega(\boldsymbol{q})$ is the phonon dispersion, and Γ_0 is the Raman linewidth. While in the ACMs, since all phonon modes in principle contribute to Raman scattering process as mentioned above, the Raman scattering intensity, I, is described by amorphous crystalline model (ACM) as follows [1]:

$$I_{\alpha\beta,,\gamma\delta}(\omega) = \sum_b C_b^{\alpha\beta,,\gamma\delta}[(1+n(\omega))/\omega]D_b(\omega) \quad (2)$$

where $D_b(\omega)$ is the PDOS of b band, implying the Raman scattering of ACMs is related to the PDOS.

Figure 1 shows the Raman spectra of polar semiconductor GaN. It has been identified for GaN bulk materials that the Raman active mode consists of discrete sharp peaks at 143 ($E_{2,\text{low}}$), 531 ($A_{1,\text{TO}}$), 559 ($E_{1,\text{TO}}$), 567 ($E_{2,\text{High}}$), 734 cm^{-1} ($A_{1,\text{LO}}$). We found that the Raman spectrum of GaN nano-particles (NPs) consisted of some broad peaks seen as Fig. 1 (a), which were totally different from the feature of corresponding bulk but very similar to that of amorphous GaN shown in Fig. 1 (b).

These observed Raman features exhibit that the GaN NPs posses of amorphous nature. The spectra with similar feature have been observed in SiC nano-rods (NRs) shown as Fig. 2(a).

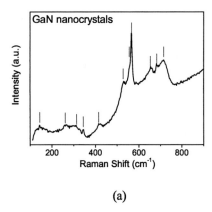
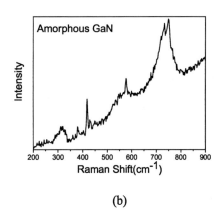

FIGURE 1. Raman spectra of NP (a) and amorphous GaN (b).

We show the observed and calculated the Raman spectra by MCM and ACM for SiC NRs in Fig. 2. The comparison of three spectra in Fig. 2 demonstrates in theory that the SiC NRs appear as amorphous materials.

Therefore, the Raman measurement indicates that the polar GaN and SiC nano-semiconductors are the amorphous ones. In contrary, X-ray diffraction (XRD) measurements showed that the samples of GaN NPs and SiC NRs are crystalline [5, 6].

It is hard to imagine in traditional sense that a material are both crystalline and amorphous. However, we found that the polar semiconductor nano-materials GaN NPs and SiC NRs can appear as crystalline and amorphous dualism.

We acknowledge the assistance from the nano-chemical group and bio-chip group of Peking University and the supports from the NSFC under grants NOs. 60290083, 50334040 and 50272017, the State Key Lab for Inferred Physics and the Beijing Key Lab for Nano-Photonics and Nano-Structure.

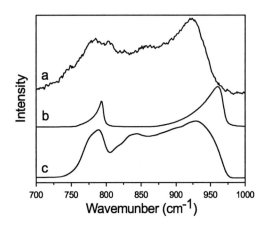

FIGURE 2. Observed Raman spectra (a) and calculated Raman spectra with MCM (b) and ACM (c) for SiC NRs.

REFERENCES

* Email: slzhang@pku.edu.cn.

1. R. Shuker and R. Gammon, Phys. Rev. Lett. **25**, 222 (1970).
2. S.-L. Zhang, Y. Hou, H.-S. Ho, B. Qian, and S. Cai, J. Appl. Phys. **72**, 4469 (1992).
3. B. Li, D. Yu, and S.-L. Zhang, Phys. Rev. B **59**, 1645 (1999u).
4. I. H. Campbell and P. M. Fauchet, Solid State Commun. **58**, 739 (1986).
5. H. D. Li, H. B. Yang, S. Yu, G. T. Zou, Y. D. Li, S. Y. Liu, and S. R. Yang, Appl. Phys. Lett. **69**, 1285 (1996).
6. H. Dai, E.W. Wong, Y.Z. Lu, S. Fan, C.M. Lieber, Nature **375**, (1995) 7613.

Strong sub-bandgap absorption in GaSb/ErSb nanocomposites attributed to plasma resonances of semimetallic ErSb nanoparticles

M. P. Hanson[1], D. C. Driscoll[1], E. R. Brown[2], A. C. Gossard[3]

[1]*Materials Department, University of California, Santa Barbara 93106*
[2]*ECE Department, University of California, Santa Barbara 93106*

Abstract. We report experimental evidence for strong surface plasma resonances on semimetallic ErSb nanoparticles grown epitaxially in semiconducting GaSb by molecular beam epitaxy. The infrared transmission spectra of ErSb/GaSb superlattices grown by molecular beam epitaxy showed a pronounced attenuation peak below the band gap of GaSb, not observed in a GaSb epitaxial control layer containing no ErSb. We attribute the attenuation peak to a surface plasmon resonance on the ErSb particles. The position of this peak shifts to longer wavelengths with increased ErSb deposition, from 2.4 μm for the smallest deposition to 3.8 μm for the largest deposition. In addition, the attenuation strength increases with the total amount of ErSb contained in the sample. The strongest effective attenuation coefficient was measured to be 1.4×10^4 cm^{-1}, which is nearly equal to the cross bandgap absorption of the bulk GaSb at 1 μm.

INTRODUCTION

Epitaxial structures containing metallic particles buried within a semiconductor matrix have been demonstrated to be effective ways of manipulating the electrical and optical properties of the host semiconductor. Previous work has shown evidence of surface-plasmon resonance on ErAs particles in GaAs below the GaAs bandgap[1].

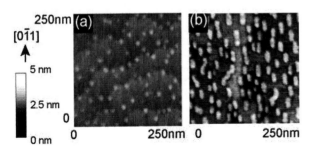

FIGURE 1. AFM images of ErSb particles on a GaSb surface for (a) 0.1 ML and (b) 1 ML of ErSb.

In this paper we present optical transmission measurements from 1 μm to 8 μm of a material consisting of layers of semimetallic ErSb nanoparticles within a GaSb matrix grown by molecular beam epitaxy (MBE). Previous work has examined the MBE growth of these superlattices and shown their ability to reduce and tune the as-grown hole concentration[2] and carrier lifetime[3] in GaSb. ErSb grows on GaSb in an island growth mode. By depositing ErSb in amounts less than that required for the islands to coalesce into a complete film, a layer of isolated particles about a nanometer in height is formed. The lateral size, shape, and density of these particles is determined by the amount of ErSb deposited. Figure 1 shows atomic force microscopy images of ErSb particles on a GaSb surface for (a) 0.1 ML and (b) 1 ML of ErSb. This layer of particles is epitaxially overgrown with GaSb and the process is repeated to form a superlattice. Superlattices examined in this paper are 620 nm thick grown on relaxed 500 nm AlSb buffer layers nucleated on semi-insulating GaAs (100) orientated substrates. Depositions of ErSb are stated in monolayers (ML) as if the ErSb grew in a layer-by-layer growth mode (1 ML = ~3.05 Å). In addition to the superlattice samples, a 620 nm thick layer of GaSb was also grown as a reference sample.

EXPERIMENT

Optical transmission measurements were made on the superlattices over wavelengths from 1 to 8 μm. For wavelengths between 1 and 3 μm a near-IR grating monochromator was used. Both reflection and transmission spectrums were obtained. An absorption signal was calculated by subtracting the sum of the percent reflected and percent transmitted from 100. For wavelengths between 3 and 8 μm a mid-IR Fourier-transform infrared (FTIR) spectrometer was used. Because the FTIR instrument geometry prohibited measurement of a reflection spectrum, the transmission data was normalized to a reference structure of the same thickness and composition but containing no ErSb.

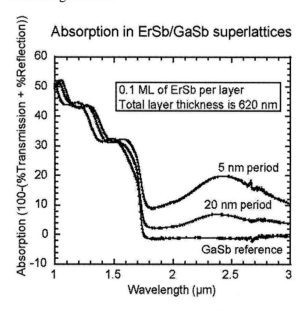

FIGURE 2. Dependence of sub-bandgap absorption on period (5 nm and 20 nm), in 620 nm thick ErSb/GaSb superlattices containing 0.1 ML of ErSb per layer relative to a reference GaSb layer.

RESULTS AND CONCLUSIONS

An absorption peak below the bandgap of GaSb is observed for all samples containing ErSb particles. No such peak is observed in the GaSb reference sample. The strength of this peak is proportional to the total amount of ErSb in the sample. Figure 2 shows absorption spectra for three 620 nm thick layers: a GaSb reference layer, and two ErSb/GaSb superlattices containing 0.1 ML of ErSb per layer, one with a 20 nm period and the second with a 5 nm period. The intensity of the absorption peak increases as the density of ErSb particles is increased through reduction of the spacing between ErSb layers. The peak position remains constant centered at about 2.4 μm.

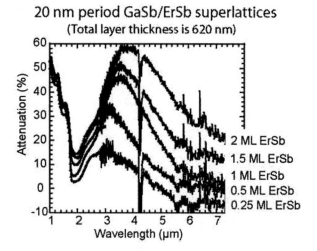

FIGURE 3. Attenuation of the transmission signal for various deposition of ErSb. Data from 1 to 3 μm from grating spectrometer, data from 3 to 7 μm from FTIR spectrometer.

As the amount of ErSb per layer is increased, in addition to the increase in attenuation, the peak position shifts to longer wavelengths from about 2.4 μm to 3.8 μm. Figure 3 shows dependence of the attenuation spectrum on the ErSb deposition. While the increase in attenuation is consistent with the increase in total ErSb in the sample, the shift in peak position is likely due to a combination of both change in particle shape and size with increased deposition.

We attribute the strong absorption peak to plasmon resonance on the ErSb particles within the GaSb matrix. The strength of the effective attenuation coefficient was measured to be 1.4×10^4 cm^{-1} for a superlattice containing 2 ML ErSb per 20 nm period. Decreasing the superlattice period could further enhance the strength of this attenuation.

REFERENCES

1. Brown, E. R., Bacher, A., Driscoll, D., Hanson, M., Kadow, C., and Gossard, A. C., *Phys. Rev. Lett.*, **90**, 077403 (2003)
2. Hanson, M. P., Driscoll, D. C., Kadow, C., and Gossard, A. C., Appl. Phys. Lett. **84**, 221 (2004)
3. Hanson, M. P., Driscoll, D. C., Zimmerman, J. D., and Gossard, A. C., submitted to Appl. Phys. Lett. (2004)

Optical Phonon Modes of InP/II-VI core-shell Nanoparticles: a Raman and Infrared Study

F. S. Manciu[1,3], R. E. Tallman[1], B. A. Weinstein[1], B. D. McCombe[1,3], D. W. Lucey[2,3], Y. Sahoo[2,3] and P. N. Prasad[2,3]

[1]*Dept. of Physics;* [2]*Dept. of Chemistry;* [3]*Intitute for Lasers Photonics and Biophotonics*
University at Buffalo, The State University of New York, Buffalo, NY 14260 USA

Abstract. Crystalline nanoparticles having ~ 3nm diameter InP cores surrounded by different II-VI cladding layers have been synthesized by colloidal chemistry. We report studies of the confined optical phonons in InP/ZnSe core-shell nanocrystals by Raman scattering and far-infrared (FIR) spectroscopy as a function or temperature. Based on comparison with a dielectric continuum model, the observed features are assigned to the surface optical (SO) modes of the InP core, and to the optical and SO modes of the ZnSe shell. Reasonable agreement is obtained between the Raman and FIR results, as well as with the calculations. Strong temperature dependent resonance Raman effects are observed.

INTRODUCTION

Interest in crystalline nanoparticles has rapidly grown because of their novel electronic and optical properties. The present studies on confined optical phonons in InP/ZnSe core-shell nanoparticles were performed by Raman scattering and infrared absorption spectroscopies. Results for other InP/II-VI nanoparticles will be given elsewhere.[1]

EXPERIMENT AND DISCUSSION

Novel colloidal chemistry methods that do not require added surfactant recently have been developed to fabricate core/shell nanocrystals with InP (zinc-blende structure) cores, and shells composed of zinc-blende ZnSe and ZnS, or wurtzite CdS and CdSe.[2] X-ray diffraction and transmission electron microscopy (TEM) confirm the crystal structures, and indicate total nanoparticle diameters distributed in the range of 5–13 nm and core diameters of ~3nm.

Raman scattering and Fourier Transform IR transmission measurements on InP/ZnSe nanoparticles press-loaded into CsI pellets were carried out at temperatures between 8 and 300K. The results in the frequency range 170–370 cm^{-1} are shown in Figures 1(a) and 1(b). A photoluminescence background has been subtracted from the Raman data.

Between T=292 K and T=190 K, the Raman spectra exhibit two features – a sharp peak at 237 cm^{-1} labeled ω_B (FWHM ~ 10 cm^{-1}) between the bulk TO(Γ) (205 cm^{-1}) and LO(Γ) (252 cm^{-1}) frequencies of ZnSe, and a much broader band in the anticipated region of the Raman active TO-like, LO-like and SO modes of the InP core. For T < 190K, two additional features appear – the sharp peaks labeled ω_C (FWHM ~ 16 cm^{-1}) and ω_A (FWHM ~12 cm^{-1}) below the bulk LO frequencies of InP and ZnSe, respectively. The three sharp Raman peaks exhibit strongly temperature-dependent intensities indicative of a resonance Raman mechanism. The upper (lower) frequency line of the ω_B(ω_A) pair is dominant at the highest (lowest) temperatures, and ω_C increases rapidly as T decreases below 110K. The three main FIR absorption features in Fig 1(b) have also been labeled ω_A, ω_B and ω_C. These labels for the Raman and FIR features refer to the theoretical results (see below) in Figures 2(a) and 2(b), respectively. The FIR spectra also contain some additional peaks, indicated by dashed arrows, which are due to residual by-products of the chemical reaction process.

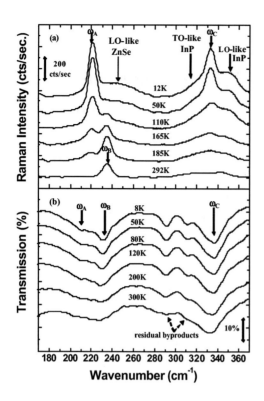

FIGURE 1. (a) Raman and (b) IR spectra at several temperatures of InP/ZnSe core-shell nanoparticles. Arrows designate assignments of observed features as indicated.

Figures 2(a) and 2(b) present calculated results for the frequencies of the confined SO modes at 50K for InP/ZnSe spherical nanoparticles having angular momenta $l = 2$ (Raman active) and $l = 1$ (IR active), respectively. These calculations employ a dielectric continuum model [3]. The parameter γ is the ratio of the core radius a to the total (core plus shell) radius b. The solid horizontal arrows indicate the frequencies corresponding to the limiting cases ($\gamma \to 1$) of an InP core embedded in an infinite CsI dielectric matrix, ($b \to \infty$) of an InP core surrounded by an infinite ZnSe shell, and ($a \to 0$) of a solid ZnSe nanoparticle surrounded by an infinite CsI dielectric material.

The frequency ranges of the calculated ω_A, ω_B, and ω_C branches in Figs. 2(a) and 2(b) correspond, within uncertainty, to the positions of the similarly labeled Raman and FIR spectral features in Figs. 1(a) and 1(b), respectively. The shaded regions in the calculated plots show the range of values of γ that best account for the experimentally observed frequencies. The FIR results are compatible with γ in the range 1.3–3.3 and the Raman results further limit this range to $\gamma = 1.7 \pm 0.3$. These findings fall within the lower range of the InP/ZnSe nanoparticle size distribution estimated by other methods.

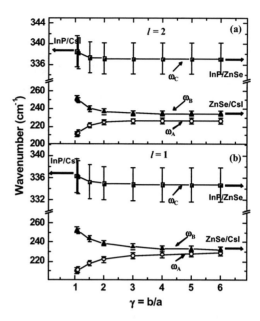

FIGURE 2. Calculated frequencies at 50 K of the SO modes vs. $\gamma=b/a$ for InP/ZnSe core-shell nanoparticles for (a) $l=2$ and (b) $l=1$. The horizontal arrows mark the limiting cases as discussed in the text.

The behavior of the Raman spectra (Fig. 1(a)) as a function of temperature is intriguing. The sharpness and strong intensity changes in the ω_A, ω_B, and ω_C Raman peaks suggest that a resonant Raman effect operates in both the InP core and the ZnSe shell materials. This could arise from the small valence band offset (~ 0.12 eV) in the InP/ZnSe system, [4][5] which would allow the hole envelope functions to extend throughout the nanoparticle, and thereby provide a temperature dependent mechanism for coupling to both the core and the shell normal vibrations. Further study of these issues is progress.[1]

ACKNOWLEDGMENTS

Supported by AFOSR/DURINT # F496200110358.

REFERENCES

1. Manciu, F. S., *et al.*, to be published.
2. Lucey, D. W., *et al.*, to be published.
3. Comas, F., and Trallero-Giner, C., *Phys. Rev. B* **67**, 115301-115307 (2003).
4. Dremel, M., *et al.*, *J. Appl. Phys.* **93**, 6142-6149 (2003).
5. Schultz, Ch., *et al.*, *J. Vac. Sci. Technol. B* **15**(4), 1260-1264 (1997).

Optical Resonances Of Single Zinc Oxide Microcrystals.

Thomas Nobis[1,3], Evgeni M. Kaidashev[1,2], Andreas Rahm[1],
Michael Lorenz[1], Jörg Lenzner[1], and Marius Grundmann[1]

[1] *Universität Leipzig, Institut für Experimentelle Physik II, Linnéstraße 5, D-04103 Germany*
[2] *on leave from Rostov-on-Don State University, 344 090 Rostov-on-Don, Russia*
[3] *corresponding author e-mail: nobis@physik.uni-leipzig.de*

Abstract. We have investigated whispering gallery modes (WGMs) of hexagonal microcavities. As a model system we used hexagonal zinc oxide microcrystals grown on *c*-plane sapphire by pulsed laser deposition. Room temperature cathodoluminescence experiments within the visible spectral range showed a series of sharp peaks accompanying the typically broad unstructured green emission of bulk material. The energetic positions of the detected peaks could be explained unambiguously by a simple plane wave model of WGMs in good agreement with the experiment.

INTRODUCTION

In the last years so-called whispering gallery resonators have attracted much interest, e.g. shaped as microdiscs [1] or microspheres [2]. Inside such a resonator light circulates around due to total internal reflection, leading to high *Q*-factors and low laser threshold power. ZnO, a promising wurtzite-material for future photonic devices within the ultra violet (UV) range, forms microcrystals that naturally exhibit a hexagonal cross section. In this paper we investigate the resonator properties, i.e. the eigenmodes, of such hexagonal ZnO microcavities, grown on *c*-plane sapphire by pulsed laser deposition [3].

THEORY

For the calculation of hexagonal whispering gallery modes (WGMs) Maxwell's equations have to be solved numerically [4]. For large mode numbers $N > 70$ a simple plane wave model has been deduced [4], allowing an approximation of the hexagonal WGMs. The main idea is that the circulating light interferes with itself when having completed one full circulation. To obtain constructive interference, the total phase shift of the wave along its path has to be an integer multiple of 2π. This leads to the following resonance condition:

$$6R_i = \frac{hc}{nE}\left[N + \frac{6}{\pi}\arctan\left(\beta\sqrt{3n^2 - 4}\right)\right] \quad (1)$$

The factor β regards polarisation, for TM-polarisation, i.e. ($\vec{E} \parallel \vec{c}$), $\beta = \beta_{TM} = n^{-1}$ has to be used; TE-polarisation ($\vec{E} \perp \vec{c}$) leads to $\beta = \beta_{TE} = n$. Due to the spectral dependence of the refractive index $n = n(E)$, eqn. 1 is an implicit equation to determine the resonance energies $E = E_N(R_i)$ in terms of the radius of the incircle R_i (inset (b) of Fig. 1), Planck's constant h and the vacuum speed of light c. The integer $N \geq 1$ characterizes the interference order of the resonance, which is identical with the WGM number [4]. Note that, since ZnO is uniaxial, $n = n_\parallel(E)$ and $n = n_\perp(E)$ have to be applied for TM- and TE-polarisation, respectively.

EXPERIMENT

ZnO is a direct semiconductor with a band gap of about 3.4 eV. Its room temperature luminescence typically consists of two bands, the first (UV) around 3.25 eV, the second (visible, VIS) around 2.35 eV. The UV band is associated with free-exciton emission;

the VIS band indicates recombination at deep levels. Since VIS emission provides an intrinsic bright and broad-band light source, it can easily be utilized in cathodoluminescence (CL) experiments to excite the microcavities and investigate their resonator properties in a wide spectral range.

Figure 1 shows two VIS emission CL spectra from different ZnO microcrystals together with a scanning electron microscopy image of the first crystal. In comparison to the broad and unstructured VIS emission of thin films or bulk material, VIS emission of the microcrystals is accompanied by a series of comparatively sharp peaks. Using the plane-wave model, those peaks can unambiguously be attributed to WGMs of a hexagonal cavity. Therefore it is necessary to determine the correct mode number N of every single detected peak. Provided that this mode number is known, eqn. 1 enables to calculate the theoretical diameter of the crystal $D = 4R_i/\sqrt{3}$ out of every single detected peak energy. Fortunately, since the peaks have to be numbered in ascending order, one only has to find the correct mode number to start. The best fitting and therefore the final peak numbering is found, if every single peak predicts *the same* radius as all the other resonance peaks, or at least if the variations in the predicted radius become smallest.

For the investigated crystals, the correct mode numbering is shown in Fig. 1, the respective theoretical diameters yield $D_1^{theory} = 2.86$ µm and $D_2^{theory} = 2.06$ µm with a small spreading of less than 20 nm. The calculations have been performed for TM polarisation, since polarisation-dependent microphotoluminescence experiments showed, that the WGMs are preferentially TM polarised. This result is consistent with former investigations of lasing in hexagonal microcavities that are reported to only emit TM modes [5]. The required data for $n_\parallel(E)$ were obtained from ellipsometry measurements on ZnO thin films [6] also grown by pulsed laser deposition.

The experimentally determined cavity diameters obtained from electron microscopy amount to $D_1^{exp} = (2.90 \pm 0.06)$ µm and $D_2^{exp} = (2.05 \pm 0.04)$ µm. Hence, the deviations between theory and experiment are in the range of 2%! This means, that the simple plane-wave model fits very well, even if N is in the range of only 15 to 30. To emphasize this fact, the theoretical values E_N calculated for two fixed diameters $D = D_{1,2}^{theory}$ are given as black arrows in Figure 1; they appear very close to the measured peaks. We note, that if $n = $ const. $\neq n(E)$ is used, the agreement between theory and experiment becomes worse.

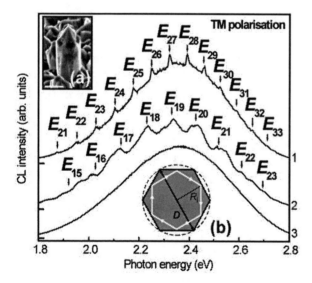

FIGURE 1. Room temperature VIS emission CL spectra of two ZnO microcrystals (curves 1, 2) and of thin film material (curve 3) scaled to the same intensity and vertically shifted for clarity. Arrows mark theoretical energetic positions of WGMs due to equation (1) for TM polarisation and different diameters ($D_1 = 2.86$ µm, $D_2 = 2.06$ µm). The inset (a) exemplarily shows a scanning electron microscopy image of the microcrystal 1. The scalebar has a length of 2 µm. The scheme (b) explains the geometry of a hexagonal cavity with radius of the incircle R_i and cavity diameter $D = 4R_i/\sqrt{3}$.

CONCLUSION

We have shown, that single hexagonal ZnO microcrystals exhibit resonant optical behavior due to whispering gallery modes of a two dimensional hexagonal cavity. The WGMs could be described theoretically by a simple plane-wave-model in a good agreement with the experiment.

ACKNOWLEDGMENTS

This work was supported by Deutsche Forschungsgemeinschaft within FOR 522.

REFERENCES

1. D.K. Armani *et al.*, *Nature* **421**, 925 (2003)
2. M. V. Artemyev *et al.*, *Appl. Phys. Lett.* **78**, 1032 (2001)
3. M. Lorenz *et al.*, *unpublished*
4. J. Wiersig, *Phys. Rev. A* **67**, 023807 (2003)
5. I. Braun *et al.*, *Appl. Phys. B* **70**, 335 (2000)
6. R. Schmidt *et al.*, *Appl. Phys. Lett.* **82**, 2260 (2003)

A New Microscopic Theory of Low Frequency Raman Modes in Ge Nanocrystals

Wei Cheng[1], Shang-Fen Ren[2] and Peter Y. Yu[3]

[1] Low Energy Nuclear Physics, Beijing Normal University, Beijing, 100875, P. R. China
[2] Department of Physics, Illinois State University, Normal, IL 61790-4560
[3] Department of Physics, University of California, Berkeley, and Materials Sciences Division, Lawrence Berkeley National Laboratory, Berkeley, CA 94720

Abstract. A bond polarizability model in combination with a lattice dynamics model has been used to compute the intensities and frequencies of low frequency Raman modes reported in Ge nanocrystals. The results indicate that the cross-polarized mode often identified as a torsional Lamb mode (which is Raman-inactive) is actually a $l=2$ spheroidal mode.

INTRODUCTION

Experimentally it has been found that nanocrystals (NC) of both metals and semiconductors often exhibit low frequency Raman modes. Typically one mode is observed for incident and scattered radiation polarized *parallel* to each other while one is observed for *cross* polarizations. The frequencies of both modes scale with the diameter d of the NC as $1/d$ with the parallel polarization mode having a higher frequency. These low frequency Raman modes have been explained successfully with a continuum model proposed by Lamb [1]. The parallel polarization mode is attributed to spheroidal vibrational mode while the cross-polarized mode is attributed to a torsional mode. These conclusions are based mainly on their Raman frequencies which can be estimated from the sound velocities of the corresponding bulk materials. However, the identification disregards the selection rule that the torsional modes are Raman inactive.

THEORY & RESULTS

To understand the microscopic origin of these modes, we have performed a lattice dynamical calculation of the vibrational modes of Ge NC containing between 47 to >7000 atoms using a Valence Force Field Model. To simplify the calculation the modes are divided according to the irreducible representations of the local T_d symmetry of the Ge atoms. From the result we have been able to obtain the phonon density-of-states, identify the surface phonons etc. The details of these calculations have been presented in several publications already and will not be repeated here [2]. In this paper we present the Raman intensity of the low frequency lattice modes calculated with a bond polarizability model. These modes obey the expected Raman selection rules in that only modes with symmetries: A_1, E and T_2 are Raman active[3]. An example of the result for the A_1 symmetry modes is shown in Fig. 1 for NC containing different numbers N of Ge atoms. To make contact with the continuum Lamb model we have also computed the amplitude of the Lamb modes for NC of different radii. The resultant spheroidal and torsional modes are then projected onto the lattice modes of a NC of the same radius. Since the Lamb modes are defined for a continuum while the lattice modes are discrete, this projection is defined only for the individual atomic positions within the NC. Using this projection technique we can calculate theoretically the Raman intensities of the Lamb modes also.

A summary of our findings are as follows: (1) Based on group theory, only the spheroidal Lamb modes with $l=0,2$ are Raman-active. Lattice model gives the same conclusion for large NC. These modes have been observed by Brillouin scattering [4].

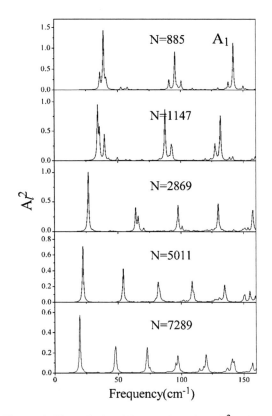

Figure 1. The calculated Raman intensity (A_i^2) of the A_1 symmetry lattice modes for Ge NC with different N.

(2) In small NC the torsional modes can contain contribution from T_2 modes which are Raman-active. However, lattice model shows that their Raman intensities are *much less* than those of spheroidal modes. (3) For large NC the decomposition of a Lamb mode into lattice modes is typically dominated by one lattice mode. However, for small NC a Lamb mode is a linear combination of a number of lattice modes of comparable strength and therefore the Lamb continuum model breaks down. Even when the Lamb model breaks down the Lamb mode frequencies still scale as (1/d) [5] where d is the NC diameter. The reason is because the frequencies of the lattice modes which contribute to the Lamb mode also scales as (1/d) until N becomes <~800 or d<3 nm. (4) Table 1 shows the Raman frequencies and intensities for some of the low frequency Raman-active lattice modes. Notice that the strongest mode with the perpendicular polarization scattering configuration (T_2) is the $n=0$, $l=2$ spheroidal modes and not the torsional mode assumed in most experimental papers [6].

ACKNOWLEDGMENTS

WC is supported by the National Natural Science Foundation of China (10075008and 10275007); SFR is supported by the National Science Foundation (0245648); and the work at Berkeley was supported in part by the Director, Office of Science, Office of Basic Energy Science, Division of Materials Sciences and Engineering, of the U.S. Department of Energy under Contract No. DE-AC03-76SF00098.

REFERENCES

1. H. Lamb, Proc. London Math. Soc. 13, 189 (1882).
2. W. Cheng, S.F. Ren, and P.Y. Yu, Phys. Rev. B 68, 193309 (2003).
3. E. Duval, Phys. Rev. B 46, 5795 (1992).
4. M. H. Kuok, H. S. Lim, S. C. Ng, N. N. Liu, and Z. K. Wang, Phys Rev. Lett.**90**, 255502 (2003).
5. A. Tanaka, S. Onari and T. Arai, Phys. Rev. B 47, 1237 (1993).
6 B. Champagnon, B. Andrianasolo and E. Duval, Materials Sci. Eng. B 9, 417 (1991). N.N. Ovsyuk, E. B. Gorokhov, V. V. Grishchenko, and A. P. Shebanin, JETP Lett. 47, 298 (1988).

TABLE 1. Calculated Raman frequencies and intensities of a few low frequency NC modes in Ge NC with N= 7289 (d=6.8 nm) and the corresponding to Lamb modes listed by symmetry

Irreducible Representations	Lamb Mode	Frequency(cm^{-1})	Raman intensity (Arbitrary unit)
A_1	$n=1, l=0$, spheroidal	19.6	0.904
A_1	$n=4, l=0$, spheroidal	47.3	0.237
A_1	$n=4, l=0$, spheroidal	48.1	0.316
E	$n=0, l=2$, torsional	11.7	0.00136
E	$n=0, l=2$, spheroidal	11.2	0.430
T_2	$n=0, l=3$, torsional	16.3	0.0197
T_2	$n=0, l=2$, spheroidal	12.9	0.686

Anomalous Stark Effect in Intraband Absorption of Silicon Nanocrystals

[1]J. S. de Sousa, [2]J.-P. Leburton, [1]V. N. Freire and [3]E. F. da Silva Jr.

[1]*Departamento de Física, Universidade Federal do Ceará, Campus do Pici, Caixa Postal 6030, 60455-970 Fortaleza, Ceará, Brazil*
[2]*Beckman Institute, University of Illinois at Urbana-Champaign, 405 N. Mathews Avenue, Urbana, IL 61801, USA*
[3]*Departamento de Física, Cidade Universitária 50670-901 Recife, Pernambuco, Brazil*

Abstract. Intraband transitions in Si/SiO_2 quantum dots (QD's) are addressed in this work. We found that external electric fields (F_{EXT}) strongly affect the absorbing properties of large QD's (> 5 nm) by creating additional absorption peaks. Moreover, the anisotropic nature and degeneracy of Si band structure couples with the shape of QD's such that the absorption coefficient response to F_{EXT} becomes highly anisotropic for non-spherical shapes.

INTRODUCTION

QD's infrared photodetectors (IP's) can overcome the insensitivity to normal incident illumination exhibited by quantum well IP's. Recently, a QD IP with operating temperature up to 260 K under normal incident illumination has been demonstrated [1]. The mechanisms ruling IP's functionality are intraband transitions. Even though the interest in such transitions in QD's has increased, investigations are pratically limited to III-V systems and almost inexistent for Silicon (Si) systems. Tight-binding calculations have shown that no-phonon and one-phonon intraband transitions in Si nanocrystals present unusual properties and high efficiency compared to III-V QD's [2]. This fact indicates that Si QD's might represent an alternative to the development of IP's that are compatible with standard Si technology. In this work, we address the theoretical investigation of intraband absorption spectra of Si/SiO_2 QD's.

MODEL

Our calculation focus on a single QD by assuming that adjacent QD's are positioned far enough to prevent the overlap among wavefunctions (WF's). The QD electronic structure is calculated by using a 3D quantum mechanical model based on the effective mass theory which takes into account the multi-valley and anisotropic characteristics of Si band structure [3]. The optical properties calculation is based on the Fermi's golden rule [4]. Moreover, phonon effects, inter-valley transitions and many-body interactions among carriers are not considered.

RESULTS

In the left panel of Fig. 1, we compare the absorption coefficient for spherical QD's with 3, 5, 7 and 10 nm of diameter. All cases exhibit two absorption peaks, one very weak and a strong one. The separation of these peaks is inversely proportional to the QD diameter, which is in agreement with the relation between the energy states separation and confinement dimension ($\Delta E_n \propto L^{-2}$), where L is the size of confinement. Under the effect of a F_{EXT}, all absorption peaks move towards higher energies. This is just a manifestation of the Stark effect. For small QD's (3-5 nm), the absorption coefficient is practically insensitive to the application of external biases, while we notice the appearance of an additional peak as F_{EXT} increases for larger QD's (7-10 nm). This is a consequence of the relaxation of some forbidden transitions. Moreover, we do not observe any anisotropy in the F_{EXT} response of the absorption spectra of spherical QD's.

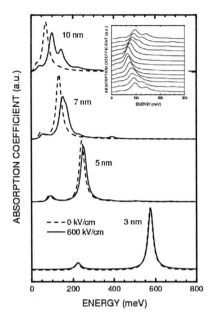 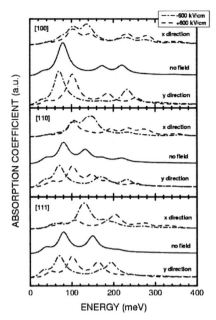

FIGURE 1. (left) Absorption coefficient of spherical QD's with 3, 5, 7 and 10 nm of diameter. The inset figure shows the variation of the absorption coefficient of a 10 nm wide QD as a function of F_{EXT}, ranging from -600 kV/cm to +600 kV/cm (top to bottom). (right) Absorption coefficient of a 10 nm wide QD with different crystalline orientation parallel to the truncation. F_{EXT} of ±600 kV/cm are applied to the x and y directions (dashed-dotted and dashed lines, respectively) and compared with to 0 kV/cm case (solid line).

For hemispherical QD's (right panel of Fig. 1), one can notice the inclusion of truncations (flat wall perpendicular to y direction) breaks the spherical symmetry and creates new transition rules, making them optically active in a broader region in comparison to spherical QD's. We also observed anisotropy with respect to the direction of F_{EXT}. For F_{EXT}'s applied to y direction, WF's have their movement restrained to a smaller dimension in comparison to the x and z directions. This constraint affects the Stark shift and the relative separation of the absorption peaks differently, whether the WF's are squeezed against the flat truncation ($F_{EXT}<0$) or the curved wall ($F_{EXT}>0$), which drastically changes the absorption spectra as shown in Fig. 1. For a F_{EXT} applied to x direction (or z direction), new peaks appear forming a different pattern because of the larger confinement dimension. Moreover, it also depends on the signal of F_{EXT}. This was not expected because of the cylindrical symmetry of the QD shape with respect to y direction. With exception of the [100] case (even though some peaks do not exhibit the same intensity), the other two cases are strongly dependent on the signal of F_{EXT} when it is applied to x direction. This effect is not so strong for the 7 nm wide QD and almost invisible for small QD's (3-5 nm).

In conclusion, we found that F_{EXT}'s create additional peaks in the absorption spectra of large Si/SiO_2 QD's that are not observed in small QD's.

Moreover, for non-spherical QD's, the F_{EXT} response becomes highly anisotropic due to the interplay among QD shape, effective mass tensor and its alignment with respect to the truncation.

ACKNOWLEDGMENTS

J. S. de Sousa is indebted to the Brazilian National Research Council (CNPq) for his research fellowship at the Physics Department of the Universidade Federal do Ceará. The authors would like to acknowledge the financial support from CNPq through the grant NanoSemiMat Project #550.015/01-9.

REFERENCES

1. L. Jiang, S. S. Li, N.-T. Yeh, J.-I. Chyi, C. E. Ross, and K. S. Jones, *Appl. Phys. Lett.* **82**, 1986 (2003).

2. G. Allan and C. Delerue, *Phys. Rev. B* **66**, 233303 (2002).

3. J. S. de Sousa, A. V. Thean, J.-P. Leburton and V. N. Freire, *J. Appl. Phys.* **92**, 6182 (2002).

4. S. L. Chuang, *Physics of Optoelectronic Devices*, Wiley Series in Pure and Applied Optics (1995), chapter 9.

Correlation between PL emission band and growth of oxide layer on surface of silicon nanocrystallites

Ikurou Umezu[*], Motohiko Koyama[*], Takayuki Hasegawa[*], Kimihisa Matsumoto[*], Mitsuru Inada[†] and Akira Sugimura[*]

[*]Konan University, Department of Physics, Kobe 658-8501, Japan
[†]Konan University, High technology Research Center, Kobe 658-8501, Japan

Abstract. A systematic study of the correlation between surface oxidation and PL properties is important to clarify the nature of silicon nanocrystal. We prepared hydrogenated silicon nanocrystallites (nc-Si:H) by pulsed laser ablation, and observed IR absorption and PL spectra. Analysis of the Si-H bond gives us information on the local configuration of Si-O bond. The PL peak wavelength shifted from 800 nm to 400 nm with increasing the Si-O bond density. The frequency resolved PL spectra indicate that the PL peak is composed of at least three frequency regions. We found that PL peak is composed of 400, 700 and 800 nm bands by comparing frequency region and PL bands. The PL peak wavelength depends not on the composition but on the thickness of surface oxide layer.

INTRODUCTION

It is well known that photoluminescence (PL) efficiency of silicon nanocrystallites is much larger than that of the bulk crystal silicon. Strong PL intensity from Si materials is important to investigate physics of nanometer sized materials and to realize Si based optelectronics. Since the PL properties of silicon nanocrystallites depend on the surface oxidation, systematic study of the correlation between surface oxidation and PL properties is important. We prepared hydrogenated silicon nanocrystallites (nc-Si:H) by pulsed laser ablation and observed IR absorption and PL spectra. Analysis of the Si-H bond gives us information on the local configuration of Si-O bond since the Si-H vibration frequency is sensitive to the number of back-bonded oxygen atoms due to the change in the electronegativity.[1] It is well known that the PL time constant of Si nanocrystallites is slow. The frequency resolved spectroscopy (FRS) is one of the methods to measure slow PL component.[2,3] A correlation between PL property and configuration of surface oxide was investigated in this paper.

EXPERIMENT

The nc-Si:H was deposited by pulsed laser ablation of Si target in hydrogen gas. The hydrogen pressure was varied from 5 to 1100 Pa. A fourth harmonic of pulsed YAG laser light (266 nm) was focused on the target and the fluence was kept at 100 J cm^2. Prepared sample was left in the atmosphere and change in the IR and PL spectra with native oxidation was monitored for 200 days.

The sinusoidally modulated excitation requires for FRS was provided by Ar laser (488nm) and a acousto-optic modulator. PL was excited in a cryostat, and detected by photomultiplier. The signal was analyzed by a lock-in amplifier, set in quadrature to the excitation modulation so that only PL with a lifetime of the order of the inverse of the modulation frequency was detected.

RESULTS AND DISCUSSION

The PL peak wavelength shifted from 800 nm to 400 nm with increasing the Si-O bond density as shown in Fig.1. Figure 2 shows FRS spectrum measured at 750 nm. This spectrum is composed of at least 3 frequency regions (time constants). The results of decomposition by assuming 3 independent time constants(~1ms, ~100μs and ~10μs) are shown in Fig.2 by solid lines. We attempted to decompose PL spectrum by using these three frequency regions. The results are shown in Fig.3. We found that PL peak is composed of 400, 700 and 800 nm bands by comparing frequency region and PL bands. This indicates that observed peak shift observed in Fig.1 is not due to shift of single emission band.

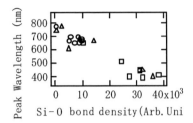

FIGURE 1. The PL peak wavelength as a function of Si-O bond density. Circle, triangle and square are the sample prepared by different gas pressure.

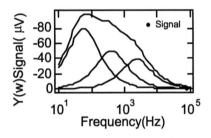

FIGURE 2. The solid squares are the FRS spectrum measured at 750 nm. The solid lines are result of decomposition.

The Si-O bond density and number of back bonded oxygen were estimate by absorption intensity of Si-O bond and frequency of the Si-H bond. They correspond to the thickness of native oxide layer and composition, x, of sub-oxide, SiO_x, respectively. We could not find remarkable correlation between Si-H vibration frequency and PL peaks. On the other hand, PL peak wavelength strongly depends on Si-O bond density. These results indicate that PL peak wavelength depends not on the composition but on the thickness of surface oxide layer. Since the relatively thin native oxide layer contribute to the 700 nm band, this band should be originated to the interface between nc-Si:H and surface oxide. Possible origins of 700 nm peak is nonbridging oxygen hole center or Si=O double bond at the interface. Since the 400 nm band depends on the thickness of oxide layer, this peak may be originated to be defects created during growth of the native oxide layer. Since 800 nm band is observed in non-oxidized sample, this band may originated to be surface self trap state or surface defects.

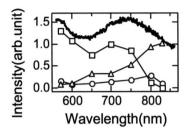

FIGURE 3. The solid line indicate PL spectrum. The circles, triangles and squares correspond to ~1ms, ~100μs and ~10μs components, respectively.

CONCLUSIONS

We compared PL and IR spectra. The PL peak wavelength changed by native oxidation. The results of FRS spectra indicate that PL spectrum is composed of at least three origins. The interface state and nonbridging oxygen hole center are possible origins of PL center.

ACKNOWLEDGMENTS

This work was partially supported by a Grant-in-Aid for Scientific Research from the Japan Society for the Promotion of Science, Nippon Sheet Glass Foundation for Materials Science and Engineering, The Kurata Memorial Hitachi Science and Technology Foundation, and The Hirao Taro Foundation of the Konan University Association for Academic Research.

REFERENCES

1. Lucovsky, G., *Solid State Commun.* **29**, 571-576(1979).
2. Depinna, S. P., and Dunatan, D. J., *Phil. Mag. B* **50**, 579-597 (1984).
3. Stachowitz, R., Schubert, M, and Fuhs, W., *Phil. Mag. B* **70**, 1219-1230 (1979).

Size Dependence Of The Optical Gap In Silicon Nanocrystals Embedded Into *a*-Si:H Matrix

V. A. Burdov[1], M. F. Cerqueira[2], M. I. Vasilevskiy[2], and A. M. Satanin[3]

[1] *Faculty of Physics, Nizhnii Novgorod State University, Nizhnii Novgorod, Russia*
[2] *Centro de Física, Universidade do Minho, 4710-057 Braga, Portugal*
[3] *Center for Computational Nanoscience, Department of Physics and Astronomy, Ball State University, Muncie, IN47306, USA*

Abstract. A simple model for Si nano-crystallites (NC's) embedded in amorphous silicon matrix is considered within the envelope function approximation. It is shown that, although the effect of the NC's on the tail states is small, there are electron and hole states localized in the NC's and separated by an energy of the order of 2eV. These states can be responsible for the experimentally observed features in the photoluminescence spectra of NC-Si/*a*-Si:H films.

INTRODUCTION

Recently, amorphous silicon (*a*-Si) films with crystalline nano-regions (nanocrystals, NC's) have attracted researcher's interest [1], based on the hope that such a system can represent luminescent quantum dots (QD's) embedded in a conductive matrix, suitable for the electrical pumping. Indeed, a size-dependent photoluminescence (PL) in the visible spectral range has been observed from Si NC's grown inside SiO$_2$ [2], indicating the quantized nature of their electronic spectra. Does a NC placed in an amorphous matrix of the same material act as a quantum dot? The answer is not obvious, because in this case there is no well-defined barrier at the NC/matrix interface. In the present work, we attempt to answer this question by calculating the electronic energy spectra and wave functions for such a system and by correlating the calculated results with PL data measured for NC-Si/*a*-Si:H films grown by magnetron sputtering.

THEORETICAL MODEL

The crystalline inclusion is assumed perfect and spherical (with radius R). The potential energy outside the NC is modeled by a random function with a non-zero mean value and dispersion estimated from the atomic radial distribution function (RDF) known from X-ray diffraction measurements and simulations [3]. Bearing in mind the hydrogenated amorphous silicon (*a*-Si:H), the mean values (V_0) of the random potential in the matrix for electrons and holes have been assumed positive and negative, respectively (100-200meV in the absolute value). The dispersion of the potential fluctuations has been assumed decreasing as r^{-2} where r is the distance from the NC center. Its value (of the order of 10eV), extrapolated to the first coordination sphere of the central atom, was estimated from the broadening of the corresponding RDF peak and the volume deformation potential of the corresponding band.

The **k·p**-method has been used to derive effective Schrodinger equations for the electrons near the X point and holes near the Γ point of the Brillouin zone. In the isotropic approximation for the random potential and dispersion relation for crystalline silicon, we found the eigenstates of the system containing a sufficiently large number of disordered atomic shells (which are assumed to have the same thickness a). The electron and hole densities of states (DS) as well as the local DS (weighted by the integrated probability to find the particle inside the NC) and the joint DS have been calculated.

RESULTS AND DISCUSSION

In our model, the crystalline inclusion is a spherical quantum well for electrons, which produces a localized state (in the absence of the random potential in the matrix) only if $R \geq R_0$, where the critical radius $R_0 = \pi\hbar/\sqrt{8mV_0}$ and m is the effective mass. Thus, for $R \leq 2-3$nm, the shallow potential well produced by the NC is not sufficient for localizing the electron. The NC-related localized state is then determined by the random potential and its wave function extends far into the matrix. The results of the DS calculations for QD's with radii close to R_0 show that the (in)existence of the electron state localized solely by the dot has insignificant effect on the (nearly Gaussian) shape of the tail below the mean conduction band bottom energy of the amorphous matrix. Therefore, the hole DS (in the tail region) has a similar shape despite the fact that the NC is supposed to be a barrier for the holes.

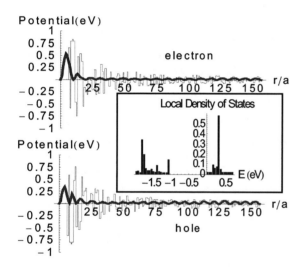

FIGURE 1. Electron and hole probability distributions calculated for a certain realization of the random potential (whose diagonal matrix elements for the conduction and valence bands are also shown) around a NC with $R=1.5$nm. They correspond to the 19-th hole ($E=0.5$eV) and 17-th electron ($E=0.3$eV) excited s-type states, respectively. The inset shows the local DS (inside the NC) for this realization.

However, we have found that there are other NC related states, lying at much higher energies and completely determined by the fluctuating part of the potential barrier between the NC and matrix. It occurs that, for any realization of the disorder, there are certain excited states for which the fluctuations constitute a potential wall almost completely localizing the particle in the vicinity of the NC. This happens both for electrons and holes (see Fig.1). The energy of such states (E) is much larger than V_0 and depends on the disorder realization. For example, for $R=1.5$nm E was found to fluctuate around 0.4eV for electrons and 0.5eV for holes. We have not yet been able to relate these values to the parameters of the problem. However, we have found that these energies decrease with the increase of the NC size. This can be understood by taking into account that increasing R means increasing the number of atoms in the first disordered shell and, therefore, decreasing the amplitude of the random potential. Notice that this amplitude decreases only as R^{-1}, therefore one should expect a weaker dependence $E(R)$ than for QD's in dielectric matrices (R^{-2}).

The resonantly localized states like those shown in Fig.1 have the largest overlap of the wavefunctions and constitute optically active electron-hole pairs. The optical gap in the considered system is dependent of the NC size but to a smaller extent than for a system with strong confinement, such as Si/SiO_2.

These findings are in qualitative agreement with the results of PL measurements for several a-Si:H/nc-Si films produced by RF sputtering. If the NC's are small, they do not manifest themselves because of the low weight of the resonant states. For samples with relatively large NC's ($R \geq 3$nm), the PL spectra show broadened peaks situated at 1.9-2eV, which can be associated with NC's since they are absent in the spectra of purely amorphous films. The above value agrees well with the calculated results. Thus, we can conclude that silicon NC's in amorphous silicon environment are similar to QD's from the point of view of their optical properties although the electronic spectra of these systems are rather different.

ACKNOWLEDGMENTS

This work was supported by FCT under project POCTI/CTM/39395/2001 and RFBR (project No 04-02-16493). MIV wishes to thank travel support from C. Gulbenkian Foundation and FLAD.

REFERENCES

1. Losurdo, M. et al, *Physica E* **16**, 414-419 (2003).
2. Pavesi, L. et al, *Nature* **408**, 440-444 (2000).
3. Ziman, J. M., *Models of Disorder,* Cambridge: Cambridge University Press, 1979, pp. 75-96.

Ab-initio Calculations Of The Electronic Properties of Silicon Nanocrystals: Absorption, Emission, Stokes Shift

Elena Degoli[1], G. Cantele[2], Eleonora Luppi[3], Rita Magri[3], Stefano Ossicini[1], D. Ninno[2], O. Bisi[1], G. Onida[4], M. Gatti[4], A. Incze[4], O. Pulci[5], R. Del Sole[5]

[1]*INFM-S3 and Universita' di Modena e Reggio Emilia-DISMI, Reggio Emilia, ITALY*
[2]*INFM-Coherentia and Universita' di Napoli "Federico II" - Dipartimento di Scienze FisicheNapoli, ITALY*
[3]*INFM-S3 and Universita' di Modena e Reggio Emilia – Dipartimento di Fisica, Modena, ITALY*
[4]*INFM and Universita' di Milano, Dipartimento di Fisica, Milano, ITALY*
[5]*INFM and Universita' di Roma "Tor Vergata", Roma, ITALY*

Abstract. The structural, optical and electronic properties of silicon nanocrystals are investigated as a function of the dimension as well as the surface passivation. Both the ground- and an excited-state configuration are studied using *ab-initio* calculations. Atom relaxation under excitation is taken into account and related with the experimentally observed Stokes shift.

INTRODUCTION

The possibility of obtaining light emission from nanosized materials and tuning their response as a function of size, has been one of the most challenging aspects of nanotechnology. Among the many, relevant results we can cite the efficient, visible light emission from porous silicon and the optical gain from silicon nanocrystals [1].

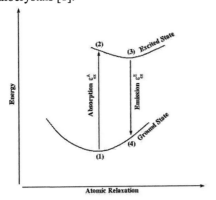

FIGURE 1. A schematic picture of the excitation-deexcitation path.

Most of the commonly observed features in the optical response of such systems are affected by excited-state properties. The investigation of excited nanocrystals by first-principle calculations can be done within several schemes; many controversial points, relative to the range of applicability of such methods as well as the meaning of their respective results, still exist [2]. The aim of this paper is showing a systematic analysis of the structural, electronic and optical properties of silicon nanocrystals, pointing out the main effects related to the system size as well as its surface passivation both in the ground- and in an excited-state configuration.

RESULTS

A pair-excitation energy model is shown in Fig. 1. Within such a model, the absorption/emission path is schematised as a four-level process that gives origin to the commonly observed Stokes shift (S.S.). The total energy calculations have been done in the framework of the density functional theory (DFT) using gradient-corrected Perdew-Burke-Ernzerhof (GGA-PBE) exchange-correlation functional (for details see ref. [2]).

The GGA-PBE calculated absorption gaps have been compared, for some cases, with those obtained within LDA using the independent particle random phase approximation (IPRPA) with or without local fields [3,4] and with GW self-energy corrections.

We consider the hydrogenated SiH_4, Si_5H_{12}, $Si_{10}H_{16}$, $Si_{29}H_{36}$ and $Si_{35}H_{36}$ and the oxidized $Si_{10}H_{14}O$ and $Si_{29}H_{34}O$ nanocrystals with the oxygen both double bonded (DB) or in a bridge (BB) configuration.

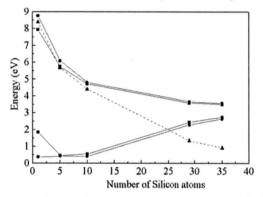

FIGURE 2. The absorption, emission energies and Stokes shift as a function of the number of Si atoms in hydrogenated nanoclusters.

Figure 2 shows, for the hydrogenated clusters, the absorption (upper curves) and emission (lower curves) energies as a function of the number of atoms in the system calculated from both pair-excitation energies (circles, corresponding to $E_{abs}=E_2-E_1$ and $E_{em}=E_3-E_4$ respectively) and HOMO-LUMO E_{H-L} gaps (squares). We see the reduction of the absorption gap with size. Moreover, it is evident that relaxation in presence of the e-h pair significantly modifies such a gap, explaining why emission occurs at lower energies than absorption. The dashed line (triangles) instead represents the S.S., obtained as the absorption-emission difference $E_{abs}-E_{em}$; it is a decreasing function of the clusters size. Being the S.S. a consequence of the different geometries in the ground- and excited-states, the decreasing trend with size is due to the fact that the excitation induces structural changes larger for small clusters.

We have found, instead, in the case of oxidized clusters, that the double bonded oxygen, contrary to the bridge-bond configurations, creates strongly localised states within the gap that induce size independent absorption and emission energies and consequently a size independent S.S. (see tab.1).

TABLE 1. Calculated absorption, emission energies and Stokes shift for the Si_{10} and Si_{29} oxidized clusters.

	$Si_{10}H_{14}O$		$Si_{29}H_{34}O$	
	DB	BB	DB	BB
E_{abs}	2.79	4.03	2.82	3.29
E_{em}	1.09	0.13	1.17	3.01
S.S.	1.70	3.90	1.65	0.28

A good agreement between the GGA-PBE and LDA HOMO-LUMO gaps for all the clusters studied has been found (see Table 2). In the case of SiH_4 and Si_5H_{12}, we have also calculated the GW quasiparticle corrections to the DFT gap [4]. This gap is, as expected, much larger of the E_{H-L} gap, since the excitonic effects are not included. In order to estimate the excitonic effects, we have also performed a TDLDA and LDA-IPRPA calculation for the optical spectra, with local fields effects included. The resulting gaps are also listed in Table 2. The agreement with the experiments is good.

TABLE 2. Calculated vs experimental values for the absorption gaps of SiH_4, Si_5H_{12} and $Si_{10}H_{16}$ clusters.

	Si_1H_4	Si_5H_{12}	$Si_{10}H_{16}$
E_{H-L} PBE	7.93	5.75	4.71
E_{H-L} LDA[3]	7.81		4.59
E_{H-L} LDA[4]	7.91	5.77	
GW	12.3	9.6	
RPA + LF	7.8	6.5	
TDLDA	7.8	6.5	
EXP.	8.8	6.5	

There is also a substantial agreement between our results and previous calculations for both the DFT ground state energies and Stokes shifts (for details see Ref. 2).

CONCLUSIONS

The ground- and excited-state properties of silicon nanocrystals have been calculated. The role of size and surface passivation have been pointed out, showing completely different features when one or more hydrogen atoms are replaced by oxygen at the surface.

ACKNOWLEDGMENTS

Financial support by INFM-PAIS CELEX and MIUR COFIN-PRIN 2002 is acknowledged. Thanks to the "Iniziativa Trasversale Calcolo Parallelo dell' INFM" for the advanced computing facilities.

REFERENCES

1. S. Ossicini, L. Pavesi. F. Priolo, *"Light Emitting Silicon for Microphotonics"*, Springer Tracts on Modern Physics 194 (Springer Verlag, Berlin 2003).
2. E. Degoli. G. Cantele, E. Luppi, R. Magri, D. Ninno, O. Bisi, S. Ossicini, Phys. Rev. B **69**, 155411 (2004)
3. M.Gatti, Tesi di Laurea Universita' di Milano.
4. DFT-LDA calculation performed through the FHI code; M. Bockstedte, A. Kley, J. Neugebauer, and M. Scheffler,Comp. Phys. Comm **107**, 187 (1997).

Preparation of surface controlled silicon nanocrystallites by pulsed laser ablation

Ikurou Umezu[*], Mitsuru Inada[†], Toshiharu Makino[†], Akira Sugimura[*]

[*]Konan University, Department of Physics, Kobe 658-8501, Japan
[†]Konan University, High technology Research Center, Kobe 658-8501, Japan

Abstract. We attempted to prepare surface controlled silicon nanocrystallites by two methods. One method is two step dry process based on plasma surface treatment and the other approach is simple one step process based on reactive pulsed laser ablation. The nanoparticles prepared in hydrogen gas are not an alloy of Si and hydrogen but Si nanocrystal covered by hydrogen or hydrogenated Si. We found that the pulsed laser ablation of Si target in hydrogen background gas is a simple one step dry process to prepare surface passivated Si nanocrystal.

INTRODUCTION

It is well known that optical properties of Si nanocrystal are governed by quantum confinement effect and surface effect. Although the effect of the surface is very important, there has been few reports on the surface effect. One of the reasons is that the control of the surface is difficult. We prepared Si nanocrystallites by pulsed laser ablation (PLA) and attempted surface control to the nanocrystallites. The sample prepared by PLA has advantages since PLA is a clean process and nanocrystal has bare surface.[1,2] Further more, PLA has unique deposition processes compared with conventional physical vapor deposition methods. Plume and shockwave induced by strong laser pulse during deposition gives unique crystallization processes.

In the present paper, we examined two approaches to modify surface of Si nanocrysal. One is post deposition two step process. This method takes an advantage of bare surface. Bare surface was modified by post deposition plasma process. The other is single step process which utilize nature of reactive pulsed laser ablation. We found that unique deposition process of PLA can be utilized to prepare surface passivated Si nanocrystallites.

EXPERIMENT

After the deposition of Si nanocrystalltes by PLA in helium gas, the surface of Si nanocrystal was irradiated by nitrogen or hydrogen plasma produced by helicon wave radical gun. The radical reached to the surface of nanocrystallites will modify the surface of nanocrystal. This is our two step dry process.

Although a concept of two step process is simple, experimental procedure is not simple. We attempted PLA in hydrogen gas as one step process and found that this is a promising method to prepare surface passivated silicon nanocrystal.

RESULT AND DISCUSSION

The number of Si-N and Si-H bonds was estimated by IR measurement after the two step surface modification. We found that number of Si-H or Si-N bonds increased with irradiation time of hydrogen and nitrogen plasma. The native oxidation was suppressed by these treatments. These results indicate that our two step process is effective technique for surface passivation.

The one step process utilize reactive PLA, that is PLA of Si target in hydrogen gas. The nanocrystallites prepared by this method may be either silicon-hydrogen alloy or silicon nanocrystallites covered by Si-H bond. The structure of the resulting nanocrystallites prepared by one step process was measured by infrared absorption, Raman scattering and TEM. The Raman scattering peak corresponding to crystalline Si was observed and amount of amorphous phase was little as shown in Fig.1. Diameter of nanocrystal observed by TEM was about 5nm. The infrared absorption peaks corresponding to Si-H_2 bonds on the Si surface were observed at 2110 cm^{-1} as shown in Fig.2.[3,4] The hydrogen content estimated by infrared absorption corresponds to the value when 5nm nanocrystal is covered by hydrogen. These results indicate that nanoparticles are not an alloy of Si and hydrogen but Si nanocrystal covered by hydrogen or hydrogenated Si; that is the surface passivated Si nanocrystal can be prepared by this one step process.

The plume excited by incident laser beam is confined by ambient hydrogen gas and form shock wave. These processes are differences with conventional physical vapor deposition techniques. We found that the pulsed laser ablation of Si target in hydrogen background gas is a simple one step dry process to prepare core-shell structured Si nanocrystal. The pulsed laser ablation in reactive gas is a promising method to prepare surface controlled nanocrystallites.

CONCLUSIONS

We succeeded to prepare surface controlled silicon nanocrystallites. One is two step process which utilize plasma treatment and the other is one step process which utilize reactive PLA. The one step process is very simple and effective method to prepare surface passivated Si nanocrystal.

ACKNOWLEDGMENTS

This work was partially supported by a Grant-in-Aid for Scientific Research from the Japan Society for the Promotion of Science, Nippon Sheet Glass Foundation for Materials Science and Engineering, The Kurata Memorial Hitachi Science and Technology Foundation, and The Hirao Taro Foundation of the Konan University Association for Academic Research.

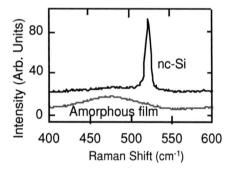

FIGURE 1. Raman spectrum of the sample prepared by one step process. This figure indicates that this material has crystalline structure. The spectra of amorphous Si film is also shown as a reference.

REFERENCES

1. Inada, M., Nakagawa, H., Umezu, I., and Sugimura, A., , *Appl. Surf. Sci.,* **197-198,** 666-669 (2002).

1. Yamada, Y., Orii, T., Umezu, I. , Takeyama, S., and Yoshida, T., *Jpn. J. Apl. Phys.* **35,** 1361-1365 (1996)

2. Shinohara, M., Seyama, A., Kimura, Y., Niwano, M., and Saito, M., *Phys. Rev. B* **65**, 075319-075326 (2002).

4. Lowe-Webb, R. R., Lee, H., Ewing, J. B., Collins, S. R., Yang, W., and Sercel, P. C., *J. Appl. Phys.* **83**, 2815-2819(1998).

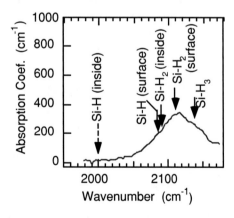

FIGURE 2. Infrared absorption spectrum of the sample prepared by one step process. The arrows in the figure are possible vibration mode.

Light Emission from GaN Microcrystals

R. Garcia[*], A. Bell[*], A. C. Thomas[§], and F. A. Ponce[*]

[*] *Department of Physics and Astronomy, Arizona State University, Tempe, AZ 85287-1504*
[§] *Rogers Corporation, Durel Division, Chandler, AZ 85224-6155*

Abstract. We have grown high quality undoped and n-type GaN crystallites by a novel technique based on direct reaction of gallium metal with ammonia using a two step method. Powders produced by this method consist of at least two differently shaped crystallites; large columnar crystals sized around 10μm and small platelets crystals between 1 and 3μm. The crystallites have a well defined wurtzite structure, with an exceptionally strong near band-edge emission at around 3.342 eV. Yellow luminescence (YL) has been observed in Si-doped and O-doped powders but not in undoped powders grown by this method.

INTRODUCTION

Because GaN is a very robust material with similar structural and electronic properties to ZnS, it is a good candidate for EL devices. An important step towards producing GaN EL powders is to achieve very well controlled n- and p-type doping[1,2,3]. In this work, we have achieved n-type Si doped GaN powders using an innovative synthesis technique. This technique has a high control over the concentration of Si in the final product based in the reactions:

2 Ga(l) + 2 NH$_3(g)$ → 2 GaN(s) + 3 H$_2(g)$ and
2 Ga-Si(l) + 2 NH$_3(g)$ → 2 GaN:Si(s) + 3 H$_2(g)$

These methods lead to highly crystalline, highly luminescent GaN powders. These superior properties are attributed to an intermediate step that dissolves ultra-high purity ammonia in the metal. The dissolved ammonia then leads to a porous melt and to the full reaction, yielding high purity crystalline GaN powders with a stoichiometric nitrogen concentration and a hexagonal wurtzite structure.

EXPERIMENTAL DETAILS

GaN powders with extraordinary luminescent characteristics have been obtained by a complete reaction between gallium metal and ammonia in a horizontal quartz tube reactor at 1100 °C. Si doped GaN powder was produced using a Ga-Si alloy as a precursor. A highly homogeneous Ga-Si alloy is prepared using ultra-high purity metals, in a ball-mill as a mechanical mixer. The alloy is placed in a boat and put inside a horizontal quartz tube reactor (see Fig. 1). When the central portion of the reactor reaches 1200°C the vacuum system is closed and a current of ammonia (~350 sccm) is flowed through the reactor. As steady-state conditions are being approached, an alloy-ammonia solution is formed, after that the boat with the solution is moved to the middle of the reactor. The reaction reaches completion in about 10 hours, then the boat with product, GaN:Si powder, is moved to the coldest part of the reactor.

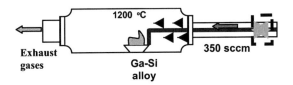

FIGURE 1. Schematic of the horizontal quartz tube reactor

CHARACTERIZATION

A typical SEM image of the powder is shown in Fig. 2, the morphology of the particles are polyhedrons presenting mainly small hexagonal platelets with a narrow size distribution between 1 and 3 μm and big columnar crystals around 10 μm. X-ray diffraction pattern shows a hexagonal (type wurtzite) crystalline structure.

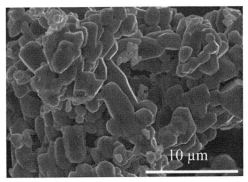

FIGURE 2. SEM image of GaN:Si powders

Room temperature PL spectra of pure GaN, GaN:O and GaN:Si powders are shown in Fig. 3. The spectra were taken using the same conditions, with a He-Cd laser (325 nm) as excitation source.

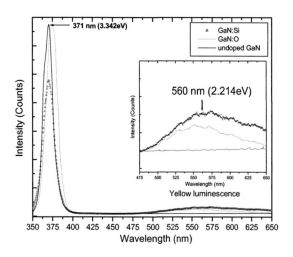

FIGURE 3. Room temperature PL spectra of undoped and n-type doped GaN powders.

Room temperature luminescence shows strong near band-edge emission at around 3.342 eV (371 nm). The FWHM of the 3.342 eV peak is 120meV which is due to impurities and phonon replicas. As shown in Fig. 3, YL is not present in our undoped GaN microcrystals, however YL is emitted by the Si- and O-doped GaN powders. This agrees with Neugebauer and Van de Walle[4] in that the origin of the YL band is best explained by a transition between shallow donors (Si_{Ga} and O_N) and a deep level located in the lower half of band gap, such as V_{Ga}-Si_{Ga} and V_{Ga}-O_N complexes. In PL measurements taken at 10K (Fig. 4), the peak at 3.47eV is the donor bound exciton typically observed in GaN thin films[5]. The 3.42eV peak has previously been attributed separately to oxygen and stacking fault emission[6, 7]. No stacking faults have been observed in these powders, so this peak is likely to be related to oxygen. CL images taken at 3.42eV show that the emission is localized at small regions in the crystallites. This suggests that the oxygen is incorporated only in specific regions of the crystallites. Also observed in the spectra of the fig. 4 is a donor acceptor pair band at 3.30eV and several phonon replicas at 3.18eV, 3.10eV, 3.02eV, 2.94eV and 2.86eV.

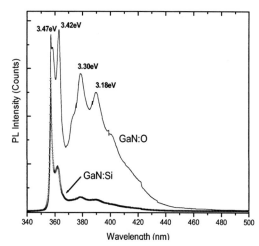

FIGURE 4. He-temperature PL spectra of GaN:Si and GaN:O powders.

CONCLUSIONS

A remarkable feature of our highly luminescent undoped GaN micro-crystals is that they do not show yellow luminescence. However the powder doped with silicon and/or oxygen show a broad yellow emission centered around 560 nm (2.214eV).

ACKNOWLEDGMENTS

The authors gratefully acknowledge technical support by David Wright of the Center for Solid State Science. This research has been partially supported by the DOE Materials Corridor Program, and Rogers Corporation, Durel Division.

REFERENCES

1. Ponce F. A., and Bour D. P., *Nature*, **386**, 351 (1997).
2. Nakamura S., *Science*, **281**, 956 (1998).
3. Strite S., and Morkoc H., *J. Vac. Technol. B.*, **10**, 1237 (1992).
4. Neugebauer J., and Van de Walle C. G., *Appl. Phys. Lett.*, **69**, 503 (1996).
5. Monemar B., *J. Phys.: Condens. Matter* **13**, 7011 (2001).
6. Chung B. –C., and Gershenzon M., *J. Appl. Phys.*, **72**, 651 (1992).
7. Rebane Y. T., Shreter Y. G., and Albrecht M., *Phys. Stat. Sol.*, **164**, 141 (1997).

Quantum Dot Emission from Selectively-Grown InGaN/GaN Micropyramid Arrays

R.A. Taylor, J.H. Rice, J.H. Na, J.W. Robinson

Clarendon Laboratory, University of Oxford, Parks Road, Oxford, OX1 3PU, UK

R.W. Martin, P.R. Edwards

Department of Physics, University of Strathclyde, 107 Rottenrow, Glasgow, G4 0NG, UK

I.M. Watson, C. Liu

Institute of Photonics, University of Strathclyde, 106 Rottenrow, Glasgow, G4 0NW, UK

Abstract. $In_xGa_{1-x}N$ quantum dots have been fabricated by the selective growth of GaN micropyramid arrays topped with InGaN/GaN quantum wells. The spatially-, spectrally-, and time-resolved emission properties of these structures were measured using cathodoluminescence hyper-spectral imaging and low-temperature microphotoluminescence spectroscopy. The presence of InGaN quantum dots was confirmed directly by the observation of sharp peaks in the emission spectrum at the pyramid apices. These luminescence peaks exhibit decay lifetimes of approximately 0.5 ns, with linewidths down to 650 μeV (limited by the spectrometer resolution).

INTRODUCTION

Most III-nitride-based quantum dot (QD) systems reported to date [1,2] have consisted of QDs that are distributed randomly across the sample area. This is a consequence of using the Stranski-Krastanov growth method. Some applications, by contrast, require the precise positioning of QDs on a substrate. Previous studies involving the selective growth of InGaN/GaN nanostructures have reported the presence of QDs [3,4], but no sharp-line QD luminescence was observed.

EXPERIMENTAL METHOD

Arrays of site-controlled GaN micropyramids were formed by MOCVD growth of GaN through an array of circular holes lithographically patterned into a silica mask on a GaN-on-sapphire substrate. The crystal symmetry of the Wurzite GaN results in the formation of hexagonal-based pyramidal structures, with facets formed by $\{10\bar{1}1\}$ planes. The structures were ~8 μm in height, with base edges of ~5 μm, and were repeated in a hexagonal array with 10 μm pitch to give an areal density of 2×10^6 cm^{-2}. A 5-period $In_xGa_{1-x}N$ quantum well (QW) structure was subsequently grown on the top surface in order to form QDs at the sharp apices.

Cathodoluminescence (CL) hyper-spectral imaging was carried out in a modified Cameca SX100 electron probe microanalyzer. Spatially-resolved time-integrated photoluminescence (PL) measurements and time-resolved single photon counting PL measurements were made using a frequency-tripled Ti:Sapphire laser at 266 nm (100 fs pulse width) and a temperature-controlled microscope cryostat. A cooled

CCD camera was employed for the time-integrated measurements. For the time-resolved PL measurements, a time-correlated single photon counting detection system based on a fast photomultiplier tube was used, giving a time resolution of 150 ps.

RESULTS

Fig. 1(a) displays a secondary electron (SE) image of the sample surface (70° tilt), showing the micropyramid array. Fig. 1(b) shows an image of the intensity of the room-temperature QW CL peak emitted by a single micropyramid, extracted from the hyper-spectral image. The peak of the QW luminescence occurs near 2.8 eV. This luminescence is seen to emanate from the entire pyramidal surface with the exception of the apex, at which position it is less intense.

The spatially-resolved PL spectrum obtained from the tip of a single pyramidal structure at 4.2 K is shown in Fig. 1(c). The PL spectrum from the pyramid apex reveals sharp peaks characteristic of localized states observed in PL from single InGaN QDs. The spectrum exhibits one strong peak and four weaker peaks; it is unclear whether the weaker peaks are due to higher-order processes or to exciton transitions in the other QDs expected to be present at the pyramid tip. The FWHM of the peaks was found to be as small as 650 μeV (limited by the resolution of the spectrometer).

Fig. 1(d) shows a micro-PL spectrum from a single InGaN QD, showing a sharp peak (marked α) superimposed upon a strong background. This background originates from excitation of the side QW. It is possible to use the time-resolved traces on each side of the QD emission [at β and γ shown in Fig. 1(d)] to correct for the QW emission collected at the same spectral position as the QD emission. Subtraction of this background enables the decay profile from the single InGaN QD to be determined, as shown in Fig. 1(e). The trace from the QD is non-exponential, and closely matches that observed from the side QW. The decay time was found to be ~0.50 ns, which is much shorter than the lifetimes of ~2 ns reported for InGaN QDs grown by self-assembly techniques [5]. The shorter exciton recombination time in selectively-grown QDs is believed to be due to the stronger coupling of the zero-dimensional QD state to the side QWs in this system, in contrast to self-assembled nanostructures where the QD is relatively decoupled from an underlying wetting layer.

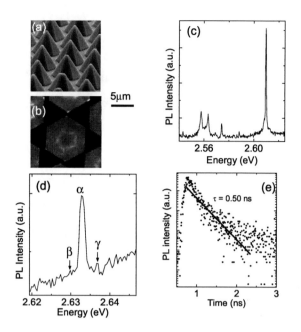

FIGURE 1. (a) SE and (b) CL intensity images of the micro-pyramids. (c) Sharp PL peaks from a single QD. (d) PL peak from a single QD and (e) its time-resolved trace following background subtraction.

ACKNOWLEDGMENTS

The authors thank H.M.H. Chong and R.M. De La Rue of the University of Glasgow for the lithographic patterning. This work was supported by EPSRC.

REFERENCES

1. K. Tachibana, T. Someya, and Y. Arakawa, *Appl. Phys. Lett.* **74**, 383 (1999).

2. R. A. Oliver, G. A. D. Briggs, M. J. Kappers, C. J. Humphreys, J. H. Rice, J. D. Smith, R. A. Taylor, *Appl. Phys. Lett.* **83**, 755 (2003).

3. J. Wang, M. Nozaki, M. Lachab, Y. Ishikawa, R. S. Qhalid Fareed, T. Wang, M. Hao, and S. Sakai, *Appl. Phys. Lett.* **75**, 950 (1999).

4. K. Tachibana, T. Someya, S. Ishida, and Y. Arakawa, *Appl. Phys. Lett.* **76**, 3212 (2000).

5. J. W. Robinson, J. H. Rice, A. Jarjour, J. D. Smith, R. A. Taylor, R. A. Oliver, G. A. D. Briggs, M. J. Kappers, C. J. Humphreys, and Y. Arakawa, *Appl. Phys. Lett.* **83**, 2674 (2003).

Considerable enhancement of PL intensity for free exciton in ZnO ultra-fine-particles and oscillatory green band emergence in PL spectrum for Cu-doped ZnO ultra-fine-particles with heat treatment

H. Kobori, T. Nanao, S. Kawaguchi, I. Umezu and A. Sugimura

Department of Physics, Faculty of Science and Engineering, Konan University
Okamoto 8-9-1, Higashinada-ku, Kobe, Hyogo 658-8501, Japan

Abstract. For the ultra-fine-particles of ZnO with the heat treatment and Cu-doping, we have observed the considerable enhancement of the PL intensity for free excitons and the oscillatory structure of the green band. These experimental results are explained by the localization model of excitons and the donor to Cu-acceptor recombination, respectively.

INTRODUCTION

Zinc oxide is a II-VI compound semiconductor with a direct wide-band-gap (3.4eV) and an wurtzite-type structure. It has aroused considerable interest [1] as an alternative to GaN for use in opto-electronic devices. The ultra-fine-particles (UFP) of ZnO, furthermore, have great potentialities as the materials of the exciton-related opto-electronic device, because of the expectation that the efficiency of the light emission increases due to the confinement of excitons. In this paper, for the ultra-fine-particles of ZnO with the heat treatment (HT) and Cu-doping, we have reported the considerable enhancement of the PL intensity for the free exciton (FE) and the oscillatory structure of the green band. These experimental results are explained by the localization model of excitons and the donor to Cu-acceptor recombination, respectively.

EXPERIMENTAL

We have employed two kinds of the UFP of ZnO whose average diameters are 70nm and 100nm, which are designated as ZnO-70nm and ZnO-100nm, respectively. For comparison, the bulk crystal of ZnO is also employed in this experiment, which is designated as ZnO-Bulk. To investigate the effect of the HT on the UFP of ZnO, they are placed in air at 800°C for 3 hours. In addition, to investigate the influence of Cu-impurity in the UFP of ZnO, we have doped these samples with Cu by the HT of them on the Cu-plate in air at 800°C for 3 hours. As the excitation source, we have used a He-Cd laser with the wavelength of 325nm (3.81eV) and the excitation power of ~70mW/cm^2. The temperature of the samples is varied from 13K to 290K by use of the closed-cycle He gas refrigerator.

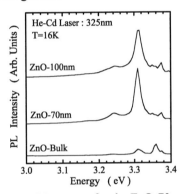

FIGURE 1. PL spectra for the ZnO-70nm, ZnO-100nm and ZnO-Bulk without the HT at 16K.

RESULTS & DISCUSSIONS

Figure 1 shows the PL spectra for the ZnO-70nm, ZnO-100nm and ZnO-Bulk without the HT at 16K. The peaks at 3.375eV, 3.311eV, 3.241eV and 3.171eV are considered to correspond to the FE, the FE associated with the one LO (FE-1LO), two LO (FE-2LO) and three LO (FE-3LO) phonons, respectively. The peaks at 3.350eV and 3.360eV are considered to correspond to the bound exciton (BE). As being pointed out in recent reports [2], we have also observed that the peak for FE-1LO remains up to the room temperature. The considerable enhancement of the PL intensity for the FE and the FE-nLO (n=1.2.3) in ZnO-70nm and ZnO-100nm has been observed as compared with that of the ZnO-Bulk. The ratio of the PL intensity of the FE-1LO for ZnO-70nm to that of ZnO-Bulk is about to ten at 16K. In addition, with the decrease of the average particle size, the increase of the PL intensity has been found. The effective Bohr radius of the FE in ZnO is about 1.5nm with the binding energy of 60meV and the dielectric constant of ~8.1. Although the exciton confinement region (ECR) in the UFP is smaller than the actual size of the UFP because of the crystal imperfection around the surface of them, the ECR is generally thought to be sufficiently larger than the effective Bohr radius of the FE in ZnO. Therefore it is difficult to consider that the considerable enhancement of the PL intensity for the FE and the FE-nLO (n=1.2.3) in ZnO-70nm and ZnO-100nm is due to the quantum confinement of the FE. As compared with ZnO-Bulk, in ZnO-70nm and ZnO-100nm, we consider that there exist more crystal defects, dislocations and impurities. Electrons and holes excited by the He-Cd laser light (3.81ev) could be localized within the extent of the localization length owing to the scatterings by the randomly configured crystal defects, dislocations and impurities. We infer that the localization length is much less than the average particle sizes (70nm and 100nm), and consequently the FE is more efficiently generated and recombined. Figure 2 shows the PL spectra at 16K for ZnO-70nm without the HT, with the HT and with Cu-doping by the HT, respectively. The several peaks between 1.6eV and 1.7eV are thought to come from deep impurities, such as the transition metals, deep defects and so on. For ZnO-70nm and ZnO-100nm, without and with the HT, we have not observed the green band, which is the broad peak frequently observed between 2.0eV and 2.7eV in ZnO. (We have, however, observed the green band for ZnO-Bulk.) As seen in this figure, by doping the ZnO-70nm and ZnO-100nm with Cu through the HT, the green band has clearly emerged for the PL spectra. Figure 3 shows the PL spectra regarding the green band at 16K for ZnO-70nm and ZnO-100nm with Cu-doping by the HT for the energy between 1.7V and 2.9eV. The oscillatory structure has been found in the green band, especially for ZnO-100nm. The energy of the interval between adjacent peaks is about 72meV, which is equal to the

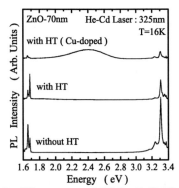

FIGURE 2. PL spectra at 16K for ZnO-70nm without the HT, with the HT and with Cu-doping by the HT.

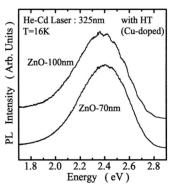

FIGURE 3. PL spectra regarding the green band at 16K for ZnO-70nm and ZnO-100nm with Cu-doping by the HT.

energy of an LO phonon. The cause of the appearance of the green band has been presented in several papers; 1) donor-acceptor pair (DAP) recombination, where Cu is given as a singly ionized acceptor in ZnO [2], and 2) the transition from deep donor level by oxygen vacancies in ZnO to valence band [3]. From our experimental results, since the PL spectrum for the UFP of ZnO with the HT without Cu-doping has not shown the green band, but those with the HT and Cu-doping have shown that, we consider that the green band results from the donor to Cu-acceptor recombination. The oscillatory structure is considered to result from the LO-phonon replica with the DAP recombination.

CONCLUSIONS

Electrons and holes in the UFP of ZnO excited by the He-Cd laser light could be localized within the extent of the localization length. We consider that the localization length is much less than the average particle sizes (70nm and 100nm) of them. Consequently the excitons are more efficiently generated and recombined. Thus the PL intensity of excitons considerably increases as compared with that of the bulk-ZnO. We consider that the green band results from the donor to Cu-acceptor recombination. The oscillatory structure is inferred to come from the LO-phonon replica with the DAP recombination.

REFERENCES

1. Reynolds D. C., Look D. C., Jogai B. and Collins T. C., *Appl. Phys. Lett.*, **93**, pp. 3994-3796 (2001).
2. Garces N. Y., Wang L., Bai L., Giles N. C., and Halliburton L. E. and Cantwell G., *Appl. Phys. Lett.*, **93**, pp. 622-624 (2002).
3. Kang, H. S., Kang, J. S., Kim J. W., and Lee S. Y., *J. Appl. Phys.*, **95**, pp. 1246-3217 (2004).

Charged excitons in modulation-doped quantum wires

T. Otterburg, D.Y. Oberli, M.-A. Dupertuis, N. Moret, A. Malko, E. Pelucchi, B. Dwir, E. Kapon

Laboratory of Physics of Nanostructures, Ecole Polytechnique Fédérale de Lausanne, 1015 Lausanne, Switzerland

Abstract. We report on the observation of negatively- and positively-charged excitons in the photoluminescence spectra of V-groove quantum wires. The charged exciton binding energy increases with the strength of the quantum confinement. We demonstrate that fluctuations of the confinement potential cause the localization of the exciton and of the charged exitons on the same location. We discover that a large fraction of the enhancement of the charged exciton "binding energies" has a kinetic origin associated with the recoil energy transferred to the remaining carrier during the emission process.

INTRODUCTION

Observation of charged excitons in the photoluminescence (PL) spectra of semiconductor quantum wires has until recently been hindered by inhomogeneous line broadening due to structural imperfections in high quality samples. A negatively charged exciton (X^-) was identified, however, in a micro-photoluminescence study of a T-shaped QWR of improved quality [1]. This observation was made possible by the reduction of the PL linewidth to about 1 meV. In order to free oneself from the inhomogeneities of the wires, which are present on a length scale of 1 to 10 µm, we prepared samples with submicron apertures made through an opaque Aluminium film. Previous optical studies, which were performed with this technique on similar and undoped QWR samples, have evidenced sharp PL lines that were ascribed to excitons recombining in the local potential minima created by structural disorder at the interface [2]. In this paper we study the PL spectra of n-type modulation-doped GaAs/AlGaAs V-groove quantum wires (QWRs), in which the carrier density was tuned by applying a potential on a Schottky gate.

RESULTS AND DISCUSSION

The samples were grown by organometallic chemical vapor deposition on [100] oriented and n-type doped substrates that were patterned. The patterns were made by electron beam lithography and consisted of lateral arrays of V-grooves with a 15 µm pitch and a depth of about 0.5 µm. For this study we designed modulation-doped structures in which the QWR was separated from the Si doped layers by spacers of 38 and 105 nm width. The thickness of the crescent-shaped QWRs (measured at the center) was either 3 nm or 7.5 nm. The samples are later referred to as *thin* or *thick* QWRs depending on this thickness. The samples were processed in mesas covered with a Schottky gate and an Al film in which submicron apertures were opened.

Typical PL spectra of the samples containing either a thin or a thick QWR are displayed in Fig. 1 for a given voltage applied on a Schottky gate. These spectra were obtained at a temperature of 10 K by exciting the sample through a 0.4 µm aperture with a laser energy of 2.41 eV and a power density of 100 W/cm^2. At a large enough negative gate voltage the QWR is completely depleted of electrons and the spectrum mainly consists of a sharp line that is assigned to a neutral exciton. At increasing gate voltage, a second sharp line emerges on the low energy side of the exciton. This line is ascribed to a negatively charged exciton, which is formed by an electron binding to a neutral exciton. At a gate voltage approaching 0V the neutral exciton line disappears from the PL spectrum and only the charged exciton line remains. The spectral separation of the doublet, often referred to as the "binding energy" of the charged exciton amounts to 4.2 meV and 2.8 meV, respectively for the thin and thick quantum wire samples. These values correspond to the average value of the binding energies that were measured on these samples over many different apertures. Consequently increasing the strength of the quantum confinement induces a significant enhancement of the "binding energies". The voltage dependence of PL spectra demonstrates that the low energy component of the doublet is a negatively charged exciton.

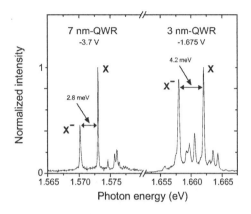

FIGURE 1. Photoluminescence spectra of the thick and thin modulation-doped QWR samples measured at 10 K through a submicron aperture. The gate voltage is tuned to observe the two transition lines associated with X and X^-.

The localized character of the charged exciton was proven by the analysis of the X^- lineshape. We found that the X^- spectral line in the temperature range of 10 to 40 K was very well fitted by assuming a Lorentzian lineshape. Theory predicts that the line shape of a charged exciton is asymmetrical when it is free [3]. From the observation of a symmetrical line shape, we conclude that the charged exciton is localized below 40 K.

The evidence for the emission of the positively-charged exciton (X^+) was obtained by choosing an applied bias that depleted the QWR of electrons and by increasing the laser excitation density. Surprisingly, the "binding energy" of the X^+ was smaller than that of the X^-, which contradicted our expectation based on the theoretical calculation of the charged excitons binding energies in QWRs. We also performed time-correlated measurements between photons emitted by the exciton (X) and the two charged excitons (X^- and X^+) in order to demonstrate that the exciton complexes were localized. We found a clear signature for photon antibunching for all three excitonic complexes in the auto-correlation curves of each associated emission line. This demonstrates that single photon emission occurs and that exciton complexes are thus localized. The cross-correlation curves also showed antibunching behavior of the photon emission, which confirmed that X, X^- and X^+ were localized on the same site along the wire.

In order to understand how interfacial disorder affects the "binding energy" of the charged exciton, we assume that the confinement potential varies slowly along the wire and thus that the center-of-mass (COM) motion of the exciton can be decoupled from its relative motion. The existence of localization introduces then an important correction to the binding energies of the charged excitons: this is a contribution of the confinement energies in the COM of X and $X^{+/-}$ and of the recoil energy of the charge carrier that remains localized once the emission of the charged exciton takes place. We evaluated this contribution by using a simple square-well description of a local minimum of the disorder potential. We calculated the Coulomb correlation energies of the two charged excitons by solving the three-particle Schrödinger equation describing the relative motion of the charged exciton. We used realistic one-dimensional Coulomb potentials for the attraction and repulsion of the carriers, which were calculated for a given structure. We then found that the contribution of the Coulomb correlation energies to the charged exciton binding energy was too small to account for the measured "binding energies" and that it could not explain the ordering of the optical transitions.

We propose that in the localization regime, the kinetic energy imparted to the remaining charged carrier contributes *for a large fraction* to the "binding energy" of a charged exciton. The optical probing of the charged exciton complexes provides thus a means to quantify the effective parameters describing a local minimum of the disorder potential in our weakly disordered semiconductor quantum wires.

REFERENCES

1. H. Akiyama, L. N. Pfeiffer, A. Pinczuk, K.W. West, *Solid State Commun.* **122**, 169 (2002).

2. F. Vouilloz, D.Y. Oberli, F. Lelarge, B. Dwir, E. Kapon, *Solid State Commun.* **108**, 945 (1998).

3. A. Esser, R. Zimmermann, E. Runge, *Phys. Status Solidi b* **227**, 317 (2001).

Influence of aspect ratio on the lowest states of quantum rods

M. Willatzen*, B. Lassen*, R. Melnik* and L. C. Lew Yan Voon[†]

*Mads Clausen Institute for Product Innovation, University of Southern Denmark, Grundtvigs Allé 150, DK-6400 Sønderborg, Denmark
[†]Department of Physics, Worcester Polytechnic Institute, 100 Institute Road, Worcester, MA 01609

Abstract. The lowest valence-band states of $In_{0.53}Ga_{0.47}As$ quantum rods with infinite barriers are studied using a four-band Burt-Foreman model. Special emphasis is given to the study of quantum-rod shape dependency and consequences for the aspect ratio at the crossing of the lowest two states. The nonseparability of the problem leads to complex ground-state envelope function (and level crossing) and demonstrates the difference between (infinite) quantum-wire structures and finite quantum-rod structures. Finally, calculations are presented for $In_{0.53}Ga_{0.47}As$ quantum-rod structures embedded in InP. It is found that the aspect ratio at crossing of the two lowest states depends on the quantum-rod radius with InP finite barriers.

INTRODUCTION

A resurgence of interest in quantum wires (QWR's) and quantum rods (QR's) has happened recently due in part to new growth techniques for making free-standing and colloidal structures, [1] and also due to the observation of linearly-polarized photoluminescence in InP nanowires [1] and CdSe QR's [2, 3] and the insensitivity of the interband transition energies on the length of the QR's. [2, 4] Evidently, an understanding of and control over these properties is necessary in order to make use of these nanostructures, e.g., as biological labeling and optoelectronic devices. Nevertheless, the theoretical study of these near-band-edge states is still at a primitive stage. Previous results by Hu *et al.* [2], using empirical pseudopotential theory revealed that a level crossing takes place between the two top valence-band states; however, they considered a rather artificial structure with cylindrical quantum rods embedded in a spherical dot. We present in this paper calculations of the energy levels of the lowest valence-band states of cylindrical QR's as a function of the aspect ratio by varying the cylinder rod length keeping the diameter constant. We also verify that a crossing of the lowest two states depends on the choice of barrier material and (for finite barriers in general) on the quantum-rod radius. This is done by computing energy levels of (a) $In_{0.53}Ga_{0.47}As$ quantum-rod structures with infinite barriers, and (b) $In_{0.53}Ga_{0.47}As$ quantum-rod structures with InP finite barriers using the Burt-Foreman model and following the method of Sercel and Vahala [5].

THEORY

The band structure of the cylindrical quantum-rod structures is calculated using a four-band $k \cdot p$ theory within the axial approximation, following the method of Sercel and Vahala. [5] It involves expressing the $k \cdot p$ Hamiltonian in terms of cylindrical polar coordinates ρ, ϕ, z and noting that a good quantum number (due to axial symmetry) is the projection of the total angular momentum (of the envelope function and Bloch state) along the rod axis (labeled z):

$$F_z = L_z + J_z, \qquad (1)$$

where the angular momentum L_z of the envelope function can only take integer values, and the J_z values belong to the $J = \frac{3}{2}$ subspace of heavy-holes (HH's) and light-holes (LH's).

There is a double degeneracy with respect to the sign of F_z due to the presence of time-reversal symmetry and inversion symmetry. Two new features we have introduced into the Sercel–Vahala theory are the use of the Burt-Foreman Hamiltonian [6, 7] (instead of the Luttinger-Kohn one originally used by Sercel and Vahala) and the implementation of confinement. The latter is done exactly as a coupled ρ–z problem, rather than the decoupling assumed by, e.g., Katz *et al.* [4]. The calculations are performed using the finite element method (FEM). Since the FEM is a variational reformulation of the problem interfaces do not need special treatment. Partial differential equations are solved in weak form by integrating equations such that slope discontinuities are captured and the correct result is found even with a fairly sparse grid [8].

NUMERICAL RESULTS AND DISCUSSIONS

Our results reveal that the two lowest energy levels converge to within 1 meV as a function of the quantum-wire length for a given diameter when the former is approximately 6 times larger than the latter. Of course, this value increases for the higher valence-band energy levels. However, if one is interested in the first state only it can be concluded that the quantum rod behaves like an infinite quantum wire above an aspect ratio of approximately 1:6.

This ratio is higher than that given by Katz *et al.* [4] since they did not require the same accuracy in convergence. In actual fact, their resolution is of the order of tens of meV only. Furthermore, the energy difference we find between the result at an aspect ratio of 1:2 and the converged value is approximately 15 meV. Hence, the variation is outside the resolution of the experiment of Katz *et al.* [4].

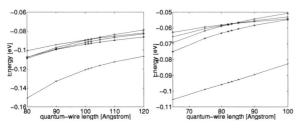

FIGURE 1. Uppermost valence-band structure of $In_{0.53}Ga_{0.47}As$ quantum rods as a function of rod length (all structures correspond to a diameter of 100 Å). Infinite barrier assumption (left) and InP barriers (right).

In Figure 1 (left), we plot the first six energy levels in the valence band near the crossing of the two lowest states slightly above a quantum-wire length of 100 Å in the case with infinite barriers. This value is computed for a diameter of 100 Å, however, our results show that the crossing does not change with the diameter if the aspect ratio is kept constant. In other words, the eigenvalues are related by a constant if the problem is scaled as a whole. In actual fact, inspection of the Hamiltonian and the boundary-conditions reveals this scaling property for infinite barriers. For the range shown [80-120 Å], the variation in energy is seen to be about 20 meV. The plotted curves with line codings: cross, circle, and plus correspond to $|F_z| = 1/2$ while the others (triangle and star) correspond to $|F_z| = 3/2$. Evidently, the nature of the groundstate is $|F_z| = 1/2$ below and above the aspect ratio at crossing.

In Figure 1 (right), we plot the first six energy levels in the valence band near the crossing of the two lowest states in the case with InP barriers for a quantum-rod of radius 50 Å. The aspect ratio at crossing now takes place near 84 Å and the range plotted is from 65 Å to 100 Å. Another significant change with finite InP barriers is that now the uppermost valence-band is a $|F_z| = 3/2$ ($|F_z| = 1/2$) state at aspect ratios below (above) the aspect ratio at crossing. Hence, the presence of a barrier region strongly affects the symmetry and structure of the eigenstates. Again, this complex behaviour is a result of the nonseparability of the problem. In addition, our calculations confirm that the aspect ratio at level crossing depends on the radius of the quantum rod with finite barriers. This is expected theoretically since the mathematical problem for the finite barrier problem does not obey the simple scaling principle applying to the corresponding infinite-barrier problem. Indeed, for a quantum rod of radius 50 (75) Å the aspect ratio at level crossing is 0.84 (0.93).

CONCLUSIONS

An aspect ratio of approximately 1:6 for the convergence of the lowest two states of a QR to within 1 meV is found using FEM calculations based on a Burt-Foreman formulation of the Sercel-Vahala method. A level crossing of the lowest two states is confirmed for our cylindrical QR based upon cubic InGaAs materials. Finite barrier calculations reveal that the aspect ratio at level crossing depends on the radius of the quantum rod and the barrier material. Also, the nature of the two uppermost levels depends strongly on the presence of a barrier material. This is verified by comparing model results obtained using infinite barriers and InP finite barriers.

ACKNOWLEDGMENTS

This work was supported by an NSF CAREER award (NSF Grant No. 9984059) and a Balslev award (Denmark).

REFERENCES

1. J. Wang, M.S. Gudiksen, X. Duan, Y. Cui, and C.M. Lieber, Science 293, 1455 (2001).
2. J. Hu, L.-S. Li, W. Yang, L. Manna, L.-W. Wang, and A.P. Alivisatos, Science 292, 2060 (2001).
3. X.-Y. Wang, J.-Y. Zhang, A. Nazzal, M. Darragh, and M. Xiao, Appl. Phys. Lett. 81, 4829 (2001).
4. D. Katz, T. Wizansky, O. Millo, E. Rothenberg, T. Mokari, and U. Banin, Phys. Rev. Lett. 89, 086801 (2002).
5. P.C. Sercel and K.J. Vahala, Phys. Rev. B 42, 3690 (1990).
6. M.G. Burt, J. Phys. Cond. Matter 4, 6551 (1992).
7. B.A. Foreman, Phys. Rev. B 48, 4964 (1993).
8. R.V.N. Melnik and M. Willatzen, Nanotechnology 15, 1 (2004).

Barrier Localization in the Valence Band of Modulated Nanowires

L. C. Lew Yan Voon*, R. Melnik[†], B. Lassen[†] and M. Willatzen[†]

Department of Physics, Worcester Polytechnic Institute, 100 Institute Road, Worcester, MA 01609
[†]*Mads Clausen Institute for Product Innovation, University of Southern Denmark, Grundtvigs Allé 150, DK-6400 Sønderborg, Denmark*

Abstract. We present evidence that the phenomenon of inversion recently discovered in a one-band model [L. C. Lew Yan Voon and M. Willatzen, J. Appl. Phys. **93**, 9997 (2003)] is much more general and is present in both multiband theories and in the excited states. A critical radius of around 15 Å (7 Å) is obtained for holes in InGaAs/InP (GaAs/AlAs) modulated nanowires.

INTRODUCTION

A new type of nanostructures, so-called nanowire superlattices or modulated nanowires, has recently been grown [1, 2, 3, 4]. Material systems studied so far include GaAs/GaP [1], Si/SiGe [2], InAs/InP [3], and ZnSe/CdSe [4], and typical dimensions have been radii of 200–400 Å. It is speculated that these structures might find applications as nano bar codes, waveguides, lasers, and LEDs, and that nanowire superlattices are an improvement over plain nanowires. Due to the large radii of the structures grown so far, there is not yet any experimental evidence of novel quantum phenomena in these structures. Here, we are reporting a unique kind of wave function localization due to the longitudinal modulation using a multiband $\mathbf{k} \cdot \mathbf{p}$ theory, whereby the bound states are localized in the barrier layer below a so-called critical radius. We had earlier predicted the phenomenon in the conduction band on the basis of a one-band model [5] and studied the impact of symmetry [6, 7, 8].

THEORY

The valence-band structures of the modulated nanowires with zincblende constituents and with an [001] wire axis were calculated using a four-band $\mathbf{k} \cdot \mathbf{p}$ theory within the axial approximation, following the method of Vahala and Sercel [9]. It involves expressing the $\mathbf{k} \cdot \mathbf{p}$ Hamiltonian in terms of cylindrical polar coordinates ρ, ϕ, z and noting that a good quantum number is the projection of the total angular momentum (of the envelope function and Bloch state) along the rod axis (labeled z):

$$F_z = L_z + J_z, \quad (1)$$

where the angular momentum L_z of the envelope function can only take integer values, and the J_z values belong to the $J = \frac{3}{2}$ subspace of heavy-holes (HH's) and light-holes (LH's). The total wave function is then written as

$$\psi(\mathbf{r}) = \sum_{J_z} f_{J_z}(\rho, z) e^{i(F_z - J_z)\phi} |\frac{3}{2} J_z\rangle, \quad (2)$$

where $f_{J_z}(\rho, z)$ are the envelope-function components. Two new features of our theory are the use of the Burt-Foreman Hamiltonian [10] (instead of the Luttinger-Kohn one) and the presence of confinement along the wire axis.

CALCULATIONS AND RESULTS

The coupled differential equations were solved using the finite-element method. We represented the superlattice structure by using a finite number of periods; it turns out that only a couple of periods are necessary for convergence.

Calculations have been carried out for GaAs/AlAs and $In_{0.53}Ga_{0.47}As$/InP modulated nanowires with layer thicknesses of 20–50 Å and a circular cross-section with radii ranging from 5 Å to 15 Å. The sum squared of the envelope function components for two different radii for GaAs/AlAs ($In_{0.53}Ga_{0.47}As$/InP) nanowires with layer widths of 50 Å (20 Å) are shown in Fig. 1 (Figs. 2–3); the occurence of the inversion is evident in both cases (the regions of highest probability densities are the small darkest ones — dark red if in color). In the case of $In_{0.53}Ga_{0.47}As/InP$, Fig. 2 is for the ground state, while Fig. 3 is for the second excited state; it turns out that the critical radius is about the same for the first excited state and the ground state.

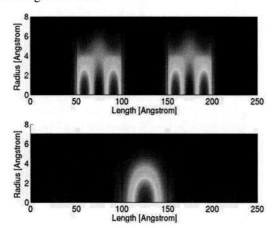

FIGURE 1. Probability density for GaAs/AlAs above and below critical radius.

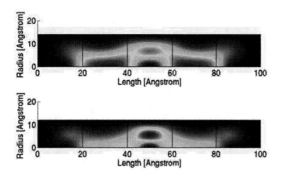

FIGURE 2. Probability densities for the ground state for $In_{0.53}Ga_{0.47}As/InP$ above and below critical radius.

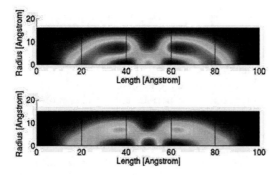

FIGURE 3. Probability densities for the second excited state for $In_{0.53}Ga_{0.47}As/InP$ above and below critical radius.

We have also calculated the critical radii using an analytical one-band model [5]; differences compared to the exact multiband results are of the order of 50%.

SUMMARY

- The existence of critical radii for the inversion of hole states in modulated nanostructures is demonstrated using a multiband calculation.
- There is a dependence of the critical radius on the layer widths.
- Excited states also display critical radii.
- The analytical one-band model is only semi-quantitatively accurate.
- It is proposed to observe the inversion via interband absorption or transport experiments in GaAs/AlAs modulated nanowires of diameters near 40 Å.

ACKNOWLEDGMENTS

This work was supported by an NSF CAREER award (NSF Grant No. 9984059).

REFERENCES

1. Gudiksen, M. S., Lauhon, L. J., Wang, J., Smith, D. C., and Lieber, C. M., *Nature*, **415**, 617–620 (2002).
2. Wu, Y., Fan, R., and Yang, P., *Nano Lett.*, **2**, 83–86 (2002).
3. Björk, M. T., Ohlsson, B. J., Sass, T., Persson, A. I., Thelander, C., Magnusson, M. H., Deppert, K., Wallenberg, L. R., and Samuelson, L., *Nano Lett.*, **2**, 87–89 (2002).
4. Solanki, R., Huo, J., Freeouf, J. L., and Miner, B., *App. Phys. Lett.*, **81**, 3864–3866 (2002).
5. Lew Yan Voon, L. C., and Willatzen, M., *J. App. Phys.*, **93**, 9997–10000 (2003).
6. Melnik, R., Willatzen, M., Lew Yan Voon, L. C., and Galeriu, C., "Finite element analysis of nanowire superlattice structures," in *Proc. ICCSA Montreal*, edited by C. T. V. Kumar, M. Gavrilova and P. L'Ecuyer, ICCS, Springer, 2003, vol. Part II, pp. 755–763.
7. Willatzen, M., Melnik, R. V. N., Galeriu, C., and Lew Yan Voon, L. C., *Mathematics and Computers in Simulations*, **65**, 387–397 (2004).
8. Galeriu, C., Lew Yan Voon, L. C., Melnik, R. N., and Willatzen, M., *Comp. Phys. Commun.*, **157**, 147–159 (2004).
9. Vahala, K. J., and Sercel, P. C., *Phys. Rev. Lett.*, **65**, 239–242 (1990).
10. Burt, M. G., *J. Phys. Cond. Mat.*, **11**, R53–R83 (1999).

High-pressure Pulsed Laser Deposition and Structural Characterization of Zinc Oxide Nanowires

Andreas Rahm[1,4], Thomas Nobis[1], Evgeni M. Kaidashev[1,3], Michael Lorenz[1], Gerald Wagner[2], Jörg Lenzner[1] and Marius Grundmann[1]

[1] *Universität Leipzig, Institut für Experimentelle Physik II, Linnéstraße 5, D-04103 Leipzig, Germany*
[2] *Universität Leipzig, Institut für Mineralogie, Kristallographie und Materialwissenschaft, Scharnhorststraße 20, D-04275 Leipzig, Germany*
[3] *on leave from Rostov-on-Don State University, 344 090 Rostov-on-Don, Russia*
[4] *corresponding author e-mail: andreas.rahm@physik.uni-leipzig.de*

Abstract. Various zinc oxide nanostructures have been grown by high-pressure PLD on gold coated sapphire substrates. Depending on growth parameters, a wide range of geometries is obtained in a controlled fashion. The whisker diameter depends on the target-to-substrate distance. The well aligned epitaxial growth of the nanowires on sapphire without rotational domains is proved by x-ray diffraction and transmission electron microscopy. Our growth mechanism seems to be different from the usually assumed vapor-liquid-solid process, because of the absence of any gold drops on top of the nanowires.

INTRODUCTION

Recently, zinc oxide (ZnO) nanostructures have gained increased interest. Due to its high exciton binding energy and the c-axis dominated growth behavior, this wide band-gap semiconductor is of particular interest for columnar UV nanolasers.

EXPERIMENT

High-pressure pulsed laser deposition (PLD) is used to grow ZnO nanostructures on a- and c-plane sapphire and silicon (111). The PLD chamber consists of a T-shaped quartz tube, which is 30mm in diameter, with a KrF excimer laser beam entering on the middle side of the T. It is focused on a rotating target that can be changed in-situ via a linear feedthrough. Our targets consist of pressed 5N powders sintered for 12 hour at 1150°C on air. Growth temperature varies between 830 and 970°C and the gas partial pressure between 1 and 200 mbar. The process is supported by an argon or oxygen flow from target to substrate. The target-to-substrate distance is varied between 5 and 40 mm. Further details will be published elsewhere.

Alternatively, we use the carbothermal evaporation method [1] at 915°C, with ZnO:C (mass ratio 1:1) pressed powders as source material. Argon is used as carrier gas to transport the zinc and zinc suboxides towards the substrate 1 cm away where they condensate [2].

Prior to the growth by PLD or evaporation, the substrates are covered by a DC-sputtered 1nm thick gold film and parts of the sample are shadowed by a TEM mask. This catalyst is thought to initiate a vapor-liquid-solid (VLS) growth process [3]. Scanning electron microscopy (SEM), high resolution x-ray diffraction (HRXRD), atomic force microscopy (AFM) and transmission electron microscopy (TEM) are used for structural characterization.

RESULTS

Figure 1 shows a uniform array of free-standing parallel nanowires obtained by PLD using a MgZnO target. Similar structures are grown with a pure ZnO source. We find that the diameter of ZnO structures changes from about 100 nm up to 3 microns with decrease of the distance from PLD target to substrate. Furthermore, the morphology of the columns can be modified in an extremely wide range by changing the PLD growth parameters. For instance, reduced laser pulse energy, lowered Ar flow rate or using a Zn enriched target, results in needle-like, branched or dodecagonally caped microcolumnar structures, respectively. A distinctive hexagonal shape is observed when a cerium oxide buffer layer is used. The nanowhiskers show random orientation when grown on Si(111). This behavior can be explained by the formation of an intermediate amorphous silicon oxide layer as observed by HRTEM.

In contrast to the highly flexible PLD process, with our carbothermal method such structural diversity is not observed. Free standing nanowires with diameters around 100 nm are typical.

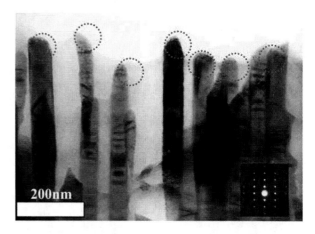

FIGURE 2. TEM bright field image of ZnO whiskers grown on a-plane sapphire (PLD). The SAD pattern which coincide for all encircled regions is shown in the inset.

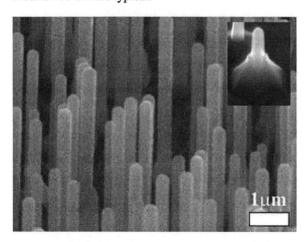

FIGURE 1. SEM image of a uniform MgZnO nanowire array grown by PLD. Inset: typical structure with hexagonal pyramid base grown at reduced laser pulse frequency.

Figure 2 shows a TEM image of a sample grown on a-plane sapphire. All nanowires have the same orientation and no gold is found on top of them. Hence our growth mechanism seems to be different from the VLS process. However, SEM and AFM images show that the nanowire growth is limited to regions originally covered with gold. The whereabouts of the Au droplets is currently under investigation. In high-resolution TEM measurements no dislocations or stacking disorder is visible. Our HRXRD data show that there are no rotational domains and the epitaxial relations are ZnO(0001)||c-sapphire(0001), ZnO[10-10]||c-sapphire[2-1-10] and ZnO[0001]||a-sapphire[11-20], ZnO<11-20>||a-sapphire[0001]. The FWHM of the ZnO(0002) reflection is as low as 0.017° for structurally optimized samples. This is comparable to state-of-the-art PLD thin films [4]. Details of the optical properties are described in [5].

CONCLUSION

ZnO nanowire and micropillar arrays with controlled size and shape are grown by high-pressure PLD and a carbothermal method. The nanowires grow epitaxially on a- and c-sapphire and exhibit excellent structural properties. Gold is required to initiate the nanowire growth. However, we do not find any gold on top of the grown structures, as described by the VLS model. Heterostructures grown by this flexible multi-target PLD technique shall pave the way to novel optoelectronic devices.

ACKNOWLEDGMENTS

This work was supported by Deutsche Forschungsgemeinschaft within FOR 522 (Project Gr 1011/11-1).

REFERENCES

1. B.D. Yao et al., Appl. Phys. Lett. **81**, 757 (2002)

2. M. Lorenz et al., Ann. Phys. **13**, No.1, 39-42 (2004)

3. M.H. Huang et al., Adv. Materials **13**, 113 (2001)

4. E.M. Kaidashev et al., Appl. Phys. Lett. **82**, 3901 (2003)

5. T. Nobis et al., Nano Letters **4**, 797 (2004)

Catalyst-free growth of semiconductor nanowires by selective area MOVPE

J. Motohisa, J. Noborisaka, S. Hara, M. Inari, and T. Fukui

Research Center for Integrated Quantum Electronics and Graduate School of Information Science and Technology, Hokkaido University, North 13, West 8, Sapporo 060-8628, Japan

Abstract. We report on a novel catalyst-free approach to form GaAs, AlGaAs and InGaAs nanowires and their arrays by selective area metalorganic vapor phase epitaxy (SA-MOVPE). At optimized growth conditions, extremely uniform array of GaAs or InGaAs nanowires with diameter d of 100 nm to 200 nm were grown on GaAs and InP substrates, respectively. It was found the shape (height h and size d) depends strongly on the growth conditions as well as the size d_0 and pitch a of the mask opening. In particular, the height h of the pillar becomes higher as d is reduced. On the other hand, h decreases as a is increased. Based on these results, we obtained hexagonal nanowires with much smaller d ($\sim$ 50 nm) and longer h (>6 μm) by doing SA-MOVPE on masked substrates with smaller d_0.

INTRODUCTION

Recently, semiconductor nanowires have been attracting interest for a new class of building blocks for nanoscale electronics and photonics in the bottom-up approach. For example, logic circuits based on the nanowires are proposed and demonstrated[1]. As other examples, ultra small light emitters[2, 3, 4] and nanoscale photodetectors[5] are proposed and attempted using the nanowires. So far, the nanowires, originally known as whiskers, have been formed by catalyst assisted vapor-liquid-solid-phase (VLS) growth[6]. We here report on a novel catalyst-free approach to form GaAs, AlGaAs and InGaAs nanowires and their arrays by elective area metalorganic vapor phase epitaxy (SA-MOVPE), which combines bottom-up and top-down technologies for the formation of nanostructures, and is expected to exhibit superior crystalline quality as well as good controllability of the growth with an atomic precision to form abrupt doping profiles and heterojunctions, including both vertical and lateral heterostructures.

EXPERIMENTAL PROCEDURE

Firstly, we prepared GaAs and InP (111)B substrates partially covered with SiO$_2$ for SA-MOVPE. Periodic array of circular mask openings was defined by using electron beam lithography and wet chemical etching. The pitch a of the mask opening was varied from 0.2 μm to 3 μm. Since the diameter d of the nanowire is directly related to the opening diameter d_0, we tried to obtain smaller d_0 as possible by controlling the amount of the EB dose as well as the designed opening size. Then, SA-MOVPE of GaAs/AlGaAs and InGaAs was carried out on the masked GaAs and InP substrates, respectively. We found the optimum growth temperature was around 750°C, 850°C, and 650°C for GaAs, AlGaAs, and InGaAs, respectively, and most of the growth was carried out at these temperature for each material. Detail of the growth conditions is reported elsewhere[7, 8, 9].

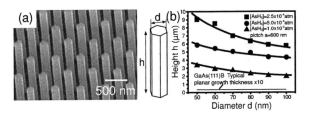

FIGURE 1. (a) Typical bird's eye SEM image of typical GaAs nanowire arrays. (b) Height h of GaAs nanowires versus nanowire diameter d grown at different AsH$_3$ partial pressures [AsH$_3$].

RESULTS AND DISCUSSIONS

Figure 1(a) shows typical SEM images of GaAs nanowires grown on (111)B. We can see an uniform array of vertically standing GaAs nanowires on the substrate. The cross section of the nanowires is hexagonal, indicating that they are surrounded by of six {110} sidewall facets normal to (111)B plane. In this result, the

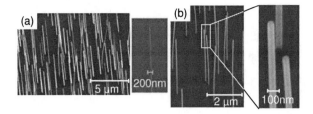

FIGURE 2. Thin ($d = 50$ nm ~ 60 nm) and tall ($h \sim 6$ μm) nanowires of (a) GaAs and (b) InGaAs.

average lateral size d of the nanowires was 0.23 μm, and their height h is about 1.4 μm.

We confirmed that the diameter d of the nanowire was almost the same as initial mask opening d_0 for both GaAs and InGaAs[9, 10]. Furthermore, h increases as d decreases, as shown in Fig.1(b). To make much thinner and higher nanowires, we attempted the growth on masked substrates with smaller circular openings. A typical result for GaAs is shown in Fig. 2(a). We can see much smaller (~ 50 nm) and longer (> 6 μm) hexagonal nanowires are formed. Figure 2(b) also shows InGaAs nanowires grown on a (111)B InP masked substrate. In this case, we obtained InGaAs nanowires with $d = 60$ nm and $h = 5.6$ μm. From these results, we conclude that semiconductor nanowires with diameter of several tens of nanometers can easily be obtained simply by reducing the initial size of the mask opening in SA-MOVPE.

Next, we measured micro-photoluminescence (PL) of GaAs-based nanowire arrays at room temperature. Figure 3 shows typical PL spectra of (a) GaAs and (b) GaAs/AlGaAs heterostructured nanowires. The average diameter d_{avr} of the nanowire is 221 nm, and h is about 0.97 μm for GaAs nanowires. For heterostructured nanowires, $h = 0.25$ μm and d_{avr} is almost the same as GaAs nanowires. PL spectrum of semi-insulating (SI) GaAs substrates is also shown for a reference in the figure. We can see the spectrum of the sample with nanowires is completely different form that of the reference, and the features in the higher energy side is attributable to the emission from nanowires both sample. We note that PL intensity is much stronger for nanowires containing GaAs/AlGaAs heterostructures than that of GaAs nanowires. We think this difference originates form the nonradiative surface recombination at the sidewall of nanowires, and stronger emission in heterostructure is due to the passivation of GaAs surface by AlGaAs.

ACKNOWLEDGMENTS

The authors thank Prof. H. Hasegawa, Dr. S. De Franceschi, Dr. M Akabori for stimulating discussions, and Mr. S. Akamatsu and A. Koike for supporting the

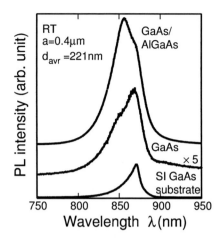

FIGURE 3. Room temperature PL spectra of (a) GaAs and (b) GaAs/AlGaAs heterotructured nanowire array with the pitch a of 0.4 μm. The average diameter of the nanowire is 221 nm, and the height $h = 0.97$ μm. PL spectrum of semi-insulating GaAs substrate (without growth) is also shown for reference.

experiment. This work was financially supported in part by a Grant-in-aid for Scientific Research supported by the Japan Society for the Promotion of Science.

REFERENCES

1. Y. Huang, X. Duan, Y. Cui, L. J. Lauhon, K.-H. Kim, C. M. Lieber, *Science* **294**, 1313 (2001).
2. M. H. Huang, S. Mao, H. Fick, H. Yan, H. Wu, H. Kind, E. Weber, R. Russo, and P. Yang, *Science* **292**, 1897 (2001).
3. N. Panev, A. I. Persson, N. Skold, and L.Samuelson, *Appl. Phys. Lett.* **83**, 2238 (2003).
4. X. Duan, Y. Huang, R. Agarwal, and C. M. Lieber, *Nature* **421**, 241 (2003).
5. J. Wang, M. S. Gudiksen, X. Duan, Y. Cui, and C. M. Lieber, *Science* **293**, 1455 (2001).
6. K. Hiruma, M. Yazawa, T. Katsuyama, K. Ogawa, K. Haraguchi, M. Koguchi, and H. Kakibayashi, *J. Appl. Phys.* **77**, 447 (1995).
7. J.Motohisa, J. Takeda, M. Inari, J.Noborisaka, T.Fukui, *Physica E* **23**, 298 (2004).
8. J. Motohisa, J. Noborisaka, J. Takeda, M. Inari, and T. Fukui, accepted for publication in J. Cryst. Growth (2004).
9. J. Noborisaka, J. Motohisa, J. Takeda, M. Inari, Y. Miyoshi, N. Ooike and T. Fukui, in *Proceedings of 2004 International Conference on Indium Phosphide and Related Materials* p.647 (2004); J. Noborisaka, J. Motohisa, and T. Fukui, submitted.
10. M. Akabori, J. Takeda, J. Motohisa, and T. Fukui, *Nanotechnology* **14**, 1071 (2003).

Synthesis and Properties of ZnO Nano-ribbon and Comb Structures

Y.H. Leung, A. B. Djurišić, and M. H. Xie

Dept. of Physics, The University of Hong Kong, Hong Kong, Hong Kong SAR, PR China

Abstract. ZnO is of great interest for photonic applications due to its wide band gap (3.37 eV) and large exciton binding energy (60 meV). A large variety of fabrication methods and nanostructure morphologies was reported up to date for this material. Obtained morphologies include nanobelts or nanoribbons, nanowires, nanorods, tetrapod nanostructures, etc. Novel nanostructures like hierarchical nanostructures, nanobridges and nanonails have also been fabricated. In this work, we report a simple method for fabrication of nanoribbon and nanocomb structures. The structures are fabricated by evaporation of a mixture of ZnO and carbon nanotubes (CNT) at 1050°C, and the deposition products have been collected on Si substrates in the temperature range 750-800°C. The growth mechanism of obtained structures is discussed.

INTRODUCTION

The wide band gap (3.37eV) and large exciton binding energy (60meV) make zinc oxide (ZnO) a promising material for photonic applications, especially in UV or blue spectral range. A wide range of nanostructures has been reported for ZnO, such as nanowires [1], tetrapod nanorods [2], nanoribbons/belts [1-3] etc. Recently, some novel ZnO nanostructures like nanobridges/nails [4] and nanosheets [5] have also been demonstrated. Among all the morphologies, ribbon/comb-like structures are of great interest for the application in nanosized devices due to the ease in manipulation, their well-defined geometry and excellent crystallinity. Recently, UV lasing from nanoribbons has been reported [6,7]. Comb-like nanostructures are also interesting for nano-cantilever arrays.

A number of ways to synthesize ZnO nanoribbon/belts have been reported. Most of the processes require high temperature (900°C~1400°C), precise pressure, gas flow and composition control [2,3,5]. In this work, we report a simple synthesis method for the fabrication of ribbon/comb-like ZnO nanostructures. ZnO nanoribbons/combs were fabricated by heating a mixture of ZnO powder (Aldrich, 99.99% purity) and single-walled carbon nanotubes (SWCNTs) (Carbolex, AP grade) in a tube furnace at 1050°C. The whole process took place in ambient atmosphere. After the desired temperature was reached, a quartz tube containing the powder mixture and the substrates (Si {111}, rough or polished side) was inserted into a tube furnace. The nanostructures were deposited on the substrates which covered a temperature range of 750-800°C. The deposited materials were examined by scanning electron microscopy (SEM) using LEO 1530 FESEM, transmission electron microscopy (TEM) and selected area electron diffraction (SAED) using Philips Tecnai 20 TEM, X-ray diffractometry (XRD) using Siemens D5000 X-ray diffractometer, and photoluminescence (PL) using a HeCd laser excitation source (325 nm).

RESULTS AND DISCUSSIONS

Figure 1 shows the representative SEM images of the ZnO nano-ribbon/comb structures grown on the rough back side of a Si substrate at source temperature of 1050°C. Bunches of nanoribbon/comb structures covered the surface of the substrate as shown in Figure 1a. The typical widths of the nanostructures are in the range of 3 to 5μm, while the lengths are about 10μm in average. Figure 1b shows a comb-like structure similar to the dentritic structures which eventually result in a nanosheet structure [5]. A growth mechanism called "1D branching and 2D filling" was proposed for the formation of sheet-like nanostructures [5]. Sidebranching on the basal 1D nanowires results in the formation of a dendritic structure. Such dendritic sidebraching is related to the supersaturation of reactant vapors [5]. It is followed by the planar filling of the interspaces between the sidebranches, which is attributed to the selective condensation of vapors on the concave corner sites between the branches [5].

FIGURE 1. Representative images of ZnO nanoribbons/combs grown at 1050°C on rough Si surface. a) bunch of nanoribbon/comb structures at low magnification, b) comb nanostructure.

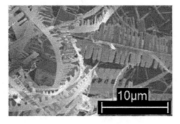

FIGURE 2. SEM image of ZnO nanoribbons/combs fabricated at 1050°C on smooth Si substrate.

Figure 2 shows the obtained nanostructures on the polished sides of Si substrates. The branches have relatively equal length and they are very broad. It can be observed that the substrate used in the synthesis has significant effect on the morphologies of the resulting nanostructures. It was also observed that higher yield of nanostructures was obtained using rough Si substrates, compared to the smooth ones. This can be attributed to the larger number of possible nucleation sites. The differences in the available nucleation sites and Zn diffusion rates on different substrates affect the subsequent growth of the nanostructures.

Figure 3 shows the photoluminescence (PL) spectra of the obtained nanostructures. Strong UV emission was observed both at room temperature (298K) and 11K, while the defect-related green emission was much weaker for all the ribbon/comb morphologies.

The shift of the high energy peak with increasing temperature is in agreement with the other works in the literature [8], while the fine structure in the UV emission at 11K can be attributed to the free and bound exciton transitions [8].

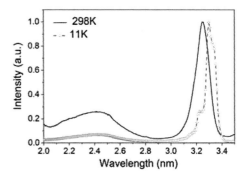

FIGURE 3. Photoluminescence (PL) spectra of the ZnO nanoribbon/combs.

CONCLUSIONS

ZnO ribbon/comb-like nanostructures were prepared at atmospheric pressure using a simple synthesis method. The morphologies were affected by the substrate type. Fabricated ribbon/comb structures exhibited strong UV but week green emission.

ACKNOWLEDGMENTS

This work is supported by the University of Hong Kong University Research Committee seed funding grant.

REFERENCES

1. B. D. Yao, Y. F. Chan, and N. Wang, Appl. Phys. Lett. 81, 757-759 (2002).
2. H. Yan, R. He, J. Pham, and P. Yang, Adv. Mater. 15, 402-405 (2003).
3. Z. W. Pan, Z. R. Dai, and Z. L. Wang, Science 291, 1947-1949 (2001).
4. J. Y. Lao, J. Y. Huang, D. Z. Wang, and Z. F. Ren, Nano Lett. 3, 235-238 (2003).
5. J.-H. Park, H.-J. Choi, Y.-J. Choi, S.-H. Sohn, and J.-G. Park, J. Mater. Chem. 14, 35-36 (2004).
6. H. Yan, J. Johnson, M. Law, R. He, K. Knuten, J. R. McKinney, J. Pham, R. Saykally, and P. Yang, Adv. Mater. 15, 1907-1911 (2003).
7. J. A. Zapien, Y. Jiang, X. M. Meng, W. Chen, F. C. K. Au, Y. Lifshitz, and S. T. Lee, Appl. Phys. Lett. 84, 1189-1191 (2004).
8. B. P. Zhang, N. T. Binh, and Y. Segawa, K. Wakatsuki, and N. Usami, Appl. Phys. Lett. 83, 1635-1637 (2003).

The Fractional-Dimensional Excitonic Absorption Theory Applied to Real V-groove Quantum Wires

K. F. Karlsson[1,2], M.-A. Dupertuis[2], H. Weman[2,1] and E. Kapon[2]

[1] *Department of Physics and Measurement Technology (IFM), Linköping University, S-581 83 Linköping, Sweden*
[2] *Laboratory of Physics of Nanostructures, Institute of Quantum Electronics and Photonics, Swiss Federal Institute of Technology Lausanne, CH-1015 Lausanne, Switzerland*

Abstract. A fractional dimensional approach is applied to a realistic V-groove quantum wire in order to calculate the excitonic absorption spectrum. An excellent agreement is obtained with much more computationally demanding models as well as with experimental photoluminescence excitation data. However, comparison with a full excitonic calculation reveal situations were the concept of fractional dimensions to some extent fails.

INTRODUCTION

In order to optimize the design of devices such as light-emitting lasers and modulators it is desired to predict the light-absorption spectrum. In the following we will present and verify a computationally simple method, which yields the full excitonic absorption spectra based on a fractional dimension approach. In order to elucidate the strengths and weaknesses of such a simplified model, the numerical results are compared with a detailed full excitonic calculation [1].

It is well known that an exciton in a real QWR is not strictly one-dimensional. A real QWR can thus be ascribed a dimension α somewhere between the ideal value one and three, depending on the strength and shape of the confinement [2]. If α is determined, all the exciton properties follow by injecting its value into general mathematical solutions. For example, the α-dimensional exciton energies is [2]

$$E_n = -\frac{Ry}{[n+(\alpha-3)/2]^2} \quad (1)$$

where Ry is the effective three-dimensional Rydberg energy and n is the principal quantum number. A corresponding expression of the the α-dimensional excitonic optical interband absorption spectrum including bound- and continuum states has been derived in Ref. 2.

THEORY

As a first step the single-particle electron and hole states are calculated by numerically solving the two-dimensional (2D) Schrödinger equation for the confined directions in the effective mass **k·p** approximation, where the valence-band mixing is taken into account by the 4×4 Luttinger Hamiltonian with spatially dependent Luttinger parameters, while the conduction bands are treated as parabolic, with a spatially dependent electron effective mass [4].

A separable form of the exciton wave function is used, written as the product of the electron and hole single particle wave functions in the confined directions, and a 1D exciton wave function along the free direction. Eventual further valence-band mixing due to the Coulomb interaction is neglected, and the exciton binding energy $(= -E_1)$ is computed by solving a 1D reduced mass Schrödinger equation [5]. Thereafter a value of α is determined by Eq 1, after which the absorption spectrum is obtained.

The fractional-dimensional results are compared with a full excitonic finite-element calculation relying on Hartree-Fock approximation [1]. The single-particle band-structure is same in both approaches [5]. Thereafter, a large number of Coulomb matrix elements are computed ($\sim 10^7$) including dielectric effects. Finally the optical absorption is obtained by computing appropriate expectation values [1]. This full calculation is essentially more computationally demanding than the simple exciton model.

NUMERICAL RESULTS

The cross-sectional Al distribution, presented in the inset of Fig. 1, is extracted from a TEM micrograph of a real V-groove GaAs/Al$_{0.33}$Ga$_{0.67}$As QWR. Given the material parameters, the Al distribution serves as the only "input parameter" into the calculations.

Figure 1 shows the computed optical density related to the first electron and hole subbands (e_1h_1). Obviously, the Coulomb interaction described by the fractional-dimension theory dramatically changes the spectrum in several ways, compared to the calculated free carrier absorption spectrum also shown in Fig, 1.: 1) the onset of the absorption is energetically down-shifted by the exciton binding energy, 2) the oscillator strength at the absorption onset is strongly enhanced, 3) additional features are present below the onset of the continuum absorption due to excited exciton states.

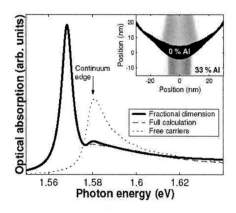

FIGURE 1. Calculated single electron and hole subbands spectra. The inset shows the used Al profile and QWR shape.

The total absorption is obtained by limiting the calculation to 9 electron and 15 hole subbands, and summing the absorption of all 135 possible combinations. The resulting excitonic spectra are shown in Fig. 2a (black lines) and the corresponding full calculation spectra are shown in Fig. 2b. It is clear that the fractional dimensional approach agrees well with the full calculation in many aspects: Strong resonances for the bound ground states with successively decreased oscillator strength for higher subband indexes, while the background absorption, on the other hand, successively increases with the photon energy. However, all resonances obtained by the full calculation are red-shifted compared with the result of the fractional dimensional model, in particular the "*lh-like*" resonance. This is an effect originating from excitonic intersubband coupling, which is strongest pronounced for the *lh-like* resonance arising due to coupling between excited states of e_1h_1 and the light-hole-like exciton e_1h_6 [1]. Thus, in situations with strong excitonic intersubband coupling the fractional dimensional theory to some extent fails. However, the excitonic effects described by fractional dimensions dramatically improve the agreement with experiment (Fig. 2c), compared to free carrier models, neglecting the Coulomb interaction (Fig. 2a, gray lines). But for best agreement, intersubband couplings (Fig. 2b) are required, not supported by fractional dimensions.

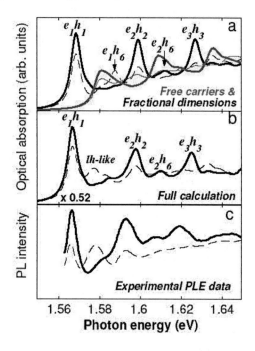

FIGURE 2. Calculated (a & b) and experimental (c) spectra. Solid (dashed) lines correspond to light polarized along (perpendicular to) the QWR axis. The labels denote involved subbands for the calculated exciton resonances (black lines).

REFERENCES

1. M.A. Dupertuis, *et al.*, Proc. of ICCN 2002, Computational Publications, Cambridge MA, USA, 227 (2002).
2. X. F. He, Phys. Rev. B **43**, 2063 (1991).
3. P. Lefebvre, *et al.*, Phys. Rev. B **48**, 17308 (1993).
4. F. Vouilloz, *et al.*, Phys. Rev. B **57**, 12378 (1998).
5. Y.C. Chang, *et al.*, Appl. Phys. Lett **47**, 1324 (1985).

Mixing of discrete and continuum excitations induced by nonperturbative Coulomb-correlations

V.M. Axt*, J. Wühr* and T. Kuhn*

Institut für Festkörpertheorie, Universität Münster, Wilhelm-Klemm Str. 10, 48149 Münster, Germany

Abstract. Taking into account all types of two-pair coherences we have calculated four-wave-mixing (FWM) spectra emitted from a quantum wire for excitation conditions where excitons and free electron-hole pairs are simultaneously excited. The spectra exhibit a mixing of discrete and continuous emissions which is shown to indicate nonperturbative Coulomb correlations. The line shapes are found to depend sensitively on the central frequency, the strength of the interaction as well as on the delay between the pulses.

While the importance of two-pair coherences for the nonlinear optics of semiconductor nanostructures is well known for selective excitations at the 1s exciton [1, 2], so far their role is widely unexplored when substantial parts of the excitation are above the band-edge. Selective excitations at the exciton are only sensitive to transitions to biexcitons or exciton-exciton scattering states but do not test for potential influences of other kinds of two-pair coherences such as transitions involving two unbound electron-hole pairs or an exciton and a free pair. In this paper we account for all types of two-pair coherences and calculate FWM spectra for broadband excitations where all these coherences are simultaneously excited. We shall concentrate on low excitation densities, i.e. the regime where the most pronounced correlation effects have been found for selective excitonic excitations [1, 2].

Our calculations are based on a microscopic density matrix theory where the hierarchy of higher order density matrices is truncated according to the dynamics controlled truncation scheme (DCT) [3]. The DCT approach is especially adapted to the low density regime where it becomes asymptotically exact. It allows us to account nonperturbatively for all relevant Coulomb correlations [4]. Concentrating on the coherent dynamics we have to follow two types of dynamical variables namely [5]: single-pair transition amplitudes $Y_\ell^j \equiv \langle \hat{Y}_\ell^j \rangle = \langle d_j c_\ell \rangle$ and correlated two-pair coherences $\bar{B}_{\ell\ell'}^{jj'} \equiv \langle \hat{Y}_\ell^j \hat{Y}_{\ell'}^{j'} \rangle - Y_\ell^j Y_{\ell'}^{j'} + Y_{\ell'}^{j'} Y_\ell^j$, where c_j (d_j) are the destruction operators for electrons (holes) in Wannier states. Y is directly related to the optical polarization while $\bar{B}$ describes the correlated dynamics of transitions to two-pair states. We consider a two-band model with two-fold degenerate bands (the valence band refers to heavy holes) representing the lowest subbands of a cylindrical GaAs quantum wire with

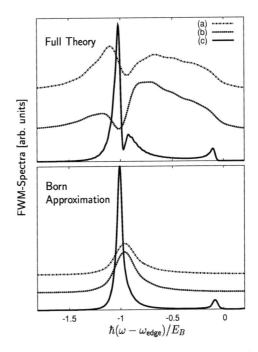

FIGURE 1. FWM spectra after collinear excitation with pulses of 20 fs duration at zero delay: (a) excitation in resonance with the exciton line for a wire with exciton binding energy $E_B = 4$ meV, (b) same as (a) but for an excitation 16 meV above the band-edge $\hbar\omega_{\text{egde}}$, (c) same as (b) but for a wire with $E_B = 18$ meV.

100 nm^2 cross-section confined by infinite barrier potentials. The static dielectric constant ε_0 which scales the Coulomb potential is treated as an adjustable parameter in order to be able to study the influence of the interaction strength. The latter is expected to be an important param-

eter for correlations effects. We have calculated standard two-pulse FWM spectra emitted in the $2\mathbf{k}_2 - \mathbf{k}_1$ direction by solving the coupled nonlinear equations for Y and $\bar{B}$ in k-space without further approximation.

The upper part of Fig. 1 shows spectra obtained for ultrafast excitation (20 fs pulse duration) with the full theory, i.e. accounting nonperturbatively for Coulomb correlations. Curves (a) and (b) correspond to a wire with weakly bound excitons, where ε_0 has been adjusted to yield a binding energy $E_B = 4$ meV. For (a) the excitation was in resonance with the exciton while for (b) it was centered 16 meV above the band-edge. In both cases the spectra exhibit complex line shapes with a dip-like structure near the $1s$ resonance and a continuous emission which is spread over the whole region between the exciton line and the band-edge. Note that the linear spetrum has a gap in this region. In contrast, when two-pair correlations are treated according to the second order Born approximation (BA) the whole emission is concentrated near the exciton as seen in the lower part of the figure. The BA represents the standard quantum kinetic scattering theory which, however, accounts only perturbatively for Coulomb correlations in the two-pair manifold [4]. Obviously, we are dealing here with nonperturbative Coulomb correlations that cannot be captured by the BA. In addition it has to be concluded that the suppression of the $1s$ emission is due to an interference between different quantum mechanical pathways involving two-pair coherences. Interestingly, it is possible to suppress a discrete resonance by mixing it with above band-edge excitations. We also see by comparing the curves (a) and (b) that even under the present broadband excitation conditions a change of the line shapes is noticeable when the central frequency is varied. Note that there is still a large suppression of the $1s$ emission even for resonant excitation [curve (a)]. For longer pulses (40 fs duration) and otherwise identical conditions the emission after resonant excitation has previously been reported to be concentrated at the $1s$ line without any dip [6].

Curve (c) in Fig. 1 has been calculated for the same conditions as in (b) but for a stronger interaction (smaller value of ε_0) corresponding to an exciton binding energy of 18 meV. Keeping the central frequency as in (b) 16 meV above the band-edge implies that we have identical band-to-band excitations in (b) and (c). However, there is less absorption at the $1s$ exciton in (c) than in (b) due to the larger binding energy in (c). Interestingly, decreasing the weight of the excition in the absorption almost removes the suppression of the $1s$ FWM emission that was seen in (b). Only a sharp dip slightly above the resonance is found in case (c). Instead of a continuous emission the spectrum is now concentrated near the $1s$ and $2s$ lines. The corresponding BA result reproduces the appearance of the $2s$ emission but still fails to describe the dip near the $1s$ line. However, there is a clear tendency that the

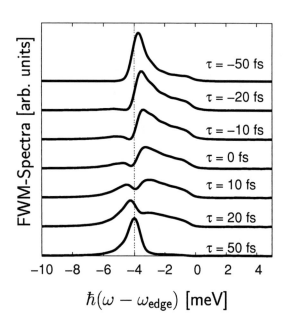

FIGURE 2. FWM spectra for different delays τ after collinear excitation 16 meV above the band-edge with pulses of 40 fs duration for a wire with $E_B = 4$ meV.

BA is closer to the full theory for stronger interactions.

Finally, Fig. 2 displays FWM spectra obtained after excitation with 40 fs pulses for different pulse delays τ. Clearly, the continuous emission in the gap region disappears for longer pulse delays and the emission concentrates at the exciton line reflecting the dephasing of continuum excitations due to destructive interference. After 50 fs delay practically all the signal is emitted at the $1s$ line but still with a noticeably asymmetric shape exhibiting more spectral weight on the lower (higher) energy side for positive (negative) delays.

In conclusion we have demonstrated a complex spectral mixing of discrete and continuum FWM emissions reflecting nonperturbative Coulomb correlations.

We thank the Deutsche Forschungsgemeinschaft for support through the grant KU 697/9-2.

REFERENCES

1. Axt, V. M., and Kuhn, T., *Rep. Prog. Phys.*, **67**, 433–512 (2004).
2. Chemla, D. S., and Shah, J., *Nature*, **411**, 549–57 (2001).
3. Axt, V. M., and Stahl, A., *Z. Phys. B*, **93**, 195–204 (1994).
4. Kwong, N. H., Takayama, R., Rumyantsev, I., Kuwata-Gonokami, M., and Binder, R., *Phys. Rev. Lett.*, **87**, 027402 (2001).
5. Axt, V. M., and Stahl, A., *Z. Phys. B*, **93**, 205–11 (1994).
6. Wühr, J., Axt, V. M., and Kuhn, T., *Phys. Status Solidi B*, **238**, 556–60 (2003).

Crossover from Excitons to an Electron–Hole Plasma in a High-Quality Single T-Shaped Quantum Wire

Masahiro Yoshita*, Yuhei Hayamizu*, Hidefumi Akiyama*,
Loren N. Pfeiffer[†], and Ken W. West[†]

*Institute for Solid State Physics, University of Tokyo and CREST, JST,
5-1-5 Kashiwanoha, Kashiwa, Chiba 277-8581, Japan
[†]Bell Laboratories, Lucent Technologies, 600 Mountain Avenue, Murray Hill, New Jersey 07974, USA

Abstract. We studied evolution of photoluminescence (PL) spectral with increasing excitation powers in an unprecedentedly high-quality quantum wire. The PL spectra obtained suggest that the transition from excitons to a dense electron–hole (e–h) plasma in the wire is a gradual crossover via bi-excitons. We also found that the continuum band edge of excitons shows no energy shift with increasing e–h densities and never crosses to the low-energy edge of the plasma PL at higher e–h densities, which is in contrast to a prevailing picture of the exciton Mott transition.

INTRODUCTION

Many-particle effects in the one-dimensional (1D) electron–hole (e–h) system of quantum wires where Coulomb correlations among the carriers become more significant have recently attracted much interest. One intriguing question is whether the transition from a dilute exciton gas to a dense 1D e–h system is described by a picture of the exciton Mott transition, which is accepted as a plausible picture in higher dimensions. In this work, we study the evolution of photoluminescence (PL) spectra from a single quantum wire with increasing excitation powers during a transition from excitons to a dense 1D e–h system.

EXPERIMENT

A single T-shaped quantum wire (T wire) was fabricated by the cleaved-edge overgrowth method with molecular beam epitaxy [1]. The quantum-wire electronic states were formed at a T-intersection of a 14-nm-thick (001) $Al_{0.07}Ga_{0.93}As$ quantum well (stem well) and a 6-nm-thick (110) GaAs quantum well (arm well). Details of the sample structure and fabrication procedures are presented elsewhere [2]. PL spectra of the wire at various excitation powers under point excitation were measured by using micro-PL setup.

RESULTS AND DISCUSSION

PL spectra of the T wire for various excitation powers at 4 K are shown in Fig. 1. At excitation powers below 5×10^{-3} mW, a dominant peak at 1.582 eV and a tiny peak on low-energy tail of the dominant peak appeared, and were respectively ascribed to free and localized excitons in the wire. Both the position and the intensity of the free-exciton PL peak were almost unchanged over 20 μm along the wire, and its linewidth was 1.3 meV, which indicate high spatial uniformity and quality of the wire. As the excitation power was increased to 10^{-2} mW, where the estimated e–h density in the wire was 2.6×10^4 cm^{-1}, a new PL peak appeared 3 meV below the free-exciton PL peak and increased its intensity superlinearly. We ascribe this peak to bi-excitons. At excitation powers above 0.1 mW where the e–h density was 1.7×10^5 cm^{-1}, the low-energy side PL peak became dominant. We ascribe the PL in this power regime to an e–h plasma. Note that the plasma PL peak is symmetrically broadened and shows no peak shift with respect to the bi-exciton peak at lower excitation powers. The evolution of the PL spectra from the wire indicates that the transition from excitons to an e–h plasma in the wire is a gradual crossover via bi-excitons. Moreover, the spectral features of the e–h plasma PL suggest that there are strong internal Coulomb correlations, in particular, bi-excitonic correlations in the 1D e–h plasma [3].

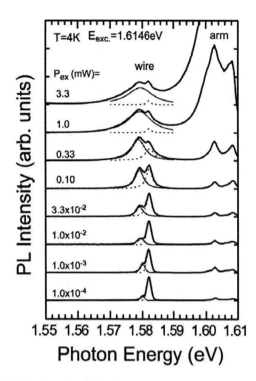

FIGURE 1. Normalized PL spectra of the T wire for various excitation powers at 4 K. Thin solid and dotted curves respectively show PL lines of the free-exciton peak and the low-energy side peak separated by line-shape analysis.

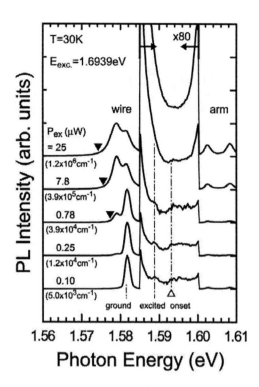

FIGURE 2. Normalized PL spectra of the T wire for various excitation powers at 30 K. Numbers in parentheses are estimated e–h densities in the wire.

Figure 2 shows the evolution of PL spectra from the T wire obtained at an elevated temperature of 30 K. In Fig. 2, a small PL peak at 1.589 eV (denoted as *excited*) due to an exciton excited state and a continuous PL band with an onset at 1.593 eV (denoted as *onset*) due to higher excited states of the excitons and 1D continuum states are clearly observed. A noticeable point in the PL spectra is that neither the *excited* peak nor the *onset* edge shows no energy shift from their initial positions. Even at a pair density of 3.9×10^5 cm^{-1} where the e–h plasma has already formed, these PL features are still observed. In the PL peaks from the wire ground state (denoted as *ground* in Fig. 2), on the other hand, low-energy edges of the plasma PL (marked by closed triangles) seen at high densities show red shift with increasing excitation powers possibly due to band-gap renormalization. However, these edges do not continuously connect to the band edge of excitons (open triangle) at the lowest density. Instead, they converge to the energy position of bi-excitons. Therefore, the level-crossing of these two edges expected in the exciton Mott transition does not occur. This result suggests that a new picture is necessary for the observed crossover from excitons to a plasma via bi-excitons in a quantum wire.

CONCLUSIONS

From micro-PL spectroscopy of the single quantum wire with increasing excitation powers, we found that the evolution from excitons to a dense e–h plasma in the wire is a gradual crossover via bi-excitons and that there are strong bi-excitonic correlations in the 1D e–h plasma. We also found that the low-energy edge of the plasma peak at high densities converges to the energy position of the bi-exciton, but never crosses to the continuum band edge of the 1D free excitons.

REFERENCES

1. L. Pfeiffer, K. W. West, H. L. Stormer, J. P. Eisenstein, K. W. Baldwin, D. Gershoni, and J. Spector, *Appl. Phys. Lett.* **56**, 1697-1699 (1990).
2. Y. Hayamizu, M. Yoshita, S. Watanabe, H. Akiyama, L. N. Pfeiffer, and K. W. West, *Appl. Phys. Lett.* **81**, 4937-4939 (2002).
3. M. Yoshita, Y. Hayamizu, H. Akiyama, L. N. Pfeiffer, K. W. West, K. Asano, and T. Ogawa, unpublished.

Photoluminescence excitation spectra of one-dimensional electron systems in an n-type doped quantum wire

Toshiyuki Ihara[1], Y. Hayamizu[1], M. Yoshita[1], H. Akiyama[1], L. N. Pfeiffer[2] and K.W. West[2]

[1]*Institute for Solid State Physics, University of Tokyo and CREST, JST, Chiba 2778581, Japan*
[2]*Bell Laboratories, Lucent Technologies, Murray Hill, NJ 07974, USA*

Abstract. Photoluminescence (PL) and PL excitation (PLE) spectra have been measured on a high-quality n-type-modulation-doped single GaAs quantum wire with a gate, where density of one-dimensional (1D) electron gas is tuned by gate voltage. We demonstrate that the absorption line shape of trions varies from symmetric to asymmetric with electron density. The high energy tail of the trion peak is analogous to a power-law anomaly in Fermi-edge singularity of 1D electron gas. The peak of trions shows no shift with electron density, while a peak of exciton shows blue-shifts. For higher electron density, the peak of trions changes to a shoulder corresponding to an absorption onset of an electron-plasma system. We found large energy gap between a high-energy cut-off in PL and a PLE onset in the wire.

INTRODUCTION

Interband optical transition spectra in the presence of a Fermi sea are expected to exhibit a power-law singularity, which reflects the final state response of Fermi-sea electrons to the attractive potential of a valence-band hole. This singularity, which is well known as the Fermi-edge singularity (FES), has been studied in detail in two-dimensional (2D) quantum wells (QWs) [1]. In 1D electron system, we previously reported this effect in photoluminescence (PL) spectra of a high-quality n-type-modulation-doped quantum wire with 1D electron densities tuned by gate voltage [2]. In this paper, we report our first study on PL excitation (PLE) spectra of the wire, which reveals many-body absorption features of the 1D system.

SAMPLES AND EXPERIMENTS

The sample was grown by the cleaved-edge overgrowth method with molecular beam epitaxy and growth interrupt annealing [3]. As schematically shown in Fig.1, the T-shaped quantum wire consisted of a 14-nm-thick $Al_{0.07}Ga_{0.93}As$ quantum well (stem well) grown on a (001) substrate, and an intersecting 6-nm-thick GaAs QW (arm well) overgrown on a cleaved (110) edge of the stem well. The electron density in the stem well was increased by Si delta-doping ($4 \times 10^{11} cm^{-2}$). By applying DC gate voltage (Vg) to the n^+ $Al_{0.1}Ga_{0.9}As$ layer relative to modulation-doped 2D electron gas in the stem well, we accumulated or depleted additional electrons in the wire and the arm well. Micro-PL and PLE measurements on the wire were performed with a cw titanium-sapphire laser with a 1μm spot size. The excitation power was 20 μW. The direction of PL detection was perpendicular to the laser excitation and their polarization were orthogonal to each other to improve signal-to-noise ratio. PL resolution was 0.15meV and PLE resolution was 0.04meV.

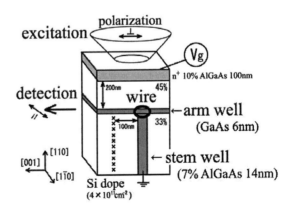

FIGURE 1. Schematic structure of T-shaped quantum wire, stem well, arm well, Si delta-doping, and a gate.

RESULTS

We measured PL and PLE spectra of the wire with various gate voltages from 0.0 to 0.7V at 5K. Figure 2 (a) shows the normalized PL (dotted lines) and PLE (solid lines) spectra. At low electron density (0.0V), the PLE spectrum is dominated by a peak of excitons (X) that shows splitting due to monolayer thickness fluctuations in the stem well.

As electron density becomes higher, X shows blue-shifts and becomes weak. Instead, a peak of trions (X^-) appears at about 2 meV below X. At $V_g = 0.2$V, PLE shape of X^- becomes asymmetric (a fast rise at low energies and a slow fall at high energies). The high-energy tail corresponds to the power-law singularity in FES for 1D electron gas. At gate voltages of 0.35-0.4V, X^- peak changes to a shoulder. Note that X^- shows no shift with electron density, while X shows blue-shifts. Thus, the energy gap between X^- and X increases with electron density. For higher gate voltages, absorption shows an onset that shows blue-shifts with state filling by a plasma.

For comparison, we measured PL and PLE spectra of the arm well. Figure 2 (b) shows the normalized PL (dotted lines) and PLE (solid lines) spectra of the arm well with various gate voltages from 0.2 to 0.8V. We observed a peak of neutral excitons (X), trions (X^-), and an electron plasma. These results are analogous to results reported for 2D QWs by another group [4], which already have supports of theoretical study on 2D electron systems [1].

The above-mentioned spectral features of the 1D wire are very similar to those of the 2D arm well. However, we found following interesting differences between them:

(a) At low electron density, energy gap between the peak of trions and that of excitons is large (2meV) in the wire, while that in the arm well is small (1.5meV).

(b) With increased carrier density, the peak of excitons decays more quickly in the wire than in the arm well. As a result, FES of the trion peak appears stronger in the wire.

(c) The peak of trions and the broad absorption onset due to the 1D plasma are more separately observed in the wire (0.3V-0.4V). In the arm well, the peak of trions changes to 2D electron plasma absorption rather continuously (0.3-0.6V). Large energy gap is observed between a high-energy cut-off in PL and a PLE onset in the wire.

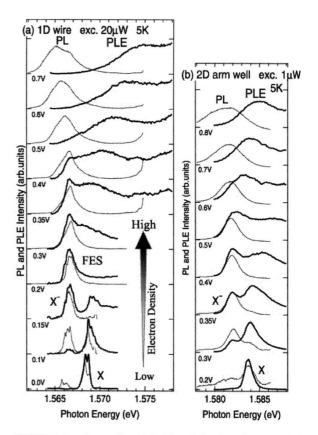

FIGURE 2. Normalized PL (dotted lines) and PLE (solid lines) spectra of the wire (a) and the arm well (b) at various gate voltages.

CONCLUSION

We measured the electron density dependence of PL and PLE spectra in an n-type-modulation-doped quantum wire. FES of trions is observed strongly in the wire. We found large energy gap between a high-energy cut-off in PL and a PLE onset in the wire.

REFERENCES

1. M. Takagiwa and T. Ogawa, *J. Phys. Chem. Solids* **63**, 1587 (2002).
2. H. Akiyama, L. N. Pfeiffer, A. Pinzuk, K. W. West, and M. Yoshita, *Solid State Commun.* **122**, 169 (2002).
3. M. Yoishita, H. Akiyama, L.N. Pfeiffer, and K. W. West, *Jpn. J. Appl. Phys.* **40**, L252 (2001).
4. G. Yusa, H. Shtrikman, and I. Bar-Joseph, *Phys. Rev. B* **62**, 15390 (2000).

Green Photoluminescence in ZnO Nanostructures

A. B. Djurišić,[1] Y. H. Leung,[1] Z. T. Liu,[1] D. Li,[1] M. H. Xie,[1] W. C. H. Choy,[2] and K. W. Cheah[3]

[1]Dept. of Physics, The University of Hong Kong, Hong Kong, Hong Kong SAR, PR China
[2]Dept. of Electric & Electronic Engineering, The University of Hong Kong, Hong Kong, Hong Kong SAR, PR China
[3]Dept. of Physics, Hong Kong Baptist University, Kowloon Tong, Hong Kong SAR, PR China

Abstract. In photoluminescence (PL) spectrum of ZnO, typically one or more peaks in the visible spectral range due to defect emission can be observed in addition to one UV peak due to band edge emission. The origin of the defect emission is controversial and several mechanisms have been proposed. In this work, we fabricated ZnO nanostructures with different methods (evaporation and chemical synthesis). We found that the preparation method influences the peak position of the defect emission. Different hypotheses for the origin of the green emission in our nanostructured samples are discussed.

INTRODUCTION

In the photoluminescence (PL) spectra of ZnO nanostructures, one UV peak and one or more peaks in the visible spectral range are observed. The UV emission from ZnO is generally believed to be due to exciton emission while emission in the visible range is due to extrinsic or intrinsic defects. Green emission [1,2] is commonly observed although other colors like yellow [3] are also reported. A number of mechanisms for the green emission has been proposed. It was suggested that the green emission originated from the transition between singly oxidized oxygen vacancies (V_o^+) and photoexcited holes [4,5]. This assignment was based on the correlation between the green PL emission and the electron paramagnetic resonance (EPR) signal at g≈1.96 on two phosphor powders [4]. However, such an assignment is quite controversial because it has been reported that V_o^+ also gives an EPR signal at g≈1.99 [6]. The signal at g≈1.96 can also correspond to shallow donors like interstitial Zn (Zn_i) [7]. Meanwhile, it was also reported that impurities like copper ions (Cu^{2+} and Cu^+) can also cause green emission in ZnO [8]. In this work, we investigated the origin of visible photoluminescence in ZnO nanostructure samples prepared by different methods. We found that the fabrication method significantly affects both the visible luminescence (peak position and relative intensity compared to the UV emission), as well as the EPR signal, but EPR and defect PL intensities were not necessarily related.

EXPERIMENTAL DETAILS

Room temperature PL and EPR studies were performed on four different ZnO nanostructure samples: tetrapods from thermal oxidation of pure Zn pellets in air and heating a ZnO:graphite mixture (1:1 molar ratio), multipods from heating a mixture of ZnO, germanium oxide (GeO_2), and graphite (1:0.1:1 molar ratio), and nanorods synthesized by a chemical method (solution of zinc nitrate hydrate and hexamethylenetetramine). The obtained nanostructures were examined by PL using a HeCd laser excitation source (325 nm), and EPR using Bruker EMX EPR Spectrometer.

RESULTS AND DISCUSSIONS

Figure 1 shows the PL spectra of all the four ZnO nanostructure samples. It is observed that tetrapod nanorod samples prepared by evaporation of pure Zn and ZnO:graphite mixture show strong UV and broad green emission. Although they exhibit similar green emissions (2.45 eV vs 2.38 eV), only one of them (samples from evaporation of ZnO:graphite mixture) shows a significantly strong EPR signal at g≈1.96, as shown in Figure 2. We may conclude that the oxygen vacancy hypothesis cannot be used to explain the green emission from all nanostructured samples [2]. It is possible that two different mechanisms are responsible for the green emission from the two tetrapod samples. In the former case (samples from evaporation of Zn in air), the green emission results from the transition between a delocalized electron and a deep trap. Such a hypothesis agrees with the mechanism proposed by Van Dijken et al. [9,10] In the latter case, the emission originates from the transition involving a shallow donor and a deep trap. The donor-acceptor hypothesis is in agreement with other works in the literature [11].

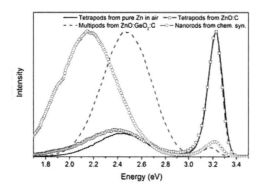

FIGURE 1. Room temperature PL spectra of the four ZnO nanostructures.

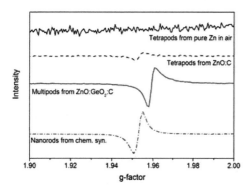

FIGURE 2. EPR spectra of the four ZnO nanostructures.

Strong green PL and g≈1.96 EPR signal are observed from the multipod samples fabricated from heating a ZnO:GeO$_2$:C mixture. The slight shift on both the EPR signal ($\Delta g \approx 0.006$) and green emission ($\Delta E \approx 0.08 eV$) compared to samples fabricated from ZnO:C mixture is likely to be attributed to the introduction of different type of intrinsic defects, since no Ge was detected. However, the reason behind this shift requires further studies.

The nanorods samples fabricated from chemical synthesis exhibit strong yellow PL emission and EPR signal at g≈1.96. This finding also contradicts the results reported by Vanheusden et al. [4]. Very likely, EPR signal corresponds to shallow donors. The visible emission (green or yellow) can originate from either conduction band-deep level or shallow donor-deep level transitions. Different types of deep level are likely to be involved in green and yellow emissions. Further studies are needed to conclusively establish the identity of these deep levels.

ACKNOWLEDGMENTS

This work is supported by the University of Hong Kong University Research Committee seed funding grant.

CONCLUSIONS

In this work, PL and EPR were performed on four different ZnO nanostructures. We found that there is no general relationship between the green emission and the g≈1.96 EPR signal, and that the existence of this signal and the type of intrinsic defects in ZnO nanostructures are strongly dependent on the synthesis conditions.

REFERENCES

1. V. A. L. Roy, A. B. Djurišić, W. K. Chan, J. Gao, H. F. Lui, and C. Surya, Appl. Phys. Lett. 83, 141-143 (2003).
2. A. B. Djurišić, Y. H. Leung, W. C. H. Choy, K. W. Cheah, and W. K. Chan, Appl. Phys. Lett. 84, 2635-2637 (2004).
3. L. E. Greene, M. Law, J. Goldberger, F. Kim, J. C. Johnson, Y. Zhang, R. J. Saykally, and P. Yang, Angew. Chem. Int. Ed. 42, 3031-3034 (2003).
4. K. Vanheusden, W. L. Warren, C. H. Seager, D. R. Tallant, J. A. Voigt, and B. E. Gnade, J. Appl. Phys. 79, 7983-7990 (1996).
5. K. Vanheusden, C. H. Seager, W. L. Warren, D. R. Tallant, and J. A. Voigt, Appl. Phys. Lett. 68, 403-405 (1998).
6. J. M. Smith, and W. E. Vehse, Phys. Lett. A 31, 147-148 (1970).
7. F. Morazzoni, R. Scotti, P. Di Nola, C. Milani, and D. Narducci, J. Chem. Soc. Faraday Trans. 88, 1691-1694 (1992).
8. N. Y. Garces, L. Wang, L. Bai, N. C. Giles, L. E. Halliburton, and G. Cantwell, Appl. Phys. Lett. 81, 622-624 (2002).
9. A. van Dijken, E. Meulenkamp, D. Vanmaekelbergh, and A. Meijerink, J. Phys. Chem. B 104, 1715-1723 (2000).
10. A. van Dijken, E. Meulenkamp, D. Vanmaekelbergh, and A. Meijerink, J. Lumin. 90, 123-128 (2000).
11. S. A. Studenikin and M. Cocivera, J. Appl. Phys. 91, 5060-5065 (2002).

Simple ideas on excitons in quantum wires

M. Combescot[*] and T. Guillet[†]

[*]*CNRS, Université Pierre et Marie Curie, Paris*
[†]*Université Montpellier 2, Montpellier*

Abstract. We show how to obtain analytical expressions for the quantum wire exciton wave functions and energies, through a "broadened" 1D potential $1/(z+d^*)$. Is also explained the way to relate the constant d^* to the wire lateral shape.

Unlike quantum wells which have ground state excitons with finite binding energy, the ground state exciton binding energy of a quantum wire is found infinite, if its thickness is taken as zero. Consequently, the properties of these quantum wire excitons crucially depend on the way this unphysical singularity is removed, i. e., on the way the wire thickness is taken into account.

One possibility is to plug the exact wire lateral shape into the exciton Schrödinger equation and to solve this 3D second order differential equation through heavy numerical calculations.

We here present an alternative, which gives not only the exciton ground state but also the lowest exciton states analytically, with an accuracy which, in most cases, is quite sufficient.

We have first reconsidered the exciton Schrödinger in D dimension and constructed a general solution of this second order differential equation, which is regular in the $D \to 1$ limit. It reads in terms of two independent hypergeometric functions $U(a,c,z)$, not in terms of $F(a,c,z)$ and $U(a,c,z)$, as the one of usual textbooks: The solution with $F(a,c,z)$ turns out to be singular for $D \to 1$.

From this solution, we derive the one for the "broadened" Coulomb potential, with $1/z$ replaced by $1/(z+d^*)$. This solution is still analytical, which is quite nice for further use in problems dealing with quantum wire excitons. However, it of course crucially depends on d^*.

It is thus necessary to find a good way to determine the appropriate d^*. For that, we first derive the formal expression of the *exact* 1D potential for the exciton relative motion along the wire. In it, enters the exact wire shape explicitly. We then show how to determine the function $d(z)$ for $1/[z+d(z)]$ to fit this potential (essentially) perfectly. From this function $d(z)$, we ultimately determine the "good" constant d^*. From it, the exciton energies and wave functions can then be obtained analytically. The precise steps of this procedure can be found in [1].

The main result of this analytical work is the fact that the energies of quantum wire excitons mainly depend on the wire area but not so much on its precise shape, provided that this shape is not too anisotropic. We in particular find that circular and square wires with same area have exactly the same energies. This should push the reader not to do heavy numerical calculations with the exact wire shape, in most cases of experimental interest.

REFERENCES

1. M. Combescot and T. Guillet, Eur. Phys. J. B **34**, 9 (2003).

Field-effect induced mid-infrared intersubband electroluminescence of quantum wire cascade structures

S. Schmult[*], T. Herrle[*], H.-P. Tranitz[*], M. Reinwald[*], W. Wegscheider[*], M. Bichler[†], D. Schuh[†] and G. Abstreiter[†]

[*]Institut für Experimentelle und Angewandte Physik, Universität Regensburg, 93040 Regensburg, Germany
[†]Walter Schottky Institut, Technische Universität München, Am Coulombwall, 85748 Garching, Germany

Abstract. Employing the Cleaved Edge Overgrowth technique, GaAs/AlGaAs quantum wire cascade structures have been fabricated. The quantum wire states are formed by the perpendicular overlap of two confinement potentials, one resulting from a strong potential modulation generated by quantum wells along the [001]-crystal direction, and a second resulting from an additional in-plane confinement generated by a silicon-δ-doping along the [110]-crystal direction. Radiative electronic transitions between discrete energy levels in coupled quantum wires are predicted in these samples. Above a threshold of 200 mA, mid-infrared electroluminescence is observed at an energy of 150 meV. The devices were temperature controlled at 20 K. The emission intensity is clearly current dependent and rises linearly with a slope efficiency of 0.1 nW/mA up to a maximum output power of 17 nW.

INTRODUCTION

Since their demonstration in 1994 [1], quantum cascade lasers (QCL) have become more and more the workhorse for the mid- [2] and far- [3] infrared spectral region. They act as compact and intensive unipolar light sources for spectroscopic applications. In 1998, the first QCL in the GaAs/AlGaAs heterosystem was presented [4]. Light emission in all these examples is caused by radiative electronic transitions between the discrete energy levels in coupled quantum wells. In the early concepts, inversion in quantum cascade structures is achieved by resonant electron-LO-phonon scattering. Even if essential for inversion, this scattering mechanism also represents a fast non-radiative decay channel in competition with the photon emission and thus compromises the performance of these devices. In our theoretical studies [5], a realistic quantum wire emitter device was introduced and we critically compared the LO-phonon transition rates of the well and wire structure. The result is a decrease of the non-radiative losses in the wire system caused by the reduced dimensionality of the underlying electron system.

SAMPLE STRUCTURE AND SET-UP

The emitter devices have been fabricated by molecular beam epitaxy (Fig. 1, [6, 7]), employing the Cleaved Edge Overgrowth technique [8]. The quantum wire states are formed due to overlap of the two one-dimensional

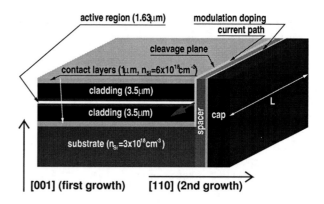

FIGURE 1. Scheme of a quantum wire cascade device. Displayed are the relevant crystal directions, and the direction and outcoupling position of the emitted light.

confinement potentials along two perpendicular crystal directions. One potential in the first growth step is the active region of a GaAs/AlGaAs quantum cascade structure similar to that published in [4] along the [001] crystal direction, but left undoped. After preparation of an atomically smooth (110) surface by cleaving, a second layer sequence is grown along the [110] direction, containing a silicon-δ-doping. This δ-doping leads to an in-plane confinement potential on the cleaved edge. The samples are then contacted with diffused indium contacts, and cleaved in devices with variable length L. Silver epoxy is used to connect these devices with wires. This is then followed by mounting the devices on the

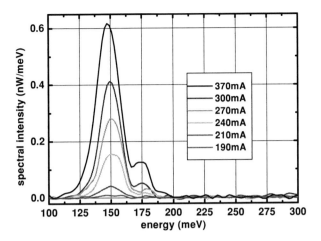

FIGURE 2. Current dependent electroluminescence spectra of a quantum wire cascade device at 20 K.

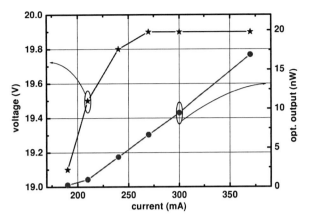

FIGURE 3. Optical output power and voltage vs. current characteristics of a quantum wire cascade device at 20 K.

cold finger of a helium flow cryostate, allowing temperatures between 4 K and room temperature. The devices are operated in pulsed current mode with pulse repetition rates of 79.1 kHz and 31.6 % duty cycle. The emitted light is collected with a lense, detected with a liquid nitrogen cooled mercurycadmiumtelluride detector and analyzed in a Fourier transform spectrometer operating in step scan mode.

RESULTS AND DISCUSSION

At cryogenic temperatures, emission spectra of the above-described devices have been recorded. Current dependent electroluminescence spectra for a device with length L=0.8 mm are displayed in Fig. 2. Above a threshold current of 200 mA, electroluminescence is observed and the detected spectral intensity of the emitted light reaches a maximum at an energy of about 150 meV, with a full width at half maximum of about 25 meV. The integrated optical output power for that device increases linearly, with a slope efficiency of about 0.1 nW/mA (Fig. 3), and reaches a maximum of 17 nW. The current-voltage characteristics for the device is also given in Fig. 3. Above emission threshold the voltage does not significantly depend on the current. Using a two-dimensional Schrödinger-Poisson solver, electronic states in the quantum wire region have been calculated. The energy difference between excited state and ground state is estimated to be 151 meV [5], which agrees reasonably well with the experimentally observed results.

SUMMARY

In conclusion, we have demonstrated field-effect induced mid-infrared electroluminescence of quantum wire cascade structures at 20 K. The emission is centered around 150 meV with a full width at half maximum of 25 meV. The observed emission peak is in good agreement with our calculations.

REFERENCES

1. J. Faist, F. Capasso, D. L. Sivco, C. Sirtori, A. L. Hutchinson, and A. Y. Cho, Science **264**, 553 (1994).
2. M. Beck, D. Hofstetter, T. Aellen, J. Faist, U. Oesterle, M. Ilegems, E. Gini, and H. Melchior, Science **295**, 301 (2001).
3. R. Köhler, A. Tredicucci, F. Beltram, H. E. Beere, E. H. Linfield, A. G. Davies, D. A. Ritchie, R. C. Iotti, F. Rossi, Nature **417**, 156 (2002).
4. C. Sirtori, P. Kruck, S. Barbieri, P. Collot, and J. Nagle, Appl. Phys. Lett. **73**, 3486 (1998).
5. I. Keck, S. Schmult, W. Wegscheider, M. Rother, and A. P. Mayer, Phys. Rev. B **67**, 125312 (2003).
6. S. Schmult, I. Keck, T. Herrle, W. Wegscheider, M. Bichler, D. Schuh, G. Abstreiter Appl. Phys. Lett. **83**, 1909 (2003).
7. S. Schmult, I. Keck, T. Herrle, W. Wegscheider, A. P. Mayer, M. Bichler, D. Schuh, G. Abstreiter, Phys. E **21**, 223 (2004).
8. L. N. Pfeiffer, K. West, H. L. Störmer, J. P. Eisenstein, K. W. Baldwin, D. Gershoni, and J. Spector, Appl. Phys. Lett. **56**, 1697 (1990).

Direct Observation of Excitonic Lasing from Single ZnO Nanobelts at Room Temperature

Kazuki Bando, Taiki Sawabe, Koji Asaka, and Yasuaki Masumoto

Institute of Physics, University of Tsukuba, Tsukuba, 305-8571, Japan

Abstract. Excitonic lasing from single ZnO nanobelts was observed at room temperature, which was due to the exciton-exciton scattering processes appearing under intense light excitation. Morphologies of the nanobelts are rectangular shapes, so that crystalline facets of the nanobelts acted as the lasing cavity mirrors. We demonstrated that mode spacings correspond to cavity lengths of the respective nanobelts, and directly observed the lasing from single ZnO nanobelts by mapping of the luminescence intensity.

INTRODUCTION

Semiconductor nanostructures are recently paid much attention for applications to electric and optical devices. In particular, the nanostructures whose dimension compare with the wavelength of light are worthy of attention, since the morphology for such nanostructures sensitively affects their optical functionalities. ZnO nanostructures are well suitable to investigate such functionalities, because it can be expected that the excitonic lasing occurs even at room temperature [1,2]. Therefore, we have studied the excitonic lasing from ZnO beltlike nanostructures (nanobelts) [3,4] whose morphology is peculiar, and tried to observe directly the lasing depending on their morphology.

EXPERIMENTAL

The ZnO nanobelts were simply grown by thermal evaporation processes of the ZnO powder without catalysts [3]. Photoluminescence (PL) from the single nanobelts under weak and intense excitation conditions were obtained by using a continuous wave (cw) He-Cd laser (325 nm), the fourth harmonic generation (266 nm) of a pulsed Nd:YAG laser and pulsed XeCl excimer laser (308 nm) as excitation sources, respectively. PL images of the single nanobelts were enlarged by using a long-distance microscope to detect the PL from only a single nanobelt from the ones dispersed on copper substrates, and the PL was detected by using a monochromator and a liquid nitrogen cooled CCD.

RESULTS AND DISCUSSION

Figure 1 shows the excitation density dependence of PL spectra at room temperature. The dimensions of the observed nanobelt are about 10 μm in length, 2 μm in width, and ~100 nm in thickness. In Fig. 1, the undermost PL spectrum was obtained under the weak excitation of the cw excitation. As seen in Fig. 1, a broad PL band appears, which consists of PL bands due to free excitons and their phonon replicas. The upper PL spectrum was obtained by the weakest excitation for the pulsed excitation. The PL peak position was observed at lower energy than that observed for the cw excitation. The PL band appears around 3.25 eV which is lower energy than the exciton resonance (~3.31 eV) by the exciton binding energy (59 meV), which indicates that spontaneous emission (P) due to the exciton-exciton scattering processes appears under intense excitation. With increasing excitation density, three sharp peaks (P_S) abruptly appear at the lower energy tail (~3.2 eV) of the P band above an excitation condition of ~350 (kW/cm^2). Considering that the PL peaks appear above a certain threshold power, it is concluded that stimulated emission occurs. Upon increasing the excitation

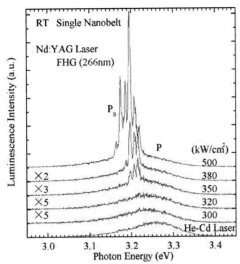

FIGURE 1. PL spectra of the single ZnO nanobelt at room-temperature.

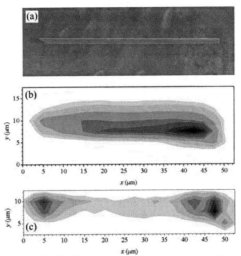

FIGURE 2. (a) Scanning electron microscopy image of the single nanobelt, (b) PL mapping for weak excitation condition, and (c) for intense excitation condition.

density further, the intensities of the sharp peaks superlinearly increase and the number of the peaks increases on the lower energy side. Then, these peaks consist of several sharp peaks whose energy spacings are constant. In addition, the energy spacings obtained from PL spectra of several nanobelts whose lengths are different from each other are also constant, depending on the lengths of the nanobelts. This clearly indicates that lasing is brought about by the self-formed cavity of the nanobelts. The cavity modes appear as the several peaks with constant energy spacings in the PL spectra. In fact, the energy spacings of the nanobelts coincided with those expected for the Fabry-Pérot cavity model [4]. It is very important that this lasing action occurs without external mirrors and the nanolaser cavity is formed by the crystalline facets alone. Although it is considered that the cavity is formed along the length of the nanobelt, the lasing was observable not only along the length of the nanobelt but also in the vertical direction of the substrate because of the scattering of the emission due to the roughness on the surface of the substrate.

In order to observe directly the lasing from the single nanobelt, mapping was performed with respect to the PL intensities. Figures 2(a), (b), and (c) show a scanning electron microcopy image of the single nanobelt, the mapping of the PL intensities performed under the cw and the pulsed excitation conditions, respectively. As seen in Fig. 2(b), the exciton PL emits throughout whole nanobelt under the cw excitation. Non-uniformity of the PL image is due to spatial distribution of the excitation. However, the mapping for the pulsed excitation condition for which the lasing occurs is obviously different from that for the cw excitation, as shown in Fig. 2(c). When the lasing occurs under the pulsed excitation, the PL due to the lasing emits only at both ends of the nanobelts. This strongly supports that the self-formed cavities of the nanobelts bring about the lasing and the cavity modes are formed along the lengths of the nanobelts.

In summary, we observed excitonic lasing from single ZnO nanobelts at room-temperature. In addition, the mapping with respect to the lasing was performed, and it was demonstrated that the luminescence emitted only at both ends of the nanobelts. It was concluded that the cavity modes were formed between both end facets of the nanobelts.

ACKNOWLEDGMENTS

This work was partly supported by the Sasakawa Scientific Research Grant from The Japan Science Society and Research Project.

REFERENCES

1. Huang M. H., Mao S., Feick H., Yan H., Wu Y., Kind H., Weber E., Russo R., and Yang P., *Science* **292**, 1897-1899 (2001).
2. Johnson J. C., Yan H., Schaller R. D., Haber L. H., Saykally R. J., and Yang P., *J. Phys. Chem.* B **105**, 11387-11390 (2001).
3. Pan Z. W., Dai Z. R., and Wang Z. L., *Science* **291**, 1947-1949 (2001).
4. Bando K., Sawabe T., Asaka K., and Masumoto Y., *J. Lumin.* **108**, 385-388 (2004).

Temperature dependent dynamics of the excitonic photoluminescence in zinc oxide nanorods

H. Priller[*], R. Hauschild[*], J. Zeller[*], C. Klingshirn[*], H. Kalt[*], F. Reuss[†], R. Kling[†], Ch. Kirchner[†] and A. Waag[**]

[*]Institut für Angewandte Physik, Universität Karlsruhe, Wolfgang-Gaede-Str.1, D-76131 Karlsruhe, Germany
[†]Abteilung Halbleiterphysik, Universität Ulm, Albert-Einstein Allee 45, 89081 Ulm, Germany
[**]Institut für Halbleitertechnik, TU Braunschweig, Hans-Sommer-Str. 66, D-38106 Braunschweig, Germany

Abstract. The temporal dynamics of the exciton photoluminescence (PL) in ZnO nanorod samples was investigated experimentally as a function of temperature and excitation intensity. Excitonic photoluminescence is observed up to room temperature. The excitation dependence of the PL dynamics reveals a saturable non-radiative recombination center. Under high excitation conditions the time-resolved photoluminescence shows two components: the ZnO M-band which decays with a temperature independent sub-100 ps time constant, and the intrinsic exciton PL with a time constant of several 100 ps increasing with temperature. Exciton-exciton scattering effects are notably absent, which is attributed to the reduced polariton phase space resulting from the small nanorod diameter of 50 nm.

Keywords: ZnO, excitons, polaritons
PACS: 71.36.+c

INTRODUCTION

Zinc oxide (ZnO) has a wide room temperature bandgap of 3.37 eV, which makes it interesting for optoelectronic applications in the near-UV[1, 2]. Especially, ZnO nanoparticles, like nanorods, are interesting as optical components integrated into nanometer-size electronic devices. The large exciton binding energy of 60 meV, in particular, has the potential for stimulated emission driven by excitons even at room temperature, promising a lower threshold and a higher gain than electron-hole plasma driven lasing processes.
In the work reported here, we investigated the dynamics of the photoluminescence spectrum in ZnO nanorods as a function of temperature and excitation intensity. In particular, the role of excitons in ZnO nanorods is explored.

SAMPLE AND EXPERIMENTAL ARRANGEMENT

The nanorods were grown by MOVPE on an a-plane sapphire substrate. They have a uniform length (5 μm) and diameter (50 nm) with an orientation perpendicular to the sapphire substrate. The hexagonal cross section is typical for the wurtzite crystal structure of ZnO.
The sample was excited by sub-100 fs pulses at 280 nm from a frequency-tripled titanium sapphire laser. The time dependent PL emission spectrum was measured at different temperatures between 9 and 290 K with a combination of spectrometer and streak camera. The spectral (temporal) resolution was 1 meV (15 ps). For high-excitation experiments, the sample was excited with a XeCl excimer laser (20 ns pulses at 308 nm).

EXPERIMENTAL RESULTS AND DISCUSSION

For spectral characterization, the cw photoluminescence spectrum under weak excitation conditions was recorded and is shown in Fig.1. At low temperatures, the PL spectrum shows the well-known lines of the transition of the donor-bound excitons (*BX*) at 3.355 eV and of the free A-excitons (*AX*) at 3.370 eV. The energies coincide with those of bulk material and prove the absence of strain. No quantum confinement effects are observed, as the nanorod diameter is larger than the exciton Bohr radius. At temperatures above 100 K the *BX* lines disappear, and one is left with the free exciton recombination and its LO-phonon replicas which merge to a broad band at room temperature. The presence of the first two LO-phonon replicas of the free excitons and the absence of the deep center emission of ZnO in the green, yellow or red prove the high crystal quality.
The normalized photoluminescence spectrum for different excitation conditions are shown in Fig.2 for a temperature of 70 K. For low excitation intensities, the spectrum is dominated by the donor-bound exciton (*BX*) at 3.35 eV and the A-exciton at 3.37 meV, accompanied by two exciton LO-phonon replica redshifted by 72 meV and 144 meV. For high excitation intensities, the well

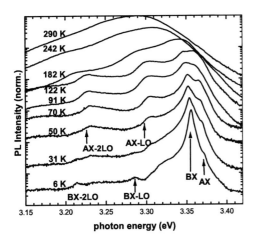

FIGURE 1. Photoluminescence spectrum under weak cw-excitation conditions at different temperatures. The spectra are shifted with respect to each other.

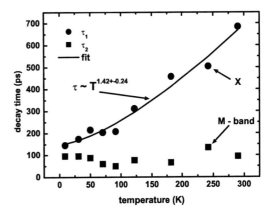

FIGURE 3. Temperature dependence of the DAS time constants for the excitonic photoluminescence (τ_2, circles) and the M-band (τ_1, squares)

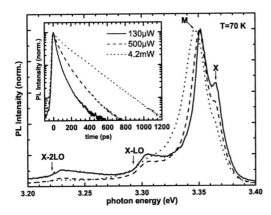

FIGURE 2. Normalized photoluminescence spectra for different excitation intensities. Inset: Time-resolved photoluminescence at the peak of the spectrum

known nonlinear M-band [5] appears and dominates both excitonic features and their LO-phonon replica.

Remarkably, the decay of the PL signal gets slower with increasing excitation intensity, as can be seen in the inset of Fig.2. This behavior is attributed to the saturation of defect states, possibly surface states, with increasing excitation intensity. They can act as non-radiative recombination centers for excitons, as the exciton radiative lifetime exceeds that for non-radiative recombination.

The spectro-temporal dynamics of the excitonic PL at different temperatures was studied under high-excitation conditions, at which non-radiative recombination centers are saturated. The measured intensity $I_{PL}(E_{Ph},t)$ (E_{Ph} :photon energy) was analyzed using the method of decay-associated spectra (DAS) [3]. This method decomposes $I_{PL}(E_{Ph},t)$ into a sum of spectral components $A_i(E_{Ph})$, which each decay exponentially with a decay time constant τ_i.

This analysis allows to separate the spectrum $A_1(E_{Ph})$ of the nonlinear M-band (τ_1 approx. 80 ps), from the slower decaying intrinsic exciton component $A_2(E_{Ph})$. The temperature dependence of the decay time constants τ_1 and τ_2 is depicted in Fig.3. While the decay of the M-band component is largely independent of temperature, the excitonic decay time constant increases from 145 to 680 ps between 9 and 290 K, following a $\tau_2 \propto T^{1.42\pm0.26}$-power law. This temperature dependence is close to the $\tau \propto T^{3/2}$ scaling expected for bulk excitons [4] and confirms the aforementioned absence of exciton confinement in our nanorod samples.

No exciton-exciton scattering (P-band [5]) was observed even under high-excitation conditions (several MW/cm^2 peak intensities) in the nanorods, unlike in ZnO epilayers [6]. This finding is attributed to the reduced phase space of exciton polaritons, which is one-dimensional in the wave-guide like nanorods, despite the bulk character for excitons just mentioned.

REFERENCES

1. M.H. Huang et al., Science **292**, 1897 (1983)
2. J.C. Johnson et al., J.Phys.Chem. **105**, 11389 (2001)
3. J.R. Knudson et al., Chem. Phys. Lett. **102**, 501 (1983)
4. L.C. Andreani et al., Solid State Comm. **77**, 641 (1991)
5. C. Klingshirn, H. Haug, Phys. Reports **70**, 315 (1981)
6. H. Priller, J. Brückner, Th. Gruber, C. Klingshirn, H. Kalt, A. Waag, H.J. Ko, T. Yao, phys. stat. sol.(b) **241**, 587 (2004)

Purely strain induced GaAs/InAlAs single quantum wires exhibiting strong charge carrier confinement

R. Schuster*, H. Hajak*, M. Reinwald*, W. Wegscheider*, D. Schuh[†], M. Bichler[†], S. Birner[†], P. Vogl[†] and G. Abstreiter[†]

*Institut für Experimentelle und Angewandte Physik, Universität Regensburg, 93040 Regensburg, Germany
[†]Walter Schottky Institut, Technische Universität München, Am Coulombwall, 85748 Garching, Germany

Abstract. Atomically precise strained quantum wire structures are obtained using the cleaved edge overgrowth technique. Strong carrier confinement is achieved purely by lateral strain variation within a single quantum well. In the first growth direction InAlAs layers serve as stressor material. Growing a GaAs quantum well directly on the cleaved (011) plane in a second growth step, one ends up with a strongly strain modulated T-shaped structure. The confinement energy rises as expected with the thickness of the stressor layer and the width of the overgrown quantum well. This is confirmed both by numerical simulations and by spatially resolved photoluminescence measurements. Experimentally, confinement energies of up to 51.5 meV with respect to the corresponding energy for the quantum well are obtained, which is approximately twice the value of $k_B T$ at room temperature. The confinement energy can be enlarged by decreasing the incident power, which can be explained by screening effects. The electron and hole wave functions are spatially separated due to the piezoelectric effect which is incorporated in the simulations. The calculations of the strain distributions and wave functions are presented as a tool for optimizing the sample layout.

INTRODUCTION

Low-dimensional systems are of interest both in fundamental research and in applied physics because of their inherent properties. Here, we discuss cleaved edge overgrowth [1] quantum wire (QWR) structures at the intersection of two perpendicular quantum wells (QWs), which have shown to provide enhanced exciton binding energies [2] and a concentration of the oscillator strength [3]. For clear one-dimensional characteristics at room temperature a large energy separation between the constituting two-dimensional QWs and the QWR is needed. The optimization of this confinement energy was sought by a variety of groups using unstrained [2, 4] and strained structures [5]. A more recent work concentrated on the improvement of the (011) GaAs interface resulting in reduced photoluminescence (PL) linewidths [6].

SAMPLE STRUCTURE

In the references cited above, the (100) QWs and the (011) QWs are of type I, thus allowing for electron and hole confinement. At the T-shaped intersection a one-dimensional QWR running along the [0$\bar{1}$1] direction is formed due to the fact that electron and hole wave functions can expand into a larger volume. In the sample

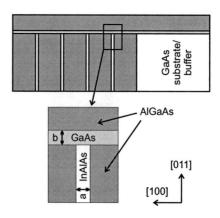

FIGURE 1. Sketch of the sample layout. QWRs are formed in the GaAs QW by the tensile strain field induced by the $In_{0.16}Al_{0.84}As$ layer.

sketched in Fig. 1, an $In_{0.16}Al_{0.84}As$ layer (width a) acts as a barrier. However, since this material has a larger lattice constant than $Al_{0.3}Ga_{0.7}As$, it is subject to tensile strain in the [100] direction, which is transmitted to the overgrown GaAs QW (width b), where a QWR is formed at the lateral positions of the stressor layers. In contrast to an early work on strained QWRs [7] the above concept, which was suggested in Ref. [8], does not need an additional barrier on the cleaved edge.

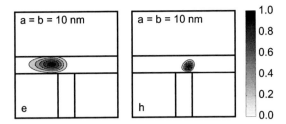

FIGURE 2. Contour plots of the electron (e) and hole (h) probability distributions in a strained GaAs/In$_{0.16}$Al$_{0.84}$As QWR structure.

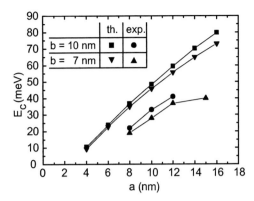

FIGURE 3. Theoretical (th.) and experimental (exp.) values for the confinement energy E_c in two series of strained GaAs/In$_{0.16}$Al$_{0.84}$As QWRs (eye guiding lines added).

RESULTS

Since this direct growth on the cleaved edge leads to strong strain modulations, large confinement energies were predicted for such structures using eight band $\mathbf{k} \cdot \mathbf{p}$ calculations [9]. Here, we present the results of computations with the device simulator next**nano**^{3} [10], which we use for the optimization of the sample layout. Although only a simple one band approach was available, the simulator has the advantage of including effects based on piezoelectricity. After the minimization of the strain energy, the Schrödinger and Poisson equations are solved self-consistently. Figure 2 depicts the electron and hole probability distributions in a strained GaAs/In$_{0.16}$Al$_{0.84}$As QWR. The separation of the electron and hole wave functions is an effect of piezoelectricity. Neglecting the piezo effect results in symmetric distributions as presented in Ref. [9].

In Fig. 3 the results of micro-PL measurements [11] are plotted and compared to the theoretical values. While the rising of the confinement energy with increasing stressor layer thickness is very well reproduced with our single band approach, the calculated absolute values do not match the experimental data very well, since no Coulomb or excitonic effects were included. The measured values are dependent on the excitation power, since screening effects have to be taken into account if an increasing number of carriers is present in the QWR. The experimental values shown in Fig. 3 represent the data for the excitation power set to 1 μW, while low excitation experiments yielded confinement energies of up to 51.5 meV. For the data shown in Fig. 3 one should note that the layer width $a = 15$ nm is close to the critical thickness as confirmed by atomic force microscopy.

In order to enhance the overlap of the electron and hole wave functions the In$_{0.16}$Al$_{0.84}$As layer can be replaced by quaternary In$_{0.16}$Al$_{0.84-x}$Ga$_x$As. With increasing x the band gap of the stressor layer gets smaller without relevantly changing the strain configuration. The simulations show that for $x = 0.70$ the electron wave function is L-shaped with the maximal value located closer to the center of the hole wave function than shown in Fig. 2.

SUMMARY

In conclusion, we demonstrated strong charge carrier confinement in strained single QWR structures. The dependence of the confinement energy on the thickness of the InAlAs and of the overgrown layer was calculated with a single band approach and verified by spatially resolved PL measurements. Thus, the applied concept of replacing the QW of the first growth direction by a stressor layer is promising for the development of room temperature devices based on strained QWR systems.

REFERENCES

1. L. Pfeiffer, K. W. West, H. L. Stormer, J. P. Eisenstein, K. W. Baldwin, D. Gershoni, and J. Spector, Appl. Phys. Lett. **56**, 1697 (1990).
2. T. Someya, H. Akiyama, and H. Sakaki, Phys. Rev. Lett. **76**, 2965 (1996).
3. H. Akiyama, T. Someya, and H. Sakaki, Phys. Rev. B **53**, R16160 (1996).
4. W. Langbein, H. Gislason, and J. M. Hvam, Phys. Rev. B **54**, 14595 (1996).
5. J. R. Jensen, J. M. Hvam, and W. Langbein, J. Cryst. Growth **227-228**, 966 (2001).
6. M. Yoshita, H. Akiyama, L. N. Pfeiffer, and K. W. West, Appl. Phys. Lett. **81**, 49 (2002).
7. D. Gershoni, J. S. Weiner, S. N. G. Chu, G. A. Baraff, J. M. Vandenberg, L. N. Pfeiffer, K. West, R. A. Logan, and T. Tanbun-Ek, Phys. Rev. Lett. **65**, 1631 (1990).
8. D. V. Regelman and D. Gershoni, *Proc. of the 24th Int. Conf. on the Physics of Semiconductors*, edited by D. Gershoni (World Scientific, Singapore, 1999), p. 1111.
9. M. Grundmann, O. Stier, A. Schliwa, and D. Bimberg, Phys. Rev. B **61**, 1744 (2000).
10. next**nano**^{3} device simulator, see websites www.wsi.tum.de/nextnano3 and www.nextnano.de
11. R. Schuster, H. Hajak, M. Reinwald, W. Wegscheider, D. Schuh, M. Bichler, and G. Abstreiter, Physica E **21**, 236 (2004).

Negative differential conductance in cleaved edge overgrown surface superlattices

T. Feil*, H.-P. Tranitz*, M. Reinwald*, W. Wegscheider*, M. Bichler†, D. Schuh†, G. Abstreiter† and S. J. Allen**

*Institut für Experimentalphysik, Universität Regensburg, 93040 Regensburg, Germany
†Walter Schottky Institut, TU München, Am Coulombwall, 85748 Garching, Germany
**Physics Department, UCSB, Santa Barbara, California 93106

Abstract. In situ overgrowth of an undoped GaAs/$Al_{0.3}Ga_{0.7}As$ superlattice with a gate electrode separated by a barrier layer produces an array of strongly coupled quantum wires. The gate allows direct control of edge channel density. The observed transport properties are very sensitive to the transport channel length. Samples with long superlattices exhibit transport characteristics dominated by an inhomogeneous density and field distribution along the channel. In shorter superlattice samples a leakage current through the bulk superlattice stabilizes the field distribution in the two-dimensional channel and allows the observation of current-voltage characteristics that exhibit strong negative differential conductance without the formation of electric field domains.

Electric field instabilities prevent the observation of the intrinsic negative differential conductance (NDC) predicted for transport in superlattice structures in a uniform electric field[1]. The lower dimensionality of the surface superlattices presented here changes the dynamics of these field instabilities and allows a greater influence by the structural surroundings. In particular, parallel bulk transport allows us, for the first time, to extract the non-equilibrium negative differential conductance without any electric field instabilities.

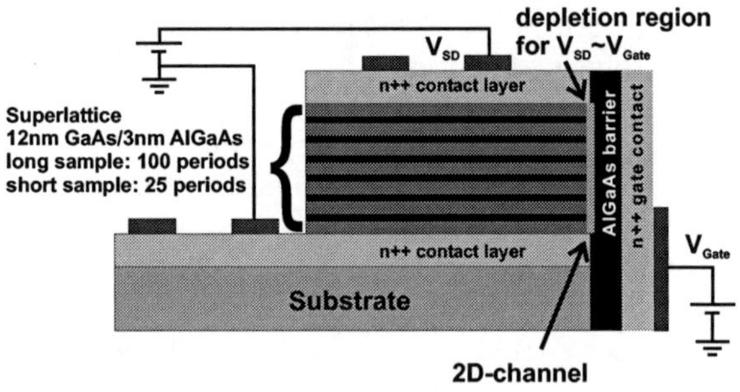

FIGURE 1. Sample structure.

THE DEVICE

The strongly coupled array of quantum wires [2] is realized by Cleaved-Edge-Overgrowth (CEO). An undoped GaAs/$Al_{0.3}Ga_{0.7}As$ superlattice, sandwiched between doped contact layers, is cleaved in situ and is overgrown with an optional thin well, a 100nm thick $Al_{0.3}Ga_{0.7}As$ barrier and a highly doped gate contact. By applying a positive gate voltage, an array of coupled electronic wires is established under the gate barrier. Current-voltage characteristics (I-V) are measured in the superlattice direction while the 2-D electron density is varied by the applied gate voltage. Here we describe results for two different channel lengths referred to as long and short channel samples (cf. figure 1). The former has 100 superlattice periods, the latter only 25. The period length in both cases is 15nm and the barrier width 3nm which results in an estimated miniband width of about 4meV.

LONG CHANNEL SAMPLES

Long channel devices [2] revealed current voltage characteristics like those shown in figure 2. For low applied gate voltages the I-V exhibits stable negative differential conductance with no hysteresis. At higher gate voltages electic field switching discontinuities appear in the I-V. Both regimes indicate transport with non-uniform electric fields. These features are expected for superlattice transport. Assuming a uniform electric field, the peak current voltage position corresponds to a Bloch frequency of only 75 GHz. If the field distribution remains uniform above this voltage, the samples should exhibit a rich terahertz photoconductivity when excited by ter-

CP772, *Physics of Semiconductors: 27th International Conference on the Physics of Semiconductors,*
edited by José Menéndez and Chris G. Van de Walle

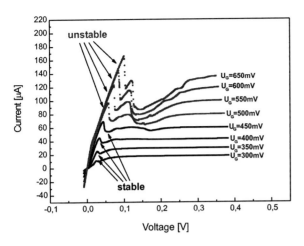

FIGURE 2. I-V-characteristics showing the initially observed NDC in the perpendicular transport through the long wire array.

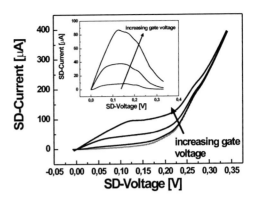

FIGURE 3. The lowest lying gray trace shows the leakage current through the superlattice of reduced thickness. The black lines show the I-V-characteristics of the combined transport of the leakage and the two-dimensional channel for different gate voltages. The inset shows the I-V-traces of the two-dimensional channel which are recovered from the overall I-V-traces by subtraction of the zero gate voltage characteristic.

ahertz radiation [3, 4]. Measurements carried out with the UCSB-Free-Electron-Laser at a variety of frequencies revealed no such spectroscopic features. Furthermore, the geometry of the long channel device is such that moderate source drain voltages are comparable to applied gate voltages. This leads to channel densities and electric fields that are non-uniform even in the absence of the superlattice instabilities. Intrinsic instabilities and extrinsic channel inhomogenities inhibit the observation of homogeneous transport in the long channel devices.

SHORT CHANNEL SAMPLES

In order to shift the NDC peak position to source drain voltages smaller than the gate voltage, short channel samples were fabricated. The I-V-characteristics of such a device are shown in figure 3. One finds, that there is a substantial current flow for zero applied gate voltage, with no edge channel present (gray line). This is caused by current flow in the bulk undoped superlattice and can be confirmed by measuring identical current flow through the bulk with the gate mechanically removed. Subtracting the zero gate voltage current from the combined bulk and edge channel transport reveals the I-V characteristics of the two-dimensional system and is shown in the inset of figure 3. To the best of our knowledge, this is the first measurement of the current-voltage characteristic of a Bloch oscillating miniband free of the electric field instabilities. It seems likely that the shunting transport in the bulk stabilizes the fields in the surface superlattice [5]. This leaves the composite conductance positive, but through gate voltage modulation we can extract the underlying superlattice I-V often predicted but never observed. Figure 3 shows the characteristic features predicted by [1]. For small channel electric fields transport is ohmic. For increasing electric fields the non-parabolic dispersion leads to a current saturation followed by a long tail of NDC. We conclude that short, gate controlled CEO surface superlattices shunted by substrate leakage offer for the first time the possibility to explore stable non-linear transport in superlattices in uniform electric fields.

ACKNOWLEDGMENTS

We gratefully acknowledge financial support by the DFG via SFB348 and by the BMBF under contract 01BM918. SJA gratefully acknowledges support by the Alexander von Humboldt Foundation and ARO.

REFERENCES

1. Esaki, L., and Tsu, R., *IBM J. Res. Dev.*, **14**, 61 (1970).
2. Deutschmann, R. A., Wegscheider, W., Rother, M., Bichler, M., and Abstreiter, G., *Physica E (Amsterdam)*, **7**, 294–298 (2000).
3. Unterrainer, K., Keay, B. J., Wanke, M. C., Allen, S. J., Leonard, D., Medeiros-Ribeiro, G., Bhattacharya, U., and Rodwell, M. J. W., *Phys. Rev. Lett.*, **76**, 2973 (1996).
4. Kroemer, H., *Cond. Mat.*, **0007482** (2003).
5. Daniel, E. S., Gilbert, B. K., Scott, J. S., and Allen, S. J., *IEEE Trans. Electron Devices*, **50**, 2434–2444 (2003).

Electronic and geometric structure of one dimensional wires: An STM Study

J. Lee, H. Kim, Y.J. Song and Young Kuk

School of Physic and CSNS, Seoul National University, Seoul, 151-747 Korea

Abstract. Many device scientists believe that current Ultra Large Scale Integration (ULSI) technology will face technical and economic difficulties in further miniaturization. It has been predicted that 1-dimensional (1-D) transistors with connecting wires in three-dimensionally stacked structures may replace current field effect transistors in planar integration structures. We propose a new scheme to fabricate and integrate 1-D active devices. As a first step, we show the way to form 1-D wires with spatially variable electronic structures and the way to characterize them.

INTRODUCTION

With the development of current silicon processing technology, a modern silicon chip has come to include more than two hundred million complementary metal oxide semiconductor (CMOS) transistors, thanks to the invention of transistor [1]. In the most recent commercial chip, the channel length of a commercial unit CMOS transistor has become as small as 0.09 µm and may continue to shrink perhaps down to 0.01 µm. However, device scientists experience many technical difficulties in this scaling-down process, such as a light source of short wavelength, accurate alignments, fine etching, thin-film deposition, and new dielectric materials. These challenges may be too difficult or expensive to achieve. In order to avoid all these processing problems, it has been suggested that we may have to replace the current unit-device structure with a new device structure, for example a one-dimensional (1-D) structure [2-4].

Scientists already discovered interesting transport properties in 1-D systems in 1974 [5]. Organic, metallic, semiconductor and biological wires have shown metallic or semiconductor properties. Straightforward candidates for functional and interconnecting wires are semiconductors and metallic wires in a nano-wire structure as shown in Fig. 1a. New tubular wires like carbon nanotubes show metallic and semiconductor property [6], therefore they can be used to produce active devices or interconnecting wires (Fig. 1b). If elemental wires are grown sequentially, one dimensional superlattice can even be produced as shown in Fig. 1c. Some groups are successful to grow these elemental wires using either catalytic seeds of nanometer-size dots [5] or zeolite-like cage structures. These wires are, however, so easily oxidized because of their dimensions. If these elemental wires are covered with other inert material as a tube form, such as carbon nanotubes, the problem of oxidation can be avoided (Fig. 1d).

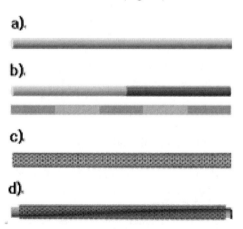

FIGURE 1. (a) Elemental wire grown in a cage structure or with catalytic seeds. (b) Superlattice 1D wires, grown sequentially. (c) 1D wire in a tube form. (d) Elemental wires grown in a tube.

CARBON NANOTUBE AS 1-D WIRE

Among the many functional 1-D systems, we chose to study carbon nanotubes (CNT) as test 1-D systems for their well-known geometrical structures, symmetric properties, electrical properties and synthesis technologie. Ever since the discovery of carbon nanotubes a decade ago [6], scientists have succeeded in synthesizing a variety of novel forms of carbon nanotubes. CNTs are known to have metallic and semiconductor properties depending on their chiralities. During growth process, various defects are formed in CNTs, such as pentagonal or heptagonal carbon rings. These defects appear as a pair or a multiple and cause the change of chiralities around them. Metal-semiconductor or semiconductor-semiconductor CNT junctions are often observed in TEM and STM images. Figure 2 shows a semiconductor-semiconductor junction [7]. Interestingly, defects at the junction appear to show a $\sqrt{3} \times \sqrt{3}$ superstructure in STM topography. In order to observe the local variation of the electronic structure, we obtained dI/dV spectra at 512 sites along the nanotube just after the topography was taken. The x-axis indicates the position along the tube, the y-axis the energy, and the z-axis the density of states, which is proportional to the dI/dV at the measured point, as shown in Fig 2d. It takes 30-60 min. to take 512 dI/dV spectra. Thermal drift is less than 0.1nm/h at this temperature, 5K. Two different band gaps are visible together with a localized donor state at the junction. In a tight binding calculation, the band structure around the junction and the donor state is explained (Fig. 2e).

Recently, it has been shown experimentally that fullerenes or endohedral metallofullerences [8] such as Gd encapsulated inside C_{82} (Gd@C_{82}: GdMF) can be inserted into single-wall nanotubes (SWNTs), forming a peapod-like structure [9-11]. When the diameter of a fullerene slightly exceeds the inner diameter of SWNT minus van der Waals bond length, the fullerene is inserted endothermally and the resultant SWNT is elastically strained. A theoretical study has predicted that the structure is severely modified, including the positions of the van Hove singularities [VHS] when an SWNT is uniaxially strained [12]. As we combine these ideas, we can perform a 'local band gap engineering' at the site where a fullerene is endothermally inserted. One can predict that the band gap of a semiconductor nanotube would be altered by the insertion of large fullerenes.

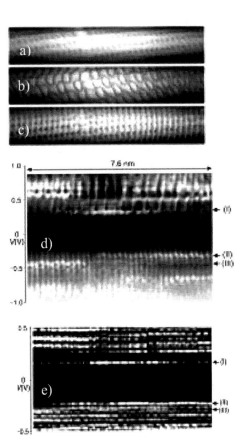

FIGURE 2. Topographic images of a semiconductor-semiconductor junction imaged at (a) 1.5 V, (b) 0.5 V and (c) -0.3 V. (d) Spatially resolved band structure: 512 composite scanning tunneling spectroscopies along the axis of the CNT in (c). (e) Calculated band structure along the axis.

As reported [11], protrusions in STM topography do not necessarily indicate the exact locations of inserted GdMFs, as the electronic structures including band edges are severely modified with the insertion as shown in Fig. 3. The chirality of the GdMF-SWNT is determined to be (11,9) from the measured diameter of 1.4 nm (nominally 1.375 nm) and the chiral angle of 27° (nominally 26.7°). In the dI/dV spectra, conduction band edge is lowered by ~0.2 eV at the position where GdMFs are inserted. There are two major causes which could account for the change of the band gap: 1) Elastic strain can change the band gap significantly. For example, a strain of 4% in the tube axis direction can induce a gap reduction of 60% for the (15,1) GdMF-SWNT. 2) Midgap states are induced with inserted GdMFs. The midgap state merges with the conduction band and results in reduction of the bandgap.

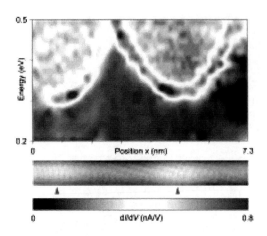

FIGURE 3. Change of conduction band with inserted Gd@C_{82} metallofullerenes. GdMFs are present at the positions of arrows.

ACKNOWLEDGMENTS

We thank J. Ihm for theoretical calculation, and H. Shinohara for their CNT samples with filled Cs and metallofullerens. This work was supported by Korean Ministry of Science and Technology through Creative Research Initiatives Program.

REFERENCES

1. Riordan, M., and Hoddeson, L., *Crystal Fire: The invention of the Transistor and the Birth of the Information Age,* New York: W.W. Norton & Co., 1990.
2. Reed, M.A., *Proc. IEEE* **87,** 652 (1999).
3. Hu, J., Ouyang, M., Yang, P., and Lieber, C.M., *Nature* **399,** 48 (1999).
4. Gudiksen, M.S., Lauhon, L.J., Wang, J., Smith, D.S., and Lieber, C.M., *Nature,* **415,** 617 (2002).
5. Aviram, A., and Ratner, M., *Chem. Phys. Lett.* **29,** 277 (1974)
6. Iijima, S., and Ichihashi, T., Nature, **363,** 603 (2000)
7. Kim, H., Lee, J., Kahng, S.-J., Son, Y.-W., Lee, S.B., Lee, C.-K., Ihm, J., and Kuk, Y., *Phys. Rev. Lett.* **90,** 2161071 (2003)
8. Shinohara, H., *Rep. Prog. Phys.* **63,** 843 (2000)
9. Smith, B.W., Monthioux, M. , and Luzzi, D.E., *Nature* **396,** 323 (1998)
10. Suenaga, K., Tencé, M., Mory, C., Colliex, C., Kato, H., Okazaki, T., Shinohara, H., Hirahara, K., Bandow, S., and Iijima, S., *Science* **290,** 2280 (2000)
11. Lee, J., Kim, H., Kahng, S.-J., Kim, G., Son, Y.-W., Ihm, J., Kato, H., Wang, Z.W., Okazaki, T., Shinohara, H. , and Kuk, Y., *Nature* **415,** 1005 (2002)
12. Yang, L. and Han, J. *Phys. Rev. Lett.* **85,** 154 (2000)

Coupling of Two Localized Magnetic Moments and its Detection

V. I. Puller and L. G. Mourokh

Department of Physics, Stevens Institute of Technology, Hoboken, New Jersey 07030, USA

A. Shailos and J. P. Bird

Department of Electrical Engineering &Center for Solid State Electronics Research Arizona State University, Tempe, AZ 85287-5706, USA

Abstract. We present results of the theoretical analysis of a device consisting of two quantum point contacts (QPCs) biased close to their pinch-off conditions, so that a local magnetic moment (LMM) is formed in each of them. Another quantum wire, running in parallel to the QPCs, serves as a detector of the states of the two LMMs. Employing the approach of our earlier theory [V. Puller *et al.*, arXiv: cond-mat/0405705] we examine the sensitivity of the linear conductance of the wire to the states of the pair of LMMs. We show that the conductance of this wire exhibits a peak only when the LMMs are in the triplet state, confirming the potential of such a device for quantum computation.

INTRODUCTION

Anomalous behavior of quantum point contacts (QPCs) biased close to their pinch-off condition has been a subject of intensive research [1]. Existing theories suggest formation of a local magnetic moment (LMM), a strongly correlated many-body state, which forms in a one dimensional channel when the energy of the exchange interaction of the two spin polarizations become substantial in comparison to the kinetic energy of the electrons passing through the QPC [2]. Recently, we have studied experimentally and theoretically the feasibility of the detection of the LMM using the resonant tunnel coupling between this local-moment and a *detector* quantum wire, in close proximity to the QPC containing the LMM [3-5].

In experiment [3], the conductance of the detector wire exhibits a peak when the QPC is biased close to the pinch off condition. In [5] we modeled the LMM as a spin-1/2 magnetic moment, coupled via the exchange interaction to electrons traveling through the QPC. Due to the possibility of tunneling between the QPC and the detector wire, electrons in the latter are also coupled to the LMM; the particular form of this exchange coupling was obtained in [5] by decoupling the Schrödinger equations for electrons propagating in the QPC and the detector wire. We showed that, for a bottleneck shape of the QPC and the detector wire, ferromagnetic coupling between electrons and the LMM may be chosen in such a way that the conductance of the QPC will have a feature similar to the 0.7-anomaly. Simultaneously, there will exist a bound state inside of the bottleneck channel of the detector wire, which will cause a peak-like increase in the conductance due to resonant tunneling of the modes propagating below threshold [7].

Here, we present the results of the theoretical analysis of a device consisting of *two* such LMMs, as shown in Figure 1. Extending the approach of [5], we examine the sensitivity of the linear conductance of the detector wire to the state of the pair of LMMs. In this way, we show that detection of the triplet or singlet state of the two LMMs is possible by all-electrical means. This result confirms the potential of this structure for implementation in quantum computation, as was suggested in [6].

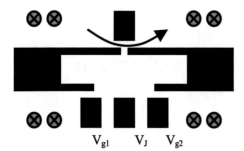

FIGURE 1. Schematic of the device. Black areas represent the metallic gates forming the *detecting* wire on the top and two QPCs formed by potentials V_{g1} and V_{g2} on the bottom. The potential on the middle gate (V_J) controls the exchange coupling between the two LMMs. Circles show the Ohmic contacts for performing various linear conductance measurements. The arrow shows the direction of the current for the conductance measurement described in the paper.

DETECTING THE STATE OF TWO LOCAL MOMENTS

Employing the approach of [5] we can show that the Schrödinger equation for electrons in the detector wire in the device shown in Figure 1 can be written in the form

$$[E - E_n - K_x - U(x) + j_1(x)\vec{\sigma}\cdot\vec{S}_1 + j_2(x)\vec{\sigma}\cdot\vec{S}_2 + J\vec{S}_1\cdot\vec{S}_2]\varphi(x) = 0, \quad (1)$$

where E is the electron energy, E_n is the energy of the bottom of the nth subband, and K_x is the electron kinetic-energy operator. The potential $U(x)$ describes the bottleneck shape of the detector wire and vanishes ($U(x) \to 0$) for $x \to \pm\infty$. $\vec{\sigma}$ and $\vec{S}_{1,2}$ are vectors of Pauli matrices for the spin of an electron passing through the detector wire and the two magnetic moments. $j_{1,2}(x)$ and J are the exchange couplings between the electron spin and the two LMMs. We consider a symmetric device with $j_1(x) = j_2(x) = j(x)$.

Algebraic manipulations allow us to express the conductance of the detector wire due to its nth mode as

$$G_n = \frac{2e^2}{h}\left[|t_0|^2 w_{\downarrow\uparrow-\uparrow\downarrow} + (w_{\uparrow\uparrow} + w_{\downarrow\downarrow} + w_{\downarrow\uparrow+\uparrow\downarrow})(2|t_P|^2 + |t_{AP}|^2)/3\right] \quad (2)$$

In this expression we assumed that electrons incident on the detector wire are not spin polarized. $w_{\uparrow\uparrow}, w_{\downarrow\downarrow}, w_{\downarrow\uparrow+\uparrow\downarrow}$, and $w_{\downarrow\uparrow-\uparrow\downarrow}$ denote the probabilities of the initial configuration of the two LMMs (three triplet and one singlet configurations, respectively). The transmission coefficients t_P, t_{AP} and t_S are obtained from the scattering solutions of the equations

$$\begin{aligned}[E - E_n - K_x - U(x) + 2j(x) + J]\varphi_P(x) &= 0, \\ [E - E_n - K_x - U(x) - 4j(x) + J]\varphi_{AP}(x) &= 0, \\ [E - E_n - K_x - U(x) - 3J]\varphi_S(x) &= 0.\end{aligned} \quad (3)$$

Here, P/AP refers to the possibilities of the incident electron spin being parallel/antiparallel to the net spin of the two LMMs when they are in their triplet state, S refers to the situation with the two LMMs in the singlet state. It is important that $t_S = t_0$, with t_0 being the transmission coefficients through the detector wire uncoupled from the LMMs. Thus, as is readily seen from Eq. (2), if the two LMMs are in their singlet state, the conductance of the detector wire will not exhibit any special features. However, if the two LMMs are in their triplet state, a conductance enhancement may be observed, similar to that of the previous experiment [3]. This is due to the fact that the singlet state has zero projection of the magnetic moment on the direction parallel to the electron spin.

In summary, the proposed device should allow routine detection of whether the two coupled LMMs are in their triplet or singlet state. The singlet state of two LMMs is an entangled state, therefore the possibility to detect it is of great importance for LMM-based quantum computation [6].

REFERENCES

1. See, e.g., A. C. Graham, K. J. Thomas, M. Pepper, N. R. Cooper, M. Y. Simmons, and D. A. Ritchie, Phys. Rev. Lett. **91**, 136404 (2003) and references therein.
2. See, e.g., S. M. Cronenwett, H. J. Lynch, D. Goldhaber-Gordon, L. P. Kowenhoven, C. M. Marcus, K. Hirose, N. S. Wingreen, and V. Umansky, Phys. Rev. Lett. **88**, 226805 (2002).
3. T. Morimoto, Y. Iwase, N. Aoki, T. Sasaki, Y. Ochiai, A. Shailos, J. P. Bird, M. P. Lilly, J. L. Reno, and J. A. Simmons, Appl. Phys. Lett. **82**, 3952 (2003).
4. V. I. Puller, L. G. Mourokh, A. Shailos, and J. P. Bird, Phys. Rev. Lett. **92**, 096802 (2004).
5. V. I. Puller, L. G. Mourokh, A. Shailos, and J. P. Bird, 2004, IEEE Transactions on Nanotechnology, accepted for publication; arXiv: cond-mat/0405705.
6. J. P. Bird et al., Proceedings of the 2004 NanoHana Workshop, IUPAP publishers, in press.
7. V. I. Puller, L. G. Mourokh, J. P. Bird, and Y. Ochiai, arXiv: cond-mat/0411126.

Electron-phonon quantum kinetics beyond the second-order Born approximation

Henning Lohmeyer*, Vollrath Martin Axt[†] and Tilmann Kuhn[†]

Institute of Solid State Physics, University of Bremen, Otto-Hahn-Allee 1, 28359 Bremen, Germany
[†]*Institute of Solid State Theory, University of Münster, Wilhelm-Klemm-Str. 10, 48149 Münster, Germany*

Abstract. For a one-band model of a quantum wire coupled to LO-phonons we present results for the dynamics of the coupled electron-phonon system calculated with quantum kinetic equations obtained in the density matrix approach. Using a correlation expansion for the truncation of the infinite hierarchy of equations of motion we go beyond the usual second-order Born approximation by taking into account two-particle correlations and twofold phonon-assisted density matrices. The different approximation levels are compared among each other and with an exact solution for a simplified case.

The emission of longitudinal optical (LO) phonons is one of the most important relaxation channels of optically excited hot carriers in semiconductors. For ultrashort timescales accessible via excitation with femtosecond pulses the semiclassical Boltzmann equation is no longer valid; a quantum kinetic theory has to be used instead to describe features such as energy non-conserving transitions and memory effects [1, 2]. Kinetic equations for the electron distribution function have been derived using different methods, in particular nonequilibrium Green's functions [3] and reduced density matrices [1]. Due to the many-body character the electron-phonon system its dynamics – apart from special cases – cannot be solved exactly and approximations are necessary.

In the density matrix approach considered here one has to truncate the infinite hierarchy of equations of motions for the reduced density matrices in a controlled way. By invoking a factorization of higher-order density matrices into products of lower-order matrices and correlation remainders and by neglecting these remainders in a certain order the second-order Born approximation (2BA) and fourth-order Born approximation (4BA) have been derived. The 2BA has been applied to the rather realistic model system of a two band semiconductor [4] whereas results for the numerically more demanding 4BA exist so far only for a one dimensional one-band model [5]. Here we limit ourself to this model system, too.

The Hamiltonian for electrons (band dispersion ε_k) and LO phonons (energy $\hbar\omega_{LO}$) interacting via the Fröhlich mechanism with coupling constant g_q^e is given by

$$H = \sum_k \varepsilon_k^e c_k^+ c_k + \sum_q \hbar\omega_{LO} b_q^+ b_q + \sum_{k,q}(g_q^e c_{k+q}^+ b_q c_k + \text{h.c.}),$$

where c_k^+ (b_q^+) denotes the creation operator of an electron (phonon) with momentum k (q). From the Heisenberg equation one obtains the equation of motion for the electron distribution function $f_k^e = \langle c_k^+ c_k \rangle$,

$$\frac{d}{dt} f_k^e = \sum_q [2\text{Re}\{s_{k+q,k}^e\} - 2\text{Re}\{s_{k,k-q}^e\}],$$

which involves the phonon-assisted density matrix $s_{k+q,k}^e = \frac{i}{\hbar} g_q^e \langle c_{k+q}^+ b_q c_k \rangle$. In the quantum kinetic description s^e is treated as an independent new dynamical variable. Its equation of motion reads

$$\frac{d}{dt} s_{k+q,k}^e = \frac{i}{\hbar}\left(\varepsilon_{k+q}^e - \varepsilon_k^e - \hbar\omega_{LO}\right) s_{k+q,k}^e$$
$$+ \frac{|g_q^e|^2}{\hbar^2}\left[f_{k+q}^e(1+n_q+B_q) - f_k^e(n_q+B_q) - f_k^e f_{k+q}^e\right]$$
$$+ \sum_{k'}\left[F_{k+q,k',q} - F_{k',k,q} + F_{k+q,k',q}^{(+)} - F_{k',k,q}^{(+)} + T_{k',k,q}\right],$$

where $n_q = \langle b_q^+ b_q \rangle$ denotes the phonon distribution function, $B_q = \langle b_q b_{-q} \rangle$ the phonon correlation, and we have defined the following two-particle correlations

$$F_{k_1,k_2,q_1} = \frac{g_q^e g_{q_1}^e}{\hbar^2}\left(\langle c_{k_1}^+ b_{q_1} b_{k_1-k_2-q_1} c_{k_2}\rangle - \delta_{k_1,k_2} f_{k_1}^e B_{q_1}\right),$$

$$F_{k_1,k_2,q_1}^{(+)} = \frac{g_{q_1}^e g_q^{e*}}{\hbar^2}\left(\langle c_{k_1}^+ b_{k_2-k_1+q_1}^+ b_{q_1} c_{k_2}\rangle - \delta_{k_1,k_2} f_{k_1}^e n_{q_1}\right),$$

$$T_{k_1,k_2,q_1} = |g_{q_1}^e|^2/\hbar^2 \left(\langle c_{k_1}^+ c_{k_2+q_1}^+ c_{k_2} c_{k_1+q_1}\rangle \right.$$
$$\left. + \delta_{k_1,k_2} f_{k_1+q_1}^e f_{k_1}^e - \delta_{q_1,0} f_{k_1}^e f_{k_2}^e\right).$$

Neglecting two particle correlations constitutes the 2BA which, in the Markov limit and by neglecting phonon

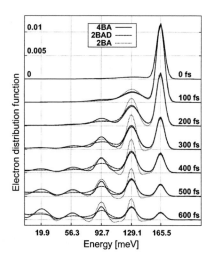

FIGURE 1. Electron distribution after excitation with a 120 ps pulse calculated on the three approximation levels.

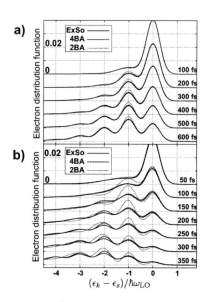

FIGURE 2. Relaxation of a given Gaussian distribution centered around ε_s in the simplified case with constant coupling $g_q = g$ and linear dispersion ε_k in 2BA and 4BA in comparison with the exact solution (ExSo) for a coupling strength of a) α=0.04 and b) α=0.16.

correlations, reduces to the well known Boltzmann equation. Analogously, the 4BA is obtained by neglecting correlations involving five operators appearing in the equations of motion for F, $F^{(+)}$ and T. It is possible to include certain self-energy like contributions from the 4BA in the 2BA by applying a formal solution and Markov approximation to the equations of motion for the two-particle correlations [4]. This results in damping terms in 2BA (2BAD) which will be compared with the 2BA and 4BA in the following, too.

Figure 1 shows the relaxation of the electron distribution calculated for a one dimensional quantum wire with GaAs material parameters ($\hbar\omega_{LO} = 36.4$ meV). The optical excitation with a 120 fs-pulse is simulated using a semiclassical generation rate obtained from the semiconductor-Bloch equations [1]. The relaxation occurs via successive emission of phonons leading to the build up of phonon replicas with increasing time. Each emission process exhibits the quantum kinetic memory effect of initial broadening of the phonon replica reflecting the energy-time uncertainty [2]. For the given excitation conditions the plain 2BA leads obviously to unphysical negative values of the distribution functions. This behavior can be corrected by including damping (2BAD) as has been seen before [4]. By comparing with the full 4BA calculation we find that the 2BAD well approximates the first replica, however it strongly overestimates the broadening of the second and following replicas. In addition, as is already known [6] it does not fully account for the conservation of total energy, in contrast to the full 2BA and 4BA levels.

A simplified case of the one-band model with linear dispersion relation $\varepsilon_k^e = v_0 k$ and constant coupling $g_q^e = g$ is exactly solvable [7]. In Fig. 2 the analytical solution for the dynamics of one electron [8] folded with an initial Gaussian distribution is compared with the results of the 2BA and 4BA for two coupling strengths characterized by the dimensionless coupling constant $\alpha = 2\pi g^2/(v_0 \omega_{LO})$. For $\alpha = 0.04$ (Fig. 2a)) the dynamics is obviously already very well described by the approximated kinetic equations in 4BA. For the broader initial distribution compared to Fig. 1 2BA gives reasonable results, too. For a four times stronger coupling (Fig 2b)) the 2BA already shows considerable deviations from the exact solution. One might expect a breakdown of the correlation expansion for higher coupling strength because of the increasing importance of higher correlations, however the 4BA shows deviations only in the second and following replica.

Financial support from DFG grant KU 697/9-1 is gratefully acknowledged.

REFERENCES

1. Rossi, F., and Kuhn, T., *Rev. Mod. Phys.*, **74**, 895 (2002).
2. Fürst, C. et al., *Phys. Rev. Lett.*, **78**, 3733 (1997).
3. Haug, H., and Jauho, A.-P., *Quantum kinetics in Transport and Optics of Semiconductors*, Springer, 1996.
4. Schilp, J., Kuhn, T., and Mahler, G., *Phys. Rev. B*, **50**, 5435 (1994).
5. Zimmermann, R. et al., *J. of Lum.*, **76 & 77**, 34 (1998).
6. Schilp, J., Kuhn, T., and Mahler, G., *phys. stat. sol. (b)*, **188**, 417 (1995).
7. Meden, V. et al., *Phys. Rev. B*, **52**, 5624 (1995).
8. Schönhammer, K., *Phys. Rev. B*, **58**, 3518 (1998).

Real Space Observation of Anisotropic Scattering in the π-Chains on Si(111)-2x1 as a Quasi 1D System

J. K. Garleff[a], M. Wenderoth[a], R. G. Ulbrich[a],
C. Sürgers[b], and H. v. Löhneysen[b,c]

[a]*IV. Phys. Institut der Universität Göttingen, Friedrich-Hund-Platz 1, D-37077 Göttingen, Germany*
[b]*Physikalisches Institut and DFG Center for Functional Nanostructures (CFN), D-76128 Karlsruhe, Germany*
[c]*Forschungszentrum Karlsruhe, Institut für Festkörperphysik, D-76021 Karlsruhe, Germany*

Abstract. The free and unperturbed surface and substitutional near-surface phosphorus donors have been investigated by scanning tunneling microscopy and spectroscopy at 8 Kelvin with atomic resolution and a spectral resolution of $\Delta E \approx$ 5 meV. The measured local density of states (LDOS) on the free surface and the calculated electronic surface structure agree very well. In the vicinity of the donors the LDOS shows a highly anisotropic contrast at the onset of the surface conduction band closely below E_F. It extends 10 nm along the reconstruction and less than 1.3 nm perpendicular. The 1D confinement along the π-bonded-chains of the reconstruction is introduced by the anisotropic band structure. For donors located in the top layer only one chain is affected. The LDOS near the donor is increased by a factor of 3-4 compared with the free surface value. There is also a strong increase of the corrugation along the affected π-chains. We discuss the observations in the context of scattering of the quasi 1-dimensional electrons by the change in potential introduced by the donor impurity. At small negative sample voltages electrons tunnel out of the surface conduction band. This indicates that the position of E_F lies within the surface conduction band.

INTRODUCTION

The Si(111) surface formed by cleavage at room temperature exhibits a 2x1 reconstruction. The surface consists of π-bonded-chains and causes a highly anisotropic surface band structure. The Fermi-Level-Pinning produces a 10 nm thick space charge layer which provides a 2D confinement for the surface states. The most important aspect of Si(111)-2x1 is the anisotropic electronic structure parallel to the surface. The group velocity parallel to the π-chains is $\sim 10^6$ m/s and reduced by at least a factor of 10 in the perpendicular direction. Theoretical as well as experimental work pointed at a 1D electronic system on this surface [1,2], but did not yet find clear evidence. Our measurements using scanning tunneling microscopy (STM) and spectroscopy (STS) at low temperature clearly show scattering exclusively along the π-chains with the P atom, which is characteristic of a 1D electronic system on Si(111)-2x1.

We prepared atomically flat and clean Si(111) surfaces by in situ cleavage at room temperature along the [$\bar{2}$11] direction of n-type Si crystals doped with $6*10^{18}$ phosphorus (P) atoms per cm^3. STM and STS was done in a home built Besocke-type STM operating in UHV at T = 8 Kelvin. The tips were electrochemically etched from polycrystalline tungsten wire. We performed I(V)-spectroscopy with a lateral resolution better than 1 Å. Numerical processing of the I(V) data to derive LDOS was done with gliding median- and cubic-spline-algorithms.

E_F PINNED WITHIN THE SURFACE CONDUCTION BAND

Typical LDOS-spectra taken on the clean Si(111)-2x1-surface and close to P atoms are shown in Fig. 1. In a first attempt the onsets of the surface valence and the surface conduction band in the spectra can be fixed at -0.6 eV and at 0 eV (see marks a,c in Fig. 1) by comparing the spectra with ab-initio calculations of the electronic surface band-structure [3,4]. From these calculations no LDOS is expected in the surface band gap in-between. Nevertheless, we find a broad peak in the STS spectra at $E - E_F \approx -0.25$ eV on the free surface as well as close to the P atom. This feature marked 'b' in Fig. 1 is strongly enhanced near the P atom.

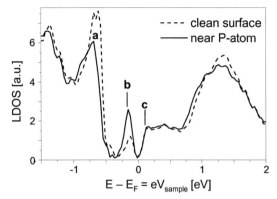

FIGURE 1. LDOS(E) on Si(111)-2x1 at T = 8 K near the P atoms (solid line) and on the clean surface (dashed line), each averaged over one surface unit cell. The positions a, b and c refer to the energies of the LDOS maps in Fig. 2.

In order to study the properties of peak (b) in the gap, the LDOS distribution relative to the underlying atomic lattice [4] is analyzed. LDOS maps at the energies of both band onsets and in the band gap are shown in Fig. 2. The valence band (a) and the conduction band (c) can be distinguished by a typical shift of the LDOS maxima from the up to the down atom in the π-bonded chains. The transition is experimentally observed at -0.5 eV. We conclude that the states in the band gap originate from the surface conduction band. Therefore the position of E_F is found closely *above* the minimum of the surface conduction band.

1D-Shaped Defect Contrast Below E_F

The contrast pattern of P atoms in Si(111)-2x1 is highly anisotropic. In LDOS maps at energies of the conduction band tail in the gap (-0.4 V < V < 0 V), the P atom introduces increased LDOS on one or two π-bonded-chains depending on the P atom substituting a Si atom in a chain or between two chains. The STM topography of a P atom (see arrow in Fig. 3) directly in the π-chain at V = -0.5 V integrates over the LDOS in the band gap. This protrusion extends over ~10 nm along the π-chain with the P atom in [0$\bar{1}$1]-direction. The STM images at voltages in the surface bands show the exact position of the P atom.

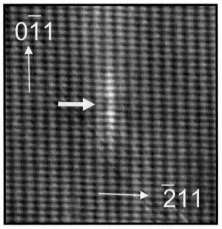

FIGURE 3. The Si(111) surface topography of (12 nm)² at V = -0.5 V and I = 0.1 nA shows the anisotropic contrast pattern along [0$\bar{1}$1] of a P atom (big arrow) in the π-chain.

ACKNOWLEDGMENTS

We thank M. Rohlfing for communicating calculated LDOS(x,y,E) of the Si(111)-2x1 surface.

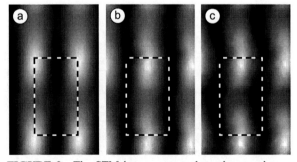

FIGURE 2. The STM images a to c show the experimentally detected LDOS-distribution relative to the surface unit cell (dashed frame) at – 0.7 eV, – 0.3 eV and + 0.1 eV.

REFERENCES

1. J. E. Northrup and M. L. Cohen, *Phys. Rev. Lett.* **49** (18) 1349 (1982)

2. R. M. Feenstra, W. A. Thompson and A. P. Fein, *Phys. Rev. Lett.* **56** (6), 608 (1986)

3. M. Rohlfing, S. G. Louie, *Phys. Rev. Lett.* **83** (4), 856 (1999),

4. J. K. Garleff, M. Wenderoth, K. Sauthoff, R.G. Ulbrich, and M. Rohlfing, submitted to *Phys. Rev. B* (2004)

Conductance of a Multiterminal Ballistic Wire

Z. D. Kvon[*], V. A. Tkachenko[*], A. E. Plotnikov[*], V. A. Sablikov[+], V. Renard[¶],
and
J. C. Portal[¶,§,#]

[*]Institute of Semiconductor Physics, Novosibirsk, 630090 Russia
[+]Institute of Radio Engineering and Electronics, Fryazino, Moscow region, 141190 Russia
[¶]Grenoble High Magnetic Fields Laboratory, MPI-FKF and CNRS, F-38042 Grenoble, France
[§]INSA-Toulouse, 31077, Cedex 4, France
[#]Institut Universitaire de France, Toulouse, France

Abstract. An experimental study of the two-, three-, and four-terminal resistance of a ballistic wire is carried out. The wire is fabricated on the basis of high-mobility 2D electron gas in an AlGaAs/GaAs heterojunction. Different behavior of mesoscopic fluctuations of multiterminal resistances is observed depending on the gate voltage and magnetic field. At weak magnetic fields the four-terminal resistance drops almost to zero and features resembling a ballistic conductance quantization are observed.

During the ten years following the discovery of conductance quantization in a ballistic wire, no experimental data on the distribution of voltage along the wire was available. When measuring this distribution by potentiometric probes, two problems arise: The first problem corresponds to the fact that the terminals connecting the probes and the wire can affect the properties of the latter. The second problem is related to the fact that, generally speaking, the potential measured by a probe is not equal to the local potential of the wire. The existing theories of ballistic transport give no universal relationship between these two quantities [1-4]. Only recently the first experiments concerning this question were made [5]. These experiments were performed using a unique cleaved-edge-grown AlGaAs/GaAs heterostrusture. The analysis of measurements suggested the conclusion [5] that the voltage applied to the wire equally drops in the near-terminal regions while the voltage drop between the probes is equal to zero. However, it is necessary to note that zero voltage between the probes cannot be considered as proof of quasi-neutrality (zero drop of potential) inside the ballistic wire [3]. In addition, in the structure studied in the cited experiment, a nonadiabatic (nonplanar) connection of the wire with the terminals was used. Therefore, the quantization steps of the two-terminal conductance of a short part of the wire had a noticeably smaller height than the quantum $2e^2/h$) (the two-terminal conductance of the whole multiterminal wire was not measured). The width of the contact of each probe with the wire was close to the length of the free part of the wire, so that the measured potential was averaged over a relatively large length. Nevertheless, the results of this experiment fit well the single-particle model of ballistic wires [1-4], and they caused a number of questions concerning electron interactions. Calculations [6, 7] showed that the interaction of electrons with the selfconsistent field of the charge arising in such a wire leads to the situation when the major part of voltage applied to the wire drops near the reservoirs while the rest of the wire remains almost equipotential.

In this work we report the results of a study of the multiterminal conductance of a ballistic wire fabricated in a more conventional way on the basis of a standard AlGaAs/GaAs heterostrusture with a high mobility two-dimensional electron gas (2DEG). The mobility of the electrons was $\mu = 10^6$ cm^2/Vs, the electron density was $N_s = 3*10^{11}$ cm^{-2} corresponding to a mean free path $l = 9$ μm. A wire of lithographic length $L = 1.4$ μm and width $W_L = 0.5$ μm was prepared by means of electron lithography and plasma etching. Along this wire two potential probes of a smaller width $W_p = 0.4$μm were prepared at the same time. The smaller width of the probes was expected to help reducing the influence of the probes on the electron motion in the wire. A metallic TiAu gate was evaporated on top of the structure. Two- three- and four-terminal resistance measurements were performed simultaneously on the wire at temperatures 0.2 – 1.5 K in zero and weak magnetic fields. First a numerical simulation of the electrostatic potential and energy spectrum of the wire including real technological parameters (the structure of the initial heterojunction, the etch depth, the presence of the upper metal layer, etc) was performed. This simulation allowed determining the number of electron modes passing through the wire and the probes at the same gate voltage. This calculation showed that triangular-shape quantum dots exist at the connection of the probes to wire (fig.1).

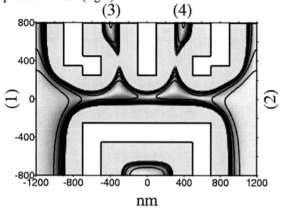

FIG. 1. 2D–Distributions of electron density in the wire; the numbers of therminals are indicated in parentheses and the etch–areas are shown by rectangles.

The analyses of two- three- and four-terminal resistance measurements (fig.2,3) allowed drawing the following conclusions.

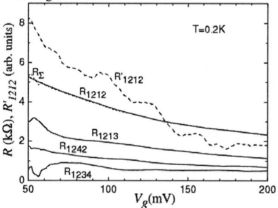

FIG. 2. Resistances of the wire versus the gate voltage at B=0. R'=|dR/dV$_g$| and R$_\Sigma$=R$_{1234}$+R$_{1242}$+R$_{1213}$.

At zero magnetic fields the main part of the voltage applied to the wire drops near the contacts. However the value of four-terminal resistance is far from zero. We show that this fact is due to scattering of electrons inside of the quantum dots mentioned above.

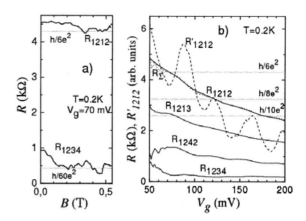

FIG. 3. Resistances of the wire versus the magnetic field and (b)–resistances of the wire at B=0.45T; the notation is the same as in Fig. 3.

It was shown that this scattering is suppressed by relatively weak magnetic field about 0.5 T. So an interesting correlation between two-terminal and four terminal resistances of multiterminal ballisctic wires controlled by magnetic field was detected: the smaller value of the four terminal resistance is observed the better defined the quantized plateaus in two-terminal resistance occurs.

ACKNOWLEDGMENTS

This work has been supported by, the programs RFBR02-02-16516, INTAS (N01-0014), PICS1577, NATO Linkage, programs "Physics and Technology of Nanostructures" of the Russian ministry of Industry and Science and "Low dimensional quantum structures" of RAS.

REFERENCES

1. H.-L. Engquist and P. W. Anderson, Phys. Rev. B **24**, 1151 (1981).
2. M. Buttiker, IBM J. Res. Dev. **32**, 317 (1988).
3. I. B. Levinson, Zh. Éksp. Teor. Fiz. **95**, 2175 (1989) [Sov. Phys. JETP **68**, 1257 (1989)].
4. M. Buttiker, Semicond. Semimet. **35**, 191 (1992).
5. R. de Picciotto *et al.*, Nature **411**, 51 (2001).
6. V. A. Sablikov, S. V. Polyakov, and M. Buttiker, Phys. Rev. B **61**, 13763 (2000).
7. V. A. Sablikov and B. S. Shchamkhalova, Physica E (Amsterdam) **17**, 189 (2003).

Dynamical response in resonance interaction in coupled quantum wires

T. Morimoto[1], T. Sasaki[1], N. Aoki[1], Y. Ochiai[1], A. Shailos[2], J. P. Bird[2], M. P. Lilly[3], J. R. Reno[3] and J. A. Simmons

[1]*Department of Materials Technology, Chiba University, 1-33 Yayoi, Inage, Chiba 263-8522, Japan*
[2]*Nanostructures Research Group, Arizona State University, Tempe, AZ 85287-5706, USA*
[3]*Nanoelectronics Group, Nanostructure and Semiconductor Physics Department, Sandia National Laboratories, MS 1415, P.O. Box 5800, Albuquerque, New Mexico 87185-1415, USA*

Abstract. We have studied the electrical transport property of the coupled Quantum Point Contacts. That resonance peak structure has appeared in conductance of the detector QPC by forming of local spin moment in swept QPC. We find that resonance peak depends on the one-dimensional density of states in detector QPC. We will discuss the consistency with theoretical model using Anderson Hamiltonian.

INTRODUCTION

Now, the "0.7 structure" [1] is great interested as many-body state formed in simple system like a Quantum Point Contact (QPC) and Quantum Wire (QW). Recently experimental [2,3] and theoretical works [4,5] focused on forming a local-spin moment vicinity or inside of QPC and possibility of occurrence the Kondo effect. To reveal the origin of "0.7 structure" is important to understand electron spin correlation in nano-scale system and control of quantum state for realization of Quantum Computing or Spintronics devices.

In this report, we show a resonance interaction appeared in transport property of the coupled QPCs [6,7]. When we changed the conductance of detector QPC, height of the resonance peaks periodically fluctuated depend on the one-dimensional density of states in the detector QPC. We will discuss a correspondence with theoretical explanation about this resonant interaction.

EXPERIMENTAL CONDITION

The device was fabricated on ultra-high mobility GaAs/AlGaAs quantum well by using split gates on substrate surface. Two-dimensional electron gas (2DEG) is 200nm below the surface and electron mobility and density are 2.0×10^6 cm/Vs and 2.7×10^{11} cm^{-2}, respectively. The sample was mounted in dilution refrigerator and measured the low temperature transports at ~80mK. Four terminal resistance measurement was performed by using a

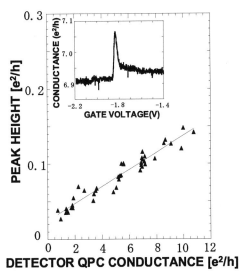

FIGURE 1. (Inset) Resonance peak line-shape of conductance of the detector QPC. Peak position is always corresponds to a gate voltage of pinch off in swept QPC. (Main panel) Peak height increases monotonically as increase of conductance of the detector QPC conductance except fluctuated small structure.

low frequency (17Hz) lock-in techniques.

EXPERIMENTAL RESULTS AND DISCUSSION

Inset of Fig. 1 shows resonance peak of conductance of the detector QPC (see [6] for a detail of measurement). This resonance peak always took placed at the swept QPC depopulation under the last conductance. Peak height of resonance enhancement is defined from the threshold of peak starting value to the top. Peak height has a dependence on the detector QPC's conductance, it increases monotonically as the gates of detector QPC open. But we can clearly observed the fluctuated the small structures having periodical behavior (shown in main panel of Fig.1).

Recently, theoretical explanation about our experimental result has been suggested that local spin moment formed in swept QPC was interacted with a conduction electron in the detector QPC by tunnel coupling through the quantum dot [8]. Their explanation is succeeded in a reproducing of the resonance peak by using Anderson Hamiltonian model. In such model, uncoupled one-dimensional density of states (DOS) in the detector QPC. So we can expect that the resonance peaks are modulated by variation of DOS in the detector QPC.

In Fig.2 (a), we show the quantized conductance in the detector QPC with upper three gates. We can roughly estimate the period of DOS in the detector QPC from transconductan (dG/dV_G) for the gate voltages.

In Fig.2 (b), fluctuated small structure appeared in the peak height of resonance interaction is almost consistent with sawtooth pattern having a similarity with shape of one-dimensional DOS. By comparison between the peak height and transconductance, we can consider that both periods take almost equal value not only in the low conductance regime but in the high conductance regime. Moreover, both phase of the peaks and dips are same to each other. It suggests that one-dimensional DOS in the fixed QPC plays an important role in our electron interaction between local spin moment formed in strong constriction and conduction electron in detector QPC.

CONCLUSIONS

We have investigated electrical transport property of coupled QPCs. We found that the resonance interaction peak in conductance of the detector QPC related to formation of the local spin moment at the swept QPC. We discussed the origin of the fluctuation like structure appeared in the peak height and found the consistency between our experimental results and theoretical work.

Recently, non-local spin control was reported in coupled dots system using a Friedel oscillation from local spin moment in quantum dot [9]. Our system also have a possibility of non-local spin control between coupled QPCs. It will be important for realization process of Quantum Computing devices.

ACKNOWLEDGMENTS

Work at ASU is supported by the Department of Energy, the National Science Foundation, and the Office of Naval Research. Work at Chiba University was supported in part by a grant-in-aid from the Japan Society for Promotion of Science (Nos. 14750005 and 13450006). Sandia is a multi-program laboratory operated by Sandia Corporation, a Lockheed Martin Com pany, for the United States Department of Energy under contract DE-AC04-94AL85000.

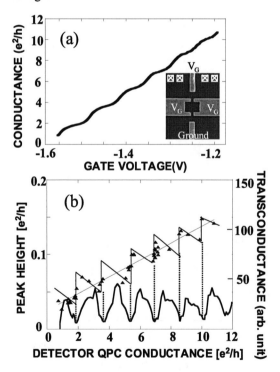

FIGURE 2. (a) Quantized conductance of upper three gates. Inset Scanning Electron micrograph shows the measurement configuration in main panel. (b) Fluctuation like structure is fitted by sawtooth pattern. Trans conductance is imported in fixed QPC conductance dependence of peak height to compare a period each other Both period and phase of peak and dip are almost same each other.

REFERENCES

1. K. J. Thomas *et al.*, Phys. Rev. Lett. **77**, 135 (1996).
2. S. M. Cronenwett *et al.*, Phys. Rev. Lett. **88**, 226805 (2002)
3. D. J. Reilly *et al.*, Phys. Rev. Lett. **89**, 246801 (2002)
4. K.-F. Berggren and I. I. Yakimenko, Phys. Rev. B **66**, 085323 (2002).
5. K. Hirose, Y. Meir, and N. S. Wingreen, Phys. Rev. Lett. **90**, 026804 (2003).
6. T. Morimoto *et al.*, Appl. Phys. Lett. **82**, 3952 (2003).
7. J. P. Bird and Y. Ochiai, Science, **303**, 1621 (2004).
8. V. I. Puller *et al.*, Phys. Rev. Lett. **92**, 96802 (2004).
9. N. J. Craig *et al.*, Science. **304**, 565 (2004)

Towards a new quantum wire structure realizable by double cleaved-edge overgrowth: Characterizing the transfer potential

S. F. Roth*, M. Grayson*, M. Bichler*, D. Schuh* and G. Abstreiter*

Walter Schottky Institut TUM, Am Coulombwall 3, Garching D-85748, Germany

Abstract. Twofold application of the well established cleaved-edge overgrowth technique can achieve an orientation of a quantum wire parallel to the growth axis of a prior grown substrate. Hence a variation of the substrate layers directly transfers the resulting potential modulation to the adjacent quantum wire. The concept of a transfer potential applied to a narrow two-dimensional system is a first step towards such a modulated one-dimesional system. First characterization data from double-cleave structures shows an unexpected appearance of two different electron densities in the wide quantum well of each sample, possibly due to charged centers in the different substrate layers.

INTRODUCTION

The cleaved-edge overgrowth [1] method makes it possible to fabricate very clean single cleave quantum wires in the GaAs/AlGaAs material system, having atomically precise control over the structural design [2]. An MBE grown heterostructure supplying a potential structure close to a narrow low-dimensional electron gas can modulate the carrier density in the system [3, 4], because the electron wave function tail is pinched by a high barrier potential. A new approach to fabricate quantum wires next to a transfer potential is double cleaved-edge overgrowth.

However, also the barrier material can influence the electron density [5], independent of the barrier potential. We present basic research on double cleave structures investigating a wide 2DEG, whose density is modulated not from quantum confinement but from they alloy content of a neighboring substrate. We propose that the density difference arises from charged defects in the AlGaAs layer.

EXPERIMENTAL

Figure 1 shows the sample structure. First a substrate is grown in [001]-direction, whose layer sequence is sandwiched by two 1μm thick n$^+$-contacts. The sequence consists of two 2μm thick GaAs layers surrounding an Al$_x$Ga$_{1-x}$As layer with an Al-content of x=0.3. After cleaving the sample in-situ, a 40nm GaAs quantum well is grown in [110]-direction separated from a n$^+$-sidegate by a 0.55μm thick Al$_x$Ga$_{1-x}$As barrier, which contains a Si modulation doping at a distance of 50nm from the quantum well, to supply electrons. A 2DEG develops in the quantum well and the density of it can be tuned by the sidegate.

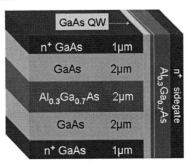

FIGURE 1. Sample after 1st CEO

The sample is completed with a second cleaved-edge overgrowth in the [1$\bar{1}$0]-direction, which in future devices is necessary to further reduce the dimension of the electron system, however, it plays no active role at the gate bias range used in this work. We characterize all samples by measuring the Shubnikov de Haas (SdH) oscillations of the quantum well 2DEG varying temperature, gate voltage and illumination conditions.

RESULTS

Typical results of the SdH-measurements are shown in Fig.2. Because the samples used are about 3mm wide, but the distance between the n$^+$-layers contacting the

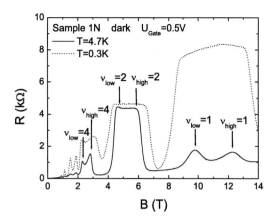

FIGURE 2. SdH-measurements of a sample with a 40nm wide quantum well

2DEG is only 6μm, the magnetoresistance has different boundary conditions than in commonly used Hall bar samples. The result is reminiscent of a Corbino geometry, with a ring shaped 2DEG with inner and outer contacts, where the quantum Hall states induce maxima rather than minima. In samples of finite width, however, the resistance at integer filling factors takes a peak value of $R_{2Pt} = R_{xy} = \frac{h}{e^2 i}$, for i filled, spin resolved Landau levels. Due to leakage effects the measured R_{2Pt} in Fig.2 is smaller than the quantized value.

Two series of peaks, equally spaced in $\frac{1}{B}$, where B is the magnetic field, can be identified in Fig.2. This observation indicates two densities, differing by a value of $\triangle$n. Biasing the sidegate positive, shifts the peak structure towards higher magnetic fields as the densities of the electron systems increase. Fig.3 shows the dependence of the densities with gate voltage and the value of $\triangle$n. The slope of the data points can be compared to the slope of the simple capacitive relation

$$\frac{dn}{dV_G} = \frac{\varepsilon_0 \varepsilon_r}{e} \cdot \frac{1}{d} \quad (1)$$

where d is the distance between the capacitor plates. The good agreement for $d = 0.55\mu$m, the expected growth value, demonstrates that both 2DEG systems are indeed located in the quantum well. The existence of a high- and a low-density 2DEG in a wide quantum well is very unexpected, because the wave function of the electrons should not be affected by the barrier potential. This suggests that the materials of the GaAs/AlGaAs heterostructure of the adjacent substrate may be responsible for the effect.

Illumination of the sample with an IR-LED affects the two systems unequally and brings forth a change of the density difference $\triangle$n. Furthermore both densities *decrease* with illumination. Hence an argument that the light simply activates the Si-donors of the modulation doping is not consistent. A possible explanation

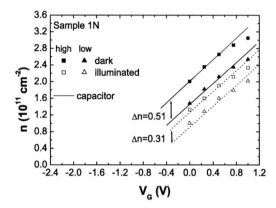

FIGURE 3. Densities of the 2DEGs depending on gate voltage and illumination condition

is provided by an earlier experiment [5] on quantum wells surrounded by $Al_xGa_{1-x}As$ barriers. Here the density of a 2DEG after illumination depended strongly on the Al-content of the back barrier. Because the growth of $Al_xGa_{1-x}As$ produces more defects with higher Al-content x, more charges can be trapped in these layers. Illumination can rearrange the charges in these traps resulting in a persistent photoconductivity effect. Accordingly the low-density 2DEG in our samples is related to the underlying AlGaAs substrate layer. This new effect will cause density modulations in addition to any wavefunction confinement effects.

ACKNOWLEDGMENTS

Financial support is given by the Bundesministerium fuer Bildung und Forschung through Project 01BM912.

REFERENCES

1. L. N. Pfeiffer, K. W. West, H. L. Stormer, J. P. Eisenstein, K. W. Baldwin, D. Gershoni and J. Spector, *Appl. Phys. Lett.* **56**, 1697 (1990)
2. A. Yacoby, H. L. Stormer, N. S. Wingreen, L. N. Pfeiffer, K. W. Baldwin and K. W. West, *Phys. Rev. Lett.* **77**, 4612 (1996)
3. H. L. Stormer, L. N. Pfeiffer, K. W. Baldwin, K. W. West and J. Spector *Appl. Phys. Lett.*, **58**, 726 (1991)
4. R. A. Deutschmann, W. Wegscheider, M. Rother, M. Bichler and G. Abstreiter *Appl. Phys. Lett.*, **79**, 1564 (2001)
5. E. P. Poortere, Y. P. Shkolnikov and M. Shayegan, *Phys. Rev. B*, **67**, 153303 (2003)

Ballistic to Diffuse Crossover in Long Quantum Wires

J. A. Seamons*[&], E. Bielejec*, M. P. Lilly*, J. L. Reno*, and R. R. Du[&]

*Sandia National Laboratories, Albuquerque, NM, 87185, USA
[&]Department of Physics, University of Utah, Salt Lake City, UT, 84112, USA

Abstract. We report a study on the uniformity of long quantum wires in the crossover from ballistic to diffuse transport with lengths ranging from 1 μm to 20 μm. For the 1 μm wire we measure 15 plateaus quantized at integer values of $2e^2/h$. With increasing length we observe plateaus at conductance values suppressed below the quantized values. With nonlinear fitting to the magnetoresistances we obtain an effective width for the quantum wires. As we find no systematic variation of the effective width as a function of sublevel index for the various length wires, we conclude that we have uniform long single quantum wires up to 20 μm.

The phenomenon of quantized conductance has been associated with ballistic transport in quantum wires and constrictions ever since its first observation in GaAs-AlGaAs heterostructures [1]. While many studies of ballistic transport exist, less is known about the electron transport as the wire length approaches the mean free path. We study the uniformity of long single quantum wires with an aim of maximizing interaction length in more complicated devices. In this paper we report a uniformity test on various length quantum wires.

The devices are fabricated on a high mobility δ doped GaAs-AlGaAs heterostructure with a 30 nm wide quantum well, nominally centered 198 nm below the surface of the semiconductor. The density of the two-dimensional electron gas (2DEG) is 3.1×10^{11} cm^{-2} with a corresponding mobility of 5.3×10^6 cm^2/Vs, after a brief illumination with a red LED. Electron beam lithography is used to pattern TiAu split gates (inset, Fig. 1) that will form the one-dimensional wire. The separation between the split gates is 0.5 μm and the lengths range from 1 μm to 20 μm. We use an ac lock-in technique with a constant voltage of 100 μV at 13-130Hz to perform 4-terminal measurements at 0.3 K. For short wires the 2-terminal measurements are consistent with the 4-terminal measurements after subtracting off a reasonable contact resistance [1,2].

In Fig. 1 we show the 4-terminal conductance as a function of the split gate voltage for quantum wires of several lengths. As negative voltage is applied on the split gates, the electrons of the 2DEG are depleted

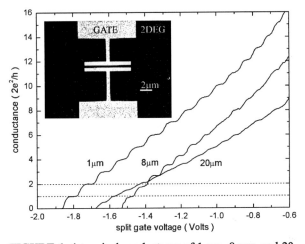

FIGURE 1. 4-terminal conductance of 1 μm, 8 μm, and 20 μm quantum wires. Inset: a scanning electron micrograph of an 8 μm quantum wire.

from under the gates and thereby form a quantum wire. With increasingly negative voltage on the split gates, the one-dimensional subbands in the quantum wire are one by one depopulated, until the conductance goes to zero. Each subband has a conductance of $2e^2/h$ in the absence of magnetic field and scattering. The 1 μm wire (left line) exhibits 15 plateaus quantized at integer values of $2e^2/h$ corresponding to 15 subbands. This high number of quantized plateaus is indicative of the high quality one-dimensional quantum wires that can be formed using this method. Short wires (0.5 μm to 5 μm) typically display quantized conductance steps, with the shorter wires having more steps. In long wires (above 5 μm to 40 μm) steps can be observed, but always at values below multiples of $2e^2/h$. In Fig. 1, the 8 μm wire (center line) has 7 steps at $0.9 \cdot i \cdot 2e^2/h$, and the 20 μm wire (right line) has oscillations at $0.5 \cdot i \cdot 2e^2/h$. An important issue for long wires such as these is the uniformity of the wire.

To study the uniformity of the wires we obtain the effective width of the wires, by employing a magnetic depopulation measurement. When the potential of the split gates is constant and a magnetic field is applied, perpendicular to the growth direction of the quantum well, the wire depopulates one sublevel at a time. This depopulation occurs as the increasing magnetic field increases the sublevel spacing. In Fig. 2a we show the 4-terminal magnetoresistance for the 1 μm wire. Each line corresponds to a fixed split gate voltage. Accumulation areas occur at the magnetic field where a subband is depopulated. For example, in the lowest line (highlighted in bold) there are 9 magnetic field values (marked with open circles) where subband depopulation is apparent as minima in the 4-terminal magnetoresistance. At zero magnetic field the conductance is just over $9 \cdot 2e^2/h$ for this split gate voltage; this is consistent with the 9 depopulation minima in the magnetoresistance. By assuming the confinement potential to be a harmonic oscillator we fit the data using equation (1) and obtain the one-dimensional density and width of the quantum wire [3]:

$$N = \left(\left(eBW / \pi n_{1D} \right)^2 + 32 / 9\pi n_{1D} W \right)^{-\frac{1}{2}} \quad (1)$$

where N is the sublevel index, e is the charge of an electron, B is the magnetic field, W is the width of the quantum wire, and n_{1D} is the one-dimensional density of the quantum wire. In Fig. 2b the fitting is shown for the highlighted magnetoresistance data in Fig. 2a. This analysis is repeated for each curve in Fig. 2a, and the results are summarized in Fig. 2c.

Although the 1 μm wire is ballistic, the 8 μm wire is on the boundary, and the 20 μm wire is clearly *not* ballistic, the effective width of each wire is the same for a given subband. If the long wires contained constricted regions, we would expect a trend to smaller widths for longer wires in Fig. 2c. Since this does not occur, we conclude that these long split gate quantum wires are uniform, and their transport is not dominated by unintentional constrictions.

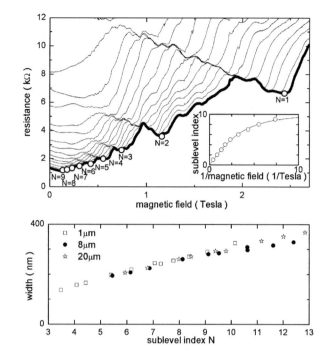

FIGURE 2. (a) 4-terminal magnetoresistance traces of 15 different split gate voltages for the 1 μm quantum wire. (b) Analysis of the lowest (bold) trace. (c) Width of the quantum wires as a function of the occupied sublevel.

We acknowledge the outstanding technical assistance from R. Dunn and D. Tibbetts. This research has been supported by The Division of Materials Sciences and Engineering, Office of Basic Energy Sciences, U.S. Department of Energy. Sandia is a multiprogram laboratory operated by Sandia Corporation, a Lockheed Martin Company, for the United States Department of Energy's National Nuclear Security Administration under contract DE-AC04-94AL85000.

REFERENCES

1. B.J. van Wees, H. van Houten, C. W. J. Beenakker, J. G. Williamson, L. P. Kouwenhoven, D. van der Marel, and C. T. Foxon, *Phys. Rev. Letters* **60**, 848 (1988).

2 K.J. Thomas, M.Y. Simmons, D.R. Mace, M. Pepper, and D.A. Ritchie, *Appl. Phys. Letters* **67**, 109 (1995).

3 K.F. Berggren, G. Roos, and H. van Houten, Phys Rev. B **37**, 10118 (1988).

Carrier Tunneling between Parallel GaAs/AlGaAs V-groove Quantum Wires

K. F. Karlsson[2,1], H. Weman[1,2], K. Leifer[1], A. Rudra[1], and E. Kapon[1]

[1]Laboratory of Physics of Nanostructures, Institute of Quantum Electronics and Photonics, Swiss Federal Institute of Technology Lausanne, CH-1015 Lausanne, Switzerland
[2]Department of Physics and Measurement Technology (IFM), Linköping University, S-581 83 Linköping, Sweden

Abstract. Carrier transfer between two asymmetric weakly coupled GaAs quantum wires separated by AlGaAs barriers ranging from 5.5 nm to 20 nm thickness are studied by photoluminescence and photoluminescence excitation spectroscopy. The chosen range includes both the regime dominated by conventional tunneling and the regime where tunneling is negligible. In contrast to the studied reference quantum wells, the charge transfer between the quantum wires can be fully explained by conventional tunneling via LO-phonon emission.

INTRODUCTION

Optical devices based on nanostructures, such as quantum wires (QWRs), normally consist of stacked layers in order to enhance the active volume. Knowledge about the carrier transfer between QWRs is thus important since the optical properties will be affected by how the injected carriers are distributed among the different layers.

Efficient electron and hole transfer between two differently sized GaAs QWRs separated by a 7 nm thick barrier has been demonstrated previously [1]. In the present study, different barrier thicknesses are chosen in a range allowing studies of the transfer due to conventional tunneling as well as alternative mechanisms, eventually active for more widely spaced QWRs. The present results can thus be directly compared with the reported efficient transfer for GaAs quantum wells (QWs) in the barrier thickness regime where conventional tunneling is negligible [2, 3].

SAMPLES, RESULTS & DISCUSSION

The QWR samples were grown by low pressure, organometallic chemical vapor deposition (OMCVD) on GaAs substrates patterned with V-grooves. One narrow and one wide GaAs/$Al_{0.3}Ga_{0.7}As$ QWR (w- and n-QWR) were grown with the nominal thicknesses 3.5 nm and 2 nm, respectively. A self-ordered AlGaAs vertical QW (VQW) is intersecting the GaAs QWRs and lower the Al concentration of the barrier. Four samples were fabricated with different nominal thickness (L_b) of the barrier: L_b = 5.5 nm, 8.5 nm, 13 nm, and 20 nm, respectively, referred to as samples QWR5.5, QWR8.5, QWR13 and QWR20. The final thicknesses of the n-QWR and the w-QWR are about 5.5 nm and 8 nm, while the thickness of the separating barrier corresponds to the nominal value. Reference GaAs QW samples were grown simultaneously on planar substrates, referred by analogous notation.

The samples were measured by conventional low-temperature (≤10 K) laser excited PL and PLE spectroscopy. A PL spectrum of QWR13 is shown in Fig. 1a (bold line) exhibiting two peaks of comparable intensities. The low- (high-) energy peak is attributed to the ground state exciton of the w-QWR (n-QWR). In contrast, only the corresponding low-energy peak dominates the PL spectrum for QWR8.5, as shown in Fig. 1b. This is a first indication of that efficient transfer takes place of the photo-excited carriers from the n-QWR to the w-QWR in QWR8.5, while such transfer is blocked in QWR13.

The PLE spectra of both the w-QWR and the n-QWR for QWR13, shown in Fig. 1a, reproduce the spectra of isolated QWRs of similar sizes. However, the PLE spectra of the w-QWR in QWR8.5, shown in Fig. 1b, is differs by two additional strong resonances at ~1.62 eV and ~1.65 eV are present. These two energies correspond to the dominating PLE resonances for the isolated n-QWR (Fig. 1a). Hence, essentially all carriers absorbed by the n-QWR are transferred to the w-QWR for QWR8.5. The experimental spectra of QWR5.5 and QWR20 resemble the spectra of QWR8.5 and QWR13, respectively. This gives the ranges of efficient transfer for $L_b \leq 8.5$ nm and blocked transfer for $L_b \geq 13$ nm.

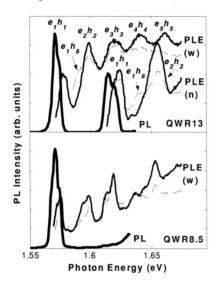

FIGURE 1. PL (bold lines) and polarized PLE (fine lines) of two QWR samples. The labels denote involved electron and hole subbands, identified in comparasion with theory. Solid (dashed) fine lines correspond to excitation light polarization parallell (perpendicular) to the QWR axis.

The experimental finding that the transfer stops in the range 8.5 nm $< L_b <$ 13 nm is in agreement with numerical results of a theoretical single particle model based on realistic shapes of the QWRs. Since the calculated electron subband separations are near the GaAs LO-phonon energy, the tunneling is assumed to occur via LO-phonon emission. This yields a value of blocked electron transfer for about $L_b > 10$ nm, corresponding to equal tunneling time and radiative lifetime. Slightly lower numerical value of L_b is obtained for the reference QWs, despite the higer barrier between the QWs due to the absence of a VQW. This difference originates from different confinement and energy level alignment in the two systems.

For comparison, the GaAs QW reference samples were subjected to analogous measurements and the results are reported in Fig. 2 (solid and dotted lines, respectively). The very weak luminescence recorded from the n-QW for $L_b \leq 8.5$ nm indicates nearly 100 % carrier transfer to the w-QW, and the transfer is still very efficient for $L_b = 13$ nm according to both the PL and the PLE spectra. Even for $L_b = 20$ nm is transfer detected in the PLE spectrum. Since conventional tunneling stops at $L_b < 10$ nm, according to the calculation, alternative mechanisms must be involved to explain the observed transfer for $L_b \geq 13$ nm.

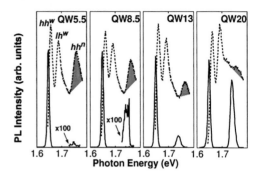

FIGURE 2. PL (solid line) and PLE (dotted line) spectra of the QW samples, the shaded areas indicate carrier transfer. The labels denote involved heavy- or light-hole subbands for the two QWRs, which confine only one electron subband.

In fact, an alternative carrier transfer mechanism with the efficiency of about 30 % have been manifested between molecular-beam-epitaxial (MBE) grown GaAs QWs separated by up to 30 nm.[2] The corresponding transfer efficiency in the presently studied OMCVD-grown QWs separated by 20 nm is estimated to about 5 %, i.e. essentially lower than the previous studies. Notably is that no alternative transfer mechanisms are observable at all for the QWRs.

A proposed photon-exchange model predicts weak transfer dependence on L_b, but vanishing transfer for strong exciton localization [5]. This is qualitatively consistent with the experimental data. The absence of photon-exchange transfer for the QWRs can be related to the strong lateral localization imposed by the two-dimensional confinement. Furthermore is the QWs very thin and more sensitive to interface roughness which enhances the effects of localization[5], and reduces the transfer compared to the results reported for thicker MBE grown QWs [2,3].

REFERENCES

1. K. F. Karlsson, *et. al.*, Phys. Rev. B **70**, 045302 (2004)
2. A. Tomita, *et al.*, Phys. Rev. B **53**, 10 793 (1996)
3. D. S. Kim, *et al.*, Phys. Rev. B **54**, 14 580 (1996)
4. B. Deveaud, *et al.*, Phys. Rev. B **42**, 7021 (1990)
5. S. K. Lyo, Phys. Rev. B **62**, 13 641 (2000)

1D-1D tunneling between vertically coupled GaAs/AlGaAs quantum wires

E. Bielejec*, J. A. Seamons*, M. P. Lilly* and J. L. Reno*

Sandia National Laboratories, Albuquerque, NM 87185

Abstract. We report low-dimensional transport and tunneling in an independently contacted vertically coupled quantum wire system, with a 7.5 nm barrier between the wires. The derivative of the linear conductance shows evidence for both single wire occupation and coupling between the wires. This provides a map of the subband occupation that illustrates the control that we have over the vertically coupled double quantum wires. Preliminary tunneling results indicate a sharp 1D-1D peak in conjunction with a broad 2D-2D background signal. This 1D-1D peak is sensitively dependent on the top and bottom split gate voltage.

The vertically coupled double quantum wire system is a simple realization of a coupled nanostructure that can be used to probe interaction effects in low-dimensional systems [1–4]. It is well known that in 1D systems electron-electron interactions play a major role leading to behavior that cannot be described by conventional Fermi-liquid theory of non-interacting particles [5]. Understanding interactions within and between the quantum wires is the first step in building more complicated interacting nanostructures.

The vertically coupled double quantum wires used in this experiment are fabricated using electron beam lithography to define metallic split gates on the top and bottom surfaces of a double quantum well GaAs/AlGaAs heterostructure. The quantum wells are 18 nm wide with an AlGaAs barrier of 7.5 nm separating the wells. The top and bottom quantum wells have individual layer densities of 1.67 and 1.79×10^{11} cm^{-2} respectively and a combined mobility of 1×10^6 cm^2/Vs. The heterostructure is thinned to approximately 0.4 μm using an epoxy-bond-and-stop-etch (EBASE) process [6]. This allows for symmetric top and bottom split gates 154 nm from the top and bottom quantum wells, aligned laterally with sub-0.1 μm resolution. The advantages of this device structure are three-fold. First, by making use of molecular beam epitaxy (MBE) growth of the tunneling barrier we have a rigid potential barrier between the layers. Second, independent contact to individual electron layers in combination with the top and bottom split gates allows for the independent formation and control of the number of occupied subbands in both the top and bottom quantum wires. Third, the close proximity of the top and bottom split gates to the electron layers leads to a well-defined confinement potential.

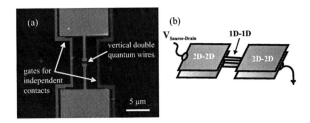

FIGURE 1. (a) Scanning electron microscope image of coupled quantum wire device. (b) Schematic diagram of the resonant tunneling measurement for this device.

In Fig. 1(a), a top view scanning electron micrograph of a coupled quantum wire device is shown. The dark areas are the GaAs/AlGaAs heterostructure described above. The active region of the device consists of six TiAu gates. The four gates in the center form pairs of split gates defining the quantum wires. Only the two gates on the top surface are visible due to the accurate alignment of the top and bottom gates. The quantum wires are formed by electrostatic confinement in the 0.5 μm wide and 1.0 μm long gap between the split gates. The remaining two gates, spanning the entire field of view, are used to independently contact the individual electron layers.

A schematic view of the resonant tunneling geometry is shown in Fig. 1(b). The active region consists of 2D-2D areas on either side of the 1D-1D quantum wire. Resonant tunneling occurs when an electron in one wire tunnels to the other wire conserving both energy and momentum [1]. Low frequency 2-probe and 4-probe measurements of the linear and tunneling conductance are made using a standard ac method with a constant excitation voltage of 100 μV at 143 Hz, with dc IV's used as a consistency check.

A gray-scale plot of the derivative of the linear conductance as a function of the top and bottom split gate voltage is shown in Fig. 2(a). While previous work concentrated on narrow barrier samples [7], we use a 7.5 nm barrier to allow separate contacts to the wires. In the top left (I) and bottom right (III) of the figure, the presence of quantized conductance steps (parallel white lines are plateaus) indicate that only one wire is occupied. In the center region (II) of the figure we observe a more complicated pattern as both of the wires are occupied and contribute to the conductance. This 1D-1D region provides a means to map out the subband occupation, as well as, to illustrate the control we have over the states of the individual wires. We define (i,j) to be the ith subband of the top wire and j^{th} subband of the bottom wire. Tunneling can be measured with any combination of the top and bottom wire subbands.

In Fig. 2(b) we show two ac tunneling conductance curves. We investigate this signal by both measuring the tunneling conductance far from the 1D-1D region where the only allowed tunneling is 2D-2D, as well as in the 1D-1D region. The dashed curve shows the tunneling conductance in region III which displays only a broad background signal with a negative dip that indicates an abrupt change in the tunneling current. Using a simple device model (circuit diagram in Fig. 2(b) inset) that consists of two parallel quantum wires each with a resistance of $R_W = 10$ kΩ, in series with a 2D-2D tunneling resistance ($R_T(\Delta)$), which corresponds to the 2D-2D tunneling areas. $R_T(\Delta)$ has a lorenzian form peaked at $\Delta = -0.5$ mV with a width of 0.1 mV and amplitude of 1.0 kΩ, values not unreasonable for 2D-2D tunneling in these samples. From the modelling results we conclude the 2D-2D signal has a broad maxima centered at $V_{Source-Drain} \sim -5$ mV as observed in the background signal. This background is a broadened 2D-2D tunneling signal. The broadening is due to the strong tunneling between our electron layers and our device geometry.

The bold curve in Fig. 2(b) shows the tunneling conductance of the (2,1) subband configuration which displays a sharp peak on top of a background signal. This sharp peak is sensitively dependent on the top and bottom split gate voltage and is not present far from the 1D-1D region. This, in conjunction with our modelling results, suggests that the sharp peak is due to 1D-1D tunneling between the wires. The 1D-1D peak is centered at - 5.9 mV. This large $V_{Source-Drain}$ indicates a substantial voltage drop across the wire exists, and equilibrium tunneling theories must be modified to understand tunneling in this system.

In conclusion we have fabricated an independently contacted vertically coupled double quantum wire device that allows for the number of occupied subbands to be varied over a wide range. We have observed a large 2D-2D background, as well as a 1D-1D peak that depends sensitively on the top and bottom split gate voltage. Future work will focus on larger barriers to eliminate the 2D-2D tunneling complications.

We acknowledge the outstanding technical assistance of R. Dunn and D. Tibbets. This work has been supported by the Division of Materials Sciences and Engineering, Office of Basic Energy Sciences, U.S. Department of Energy. Sandia is a multiprogram laboratory operated by Sandia Corporation, a Lockheed Martin Company, for the United States Department of Energy under contract DE-AC04-94AL85000.

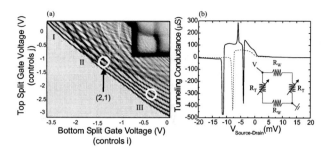

FIGURE 2. (a) Gray scale plot of the derivative of the linear conductance with respect to the top and bottom split gate voltage. Darker color stands for higher differential conductance. The parallel white lines in the top left (I) and bottom right (III) of the figure are regions of uniform conductance steps. The center (II) crossing region is where both wires are occupied and contribute to the conductance. The (2,1) subband configuration is indicated on the figure. (b) Ac IV tunneling conductance for the (2,1) configuration (bold), and for region III (dashed) as indicated in (a). Inset: A circuit diagram of the model.

REFERENCES

1. U. Zulicke, Science **295**, 810 (2002).
2. O. M. Auslander, A. Yacoby, R. de Picciotto, K. W. Baldwin, L. N. Pfeiffer and K. W. West, Science **295**, 825 (2002).
3. Y. Tserkovnyak, B. L. Halperin, O. M. Auslander and A. Yacoby, Phys. Rev. Lett. **89**, 136805 (2002).
4. Y. Tserkovnyak, B. L. Halperin, O. M. Auslander and A. Yacoby, Phys. Rev. B **68**, 125312 (2003).
5. F. D. M. Haldane, J. Phys. C **295**, 810 (2002).
6. M. V. Weckwerth, J. A. Simmons, N. E. Harff, M. E. Sherwin, M. A. Blount, W. E. Baca, and H. C. Chui, Supperlatt. Microstruct. **20**, 561 (1996).
7. M. A. Blount, J. S. Moon, J. A. Simmons, S. K. Lyo, J. R. Wendt, and J. L. Reno, Physica E **6**, 689 (2000).

Magnetotransport spectroscopy of mode coupling in electron wave guides

G. Apetrii*, S.F. Fischer*, U. Kunze*, D. Schuh[†] and G. Abstreiter[†]

*Werkstoffe und Nanoelektronik, Ruhr-Universität Bochum, Universitätsstr. 150, D-44780 Bochum, Germany
[†]Walter Schottky Institut, Technische Universität München, Am Coulombwall, D-85748 Garching, Germany

Abstract. High-quality one-dimensional electron wave guides were fabricated by atomic-force microscope lithography from a AlGaAs/GaAs 30 nm wide quantum well with two occupied subband. In the case of strong coupling the quantized conductance at 4.2 K shows missing or ill-defined steps with otherwise regular increase in multiples of $2e^2/h$ which are caused by coincidence of 1D subbands of different vertical modes. Magnetotransport in transverse fields show nearly regular patterns of level crossings and anticrossings.

Coupling of nearly degenerate levels in confined electron systems not only depends on the spatial overlap of the corresponding wave functions but also on the detailed symmetry of the perturbing potential [1]. This is of particular importance if the electron states are confined to the same spatial region. The present work deals with a ballistic one-dimensional (1D) constriction cut from a two-dimensional electron gas (2DEG) with two occupied subbands. The resulting 1D system contains two series of 1D subbands with spatially nearly coincident wave functions, in contrast to stacked and tunnel-coupled electron waveguides [2, 3, 4, 5, 6]. Since our system is based on a GaAs/AlGaAs heterostructure with modulation doped square quantum well (QW) it is fundamentally different from the vertically harmonic well approximation [6], where ideally wave function mixing is absent.

The 30 nm wide GaAs quantum well embedded by $Al_{0.68}Ga_{0.32}As$ barriers is situated 60 nm below the semiconductor surface. The 2DEG carrier mobility and density are 8.5×10^5 cm^2/Vs and 4.2×10^{11} cm^{-2}, respectively, measured at 4.2 K in the dark. A numerical simulation indicates a symmetric QW at top-gate voltage of V_g = +0.1 V. At larger V_g the QW is tilted resembling a triangular shape. From Shubnikov-de Haas measurements as a function of V_g we determined a threshold voltage of -0.4 V and the onset of second subband population at +0.02 V [7].

Quantum point contacts (QPCs) with 1D-level spacing in excess of 10 meV were prepared by nanolithography with an atomic force microscope [8]. Hereby, 1D constrictions of 30-260 nm width and 100 nm length are formed by deeply (70-85 nm) etched grooves as depicted in the inset of Fig. 1. The constriction and the surrounding reservoirs are covered by a metallic top gate elec-

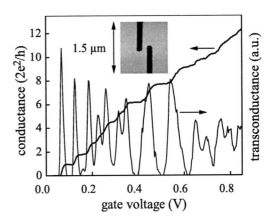

FIGURE 1. Conductance and transconductance versus gate voltage of a QPC with 165 nm geometric width show characteristica of two vertical modes at 4.2 K. The inset presents an AFM top view of the groove pattern (black) defining the 1D constriction.

trode.

At a temperature of T = 4.2 K the two-terminal differential drain conductance $g_d = I_d/V_d$ was measured using standard lock-in technique, where I_d denotes the drain current and V_d the drain voltage. The rms drain excitation was 0.3 mV at 433 Hz. The transconductance $g_m = \partial^2 I_d/\partial V_d \partial V_g \approx \partial g_d/\partial V_g$ has been determined from the second harmonic signal in I_d when additionally the gate voltage is modulated with a 3 mV rms. Transport data under high magnetic field are taken at T = 2 K and at fixed magnetic field up to B = 8 T.

The conductance characteristics show quantum steps at multiples of $2e^2/h$ height (Fig. 1). Double steps of different slope quality indicate close or coincident 1D lev-

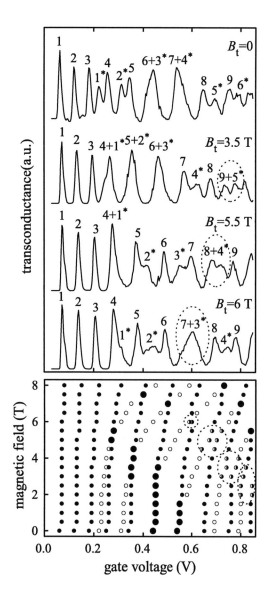

FIGURE 2. Top: Transconductance at different transversal magnetic fields of a 165 nm wide QPC at 2 K. The first vertical mode is labelled by 1,2,3... and the second by 1^*, 2^*, 3^*,... Bottom: Magnetic field and gate voltage dependence of transconductance maxima. Maxima of the first vertical mode are identified by black filled circles, those of the second vertical mode by black open circles. Larger black filled circles depict maxima enhancements. Maxima repulsions are encircled by dashed lines.

els belonging to different vertical wave functions (vertical modes). The corresponding transconductance vs. gate voltage characteristics show two distinctively different series of maxima, each signaling the subsequent population of 1D levels within a vertical mode. For a single vertical mode a sequence consists of nearly regularly spaced transconductance maxima which decrease monotonously with increasing gate voltage. Such transconductance characteristic is significantly modified at the onset of the 1D level population of a second vertical mode. Maxima of lower amplitude appear in between the first peak series. Furthermore, if maxima of the two vertical modes coincide two cases are observed: amplitude superposition or repulsion. In order to verify the latter, we applied magnetic fields transverse to the 1D current flow. Differing magnetodispersions are expected due to a distinctly different diamagnetic shift which relates to the increasing spatial extension of the wave function in growth direction for higher vertical modes. Hence, increasing transverse magnetic fields shift the transconductance maxima of the two vertical modes relative to each other (Fig. 2 (top)). We label transconductance maxima assigned to the first vertical mode by 1,2,3... and those of the second vertical mode by $1^*, 2^*, 3^*$,.... For example, the relative position of 1^* and 4 to each other is changed as the following: two distinct peaks ($B_t = 0$ T) $1^*, 4$ merge at $B_t = 5.5$ T to a single maximum with increased amplitude $4+1^*$ and reappear at $B_t = 6$ T as two peaks $4, 1^*$. Fig. 2 (top) details the superposition peak structures of $4+1^*, 5+2^*, 6+3^*, 7+4^*$ and $7+3^*$. The evolution of peak superpositions in the magnetic field vs. gate voltage plane is shown in Fig. 2 (bottom), which clearly reflects 1D level crossing of two vertical modes. However, for coincidences of $8+4^*, 9+5^*$ and $10+6^*$ 1D level degeneracy is not observed. Level repulsions instead are visible as anticrossings of transconductance peaks which are encircled by dashed lines. As a result of the two different lateral extensions of the 1D electron systems which are formed from an asymmetric vertical confinement we find that anticrossings are only observed between levels of equal parity of the lateral wave function.

ACKNOWLEDGMENTS

GA gratefully acknowledges the financial support of the Deutsche Forschungsgemeinschaft via Graduiertenkolleg 384. Part of this work was supported by the Bundesministerium für Bildung und Forschung under grant no. 01BM920.

REFERENCES

1. R. Ferreira, G. Bastard, Rep. Prog. Phys. **60**, 345 (1997).
2. I.M. Castleton et al., Physica B 249-251, 157 (1998).
3. K.J. Thomas et al., Phys. Rev. B 59, 12252 (1999).
4. M.A. Blount et al., Physica E 6, 689 (2000).
5. S. Roddaro et al., J. Appl. Phys. 92, 5304 (2002).
6. G. Salis et al., Phys. Rev. B 60, 7756 (1999).
7. G. Apetrii et al., Physica E 22, 398 (2004).
8. G. Apetrii et al., Semicond. Sci. Technol. 17, 735 (2002).

Tunneling between Parallel Quantum Wires

M. Yamamoto[1], Y. Tokura[2,3], Y. Hirayama[2,3], M. Stopa[4], K. Ono[1,3], S. Tarucha[1,4]

[1]Department of Applied Physics, University of Tokyo, Bnkyo-ku, Tokyo, 113-0033, Japan,
[2]NTT Basic Research Laboratories, Atsugi-shi, Kanagawa, 243-0198, Japan,
[3]SORST-JST, Atsugi-shi, Kanagaw, 243-0198, Japan,
[4]ERATO-JST, Atusgi-shi, Kanagawa, 243-0198, Japan,

Abstract. We measure electron tunneling between parallel coupled two split-gate quantum wires to investigate the microscopic electronic properties of one-dimensional (1D) electron systems. The differential tunneling conductance dI/dV has peaks when the subband bottoms align between the wires. By tracing these peaks as a function of voltage V between the wires and gate voltages, we characterize the 1D subband structures and inter-subband interactions. We also observe charging and filling of the 1D energy bands.

INTRODUCTION

A number of transport experiments have been performed on quantum wires made in a two-dimensional electron gas (2DEG) to study the 1D characteristics [1]. In a conventional transport experiment using a single quantum wire with finite length, the conductance usually shows neither well-defined quantized steps or interaction effects. For a coupled-wire system, on the other hand, specific measurements like Coulomb drag [2] and inter-wire tunneling are feasible, and these measurements can provide microscopic nature of 1D electrons including electron-electron interaction. In this work, we measure tunneling between parallel split-gate quantum wires to study the microscopic electronic properties of coupled 1D system.

EXPERIMENT AND DISCUSSION

Our device consists of parallel coupled two split-gate wires defined in an n-AlGaAs/GaAs 2DEG using a Schottky gate technique (See Fig. 1). The two wires are separated by a 30 nm wide line Schottky gate. The length of the wire is 4 μm. The tunneling current I between the wires is varied as a function of the line gate voltage, and the electron densities of the two wires are both varied as a function of the two side gate voltages. The line center gate is sufficiently negatively biased to probe the transport properties in the regime of weak tunneling [4], where $dI/dV \ll 2e^2/h$. The device is placed in a dilution refrigerator with a base temperature of 10 mK.

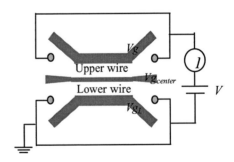

FIGURE 1. Experimental setup for the tunneling measurement between two quantum wires.

We firstly explain possible tunneling mechanism in our device. When positive V is applied in Fig. 1, tunneling occurs from the occupied state in the upper wire subband to the Fermi surface or unoccupied state in the lower wire subband, and vise versa for negative V. Transverse electron momentum is conserved in the tunneling process. After the tunneling, electrons with excess energy finally relax onto the Fermi surface. Here we assume that applied bias V mainly contributes to shift the energy bands rather than to fill them [4], neglect the electron-electron interaction, and only consider a single occupied mode for each wire. As V is initially increased, the tunneling current begins to flow when the electrons can directly tunnel onto the Fermi surface of the opposite wire with no energy cost (See band diagrams in Fig. 2). When Vg_{center} and Vg_L are fixed, the Fermi wave vector in the lower wire is almost constant. Then above mentioned tunneling occurs along the solid line in the plane of (V, Vg_U) in Fig. 2. Note along this line, the differential conductance dI/dV has a peak as a function of V and Vg_U.

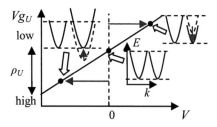

FIGURE 2. dI/dV peak position as a function of V and Vg_U for fixed Vg_{center} and Vg_L when each wire has a single occupied subband.

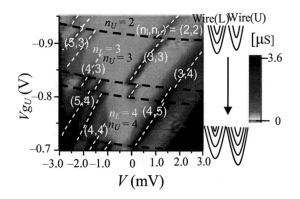

FIGURE 3. Gray scale plot of dI/dV for $Vg_{center} = -0.805$ V and $Vg_L = -0.90$ V. $n_{L(U)}$ denotes the number of the occupied subbands in the lower (upper) wire and (n_l, n_r) denotes the alignment of the subband bottoms between n_l-th subband in the lower wire and n_r-th subband in the upper wire.

Fig. 3 shows a typical dI/dV data obtained in our experiment. We can see many peak lines associated with alignment between various subbands of the two wires. The number of the occupied subbands n_L and n_U are decided from measurements of wire conductance and its magnetic depopulation. The dotted black lines in Fig. 3 indicate the boarder between different n_U's or different n_L's. Note that Vg_U also changes the chemical potential in the lower wire, resulting in the change of n_L, for example, between $n_L = 3$ and 4 at $Vg_U \approx -0.8$ V. Hence one of the two dotted lines around $Vg_U \approx -0.8$ V corresponds to the boarder between $n_L = 3$ and 4 and the others correspond to the boarder between different n_U's. Note also that these lines are not parallel to the horizontal axis. The inclination represents the ratio of filling (charging) and shifting of 1D subbands with bias V. The peak lines finally identified are indexed by (n_l, n_U) in Fig. 3, although it is difficult to index all of the peak lines because there are so many subbands. We derive the subband spacing or parabolic confining energy of the lower wire from the distance between the neighboring peak lines, (n,m) and $(n-1, m)$. For example, the value is 1.5 meV for $Vg_U = -0.90$ V and 1.3 meV for $Vg_U = -0.80$ V. Jumps of the peak positions are also observed along the black dotted lines. These jumps arise from the inter-subband interaction: Occupation of a new subband raises electrostatic enegy of the subbands. This is a charging effect of the wire, and occurs when the bottom of the subband having a high density of states is just occupied. Also note that the tunneling current is suppressed around $V = 0$, probably due to the interaction effect.

In summary, we observe electron tunneling between parallel split-gate quantum wires. The obtained results reflect the microscopic electronic nature, including inter-subband interaction and ratio between the charging and shifting of 1D energy bands.

ACKNOWLEDGMENTS

We acknowledge financial support from the Grant-in-Aid for Scientific Research A (No. 40302799).

REFERENCES

1. See for example, S. Tarucha *et al.,* Solid State Commun. **94**, 413 (1995).

2. P. Debray *et al.*, Physica E **6** (2000) 694; P. Debray *et al.*, J. Phys.: Condense. Matter **13** (2001) 3389; M. Yamamoto *et al.*, Proc. of the 25th ICPS (2001); M. Yamamoto *et al.*, Physica E **12** (2002) 726.

3. O.M. Auslaender *et al.*, Science **295**, 825 (2002); Y. Tserkovnyak *et al.*, Phys. Rev. Lett. **89**, 136805 (2003).

4. D. Boese *et al.*, Phys. Rev. B **64**, 085315 (2001).

Optical Spectroscopy on Single Localized States in an InGaN/GaN Structure

S. Halm [1], G. Bacher [1], H. Schömig [2], A. Forchel [2]
J. Off [3], F. Scholz [3,4]

[1] *Werkstoffe der Elektrotechnik, Universität Duisburg-Essen, 47057 Duisburg, Germany*
[2] *Technische Physik, Universität Würzburg, 97074 Würzburg, Germany*
[3] *4. Physikalisches Institut, Universität Stuttgart, 70550 Stuttgart, Germany*
[4] *Abteilung Optoelektronik, Universität Ulm, 89069 Ulm, Germany*

Abstract. By employing photoluminescence (PL) spectroscopy with subwavelength lateral resolution we succeeded in resolving individual localization centers in a thin InGaN/GaN quantum well. Spectrally narrow PL lines with a linewidth as small as 0.8meV could be observed and attributed to the recombination of an electron-hole pair occupying a single localized state. These individual emission lines show virtually no energy shift with increasing excitation density. Instead new high-energetic lines appear and cause a blue shift of the overall spectrum.

INTRODUCTION

Although the crystalline quality of InGaN/GaN quantum films is generally poor due to a high density of threading dislocations the internal quantum efficiencies are nevertheless large enough to fabricate high brightness light-emitting diodes and diode lasers for the blue spectral range. Localization effects possibly resulting from In-rich nanoislands [1,2] and monolayer and compositional fluctuations [3] as well as the strong internal electric fields [4] are thought to play an important role for the emission characteristics of such structures but the details are still under controversial debate. In this paper we present highly spatially resolved micro-photoluminescence (μPL) spectroscopy measurements on single localization centers in an InGaN/GaN quantum well using a metal mask nanoaperture technique.

EXPERIMENTS

The samples were grown by metal organic vapor phase epitaxy (MOVPE) on a 6H-SiC substrate. After a 300nm AlGaN buffer layer and a 700nm GaN barrier, a 3nm-thick $In_xGa_{1-x}N$ quantum well with a nominal In content of x=0.15 was grown, capped by 40nm GaN. Subwavelength lateral resolution was achieved by lithographically defining an opaque metal mask on top of the semiconductor with nanoapertures down to 100nm in diameter (see inset Fig. 1). For the μPL measurements the sample was mounted in a helium-flow cryostat (T≥3.5K) and excited by the UV-lines of an argon ion laser. The laser was focused onto the sample by a 50x microscope objective with a high numerical aperture of N.A.=0.6. The approximate diameter of the laser spot on the sample was measured to be around ~600nm. The photoluminescence was collected through the same microscope objective and subsequently dispersed by a 0.55m-monochromator and detected by a charge-coupled-device camera.

Figure 1 shows μPL spectra from the InGaN/GaN quantum well measured beneath two apertures of different sizes. In case of the 5μm aperture a large number of localization centers are excited simultaneously. Thus, the resulting spectrum is inhomogeneously broadened with a FWHM (full width at half maximum) of 40meV. This situation changes drastically as we reduce the aperture size and thereby the number of excited localization centers. The broad spectrum splits into discrete lines. For an aperture size of 175nm only a single spectrally narrow emission line remains which we attribute to the recombination process of an electron-hole pair occupying a single localized state.

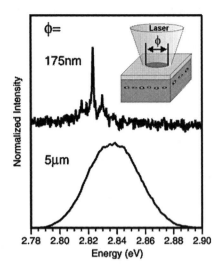

FIGURE 1. Low temperature micro-photoluminescence spectra for two different aperture sizes ϕ. The inset schematically demonstrates the mask technique.

Apart from enabling us to confirm three-dimensional carrier localization in a InGaN/GaN quantum well in a direct optical experiment, the µPL measurements also shed light on details of the recombination mechanisms that so far were hidden by the inhomogeneous broadening effects inherent for macro-PL studies. It is well known for InGaN/GaN quantum wells that macro-PL spectra exhibit a pronounced and continuous blue shift for increasing excitation power. This blue shift is commonly believed to result from a screening of the strong internal electric fields by optically generated charge [4], as well as from filling of higher energy states (called 'band tail filling') [5,6].

We analyzed the exitation density dependence of a small set of localization centers (ϕ=1000nm) and obtained the µPL spectra shown in Fig. 2. One observes that the single emission lines show virtually no energy shift although the excitation density is varied by about three orders of magnitude. Instead the relative intensity of the higher energy lines increases and new lines appear on this side of the spectrum. This causes the spectral weight to shift to the blue. These findings suggest the following microscopic picture: At low excitation densities, single electron-hole pairs occupy mainly the ground state of low-energetic localization centers. With increasing excitation density the low energy localization centers become saturated and higher energy centers are filled. As a consequence new high energetic lines emerge in the spectrum. Furthermore, several electron-hole pairs may be trapped in the same localization center giving rise to biexciton as well as other multiexciton states. Indeed we found experimental evidence for a biexciton state formation [7]. Surprisingly we found a negative binding energy of about -5meV. We believe that it is caused as an effect of spatial separation of the excited charge carriers by the strong internal electric fields.

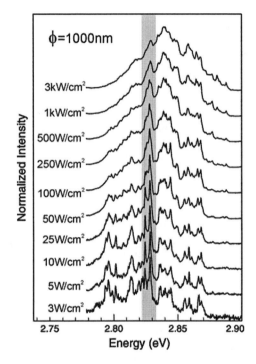

FIGURE 2. µPL spectra at varying excitation densities and a temperature of T=4K

ACKNOWLEDGEMENTS

The authors would like to thank M. Emmerling for expert technical assistance.

REFERENCES

1. I. L. Krestnikov et al., *Phys. Rev. B* **66**, 155310 (2002).
2. A. Morel et al., *Phys.Rev B* **68**, 045331 (2003).
3. S. Chichibu et al., *J. Vac. Sci. Technol. B* **16**, 2204 (1998)
4. T. Kuroda et al., *Appl. Phys. Letters* **76**, 3753 (2000)
5. E. Kuokstis et al., *Appl. Phys. Letters* **80**, 977 (2002)
6. U. Jahn et al., *J. Appl. Phys.* **93**, 1048 (2003)
7. H. Schömig et al., *Phys. Rev. Letters* **92**, 106802 (2004)

The effect of inter-wall correlations and electric field on inhomogeneous broadening of excitons in quantum wells

Ilya V. Ponomarev*, Lev I. Deych* and Alexander A. Lisyansky*

Queens College of City University of New York, Flushing, NY 11367, USA

Abstract. We show that the vertical correlations between rough interfaces in narrow quantum well reduce strongly the interface roughness contribution to exciton line width. We also show that the applied uniform electric field has different effects on alloy disorder and interface roughness contributions to inhomogeneous exciton broadening in QW. In some range of parameters the electric field induced narrowing of the exciton line-width is possible. It is possible effectively distinguish two different disorder mechanisms. This can be employed for applications in monitoring of quality in semiconductor nanostructures.

It is well known that the exciton line width in absorption and luminescence spectra in quantum wells (QW) is predominantly inhomogeneous at low temperatures. The compositional disorder and interface roughness are two main contributions to this broadening. From the earliest days in the history of semiconductor nanostructures the inhomogeneous line width was used to characterize the quality and structure of the material. It is also a common belief that exciton line width is a quick way to determine the interface quality: narrow line widths imply uniform interfaces, and a good QW[1]. Estimations show however that usually both disorder yield comparable contributions to the total width. This imposes an additional difficulty for experimental identification of the interface quality in QW from optical spectra, since absolute values of each contributions are unknown.

In this paper we demonstrate that the presence of vertical inter-wall correlations strongly reduce the interface roughness contribution to the inhomogeneous broadening. Therefore the correspondence between QW thickness fluctuations and exciton line width should be used with caution, especially for narrow quantum wells. We also show that application of the uniform electric field perpendicular to the QW growth direction allows one to separate different disorder mechanisms. Another surprising counterintuitive effect of applied electric field is that in certain range of QW widths the motion narrowing of original exciton line-width is possible.

In the presence of disorder, the center-of-mass (C.O.M) motion of the exciton in a QW structure, is no longer described by a plane wave, but is determined by Schrödinger equation:

$$\left[-\frac{1}{2M}\Delta_R + V_{eff}(\mathbf{R})\right]\Phi(\mathbf{R}) = E_X\Phi(\mathbf{R}) \quad (1)$$

where $M = m_h + m_e$ is a total mass of the exciton. The random effective potential, $V_{eff}(\mathbf{R}) = V_e(\mathbf{R}) + V_h(\mathbf{R})$, is derived by averaging the atomic scale electron and hole disordered potentials over all but **R** coordinates with an exciton wave function. This function is presented as a product of the electron and hole confinement wave functions, $\chi(z_{e,h})$, in the growth direction, z, and the in-plane wave function, $\phi(r)$, for relative and C.O.M exciton motion. The variance, $\sigma^2 = \langle V_{eff}(\mathbf{R})^2 \rangle = \sigma_{ee}^2 + \sigma_{eh}^2 + \sigma_{hh}^2$, of the effective potential is an important characteristic determining the total inhomogeneous width of exciton spectra. It is obtained by the statistical averaging over different random configurations of microscopic random potentials. For QW with a heavy hole and light electron ($m_h \gg m_e$) the main contribution stems from hole-hole variance, σ_{hh}^2. In general, we may split the variance in two the sum of contributions from compositional disorder and interface roughness, $\sigma_{hh}^2 = \sigma_{hh}^{(comp)2} + \sigma_{hh}^{(int)2}$. One can show that for the case of compositional disorder

$$\sigma_{hh}^{(comp)2} \propto \int \chi^4(z_h) dz_h \quad (2)$$

while for the interface disorder the contribution is

$$\sigma_{hh}^{(int)2} \propto V_h^2 \left[\chi_L^4 \langle \xi_1^2 \rangle + \chi_R^4 \langle \xi_2^2 \rangle - 2\chi_L^2 \chi_R^2 \langle \xi_1 \xi_2 \rangle\right] \quad (3)$$

where V_h is the height of hole potential in a QW, $\xi_{1,2}$ describe random fluctuations of the positions of two interfaces forming the well, and $\chi_{L,R} = \chi(\mp L/2)$. The statistical properties of the interfacial roughness in multi-layered systems are characterized by height-height correlation functions

$$\langle \xi_i(\boldsymbol{\rho}_1)\xi_j(\boldsymbol{\rho}_2) \rangle = \Delta^2 f_{ij}\zeta(|\boldsymbol{\rho}_1 - \boldsymbol{\rho}_2|), \quad (4)$$

where Δ is an average height of interface inhomogeneity. We assume here that the dependence of both diagonal and non-diagonal correlations on the lateral coordinates $\boldsymbol{\rho}$ is described by the same function $\zeta(\boldsymbol{\rho})$. Diagonal elements f_{ii} are in general different constants that describe a possibility for a different vertical size of corrugations on different interfaces (for example, due to the growth interruption technique for one of walls) and the respective functions describe lateral correlation properties of a given interface (*self-correlation functions*). Nondiagonal elements introduce correlations between different interfaces; the respective quantity $f_{12}(L/\sigma_\parallel)$, which can be called a *cross-* or *vertical-correlation function*, is a function of the average width of the well and is characterized by the vertical correlation length $\sigma_\parallel$.

To the best of our knowledge, the inter-wall correlator f_{12} was completely neglected in all previous studies. This might be justified for wide QWs, but it cannot be neglected in narrow QW. Experimental techniques such as cross-sectional scanning tunnel microscopy[2] and X-ray reflection measurements[3] clearly confirm that vertical correlations do certainly exist. Another relevant example is vertical stacking of quantum dots[4], where vertical correlation length is observed up to 80ML thickness. In general, $f_{12}(L/\sigma_\parallel)$ is some monotonic function which tends to unity at small L, and it has zero asymptote at large QW's widths.

We show that the presence of these correlations strongly suppresses the interface disorder contribution into inhomogeneous broadening. At small QW thicknesses the suppression factor has a form: $\sigma_{hh}^{(int)^2} \propto L^{2+\alpha}$, where α depends on the form of the inter-wall correlator $f_{12}(L/\sigma_\parallel)$, and an additional power of 2 comes from L-dependence of the wave-function $\chi(z)$ at the QW interfaces. In the limit of very narrow symmetric QWs, the interface-roughness contribution to the inhomogeneous width totally vanishes due to symmetry. Schematically, it can be understood in the following way: if one random interface repeats the pattern of the second one, then the disorder contribution tends to cancel in the first order of perturbation theory.

Application of the electric field, E, in z-direction increases the average separation between the electron and the hole and thus considerably modifies the excitonic wave functions, which in turn change the properties of the effective random potential. We show that in the Stark regime of not very strong fields these changes for compositional disorder and interface roughness contributions dominate and have the following forms:

$$\sigma_{hh}^{(int)^2}(E) = \sigma_{hh}^{(int)^2}(0) + \nu_1 (f_{22} - f_{11}) E + \nu_2 E^2, \quad (5)$$

$$\sigma_{hh}^{(comp)^2}(E) = \sigma_{hh}^{(comp)^2}(0) - \mu E^2, \quad (6)$$

where $\nu_1 > 0$, and ν_2, μ are some functions of L.

First of all, there is a remarkable linear in field term in the interface roughness mechanism, which is nonzero for the interfaces with different size of corrugations ($f_{11} \neq f_{22}$). This means that one can broaden or narrow the line width by reversing the direction of the electric field. This fact has a simple explanation: the field affects a shape of QW wave function, pushing its maximum towards one of the interfaces. If this push directed away from the roughest interface then the overall result will be decreasing of exciton line-width.

Even if the linear term is absent for equally corrugated walls, the contributions from alloy disorder and interface roughness broadening still can be distinguished due to different behaviors of the respective factors $\nu_2(L)$ and $\mu(L)$. Both these functions are monotonic. They are negative for very narrow QW widths and positive for wide QWs. However, the values L_{cr} of the critical widths, where $\nu_2(L)$, $\mu(L)$ change signs (*i.e.* vanish) are different. This means that we can effectively turn off the quadratic contribution from one of two sources of the inhomogeneous broadening by growing a QW with width close to the respective critical value, L_{cr}. In this case, all the field induced changes in the exciton broadening will be cause mostly by the other broadening mechanism. In principle, this can allow for unambiguous discrimination between alloy and interface disorder contributions to the exciton line width. Qualitatively a different sign of ν_2, μ for shallow and deep QWs can be explain by a competition of two processes. On one hand, electric field pushes part of the wave function outside of the well and away from the influences of the disorders, promoting narrowing of the exciton line. On the other hand, the field changes a shape of the wave function, pushing it towards an interface and slightly squeezing. The latter results in greater localization of the wave function and, hence, broadens the exciton line. It is clear that the first process dominates for shallow QWs, while the second one prevails for QW with larger widths.

Eqs. (6) and (5) show that studying exciton line width as a function of electric field one can determine the main disorder mechanism contributing to the inhomogeneous broadening of excitons. The described effect can also be used for more reliable monitoring the interface quality in the process of growth of the QW structures. This work was supported by AFOSR grant F49620-02-1-0305.

REFERENCES

1. C. Weisbuch, R. Dingle, A.C. Gossard, and W. Wiegmann, *Solid State Commun* **38**, 709 (1981).
2. Y. Yayon et al. *Phys. Rev. Lett.* **89**, 157402 (2002).
3. E.A. Kondrashkina et al, *Phys. Rev. B* **56**, 10469 (1997).
4. Q. Xie, A. Madhukar, P. Chen, and N. Kobayashi, *Phys. Rev. Lett.* **65**, 2542 (1995).

Low Temperature Photoluminescence Of GaAs/GaInP Heterostructures Measured Under Hydrostatic Pressure

Toshihiko Kobayashi[1], Atsushi Nagata[1], Andrew D. Prins[1,2], Yasuhiro Homma[1], Kazuo Uchida[3], and Jun-ichiro Nakahara[4]

[1] *Department of Electrical and Electronics Engineering, Kobe University, Kobe 657-8501, Japan*
[2] *Department of Physics, Queen Mary, University of London, London E1 4NS, UK*
[3] *Department of Electronic Engineering, University of Electro-Communications, Chofu 182-8585, Japan*
[4] *Division of Physics, Graduate School of Science, Hokkaido University, Sapporo 060-0810, Japan*

Abstract. A study of 11 K photoluminescence measurements of metalorganic vapor phase epitaxy grown GaAs/GaInP quantum wells is reported for the first time at pressures up to ~5 GPa. The use of low temperature allows us to study the true nature of a peak at ~1.46 eV, which dominates instead of the GaAs QW emission, even at very low excitation intensities to pressures well above the Γ-X crossover in GaInP. Our results suggest that the ~1.46 eV emission is a spatially indirect transition of electrons and holes separated at the interface in a type II band alignment.

INTRODUCTION

Heterojunctions in the GaAs/GaInP system have attracted recent attention. However, because of the imperfect CuPt-type ordering in currently available GaInP grown by metalorganic vapor phase epitaxy (MOVPE), discrepancies exist between measurements such as band offsets even for similar growth conditions. Theoretical calculations show the band alignment to be dependent on degree of ordering [1]. The insertion of thin layers between the two material interfaces has been shown to drastically improve interface characteristics. In the GaAs/partially ordered GaInP quantum well (QW) structures without insertion of two thin GaP layers, an undesirable, intense emission band at ~1.46 eV often dominates instead of QW emission [2]. We report for the first time 11 K photoluminescence (PL) measurements of MOVPE grown GaAs/GaInP quantum wells at pressures up to ~5 GPa. We examine the distinct effects of laser excitation photon energy and excitation intensity. The various PL features observed at 11 K clearly show that the true nature of this emission and the interface properties are only revealed at low excitation powers and arise from the GaInP layer. This is in contrast to the earlier high-pressure study at 77 K [3].

EXPERIMENT

Four types of GaAs/GaInP single QW structures were grown by MOVPE on (100) GaAs substrates at 550 °C under the same conditions [2]. Sample No. 1 consists of a 100-Å-thick GaAs well and 1100 Å partially ordered GaInP barriers. Sample No. 2 has a 14-Å-thick GaP layer between the lower GaAs/GaInP interface (i.e., nearer the substrate) only, while sample No. 3 has a 28 Å GaP layer between the upper interface. Sample No. 4 contains GaP layers sandwiching the GaAs well. Liminescence spectra at 10–12 K were excited by the 488 nm, 532 nm, 633 nm, 682 nm, and 783 nm lines of lasers. The high-pressure PL measurements were performed at 11 K using a diamond anvil cell.

RESULTS AND DISCUSSIONS

Figure 1 shows the comparison of the typical PL spectra of sample No.2 under several different excitation photon energies, measured at 10–12 K and at atmospheric pressure. Excitation intensities are in the range 0.7–3 mW. Among all the emission peaks in the spectra, only a below-band gap PL spectrum (denoted as peak A) with a peak energy (~1.46 eV) less than either band gap of GaInP and GaAs shows a significant blueshift of its peak energy with excitation laser power. Similar intense PL signal is also observed from samples No.1 and No.3, while the quantum well emission from the GaAs QW whose peak energy (~1.53 eV) is slightly higher than the sharp excitonic transitions of GaAs at ~1.51 eV (denoted as P2) or the impurity-related peak at 1.49 eV (denoted as P3) is obtained only for sample No.4, where GaP layers are present at both GaAs/GaInP interfaces. Selecting a photon energy that does not excite GaInP layers apparently weakens the emission peak at ~1.46 eV relative to near band-edge transitions P2 and P3 in the spectra. The presence of the photoexcited electrons in the partially ordered GaInP layers at the GaAs/GaInP interface with no thin GaP intermediate layer is greatly responsible for the observed ~1.46 eV emission. These results strongly suggest the peak A is attributed to the spatially indirect recombination of the electrons and holes confined at the type II heterojunction, GaInP(Γ_C)–GaAs(Γ_V).

clearly exhibits the shift towards lower energies and a drop of its intensity. This is again strongly dependent on excitation intensity. Since the Γ-X crossover in partially ordered GaInP in sample No.2 occurs at 3.2–3.5 GPa, the transition, GaInP(Γ_C)–GaAs(Γ_V), is replaced by the indirect GaInP(X_C)–GaAs(Γ_V) transition both in real space and momentum space at the type II heterojunction.

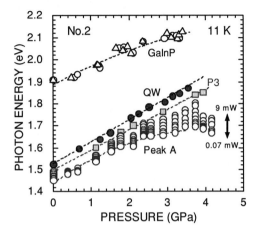

FIGURE 2. PL peak energies of sample No.2 at 11 K as a function of pressure. The excitation intensities at 2.54 eV are in the range 0.07–9 mW. Results for QW emission in sample No. 4 are also shown for comparison.

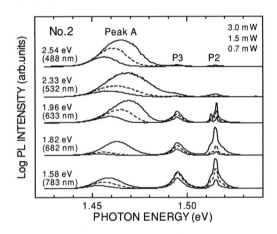

FIGURE 1. Comparison of typical PL spectra of sample No.2 at 10–12 K for five excitation energies 2.54 eV, 2.33 eV, 1.96 eV, 1.82 eV and 1.58 eV.

Figure 2 shows the pressure dependence of the PL peak energies of sample No. 2 at 11 K, measured under relatively lower excitation intensities below 9 mW at 2.54 eV. The ~1.46 eV emission band (peak A) shows a *sublinear* shift towards higher energies up to ~3.5 GPa depending strongly on the excitation intensity. It should be noted that its peak shift measured at sufficiently lower excitation is rather similar to that obtained for partially ordered GaInP layers [4]. In addition, it is found that the energy shifts of peak A with increased excitation intensity at higher pressures are larger than those at lower pressures. These features are in contrast to the QW emission at 1.53 eV in sample No. 4. Beyond ~3.5 GPa, the emission peak

According to recent theoretical calculations by Zhang et al. [1], band alignment between GaAs and partially ordered $Ga_xIn_{1-x}P$ is found to change from type I to type II at the order parameter $\eta = 0.46$ for a GaAs lattice matched composition. Furthermore, since the pressure coefficient of the GaAs band gap energy is much larger than that of the partially ordered GaInP, high pressure may convert the band alignment from type I to type II. Thus, the luminescence due to a spatially indirect recombination of electrons and holes separated at the interface is made possible by a type II band alignment.

REFERENCES

1. Zhang, Y., Mascarenhas, A., and Wang, L-W., *Appl. Phys. Letters* **80**, 3111-3113 (2002).

2. Uchida, K., Arai, T., and Matsumoto, K., *J. Appl. Phys.* **81**, 771-776 (1997).

3. Kwok, S. H., Yu, P. Y., Uchida, K., and Arai, T., *J. Appl. Phys.* **82**, 3630-3632 (1997).

4. Kobayashi, T., and Deol, R. S., *Appl. Phys. Lett.* **58**, 1289-1291 (1991).

Red to Blue Excitonic Emission with Ultra-Thin Quantum Wells and Fractional Monolayer Quantum Dots of II-VI Semiconductors

I. Hernández-Calderón, M. García-Rocha, and P. Díaz-Arencibia

Physics Department, CINVESTAV-IPN, Ave. IPN 2508, 07360, Mexico, DF, Mexico.

Abstract. Ultra-thin quantum wells of CdTe and CdSe and fractional monolayer quantum dots of CdSe present very strong excitonic emission covering the red-blue spectral range. They are grown by atomic layer epitaxy and the thickness fluctuations can be completely avoided in many cases by choosing the appropriate substrate temperature. These systems appear as promising candidates for the elaboration of red to blue light emission devices.

INTRODUCTION

A broad range of II-VI materials has been employed in order to produce light in the full visible range [1]. Since CdTe/ZnTe and CdSe/ZnSe quantum wells (QWs) could be tuned, in principle, from the infrared to the green and to the blue, respectively, the approach that we present in this work is to cover the full red-blue range employing a combination of CdTe/ZnTe and CdSe/ZnSe QWs. The calculation of excitonic transitions indicates that no more than 4 monolayers (1 ML = $a/2$, where a is the lattice constant) of CdTe or CdSe are required to cover that range. Although there are many efforts dealing with the growth of submonolayers or few monolayers of CdTe or CdSe, most of them are focused on quantum dot (QD) formation; there are only few efforts whose aim is the growth of high quality ultra-thin quantum wells (UTQWs) [2]. We are interested in the growth of ultra-thin 2D layers and we look for the optimum growth parameters that avoid island growth or the Stranski-Krastanow growth mode.

EXPERIMENTAL DETAILS

The samples were prepared in an epitaxial growth system with a basis pressure of 5×10^{-11} Torr on GaAs(001) substrates. More details about substrate preparation and buffer layer growth can be found in ref. [3]. Considering the need of a high sample reproducibility and precise control of the QW thickness, we employed the atomic layer epitaxy (ALE) growth method for the production of the UTQWs. The 2D ALE growth was verified by monitoring the intensity of reflection high energy electron diffraction (RHEED) oscillations. It is well known that in the 260 – 290 °C range the self-regulated ALE growth results in a saturation coverage of Cd or Zn of ~ 0.5 ML of the (001) surfaces of ZnSe, CdSe, ZnTe and CdTe, this coverage is limited by the surface reconstruction. Se or Te finished surfaces present a coverage ~ 1 ML. Then, an ALE cation-anion cycle will result in the deposition of a ~ 0.5 compound monolayer. A typical photoluminescence (PL) setup with Ar and HeCd lasers was employed.

RESULTS AND DISCUSSION

Figure 1 presents the low temperature PL spectra of CdTe UTQWs grown at a substrate temperature T_s = 290 °C. The 1 ML UTQW presents a blue shift of 0.74 eV in respect to the infrared band gap of CdTe. The 4 cycles CdTe UTQW presented only one intense and well defined peak with very high intensity. However, in the case of the 2 and 8 cycles

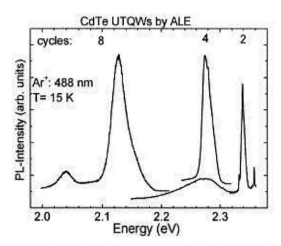

FIGURE 1. Photoluminescence spectra of UTQWs of CdTe grown at a substrate temperature of $T_s = 290$ °C.

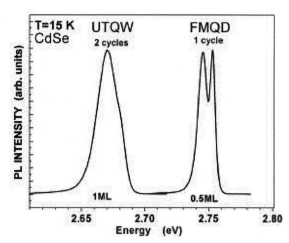

FIGURE 3. PL spectra of the 1ML CdSe UTQW and the FMQD of 0.5 ML, both samples grown at 260 °C. The FMQD has the same thickness of the 1ML UTQW.

we can observe a more complex structure. In the case of the 2 cycles CdTe UTQW we observe the formation of QDs with an average thickness of 2 ML. In the case of the 8 cycles sample the QW presents an average lower thickness due to substitution of Cd atoms by Zn atoms [4]. UTQWs of CdTe grown at T_s=270 °C presented narrower single peaks for thickness above 1ML [3]. Figure 2 presents the spectra of CdSe/ZnSe UTQWs with thickness in the range from 1 to 4 ML. In contrast with the CdTe QWs, these samples exhibited only one peak in the whole spectral range, demonstrating their very high thickness homogeneity; fluctuations of ± 1 ML would be clearly visible. If we deposit only one Cd-Se ALE cycle we obtain a 0.5 ML coverage and then we have ultra-thin islands with 3D confinement, this fractional monolayer quantum dots (FMQD) exhibit very high intensities and very narrow, reproducible,

peaks. The lateral confinement is exhibited as a blue shift related to the 1ML UTQW. In all cases the UTQWs and FMQDs grown at higher temperatures presented clear blue-shifts related to lower Cd incorporation. This effect can be advantageously employed for fine tuning of the emission of the QWs. In summary, we have demonstrated that employing UTQWs and FMQDs of CdTe and CdSe we can cover the full red-blue spectral range. These systems present very intense excitonic emission and could be employed for the elaboration of LEDs and lasers in the red to blue spectral range.

ACKNOWLEDGMENTS

This work was partially supported by Conacyt-Mexico. We thank tha technical assistance of H.Silva, Z. Rivera and A. Guillén.

REFERENCES

1. See, for example, M. C. Tamargo, W. Lin, S. P. Guo, Y. Guo, Y. Luo, and Y. C. Chen, J. Cryst Growth 214-215, 1058 (2000).
2. See, for example, H. Zajicek, P. Juza, E. Abramof, O. Pankratov, H. Sitter, M. Helm, G. Brunthaler, W. Faschinger, and K. Lischka, Appl. Phys. Lett. **62**, 717 (1993).
3. I. Hernández-Calderón, M. García-Rocha, and P. Díaz-Arencibia, Physica Status Solidi (b) **241**, 558 (2004) and references therein.
4. E. López-Luna, P. Díaz-Arencibia, I. Hernández-Calderón, Phys. Status Solidi (c) **1**, 819 (2004).

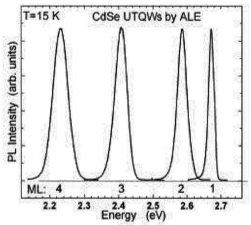

FIGURE 2. Photoluminescence spectra of UTQWs of CdSe. The 3 ML QW was grown at Ts = 275 °C, the others at 260 °C. Each spectrum presented a strong, single peak.

Characterization of the emitting states in quantum wells with planar nano-islands by polarization spectroscopy

A. Reznitsky[1], A. Klochikhin[1,2], S. Permogorov[1], L. Tenishev[1], S. Yu. Verbin[1], H. Kalt[3], and C. Klingshirn[3]

[1]*A.F.Ioffe Physical-Technical Institute, RAS, St.Petersburg, Russia,* [2] *Petersburg Nuclear Physics Institute, RAS, St.Petersburg, Russia,* [3]*Institut für Angewandte Physik, Universität Karlsruhe, and Center für Functional Nanostructures (CFN), Karlsruhe, Germany*

Abstract. he temperature dependence of photoluminescence spectra of ultrathin ZnCdSe/ZnSe quantum wells with inhomogeneous Cd distribution indicates that there are two types of emitting states (metastable and ground), which contribute to the high-energy and low energy part of the PL band, respectively. Optical orientation and optical alignment experiments at resonant excitation show that considerable part of islands contains an extra electron. As a result, the deepest states of islands are formed mostly by the ground states of trions while the metastable states in charged island correspond to the exciton states spatially isolated from localized electron.

We have studied low-temperature photoluminescence (PL) and excitation of PL (PLE) spectra of ultra narrow quantum wells (QWs) with planar nanoislands formed by MBE insertion of a few monolayers (ML) of CdSe into the ZnSe matrix. The CdSe/ZnSe samples with different nominal CdSe thickness (1 - 3 ML) were grown at 280C by (i) migration enhanced epitaxy using the multi-cycled deposition of CdSe on the ZnSe surface, or (ii) conventional MBE with CdS compound as Cd source and elemental Se source (for details see Ref. [1]). High resolution transmission electron microscopy (HRTEM) shows that the typical halfwidth of CdSe sheet in samples is 2 - 2.5 nm (Ref. [2]).

PLE spectra for different detecting energies, being normalized at excitation energies slightly below the barrier exciton, are diverging below the energy E_{ME} (Fig.1)identified as exciton mobility edge in the QW with nano-islands [3]. The depth of potential well of a quantum island, which is the difference between the emission energy and E_{ME}, has an order of 0.2-0.3 eV. Obtained potential well depths and the island sizes estimated from HRTEM allow to calculate the exciton wave-functions.

The characteristic feature of PLE spectra of the states on the high energy side of PL band is the oscillating structure with the period of the order of optical phonon (Fig.1). Such behavior shows that this side of PL band is due to the emission of excited states of the islands. Relaxation rates of these states strongly depend on the temperature, at low temperatures they are metastable [1]. The PLE spectra of the states forming the low energy part of the PL band do not depend on the detection energy, which indicates that these states have no ways for further relaxation and represent the ground states of islands.

An increase of relaxation rate of the metastable states with temperature leads to depopulation of these states and to the low-energy shift of PL band (see insert to Fig.2). Therefore, the PL spectrum at the temperature corresponding to the maximum Stokes shift represents mainly the spectrum of ground states (curve 2 in Fig.2). Further rise of temperature and establishment of equilibrium between metastable and ground states causes the high-energy shift of the PL band maximum resulting in the anomalous ("S-shape") behavior of the PL band. Then, the difference between the low temperature PL band (curve 1 in Fig.2) and spectrum 2 gives the spectrum of metastable states (curve 3 in Fig.2).

In order to elucidate the nature of metastable and ground states of islands, the optical orientation and optical alignment experiments at resonant excitation at low temperature were performed. It was found that at resonant excitation of the island states both by linear and circular polarized light the resulting emission of the metastable states shows a considerable degree of corresponding polarization (Fig.3a and 3b, respectively). This indicates that the metastable states have an exciton nature and are populated as a result of energy relaxation of localized excitons originally excited within the island. In distinction, Fig.3c demonstates that at resonant excitation of the ground states the linear polarization of the emission at linearly polarized excitation (optical align-

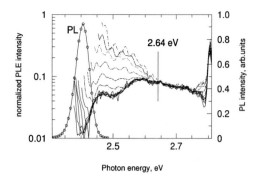

FIGURE 1. PL spectrum at above ZnSe barrier excitation (solid line with symbols) and normalized PL excitation spectra for different detection positions with step 10 meV: six spectra for detection in spectral range 2.34 - 2.39 eV (solid lines) and six spectra in the range 2.40 - 2.45 eV (dashed-dotted lines). All spectra are obtained at T=5K. Exciton mobility threshold E_{ME} at 2.64 eV is indicated by vertical solid line.

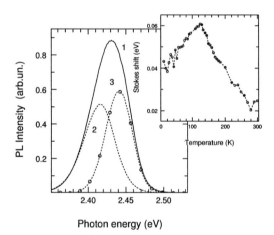

FIGURE 2. PL spectra at above ZnSe barrier excitation at T=2K and 100K (solid line 1 and dashed line 2, respectively). Line 3 with symbols is the difference between bands 1 and 2. Insert: Temperature dependence of the Stokes shift between PL and PLE maxima for the same sample.

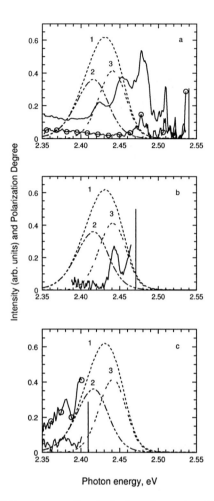

FIGURE 3. Spectra of optical orientation and optical alignment (solid lines with and without open circles, respectively) for resonant excitations in different spectral range (shown by solid vertical lines): 2.540 eV (a), 2.471 eV (b) and 2.409 eV (c). Dashed lines: PL band at above ZnSe band gap excitation (curve 1) and subbands due to recombination through ground (curve 2) and metastable (curve 3) states are shown for the convenience.

ment) is not observed, whereas the circular polarization at circularly polarized excitation (optical orientation) is still detectable like in the former case.

The obtained polarization characteristics of the emission can be explained if we assume that a considerable part of the CdSe islands in ZnSe matrix contains extra electrons (both compounds are unintentionally weakly n-doped) and that the deepest island states represent mostly the ground states of trions. In turn, the metastable states in the charged island correspond to the exciton states in local potential island minima spatially separated from the absolute minimum occupied by excess electron.

In conclusion, we have demonstrated that in nominally undoped CdSe nanoislands in ZnSe matrix the charged and neutral exciton states can coexist within one and the same island. This situation agrees well with HRTEM data indicating a complex topological shape of islands, which, as a rule, consist of weakly connected parts.

This work was partly supported by Deutsche Forschungsgemeinschaft, by Russian Program "Physics of Solid State Nanostructures" and by Russian Foundation for Basic Research.

REFERENCES

1. A. Klochikhin, et al., Phys.Rev. B **69**, 085308 (2004).
2. N. Peranio et al., Phys.Rev. B **61**, 16819 (2000).
3. A. Reznitsky, et al., phys.stat.sol.(b) **229**, 509 (2002).

Imaging of the Electric Fields and Charge Associated with Modulation-Doped 4H/3C/4H Polytypic Quantum Wells in SiC

M. K. Mikhov[*], G. Samson[*], B. J. Skromme[*], R. Wang[†], C. Li[†], and I. Bhat[†]

[*]*Department of Electrical Engineering and Center for Solid State Electronics Research, Arizona State University, Tempe, AZ, 85287-5706, USA*
[†]*ECSE Department, Rensselaer Polytechnic Institute, Troy, NY 12180-3590, USA*

Abstract. Polytypic 3C quantum wells (double Shockley stacking faults) are spontaneously generated during thermal processing of moderately doped 4H-SiC epilayers grown on substrates with heavy *n*-type doping above ~3×10^{19} cm^{-3}. They intersect the wafer surface as straight lines, due to the 8° misorientation of the wafer from the *c*-axis. We describe observations of electric fields and charge associated with these intersections using electrostatic force microscopy (EFM) and scanning Kelvin probe microscopy (SKPM). The results are compared to two-dimensional electrostatic simulations.

Polytypic quantum wells are a novel system where variations in stacking order alone can lead to quantum confinement, modulation doping, and related effects within a single material system such as SiC. Here, we study the case of thin (six bilayer) 3C lamellae formed by spontaneous nucleation during thermal processing of *n*-type 4H-SiC with doping levels above ~3×10^{19} cm^{-3} [1-5]. They consist of double Shockley stacking faults formed on the (0001) <11-20> hexagonal slip system by dissociation of perfect dislocations into two partial dislocations, according to the reaction $a/3$ <2-1-10> → $a/3$ <1-100> + $a/3$ <10-10>. Electron transfer from the 4H matrix into the 3C layer, which acts like a quantum well (QW), has been proposed to drive the transformation [3].

If this modulation doping effect occurs, the wells should become sheets of negative charge surrounded by positively-charged depletion layers. Where they intersect the surface, fringing electric fields should exist. Here, we report direct observation of these fields and charges using the EFM and SKPM methods, and compare the observations to two-dimensional electrostatic simulations.

The sample is a moderately doped ([N]~1.3×10^{17} cm^{-3}) epitaxial layer on a heavily doped ([N]~3×10^{19} cm^{-3}) 4H-SiC substrate tilted 8° off (0001). A Digital Instruments NanoScope IIIa MultiMode atomic force microscope (AFM) and Pt/Ir coated Si tips (Nanosensor Tech.), was used for AFM, EFM, and SKPM. Further details and related secondary electron imaging results can be found in [5].

Figure 1 shows a tapping mode AFM image of the surface morphology along with a simultaneously

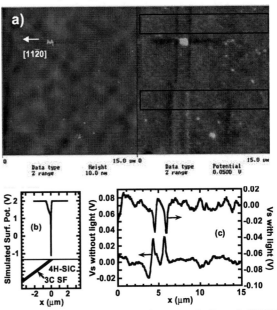

FIGURE 1. (a) Tapping-mode AFM (left) and SKPM (right) images of two parallel faults intersecting the surface. A D$_2$ lamp illuminated the surface during the upper portion of the scan (only). (b) Simulated electrostatic potential profile for the inclined fault. (c) Surface potential profiles averaged over the boxes in the SKPM image (upper curve = upper box, lower curve = lower box)

acquired SKPM image. The intersections of two separate double faults with the surface are visible as vertical stripes in the SKPM image (a third fault is partially visible at the extreme left). The stripes are clearly not related to the surface topography. The upper portion (only) of the SKPM image was acquired under UV irradiation from a D_2 lamp; the lower portion of the image was recorded without such illumination. Profiles of the SKPM signal (averaged parallel to the fault direction to reduce noise) are shown for each portion of the scan in Fig. 1(b). Under UV illumination, the profile becomes narrower and shifts to the right of the profile without illumination.

To understand the observed profiles, we performed a two-dimensional simulation of the electrostatic properties of the inclined fault intersecting the surface, using ISE Dessis software. The spontaneous polarization in the 4H-SiC [6] was included as charge sheets at the interfaces. The calculated electrostatic potential profile along the semiconductor surface is shown in Fig. 1(c). A depletion region is formed along each side of the fault due to modulation doping. Away from the surface, it extends farther from the "bottom" side of the fault due to the asymmetry caused by the polarization charge, which is positive on the top (+c) interface and negative on the bottom (–c) interface. Where the depletion region intersects the surface, this asymmetry is reversed by the interaction with the surface, and even more so due to the negatively-charged region where the negative charge sheet extends beyond the edge of the positive charge sheet. At the surface, the depletion region therefore becomes extended on the left side of the inclined fault and shrinks on its right side. An accumulation layer is formed everywhere within the QW. The calculated surface potential profile extends about 0.7 μm laterally, much wider than the ~11 nm width of the intersection of the ~1.5 nm thick 3C region with the surface. The SKPM profile of each fault without illumination is even wider (~1.5 μm), which is not yet understood.

Under illumination, the photogenerated holes are swept into the potential well formed by the negatively charged fault, and partially neutralize the electrons. The depletion region shrinks as a result. The left side of the potential profile should therefore be flattened and made shorter, causing a shift of the imaged potential profile to the right. This effect explains the experimentally observed shift and apparent sharpening of the image. The apparent change in the phase of the profile is not yet understood, however, and requires further study.

We also employed the complementary EFM method to study the electrostatic field gradients due to the faults. A tapping mode surface topology image and phase-contrast EFM image are shown in Fig. 2. The

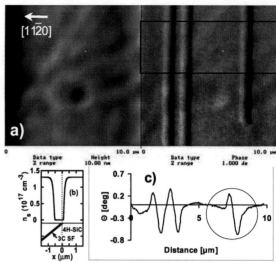

FIGURE 2. a) Tapping mode AFM image of the surface morphology (left) and EFM image (right) of the 3C SF (QW) intersecting the wafer surface. b) Simulated free electron concentration along the surface around the fault. c) Phase profile of the EFM image, averaged parallel to the SF over the indicated rectangular area.

EFM image shows clear stripes due to the faults intersecting the surface, which start and stop at various points along the surface due to partial dislocations at those locations. They are not related to surface topography. The phase variations are dominated by "fixed" surface charges (i.e., charges not capacitively induced by the tip), because the contrast is reversed when the applied bias is reversed in sign [5]. The simulated free electron concentration along the surface is shown in Fig. 2(c). The positively-charged depletion region extends mainly to the left side of the fault. This asymmetry is likely related to the asymmetry in the EFM profile. Further work is in progress to understand these images in detail, but it is clear that they prove the existence of electric fields associated with modulation doping, and they should therefore be useful to investigate this effect in more detail.

The ASU work was supported by the National Science Foundation under Grant Nos. ECS 0080719 and ECS 0324350, and by Motorola. Work at RPI is supported by DARPA Contract #DAAD19-02-1-0246.

REFERENCES

1. B. J. Skromme et al., *Mater. Sci. Forum* **389-393**, 455 (2002).
2. R. S. Okojie et al., *Appl. Phys. Lett.* **79**, 3056 (2001).
3. J. Q. Liu et al., *Appl. Phys. Lett.* **80**, 2111 (2002).
4. B. J. Skromme et al., *Mater. Sci. Forum* **457-460**, 581 (2004).
5. M. K. Mikhov et al., *MRS Proc.*, **799**, Z5.21.1 (2004).
6. A. Qteish et al., *Phys. Rev. B* **45**, 6534 (1992).

Study of InGaAs band structure for nonparabolic conduction band and parabolic heavy-hole band using InGaAs/InAlAs multi-quantum well structure

Nobuo Kotera* and Koichi Tanaka[†]

Kyushu Institute of Technology, Iizuka City, Fukuoka 820-8502, Japan
[†]Hiroshima City University, Asa-Minami, Hiroshima 731-3194, Japan

Abstract. Experimental study of nonparabolic band is difficult because electron kinetic energy cannot be controlled widely. We propose here to use quantization energy in quantum well (QW) where eigen energy is changed by controlling QW thickness. Thirteen eigen energies from four QW specimens were obtained. Effective mass was determined as a function of QW eigen energy or electron kinetic energy. This gave a nonparabolic effective mass of bulk InGaAs. Heavy-hole band was also checked and found that the band was parabolic.

INTRODUCTION

In 1961, nonparabolic increase of cyclotron effective mass as a function of magnetic field was reported in InSb conduction band (CB). Experimental study of nonparabolic band is still difficult because electron kinetic energy cannot be controlled widely from band edge. Use of quantization energy in QW is promising because energy can be controlled by QW thickness. Square well model for QW potential fit well in experiments. InAlAs barrier thickness of 10 nm or more will suffice to separate each InGaAs QW without coupling. Mixed crystal of $In_{0.53}Ga_{0.47}As$ has a bandgap energy of 0.81 eV at low temperature and the CB nonparabolicity could be measured. Mixing of heavy-hole (HH) band and light-hole band in QW might be expected.

In this paper, CB effective mass was determined as a function of eigen energies in QW. Obtained effective mass vs. energy curve was compared with a simple three-level band theory. Obtained CB mass was 5-7% larger than band theory. The theoretical curve agreed fair with experiments. HH band was checked and effective mass was determined in QW.

EXPERIMENTAL RESULTS AND ANALYSES

Specimen Structure and Measurements

Specimen was fabricated by MBE on sliced InP (100) substrate surface and specimen structures were checked by double-crystal X-ray diffraction. An $In_{0.52}Al_{0.48}As$ p-i-n junction including undoped MQW structures in undoped $In_{0.52}Al_{0.48}As$ i-layer was fabricated. Substrate was n-type-doped. On top of the surface, highly p-type doped $In_{0.52}Al_{0.48}As$ layer was fabricated to form an Ohmic contact. Above the contact layer, metal needle was pressed to measure the photocurrent. Photocurrent caused by interband optical absorption effect were observed between $In_{0.53}Ga_{0.47}As$ and $In_{0.52}Al_{0.48}As$ band gap energies. A reference specimen without MQW was needed to correct the light intensity introduced. Step-like optical absorption in QW was observed clearly, especially in 10-nm-thick $In_{0.53}Ga_{0.47}As/In_{0.52}Al_{0.48}As$ QW [1]. Under a negative bias electric field, additional forbidden transitions were observed besides the allowed ones.

Conduction Band Analysis

Thirteen eigen energies from specimens with well thickness of 5, 9.4, 10 and 20 nm were obtained from spectrum analysis at room temperature. Corresponding specimen name was V720, V706, U10, and V748, respectively. CB eigen energies in $In_{0.53}Ga_{0.47}As$ QW depended sublinearly on the square of assigned quantum numbers ≤ 7 in the energy range ≤ 0.5 eV. If a square-well potential model with a finite step difference is applicable, one CB eigen energy leads to one electron effective mass of $In_{0.53}Ga_{0.47}As$ if band-offset of CB in QW and effective mass of $In_{0.52}Al_{0.48}As$ barrier were assumed 0.52 eV and $0.075m_0$, respectively. The band offset was obtained consistently from our previous mea-

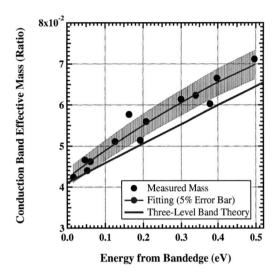

FIGURE 1. Conduction band effective mass of InGaAs as a function of electron kinetic energy, determined from QW eigen state energies (solid mark). The least-square fitting curve with 5%-error bars is also shown.

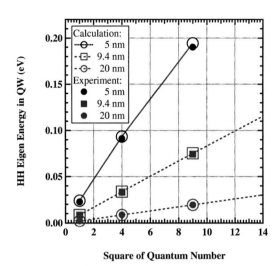

FIGURE 2. Heavy-hole eigen energies vs. square of quantum number. Band offset of 0.22 eV and effective mass of $0.38m_0$ was used in calculation. QW thickness, 5, 9.4, and 20 nm, corresponds to specimen name, V720, V706, and V748.

surement. CB effective mass of $In_{0.53}Ga_{0.47}As$ was thus determined as a function of electron kinetic energy as shown in Fig. 1. Least-square fitting of mass vs. energy curve by thirteen points at room temperature gave the following empirical mass formula; $m_e^*(E)/m_0 = 0.042 + 0.070E - 0.029E^2$ for $In_{0.53}Ga_{0.47}As$ QW, where E is energy in eV. The formula gave a bandedge mass, $0.042m_0$, which agreed with the conventional electron effective mass of $In_{0.53}Ga_{0.47}As$, $0.041m_0$, at low temperatures within 3%.

Obtained mass vs. energy curve was compared with a three-level band theory given by $m_e^*(E)/m_e^*(0) = (E + \varepsilon_g)(E + \varepsilon_g + \Delta)(\varepsilon_g + 2\Delta/3)/\varepsilon_g(\varepsilon_g + \Delta)(E + \varepsilon_g + 2\Delta/3)$ where E, ε_g, and Δ are an electron energy, a bandgap energy, and energy of spin-orbit splitting, respectively, where $\varepsilon_g = 0.81$ eV and $\Delta = 0.36$ eV were used. Experimentally obtained CB mass, $m_e^*(E)$, was more-than-5% larger than the theory as shown in Fig. 1. Inclusion of higher bands was needed to fit both, however, the simple band theory was found almost applicable within 5-to-7% error in wide energy range.

Heavy-Hole Band Analysis

About CB, eigen-energies depended sublinearly on the square of quantum number because of the nonparabolicity. About HH band, eigen-energies depended linearly as shown in Fig. 2. The heavy-hole band was found parabolic and effective mass was determined to be $0.38m_0$. This was the same with the conventional bulk value. Evidence on mixing of heavy- and light-hole bands was not observed below 0.19 eV.

CONCLUSION

Nonparabolic conduction band in bulk InGaAs has been analyzed using quantum well structures. Effective mass varying between $0.04m_0$ and $0.07m_0$ has been obtained as a function of energy between 0 (bulk bandedge) and 0.51 eV. This was 5-to-7% larger than the three-level band theory in QW. Heavy-hole band was found parabolic below 0.19 eV. The mass could fit at $0.38m_0$.

REFERENCES

1. K. Tanaka, N. Kotera, and H. Nakamura, *Elec. Lett.* **34** 2163 (1998).

Wavefunction Engineering for GaN-Based Quantum Wells and Superlattices

L. R. Ram-Mohan*, A. M. Girgis*, J. D. Albrecht[†], C. W. Litton[†], T. D. Steiner[¶]

Worcester Polytechnic Institute, Worcester, Massachusetts
[†]*Air Force Research Laboratory, Wright-Patterson Air Force Base, Ohio*
[¶]*Air Force Office of Scientific Research, Arlington, Virginia*

Abstract. The electronic band structure of GaN-based heterostructures such as those comprised of AlGaN/GaN epitaxial layers is investigated using envelope function **k•P** theory. We obtain the valence states through minimization of a Lagrangian describing the electronic interaction with the lattice potential including spin-orbit effects. This construction yields transparent derivative operator ordering appropriate for arbitrary allowed crystallographic orientations. The issues of derivative operator ordering and boundary conditions at material interfaces can be significant given the strongly anisotropic character of the wurtzite crystal structure and we explore these effects on the resulting in-plane dispersions in detail. These results are of interest, for example, in treating A-plane wurtzite heterostructures such as GaN/AlGaN quantum wells grown on R-plane sapphire. Numerical examples are obtained using finite-element discretization in order to obtain the valence band electronic states for quantum wells and superlattices.

INTRODUCTION

It is well known that AlGaN/GaN heterostructures can exhibit strong or weak polarization field effects depending on growth direction. In transistor structures grown along [0001] on sapphire or SiC substrates, for instance, large spontaneous and piezoelectric polarization fields are induced and can be tailored to optimize the channel properties. For designing optical transitions these built-in fields are often less desirable. For [0001]-oriented quantum wells the presence of polarization fields spatially separates the electrons and holes giving rise to smaller wavefunction overlap and weaker transitions. This effect can be eliminated through the growth of A-plane wurtzite structures such as GaN structures on R-plane sapphire substrates.

We have developed an envelope function (EF) formalism for representing the valence eigenstates of wurtzite heterostructures within **k•P** theory. The EF treatment for growth along the hexagonal axis is a straightforward extension of the standard method used for cubic systems. For the z-direction along the c-axis, the 6 x 6 Rashba-Sheka-Pikus Hamiltonian in the basis of p-states can be block diagonalized with a basis transformation that is independent of k_z [1]. Then the EF theory is derived by substituting spatial derivative operators for k_z, etc. in direct analogy to cubic EFs [2].

The situation for EFs in the [11$\bar{2}$0] direction (z') that corresponds to A-plane structures is substantially more complicated. A convenient block diagonal form for the valence band Hamiltonian does not exist in this case because the original off-diagonal terms involving k_x and k_y now involve a spatial variation after rotation for EFs along z'. In addition, the operator ordering of the derivatives at material interfaces is no longer obvious by inspection. An extensive investigation of these and other associated issues has been carried out and the details will be published in a longer format. Presently, we illustrate our theory by reporting a comparison of the EF calculations made for [0001] and [11$\bar{2}$0] quantum wells by examining the in-plane dispersions of the lowest energy hole eigenstates.

THEORY AND RESULTS

We favor a theoretical approach to the EF problem based on minimization of an energy functional, in this case the integrated Lagrangian density for the valence states. The interactions with remote bands are treated

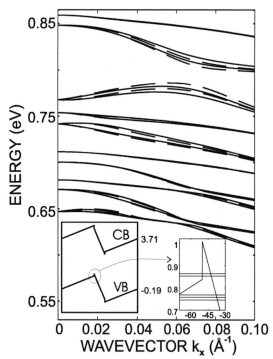

FIGURE 1. Energy dispersion of hole eigen-states for a [0001] $Al_{0.2}Ga_{0.8}N/GaN/Al_{0.2}Ga_{0.8}N$ 100 Å QW.

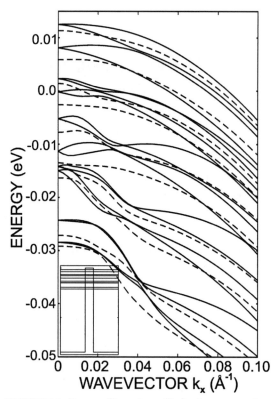

FIGURE 2. Energy dispersion of hole eigen-states for a [11$\underline{2}$0] $Al_{0.2}Ga_{0.8}N/GaN/Al_{0.2}Ga_{0.8}N$ 100 Å QW.

within the framework of Löwdin perturbation theory and parameterized, resulting in a six-band description. The details of the formal mathematical approach are beyond the scope of this report. We can point out that the resulting EF theory differs slightly from previous derivations for wurtzite semiconductors [3]. Also, the Lagrangian derivation is general and leads to the correct operator ordering for layered structures grown along any allowed crystallographic direction. We solve for the eigenstates by finite-element analysis based on the variation of the discretized Lagrangian. The numerical results include strain and polarization (spontaneous and piezoelectric) effects. The parameters are tabulated elsewhere [4].

Figures. 1 and 2 show the in-plane energy dispersions of the lowest energy hole states in quantum wells of two common orientations. The solid (dashed) curves are for $A_7=0$ ($A_7 \neq 0$). The obvious difference in the two structures is the presence of large polarization electric fields for the [0001] case which dominates the spectrum. From the inset in Fig. 1 it is seen that we compute only two bound states with strong confinement to the GaN well layer. This is in sharp contrast to Fig. 2 where the first ten states are in the well. This corresponds to the AlGaN-like character of the higher energy states in Fig. 1 and the more well-like states of Fig. 2. The second major point is that the inversion asymmetry effects parameterized in A_7 are more pronounced for the [11$\underline{2}$0] case in Fig. 2. Not only is the band degeneracy lifted, but the energy spectrum is shifted at **k**=0. These shifts are significant and could give rise to measurable shifts in the energies of optical interband and intersubband transitions.

In summary, we have developed an EF theory appropriate for wurtzite heterosturctures with flexibility to handle cases of arbitrary orientantation.

ACKNOWLEDGMENTS

The work at WPI is supported by AFOSR, NSF, and the DARPA SpinS program.

REFERENCES

1. G. B. Ren, Y. M. Liu, and P. Blood **74**, 1117 (1999).

2. G. Bastard, *Wave Mechanics Applied to Semiconductor Heterostructures* (Les Editions de Physique, Les Ulis, France, 1988).

3. F. Mireles and S. E. Ulloa, Phys. Rev. B **60**, 13659 (1999).

4. I. Vurgaftman, J. R. Meyer, and L. R. Ram-Mohan, J. Appl. Phys. **89**, 5815 (2001); **94**, 3675 (2003).

Blue-light Emission from GaN/AlGaN Multiple Quantum Wells with an $Al_{0.5}Ga_{0.5}N$ Perturbation Monolayers Grown by Molecular Beam Epitaxy

Y. S. Park*, S. H. Lee†, J. E. Oh†, C. M. Park*, and T. W. Kang*

*Quantum-functional Semiconductor Research Center, Dongguk University, Seoul 100-715, Korea
†Center for Electronic Materials and Components, Hanyang University, Ansan 425-791, Korea

Abstract. We have studied the influence of AlGaN inserting layer into GaN well region on the light emission from a strained GaN/AlGaN multiple-quantum-well system. We have found that, by simply inserting thin AlGaN layer, the luminescence is dramatically red-shifted with respect to that of the normal GaN/AlGaN quantum well, which is centered at 2.96 eV, nearly 0.52 eV below the bulk GaN band gap. We attribute this enormous redshift to an additional 0.7 MV/cm field present in the well due to the perturbation of well region by inserting AlGaN layer. The result reveals to be of great importance in the design and analysis of nitride heterostructure devices and which can be exploited to advantage in nitride materials and device engineering.

INTRODUCTION

In biaxially strained III-nitride heterostructures grown in the wurtzite structure with the *c*-axis parallel to the growth direction, giant piezoelectric and spontaneous polarization effects are expected to be present as a consequence of the non-centrosymmetry of the wurtzite structure. Actually, for fully strained GaN on AlN ($\Delta\varepsilon=2.4\%$) piezoelectric fields as high as several MV/cm are expected. These values are more than an order of magnitude larger than the piezoelectric fields that can be found in zinc blende semiconductors for the same amount of strain. If spontaneous polarization effects are present and can be engineered[1], then a new degree of freedom in engineering the electric fields within a structure can be envisaged.

We have found that, by simply inserting thin AlGaN layer into GaN well region, the luminescence is dramatically red-shifted with respect to that of the normal GaN/AlGaN multiquantum wells (MQW), which is centered at 2.96 eV, nearly 0.52 eV below the bulk GaN band gap. The result reveals to be of great importance in the design and the analysis of nitride heterostructure devices and which can be exploited to advantage in nitride materials and device engineering.

Finally, we suggest that a possible application of these results could be achievement of high efficiency blue light emitting devices with no indium containing compounds.

EXPERIMENTALS

The samples used in the present study were grown on (0001) sapphire substrate using rf-plasma assisted molecular beam epitaxy. A high-temperature AlN layer, about 10 nm thick, was deposited to assure the Ga-face growth condition. Following the growth of a 0.8 μm thick GaN layer, 50 periods of GaN/$Al_{0.5}Ga_{0.5}N$ MQWs were grown. The thickness of $Al_{0.5}Ga_{0.5}N$ barrier was kept to be 100 Å. Each well consists of the first 20 Å of GaN, thin $Al_{0.5}Ga_{0.5}N$ layer, and the rest 10 Å GaN. The structural properties of MQW have been measured by high-resolution x-ray diffraction (HRXRD). The surface morphology and the cross sectional images have been investigated by a field emission scanning electron microscopy (FESEM) measurement. Cathodoluminescence (CL) measurements were carried out at 300 K.

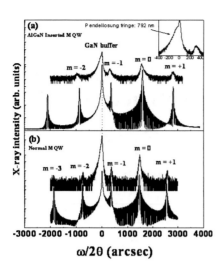

FIGURE 1. HRXRD spectra and simulation of (a) the AlGaN layer inserted MQW and (b) the normal MQW for (0002) plane. As shown in the inset, Pendellogsung fringes with a period of 21 arcsec is observed, indicating a high quality.

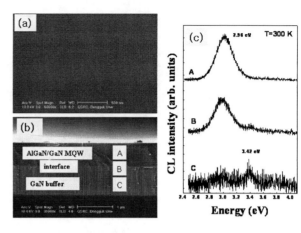

FIGURE 2. (a) The SEM surface morphology and (b) cross-sectional view of the MQW sample A. (c) The comparison of room-temperature CL spectra obtained at different points, A, B, and C is shown.

RESULTS AND DISCUSSION

Detailed HRXRD ω-2θ scans of the AlGaN layer inserted MQW and the normal MQW structure around (0002) Bragg peak of GaN are shown in Fig. 1(a) and (b), respectively. Due to the highly-organized structure, the peak positions of the Bragg-reflexes of superlattice peaks as well as wurtzite GaN buffer layer are clearly resolved. In order to confirm the peak assignment a simulation using a standard kinematical scattering is performed and also shown in Fig. 2. Indexing and positions of superlattice peaks are identical for both samples with and without inserting AlGaN layers. As shown in the inset of Fig. 2(a), a clear Pendellogsung fringe with a period of 21 arcsec is observed in the left shoulder of GaN buffer peak, which implies the GaN buffer thickness of 792 nm.

The spatial localization of the QW emission is unambiguously determined by mono-chromatic CL measurements. The surface morphology and cross-sectional view of the MQW with inserting layer are shown in Fig. 2(a) and (b), respectively. The surface is flat and featureless, and no distinct dislocation network is observed in the well region. Figure 2(c) compares room-temperature CL spectra obtained at different points, A, B, and C. Strong and broad luminescence at 2.96 eV is observed at point A of MQW region, while a bound exciton line of wurtzite GaN at 3.42 eV dominates at a point C, GaN buffer region. The observation shows that the blue emission observed in sample the MQW with inserting layer originates from the MQW region, rather than from the deep level in GaN buffer layer.

CONCLUSIONS

It has been experimentally demonstrated that anomalously strong red-shift of luminescence energy of GaN/AlGaN MQWs can be achieved by simply inserting a thin layer of AlGaN into the GaN well region. The additional piezoelectric and spontaneous polarization field induced by potential perturbation create huge localized giant fields responsible for the blue emission. Finally, we suggest that a possible application of these results could be achievement of high efficiency blue light emitting devices with no indium containing compounds.

ACKNOWLEDGMENTS

This work was supported by Tera-Bit Nano Devices Program as a part of Frontier-21 Project and Quantum Functional Semiconductor Research Center at Dongguk University sponsored by Korea Science and Engineering Foundation and partially by Center for Electronic Materials and Components at Hanyang University.

REFERENCES

1. G. Vaschenko, D. Patel, C. S. Menoni, H. M. Ng, and A. Y. Cho, *Appl. Phys. Lett.* **80**, 4211-4213 (2002).

Theory of Raman Lasing due to Coupled Intersubband Plasmon-Phonon Modes in Asymmetric Coupled Double Quantum Wells

S. M. Maung and S. Katayama

School of Materials Science, Japan Advanced Institute of Science and Technology, 1-1 Asahidai, Tatsunokuchi, Ishikawa 923-1292, Japan

Abstract. A theory of Raman laser gain due to coupled intersubband (ISB) plasmon-optical phonon modes in asymmetric coupled double quantum wells (ACDQWs) is presented. Based on the charge-density-excitations (CDE) mechanism, we take into account the electron-electron and electron-phonon (confined LO phonon and interface (IF) phonons) interactions in the scattering cross-section. For $Al_{0.35}Ga_{0.65}As/GaAs$ ACDQWs the calculated coupled mode energies which are responsible for the lasing Stokes emission are well consistent with recent experiments.

INTRODUCTION

Recently, optically pumped unipolar three-level intersubband Raman laser (IRL) has been realized [1] for mid-infrared radiation in modulation-doped GaAs/$Al_{0.35}Ga_{0.65}As$ asymmetric coupled double quantum wells (ACDQWs). By using various samples with CO_2 laser, a new evidence associated with coupled intersubband (ISB) plasmon-optical phonon modes has been found [2]. It has been clarified theoretically based on the charge-density-excitation (CDE) mechanism that the coupled ISB plasmon-LO phonon mode with the largest Raman cross-section dominates the lasing action [3]. In this paper we further develop a theory of Raman lasing due to coupled ISB plasmon-phonon modes by considering confined nature of LO phonon and including the interface (IF) phonons in ACDQWs.

THEORY

The schematic view of IRL is shown in Fig. 1a. In our CDE mechanism Stokes emission can be understood by adopting the 2-step scattering process as illustrated in Fig. 1b [3] so that the light scattering process contains the ISB excitations (1→2). Further, owing to electron-electron and electron-phonon interactions, the ISB excitation energy is shifted from E_{21} (=E_2-E_1) to the coupled ISB plasmon-phonon mode energy. The Raman laser gain (G_R) in IRL has been given in terms of the scattering cross-section ($d^2\sigma/d\omega d\Omega$), subband populations ($N_{1(2)}$) and pump intensity ($I_{pump}$) in ref. 3 as

$$G_R \propto (d^2\sigma/d\omega d\Omega)(N_1 - N_2)I_{pump}. \quad (1)$$

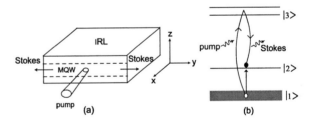

FIGURE 1. (a) Schematic view of IRL device, (b) 2-step CDE Raman scattering process.

Within the RPA we obtain the scattering cross-section with use of dynamical susceptibility function χ_{11} related to the virtual (1,2)→(2,1) transitions as

$$\frac{d^2\sigma}{d\omega d\Omega} = -\frac{r_0^2 \hbar}{\pi}\left(\frac{\omega_s}{\omega_i}\right)|R|^2(n(\omega)+1)\,\mathrm{Im}\,\chi_{11}(q,\omega), \quad (2)$$

where r_0 is the classical radius of an electron, R is the intersubband Raman tensor and $n(\omega)$ is the Bose-Einstein factor. Poles of $\chi_{11} = \chi^0/[1-V_{11}^{eff}\chi^0]$ are determined by the linear response function χ^0 and the effective electron-electron interaction V_{11}^{eff} given by

$$V_{11}^{\text{eff}} = \frac{4\pi e^2 L_{11}}{\varepsilon_\infty} + \sum_\nu \frac{2\hbar\omega_{q,\nu}|V_{q,\nu}|^2}{\hbar^2(\omega^2 - \omega_{q,\nu}^2)}, \quad (3)$$

where L_{11} and $V_{q,\nu}$ denote the matrix element of Coulomb interaction and the electron-phonon coupling strength for mode ν.

CALCULATIONS AND DISCUSSION

To evaluate the scattering cross-section of GaAs/Al$_{0.35}$Ga$_{0.65}$As ACDQWs with $N_s = 3 \times 10^{11}$ cm^{-2}, we calculate the subband states as well as $V_{q,\nu}$ and $\omega_{q,\nu}$. Figure 2 shows the calculated cross-section for three ACDQWs. It is seen in Figs. 2a and 2b that the peak

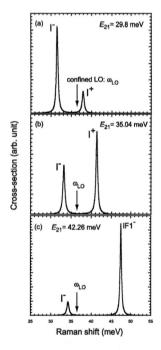

FIGURE 2. Calculated scattering cross-section for three ACDQWs samples. In (a) $E_{21} < \hbar\omega_{\text{LO}}$ (confined GaAs), (b) $\hbar\omega_{\text{LO}} \sim E_{21} < \hbar\omega_{\text{IF1}}$, and (c) $\hbar\omega_{\text{LO}} < E_{21} < \hbar\omega_{\text{IF1}}$.

(I$^-$) corresponding to lower coupled ISB plasmon-confined LO phonon mode dominates the cross-section for $E_{21} < \hbar\omega_{\text{LO}}$, while the upper peak (I$^+$) becomes dominant in spectrum for $\hbar\omega_{\text{LO}} \sim E_{21} < \hbar\omega_{\text{IF1}}$. More important fact is seen in Fig. 2c in which the peak (IF1$^-$) corresponding to coupled ISB plasmon-IF1 phonon mode has the strongest intensity for $\hbar\omega_{\text{LO}} < E_{21} < \hbar\omega_{\text{IF1}}$. These results can be explained by the two factors appearing in V_{11}^{eff} of Eq. (3). The first one is the resonance between E_{21} and $\omega_{q,\nu}$ and the second is the coupling strength $V_{q,\nu}$. To compare with the IRL experiments [2], we have calculated the density of Stokes photons associated with coupled modes using laser rate equations [3].

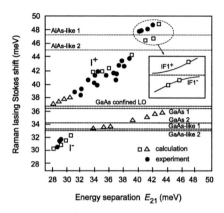

FIGURE 3. Raman lasing Stokes shift versus E_{21}. All calculated phonon mode-energies for wave vector $qa = 0.016$ (a being the wide barrier width) in ACDQWs samples are also shown.

In Fig. 3 we plot the calculated Raman Stokes shifts (coupled mode energies) versus E_{21} by solving the laser rate equations for several ACDQWs. The horizontal dashed lines show the calculated six IF phonon modes and confined GaAs LO phonon mode. The open squares and open triangles indicate the calculated Raman lasing Stokes shifts and non-lasing Stokes shifts corresponding to coupled ISB plasmon-phonon modes (I$^-$, I$^+$, IF1$^-$ and IF1$^+$). The Solid circles represent the experimental Raman lasing Stokes shifts. Calculated coupled mode energies are well consistent with the experimental results [2].

In conclusion, we have developed a theory of Raman laser in ACDQWs, based on the differential scattering cross-section due to the coupled ISB plasmon-optical phonon modes. The theory can predict the correct lasing mode frequencies observed in recent experiments.

REFERENCES

1. H. C. Liu et al., *Appl. Phys. Letters* **78**, 3580-3582, (2001).
2. H. C. Liu et al., *Phys. Rev. Letters*. **90**, 077402-1-4, (2003).
3. S. M. Maung and S. Katayama, *Physica E* **21**, 774-778, (2004).

A^+-Centers and "Barrier-Spaced" A^0-Centers in Ge/GeSi MQW Heterostructures

Anton Ikonnikov,[*] Irina Erofeeva,[*] Dmitry Kozlov,[*] Oleg Kuznetsov,[*] Vladimir Aleshkin,[*] Vladimir Gavrilenko,[*] Dmitry Veksler,[**] and Michael S. Shur[**]

[*]*Institute for Physics of Microstructures RAS, Nizhniy Novgorod, Russia*
[**]*Rensselaer Polytechnic Institute, Troy, NY, USA*

Abstract. New shallow acceptor magnetoabsorption lines in THz range have been discovered under bandgap photoexcitation in strained Ge/GeSi quantum well (QW) heterostructures. The magnetoabsorption lifetime was as high as 10^{-4} s, indicating that the absorption is due to the transitions from the ground states of very shallow acceptors. The observed absorption resonances are shown to result from the photoexcitation of A^+-centers and $1s \to 2p_+$ type transitions from the ground state of the barrier-situated A^0-centers into excited states in the 1st and 2nd electric subbands. The shallowest discovered ground acceptor states ($E_B \leq 0.5$ meV) are attributed to the "barrier-spaced" acceptors (a hole bound with an acceptor ion in the *neighboring* Ge QW).

INTRODUCTION

Recent interest in the THz range has stimulated investigations of shallow impurities in strained Ge/Ge$_{1-x}$Si$_x$ multiple-quantum-well (MQW) heterostructures. The strain induced valence band splitting decreases the effective masses of two-dimensional holes and the binding energies of confined acceptors compared to those in bulk Ge [1]. In this paper, we employ the newly developed technique [2] to distinguish between different type shallow acceptor centers contributing to the magnetoabsorption in the THz range in Ge/GeSi heterostructures.

RESULTS AND DISCURSION

The structures under study were grown by vapor-phase epitaxy on Ge(111) substrate. The sample parameters are given in the captions to Figures 1 and 2. The residual shallow acceptor concentration was estimated to be $\sim 10^{14}$ cm^{-3} [1]. Magnetoabsorption of THz radiation (0.3 to 1.25 THz) was measured at $T =$ 4.2K sweeping the magnetic field (parallel to the structure axis) up to 3.5 T. Free carriers were excited by the modulated band-gap light. The differential impurity absorption was observed due to the carrier capture by compensated ionized impurities. The elliptical polarization of the THz radiation was employed to distinguish between the acceptor and donor absorption.

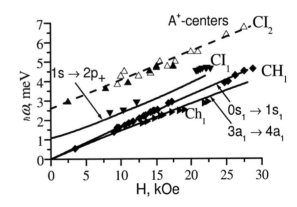

FIGURE 1. Spectral positions of absorption (closed symbols) and photoconductivity (open symbols) [1,3] resonances in sample #306a ($x = 0.12$, $d_{Ge} = 20$ nm, $d_{GeSi} = 26$ nm, elastic deformation of Ge layers $\varepsilon_{xx} = 2.2 \times 10^{-3}$, $N_{period} = 162$). Solid lines - calculated $\omega(H)$ dependences for hole CR lines CH$_1$ and Ch$_1$ and $1s \to 2p+$ transitions for barrier-centred A^0 acceptors (CI$_1$). Dashed line - resonances attributed to photoionization of A^+-centres (CI$_2$).

Figure 1 shows the magnetoabsorption resonant line positions in sample #306a with narrow Ge QWs. Two spectral lines CH_1 and Ch_1 are known to result from cyclotron resonance (CR) transitions from two lowest Landau levels of holes in the QWs [2]. In contrast to the CR lines, the positions of two other lines CI_1 and CI_2 are not extrapolated to the origin of coordinates. For uniformly distributed residual impurities, only QW-centered neutral acceptors (with the maximal binding energy about 7 meV) and the barrier-centered ones (with the minimal binding energy about 2 meV) give rise to the resonant absorption. Earlier we had observed the CI_2 line in the impurity photoconductivity spectra [1,3] (see Fig. 1) and attributed this line either to the photoionization of the A^+-center (the A^+-center is the. QW situated acceptor ion binding two holes) or to the $1s \rightarrow 2p^+$ transitions of the barrier situated neutral acceptors (A^0-centers). To distinguish between these two possibilities, we calculated the energies of $1s \rightarrow 2p^+$ transitions for the barrier situated A^0-centers in the magnetic fields using the expansion of the acceptor wavefunction in terms of the hole wavefunctions in the Ge QW at $H = 0$ [4]. Comparing the calculation results with the experimental data (see Fig. 1), we attributed the CI_1 line to the $1s \rightarrow 2p^+$ transitions. The CI_3 line seems to result from the photoionization of the A^+-centers. We estimate the corresponding binding energy to be about 2 meV at $H = 0$ [5].

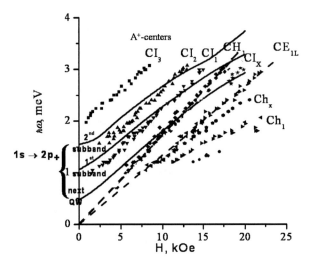

FIGURE 2. Spectral positions of the observed absorption resonances in the sample #308a (x = 0.09, d_{Ge} = 35 nm, d_{GeSi} =16 nm, ε_{xx}= 4.4×10^{-4}, N_{period} = 162). Solid lines - calculated $\omega(H)$ dependences for transitions from $1s$ state to $2p+$ states pertained to the 1st and the 2nd electric hole subband in the QW (CI_1 & CI_2) and for "barrier spaced" A^0-center (CI_x).

In the sample with broader Ge QWs, we observed three acceptor related absorption lines $CI_1 - CI_3$ (Fig. 2). The time-resolved measurements show that typical relaxation times for all impurity lines are as high as 10^{-4} s [4]. Therefore, these lines result from the optical transitions from the *ground* state of shallow acceptors rather than from their excited states (cf. [6]). Just as for sample #306a, the highest frequency line (CI_3) can be attributed to the photoionization of the A^+-centers. The calculation results for sample #308a with broad QWs predict two optical transitions of comparable intensity from the ground $1s$ state of barrier-situated A^0-center into $2p^+$ states related to the 1st and the 2nd hole electric subbands [4]. This results from a lower (compared to sample #306a) energy of the size quantization and from the state mixing in the 1st and the 2nd subbands.

In conclusion, we discovered the new acceptor related magnetoabsorption line CI_x that becomes discernible on the higher H side of the main hole CR line CH_1 with the frequency increase (see Fig.2). We speculate that this line results from the $1s \rightarrow 2p^+$ transitions for very shallow "barrier-spaced" A^0-center consisting of the hole bound with an acceptor ion in the *neighboring* Ge QW. According to our calculations, the energy of such transition only slightly exceeds that of the main hole CR transition in the magnetic fields up to 1 T, and this line is typically masked by the powerful absorption line CH_1. However, this transition becomes distinguishable in H > 17 kOe.

ACKNOWLEDGMENTS

This work was financially supported by RFBR (Grants #03-02-16808, #04-02-17178) and by the National Science Foundation (Project Monitor Dr. Rao Mulpuri). The authors are thankful to M.D. Moldavskaya for collaboration.

REFERENCES

1. Gavrilenko V. I. et al., *JETP Lett.* **65**, 209-214 (1997).
2. Aleshkin V. Ya. et al., *Phys. Solid State* **46**, 126-130 (2004).
3. Aleshkin V. Ya. et al., *Physica E* **7**, 608-611 (2000).
4. Aleshkin V. Ya. et al., *Phys. Solid State* **47** (2005) (submitted)
5. Aleshkin V. Ya. et al., *Proc. Workshop on Nanophotonics*, N. Novgorod, Russia, 2003, pp.318-321.
6. Meshov S. V. and E. I. Rashba, *Sov. Phys. JETP* **49**, 1115-1121 (1979).

Intersubband Hole Cyclotron Resonance in Strained Ge/GeSi MQW Heterostructures

Anton Ikonnikov,[*] Irina Erofeeva,[*] Dmitry Kozlov,[*] Oleg Kuznetsov,[*] Vladimir Aleshkin,[*] Vladimir Gavrilenko,[*] Dmitry Veksler,[**] and Michael S. Shur[**]

[*]*Institute for Physics of Microstructures RAS, Nizhniy Novgorod, Russia*
[**]*Rensselaer Polytechnic Institute, Troy, NY, USA*

Abstract. Hole cyclotron resonance lines due to transitions between different electric subbands were discovered in the magnetoabsorption spectra of strained Ge/GeSi heterostructures with wide (~80 nm) QWs. The measurements were performed on the undoped samples in the frequency range from 0.35 to 1.25 THz at T = 4.2 K with optical band-gap excitation.

INTRODUCTION

Investigations of Ge/GeSi multiple-quantum-well (QW) heterostructures with different QW size allow tracing the charge carrier spectra transformation from the bulk material to strained quantum-size layers. A small value of the hole cyclotron resonance (CR) mass of $0.07m_0$ has been observed in relatively narrow strained Ge QWs (d_{Ge} ~ 20 nm) [1]. In quantizing magnetic fields ($\hbar\omega_C \gg k_BT$), the hole CR line splits into two components corresponding to the CR transitions from the two lowest Landau levels into upper ones in the 1st electric subband [2]. In heterostructures with thicker Ge layers, the energy separation between the hole electric subbands in QWs and the CR energies are of the same order (up to 5 meV). Hence, the interaction between the Landau levels from different electric subbands allows us to observe the intersubband hole CR.

EXPERIMENT

The structure under investigation was grown by vapor-phase epitaxy on a lightly doped Ge(111) substrate. The parameters are given in the caption to Fig. 1. The total thickness of the structure exceeds the critical value, thus leading to the stress relaxation at the heterostructure–substrate interface. As a result, the Ge layers are biaxially strained and GeSi layers are biaxially stretched. The structure was not intentionally doped, with the residual shallow acceptor concentration being estimated as 10^{14} cm^{-3} [1]. The sample was placed in the center of a superconductive solenoid and cooled to the liquid helium temperature. We measured THz magnetoabsorption spectra (from 0.3 to 1.25 THz) in the Faraday configuration with the magnetic field being parallel to the structure axis. Free carriers were excited by modulated band-gap light. The transmitted radiation was detected by an n-InSb crystal. The signal proportional to the differential absorption of the THz radiation was measured as a function of the magnetic field up to 3.5 T.

RESULTS AND DISCURSION

Figure 1 shows the typical absorption spectra of the sample for elliptic polarization of THz radiation. The measurements for two opposite directions of the magnetic field allow us to identify the electron and hole related features in the spectra. Contrary to [1,2] which report only hole CR lines to be observable in Ge/GeSi heterostructures with narrow (20 nm) Ge layers, we discovered the CR line CE$_{1L}$ of electrons in 1L valley with a light effective mass (m$_C$ = 0.083m$_0$). 1L valley forms the conduction band bottom in stretched GeSi layers (QWs for electrons) [6]. Thus, the Ge/GeSi heterostructures with thick Ge layers (such as #309a) proved to be type II heterostructures.

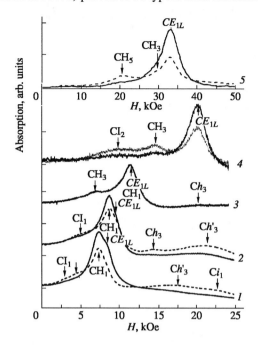

FIGURE 1. Magnetoabsorption spectra of sample #309a (d_{Ge} = 85 nm, d_{GeSi} = 20 nm, ε_{xx} = 4.6x10^{-4}, N_{period} = 83) $\hbar\omega$ = (1) 1.16, (2) 1.34, (3) 1.72, (4) 2.81, and (5) 4.73 meV. Solid and dashed lines correspond to different directions of the magnetic field. **CH** and **Ch** specify the hole CR peaks related to the different ladders of Landau levels. **CI** and **Ci** designate the impurity related absorption peaks. **CE** indicates the electron CR.

A lateral DC electric field was used to distinguish between the CR and impurity assisted transitions. The intensities of the impurity lines decrease with the electric field (probably due to the impact ionization of the shallow acceptors), while those corresponding to the CR lines persist. The energy positions of the resonance lines in the magnetoabsorption spectra are shown in Fig. 2. According to the calculation results of the hole Landau levels and CR transition matrix elements (using the Luttinger Hamiltonian, see [3,4] for details), we assign the lines CH_1, CH_3 and CH_5 to the transitions from the lowest Landau levels of the 1st electric subband $0s_1$ into the upper Landau levels of the 1st, 3rd and 5th subbands $1s_1$, $1s_3$ and $1s_5$, accordingly. The intersubband hole CR transitions become possible due to a strong interaction of the hole Landau levels from different electric subbands of the same parity resulting in the state anticrossing and mixing. The observation of the intersubband CR makes it possible to determine not only in-plane (cyclotron) hole mass but also the transverse hole mass responsible for the electric quantization, i.e. we can determine the parameters of the valence band dispersion law, that seems to be slightly different from those in bulk Ge (see caption to Fig. 2).

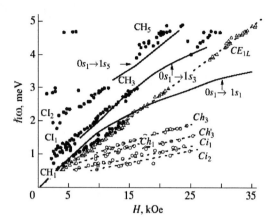

FIGURE 2. Energy positions of the resonant lines in the magnetoabsorption spectra of sample #309a (points). Solid lines indicate the calculated magnetic-field dependences of the transition energies between different hole Landau levels in the spectral regions, where these transitions occur with a sufficiently high intensity. The following parameters have been extracted from the comparison between the theory and experimental data: A = 0.99γ_1, B = 1.98γ_2, and D = 2.06$\gamma_3/3^{1/2}$, where γ_1, γ_2, and γ_3 are the parameters of the Luttinger Hamiltonian for holes in bulk Ge [5].

ACKNOWLEDGMENTS

This work was financially supported by RFBR (Grants #03-02-16808, #04-02-17178) and by the National Science Foundation (Project Monitor Dr. Rao Mulpuri). The authors are thankful to M.D. Moldavskaya for collaboration.

REFERENCES

1. Gavrilenko V. I. et al., *JETP Lett*. **59**, 348-352 (1994).
2. Aleshkin V. Ya. et al., *Proc. 6th Int. Symp. "Nanostructures: physics and technology"*, Ioffe Institute, St. Petersburg, 1999, pp. 356-359.
3. Aleshkin V. Ya., Vaks V. L., Veksler D. B., et al., *Izv. Akad. Nauk, Ser. Fiz.* **64**, 308-312 (2000).
4. V. Ya. Aleshkin, V. I. Gavrilenko, D. B. Veksler, and L. Reggiani, *Phys. Rev. B* **66**, 155336-155347 (2002).
5. J. C. Hensel and K. Suzuki, *Phys. Rev. B* **9**, 4219-4257 (1974).
6. V. Ya. Aleshkin and N. A. Bekin, *Semiconductors* **31**, 132-138 (1997).

Dielectric properties of ultra-thin films

Jun Nakamura, Shunsuke Ishihara, Hideki Ozawa, and Akiko Natori

Department of Electronic-Engineering, The University of Electro-Communications, 1-5-1 Chofugaoka, Chofu, Tokyo 182-8585, Japan

Abstract. A novel evaluation method of a dielectric constant is proposed with the use of first-principles calculations for the ground states in external eletrostatic fields, which is applicable to ultrathin films. The optical dielectric constant evaluated at the innermost region of the film approaches a constant value near to the experimental bulk dielectric constant with increasing the thickness of Si(111) ultra-thin films up to only 8 bi-layers, while the energy gap of the film is much larger than that of bulk Si. The theoretical value of the optical dielectric constant for the Si (111) film, which is converged at a large thickness, well reproduces the experimental one for bulk Si.

INTRODUCTION

Current device technology requires a drastic reduction in the size of metal-oxide-semiconductor field-effect transistors (MOSFET's). Indeed, really 15 nm gate lengths MOSFET has been realized and its physical gate oxide thickness is scaled down to 0.8 nm which corresponds to four Si atoms across [1]. Therefore, much interest has been devoted to the structural and electronic properties of such ultra-thin films. In particular, it has been recognized that one of the important physical parameter to characterize the gate film is the dielectric constant of the film. Here, we propose a novel method to theoretically evaluate the dielectric constant of ultra-thin films. Before turning to an actual gate dielectric, we begin, in this study, with the thin-film of silicon typical of semiconductors in order to asses the capability of our method.

CALCULATIONS

We performed first-principles total energy and band calculations based on the density functional theory [2, 3] by using the norm conserving pseudopotential suggested by Troullier and Martins [4]. Exchange and correlation were treated with a generalized gradient approximation [5]. The wave functions were expanded in a plane-wave basis set with a kinetic-energy cutoff of 25 Ry. Brillouin zone integration was done at 16 k points in the two-dimensional zone, and structures were optimized with the use of a conjugate gradient method. A repeated slab geometry was used for the simple calculation, which has a supercell consisting of n bi-layers (BL) of Si(111) (n =1 ~ 20) and of a vacuum region corresponding to about 6 BL in thickness. The lattice constant for the slab was assumed to be that calculated for the bulk Si crystal, and the (1 × 1)-unit cell was adopted parallel to Si(111). The frontside and backside of the slab were terminated with H atoms that eliminate artificial dangling bonds. The bond length of Si-H was set to 1.51 Å for all the models, which was optimized for the Si(111)-10BL model.

RESULTS AND DISCUSSION

In order to apply an external electrostatic field perpendicular to the surface of the slab, we introduce a planar dipole layer in the middle of the vacuum region [6, 7], as shown in Fig.1(a). Potential jumps in the vacuum region are due to the dipole layer. We evaluated the dielectric constant, ε_{slab}, as the ratio of the applied field E_z^{ext} to the planar averaged internal electric field E_z^{in},

$$\varepsilon_{slab} = \frac{E_z^{ext}}{<E_z^{in}>} \quad (1)$$

in cgs Gaussian system of units. This relationship is derived from the continuity of the normal dielectric displacement fields between the outside and the inside of the slab. The planar averaged internal field is evaluated as follows:

$$\left\langle E_z^{in}(z') \right\rangle \equiv \left\langle \frac{d(\Delta v(z))}{dz} \right\rangle_{1BL} \quad (2)$$

$$= \frac{\Delta v\left(z' - d_{\frac{1}{2}BL}\right) - \Delta v\left(z' + d_{\frac{1}{2}BL}\right)}{d_{1BL}} \quad (3)$$

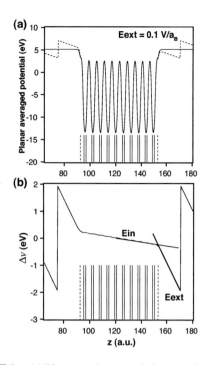

FIGURE 1. (a) Planar (x-y) averaged electrostatic potentials with (solid) and without (dashed) external field, (b) difference between planar (x-y) averaged effective potentials with and without external field for the Si(111)-10BL model. Vertical solid and dashed lines show the z-positions of Si(111) atomic layers and terminated H layers of the slab, respectively.

E_z^{in} is evaluated at the deepest bi-layer of the slab. Here, $\Delta v(z)$ is the difference between planar (x-y) averaged electrostatic potentials with and without an external field. The mean value theorem assures the equal sign between Eq.(2) and Eq.(3). In Fig.1(b) is shown $\Delta v(z)$ for the model of ten bi-layers of Si(111)-H. As shown in this figure, the external electric field penetrates inside the Si(111) slab, which reflects the semiconducting nature of this system. It is clear that the planar averaged internal electric field approaches a constant value within a few bi-layer depth from the surface. Figure 2 shows the dielectric constant evaluated using Eq.(1). When a slab thickness increases, the dielectric constant evaluated at the innermost region of the slab approaches a constant value near to the experimental bulk dielectric constant at a thickness of about 8 bi-layers, while the energy gap is much larger than the bulk value (+25 %). Converged dielectric constant successfully reproduces experimental one for bulk Si; the dielectric constant for the Si(111)-20BL (12.85) is only 6.2 % higher than the experimental one (12.1). Further, it is noted that the dielectric constant lowers with decreasing Si layer thickness, because of the enhancement of a quantum confinement effect.

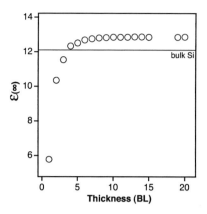

FIGURE 2. Dielectric constant evaluated from the relationship between the external and internal electric fields. Horizontal line indicates the experimental dielectric constant for bulk Si (12.1).

CONCLUSIONS

We have succeeded in evaluating the dielectric constant of Si(111)-H slab by using the first-principles calculations in external electrostatic fields. These calculations are performed within the framework of the ground states calculations. The theoretical value of the optical dielectric constant for the Si (111) film is 12.85, which agree well with the experimental one for bulk Si. We can develop this method to the evaluation of the static dielectric constant by considering the lattice relaxation of slabs in external electric fields.

ACKNOWLEDGMENTS

This research is conducted as a program for the "Promotion of Leading Researches" in Special Coordination Funds for Promoting Science and Technology by Ministry of Education, Culture, Sports, Science and Technology.

REFERENCES

1. B.Doyle, Arghavani, R., D.Barlage, S.Datta, M.Doczy, J.Kavalieros, A.Murthy, and R.Chau, *Intel Technolygy J.*, **6**, 42 (2002).
2. P.Hohenberg, and W.Kohn, *Phys. Rev.*, **136**, B864 (1964).
3. W.Kohn, and L.J.Sham, *Phys. Rev.*, **140**, A1133 (1965).
4. N.Troullier, and J.L.Martins, *Phys. Rev. B*, **43**, 1993 (1991).
5. J.P.Perdew, K.Burke, and M.Ernzerhof, *Phys. Rev. Lett.*, **77**, 3865 (1996).
6. J.Neugebauer, and M.Scheffler, *Phys. Rev. B*, **46**, 16067 (1992).
7. B.Meyer, and D.Vanderbilt, *Phys. Rev. B*, **63**, 205426 (2001).

Localized and extended states in semiconductor quantum wells with wire-like interface islands

N. Shtinkov*, P. Desjardins* and R. A. Masut*

Département de Génie Physique and Regroupement Québécois sur les Matériaux de Pointe (RQMP), École Polytechnique de Montréal, C.P. 6079, Succursale "Centre-Ville", Montréal (Québec), H3C 3A7, Canada

Abstract. We study theoretically the effect of lateral carrier confinement in quantum wells (QWs) with one-dimensional (wire-like) islands at the interfaces. Using a newly developed efficient Green's function approach and the sp^3s^* tight-binding model, we calculate the conduction band electronic structure of 14 monolayers thin $In_{0.32}Ga_{0.68}As/GaAs$ (001) QWs. The interface islands induce weak carrier localization with a negligible effect on the subband energies. However, the extended states exhibit changes in their orbital character that can have an important impact on the optical properties of the structures.

The device properties of nanostructures are affected by the interface morphology and its impact on the electronic structure, yet precise calculations of these effects are scarce due to the numerical complexity of the problem. Most of the attention has been focused on investigating interface interdiffusion, segregation, and roughness [1], which require relatively small supercells. Although experimental results have demonstrated the important effect of interface steps on both the extended and the localized states of quantum well (QW) structures [2], only the behavior of localized states has been theoretically studied [3]. We have recently developed an efficient Green's function method for the calculation of the electronic structure of QWs with one-dimensional (1D) interface islands, and have used it to study the localized and extended states in ultrathin 2 monolayers (ML) wide InAs/InP (001) QWs with 1 ML interface steps [4, 5]. In these structures a change of the QW width by only 1 ML modifies the states' energies by several tens of meV; therefore, their electronic structure is drastically affected by the interface morphology. The principal effects of this large lateral confining potential have been identified as: i) carrier localization, leading to a large band gap reduction and decrease of the heavy-light hole splitting; and ii) significant changes in the symmetry of the extended (QW) states, which have important implications on the electronic and optical properties of the structures [4, 5]. In wider QWs such as those often used in optoelectronic devices the fluctuations in the QW width give rise to a lateral confining potential of only several meV, therefore their electronic properties should be much less sensitive to the interface morphology. However, to the best of our knowledge, no exact calculations of these effects have been carried out until now.

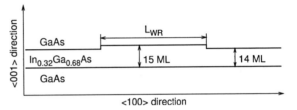

FIGURE 1. Cross-section of the structures under study. The GaAs/InGaAs interfaces are shown with a thick solid line.

In the present work, we study the effect of the interface islands on the conduction band states of 14 ML $In_{0.32}Ga_{0.68}As/GaAs$ (001) QWs, designed for infrared photodetectors (QWIPs) in the 8 μm range. The structures under study have one planar interface and the other exhibits two 1 ML steps, delimiting a wire-like interface island forming a region that contains a 15 ML QW (see Fig. 1). The width of the wire-like region L_{WR} is varied between 10 and 50 ML with a step of 10 ML. The structure is infinite in the (001) plane and periodic in the $\langle 001 \rangle$ direction with a period of 40 ML, which is sufficient to ensure minimal coupling between neighboring QWs. The islands are oriented along the $\langle 010 \rangle$ direction; because of the isotropy of the conduction band (CB), the dependence on the orientation is negligible [4, 5]. The local density of states (DOS) is calculated using a recently developed Green's function method [4] and the semi-empirical nearest-neighbors sp^3s^* TB model including spin and strain effects [6]; more details can be found in Ref. [4]. The InAs/GaAs valence band (VB) offset of 0.21 eV [7] is added to the on-site energies of InAs.

The n-doped 14 ML $In_{0.32}Ga_{0.68}As/GaAs$ (001) QWs under study have been designed for QWIP applications

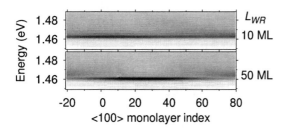

FIGURE 2. CB DOS for island widths 10 and 50 ML. Each horizontal line represents the lateral variation of the local DOS in the well center for a given energy value. The wire-like region starts at monolayer index 1. The intensity scale is linear.

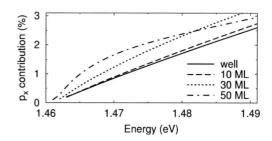

FIGURE 3. Percentage contribution of the p_x orbital to the CB DOS in the wire center. The solid line shows the dependence in the absence of a wire-like region (perfect QW).

in the 8 μm range, based on transitions between the localized 2D electron subband and resonant electron states in the GaAs CB. We investigate the initial states of these transitions, lying in the energy region from the QW ground state to above the Fermi level, which for typical doping concentrations is $\approx$ 10 meV above the CB edge. The calculated lateral spatial dependence of the local DOS in this energy region for L_{WR} 10 and 50 ML is shown in Fig. 2. The QW subband edge is at 1.462 eV, using an energy reference with zero at the GaAs VB maximum. For energies above 1.462 eV one observes the continuum of 1D extended states, originating from the projection of the QW electron subband to the center of the 1D Brillouin zone. The 1D carrier localization is manifested by an increase of the DOS in the island region and a shift of the CB edge to lower energies. The effect is negligible for $L_{WR} = 10$ ML and becomes more pronounced with increasing L_{WR}. However, even for $L_{WR} = 50$ ML the localization remains weak and the energy shift is less than 2 meV. This is due to the small (6 meV) energy difference between the ground states of the 14 and the 15 ML QWs, which acts as a lateral confining potential.

Apart from inducing lateral localization, the interface islands also affect the extended states of the structure by altering their orbital character [4]. The calculated contributions of the p_y and p_z orbitals (oriented along the 1D island and in the growth direction, respectively) to the CB DOS remain below 0.5% in the energy range 1.46eV $\leq E \leq$ 1.5eV. The contributions of the p_x orbital are plotted in Fig. 3 for several values of L_{WR}, as well as for a 14 ML QW with perfectly abrupt interfaces. Again, for $L_{WR} = 10$ ML the changes in the orbital character introduced by the presence of an interface island are negligible; however they increase with L_{WR} and, for a 50 ML wide island, a twofold increase in the p_x contribution to the extended states' DOS is observed. The nonzero contributions of in-plane p orbitals have been identified as the main source of the nonzero normal-incidence optical absorption in n-doped QWs [4]. Thus, the observed increase of the p_x contribution to the extended states below the Fermi level will increase the (usually very small) absorption coefficient for normal incidence light, having a direct impact on the device properties of the structure. The calculated p_x contribution is comparable to that in ultrathin QWs [4]; therefore, although in wide QWs the 1D localization caused by the interface islands is small, the changes in the extended states can be significant.

In conclusion, we have studied the effect of lateral confinement on the conduction band states of 14 ML In$_{0.32}$Ga$_{0.68}$As/GaAs (001) QWs with wire-like interface islands. The island-induced weak lateral localization leads to negligible shifts in the states energies. The extended states, however, can exhibit significant modifications in their orbital character with respect to the QW states. The observed modifications of the electronic structure may cause anisotropic behavior, relax some of the selection rules, and enhance the oscillator strengths of the optical transitions relevant in devices such as QWIPs.

We acknowledge the financial support of the Natural Sciences and Engineering Research Council of Canada and the Canada Research Chair Program.

REFERENCES

1. Capaz, R. B., Dargam, T. G., Martins, A. S., Chacham, H., and Koiller, B., *Phys. Stat. Sol. (a)*, **173**, 235 (1999).
2. Paki, P., Leonelli, R., Isnard, L., and Masut, R. A., *J. Vac. Sci. & Tech.*, **A 18**, 956–9 (2000).
3. Dargam, T. G., Capaz, R. B., and Koiller, B., *Phys. Rev. B*, **64**, 245327 (2001).
4. Shtinkov, N., Desjardins, P., Masut, R. A., and Vlaev, S. J., submitted to *Phys. Rev. B* (2004).
5. Shtinkov, N., Desjardins, P., and Masut, R. A., *Microelectronics Journal*, **34**, 459 (2003).
6. Priester, C., Allan, G., and Lannoo, M., *Phys. Rev. B*, **37**, 8519 (1988).
7. Vurgaftman, I., Meyer, J. R., and Ram-Mohan, L. R., *J. Appl. Phys.*, **89**, 5815 (2001).

Photoluminescent and kinetic properties of A(+) centers in quantum well

K.S. Romanov*, N.V. Agrinskaja*, N.S. Averkiev*, Yu.L. Ivanov*, P.V. Petrov* and V.M. Ustinov*

*A. F. Ioffe Physico-Technical Institute, Russian Academy of Sciences, 26 Politekhnicheskaya, 194021, St. Petersburg, Russia

Abstract. Theoretical and experimental studies of photoluminescence and hopping conductivity of A(+) centers in GaAs/AlGaAs based quantum well (QW) samples were performed. The binding energy of hole localized at A(+) center in the QW as a function of the width of the QW was calculated. The comparison of calculated binding energies for various well widths showed good agreement with experimental data. The calculated characteristic size of wavefunction is consistent with results of our analysis of the hopping conductivity. The model developed requires only one fit parameter.

INTRODUCTION

Quantum structures containing impurities are extensively studied not only in view of their promising applications but also because they have some new physical properties.

Quantum wells containing donor or acceptor were studied both theoretically [1] and experimentally [2]. The models used are quite sophisticated and require more than one fitting parameter.

The goal of the present work is the experimental and theoretical study of of the A(+) center in $A^{III}B^{V}$ quantum wells. The comparison with experimental data shows that the developed theoretical model describes well two particle complexes in low dimensional heterostructures.

THEORY

The calculation of A(+) center binding energy is quite difficult due to several reasons. First of all it is necessary to take into account interaction between two holes localized at acceptor. Secondly the exact acceptor potential is unknown due to the presence of its short-range part.

One of the ways to overcome the difficulties with the A(+) potential is the usage of zero-radius potential method. It is possible because hole coupling by A(+) center is relatively weak. The corresponding binding energy is about 5 meV while the energy of hole localized at acceptor is about 10 times larger. This weak coupling leads to quite large localization range of hole bounded on A(+) center.

Another reason to use zero-radius potential model is the screening of acceptor potential by the potential of the first hole. This hole potential compensates the long-range Coulomb part of the acceptor so that resulting potential attracting the second hole (i.e. A(+) potential) is relatively short-range.

In the present work the effective mass approach for studying A(+) states was used. Luttinger Hamiltonian was used for degenerated valence band description. The quantum well with infinite borders was considered for simplicity.

Following the zero-range potential model the potential of a defect is described by modifying the Schredinger equation for a free particle. Namely solutions to the Schredinger equation in the absence of attractive potential, which decay at infinity, are constructed. The attractive potential is taken into account by introducing the special boundary condition that determines the asymptotic behavior of the spherically symmetric part of the hole wave function near a defect:

$$\psi(r)|_{r\to 0} = C\left(\frac{1}{r} - \alpha\right) + o(r), \quad (1)$$

where α - is the parameter of the A(+) potential, r denotes the distance from A(+) center, ψ is the spherically averaged wave-function.

It is more convenient [3] to find the Green function of Schredinger equation rather than eigenfunctions.

The mixed coordinate-momentum representation has been used for calculations. In this representation the Green function $\psi(\vec{q},z)$ depends on $\vec{q}$ - 2D wave vector parallel to the well plane and z - coordinate perpendicular to the well plane. The Green functions of consideration can be found analytically in this representation.

To find the energy levels we have to use boundary condition (1) which characterizes the potential. Therefore the Green functions should be transformed into the coordinate representation and then averaged over the polar and azimuth angles - $\bar{\psi}$. It is rather difficult to find analytical expressions for these functions but in order to calculate the energy levels we only need to know the asymptotic behavior of these functions at $r \to 0$. Specifically only the first two terms of asymptotic expansion

$$\bar{\psi}(r,\varepsilon) = a(\varepsilon)_{-1} \cdot \frac{1}{r} + a_0(\varepsilon) + a_1(\varepsilon) \cdot r \ldots \quad (2)$$

are to be found: $a_{-1}(\varepsilon)$ and $a_0(\varepsilon)$ where ε is the A(+) binding energy.

The resulting dispersion equation binds the energy of the localized state ε with the power of attracting potential α:

$$\alpha = -\frac{a_0(\varepsilon)}{a_1(\varepsilon)}. \quad (3)$$

There are two twice-degenerated solutions of this equation. The 2D localization range of these states is

$$\rho = 1/\sqrt{2m_l \varepsilon}, \quad (4)$$

where m_l is the light hole effective mass.

The only fitting parameter of this model is α (1). But usually it is more convenient to use the energy of A(+) center in bulk material. The parameter α can be derived from this energy using (3).

RESULTS AND DISCUSSION

The comparison of the A(+) center energy dependence on the thickness of the quantum well computed using (3) with our experimental data was performed. The theoretical curves for two A(+) center energies in bulk semiconductor with experimental data points are shown on figure 1.

Experiments were performed with the samples prepared by molecular-beam epitaxy and selectively doped with beryllium; the bulk hole concentration in GaAs amounted to $10^{17} cm^3$. Stable A(+) centers were formed in quantum wells by the so-called double selective doping method. Photoluminescence spectra were measured from the samples immersed directly in liquid helium. A glass optical fiber was used to feed light from a HeNe laser to a sample and to output a luminescence signal. The radiation was recorded by a diffraction spectrometer and a photomultiplier in the photon-counting mode.

It is seen that there is a good agreement between experimental and theoretical data when in the theoretical

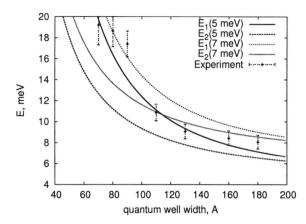

FIGURE 1. The energy of the transition between the localized levels of A(+) center and the first level of heavy holes in the quantum well versus the quantum well thickness. Two different energies of A(+) center in bulk material were taken: 7 meV and 5 meV.

calculation the binding energy in bulk semiconductor is about 5meV.

We've estimated the size of an A(+) center by analyzing the hopping conductivity [4]. According to the formula (4) the characteristic size of the wave function in quantum wells is about 100Å. This result is consistent with the data of [4] (80 Å).

Thus it has been shown that the method developed describes A(+) centers quite well. So it can be used for identification of impurities in quantum wells.

ACKNOWLEDGMENTS

This work was supported in part by grants from the Russian Foundation for Basic Research; INTAS; the European Office of Aerospace Research and Development; ISTC (grant no. 2206); the programs of the Ministry of Industry, Science, and Technology of the Russian Federation; the programs of the Presidium and the Department of Physics of the Russian Academy of Sciences; and the Foundation for the Support of Russian Science.

REFERENCES

1. Belyavskii, V. I., Gol'dfarb, M. V., and Kopaev, Y. V., *Semiconductors*, **31**, 936–940 (1997).
2. Masselink, W. T., Chang, Y.-C., and Morkoc, H., *Phys. Rev. B*, **28**, 7373–7376 (1983).
3. Pahomov, A. A., Halipov, K. V., and Yassievich, I. N., *Semiconductors*, **8**, 1387–1394 (1996).
4. Agrinskaya, N. V., Ivanov, Y. L., Ustinov, V. M., and Poloskin, D. A., *Semiconductors*, **35**, 550–553 (2001).

How Do Electrons, Excitons and Trions Share The Reciprocal Space In A Quantum Well?

M. T. Portella-Oberli[1], J. H. Berney[1], V. Ciulin[2], M. Kutrowski[3], T. Wojtowicz[3], and B. Deveaud[1]

[1]*IPEQ-Ecole Polytechnique Fédérale de Lausanne (EPFL), CH1015 Lausanne, Switzerland*
[2]*IQUEST-UC-Santa Barbara, Ca93106, USA*
[3]*Institute of Physics-Polish Academy of Science, Warsaw, Poland*

Abstract. We report on nonlinear optical dynamical properties of the excitonic complexes in CdTe modulation-doped quantum wells. The results reveal correlated effects of excitons and trions. We propose that the main source of these correlations is due to the presence of electrons in the quantum well and that its physical origin is the Pauli exclusion-principle.

In undoped semiconductor quantum wells, Coulomb interaction gives rise to the electron-hole bound states (excitons) that plays a crucial role in determining the linear and nonlinear optical properties near the band edge [1]. The optical spectrum of moderately doped quantum wells also feature a (trion) charged exciton resonance [2]. Then, modulation-doped quantum wells provide a system in which electrons, excitons and trions cohabit. A very fundamental question immediately rises: How do they share the reciprocal space in this quantum well? To probe this, we investigated the many-body interactions among excitons, trions and electrons through the nonlinear dynamical properties in the excitonic complexes in modulation doped CdTe quantum wells. We report the optical nonlinearities of excitons and trions as either excitons or trions are created in the quantum well. The results reveal that the nonlinearities induced by trions are different from those induced by excitons, moreover they are mutually correlated.

The sample [3] is a one-side modulation-doped CdTe/Cd$_{0.73}$Mg$_{0.27}$Te heterostructure containing one single quantum well of 8 nm with an electron concentration of about 4×10^{10} cm^{-2}. The reflectivity and luminescence spectra present the trion line situated about 3 meV below the neutral exciton [4]. We use temporally and spectrally resolved pump and probe experiments to investigate the nonlinear behaviour of the excitonic complex spectrum. The sample is excited with a narrow tunable 1.3 ps pulse and the modifications in the optical spectrum are probed by a broad 100 fs probe pulse. The pump-probe nonlinear spectra are detected at different delay times between pump and probe pulses. The differential reflectivity spectrum $\Delta R = (R-R_0)/R_0$ is plotted for each delay time, R_0 and R being the unexcited and excited reflectivity spectra, respectively. The experiments are performed at 5 K with linearly polarized pulses.

In Figure 1, we plot the differential reflectivity spectra obtained when pumping selectively at exciton (Fig. 1a) and, at trion (Fig. 1b) resonances, at zero delay time. These results evidence the distinct nonlinear optical effects induced by an exciton and a trion populations. In Fig. 1a, we observe a blue-shift of the exciton transition, a bleaching and a red-shift of the trion resonance induced by the presence of excitons in the quantum well. The renormalization of the exciton line is due to repulsive short-range fermion exchange.[5] The red-shift of the trion line, observed only at low temperatures [4, 6], is a consequence of the exciton-electron interaction, which induces a changing in the electron distribution. This results in an increase of the electron occupation at higher energies favoring the trion transitions at lower energies [4]. The red-shift is only observed at short times and, at later times, it is masked by the bleaching effect due to trion formation from the exciton population (insert of Fig. 1a).

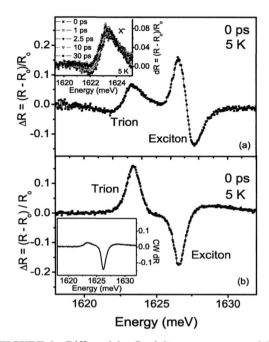

FIGURE 1. Differential reflectivity spectra, at zero delay, when pumping at (a) exciton (3×10^{10} excitons/cm^2) and (b) trion (2×10^{10} trions/cm^2) resonances. Inset in (a): Time evolution of the nonlinear trion signal. Inset in (b): differential reflectivity spectrum obtained from cw relectivity spectra with a difference in the electron density of about 2×10^{10} cm^{-2}.

It is important to note that the trion bleaching signal does not increase with time. This almost constant value over about 10 ps, the time that trions are formed from exciton population [7], evidences that this bleaching evolves according to two opposite contributions of the same order of magnitude: the contribution due to the trion population which increases with time, and that due to the exciton population which decreases with time. Thus we attribute the trion bleaching, at short times, to the phase-space filling of the optically-accessible k-space by the photogenerated excitons, resulting into a blocking of the trion transition. This result evidences that excitons and trions share the same k-space and originates from the same ground state.

The spectra in Fig. 1b, evidence that a trion population induce a bleaching of the trion resonance and an induced absorption of the exciton line, effects with similar amplitudes. We find that the decreasing of the electron concentration in the well induces these same effects, but with dissimilar amplitudes (inset in Fig. 1b). We compare those signals, for similar density of generated trions and decreased electron concentration. In contrast, we observe that the trion bleaching is much more important when trions are generated. This difference comes from phase-space filling of trions. Therefore, the trion bleaching in the pump-probe signal has two contributions originate from: phase-space filling of trions and phase-space unfilled by electrons. The induced absorption of excitons is already a very large effect as the electron gas concentration is decreased and is attributed as resulting from reduction of the screening of excitons by electrons [4, 6]. This effect is less pronounced in the pump-probe experiments, when a trion population is generated. This difference can be attributed to phase-space filling by trions: in the same way excitons have been observed to block the trion transition due to phase-space filling, the presence of trions should block the exciton transition by the same argument. A screening of excitons by trions, much less important that by electrons can also be a contribution to the reduced signal.

In conclusion, the results reveal that excitons and trions share the same k-space in a modulation-doped quantum well. Furthermore, the electron gas is perturbed in the presence of either excitons or trions, inducing correlated effects of excitons and trions.

This work was partially supported by the Swiss National Science Foundation.

REFERENCES

1. Chemla, D. S., and Miller, D. A. B., J. Opt. Soc. Am. B **2**, 1155 (1984).
2. Keng, K., Cox, R. T., Merle d'Aubigné, Y., Bassani, F., Saminadayar, K., and Tatarenko, Phys. Rev. Lett. **71**, 1752 (1993).
3. Wojtowicz, T., Kutrowski, M., Kaczewski, G., and Kossut, J., Appl. Phys. Lett. **73**, 1379 (1998).
4. Portella-Oberli, M. T., Ciulin, V., Berney, J. H., Deveaud, B., Kutrowski, M., and Wojtowicz, T., Phys. Rev. B **69**, 235311 (2004).
5. Schimitt-Rink, S., Chemla, D. S., and Miller, D. A. B., T., Phys. Rev. B **32**, 6601 (1985).
6. Portella-Oberli, M. T., Ciulin, V., Berney, J. H, Kutrowski, M., Wojtowicz, T., and Deveaud, B., Phys. Status Solidi C **1**, 484 (2004).
7. Portella-Oberli, M. T., Ciulin, V., Deveaud, B., Kutrowski, M., and Wojtowicz, T., Phys. Status Solidi B **238**, 513 (2003).

Effect of the Electron and Hole Scattering Potentials Compensation on Optical Band Edge of Heavily Doped GaAs/AlGaAs Superlattices

Yu.A.Pusep[1], F.E.G.Guimarães[1], M.B.Ribeiro[1], H. Arakaki[1], C.A. de Souza[1], S.Malzer[2] and G.H.Döhler[2]

[1] *Instituto de Fisica de São Carlos, Universidade de São Paulo, 13560-970 São Carlos, SP, Brazil.*

[2] *Institut für Technische Physik I, Universität Erlangen, D-91058 Erlangen, Germany*

Abstract. The optical broadenings studied by the photoluminescence in the intentionally disordered GaAs/AlGaAs superlattices were compared with the broadenings of the individual electron states measured by the Shubnikov-de Haas oscillations. It was shown that the combined effect of the electron and hole energy blurrings is to decrease the optical broadening with respect to the individual state broadenings resulting in very sharp optical edges even in highly disordered superlattices.

INTRODUCTION

The great advantage of the intentionally disordered superlattices (SLs) is a controlled nature of the disorder which allows distinguishing the impact of the randomness on their properties. The optical measurements, which are determined by the joint density of states, present a powerful method to study the disorder effects. The disorder affects the joint density of states through the modification of the energy and broadening of the electron conduction and valence band states. It is well known that in disordered materials the indirect optical transitions form the broadened band edge. In this work we demonstrate that actually, the shape of the optical band edge drastically depends on the character of the optical transitions. Namely, a sharp band edge emerges even in the presence of strong disorder when the direct interband transitions dominate.

THEORETICAL CONSIDERATIONS

The optical broadening studied by the photoluminescence (PL) in the intentionally disordered Si-doped GaAs/AlGaAs SLs as a function of disorder was compared with the broadenings of the individual electron states measured by the Shubnikov-de Haas oscillations. The interaction of the carriers with imperfections causes the blurring of their energy [1]:

$$\frac{\hbar}{\tau_{e(h)}} = 2\pi u_{e(h)}^2 N_i g_F \quad (1)$$

where $\tau_{e(h)}$ is the electron (hole) single-particle relaxation time, $u_{e(h)}$ is the electron (hole) scattering potential, N_i is the concentration of imperfections and g_F is the density of states on the Fermi surface. This individual electron (hole) blurring results in a broadening of the Landau levels which can be obtained by the magnetotransport measured in the range of the Shubnikov-de Haas oscillations.

On the other hand, both the electron states of the valence and conduction bands contribute to the optical interband transitions. Therefore, the electron and hole blurrings together give rise to the broadening of the PL edge. In such a case, according to [2], the optical band edge broadening is determined by the value:

$$\frac{\hbar}{\tau} = 2\pi(u_e - u_h)^2 N_i g_F \quad (2)$$

This means that the relaxation time τ, which characterizes the broadening of the optical edge cannot be expressed in terms of the mean free times τ_e and τ_h. the optical broadening depends on the difference of the electron and hole scattering potentials and therefore, may be considerably smaller than the individual broadenings.

RESULTS AND DISCUSSION

The disorder in the studied here SLs was produced by a random variation of the well thicknesses.[3]. The disorder strength was characterized by the disorder parameter $\delta_{SL}=\Delta/W$, where Δ is the full width at half maximum of a Gaussian distribution of the electron energy calculated in the isolated quantum wells and W is the miniband width of the nominal SL in the absence of disorder.

Some of the PL spectra of the disordered SLs are shown in Fig.1(a). They show that the disorder leads to a considerable red shift and a relatively weak broadening of the PL edge. The magnetoresistance traces measured with different orientations of the magnetic field are shown for selected SLs in Fig.1(b).

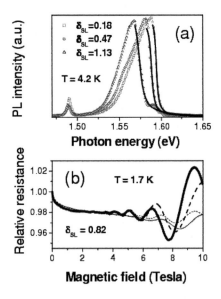

FIGURE 1 Photoluminescence spectra (a) and magnetoresistance traces (b) measured in the $(GaAs)_m(Al_{0.3}Ga_{0.7}As)_6$ SLs with n = 7×10^{17} cm^{-3} and with different disorder strengths δ_{SL}. The full lines in (a) are the PL intensities calculated according to [2]. The thick (thin) magnetoresistance lines were measured at the magnetic fields parallel (perpendicular) to the growth direction. The dash lines are the calculated magnetoresistances.

The characteristic broadenings of the PL edges shown in Fig.2(a) were determined by the fit of the PL spectra calculated according to [2]. The electron energy broadenings associated with the broadenings of the Landau levels were obtained by the fits of the calculated magnetoresistances. The dependences of the vertical and parallel electron broadenings on the disorder strengths obtained in the disordered SLs are shown in Fig.2(b). The vertical electron energy broadening and the optical broadening display similar behaviors - the noteworthy enhancement with the increasing disorder. While, as expected, the parallel electron energy broadenings were not affected by the vertical SL disorder. The broadenings of the individual electron states were found considerably higher than the characteristic broadenings of the PL edges. This discrepancy is explained by the effect of the partial compensation of the electron scattering potential by the hole scattering potential, which is demonstrated by Eq.(2).

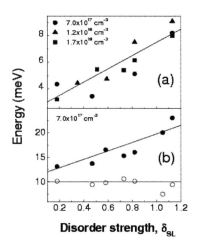

FIGURE 2. Characteristic broadenings of the PL edge $\hbar/\tau$ (a) and the electron broadenings $\hbar/\tau_e$ (b) obtained in disordered SLs $(GaAs)_m(Al_{0.3}Ga_{0.7}As)_6$. The open and full circles on panel (b) correspond to the parallel and vertical electron energy broadenings respectively. The lines are guides for eyes.

ACKNOWLEDGMENTS

The financial support from FAPESP is gratefully acknowledged.

REFERENCES

1. A.A.Abrikosov, L.P.Gor'kov, and I.E.Dzyaloshinskii, Quantum Field Theoretical Methods in Statistical Physics, Pergamon Press, Oxford (1965).

2. E.G.Batyev, Yu.A.Pusep, and M.P.Sinyukov, Sov.Phys.Solid.State **27**, 708 (1985).

3. Yu.A.Pusep, W.Fortunato, P.P.Gonzalez-Borrero, A.I.Toropov, J.C.Galzerani, Phys. Rev.B **63**, 115311 (2001).

Polarization Dependence of Photosensitivity of the Schottky Barrier Diodes Based on the InGaAs/GaAs Quantum Well and Quantum Dot Structures

D. O. Filatov, I. A. Karpovich, V. Ya. Demikhovskii, D. V. Khomitskiy and V. V. Levichev

University of Nizhni Novgorod, 23 Gagarin Ave., Nizhni Novgorod 603 950 Russia

Abstract. The polarization dependence of the photosensitivity (PS) of the Schottky barrier diodes based on the InGaAs/GaAs quantum well (QW) and quantum dot (QD) structures has been studied. The PS polarization dependence follows the theoretical one with good agreement. The effect of the hole state mixing in the QW on the PS polarization dependence has been studied. The asymmetry of the PS polarization dependence of the QDs in the [110] and [1$\bar{1}$0] directions has been observed attributed to the effect of piezoelectric field induced by the elastic strain.

Optical spectroscopy of the quantum semiconductor structures is an important area of the semiconductor physics. The optical absorption spectra, photoluminescence, etc. reveal the characteristics of size quantization, density of states, excitonic effects, etc. The polarization dependencies reveal the optical transition selection rules, which are related to the nature of the initial and final states.

While a lot of work was devoted to the optical spectroscopy of the low dimensional systems, the polarization dependencies were much less studied. In this work we studied the polarization dependence of the photosensitivity (PS) of the Schottky barrier diodes with built-in InGaAs/GaAs quantum wells (QWs) and InAs/GaAs quantum dots (QDs) at linearly polarized excitation.

The samples were grown by Atmospheric Pressure Metal Organic Vapor Phase Epitaxy (AP-MOVPE) on n^+ GaAs (001) substrates. The GaAs buffer and cap layers in the QW structure (thickness 1.0 and 0.4 μm respectively) were doped by Si ($n_0 = 3 \times 10^{16} \text{cm}^{-3}$). The QW thickness was 4.9 nm, and In molar fraction was 0.21. The InAs nominal thickness in the QD structure was 1.5 nm, the cap layer thickness was 30 nm. The lateral size of the QDs according to AFM data was 35-40 nm, average height 6 nm, the surface density $\approx 10^9 \text{cm}^{-2}$.

The Au Schottky contacts were deposited by the thermal evaporation. The chips were cleaved across the Schottky contacts (see inset in Fig.1). The diodes could be excited either normally to the QW or through the cleaved facet so the light inside the structure propagates almost parallel to the QW. Although a very small fraction was absorbed by the QW, the PS was enough to measure

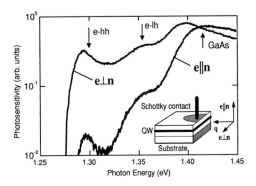

FIGURE 1. PS spectra (300K) of a Schottky barrier diode with InGaAs/GaAs QW at in-plane linearly polarized excitation. In the inset the measurement scheme in shown.

its polarization dependence.

The open circuit photovoltage spectra were measured at 300K by standard lock-in technique using a grating monochromator and Glan-Taylor prism as a polarizer.

Figure 1 presents the PS spectra of the Schottky barrier diode measured with in-plane polarized excitation for $\mathbf{e} \parallel \mathbf{n}$ and $\mathbf{e} \perp \mathbf{n}$. Two PS bands attributed to the ground state transitions involving the heavy hole (*hh*) states and the light hole (*lh*) ones were observed. Identification of the transitions was based on calculations of the QW energy spectrum taking into account the elastic strain according to [2]. Usually, the *lh* subband in the InGaAs/GaAs QWs is considered to be pushed out of the QW due to elastic strain [2]. However, band bending near a QW due to its charging makes an additional barrier for holes which can lead to *lh* confinement [3].

The PS of a quantum confined structure is proportional to the interband optical transition matrix element M_{e-hh}. For a QW at $k=0$ (k is the module of the electron (hole) wave vector) it is given by [1]

$$|M_{e-hh}(\mathbf{e})|^2 \sim [1-(\mathbf{e}\cdot\mathbf{n})^2] \quad (1)$$

for the heavy holes and

$$|M_{e-lh}(\mathbf{e})|^2 \sim [1+3(\mathbf{e}\cdot\mathbf{n})^2] \quad (2)$$

for the light holes. Here $\mathbf{e} = \mathbf{E}/|\mathbf{E}|$ is an unit vector parallel to the electric field $\mathbf{E}$ in the incident electromagnetic wave and $\mathbf{n}$ is an unit vector normal to the QW.

The interband PS near the QW absorption edge is polarized almost completely (Fig.1). The PS polarization degree $P = (S_\perp - S_\parallel)/(S_\perp + S_\parallel)$ where S is the photosensitivity was 0.87 at $h\nu = 1.32$ eV in agreement with (1). At higher $h\nu$ P decreases with increasing $h\nu$. This effect was attributed to the mixing of the *hh* and *lh* states in the QW at $k \neq 0$. To account for this effect the polarization dependence of M_{e-h} at $k \neq 0$ was calculated. The electrons were described by the effective mass approximation and the hole states were described by Luttinger Hamiltonian. The results of calculations and the experimental results are presented in Fig.2. The shape of the M_{e-h} polarization dependence is intermediate between those given by (1) and (2). The experimental data is consistent with the theory. However, the measured value of P is lower than the calculated one that can be attributed to the QW imperfections and/or to the stray light.

In the PS spectrum of the QD structure (Fig.3) two well resolved Gaussian peaks at $h\nu = 0.97$ eV and 1.06

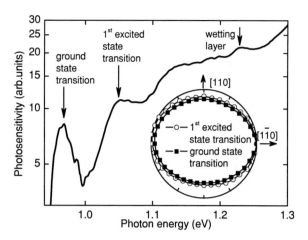

FIGURE 3. PS spectrum (300K) of the Schottky barrier diode with InAs/GaAs QDs. Inset: PS polarization dependencies of the ground state transition in the QDs and of the first excited state one measured at normal excitation.

eV attributed to the ground state transition in the QDs and to the first excited state, respectively are present. In the inset to Fig.3 the polarization dependencies of the PS in the respective peaks measured at the normal photoexcitation are presented. The asymmetry of the PS polarization dependencies in the [110] and [1$\bar{1}$0] directions was attributed to the effect of the piezoelectric field induced by the elastic strain [4]. The polarization degree observed ($P = 0.10$ and 0.07 for the ground state transition and the first excited one, respectively) was lower then the one predicted by the theory [4] ($P = 0.18$ for the ground state interband transition). This can be explained by deviation of the actual shape of the QDs in the investigated structure from the pyramidal one considered in [4].

ACKNOWLEDGMENTS

The work was supported by Russian American Program "Basic Research and Higher Education" (REC-NN-001).

REFERENCES

1. F. T. Vasko and A. V. Kuznetsov, *Electronic states and optical transitions in semiconductor heterostructures*, Springer, New York, 1999, p.93.
2. G. Ji, D. Huang,, U. K. Reddy, T. S. Henderson, R. Houdre and H. Morko, *J.Appl.Phys.* **62**, 3366-3373 (1987).
3. D. O. Filatov, I. A. Karpovich and T. V. Shilova, *Phys. Low-Dim. Struct.*, 2003, **1/2**, pp.143- 149.
4. O. Stier, M. Grundmann and D. Bimberg, *Phys. Rev. B* **59**, 5688-5071 (1999).

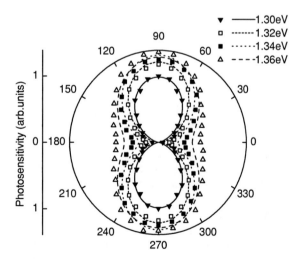

FIGURE 2. PS polarization dependencies of the Schottky barrier diode with InGaAs/GaAs QW at different photon energies. Lines - theory, symbols - experiment. The experimental data was normalized to unity for $\mathbf{e} \perp \mathbf{n}$ and $h\nu = 1.30$ eV.

Temperature-dependent conduction band structure of GaNAs and GaInNAs / GaAs quantum wells

M. Hetterich[*], A. Grau[*], T. Passow[*], A. Yu. Egorov[†] and H. Riechert[†]

[*]*Institut für Angewandte Physik and Center for Functional Nanostructures (CFN), Universität Karlsruhe (TH), D-76131 Karlsruhe, Germany*
[†]*Infineon Technologies, D-81730 Munich, Germany*

Abstract. We use photoreflectance spectroscopy to investigate the temperature-dependence of the electronic transitions in a $GaN_{0.012}As_{0.988}$ epilayer and a $Ga_{0.7}In_{0.3}N_{0.017}As_{0.983}$ (6.2 nm) / GaAs multiple quantum well. From fits to our experimental data using the band anticrossing model we find evidence, that the nitrogen level E_N in $Ga_{1-x}In_xN_yAs_{1-y}$ shifts significantly to higher energies with decreasing temperature. Our results also suggest a simultaneous rise in the coupling parameter C_{NM}.

INTRODUCTION

It is well known that the incorporation of nitrogen into $Ga_{1-x}In_xAs$ leads to a strong bandgap reduction, a highly nonparabolic conduction band $E_-(k)$ as well as the appearance of a new band E_+ above the conduction band edge (see, e.g., [1, 2]). Within the so-called band anticrossing (BAC) model these effects are interpreted to arise from a repulsive interaction between the conduction band and a higher-lying nitrogen-related energy level E_N. At $k = 0$ one obtains [1]

$$E_\pm = \frac{1}{2}\left[(E_N + E_M) \pm \sqrt{(E_N - E_M)^2 + 4C_{NM}^2 y}\right], \quad (1)$$

where E_M is the bandgap of $Ga_{1-x}In_xN_yAs_{1-y}$ neglecting the anticrossing with E_N, and C_{NM} is a coupling parameter depending on the indium concentration x.

For the BAC model to be useful in practice, accurate values for E_N and C_{NM} are needed. In the following we mainly focus on the temperature dependence of these parameters. This aspect is of particular interest, because the introduction of nitrogen into $Ga_{1-x}In_xAs$ is known to significantly reduce the temperature dependence of the bandgap, which has been attributed to the repulsive interaction between the conduction band and E_N in the literature [3, 4, 5].

EXPERIMENTAL DETAILS

The (strained) $GaN_{0.012}As_{0.988}$ / GaAs epilayer and $Ga_{0.7}In_{0.3}N_{0.017}As_{0.983}$ / GaAs multiple quantum well (MQW) samples discussed below were grown by solid-source molecular-beam epitaxy on undoped GaAs (001), using a RF-coupled plasma source for nitrogen. Afterwards, thermal annealing was used to improve their optical quality. To measure the different transition energies in our samples we applied photoreflectance (PR) spectroscopy. A He–Cd laser (442 nm line) provided the photomodulation in case of the $GaN_{0.012}As_{0.988}$ layer. For the MQW a 670 nm laser diode was used instead.

RESULTS AND DISCUSSION

The PR spectra of the investigated $GaN_{0.012}As_{0.988}$ / GaAs sample comprises features due to E_-, $E_- + \Delta_0$ (Δ_0 is the spin–orbit split-off energy) and E_+ transitions in the layer as well as the bandgap of the GaAs substrate E_0^{GaAs}. Fig. 1 shows the measured temperature dependence of E_-, E_+ and E_0^{GaAs}. Based on this data the ener-

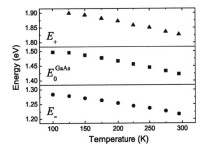

FIGURE 1. Measured temperature dependence of E_-, E_+ and E_0^{GaAs} in a strained $GaN_{0.012}As_{0.988}$ / GaAs sample.

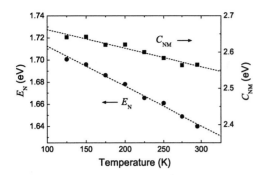

FIGURE 2. Temperature dependence of E_N and C_{NM} for a strained $GaN_{0.012}As_{0.988}$ / GaAs epilayer.

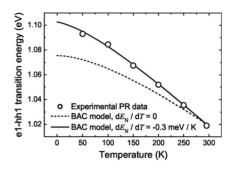

FIGURE 3. Temperature dependence of the e1–hh1 transition energy in a $Ga_{0.7}In_{0.3}N_{0.017}As_{0.983}$ (6.2 nm) / GaAs MQW. The data points have been extracted from PR measurements, the curves are theoretical predictions of the BAC model.

getic position of the nitrogen level E_N and the coupling parameter C_{NM} as a function of temperature can be determined. From (1) we find

$$E_N = E_+ + E_- - E_M \quad (2)$$

$$C_{NM} = \sqrt{\frac{(E_+ - E_-)^2 - (E_N - E_M)^2}{4y}}. \quad (3)$$

Following [6] we set $E_M(y) = E_0^{GaAs} - \alpha y$, where the correction factor $\alpha = 1.55$ eV is assumed to be temperature-independent. The results of this procedure for the $GaN_{0.012}As_{0.988}$ layer discussed above are summarized in Fig. 2. At room-temperature we obtain $E_N = 1.64$ eV and $C_{NM} = 2.57$ eV, in quite good agreement with previously published data [1, 2, 3, 7]. Most importantly, a strong thermal shift of E_N relative to the valence band edge is found. This qualitatively confirms the results in [2], where the author deduced $dE_N/dT = -0.25$ meV/K for free-standing $Ga_{0.93}In_{0.07}N_{0.004}As_{0.996}$ and $Ga_{0.96}In_{0.04}N_{0.01}As_{0.99}$. However, we obtain the somewhat higher value $dE_N/dT \approx -0.36$ meV/K for our $GaN_{0.012}As_{0.988}$ epilayer. This may partly be due to the fact that C_{NM} was assumed to be temperature-independent in [2], while our analysis suggests a decrease with increasing temperature ($dC_{NM}/dT \approx -0.5$ meV/K).

To extend our measurements to $Ga_{1-x}In_xN_yAs_{1-y}$ with high In concentrations we also analyzed the temperature dependence of the lowest electronic transition (e1-hh1) in a $Ga_{0.7}In_{0.3}N_{0.017}As_{0.983}$ (6.2 nm) / GaAs MQW. These results are discussed in detail in [5] and summarized in Fig. 3. In the theoretical modelling of our PR data we calculate the bound electron states using a special technique based on the boundary conditions derived for the conduction band BAC Hamiltonian wavefunction [5]. The influence of temperature and strain on the band structure of the Ga(In)As host material are taken into account, however, following [2] we neglect the temperature dependence of C_{NM} here, because the latter is found to affect the bandgap significantly less than the shift in E_N which we want to estimate. As can be seen in Fig. 3, good agreement of the BAC model prediction with experiment can again only be achieved if a strong thermal shift of the nitrogen level relative to the unstrained valence band ($dE_N/dT \approx -0.3$ meV/K) is assumed.

In conclusion we have presented evidence that the nitrogen level E_N in GaN_yAs_{1-y} and high In content $Ga_{1-x}In_xN_yAs_{1-y}$ shifts significantly to higher energies with decreasing temperature. For GaN_yAs_{1-y} our results also suggest a simultaneous rise in C_{NM}.

ACKNOWLEDGMENTS

This work has been financially supported by the Deutsche Forschungsgemeinschaft (DFG) within the Emmy Noether program and the Center for Functional Nanostructures (CFN) at the University of Karlsruhe (project A2).

REFERENCES

1. Shan, W., Walukiewicz, W., Yu, K. M., Ager III, J. W., Haller, E. E., Geisz, J. F., Friedman, D. J., Olson, J. M., Kurtz, S. R., Xin, H. P., and Tu, C. W., *Phys. Status Solidi B*, **223**, 75–85 (2001).
2. Skierbiszewski, C., *Semicond. Sci. Technol.*, **17**, 803–814 (2002).
3. Suemune, I., Uesugi, K., and Walukiewicz, W., *Appl. Phys. Lett.*, **77**, 3021–3023 (2000).
4. Polimeni, A., Capizzi, M., Geddo, M., Fischer, M., Reinhardt, M., and Forchel, A., *Phys. Rev. B*, **63**, 195320 (2001).
5. Hetterich, M., Grau, A., Egorov, A. Y., and Riechert, H., *J. Phys.: Condens. Matter*, **16**, S3151–S3159 (2004).
6. O'Reilly, E. P., Lindsay, A., Tomić, S., and Kamal-Saadi, M., *Semicond. Sci. Technol.*, **17**, 870–879 (2002).
7. Hetterich, M., Grau, A., Egorov, A. Y., and Riechert, H., *J. Appl. Phys.*, **94**, 1810–1813 (2003).

Photoluminescence from Ion Implanted and Low-Power-Laser Annealed GaAs/AlGaAs Quantum Wells

A. Sonkusare, D. Sands, S. I. Rybchenko*, I. E. Itskevich*

Dept Physical Sciences (Physics), University of Hull, Kingston-upon-Hull, HU6 7RX, UK
* Dept Engineering, University of Hull, Kingston-upon-Hull, HU6 7RX, UK

Abstract. A triple-quantum-well $Al_{0.3}Ga_{0.7}As$/GaAs structure was implanted with Zn at relatively low doses to disrupt the lattice and destroy the photoluminescence (PL). A single 10-ns pulse from an excimer laser was used to anneal the damage. It was found that there is an optimum energy for annealing, at which strong PL signals from all the quantum wells are recovered. After implantation, the lines are blue-shifted and the relative intensities also change. The blue shift is ascribed to ballistic intermixing of Al during the implantation.

Pulsed laser annealing of semiconductors has been investigated since the late 1970s. At that time the laser was seen as an ideal tool for activating dopants which were implanted into semiconductors. Early work concentrated on melting and recrystallisation [1], but this resulted in a number of undesirable effects in compound semiconductors such as GaAs. Not least among them is the preferential loss of the most volatile element from the surface and the concomitant creation of vacancies in the sub-surface region. Hence, in 1991 Vitali *et al* reported low power laser annealing (LPLA) of GaAs using a Q-switched ruby laser [2]. Low power is not an absolute term but is defined relative to the melting point of the semiconductor. LPLA causes a re-ordering of the implantation-damaged lattice within the solid phase.

All the work to date on LPLA was performed using bulk material. Here we present the results of XeCl-excimer-laser LPLA performed on GaAs/$Al_{0.3}Ga_{0.7}As$ quantum wells (QWs) implanted with Zn at doses ranging from 10^{11} cm^{-2} to 10^{13} cm^{-2}. Attention will be focused on the lowest implantation dose, though similar results are found for all doses. The samples incorporated three QWs, 4 nm, 7 nm, and 10 nm wide (4-nm well on top), separated by 50 nm barriers. The implant energy (200 keV) was chosen so that the implantation tail just straddles the lower (10 nm) well. Annealing was carried out in Ar at a pressure of 4 bars using fluences in the range 180 to 250 mJ cm^{-2}. Photoluminescence (PL) was measured using an Ar$^+$-ion laser and a liquid nitrogen immersion cryostat.

Figure 1a shows the effect of implantation at 10^{11} ions cm^{-2} on the PL spectrum. Most noticeably, the intensity is strongly reduced (by over 10^4) and the emission energies from the wells are blue-shifted. Also, other broad emission lines emerge in the spectral region of the 4-nm well. The emission extends down to 1.60 eV on the low energy side and up to 1.68 eV on the high energy side.

Figure 1b shows the spectra after a single 10-ns excimer laser pulse was applied. The positions of the principal lines are only slightly shifted relative to the implanted material. After annealing at 185 mJ cm^{-2} there is little change to both the intensity and the shape of the spectrum, apart from a transition at 9 meV above the line from the topmost (4-nm) well. At 250 mJ cm^{-2} the intensity decreases even more relative to the implanted sample and the broad line at ≈ 1.63 eV becomes stronger. After annealing at 209 mJ cm^{-2}, however, not only does the intensity of the signal rise by three orders of magnitude, but there are also no defect signals present.

The effect of annealing is summarized in Fig. 2. It demonstrates a very narrow annealing recovery "window" as a function of fluence.

Fluences below the "window" are apparently not sufficient for recovery. We find in general that the peaks observed in the spectra after annealing at 185 mJ cm^{-2} are those arising after implantation. Their relative intensities are altered, however, so that the heavy-hole exciton transition from the 4-nm QW appears as a low energy tail to a dominating "defect" line. On the other hand, fluences above the "window" cause additional damage to the sample, with the broad defect line at 1.63 eV becoming more prominent. The origin of the defects is unclear. The substitutional Zn

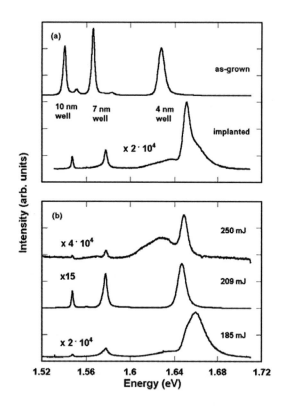

FIGURE 1. PL spectra for as-grown and Zn-implanted sample (a) and for samples annealed using a single excimer laser pulse at the energies (per sq cm) shown. The spectra are normalized and offset from each other for clarity.

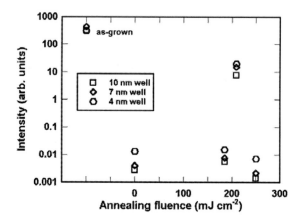

FIGURE 2. The intensity of the PL lines as a function of laser fluence. Zero fluence relates to spectra after implantation. Data for the as-grown sample are also shown.

level lies 30 meV above the valence band edge [3] and vacancies and anti-site defects lie even deeper in the band gap [4,5]. Meanwhile, defect lines on the low energy side are observed between 5 and 25 meV from the QW line. These lines are clearly associated with implantation and are probably due to point defects either in the QW or in the barrier.

Simple calculations of the temperature rise during the laser pulse [6] suggest that for all three fluences the surface temperature may reach the melting point. However, the surface does not melt, except perhaps at 250 mJ cm^{-2}, because there is not enough energy within the pulse to overcome the latent heat. Even at the highest fluence it is only the first few nanometers that melt. This surface damage is probably responsible for the low intensity of the luminescence.

Similar thermal calculations of the total diffusion during the laser pulse using macroscopic diffusion coefficients have been shown to be accurate [7]. Using accepted values for the inter-diffusion coefficients in the AlGaAs system [8] we derive negligibly small total Al diffusion lengths of $\approx 3 \times 10^{-4}$ nm. This agrees with our observation that no further blue-shift occurs upon laser annealing. It is possible therefore to ascribe the blue-shifting entirely to ballistic intermixing during implantation. The blue shift is independent of implant dose and corresponds to an effective diffusion length of 0.5 - 0.7 nm, depending on the well position.

To conclude, we have shown that a single pulse from an excimer laser can anneal implantation induced damage in GaAs QWs. The damage causes 1) a strong reduction in luminescence intensity; 2) the appearance of broad "defect" emission lines either side of the main QW transition; and 3) blue-shift in QW emission which is due to ballistic intermixing. Annealing restores the intensity and decreases the "defect" emissions without adding further blue-shifting.

The authors acknowledge support from the EPSRC III-V growth facility in Sheffield and the Ion Beam Analysis center (SCRIBA), Surrey. AS acknowledges the support of the Indian Government.

REFERENCES

1. R. F. Wood, C. W. White, R. T. Young, "Pulsed Laser Processing of Semiconductors", in *Semiconductors and Semimetals*, vol 23, Academic Press, 1984.
2. G. Vitali et al, *J. Appl. Phys.* **69**, 3882 (1991)
3. D. W Kisker, H Tews, W Rehm, *J. Appl. Phys.* **54**, 1332 (1983)
4. H. K. Nguyen et al, *J. Appl. Phys.*, **69**, 7585 (1991)
5. T S Shamirzaev et al, *Semicond. Sci. Technol.* **13**, 1123 (1998)
6. M. K. Eladawi, M. A. Abdelnaby, S.A. Shalaby, *Int. J. Heat Mass Tran.*, **38**, 947 (1995)
7. H Howari et al, *J. Appl. Phys.* **88**, 1373 (2000)
8. W P Gillin, D J Dunstan, *Computational Materials Science* **11**, 96 (1998)

Vertical correlation of highly organized exciton complex at ZnSe / BeTe type-II asymmetric superlattices

A. Fujikawa, H. Mino, K. Oto, R. Akimoto [A] and S. Takeyama [B]

Graduate School of Science and Technology, Chiba University, Chiba 263-8522, Japan
[A] *AIST Photonics Research Institute, Ibaraki 305-8568, Japan*
[B] *Institute for Solid State Physics, University of Tokyo, Kashiwa 277-8581, Japan*

Abstract. The linearly polarized PL spectra of the asymmetric superlattice ZnSe / BeTe have been investigated at various exciting light intensities. The ZnSe layers have two kinds of well widths. It has been shown that the typical PL approximately consists of two peak components reflecting in the deference of the recombination energies by size effect, and the polarization degrees of these peak components invert each other under strong excitation. This behavior is probably an important clue that V-shape electron-hole-electron complexes are formed with multi-interlayer correlation at high-density regime.

INTRODUCTION

In ZnSe/BeTe type-II superlattice the spatially indirect photoluminescence (PL) was strongly observed below about 10 K. In a former work we had suggested an existence of charged excitons, biexcitons or excitonic polymers in which electrons and holes are combined over more than two heterointerfaces at high carrier density [1].

In a symmetric superlattice the eigenvalues of electrons in each ZnSe layer and holes in each BeTe layer are nearly equal respectively. The energy level of electrons and holes might be very important to produce an electron-hole complex. In order to make this point clear, we have prepared an asymmetric superlattice, and investigated the PL excitation (PLE) spectra and the linearly polarized PL spectra at various exciting light intensities.

EXPERIMENT

The investigated sample was an asymmetric superlattice (ZnSe/BeTe/ZnSe/BeTe : 7.9 nm / 2.8 nm / 4.0 nm / 2.8 nm) ×12 periods, which was grown on a (001)-GaAs substrate by molecular beam epitaxy. We have performed PL measurements at exciting light intensities (40 mW/cm^2 ~ 40 W/cm^2) and 6 K to avoid the distraction of the 'Helium' bubbles. Linear polarization dependence of this PL in the direction parallel to the heterointerfaces has also been measured. Frequency-doubled mode-locked Ti:Sapphire laser was used as an excitation light source. At the linearly polarized PL measurement, the excitation energy was adjusted to 3.1eV (400nm) to excite all ZnSe layers.

RESULT AND DISCUSSION

In this sample typical PL approximately consists of two peak components around 1.9 eV and 2.0 eV. The recombination energies between spatially separated electrons and holes should depend on two kinds of electron states that are affected by size effect as shown in INSET of FIG. 1. So we have identified the peak component at about 2.0 eV as the irradiative recombination of electrons in 4.0 nm ZnSe and holes in 2.8 nm BeTe. In a similar manner, we have done that at about 1.9 eV as the irradiative recombination of electron in 7.9 nm ZnSe and holes in 2.8 nm BeTe. We have confirmed above-mentioned items by the PLE spectra.

In ZnSe / BeTe type-II superlattice PL occurs in an extremely narrow region at the interface, so that we can expect the observation of the PL that depends on the direction of the chemical bonds [2]. So we have investigated the linear polarization dependence of this PL. FIG. 1 shows the linear polarization dependence of the PL spectra at various exciting light intensities. In our sample, all ZnSe and BeTe layers are terminated Zn and Te respectively. In this case the chemical bonds lie in mutually orthogonal planes (110) and ($1\bar{1}0$) at a normal and an inverted interface respectively, and their contributions to the anisotropy should cancel each other [2]. We labeled these interfaces as INSET of FIG. 1 on this occasion [3].

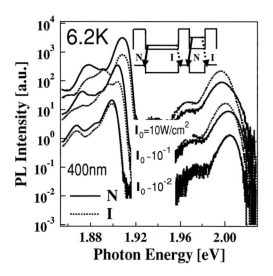

FIGURE 1. The linear polarization dependence of the PL spectra at various exciting light intensities, **INSET.** Band structure of this asymmetric superlattice.

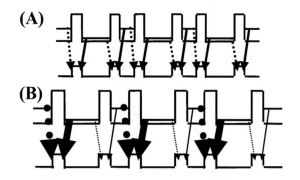

FIGURE 2. The image of the PL at each interface at **(A)** weak and **(B)** strong exciting light intensity,

At weak exciting light intensity, no polarization degree has been detected. This result is reflected in the cancellation of the anisotropy. FIG. 2(A) shows the image of the PL at each interface at weak exciting light intensity. While under strong excitation, the PL of electrons in 7.9 nm ZnSe and holes in 2.8 nm BeTe was dominant at N-interfaces, and that of electrons in 4.0 nm ZnSe and holes in 2.8 nm BeTe was dominant at I-interfaces, respectively. FIG. 2(B) shows the image of the PL at each interface at strong exciting light intensity. This result indicates that if we notice one BeTe layer and the PL at both interfaces in this layer has been stronger; the PLs at both interfaces in the next BeTe layers have been weaker. This behavior is probably an important clue that V-shape electron-hole-electron complexes are formed with multi-interlayer correlation at high-density regime. As seen in FIG. 1, the peak energy of the PL of electrons in 7.9 nm ZnSe and holes in BeTe has been higher, while that of electrons in 4.0 nm ZnSe and holes in BeTe lower. This result suggests existence of the vertical correlation that works to fill the gap in two kinds of electrons' energy levels when V-shape complexes are formed. There is the effect of internal electric field under strong excitation, so that we need to get more in-depth data on the vertical correlation.

CONCLUSION

The linearly polarized PL spectra of asymmetric superlattice ZnSe / BeTe have been investigated at various exciting light intensities. It has been shown that the typical PL approximately consists of two peak components reflecting in the deference of the recombination energies, and the polarization degrees of these peak components invert each other under strong excitation. This behavior is probably an important clue that V-shape electron-hole-electron complexes are formed with multi-interlayer correlation at high-density regime.

ACKNOWLEDGEMENTS

We thank K. Muro and T. Matsusue for the experimental and theoretical advice.

REFERENCES

1. H. Mino et al., in Proceeding of *the 26 th International conference on the Physics of Semiconductors, (H67)*
2. A.V.Platonov et al., *Phys.Rev.Letters* **83**, 3546 (1999).
3. S.V.Zaitsev et al., *J.Appl.Phys.* **91**, 652 (2002).

New efficient approach to calculation exciton resonance position and width for quantum-confined Stark effect in shallow quantum wells

Ilya V. Ponomarev*, Lev I. Deych* and Alexander A. Lisyansky*

*Queens College of City University of New York, Flushing, NY 11367, USA

Abstract. A computationally efficient approach to calculating characteristics of quantum well excitons in an external electric field is introduced. The non-stationary nature of eigenstates in the presence of electric field is taken into account with the help of the complex scaling approach. The method allows one to obtain simultaneously the filed-induced shift and broadening of the exciton resonance. The method is applied to a shallow quantum well in the regime of strong confinement. It is shown that in this case the field induced broadening is strongly affected by the effective electron-hole interaction at small to moderate electric fields.

The quantum-confined Stark effect has received considerable attention since its discovery[1]. However, even in a case of a single quantum well it is not possible to express a value of the exciton resonance's position and width as a function of quantum well parameters. Since variables in this problem cannot be separated, usually a variational method is employed from which the exciton characteristics can only be obtained with extensive numerical calculations. Moreover, all contemporary methods do not take into account the field-induced exciton width itself. At most, imaginary corrections to the quasi-bound state energy are considered for single-particle electron- and hole wave functions in the direction of the quantum well growth, z (which is also the direction of the electric field). We develop a new computationally efficient approach to this problem that provides a deeper physical insight, significantly reduces required computational resources and yields some analytical results for a case of shallow quantum wells. The approach is based on a complex-coordinate exterior-scaling procedure[2] that allows one to find the resonance position and the broadening for the exciton in electro-absorption spectra in quantum wells.

The main idea of the method is to introduce such a transformation of an original Hamiltonian that would produce an non-Hermitian Hamiltonian, whose complex eigenvalues would give the positions and the widths of the resonances. It is possible to do with the help of complex scaling of coordinates in an original Hamiltonian, $\hat{H}$, $r \to r\exp(i\theta)$, which can be described as an similarity transformation of the Hamiltonian

$$\hat{\tilde{H}}(\theta) = \hat{S}(\theta)\hat{H}\hat{S}^{-1}(\theta) \qquad (1)$$

where the complex scaling operator $\hat{S}$ is defined as $\hat{S}f(r) = f(re^{i\theta})$. The physical meaning of this transformation can be illustrated by applying it to a typical asymptotic form of a scattering wave function for a vanishing at infinity potential

$$\psi(r \to \infty) \propto \exp(ikr), \qquad (2)$$

where r is a radial coordinate and k is a wave number. If one tries to describe a resonance as an eigenvalue problem with a boundary condition given by Eq.(2), the value of k will come out complex, $k = |k|e^{-i\phi}$ and the respective wave function will exponentially diverge at infinity. If, however, we apply complex rotation of coordinates to this wave function, it will become $\hat{S}\psi(r \to \infty) \propto e^{i|k|\exp[i(\theta-\phi)]}$, which with the proper choice of θ can be made square integrable. This example illustrates the main idea of the complex scaling: with an appropriate choice of the transformation parameter θ, the transformed Hamiltonian, $\hat{\tilde{H}}(\theta)$ can be made to have square integrable (L^2) eigenfunctions with complex eigenvalues. Their real and imaginary parts are interpreted as the energies and widths of the resonance respectively. Since the transformed eigenfunctions belong to L^2 Hilbert space, one can calculate the value of energy following the standard variational procedure

$$E = \frac{\langle \psi(r)|\hat{\tilde{H}}(\theta)|\psi(r)\rangle}{\langle \psi(r)|\psi(r)\rangle} \equiv \frac{\langle \psi(re^{-i\theta})|\hat{H}|\psi(re^{-i\theta})\rangle}{\langle \psi(re^{-i\theta})|\psi(re^{-i\theta})\rangle}, \qquad (3)$$

where, however, the traditional definition of the scalar product must be modified because of the non-Hermitian nature of the Hamiltonian. While the standard definition requires complex conjugation of the function appearing at the left side of the product, the new rule requires to conjugate only those parts of the wave-function, which would have been complex without the scaling transformation. Since exact eigenfunctions of the transformed Hamiltonian are usually not known, any kind of an approximate representation of these functions would result in θ-dependence of the resulting resonance energies. It was suggested, therefore, that the approximate values must be stabilized with respect to the changes of θ. Thus θ play a role here as an additional variational parameter.

In our approach the exciton envelop function is presented as a product of three functions:

$$\Psi_{\text{trial}}(r, z_e, z_h) = \chi_e(z_e e^{-i\theta_e}) \chi_h(z_h e^{-i\theta_h}) \psi(r e^{-i\theta_r}), \quad (4)$$

where each function is assumed to be normalized. First two functions here obey single particle equations for electrons and holes respectively

$$H_e(z_{e,h}) \chi_{e,h}(z_{e,h} e^{-i\theta_{e,h}}) = W_{e,h} \chi_{e,h}(z_{e,h} e^{-i\theta_{e,h}}), \quad (5)$$

where $H_{e,h}$ describe single particle motions of electrons and holes in QW potential and electric field. These equations solved with rotated Siegert boundary conditions yield complex energy values, $W_{e,h}$, but because of the complex scaling transformation the respective wave functions are square integrable despite of the presence of electric field. This fact makes these functions suitable for calculating effective potential, $\overline{V}_r(r)$, that enter an equation determining the last of the function in Eq.(4):

$$\left[K_r(r) + \overline{V}_r(r) \right] \psi(r e^{-i\theta_r}) = W_X \psi(r e^{-i\theta_r}) \quad (6)$$

where $\overline{V}_r(r)$ is defined according to

$$\overline{V}_r(r) = \langle \chi_e \chi_h | - \left[r^2 + (z_e - z_h)^2 \right]^{-1/2} | \chi_e \chi_h \rangle = V_{Re} + i V_{Im}. \quad (7)$$

Solving the equations (5) and (6) we obtain the best approximation for the entire wave function in the form of a product (4). The corresponding value of the total quasi-bound energy $W = E - i\Gamma/2$ can be found as

$$W = \langle \Psi | \hat{H} | \Psi \rangle = W_e + W_h + W_X \quad (8)$$

The main advantage of the approach described above is that it allows one to calculate, in principle, not only field-induced single-particle widths $\Gamma_{e,h}$ but the *exciton width* Γ_X as well, which describes a renormalization of the electron-hole pair life-time by the effective interaction. Below we consider solutions of Eqs. (5) and (6) for the a shallow quantum well, where some analytical results can be derived.

For strongly confined excitons the effective exciton potential has two different regimes of behavior as a function of the distance, r. It is governed by the complex parameter d, which real part is an average distance between

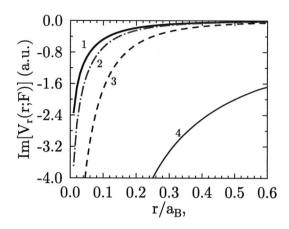

FIGURE 1. The profile of the imaginary part of the effective exciton potential $\overline{V}_r(r; F)$. The labels for the curves correspond to, curve 1, $F = 10^5$ V/cm, curve 2, $F = 2. \times 10^5$ V/cm, curve 3, $F = 4 \times 10^5$ V/cm. The thick solid line (curve 4) below the other curves represents $-1/r$ behavior, and is given for comparison. All data are for $L = 20$ finite quantum well in AlGaAs/GaAs materials.

an electron and a hole inside the quantum well:

$$d^2 = \langle \chi_e \chi_h | (z_e - z_h)^2 | \chi_e \chi_h \rangle. \quad (9)$$

When $r > d$ the potential V_r has a form $-1/\sqrt{r^2 + d^2}$ with the Coulomb tail, while at small but finite distances it has a logarithmic shape similar to the true Coulomb potential of a point charge in two dimensions. For a shallow quantum well a further analytical treatment is possible. We use a δ-functional model for the quantum well that allows us to obtain an explicit form of the effective potential. Figure 1 shows the profile of the imaginary part of this potential calculated for several different electric fields. As a result, we derive a simple analytical formula for the exciton's resonance position and broadening.

REFERENCES

1. D. A. Miller, D.S. Chemla, T.C. Damen, A. C. Gossard, W. Wiegmann, T.H. Wood, and C.A. Burrus, *Phys. Rev B* **32**, 1043 (1984).
2. W. P. Reinhart, *Annu. Rev. Phys. Chem*, **33**, 223 (1982).

Exciton complexes in ZnSe/BeTe type-II single quantum wells

H.Yamamoto, Ziwu Ji, H.Mino, R.Akimoto [1], and S.Takeyama [2]

Graduate School of Science and Technology, Chiba University, Chiba, Japan
[1] *AIST Photonics Research Institute, Tsukuba, Japan*
[2] *Institute for Solid State Physics, University of Tokyo, Kashiwa, Japan*

Abstract. We have attempted spatially-resolved photoluminescence (PL) measurements to ZnSe/BeTe type-II single quantum wells. We have found anomalous PL extended spatially to an order of 300 μm. The spatial extension grew with decreasing temperature. The results were unable to explain by a simple exciton diffusion model. The exciton complexes as a result of high density excitation could be responsible for the striking spatial propagation.

INTRODUCTION

ZnSe/BeTe QW structures with a type-II band alignment recently have a much attention due to large band offsets and an in-plane optical anisotropy. A number of studies have been carried out experimentally and theoretically in such systems [1-2]. However, up to now there are few reports stated the origin of the spatially indirect excitonic transition in II-VI semiconductors. In ZnSe/BeTe QWs, the spatially indirect PL shows high intensity, regardless of fairly small electron-hole wave function overlap. The long recombination time owing to the small electron-hole wave function overlap, leads to easy realization of a high density excitation state. Recently in biased GaAs/AlGaAs double quantum well, a macroscopically ordered exciton state has been observed [3]. Its PL transition is indirect in a real space like type-II QWs, realized by an applied electric field. Therefore, we have also paid attention to the spatial behavior of the indirect PL in ZnSe/BeTe QWs. In Ref.4, we have performed the magneto-PL measurement in a Voigt configuration. We found that exciton complexes probably become the indirect PL origin at a high density regime. In this paper, we have applied the spatially-resolved PL to the spatially indirect PL spectra in ZnSe/BeTe type-II single QWs, on attempt to clarify the underlying mechanism governing the strong optical recombination of our indirect type PL.

EXPERIMENTS

A non-doped asymmetric ZnSe/BeTe/ZnSe (7.9nm/2.8nm/4nm) quantum well structure was grown on (001)-oriented GaAs substrates by molecular beam epitaxy. For details of this sample, see Ref.5. For the spatially-resolved PL spectra the exciting laser was focused on a sample surface with 40 μm in diameter. The PL image at the surface was enlarged and projected on the screen by a camera lens. The spatially-resolved PL component was picked up through a pinhole-terminated glass fiver attached on the screen whose position is finely adjustable. The spatial resolution of this system is about ~70 μm. The temperature dependence of PL spectra have been studied at 5K, 10K, and 20K. A frequency double of a mode-lock Ti-Sapphire laser with a repetition rate of 76 MHz, pulse duration of ~100 fs and the wavelength of 370 nm (3.35 eV) was used for photo-excitation, but used as a steady excitation.

RESULTS AND DISCUSSION

A typical spatially indirect PL spectrum in ZnSe/BeTe type-II single QWs is shown in Fig.1 (*T*=

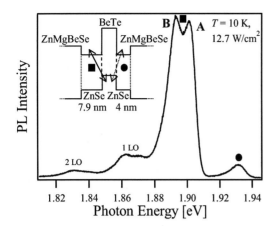

FIGURE 1. Typical PL spectra in ZnSe/BeTe type-II single QWs at 10 K. Inset: The sample structure and the indirect transition configuration expressed by ■ and ●.

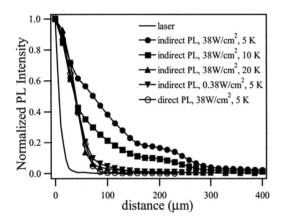

FIGURE 2. Spatial PL intensity profiles in various temperature: indirect transition PL at 5 K (●, ▼), 10 K (■), and 20 K (▲) with excitation intensity 38 W/cm^2 except ▼ with 0.38 W/cm^2 and direct transition PL at 5 K (○) with 38 W/cm^2. Note that spatial laser profile is measured by another method with a knife edge. Accordingly it is not indicate the spatial resolution (~70μm) of this system.

10 K), where some PL peaks are observed. We interpreted two peaks, at 1.90 eV and at 1.93 eV, as emitting from different ZnSe layer, which is illustrated by the markers in the inset. Here we remark the PL peak around 1.90 eV, which is attributed to the recombination of an electron at the ZnSe layer of 7.9 nm and a hole at the BeTe. Its PL peak splits into two components (A and B in Fig.1) at 10 K. The B peak was assumed to be an exciton complex from the magneto-PL measurement reported in a different work [4]. Its intensity increased nonlinearly against that of A peak and never saturated at a high density excitation regime. The PL intensity of B rapidly decreased with increasing temperature. Figure.2 shows the spatially-resolved PL at various temperatures. At 5 K, the spatial distribution of PL intensity under high density excitation, where the B peak is dominant, extended more than 300μm. With increasing temperature, the PL spatial expansion becomes small and then cannot be resolved within spatial resolution at 20K. When the A peak is dominant in the low excitation density at 5 K, the spatial-resolved PL showed a smaller spatial profile than the resolution of this system. These results indicated that the B component anomalously expanded in comparison with that of A. The large B PL expansion under lower temperature and more intense laser excitation over macroscopic region is not explained by a simple diffusion model. Taking account of the fact that the B PL component kept its spectral line shape even at far point from an excitation center, the high density exciton complex propagated coherently in the real space to the extent of the macroscopic scale. These results meant that the high density exciton complex state might become a condensed state.

CONCLUSION

We have performed spatially-resolved PL measurements on a ZnSe/BeTe quantum well (QW). The spatially indirect PL peak split into A and B peak at 10 K. The B peak showed unusual spatial macroscopic expansion. We have suggested the possible occurrence of the condensed state of the high density exciton complexes.

ACKNOWLEDGMENTS

We gratefully acknowledge fruitful discussions with Prof. K.Muro and Asis. Prof. K.Oto.

REFERENCES

1. D.R.Yakovlev, A.V.Platonov, E.L.Ivchenko, V.P.Kochereshko, C.Sas, W.Ossau, L.Hansen, A.Waag, G.Landwehr, and L.W.Molenkamp, *Phys. Rev. Lett.* **88**, 257401 (2002).
2. E.L.Ivchenko, and M.O.Nestoklon, *J.Exp.Theor.Phys.* **94**, 644 (2002).
3. L.V.Butov, A.C.Gossard, and D.S.Chemla, *Nature*, vol418. 751 (2002).
4. H.Yamamoto, Ziwu Ji, H.Mino, R.Akimoto, and S.Takeyama, the 16th International Conference on HighMagnetic Fields in Semiconductor Physics *submitted* (2004).
5. Ziwu Ji, H.Yamamoto, H.Mino, R.Akimoto, and S.Takeyama, Physica, E22. 632-635 (2004).

Singlet And Triplet Trion States In QW Structures

D. Andronikov[1], V. Kochereshko[1], A. Platonov[1], S.A. Crooker[2], T. Barrick[2], G. Karczewski[3]

[1]*Ioffe Physico-Technical Institute RAS, St-Petersburg, Russia*
[2]*National High Magnetic Field Laboratory, Los Alamos, USA*
[3]*Institute of Physics Polish Academy of Sciences, Warsaw, Poland*

Abstract. Photoluminescence (PL) spectra of modulation-doped CdTe/CdMgTe quantum well structures containing two dimensional electron gases different electron concentrations have been studied in magnetic fields up to 45T. Recombination line of triplet trion state was found in the spectra. A model calculation of PL spectra in magnetic fields, which takes into account singlet and triplet trion states, was carried out. It was shown that the dark triplet becomes observable in PL spectra because it becomes the only recombination channel when the formation of the singlet trion state is suppressed by magnetic fields.

INTRODUCTION

Charged exciton-electron complexes (trions) are bound states of three particles. They consist of two electrons and one hole or two holes and one electron [1]. The ground state of a trion in zero magnetic field is a singlet. The triplet state of a trion is not bound in the absence of magnetic field.

In the present work the triplet trion states were studied in CdTe-based quantum well structures. The energetic structure of the exciton and trion states in the magnetic field in these materials is different from that in previously studied ZnSe [2]. This allows us to check the idea that the anti-crossing of the singlet and triplet states affects the oscillator strength of the triplet trion and to identify the spin structure of the observed triplet trion state.

EXPERIMENT

We studied modulation doped CdTe/(Cd$_{0.7}$Mg$_{0.3}$)Te QW structures containing a 2DEG with electron density varied from $n_e<10^{10}$ up to 10^{12} cm^{-2}. These structures contained a 100 Å single quantum well grown on GaAs (100) substrates, and were δ-doped in the barriers at 100 Å distance from the QW.

Polarized photoluminescence from these samples was measured with magnetic fields up to 45 T applied in the Faraday configuration at 1.6K. The emitted light was detected in both circular polarizations.

In the PL spectra in low magnetic fields a bright PL line (T_s) is observed at ∝1.614 eV (see Fig. 1). It is attributed to the singlet state of a trion [1, 3]. With the growth of the magnetic fields a PL line of the neutral exciton appears in the spectra.

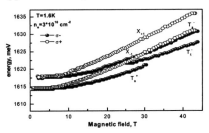

FIGURE 1. Magnetic field dependences of the energy positions of the maxima of X, T_s and T_t PL lines

In magnetic fields higher than 25T a new line (T_t) appears. We identify it as the triplet trion state.

As seen from Fig. 2 in low magnetic fields the intensity of the trion line is much larger than that of the exciton, which is due to the rapid rate of trion formation. With increasing magnetic field the intensity of the trion line T_s decreases in both circular polarizations, meanwhile the intensity of the exciton line X increases in σ^- and remains constant in σ^+ polarization. These peculiarities of the Pl behavior are analyzed in detail in [4], where it is shown that they are linked to the fast spin-dependant processes of the trion formation.

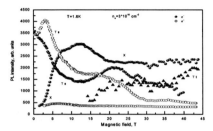

FIGURE 2. Intensity of X, T_s and T_t PL lines as a function of magnetic fields. Experiment

RESULTS AND DISCUSSION

A model calculation using the system of kinetic equations describing the PL of the exciton-trion system was performed. We used the following parameters [4, 6]: exciton and singlet trion radiative lifetime τ_{Ex}= 60ps , τ_{Ts}= 60ps, electron and hole spin relaxation time τ_e^s= 60ps , τ_h^s= 30ps and the trion formation time τ_{form}= 11ps . The lifetime of the trion triplet state was taken $\tau_{Tt}^0 = 50\tau_{Ts}^0$.

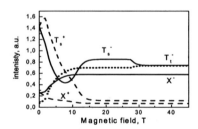

FIGURE 3. Intensity of X, T_s and T_t PL lines as a function of magnetic field. Calculation. $N_e = 3 \times 10^{10}$ cm^{-2}

As it is seen from Fig. 3 the triplet trion PL line becomes brightly observable in rather high magnetic fields and its intensity is comparable with the intensity of the exciton recombination line. This is due to the preferable population of exactly this triplet trion state in the magnetic field.

The upper Zeeman sublevel of the singlet trion becomes populated when the electron from the lower Zeeman sublevel (spin −1/2) and the exciton from the upper Zeeman sublevels (+1/2, -3/2) and (+1/2, +3/2) become bound. It is evident that in high enough magnetic field along with the low temperature the concentration of such excitons is low and thus the concentration of the trions on the upper sublevels is also low. The lower Zeeman sublevel of the singlet trion becomes populated when the electron from the upper sublevel (+1/2) and the exciton from the lower sublevels (-1/2, +3/2) and (-1/2, -3/2) become bound. In high magnetic field the concentration of such trions is also low due to low population of the upper electron sublevel [4, 5].

The triplet trion is formed when the electron and the exciton from the lower Zeeman sublevels become bound. The population of these sublevels rises with the magnetic field and thus the concentration of this trions can appear to be rather high. Thus when the magnetic field rises, the intensity of the singlet trion luminescence line T_s should fall in both circular polarizations and the intensity of the exciton recombination from the lower Zeeman sublevel should rise even when the time of trion formation is smaller compared to the lifetime of the exciton. At the same time the intensity of the triplet line T_t can rise. This growth of the intensity of triplet recombination is exactly due to the fact that the recombination channel through the singlet trion state is suppressed in the magnetic field.

This conclusion is confirmed by the temperature dependence of the PL spectra. At 4.2K the triplet trion PL line appeared only at 40T and at 15K it was not observed even at 45T while the intensity growth of the recombination line of the singlet trion was observed.

ACKNOWLEDGMENTS

The work was supported in part by grants of RFBR, the program "Nanostructures" of the Russian Ministry of Sciences and the programs of the Presidium of RAS.

REFERENCES

1. K.Khunteak, R.Cox et al. Phys.Rev.Lett. **71**, 1752 (1993)
2. J.Homberg, K.Sebald, et al. Phys. Rev. **B62**, 7413 (2000)
3. D.R.Yakovlev, V.P.Kochereshko, W.Osau, et al. J. Crystal Growth **184/185**, 818, (1998)
4. C.R.L.P.N.Jeukens et al. Phys.Rev.**B66**, 235318 (2002)
5. D.R.Yakovlev, V.P.Kochereshko, W.Ossau, et al. Proc. ICPS-24, Jerusalem 1998
6. E. Vanelle, et al. Phys. Rev **B62**, 2696, (2000)

Quantum theory of spatially resolved photoluminescence in semiconductor quantum wells

G. Pistone, S. Savasta, O. Di Stefano, and R. Girlanda

*INFM and Dipartimento di Fisica della Materia e Tecnologie Fisiche Avanzate,
Università di Messina, Salita Sperone 31, I-98166 Messina, Italy*

Abstract. We present a microscopic quantum theory of spatially resolved photoluminescence in quantum structures with interface fluctuations that includes light quantization, acustic phonon scattering, and inhomogeneous sample-excitation and/or light-detection. The obtained numerically calculated images agree with images from near-field photoluminescence experiments and put forward the potentials of the method for the understanding of near-field light emission from semiconductor quantum structures.

INTRODUCTION

Near-field optical microscopy and spectroscopy, which uses optical interaction in the visible or near-infrared range has demonstrated its ability to image optical fields and surface structures at a sub-wavelength scale. In recent years, measurements based on spatially-resolved photoluminescence (PL) provided direct information on the spatial and energy distribution of light emitting nanometric centers of semiconductor quantum structures, thus opening a rich area of physics involving spatially resolved quantum systems in a complex solid state environment [1, 2]. In particular, the ability of this kind of optical microscopy and spectroscopy to identify the individual quantum constituents of semiconductor quantum structures has been widely demonstrated. Among the most investigated quantum structures are quantum dot (QD) systems formed naturally by interface fluctuations in narrow quantum wells. Detailed simulations of Zimmermann et al. [3] have clarified many aspects of the intrigued non-equilibrium dynamics giving rise to photoluminescence spectra in these quantum structures. However, theoretical simulations of near-field imaging spectroscopy of semiconductor quantum structures focus on calculations of local absorption. In contrast, as a matter of fact, almost all experimental images are obtained from PL measurements. The theory here presented can model PL and PLE experiments performed both in illumination and collection mode or diffusion experiments where the spatial positions of excitation and collection can all independently be scanned [4]. Numerical results on disordered GaAs QWs are able to reproduce the observed transition from an inhomogeneous 2D quantum system to a system of QDs when lowering the temperature and to explain the differences between collection-mode and illumination-mode near-field images.

THEORY

The positive frequency components of the operator describing the signal that can be detected by a general near-field setup can be expressed as $\hat{S}_t^+ = \hat{A}_{bg}^+ + \hat{S}^+$, where $\hat{A}_{bg}^+$ is the elastic background signal (largely uniform along the $x-y$ plane) proportional to the input electric-field operator, and $\hat{S}^+$ is related to the sample polarization density operator $\hat{\mathbf{P}}^+(\mathbf{r})$: $\hat{S}^+ = \mathscr{A} \int d\mathbf{r}\, \hat{\mathbf{P}}^+(\mathbf{r}) \cdot \mathbf{E}_C(\mathbf{r})$, where $\mathscr{A}$ is a complex constant and $\mathbf{E}_C(\mathbf{r})$ is the signal mode delivered by the tip. Photoluminescence can be defined as the incoherent part of the emitted light intensity. The PL that can be measured by a photodetector after the collection setup (broadband detection) is proportional to $I_{PL} = I - I_{coh} = \langle \hat{S}^-(\tau)\hat{S}^+(\tau)\rangle - |\langle \hat{S}^+(\tau)\rangle|^2$. Analogously, the steady-state spectrum of incoherent light emitted by the semiconductor quantum structure and detected by the SNOM setup can be expressed as

$$\mathscr{I}(\omega_C) = \frac{1}{\pi}\int_0^\infty d\tau\, \langle \hat{S}^-(0)\hat{S}^+(\tau)\rangle e^{i\omega_C \tau}. \quad (1)$$

The interband polarization density operator can be expressed in terms of exciton operators as $\hat{\mathbf{P}}^+(\mathbf{r}) = \mu_{eh}f(z)\Psi_\alpha^{eh}(\rho=0,\mathbf{R})\hat{B}_\alpha$. The operator $\hat{B}_\alpha^\dagger$ creates an exciton state $\hat{B}_\alpha^\dagger |0\rangle \equiv |E_{1,\alpha}\rangle$ with energy $\omega_{1,\alpha}$. $\Psi_\alpha^{eh}(\rho,\mathbf{R})$ is the exciton wavefunction with ρ indicating the relative in-plane electron-hole eh coordinate and $\mathbf{R}$ describes the centre of mass motion, while $f(z) = u_e(z)u_h(z)$ is the product of the electron and hole envelope functions along the confinement direction. If the disorder induced broadening is small compared to the exciton binding energy, only the lowest bound state 1s at the fundamental sublevel transition has to be considered and the exciton wave function can be factorized as $\Psi_\alpha^{eh}(\rho,\mathbf{R}) = \phi_{1s}(\rho)\psi_\alpha(\mathbf{R})$. The kinetic equation for diagonal terms of the exciton density matrix $N_\alpha = \langle \hat{B}_\alpha^\dagger \hat{B}_\alpha \rangle$, can be derived starting from the Heisenberg equation of motion under the influence of an Hamiltonian describing the interaction with acoustic phonons and with a (possibly inhomogeneous) light field. Assuming a low density regime, one obtains

$$\partial_t N_\alpha = G_\alpha(\omega_I) + \sum_\beta \gamma_{\alpha \leftarrow \beta} N_\beta - 2\Gamma_\alpha N_\alpha, \quad (2)$$

where $\gamma_{\beta \leftarrow \alpha}$ are the resulting phonon-assisted scattering rates [3, 4]. $2\Gamma_\alpha = r_\alpha + \sum_\beta \gamma_{\beta \leftarrow \alpha}$ is the total out-scattering rate, being r_α the rate for spontaneous emission proportional to the exciton oscillator strength: $r_\alpha = r_0 \left| \int d^2\mathbf{R}\, \psi_\alpha(\mathbf{R}) \right|^2$. In this equation the generation term depends on the spatial overlap between the illuminating beam and the exciton wavefunctions corresponding to exciton levels resonant with the input light, and can thus be function of the tip position and shape: $G_\alpha(\mathbf{R}_I) = r_0 \left| o_\alpha^I(\mathbf{R}_I) \right|^2 \mathcal{L}_\alpha(\omega_I)$ with $\pi\mathcal{L}_\alpha(\omega) = \Gamma/[(\omega - \omega_\alpha)^2 + \Gamma^2]$ and $o_\alpha^I(\mathbf{R}_I) = \int d^2\mathbf{R}\, \tilde{E}_I(\mathbf{R})\psi_\alpha(\mathbf{R})$ where $\tilde{E}_I(\mathbf{R}) = \int E_I(\mathbf{r}) f(z) dz$. In numerical calculations concerning the illumination-mode we will assume an input light field with a given Gaussian profile centered around the tip position $\mathbf{R}_I$: $\tilde{E}_I(\mathbf{R}) = E_I^0 g(\mathbf{R} - \mathbf{R}_I)$. In this case the generation term becomes function of the beam position and shape (spatial resolution). Once the exciton densities have been derived, the PL spectrum can be readily obtained. According to the quantum regression theorem, $\langle \hat{S}^-(0)\hat{S}^+(\tau) \rangle$ has the same dynamics of $\langle \hat{S}^+(\tau) \rangle$ (proportional to the exciton operator), but with $\langle \hat{S}^-(0)\hat{S}^+(0) \rangle$ as initial condition. Following this procedure we obtain

$$\mathscr{I}(\mathbf{R}_C,\omega_C) = r_0 \sum_\alpha \left| o_\alpha^C(\mathbf{R}_C) \right|^2 \mathcal{L}_\alpha(\omega_C) N_\alpha \quad (3)$$

NUMERICAL RESULTS

We consider a system of QDs arising from interface fluctuations of GaAs quantum wells. The effective disordered potential $V(\mathbf{r})$, used in our simulations, is obtained summing up two different contributions. They are both modelled as a zero mean, Gauss distributed and spatially correlated process with prescribed amplitude v_0 and correlation length ξ. For the two contributions we have chosen $\xi = 16$ nm, $v_0 = 1.5$ meV; and $\xi = 8$ nm and $v_0 = 0.5$ meV respectively. Fig. 1(a) displays the specific realization of the effective disordered potential used for all the calculations. Fig. 1(b-d) shows energy-integrated PL images obtained after uniform illumination of the sample

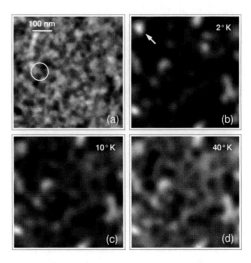

FIGURE 1. (a): Disorder potential; (b-d): energy-integrated PL images obtained after uniform illumination of the sample at energy $\omega_I = 1$ meV and collecting locally the emitted light.

at energy $\omega_I = 1$ meV (the zero of energy is fixed at the energy of the 1s exciton in absence of disorder) and collecting locally the emitted light (C mode with spatial resolution $FWHM = 47$ nm). It is worth noting that the energy integrated excitonic local density of states does not depend on position. So the observed structures are a direct consequence of the increasing ratio between radiative and nonradiative scattering rates for exciton states localized at the potential minima (compare e.g. images a and c). As the temperature of the structure is lowered, a transition from a broad and fairly continuous PL to an intense set of few spatially localized luminescence centers can be observed. Another interesting feature is the non-monotonous brightness of some luminescence centers when the temperature is increased (see e.g. the location indicated by an arrow in image b), due to excitons which can overcome shallow local minima by thermal activation and fall down in still deeper states.

REFERENCES

1. H. F. Hess, E. Betzig, T. D. Harris, L. N. Pfeiffer, and K. W. West, Science **264**, 1740 (1994).
2. D. Gammon, E. S. Snow, B. V. Shanabrook, D. S. Katzer, and D. Park, Phys. Rev. Lett. **76**, 3005 (1996); Science **273**, 87 (1996).
3. R. Zimmermann and E. Runge, Phys. Status Solidi (a) **164**, 511 (1997).
4. G. Pistone, S. Savasta, O. Di Stefano, and R. Girlanda, Appl. Phys. Lett. **84**, 2971 (2004).

Determination Of InAsP/InP And InGaAs/InP Band Offsets Using Blue Shifting Type II Asymmetric Multiple Quantum Wells

A. C. H. Lim [a], R. Gupta [a], S. K. Haywood [a], P. N. Stavrinou [b], M. Hopkinson [c], G. Hill [c]

[a]*Dept. of Engineering, University of Hull, Hull, HU6 7RX, UK*
[b]*The Blackett Laboratory, Imperial College, London, SW7 2BZ, UK*
[c]*Dept. of Electronic and Electronic Engineering, University of Sheffield, Sheffield, S1 3JD, UK*

Abstract. Photocurrent measurements on two molecular beam epitaxy-grown $InAs_xP_{1-x}$/ $In_{0.53}Ga_{0.47}As$ asymmetric quantum wells with InP barriers, were carried out at various temperatures. A significant electric field-induced blue shift of this transition was observed at all temperatures. Higher energy transitions were also observed at 77K. These higher energy transition were used to determine the band offsets in these structures. The conduction band offset (ΔE_c) for InAsP/InP was found to be ~65% of the bandgap difference (ΔE_g) and for InGaAs/InP heterostructures was found to be ~55% of ΔE_g.

INTRODUCTION

Self-electro-optic effect devices based on Stark shifts in multiple quantum wells (MQWs) are used for switching applications. Normally MQWs exhibit a red shift with applied electric field. However, MQWs exhibiting a field-induced blue shift could offer better performance in terms of on/off ratio, operating voltage and insertion loss. Various routes to achieve this blue shift have been proposed but a large enough shift for practical applications has not yet been realised.

In this paper, we present low temperature results on InP/InAsP/InGaAs asymmetric MQW structures with a type II band alignment in which the stepped well causes separation of the electron and hole wave functions at zero bias. A blue shift is realised when an electric field is applied in a direction reducing this separation. [1]

SAMPLES DESIGN

A three-band Kane **k.p** model within the envelope function approximation and a tunnel resonance method were used to model the structure. The model successfully predicted band structure and transition energies for a variety of strained InGaAs structures. [2]

Two structures, namely 1866 and 1867 with different layer widths and arsenic compositions were designed in order to confine the e1 level in the InAsP layer. These structures were grown on a (100) InP substrate using molecular beam epitaxy. The layer width and composition of both samples are shown in Table 1. Both samples consist of 10 InAsP/InGaAs/InP well/barrier periods embedded in the i-region of a p-i-n diode. The samples were fabricated into mesa diodes with circular optical access of 200 or 400 μm radius to enable photocurrent measurements.

TABLE 1. Composition and layer thickness.

Samples	$InAs_xP_{1-x}$		$In_yGa_{1-y}As$		InP
	x	L_1(Å)	y	L_2(Å)	L_b
1866	0.6	36	0.53	54	300
1867	0.4	55	0.53	35	300

RESULTS AND DISCUSSION

At 300K a field-induced blue shift of the lowest exciton peak was observed for both samples. A net blue shift of 5.3meV at 6V applied field was found in 1867 and 2meV at 2V applied field in 1866. Further increase in field resulted a red shift. Both samples are believed to have good sample quality since a full width at half maximum of 30 meV and 35 meV respectively was obtained from the photoluminescence measurement at room temperature. The lowest transition energy, e1-hh1 (electron 1 – heavy hole 1) of both measurements was found to be 1.545 μm (0.8 eV) for 1866 and 1.465 μm (0.85 eV) for 1867.

Experimental and modelled results agreed to within 50kV/cm which can be attributed to the internal field arising from the unintended doping in the samples, a percentage of layer intermixing in the well and the band offset ratio used. [1]

A further study of these results was carried out by varying the temperature (77K- 300K) of the photocurrent measurement. The photocurrent results at different temperature are shown in Figures 1 and 2.

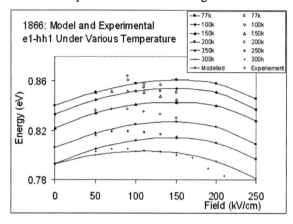

FIGURE 1. Experimental and modelled transition energies for 1866 at different temperature.

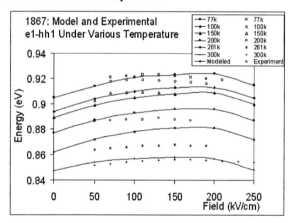

FIGURE 2. Experimental and modelled transition energies for 1867 at different temperature.

The photocurrent spectra at 77k and 300K at 0V are shown in figure 3. Although a variation in temperature shifts the e1-hh1 transition, the net field-induced blue shift remains the same. High energy transitions, e2-hh2 (electron 2 – heavy hole 2) and e2-lh2 (electron 2 – light hole 2) at 77K for sample 1867 confirmed the band offsets used in the model for these structures.

The results could only be modelled by a conduction band offsets ratio ~0.65 ΔE_g for InAsP/InP and 0.55 ΔE_g for InGaAs/InP. The band offsets ratio for both layers suggested in literature would lead to a large discrepancy between the experimental and modelled results.

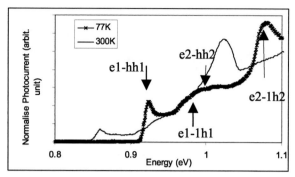

FIGURE 3. Photocurrent results at 77K and 300K for sample 1867 (The assignment for various transition are only shown for 77K).

The observed transition energies and modelled values at 77K with various band offsets values are shown in Table 2.

TABLE 2. Experimental and modelled high energy transitions for 1867 at 77K.

Transition Energy	Experimental Result (eV)	Modelled Results (eV)
e1hh1	0.922	0.906
e1lh1	0.98	0.965
e2hh2	1.01	0.996
e2lh2	1.07	1.05

In conclusion a net field induced blue shift of >5meV for the lowest exciton peak was demonstrated at various temperatures (77K- 300K) in type II band line up structures. The results also suggest that the conduction band offsets ratio in the InAsP/InP heterostructures to be 0.65 of ΔE_g and InGaAs/InP heterostructures to be 0.55 of ΔE_g.

ACKNOWLEDGEMENT

The authors would like to acknowledge funding of a research studentship (A.C.H.Lim) by Keithley Instruments in support of this work.

REFERENCE

1. S.K.Haywood, A.C.H.Lim, R.Gupta, S.Emery, J.H. C. Hogg, V.Hewer, P.N.Stavrinou, M.Hopkinson and G.Hill, J.Appl. Phys. **94**, 322 (2003).
2. R. Mottahedeh, D. Prescott, S. K. Haywood, D. A. Pattison, P. N. Kean, I. Bennion, M. Hopkinson, M. Pate and L. Hart, J.Appl. Phys. **83**, 306 (1998).

Van Hove singularities detected by photoluminescence in doped AlGaAs/GaAs superlattices

A.B.Henriques, R.F.Oliveira, T.E.Lamas, and A.A.Quivy

Instituto de Física, Universidade de São Paulo, Caixa Postal 66318, 05315-970 São Paulo, Brazil

Abstract. The photoluminescence (PL) of a doped 5/5nm GaAs/Al$_{0.21}$Ga$_{0.79}$As superlattice was studied. A doublet of frequencies, which can be associated with Van Hove singularities in the density of states of the superlattice, and allows us to deduce characteristic parameters of the superlattice, characterizes these oscillations.

Direct measurement of miniband parameters in undoped semiconductor superlattices can be done using intraband optical techniques in the far infrared [1]. Interband optical methods, however, could not be used successfully, because saddle point excitons at the van-Hove singularities M_0 and M_1 have different binding energies [2]. In heavily doped superlattices, exciton formation is suppressed, due to Coulomb screening and phase-space filling. With sufficient doping, the Fermi level will lay above the energy at top of the electronic miniband, and luminescence from miniband states below the Fermi level should be observed. In this work we take advantage of doping in order to observe free carrier recombination that is not shifted due to the exciton binding energy.

The superlattice sample was grown by molecular beam epitaxy (MBE) and consisted of 20 GaAs quantum wells of 5nm thickness, separated by 19 Al$_{0.21}$Ga$_{0.79}$As inner barriers of thickness 5 nm. The internal AlGaAs barriers were delta-doped with Si at their center, with an areal density of 1.5 x10^{12} cm^{-2}. The outer AlGaAs layers were also delta-doped, but with half of the same areal concentration, at a distance of 2.5 nm from the adjacent GaAs layer. Doping of the outer layers is necessary in order to avoid the formation of Tamm states, which otherwise will dominate completely the PL of heavily doped superlattices [3]. Shubnikov-de Haas (SdH) measurements of in-plane conductivity in tilted fields showed that the sample had electrons of in-plane mass of m_e=0.068 m_0, occupying a miniband with an energy width of Δ=19.6 meV, a Fermi level Φ=59.4 meV above the miniband threshold, and a level broadening at the Fermi level of γ=8 meV. PL measurements were done in the Faraday geometry, at T=2K, using optical fibers and *in situ* miniature focusing optics.

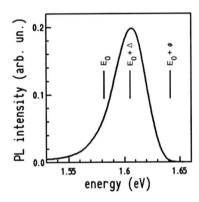

FIGURE 1. Superlattice PL at B=0T and T=2K.

The broad superlattice PL (Fig. 1) showed weak intensity oscillations as a function of the magnetic field applied perpendicular to the epitaxial layers (see Fig. 2).

We assume that the PL oscillations are proportional to the combined density of states, and use the tight-binding approximation to describe the miniband dispersion. If we ignore the very small energy width of the hole miniband (less than 0.5 meV), then the PL intensity at the photon energy hν and field B will be given by [4]

$$I(h\nu,B) \sim -e^{-\frac{2\pi\mu\gamma}{\hbar eB}} J_0\left(\frac{\pi\mu\Delta}{\hbar eB}\right)\cos\left(2\pi\mu\frac{h\nu-E_0-\frac{\Delta}{2}}{\hbar eB}\right) \quad (1)$$

where μ is the reduced mass of the electron-hole pair.

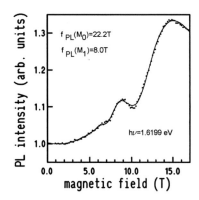

FIGURE 2. Oscillatory PL at $h\nu$=1.6199 eV.

Figure 2 shows the oscillations in the PL intensity as a function of magnetic field for a photon energy of $h\nu$=1.6199 eV, and a least square fit to equation (1), including a monotonous parabolic background. To avoid local minima, the minimization was done in two stages. In the first stage, parameters Δ, γ and μ where fixed at the Shubnikov-de Haas values. In the second stage, these parameters, and the remaining parameter E_0, were set free. The fitting procedure was repeated for all photon energies measured. From the fit, the frequencies of PL oscillation associated with the M_0 and M_1 Van Hove singularities could be deduced:

$$f_{PL}(M_0) = \mu\frac{h\nu-E_0}{\hbar e}, \quad f_{PL}(M_1) = \mu\frac{h\nu-E_0-\Delta}{\hbar e} \quad (2)$$

The straight lines depicted in figure 3 were obtained from a simultaneous linear fit of the two sets of data points with eq. (2). The fit yields the intersection of the lines with the energy axis, which according to eq.(2) will occur at the energies $h\nu=E_0$=1.581eV and $h\nu = E_0+\Delta$=1.604eV, giving Δ=23 meV. The latter parameter is in reasonable agreement with the electronic miniband width estimated from the Shubnikov-de Haas measurements. The slope of the line determines the reduced effective mass of the electron-hole pair, i.e. μ=0.0612 m_0.

The estimated superlattice band gap, E_0=1.581 eV, can be compared to the theoretical value of the band gap, obtained from the **kp** equation for this structure, E_0^{th}=1.605 eV, which does not take into account the lowering of the band gap due to many body effects. By equating $E_0= E_0^{th}$ +BGR, we obtain BGR=-24 meV. This is greater than the BGR estimated for a three-dimensional electron system (-21.9 meV) [5], but less than for electrons confined in a 5nm quantum well (-28 meV) [6].

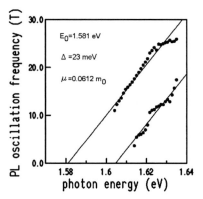

FIGURE 3. PL oscillation frequencies, $f_{PL}(M_0)$ and $f_{PL}(M_1)$, as a function of the photon energy $h\nu$.

In conclusion, superlattice structures of GaAs/AlGaAs composition were produced, with a modulation doping profile that prevents the formation of Tamm states. PL was studied in high magnetic fields, and a broad luminescence band was observed above the GaAs band gap. For fixed photon energy, PL oscillates as a function of the magnetic field. From the PL oscillations we estimated the superlattice parameters - the electronic miniband width Δ, the electron-hole pair reduced mass, μ, and the band gap renormalization.

CNPq and FAPESP – Brazilian agencies, supported this work.

REFERENCES

1. H.T.Grahn, *Semiconductor Superlattices - Growth and Eletronic Properties*, (World Scientific, Singapore, 1995).
2. R. H. Yan *et al*, *Appl. Phys. Lett.* **54**, 1549 (1989).
3. A. B. Henriques, *Appl. Phys. Lett.* **78**, 691 (2001).
4. A. B. Henriques, *Phys. Rev. B* **50**, 8658 (1994).
5. S. Schmitt-Rink *et al*, *Sol. St. Comm.* **52**, 123 (1984).
6. T. L. Reinecke and S. C. Ying, *Phys. Rev. Lett.* **43**, 1054 (1979).

Superradiance of excitons in multi-quantum-well structures based on InAs-GaSb coupled quantum wells

Boris Laikhtman, Leonid D. Shvartsman

The Racah Institute of Physics, The Hebrew University, Jerusalem 91904, Israel

Abstract. We suggest the system of excitons in multiple InAs-GaSb coupled quantum well as a new source of superradiance. Coupled InAs-GaSb quantum well can be tuned to a very large wavelength that makes possible to fabricate a sample with a large number of quantum wells and in this way to increase the superradiant effect

Superradiance considered first by Dicke in 1954 [1] does not loose its attractiveness not only because it is a remarkable macroscopic coherent phenomena but also as a way to obtain ultra short (for solids, especially for QW excitons) radiative pulses from an ensemble of identical two-level systems (TLS) [2]. A specific interest in this respect present excitons in quantum wells due to a rather high flexibility of this system. First of all, the number of excitons is not fixed and is controlled by the intensity of pumping. Then, the wavelength of the exciton radiation can be tuned not only by the choice of the material but also by the change of the quantum well width.

Since the first papers about the superradiance [1,2] this term acquired also another meaning. Lee et al. used this term to describe a significant enhancement of the exciton radiative recombination rate. [3] Thus, the term "superradiance" applying to ensemble of excitons can be used in both meanings.

A natural way to make the Dicke exciton superradiance stronger is to increase the number of radiators, i.e., excitons. This immediately raises a question about effect of statistics on the exciton superradiance. The superradiance in the sense of Dicke is based on a remarkable property of the radiation matrix elements. An ensemble of TLS can be in such states that the radiation matrix element between them is proportional to the number of TLS. This property is inseparably connected to the excitation statistics. In an ensemble of excitons that with a very good accuracy can be considered as Bose particles the situation is different from TLSs. The recombination matrix element is proportional to the matrix element of the Bose annihilation operator that is equal to the square root of the number of excitons. That is, in general, the ensemble of excitons in the same quantum well is not superradiant in the Dicke sense. This is similar to the absence of such superradiance in an ensemble of classical oscillators. [4] But similar to the classical oscillators, the increase the number of radiators enhances the radiation recombination rate. That is the system of exciton in a quantum well reveals the Lee superradiance not only due to a large coherent volume of each exciton [3] but also due to a large number of excitons that can be pumped in the same state.

There are also ways to construct ensemble of excitons that reveals the Dicke superradiance. The first one is the situation when excitons are localized in fluctuations of the quantum well width and their wave functions don't overlap. (compare Ref. [5]) Another, more controllable way is to generate an ensemble of excitons in multi-quantum well samples. Both ways, however, have disadvantages. Interaction between a few excitons pumped in the same destroys their coherence with excitons in other fluctuations. Multi-quantum well sample can be longer than the wavelength that affects the shape of superradiance pulse and makes it weaker. [4,6] The problems can be avoided for excitons in InAs/GaSb coupled quantum wells. Due to a unique band structure of this material, by the choice of appropriate well widths it is possible to tune the exciton optical energy in the interval between zero and more than a few hundred of meV (the wavelength of about 3 μm and larger). For such a radiation the number of radiators concentrated in a sample shorter than the wavelength is limited only by the growth limitations, and the superradiance can show up is a much larger scale than in a short wavelength radiation. Excitons in InAs-GaSb coupled QW have dipole moments. The dipole – dipole interaction is rather strong and prevents the trapping of two or more excitons in the same well width fluctuation.

An advantage is that in a multi-quantum well sample a large area excitons form a nearly periodic system. In such a system the dephasing coming from the dipole - dipole interaction [6] can modify the superradiant pulse but does not destroy it. [7]

In order the exciton – exciton interaction in the same quantum well does not destroy the superradiance, the exciton concentration has to be small. In InAs/GaSb coupled quantum wells each exciton has a dipole moment p$\approx$eL where L is the distance between the centers of the wells. The dipole moments of all excitons are parallel and directed along the growth direction so that the exciton - exciton interaction energy $\sim p^2 n^{3/2}/\kappa$, where n is the two-dimensional exciton concentration and $\kappa \approx 15$ is the dielectric constant. This interaction does not play any role if it is smaller than the energy uncertainty coming from the exciton recombination. Theoretical estimates give for the recombination time the values 5×10^{-11}-5×10^{-10} s, depending on the well widths. [8] The interaction energy between exciton in different coupled wells is even smaller. That is the dipole - dipole interaction can be neglected if the exciton concentration is of the order of or smaller than 10^9 cm^{-2}.

This estimate leads to the following qualitative dependence of the superrandiant pulse on the pumping intensity. The exciton concentration grows with the pumping intensity. But when it reaches 10^9 cm^{-2} the interaction destroys the coherence. That is the width of the pulse first decreases with the pumping intensity and then saturates. The saturation level of the pumping and the pulse width grow with the number of quantum wells. In experiment these qualitative dependences can be considered as a signature of the exciton superradiance.

There is a rather strong limitation on the temperature of the exciton superradiance. It is natural to expect that only excitons in the ground state can recombine coherently. At the concentration of 10^9 cm^{-2} the probability to find excitons in the ground state is of the order of unity at temperatures about 0.5 K and below.

In quantum wells of 70 Å/70 Å width the exciton recombination time $\tau_0 \approx 10^{-10}$ s and the exciton optical energy is around 57 meV. [8] This corresponds to the wave length 21 μm. The period of a multi-well structure with 200 Å barrier of AlSb between the wells is 340 Å. So at one half of the wavelength it is possible to grow around N=300 periods. The superradiance of such a sample can produce pulses with duration of $\tau=\tau_0/N \approx 3\times10^{-13}$ s. The pulse can be done more narrow if more than one exciton in each period participates in the superradiance.

In conclusion, the system of excitons in multiple InAs-GaSb coupled quantum wells is very promising for generation of short superradiance pulses. The system is flexible that allows a researcher clearly distinguish superradiance from other accompanying phenomena.

We acknowledge the support of the Israel Science Foundation founded by the Israel Academy of Sciences and Humanities

REFERENCES

1. R. H. Dicke, Phys. Rev. **93**, 99-110 (1954).

2. J. H. Eberly and N. E. Rehler, *Phys. Lett.* A **29**, 142-143 (1969); N. E. Rehler and J. H. Eberly, *Phys. Rev.* A **3**, 1735-1751 (1971); R. Bonifacio, P. Schwendimann, and F. Haake, *Phys. Rev.* A **4**, 302-313 (1971).

3. Y. C. Lee and K. C. Liu, J. Phys. C: Solid State Phys., **14**, L281-L285 (1981); E. Hanamura, Phys. Rev. B **38**, 1228-1234 (1988); M. Hirasawa, T. Ogawa, and T. Ishihara, Phys. Rev. B **67**, 075310 (2003).

4. A. V. Andreev, V. I. Emel'yanov, and Yu. A. Il'inskii, *Cooperative Effects in Optics*, IOP Publishing, London, 1993.

5. G. Björk, S. Pau, J. Jacobson, Y. Yamamoto, *Phys. Rev.* B **50**, 17336-17348 (1994).

6. R. Friedberg, S. R. Hartmann, and J. T. Manassah, *Phys. Lett.* A **40**, 365-366 (1972); R. Friedberg and S. R. Hartmann, *Phys. Rev.* A **10**, 1728-1739 (1974).

7. T. Tokihiro, Y. Manabe, and E. Hanamura, Phys. Rev. B **47**, 2019-2030 (1993); **48**, 2773-2776 (1993); **51**, 7655-7668 (1995).

8. S. de-Leon and B. Laikhtman, Phys. Rev. B **61**, 2874-2887 (2000).

Excitons and Trions in Heavily Doped QWs at High Magnetic Fields

V. Kochereshko[1], D. Andronikov[1], G. Karczewski[2], S.A. Crooker[3],

[1]Ioffe Physico-Technical Institute RAS, St-Petersburg, Russia
[2]Institute of Physics Polish Academy of Sciences, Warsaw, Poland
[3]National High Magnetic Field Laboratory, Los Alamos, USA

Abstract. Modification of photoluminescence spectra taken from modulation doped CdTe/CdMgTe QW structures have been studied in magnetic fields up to 45T at high concentrations of 2D electrons. The following peculiarities were found in relatively low magnetic fields when filling factors ν was higher than 2: (*i*) linear energy shift of the photoluminescence (PL) peak with increasing magnetic fields, (*ii*) periodic variation of PL intensities for σ⁻ and σ⁺ circular polarizations, (*iii*) smooth jumps of the PL peak position at integer filing factors. In high magnetic fields the behavior of the PL specter of the heavily doped structure coincided with the behavior of the lightly doped one.

INTRODUCTION

After the first experimental detection of the trion in modulation doped CdTe based quantum wells (QW) containing 2DEG [1] the trion states had been studied in various A_2B_6 or A_3B_5 heterostructures. At present time the properties of the trion states in QW optical spectra with low electron concentrations are studied rather well. However the optical spectra of the structures with high densities of the 2D electron gas (2DEG) are not well studied and the phenomena observed in these conditions are subjects to speculation. The main peculiarity of the PL spectra behavior with high concentration of 2DEG in magnetic field is that the maximum of the PL line shifts linearly with the magnetic field towards high energies when the value of the filling factor is more than 2. When value of the filling factor is less than 2 the maximum of the PL line experiences a usual diamagnetic shift. In order to explain the observed behavior different conceptions were used in literature: renormalization of the band gap, Fermi edge singularity, uncompressible electron liquid, [2] and other. In the present work we suggest a different explanation of this behavior.

EXPERIMENT AND DISCUSSIONS

We studied modulation doped $CdTe/(Cd_{0.7}Mg_{0.3})Te$ QW structures containing a 2DEG with electron density varied from $n_e<10^{10}$ up to 10^{12} cm⁻². These structures contained a 100 Å single quantum well (SQW) grown on GaAs (100) substrates, and were δ-doped in the barriers at 100 Å distance from the QW.

Polarized photoluminescence from these samples was measured with magnetic fields up to 45 T applied in the Faraday configuration at 1.6K. The emitted light was detected in both circular polarizations.

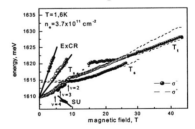

FIGURE 1. Magnetic field dependences of the energy positions of the T_s and T_t PL lines. *SU* –Shake-up lines. Arrows indicate integer filling factors.

Figure 1 presents the dependences of the energies of all the observed PL lines in a structure with electron concentration 3.7x10^{11} cm^{-2} on the magnetic fields. For comparison the dashed lines show the dependences of the PL lines of the lightly doped structure with electron concentration n_e=3x10^{10} cm^{-2}.

It is evident that the position of the line in heavily doped structure coincides with the position of the line in lightly doped ones when the value of the filling factor is less than 2. When the value of the filling factor is more than 2 a significant difference of the positions of the PL lines in these structures is observed. In heavily doped structure in zero magnetic fields the maximum of the main PL band shifts towards lower energies compared to the lightly doped structure. This shift is of the order of magnitude of the 2DEG Fermi energy. As the magnetic field grows this maximum moves towards higher energies.

Besides in low magnetic fields the lines ExCR are observed in the PL. These lines also shift linearly towards high energies. The slope of this dependence is divisible by the cyclotron energy of the electron $N\hbar\omega_c^e$. Such transitions were previously observed in PLE and reflectivity spectra and had been identified as Combined Exciton Cyclotron Resonance [3]. This transitions reveal on combined frequencies equal to the sum of the exciton transition frequency and electron cyclotron frequency and are connected with the following process: an incident photon creates an exciton and simultaneously excites a transition of an additional electron between Landau levels.

The main feature of the PL spectra of the heavily doped structures is linear on the magnetic field shift of the maximum of the PL line in low magnetic field equal to $(1/2)\hbar\omega_c^e$.

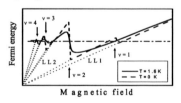

FIGURE 2. Fermi energy as a function of magnetic fields.

The presence of such linear shift in the magnetic field definitely indicates that besides the exciton a free electron should take part in the PL process. Indeed a free electron remains in the final state after the recombination of the trion. This electron can be found only in vacant positions above the Fermi level. Thus, $E_{tr} = \hbar\omega + E_F \Rightarrow \hbar\omega = E_{tr} - E_F$, the trion PL line should be shifted towards lower energies. This shift is of the order of magnitude of the 2DEG Fermi energy.

As the magnetic field grows this shift decreases as $(1/2)\hbar\omega_c^e$. (see Fig.2) At the same time small jumps of the PL line position owing to the non-monotonous dependence of the Fermi level on the magnetic field should be observed. These jumps can be smoothed by non-zero electron temperature and the effects of localization and disorder.

Periodical variations of the intensity of the trion PL are linked to the periodical population of the upper Zeeman components on each Landau level when the value of the filling factor is more than 2. This periodical population of the electron Zeeman components leads to periodical changes in efficiency of the trion formation and hence to periodical changes of the PL intensity [4].

CONCLUSION

Modification of photoluminescence spectra taken from modulation doped CdTe/CdMgTe QW structures have been studied in magnetic fields up to 45T at high concentration of 2D electrons. The following peculiarities were found in relatively low magnetic fields when filling factors ν was higher than 2: *(i)* Low energy shift of the trion PL band at zero magnetic fields, *(ii)* Linear energy shift of the photoluminescence (PL) peak with increasing magnetic field increases equal to $(1/2)\hbar\omega_c^e$, *(iii)* Periodic variation of PL intensities for σ$^-$ and σ$^+$ circular polarizations, *(iv)* Smooth jumps of the PL peak position at integer filing factors. All this peculiarities are explained by a model that takes into account the initial and final state of the trion recombination.

ACKNOWLEDGMENTS

The work was supported in part by grants of RFBR, the program "Nanostructures" of the Russian Ministry of Sciences and the programs of the Presidium of RAS.

REFERENCES

1. K.Khunteak, R.T.Cox, Y.Merle d'Aubigne, et al. Phys.Rev.Lett. **71**, 1752 (1993)
2. D.Gekhtman, E.Cohen, Arza Ron et al., Phys. Rev. **B54**, 10320 (1996); F.J.Teran, M.L.Sadowski, M.Potemski et al. Physica **B256-258**, 577, (1998); .E.I.Rashba, M.Sturge Phys. Rev. **B63**, 045305, (2001)
3. D.R.Yakovlev, V.P.Kochereshko, R.A.Suris et al. Phys.Rev.Lett. **79**, 3974, (1997)
4. D.R.Yakovlev, V.P.Kochereshko, W.Ossau et al. Proc. ICPS 24 Jerusalem 1998

Investigation of Carrier Recombination Processes and Transport Properties in GaInAsN/GaAs Quantum Wells

R. Fehse[1], S.J. Sweeney[2], A.R. Adams[2], E.P. O'Reilly[1],
D. McConville[2], H. Riechert[3] and L. Geelhaar[3]

[1]NMRC, University College, Lee Maltings, Prospect Row, Cork, Ireland
[2]Advanced Technology Institute, University of Surrey, Guildford, Surrey, GU2 7XH, UK
[3]Infineon Technologies AG, Corporate Research, 81730 Munich, Germany

Abstract. It is shown that the dramatic changes in threshold current density with changing active region growth temperature in 1.3μm GaInNAs-based lasers can be attributed almost entirely to changes in the defect related monomolecular recombination current in the optically active material. In addition, growth temperature dependent changes in the QW morphology are shown to have a significant influence on the transport properties of the structure.

INTRODUCTION

The dilute III-V alloy GaInAsN is of great interest for its use in optoelectronics. It can be grown lattice-matched on GaAs substrates and due to its unusually strong bowing the band gap can cover the range $0.8\ eV < E_g < 1.4\ eV$, even with N content less than 4%. This makes GaInNAs-based devices ideal for application in the 1.3μm-1.55μm communication window. However, the detailed band structure of GaInAsN remains relatively unexplored and is the subject of some controversy.

To understand and improve the performance of opto-electronic devices, knowledge of the magnitudes of the various recombination processes is a key factor. In this paper we show that an operating semiconductor laser provides an ideal tool to experimentally determine the required information right up to the carrier density, n_{th}, required for lasing operation.

EXPERIMENTAL

The laser structures used in this study are GaInNAs-based SQW devices grown by molecular beam epitaxy (MBE) on n$^+$- GaAs substrates. The GaInNAs QWs have an In content of about 30%, an N content of about 1.6% and a width of 6.5 nm. Two nominally identical series were grown differing only in the growth temperatures of the active region, T_g, which were set at 422°C and 456°C respectively. The main experimental analysis utilizes a method, successfully applied before [1], which enables the determination of the monomolecular-, radiative- and Auger-related current contributions to the threshold current by studying the integrated spontaneous emission rate from a window milled into the n-contact of the devices.

RESULTS AND DISCUSSION

The structural homogeneity of the QWs in the two laser structures is estimated on the basis of transmission electron microscopy (TEM) images that were acquired from simpler reference structures. These reference structures were grown at the same temperatures, but the respective QWs contained somewhat more In (37%). The measurements suggest for the low T_g laser structure a very even and homogeneous QW, and for the high T_g laser a clear non-uniformity in composition. This has a measurable effect on the transport properties of the devices. In an ideal QW laser the carrier density and hence the integrated spontaneous emission rate becomes pinned at its threshold value, n_{th}, for pump currents larger than the threshold current. However, in strongly inhomogeneous devices, carriers can localise in potential minima above the lasing energy and cannot take part directly in the lasing process. The transfer rate between the minima is slow compared to the depletion rate by stimulated emission and a quasi Fermi-Dirac electron distribution is not maintained. This leads to an increasing carrier density above threshold and, hence, an increasing spontaneous emission rate. With rising temperature, thermal excitation out of the potential minima provides faster transport and improved pinning of the carrier density. These effects are observed in the low T_g sample, with the more homogeneous QW morphology, where we observe a poor pinning for temperatures lower than 200K. For T>200K the pinning improves and for T>250K the carrier density pins perfectly (Fig 1a).

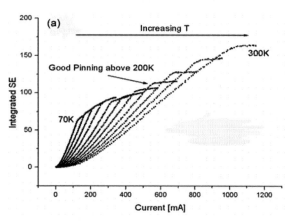

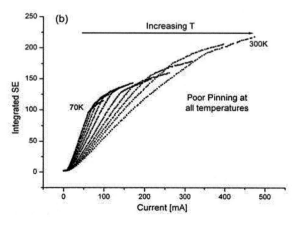

FIGURE 1. Integrated spontaneous emission rate versus current for increasing temperatures (70K – 300K) for QW growth temperatures $T_g = 422°C$ (a) and $T_g = 456°C$ (b). A good pinning for T>200K can be observed for the low T_g samples.

In contrast to this in the high T_g samples (Fig 1b), where significant compositional fluctuations occur, the carrier density does not pin over the whole temperature range measured (70K – 300K), indicating the strong influence of the QW morphology on the carrier transport properties.

Comparing the threshold current densities, J_{th}, of the two device types around room temperature (T=290K) we observe that $J_{th} \sim 970$ A/cm^2 for the low T_g lasers, while $J_{th} \sim 330$ A/cm^2 for the higher T_g devices. This is a surprising result, as the structure with the more pronounced structural inhomogeneity exhibits the lower threshold current. To explain the origin of the threefold difference in J_{th} value, the current densities at threshold as a function of temperature for the monomolecular current, J_{mono}, radiative current, J_{rad}, and Auger current, J_{Aug}, were determined experimentally for the two laser structures. For the low T_g devices we find that the monomolecular current contributes ~61% to the total threshold current at 290K and ~48% for the high T_g devices. The total monomolecular current density is reduced by a factor of 3.8 from $J_{mono} \sim 600$ A/cm^2 down to 160 A/cm^2 when increasing T_g from 422 °C to 456°C. However, the reduction observed for the Auger ($\propto n^3$) and radiative ($\propto n^2$) component is much less significant and can be attributed to a slight variation in the threshold carrier density, n_{th}, between the two structures. However a slightly smaller n_{th} cannot explain the strong reduction in the monomolecular recombination ($\propto n$), which therefore requires a strongly decreased monomolecular recombination coefficient with increasing T_g.

The results suggest that in term of carrier transport homogeneous 2D QWs are desirable. However, these structures do not necessarily show the lowest threshold current densities, as the defect density and hence the threshold current density seems to be more dependent on the growth temperature than on the QW morphology.

ACKNOWLEDGMENTS

The authors gratefully acknowledge EPSRC (UK) and Infineon (Munich) for financial support and for providing a studentship for DM. The authors would also like to thank M. Albrecht, T. Remmele and H. Strunk for providing the TEM images of the reference structures.

REFERENCES

1. Fehse, R., Tomic, S., Adams, A.R., Sweeney, S.J., O'Reilly, E.P., Andreev, A., Riechert, J. of Selected Topics in Quantum Electronics, Vol.8, No.4, pp801-810, Jul/Aug, 2002

Inhibition Of Exciton Formation In Iron Doped InGaAs/InP Multiple Quantum Wells

M. Guezo, S. Loualiche, J. Even, A. LeCorre, O. Dehaese, C. Labbe

LENS, INSA Rennes, 20, Avenue des buttes de Coesmes - 35043 Rennes, France
Mail : slimane.loualiche @insa-rennes.fr

Abstract. Saturable absorber based on GaInAs/InP multi-quantum well structures are studied. Iron doping is used to control their absorption recovery time from ns down to sub ps regime reaching a record value of 0.29 ps. However at high doping level (10^{19} cm^{-3}) and sub ps regime, the non-linear amplitude variation is reduced by Fe-exciton interaction.

INTRODUCTION

Future ultra high-bit-rate wavelength-division-multiplexing communication systems require ultrafast all-optical switching devices with low switching fluence (SF) and high contrast ratio (CR) on a large spectral bandwidth in the 1.55 µm-range. Multiple quantum wells (MQWs) structures are materials of choice to obtain these properties thanks to their large optical absorption nonlinearity and low saturation fluence enhanced by excitonic (X) effects. These structures are very efficient in increasing long distance transmissions (+4000 km at 20Gb/s, [1]) when used inside resonant microcavities or for ultra fast and ultra high repetition rate optical pulse generation when used as intracavity or end cavity elements.

EXPERIMENTAL RESULTS

The studied GaInAs/InP MQWs (40 wells) structures are grown by molecular beam epitaxy and their exciton peak wavelength is at 1.55µm. As the photoresponse time constant of these MQWs structures is in the nanosecond (ns) range, their interest for applications in the ultrafast optical field is greatly reduced. Several approaches have been used in order to minimize this photoresponse time. In our way, iron doping is a simple method which takes place during the molecular beam epitaxy growth of the MQWs. Fe is well-known to be a deep acceptor level, which acts as a fast recombination center carriers trap and in InGaAs, as well as in InP [2,3]. Iron doping concentrations ranging from 10^{17} cm^{-3} to few 10^{19} cm^{-3} have been homogeneously incorporated. Nonlinear absorption dynamic is investigated in pump-probe experiments (PPE) using femtosecond pulses at 1.55µm [4]. In the present work we demonstrate that such iron concentrations reveals absorption decay times from ns down to sub-picosecond (0.29ps, figure 1).

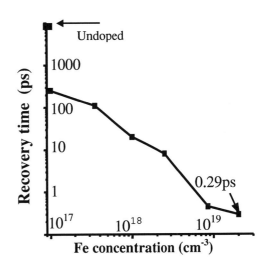

FIGURE 1. MQW decay time with Fe doping.

Absorption decay usually presents two characteristic time constants. The fastest component (τ_X <0.3 ps) corresponds to the exciton ionization as a consequence of interactions with LO phonons at room temperature. The second characteristic time constant is controlled by the iron doping concentration. Picosecond photoresponse time confirms the efficiency of such recombination centers. However, for ultrafast sub-picosecond decay time, a reduction of the differential absorption amplitude variation for transmitted intensity (figure 2) is additionally observed. At the corresponding high doping concentration, the mean distance between iron traps is comparable to the exciton Bohr radius. Exciton is known to be sensitive to Coulomb interactions (fixed or free charges), so exciton formation can be perturbed by the presence of these iron related defects. Inhibition of exciton formation seems to be responsible of the observed attenuation in absorption change amplitude.

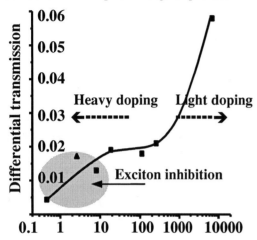

FIGURE 2. Differential transmission with decay time.

Interpretation Of Experimental Results

A very simple carrier dynamic model[5] has been used to explain the observed results. For the lightly Fe-doped structures, the optical non linearity dynamics presents two distinct components. The first sub ps component corresponds to the usual exciton ionization time of InGaAs MQW ($\tau_X \approx 0.22$ ps). The subsequent ps decay component is related to the electron-hole recombination ($\tau_{e/h}$), controlled by the incorporated iron concentration. The excitonic absorption dynamics is also governed by the evolution of exciton population (N_x) and electron-hole plasma ($N_{e/h}$). However, for the highest Fe-doped samples, only one ultrafast component (0.29 ps) is experimentally observed in the absorption recovery dynamics. This phenomenon is attributed to exciton formation inhibition and has been reported previously[3]. In the limit case, the dynamics of MQWs non-linear absorption is governed by the unique population of electron-hole plasma ($N_{e/h}$). The absorption amplitude can be represented by the sum of two non linear absorption contributions[6] : bound states and continuum states. The excitonic bound states absorption saturation threshold is lower (~ factor 10) than the corresponding value for continuum states absorption. In the case where the number of excitons is strongly reduced by the presence of ionized Fe atoms through coulomb interactions, the excitonic contribution to the absorption non linearity follows the same trend. As a consequence the optical threshold fluence needed to reach the absorption saturation which is proportional to the number of carriers is also increased. The experimental value found in the case of very heavy Fe doping samples (subpicosecond time constants) is an order of magnitude higher than on undoped structures. This trend has been also observed in irradiated[7] as well as on low temperature grown Be doped[8] structures.

CONCLUSION

Fe doping is used at high concentration to enhance saturable absorber time constants. Record decay time as low as 0.29ps is reached. However, differential amplitude variation of the optical absorption is reduced at high Fe doping. This is interpreted as Fe-X Coulomb interaction which reduces the excitonic part of the MQW absorption amplitude.

REFERENCES

1. J. L. Oudar, G. Aubin, J. Mangeney, S. Loualiche, J.C. Simon, A. Shen, O. Leclerc, *Annals of telecommunications* **58**, 1667, (2003)

2. D. Söderström, S. Marcinkevicius, S. Karlsson, S. Iordudoss, *Appl. Phys. Lett.* **70**, 3374 (1997)

3. M. Guezo, S. Loualiche, J. Even, A. LeCorre, H. Folliot, C. Labbé, O. Dehaese, C. Dousselin, *Appl. Phys. Lett.* **82**, 1670 (2003)

4. M. Guezo , A. Marceaux, S. Loualiche, J. Even, A. LeCorre, H. Folliot, O. Dehaese, Y. Pellan, *J. Appl. Phys.* **94**, 2355 (2003).

5. M. Guezo, S. Loualiche, J. Even, C. Labbé, O. Dehaese, A. Le Corre, H. Folliot, Y.Pellan, *Appl. Phys. Lett.* 2004, in press.

6. A.M. Fox, A.C. Maciel, M.G. Shorthouse, J.F Ryan, M.D. Scott, J.I. Davies, A. Riffat, *Appl. Phys. Lett.* **51**, 30, (1987)

7. J. Mangeney, J.L. Oudar, J.C. Harmand, C. Meriadec, G. Patriarche, G. aubin, N. Stelmakh, J.M. Lourtioz, *Appl. Phys. Lett.* **76**, 1372, (2000)

8. R. Takahashi, Y. Kawamura, T. Kagawa, H. Iwamura, *Appl. Phys. Lett.* **65**, 1790, (1994)

Optical Spectroscopy of Polytypic Quantum Wells in SiC

G. Samson*, L. Chen*, B.J. Skromme*, R. Wang[†], C. Li[†], and I. Bhat[†]

*Department of Electrical Engineering and Center for Solid State Electronics Research,
Arizona State University, Tempe, AZ, 85287-5706, USA
[†]ECSE Department, Rensselaer Polytechnic Institute, Troy, NY 12180-3590, USA

Abstract. Optical characterization is used to study spontaneously formed polytypic quantum well structures in lightly doped epilayers on heavily doped ([N] > 3×10^{19} cm^{-3}) 4H-SiC substrates. Low temperature (1.8 K) photoluminescence (PL) shows emission from the wells in both epilayer and substrate, the latter occurring at higher energy. A self-consistent model of the quantum wells including polarization charge is used to explain the results. Raman scattering data provide direct evidence of depletion of the 4H barriers in the substrate due to modulation doping into these wells.

Device degradation and reliability issues associated with stacking faults have driven the need to understand their formation and the resulting quantum well (QW) structures in SiC. The low stacking fault (SF) energy of 4H-SiC can lead to spontaneous SF formation at typical processing temperatures. Optical and structural studies by ourselves and others revealed the formation of I$_2$ stacking faults (SFs, displacement vector **R** = 1/3<10–10>) during thermal processing of heavily n-doped ([N] > 3×10^{19} cm^{-3}) SiC wafers [1-4]. Liu et al. [3] modified a model originally proposed by Miao et al. [5] for single fault formation and suggested that the energy gain from modulation doping of the 3C quantum well inclusions could drive the motion of the Shockley partials that forms these SFs. Here we present PL and Raman scattering data, which provide evidence of depleted barriers due to modulation doping, and polarization fields in the wells. A self-consistent model is used to explain the experimental results and to estimate the spontaneous polarization.

The 4H-SiC sample used in this study has a heavily n-doped substrate ([N]~3×10^{19} cm^{-3}) with a 2 μm thick epilayer ([N]~1.3×10^{17} cm^{-3}). The wafer is oriented 8° off [0001]. Faults were generated by thermal oxidation at 1150 °C for 90 min. in dry oxygen [2].

The upper spectrum in Fig. 1 is PL data recorded from the epilayer at 1.8 K. In addition to the normal 4H donor-bound exciton peaks and their phonon replicas, it reveals well-defined new peaks at 2.408, 2.433 and 2.464 eV, identified as LO (X) (104 meV), LA (X) (77 meV) and TA (X) (46 meV) momentum conserving phonon replicas of an exciton with an energy of ~ 2.510 eV. (Other spectra resolved the TO (X) mode at 96 meV as well.) These peaks, just above the bulk 3C band gap, are attributed to emission from triangular 3C QWs.

The PL spectrum from the substrate, recorded after polishing away the epilayer, reveals a much broader feature (bottom, Fig. 1). Comparing these spectra, we estimate that the substrate peaks are shifted above those of the epilayer by very roughly 52 meV. The spectral broadening in the substrate is attributed to band-filling, band-tailing, and fluctuations in well spacing. Many-body effects are likely to lower the peak position in the substrate.

The Raman scattering data in Fig. 2 show an essentially unscreened A$_1$ (LO) (967 cm^{-1}) mode in the transformed regions of the substrate, whereas it is

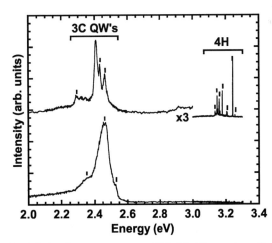

FIGURE 1. Low temperature (1.8 K) PL spectra of an oxidized 4H-SiC wafer. Upper spectrum: epilayer (excited at 4.127 eV). Lower spectrum: substrate after removing epilayer (excited at 3.691 eV).

strongly broadened and exhibits a larger Stokes shift in the untransformed regions. The latter effect is due to phonon-plasmon interaction due to the high electron concentration [6]. Single particle excitations of free carriers also cause a characteristic Fano interference distortion of the E_2(TA) mode profile [6]. In regions with a high density of stacking faults, ~all of the free carriers dump into the QWs by modulation doping, leaving depleted 4H barriers. Even though the high carrier density in the wells screens the 3C LO mode at (972 cm^{-1}), the reduced phonon-plasmon interaction in the depleted barriers explains the sharper, much less shifted LO mode characteristic of low-doped 4H-SiC in the transformed regions. The Fano distortion is also eliminated. Therefore, Raman scattering provides direct evidence for the modulation doping process.

We performed self-consistent solutions of the Schrödinger and Poisson equations for the QW structures in the epilayer and substrate to interpret the PL and Raman data. We used the experimental doping values of [N]=1.3x10^{17} cm^{-3} in the epilayer and [N]=3x10^{19} cm^{-3} in the substrate, and well spacings (77 nm in the epilayers and 10 nm in the substrate) based on our structural studies [4]. The spontaneous polarization in 4H-SiC [7] is included as charge sheets at the 3C/4H interfaces. Details of the materials parameters we used are discussed elsewhere [8].

The appropriate width to assume for the 3C QW is not unambiguously clear from the stacking sequence, so its value was varied from 4 to 6 bilayers, using trial spontaneous polarization values of 0.0054, 0.0108, and 0.0216 C/m^2. All simulations were performed at 2 K and assumed a type-II band alignment with offsets of ΔE_c = 0.919 eV and ΔE_v = –0.05 eV [7]. Good agreement with the experimental PL emission energy of 2.510 eV in the epilayer is obtained for a QW width

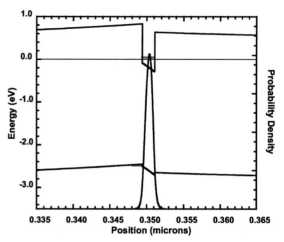

FIGURE 3. Self-consistent simulation of a 3C inclusion in the epilayer (only the vicinity of one well is shown). The probability density shown is that of electrons.

of 15 Å (6 bilayers) and a spontaneous polarization of 0.0108 C/m^2 (half the theoretical value of 0.0216 C/m^2 [7]), for which we predict an energy of 2.500 eV (neglecting the type II exciton binding energy). Figure 3 shows the corresponding simulation results. The calculations predict a band-to-band recombination energy of 2.620 eV in the heavier-doped substrate. The shift to higher energy is due to screening by the higher carrier densities in the wells. This value is only in fair agreement with experiment (2.562 eV), but the experimental value is difficult to determine accurately because of broadening and many-body effects.

In conclusion, our results are consistent with a spontaneous polarization of ~0.01 C/m^2 for 4H-SiC, in agreement with a recent determination by Bai *et al.* [9]. The simulations predict (1) large polarization fields in the QWs; (2) fully depleted 4H barriers between the wells; and (3) doping-dependent recombination energies, which are verified by spectroscopy.

The work at ASU was supported by the National Science Foundation under Grant Nos. ECS0080719 and ECS0324350 and by Motorola. Work at RPI is supported by DARPA Contract #DAAD19-02-1-0246.

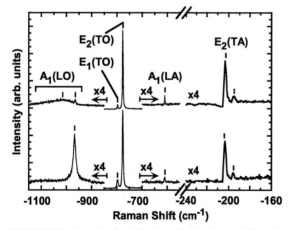

FIGURE 2. Raman scattering data in $z(x,x)$–z configuration for a thermally oxidized 4H-SiC wafer (excited at 2.409 eV). Upper spectrum: untransformed regions (without SFs, lower substrate doping). Lower spectrum: transformed regions (with SFs, due to locally higher substrate doping).

REFERENCES

1. R. S. Okojie *et al.*, *Appl. Phys. Lett.* **79**, 3056 (2001).
2. B. J. Skromme *et al.*, *Mater. Sci. Forum* **389-393**, 455 (2002).
3. J. Q. Liu *et al.*, *Appl. Phys. Lett.* **80**, 2111 (2002).
4. B. J. Skromme *et al.*, *MRS Proc.* **742**, K3.4.1 (2003).
5. M. S. Miao *et al.*, *Appl. Phys. Lett.* **79**, 4360 (2002).
6. S. Nakashima *et al.*, *Phys. Stat. Sol.* (*a*) **162**, 39 (1997).
7. A. Qteish *et al.*, *Phys. Rev. B* **45**, 6534 (1992).
8. B. J. Skromme *et al.*, *Mater. Sci. Forum* **457-460**, 581-584 (2004).
9. S. Bai *et al.*, *Appl. Phys. Lett.* **83**, 3171 (2003).

Excitonic Properties of ZnO Films and Nanorods

A. A. Toropov [a], O. V. Nekrutkina [a], T. V. Shubina [a], S. V. Ivanov [a], Th. Gruber [b], R. Kling [b], F. Reuss [b], C. Kirchner [b], A. Waag [c], K. F. Karlsson [d], J. P. Bergman [d], and B. Monemar [d]

[a] *Ioffe Physico-Technical Institute, St. Petersburg 194021, Russia*
[b] *Department of Semiconductor Physics, Ulm University, 89081 Ulm, Germany*
[c] *Institute of Semiconductor Technology, Braunschweig Technical University, 38106 Braunschweig, Germany*
[d] *Linköping University, S581 83 Linköping, Sweden*

Abstract. We report on the comparative studies of linearly polarized photoluminescence (PL) in a ZnO epitaxial film and ZnO nanorods. At low temperatures the PL spectrum of both samples included a number of narrow lines attributed to donor-bound excitons and a peak of free A excitons. An additional line observed in the nanorod sample was assigned to the excitons bound to some defects introduced during the sample post-growth history and located near the nanorod tips. The emission of mixed longitudinal – transverse exciton polariton modes was observed at elevated temperatures in both samples.

Remarkable excitonic properties of ZnO (60 meV exciton binding energy and ~10 meV longitudinal-transverse splitting of B and C excitons) have allowed one to consider this material as a polaritonic medium suitable for sophisticated optoelectronic applications, such as a microcavity polariton laser [1]. Especially interesting are ZnO nanorods, which are one-dimensional structures fabricated by a self-organization process [2]. The typical nanorods diameter (10-100 nm) is too large to ensure the electronic quantum confinement effect in ZnO, where the exciton radius is ~ 14 Å. Nevertheless, these nanostructures are promising for the fabrication of devices based on confined exciton polaritons.

In this work we have performed comparative optical studies of excitons in a ZnO film and ZnO nanorods, grown by metalorganic vapor phase epitaxy. The 1.2 µm thick layer was grown on the (0001) plane of a GaN template deposited on a sapphire substrate. The vertically aligned nanorods with the average height ~7000 nm and the diameter in the range of 40-80 nm were grown by a catalyst-free process on (11-20) sapphire. Cross-section scanning electron microscopy (SEM) images of both the nanorod sample (a) and the bulk layer (b) are illustrated in Fig 1.

Photoluminescence (PL) studies were performed either from the sample surface or from the cleaved edge. A microphotoluminescence (µ-PL) set up was used in the latter case. Figure 1 (b) explains the measurement geometry. The exciting beam (a *cw* 266 nm laser line) was focused by a reflective microobjective into a spot of ~ 1 µm in diameter, as illustrated in Fig 1 (b) by a white circle. The PL signal was collected by the same objective, passed through a monochromator and detected by a nitrogen-cooled charge-coupled device camera. The numerical aperture of the objective (NA=0.5) led to the collection of light within ±30° from the normal to the surface. A linear

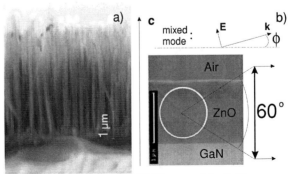

FIGURE 1. Cross-section SEM images of the ZnO nanorods (a) and the bulk ZnO layer (b).

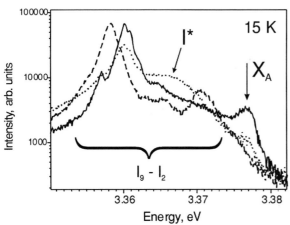

FIGURE 2. 15 K surface PL spectra measured in the bulk layer (dashed curve), as grown nanorod sample (solid curve), and the same nanorod sample after a few cryogenic measuring cycles (dotted curve).

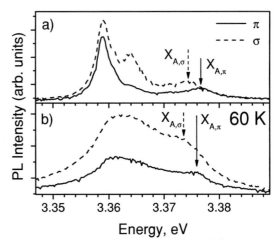

FIGURE 3. 60 K μ-PL spectra of the bulk layer (a) and nanorod sample (b) cleaved edge emission, measured in σ geometry (dashed curves) and geometry of an extraordinary beam (solid curve).

polarizer was mounted between the objective and monochromator. Depending on the polarizer orientation, one could detect either the light with electric field vector $E \perp c$ (σ geometry), where c is the principle axis of the wurtzite crystal, or the light with $E \| c$. The latter geometry corresponds to the detection of extraordinary beams in a uniaxial crystal.

Low-temperature PL spectra measured from the sample surface are shown in Fig. 2. The solid curve corresponds to the measurements carried out with a fresh as-grown nanorod sample. A number of narrow lines are visible, corresponding to bound excitons ($I_2 - I_9$) and free A excitons (X_A) in a practically unstrained ZnO crystal. The dotted curve in Fig. 2 represents the spectrum measured in the same nanorod sample after a few cryogenic cooling cycles. An additional broad PL line (I*) appears in the spectrum. The spatially resolved edge emission spectra showed that this line corresponds to the emission of excitons bound to some defects introduced near the nanorod tips. The dashed curve represents the spectrum measured in the bulk layer. It exhibits PL lines of nearly the same width, which are somewhat shifted due to the action of strain. No ageing effects were observed in this sample.

Figure 3 shows the edge spectra of the bulk layer (a) and nanorod sample (b) measured in the geometry of an extraordinary beam (solid curves) or in a σ geometry (ordinary beam, dashed curves) at the elevated temperature of 60 K. The two linearly polarized PL lines relevant to free A excitons ($X_{A,\sigma}$ and $X_{A,\pi}$) are visible in the spectra of the bulk layer. The gap between the lines is ~2-3 meV. Previously, observation in ZnO of the split PL line with the polarization of an extraordinary beam was attributed to the emission of mixed longitudinal-transverse exciton polariton modes [3]. Surprisingly, this line is also visible in the edge emission spectra of the nanorod sample (see Fig. 3 (b)), while not so clearly as in the bulk layer.

This observation confirms the emergence of nearly bulk exciton polariton modes in the nanorod sample. This is in general agreement with the expected dominance of the nanorod diameter over the exciton mean free path. Since the nanorod diameter is larger, the majority of excited excitonlike polaritons cannot reach the sample surface and radiate due to the scattering into the photonlike modes [4].

This work was supported by RFBR Gr. 04-02-17652 and by the "Landesschwerpunkt Funktionelle Nanostrukturen Baden-Württemberg".

REFERENCES

1. M. Zamfirescu, A. Kavokin, B. Gil, G. Malpuech, and M. Kaliteevski, *Phys. Rev. B* **65**, 161205 (2002).

2. M. H. Huang, S. Mao, H. Feick, H. Yan, Y. Wu, H. Kind, E. Weber, R. Russo, and P. Yang, , *Science* **292**, 1897 (2001).

3. A. A. Toropov, O. V. Nekrutkina, T. V. Shubina, Th. Gruber, C. Kirchner, A. Waag, K. F. Karlsson, P. O. Holz, and B. Monemar, *Phys. Rev. B* **69**, 165205 (2004).

4. C. Benoit a la Guillaume, A. Bonnot, and J. M. Debever, *Phys. Rev. Lett* **24**, 1235 (1970).

Observation of current resonances due to enhanced electron transport through stochastic webs in superlattices

S. Bujkiewicz, D. Fowler, T.M. Fromhold, A. Patanè, L. Eaves, A.A. Krokhin, P.B. Wilkinson, S.P. Stapleton, D. Hardwick, M. Henini, and F.W. Sheard

School of Physics and Astronomy, University of Nottingham, Nottingham NG7 2RD, UK

Abstract. We present experimental and theoretical studies of a unique type of stochastic electron motion in superlattices (SLs) with a magnetic field **B** tilted at an angle to the SL axis. The magnetic field couples electronic Bloch oscillations along the SL axis to cyclotron orbits in the plane of the layers. At discrete values of the applied voltage, this coupling transforms the localised Bloch trajectories into unbounded chaotic electron paths, which propagate through intricate "stochastic web" patterns in phase space. This abrupt metal-insulator-like transition reveals itself in our experiments as a large resonant increase in the current flow and conductance, observed at critical values of the applied voltage.

INTRODUCTION

Chaotic electron transport has been explored in experiments performed on a variety of semiconductor nanostructures [1]. Despite the diversity of these experiments, they all involve systems in which the transition to chaos occurs by the gradual and progressive destruction of stable orbits in response to an increasing perturbation. But there is also a much rarer type of chaos, known as non-KAM dynamics [2], which switches on and off abruptly when the perturbation reaches certain critical values. The theory of non-KAM chaos is of great interest due to applications in tokamaks, turbulence, and quasicrystals [2]. But, to our knowledge, such dynamics have previously been inaccessible to experiment. In this paper, we show that electrons in a SL miniband with a tilted magnetic field provide an experimentally-accessible non-KAM system. As predicted in our previous theoretical work [3], the onset of this unique type of chaotic dynamics produces a sharp increase in the current flow, which we observed as a strong resonant peak in the conductance-voltage, $G(V)$, curves.

To observe the chaos-induced conductance resonance, we use a novel type of InAs/GaAs/AlAs SL with a magnetic field **B** tilted at an angle θ to the SL (x) axis. Details of the sample composition are given elsewhere [4,5].

EXPERIMENT AND THEORY

When $\theta = 0$, the experimental $G(V)$ plot [Fig. 1(a), lower trace] exhibits the decrease with increasing V (arrow MB) associated with Bloch electron localisation. The curve for $\theta = 0$ also reveals the Stark-cyclotron resonance (arrow SC), which occurs when the quantised Wannier Stark and Landau levels are equally spaced [6]. As θ increases from 0, a much stronger resonant peak emerges (arrow SW), which has a fundamentally different origin from the Stark-cyclotron resonance, as we now explain.

To interpret the experimental data, we investigate the dynamics of electrons injected into the lowest SL miniband, using the semiclassical equations of motion $\mathbf{v} = \partial E(\mathbf{p})/\partial \mathbf{p}$ and $d\mathbf{p}/dt = -e[\mathbf{F} + (\mathbf{v} \times \mathbf{B})]$, where **v** and **p** are, respectively, the electron's velocity and crystal momentum, e is the magnitude of the electronic charge, **F** is the electric field produced by the applied voltage V, **B** is the applied magnetic field, and $E(\mathbf{p})$ is the energy versus crystal momentum dispersion relation for an electron moving in the first miniband of the SL [3], which is 19 meV wide. We consider motion in a tilted magnetic field which lies in the x-z plane at an angle θ to the SL axis (Fig. 2 inset).

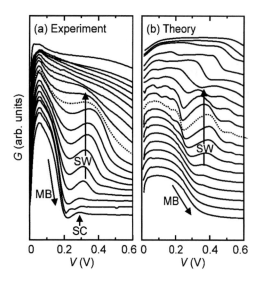

FIGURE 1. $G(V)$ curves (a) measured and (b) calculated for the SL [4] with $B = 11$ T, and $\theta = 0$ (bottom trace) to $90°$ (top trace) at $5°$ intervals. Curves for $\theta = 45°$ are shown dotted.

Remarkably, it can be shown that the dynamics of a miniband electron in tilted **B** reduce to a one-dimensional simple harmonic oscillator, of angular frequency $\omega_C \cos\theta$, where ω_C is the cyclotron frequency, driven by a time-dependent plane wave whose angular frequency equals the Bloch frequency ω_B [3,4]. So, even though the voltage and magnetic field are stationary, they act like an effective THz source. At V values for which $\omega_B = n\omega_C \cos\theta$, where n is an integer, the electron orbits change from localised Bloch-like trajectories [Fig. 2(a)] to *unbounded* stochastic orbits [Fig. 2(b)], which diffuse rapidly through intricate web patterns in phase space (Fig. 2 inset). To quantify how these webs affect the experimental data, we make drift-diffusion calculations of $G(V)$ including the effects of space-charge build up [4,7]. Our theoretical $G(V)$ curves [Fig. 1(b)] simulate all the key features of our experimental data and confirm that the large resonant peak observed in tilted **B** (arrow SW) originates from stochastic delocalisation of the electron orbits. Quantum-mechanical calculations of $G(V)$ in a tilted **B**-field, based on the non-equilibrium Green function formalism [7], closely resemble the semiclassical calculations shown in Fig. 1(b) [8]. The quantised eigenstates change discontinuously from a highly localised character when the system is off resonance [Fig. 2(a)] to a fully delocalised form when the resonance condition is satisfied [Fig. 2(b)]. Further work is required to relate the resonances observed here for hot electrons in a tilted **B**-field with those observed for electrons close to equilibrium in weakly-coupled SL structures [9].

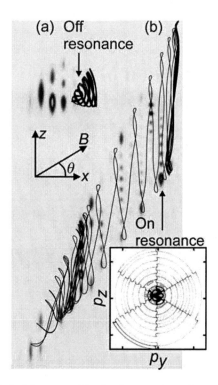

FIGURE 2. Electron orbits in the x-z plane (axes inset) superimposed on probability density plots (black = high) of quantised eigenstates. (a) A highly localised orbit (offset to right of wavefunction for clarity) *off* resonance at $\theta = 30°$. (b) An unbounded chaotic orbit *on* the $n = 1$ resonance at $\theta = 60°$. Inset: Stochastic web constructed by plotting (p_y, p_z) at discrete times separated by $2\pi/\omega_C \cos\theta$ [4]. $B = 11$ T.

ACKNOWLEDGMENTS

This work is funded by EPSRC and the EU.

REFERENCES

1. For a review see Stöckmann, H. -J., *Quantum Chaos: an Introduction*, Cambridge: Cambridge University Press, 1999.
2. Zaslavsky, G. M. et al, *Weak Chaos and Quasi-Regular Patterns*, Cambridge: Cambridge University Press, 1991.
3. Fromhold, T. M. et al, *Phys. Rev. Lett.* **87**, 046803 (2001).
4. Fromhold, T. M. et al, *Nature* **428**, 726-730 (2004).
5. Patanè, A. et al, *Appl. Phys. Lett.* **81**, 661-663 (2002).
6. Canali, L. et al, *Phys. Rev. Lett.* **76**, 3618-3621 (1996).
7. Wacker, A., *Phys. Rep.* **357**, 1-111 (2002).
8. Mori, N. et al, to be published.
9. Osada, T. et al, *Proc. 26th Int. Conf. on Physics of Semiconductors*, edited by A.R. Long and J.H. Davies, Institute of Physics Conference Series Number 171, Bristol: IoP Publishing, 2003, P155.PDF (CD-ROM).

Mobile potential dots in GaAs quantum wells

J.A.H. Stotz*, T. Sogawa†, F. Alsina*, H. Hey* and P.V. Santos*

*Paul-Drude-Institut für Festkörperelektronik, Hausvogteiplatz 5–7, 10117 Berlin, Germany
†NTT Basic Research Laboratories, 3-1, Morinosato-Wakamiya, Atsugi-shi, Kanagawa 243-0198, Japan

Abstract. Confined and mobile potential dots (dynamic dots, DDs) are created using two orthogonally propagating surface-acoustic-wave beams. Using spatially and time-resolved photoluminescence measurements, the compressive and tensile strain fields at the DD centers have been imaged by analyzing the polarization-dependent luminescence from charge carriers transported by the DDs.

There has been a growing interest to transport and manipulate carriers in confined dynamic potentials created by surface acoustic waves (SAWs). Earlier work has focused on transport through quantum point contacts [1] or along quantum wires.[2] Recently, we have realized a third geometry to create dynamic dots (DDs) using two orthogonally travelling SAWs.[3] The creation of DDs using this two-beam system is unique in that the mobile DD array is perfectly periodic and its formation does not require physical structuring of the interaction region.

The absence of a strain-induced splitting in the photoluminescence (PL) spectral lines from the DDs indicates a distinct separation of the hydrostatic strain s_0 and piezoelectric potential Φ_{SAW} in the SAW interference region.[3] While s_0 vanishes in the DDs, they are still subjected to a pure shear strain that leads to a polarization anisotropy of the optical properties. In this contribution, we investigate this anisotropy using polarization-dependent PL measurements.

The SAWs were generated by focussing interdigital transducers (IDTs) patterned on the (001) surface of a GaAs/Al$_{0.3}$Ga$_{0.7}$As QW sample grown by molecular-beam epitaxy. The IDTs generate two narrow SAW beams (f_{SAW} = 520 MHz) that propagate along the x = [110] and y = [1$\bar{1}$0] axes and interfere to create a square array of DDs. The periodicity of the DD array is defined by the SAW wavelength λ_{SAW} = 5.6 µm. The sample contains three QWs with thicknesses of 20, 15, and 12 nm placed between 100 and 170 nm below the surface of the sample. The QW PL signal from the DD region was measured at 12 K using a cw Ti:sapphire laser beam for excitation (λ_L = 768 nm). The excitation spot was polarized perpendicular to the DD propagation direction and expanded to cover the entire collection area. After being collected by a microscope objective, the QW luminescence passed through a polarizing beam displacer to spatially separate the PL into its x ($I_{PL,x}$) and y ($I_{PL,y}$) components. The two polarizations were simultaneously imaged onto an intensified charge coupled device (ICCD) camera. Stroboscopic, time-resolved PL measurements with a temporal resolution of 300 ps were performed by gating the intensifier with pulses synchronized with the SAW.

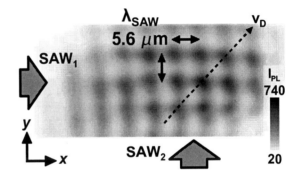

FIGURE 1. Spatially resolved PL snapshot from DDs created by orthogonally propagating SAWs for a particular DD phase ϕ_D. The dark regions correspond to high PL intensities.

Figure 1 shows a time-resolved snapshot of the integrated PL from the DD region including the electron-heavy hole emission from all 3 QWs. Note that the DDs are not static, and images recorded with different time delays demonstrate the DD movement along well-defined channels in the direction indicated by the arrow with velocity $v_D = \sqrt{2}v_{SAW}$, where v_{SAW} denotes the SAW phase velocity. The areas of strongest PL appear at the maxima of Φ_{SAW}, where the electrons are stored — described here as negative DDs (DD$^-$s). When electrons and holes are photoexcited, the electrons quickly relax to the DD$^-$s. Holes are less mobile, and those generated close to the DD$^-$s have a higher probability of recombi-

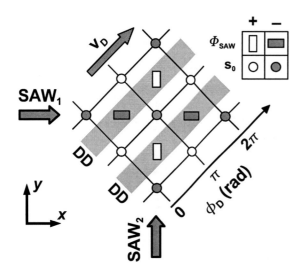

FIGURE 2. Schematic representation of the piezoelectric potential Φ_{SAW} (rectangles) and hydrostatic strain s_0 (circles) within the DD region. The DDs coincide with saddle-points in s_0, corresponding to zero hydrostatic strain but possessing anisotropic strain profiles. The s_0 dots, in contrast, consist of large, isotropic hydrostatic strain fields.

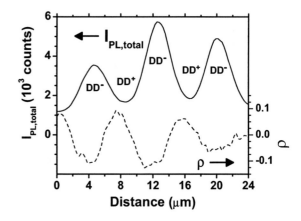

FIGURE 3. Total PL intensity $I_{PL,total}$ and polarization anisotropy ρ along the DD channel (shaded region in Fig. 2 and dotted line in Fig. 1) showing the positive (DD$^+$s) and negative (DD$^-$s) DDs.

nation with the stored electrons before they migrate towards the lower potential regions (positive DDs, DD$^+$s). The highest PL intensities are, therefore, found around the DD$^-$s, which form a square array with a periodicity defined by the SAW wavelength λ_{SAW}.

As we have discussed earlier,[3] the intersection of two orthogonally propagating SAWs on GaAs (001) produces the unique distribution of strain and piezoelectric fields outlined in Fig. 2. Essentially, SAWs propagating along the x and y directions of the GaAs (001) surface generate Φ_{SAW} and s_0 fields, which are in-phase and out-of-phase, respectively.[4] Summing the respective modulations in the SAW interference region yields Φ_{SAW} channels (corresponding to the DD$^+$s and DD$^-$s) separated by s_0 channels. As a result, while the strain at the maxima and minima of s_0 is purely hydrostatic (circles), the DDs (rectangles) reside on saddle points of the strain field, where the compressive strain in one direction is compensated by a tensile strain in the orthogonal direction. This pure shear strain is expected to lead to polarized PL from the DDs.

Figure 3 compares the total PL intensity ($I_{PL,total} = I_{PL,x} + I_{PL,y}$) and the degree of PL polarization ($\rho=(I_{PL,x}-I_{PL,y})/(I_{PL,x}+I_{PL,y})$) measured along the DD channel (indicated by the shaded region in Fig. 2 and dotted line in Fig. 1). Figure 3 shows the expected polarization oscillation between the DD$^+$s (minima of $I_{PL,total}$) and DD$^-$s (maxima of $I_{PL,total}$). For example, the DD$^-$s (Fig. 2) are compressed in the x direction and extended in the y direction — in agreement with band structure calculations (not shown), this strain configuration yields a PL preferentially polarized along y. The inverse holds for the positive DDs. The shift of the ρ extrema relative to the exact center of the DDs is presently ascribed to polarization dependences related to electro-optic effects, and it is an issue of continuing investigation. Nevertheless, polarization effects around 10% are observed illustrating the anisotropic nature of the DDs.

We have used spatially, time- and polarization-resolved PL measurements to map the strain and piezoelectric fields from DDs produced by SAW interference. In agreement with earlier predictions,[3] the positive and negative DDs are located at saddle-points of the strain field where the strain is of pure shear nature and leads to a significant PL polarization anisotropy.

ACKNOWLEDGMENTS

The authors thank H.T. Grahn for a critical reading of the manuscript. J.A.H. Stotz also thanks the Alexander von Humboldt Foundation for financial support.

REFERENCES

1. Talyanskii, V. I., Shilton, J. M., Pepper, M., Smith, C. G., Ford, C. J. B., Linfield, E. H., Ritchie, D. A., and Jones, G. A. C., *Phys. Rev. B*, **56**, 15180 (1997).
2. Alsina, F., Santos, P. V., Schönherr, H.-P., Seidel, W., Ploog, K. H., and Nötzel, R., *Phys. Rev. B*, **66**, 165330 (2002).
3. Alsina, F., Stotz, J. A. H., Hey, R., and Santos, P. V., *Solid State Commun.*, **129**, 453 (2003).
4. Alsina, F., Santos, P. V., and Hey, R., *Phys. Rev. B*, **65**, 193301 (2002).

Tuning of transmission function and tunneling time in superlattices

C. Pacher[*], U. Merc[†] and E. Gornik[*]

[*]*Institut für Festkörperelektronik, Technische Universität Wien, A-1040 Wien, Austria and ARC Seibersdorf research GmbH, Tech Gate Vienna, A-1220 Wien, Austria*
[†]*Faculty of Electrical Engineering, University of Ljubljana, 1000 Ljubljana, Slovenia*

Abstract. The coherent transmission function and the tunneling time of semiconductor superlattices can be adjusted without changing the resonance energies and therefore the miniband positions. The increase of the transmission comes along with a faster tunneling process through the transmission resonances. Using a hot-electron transistor we give experimental evidence that the faster tunneling process leads to a reduction of the incoherent current.

Semiconductor superlattices are artificial crystals, having a much larger unit cell than natural crystals.

Several experiments performed with three terminal devices [1, 2] showed that coherent transport dominates in short AlGaAs/GaAs superlattices (in minibands smaller than the LO phonon energy).

Here we give experimental evidence for a method that increases the coherent transmission function and reduces at the same time the phase tunneling time of periodic superlattices. The special feature of this method is to keep the energetic positions of the transmission resonances and also the miniband edge energies constant [3].

Together with the envelope function approximation the transfer matrix formalism allows for a very simple treatment of coherent transport in GaAs/AlGaAs heterostructures below the energy of the Γ and L-valley of AlGaAs.

Then in good approximation the transfer matrix $M_{SL}^{(1)}$ of the superlattice (SL) unit cell is a 2x2 matrix:

$$M_{SL}^{(1)} = \begin{pmatrix} a_{SL}^{(1)} & b_{SL}^{(1)} \\ b_{SL}^{(1)*} & a_{SL}^{(1)*} \end{pmatrix}. \quad (1)$$

The resonance energies inside the minibands of a SL with n periods are then given by

$$\text{Re}\{a_{SL}^{(1)}\} = \cos(k\pi/n), \quad k = 1,\ldots,n-1, \quad (2)$$

and the coherent superlattice transmission reads

$$T_{SL}^{(n)} = \left(1 + |b_{SL}^{(1)}|^2 U_{n-1}^2(\text{Re}\{a_{SL}^{(1)}\})\right)^{-1}. \quad (3)$$

The envelope of the minima of $T_{SL}^{(n)}$ does not depend on the number of periods:

$$T_{SL}^{\min} = \left(1 + \frac{|b_{SL}^{(1)}|^2}{1 - \text{Re}^2\{a_{SL}^{(1)}\}}\right)^{-1}. \quad (4)$$

The necessary unit cell functions $\text{Re}\{a_{SL}^{(1)}\}$ and $|b_{SL}^{(1)}|$ can be calculated from

$$\text{Re}\{a_{SL}^{(1)}\} = \cosh \kappa L_b \cos k L_w - c_2 \sinh \kappa L_b \sin k L_w, \quad (5)$$

$$|b_{SL}^{(1)}| = c_1 \sinh \kappa L_b. \quad (6)$$

Here $c_{1,2} = \frac{1}{2}\left(\frac{k}{\kappa}\frac{m_b}{m_w} \pm \frac{\kappa}{k}\frac{m_w}{m_b}\right)$. The electron wave vector in the wells and the decaying electron wave vector in the barriers are given by $k = (2m_w E)^{1/2}/\hbar$ and $\kappa = [2m_b(V_b - E)]^{1/2}/\hbar$, respectively, m_b (m_w) is the effective electron mass in the barriers (wells), V_b is the potential energy of the barriers, and E is the energy.

A unit cell of a normal superlattice consists of the sequence "barrier | well". By shifting the origin in the SL unit cell to the middle of the barrier a new unit cell that is formed by the sequence "barrier/2 | well | barrier/2" arises. That origin shift is reflected in a similarity transformation of the corresponding transfer matrix $M_{SL}^{(1)'}$ [3]. The trace of $M_{SL}^{(1)}$, $\text{Re}\{a_{SL}^{(1)}\}$ and henceforth [see Eq. (2)] the SL transmission resonances are invariant under this similarity transformation [3]. The matrix element $b_{SL}^{(1)'}$ of the transformed unit cell is given by

$$b_{SL}^{(1)'} = 2ic_1 \sinh(\kappa L_b/2) \quad (7)$$
$$\times [\cosh(\kappa L_b/2) \cos k L_w - c_2 \sinh(\kappa L_b/2) \sin k L_w].$$

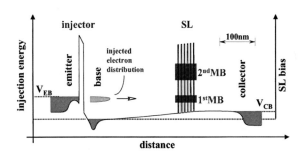

FIGURE 1. Schematic bandstructure of a hot-electron transistor used to measure the transfer ratio of a superlattice.

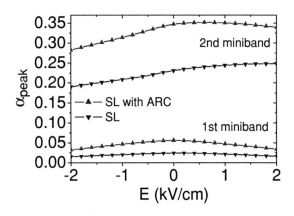

FIGURE 2. Peak values of the measured transfer ratios $\alpha = I_C/I_E$ at T=4.2K vs. electric field across the superlattice (SL) and the SL with anti-reflection coating (SL+ARC).

Since $|b_{SL}^{(1)'}| < |b_{SL}^{(1)}| \Leftarrow E < V_b$, this leads to an increase of the average transmission in a periodic sequence of the new unit cells [3]. The new SL structure can be seen as a SL with an additional anti-reflection coating (ARC) [2].

At the same time the phase tunneling time at resonance, which is given by

$$\tau_{\text{res}}^{(n)} = \hbar n \frac{\text{Im}\{a_{SL}^{(1)}\}}{1-\text{Re}^2\{a_{SL}^{(1)}\}} \frac{\partial \text{Re}\{a_{SL}^{(1)}\}}{\partial E} \qquad (8)$$

is reduced [3]. This time is equal to the dwell time at resonance [4].

Since the ratio of scattering time to tunneling time determines the ratio of incoherent to coherent transmission, a decrease of the tunneling time should reduce the incoherent current.

Incoherent transmission can be distinguished from coherent transmission by applying an electric field across the SL. Incoherent transmission is mainly in the direction of the applied field (ohmic current), whereas coherent transmission does not depend on the direction of the field.

In a hot-electron transistor [1, 5] it is possible to measure the transmission as a function of the applied electric field and independently as a function of the electron energy. Figure 1 shows a band lineup for the Γ-valley in the conduction band. The SL parameters are: 2.5 nm $Al_{0.3}Ga_{0.7}As$ barriers, 6.5 nm GaAs wells, $n = 6$ periods. A more detailed description is given in Ref. [2]. The static transfer ratio $\alpha = I_C/I_E$ at T=4.2K is a measure for the transmission.

In Fig. 2 we show the measured peak transfer ratios of a normal SL and a SL with ARC for the first and second miniband as a function of the applied electric field across the SL. The amplitudes show clearly the increase in the transmission, which is in good agreement with theory [2].

Now we turn to the question coherent vs. incoherent current. In the first miniband ($\Delta_1 = 20$meV) the transfer ratios do not depend on the field direction for both samples. The maximum transfer ratio is at zero field for both samples. This is a clear indication that the transport is predominantly coherent in the first miniband.

The second miniband is broader ($\Delta_2 = 60$ meV) than the LO phonon energy ($\hbar\omega_{LO} = 36$ meV). Now incoherent (LO phonon induced) current shows up in an asymmetry of the transfer ratio α of both samples. For the normal SL α_{peak} is monotonic increasing for fields at least up to 2 kV/cm. In contrast we observe a decrease of α_{peak} for fields $\gtrsim 0.7$ kV/cm in the ARC sample. This shows that in the second miniband in the SL with ARC incoherent current gets suppressed by the shorter tunneling time, compared to the normal SL.

In conclusion, a hot-electron transistor was used to show that a strong increase in the overall superlattice transmission and a reduction of the fraction of incoherent current can be obtained simultaneously. The reduction of the incoherent current is a result of the shorter tunneling time in the modified SL structure.

This work was supported by the Gesellschaft für Mikroelektronik (GMe) and the FWF Grant No. Z24.

REFERENCES

1. Rauch, C., Strasser, G., Unterrainer, K., Boxleitner, W., Gornik, E., and Wacker, A., *Phys. Rev. Lett.*, **81**, 3495 (1998).
2. Pacher, C., Rauch, C., Strasser, G., Gornik, E., Elsholz, F., Wacker, A., Kießlich, G., and Schöll, E., *Appl. Phys. Lett.*, **79**, 1486 (2001).
3. Pacher, C., and Gornik, E., *Phys. Rev. B*, **68**, 155319 (2003).
4. Pacher, C., Boxleitner, W., and Gornik, E. (2004), to be published.
5. Kast, M., Pacher, C., Strasser, G., and Gornik, E., *Appl. Phys. Lett.*, **78**, 3639 (2001).

Pekar Mechanism of Electron-Phonon Interaction in Nanostructures

B.A. Glavin[1], V.A. Kochelap[1], T.L. Linnik[1], K.W. Kim[2], and V.N. Sokolov[1,2]

[1]*V.E. Lashkar'ov Institute of Semiconductor Physics, Kiev, Ukraine*
[2]*Department of Electrical and Computer Engineering, North Carolina State University, Raleigh, USA*

Abstract. We demonstrate that in nanostructures an additional mechanism of electron-acoustic phonon interaction has to be considered along with traditional mechanisms. It was addressed previously by Pekar for bulk crystals with extremely high dielectric permittivity. The additional coupling is due to macroscopic electrical potential which is caused by modulation of the dielectric permittivity by a phonon in the presence of external electric field. We show that in nanostructures this mechanism is important because of strong confining electric fields. In this case both absolute value and spatial distribution of the electric field influence strength of interaction. We consider two specific examples where Pekar interaction is important: low-temperature energy relaxation of 2D-electrons in silicon inversion layers and phonon emission from nitride multiple quantum well structures.

INTRODUCTION

In bulk semiconductors, two mechanisms of electron-acoustic phonon coupling are usually considered: deformation potential and piezoelectric interactions. An additional mechanism was considered by Pekar for the problem of sound amplification [1]. It originates in the presence of external electric field E and is caused by phonon-strain-induced modulation of the crystal dielectric permittivity. The phonon wavevectror dependence of the Pekar potential is similar to that of piezoelectric effect, and an effective "piezo"-constant can be introduced: $e_{eff} = \varepsilon_0 \varepsilon^2 pE$, where ε is dielectric permittivity of unstrained crystal, p is photoelastic coefficient, and ε_0 is absolute dielectric constant. Unlike piezoelectric effect, Pekar interaction exists in crystals of arbitrary symmetry. For bulk crystals, Pekar mechanism was found important only for materials with extremely high ε. For nanostructures e_{eff} also becomes important due to the presence of very strong confining electric fields. We have found that in this case spatial distribution of the electric field is also important, and we provide corresponding examples. In particular, for structures where electric field is localized in some region, as in silicon inversion layers, the phonon wavelength has to be smaller than dimensions of that region for strong Pekar interaction. On the other hand, for periodic field distributions, as in nitride multiple quantum wells (MQW), Pekar interaction is strongly enhanced for resonant wavevectors $2\pi n/d$, where d is MQW period and n is integer.

LOW-TEMPERATURE ENERGY RELAXATION IN SILICON INVERSION LAYERS

In silicon inversion layers confinement of electron is achieved due to the strong electric field at the silicon-silicon oxide interface. The absolute value of the field at the interface, E^*, can be as high as 10^7 V/m. Solution of the Poisson equation for the Pekar potential suggests that strong interaction is possible for structures with thick enough oxide layers (for typical parameters its thickness should be greater than a few tenth of a micron). In that case, at low temperatures the power dissipated by heated electron to the lattice is $Q = a(T_e^3 - T_{lat}^3)$ with coefficient a proportional to $(E^*)^2$. Similar temperature dependence would be due to piezoelectric coupling, which, however, is not expected for silicon. Since "traditional" deformation potential interaction provides T^6 dependence, Pekar contribution can be easy detected in experiments. Details of calculations and numerical results will be published elsewhere. Note, that a number of energy losses and thermopower measurements suggest piezo-like interaction in silicon inversion layers as well as in SiGe quantum wells [2-4]. Since strong electric field is present in all these structures, such observations can be attributed to the Pekar interaction.

ACOUSTIC PHONON EMISSION FROM NITRIDE MULTIPLE QUANTUM WELLS

Nitride MQW structures are characterized by strong periodic built-in electric field, up to 10^8 V/m, which is due to piezoelectric and pyroelectric effects in nitride semiconductors. For such system the Poisson equation for the Pekar potential φ is

$$\frac{d^2\varphi}{dz^2} - q_\parallel^2 \varphi = \Phi(z)\exp(iq_z z), \quad (1)$$

where $\Phi(z)$ is periodic function with zero average value and the period of MQW, d, and $q_\parallel$ and q_z are in-plane and perpendicular components of the phonon wavevector, respectively. One can see that if $q_z = 2\pi n/d$ with integer n, then the potential φ has close-to-constant component even though the phonon is characterized by large wavevector. In addition, for small $q_\parallel$ the amplitude of φ is large. As a result, Pekar interaction provides resonance-like electron-phonon coupling for large-wavevector phonons which practically do not interact with 2D electrons via deformation potential and piezoelectric mechanisms due to small form-factor. We demonstrate that this leads to specific features of acoustic phonon emission by hot electrons from MQW structures. For small angles between the phonon wavevector and the normal to the MQW layers, the energy spectrum of emitted phonons consists from the sequence of sharp peaks corresponding to the resonances of Pekar interaction. This is illustrated in Fig. 1. As we see, the emission spectra are characterized by very sharp peaks with a small background due to deformation potential and piezoelectric interactions. This indicates that nitride MQWs can be used as a source of highly collimated quasi-monochromatic beams of terahertz acoustic phonons. Such beams can be detected in phonon imaging experiments.

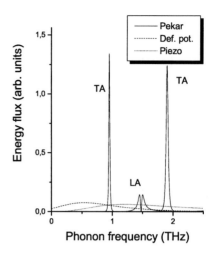

FIGURE 1. Energy spectrum of the emitted phonons for GaN/AlGaN MQW of period d=4 nm for electron temperature 300K and angle between the phonon wavevector and the normal to MQW 0.01. "LA" and "TA" marks correspond to the resonances of LA and TA phonons, respectively. The data for deformation potential and piezoelectric interaction are given for comparison.

ACKNOWLEDGMENTS

We are indebted to V.N. Piskovoi for many valuable comments and discussions. The work was supported in part by AFOSR and CRDF (grant UE2-2439-KV-02).

REFERENCES

1. Pekar S.I., *Zh. Eksp. Teor. Fiz.* **49**, 621-629 (1965).

2. Prus O., Reznikov M., Sivan U., Pudalov V., *Phys. Rev. Lett.* **88**, 1-4 (2002).

3. Leturcq R., L'Hote D., Tourbot R., Senz V., Gennser U., Ihn T., Ensslin K., Dehlinger G., and Grutzmacher D., *Europhys. Lett.* **61**, 499-505 (2003).

4. Possanzini C, Fletcher R, Tsaousidou M, Coleridge P.T, Williams R.L, Feng Y, Maan J.C., *Phys. Rev. B* **69**, 195306 (2004).

Electronic Band Structure and New Magneto-transport Properties in p-type Semiconductor Medium-infrared HgTe / CdTe Superlattice

Ab. Nafidi*, A. EL Abidi, A. El Kaaouachi, and Ah. Nafidi

Condensed Matter Physics Laboratory, Faculty of Sciences, B.P 8106, City Dakhla, University Ibn Zohr, 80000 Agadir, Morocco

Abstract: We report here the band structure and new magneto-transport results for HgTe (56 Å) / CdTe (30 Å) superlattice grown by molecular beam epitaxy (MBE). The angular dependence of the transverse magnetoresistance follows the two-dimensional (2D) behaviour. At low temperature, the sample exhibits p type conductivity with a concentration of 1.84×10^{12} cm^{-2} and a Hall mobility of 8200 cm^2/Vs. The observed Shubnikov-de Haas effect gives a carrier density of 1.80×10^{12} cm^{-2}. The superlattice heavy holes dominate the conduction in plane with an effective mass of 0.297 m_0 and Fermi energy (2D) of 14 meV. In intrinsic regime, the measured gap E_g =190 meV agree well with calculated $E_g(\Gamma, 300\ K)$ =178 meV. The formalism used here predicts that the system is semiconductor, for our HgTe to CdTe thickness ratio d_1/d_2 = 1,87, when d_2 < 140 Å. In our case, d_2=30 Å and $E_g(\Gamma, 4.2\ K)$ = 111 meV. In spite of it, the sample exhibits the features typical of a p type semiconductor and is a medium-infrared detector (7 µm< λ< 11µm).

INTRODUCTION

The development of MBE was successfully applied to fabricate different quantum wells and superlattices. Among them, III-V superlattices (Ga$_{1-x}$Al$_x$As-GaAs - type I), IV-VI (InAs/GaSb - type II) and later II-VI superlattice (HgTe/CdTe - type III). It has been predicted that HgTe/CdTe superlattices could be a stable alternative for application in infrared optoelectronic devices than the random alloy Hg$_{1-x}$Cd$_x$Te. HgTe is a zero gap semiconductor when it is sandwiched between the wide gap semiconductor CdTe layers yield to a small gap HgTe/CdTe superlattice (SL) which is the key of an infrared detector. A number of papers have been published devoted to the band structure of this system as well as its magnetooptical and transport properties [1]. We report here the band structure and new magneto-transport results for a HgTe/CdTe SL grown by MBE.

THEORY AND EXPERIMENTAL

The sample (90 layers), with a period d of 56Å (HgTe) / 30Å (CdTe), was grown on a CdTe (111) substrate at 180 °C. Calculations of the specters of energy $E(k_z)$ and $E(k_p)$, respectively, in the direction of growth and in plane of the SL "Fig. 1"; were performed in the envelope function formalism [2]; with a valence band offset between heavy hole band edges of HgTe and CdTe of Λ= 40 meV determined by magneto-optical absorption experiments [3]. This offset agrees well with our experimental results contrary to 350 meV given by [3]. Carrier transport properties were studied in the temperature range 1,5 -

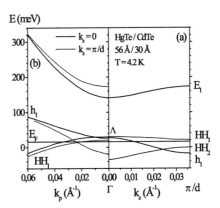

FIGURE 1. Electronic band structure along the wave vector k_z (a) and in plane $k_p(k_x,k_y)$ (b) of HgTe/CdTe SL at 4,2 K.

300 K in magnetic field up to 18 Tesla. Conductivity, Hall Effect "Fig. 4" and angular dependence of the transverse magnetoresistance ΔR with respect to the magnetic field were measured "Fig. 2".

RESULTS AND DISCUSSION

ΔR vanishes, when the field is parallel to the plane of the SL, indicating a 2D behaviour "Fig. 2".

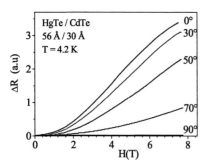

FIGURE 2. Variation of magnetoresistance of the sample with various angles between the magnetic field and the normal to the HgTe/CdTe superlattice surface.

At low temperatures, the sample exhibits p type conductivity (confirmed by thermoelectric power measurements) with a concentration $p=1.84 \times 10^{12}$ cm^{-2} and a Hall mobility $\mu_p = 8200$ cm^2/Vs "Fig. 4". The decrease of R_H (1/T) at 40 K shown in "fig. 4 b" can be due to coupling between HgTe well (small d_1/d_2 and d_2), to the widening of carrier sub-bands under the influence of the magnetic field and/or to the overlap between involved carriers sub-bands (HH$_1$) and (h$_1$) at (k_z; k_p) = (π/d; 0,023 Å^{-1}) along E(k_p). In intrinsic regime, the measure of the slope of the curve $R_H T^{3/2}$ indicates a gap $E_g = E_1 - HH_1 = 190$ meV witch agree well with calculated E_g (Γ, 300 K) = 178 meV. This relatively high mobility allowed us to observe the

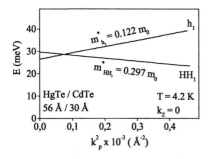

FIGURE 3. Determination of electronic effectives masses m^*_{HH1} and m^*_{h1}, respectively, of heavy-hole HH$_1$ and light-hole h$_1$ subbands at 4.2 K in the center Γ of the first Brillion zone of the HgTe/CdTe superlattice.

Shubnikov-de Haas effect until 18 Tesla. The measured period of the magnetoresistance oscillations (2D) gives $p = 1.80 \times 10^{12}$ cm^{-2} in good agreement with weak field Hall effect. At low temperature, the superlattice heavy holes dominate the conduction in plane. The HH$_1$ (and h$_1$) band is parabolic "Fig. 3". That permits us to estimate the Fermi energy (2D) at
$$\left|E_F - E_{HH_1}\right| = \left|p\pi\hbar^2/m^*_{HH1}\right| = 14 \text{ meV}.$$

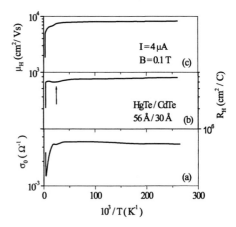

FIGURE 4. Temperature dependence of the conductivity (a), weak-field Hall coefficient (b) and Hall mobility (c) in the investigated HgTe/CdTe superlattice.

CONCLUSIONS

The formalism used here predicts that the system is semiconductor, for our HgTe to CdTe thickness ratio $d_1/d_2 = 1,87$, when $d_2 < 140$ Å. In our case, $d_2 = 30$ Å and E_g (Γ, 4.2 K) = 111 meV. In spite of it, the sample exhibits the features typical for the semiconductor type p conduction mechanism. In the used temperature range, this simple is a medium-infrared detector (7 μm<λ<11 μm). Note that we had observed a semimetallic conduction mechanism in the 2D p type HgTe/CdTe (180 Å / 44 Å) superlattice [4].

REFERENCES

1. Y. C. Chang et al, *Phys. Rev. B* **31**, 2557 (1985). F. J. Boero et al, *Solid State Commun.* **54**, 35 (1985).
2. Ab. Nafidi et al, *the proceedings of the International Conference on theoretical Physics (HT 2002)*, Paris, France July 22-27, 2002, p. 274.
Bastard, *Phys. Rev. B* **24** (10), 5693 (1981).
3. N. F. Johnson et al, *Phys. Rev. Lett.* **61** (17), 1993 (1988).
4. Ab. Nafidi, et al, *the proceedings of The 11th Intern. Conference on Modulated Semiconductor Structures (MSS11)*, Nara, Japan July 14-18, 2003, pp. 280-281.

Spin Effects in Magnetotransport of an n-In$_x$Ga$_{1-x}$As/GaAs Double Quantum Well Under Parallel Magnetic Fields

M. V. Yakunin[*], G. A. Alshanskii[*], Yu. G. Arapov[*], G. I. Harus[*], V. N. Neverov[*], N. G. Shelushinina[*], B. N. Zvonkov[†], E. A. Uskova[†], A. de Visser[¶], and L. Ponomarenko[¶]

[*]*Institute of Metal Physics RAS, Ekaterinburg 620219, Russia*
[†]*Physical-Technical Institute at Nizhnii Novgorod State University, Nizhnii Novgorod, Russia*
[¶]*Van der Waals - Zeeman Institute, University of Amsterdam, the Netherlands*

Abstract. In a magnetic field configured parallel to the layers of the n-In$_x$Ga$_{1-x}$As/GaAs double quantum well, peculiarities are revealed in the magnetoresistance (MR) caused by passing of the tunnel gap edges through the Fermi level. Shown is that to reach coincidence of the calculated MR maximum position with the corresponding experimental one, spin splittings in the energy spectrum should be considered, and only the lower spin-split subband manifests in the experiment while the peculiarity due to the upper one is suppressed because of its lower population and vanishing probability of the spin-flip transitions. Spin splittings definitely manifest in the same samples in the quantum Hall effect.

INTRODUCTION

Two-dimensional structures of materials with a considerable g-factor are promising candidates for spintronic phenomena and devices [1]. Usually magnetic fields are used to explore the spin splitting, but under high fields it may vary considerably due to exchange correlation effects [2]. Although it is often assumed that the spin splitting is primarily determined by the perpendicular component of the field [3], there are indications that the effective g-factor depends on the parallel field too [4]. In this context we try to use the magnetoresistivity (MR) of a double quantum well (DQW) induced by the parallel magnetic field to estimate the spin splitting. A supposed advantage of DQW is in that the MR maximum caused by a passing of the lower tunnel gap edge through the Fermi level E_F may be tuned to occur in a high magnetic fields, where the spin splitting resolves easier.

We investigate DQW created in n-In$_x$Ga$_{1-x}$As/GaAs ($x \approx 0.18$) heterosystem. In case of a rather thick barrier, the Fermi level initially, without a magnetic field, is located above the tunnel gap, and under the parallel magnetic field two kinds of peculiarities are observed: a minimum in weak fields, when the upper edge goes beyond E_F, and a maximum, when the saddle point in the energy spectrum connected with the lower gap edge goes beyond E_F [5]. On the other hand, in a sample (#2984 [5]) with thin enough barrier (3.5 nm) the tunnel gap is wide (~23 meV) and E_F initially lies within it. As a result, only a maximum is observed experimentally in fields around 30 T that confirms our approach to the interpretations. The topic of this paper is a deeper analysis of the spin effects in the nature of the observed MR maximum.

A SADDLE POINT IN THE DQW ENERGY SPECTRUM AND A MAGNETORESISTIVITY MAXIMUM

As was revealed in Ref. 5, the experimental MR maximum position is ~5 T higher than the calculated one, well beyond the errors, and calculations in higher approximations don't remove this discrepancy (Fig. 1 at $g = 0$). The only way found was to consider the spin splittings in the energy spectrum. This was a novel approach since so far this kind of MR peculiarity in a DQW has only been observed in the GaAs/AlGaAs heterosystem where the spin splitting in the GaAs

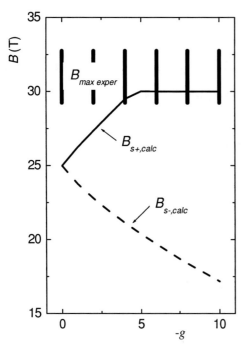

FIGURE 1. MR maximum position: experimental (bars) and calculated for the lower (solid curve) and upper (dashed) spin-split saddle points.

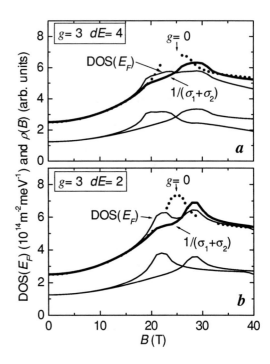

FIGURE 2. Calculated DOS(E_F) for two spin-split subbands (lower curves), its sum (upper thin curves), and MR= $1/(\sigma_1+\sigma_2)$ (bold solid curve). $|g| = 3$. Compared with $g = 0$ (dotted curves). **a** and **b** – for two level widths (indicated).

conduction band certainly could be ignored due to a very small $|g| \approx 0.4$. In our case the MR maximum position for the lower spin-split subband do agree with experiment at $|g| > 3$ (Fig. 1). This value seems quite realistic from interpolation for the given component relation in $In_xGa_{1-x}As$, and experiments in perpendicular magnetic fields reveal quantum Hall peculiarities at odd filling factors $\nu = 3$ and 5 that definitely indicate resolved spin sublevels at fields $B > \sim 3$ T. But the question arises: why in this case the second spin-split component doesn't manifest at lower fields (the lower branch in Fig. 1)?

To answer it we attempted to estimate the relative amplitudes of these two MR components. In the first step we estimated the total density of states at E_F, DOS(E_F), as a function of parallel magnetic field, assuming that it reflects MR if all the transitions are equally probable within k_BT around E_F: see the two curves in Fig. 2a,b for spin subbands and its sum. In this approach, the upper spin subband yields a peak (at lower fields) comparable in amplitude to that for the lower spin subband.

In the second step, we supposed that inter-spin-subband transitions may be neglected, then all the scatterings occur within each subband and MR is proportional to $1/(\sigma_1 + \sigma_2)$ with $\sigma_i \sim n_i/\mathrm{DOS}(E_F)$, n_i – the electron gas density in the i-subband. In these estimations, the upper spin subband yields much smaller contribution to MR since it is less populated, and the lower field MR spin component is suppressed considerably (Fig. 2).

According to this, the upper spin subband contribution is still more suppressed for higher $|g|$-factor values. Then, the experimentally unobserved second peak may indicate that $|g| > 3$ (while the motion of higher field spin peak saturates at $|g| > \sim 3$).

ACKNOWLEDGMENTS

The work is supported by RFBR, projects 02-02-16401, 04-02-16614 and the Program "Solid state nanostructures".

REFERENCES

1. Zutic, I., Fabian, J, and Das Sarma, S., *Rev. Mod. Phys.* **76**, 323-413 (2004).
2. Ando, T., and Uemura, Y., *J. Phys. Soc. Japan* **37**, 1044-1052 (1974).
3. Nicholas, R. J., Haug, R. J., v. Klitzing, K., and Weimann, G., *Phys. Rev. B* **37**, 1294-1302 (1988).
4. Tutuc, E., Melinte, S., and Shayegan, M., *Phys. Rev. Lett.* **88**, 036805 (1-4) (2002).
5. Yakunin, M. V., Alshanskii, G. A., Arapov, Yu. G., Harus, G. I., Neverov, V. N., Shelushinina, N. G., Kuznetsov, O. A., Zvonkov, B. N., Uskova, E. A., Ponomarenko, L., de Visser, A., *Physica E* **22**, 68-71 (2004).

Z-shaped current-voltage characteristics of disordered semiconductor superlattices

Olga Pupysheva*, Alexey Dmitriev* and Alexander Kozhanov*

Department of Low Temperature Physics, Moscow State University, Moscow 119992, Russia

Abstract. Vertical electron transport in weakly-coupled semiconductor superlattices is studied theoretically. The Poisson equation and kinetic equation are solved self-consistently, taking into account the global charge conservation of the whole system including the contact layers. The current-voltage characteristics obtained numerically exhibit Z-shaped portions in addition to the conventional N-shaped maxima. We discuss the origin of N- and Z-shaped current maxima, and of the transitions between them. It is shown that the superlattice transport properties can be controlled by introducing the disorder into the layer parameters.

INTRODUCTION

Semiconductor superlattices (SLs) are attractive for various practical applications, in particular, due to the negative differential conductivity region, often possessed by their current-voltage characteristics along the growth axis. The conventional N-shaped maxima of *I-V* curves can be also deformed into more unusual Z-shaped portions, a phenomenon known as intrinsic bistability [1]. It arises due to a tunneling resonance between the emitter and the quantum well. The continuous Z-shaped *I-V* curves can be observed directly by means of the negative output resistance technique, originally proposed for semiconductor double-barrier structures [2]. The overview of various approaches to the transport problem in semiconductor superlattices can be found in [3, 4].

We theoretically study the vertical electron transport in weakly-coupled semiconductor SLs. We use the sequential tunneling model described in details in [5]. The stationary *I-V* curves are calculated for the periodic and disordered SLs. We give a simple explanation of the observed N- and Z-shaped current maxima. The possibility to control the vertical transport in SLs by introducing a disorder into the layer parameters is discussed.

MODEL

An n-type SL with thick quantum barriers is considered. Its potential profile under electric field is shown schematically in Fig. 1. The electrons are considered as localized at the size-quantization levels in the quantum wells. Only the transitions between the states localized in adjacent quantum wells are taken into account.

The non-uniformity of electric field along the superlattice axis is essential in the weakly-coupled structures. We take into account the charge redistribution among the inner quantum wells, as well as the contact layers (emitter and collector). The latter are simulated by two "fictitious" quantum wells marked by indices $k = 1$ and $k = N$ (see Fig. 1). The charge of a quantum well is expressed through the electric field in surrounding barriers by integrated Poisson equation. The global charge conservation of the whole system, including the contacts, is adopted as the boundary condition.

The kinetic equation for the current density between the adjacent wells can be written using the integral rates P_k and R_k of the direct and reverse transitions:

$$I = e(P_k n_k - R_k n_{k+1}),$$

where n_k is 2D electron concentration in the k-th well. The transition probabilities depend on the electrostatic potential drop $\Delta u_k = u_k - u_{k+1}$ in the corresponding quantum barrier. These dependences are of resonant character and are determined by the dominant mechanism of scattering accompanying the interwell transition [5].

This nonlinear system of equations is solved self-consistently regarding the variables n_k and Δu_k at fixed I value. The voltage on the SL is given by the sum of the potential drops Δu_k in all the barriers.

N- AND Z-SHAPED *I-V* CURVES

The *I-V* curves, calculated as described above, exhibit two different types of maxima: conventional N- and Z-

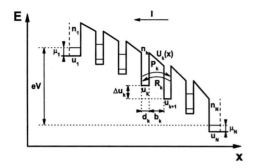

FIGURE 1. Superlattice potential profile in electric field.

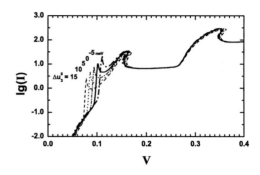

FIGURE 2. I-V curves of periodic ($\Delta u_3^0 = 0$) and disordered structures.

shaped. Moreover, a transition between them occurs with the increase of the number of wells N.

This has a simple explanation. Let us consider the resonance in some pair of adjacent wells, taking place when the corresponding transition probability $P_k(\Delta u_k)$ is near maximum. If this maximum is sharp enough, the current density also passes through a maximum as Δu_k increases. The total potential drop on the non-resonant part of the SL is proportional to I, while Δu_k increases monotonously. This leads either to an increase, or to a decrease of the voltage on the whole structure. The former corresponds to a conventional N-shaped portion of the I-V curve, and the latter corresponds to a Z-shaped portion. The resistance of the non-resonant part of the structure, mentioned above, grows with N. This makes the Z-like shape of the corresponding current maximum more pronounced in the longer SLs.

EFFECT OF THE DISORDER

The I-V curves for periodic and disordered structures are depicted in Fig. 2.

There are various types of the disorder, which can be introduced into the layer parameters of the SL, and all of them would affect the transport properties. Fluctuations of the well parameters change the transition probabilities via the size-quantization energies, while disorder in the barrier parameters affects, first of all, the tunneling component of the transition probabilities.

As a specific example, here we will concentrate on the disorder in the well compositions, i.e., in their potentials. Their effect on the I-V curves of the triple-barrier structures is shown in Fig. 2. One should note the shift of the first N-shaped current maximum, corresponding to the coincidence of the energy levels in the inner quantum wells. At the same time, the other two Z-shaped maxima at higher voltages, which arise due to the resonances between the emitter and the adjacent inner well, are almost not affected. This is in agreement with the natural fact that only the functions $P_k, R_k(\Delta u_k)$ and $P_{k+1}, R_{k+1}(\Delta u_{k+1})$ are changed by changing the single parameter Δu_k^0.

For the SLs with larger N this gives us the possibility to shift any particular pair of current maxima, without affecting the other features of the curve. As a result, we can control the transport properties of semiconductor SLs by introducing the intentional disorder into the layer parameters.

ACKNOWLEDGMENTS

This work was partially supported by the Russian Foundation for Basic Research (Grant No. 1786.2003.2).

REFERENCES

1. Goldman, V.J., Tsui, D.C., and Cunningham, J.E., *Phys. Rev. Lett.*, **58**, 1256–1259 (1987).
2. Martin, A.D., Lerch, M.L.F., Simmonds, P.E., and Eaves, L., *Appl. Phys. Lett.*, **64**, 1248–1250 (1994).
3. Wacker, A., *Phys. Rep.*, **357**, 1–111 (2002).
4. Bonilla, L.L., *J. Phys.: Condens. Matter*, **14**, R341–R381 (2002).
5. Pupysheva, O.V., Dmitriev, A.V., Farajian, A.A., Mizuseki, H., and Kawazoe, Y., to be published.

Quantum Interference and Localization in Disordered GaAs/AlGaAs Superlattices

Yu.A.Pusep

Instituto de Física de São Carlos, Universidade de São Paulo, 13560-970 São Carlos, SP, Brazil

Abstract. The coherence and the dimensionality of electrons were studied in the intentionally disordered GaAs/AlGaAs superlattices where the disorder was produced either by the random variation of the well thicknesses or by the interface roughness. Depending on relative values of the disorder energy and the Fermi energy the coherent and incoherent diffusive transport regimes were distinguished and the disorder driven metal-to-insulator transition was achieved. The qualitatively different magnetoresistances with the positive and negative concavities were observed in the metallic and insulating samples respectively. Good agreements were found with the theories developed in the limits of the weak and strong disorder. The localization properties of the single-particle excitations were compared with those of the collective excitations.

INTRODUCTION

The electron systems built up the weakly coupled metallic layers exhibit many interesting properties due to their electric anisotropy, when the conductivities perpendicular and along the layers are considerably different. The interplay between the interlayer and intralayer conductivities is essentially linked to the dimensionality of the system: the vanishing interlayer conductivity results in two-dimensionalization of the electronic properties. The interlayer transport is also linked to the coherence of quasiparticles responsible for the parallel metallic conductivity [1]. The anisotropy of the layered materials naturally sets two temporal scales: the decoherence time τ_φ and the time an electron needs to change a plane τ_0. The relationship between these two characteristic times establishes the coherence-incoherence crossover in the layered electron systems. Moreover, a possibility to produce intentional disorder in the layered electron systems permits to realize the insulating regime and thus, to study the electron coherence when crossing the metal-to-insulator transition.

RESULTS AND DISCUSSION

The weak-field magnetoresistance was explored in the intentionally disordered doped GaAs/AlGaAs superlattices (SLs) with different strengths of disorder exhibiting both the metallic and the insulating behavior. Depending on the disorder strength, three different regimes of the quantum transport were distinguished [2]: two regimes of the weak localization ($k_F l \gg 1$) identified as (i) the propagative regime ($t_z \tau > \hbar$) and (ii) the diffusive one ($t_z \tau < \hbar$) and (iii) the strong localized insulating regime ($k_F l \ll 1$), where k_F, l, t_z and τ are the Fermi wave number, the mean free path length, the vertical coupling energy (which is linked to the interlayer tunneling time τ_0) and the elastic scattering time respectively.

The coherence of the electrons was studied in the periodic $(GaAs)_m(Al_{0.3}Ga_{0.7}As)_6$ SLs, where thicknesses are expressed im monolayers (ML), as a function of the well thickness (m) and consequently, the disorder induced by the interface roughness. In such SLs the disorder driven metal-to-insulator transition was achieved with the decreasing well thickness. In this case the quantum interference was found to be responsible for the weak-field negative magnetoresistances observed on both sides of the metal-to-insulator transition (MIT). The dependences of the resistances on the magnetic field revealed different behavior on the metallic and insulating sides of the transition. It demonstrated the positive and negative concavities in the metallic-type (m = 50 ML) and insulating-type (m = 10,15 ML) samples respectively, which were found in good agreements with the weak [3] and strong [4] localization theories in SLs (Fig.1). The different temperature dependences of the phase breaking times shown in Fig.2 were observed on different sides of the MIT reflecting dissimilar inelastic scattering mechanisms. Thus, the MIT was found to manifest itself in the corresponding modification of the quantum interference [5].

The diffusive transport regime was realized in the SLs there the disorder was produced by a random variation of the well thicknesses. In this case the disorder strength was characterized by the disorder parameter $\delta_{SL}=\Delta/W$, where Δ is the full width at half maximum of a Gaussian distribution of the electron energy calculated in the isolated quantum wells and W is the miniband width of the nominal SL in the absence of disorder.

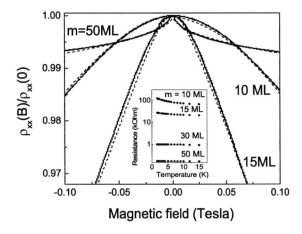

FIGURE 1. Relative magnetoresistivities measured in the $(GaAs)_m(Al_{0.3}Ga_{0.7}As)_6$ SLs with different well thicknesses at T = 1.6 K. The insertion shows the temperature dependences of the zero-field resistivities.

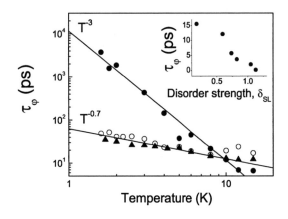

FIGURE 2. Temperature dependences of the dephasing times τ_φ obtained in the metallic-type $(GaAs)_m(Al_{0.3}Ga_{0.7}As)_6$ SL with m = 50 ML (full circles) and in the insulating-type SLs with m = 15 ML (open circles) and m = 10 ML SLs (triangles). The insertion shows the dependence of the in-plane phase-breaking time on the strength of the vertical disorder obtained in the SLs with the intentional vertical disorder.

Such anisotropic disorder was found to result in the anisotropy of the phase-breaking time.

Moreover, in the diffusive transport regime the disorder driven coherence-incoherence crossover was detected and the dependence of the in-plane decoherence time was obtained as a function of the vertical disorder. The data presented in the insertion to Fig.2 show this time decreasing with the increasing disorder strength. This demonstrates that the disorder induced break-down of the interlayer coherence of quasi-particles drastically affected their intralayer coherence.

The phase coherency length obtained by the magnetoresistance measurements determines the minimum width of an electron wave packet and therefore, it may serve as the lower cutoff for the localization length [6]. On the other hand, the localization length associated with the indetermination of the quasi-momentum of the plasmons in the presence of disorder can be measured by Raman scattering [7]. It was demonstrated that in the presence of disorder the localization length of the individual electron is considerably longer than that of the collective excitations. This indicates that the disorder has weaker effect on the individual electrons than on their collective motion and that the interaction, which gives rise to the collective effects, enhances localization.

ACKNOWLEDGMENTS

H.Arakaki, C.A.Souza, A.I.Toropov and S.Malzer are thanked for growth of the samples, the assistance in the experiments from M.B.Ribeiro and P.Zanello is appreciated and the financial support from FAPESP is gratefully acknowledged.

REFERENCES

1. T.Valla, et al., Nature (London), **417**, 627 (2002)

2. W.Szott, et al., Phys.Rev. B **40**, 1790 (1989)

3. V.V.Bryksin and P.Kleinert, Z.Phys. B **101**, 91 (1996)

4. Yu.A.Pusep, et al., Phys.Rev.B **68**, 195207 (2003)

5. Yu.A.Pusep, et al., Phys.Rev.B **68**, 205321 (2003)

6. B.L.Altshuler, et al, J.Phys.C **15**, 7367 (1982)

7. Yu.A.Pusep, et al., Phys.Rev.B **61**, 4441 (2000)

Individual band mobilities in a double quantum well

R. Fletcher*, M. Tsaousidou[†], T. Smith*, P. T. Coleridge**, Z. R. Wasilewski** and Y. Feng**

*Physics Department, Queen's University, Kingston, Ontario, Canada, K7L 3N6.
[†]Materials Science Department, University of Patras, Patras 26 504, Greece.
**Microstructural Sciences, National Research Council, Ottawa, Canada K1A OR6.

Abstract. The transport mobilities at 0.3K have been measured for the two bands in a GaAs double quantum well as a function of top-gate voltage. The results were obtained by fitting the experimental semi-classical low-field magneto-resistance to the expected 2-band behaviour with inter-subband scattering taken into account. The mobilities lie on two smooth curves which cross at the resonance point. They are in good agreement with theoretical values obtained by solving the Boltzmann equation when two subbands are occupied.

When the two wells of a double quantum well (DQW) are adjusted so that the energy levels coincide, the wave functions in the wells are strongly mixed to form symmetric and antisymmetric states separated by an energy gap which depends on the barrier between the wells, and the electrons have equal probability of being in either well; this is the resonance condition. As the system moves away from resonance (in the present case by the use of a top-gate voltage) the two bands become progressively localized in one or other of the wells. We describe experiments to determine the transport mobilities at 0.3K for the symmetric and the anti-symmetric states as a function of top–gate voltage. We also present theoretical results of the transport mobilities obtained by solving the Boltzmann equation when two subbands are occupied and assuming that the electron mobility is limited by scattering due to ionized impurities.

The present experiments made use of a sample with two 18nm wide GaAs wells separated by a 3.4nm barrier of $Al_{0.67}Ga_{0.33}As$. Electrons were provided by a δ-doped Si layer on each side of the wells and separated from them by 120nm of $Al_{0.67}Ga_{0.33}As$. The relative energies of the wells were adjusted by a top gate which comprised a gold film insulated by 30nm of SiO_2. Examples of the measured semi-classical magneto-resistivity around $B = 0$ and the fitted curves (see below) for various gate voltages are shown in Fig. 1.

The standard two-band model for the semi-classical magneto-resistivity in a perpendicular magnetic field assumes that two groups of carriers exist with densities n_i ($i = 1, 2$, with 1 corresponding to the symmetric state in the present context) and transport mobilities μ_i and that the two groups carry current independently. For the two

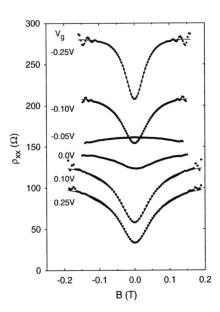

FIGURE 1. Semi-classical magneto-resistivities at various fixed values of V_g. The solid dots are experimental data and the lines are the fitted curves using the sum of Eq. (1) and Eq. (2).

bands in our DQW the last assumption is not satisfied around the resonance point where interband scattering is strong. Zaremba [1] has shown that interband scattering can be taken into account and the resulting resistivity can be written

$$\rho_{xx} = \rho_0 \left[1 + \frac{rn_1 n_2 \mu_1 \mu_2 (\mu_1 - \mu_2)^2 B^2}{(n_1\mu_1 + n_2\mu_2)^2 + (rn_T \mu_1 \mu_2)^2 B^2} \right] \quad (1)$$

where $\rho_0 = 1/(n_1\mu_1 + n_2\mu_2)e$ is the zero–field resistivity, $n_T = n_1 + n_2$ and the dimensionless parameter r

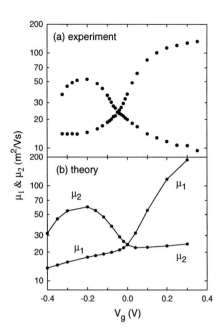

FIGURE 2. The top panel shows the experimental mobilities, the bottom panel the calculated mobilities.

takes into account interband scattering. When $r = 1$ this equation reduces to the usual result for two independent bands.

We analyzed ρ_{xx} using the full expression Eq. (1), i.e., including ρ_0. The values of n_i for the two bands were taken from the deHS oscillations. To obtain the best fits it was necessary to include an extra resistivity contribution of the form

$$\rho_{xx}^{ex} = aB + bB^2 \qquad (2)$$

The first term was very small ($\leq 1\%$ of ρ_0) and arises because the data were not precisely even in B, presumably due to a small admixture of the Hall resistivity. The coefficient b was always negative. A negative quadratic behaviour is often seen in magneto-resistance experiments and has been ascribed to Coulomb interactions, e.g., Beenakker and van Houten [2]. Eq. (1) is also approximately quadratic (and positive) near $B = 0$, but the factor bB^2 gave a much smaller contribution and was clearly visible only near resonance.

The fit had five free parameters comprising μ_1, μ_2, r, a and b. The fitted curves are shown in Fig. 1 and the resulting values for μ_i as a function of V_g are shown in Fig. 2(a). The values of μ_i lie on two smooth curves which cross at the resonance point ($V_g \approx -0.05$V). The resulting values of μ_i were hardly changed if we set $a = 0 = b$, though the fitted curves were visibly worse. At resonance the two bands have identical mobilities and the minimum around $B = 0$ from Eq. (1) disappears in Fig. 1. At this point only Eq. (2) makes a contribution and the negative quadratic term bB^2 can be seen. At ≈ -0.4V

(not shown) the upper well empties and the magneto-resistivity again shows no minimum. The width of the resonance observed in the mobilities in Fig.2 correlates well with that in the measured zero-field resistivity for this sample [3].

The mobility of each subband is $\mu_i = e\tau_i/m^*$ where τ_i is the elastic (transport) relaxation time due to electron scattering by ionized impurities. τ_i was calculated within the Boltzmann framework [4, 5] taking into account screening effects and interband scattering. In our DQW, carriers in the top well are scattered much more weakly than those in the lower well and to take this into account in the calculations the concentration of the impurities near the lower well was taken to be almost 10 times larger than the concentration near the upper well. The results are shown in Fig. 2(b). Note that in the experiments the resonance appears at $V_g \approx -0.05$V, but in the calculations it occurs at 0V. At resonance the carriers spend equal times in each well (assuming the tunneling time is short compared to the scattering time) so that the impurity scattering rate for all carriers is the same. Well away from resonance the carriers in the each state are essentially localized in one or other of the wells and act almost independently. In this limit the factor r in Eq. (1) is unity, as expected.

In conclusion, the experimental and calculated mobilities of the two bands of the DQW are in good agreement. The mobilities fall on two smooth curves which cross at the resonance point.

ACKNOWLEDGMENTS

This work was supported by a grant from the Natural Sciences and Engineering Research Council of Canada. We thank Jean Beerens at the University of Sherbrooke for providing access to pieces of the wafer. MT would also like to thank the University of Warwick for providing access to its computer facilities.

REFERENCES

1. E. Zaremba, Phys. Rev. B **45**, 14143 (1992).
2. C. W. Beenakker and H. van Houten in *Solid State Physics*, edited by H. Ehrenreich and D. Turnbull (Academic Press, New York, 1991) p 1.
3. T. Smith, M. Tsaousidou, R. Fletcher, P. T. Coleridge, Z. R. Wasilewski and Y. Feng, Phys. Rev. B **67**, 155328 (2003).
4. S. Mori and T. Ando, Phys. Rev. B **19**, 6433 (1979).
5. G.-Q Hai, N. Studart, F.M. Peeters, Phys. Rev. B **52**, 8363 (1995).

An External ac Bias Induced Expansion of Dynamic Voltage Bands in a Weakly Coupled GaAs/AlAs Superlattice

H.T. He, Z.Z. Sun, X.R. Wang, Y.Q. Wang, W.K. Ge, J.N. Wang[1]

Physics Department, Hong Kong University of Science and Technology, Clear Water Bay, Hong Kong, China

Abstract. In the intermediate temperature region, an external ac bias induced expansion of dynamic voltage band (DVB) is investigated experimentally in a weakly coupled GaAs/AlAs superlattice. The observed results are explained based on an analysis of general nonlinear dynamics. It suggests that the effect of an external ac bias is to tighten the unstable limit cycle onto the stable fixed point. Beyond a certain critical value, the stable limit cycle is the only stable solution of the system resulting in the observation of self-sustained current oscillation.

Since the pioneering work of Esaki and Chang [1], the vertical transport of weakly coupled superlattice (SL) has been shown to exhibit many interesting behaviors. It includes the saw-tooth-like current-voltage (I-V) characteristics [2], self-sustained current oscillation (SSCO) [3-4] and chaos [5]. It's understood that the formation of stable electric field domain (EFD) in SL is responsible for the occurrence of saw-tooth like I-V characteristics and the SSCO is attributed to the traveling of domain boundary within SL [6], which can be induced by varying the doping density (N_D) [3], changing temperature (T) or applying transverse magnetic field (B) [4]. In 1999 J.N.Wang et.al. [4] found that in the transition from static to dynamic EFD, a so-called dynamic voltage band (DVB) emerged at each saw-tooth like current branch. In general, the control parameter space (N_D, T or B) can be divided into three regions. One corresponds to static EFD, in which only saw-tooth-like I-V characteristics is observed, the other dynamic EFD, in which the whole sequential tunneling plateau exhibits SSCOs. In between, there's an intermediate region, where alternating DVB and static voltage band (SVB) are observed. In this region the applied dc bias is an additional control parameter for the system stability. In this work, we focus on the influence of an external ac bias on the DVB in the intermediate region of temperature. It is found that an external ac bias can induce the DVB and increase the width of DVB as the ac amplitude increases. A possible mechanism is given in terms of general nonlinear dynamic analysis.

The GaAs/AlAs SL sample used in this work is grown by molecular beam epitaxy. It consists of 30 periods of 14nm GaAs well and 4nm AlAs barrier and is sandwiched between two n^+-GaAs layers. The central 10nm of each GaAs well is doped with Si ($n=2\times10^{17} cm^{-3}$). The sample is fabricated into $0.2\times0.2 mm^2$ mesas. The sample temperature is fixed at 97K, at which the temperature-induced formation of DVB is observed in the I-V. For clarity, only part of I-V from the first sequential tunneling plateau is given in Fig.1. It shows the presence of one DVB, along with two complete saw-tooth-like branches. The SSCOs from this DVB is given in the inset of Fig.1 with dc bias at 537mV and the SSCO frequency f_0 at 2.89 KHz.

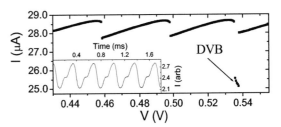

FIGURE 1. I-V characteristics show the coexistence of SVB and DVB. Inset: SSCOs obtained in the DVB indicated.

The influence of a small external ac bias on the I-V characteristics is investigated using a fixed ac frequency (3 KHz) but varying amplitude V_{ac} (0–

---

[1] Corresponding author email: phjwang@ust.hk

14mV). Figure 2 shows the obtained results in grey-scale plot. The measured time-average current is shown in density plot where darker area corresponds to lower current value. For clarity, the dc bias range in Fig.2 is the same as in Fig.1. Referring to Fig.1 it is clear that the dark area in Fig.2 corresponds to DVB, where the current is smaller than that at SVB. In Fig.2, two interesting features are observed. One is that the DVB is induced at the first and second saw-tooth-like branches when V_{ac} is larger than about 2 mV. The other is the expansion of each DVB with increasing V_{ac}. At V_{ac}=12mV, the three DVBs join up together turning the whole dc voltage region into dynamic.

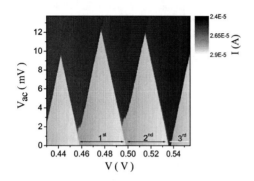

FIGURE 2. The grey scale plot showing the influence of an ac signal amplitude on the DVB.

In our previous work, it has been shown that the transport dynamics of a weakly coupled SL can be well described in the framework of nonlinear dynamics. The saw-tooth-like I-V characteristics corresponds to a stable fixed point (SFP) in phase space (see Fig.3(c)), but the formation of a stable limit cycle (SLC) (see Fig.3(a)) results in SSCO [7]. Based

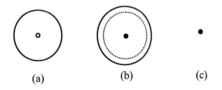

FIGURE 3. (a)-(c) show the phase portraits corresponding to the three different regions of control parameter space.

on these, we propose that in the intermediate region of the control parameter space the phase portrait of the system consists of a SLC (solid circle), an unstable limit cycle (ULC) (dot circle) and a SFP (solid dot) as illustrated in Fig. 3 (b). In this region, when the control parameters (N_D, T, B, V) are varied in the direction to bring the system from SLC to SFP, the SLC and ULC are forced to get closer and closer. At a critical value, the two limit cycles collide and annihilate leaving only a SFP in the phase space (see Fig.3 (c)). On the other hand, when the control parameters are varied to bring the system from SFP to SLC, the ULC shrinks and finally swallows the SFP, rendering it unstable (see the open dot in Fig.3(a)). As a result, a SLC becomes the only stable solution of the system. To explain our experimental results, the influence of an external ac bias applied in the intermediate region of control parameters can be considered as to further separate the SLC and ULC and to force ULC to tighten onto the SFP. As V_{ac} increases and reaches a critical value the ULC annihilates the SFP turning it into an unstable fixed point. As a result, the system changes from SFP to SLC solution and the SSCO is observed. However, the critical value of V_{ac} depends on the other control parameters. In this experiment, only the applied dc bias V is varied. Different V corresponds to different critical values of V_{ac}. Within a saw-tooth-like current branch, the critical V_{ac} increases with increasing V. As applied V_{ac} is gradually increased the DVB emerging first at the lower voltage side in the branch is becoming wider and wider. So as shown in Fig.2, the DVB at the first and second branches is induced and that at the third branch is expanded with increasing V_{ac}.

In conclusion, the influence of an ac bias on the emergence and expansion of DVB is studied experimentally. It reveals that the width of DVB increases with the ac amplitude and finally the whole plateau exhibits SSCOs. The underlying mechanism is given based on the general nonlinear analysis of phase portraits of the system.

ACKNOWLEDGMENT

This work is supported by UGC, Hong Kong, via grants HKUST6149/00P and HKUST6162/02P.

REFERENCES

1. L. Esaki et al., IBM J.Res.Develop. 14, 61 (1970); R. Tsu et al., Appl.Phys.Lett. 22, 562 (1973); L. Esaki et al., Phys.Rev.Lett. 33, 495 (1974).
2. K. K. Choi et al., Phys. Rev. B 38, 12362 (1986); H. T. Grahn et al., Phys. Rev. Lett. 67, 1618 (1991); P. Helgesen et al., J. Appl. Phys. 69, 2689 (1991); Y. Zhang et al., Appl. Phys. Lett. 64, 3416 (1994); S. H. Kwok et al., Phys. Rev. B. 50, 2007 (1994).
3. S. H. Kwok et al., Phys. Rev. B. 51, 10171 (1995); J. Kastrup et al., Phys. Rev. B. 52, 13761 (1995); H. T. Grahn et al., Jpn. J. Appl. Phys. 34, 4526 (1995).
4. J. N. Wang et al., Appl. Phys. Lett. 75, 2620 (1999); X. R. Wang et al., Phys. Rev. B. 61, 7261 (2000); J. N. Wang et al., Solid State Commun. 112, 371 (1999).
5. Y. Zhang et al., Phys. Rev. Lett. 77, 3001 (1996).
6. J. Kastrup et al., Phys. Rev. B. 55, 2476 (1997); A. Wacker et al., Phys. Rev. B. 55, 2466 (1997).
7. Z. Z. Sun et al., Phys. Rev. B. 69, 045315 (2004).

A new class of small low-field magnetoresistance oscillation in unidirectional lateral superlattice

Akira Endo* and Yasuhiro Iye*

Institute for Solid State Physics, University of Tokyo, 5-1-5 Kashiwanoha, Kashiwa, Chiba 277-8581, Japan

Abstract. A new class of small-amplitude magnetoresistance oscillation has been unveiled in unidirectional lateral superlattice (ULSL) samples in a low magnetic field regime dominated by positive magnetoresistance. The oscillation is ascribed to geometric resonance of Bragg-reflected cyclotron orbits. The predicted approximate $\propto a^{-3} k_F^{-1}$ dependence of the positions of maxima, where a and k_F represent the period and the Fermi wavenumber, respectively, has been confirmed by comparing ULSL samples with a=184 and 207 nm.

Unidirectional lateral superlattice (ULSL) samples are well known to display two types of magnetotransport characteristics: positive magnetoresistance (PMR) [1] at magnetic fields around zero, and commensurability oscillation (CO) [2] above the field range for PMR. Both PMR and CO can be basically understood by the semiclassical motion of electrons under periodic potential modulation $V_0 \cos(2\pi x/a)$ and a perpendicular magnetic field B. Very recently we have uncovered a small-amplitude magnetoresistance oscillation superposed on the slope of PMR ([3], referred to as paper I, hereafter). The new oscillation is observed for four ULSL samples with the period a ranging from 138 to 184 nm, and from the a- and the electron density n_e- dependences, is attributed to geometric resonance between a and the width $b_{j,k}$ of an open orbit resulting from the miniband structure of the superlattice. The resonant condition where magnetoresistance $\Delta \rho_{xx}/\rho_0$ takes local maximum is given by equating $b_{j,k}$ with the multiple of the period, na. The width $b_{j,k}$ of the open orbit between j-th and k-th reflection points is given by (assuming $\eta \equiv V_0/E_F \ll 1$, E_F being the Fermi energy),

$$b_{j,k} = \frac{\hbar k_F}{e|B|} \left[\sqrt{1 - \left(\frac{j\pi}{a k_F}\right)^2} - \sqrt{1 - \left(\frac{k\pi}{a k_F}\right)^2} \right], \quad (1)$$

where $k_F = \sqrt{2\pi n_e}$ denotes the Fermi wavenumber. The positions of local maxima (or minima in $d^2(\Delta\rho_{xx}/\rho_0)/dB^2$ to be explained later), therefore, read, up to the first order of $(\pi/a k_F)^2$,

$$|B_{\min}^{j,k,n}| \simeq \frac{\pi^2 \hbar}{2 n e a^3 k_F} (k^2 - j^2). \quad (2)$$

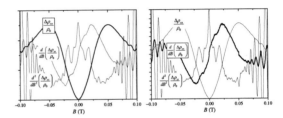

FIGURE 1. Left: Magnetoresistance with its numerical first and second derivative. Right: Directly measured first derivative with its numerical integral and derivative. Thick curves represent raw data. The data were taken on Sample A at 4.2 K.

Because of the a^{-3} dependence, $|B_{\min}^{j,k,n}|$ shifts rapidly with a, especially for smaller a. However, the oscillation is limited to a narrow window $|B| < \sim 0.04$ T, since the open orbits become demolished by the dominance of the magnetic breakdown at higher fields. In paper I, $|B_{\min}^{j,k,n}|$'s that appeared in the window belonged to different sets of parameters (j,k,n) for different samples having different a, which made the verification of the a-dependence rather unclear. In the present paper, we measure another ULSL sample with larger period, a=207 nm, and compare it with the result for the a=184 nm sample. The two samples show directly corresponding $|B_{\min}^{j,k}|$'s (n=1) which actually scales in accordance with Eq. (2), furnishing another support for our interpretation.

Two ULSL samples with periods a=184 nm (Sample A) and 207 nm (Sample B) were prepared from the same GaAs/AlGaAs wafer and had the same Hall-bar pattern (64×37 μm^2). Potential modulation was introduced by strain-induced piezoelectric effect [4] from a surface grating. Modulation amplitudes were measured from the amplitude of CO [5] and were $V_0 \sim 0.25$ and

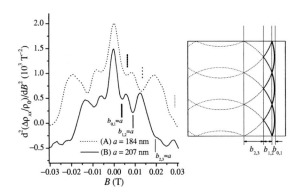

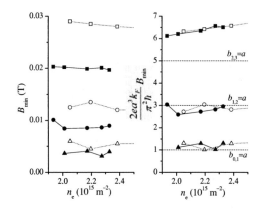

FIGURE 2. Left: $d^2(\Delta\rho_{xx}/\rho_0)/dB^2$ traces for Samples A and B for similar n_e; $n_e=2.38\times10^{15}$ m^{-2} and 2.33×10^{15} m^{-2} for dotted (A) and solid (B) curves, respectively. The former is vertically offset for clarity. Vertical lines mark the positions of minima. Right: Sketch of the open orbits (Bragg-reflected cyclotron orbits).

FIGURE 3. Left: Plot of minima position $B_{min}(>0)$ versus n_e. Open and solid symbols represent Samples A and B, respectively. Right: B_{min} normalized by $\pi^2\hbar/2ea^3k_F$ [see Eq. (2)].

0.22 meV for Samples A and B, respectively, and did not vary much with n_e. The measurement of ρ_{xx} was done by standard low-frequency lock-in technique. In order to pick out the small-amplitude oscillation from much larger PMR background, we numerically differentiated magnetoresistance with respect to B. Alternatively, $d\rho_{xx}/dB$ was directly measured by picking up the resistance that follows a small-amplitude (rms 1 mT) ac component of magnetic fields superposed on an ordinary dc sweep. Fig. 1 demonstrates the consistency of both types of measurements. We use the data obtained by standard measurements in the following. Close inspection of the traces in Fig. 1 reveals that minima in $d^2\rho_{xx}/dB^2$ correspond to local maxima in ρ_{xx}. To see the dependence of the oscillation on k_F, n_e was varied by employing persistent photoconductivity effect.

Fig. 2 compares second derivative traces for the two samples at nearly the same n_e. Straightforward correspondence of the minima positions (marked by dotted and solid vertical lines) can be observed between the two traces. The minima are identified as the condition that the width $b_{j,k}$ (right panel in Fig. 2) satisfies $b_{j,k}=a$ ($n=1$). It can readily be seen that the minima for Sample B appears at smaller $|B|$. In the left panel of Fig. 3, positions of minima are plotted as a function of n_e. The figure clearly shows the trend of smaller B_{min} for Sample B having larger a, in the ranges of n_e. To be more quantitative, B_{min} is normalized by the prefactor $\pi^2\hbar/2ea^3k_F$ of Eq. (2) and plotted in the right panel of Fig. 3. The normalized values coincide between the two samples, unambiguously confirming the a^{-3} dependence. Moreover, the normalized value is expected from Eq. (2) to be equal to k^2-j^2 (shown as horizontal dashed line), which is also evident in the figure, at least for the two smaller B_{min}'s. The deviation from $3^2-2^2=5$ for the largest B_{min} will partly be explained by the lesser applicability of the first order approximation in Eq. (2) for larger j and k. Owing to rather narrow n_e range, k_F^{-1} dependence is less evident, which has been unequivocally shown in paper I.

To summarize, we have confirmed the expected a-dependence of the new low-field magnetoresistance oscillation we have found in ULSL, supporting our interpretation that the oscillation originates from the geometric resonance of open orbits. This provides another piece of evidence of the formation of miniband structure in ULSL [6].

ACKNOWLEDGMENTS

This work was supported by Grant-in-Aid for Scientific Research (C) (15540305) and (A) (13304025) and Grant-in-Aid for COE Research (12CE2004) from Ministry of Education, Culture, Sports, Science and Technology.

REFERENCES

1. Beton, P. H., Alves, E. S., Main, P. C., Eaves, L., Dellow, M. W., Henini, M., Hughes, O. H., Beaumont, S. P., and Wilkinson, C. D. W., *Phys. Rev. B*, **42**, 9229 (1990).
2. Weiss, D., v. Klitzing, K., Ploog, K., and Weimann, G., *Europhys. Lett.*, **8**, 179 (1989).
3. Endo, A., and Iye, Y. (2004), cond-mat/0406667.
4. Skuras, E., Long, A. R., Larkin, I. A., Davies, J. H., and Holland, M. C., *Appl. Phys. Lett.*, **70**, 871 (1997).
5. Endo, A., Katsumoto, S., and Iye, Y., *Phys. Rev. B*, **62**, 16761 (2000).
6. Deutschmann, R. A., Wegscheider, W., Rother, M., Bichler, M., Abstreiter, G., Albrecht, C., and Smet, J. H., *Phys. Rev. Lett.*, **86**, 1857 (2001).

Interlevel Crossings In Double Period Superlattices

M. Coquelin, C. Pacher, M. Kast G. Strasser and E. Gornik

Institut für Festkörper Elektronik, Technische Universität Wien, Floragasse 7, A 1040 Wien, Austria

Abstract. The transmission properties of a biased GaAs/AlGaAs-superlattice with two different alternating well widths were theoretically and experimentally studied. The minibands of this structure split into two separated subminibands. Both subminibands show Wannier Stark splitting due to the applied bias. At special bias values crossings between the stark levels of both subminibands can be induced. A three terminal device was used to derive the transmission spectra. It is shown that the strength of the level crossing can be engineered.

INTRODUCTION

Since the idea of polytype superlattices was proposed by Esaki et al. [1], several studies have been made on superlattices with multilayer bases. The Stark lokalisation in superlattices with alternating well widths was studied by photocurrent measurements [2]. Resonant tunneling due to interlevel crossing was measured by Kristensen et al. [3].

In this work the transmission properties of a biased finite superlattice with two different alternating wells are studied, using a three terminal device (3TD), which is based on the THETA-device of Heiblum et al. [4]. The 3TD is a transistor like structure which enables to perform hot electron spectroscopy on arbitrary biased superlattice structures. Kast et al. described a 3TD with an advanced resolution of approximately 10meV [5], which was used to map the Wannier Stark splitting in superlattices with a simple basis [6].

THREE TERMINAL DEVICE (3TD)

The experiments have been performed using a 3TD at a temperature $T=4.2K$. In a 3TD hot electrons are injected through an AlGaAs-tunnel barrier between two contacts called emitter and base into a GaAs-drift region. The energy of these hot electrons is tunable by the applied voltage U_{eb} between these contacts. The Fermi energy $E_f=eU_{eb}$ can be assumed as the highest possible energy of the injected electrons. After traversing the drift region, the hot electrons reach the superlattice structure, which is placed between the base and a third contact (collector). Therefore the electrons, which are transmitted through the superlattice, are detected as collector current I_C. The three contacts are highly doped (n=1*10^{18}cm^{-3}) n-GaAs-layers. By dividing collector and emitter current one gets the transfer ratio $\alpha=I_C/I_E$ which equals the convolution of the transmission function of the superlattice and the energy distribution of the electrons which reach the superlattice. Therefore miniband transport leads to peaks in α. If the energy of the injected electrons is much higher than the position of the minibands the transfer ratio is still nonzero. This is caused by scattering processes in the drift region. The emission of LO-phonons is the most relevant scattering mechanism.

WANNIER STARK SPLITTING

The investigated structure consists of 5 GaAs-wells between $Al_{0.3}Ga_{0.7}As$-barriers with a constant width $b=3.6nm$. The width of the wells is $w_1=4.1nm$ for the first, third and fifth layer and $w_2=3.6nm$ for the second and fourth. Without an applied bias the energy levels of wells of the same width are coupled. This leads to a splitting of the first miniband into two subminibands at 99.5meV 122meV. If a strong electric field is applied to the superlattice the energy levels (E_i) of both subminibands localize. This leads to a splitting of the levels, which increases with increasing bias. At special bias values the energy levels of wells with different well width align (Figure 1).

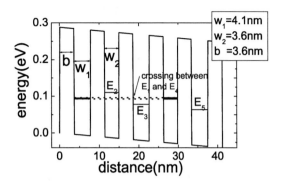

FIGURE 1. Conduction band of the superlattice with applied field (10kVcm^{-1}).

While the transfer ratio through the subminibands decreases with increasing bias due to the localization of the subminiband states one finds an increase of the transfer ratio due to interlevel crossings. Figure 2 shows the transfer ratio the versus the V_{eb} for different biases between collector and base. The first peak of the transfer ratio at $V_{cb}=216mV$ shows a strong increase. This increase of the transmission is caused by the alignment of E_1 and E_4. One can also find an increase in the transfer ratio for $V_{cb}>500mV$.

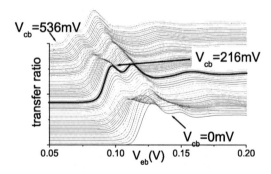

FIGURE 2. Transfer ratios versus bias voltage between emitter and base for different collector basis bias voltages.

The position of the transmission peaks can be derived from the second derivative of the transfer ratio. Due to undoped GaAs layers around the superlattice V_{cb} does not equal the bias at the superlattice. Nevertheless it can be assumed, that these values are proportional. The energy positions of the transmission resonances can only be analyzed in relation to other resonances. Therefore Figure 3 shows the measured difference between the peak position of the highest level E_2 and the lower levels (crosses) compared to the calculated values derived by using the transfer matrix method (lines). Up to $V_{cb}=300meV$ the peak positions of the levels E_1–E_4 can be found. At higher biases the peaks of E_3 and E_4 vanish. Due to the low transmission properties no peak, indicating transmission through the lowest level E_5, can be found.

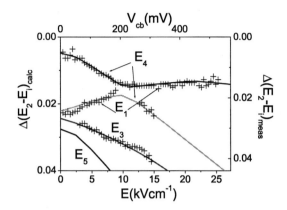

FIGURE 3. Measured and calculated energy positions of the transmission resonances in relation to the highest resonance.

The measurements are in good agreement to the calculated values. Between $V_{cb}=192mV$ and $V_{cb}=256mV$ one finds interlevel crossing between E_1 and E_4. Due to the coupling between the two energy levels the calculations show the formation of a subminiband consisting of two states. The minimal energy difference between these two states is $\Delta E=3meV$ (anticrossing behavior). ΔE_{14} indicates a Rabi frequency of $f_r=725GHz$. At $V_{cb}=450mV$ the same behavior can be found for the levels E_1 and E_2. The splitting of $\Delta E_{12}=13meV$ indicates a Rabi frequency of $f_r=3.14THz$. These two states are represented by the two peaks in the transfer ratios at high biases between collector and base (see Figure 2). This demonstrates that the level crossing splitting can be engineered in a wide energy range.

REFERENCES

1. Esaki, L, Chang, L. L., and Mendez, E. E., *Jpn. J. Appl. Phys.* **20**, L529-532 (1981).
2. Hinooda, S., and Fujiwara, K., *Solid State Comm.* **94**, 817-820 (1995).
3. Heiblum, M., Nathan, M. I., Thomasv, D. C., and Knoedler, C. M., *Phys. Rev. Lett.* **55**, 2200-2203 (1997).
4. Kristensen, A., Lindelof, P.E., Sorensen, C. B., and Wacker, A., *Semicond. Sci. Technol.* **13**, 910-914 (1998)
5. Kast, M., Pacher, C., Strasser, G., and Gornik, E., *Appl. Phys. Lett.* **78**, 3639-3641 (2001).
6. Kast, M., Pacher, C., Strasser, G., Gornik, E., and Werner, W.S.M., *Phys. Rev. Lett.* **89**, 136803 (2001).

Spatiotemporal dynamics of carriers in quantum wells modulated by surface acoustic waves

A. García-Cristóbal*, A. Cantarero* and P. V. Santos[†]

Materials Science Institute, University of Valencia, P. O. Box 22085, E-46071 Valencia, Spain
[†]*Paul-Drude-Institut für Festkörperelektronik, Hausvogteitplatz 5-7, 10117 Berlin, Germany*

Abstract. We present a theoretical model together with numerical simulations of the carrier dynamics in a quantum well under the influence of the piezoelectric field of a surface acoustic wave. We also analyse the associated spatiotemporal modulation of the radiative recombination under conditions corresponding to different experimental configurations.

The dynamics of photogenerated carriers in semiconductor structures modulated by periodic potentials produced by interdigitated metal gates and surface acoustic waves (SAWs) has received considerable attention in the last years [1, 2, 3, 4, 5]. The potential modulation spatially separates the electrons and holes, thus dramatically enhancing their radiative lifetime. This dynamic control of the radiative recombination can be applied for the design of optoelectronic devices based on the storage and transport of electron-hole (e-h) pairs [1, 2]. In this work, we present a theoretical model together with numerical simulations of the dynamics of two-dimensional photogenerated carriers under the influence of the space- and time-dependent electric field associated with a SAW.

We assume that the carriers are constrained to move in the plane of a quantum well (XY plane). The propagation direction of the acoustic wave is taken as the X-axis, and only the longitudinal component of the associated piezoelectric field is considered:

$$E_{SAW}(x,t) = E_0 \cos\left[2\pi\left(\frac{x}{\lambda_{SAW}} - \frac{t}{T_{SAW}} + \phi\right)\right], \quad (1)$$

E_0 being the amplitude of the oscillating field, and λ_{SAW}, T_{SAW} and $v_{SAW} = \lambda_{SAW}/T_{SAW}$ the SAW wavelength, period and velocity. The e-h pair generation rate G is taken to be uniform in the Y-direction so that the problem becomes effectively one-dimensional. The dynamics of the electron and hole concentrations, $n(x,t)$ and $p(x,t)$, is governed by the drift-diffusion equations:

$$\frac{\partial n}{\partial t} - \mu_n \frac{\partial (nE)}{\partial x} = D_n \frac{\partial^2 n}{\partial x^2} + G - (r+c)np \quad , \quad (2)$$

$$\frac{\partial p}{\partial t} + \mu_p \frac{\partial (pE)}{\partial x} = D_p \frac{\partial^2 p}{\partial x^2} + G - (r+c)np \quad , \quad (3)$$

coupled with the Poisson equation, which determines the internal field E_{ind} induced by the space charge:

$$\frac{\partial E_{ind}}{\partial x} = \frac{e(p-n)}{\varepsilon} \quad . \quad (4)$$

In the above equations, μ_n and μ_p (D_n and D_p) are the electron and hole mobilities (diffusivities), and ε is the background dielectric constant. The total electric field is given by $E = E_{SAW} + E_{ind}$. The parameters r and c are the free-carrier radiative-recombination and exciton formation coefficients, respectively. In addition, we consider the spatiotemporal dynamics of the exciton concentration $N(x,t)$ as described by:

$$\frac{\partial N}{\partial t} = D_X \frac{\partial^2 N}{\partial x^2} + cnp - \frac{N}{\tau_X} \quad , \quad (5)$$

where D_X is the exciton diffusivity and τ_X the exciton radiative lifetime.

In the following we present some illustrative numerical results obtained from Eqs. (1)-(5). Further details of the theoretical model and the values of the parameters used in the simulations can be found in Ref. [6].

We first consider a simplified form of the theoretical model, disregarding for the moment the generation-recombination processes and the space-charge induced field. Under these conditions, the hole transport equation can be written in dimensionless form as:[1]

$$\frac{\partial p(\tilde{x},\tilde{t})}{\partial \tilde{t}} + \tilde{v}\frac{\partial}{\partial \tilde{x}}[u(\tilde{x},\tilde{t})p(\tilde{x},\tilde{t})] = \tilde{d}\frac{\partial^2 p(\tilde{x},\tilde{t})}{\partial \tilde{x}^2} \quad , \quad (6)$$

with $\tilde{x} = x/\lambda_{SAW}$, $\tilde{t} = t/T_{SAW}$ and $u(\tilde{x},\tilde{t}) = \cos[2\pi(\tilde{x} - \tilde{t} + \phi)]$. This simplified model is governed by two dimensionless parameters, namely $\tilde{v} = \mu_p E_0/v_{SAW}$ and $\tilde{d} =$

---

[1] An equivalent equation results for the electron concentration.

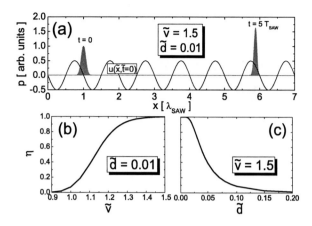

FIGURE 1. (a) Time evolution of an initial Gaussian concentration as given by Eq. (6). (b) and (c) Dependence of the efficiency parameter η (see text) on $\tilde{v}$ and $\tilde{d}$.

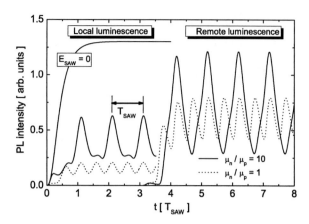

FIGURE 2. Time evolution of the PL intensity recorded at the illumination spot and at a remote position along the SAW path.

$D_p/(\lambda_{SAW} v_{SAW})$. We present in Fig. 1(a) the result of a typical simulation, in which an initial Gaussian concentration is evolved according to Eq. (6). The whole concentration is transported with the SAW velocity as a narrow and stable charge package. In order to characterize the efficiency of this SAW-induced charge transport, we introduce the parameter η defined as the ratio of the concentration at $t = 5T_{SAW}$, integrated around $x = 6\lambda_{SAW}$, to the total initial concentration. In Fig. 1(b) we show the dependence of η on the normalized drift velocity $\tilde{v}$. It is apparent that the acoustically induced transport is established when the parameter $\tilde{v}$ exceeds a threshold value. In the ideal diffusionless case ($\tilde{d} = 0$) this value would be exactly $\tilde{v} = 1$, since then the drift velocity $v_p = \mu_p E_0$ exactly matches the SAW velocity and the carriers are able to follow the wave. If the diffusivity is finite, as in Fig. 1(b), some of the transported carriers fall back to the preceding periods, thus reducing the value of η, and a higher value of $\tilde{v}$ (~ 1.5) is needed to compensate the diffusion effect. This is supported by the results in Fig. 1(c), where it can be seen that a further increase in $\tilde{d}$ (i.e., in the diffusivity) can destroy the SAW-induced transport. In practice, $\tilde{v}$ can be adjusted externally by changing the SAW power (and therefore E_0) and $\tilde{d}$ can be governed to some extent by controlling the temperature.

The SAW-induced transport can be studied experimentally by spatially monitoring the photoluminescence (PL) originated from the recombination of photogenerated e-h pairs [3, 4, 5]. Since the electrons and holes accumulate at different positions within the SAW potential profile, their recombination is inhibited as they are transported by the wave. In practice, the PL can be recovered at a remote position from the generation point by deliberately screening the SAW field [3, 4, 5]. To illustrate these issues we have performed numerical simulations based on Eqs. (1)-(5) of the PL spatiotemporal dynamics resulting from continuous generation of e-h pairs in a localized spot. Figure 2 displays the time evolution of the PL intensity (assumed to be proportional to the exciton population) spatially integrated around the generation spot position (local luminescence) and around a distant point where the SAW field has been forced to vanish (remote luminescence). The quenching of the local luminescence with respect to the no-field case ($E_{SAW} = 0$) is a consequence of the onset of the SAW-induced transport [3]. This is also recognized by the appearance at later times of a stronger remote luminescence. Moreover, both PL traces in Fig. 2 (solid lines) exhibit a characteristic temporal modulation, with the period of the SAW [4, 5]. This modulation is associated with the passage of the higher mobility carriers (electrons) through the detection point. In the hypothetical case of equal mobilities for electrons and holes, the PL would be temporally modulated with the period $T_{SAW}/2$, as shown by the dotted lines in Fig. 2.

REFERENCES

1. S. Zimmermann, A. Wixforth, J. P. Kotthaus, W. Wegscheider, and M. Bichler, Science **283**, 1292 (1999).
2. S. K. Zhang, P. V. Santos, R. Hey, A. García-Cristóbal, and A. Cantarero, Appl. Phys. Lett. **77**, 4380 (2000).
3. C. Rocke, S. Zimmermann, A. Wixforth, J. P. Kotthaus, G. Böhm, and G. Weimann, Phys. Rev. Lett. **78**, 4099 (1997).
4. F. Alsina, P. V. Santos, R. Hey, A. García-Cristóbal, and A. Cantarero, Phys. Rev. B **64**, 041304 (2001).
5. F. Alsina, P. V. Santos, H. P. Schönherr, W. Seidel, K. H. Ploog, and R. Nötzel, Phys. Rev. B **66**, 165330 (2002).
6. A. García-Cristóbal, A. Cantarero, F. Alsina, and P. V. Santos, Phys. Rev. B **69**, 205301 (2004).

High field magnetoresistance of strongly coupled InAs/GaSb superlattices

R. S. Deacon*, A. B. Henriques†, R.J. Nicholas* and P. Shields*

*Department of Physics, Oxford University, Clarendon Laboratory, Parks Rd., Oxford, OX1 3PU, U.K.
†Instituto de Fisica, Universidade de Sao Paulo, Caixa Postal 66318, 05315-970 Sao Paulo, Brazil

Abstract. We investigate vertical magnetotransport measurements in strongly coupled InAs/GaSb superlattices within the miniband transport regime. Sample magnetoresistance curves display oscillations due to the successive un-nesting of Landau level minibands. This qualitative picture is strongly supported by Monte Carlo simulations, which include acoustic and optic phonon scattering as well as Umklapp processes, providing a semi-classical description of the miniband transport.

Since Esaki and Tsu introduced the concept of the semiconductor superlattice [1] extensive studies of short-period superlattices have been performed, primarily in GaAs-AlAs. Strong coupling between energy levels in adjacent wells results in delocalised wave functions leading to the formation of minibands. Miniband transport is characterised by the onset of negative differential conductivity (NDC) where the electron group velocity reaches a peak then decreases for higher applied field.

A magnetic field parallel to the growth axis quantizes the electronic motion in the plane of the superlattice layers. The magnetically quantized states are separated in energy by $(n+1/2)\hbar\omega_c$, where n is an integer Landau level index and $\hbar\omega_c$ is the cyclotron energy. At low magnetic fields the miniband dispersion curve breaks up into a series of 'nested' Landau level minibands (LLMB's) separated by the cyclotron energy.

Eaves et al.[3] investigated the quenching of 1D miniband conduction in $GaAs/Al_xGa_{1-x}As$ superlattices at low temperatures and at high magnetic fields. A supression in conduction is observed when the cyclotron energy exceeds the miniband width Δ. When the cyclotron energy is greater than the miniband width the LLMB's become resolved or 'un-nested' and the superlattice enters a 1D transport or Quantum Box Superlattice (QBSL) regime. This results in true one dimensional conduction through the Landau miniband. Conduction is quenched in the 1D regime due to the suppression of inter-Landau level transitions as electrons are constrained to scatter within a single Landau level[4].

Type-II superlattices have the interesting property that the strength of electron tunnelling between wells is controlled by the proximity to the barrier valence band states. This results in higher index Landau levels exhibiting smaller miniband widths due to decreased coupling with adjacent well states [2]. LLMB narrowing is greater for higher landau indexes where the equivalent in-plane wavevector is higher.

Inset figure 1 displays the allowed scattering mechanisms above and below the QBSL transition. When the Landau levels are un-resolved the carriers may climb the Landau level 'ladder' through quasi-elastic scattering. At high energy electrons are able to efficiently relax via LO phonon emission. Above the QBSL transition there is no scattering mechanism to allow carriers to reach the higher landau levels. Carriers are constrained to move and scatter between states in a single Landau level and conductivity is suppressed.

We have studied 100 period undoped InAs/GaSb superlattices grown by MOVPE on InAs substrates. Sam-

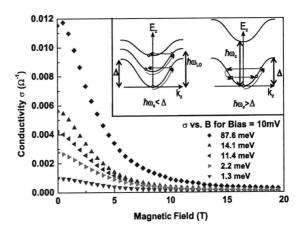

FIGURE 1. Magneto-conductivity curves for a range of miniband widths taken at a bias of 10 mV across the superlattice structure. *Inset*, Schematic of energy dispersion curves in the presence of a quantising magnetic field.

ple periods are determined with X-ray scattering allowing **k.p** calculations of the miniband dispersion relation. Strong interband coupling gives approximate ground state miniband widths ($\triangle$) between 1 meV and 88 meV. The Miniband gap in all samples is large enough to suppress inter-miniband tunnelling so that only transport in the lowest miniband need be considered. All results were taken in the electric field range below the miniband transport peak such that the degree of Wannier-Stark localisation is small and the transport may be analysed with the semi-classical Drude model.

Magneto-conductivity curves for a a range of samples are displayed in figure 1. The zero field conductivity falls with miniband width and a strong quenching of miniband conduction is apparent with increasing B field. Large miniband width samples display a significantly larger magnetoresistance attributed to the transition to the QBSL regime.

Low bias miniband transport has been modelled with semiclassical Monte Carlo simulations[5] in which Wannier-Stark quantisation (WSQ) is omitted. The simulation expands on the Esaki-Tsu model to include acoustic, optical and Umpklapp scattering processes. Characteristic parameters for the LLMB's were obtained from 8-band **k.p** calculations and were used as input for the Monte Carlo simulation.

Typical Monte-Carlo simulation results are presented in figure 2. For low electric fields IB curves reveal a weak magnetoresistance. Electrons remain close to the minibrillouin zone centre Γ and un-nesting effects are not observed. At higher electric fields carriers can gain sufficient energy to reach the top of the LLMB and a series of conductivity steps become apparent at fields for which successive Landau levels shift out of energy resonance with the fundamental LLMB. The un-nesting condition is given by $n\hbar\omega_c = \triangle_0(B)$, where the miniband width is a function of B due to strong $k_{x,y}$ dependance of the inter-well coupling. Un-nesting features increase in strength at higher electric fields. However, in practice the WSQ breaks up the miniband above F_c complicating observation of strong un-nesting features.

Un-nesting features are experimentally identified in $\triangle = 75$ meV and $\triangle = 48$ meV samples in figure 3 by taking the second derivative of the sample resistivity. Minima in the d^2I/dB^2 plot identify resistivity peaks and are equivalent to un-nesting steps in figure 2. Low magnetic field features are attributed to 3D Shubnikov de Haas oscillations in the substrate. For the $\triangle = 78$ meV sample un-nesting features are observed at 8 T and 12.5 T, with the final Landau level expected to un-nest at a B field above the range studied at ~ 25 T. In the $\triangle = 48$ meV data a single un-nesting feature is observed at 10.5 T with the final en-nesting beginning to emerge at $\sim 20 T$. We conclude therefore that we find the first evidence for resonant unnesting of superlattice Landau levels.

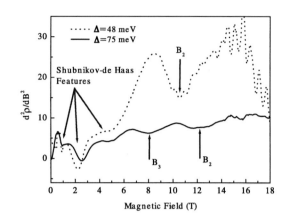

FIGURE 3. Plot of $d^2\rho/dB^2$ for $\triangle = 75$ meV and $\triangle = 48$ meV samples with a bias of 800 mV and 500 mV respectively. Un-nesting transitions for the $n = 2$ and $n = 3$ LLMB's are indicated.

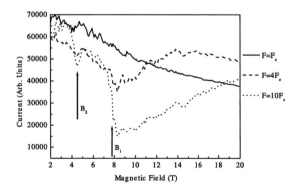

FIGURE 2. Simulated IB curves for a $\triangle = 28$ meV sample with $m\star = 0.05$, 53 Å InAs and 64 Å GaSb.

ACKNOWLEDGMENTS

This work is supported by the Engineering and Physical Science Research Council (EPSRC). A. B. Henriques acknowledges support from FAPESP (Pro-jeto No. 99/10359-7) and CNPq (Projeto No. 306335/88-3).

REFERENCES

1. Esaki, L., and Tsu, R., *IBM J. Res. Dev.*, 1970, 14, pp. 61.
2. Hales, V., *Physica E*, 2000, 7, pp. 84.
3. Eaves, L., *Physica B*, 1999, 272, pp. 190.
4. Mori, M., *Physica B*, 2001, 298, pp. 329.
5. Henriques, A. B., *BrJPhys.*, 2004, 34, pp. 605.

Magnetoresistance over 1000 % in CoFe/carbon nanotube/CoFe junctions

Yoichi Ishiwata[1], Hideyuki Maki[1], Daiju Tsuya[1,2], Masaki Suzuki[1,3], and Koji Ishibashi[1,3]

[1]*Advanced Device Laboratory, The Institute of Physical and Chemical Research (RIKEN), Wako, Saitama 351-0198, Japan*
[2]*Department of Information Processing, Tokyo Institute of Technology, Yokohama, Kanagawa 226-8503, Japan*
[3]*CREST, Japan Science and Technology (JST), Kawaguchi, Saitama 332-0012, Japan*

Abstract. We have studied spin-dependent transport through a single-wall carbon nanotube, contacted by two CoFe electrodes with a gap of about 300 nm. At 1.8 K, the junction exhibits extraordinarily high magnetoresistance ratio over 1000 %.

INTRODUCTION

Recently, extremely high magnetoresistance (MR) ratios have been observed in nanocontacts formed between two ferromagnets (FMs) [1]. This suggests the nanocontacts realize unusual conduction with very high spin polarization (*P*). In this case, the insertion of a spin mediator into magnetic nanocontacts should preserve the large MR. Carbon nanotubes (CNTs) with a small diameter of 1 – 80 nm operate as good spin mediators [2]. There is generally a weak interaction between the CNT and the contact metal, resulting in the contacts with discrete metal nanoparticles [3]. Hence FM/CNT/FM junctions are expected to introduce the highly spin-polarized current from the formation of nanocontacts.

EXPERIMENT

The device is composed of a single-wall CNT (SWNT) and two CoFe electrodes with the separation of about 300 nm on a SiO_2/Si substrate. The thicknesses of electrodes are about 80 nm. The device is covered with PMMA resist. We measured the dc current *I* with a bias voltage *V* applied to the two electrodes in a helium bath cryostat. Also, the magnetic field was applied perpendicular to the SWNT axis and parallel to the substrate.

RESULTS AND DISCUSSION

Figure 1 shows the resistance $R = V/I$ as a function of applied magnetic field *B* for a CoFe-contacted SWNT at 1.8 K with varying *V*. As the *B* field swept from negative to positive, the MR steeply reduces at about 100 mT (solid line). When it is swept from negative to positive, the steep rise in the MR appear at about -100 mT (dotted line). A pair of the MRs swept in opposite directions forms a hysteresis loop, showing the emergence of spin-dependent transport for the CoFe/SWNT/CoFe junction. Here, the lower and higher resistances correspond to the parallel and antiparallel alignments of the contact magnetizations, and are denoted by R_p and R_{ap}, respectively. One CoFe contact reverses the magnetization direction at ±100 mT, while the other remains fixed in the positive *B* direction. The junction shows very high remanence in the MR, and the pinned contact keeps the magnetization direction even up to the *B* field of ±1 T (not shown). Such a large coercivity should originate from the magnetic nanograin of CoFe which is in weak contact with the SWNT [4]. This suggests the formation of nanocontacts at SWNT/CoFe interfaces. In addition, the steep changes in the MR suggest that a few nanocontacts contribute to electrical conduction. Note that the MR characteristics do not involve the shift of a hysteresis loop (exchange bias [6]). MR characteristics measured at V = 15, 30 and 50 mV are scaled to indicate MR ratios defined as $\Delta R/R_p = (R_{ap} - R_p)/R_p$. At V = 50 mV, the MR ratio is 90 %, larger than deduced from the known P of CoFe (~0.5) [5]. At V = 30 mV, the MR ratio further increases to 155 %. As V is decreased to 15 mV, the MR ratio reaches to 1200 %. Therefore, the experimental observations indicate the emergence of the highly spin-polarized current through a SWNT.

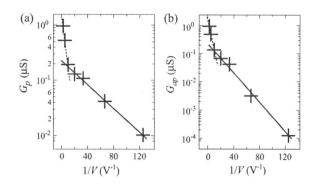

FIGURE 2. The log plots of conductances as a function of V^{-1} for the CoFe-contacted SWNT in (a) parallel and (b) antiparallel configurations at 1.8 K, which are denoted by G_p and G_{ap}, respectively.

In summary, we have investigated spin-dependent transport through a CoFe-contacted SWNT. At 1.8 K, the junction induces the high MR ratio over 1000 %.

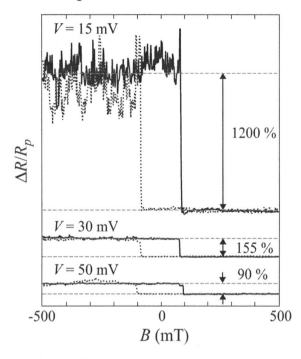

FIGURE 1. Resistance as a function of applied magnetic field B for the CoFe/SWNT/CoFe junction at 1.8 K with V of 15, 30, and 50 mV. The solid and dotted lines correspond to the positive and negative sweep directions, respectively. MR characteristics are scaled to indicate MR ratios defined as $\Delta R/R_p$.

The MR ratio has V dependence as shown in Fig. 1. We have also measured the MR at 1.8 K with V of 8, 100, 200, and 400 mV. As a result, we have found the unique V^{-1} dependence of the log of conductances I/V for the CoFe-contacted SWNT in (a) parallel and (b) antiparallel magnetization configurations at 1.8 K, which are denoted by G_p and G_{ap}, respectively. In addition, the plots are classified as two regions of 8 ~ 100 mV and 100 ~ 400 mV in both configurations. The difference between linearity for parallel and antiparallel configurations should relate to the spin-dependent transport, although the origin has not been clear.

ACKNOWLEDGMENTS

We would like to thank J. Inoue, G. Tatara, K. Kusakabe, S. Uryu, S. Kokado, T. Kimura, and K. Arai for helpful discussions, and S. Moriyama and T. Fuse for technical support.

REFERENCES

1. Hua, S. Z., and Chopra, H. D., *Phys. Rev. B* **67**, 060401 (2003).

2. Tsukagoshi, K., Alphenaar, B. W., and Ago, H., *Nature* **401**, 572 (1999).

3. Zhang, Y., and Dai, *Appl. Phys. Lett.* **77**, 3015 (2000).

4. Shima, T., Takahashi, K., Takahashi, Y. K., and Hono, K., *Appl. Phys. Lett.* **81**, 1050 (2002).

5. Moodera, J. S., Kinder, L. R., Wong, T. M., and Meservey, R., *Phys. Rev. Lett.* **74**, 3273 (1995).

6. Nogués, J., and Schuller, I. K., *J. Magn. Magn. Mater.* **192**, 203 (1999).

Laser-Resonance Chirality Selection in Single-Walled Carbon Nanotubes

Kenzo Maehashi, Yasuhide Ohno, Koichi Inoue and Kazuhiko Matsumoto

*The Institute of Scientific and Industrial Research, Osaka University,
8-1 Mihogaoka, Ibaraki, Osaka 567-0047, Japan*

Abstract. Laser Resonance Chirality Selection method has been proposed to select single-walled carbon nanotubes (SWNTs) with specific chirality from mass of SWNTs by intense laser irradiation. Absorption of the exciting laser beam is strongly enhanced for SWNTs with specific chirality when the laser excitation energy is close to the energy separation in the density of states. As a result of the resonance effect, at a certain threshold excitation power density, only these SWNTs can be oxidized and be removed selectively. We have demonstrated with Raman scattering measurements that SWNTs with specific chirality are selectively removed after intense laser irradiation.

INTRODUCTION

Single-walled carbon nanotubes (SWNTs) are one of the best candidates for nano-scale electronic devices. Depending on the nanotube symmetry and diameter, two-thirds of SWNTs are expected to be semiconductors, the other being metals. Semiconducting SWNTs have been applied to field-effect transistor, and metallic SWNTs to single-electron transistors [1]. Their performances strongly depend on the chirality of SWNTs. Therefore, for the precise control of the device performance, it is indispensable to control the chirality of SWNTs. In this letter, we propose a "Laser Resonance Chirality Selection (LARCS)" method to select SWNTs with specific chirality by intense laser irradiation. We have investigated the effect of the LARCS method with Raman scattering.

PRINCIPLE AND EXPERIMENTAL

The SWNT density of states (DOS) is characterized by multiple van Hove singularities since SWNTs are one-dimensional materials. The DOS of SWNTs is strongly dependent on their chirality. Figure 1 shows a schematic illustration of the LARCS method. When a sample is irradiated by the laser beam with the energy ($h\nu$), as shown in Fig. 1(a), absorption of the exciting beam can be strongly enhanced for the SWNTs with specific chirality, whose energies of the allowed electronic transitions match the energy ($h\nu$) of the incident photon. As a result of the resonant effect, when the excitation power density of the laser beam increases in an atmosphere of air, at a certain threshold excitation power density, only these SWNTs are oxidized and are removed selectively, as shown in Fig. 1(b). The LARCS is considered to be a simple and easy method to select the SWNTs with specific chirality.

SWNTs were formed by a thermal chemical vapor deposition method on Si substrates. After the catalyst, which consists of Fe, Mo and alumina nanoparticles in the liquid phase, was deposited on the surfaces, ethanol vapor was supplied for 30 min at 900°C. For Raman scattering measurements, the excitation lasers were focused to about 20 μm spot size in diameter.

RESULTS AND DISCUSSION

Figure 2 shows the low frequency region of typical Raman spectra, which were measured with the 457.9, 488.0 and 514.5 nm lines of Ar-ion laser. The excitation density of the Ar-ion laser for Raman scattering measurements is low power of about 10 W/cm^2. For all spectra, the peaks (about 520 cm^{-1}) from Si substrates and the nanotube radial breathing modes (RBMs) (150 - 300 cm^{-1}) that are specific to SWNTs are clearly observed, as shown in Fig. 2. From the RBM signals, the diameter of SWNTs is

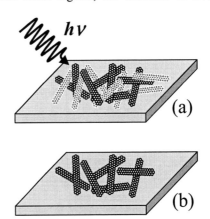

FIGURE 1. Schematic illustration of the LARCS method.

estimated to be about 1 - 2 nm. In addition, the strong *G*-band peak was observed for three kinds of laser excitation energy. The *D*-band peak, which is related to structural disorder, was quite weak. These results indicate that high-purity and high-quality SWNTs are formed on Si substrates.

The Raman peak position from Si substrates is independent from the laser excitation energy, as shown in Fig. 2. However, the peak positions of RBM signals from SWNTs strongly depend of the laser excitation energy. This phenomenon is attributed to the resonant effects. Each RBM signal corresponds to the SWNT with specific chirality for which the absorption of the exciting beam is strongly enhanced, when the energy separations between van Hove singularities in DOS of SWNTs are close to the laser excitation energy.

The following experiment has been carried out to investigate the effect the LARCS method. First, Raman spectra of SWNTs were measured with the 457.9, 488.0 and 514.5 nm lines of Ar-ion laser with low excitation power density at the same spot. Second, the samples were irradiated with the 514.5 nm line of Ar-ion laser with high excitation power density at the same spot. Finally, Raman spectra of SWNTs with the three excitation wavelengths were measured again at the same spot. The intensities of RBM signals of SWNTs before and after intense laser irradiation were compared since they depend on the amount of the specific SWNTs.

After the sample is irradiated with the 514.5 nm line at high power density of 20 kW/cm^2 for 30 min, the intensities of RBM signals from SWNTs decrease for three different Raman excitation energies. Figure 3 shows the average intensity ratio of the RBM signals from SWNTs after intense laser irradiation of 514.5 nm line to those before intense laser irradiation as a

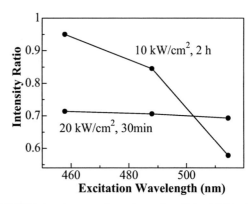

FIGURE 3. Average intensity ratio of the RBM signals from SWNTs after laser irradiation to those before the laser irradiation as a function of Raman excitation wavelength.

function of Raman excitation wavelength. The average intensity ratio of the RBM signals from SWNTs is about 0.7 for three different Raman excitation energies, revealing that it is independent of Raman excitation energy. This indicates that all kinds of SWNTs are reduced simultaneously because of too high power density of the laser irradiation.

After the sample is irradiated with the 514.5 nm line at mild power density of 10 kW/cm^2 for 2 hours, the intensities of RBM signals are almost constant for 457.9 nm excitation, and the RBM signals decrease slightly in intensity for 488.0 nm excitation. On the other hand, the intensities of RBM signals are significantly reduced for 514.5 nm excitation. The average intensity ratio of the RBM signals from SWNTs for 457.9, 488.0 and 514.5 nm excitations is about 0.95, 0.85 and 0.57, respectively, as shown in Fig. 3. This indicates that the SWNTs that are at resonance with 514.5 nm excitation are selectively removed after the intense laser irradiation. In contrast, the SWNTs that are at resonance with 457.9 and 488.0 nm excitations are relatively stable even after the intense laser irradiation. Therefore, we have succeeded in selecting SWNTs with specific chirality using LARCS method.

CONCLUSION

We have proposed a "LARCS" and succeeded in selecting SWNTs with specific chirality by intense laser irradiation. The effect of the method has been investigated with Raman scattering measurements. By changing the energy of the intense laser beam, SWNTs with specific chirality, which are not necessary, will be removed selectively by LARCS method.

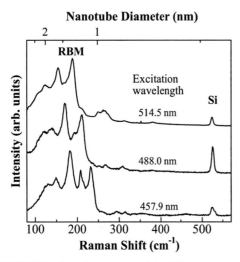

FIGURE 2. Low frequency region of Raman spectra.

REFERENCE

1. K. Matsumoto *et al*, Jpn. J. Appl. Phys. **42**, 2415 (2003).

A symmetrized-basis approach to excitons in carbon nanotubes

Giovanni Bussi*, Eric Chang*, Alice Ruini* and Elisa Molinari*

INFM National Center on NanoStructures and BioSystems at Surfaces (S3) and Dipartimento di Fisica, Università di Modena e Reggio Emilia, Via Campi 213/A, 41100 Modena, Italy

Abstract. We calculate from first-principles the optical spectrum of a (4,2) single-wall carbon nanotube including quasi-particle corrections and excitonic effects. We expand every quantity on a special basis sets which completely exploits the symmetries of the nanotube, allowing calculations for tubes with a long unit cell. The results indicate that the excitonic effects are crucial and a strong peak in the absorption spectrum is predicted at 2.2 eV. This value is compared with experimental results, with excellent agreement.

Carbon nanotubes [1] are promising materials for novel opto-electronic applications [2], and show important dimensionality effects. The nature of optical excitations however remains an open question. Optical spectra have been measured for a set of tubes with equal radius [3, 4], but an *ab initio* investigation is required to assign the peaks to the different possible tubes. In this work we calculate the electronic structure and the optical spectrum of a (4,2) single-wall carbon nanotube (SWCN) using a first-principles approach based on the Bethe-Salpeter equation (BSE) scheme [5]. Traditional plane-wave implementations of this method are quite demanding even for tubes with a small number of atoms per unit cell [6]. For this reason, we adopt a symmetrized basis set which allows us to generate the whole tube from a pair of independent atoms.

The first step in our approach is the calculation of the self-consistent single-particle Hamiltonian of the nanotube, based on the density-functional theory in the local density approximation (LDA) [7]. We adopt here a conventional plane-wave basis set with pseudopotentials [8]. The Hamiltonian is then projected on a new basis set and diagonalized to obtain the LDA band-structure. The tube is invariant with respect to two independent symmetries: (a) a translational symmetry $\hat{G}_1$ of length T in the axial direction; (b) a screw axis operation $\hat{G}_2$, obtained applying a rotation around the tube axis of an angle $\frac{2\pi}{N}$, where N is the number of hexagons per unit cell, and a translation in the axial direction with length $\frac{MT}{N}$, where M is an integer. All the parameters (T, N and M) can be obtained starting from the (n,m) pair and the lattice parameter of the graphene sheet [1]. We choose the new basis set as a linear combination of atomic-centered Gaussian orbitals, symmetrized so that they are eigenstates of both

TABLE 1. Optical transitions (eV) in 4 Å diameter carbon nanotubes. The number of atoms in the translational unit cell is also shown.

	N_{atoms}	Theory			Experiment
		LDA	GW	BSE	
(3,3)	12	2.8[a]	3.3[a]	3.2[a]	3.1[b]
(5,0)	20	1.2[a]	1.3[a]	1.3[a]	1.4[b]
(4,2)	56	1.7[c]	**3.1[d]**	**2.2[d]**	2.1[b]

[a] Reference [6].
[b] Reference [3].
[c] Our previous work [10].
[d] **This work.**

the $\hat{G}_1$ and the $\hat{G}_2$ operations. Since $\hat{G}_1^M = \hat{G}_2^N$, it is possible to relate the eigenvalue with respect to $\hat{G}_2$ to an integer quantum number ranging from 1 to N which we call h. The eigenrelations $\hat{G}_1|\chi_{\sigma nlm}^{k,h}\rangle = e^{-2\pi i k}|\chi_{\sigma nlm}^{k,h}\rangle$ and $\hat{G}_2|\chi_{\sigma nlm}^{k,h}\rangle = e^{-2\pi i \frac{Mk+h}{N}}|\chi_{\sigma nlm}^{k,h}\rangle$ are then satisfied. Using this *ad hoc* basis set, we developed a standard GW approach with a plasmon-pole model for the screening similar to the model described in Ref. [9]. We then include, on top of the quasiparticle structure, the BSE scheme, taking advantage of the selection rules on the quantum number h. The matrix elements needed for the GW and the BSE calculations can be easily obtained on our basis set [10], allowing us to study with a reasonable computational effort a tube with a quite large number of atoms per unit cell (see Table 1), such as the (4,2) SWCN.

We calculate the self-energy correction for three inequivalent k points in the Brillouin zone and interpolate the shift as a function of the LDA energies. The resulting correction is a combination of a shift of the conduction

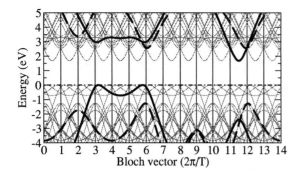

FIGURE 1. Band structure of (4,2) nanotube, including GW corrections. The grey lines represent the repeated zone scheme, while the black lines show the unfolded band structure (see text). The solid lines refer to odd values of character h, and the dashed ones to even values.

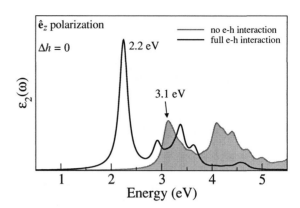

FIGURE 2. Calculated absorption spectra of a (4,2) carbon nanotube for light polarized parallel to the tube axis. The shaded plot is obtained neglecting excitonic effects. The solid line is obtained including the full electron-hole interaction.

bands and a stretching of the valence bands. The resulting quasi-particle energies depend on the Bloch vector k and on the character h. In Fig. 1 the traditional band-structure, depending on the Bloch vector only, is plotted with grey lines. We now unfold the structure connecting all the bands with the same character h. Since increasing the k vector by a lattice constant is equivalent to decreasing the character h by M, we can group together all states with odd h (solid lines) and all states with even h (dashed lines). This band structure can be compared to the unfolded LDA band structure in Ref. [10].

In Fig. 2 we show the predicted absorption spectrum for the relevant polarization, i.e. for a field polarized along the tube axis. The shaded line is obtained in the independent quasi-particle scheme, where the electron-hole interaction is neglected. When we switch on the full electron-hole coupling (solid plot), the spectrum is crucially modified: a major peak arises at 2.2 eV, due to a bound exciton, and other structures appear at energy higher than 2.8 eV. Comparing our results with the experiment described in Ref. [3] and with the theoretical predictions for tubes (3,3) and (5,0) described in Ref. [6], we can complete the assignment of the experimental peaks, see Table 1. In addition, the peaks that we find in the range between 2.9 and 3.6 eV may contribute to the experimental structure at approximately 3.1 eV in Ref. [3]. It can be also observed that the agreement between the peak for absorption spectrum of the (4,2) nanotube in Ref. [11] and the experiment is due to a cancellation of errors. In fact, the blue-shift induced by the self-energy corrections is partially cancelled by the red-shift due to the binding of the exciton. Including both effects we find the correct peak position.

In conclusion, we implemented the GW plus BSE approach, which is the state of the art for optical properties of materials, in a symmetrized Gaussians basis. This basis set allows for the description of a SWCN in term of two generating atoms, and calculations for a (4,2) nanotube are feasable. For this tube, we calculate the quasi-particle band structure and the optical spectrum, including excitonic effects, which prove to have a large impact on the peak shapes and positions. We find a peak in the absorption spectrum at 2.2 eV, in agreement with experimental results.

We are grateful to G. Goldoni, and M. Rohlfing for fruitful discussion. Computer time was partly provided by CINECA through INFM Parallel Computing Projects. The support by the RTN EU Contract "EXCITING" No. HPRN-CT-2002-00317, and by FIRB "NOMADE" is acknowledged.

REFERENCES

1. Saito, R., Dresselhaus, G., and Dresselhaus, M. S., *Physical Properties of Carbon Nanotubes*, Imperial College Press, London, 1998.
2. Misewich, J. A., et al., *Science*, **300**, 783 (2003).
3. Li, Z. M., et al., *Phys. Rev. Lett.*, **87**, 127401 (2001).
4. Guo, J., et al., *Phys. Rev. Lett.*, **93**, 017402 (2004).
5. Benedict, L. X., et al., *Phys. Rev. Lett.*, **80**, 4514 (1998); Albrecht, S., et al. *Phys. Rev. Lett.*, **80**, 4510 (1998); Rohlfing, M., and Louie, S. G., *Phys. Rev. Lett.*, **81**, 2312 (1998).
6. Spataru, C. D., Ismail-Beigi, S., Benedict, L. X., and Louie, S. G., *Phys. Rev. Lett.*, **92**, 077402 (2004).
7. Kohn, W., and Sham, L. J., *Phys. Rev.*, **140**, A1133 (1965).
8. Baroni, S., Dal Corso, A., de Gironcoli, S., and Giannozzi, P. (2001), http://www.pwscf.org.
9. Rohlfing, M., Krüger, P., and Pollmann, J., *Phys. Rev. B*, **52**, 1905 (1995).
10. Chang, E., Bussi, G., Ruini, A., and Molinari, E., *Phys. Rev. Lett.*, **92**, 196401 (2004).
11. Marinopoulos, A. G., Reining, L., Rubio, A., and Vast, N., *Phys. Rev. Lett.*, **91**, 046402 (2003).

Chirality Assignment of Single-Walled Carbon Nanotubes with Strain

Lain-Jong Li*, R. J. Nicholas*, R. S. Deacon*, P. A. Shields*, Chien-Yen Chen[†], R. C. Darton[†] and S. C. Baker**

*Clarendon Laboratory, Physics Department, Oxford University, Parks Road, Oxford OX1 3PU, U.K.
[†]Department of Engineering Science, Oxford University, Parks Road, Oxford, OX1 3PJ, U.K.
**School of Biological and Chemical Sciences, Birkbeck College, Malet Street, London, WC1E 7HX, U.K.

Abstract. Strain-induced band gap shifts that depend strongly on the chiral angle have been observed by optical spectroscopy in polyvinylpyrrolidone(PVP)- or Surfactin micelle-wrapped single-walled carbon nanotubes (SWCNTs). Uniaxial and torsional strains are generated by changing the environment surrounding the SWCNTs, using the surrounding D_2O ice temperature or the hydration state of a wrapping polymer. These methods can be used as diagnostic tools to determine the quantum number, q, and examine chiral vector indices for specific nanotube species.

INTRODUCTION

SWCNTs possess unique one-dimensional properties that depend on the tube chiral indices. Modifications of the band gap by mechanical deformation offer a route to characterize the tubes. It has been reported theoretically[1] that the sign of the quantum number, q, is critically important in determining the strain-induced change in band gap for semiconducting tubes. Tubes with q = 1 and q = -1 represent cases where the relative positions in k-space of the folded dispersion relations relative to the Brillouin zone boundary are opposite, so that perturbations to the Fermi wavevector k_F act in opposite relative directions and consequently cause shifts in band gap of opposite signs.

RESULTS AND DISCUSSIONS

We have carried out photoluminescence(PL) measurements for the SWCNTs wrapped with Sodium Dodecyl Sulfate(SDS) micelle, surfactin micelle and PVP in aqueous solutions as a function of temperature (4.2K to 353K). Surfactin is an anionic lipopeptide biosurfactant and forms rod-shaped micelles in aqueous solution[2]. The PL profiles of the SWCNTs wrapped with surfactin or PVP show very little change with solution temperature from 353 to 278K. In contrast, the PL profile of SDS-wrapped SWCNTs significantly vary with solution temperature, which is attributed to the drastic solubility change of SDS with temperature and which also indicates a relatively poor wrapping of SDS around nanotubes[3]. When the temperature was further lowered the overall PL intensities of PVP- and surfactin micelle-wrapped SWCNTs were enhanced while the PL peaks of SDS micelle-wrapped SWCNTs were broadened, suppressed and strongly dependent on their cooling rate. Therefore PVP and surfactin were used for low temperature PL study of SWCNTs in our research. It has been calculated that the linear thermal contraction of a (10,10) SWCNT is estimated to be 0.1 % from 80K to 260K[4]. In contrast to the SWCNT, D_2O ice expands by about 0.6 % over the same temperature range[5]. It is likely therefore that a combination of uniaxial tension and torsional stretching can be produced in ice-surrounded nanotubes by simply lowering the temperature.

Figure 1 shows the effects of temperature on the band gap luminescence for PVP- and surfactin-wrapped SWCNTs excited with 633nm radiation. Both spectra show a similar trend of peak-shifts for the identified peaks P1-P10. In addition, variation of the excitation wavelength allows us to induce preferential luminescence from specific nanotubes by selective excitation into specific E_{22} band gaps[6, 7].

Yang's theory[8] suggests that the sign of d(Band gap)/d(Strain) is the same as that of the q value. We expect that the dominant contraction of the ice once the solution has frozen will result in a negative strain for the nanotubes at low temperatures. Therefore, the change in bandgap on re-heating will result in a decrease in the (negative) strain and thus the signs of the peak shifts allow us to directly assign the peaks to be q=+1 and q=-1 respectively, as also shown in Fig.2. The assignments

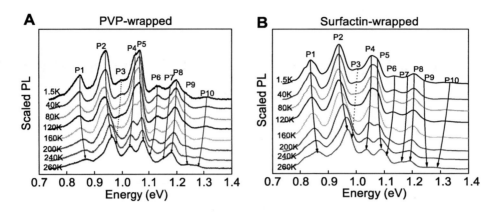

FIGURE 1. The temperature effects on the band gap scaled luminescence for (A)PVP- (B) Surfactin-wrapped SWCNTs solution excited with 633 nm radiation.

from Bachilo [6] et al. and Hagen and Hertel [9] Scenario I and II are shown in Fig.2.

The power of the analysis can be seen by comparing the results with the q assignments in Hagen and Hertel's Scenario II which clearly demonstrate that this scheme disagrees strongly with the q values deduced from the strain dependence. Although both the q values from the Bachilo et al and Hagen and Hertel's Scenario I assignments match well with the strain dependence, we find that a detailed study of bandgap shift vs. chiral angle leads us to favour the Bachilo et al assignment. According to Yang and Han's theory [1], the band gap change (ΔE_g) of SWCNTs under small strain is given by,

$$\Delta E g = sgn(2q+1)3t_0[(1+\nu)\sigma cos3\theta + \gamma sin3\theta] \quad (1)$$

where t_0 and ν denote the carbon-carbon transfer integral and Poisson's ratio, and θ is the tube chiral angle. σ and γ are strains along the tube axis and circumference, corresponding to uniaxial and torsional strains on the nanotubes. Fittings of the heating induced band gap shifts to Equation (1) using the uniaxial and torsional strains as fitting parameters strongly suggests that the Bachilo et al assignment is preferable [7]. Straining the nanotubes is also possible at room temperature by changing the hydration state of a wrapping polymer [7]. Our measurements show that more than 0.2 % uniaxial and 0.08 % torsional strain can be generated on SWCNTs simply by changing the surrounding D_2O ice temperature or by polymer swelling.

In summary, the application of strain to carbon nanotubes by changing temperature or hydration state of a wrapping polymer can provide a clear signature to enable quick and accurate q value assignments to be made. In addition, the use of polymer wrapping to strain the nanotubes at room temperature offers new routes to potential nanostructure fabrication.

We acknowledge with thanks the support from Dr. C. Pears and Mr. D.W. Hsu for the provision of the ultracentrifuge facilities. The author L.J.Li would thank the Swire Group for providing his graduate scholarship.

Photoluminescence spectral data and assignments

PVP-wrapped SWCNTs	Bachilo et al				Hagen and Hertel I				Hagen and Hertel II			
Peak	ΔE_g (meV)	q	n	m	q	Chiral angle	n	m	q	Chiral angle	n	m	q	Chiral angle
P1	26.46	1	12	2	1	0.132	10	6	1	0.380	8	7	1	0.485
P2	19.73	1	10	3	1	0.222	9	5	1	0.360	11	3	-1	0.205
P3	(-)#	-1	10	5	-1	0.333	8	6	-1	0.442	12	1	-1	0.069
P4	(-)	-1	8	6	-1	0.441	12	1	-1	0.069	9	4	-1	0.305
P5	10.49	1	7	6	1	0.479	9	2	1	0.172	7	5	-1	0.428
P6	(-)	-1	9	4	-1	0.305	9	4	-1	0.305	10	2	-1	0.156
P7	(-)	-1	10	2	-1	0.156	10	2	-1	0.156	11	0	-1	0.000
P8	-13.79	-1	7	5	-1	0.428	11	0	-1	0.000	8	3	-1	0.267
P9	9.69	1	6	5	1	0.471	7	3	1	0.297	6	4	-1	0.409
P10	-33.47	-1	8	3	-1	0.267	8	3	-1	0.267	9	1	-1	0.091

Unresolved peaks; (-) negative shift

FIGURE 2. PL Spectral data and chirality assignments. The peak shifts (ΔE_g, from 80 to 260K) were calculated from PVP-wrapped SWCNTs excited by laser wavelengths of 633nm.

REFERENCES

1. L. Yang and J. Han, *Phys. Rev. Lett.* **85**, 154 (2000).
2. Y. Ishigami, M. Osman, H. nakahara, Y. Sano, R. Ishiguro and M. Matsumoto, *Colloids. Surfaces B: Biointerfaces* **4**, 341 (1995).
3. L.J. Li, R. J. Nicholas, to be published.
4. Y. K. Kwon, S. Berber, and D. Tomanek, *Phys. Rev. Lett.* **92**, 015901 (2004).
5. K. Rottger, A. Endriss, J. Ihringer, S. Doyle, and W. F. Kuhs, *Acta Cryst.* **B50**, 644 (1994).
6. S. M. Bachilo, M. S. Strano, C. Kittrell, R. H. Hauge, R. E. Smalley and R. B. Weisman, *Science* **298**, 2361 (2002).
7. L. J. Li, R. J. Nicholas, R. S. Deacon and P. A. Shields, to be published.
8. L. Yang, M. P. Anantram, J. Han, and J. P. Lu, *Phys. Rev. B* **60**, 13874 (1999).
9. A. Hagen and T. Hertel, *Nano Lett.* **3**, 383 (2003).

Inter-tube transfer of electrons in various double-wall carbon nanotubes

Seiji Uryu and Tsuneya Ando

Department of Physics, Tokyo Institute of Technology, 2-12-1 Ookayama, Meguro-ku, Tokyo 152-8551, Japan

Abstract. Conductance for inter-tube transfer in double-wall carbon nanotubes is calculated in a tight-binding model. The conductance remains extremely small with fluctuations around an averaged value independent of the tube length. This can be regarded as a direct manifestation of quasi-periodic nature of the system in the conductance.

Carbon nanotubes are often self-assembled to be complex systems such as multi-wall tubes and tube bundles. In these systems, inter-tube transfer of electrons can occur and become an important factor to modify electronic properties. The purpose of the present study is to clarify effects of inter-tube transfer on electronic transport properties of multi-wall tubes.

Using double-wall tubes, some experiments revealed a structural feature that lattices of tubes are incommensurate with each other and therefore, the system is quasi-periodic [1, 2]. Some theoretical studies suggested that effect of inter-tube transfer on electronic transport properties is small [3, 4]. However, the full understanding of inter-tube transfer effects has not been achieved yet. In this paper the conductance between the outer and inner tubes of a double-wall tube is calculated, because it is the quantity dominated by inter-tube transfer.

Figure 1 is a schematic illustration of our two-terminal system. A semi-infinite long outer tube is attached to a reservoir on the right-hand side, while a semi-infinite long inner tube is attached to another reservoir on the left-hand side. The two tubes are overlapped with each other in the middle double-wall-tube region with length A where inter-tube transfer occurs. Each tube plays a role of an ideal lead outside this region.

In numerical calculations we use a tight-binding model including only π orbitals and take into account transfers between nearest-neighbor sites in each wall and inter-tube transfers between all sites within the hopping range [5]. Conductance is calculated by using recursive Green's function method [6] and the Landauer formula.

In this paper conductance of (4,7)-(4,16) tube is presented as a typical result of incommensurate double-wall tubes with metallic inner and outer tubes. Energy is chosen as a value close to the Fermi energy throughout the paper.

Figure 2 shows length-dependence of conductance G in the lower panel and averaged conductance $\tilde{G}$ and conductance fluctuation ΔG in the upper panel. The result is in the case of short tubes $0 < A/L_{\text{out}} \leq 0.8$ where L_{out} is the circumference of the outer tube. In the upper panel averaged values and fluctuations are calculated from the conductances of the lower panel in the ranges $0.2n \leq A/L_{\text{out}} < 0.2(n+1)$ with $n=0, 1, 2$, and 3.

The conductance is extremely small in comparison with the quantum conductance and rapidly fluctuates as shown in the lower panel. In the upper panel the averages and fluctuations are essentially independent of the length.

A similar result in the case of long tubes is shown in Fig. 3. In the upper panel averages and fluctuations are calculated from the conductances in the ranges $500n \leq A/L_{\text{out}} < 500(n+1)$ with $n=0, 1, 2$, and 3.

Averaged values and fluctuations are slightly enhanced as compared to those in the previous result. Similarly to the previous case, however, they are almost independent of the length in spite of the huge difference in the length scale (2500 times larger than that of the previous case), indicating that all the features of the conductance are independent of the length.

This result shows that the conductance due to inter-tube transfer exhibits quasi-periodic (and therefore fractal-like) behavior as a function of length. It is to be expected because double-wall tubes themselves are quasi-periodic except in some commensurate tubes. However, closer investigation of the length dependence of the conductance is needed in order to clarify more details on the quasi-periodicity and self-similarity. Details of the study will be published elsewhere.

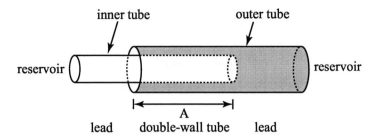

FIGURE 1. Schematic illustration of two-terminal double-wall tube.

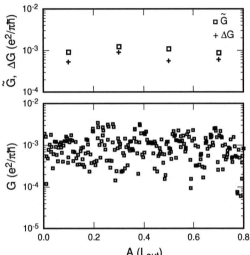

FIGURE 2. In lower panel length-dependence of conductance and in upper panel averaged conductance and fluctuation are plotted by squares and crosses, respectively, in the case of short tubes. The length is changed as finely as possible. Energy E is chosen as $EL_{out}/2\pi\gamma = -0.04$ where the Fermi energy is set to zero and $\gamma = \sqrt{3}a\gamma_0/2$ with a being the lattice constant and γ_0 being the overlap integral between nearest-neighbor sites.

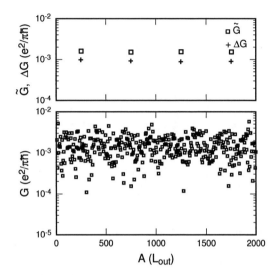

FIGURE 3. In lower panel length-dependence of conductance and in upper panel averaged conductance and fluctuation are plotted by squares and crosses, respectively, in the case of long tubes. In the lower panel the conductance is calculated at every $12T_{out}$ length with T_{out} being the period of the outer tube in the tube axis direction, $T_{out}/L_{out} = 1/4\sqrt{3}$, and the conductances are plotted at every $36T_{out}$ length for simplicity. The maximum length $2000L_{out}$ is about 9 μm.

ACKNOWLEDGMENTS

This work was supported in part by a 21st Century COE Program at Tokyo Tech "Nanometer-Scale Quantum Physics" and by Grants-in-Aid for Scientific Research and for COE (12CE2004 "Control of Electrons by Quantum Dot Structures and Its Application to Advanced Electronics") from the Ministry of Education, Culture, Sports, Science and Technology, Japan. Numerical calculations were performed in part using the facilities of the Supercomputer Center, Institute of Solid State Physics, University of Tokyo and the Advanced Computing Center, RIKEN.

REFERENCES

1. M. Kociak, K. Suenaga, K. Hirahara, Y. Saito, T. Nakahira, and S. Iijima, Phys. Rev. Lett. **89**, 155501 (2002).
2. J. M. Zuo, I. Vartanyants, M. Gao, R. Zhang, and L. A. Nagahara, Science **300**, 1419 (2003).
3. S. Roche, F. Triozon, A. Rubio, and D. Mayou, Phys. Rev. B **64**, 121401 (2001).
4. Y.-G. Yoon, P. Delaney, and S. G. Louie, Phys. Rev. B **66**, 073407 (2002).
5. S. Uryu, Phys. Rev. B **69**, 075402 (2004).
6. T. Ando, Phys. Rev. B **44**, 8017 (1991).

First-Principles Band Offsets of Carbon Nanotubes with III-V Semiconductors

Yong-Hyun Kim*, M. J. Heben* and S. B. Zhang*

National Renewable Energy Laboratory, 1617 Cole Boulevard, Golden, Colorado 80401, USA

Abstract. Band offsets for carbon nanotubes assembled with III-V semiconductors are calculated within first-principles density functional formalism. It turns out that in many cases the calculated mid-gap levels of the semiconducting nanotubes, corresponding to the original graphene Fermi level, are located at 0.53±0.08 eV above the valence band maximum of bulk InAs, regardless of the tube diameter and chirality. Direct calculations for the nanotube/GaAs systems confirm the applicability of the above graphene Fermi level theory, as well as the transferability of the band offsets among III-V semiconductors.

INTRODUCTION

Incorporation of carbon nanotubes into conventional semiconductor fabrication techniques is a subject of intensive study [1, 2, 3] with the potential for eventual nano-electronic and optoelectronic applications. Understanding the physics and chemistry of the nanotube/semiconductor assemblies is important for the realization of such a potential and ultimately for the design of desirable quantum functional devices.

In a previous report [4], we have dealt with the issues regarding binding energy and mechanism, band offsets, charge transfer, and surface dipole effects on the electronic structure of carbon nanotubes on one particular semiconductor surface, namely the commensurate InAs surface. Here we will focus on the more general question of the transferability of the band offsets among different semicondcutors, which provides a principal guideline for designing heterostructures of carbon nanotube with conventional semiconductors. In cases when a strong surface dipole presents or when a significant charge transfer takes place across the interface, strictly speaking, band offset will subject to change. Even in such cases, however, our model can qualitatively account for the direction of the changes.

MATCHING THE LATTICES

When placing a nanotube on a surface, it is natural to first consider if the irreducible tube lattice parameters commensurate or incommensurate with those of the surface. For tubes that are incommensurate, the binding energies are expected to be smaller than their commensurate counterparts. In III-V semiconductors, the smallest (periodic) length scale is the lattice parameter of face-centered cubic (fcc) unit cell, i.e., $d_{[110]}$ as shown in Fig. 1(a) for InAs. For carbon nanotubes, the two smallest (periodic) length scales are the lattice constants for armchair and zigzag nanotubes, as indicated in Fig. 1(b). By surveying the $d_{[110]}$ values for binary semiconductors [see Fig. 1(c)], we find that the zigzag tubes could match well with InAs and CdSe in the [110] direction. On the other hand, armchair tubes are hardly commensurate with any semiconductor surfaces.

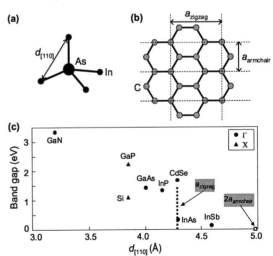

FIGURE 1. (a) Lattice parameter of the fcc unit cell, $d_{[110]}$, as defined for InAs. (b) The two lattice parameters for carbon nanotubes, a_{zigzag} and $a_{armchair}$. (c) Band gap vs. $d_{[110]}$ for eight different semiconductors [5]. Dotted line and open square indicate the lattice parameter and bang gaps for zigzag and armchair carbon nanotubes, respectively. Note that the gap for nanotube is quasicontinuous from 0 to 1.5 eV [6].

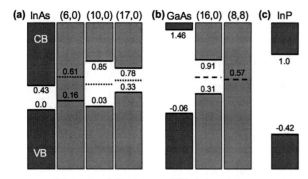

FIGURE 2. Calculated band offsets of various carbon nanotubes for (a) InAs and (b) GaAs. In (b)-(c), the band offsets of III-V semiconductors to InAs from Ref. [7] are used.

BAND OFFSETS

Commensurate InAs. Band offsets of several carbon nanotubes with lattice-matched InAs have been calculated by aligning the average potential in the bulk region of the InAs(110) slab with that of bulk InAs. Details on the first-principles density functional calculations can be found in Ref. [4]. Figure 2(a) shows the calculated band offsets of (6,0), (10,0), and (17,0) nanotubes with respect to the valence band maximum (VBM) of bulk InAs. The mid-gap levels of semiconducting nanotubes (and the Fermi level of metallic tubes) correspond to the original graphene Fermi level, denoted by the dotted lines in Fig. 2. We note that these levels are well aligned at 0.53 ± 0.08 eV above the VBM of InAs.

Incommensurate GaAs. In this case, we need a bigger supercell than InAs to match reducible lattice parameters. By inspecting the lattice parameters of GaAs(110) surface, we find that a 5 x 3 surface unit cell matches with $8a_{armchair}$ and $4a_{zigzag}$. We have calculated the band offsets of (8,8) and (16,0) tubes with respect to the VBM of bulk GaAs the same way as for InAs. In Fig. 2(b), we align the VBM of GaAs with respect to that of InAs according to Ref. [7], i.e., the VBM of GaAs is 0.06 eV below that of InAs. The figure shows clearly that the graphene Fermi level, denoted by dashed lines here (0.59 ± 0.02 eV), matches well to the dotted lines in Fig. 2(a) for InAs. This confirms the transferability of the band offsets of carbon nanotubes to different III-V semiconductors beyond InAs, just like the ordinary semiconductors do.

From the transferability of the band offsets, we can easily predict band offsets of carbon nanotubes with other III-V semiconductors whose band offsets to InAs and GaAs are known. For example, InP should form a type-I junction with (16,0) nanotube, as shown in Fig. 2(c).

Strickly speaking, Figure 2 applies only to cases where the interfacial dipole and the degree of interfacial charge transfer are small. It does not apply to cases where the above interfacial effects are large. This is because nanotubes are one dimensional periodic molecules whose density of states (DOS) is not nearly as large as a two dimensional periodic surface. Even in the latter cases, however, Figure 2 is still very useful as it provides a convenient guidance in terms of the directions of the changes, as detailed in Ref. [4].

CONCLUSION

Band offsets between carbon nanotubes and III-V semiconductors have been calculated by first-principles. The mid-gap levels of semiconducting nanotubes, corresponding to the original graphene Fermi level, are located at 0.53 ± 0.08 eV above the valence band maximum of bulk InAs, regardless of the tube diameter and chirality. Graphene Fermi level is transferable among different tube systems, so does the band offsets between conventional semiconductors. Hence, our theory is capable of predicting band offsets of carbon nanotubes with any III-V semiconductors.

ACKNOWLEDGMENTS

We thank Yufeng Zhao and Mao-Hua Du for stimulating discussions. YHK and MJH were supported by the U.S. Department of Energy, Office of Basic Energy Sciences, Division of Chemical Sciences, under contract No. DE-AC36-99GO10337.

REFERENCES

1. Albrecht, P. M., and Lyding, J. W., *Appl. Phys. Lett.*, **83**, 5029 (2003).
2. Ruppalt, L. B., Albrecht, P. M., and Lyding, J. W., *J. Vac. Sci. Tech. B*, to be published (2004).
3. Jensen, A., Hauptmann, J. R., Nygard, J., Sadowski, J., and Lindelof, P., *NanoLett.*, **4**, 349 (2004).
4. Kim, Y.-H., Heben, M. J., and Zhang, S. B., *Phys. Rev. Lett.*, **92**, 176102 (2004).
5. Sze, S. M., *Physics of Semiconductor Devices*, Wiley-interscience publication, New York, 1981, pp. 848–849.
6. Bachilo, S. M., Strano, M. S., Kittrell, C., Hauge, R. H., Smalley, R. E., and Weisman, R. B., *Science*, **298**, 2361 (2002).
7. Wei, S.-H., and Zunger, A., *Appl. Phys. Lett.*, **72**, 2011 (1998).

Electron Transport in Carbon Nanotubes using Superconductiong Electrodes

S. Ishii[a,d], J-F Lin[c], E. S. Sadki[d], K. Kida[a], T. Sasaki[a,b], N. Aoki[a,c], S. Ooi[d], J. P. Bird[c], K. Hirata[d], Y. Ochiai[a,b]

[a]*Graduate School of Science and Technology, Chiba University, 1-33 Yayoi, Inage, Chiba 263-8522, Japan*
[b]*Center for Frontier Electronics and Photonics, Chiba University, 1-33 Yayoi, Inage, Chiba 263-8522, Japan*
[c]*Nanostructures Research Group, Arizona State University, Tempe, AZ 85287-5706, USA.*
[d]*Superconducting Materials Center, National Institute for Materials Science, 1-2-1 Sengen, Tsukuba 305-0047, Japan*

Abstract. Superconducting properties in multi-walled carbon nanotubes (MWNTs) have been studied due to an interest on the flux field effect transistor applications. We have prepared a sample of a lope or bundle of MWNTs contacted with Indium metal at the both ends and observed a clear decrease in the magneto-resistance near the transition temperature of the Indium metal. This behavior has been expected to arise from superconducting precursor or the fluctuation effects near the transition temperature.

INTRODUCTION

Superconductivity in single-walled carbon nanotubes (SWNTs) and multi-walled carbon nanotubes (MWNTs) has been started to explore its possibility on device applications for quantum computing. While in most literatures the superconductivity of carbon nanotubes has been reported to be due to the so-called proximity effect [1,2], there has been few claiming it is due to the intrinsic properties of carbon nanotubes. In order to clarify this problem, we study electrical transport properties of MWNTs with superconducting electrodes at low temperatures.

EXPERIMENTS

The samples were prepared by the following procedures. The MWNTs, which were sonicated in dichloroethane, were dispersed onto SiO_2 substrates. The superconducting contacts were then defined by photolithography, followed by Indium metal lift-off. Typical diameter of MWNT bundle in our study is ~ 1 μm and the spacing between two pads is ~ 5 μm. The bundle between two pads includes several MWNTs. Temperature dependence measurements of magneto-resistance have been performed in the cryostat at magnetic field of 0, 20, and 50 Oe, and at temperatures from 1.7 to 290K.

RESULTS AND DISCUSSION

Before starting our superconducting measurements, we performed ESR and resistance measurements for the same type of the MWNTs [3]. The transport properties of these MWNTs show metallic behavior. Temperature dependence of the resistance and I-V characteristics at several temperatures have been also measured in MWNT samples and the resistance is slightly higher than the sample only consisting in MWNT. It suggests that weak localization plays an important role in these superconducting electrodes.

The low temperature resistance of MWNTs with Indium contacts decreases with lowering temperatures from room temperature, and then it begins to turn to the increase between 150 and 200 K. These are a typical semiconductor behavior for MWNTs as reported in previous studies.

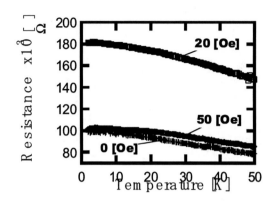

FIGURE 1. Temperature dependence (Wide Range) of the resistance for the MWCNT having Indium electrodes under the magnetic fields, 0, 20 and 50 Oe.

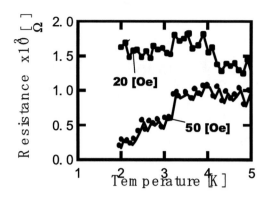

FIGURE 2. Temperature dependence (Narrow Range) of the resistance for the MWNTs with Indium electrodes under the magnetic fields, 0, 20 and 50 Oe, which is shifted for comparing in this graph.

But, at around 3.41 K, the drastic drops of the resistance have been observed in the case of the magnetic field of 20 and 50 Oe as shown in Figs.1 and 2. However, such a decrease has not been observed in the case of 0 Oe. The temperature where we observed the resistance drop is almost equal to the critical temperature of Indium. Probably, these drops can be considered to relate to superconducting precursor or the fluctuation effects near the transition temperature. However, we need to study further more on the magneto-resistance measurements for the wire samples of MWNTs having superconducting metal leads.

SUMMARY

Superconducting behaviors in MWNTs have been studied with an interest on the flux field effect transistor applications. In the case of the magnetic field of 20 and 50 Oe the drastic drops of the resistance have been observed in the magneto-resistance near the transition temperature of the Indium metal. We suggest that the decrease should come from superconducting precursor or the fluctuation effects near the transition temperature.

ACKNOWLEDGMENTS

Work at Chiba University was supported in part by a grant-in-aid from the Japan Society for Promotion of Science (Nos. 14750005 and 13450006). Also, This work is supported in part by 2003 Academic Frontiers Student Exchange Promotion Program scholarship in Minister of Education, Culture, sports, Science and Technology, Japan.

REFERENCES

1. A. Yu. Kasumov, R. Deblock. M. Kociak, B. Resulet, H. Bouchiat, I. I. Khodos, Yu. B. Gorbatov, V. T. Volkov, C. Journet, and M. Burghard, Science, **284**, 1508 (1999).
2. J. Haruyama, K. Takazawa, S. Miyadai, A. Takeda, N. Hori, I. Takesue, Y. Kanda, N. Sugiyama, T. Akazaki, and H, Takayanagi, Phys. Rev. B., **68**, 165420 (2003).
3. S. Ishii, K. Miyamoto, N. Oguri, K. Horiuchi, T. Sasaki, N. Aoki and Y. Ochiai, Physica E., **19**, 149 (2003).

The transport property at cross-junction of multiwall carbon nanotubes

Nobuyuki Aoki, Takahiko Mihara, Michio Kida, Koichiro Miyamoto, Takahiko Sasaki, and Yuichi Ochiai

Department of Electronics and Mechanical & Center for Frontier Electronics and Photonics, Chiba University, 1-33 Yayoi-cho, Inage-ku, chiba-shi, Chiba 263-8522, Japan

Abstract. Transport property of crossed two multiwall carbon nanotubes has been investigated through low temperature measurement. We have succeeded to obtain the essential transport characteristics at the junction by means of the four-terminal measurement, where the conductance showed no Tomonaga-Luttinger liquid behabior, but showed almost metallic transport corrected by weak localization.

INTRODUCTION

Carbon nanotubes (CNTs) have been proposed as one of a possible material for building up a nano-scale integrated circuit and a molecular device [1]. In order to realize such a nano-device, it is very important to clear the transport properties at the junction of CNTs. Especially, multi-wall carbon nanotube (MWNT) is considered as a candidate for electrical wiring because of the high conductivity and the high maximum current density [2]. Recently, transport properties of cross junctions of CNTs have been studied and shown very characteristic transport behavior [3-6]. However, these studies have discussed using two-terminal (2-t) measurement which includes the properties of the contact region, the CNT and the junction, or have not shown the local transport at the junction. In this study, we have succeeded to measure the real conductance at the junction of MWNTs using a four-terminal (4-t) method. In order to fabricate the crossed structure, we have established our original manipulation technique [7] and the transport properties have been studied by multi-terminal measurements.

SAMPLE PREPARATION

The MWNT used here is synthesized by arc discharge method, and the typical diameter and length are 150 nm and 20 μm, respectively. It is purified only by a centrifugal separator without thermal oxidation. It is dispersed in a solvent with ultrasonic and is coated on a SiO_2 layer on a Si wafer. Under a high resolution CCD microscope, we selected suitable MWNTs and set them to form a crossed structure by manipulating a glass micro-capillary. The electric contacts are performed by Au/Ti, and the sample is rapidly annealed in reduction environment in order to reduce the contact resistance. Figure 1 shows a scanning electron microscope (SEM) image of the sample.

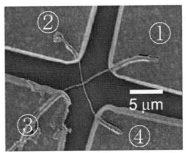

FIGURE 1. SEM image of the sample. The metal pads consists of Au/Ti.

RESULT AND DISCUSSION

In the 2-t measurement using electrodes 1 and 3, the transport properties of the individual MWNT was obtained. Figure 2 (A) shows the temperature dependence of the 2-t conductance (G_{1-3}) on a log-log scale, which shows power law dependence ($G \propto T^{\alpha}$) with $\alpha = 0.71$. It indicates that this MWNT shows a one-dimensional Tomonaga-Luttinger liquid (TLL) behavior [8]. On the other hand, if we measure electrodes 2 and 3, we obtain the curve resistance including the properties of these two MWNTs and the junction. Figure 2 (B) shows the temperature dependence of G_{2-3} on a log-log scale, which also shows a power law dependence at higher temperature with $\alpha = 1.7$. The power is more than twice of that of G_{1-3}. It has theoretically predicted that the transport at a junction of CNTs would be a tunneling between TLL, consequently the power would be increased as the summation of α of the individual CNTs [9].

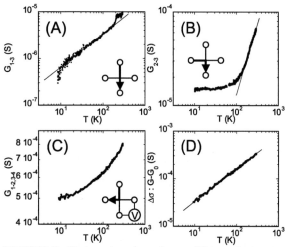

FIGURE 2. Temperature dependence of 2-t conductance at the straight (A; G_{1-3}) and the curve configuration (B; G_{2-3}). Temperature dependence of 4-t conductance at the junction (C; $G_{1-2, 3-4}$) and the conductance correction (D; $\Delta\sigma$). The insets schematically indicate the current pass. Solid lines are guide for eyes.

However, in order to obtain the local conductance at the junction of MWNTs, 4-t measurement was performed. The local conductance at the junction ($G_{1-4,2-3}$) was measured by passing a current from 1 and 2, and by measuring the voltage between electrodes 3 and 4. Then the $G_{1-2,3-4}$ and $G_{1-4,3-2}$ showed almost same value of 1.19 kΩ and 1.21 kΩ at room temperature (RT), respectively, moreover they almost coincide at all temperature range. Figure 2 (C) shows the temperature dependence of $G_{1-2,3-4}$ on a log-log scale. It monotonically decreases with decreasing the temperature however there is no straight dependence in the plot. It is revealed that the transport at the junction does not show TLL behavior contrary to the prediction [9]. In order to understand the transport characteristic at the junction, we tried to fit the data with some of transport regimes. The most accurate fit is shown in Fig. 2 (D). The $\Delta\sigma$ that is obtained by the deference from the conductance at 0 K, shows well-fitted straight dependence on temperature on a log-log scale. It indicates that the transport at the junction is essentially metallic but it is corrected by weak localization at finite temperature [10]. It indicates that quantum interference exists at the junction of MWNTs.

SUMMRY

Transport properties at a junction of MWNTs have been investigated. TLL behavior was observed in the 2-t conductance, and the value of α in the curve configuration showed more than twice value than in the case of the individual MWNT. However, the 4-t conductance showed no trace of TLL behavior, which showed a metallic transport corrected by weak localization.

ACKNOWLEDGMENT

This work is supported in part by Grant-in-Aid for Scientific Research of Japan Society for the Promotion of Science (JSPS), No.14750005 and 13450006.

REFERENCE

1. Dekker, C., *Phys. Today* **52**, pp.22 (1999).
2. Aoki, N., Takayama, J., Kida, M., Horiuchi, K., Yamada, S., Ida, T., Ishibashi, K., Ochiai, Y., *Jpn. J. Appl. Phys.* **42**, pp.2419-2421 (2003).
3. Postma, H.W.C., Jonge, M., Yao, Zhen, Dekker, C., *Phys. Rev. B* **62**, pp.10653-10656 (2000).
4. Kim, J., Kang, K., Lee, J-O., Yoo, K-H., Kim, J-R., Park, J.W., So, H.M., Kim, J-J., *J. Phys. Soc. Jpn.* **70**, pp.1464-1467 (2001).
5. Park, J.W., Kim, J., Yoo, K.-H., *J. Appl. Phys.* **93**, pp.4191-4193 (2003).
6. Kim, J., Kim, J-R., Lee, J-O., Park, J.W., So, H.M., Kim, N., Kang, K., Yoo, K-H., Kim, J.-J., *Phys. Rev. Lett.* **90**, pp.166403-1-4 (2003).
7. Mihara, T., Miyamoto, K., Kida, M., Sasaki, T., Aoki, N., Ochiai, Y., *Super Latt. and Microstruct.*, now printing.
8. Bockrath, M., Cobden, D.H., Lu, J., Rinzler, A.G., Smalley, R.E., Balents, L., McEuen, P.L., *Nature* **397**, pp.598-471 (1999).
9. Egger, R., Gogolin, A.O., *Phys. Rev. Lett.* **79**, pp.5082-5085 (1997).
10. Anderson, P.W., *Physica* **117B & 118B**, pp.30-36 (1983).

Quasiparticle Band Structure of Carbon Nanotubes

Takashi Miyake and Susumu Saito

Department of Physics, Tokyo Institute of Technology, 2-12-1 Oh-okayama, Meguro-ku, Tokyo 152-8551, Japan

Abstract. The electronic states of small-diameter carbon nanotubes are investigated using the *ab initio GW* approximation. It is found that quasiparticle excitation energies are increased significantly in semiconducting nanotube due to the many-body correction. Lifetimes of the quasiparticles are also calculated and turned out to be much longer in semiconducting tube than in metallic tube.

INTRODUCTION

Since their discovery in 1991 [1], the electronic properties of carbon nanotubes have been a topic of intensive researches. Carbon nanotubes are either metallic or semiconducting depending on their diameter and helical arrangement [2, 3]. This is understood by imposing appropriate boundary condition to the electronic band structure of a graphite sheet. However, this simple zone-folding analysis would be invalid for small-diameter nanotubes since curvature effect should be important. Local density approximation (LDA) in the framework of density functional theory [4, 5] is a standard and powerful tool to handle this kind of problems. However, when it comes to the electron excitation energies, LDA underestimates the band gaps of semiconductors significantly.

As an alternative procedure, the *GW* approximation [6] has been applied to many semiconductors for the last two decades and turned out to give good description of band gaps. In the present work, we study the (7,0) and (5,0) nanotubes within the *GW* approximation. We discuss the many-body effect on quasiparticle excitation energies and their lifetimes.

METHODS

We first perform LDA calculation with the full-potential LMTO method. The self-energy is then estimated in the *GW* approximation using the LDA wave functions and eigenvalues. The frequency-dependence of the dielectric function is evaluated numerically, and plasmon-pole approximation is not applied. More details of numerical techniques are found elsewhere [7]. The test calculation for diamond shows that the fundamental gap is 5.6 eV, to be compared to 5.48 eV (experiment) and 4.1 eV (LDA).

The geometry of the nanotube is optimized by means

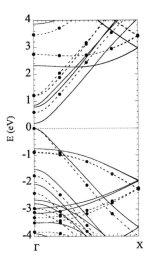

FIGURE 1. Quasiparticle band structure of the (7,0) tube.

of LDA. A nanotube is placed in the supercell in which the wall-to-wall distance is set to 6.5 Å. The nanotube is periodic in the tube direction, and 8 k points are sampled in this direction. In order handle a singularity of the Coulomb interaction, the $k = 0$ point is shifted slightly [8].

RESULTS AND DISCUSSIONS

Figure 1 shows the electronic structure of the (7,0) nanotube. The solid lines are the LDA results and closed circles are the *GW* quasiparticle energies. Dashed lines are the guide for the eyes. The energy is measured from the top of the valence band. The zone-folding analysis predicts that the (7,0) tube is a moderate-gap semiconductor with the fundamental gap formed at the Γ point.

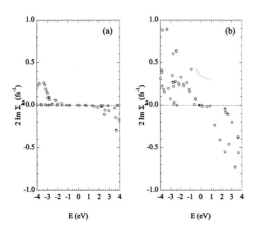

FIGURE 2. The imaginary part of the self-energy. (a) (7,0) tube. (b) (5,0) tube.

The LDA band structure shows that it is actually a semiconductor with both the top of the valence band and the bottom of the conduction band at the Γ point. However, the gap is quite small and the tube is a narrow-gap semiconductor. This is because the lowest conduction band is pulled down by the curvature effect [9]. We found the geometry optimization enhances this effect. The LDA gap before geometry optimization is 0.5 eV, while it is reduced to 0.2 eV by optimization [7]. The curvature effect is more significant in the thinner (5,0) tube. It is metallic in contrast with the zone-folding analysis.

When we include the GW many-body correction, the conduction bands in the (7,0) tube are shifted up as a whole, so that the band gap is increased to 0.6 eV. Hence the GW gap for the optimized geometry is accidentally close to the LDA gap for the non-optimized geometry. On the other hand, the many-body correction is of minor importance in the (5,0) tube, and it remains metallic even in the GW level.

The imaginary part of the self-energy, $2\mathrm{Im}\langle \psi_{\mathbf{k}n}|\Sigma(E_{\mathbf{k}n})|\psi_{\mathbf{k}n}\rangle$, is plotted in Fig.2, where $\psi_{\mathbf{k}n}$ and $E_{\mathbf{k}n}$ are LDA wave functions and eigenvalues, with $\mathbf{k}$ being the wave vector and n the band index. Each open circle corresponds to a set of $(\mathbf{k},n)$. Following the general properties, the imaginary part is positive for $E<0$, whereas it is negative for $E>0$. The inverse of the absolute value gives the lifetime of the excited electrons for $E>0$ [10] and holes for $E<0$ [11]. The lifetime is an inelastic one which arises from the electron-electron interactions. The imaginary part has larger values as the energy is away from the Fermi level, which corresponds to shorter lifetime of the quasiparticles at higher energies. Although more careful analysis is necessary, it seems that the lifetime does not obey a simple E^2 scaling.

Comparing Fig.2(a) which is the result for the (7,0) tube, with (b) for the (5,0) tube, we find the latter has much larger imaginary part. This strongly suggests that the quasiparticle decays faster in metallic (5,0) tube than in semiconducting (7,0) tube.

CONCLUDING REMARKS

The quasiparticle excitation of the small-diameter zigzag carbon nanotubes has been investigated using the GW approximation. It is found that the many-body effect significantly affects the excitation energies in the semiconducting tube. The lifetime of hot electrons (holes) were also studied. Quasiparticles have much longer lifetime in semiconducting tube than in metallic tube. Applications of the present method to larger diameter tubes as well as tubes with various chirality are highly anticipated for comparison with experiments.

ACKNOWLEDGMENTS

We wish to acknowledge Dr. T. Kotani and Prof. M. van Schilfgaarde for providing us with the GW code. We also thank Prof. A. Oshiyama, Prof. T. Nakayama, Dr. M. Saito, and Prof. O. Sugino for the pseudopotential program, and Prof. N. Hamada and Prof. S. Sawada for the tight-binding code. Numerical calculations were performed partly on Fujitsu VPP5000 at the Research Center for Computational Science, Okazaki National Institute. This work was supported by Grant-in-Aid from the Ministry of Education, Science and Culture of Japan, and the 21st Century Center-of-Excellence Program at Tokyo Institute of Technology 'Nano-Scale Quantum Physics' from the Ministry of Education, Culture, Sports, Science, and Technology of Japan.

REFERENCES

1. S. Iijima, Nature **354**, 56 (1991).
2. N. Hamada, S. Sawada, and A. Oshiyama, Phys. Rev. Lett.**68**, 1579 (1992).
3. R. Saito et al., Appl. Phys. Lett. **60**, 2204 (1992).
4. P. Hohenberg and W. Kohn, Phys. Rev. B**136**, 864 (1964).
5. W. Kohn and L. J. Sham, Phys. Rev. A**140**, 1133 (1965).
6. L. Hedin and S. Lundqvist, *Solid State Physics* vol. 23, edited by H. Ehrenreich, F. Seitz, and D. Turnbull (New York, Academic 1969).
7. T. Miyake and S. Saito, Phys. Rev. B **68**, 155424 (2003).
8. T. Miyake and S. Saito, Transactions of the Materials Research Society of Japan **29**[2], 553 (2004).
9. X. Blase et al., Phys. Rev. Lett. **72**, 1878 (1994).
10. I. Campillo et al., Phys. Rev. Lett. **83**, 2230 (1999).
11. I. Campillo et al., Phys. Rev. Lett. **85**, 3241 (2000).

Magneto-Optical Properties Of Aligned Single-Walled Carbon Nanotubes

Masao Ichida[1,2], H. Wakida[1], H. Kataura[3], Y. Achiba[3], and H. Ando[1]

[1]*Faculty of Science and Engineering, Konan University, Kobe 658-8501, Japan*
[2] *PRESTO, JST, Kawaguchi, 332-0012, Japan*
[3] *Faculty of Science, Tokyo Metropolitan University, Hachioji 192-0397, Japan*

Abstract. We report on the magneto-optical properties of aligned single-walled carbon nanotubes in polymer. With appling magnetic fields parallel to the aligned tube axis, the optical absorption spectra significantly change. The observed spectral change is explained by the magnetic-field-induced band splitting. The physical origin of energy splitting is discussed in relation to Aharanov-Bohm and Zeeman effects.

INTRODUCTION

The fundamental structure of single-walled carbon nanotube (SWNT) can be regard as a cylinder. Since the motion of electron in SWNT is confined on the cylinder surface, we can expect interesting magnetic effects in the electronic and optical properties due to Aharonov-Bohm (AB) effect[1]. The theoretical calculations have predicted that with increasing magnetic flax in the cross section of SWNT, the gap will be opened at the Fermi energy in metallic tube and the energy gap split into two in semiconducting SWNT[1]. This effect should be observed in the optical spectra in SWNTs. Quite recently, magneto-optical spectra for micelled SWNTs in water have been measured[2]. However, since the micelled SWNTs in water were aligned by magnetic field, the observed magnetic effects were mixed magnetic induced tube alignment and electronic effects. In this paper, we report on the magnetic effects in optical transitions of aligned SWNTs in polymer. We discuss the origin of magnetic induced absorption change in relation to AB and Zeeman effects.

EXPERIMENTAL

We prepared aligned SWNTs thin film samples by mechanically stretching with polymer[3]. SWNTs in the sample are aligned to the stretching axis. Using a tungsten lamp as a light source, Si and InGaAs photodiodes as detectors, the polarization dependence of absorption spectra were measured under the magnetic fields with two different configurations, magnetic field parallel and perpendicular to the aligned axis of SWNTs. The magnetic fields up to 10 T were applied by using super conducting magnet.

RESULTS AND DISCUSSION

Solid curve in Fig. 1 shows the absorption spectrum of aligned SWNTs sample measured with the polarization of incident light parallel to the alignment axis. Broad absorption band is observed at 0.8 eV. This band is attributed to the excitonic transition from the first valence band to the first conduction band in semiconducting SWNTs[4]. Broken curve in Fig. 1 displays the absorption spectrum measured under magnetic field of 10 T parallel to the tube axis. The absorption spectrum significantly changes by applying the magnetic field. Figure 1(b) displays the difference in absorption spectra B=0 T and B=10 T. The optical absorption decreases in the vicinity of the band peak and increases below the band. On the other hand, in the case of magnetic field perpendicular to the tube axis, no absorption change is observed. In order to investigate the physical origin of the absorption change, we have conducted the line shape analysis. The absorption spectrum without magnetic field is well fitted to the Gaussian type line shape. We assume that the observed absorption change under the magnetic field in the case of parallel to the tube axis is caused by a band splitting. The absorption spectrum under the magnetic field is analyzed by two Gaussian line shape function; one located at lower energy side and another at higher energy side by ΔE than that for without magnetic field. Dotted curve in Fig. 1 (b) shows the best fit result for the spectrum of absorption change. The estimated energy splitting $2\Delta E$ is about 54 meV at 10 T.

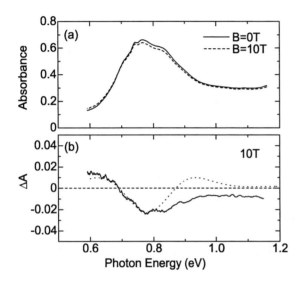

FIGURE 1. (a) Absorption spectra of aligned SWNTs in polymer with and without magnetic field. The magnetic field of 10 T is applied parallel to the aligned axis. (b) Absorption change due to the magnetic field.

For the magnetic field parallel to the tube axis, both Zeeman and AB effects cause the energy splitting in the first transition band A of semiconducting SWNT[1]. We estimate the energy splitting for these effects at 10 T. In the SWNT, the absorption spectra can be calculated by the tight-binding model with periodic boundary condition in the circumference on the tube. Under the magnetic field, the boundary condition is expressed as,

$$\mathbf{k} \cdot \mathbf{L} = 2\pi m + 2\pi(\phi/\phi_0), \quad (1)$$

where $\mathbf{k}$ is a wavevector, $\mathbf{L}$ is a chiral vector (circumference), m is a integer corresponding to the quantized angular momentum, ϕ is a magnetic flux in the cross section of SWNT, and $\phi_0 = h/e$ is the magnetic flux quantum[1]. Solid and broken curves in Fig. 2 show the calculated absorption spectra of SWNT with tube diameter of 1.22 nm for chiral index of (10,8) under B=0 T and B=10T, respectively. Under the magnetic field, the absorption peak splits into two peaks. The splitting energy $2\Delta E$ is about 12 meV. This value is smaller than the measured value of 54 meV. On the other hand, the splitting caused by the spin Zeeman effects is estimated as about 6 meV. When we consider the orbital Zeeman effect semiclassically, the splliting energy can be estimated from $2m\mu_B B$. A simple tight-binding and band-folding model predicts the angular momentum of 27 for the first valence and conduction bands of (10,8) tube. The energy splitting for this transition amounts to about 30 meV at B=10 T. Thus, we conclude that not only AB effect but Zeeman effects play an important role in electronic and optical properties of SWNT under the high magnetic field. We note that, in the SWNT based on the graphene 2D electronic band, the energy has k-linear dependence instead of k^2 dispersion. This may cause reduction in orbital Zeeman effect. We may need further study to investigate the magneto-optical properties in SWNTs taking into account the exciton effects.

This work was partially supported by the grant to private university from the Ministry of Education, Culture, Sports, Science and Technology of Japan.

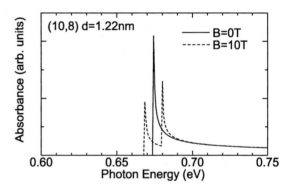

FIGURE 2. Calculated absorption spectra of a SWNT with tube diameter of 1.22 nm and chiral index of (10,8). Solid and broken curves show the spectra at B=0 T and B=10 T, respectively.

REFERENCES

1. H. Ajiki and T. Ando, *Physica B* **201**, 349 (1994).

2. S. Zaric, G. N. Ostojic, J. Kono, J. Shaver, V. C. Moore, M. S. Strano, R. H. Hauge, R. E. Smalley, X. Wei, *Science* **304**, 1129 (2004).

3. M. Ichida, S. Mizuno, H. Kataura, Y. Achiba, and A. Nakamura, *Appl. Phys. A* **78**, 1117 (2004).

4. M. Ichida, S. Mizuno, Y. Tani, Y. Saito, and A. Nakamura, *J. Phys. Soc. Jpn.* **68**, 3131 (1999).

Transport Properties of Charge Carriers in Single-Walled Carbon Nanotubes by Flash-Photolysis Time-Resolved Microwave Conductivity Technique

Yasuhide Ohno, Kenzo Maehashi, Koichi Inoue, Kazuhiko Matsumoto, Akinori Saeki, Shu Seki, Seiichi Tagawa

The Institute of Scientific and Industrial Research, Osaka University, 8-1 Mihogaoka, Ibaraki, Osaka 567-0047, Japan

Abstract. Transport properties of charge carriers in single-walled carbon nanotubes (SWNTs) were investigated by flash-photolysis time-resolved microwave conductivity (FP-TRMC) technique. With this technique, it is possible to monitor the change in conductivity on pulsed laser excitation on a nanosecond timescale, without contacting layer with electrode. The FP-TRMC signals obtained by SWNT sample is drastically larger than that of only catalyst. The dependence of excitation wavelength on $\phi\Sigma\mu$, which is a product of a quantum yield and sum of mobility, was obtained, indicating variety in transport property of size-distributed SWNTs.

INTRODUCTION

Single-walled carbon nanotubes (SWNTs) have been investigated eagerly from the viewpoint of their ideal one-dimensional materials. Carbon nanotube nanoelectronics applications, such as field-effect transistors (FETs) have been proposed for chemical and biochemical sensors [1]. Transport properties like mobility of charge carriers in SWNTs are very crucial to these applications. Mobility in semiconducting SWNTs, however, has not been evaluated yet in detail. Estimates of mobility by FET properties were considerably scattering, which was 20 cm^2/Vs to infinite [2], 79,000 cm^2/Vs [3] or ballistic. It can be considered that these estimated mobilities contain the effect of contact resistivity between SWNTs and electrode metal. In this letter, we investigated transport properties of charge carriers in SWNTs without electrode by flash-photolysis time-resolved microwave conductivity (FP-TRMC) technique.

EXPERIMENT

SWNT samples were grown by thermal chemical vapor deposition (CVD) method like follows; (1) deposition of Fe(NO$_3$)$_3$·9H$_2$O, MoO$_2$(acac)$_2$ and alumina nanoparticles in the methanol on Si or quartz substrate, (2) heating the substrate in a quartz tube furnace, which evacuated by rotary pump, to reach 900°C with Ar flow rate of 1 SLM (standard liter/min) due to maintain the pressure at 0.4~0.5 Torr, (3) the Ar flow is replaced by ethanol vapor from reservoir for 10 min at pressure of 1~2 Torr, (4) electric furnace was turned off, and 1 SLM Ar was flowed while it cooled at room temperature.

In FP-TRMC method [4] (shown in Fig. 1), which probe the change in conductivity on photoexcitation of SWNTs, neither electrical contacts nor electrolyte is necessary, and the change in reflected microwave is directly proportional to the conductivity which comes from charge carriers generated by pulsed laser.

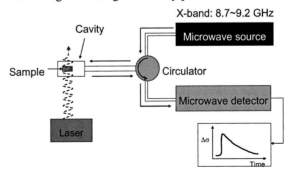

FIGURE 1. Schematic illustration of FP-TRMC method.

The change in microwave power reflected by the cell on flash-photolysis was monitored using microwave circuitry and detection equipment. The change in reflected power (ΔP) is related to the change in conductivity of the sample ($\Delta \sigma$) by the sensitivity factor A, via

$$\frac{\Delta P}{P} = -A\Delta\sigma. \quad (1)$$

The sensitivity factor A is dependent on the frequency of the microwave field and on the material properties of the measured layers. The value of A can be determined independently by an electrostatic analysis of the resonant cavity and an optical property of the material. The change in the conductivity ($\Delta \sigma$) is related to the average concentration of negative (n_n) and positive (n_p) charge carriers present, and their respective mobilities, μ_n and μ_p. Therefore, the change in conductivity can be written as

$$\Delta\sigma = e(n_n\mu_n + n_p\mu_p) \quad (2)$$

where e is an elementary charge. The number of charge carriers ($n \sim n_n \sim n_p$) per volume which is a function of the amount of the incident flux in the pulse, I_0 (photons/m^2), correction factor F_{Light}, which is determined by sample film thickness and position, and the quantum efficiency for charge carrier generation, ϕ, can be determined by the equation

$$n = \phi I_0 F_{Light} \quad (3)$$

By combining Eqs. (2) and (3), the $\Delta\sigma$ measured, can be related to the product of ($\mu_n + \mu_p$) and ϕ by,

$$\phi(\mu_e + \mu_p) = \phi\sum\mu = \frac{\Delta\sigma}{eI_0 F_{Light}} \quad (4)$$

RESULTS

FP-TRMC transients obtained from SWNTs and only catalyst on quartz substrate, which were excited by 355 nm, 0.1 W (12 Hz) of third-harmonic generation in Nd-YAG laser, are shown in Fig. 2. The solid line shows the FP-TRMC transient from SWNTs and the dashed line shows that of catalyst. The $\Delta\sigma$ values measured from SWNTs is drastically larger than that of only catalyst sample, which indicates the FP-TRMC system can observe the change in conductivity in the SWNTs.

Then we measured the dependence of excitation wavelength on $\phi\Sigma\mu$ in SWNTs (shown in Fig. 3). The $\phi\Sigma\mu$ obtained from the sample that excited at 532 nm is larger than that from excited at 355 nm.

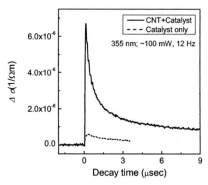

FIGURE 2. FP-TRMC transients obtained from SWNTs and catalyst.

It is probably due to the band gap difference of each SWNT. Generally, band gap and diameter of SWNTs depend on their chirality (m, n). SWNTs excited at 532 nm and 355 nm differ from each other because of their resonance absorption. Therefore, if we can estimate the value of ϕ, we can obtain mobility of charge carrier in SWNTs. Now, we are investigating the ϕ value of SWNTs. These results indicate that the FP-TRMC method is very useful to investigate the transport properties in SWNTs.

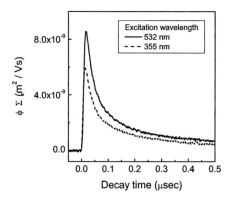

FIGURE 3. Excited wavelength dependence of FP-TRMC transients.

REFERENCES

[1] J. Kong, N. R. Franklin, C. Zhou, M. G. Chapline, S. Peng, K. Cho and H. Dai, *Science* **287**, 622-625 (2000).

[2] R. Martel, T. Schmidt, H. R. Shea, T. Hertel and Ph. Avouris, *Appl. Phys. Lett.* **73**, 2447-2449 (1998).

[3] T. Durkop, S. A. Getty, E. Cobas and M. S. Fuhrer, *Nano Lett.* **4**, 35-39 (2004).

[4] J. M. Warman, G. H. Gelinck and M. P. de Haas, *J. Phys.: Condens. Matter.* **14**, 9935-9954 (2002).

Modification of the Band Gaps and Optical Properties of Single-Walled Carbon Nanotubes

J.G. Wiltshire*, L.J. Li*, A.N. Khlobystov†, M. Glerup**‡, P. Bernier** and R.J. Nicholas*

Clarendon laboratory, Department of Physics, University of Oxford, Parks Road, Oxford OX1 3PU, UK
†Department of Materials, University of Oxford, Parks Road Oxford OX1 3PH, UK
**GDPC (UMR 5581), Université Montpellier II, Pl. E. Bataillon, 34095, Montpellier, Cedex 5, France
‡Grenoble High Magnetic Field Laboratory, MPI-FKF/CNRS, 25 Avenue des Martyrs, 38042 Grenoble, Cedex 9, France

Abstract. We report a study of the effects of various sample purification and modification techniques on the band gaps and optical properties of bulk single walled carbon nanotubes (SWNTs). These include a variety of thermal and chemical oxidation treatments and a study of nitrogen and boron doping.

Single-walled carbon nanotubes were first discovered in 1993 by Iijima et al [1] and since then they have attracted a large amount of interest within the semiconductor industry due to their novel band structure which is dependent on the geometry of the carbon nanotubes. However the bulk production of SWNTs through methods such as electric arc discharge, CVD or high-pressure CO conversion (HiPCO) produces nanotubes with a range of diameters and chiralities. We are interested in studying methods that modify the diameter distribution of nanotubes produced by various methods. Here we present two approaches to alter this distribution either during the growth stage or as a post production treatment.

For absorption measurements the SWNTs were deposited as thin films on ZnSe substrates using a modified Badger 200NH airbrush. For use with the airbrush the sample material was first suspended in MeOH and sonicated for ~15-30min at room temperature, evaporation of the solvent was aided by heating the substrate during airbrushing using a Corning PC-35 hotplate at either 150 or 250°C. Absorption spectra were obtained using a Perkin Elmer Lambda 9 Spectrophotometer.

Liquid and Gas Phase Oxidative Treatments

Ultrasonically assisted liquid phase purification was conducted in a glass test tube filled with 10mg of raw HiPCO SWNTs in 40ml of $H_2SO_4:HNO_3=$ 3:1 solution and sonicated in an ultrasonic bath for various times at 35-50°C [2]. Following ultrasonication the resultant suspension was diluted with 200ml of water and collected on filter paper before washing in NaOH.

Thermal purification was performed by placing the raw HiPCO material in an alumina crucible and annealing it in air at various temperatures and times. The SWNTs were then washed in HCl, collected on a PTFE filter, then washed with water and MeOH and finally dried in air at room temperature.

Background adjusted absorbance spectra for the chemically and thermally treated samples are presented in Fig. 1, the broad peak centered at 0.9eV is a superposition of individual semiconducting nanotube transitions between the first van Hove singularities. In Fig. 1a and 1b the integrated area of the E_{11} mode has been normalized to the appropriate sample weight loss.

Figure 1a shows that with increasing oxidation time the higher energy features of E_{11} are depressed relative to the lower energy side. The same trend is seen in Fig. 1b with increased oxidation temperature. In both cases this is consistent with a preferential suppression of the contribution from smaller diameter nanotubes [3], which may react with the acid at a faster rate possibly because of their higher C-C bond strain. TEM and Raman spectroscopy of these samples [4] demonstrated that the nanotubes are substantially shortened by both processes. However the chemically treated nanotubes also developed a higher degree of damage with increased exposure to the acid, while no increase in damage to the thermally treated nanotubes was observed.

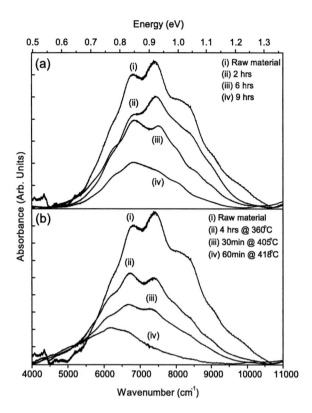

FIGURE 1. Background adjusted absorbance spectra of (a) the E_{11} mode for (i) the as purchased raw material, (ii) 2hrs treatment, (iii) 6hrs treatment and (iv) 9 hrs treatment, (b) the E_{11} mode for (i) the as purchased raw material, (ii) 4hrs@360°C (iii) 30min@405°C and (iv) 1hr@418°C.

Nitrogen and Boron Doping

Nitrogen and boron doped nanotubes were also studied in an attempt to understand their influence on the electronic structure and diameter distribution produced. These samples were grown using an arc discharge method where boron and nitrogen were introduced into the chamber at a ratio of 1:1 [5]. Typically arc-discharge nanotubes have a larger diameter than HiPCO produced nanotubes and have a smaller initial diameter distribution.

Background adjusted absorbance spectra for the raw and undoped materials are shown in Fig. 2a. For the raw undoped SWNTs the E_{11} transition is measured as ~0.71eV and the second transition (E_{22}) is observed at ~1.25eV. Figure 2b shows that both the E_{11} and E_{22} transitions are shifted up by 0.07eV and 0.15eV respectively for the doped sample which is thought to contain nominally 1% doping.

The similar proportional shifts in the E_{11} and E_{22} transitions suggests that this is caused by changes in the relative proportions of the diameter species present

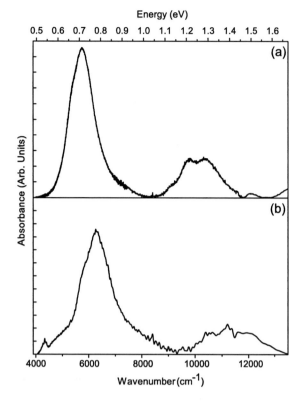

FIGURE 2. Background adjusted absorbance spectra of (a) the E_{11} (~0.71eV) and E_{22} (~1.25eV) modes for raw undoped material, (b) doped sample with approximately 1% nitrogen and boron doping.

[6], which is also supported by Raman spectroscopy. Therefore doping may provide a new route to controlled nanotube diameter growth.

ACKNOWLEDGMENTS

Parts of this work were funded by Foresight LINK Award (Nanoelectronics at the Quantum Edge), funded by the DTI, EPSRC and Hitachi Europe Ltd. We would like to thank the Swire group for the provision of a graduate scholarship for L.J. Li.

REFERENCES

1. Iijima, S., *Nature*, **363**, 603 – 605 (1993).
2. Liu, J., *Science*, **280**, 1253–1256 (1998).
3. Chiang, I. W., *J. Phys. Chem B*, **105**, 8297–8301 (2001).
4. Wiltshire, J. G., *Chem Phys Lett*, **386**, 239–243 (2004).
5. Glerup, M., *Chem Phys Lett*, **387**, 193–197 (2004).
6. *To be published* (2004).

A new type of superlattice based on carbon nanotubes

O.V. Kibis*, D.G.W. Parfitt[†] and M.E. Portnoi[†]

Novosibirsk State Technical University, Novosibirsk 630092, Russia
[†]*School of Physics, University of Exeter, Stocker Road, Exeter EX4 4QL, United Kingdom*

Abstract. We show that electron motion in a $(n,1)$ carbon nanotube corresponds to a de Broglie wave propagating along a helical line on the nanotube wall. This helical motion leads to periodicity of the electron potential energy in the presence of an electric field normal to the nanotube axis. The period of this potential, which is proportional to the nanotube radius, is always greater than the interatomic distance in the nanotube. Therefore the behavior of an electron in such a nanotube subject to a transverse electric field is similar to that in a semiconductor superlattice.

The electronic energy spectrum of a (n,m) single-wall carbon nanotube (CNT) is usually described in terms of the two components of the electron wave vector $\mathbf{k}$ along the translation vector $\mathbf{T}$ and chiral vector $\mathbf{C}_h$ as $\mathbf{k} = k_\| \mathbf{T}/T + k_\perp \mathbf{C}_h/C_h$, where $k_\|$ and $k_\perp$ are subject to the following constraints: $-\pi/T < k_\| \leq \pi/T$ and $k_\perp = 2\pi l/C_h$, with the integer l representing the electron angular momentum along the nanotube axis [1].

The energy spectrum obtained in terms of $k_\|$ and $k_\perp$ in the tight-binding approximation [1] can be rewritten as

$$\varepsilon = \pm \gamma_0 \left[1 + 4\cos\left(\frac{k_s a}{2}\right) \cos\left(\frac{2n+m}{2m} k_s a - \frac{2\pi l}{m}\right) + 4\cos^2\left(\frac{k_s a}{2}\right) \right]^{1/2}, \quad (1)$$

where $k_s = k_\perp \cos\theta + k_\| \sin\theta$, and θ is the chiral angle, $\tan\theta = \sqrt{3}m/(2m+n)$. In the energy spectrum (1), the plus and minus signs correspond to the conduction and valence bands, respectively, $\gamma_0 \approx 3$ eV is the transfer integral between π-orbitals of neighboring carbon atoms, and the lattice constant of graphite $a = 2.46$ Å. For $m = 1$, Eq. (1) becomes independent of l, and we obtain the electron energy spectrum of a $(n,1)$ CNT in the form

$$\varepsilon = \pm \gamma_0 \left[1 + 8\cos\left(\frac{n+1}{2} k_s a\right) \right.$$
$$\left. \times \cos\left(\frac{n k_s a}{2}\right) \cos\left(\frac{k_s a}{2}\right) \right]^{1/2}. \quad (2)$$

It should be noted that the spectrum (2) depends on the parameter k_s alone, in contrast to the general case of a (n,m) CNT, for which the electron energy spectrum depends on two parameters ($k_\|$ and $k_\perp$ are conventionally used). This peculiarity of a $(n,1)$ CNT is a consequence of its special crystal symmetry: the $(n,1)$ CNT lattice can be obtained by translation of an elementary two-atom cell along a helical line on the nanotube wall. As a result, the parameter k_s has the meaning of an electron wave vector along the helical line, and so any possible electron motion in a $(n,1)$ CNT can be described by a de Broglie wave propagating along such a line. Thus, $(n,1)$ CNTs represent a previously overlooked distinctive class of nanotubes, which may be termed 'helical' nanotubes. Both descriptions of the energy spectrum of a $(n,1)$ CNT — by two parameters, $k_\|$ and $k_\perp$, or a single parameter k_s — are physically equivalent. However, the second description is more convenient for studies of electron processes determined by the above-mentioned helical symmetry of electron motion, and allows one to discover new physical effects (e.g. the electron-electron interaction should be strongly modified for helical one-dimensional motion [2]). We shall now show that such helical symmetry results in superlattice behavior of a $(n,1)$ CNT in the presence of an electric field oriented perpendicular to the nanotube axis.

The potential energy of an electron on a helix subject to a transverse electric field takes the form

$$U = eER\cos(2\pi s/l_0), \quad (3)$$

where e is the electron charge, E is the electric field strength, $R = C_h/2\pi$ is the radius of the CNT, s is the electron coordinate along the helical line,

$$l_0 = 2\pi R/\cos\theta = 2a(n^2+n+1)/(2n+1) \quad (4)$$

is the length of a single coil of the helix, and the electric potential is assumed to be zero at the axis of the CNT. The potential energy (3) is periodic in the electron coordinate s along the helical line and the period of the potential is equal to l_0. Since this period (4) is proportional to the CNT radius R and is greater than the interatomic distance, the CNT assumes typical superlattice properties. In particular, Bragg reflection of electron waves with wave vectors $k_s = \pm \pi/l_0$ results in energy splitting within the conduction and valence bands of the CNT. It can be shown that for sufficiently weak fields, $E \ll \gamma_0 a/(eR^2)$, the energy of Bragg band splitting

$$\Delta \varepsilon \sim eER. \quad (5)$$

Thus, even a small electric field results in a superlattice-like change of the electron energy spectrum in $(n,1)$ CNTs, with the appearance of Bragg energy gaps proportional to the field amplitude E and the nanotube radius R. Notably, this dependence of the Bragg gaps on the external field and radius applies to any helical quasi-one-dimensional nanostructure in a transverse electric field: this generic feature arises from the symmetry of the nanostructure, and is independent of the parameters of the tight-binding model used to derive Eq. (5). It should be emphasized that the discussed superlattice behavior is a unique feature of $(n,1)$ CNTs. For the general case of a (n,m) CNT with $m \neq 1$, the energy spectrum (1) depends on the quantum number l in addition to k_s. As already mentioned, l represents the projection of the electron angular momentum on the nanotube axis, and it follows from the corresponding selection rule that the transverse electric field only mixes electron states with angular momentum l and $l \pm 1$. For $m \neq 1$, however, states with l differing by one correspond to different subbands, and in general have different energies for $k_s = \pm \pi/l_0$, so that there is no Bragg scattering between these states. The only effect of the electric field, therefore, is to mix electron states with different energies, which does not lead to noticeable modification of the dispersion curves for weak electric fields [3].

For the particular case of a $(1,1)$ CNT the energy spectrum can be obtained in analytic form for any electron state. This spectrum is shown in Fig. 1 (solid lines) for a range of wave vectors $-\pi/a \leq k_s \leq \pi/a$. In the Figure, positive energies correspond to the conduction band and negative energies to the valence band. The energy spectrum in the absence of the field is shown for comparison (dashed lines). According to Eq. (4), the superlattice period l_0 for a $(1,1)$ CNT is equal to twice the lattice constant a. Therefore, as can be seen in Fig. 1, the width of the first Brillouin zone in the presence of a transverse electric field is half that without the field. It can also be seen that the electric field opens gaps in the dispersion curve at $k_s = \pm \pi/(2a)$ due to the aforementioned Bragg reflection of electron waves.

In the general case of a $(n,1)$ nanotube, for external electric field intensities attainable in experiment ($E \sim 10^5$ V/cm) and for a typical nanotube of radius $R \sim 10$ Å,

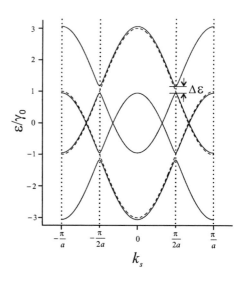

FIGURE 1. Electron energy spectrum of a $(1,1)$ CNT in the presence of a transverse electric field $E = \sqrt{3}\gamma_0/(ea)$ (solid lines) and without the electric field (dashed lines).

the value of the Bragg gap given by (5) is $\Delta \varepsilon \sim 10^{-2}$ eV, which is comparable to the characteristic energy of band splitting in conventional semiconductor superlattices. As a consequence, the discussed superlattice effects generated by the transverse electric field in $(n,1)$ CNTs should be observable in experiments, and may take place in existing CNT field-effect devices [4]. The inherent regularity of a nanotube-based superlattice, with the superlattice period determined by the CNT radius, presents a distinct advantage over semiconductor superlattices, in which monolayer fluctuations are unavoidable. A whole range of new nanoelectronic devices based on the discussed superlattice properties of $(n,1)$ CNTs can be envisaged, including Bloch oscillators [5] and quantum cascade lasers [6]. An evaluation of the feasibility of these novel devices and selection of their optimal parameters will undoubtedly form the subject of extensive future research.

This work is supported by INTAS, the Royal Society, the Russian Foundation for Basic Research, and the 'Russian Universities' program.

REFERENCES

1. Saito R., Dresselhaus G., and Dresselhaus M.S, *Physical Properties of Carbon Nanotubes*, Imperial College Press, London, 1998.
2. Kibis O.V., *Phys. Lett. A*, **166**, 393 (1992).
3. Li Y., Rotkin S.V., and Ravaioli U., *Nano Lett.*, **3**, 183 (2003).
4. Appenzeller J. et al., *Phys. Rev. Lett.*, **89**, 126801 (2002).
5. Esaki L. and Tsu R., *IBM J. Res. Dev.*, **14**, 61 (1970).
6. Faist J. et al., *Science*, **264**, 553 (1994).

Temperature and Hydrostatic Pressure Effects on the Band Gap of Semiconducting Carbon Nanotubes

Rodrigo B. Capaz[1,2,3], Catalin D. Spataru[2,3], Paul Tangney[2,3], Marvin L. Cohen[2,3], and Steven G. Louie[2,3]

[1] *Instituto de Física, Universidade Federal do Rio de Janeiro, Rio de Janeiro, RJ 21941-972, Brazil*
[2] *Department of Physics, University of California at Berkeley, Berkeley, CA 94720, USA*
[3] *Materials Science Division, Lawrence Berkeley National Laboratory, Berkeley, CA 94720, USA*

Abstract. We describe the temperature and pressure dependences of the band gap of semiconducting single-wall carbon nanotubes (SWNTs). At low pressures and temperatures, the band gaps display an unusual behavior, showing either positive or negative pressure and temperature shifts depending on the value of (n,m), with a distinct "family" (same $n-m$) behavior. We find that both pressure and temperature family behaviors have the same qualitative origin, unveiling an interesting and unsuspected connection between static and dynamical electron-lattice couplings.

The pressure and temperature dependences of the band gap (E_g) are two of the most important characteristics of a semiconductor. Recently, it has become possible to measure such properties for isolated semiconducting single-wall carbon nanotubes (SWNTs) [1, 2]. In this work, we describe these effects within a single-particle theory.

For low enough pressures P, the structural properties of isolated carbon nanotubes can be described by radial and axial strains: $\varepsilon_r = -P/C_r$ and $\varepsilon_z = -P/C_z$, where C_r and C_z are radial and axial elastic constants, respectively. First-principles calculations [3, 4] show a clear anisotropy in these elastic constants ($C_z > C_r$). Using the calculated values of ε_z and ε_r, we compute the pressure coefficients for the gap, dE_g/dP, for a variety of semiconducting SWNTs using a simple one-orbital tight-binding scheme [3]. The results are shown in Fig. 1. Values of dE_g/dP seem to follow trends according to the specific values of $(n-m)$, a so-called "family behavior". Also, there is a tendency for negative or positive dE_g/dP for $q = (n-m)$ mod 3 equals to 2 or 1, respectively. Similar types of family behavior have been found in the study of band-gap changes in SWNTs under uniaxial stress [5]. Using an analogy between the two situations, we can derive an expression for the pressure coefficient:

$$\frac{dE_g}{dP} = \frac{4|\gamma|a_{C-C}}{C_r d} - 3|\gamma|\left(\frac{1}{C_r} - \frac{1}{C_z}\right)(-1)^q \cos(3\theta), \quad (1)$$

where a_{C-C} is the C-C bond length, θ is the chiral angle and $\gamma = -2.89$ eV is the tight-binding hopping.

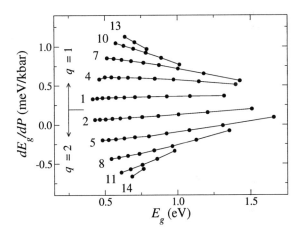

FIGURE 1. Band-gap pressure coefficient as a function of the gap for a large number of semiconducting SWNTs. Each dot corresponds to a particular nanotube. Tubes are grouped according to their $(n-m)$ family. The values of $(n-m)$ for each family are also shown in the figure.

Equation (1) has a very clear physical meaning. Hydrostatic pressure causes an overall shortening of C-C bonds, therefore increasing the magnitude of the hopping. In a graphene sheet, this would lead to an increase in the Fermi velocity. Because, in a simplified picture, energy gaps of semiconducting SWNTs are obtained by slicing the graphene bands, this would lead to an overall tendency of E_g to increase upon applying pressure, for all SWNTs. This is the meaning of the first term in Eq.(1), a chirality-independent positive constant. In addition to that, hydrostatic pressure breaks the triangular lattice symmetry of the parent graphene sheet due to the difference in radial and axial strains. This leads to a relative shift of the slicing planes with respect to the graphene Fermi point. This shift depends on chirality: For $q = 1$ (2), the planes move away from (closer to) the Fermi point, therefore increasing (decreasing) the gap [6]. That is the meaning of the second, chirality-dependent term in Eq. (1).

Experimentally, however, optical gaps of micelle-wrapped SWNTs in aqueous solution show an overall *decrease* with pressure, with magnitudes of dE_g/dP almost ten times higher than our calculated values for similar diameters [1]. Although excitonic effects (not included here) are crucial for a quantitative description of such optical experiments [7], trends are often well described by a single-particle picture. Therefore, this *qualitative* disagreement between theory and experiment is puzzling and should motivate further work.

We now focus on temperature effects on the band gap. The harmonic contribution to the gap shift is given by [8]:

$$\Delta E_g = \sum_j \frac{\partial E_g}{\partial n_j} \left(n_j + \frac{1}{2} \right), \quad (2)$$

where n_j is the Bose-Einstein occupation number of phonon mode j. We calculate the electron-phonon coupling coefficient $\partial E_g/\partial n_j$ using a "frozen-phonon" scheme in a supercell. A multiple-neighbor, non-orthogonal tight-binding method is used [9, 10].

Figure 2 shows $\Delta E_g(T) = E_g(T) - E_g(0)$ for the (10,0) and (11,0) nanotubes. The (10,0) SWNT shows the usual monotonic decrease of $E_g(T)$, non-linear at low T and linear at sufficiently high T. Interestingly, however, $\Delta E_g(T)$ for the (11,0) SWNT shows a non-monotonic behavior, being positive at low T. A more complete analysis [3] for a variety of tubes shows, at low temperatures, a tendency for negative or positive ΔE_g for $q = (n - m)$ mod 3 equals to 1 or 2, respectively, precisely the opposite trend as dE_g/dP.

The understanding of this result involves the following steps: (i) It can be shown, that the major contributions to $E_g(T)$ at low temperatures come from the lowest-energy optical modes, the so-called "shape-deformation modes" (SDMs); (ii) The contribution of these modes to the electron-phonon coupling can be described by a Debye-Waller theory, i.e., weakening of pseudopotential form factors due to phonon displacements. In the language of tight-binding theory, this can be translated into a reduction of the magnitude of hopping matrix elements along certain bonds; (iii) Since the SDMs involve only radial displacements, only bonds with circumferential components are affected; (iv) As far as the gap response is concerned, this is *formally equivalent to a radial expansion of the nanotube, i.e., negative pressure*; This is

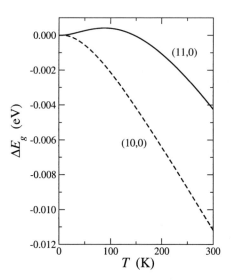

FIGURE 2. Temperature dependence of the band-gap in (10,0) and (11,0) nanotubes, prototype examples of $q = 1$ and $q = 2$ SWNTs.

the reason for the opposite family behavior in $E_g(P)$ and $E_g(T)$ of SWNTs at low pressures and temperatures.

In conclusion, semiconducting SWNTs show opposite family behavior in $E_g(P)$ and $E_g(T)$. This intriguing correlation is explained by a fascinating and unsuspected connection between dynamical and static deformations. A more detailed description of these results can be found elsewhere [3].

RBC acknowledges financial support from the John Simon Guggenheim Memorial Foundation and Brazilian funding agencies CNPq, FAPERJ, Instituto de Nanociências, FUJB-UFRJ and PRONEX-MCT. Work partially supported by NSF Grant No. DMR00-87088 and DOE Contract No. DE-AC03-76SF00098. Computer time was provided by the NSF at the National Center for Supercomputing Applications and by the DOE at the Lawrence Berkeley National Laboratory (LBNL)'s NERSC center.

REFERENCES

1. Wu, J. et al., Phys. Rev. Lett. **93**, 017404 (2004).
2. Lefebvre, J. et al., cond-mat/0403715 (2004).
3. Capaz, R. B., Spataru, C. D., Tangney, P., Cohen, M. L., Louie, S. G. to be published.
4. Reich, S. et al., Phys. Rev. B **65**, 153407 (2002).
5. Gartstein, Yu. N. et al, Phys. Rev. B **68**, 115415 (2003).
6. Yang, L. and Han, J., Phys. Rev. Lett. **85**, 154 (2000).
7. Spataru, C. D. et al, Phys. Rev. Lett. **92**, 077402 (2004).
8. Allen, P. B. and Cardona M., Phys. Rev. B **23**, 1495 (1981).
9. Hamada, N. et al., Phys. Rev. Lett. **68**, 1579 (1992).
10. Okada, S. and Saito, S., J. Phys. Soc. Japan **64**, 2100 (1995).

Ensemble Monte Carlo Transport Simulations for Semiconducting Carbon Nano-Tubes

A. Verma[*], M.Z. Kauser[✢], B.W. Lee[*], K.F. Brennan[*], and P.P. Ruden[✢]

[*] School of ECE, Georgia Institute of Technology, Atlanta, GA 30332, USA
[✢] Department of ECE, University of Minnesota, Minneapolis, MN 55455, USA

Abstract. In this study we present results of ensemble Monte Carlo simulations for carbon nano-tubes, focusing particularly on semiconducting, single wall, zigzag (n,0) structures of relatively small diameter. The basis for the Monte Carlo simulations is provided by electronic structure calculations in the framework of a tight binding model. The principal scattering mechanism considered is due to the electron-phonon interaction involving longitudinal acoustic and optical phonons. From the ensemble Monte Carlo simulation we determine the electron distribution function as a function of position and time. We explore steady state and transient characteristics. An intriguing result is the oscillatory behavior of the average drift velocity, which we attribute to the optical phonon scattering within the limited phase space of the dynamically one-dimensional band structure.

INTRODUCTION

Carbon nano-tubes (CNTs) have excellent potential as basic building blocks for nanometer scale electronic and optoelectronic devices. Recent experiments demonstrating light emission from relatively long semiconductor CNTs, subject to injection of electrons at one end and holes at the other, have stimulated interest in high field charge carrier transport and thermalization phenomena [1]. Ensemble Monte Carlo simulations constitute a flexible and accurate tool for the theoretical exploration of high field charge carrier transport in the steady state and transient regimes. The semiclassical transport model forms the basis of these simulations: charge carriers described by wavepackets composed of states belonging to a band are accelerated by an applied electric field and may be scattered through various mechanisms either in the same band or to other bands. Numerical bandstructures and scattering rates can be used in order to avoid additional approximations [2]. In the case of CNTs, we find that the very restricted (one-dimensional) phase space may lead to novel transient phenomena not encountered in conventional semiconductors. Here we focus on single wall zigzag CNTs of relatively small diameter.

MODEL DESCRIPTION

The electronic bandstructures for zigzag CNT are calculated based on the tight binding (TB) description of the π and π^* bands of grapheme. Cyclic boundary conditions around the periphery of the CNT are imposed, implying a folding of the two-dimensional graphene Brillouin zone (BZ) into a series of one-dimensional BZs [3]. The effect of the curvature is incorporated by allowing for two different effective bond lengths, distinguishing the bond parallel to the CNT axis from the zigzag bonds. Based on *ab initio* electronic structure calculations, an empirical relation for the bond lengths has been reported [4]. That relationship and the universal TB scaling scheme [5] are used here to relate the two different hopping matrix elements for the TB calculations. The remaining parameter is determined by comparing the resulting energy gap with experimental data [6].

Longitudinal acoustic and optical phonons are the dominant scattering mechanisms in simple, undoped, non-polar semiconductors. However, along a general direction in k-space, longitudinal and transverse modes may be mixed. For the simulations reported here, we simplify the phonons of the CNTs by fitting simple polynomials to the calculated dispersion curves of the longitudinal acoustic and optical modes of graphene (neglecting the coupling to transverse modes). We then apply the zone folding scheme to generate the corresponding CNT phonons.

Scattering of electrons by longitudinal phonons is treated in the framework of deformation potential coupling. For the acoustic deformation potential constant the same value is used as in previously reported Monte Carlo calculations for CNTs [7], while the optical deformation potential was chosen to agree with recent experimental data [8].

RESULTS AND DISCUSSIONS

The calculated conduction band structures for (10,0) and (14,0) CNTs are shown in Fig. 1. For the present work, only the lowest three conduction subbands are populated and need be considered. As expected, the curvature correction is small, but not negligible in the case of the (10,0) CNT.

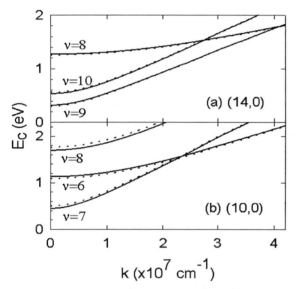

FIGURE 1. Lowest three conduction bands for (a) (14,0) and (b) (10,0) CNTs. Solid and dotted lines represent bands with and without considering the curvature effect, respectively.

Similarly to previous reported Monte Carlo results for CNTs [7], we find regions of negative differential mobility in the steady state velocity vs. field curves. We also examine the transient behavior and find that the mean free path is on the order of 10nm for field strengths in the range of kV/cm.

Perhaps most intriguing is the result shown in Fig. 2 depicting the average drift velocity as a function of time after injecting thermal electrons into a CNT. The velocity shows oscillations that we attribute to the strong scattering through optical phonon emission, which involves predominantly backscattering in the one-dimensional BZ.

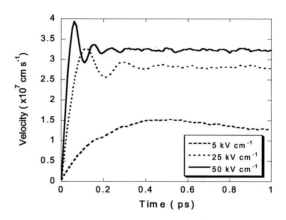

FIGURE 2. Average drift velocity transient for a (10,0) CNT subject to different electric fields.

ACKNOWLEDGEMENTS

We thank Dr. Phaedon Avouris for stimulating discussions and Dr. John Albrecht for his assistance. This work is partially supported by NSF-ECS.

REFERENCES

1. J. A. Misewich, R. Martel, Ph. Avouris, J. C. Tsang, S. Heinze, J. Tersoff, Science **300**, 783-786 (2003).

2. *'Monte Carlo Device Simulation: Full Band and beyond'*, K. Hess, ed., Kluwer Academic Publishers, Dordrecht, Netherlands (1991).

3. R. Saito, M. S. Dresselhause and G. Dresselhause, *Physical properties of carbon nanotubes*, Imperial College Press, London (1998).

4. K. Kanamitsu and S. Saito, *J. Phys. Soc. Jpn.* **71**, 483-486 (2002).

5. Walter A. Harrison, *Electronic structure and the properties of solids : the physics of the chemical bond*, Freeman, San Francisco (1980).

6. T. W. Odom, J. L. Huang, P. Kim and C. M. Lieber, *Nature* **391**, 62-64 (1998).

7. G. Pennington and N. Goldsman, Phys. Rev B **68**, 045426-1-11 (2003).

8. A. Javey, J. Guo, M. Paulsson, Q. Wang, D. Mann, M. Lundstrom, and H. Dai, Phys. Rev. Lett. **92**, 106804-1-4 (2004), and, M. Freitag, V. Perebeinos, J. Chen, A Stein, J.C. Tsang, J.A. Misewich, R. Martel, and Ph. Avouris, Nanoletters **4**, 1063-1066 (2004).

Growth and Characterization of SnO$_2$ micro- and nanotubes

D. Maestre, A. Cremades[1] and J. Piqueras

Departamento de Física de Materiales, Facultad de Ciencias Físicas, Universidad Complutense de Madrid, 28040 Madrid, SPAIN

Abstract. Micro- and nanotubes, and other elongated structures of SnO$_2$ as wires and rods, were grown after sintering in argon flow at temperatures ranging from 1350°C to 1500°C. The morphology and luminescence properties of these structures have been investigated by means of the secondary electron and the cathodoluminescence (CL) modes of the scanning electron microscope (SEM).

INTRODUCTION

Fabrication and characterization of semiconductor structures in form of nanowires, nanorods and nanotubes is a subject of increasing interest due to the potential application of the nanostructures in future nanoelectronic systems. In the case of semiconductors with gas sensing applications, as SnO$_2$, the high surface to volume ratio of the nanostructures provides a high sensitivity to the interaction with gases. SnO$_2$ is also a promising material for optoelectronic devices. Recently SnO$_2$ nanowires and belts, and discontinous tubular structures have been grown by thermal evaporation of powder on a substrate [1].

In this work the growth of micro and nanotubes and other elongated structures of SnO$_2$ during sintering treatments is investigated. The samples were characterized by means of the secondary electron and cathodoluminescence (CL) modes in the scanning electron microscope (SEM).

EXPERIMENTAL

The starting material used was commercial SnO$_2$ powder, formed by particles and aggregates of rounded particles with sizes of about 200 nm. Samples were prepared by compacting the powder under a compressive load of 2 tons to form disks. The powder was either untreated or mechanical ball milled, which leads to smaller and more homogeneous particle size. The samples were sintered in argon flow at temperatures from 1000 °C to 1500 °C. CL measurements were performed with a LEICA 440 SEM at a temperature of 80 K and accelerating voltages ranging from 12 KV to 18 KV. Spectra were recorded with a charge coupled device (CCD) camera with a built in spectrograph (Hamamatsu PMA-11).

RESULTS

Sintering under argon flow at temperatures between 1350 °C and 1500 °C leads to the formation of wires, rods and tubes with lateral dimensions ranging from tens or hundreds of nanometers to some micrometers, as shown in Fig. 1.

FIGURE 1. Micro and nanostructures grown after treatments in argon flow at different temperatures.

---

[1] cremades@fis.ucm.es

Sintering at 1500 °C causes the formation of a wire-like structure on the sample surface. Typical cross-sectional dimension of the wires is several microns and the length is of hundreds of microns or even reach the millimeter range. The wires show a higher luminescence than the sample background (Fig. 2). The CL spectra reveal the presence of a band at 1.94 eV in the wires, related to the oxygen vacancies [2].

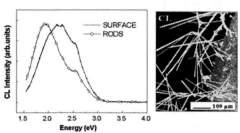

FIGURE 2. CL spectra and image of the surface of a sample sintered at 1500°C in Ar flow

In the samples prepared at temperatures between 1350 °C and 1450 °C micro and nanorods with rectangular cross-section are observed. Some of the rods have a tubular structure with a hollow and well defined lateral faces (Fig. 1a, 1c and Fig. 3).

In samples prepared from ball milled powder, a more regular distribution of the structures is obtained.

The external faces of the tubes are normally flat, while the interior can present different appearances (scaled as seen in Fig. 3a, or oriented layered as Fig. 3b shows), depending on the stage of the growth process.

FIGURE 3. SEM images of tubes showing inner scaled (a) and oriented layered nanostructure (b).

According to ref. [3], the most probable growth direction seems to be the [101]. Lateral surfaces of the rectangular tubes would be the (010) and the (10-1).

The presence of tubes with high dimensions enables to record spectra in different crystal faces as well as in the tube interior. The inner surfaces of the tubes show a more intense CL emission with a peak at 2.58 eV, while the spectra of the external surfaces reveal in addition the presence of the 1.94 eV band (Fig. 4).

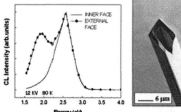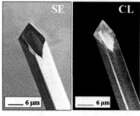

FIGURE 4. CL spectra recorded in different tube faces (internal and external) and SE and CL images.

Some samples sintered at 1350 °C show a surface structure (Fig. 5) which is thought to be the first steps of the growth of the elongated structures.

FIGURE 5. First steps of the growth process

CONCLUSIONS

Sintering SnO_2 compacted disks under argon flow in the temperature range from 1350°C to 1500°C leads to the formation of wires, rods and tubes on the sample surface. Since not a catalytic process or a foreign substrate is involved, it appears that the formation of the structures takes place by a vapor-solid process.

Enhanced luminescence emission arises from the internal region of the tubes, which is of potential application in high resolution displays.

ACKNOWLEDGMENTS

This work has been supported by MCYT (Project MAT 2003-00455).

REFERENCES

1. Pan,Z.W, Dai, Z.R Wang,Z.L, *Science* **291** 1947-1949 (2001)

2. Maestre, D., Cremades, A. and Piqueras, J. *J.Appl.Phys.* **95**, 3027-3030 (2004)

3. Beltrán, A. Andrés, J., Longo, E. and Leite, E.R. *Appl Phys. Lett.* **83**, 635-637 (2003)

Quantum interference in cross-linked carbon nanotube

Michio Kida, Sigeki Harada, Takahiko Mihara, Takahiro Morimoto, Takahiko Sasaki, Nobuyuki Aoki, Yuichi Ochiai

Department of Materials Technology, Chiba University, Japan
1-33 Yayoi-cho, Inage-ku, Chiba-shi, Chiba 263-8522, Japan

Abstract. We have fabricated a cross-linked structure of multi-walled carbon nanotube(MWNT) in order to observe Fano-resonance in quantum coherent nano-devices. We can observe a certain distorted line shape in the zero-bias anomaly of the I-V characteristics for the cross-linked MWNT in wide temperature range. Especially, at low temperatures near 4.2K, the resonance peak can be observed to be similar to Fano resonance line shape.

INTRODUCTION

In recently, there are many experimental studies related to quantum interference effects in small semiconductor devices. Also such coherent transport behavior have been expected even in novel nano-structure devices as well as carbon nanotubes(CNTs). Above coherent transport properties are strongly connected to quantum computing device application by means of nano-scale circuits using CNT [1]. Fano resonance arises from an interference between discrete state and continuum in the coherent transport system. This effect has been observed in a semiconductor AB ring constructed with a semiconductor quantum dot [2]. However, this effect can be observed only at very low temperatures(~mK) because of low charging energy and broad linewidth of the ring. In this work, we are trying to observe coherent interference, such as Fano-resonance in the multiwalled carbon nanotubes(MWNTs) in order to exploit novel nano-structure devices for application of quantum computing.

SAMPLE PREPARATION

The MWNT employed here is synthesized by arc discharge method and purified only by a centrifugal separator without thermal oxidation. It is dispersed in dichloro-ethane with ultrasonic and is coated on a SiO$_2$

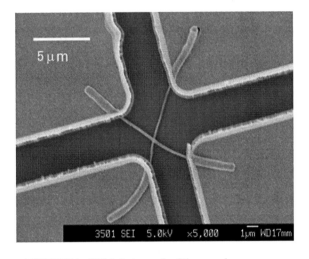

FIGURE 1. SEM photograph of the sample.

layer on a Si wafer. Under a high resolution CCD microscope, we selected suitable MWNTs and set them to form a crossing structure by manipulating a glass micro-capillary [3]. The electrical-lead contacts are set up by photo lithography and liftoff process with sequential deposition of Ti and Au. Moreover, in order to reduce the contact resistance between MWNT and Ti, the device sample has been rapidly heated in acceleration of alloying at the contact. Figure 1 shows an scanning electron microscope (SEM) image of the cross junction device.

RESULTS AND DISCUSSION

In the two terminal resistance measurement of a lope of MWNTs, the transport property behavior of such an individual CNT clearly shows a power law dependence ($G \propto T^\alpha$) with $\alpha \leq 1$.[4] This may be explained by a tentative transport based on Tomonaga-Luttinger Liquid (TLL) model [5]. However, if we measure the trans-conductance such as a cross over passing through between two lopes of MWNT, another transport behavior has been observed. For the four terminal(4t) resistance measurement of the trans-conductance in the cross junction, the temperature dependence shows metallic transport explained by weak localized correction of quantum interferences in the scattering of impurities or lattice defects in the wire. The difference between above two type transports has been discussed in our presentation of this conference with parallel.

On the other hand, we have observed a distorted peaks structure in the zero bias anomaly of the I-V characteristics of the 2t trans-conductance as shown in Fig.2. The shape of the distorted peak structure almost corresponds to the line shape of Fano resonance. It is noted that the shape of the peak can observe even at room temperature. Recently, Fano resonance peak in the zero bias anomaly has been also reported in the MWNT junction device as a similar configuration in our junction device [6,7]. If our results also come from Fano effect in the cross junction of the wire, the resonance peak would exist up to a higher temperature than that in the above reported. Here, there exists two important problems for our result. The first is whether this distorted line shape comes from Fano effect or not. Another possibility for this distortion can be considered to be due to an anomaly from Kondo effect in a single dot as well as widely discussed in semiconductor quantum dots. The second is an exact location where coherent interferences are prominent in electron waves of the MWNT. It should be a certain single dot structure constructed between two lopes of MWNT. However, in order to determine a possible mechanism of this distorted line shape in the I-V characteristic, we must make further experimental works on the low temperature transports of the cross-linked of MWNT.

ACKNOWLEDGEMENT

This work is supported in part by Grant-in-Aid for Scientific Research of Japan Society for the Promotion of Science (JSPS), No14750005 and 13450006.

REFERENCES

1. C. Dekker, *Phys Today* **52**, 22 (1999)

2. K. Kobayashi, H. Aikawa, S. Katsumoto, Y. Iye, *Phys. Rev. Lett.* **88,** 256806 (2002)

3. T. Mihara, K. Miyamoto, M. Kida, T. Sasaki, N. Aoki, Y. Ochiai, *Superlattices and Microstructures* Now in press

4. M. Kida, T. Mihara, K. Miyamoto, S. Harada, T. Sasaki, N. Aoki, Y. Ochiai, to be published in IPAP special issue.

5. M. Bockrath, D. H. Cobden, J. Lu, A. G. Rinzler, R. E. Smalley, L. Balents, P. L. McEuen, *Nature* **397**, 598 (1999)

6. J. Kim, K. Kang, J.-O Lee, K.-H. Yoo, J.-R. Kim, J. W. Park, H. M. So, J.-J. Kim, J. phys. Soc. Jpn. **70**, 1464 (2001)

7. J. Kim, J.-R. Kim, J.-O Lee, J. W. Park, H. M. So, N. Kim, K. Kang, K.-H. Yoo, J.-J. Kim, *Phys. Rev. Lett.* **90**, 166403 (2003)

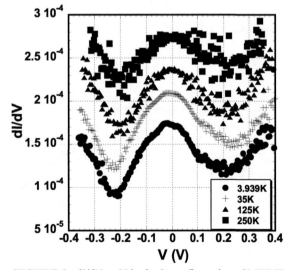

FIGURE 2. dI/dV vs V in the 2t configuration of MWNT.

Effect of Interwall Interaction on the Transport Properties of Mulit-wall Carbon Nanotubes

Takanori Matsumoto and Susumu Saito

*Department of Physics, Tokyo Institute of Technology,
2-12-1 Oh-okayama, Meguro-ku, Tokyo 152-8551, Japan*

Abstract. We study the electronic transport properties of the finite-length double-wall nanotubes, in which (10,10) tube encapsulates a finite-length (5,5) tube. Using the Landauer-Buttiker formula combined with the tight-binding model, we calculate the conductance $G(l,E)$ as a function of the length l of an inner tube as well as energy E. Our result shows the strong dependence on the inner-tube length due to the interwall interaction. We find two kinds of oscillations in the inner-tube length dependence. One is the rapid oscillation with the wavelength corresponding to the Fermi wavelength, and the other is the long-length oscillation which envelopes the maxima of the rapid oscillation. We also discuss the energy dependences of conductance. The conductance exhibits dips which are attributed to the quasi-bound states. In the whole conductance $G(l,E)$, we find that, for l=61.40A G takes the minimum value independent of E due to strong localization in the inner tube.

INTRODUCTION

Carbon nanotubes have one-dimensional tubule network of graphitic cylinders [1]. From band theory and experiments, it is known that the electronic properties of single-wall nanotubes (SWNTs) can be either conducting or insulating depending sensitively on their geometries, which are characterized by tube indices (n, m). In addition, the electronic transport in SWNTs is theoretically predicted to be ballistic, implying the absence of the back scattering. On the other hand, multi-walled carbon nanotubes (MWNTs) comprise several concentric SWNTs nesting one another, so that their structures are much more complicated than those of SWNTs. It is suggested that in such multi-wall systems, the effect of interwall interaction plays an important role in the electronic properties. For the electronic transport in MWNTs, the quantized conductance has been reported in the recent experiment [2]. However, the observed values turned out to be half of the values given by theoretical prediction. The difference in the conductance has triggered the interest in the transport properties of multi-wall systems with the interwall interaction.

In this work, we investigate the conductance of double-wall structures, in which the inner tube has various finite lengths. Such structures have been observed in the experiments on the nanotubes encapsulating fullerene linear chain inside. These double-wall systems can be synthesized by fusion of inside fullerenes. It is likely to occur that the synthesized inner tubes have various lengths. The calculated conductance as a function of inner-tube length (l) as well as energy (E), $G(l, E)$, indicates that the inner-tube length should be an important parameter of MWNT systems.

METHODS AND MODELS

We use the tight-binding (TB) model to calculate the electronic properties for finite-length double-wall nanotubes. In this TB model proposed by Hamada and Sawada, the basis set consists of the 2s and 2p orbitals of an isolated carbon atom. Not only the transfer integrals but also overlap integrals are introduced. The parameters have been chosen so that this model can reproduce well the electronic structures of graphite and fullerene C_{60}. By calculating the Green function from this TB Hamiltonian, we obtain the transmission functions. We next use the Landauer-Buttiker formula to calculate the conductance.

To consider the systems similar to the structures observed in the experiments, we focus (cap+$(5,5)_m$)@(10,10) finite double-wall structures shown in Fig. 1. This system consists of the finite (5,5) tube with length $l = ma$ (m is the number of unit cells and a is lattice constant, 2.456 A) inside the infinite (10, 10) tube. This inner tube has end-geometry corresponding to the half of fullerene C_{60} whereas the outer tube is in contact with electrodes.

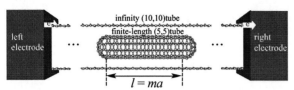

FIGURE 1. Schematic view of $(cap+(5,5)_m)@(10,10)$ finite-length double-wall system. l is the length of inner (5,5) tube part, i.e., m times the lattice constant a (=2.456 A).

RESULTS AND DISCUSSIONS

Figure 2 shows the conductance G as a function of both inner-tube length l (=ma) and energy E. In this system, the conductance depends strongly on the inner tube length l although the inner tube part is not "chemically" connected to the other outer tube. The l dependence of conductance is shown in Fig. 3(a). The conductance values are calculated at the Fermi energy of an isolated (10,10) tube. In this figure, we find two kinds of conductance oscillations. One is the rapid oscillation with the wavelength of $3a$. The other is a slower oscillation shown by a dashed line, which envelops the maximum values in the rapid oscillations. The wavelength of this envelope oscillation is near $50a$. The minimum G values are about a half of conductance of an isolated (10,10) tube, i.e., l=0. These dependences are associated with the interwall mixing in the double-wall region. Near the Fermi level, there are two conducting channels are the outer tube. By calculating each transmission rate for two channels, we find that one of two channels is reflected due to the interwall mixing whereas the other can transmit without scattering. Therefore, the conductance corresponding one channel is reduced. In Fig. 3(b), we also show the conductance curves as a function of energy for three structures with different inner-tube lengths. The filled circles, crosses, and open circles denote the case with m=10, 25, and 40, respectively. In this energy region near the Fermi level of isolated (10,10) tube, the periodic conductance dips exist at m=10 and 40. The number of conductance dips increases with the inner-tube length. The behavior of the conductance dips is connected with the formation of the quasi-bound states in double-wall region. In the particular case of m=25 where the conductance takes the minimum value, E dependence is found to be depressed. The electron wave is strongly localized in the inner tube due to the large interference between inner and outer tubes [3].

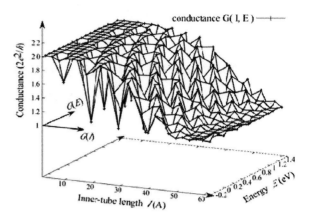

FIGURE 2. Conductance $G(l,E)$ of $(cap+(5,5)_m)@(10,10)$ as a function of both inner-tube.

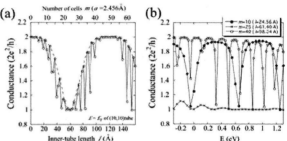

FIGURE 3. Conductance dependence on (a) inner-tube length l at the Fermi energy and (b) Energy for three structures with m=10 (l=24.56A), 25 (61.40A), 40 (98.24A).

ACKNOWLEDGMENTS

We would like to thank Dr. T. Miyake, Dr. S. Okada, and Mr. Y. Akai for useful discussion. Numerical calculations were performed on Fujitsu VPP and computers at the Research Center for Computational Science, Okazaki Research Facilities, National Institute of Natural Sciences. This work was supported by Grant-in-Aid from the Ministry of Education, Science and Culture of Japan, and the 21st Century Center-of-Excellence Program at Tokyo Institute of Technology 'Nanometer-Scale Quantum Physics' from the Ministry of Education, Culture, Sports, Science, and Technology of Japan.

REFERENCES

1. S. Iijima, Nature **354,** 56 (1991)
2. S. Frank, P. Poncharal, Z. L. Wang, and W. A. de Heer: Science **280**, 1744 (1998)
3. T. Matsumoto and S. Saito, to be published.

Systematic First-Principles Electronic-Structure Study of Carbon Nanotubes

Yoshio Akai and Susumu Saito

Department of Physics, Tokyo Institute of Technology,
2-12-1, Oh-okayama, Meguro-ku, Tokyo, 152-8551, Japan

Abstract. By using the local density approximation (LDA) in the framework of the density-functional theory, we study the electronic structure of all the chiral carbon nanotubes with the diameters between 0.8 nm and 2.0 nm. We find that the LDA result gives the reasonable energy gap and the density of states (DOS) by comparing with the STM/STS experimental result. We plot the peak-to-peak energy separations of the DOS as a function of the nanotube diameter. For the semiconducting nanotubes, we find the peak-to-peak separations can be classified into two types according to the chirality. This chirality dependence of the LDA result is opposite to that of the simple θ tight-binding result.

INTRODUCTION

It is well known that carbon nanotubes can be metallic or semiconducting depending on their chirality [1-3]. Moreover, semiconducting nanotubes even with similar diameters show different electronic properties if their chiral angles are different. As recent progress of the experiment, such as STM/STS [4] or spectrofluorimetric measurement [5], makes it possible to investigate the individual single-walled carbon nanotube, a detailed systematic theoretical prediction of their electronic structure from first principles has been awaited to understand these experimental results. So far, systematic study of their electronic properties has been performed by using the simple θ tight-binding (θ TB) method because one can easily obtain the electronic structure for any kind of carbon nanotube. However, this result is obtained by a zone folding of the graphite-sheet θ and θ^* bands and does not take into account the effect of the hybridization of θ and θ states induced by the curvature of the nanotube sp^2 surface.

Here, we report the detailed electronic structure of chiral carbon nanotubes by using the local density approximation (LDA) in the framework of the density-functional theory.

METHOD

We adopt the real space method, i.e. higher-order finite-difference pseudopotential method. In general, first principles calculations of carbon nanotubes has been limited to so-called armchair and zigzag nanotubes, because one has to consider a large number of atoms in a translational unit cell for chiral nanotubes. In this study, we adopt the helical and rotational symmetries method [6], by which the number of atoms in a unit cell becomes two for any type of nanotube.

RESULTS AND DISCUSSIONS

In Fig. 1, LDA DOS for two semiconducting chiral nanotubes with slightly different (n,m) indices, (a) (16,2) tube (Dt=1.34 nm, θ=5.83 deg) and (b) (18,2) tube (Dt=1.49 nm, θ=5.21 deg), are shown, where Dt is nanotube diameter and θ is chiral angle. STS experimental DOS of a nanotube with Dt=1.4 nm, θ=5 deg [4] is also shown in Fig. 1. The shapes of LDA DOS are similar to each other near the Fermi level. On the other hand, third valence and conduction peak positions of these two nanotubes are quite different. This difference of DOS is the key to assign the tube index (n,m). Hence, the present systematic

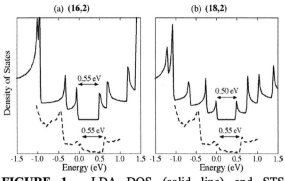

FIGURE 1. LDA DOS (solid line) and STS experimental DOS (broken line).

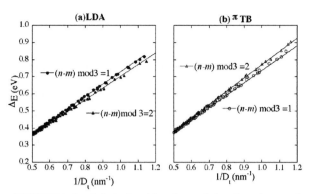

FIGURE 3. Mod classification of Semiconductor 1 for (a) LDA and (b) θ TB.

study should be very important as a reference for the experimental electronic and geometric structure study.

We plot energy separations of DOS for all chiral carbon nanotubes with the diameters between 0.8 nm and 2.0 nm as a function of D_t in Fig. 2. For the semiconducting nanotubes, we find that both Semiconductor 1 and 2 can be classified into two types according to $(n-m)$ mod 3 value, 1 or 2. In Fig. 3, this classification for Semiconductor 1 is shown together with the θ TB result for comparison. The "mod=1" group has wider separation than the "mod=2" group in the LDA whereas the trend is opposite in the θ TB. The LDA result obtained by the zone folding method for the graphite sheet also gives the same result with θ TB. Hence, our result shows the importance of the nanotube curvature. We performed the realistic θ and θ TB study, and it also gives the same trend with the LDA result. On the other hand, in the case of Semiconductor 2, the "mod=1" group has wider separation than the "mod=2" group for both LDA and θ TB. In addition to the energy separation vs. D_t plot, Semiconductor 1 vs. 2 plot is required to analyze the spectrofluorimetric measurement result. This plot for the present LDA result is found to form two parallel lines corresponding to the above mod classification [7], while the realistic θ and θ TB result fails to give mod classification in this plot.

ACKNOWLEDGMENTS

We thank Dr. J.-I. Iwata and Prof. K. Yabana for the LDA code. Numerical calculations performed partly on NEC SX-7 at Research Center for Computational Science, Okazaki National Facilities, National Institutes of Natural Science and on NEC SX-5 at Global Scientific Information Center, Tokyo Institute of Technology. This work was supported by a 21st Century COE Program at Tokyo Institute of Technology "Nanometer-Scale Quantum Physics" and Grant-in-Aid by the Ministry of Education, Culture, Sports, Science and Technology of Japan.

REFERENCES

1. N. Hamada, S. Sawada and A. Oshiyama: Phys. Rev. Lett. **68**, 1579 (1992).
2. K. Tanaka, K. Okahara, M. Okada and T. Yamabe: Chem. Phys. Lett. **191**, 469 (1992).
3. R. Saito, M. Fujita, G. Dresselhaus and M. S. Dresselhaus: Appl. Phys. Lett. **60**, 2205 (1992).
4. J. W. G. Wildöer, L. C. Venema, A. G. Rinzler, R. E. Smalley, C. Dekker: Nature **391**, 59 (1998).
5. S. M. Bachilo, M. S. Strano, C. Kittrell, R. H. Hauge, R. E. Smalley, R. B. Weisman: Science **298**, 2361 (2002).
6. C. T. White, D. H. Robertson and J. W. Mintmire: Phys. Rev. B **47**, 5485 (1993).
7. Y. Akai, S. Saito, J.-I. Iwata and K. Yabana: *to be published*.

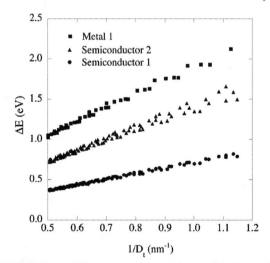

FIGURE 2. LDA energy separations of DOS peaks for the first valence-band peak to the first conduction-band peak (Semiconductor 1) and the second peak-to-peak (Semiconductor 2) of semiconducting tubes and the first peak-to-peak of metallic tubes (Metal 1).

Effects of Symmetry Breaking on Perfect Channel in Metallic Carbon Nanotubes

Tsuneya Ando and Koichi Akimoto

Department of Physics, Tokyo Institute of Technology, 2–12–1 Ookayama, Meguro-ku, Tokyo 152-8551, Japan

Abstract. Numerical calculations of the conductance are performed to study symmetry breaking effects on a perfectly conducting channel in metallic carbon nanotubes existing unless the potential range of scatterers is smaller than the lattice constant. The perfect channel is fragile against perturbations when there are several bands at the energy, while it is robust in the energy range of metallic linear bands.

Transport properties of carbon nanotubes are extremely interesting. In metallic nanotubes there is no backscattering even in the presence of scatterers unless their potential range is smaller than the lattice constant of a two-dimensional graphite [1]. When there are several bands at the Fermi level, interband scattering appears, but a perfectly conducting channel transmitting through the system without being scattered back exists and the conductance is always larger than $2e^2/\pi\hbar$ [2].

In this paper we consider effects of various kinds of perturbations breaking the symmetry leading to such unique transport properties and show that the perfect channel is very fragile against such perturbations when there are several bands at the energy, while it is much more robust in the energy range of metallic linear bands.

In a graphite sheet the conduction and valence bands consisting of π states cross at K and K' points of the Brillouin zone. Electronic states near a K point are described by the $\mathbf{k}\cdot\mathbf{p}$ equation $\gamma(\vec{\sigma}\cdot\hat{\mathbf{k}})\mathbf{F}(\mathbf{r}) = \varepsilon\mathbf{F}(\mathbf{r})$, where γ is the band parameter, $\vec{\sigma} = (\sigma_x, \sigma_y)$ is the Pauli spin matrix, $\hat{\mathbf{k}} = (\partial/i\partial x, \partial/i\partial y)$, and $\mathbf{F}(\mathbf{r})$ is the envelope function with two components representing the amplitude at two carbon atoms in a unit cell. In a metallic nanotube, states can be obtained by imposing periodic boundary conditions in the circumference direction $\mathbf{F}(\mathbf{r}+\mathbf{L}) = \mathbf{F}(\mathbf{r})$. Then, the energy bands are given by $\varepsilon_n(k) = \pm\gamma\sqrt{\kappa(n)^2 + k^2}$, where $n = 0, \pm 1, \cdots$, $\kappa(n) = 2\pi n/L$, k is the wave vector along the axis, and $+$ and $-$ for the conduction and valence bands, respectively. They are shown in Fig. 1 and independent of the structure such as armchair or zigzag as long as the nanotube is metallic.

This equation is the same as that of a massless neutrino and possesses a topological singularity at $\mathbf{k} = 0$ giving rise to Berry's phase under a rotation around the origin in the wave-vector space [3]. It has special time reversal

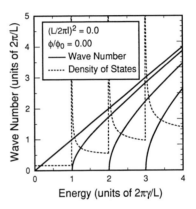

FIGURE 1. The energy bands and density of states of a metallic nanotube obtained in the $\mathbf{k}\cdot\mathbf{p}$ scheme. They are doubly degenerate except those with a linear dispersion.

symmetry which leads to various exotic transport properties mentioned above if being combined with Berry's phase. This symmetry is destroyed by a magnetic field H perpendicular to the axis or a flux ϕ due to a field parallel to the axis. The dimensionless quantity characterizing the strength is $(L/2\pi l)^2$ for the magnetic field and ϕ/ϕ_0 for the flux, where $L = |\mathbf{L}|$, $l = \sqrt{c\hbar/eH}$, and $\phi_0 = ch/e$.

The symmetry is affected also by scatterers with range smaller than the lattice constant a. Its strength is characterized by δ which denotes the amount of short-range scatterers relative to the total amount of scatterers ($0 \leq \delta \leq 1$). It is destroyed also by the presence of trigonal warping of the bands around the K and K' points. The effect of warping can be described by higher order $\mathbf{k}\cdot\mathbf{p}$ terms [4] and is characterized by $\beta a/L$ where β is a dimensionless quantity of the order of unity.

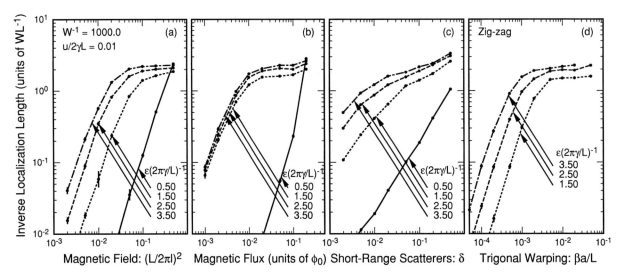

FIGURE 2. The inverse localization length as a function of (a) the magnetic field, (b) the magnetic flux, (c) the amount of short-range scatterers, and (d) the trigonal warping of the band in a logarithmic scale for a zigzag nanotube. The results are independent of the nanotube structure in (a)–(c) and weakly dependent in (d).

For actual calculations, we consider a system with randomly distributed scatterers with strength $\pm u$ and concentration n_i. The scattering strength is characterized by the dimensionless quantity $W = n_i u^2 / 4\pi\gamma^2$. Using numerical results of the geometric average of the conductance, i.e., $\bar{G} \propto \exp(-2\alpha A)$, we can determine the inverse localization length α, where A is the length. This inverse localization length is a good measure describing the magnitude of effective backward scattering.

Resulting α's at different values of energy ε are shown in Fig. 2. The number of the bands are 1, 3, 5, and 7 for $\varepsilon(2\pi\gamma/L)^{-1} = 0.5, 1.5, 2.5$, and 3.5, respectively, as can be seen in Fig. 1. When there are several bands at the energy, the inverse localization length exhibits a near-plateau behavior above a small critical value. The plateau values are nearly same in all the cases and given approximately by the inverse of the sum of the mean free path of the bands [2] at the energy. When the energy is in the range of linear bands, the inverse localization length is much smaller unless the symmetry breaking parameter becomes appreciable. Therefore, the perfect channel can easily be destroyed by a very small perturbation when there are several bands at the energy, while it is very robust in the energy range of linear bands.

It should be noted that various effects can be incorporated as an effective magnetic flux such as the nonzero curvature present in actual nanotubes [5, 6] and the lattice distortion [6, 7, 8]. It is highly likely, therefore, that an effective flux of the order of $\phi/\phi_0 \sim 10^{-2}$ is always present in actual nanotubes. It turns out further that effects of the trigonal warping are surprisingly strong among those symmetry breaking effects. Even considering the ambiguity in the exact value of β, we can expect that its effective strength is well in the near-plateau region for typical tubes. In actual nanotubes, therefore, the perfectly conducting channel is likely to be destroyed almost completely when there are several bands at the energy. The absence of backward scattering when the Fermi level lies in the region of metallic linear bands remain quite robust, however.

This work was supported in part by a 21st Century COE Program at Tokyo Tech "Nanometer-Scale Quantum Physics" and by Grants-in-Aid for Scientific Research and for COE (12CE2004 "Control of Electrons by Quantum Dot Structures and Its Application to Advanced Electronics") from Ministry of Education, Culture, Sports, Science and Technology Japan. Numerical calculations were performed in part using the facilities of the Supercomputer Center, Institute for Solid State Physics, University of Tokyo.

REFERENCES

1. T. Ando and T. Nakanishi, J. Phys. Soc. Jpn. **67**, 1704 (1998).
2. T. Ando and H. Suzuura, J. Phys. Soc. Jpn. **71**, 2753 (2002).
3. T. Ando, T. Nakanishi, and R. Saito, J. Phys. Soc. Jpn. **67**, 2857 (1998).
4. H. Ajiki and T. Ando, J. Phys. Soc. Jpn. **65**, 505 (1996).
5. T. Ando, J. Phys. Soc. Jpn. **69**, 1757 (2000).
6. C. L. Kane and E. J. Mele, Phys. Rev. Lett. **78**, 1932 (1997).
7. C. L. Kane, E. J. Mele, R. S. Lee, J. E. Fischer, P. Petit, H. Dai, A. Thess, R. E. Smalley, A. R. M. Verschueren, S. J. Tans, and C. Dekker, Europhys. Lett. **41**, 683 (1998).
8. H. Suzuura and T. Ando, Phys. Rev. B **65**, 235412 (2002).

Excitonic Effects and Optical Spectra of Single-Walled Carbon Nanotubes

Catalin D. Spataru[*][†], Sohrab Ismail-Beigi[¶], Lorin X. Benedict[‡], and Steven G. Louie[*][†]

[*]*Department of Physics, University of California at Berkeley, Berkeley, California 94720, USA*
[†]*Materials Sciences Division, Lawrence Berkeley National Laboratory, Berkeley, California 94720, USA*
[¶]*Department of Applied Physics, Yale University, New Haven, Connecticut 06520, USA*
[‡]*H Division, Physics and Advanced Technologies Directorate, Lawrence Livermore National Laboratory, University of California, Livermore, California 94550, USA*

Abstract. Recent optical measurements on single-walled carbon nanotubes (SWCNT) showed highly anomalous behaviors indicating strong many-electron effects. To understand these data, we performed *ab initio* calculation of electron-hole interaction (excitonic) effects on the optical spectra of several SWCNTs. We employed a many-electron Green's function approach that determines both the quasiparticle and optical excitations from first principles. For both semiconducting and metallic SWCNTs, we found that their optical spectra are strongly affected by excitonic effects. These large many-electron effects explain the unusual optical response of these quasi-one-dimensional systems.

Recent experimental advances allowed the measurement of optical response of well-characterized, individual SWCNTs [1,2]. However, the response is quite unusual and cannot be explained by conventional, one-particle interband theories. Our *ab initio* results show that many-electron interaction (excitonic) effects are very important and can change qualitatively the optical spectra of both metallic and semiconducting SWCNTs [3,4].

We have computed the optical absorption spectra of the four SWCNTs: (3,3), (5,0), (8,0) and (11,0). Our first-principles calculations are done in three stages [5]: (i) we treat the electronic ground-state with *ab initio* pseudopotential density-functional theory [6], (ii) we obtain the quasiparticle energies within the *GW* approximation for the electron self-energy [7], and (iii) we calculate the coupled electron-hole excitation energies and spectrum by solving the Bethe-Salpeter (BS) equation of the two-particle Green's function [5].

Our results for the metallic SWCNTs (3,3) and (5,0) can be compared with those from the experiments by Li, et al. [1]: by growing 4 Å SWCNTs inside zeolite channels, three main peaks were found in the measured absorption spectra (see Table 1). While 4 Å diameter SWCNTs come in only three chiralities, (3,3), (5,0) and (4,2), it was not possible to assign specific peaks to specific tubes experimentally. As shown in Table 1, our results for the (3,3) and (5,0) tubes are in excellent quantitative agreement with experiment and provide a concrete identification for two of the observed peaks. In particular the 3.1 eV is shown to be due to a bound excitonic state of the metallic (3,3) tube. We conclude that the remaining peak at 2.1 eV is due to the (4,2) tube. This assignment is similar to that deduced from DFT calculations neglecting excitonic effects [1].

TABLE 1. Peak positions and optical transitions in 4 Å SWCNTs.

Nanotube	Observed experimental peaks[a]	Theoretical prominent excitation energy
(5,0)	1.37 eV	1.33 eV
(3,3)	3.1 eV	3.17 eV
(4,2)	2.1 eV	-

[a]See reference [1].

TABLE 2. Lowest two optical transition energies of the (8,0) and (11,0) semiconducting SWCNTs.

	(8,0)		(11,0)	
	Expt.[a]	Theory	Expt.[a]	Theory
E_{11}	1.6 eV	1.6 eV	1.2 eV	1.2 eV
E_{22}	1.9 eV	1.8 eV	1.7 eV	1.7 eV
E_{22}/E_{11}	1.19	1.13	1.42	1.42

[a]See references [2,8].

Our results for the semiconducting SWCNTs (8,0) and (11,0) can be compared with those from the experiments by Bachilo et al. [2,8]: by performing spectrofluorimetric measurements on various semiconducting SWCNTs, they provided results for the first and second optical transition energies (E_{11} and E_{22}) of the (8,0) and (11,0) SWCNTs. These are shown to be strong excitonic transitions in the theory.

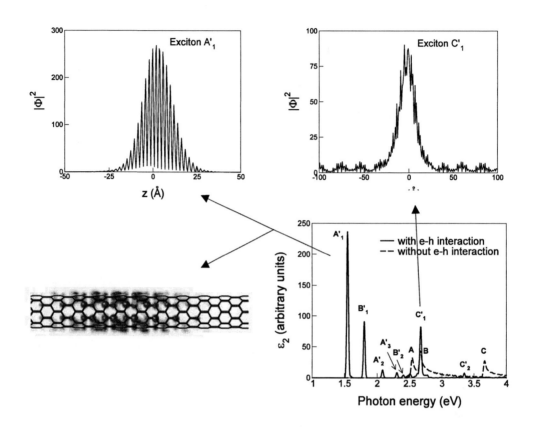

FIGURE 1. Electron-hole amplitude (or exciton wavefunction) and absorption spectra in (8,0) SWCNTs.

Theory predicts two varieties of excitons in semiconducting tubes: bound and resonant. Figure 1 shows the electron-hole pair probability distribution as a function of the electron position (z is the coordinate along the tube axis), for a bound (upper left panel) and resonant (upper right panel) photoexcited excitons in the (8,0) semiconducting SWCNTs; the position of the hole is fixed on a carbon π orbital, as indicated by the star in the bottom left panel. The bottom right panel shows the optical absorption spectra of the (8,0) SWCNT, calculated with and without the electron-hole interaction when solving the BS equation.

In conclusion, we have studied the optical absorption spectra of metallic and semiconducting small-diameter SWCNTs and shown that there are strong excitonic effects in both types of tubes.

Work supported by NSF Grant No. DMR00-87088 and DOE Contract No. DE-AC03-76SF00098. Computer time was provided by the NSF at the NCSA and by DOE at LBNL's NERSC center.

REFERENCES

1. Li Z.M. et al., *Phys. Rev. Lett.* **87**,127401 (2001).
2. Bachilo S.M. et al., *Science* **298**, 2361 (2002).
3. Spataru C.D., Ismail-Beigi S., Benedict L.X., Louie S.G., *Phys. Rev. Lett.* **92**, 077402 (2004); *Appl. Phys. A* **78**, 1129 (2004).
4. Chang E. et al., *Phys. Rev. Lett.* **92**, 196401 (2004).
5. Rohlfing M. and Louie S.G., *Phys. Rev. Lett.* **81**, 2312 (1998); *Phys. Rev. B* **62**, 4927 (2000).
6. Kohn W. and Sham L.J., *Phys. Rev. B* **140**, A1133 (1965).
7. Hybertsen M.S. and Louie S.G., *Phys. Rev. B* **34**, 5390 (1986).
8. Weisman R.B. and Bachilo S.M., *Nano Lett.* **3**, 1235 (2003).

Magneto-Transport Studies Of Fe/Alq$_3$/Co Organic Spin-Valves

F. J. Wang, Z. H. Xiong, D. Wu, J. Shi and Z. V. Vardeny

Department of Physics, University of Utah, Salt Lake City, Utah 84112, USA

Abstract. We have studied vertical organic spin-valves made of evaporated π-conjugated molecule, namely 8-hydroxy-quinoline aluminum (Alq$_3$) semiconductor spacer sandwiched between two ferromagnetic (FM) electrodes with spin injecting capability. Recently we have communicated on our work with organic spin-valve devices using half-metallic manganites as one of the spin injecting FM electrode that showed giant magnetoresistance (GMR) of about 40% at 11 K. However, we found that the GMR decreases at high temperatures and actually disappears at 300 K, partially because the FM manganite loses its magnetic properties at ambient temperature. To realize room temperature organic spin-valve devices, in the present work we report on our investigations of Fe/Alq$_3$/Co spin-valve devices, where both spin-injecting FM electrodes (Fe and Co) have high Curie temperatures and thus maintain their magnetic properties at ambient temperature. We found that these devices show GMR of about 5% at 11K. However at elevated temperatures the GMR value steeply decreases, and in fact vanishes at 90K. We attribute this decrease to the increase of the spin relaxation rate of the injected carriers in the Alq$_3$ organic spacer at elevated temperatures, and the carrier injection across the Fe/Alq$_3$ interface that is dominated by thermo-ionic emission at high temperatures.

INTRODUCTION

A spin-valve device is a layered structure of magnetic and non-magnetic (spacer) materials of which electrical resistance depends on the spin state of electrons passing through the device. The discoveries of giant magnetoresistance (GMR) [1] in metallic spin-valves have promoted the new field of Spintronics [2], or spin-electronics [3], which has revolutionized applications such as magnetic recording and magnetic memory [4, 5]. Intense research efforts have been devoted recently to extending these spin-dependent effects to semiconductors that are sandwiched between the two spin-injecting ferromagnet (FM) electrodes [6-8]; however spin-valves with inorganic semiconductors have not as yet been achieved. Due to their inherent lattice flexibility and long spin coherence [9], π-conjugated organic semiconductors (OSEC) may offer a promising alternative to incorporate semiconductor materials in spintronics devices, having novel functionalities such as light emitting capabilities.

Recently there has been substantial research progress in the field of organic spintronics, which include horizontal and vertical spin-valves [10, 11], spin-organic light emitting diodes (OLED) [12], and spin diodes. Recently we reported the fabrication and magneto-transport studies of the π-conjugated molecule 8-hydroxy-quinoline aluminum (Alq$_3$)-based spin valves, where one of the FM electrodes was a half-metallic manganite, for which a GMR value of 40% was measured at 11 K [11]. However due to weakening of the manganite ferromagnetic properties at ambient temperature, it is desirable to fabricate spin-valve devices where more regular metallic FM's, such as d-band metals (Fe, Ni, and Co) with high Curie temperatures are used as the spin-injecting electrodes.

Here we report the fabrication and magneto-transport studies of Fe/Alq$_3$/Co organic spin-valves. We obtained GMR of about 5% at 11K; however the GMR decreases at elevated temperatures and actually vanishes at 90K. Two effects contribute to the GMR steep decrease with the temperature: (i) the increase of the spin relaxation rate of the injected carriers in the Alq$_3$ organic spacer, and (ii) the change in the carrier injection process across the Fe/Alq₃ interface with temperature, which is dominated by thermo-ionic emission at elevated temperatures.

EXPERIMENTAL

The spin-valve devices that we have fabricated are in the form of a vertical sandwiched structure, which consists of two FM electrodes and an evaporated layer of Alq₃ OSEC. The electrical current through our simple two-terminal devices was perpendicular to the deposited films and the direction of the externally applied magnetic field, H was parallel to the films. For the spin-valve devices described here we have chosen iron (Fe) as the bottom FM electrode and cobalt (Co) as the top FM electrode. Iron and cobalt are metallic FM having spin-polarization injection capability, P of about 40% and -34%, respectively [13, 14].

The spin-valve device studied here was fabricated on a silicon substrate. Following a cleaning procedure using acetone, the silicon substrate was introduced into the evaporation chamber with a base pressure of 5×10^{-7} torr. Firstly the bottom iron electrode was thermally evaporated with a film thickness of about 25 nm. Without breaking the vacuum, we then deposited the OSEC layer (Alq₃) with thickness, d of about d = 140 nm; the OSEC layer completely covered the iron electrode to protect it from oxidation. Then we opened the vacuum chamber and changed the shadow mask for the deposition of the top Co electrode. For protecting the deposited Co layer (3.5 nm thick) from oxidation and water contamination we covered it with an aluminum (Al) contacting film of about 100 nm thick. The obtained active device area was about 2×3 mm². We have used a thickness monitor to measure the film thickness of the electrodes and OSE spacer.

Figure 1 shows the SEM photograph of our spin-valve device. From the photograph the sandwiched structure of the device can be clearly seen. It is composed of four layers: the silicon substrate, Fe bottom layer, OSEC spacer, and the capped layer that is composed of a thin Co film and a thicker Al film.

The magneto-resistance (MR) properties of the fabricated device were measured in a closed-cycle refrigerator where the temperature, T varied from 11 to 300 K. The MR measurements were done by applying a constant voltage between the electrodes while measuring the current, I as a function of the in-plane external magnetic field, H. The magnetization properties of the FM electrodes versus H were measured as a function of temperature using the magneto-optic Kerr effect (MOKE). We found that the coercive fields, H_c of the two FM electrodes are different from each other, and this allows us to change their relative magnetization directions to be either parallel or anti-parallel to each other, when H was changed.

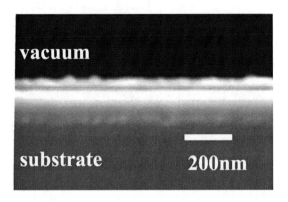

FIGURE 1. Scanning electron micrograph of a functional organic s01pin-valve consisting of a 25-nm-Fe film, 140-nm-thick Alq₃ spacer, a 3.5-nm thick Co electrode and Al covering layer; the device is grown on a silicon substrate.

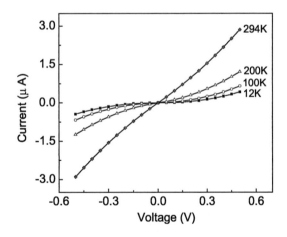

FIGURE 2. The I-V characteristics of the Fe/Alq₃/Co spin valve device at various temperatures.

RESULTS AND DISCUSSION

At low applied bias voltages V, we found that the device I-V characteristic was nonlinear, where the electrical resistance, R increases as the temperature decreases (Fig.2). The rather steep temperature dependence and the nonlinear I-V behavior indicate that carrier injection from the FM electrodes to the OSEC layer is composed from a combination of both tunneling and thermo-ionic emission, of which contribution depends on T. At 12K (the lowest T measured) we expect that injection by tunneling is dominant. However carrier injection becomes more dominant by thermo-ionic emission at elevated T.

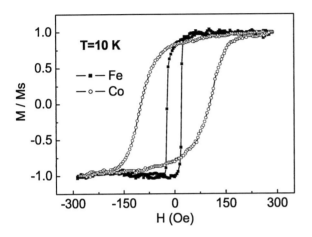

FIGURE 3. Magnetic hysteresis loops of the iron bottom electrode and the cobalt top electrode that form the organic Fe/Alq$_3$/Co spin-valve device, measured using the MOKE technique.

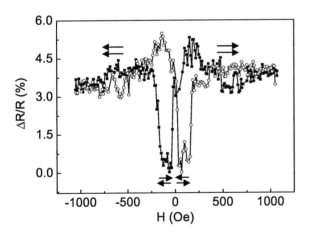

FIGURE 4. Giant magnetoresistance loop of the spin-valve device Fe(25 nm)/Alq$_3$(140 nm)/Co(3.5 nm) measured at 11 K and 50 mV. The arrows represent the mutual magnetization directions of the two FM electrodes.

Moreover, we found that devices with d <100 nm showed a linear I-V behavior and lack of electroluminescence at high current. This has led us to believe that due to the softness of the Alq$_3$ layer, the deposited cobalt layer may penetrate into the OSEC spacer in the form of inclusions. We have therefore concluded that such devices have an "ill-defined" layer with thickness $d_0 = 100$ nm that may contain a layer of the OSEC with pinholes and cobalt inclusions [11].

MOKE measurements performed separately on the bottom Fe electrode and the top Co electrode are shown in Fig. 3. These measurements indicate that at 11 K H_C of the Fe electrode is about 25 Oe, and H_C of the Co electrode is about 150 Oe. Therefore when the external magnetic field H is between 25 and 150 Oe, the magnetization directions of the two FM electrodes in the device are anti-parallel to each other; whereas when $H > 150$ Oe the magnetization directions of the electrodes are parallel to each other, as schematically represented by the arrows in Fig.4.

Figure 4 shows a typical MR hysteresis loop of the Fe/Alq$_3$/Co spin-valve device with d = 140 nm. The electrical resistance, R in the anti-parallel magnetizations alignment is smaller than that in the parallel alignment. This is opposite to many other metallic spin-valves, and is caused by the different signs of the electrode spin polarizations [15]. Whereas the spin polarization, P is negative for the Co electrode, the Fe spin polarization is in fact positive.

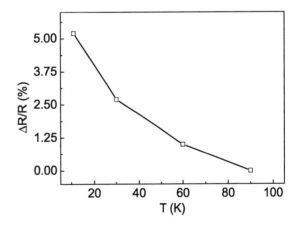

FIGURE 5. Giant magnetoresistance $|\Delta R/R|$ of the spin-valve device Fe(25 nm)/Alq$_3$(140 nm)/Co(3.5 nm) as a function of temperature.

The maximum relative GMR, $\Delta R/R$, where ΔR is the change in R between parallel and anti-parallel magnetization alignments of the FM electrodes, is seen to be about 5% at 11 K. $\Delta R/R$ may be in fact calculated using a modified Julliére equation [11, 16]; $\Delta R/R = 2P_1P_2D/(1 + P_1P_2D)$, where P_1 and P_2 are the spin polarizations of the two FM electrodes, and D =

exp$[-(d-d_o)/\lambda_s]$, where λ_s is the spin diffusion length in the OSEC, d is the OSEC thickness and $d_o = 100$ nm is the "ill-defined" OSEC layer thickness mentioned above. Using the spin polarization of iron and cobalt FM electrodes ($P_1 = 0.4$ and $P_2 = -0.34$) and $\lambda_s = 45$ nm in Alq$_3$ at 11K [11], we calculated $\Delta R/R = 11\%$. This is in fair agreement with the experimental GMR value of about 5%, and indicates that the device structure may be improved in the near future to give a higher GMR value.

We found that $\Delta R/R$ strongly depends on the temperature. Figure 5 shows that the GMR response steeply decreases with increasing temperature; in fact $\Delta R/R \approx 0$ at about 90K. Since the Curie temperatures of the iron and cobalt are 1043 K and 1388 K, respectively, much higher than 90K, we thus conclude that the steep $\Delta R/R$ decrease with temperature cannot be due to the loss of spin injection capability of the FM electrodes. We attribute the GMR decrease with temperature mainly due to the temperature dependence of λ_s, which steeply decreases at elevated temperature because of an increase of the spin relaxation rate [11].

However we note that $\Delta R/R$ decreases more steeply with the temperature in the Fe/Alq$_3$/Co spin-valves compared with the LSMO/Alq$_3$/Co spin-valve devices measured previously [11]. In the later devices we also measured I-V characteristic that are less dependent with the temperature compared with that of the Fe/Alq$_3$/Co devices, and this indicates that tunneling injection is more dominant at the LSMO/Alq$_3$ interface compared to the Fe/Alq$_3$ interface. Since carrier injection by tunneling is a prerequisite for effective spin injection from metals into semiconductors [17], we therefore conclude that the steep GMR decrease with temperature that we found in the Fe/Alq$_3$/Co devices may be related with the carrier injection process across the Fe/Alq$_3$ interface.

CONCLUSION

In conclusion, we have fabricated and studied a novel organic spin-valve device, namely Fe/Alq$_3$/Co where the OSEC is sandwiched between two d-band FM electrodes with high Curie temperatures. This device shows a GMR value of about 5% at 11 K, in agreement with a theoretical estimate; but decreases to zero value at about 90 K. The GMR decrease with temperature is mainly caused by an increase in the carrier spin relaxation rate with temperature, but is also influenced by carrier injection at the Fe/Alq$_3$ interface. We anticipate that increased purity and other precautions taken with the Alq$_3$ OSEC, as well as careful growth of the OSEC at the Fe/Alq$_3$ interface may push the device working temperature to reach 300 K.

ACKNOWLEDGMENTS

We thank Drs. Vladimir Burtman and Matt Delong, and C. Yang for help with the device fabrication, measurements and useful discussion. This work was supported in part by the DOE grant 03 ER-45490, ONR grant N00014-02-10595, and the University of Utah TCP program.

REFERENCES

1. Baibich M., et al., *Phys. Rev. Lett.* **61,** 2472-2475 (1988).
2. Wolf, S. A., et al., *Science* **294**, 1488-1490 (2001).
3. Prinz, G. A., *Science* **282**, 1660-1661 (1998).
4. Tsang, C. et. al., *IEEE Trans. Magn.* 30, 3801 (1994).
5. Dax, M., *Semicond, Int.* **20**, 84 (1997).
6. Kikkawa, J. M. and Awachalom, D.D., *Nature* **397**, 139-141 (1999).
7. Ohno, Y. et al., *Nature* **462**, 790-792 (1999).
8. Hanbicki, A. T., *Appl. Phys. Lett.* **80**, 1240-1242 (2002).
9. Krinichnyi, V. I., *Synth. Met.* **108**, 173 (2000).
10. Dediu, V., Murgia, M., Matacotta, F. C., Taliani, C. and Barbanera, S., *Solid State Commun.* **122**, 181-185 (2002).
11. Xiong, Z. H., Wu, D., Vardeny, Z. V., and Shi, J., *Nature* **427**, 821-824 (2004).
12. Arisi, E., et al., *J. Appl. Phys.* **93**, 7682-7684 (2003).
13. Soulen, R. J., et al., *Science* **282**, 85-88 (1998).
14. Fulde, P., *Adv. Phys.* **22**, 667 (1973).
15. Dieny, B., et al., *Phys. Rev. B* **45**, 806 (1992).
16. Jullière, M., *Phys. Lett.* A **54**, 225-228 (1975).
17. Rashba, E. I., *Phys. Rev. B* **62**, R16267-R16270(2000).

Saturated carboxylic acids on silicon: a first-principles study

Clotilde Cucinotta*, Alice Ruini*, Marilia J. Caldas[†] and Elisa Molinari*

*INFM National Center on NanoStructures and BioSystems at Surfaces (S3) and Dipartimento di Fisica,
Università di Modena e Reggio Emilia, Via Campi 213/A, 41100 Modena, Italy
[†]Instituto de Física, Universidade de São Paulo, Cidade Universitária, 05508-900 São Paulo, Brazil

Abstract. We present a first-principles calculation of the energetics of different possible dissociative chemisorption reactions leading to the attachment of organic acids with a functional carboxylic group to a hydrogenated silicon surface. Our study allows us to understand the role of oxygen atoms in the stable anchoring of the organic layer to the surface.

INTRODUCTION

Organic functionalization is emerging as a key advance for semiconductor-based material science [1]. A particularly promising new class of hybrid devices involves the functionalization of silicon surfaces [2, 3], which may allow for biocompatibility. The control of surface characteristics –polarity, hydrophobicity or affinity, and so forth– is a clear issue, and depends strongly on the possibility of exploiting intrinsic self-assembling properties of the chosen organic on Si. This is usually attempted through radical chain reactions occurring in typical hydrosilation mechanisms [2], where the pre-existence of a dangling bond in a silicon-hydride surface group is needed to initialize the chemisorption reaction. The required surface dangling bond can be created by means of proper activation techniques, such as thermal energy, UV irradiation, reaction involving a radical initiator; however, it is known that even a modest presence of oxygen-containing impurities in the reaction environment hampers the formation of stable and ordered organic layers on silicon. On the other hand, it was recently shown that the presence of oxygen in the functional group of the molecule itself can be conducive to a stable anchoring of the molecule to the surface [4, 5].

In order to understand the role of oxygen in the formation of a stable functionalized silicon surface, we consider organic acids with a functional carboxylic group (COOH), where two oxygen atoms are involved in the carbonyl (C=O) and in the hydroxyl (O-H) group. In particular, we consider here propionic acid, a saturated (methyl-terminated) carboxylic acid that allows us to rule out the influence of any other possible competitive reaction mechanism not involving the COOH group.

The selected substrate for our study is the hydrogen-terminated (100) silicon surface: the H-saturation makes silicon resistant to oxide formation, and allows us to reproduce wet-chemistry reactions, that are easy and cheap compared to the ultra-high vacuum techniques needed to functionalize a clean silicon surface. In particular we consider the (1x1) H:Si(100) [6], corresponding to a dihydride termination, since this surface is technologically relevant and can be produced with atomic flatness and environmental robustness [7]; moreover its structure allows a molecular anchoring involving both oxygen atoms of the COOH group, thereby leading to a number of different possible attaching chemistries.

THEORETICAL APPROACH

We work within an *ab initio* framework based on the density-functional theory [8]; the Kohn-Sham equations are solved [9] in a plane-wave basis set and ultrasoft pseudopotential [10] are employed to describe the electron-ion interaction. We checked that our results do not depend on the specific choice of the exchange-correlation kernel in the energy functional (to this end, we tried both the local-density approximation and generalized-gradient approximations such as PW91). The functionalized system is described using a slab geometry in the supercell approach: this enables us to study different coverage regimes and to fully include the effect of the extended surface electronic structure on the energetics, which is neglected in small-cluster models [5] previously used to investigate this class of systems.

RESULTS

We show in Fig. 1 the most relevant configurations resulting from our calculations. We simulated different

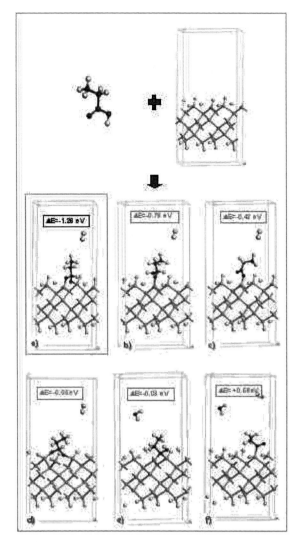

FIGURE 1. Relaxed structure of the most stable configurations formed after the chemisorption through the carboxylic group of propionic acid on H:Si(001) in the 50% coverage regime. The top panel shows the initial reagents. In panel (a), the favored bidentate oxygen configuration is depicted, while panels (b) and (c) report two possible single-oxygen-bridge attachments (in (b) the oxygen involved in the surface bond comes from the molecular hydroxyl group and in (c) from the carbonyl group); panel (d) shows a less favorable bidentate oxygen anchoring, while two Si-C linked configurations are depicted in panels (e) and (f).

possible reactions leading therefore to the formation of oxygen-bridged Si-O-C (Fig. 1(a-d)) and direct Si-C bonded structures (Fig. 1(e-f)). The first class is characterized by the production of a residual H_2 molecule, while the Si-C linked configurations release an H_2O molecule.

An important result of our calculation is that the most energetically favorable configuration is bidentate, with both O atoms of the COOH group involved in the bond, in a bridge position between the carbon atom of the carboxylic group and two surface silicon atoms. As a result, the oxygen in the reactive group of the molecule plays a key-role for the attachment, which exploits the remarkable strength of the resulting Si-O bond.

It is also relevant to observe that none of the considered reactions necessarily require any surface activation, the driving mechanism of the reactions proposed in this work can be identified as the mutual electrostatic attraction of the carboxylic reactive atoms and the silicon surface groups.

The supercell-based scheme employed in our calculation enabled us to study the stability of different coverage regimes for the organic moiety on the substrate. We studied different packing densities corresponding to 1 ML, 0.5 ML and 0.25 ML coverages: our results show no significant difference between the three regimes (the high molecular packing corresponding to 1 ML coverage is slightly preferred by 0.03 eV/mol over the 0.5 ML coverage, which is in turn favored over the 0.25 ML case by only 0.02 eV/mol). These results allow us to argue that the Si-O-C linked configuration is highly compatible with a full monolayer coverage. This is further evidence in favor of high packing densities for carboxylic acid chemisorbed layers, and since the 1 ML coverage hampers the intrusion of water molecules, this suggests also higher stability for these functionalized surfaces.

We are grateful to M. Stutzman and S. Corni for fruitful discussion. Computer time was partly provided by CINECA through INFM Parallel Computing Projects. The support by FIRB "NOMADE" is also acknowledged.

REFERENCES

1. Joachim, C., Gimzewski, J. K., and Aviram, A., *Nature*, **408**, 541–548 (2000).
2. Buriak, J. M., *Chem. Rev.*, **102**, 1271–1308 (2002).
3. Bent, S. F., *Surf. Sci.*, **500**, 879–903 (2002).
4. Boukherroub, R., Morin, S., Sharpe, P., and Wayner, D. D. M., *Langmuir*, **16**, 7429–7434 (2000).
5. Pei, Y., Ma, J., and Jiang, Y., *Langmuir*, **19**, 7652–7661 (2003).
6. Northrup, J. E., *Phys. Rev. B*, **44**, 1419–1422 (1991).
7. Cerofolini, G. F., Galati, C., Reina, S., and Renna, L., *Semicond. Sci. Technol.*, **18**, 423–429 (2003).
8. Kohn, W., and Sham, L. J., *Phys. Rev.*, **140**, A1133–A1138 (1965).
9. Baroni, S., Dal Corso, A., de Gironcoli, S., and Giannozzi, P. (2001), http://www.pwscf.org.
10. Vanderbilt, D., *Phys. Rev. B*, **41**, 7892–7895 (1990).

Interchain Effects on the Vibrational Properties of PPP and PPV

R. L. de Sousa and H. W. Leite Alves

Departamento de Ciências Naturais, Universidade Federal de São João del Rei
C. P.: 110, CEP: 36.300-000 São João del Rei, MG, Brazil

Abstract. In this work, we have calculated the vibrational modes of the poly-paraphenylene (PPP) in both Pbam and Pnnm structures, and of the poly-paraphenylene-vinylene (PPV) in the $P2_1/c$ structure. Our results agree well (within 3 % of error) with the available experimental data, whenever these comparison is possible. Based on our calculations, we show that the acoustical longitudinal mode stretches the π-bonds, while the acoustical transversal ones leads only to their torsion.

INTRODUCTION

The π-conjugated polymers have become the most promising materials for optoelectronic device technology in the last decade. However, while the structural and electronic properties of these materials are well known [1,2], the experimental and theoretical data on the phonon modes available in the literature are very scarce [3-6]. To supply the missing information on the vibrational properties of both PPP and PPV, and to check the validity of the current adopted models, we have calculated, *ab initio*, their principal phonon modes by using the density-functional theory within the local density approximation (LDA), plane wave expansions and the pseudopotential method (ABINIT code) [7]. In our calculations, we have used the Troullier-Martins pseudopotentials (evaluated by the fhi98PP code [8]). Details about the structural properties of these systems, as well as the vibrational modes are described in our previous work [9].

VIBRATIONAL MODES

Our calculated vibrational modes for PPP (in both Pbam and Pnnm structures) show narrow flat bands, located between 3053.8 and 3076.6 cm^{-1}, corresponding to C-H bond stretching vibrations. Also, the calculated phonon dispersion has the same features in both symmetries. The dispersion curves show some frequency gaps located between 1642-3053.8 cm^{-1}, 1372-1440 cm^{-1} and 1180-1270 cm^{-1}. In Fig. 1, we plot our calculated phonon dispersion for PPP along the chain direction (full lines), compared with the phonon modes of a single chain of PPP (dotted lines). We have excluded the flat bands at 3053.8 to 3076.6 cm^{-1} in order to see the details of only the acoustic bands. It is interesting to note, that the single-chain approach can describe well the main features of the phonon dispersion. From the analysis of calculated oscillator strenghts for the modes at Γ point of the Brillouin zone, conjugated with the peaks of the calculated phonon density of states, the most intense ones are located at 423.0(B_{2g}), 424.3(B_{3g}), 506.0(B_{2g}), 620.6(B_{3g}), 669.1(B_{3g}), 799.1(A_g), 820.6(A_g), 832.3(B_{1g}), 1044.5(A_u), 1176.5(A_g), 1322.0(B_{1g}), 1372.9(B_{3u}), 1600.0(B_{3g}) and 3084.5(A_g+B_{1g}) cm^{-1}. Only a few of these assignments agree well with the single chain calculations by Capaz and Caldas [3], but agree well with the Raman and infrared data [6]. It is interesting to remark that the 620.6 cm^{-1} agrees well with the Raman active B_{3g} mode at 623 cm^{-1}. This is a longitudinal mode that stretches the phenyl π-bonds.

In Fig. 2, we show our calculated phonon dispersion for PPV, in the same way as in Fig. 1. Also, the excluded narrow flat band corresponds to C-H bond stretching vibrations. From this Figure, we observe that the single chain approach describes well only the modes at Γ. So, for PPV, excluding the interchain effects can be a good approximation to analyze the Raman as well as infrared results, that can only measure the zone-center modes.

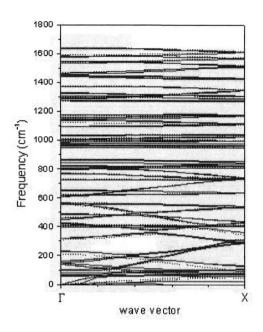

FIGURE 1. Calculated phonon dispersion for PPP along the chain direction (full lines), compared with the phonon modes of their single chain (dotted lines).

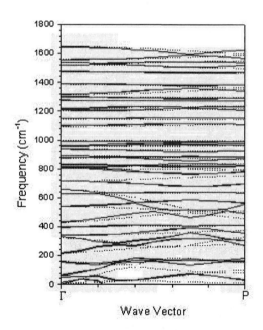

FIGURE 2. Same as Figure1 for PPV.

From the analysis of calculated oscillator strenghts for the zone-center modes, the most intense ones are located at $431.2(B_u)$, $540.1(B_u)$, $807.9(B_u)$, $832.5(A_u)$, $939.2(B_u)$, $957.8(B_u)$, $993.1(A_u)$, $1098.8(A_u)$, $1100.3(B_u)$, $1208.1(B_u)$, $1315.2(B_u)$, $1390.6(A_u)$, $1471.0(B_u)$, and $1556(A_g)$ cm^{-1}, and the modes with some intensity were located at $328.5(B_g)$, $627.7(A_g+B_g)$, $715.9(A_g)$ and $1643.3(A_g)$ cm^{-1}, respectively. These assignments agree well with the single chain calculations by Capaz and Caldas [3], as well as with Raman data [5]. Analyzing the displacement patterns for some modes, we found a longitudinal one at 259 cm^{-1}, which stretches the π-bonds(mainly in the vinyl group), and other longitudinal mode at 1643.3 cm^{-1}, which leads only to π-bonds torsion.

In summary, we have presented the vibrational modes of PPP and of PPV in their full structure. Our results agree well with the available experimental data, whenever these comparison is possible. From the obtained results, we detected that single-chain models only describe well the zone-center modes. Complete description of our obtained results for the dynamical properties of PPP will be published soon elsewhere.

ACKNOWLEDGMENTS

This work supported by MEC-CAPES scientific program, Brazil.

REFERENCES

1. C. Ambrosch-Draxl, J. A. Majewski, P. Vogl and G. Leising, *Phys. Rev.* **B51**, 9668(1995), and references therein.

2. P. Gomes da Costa, R. G. Dandrea and E. M. Conwell, *Phys. Rev.* **B47**, 1800(1993), and references therein.

3. R. B. Capaz and M. J. Caldas, *Phys. Rev.* **B67**, 205205(2003).

4. J. Gierschner, H.-G. Mack, L. Lüer and D. Oelkrug, *J. Chem. Phys.*, **116**, 8596(2002).

5. E. Mulazzi, A. Ripamonti, J. Wery, B. Dulieu and S. Lefrant, *Phys. Rev.* **B60**, 16519(1999).

6. L. Cuff and M. Kertesz, *Macromol.* **27**, 762-770(1994).

7. X. Gonze, et al., *Comput. Mater. Sci.* **25**, 478(2002), and references therein.

8. M. Fuchs and M. Scheffler, *Comp. Phys. Commun.* **119**, 67(1999).

9. R. L. Sousa, J. L. A. Alves, and H. W. Leite Alves, *Mater. Sci. Eng. C* (2004) to be published.

Fabrication Of Atomic-scale Gold Junctions By Electrochemical Plating Technique Using A Common Medical Disinfectant

Akinori Umeno and Kazuhiko Hirakawa

Institute of Industrial Science, University of Tokyo, 4-6-1 Komaba, Meguro-ku, Tokyo 153-8505, Japan

Abstract. Iodine tincture, a medical liquid familiar as a disinfectant, was introduced as an etching/deposition electrolyte for the fabrication of nanometer-separated gold electrodes. In the gold dissolved iodine tincture, the gold electrodes were grown or eroded slowly in atomic scale, enough to form quantum point contacts. The resistance evolution during the electrochemical deposition showed plateaus at integer multiples of the resistance quantum, $(2e^2/h)^{-1}$, at the room temperature. The iodine tincture is a commercially available common material, which makes the fabrication process to be the simple and cost effective. Moreover, in contrast to the conventional electrochemical approaches, this method is free from highly toxic cyanide compounds or extraordinary strong acid. We expect this method to be a useful interface between single-molecular-scale structures and macroscopic opto-electronic devices.

Keywords: quantum point contact; electrochemical plating; iodine tincture; gold
PACS: 81.07.Lk

INTRODUCTION

Measuring transport properties at the level of single molecules opens new possibilities to future device technologies. In single-molecule devices, new ingredients, such as quantum mechanically well-defined molecular orbitals, localized electron spins, nanomechanical oscillations, etc., can be incorporated into the functionalities of optoelectronic devices. However, the key issue is how to reproducibly interface single molecules to macroscopic device structures. In this work, we have developed a simple and nontoxic fabrication technique for atomic-scale metallic electrodes with high controllability.

EXPERIMENTAL PROCEDURE

The experimental procedure involves two stages, the preparation of the electrolyte and the electrochemical deposition/dissolution of gold electrodes. In the first stage, enough amounts of gold films (Kanazawa Katani Sangyo Co., ltd) were dissolved in the commercially available iodine tincture (Ken-ei Co., ltd). Typical constituents of iodine tincture are iodine and potassium iodide dissolved in ethanol. Such halogen-halide-organic solutions had been reported to serve as solvents for noble metals [1] and recently used as a teaching material at high schools for electroplating gold onto bulk copper surface [2]. A proper quantity of L(+)-ascorbic acid (Wako Co., ltd), known as vitamin C, was added until the color of the solution turned into transparent.

In the second stage, with a counter electrode the lithographically patterned gold electrodes were dipped in the electrolyte. The gaps of starting electrodes were about 200 nm. The patterned electrodes were then grown or eroded by the electrochemical reaction. During the electrochemical reactions, the time evolution of the resistance across the electrodes was monitored with a lock-in amplifier.

RESULTS AND DISCUSSION

With constant deposition voltage of 650 mV, the resistance across the electrodes was gradually decreased down to single resistance quantum, $R_0 = (2e^2/h)^{-1}$. This could be interpreted that the separation of the electrodes was slowly decreased in atomic scale down to quantum point contact (Fig. 1). The observed exponential decrease in resistance indicates the enhancement of tunneling probability between the tip of the electrodes since the deposition speed expected to be constant under the constant deposition voltage. The resistance, then, jumped down abruptly, which indicates the macroscopic electrical contact was formed between the electrodes. Under this deposition voltage, however, no clear conductance plateaus were observed at integral multiples of the conductance quantum, $G_0 = 2e^2/h$, with larger integers.

By decreasing deposition voltage down to 30 mV, conductance steps with larger integral multiple of G_0 were observed (Fig. 2). In this region, 2nd or 3rd multiples of the G_0 were not observed under the time resolution of 500 msec., which was limited by the time constant of our lock-in measurement. Off-integral plateaus were also observed around $5G_0$, $10G_0$, $18G_0$, which could be interpreted as a series connection of 1D channels with different cross sections [3].

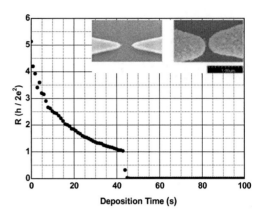

FIGURE 1. Time evolution of resistance across the electrodes during electrochemical deposition in the gold dissolved iodine tincture. The bias voltage was 650 mV and longitudinal axis is normalized by single resistance quantum. Insets are scanning electron micrographs of electrode pair before (left) and after (right) the electrochemical deposition for a certain amount. The bar corresponds to 1 μm.

SUMMARY

In this work, commercially available medical liquid, iodine tincture, was introduced as an effective electrolyte for the atomic-scale electroplating. With a gold dissolved iodine tincture, a pair of gold electrode was grown or etched slowly in atomic scale enough to exhibit conductance quantization. With an appropriate feedback control, the distance between the electrodes can be controlled below 1 nm, which meets the requirements for realizing single-molecule devices. Stable metallic 1D channels can also be fabricated for chemical sensing applications. Furthermore, this simple nanofabrication technique is compatible with microfluidic processes. This method is free from highly toxic cyanide materials or extraordinary strong acid, which are used in conventional electrochemical fabrication techniques with gold electrodes. We expect this simple nanofabrication method, as other reported electrochemical approaches are [4], could be a useful connection between single-molecular-scale structures and macroscopic opto-electronic devices.

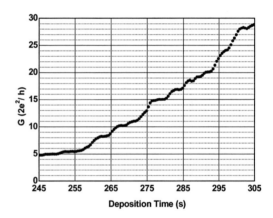

FIGURE 2. Conductance time evolution across the electrodes during the electrochemical deposition with 20 times slower speed than Figure 1 case. The longitudinal axis is normalized by single conductance quantum of 77μS.

ACKNOWLEDGMENTS

We thank H. Sakaki, Y. Arakawa and S. Komiyama for their continuous encouragement and technical supports. This work has been supported in part by the 21st Century COE Program of Japan Society for the Promotion of Science (JSPS).

REFERENCES

1. Y. Nakao and K. Sone, *Chem. Commun.*, **897** (1996).
2. H. Takagi, *Japan Society of Physics and Chemistry Education Meeting* (2002).
3. W. A. de Heer, S. Frank, D. Ugarte, *Z. Phys. B* **104**, 469 (1997).
4. A. F. Morpurgo, C. M. Marcus, and D. B. Robinson, *Appl. Phys. Lett.*, **74**, 2084 (1999).

Room Temperature Polariton Photoluminescence in a Two-dimensional Array of Inorganic-organic Hybrid-type Quantum-wells

Junko ISHI-HAYASE and Teruya ISHIHARA

Institute of Physical and Chemical Research (RIKEN), 2-1 Hirosawa, Wako, Saitama 351-0198, JAPAN

Abstract. Photoluminescence (PL) properties of polaritons at room temperature were investigated in a two-dimensional array of inorganic-organic hybrid-type semiconductor quantum-wells. In angularly-resolved PL spectra, PL peaks attributed to the lower branch of polariton shift with a detection angle and show a great change of its intensities. The central energy of PL peak at a certain detection angle is connected by the dispersion relation of the polariton, which is much different from polaritons in a bulk semiconductor. PL intensity of polariton is significantly enhanced at a half of Rabi splitting energy lower than exciton resonant energy due to the polariton bottleneck effect. PL intensity around Γ point is found to be strongly modulated depending on material parameters of the samples.

INTRODUCTION

The strong coupling between an exciton and a photon results in the formation of cavity polaritons which leads to a dramatic change of optical responses of semiconductors[1]. So far, most of investigations of cavity polaritons have been performed in semiconductor-embedded Fabry-Perót (FP) microcavities. The concept of the cavity polariton in a FP microcavity can be extended to the exciton-photon coupled system in a photonic crystal (PhC). Though such system is expected to give us new environment for light-matter interaction, there are few reports about polaritons in PhCs. Recently, our group found the clear evidence of polaritons in one-dimensional and two-dimensional (2D) photonic crystal slabs composed of inorganic-organic hybrid-type quantum-wells[2]. Due to the large oscillator strength of excitons in these materials[3], Rabi splitting Ω reaches to over 130 meV in our system which enables us to observe polaritons even at room temperature. In this paper, we report the photoluminescence properties of room temperature polaritons in a similar system under nonresonant excitation.

EXPERIMENT

The sample used in our experiments is schematically shown in Fig 1. An inorganic-organic hybrid-type semiconductor $(C_6H_{13}NH_3)_2PbI_4$ was incorporated into 2D periodic holes on a quartz substrate by spin-coating method. The substrate was covered by polystyrene film so that the electromagnetic field is well confined. Such system can be regarded as a 2D PhC slab. We fabricated various samples with different pitch of holes (Λ) and filling factor $\eta = a / \Lambda$.

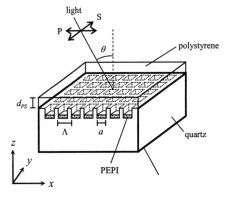

FIGURE 1. Schematic structure of our sample (not to scale).

RESULTS AND DISSCUTTIONS

Figure 2(a) shows S-polarized PL spectra for various detection angle θ_d measured in a sample with $\Lambda = 0.76$ μm and $\eta = 0.63$. The central energy of excitation laser was tuned to 3.1 eV which is much higher than exciton resonant energy E_{ex} in $(C_6H_{13}NH_3)_2PbI_4$, 2.409 eV. Each spectrum was subtracted the PL spectrum for bulk $(C_6H_{13}NH_3)_2PbI_4$ to extract only the contribution of polaritons. We observed strong PL peaks shifting with θ_d at the lower energy side of E_{ex}. This PL peaks were attributed to the lower branch of polariton which is strongly-coupled state between an exciton and a quasi-guided mode. On the other hand, upper polariton PL cannot be found because of the large Rabi splitting. The dispersion curves estimated from the result in Fig 2(a) is shown in Fig 2(b). In our system, the energy of polariton PL at a certain θ_d is directly connected by the dispersion relation of the polariton. This result is much different from the situation of polariton in bulk semiconductor where the polariton PL is strongly modulated by the propagation effect of polaritons. Figure 2(c) plots the PL peak intensity as a function of photon energy calculated from the corresponding PL intensities. The PL intensity is significantly enhanced at $E_{ex} - \Omega/2$. This result demonstrated that polariton has the largest population at the point (polariton bottleneck effect). Moreover, we found small energy gap (~ 15 meV) around $\theta_d = 0°$ (Γ poiont) which corresponds to the photonic band bap (PBG). PL intensity around the edge of PBG shows anomalous behavior which reflect the symmetry of polariton mode at the edge of PBG[4]. In Fig 3(c) the polariton mode at the top of PBG is optically inactive, while the bottom is optically active.

Figure 3 shows similar graphs to the Fig 2(c) in samples with various η and the same Λ. This result demonstrates that PBG can be tuned by changing η. It is also found that the symmetry of polariton modes at the top and bottom of PBG is changed depending on η. Thus, the optical properties of polaritons can be controlled by material parameters of PhC slabs.

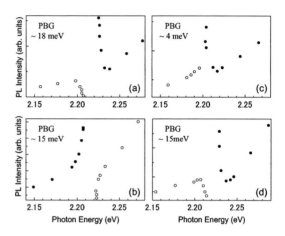

FIGURE 3. The intensity vs. the central energy of PL peak for S-polarization measured in samples with various η ($\Lambda = 0.76$ mm). (a) $\eta = 0.40$, (b) $\eta = 0.63$, (c) $\eta = 0.76$, (d) $\eta = 0.86$.

ACKNOWLEDGEMENTS

Patterned-substrates used in our experiment were fabricated by means of e-beam lithography at advanced device laboratory, RIKEN.

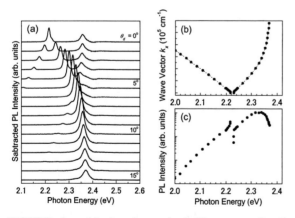

FIGURE 2. (a) Angular-resolved PL spectra for S-polarization ($\Lambda = 0.76$ mm, $\eta = 0.63$). (b) Dispersion curves, and (c) PL intensity vs. photon energy estimated from the result in (a).

REFERENCES

1. Weisbuch,, C., Nishioka, N., Ishikawa, A., and Arakawa. Y., *Phys. Rev. Lett.* **69**, 3314 (1992).
2. Fujita, T., Sato, Y., Kuitani, T., and Ishihara, T., *Phys. Rev.* **B 57**, 12428 (1998); Shimada, R., Yablonskii, A.L., Tikhodeev, S.G. ,and Ishihara, T., *IEEE J. Quantum Electron.* **38**, 872 (2002); Ishi-Hayase, J. and Ishihara, T., *Semicond. Sci. Technol* **18**, S411 (2003).
3. Ishihara, T., in *Optical Properties of Low-Dimensional Materials*, edited by T. Ogawa and Y. Kanemitsu (World Scientific, Singapore, 1995), Chap. 6.
4. Tikhodeev, S.G., Yablonskii, A.L., Muljarov. E.A., and Ishihara T., *Phys. Rev.* **B 66**, 045102 (2003)

Optical Absorption, Photoluminescence and Photoconductivity of Organic-Inorganic One-Dimensional Semiconductors $C_5H_{10}NH_2PbI_3$ and $[NH_2C(I)=NH_2]_3PbI_5$

K. Matsuishi, Y. Kubo, T. Ichikawa, and S. Onari

Institute of Materials Science, University of Tsukuba
Tsukuba, Ibaraki 305-8573, Japan

Abstract. The optical properties of the semiconductive multi-quantum wire materials, $C_5H_{10}NH_2PbI_3$ and $[NH_2C(I)=NH_2]_3PbI_5$, have been investigated by optical absorption, photoluminescence and photoconductivity measurements. Comparison in the electronic and excitonic states between the two materials is presented and discussed in the light of the connectivity of constituent octahedra in the inorganic wires.

INTRODUCTION

A series of lead-halide based inorganic-organic perovskite semiconductors forms self-organized quantum structures composed of corner- or face-sharing $[PbX_6]^{4-}$ (X: halogen) octahedra separated by insulating organic molecular parts [1,2]. The octahedral network can be designed and controlled to act as multi-quantum wells, wires and dots for carriers confined by the dielectric barriers of the organic parts. The multi-quantum wire crystals, $C_5H_{10}NH_2PbI_3$ and $[NH_2C(I)=NH_2]_3PbI_5$ (hereafter, referred to as pp-PbI3 and cy-PbI5, respectively), are composed of one-dimensional (1D) semiconductive chains of face- and corner-sharing octahedra, respectively, separated by the insulating organic molecular parts. In the present study, we have investigated the differences in the electronic and optical properties between the face- and corner-sharing octahedral multi-quantum wire materials. We report the optical absorption (OA), photoluminescence (PL) and photocurrent (PC) spectra associated with the excitonic states and their anisotropy in these 1D semiconductors.

EXPERIMENTAL

Needle-like single crystals with an orthorhombic structure for pp-PbI3 and a monoclinic structure for cy-PbI5 were grown by a solution method. PL measurements were performed at 10 to 300 K using the 325, 442 and 458 nm laser lines, and PC measurements at 90 to 370 K using an intense monochromatized light (300 to 600 nm) from a Xe lamp chopped at frequencies of 20 to 500 Hz.

RESULTS AND DISCUSSION

The OA spectra of pp-PbI3 and cy-PbI5, which were obtained from reflection spectra at room temperature using the Kramers-Kronig relation, are shown in Fig. 1(a) and 1(b), respectively. The intense peaks at 3.25 eV for pp-PbI3 and at 2.8 eV for cy-PbI5 can be assigned to free exciton absorption. The rises in the high energy region are considered due to the edge of the 1D interband transitions, reflecting the 1D density of state. The differences in the exciton and band gap energies between the two materials can be explained by the overlapping between neighboring Pb 6p orbitals along the semiconductive wire to form partially the conduction band for pp-PbI3.

Representative PL spectra of pp-PbI3 and cy-PbI5 are shown in Fig. 1(a) and 1(b) together with the OA spectra. The broad PL band at 1.95 eV for pp-PbI3 can be assigned to emission from self-trapped excitonic states with a Stokes shift of 1.3 eV [3,4]. This is an indication of significant electron-lattice interaction due to a narrow band width as a low-dimensional band characteristic. On the other hand, four distinct PL bands are observed at 1.6, 2.1, 2.4, and 2.6 eV for cy-PbI5. All of them are sharper than that of pp-PbI3. The PL band at 1.6 eV, which grows with laser irradiation, can be assigned to emission from excitons bound by defects. The bands at 2.1 and 2.4 eV, which are more significant at lower temperatures, would be from trapped excitons, and the band at 2.6 eV from free excitonic states. It is noteworthy that emission from free excitonic states has not been observed for pp-PbI3. We estimated the exciton binding energy to be 300 meV for pp-PbI3 and 430 meV for cy-PbI5.

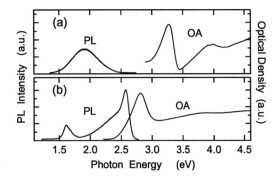

FIGURE 1. OA and PL spectra for (a) pp-PbI3 and (b) cy-PbI5 at room temperature. The PL was excited by the 441.6 nm laser line.

Figure 2 shows the PC spectra $j_{p,//c}$ and $j_{p,\perp c}$ of pp-PbI3 with an applied electric field parallel and perpendicular to the semiconductive wire (c-axis), respectively. The PC $j_{p,//c}$ was about 40 times larger than $j_{p,\perp c}$ at 2.85 eV, while the dark current, $j_{d,//c}$, was about 10 times larger than $j_{d,\perp c}$ at 3 kV/cm. The electrical anisotropy in j_p and j_d is evidence of the anisotropy of the 1D electronic band structure. For both pp-PbI3 and cy-PbI5, the PC spectra exhibit a strong peak due to the thermal dissociation of excitons confined in the quantum wire, and a small hump in the higher energies due to the lowest interband transition reflecting the 1D density of states. The temperature dependence of photocurrent above 250 K followed an Arrhenius form at each excitation energy, E_{ph}, above 2.8 eV. We determined the activation energy, E_a, as a function of E_{ph} by Arrhenius plots for $j_{p,//c}$, as shown in the inset of Fig. 2. E_a decreases almost linearly as E_{ph} increases from 2.8 to 3.3 eV and becomes almost independent of E_{ph} above 3.3 eV. The result suggests that the thermal dissociation of excitons is responsible for photocurrent in the low energy excitation region, while the hopping conduction dominates in the higher energies. The variation of E_a in the low energy excitation region implies that the free excitonic states

have distribution in the exciton binding energy, and excitons could be weakly localized or trapped due to random potentials induced by perturbation from lattice due to imperfection. We observed a similar behavior in the photoconductivity of cy-PbI5.

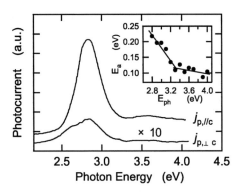

FIGURE 2. PC spectra of pp-PbI3 at 10 V/cm at room temperature. The inset indicates the activation energy, E_a, determined by Arrhenius plots from the temperature dependence of $j_{p,//c}$ at each excitation energy, E_{ph}.

CONCLUSION

We have investigated the electronic and excitonic properties of the semiconductive multi-quantum wire materials, $C_5H_{10}NH_2PbI_3$ and $[NH_2C(I)=NH_2]_3PbI_5$, by measuring optical absorption, photoluminescence and photoconductivity. The optical spectra exhibit large anisotropy due to the 1D electronic band structures. We have found that the optical properties and their temperature dependences are intriguingly different between the two materials, indicating that the connectivity of constituent octahedra affects significantly their electronic and excitonic states.

This work was supported in part by the Grant-in-Aid for Scientific Research from the Ministry of Education, Culture, Sports, Science and Technology in Japan.

REFERENCES

1. Ishihara T., "Optical Properties of Low-Dimensional Materials" edited by T. Ogawa and Y. Kanemitsu, Singapore: World Scientific, 1995, chapter 6.
2. Calabrese J., Jones N. L., Harlow R. L., Herron N., Thorn D. L., and Wang Y., *J. Am. Chem. Soc.* **113**, 2328-2330 (1991).
3. Nagami A., Okamura K. and Ishihara T., *Physica B* **227**, 346-348 (1996).
4. Fukumoto T., Hirasawa M. and Ishihara T., *J. Lumin* **87-89**, 497-499 (2000).

Linear And Nonlinear Optical Spectroscopies Of PPE/PPV Copolymer Semiconductors

M. H. Tong[a], Z. V. Vardeny[a] and Y. Pang[b]

[a]*Department of Physics, University of Utah, Salt Lake City, Utah 84112-0830*
[b]*Department of Chemistry & Center for High Performance Polymers and Composites, Clark Atlanta University Atlanta, Georgia 30314*

Abstract. We used a variety of picosecond transient and cw spectroscopies to investigate the linear and nonlinear optical properties of films of PPE/PPV copolymer semiconductor. The spectroscopies used include absorption, electro-absorption (EA) and two-photon-absorption (TPA). In addition to the odd parity excitons at ca. 3.0 eV, we also found a broad two-photon absorption band centered at about 4 eV, which is due to strongly coupled even parity excitons.

INTRODUCTION

π-conjugated semiconductor polymers have become important materials because of their potential applications as active media in organic light emitting diodes (OLED's) [1], and other optoelectronic devices, such as field effect transistors[2] and photovoltaic cells. Recently a PPE/PPV hybrid polymer has been successfully synthesized and was shown to be an excellent active medium for OLED applications [3]. This copolymer contains both PPE and PPV chromophores with a nearly perfect overlap between the emission band of the PPE chromophores and the absorption band of the PPV chromophores that leads to a very efficient energy migration from PPE to PPV excited states. We have already studied and reported the photoexcitation dynamics and laser action in solutions and thin films of PPE-PPV copolymer using a variety of ultrafast and steady-state spectroscopy techniques [4]. Those studies are considered to be the first laser action reported in PPE-type polymers. In the present paper we report our studies concerning the even parity excited states of PPE/PPV copolymer.

EXPERIMENT

The two-photon absorption (TPA) spectrum was measured using the pump-probe correlation technique. The pump beam (from a homemade Ti-Sapphire regenerative amplifier) was set at 1.55 eV, below the copolymer absorption band, whereas the probe beam (from a super-continuum generated in a sapphire crystal) covers the spectral range from 1.5 to 2.6 eV. The temporal and special overlap between the pump and probe beams leads to a photoinduced absorption (PA) signal that peaks at t = 0. This PA has a temporal profile similar to the cross-correlation function of the pump and probe pulses, which we interpret here as due to TPA of one pump and one probe photons.

For the electro-absorption (EA) measurements, the light source was derived from a Xe lamp, with broadband visible and ultraviolet spectral range. The light was dispersed through a monochromator, focused on the sample, and detected by a UV-enhanced silicon photodiode. A small sine-wave source was connected to a custom-built transformer, the output of which was connected to the electrode (40 μm spacing). The electrode was placed in a cryostat for low temperature measurements. For each EA spectrum, the transmission (T) was measured with a mechanical chopper while the electric field was turned off. The differential transmission (ΔT) was measured without the chopper, with the electric field on, and with the lock-in amplifier set at twice (2f) the electric field modulation frequency, f.

RESULTS AND DISCUSSION

The room-temperature optical absorption spectrum of the PPE/PPV copolymer film is shown in the inset of the Fig. 1. Figure.1 shows the TPA spectrum of a PPE-PPV free-standing film. It is seen that the threshold of the TPA spectrum is blue-shifted respect to that of the linear absorption. The TPA spectrum also displays a peak at about 4eV, which is consistent with a similar spectral feature found in the EA spectrum (Fig. 2). As is well known, most of semiconductor polymers belong to the C_{2h} symmetry group. In this case the excited energy states in the correlated electron picture belong to A_g (even) or B_u (odd) irreducible representations. The ground state is obviously $1A_g$. For the C_{2h} symmetry group a one-photon transition is electric dipole allowed between A_g and B_u states. Consequently even parity excitons (A_g) are forbidden in linear absorption, but become allowed in a two-photon absorption process, which is a $\chi^{(3)}$ non-linear optical process. It follows that strong resonances in the TPA spectrum peak at A_g excited states. We thus can assign the strong TPA band at 4eV to the two-photon allowed mA_g state, which is about 0.8eV above the lowest, $1B_u$ exciton energy.

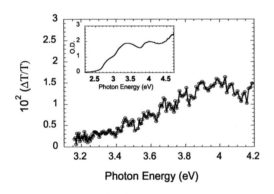

FIGURE 1. Two-photon absorption spectrum of a PPE/PPV film. The inset shows the linear absorption spectrum.

Figure. 2 shows the PPE/PPV EA spectrum. Below 4eV the EA spectrum is composed of strong features in the range of band I in the linear absorption spectrum. It exhibits a derivative-like feature with zero crossing at 2.97eV. In addition, there are two vibrational satellites at 3.2eV and 3.7eV, respectively. These features are formed due to an electric field induced red-shift (Stark shift) of the $1B_u$ exciton and its phonon sidebands. In addition to the Stark shift at low energy there is an electric field induced absorption band at 4.17eV. In agreement with the TPA spectrum we assigned this EA band to the mA_g exciton absorption, which is forbidden in linear absorption. This state is observable in EA since the external electric field transfers some oscillator strength from the allowed $1A_g \to 1B_u$ transition to the $1A_g \to mA_g$ forbidden transition.

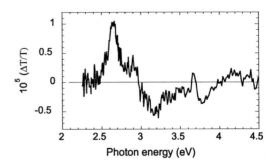

FIGURE 2. EA spectrum of PPE/PPV copolymer measured at 30K with field strength $F=5\times10^4$V/cm.

CONCLUSION

In this work we studied in films of PPE/PPV copolymer exciton states with even (A_g) and odd (B_u) parity using absorption, TPA and EA spectroscopies. From the EA spectrum we determined the $1B_u$ exciton to be at about 3 eV, following by two phonon side bands, and a strong mA_g state at $\approx$ 4 eV, about 1 eV above the $1B_u$ state.

ACKNOWLEDGMENTS

At the University of Utah this work was supported in part by NSF-DMR Grant No. 02-02790 and DOE Grant No. 02-04ER46109. The work at Clark Atlanta University was supported in part by the Air Force Grant No. F49620-00-1-0090.

REFERENCES

1. J. H. Burroughes, D. D. C. Bradley, A. R. Brown, R. N. Marks, K. Mackay, R. H. Friend, P. L. Burn, and A. B. Holmes, *Nature* (London) **347**, 539 (1990)
2. Y. Yang and A. J. Heeger, *Nature* (London) **372**, 344 (1994).
3. Q. Chu, Y. Pang, L. Ding, and F. E. Karasz, *Macromolecules* **36**, 3848 (2003).
4. M. Tong, C. X. Sheng, C .Yang, Z.V.Vardeny, and Y. Pang, *Phys. Rev. B* **69**, 155211 (2004) .

Raman Scattering from Organic Light Emitting Diodes

S. Guha[1], M. Arif[1], J.G. Keeth[2], T.W. Kehl[2], K. Ghosh[2] and R. E. Giedd[2]

[1] *Department of Physics, University of Missouri, Columbia, MO65211*
[2] *Department of Physics, Astronomy & Materials Science, Southwest Missouri State University, Springfield, MO 65804*

Abstract. We present Raman studies from working organic light emitting diodes (OLEDs) fabricated from a single layer of *para*-hexaphenyl (PHP). The OLEDs were fabricated on patterned ITO substrates with Al cathode. The Raman intensity ratio, I_{1280}/I_{1220}, increases with current injection. This is interpreted in terms of a conformational change where the phenyl rings become more aromatic (benzenoid) upon charge injection. Further, capacitance vs. voltage measurements indicate the presence of interface states.

INTRODUCTION

π-conjugated organic semiconductors such as short-chain oligomers and long-chain polymers continue to attract a lot of attention because of their great promise for low-cost, large-area optoelectronic and photonic device applications.[1] Semiconducting properties are defined by the ability of these materials to efficiently transport charge along the chain or between adjacent chains due to their π-orbital between neighboring molecules. Devices such as organic light-emitting diodes (OLEDs), transistors, and photodiodes are currently attracting much attention. Blue photoluminescence with a quantum yield of 30% has been the motivation for using *para*-hexaphenyl (PHP) as the emitting layer in OLEDs. PHP is characterized by a torsional degree of freedom between neighboring phenyl rings. In the crystalline state the molecules are arranged in layers, with a herringbone type arrangement found in each layer. PHP, on the average, at room temperature is more planar than at lower temperatures, and can be planarized further by the application of hydrostatic pressure. [2]

The Raman spectrum of PHP is characterized by three strong peaks at 1220 cm^{-1}, 1280 cm^{-1}, and 1600 cm^{-1}. They have been assigned to the C-H in plane bend, C-C inter-ring stretch, and the C-C intra-ring stretch modes, respectively. The Raman intensity ratio of the 1280 cm^{-1} to the 1220 cm^{-1} (I_{1280}/I_{1220}) is an indication for the number of π-conjugated phenyl rings and hence an indicator of planarity of the molecule. The Raman intensity ratio, I_{1280}/I_{1220}, typically decreases upon planarization. [2] In shorter oligophenyls the 1600 cm^{-1} peak is a double peak structure that has been suggested to originate from a Fermi resonance. [3] The higher energy peak in this region is an overtone mode.

EXPERIMENTAL DETAILS

Highly purified PHP was obtained from Tokyo Chemical Industries Ltd. OLEDs were prepared by depositing a 100 nm PHP film by vacuum evaporation (10^{-6} mbar) on patterned ITO coated glass substrates. Al (capped with Ag) was deposited in strips across the PHP layer using an evaporation mask. Each glass slide supported 25 devices. The IV and CV measurements were measured with a Keithley 2400 Sourcemeter and HP4284A LCR meter, respectively. The Raman measurements were carried out using a fiber optically coupled confocal micro-Raman system (TRIAX 320).

RAMAN SCATTERING

The Raman spectrum of PHP from working OLEDs was measured from the transparent ITO side in a perfect backscattering geometry. Figure 1 shows the Raman spectra of PHP film from an OLED under the application of an electric field for various values of current. The Raman intensity ratio, I_{1280}/I_{1220}, increases with increasing current and the changes are completely reversible. At the two non-zero values of current in Fig. 1, the device showed a blue electroluminescence. The changes observed in the Raman spectra are unlikely due to any heating effects since at lower temperatures, where the molecule is more non planar, the Raman intensity ratio I_{1280}/I_{1220} is enhanced compared to room temperature, where the molecule on the average is more planar. [4]

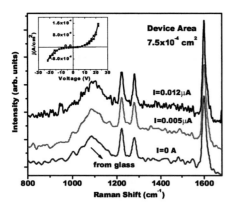

FIGURE 1. Raman spectra from a working PHP OLED. The inset shows the current density vs. voltage.

The doublet in the 1600 cm^{-1} region also shows subtle changes upon current injection. With increasing current, the overtone mode at 1615 cm^{-1} gains intensity compared to the 1600 cm^{-1} peak. These changes in the Raman spectrum of PHP with charge injection can be interpreted in terms of a more benzenoid conformation of the phenyl rings. This is in contrast to chemical doping where the phenyl rings tend toward a more planar quinoid conformation. [5]

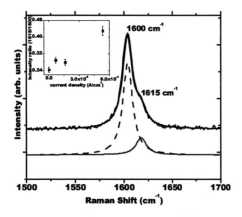

FIGURE 2. The 1600 cm^{-1} region is fit to two peaks at 0V. The inset shows the relative intensity ratio of the 1615 to 1600 cm^{-1} peak vs. current density.

CAPACITANCE MEASUREMENT

Figure 3 shows 1/capacitance^2 and conductance vs. voltage from the above device. The peak at 5 V signifies presence of interface states. The Fermi level is resonant with the interface states for a bias voltage of 5 V and is partly filled. At this voltage the Fermi level is modulated and charge flows into and out of the interface states. This is equivalent to a capacitance. When the bias is further reduced, the Fermi level is completely above the interface states and modulating it has no effect on the amount of charge present.

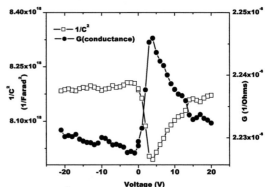

FIGURE 3. 1/C^2 and conductance vs. voltage from a PHP OLED.

In summary, increasing I_{1280}/I_{1220} Raman intensity ratio with current injection in a PHP OLED shows a more non planar benzenoid conformation of the phenyl rings. CV measurements show the presence of interface states.

ACKNOWLEDGMENTS

SG acknowledges the financial support of the Petroleum Research Fund Grant No. 35735-GB5 and the University of Missouri Research Board. Work at SMSU is funded by the Office of Naval Research Award # N00014-03-0893.

REFERENCES

1. For a review see *Handbook of Conducting Polymers*, edited by T.A. Skotheim, R.L. Elsenbaumer and J.R. Reynolds, Marcel Dekker, Inc. (1997).
2. S. Guha *et al.*, *Phys. Rev. Lett.* **82**, 3625 (1999).
3. M. Rumi and G. Zerbi, *Chem. Phys.* **242**, 123 (1999).
4. S. Guha *et al.*, *J. Phys. Chem. A* **105**, 6203 (2001).
5. J.L. Bredas *et al.*, *Phys. Rev. B* **29**, 6761 (1984).

Infrared Ultrafast Optical Probes of Photoexcitations in π-Conjugated Organic Semiconductors

C. X. Sheng and Z. V. Vardeny

Department of Physics, University of Utah, Salt Lake City, Utah 84112

Abstract: We measured the ultrafast dynamics of photoexcitations in a variety of semiconductor π-conjugated polymer films and solutions, in the spectral range from 0.13 eV to 1.05 eV and time domain from 100 fs to 800 ps. The measurements were made in the low signal regime, where the relative changes in transmission, $\Delta T/T < 10^{-5}$. In pristine poly [2-methoxy-5(2'-ethyl-hexyloxy)-p-phenylene vinylene] (MEH-PPV) solution we found that the primary photoexcitations are excitons with a photoinduced absorption (PA) band PA_1 at 1.0 eV. In pristine MEH-PPV film, however we found two PA bands, PA_1 at 1.0eV and P_1 at 0.3eV. We consider P_1 to be due to a polarons that are photogenerated intrinsically via defects and/or impurities. When following the transient dynamics of PA_1 in C_{60}-doped PPV-based films we could easily measure the dynamics of exciton dissociation from the polymer chains onto the C_{60} molecules. We found the charge transfer time to range from about 100 fs to 100 ps, depending on the concentration of C_{60} molecules in the films.

INTRODUCTION

Photoexcitations in π-conjugated polymers have been extensively investigated both theoretically and experimentally because of the surge of potential optoelectronic applications for these materials [1,2]. In addition, films of polymers/C_{60} mixtures, and especially MEH-PPV/C_{60} composites have been also extensively studied, driven by their possible use in photovoltaic devices [3,4]. This interest calls for detailed ps transient spectroscopy measurements, especially in the mid-IR spectral range, to study the role of excitons and polarons in the earliest events following photon absorption.

Here we report on the fs time-resolved transient spectroscopy in the mid IR range from 0.13 to 1.05 eV, in pristine and C_{60}-doped semiconductor polymer films and solutions that include MEH-PPV and MEH-PPV/C_{60} composites. We found that excitons are the primary photoexcitations. However polarons can also be generated at early times due to imperfections. In C_{60}-doped films we measured the charge transfer time from the polymer chains onto the C_{60} molecules, which we found to depend on the concentration of the C_{60} molecules in the polymer/C_{60} mixture.

EXPERIMENTAL

For our transient photomodulation (PM) measurements we used the fs two-color pump-probe correlation technique with linearly polarized light beams. The ultrafast laser system used was a 100 fs titanium-sapphire oscillator operating at a repetition rate of about 80 MHz, which pumped an optical parametric oscillator (OPO). The OPO generates signal and idler beams that were used as probes at photon energy ω_S and ω_I ranging between 0.55 and 1.05 eV. These two beams were mixed in a nonlinear crystal (AgGaS$_2$) to generate probes at $\omega = \omega_S - \omega_I$ in the spectral range of 0.13 to 0.43 eV. The pump beam was the second harmonic of the fundamental at 3.2eV with energy/pulse of about 0.1 nJ.

The semiconductor polymer films were obtained by dissolving the polymers in toluene solution and subsequently drop casting onto CaF$_2$ substrates. MEH-PPV/C_{60} composites were obtained by dissolving the polymer and C_{60} powders (1:1 by weight) in toluene.

RESULTS AND DISCUSSION

The transient PM spectra of pristine MEH-PPV film and solution are shown in Fig. 1. The PM spectrum in Fig. 1 contains two PA bands for the film but only one PA band for the solution. Since polarons cannot be generated in solution, where the polymer chains are isolated, we thus conclude that the primary excitations in semiconductor polymers are excitons, with PA_1 at 1.0 eV. It is noteworthy that excitons in MEH-PPV have another PA band in the near ir range (ca. 1.6 eV), which, however cannot be reached with our present set-up. In solution, all probe wavelengths share the same dynamics that are longer than in films. In films, however there is a new PA band at 0.3eV that does not have the same dynamics as PA_1, which according to steady state PA measurements [4] can be attributed to the lower polaron PA band, P_1. Since the optical properties of MEH-PPV film are extremely sensitive to imperfections such as impurities and defects[4], we attribute the difference in the early-time photoexcitations between solution and film to a defect- and/or impurity- induced charge transfer in the film.

by following the transient dynamics of PA_1. This band disappears in about 1 ps (Fig. 2(b) inset) when the charge transfer process is completed. In C_{60}-doped MEH-PPV the lower polaron band P_1 is at ca. 0.4 eV [4], and its lifetime is much longer (Fig. 2(a)) compared with PA_1 of excitons in pristine MEH-PPV (Fig. 1). We also note that in 10% C_{60}-doped film (not shown here) the charge transfer time is longer than in 50% C_{60} doped film, and occurs within 10 ps [5]. The other noticeable spectral feature in Fig. 2(a) is the dip at ca. 0.15 eV that is due to infrared active vibrational (IRAV) modes [1,4] that are a unique signature of charges induced on the polymer chains; this also support the ultrafast charge transfer in MEH-PPV/C_{60} composites.

CONCLUSIONS

We report the ultrafast spectroscopy of photoexcitations in pristine MEH-PPV film and solution, and in MEH-PPV/C_{60} composites, in the spectral range from 0.13 to 1.05 eV. We found that excitons are the primary photoexcitations in single polymer chain. However polarons can also be photogenerated at early time in films, but we consider this process to be extrinsic in nature. In C_{60}-doped PPV-based films we found that the charge transfer dynamics depends on the concentration of the C_{60} molecules in the composite.

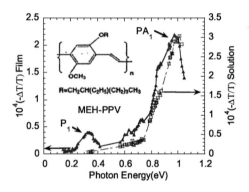

FIGURE 1. Transient PM spectrum of pristine MEH-PPV film (triangles) and solution (squares) at t = 0. MEH-PPV repeat unit is shown in the inset.

In Fig. 2 we show that in 50% C_{60}-doped MEH-PPV film we can measure the dissociation time of excitons from the polymer chains onto C_{60} molecules

ACKNOWLEDGMENTS

We thank Matt DeLong for synthesizing some of the polymer used here. This work was supported in part by the NSF (DMR 02-02790).

REFERENCES

1. D. Moses et al., Phys. Rev. B **54**, 4748 (1996); P. B. Miranda et al. Phys. Rev. B **64**, 181201(2001); D. Moses et al., Phys. Rev. B **61**, 9373 (2000)

2. S.V. Frolov, et al., Phys. Rev. B **65**, 205209 (2002); E. Hendry, et al., Phys. Rev. Lett. **92**, 196601 (2004)

3. S. Alem, et al., Appl. Phys. Lett. **84**, 2178 (2004); U. Mizrahi, et al. Synth. Metals **102**, 1182(1999)

4. X. Wei, et al., Phys. Rev. B **53**, 2187 (1996)

5. S. V. Frolov, et al., Chem. Phys. Lett., **286**, 21(1998)

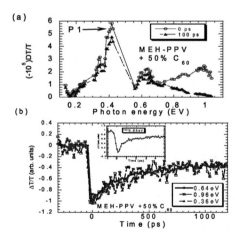

FIGURE 2. (a) Transient PM spectra of C_{60}/MEH-PPV film at t = 0 (squares) and t = 100 ps (triangles). (b) PA decay dynamics at various probe energies. The inset to (b) is the ultrafast PA_1 decay measured at 0.96 eV.

Nano-scale organic FET fabricated with carbon nanotubes

Kazunaga Horiuchi*, Tomohiro Kato, Michika, Mochizuki, Shinobu Hashii, Akira Hashimoto, Takahiko Sasaki, Nobuyuki Aoki and Yuichi Ochiai

Advanced Research Lab., Corporate Research Center, Fuji-Xerox Co. Ltd., 430 Sakai, Nakai, Ashigarakami-gun, Kanagawa 259-0157, Japan
Department of Materials Technology, Chiba University, 1-33 Yayoi-cho, Inage-ku, Chiba-city, Chiba 263-8522, Japan

Abstract. A nano-scale C_{60} field-effect transistor has been fabricated with carbon nanotubes (C_{60}CN-FET). A rope of multi-walled carbon nanotubes has been anchored by metal pads and cut by a focused Ga^{2+} ion beam ablation. The ablated ends of the rope have been integrated as electrodes into the C_{60}CN-FET, which exhibits an excellent performance of a low-voltage operation, even as small as 100 nm of channel length. The electrodes have been applied also for two-terminal conductance of salmon's DNA.

A size-reduction of device structure is sometimes needed to improve device performance of organic field-effect transistor (OFET), since a large crystal is not always available because of its thermal instability. Some preliminary results have been reported for the size-reduction, where a comb-type [1] or a wide-type [2] of electrodes has been used in order to collect enough signals. However, in our opinion, their electrodes are too elaborative to fabricate and a facile simple structure is favored for future device integrations.

In this report, carbon nanotubes (CNTs) have been applied as channel electrodes in the OFET (C_{60}CN-FET). The CNTs were cut so as to form a channel gap by focused-ion beam (FIB) ablation. Some other cutting methods have been reported, such as O_2 plasma ablation and current breakdown [3,4]. However, in the former case, it needs a lithographic patterning and chemical removal process of a photo-polymer covering. In the latter case, it is hard to control the position and width of cutting when the breakdown occurs.

A single rope of several multi-walled CNTs was placed on a substrate, which was very conductive Si wafer of about 0.005 Ω cm, but surface-insulated by a thermally grown SiO_2 of about 500 nm thick. The CNTs synthesized by an arc-discharge, were purchased from MTR Inc. in US. Au of 300 nm and Ti of 100 nm thick were deposited as metal pads, on both ends of the rope in order for a physical anchorage and an electric contact. Thermal annealing at 873 K has been also applied to them in order to reduce contact resistance, in a H_2/Ar_2 gas for a short period of 30 sec and then, a resistance of 5 kΩ for the CNT rope connected to the metal pads was obtained.

The CNT rope has been cut by ablation with a focused beam of Ga^{2+} ion using JFIB-2300 (JEOL). A precise position of beam exposure has been determined under SEM observations. The ablation process has been executed until total dosage of the ion was reached at 300×10^5 cm^{-2}, which enables us to avoid a damage of silicon surface. Separation length between the ablated ends of CNT rope could be controlled as narrow as 50 nm, which was slightly broadened in comparison with the beam diameter of about 22 nm. Non-ablated potion of the CNT rope seemed no damage without any deformation, such as curl or split.

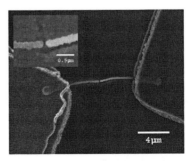

FIGURE 1. SEM image of C_{60}CN-FET. Insertion is a magnification around channel gap between the ablated ends of CNT rope and closely packed C_{60} clusters are also seen.

SE image of C_{60}CN-FET is shown in Fig.1, where C_{60} thin-film has been deposited by the thickness of about 50 nm in a high vacuum at 10^{-4} Pa, after the ablation process. Inserted image of Fig.1 clearly shows a channel gap between the ablated ends of CNT rope and the channel length seems less than 100 nm.

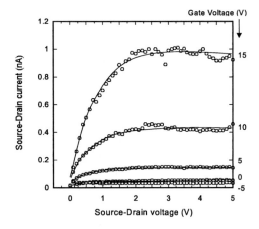

FIGURE 2. Device performance of C_{60}CN-FET. Carrier mobility can be roughly estimated at about 1×10^{-3} cm^{2}V^{-1}s^{-1}, assuming width and length of the CNT channel are 150 and 100 nm, respectively.

Device performance of the C_{60}CN-FET was shown in Fig.2. Conductance between the metal pads should be negligible, because separation length between the metal pads is much longer than that between the ablated ends of CNT rope. A distinctive plateau of saturation current has been observed in each gate voltage, which seems no short-channel effect in our C_{60}CN-FET. According to a classical FET theory, carrier mobility of this C_{60}CN-FET can be roughly estimated at 1×10^{-3} cm^{2}V^{-1}s^{-1}. Although the result is actually disappointing in comparison with that in previous C_{60}FETs [5,6], effective contact area in our estimation might be wrong due to the morphology of C_{60} crystals or shapes of the ablated ends of CNT rope.

It is worth to mention that the C_{60}CN-FET, in Fig.3, can operate by only several source-drain voltages. It have not seen in the previous C_{60}FETs [5,6,7], where considerable contact barriers [8,9] could be existed at interfaces of metal electrodes and organics. If this result comes from our usage of the CNTs, it is possibly explained that the contact barrier between the ablated ends of CNT rope and C_{60} thin-film may be negligibly smaller than that with the metal electrodes.

In Fig. 3, the ablated ends of CNT rope have been applied for salmon's DNA. Although measurement of DNA conductance has been so complicated [10,11], our fabrication offers a straightforward way for it.

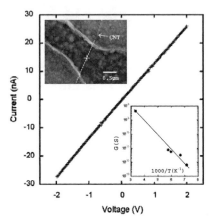

FIGURE 3. Two terminal conductance of salmon's DNA between the ablated ends of CNT rope, in various gate voltages at room temperature. Top insertion is a SEM image of the DNA thin-film covered on the CNT rope. Bottom insertion is temperature dependence in the conductance of salmon's DNA.

Unfortunately, there has been still no field-effect in the DNA conductance. Carrier transport mechanism of DNA, in turn, is under investigation in terms of temperature dependence (see bottom insert) and size-effect of channel length between the ablated ends.

REFERENCES

1. M. D. Austin and S. Y. Chou, Appl. Phys. Lett., **81**, 4431 (2002).
2. J. Collet, O. Tharaud, A. Chapoton and D. Vuillaume, Appl. Phys. Lett., **76**, 1941 (2000).
3. N. Yoneyama, E. Watanabe, K. Tsukagoshi and Y. Aoyagi, Appl. Phys. Lett., **79**, 1465 (2001).
4. P. G. Collins, M. S. Arnold, and P. Avouris, Science **292**, 706 (2001).
5. R. C. Haddon, A. S. Perel, R. C. Morris, T. T. M. Palstra A. F. Hebard and R. M. Fleming, Appl. Phys. Lett., **67**, 121 (1995).
6. K. Horiuchi, S. Nakada, S. Uchino, S. Hashii, N. Aoki, Y. Ochiai and M. Shimizu, Appl. Phys. Lett., **81**, 1911 (2002).
7. K. Horiuchi, S. Uchino, K. Nakada, N. Aoki, M. Shimizu and Y. Ochiai, Physica B **329-333**, 1538 (2003).
8. R. A. Street and A. Salleo, Appl. Phys. Lett., **81**, 2887 (2002).
9. K. Seshadri and C. D. Frisbie, Appl. Phys. Lett., **78**, 993 (2001).
10. H.W. Fink and C. Schönenberger, Nature **398**, 407 (1999)
11. D. Porath, A. Bezryadin, S.D. Vries and C. Dekker, Nature **403**, 635 (2000)

Light-Emitting Polymers: a First-Principles Analysis of Singlet-Exciton Harvesting in PPV

Marília J. Caldas[*], Giovanni Bussi[†], Alice Ruini[†] and Elisa Molinari[†]

[*]*Instituto de Física, Universidade de São Paulo, Cidade Universitária, 05508-900 São Paulo, Brazil*
[†]*INFM National Center on nanoStructures and bioSystems at Surfaces (S^3) and Dipartimento di Fisica, Università di Modena e Reggio Emilia, Via Campi 213/A, 41100 Modena, Italy*

Abstract. We study poly(*para*-phenylene-vinylene) PPV in a π-stacked crystalline configuration, through ab initio density functional techniques for the electronic structure and optical properties. We find that interchain interactions, while maintaining the quasi-1D characteristics of the lowest singlet and triplet excitons, introduces other bound excitons that should favor intersystem crossing and enhance the singlet exciton yield.

Polymer light-emitting diodes are good candidates for use in active-matrix displays, due to the possibility of blue-green light emission coupled to mechanical properties, and ease of manufacture. One of the most studied materials is poly(*para*-phenylene-vinylene) PPV, a ring-structured polymer with a natural gap in the green, a reasonably good conductor, and efficient photoluminescence. The electroluminescence EL of an organic layer faces however different problems compared to conventional semiconductors. In simple independent-particle spin statistics, neglecting electron-electron correlation, the efficiency of singlet-exciton formation from electron-hole pairing is limited to 25%. In inorganic semiconductors, where the large exciton radii and small binding energies result in quasi-degenerate singlet and triplet states, this upper limit can easily be exceeded due to efficient inter-system (triplet-singlet) crossing ISC. In organic films the charge carriers are strongly localized on one chain, and the lowest excitons (singlet and triplet) also show strongly localized nature with large binding energies of the order of tenths of an eV, thus in principle one would expect very limited EL efficiency, as indeed seen for molecular films [1]. For PPV and other polymeric films, however, the EL internal efficiency[2] $\eta_{int} = \gamma q \eta_{ST}$ has been found to be as high as 50% (here γ is the capture ratio, created excitons per injected carrier pair, η_{ST} is the singlet/triplet ratio, and q is the singlet exciton optical efficiency). Several models have been proposed to account for the excess singlet harvesting, either by predicting higher ISC crossing [3], or higher η_{ST} [4, 5, 6]. Most of the previous theoretical works focused on isolated-chain models, valid in the disordered regions of the polymer films. We here treat the particular π-stacking morphology characteristic of some highly

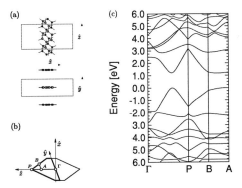

FIGURE 1. Representation of the crystalline arrangement for πS PPV, (a) real space unit cell, and (b) first Brillouin Zone BZ. In (c) we show the band structure close to the energy gap, for relevant directions in the BZ.

luminescent PPV derivatives, and thus we are able to extract information useful for this kind of modelling. We find that the interchain interaction in the stack is sufficient to introduce in the gap a second pair of singlet-triplet states, degenerate in energy, which strongly points to an ISC mechanism very similar to that proposed by Hong and Meng [6].

We show in Fig. 1 the unit cell of the π-stacked PPV chains, the associated first Brillouin zone, and the single-particle band structure. We work within density functional theory in the local approximation, with norm conserving pseudo-potentials and plane-waves basis set [7]. We see that the electronic structure is highly anisotropic, with the largest dispersion occurring along the long axis of the chain z, but there is also a noticeable dispersion along the stacking direction y. The optical properties are calculated from this LDA band states, including electron-hole correlation within a Bethe-Salpeter approach as described in [8].

The excitonic structure shown in Fig. 2 has several remarkable characteristics. The lowest singlet exciton DS is optically active, in contrast with the dark state found in herring-bone packing [8, 9]. The DS has a large binding energy of $E_B \approx 0.4$eV, is polarized along z and is direct, that is, electron and hole are localized on a single chain with a localization length characterized by FWHM of $L_{DS} \approx 22$Å; the twin state DT, the lowest triplet, is found at $E_B \approx 0.9$eV. This singlet-triplet splitting $\Delta_{ST} = 0.5$eV is large compared to the optical phonon energy ~ 0.2eV in the PPV chains [10]. The DT localization length is, as expected, smaller $L_{DS} \approx 12$Å. Most interestingly, there is also structure in the y-polarization through twin charge-transfer singlet and triplet excitons, with electron and hole are localized on different, nearest-neighbor chains; the CTS and CTT are degenerate, and $L_{CTT} \approx L_{CTS}$.

In the polymer system, ISC is usually proposed by invoking impurities or defects [3] which should not be relevant for high-purity crystalline samples. On the other hand, even at very low temperature the soft torsional phonons are active [10]. Hong and Meng [6] have pointed out, for the isolated PPV chain, that a very small torsion angle of $\sim 7°$ of the phenyl ring can bring into action sufficient ISC between DS and DT to enhance η_{ST}. Furthermore, it is reasonable to suppose that capture occurs not at the lowest DS or DT states, but at higher lying states, so that they propose capture at a higher triplet, with two competing decay channels: ISC to the lowest singlet (and optical recombination) or phonon emission to the lowest triplet (and from there competition between ISC to the lowest singlet, and non-radiative decay). However, if Δ_{ST} is large compared to the optical phonon energy, as is the case here, there occurs a phonon bottleneck virtually closing the channel from higher to lowest triplet. In the crystalline case we treat here, even supposing that capture occurs at higher, band-resonant excitons, followed by progressive decay to the lowest DS and DT, there will be another ISC channel from the CTT to the CTS, highly efficient due to the twin's degeneracy; moreover, phonon decay from the CTS to the DS should also be highly efficient due to the large localization length of the DS.

In conclusion, we find that interchain interaction, while maintaining the one-dimensional characteristics of light emission, interferes with the exciton structure at higher energies, resulting in overall increase of light harvesting for EL in packed polymer films.

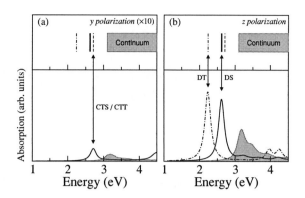

FIGURE 2. Excitons and absorption spectra for πS PPV, for polarization along the direction of the stack(a) and along the axis of the chains (b). The upper-lying graphs are energy diagrams for the lowest excitons in each system, while the lower-lying graphs show the calculated absorption spectra (obtained using a 0.1 eV lorentzian broadening. Shaded regions represent spectra calculated without inclusion of electron-hole interaction, lines show the results for the two-particle calculation. The grey box represents the continuum of dissociated transport states. Excitons and peaks are labelled as direct singlet or triplet (DS, DT), and charge transfer singlet or triplet (CTS, CTT).

ACKNOWLEDGMENTS

Computer time was partly provided by CINECA through INFM Parallel Computing Projects. We acknowledge support by the RTN EU Contract "EXCITING" No. HPRN-CT-2002-00317 and by FIRB "NOMADE", MJC also thanks INFM-S³, Italy, FAPESP and CNPq, Brazil.

REFERENCES

1. Wilson, J. S., Dhoot, A. S., Seeley, A. J. A. B., Khan, M. S., Köhler, A., and Friend, R. H., *Nature*, **413**, 828–831 (2001).
2. Friend, R. H., Gymer, R. W., Holmes, et. al., *Nature*, **397**, 121–128 (1999).
3. Frolov, S. V., Liess, M., Lane, P. A., Gellermann, W., Vardeny, Z. V., Ozaki, M., and Yoshino, K., *Phys. Rev. Lett.*, **78**, 4285–4288 (1997).
4. Shuai, Z., Beljonne, D., Silbey, R. J., and Brédas, J. L., *Phys. Rev. Lett.*, **84**, 131–134 (2000).
5. Wohlgenannt, M., Tandon, K., Mazumdar, S., Ramasesha, S., and Vardeny, Z. V., *Nature*, **409**, 494–497 (2001).
6. Hong, T.-M., and Meng, H.-F., *Phys. Rev. B*, **63**, 075206–075211 (2001).
7. Baroni, S., Dal Corso, A., de Gironcoli, S., and Giannozzi, P. (2001), http://www.pwscf.org.
8. Ruini, A., Caldas, M. J., Bussi, G., and Molinari, E., *Phys. Rev. Lett.*, **88**, 206403–206406 (2002).
9. Ruini, A., Ferretti, A., Bussi, G., Molinari, E., and Caldas, M. J., *Semicond. Sci. Technol.*, **19**, S362–S364 (2004).
10. Capaz, R. B., and Caldas, M. J., *Phys. Rev. B*, **67**, 205205–205214 (2003).

Probing Nanoscale Pentacene Films by Resonant Raman Scattering

Rui He*, Irene Dujovne*†, Liwei Chen*, Qian Miao*, Cyrus F. Hirjibehedin*†, Aron Pinczuk*†, Colin Nuckolls*, Christian Kloc† and Graciela B. Blanchet**

*Columbia Nanoscale Science and Engineering Center, Columbia University, New York, New York 10027, USA
†Bell Labs, Lucent Technologies, Murray Hill, New Jersey 07974, USA
**DuPont, Central Research and Development, P.O. Box 80356, Wilmington, Delaware 19880, USA

Abstract. Resonant enhancements of Raman scattering intensities offer the sensitivity required to study nanoscale pentacene films that reach into monolayer thickness. In the results reported here structural characterization of ultra-thin layers and of their fundamental optical properties are investigated by resonant Raman scattering from intra-molecular and inter-molecular vibrations. In this work Raman methods emerge as ideal tools for the study of physics and characterization of ultra-thin nanoscale films of molecular organic materials fabricated on diverse substrates of current and future devices.

INTRODUCTION

Organic molecular semiconductors have been receiving growing attention for their promising applications in field effect transistors. Transistors made of pentacene have reached mobilities of more than $0.1 cm^2/Vs$ [1, 2], which makes organic semiconductors potential substitutes for amorphous Si. Performance of organic devices strongly depends on the material perfection and structural ordering of the organic thin films grown on dielectric layers [3]. We have recently shown that resonance Raman scattering is an ideal tool for thin film characterization and the study of fundamental optical properties [4].

Raman scattering has several advantages in thin film characterization. It probes vibrational modes from the films almost regardless of the nature of the substrates. Furthermore, resonance enhancements of Raman intensities give insights of optical and electronic properties of materials.

In this paper we consider Raman scattering from pentacene single crystals and nanoscale films that exhibit large resonance enhancements when the outgoing photon energy overlaps the free exciton optical transitions observed in luminescence. The resonance enhancements provide the sensitivity required to study ultra-thin films with thickness reaching a single monolayer. Spectra of intra-molecular and inter-molecular (phonon) modes of the nanoscale thickness films display major differences with those measured in single crystals.

EXPERIMENTS

Two pentacene thin films with different average thickness, 65nm (**F1**) and 1.9nm (**F2**), and a pentacene single crystal (**CR**) were studied. The crystal was made by physical vapor deposition [5]. The films were grown by thermal evaporation in high vacuum on oxidized Si substrates that are kept at room temperature. A deposition rate of about 0.05nm/sec was used to grow sample **F1**. Monolayer film **F2** was grown by an evaporation rate of 0.1monolayer/min for a typical run time of 2 minutes. The AFM image shows that in **F2** the substrate is not completely covered by the pentacene islands. The average thickness of those islands is about 1.9nm, which is approximately the length of a pentacene molecule along its long axis.

Inelastic light scattering experiments were conducted at 77K. Samples were mounted in a cryostat and immersed in liquid nitrogen. A dye laser was used with the tuning range from 585 to 685nm which overlaps the fundamental optical transition in pentacene. The incident power density was kept to less than $5W/cm^2$, and the scattered light was collected and dispersed by a Spex 1404 double spectrometer with CCD multichannel detection.

RESULTS AND DISCUSSIONS

Figure 1 (a), (b), and (c) displays intra-molecular Raman scattering signals in the range of $1150-1190 cm^{-1}$

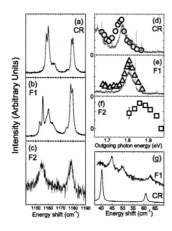

FIGURE 1. Panel (a), (b) and (c): Raman scattering of C-H in-plane bending modes from samples **CR**, **F1**, and **F2**, respectively. Panel (d), (e) and (f): The continuous traces are luminescence from **CR** and **F1**. The circles, triangles, and squares are Raman intensities of the modes at 1179, 1178, and $1178 cm^{-1}$ in **CR**, **F1**, and **F2**, respectively. Panel (g): Comparison of lattice modes in sample **CR** and **F1**. All data are at 77K.

from samples **CR**, **F1**, and **F2**. Vibrational modes in this energy range are assigned to C-H in-plane bending [6]. Sample **F1** displays more Raman modes between 1150 and $1170 cm^{-1}$ than the crystal **CR**. This could be due to the emergence of different polymorphs in the film [7]. Inhomogeneity and disorder of local environments could break the intrinsic symmetry of pentacene molecules, and may also contribute to the appearance of additional Raman lines in sample **F1**. For sample **F2** the signal is much weaker and noisy. Peak doublets are not resolved in **F2**. The larger mode broadening in **F2** could be due to disorder and interaction between pentacene molecules and the substrate.

Resonance profiles of Raman scattering intensities as functions of outgoing photon energies by C-H in-plane bending modes are shown in Fig. 1 (d), (e), and (f) for the three samples. The continuous traces are luminescence from **CR** and **F1**. It can be seen that the strongest enhancements of Raman intensities occur when the outgoing photon energies overlap the free exciton emissions observed as luminescence bands in sample **CR** and **F1**. The free exciton transition exhibits a blue-shift of about 0.05eV in sample **F1** compared to crystal **CR**. The blue-shift could be linked to the different crystalline structures in the thin film [7]. In the monolayer sample **F2**, the free exciton luminescence band is not observed, but there is still a well-defined peak in the resonance Raman profile. One can conjecture that the free exciton transition in the monolayer film could also happen at the energy where the Raman intenstiy has the strongest enhancement. This implies that resonance Raman scattering is an alternative way to explore material electronic properties.

By using a wavelength close to the strongest enhancement in the resonance profile (free exciton transition energies), one can observe the inter-molecular (lattice) modes in nanoscale thin film **F1**. Figure 1 (g) displays lattice modes of sample **F1** and **CR** in the region of 35-$70 cm^{-1}$. One can see that thin film **F1** shows additional lattice vibrational modes that are absent in the crystal **CR**. The mode at $63 cm^{-1}$ shifts by about $2 cm^{-1}$ to higher energy compared to **CR**. The lattice modes in **F1** are significantly broadened in comparison to the single crystal phonon modes. More disorder, inhomogeneity, and grain boundaries effects in the thin film could be responsible for the broadening of the phonon modes.

Resonance enhancements of Raman scattering signals provide the sensitivity required to study vibrational modes in nanoscale films with thickness reaching one monolayer. The sensitivity required does not seem to depend on the choice of substrates or dielectric layers. Under the resonance condition, lattice modes can be observed in thin films less than 100nm thick. Resonance Raman scattering is an ideal method for thin film characterization, device characterization, study of low dimensional behavior and interface effects.

ACKNOWLEDGMENTS

This work was supported primarily by the Nanoscale Science and Engineering Initiative of the National Science Foundation under NSF Award Number CHE-0117752 and by the New York State Office of Science, Technology, and Academic Research (NYSTAR), and by a research grant of the W. M. Keck Foundation. L. Chen thanks A. Schrott for help with sample preparation at the IBM T. J. Watson Research Center.

REFERENCES

1. Nelson, S. F., Lin, Y.-Y., Gundlach, D. J., and Jackson, T. N., *Appl. Phys. Lett.*, **72**, 1854–1856 (1998).
2. Butko, V. Y., Chi, X., Lang, D. V., and Ramirez, A. P., *Appl. Phys. Lett.*, **83**, 4773–4775 (2003).
3. Blanchet, G. B., Fincher, C. R., and Malajovich, I., *J. Appl. Phys.*, **94**, 6181–6184 (2003).
4. R. He, *et al.*, *Appl. Phys. Lett.*, **84**, 987–989 (2004).
5. Laudise, R. A., Kloc, C., Simpkins, P. G., and Siegrist, T., *J. Cryst. Growth*, **187**, 449–454 (1998).
6. Jentzsch, T., Juepner, H. J., Brzezinka, K.-W., and Lau, A., *Thin Solid Films*, **315**, 273–280 (1998).
7. C. C. Mattheus, *et al.*, *Synthetic Metals*, **138**, 475–481 (2003).

Control of amino-acid electronic structures on semiconductor surfaces

Masato Oda and Takashi Nakayama

Department of physics, Chiba University, 1-33 Yayoi, Inage, Chiba 263-8522, Japan

Abstract. Electronic structures of amino acids on the Si(111) surfaces are investigated by using ab-initio Hartree-Fock calculations. It is shown that among various ionic amino acids a histidine is the only one that can be positively ionized when hole carriers are supplied in the Si substrate, by transferring the hole charge from the Si substrate into an amino acid. This result indicates that the ionization of a histidine, which will activate the protein functions, can be controlled electrically by producing amino-acid/Si junctions.

INTRODUCTION

Recent developments of nanotechnology have enabled to grow bio-molecules on semiconductor surfaces and opened the possibility of bio-devices[1,2]. Among various bio-molecules, proteins are located at the special position because they possess catalysis functions that induce almost all of chemical reactions in living bodies. These functions are revealed when some amino-acid residues inside the protein are ionized by changing the solvent pH and when such ionization induces the transformation of protein shapes by way of the Coulomb interaction. As shown in Fig.1, if we arrange the protein on the semiconductor surface and inject the charged carriers into the protein by doping the semiconductor, we can electrically control the ionization of amino acids and thus the conformation of proteins, which might produce new types of bio-devices. The purpose of this study is to theoretically investigate the possibility of such carrier injections, by considering joint systems of proteins and semiconductors.

METHODS

To model the joint system of a protein and a semiconductor, we selected a single amino-acid branch in the protein and arranged it on the Si(111) surface. In this study, we considered five ionic amino acids; aspartic acid (D), glutamic acid (E), lysine (K), arginine (R), and histidine (H). We assumed that there is a covalent bond between the amino acid and the surface-top Si atom. The Si_8H_{14} cluster was adopted as a silicon surface and the joint system was placed in the vacuum. All calculations were performed by the ab-initio Hartree-Fock method using the Gaussian-98 package. STO-3G and 6-31G** bases were employed for the geometry optimization and the electronic structure calculation, respectively.

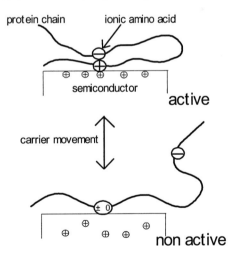

FIGURE 1. Schematic picture of the electric control of protein's activity and conformation, by arranging a protein on the semiconductor surface and injecting carriers from the semiconductor into the amino acid part of the protein.

RESULTS AND DISCUSSIONS

First we compare electronic structures of isolated neutral amino acids with that of an isolated neutral Si cluster and estimate the possibility of carrier injection. Figure 2 shows calculated energy levels of the highest occupied molecular orbitals (HOMOs) and the lowest unoccupied molecular orbitals (LUMOs). It is seen that the LUMO energies of all amino acids are higher than that of the Si cluster. This result indicates that it is difficult to negatively ionize the present amino acids by connecting to the n-type-doped Si substrate. On the other hand, the HOMO energies of R and H acids are higher than that of the Si cluster. Thus, when the hole carrier is supplied into the Si substrate and such Si substrate is connected to the R or H acid, the hole carrier is expected to move from Si to the acid.

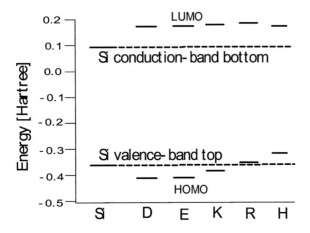

FIGURE 2. Calculated HOMO and LUMO energy levels of Si cluster and five ionic amino acids. D,E,K,R and H indicate an aspartic acid, a glutamic acid, a lysine, an arginine, and a histidine, respectively.

To check the above-mentioned estimate of the carrier injection, we then constructed the joint systems of amino acids and the Si cluster as shown in Fig.3, and calculated the hole-charge distribution in these joint systems by using the Mulliken-charge analysis. In the case of a histidine/Si joint system, 60% of hole charge is localized around the imidazole ring and 27% is localized around Si cluster. On the other hand, in the case of an isolated histidine, the corresponding hole charge around the imidazole ring is 85%. Thus, we can see about 25% of hole charge is transferred from the histidine to the surface-top Si atom by producing the histidine-Si covalent bond. However, the supplied hole charge is mainly localized around the imidazole ring of a histidine for both an isolated histidine and a joint histidine/Si system, which indicates that one can control the ionization of a histidine by injecting hole carriers from the Si substrate. On the other hand, in the case of an arginine/Si joint system, the hole charges of about 62% is extended into the Si substrate. This is because the difference of HOMO energies between an arginine and the Si substrate is too small, around 0.01 Hartree. Therefore, one can conclude that when the ionic amino acids are arranged on the Si surface, the ionization is only possible for a histidine.

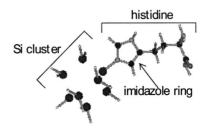

FIGURE 3. Schematic view of the joint system of a histidine amino acid and the Si cluster.

CONCLUSION

We investigated the carrier injection into the ionic amino acids when the amino acids are arranged on the Si(111) substrate and the carriers are supplied into the Si substrate. By comparing the HOMO energies and analyzing the hole-charge densities, we found that only the positive ionization of a histidine can be electrically controlled when a histidine is adsorbed on the Si(111) surface.

ACKNOWLEDGMENTS

This work was supported by the CREST program of JST, Japan, the Futaba Memorial Foundation, and the 21COE program of Chiba University. We also thank the Super Computer Centers, ISSP, University of Tokyo and Chiba University for the use of facilities.

REFERENCES

1. E. R. Goldman, E. D. Balighan, H. Mattoussi, M. K. Kuno, J. M. Mauro, P. T. Tran, and G. P. Anderson, J. Am. Chem. Soc. **124** 6347 (2002)

2. F. Pinaud, D. King, H. P. Moore, and S. Weiss, J. Am. Chem. Soc. **126** 6115 (2004)

Transport measurements of DNA molecules by using carbon nanotube nano-electrodes

Takahiko K. Sasaki, Asato Ikegami, Michika Mochizuki, Nobuyuki Aoki, and Yuichi Ochiai

Center for Frontier Electronics and Photonics & Department of Materials Technology, Chiba University
1-33 Yayoi, Inage, Chiba 263-8522, Japan

Abstract. We have studied to try on direct measurement of electron transport between DNA molecules by means of carbon nanotube (CNT) as a nano-electrode. The CNT electrodes were fabricated by focused ion beam bombardment (FIBB). Very short channel having about 50nm gap was easily formed between the severed CNT. Poly (dG)-poly (dC) DNA molecules containing identical base pairs were positioned between the CNT nano-electrodes by electrostatic trapping method. The current-voltage (I-V) characteristics showed a very short hopping distance of 3.8 Å and were similar to the distance between general base pairs (~3.4 Å) of poly (dG)-poly (dC) DNA molecules. We also measured the gate-voltage dependence in the I–V characteristics and found that poly (dG)-(dC) DNA molecules exhibited p-type conduction. It confirms that the current seems to flow really through the DNA molecules.

INTRODUCTION

In present, many investigations for DNA molecules by use of nanotechnology methods give a significant implication for the application to the electronic devices and DNA-based electrochemical biosensors [1-4]. On the other hand, Carbon nanotube (CNT) has been proposed as a candidate transducer material for application into biosensors [5-7]. Here, we want to propose a planar biological sensor using CNT nano-electrodes and a standard semiconductor integrated fabrication technique, in which the detection can be monitored electrically. In this study, we have performed a direct measurement of the electrical transport between DNA molecules using CNT nano-electrodes.

SAMPLE PREPARATION

The studied CNTs were multiwalled type (the diameter: about 50 nm) which is synthesized by arc discharge method. CNT nano-electrodes were obtained in the process by using focused ion beam bombardment (FIBB) as shown in Fig. 1 (a). Firstly, the solution containing CNTs was dispersed on SiO₂ substrate using spin rotational method. The electrical contacts were made using Ti/Au pads (15 nm / 65 nm) and a lift-off process by means of a high alignment photolithography system. The distance of the electrodes was 4 μm. The FIBB has been carried out at the central part of CNT as shown in Fig. 1 (a). It was performed by Ga ion beam.

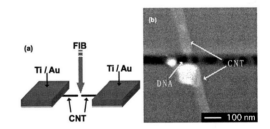

FIGURE 1. (a) A schematic image of the fabrication of CNT nano-electrode. (b) The SEM image of trapped DNA molecules between CNT nano-electrodes.

A very short channel, 50nm gap, was easily formed between sharply cut edges of the electrode CNT as shown in Fig. 1 (b). This structure corresponds to two closely spaced nano-electrodes. DNA molecules were located between CNT nano-electrodes by electrostatic trapping [3, 8] from a dilute solution of DNA containing about 1mg/mℓ. In order to trap DNA molecules by placing a drop of solution of suspended DNA molecules on top of the CNT nano-electrodes, we applied radio waves of the frequency 2MHz, near the electrodes. In this way, DNA molecules can be trapped between the electrodes as shown in Fig. 1 (b). The measured DNA molecules were poly (dG)-(dC) containing identical base pairs. After trapping the DNA molecules between CNT electrodes, the samples were dried in vacuum by pumping.

Results and discussion

I–V characteristics of the DNA molecules have been studied at room temperature. Transport measurement has been carried out in vacuum for all cases. The solid line of Fig. 2 (a) shows the *I–V* curve measured at room temperature on poly (dG)-(dC). The *I-V* curve shows clearly a nonlinear dependence. We assume a contribution of a long-distance charge transport [9], which comes from a *G* radical cation to a *G* -rich sequence in poly (*dG*)-poly (*dC*) DNA molecules. In order to discuss the *I–V* characteristic of poly (d*G*)-(d*C*) DNA molecules, we tentatively carried on curve fitting based on small polaron hopping model [3, 10] as shown in dotted curve in Fig. 2. In this case, the activation energy (0.12eV) was referred from [3] related for our curve fitting. The hopping distance can be estimated to be 3.8Å at room temperature. It is almost corresponding to the distance between general base pairs (∼3.4 Å). Furthermore, DNA molecules, poly (*dG*)-poly (*dC*), have identical base pairs, it should be clear that our result is applicable to the long-distance charge transport mechanism from a guanine (*G*) radical cation to a *G*-rich sequence [9] based on the model of small polaron hopping. More details are required to discuss this analysis, if we can obtain clear results on an oxidative damage and on temperature dependence of the *I–V* characteristic. Figure 2 (b) shows the *I–V* curves measured at room temperature with various back-gate voltages, V_{gate}, for poly (*dG*)-poly (*dC*). The inset is the schematic diagram of the same measurement. The current seems to be a small depletion by applying a positive V_{gate} or a slightly enhancement by a negative V_{gate}, implying that poly (*dG*)-poly (*dC*) acts as a *p*-type semiconductor. This supports the assumption that there are many positive charge carriers in *G*-rich sequences. From above results, we can conclude that the direct transport measurements of DNA molecules have been successfully achieved using nano-electrodes of CNTs. Thus, it may be needed on further electrical measurements for DNA molecules using this method based on CNT nano-electrodes and we must make more discussions or a development of new ultra-high sensitive electronic biosensors utilizing maskless fabrication technique and controlling an interface between DNA and CNT.

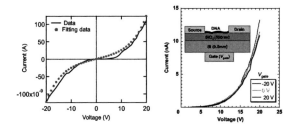

FIGURE 2. (a) The solid curve is *I–V* characteristic of poly (d*G*)-(d*C*) DNA molecules between two CNT electrodes at room temperature. Doted curve is the calculated one based on the small polaron hopping model [3, 10]. The parameter was referred from [3] in this study. (b) The *I-V* curves measured at room temperature for various values of the back-gate voltage (V_{gate}) for poly (d*G*)-(d*C*). The inset is the schematic diagram of electrode arrangement for gate dependent transport experiments.

This work is supported in part by Grant-in-Aid for Scientific Research, No. 13450006 and 14655003.

REFERENCES

1. Denny Porath, Alexey Bezryadin, Simon de Vries, and Cees Dekker, *Nature* **403**, 635 (2000).

2. P. Tran, B Alavi and G. Gruner. *Phys. Rev. Lett.* **85**, 1564 (2000).

3. K. –H. Yoo, D. H. Ha, J. –O. Lee, J. W. Park, Jinhee Kim, H –Y. Lee, T. Kawai, and Han Young Choi, *Phys. Rev. Lett.* **87**, 198102-1 (2001).

4. Lintao Cai, Hitoshi Tabata, and Tomoji Kawai, *Appl. Phys. Lett.* **77**, 3105 (2000).

5. Y. Cui, Q. Wei, H. Park, C. M. Leiver, *Science* **293**, 1289 (2001).

6. Yuehe Lin, Fang Lu, Yi Tu, Zhifeng Ren, *Nano Letters* **4**, 191 (2004).

7. C. R. Martin, *Science* **293**, 1289 (2001).

8. A. Bezryyadin, C. Dekkaer, and G. Schmid, *Apple. Phys. Lett.* **71**, 1273 (1997).

9. Brend Giese, *Acc. Chem. Res.* **33**, 631 (2000).

10. H. Böttger and V. V. Bryksin: Hopping Conduction in Solids (Akademie-Verlag, Berlin, 1985).

Electronic Transport in DNA — the Disorder Perspective

Daphne K. Klotsa, Rudolf A. Römer, Matthew S. Turner

Physics Department and Centre for Scientific Computing, University of Warwick, Coventry CV4 7AL, U.K.

Abstract. We are investigating the electronic properties of DNA by looking at a tight-binding model and four DNA sequences. The charge transfer is studied in terms of localisation lengths as a function of Fermi energy and backbone disorder, the latter accounting for different environments. We have performed calculations on poly(dG)-poly(dC), telomeric DNA, random-ATGC DNA, λ-DNA and find that random-ATGC and λ-DNA have localization lengths allowing for electron motion among a few dozen base pairs only, whereas for telomeric and poly(dG)-poly(dC) DNA they are much larger. Enhancement of localisation lengths is observed at particular energies for an increasing binary backbone disorder.

The question on whether DNA can conduct electricity, and if so how this can be utilized, has been a subject of discussion particularly since direct experimental results became available [1]. Part of the motivation for such studies is the potential use of DNA in nanotechnology and also the possibility of DNA damage-repair mechanisms via electron transfer [2]. Various experiments, models and ideas exist that aim to describe its electronic transport properties and these have recently been reviewed in Refs. [3, 4]. Despite the enhanced activity in both experimental and theoretical studies, the complexity of DNA is still preventing us from forming a consistent understanding.

In most models [5, 6] it has been assumed that electronic transport takes place along the long axis of the DNA molecule and that the conduction path is due to π-orbital overlap between consecutive bases; density-functional calculations [7] have shown that the bases, especially Guanine, are rich in π-orbitals. Quantum mechanical approaches to the problem use standard one-dimensional (1D) tight-binding models [8]. Of particular interest to us is a 1D model [5] which includes the backbone structure of DNA explicitly and exhibits a semiconducting gap.

In this paper, we extend this 1D model to a biologically more relevant ladder structure, as shown in Fig. 1 and study its electronic properties for various DNA sequences, including poly(dG)-poly(dC), telomeric, random-ATGC and λ- (bacteriophage) DNA. Our approach uses standard transfer-matrix techniques [9] which give estimates of the localisation lengths of a single electronic excitation averaged over the DNA strand at zero temperature. The ladder model is a planar projection of the structure of the DNA-the helix being unwound. There are two connected central branches with sites that represent the DNA bases. These central branches are the

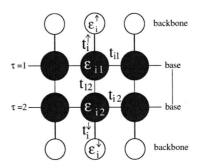

FIGURE 1. Ladder model of DNA corresponding to the Hamiltonian (1).

π-orbital conduction pathways for the electrons and their sites are additionally linked to upper and lower sites corresponding to the upper and lower backbones; backbone sites are not interconnected. The Hamiltonian for the ladder model is given by:

$$H = \sum_{i=1}^{L} \left(\sum_{\tau=1,2} t_{i,\tau}|i,\tau\rangle\langle i+1,\tau| + \varepsilon_{i,\tau}|i,\tau\rangle\langle i,\tau| \right.$$
$$\left. + \sum_{q=\uparrow,\downarrow} t_i^q|i,q\rangle\langle i,q| + t_{1,2}|i,1\rangle\langle i,2| + \varepsilon_i^q|i,q\rangle\langle i,q| \right)$$
$$+ h.c. \qquad (1)$$

where $t_{i,\tau}$, $\tau=1,2$, is the hopping amplitude along the τ-branch, t_i^q gives the hopping to the upper ($q=\uparrow$) and lower ($q=\downarrow$) backbone, $t_{1,2}$ represents the hopping between the two central branches; $\varepsilon_{i,\tau}$ is the onsite potential energy on each site along the two central branches and ε_i^q gives the onsite energy at the sites of the backbone.

In Ref. [5], it has been argued that $t_{i,\tau} \approx t_i^q/2 \sim 0.5\text{eV}$ can describe experimental results in poly(dG)-poly(dC) DNA for a simplified one-chain version of the ladder

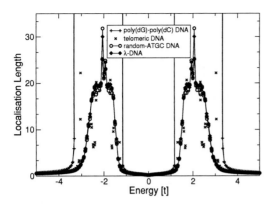

FIGURE 2. Localisation lengths as a fucntion of energy for poly(dG)-poly(dC), telomeric, random-ATGC, and λ-DNA as described in the text. The energy is measured in units of hopping strength between like base pairs. The spectrum is symmetric in energy. Lines are guides to the eye only.

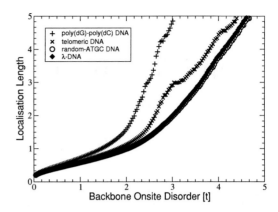

FIGURE 3. Localisation lengths as a function of binary backbone disorder for poly(dG)-poly(dC), telomeric, random-ATGC, and λ-DNA at energy $E = 0$. The disorder corresponds to a situation when 50% of all backbone sites are occupied, e.g., by a salt ion.

model. Semi-empirical calculations on DNA base pairs using the SPARTAN package have shown that the wavefunction overlap between the individual bases of a base pair is weak and therefore we take it to be $t_{1,2} = t_{i,\tau}/10$. We make the further assumption that the wavefunction overlap between consecutive bases along the DNA strand is weaker between unlike and non-matching bases, for which we thus choose $t_{i,\tau} = t_i^q/4$, whereas $t_{i,\tau} = t_i^q/2$ between identical and matching bases (i.e. AT/TA, GC/CG, AA, TT, GG, CC). Initially, all small onsite potential fluctuations due to differences in base-ionization energies are ignored. In Fig. 2, the energy dependence of the localisation lengths computed using model (1) is shown for four different sequences of DNA, namely, poly(dG)-poly(dC) DNA with 10,000 base pairs, a telomeric DNA (repetitions of pattern TTAGGG truncated at 6000 base pairs), random-ATGC DNA with 10,000 bases and λ-DNA (bacteriophage) with 48,502 bases. As expected, poly(dG)-poly(dC) and telomeric DNA give rise to perfect conductivity, due to their periodic electronic structure, within small minibands centered around $E = 0$. On the other hand, random-ATGC and λ-DNA give two bands with finite localisation lengths in the energy regions $(-4, -1)$ and $(1, 4)$. The localisation lengths, which roughly equal the average distance an electron would be able to travel (conduct), are close to the distance of 20 bases within the band, with a maximum of ~ 30 bases at the centre of each band.

In vivo and most experimental situations, DNA is exposed to diverse environments. The solution, the thermal effects and the available space (causing the DNA to bend) are factors that alter the structure and properties that one is measuring [10, 11]. Here, we model this by introducing various types of disorder into the hopping (t) and onsite-energy parameters (ε). In general, this leads to a reduction of the localisation lengths and a gradual filling of the gap. A particularly intriguing result emerges when we use *binary* onsite disorder at the backbone in order to model the random adhesion of saline solvents [11]. In Fig. 3, we show that at energy $E = 0$, the localisation length increases with increasing disorder. This is also true for the model of Ref. [5]. Thus, it appears that adding binary disorder leads to *partial delocalisation*.

REFERENCES

1. H.-W. Fink and C. Schonenberger, Nature **398**, 407 (1999); D. Porath, A. Bezryadin, S. Vries, and C. Dekker, Nature **403**, 635 (2000).
2. H. Park, S. Kim, A. Sancar, and J. Deisenhofer, Science **268**, 1866 (1995); P. O'Neil and E. M. Fielden, Adv. Radiat. Biol. **17**, 53 (1993); J. Retel et al., Mutation Res. **299**, 165 (1993);
3. D. Porath, G. Cuniberti, and R. Di Felice, Topics in Current Chemistry **237**, 183 (2004).
4. C. Dekker and M. A. Ratner, Physics World **14**, 29 (2001).
5. G. Cuniberti, L. Craco, D. Porath, and C. Dekker, Phys. Rev. B **65**, 241314 (2002).
6. J. Zhong, (private comminucation); R. Bruinsma, G. Gruner, M. R. D'Orsogna, and J. Rudnick, Phys. Rev. B **85**, 4393 (2000); S. Priyadarshy, S. M. Risser, and D. N. Beratan, J. Phys. Chem. **100**, 17678 (1996).
7. P. J. Pablo et al., Phys. Rev. Lett. **85**, 4992 (2000).
8. S. Roche, Phys. Rev. Lett. **91**, 108101 (2003); W. Zhang and S. E. Ulloa, Phys. Rev. B **69**, 153203 (2004); Microelectronics Journal **35**, 23 (2004); S. Roche, D. Bicout, E. Macia, and E. Kats, Phys. Rev. Lett. **91**, 228101 (2003); H. Wang, J. P. Lewis, and O. F. Sankey, Phys. Rev. Lett. **93**, 016401 (2004).
9. B. Kramer and A. MacKinnon, Rep. Prog. Phys. **56**, 1469 (1993).
10. Z. Yu and X. Song, Phys. Rev. Lett. **86**, 6018 (2001).
11. R. N. Barnett et al., Science **294**, 567 (2001).

Molecular Signature in the Photoluminescence of α-Glycine, L-Alanine and L-Asparagine Crystals: Detection, *ab initio* Calculations, and Bio-sensor Applications

E. W. S. Caetano[1], J. R. Pinheiro[1], M. Zimmer[1], V. N. Freire[1], G. A. Farias[1], G. A. Bezerra[2], B. S. Cavada[2], J. R. L. Fernandez[3], J. R. Leite[3], M. C. F. de Oliveira[4], J. A. Pinheiro[4], J. L. de Lima Filho[5] and H. W. Leite Alves[6]

[1]*Departamento de Física, UFC, C. P.: 6030, CEP: 60.455-900, Fortaleza, CE, Brazil*
[2]*Departamento de Bioquímica, LabMol, UFC, CEP: 60.455-900 Fortaleza, CE, Brazil*
[3]*Instituto de Física, USP, C.P.: 66318, CEP: 05.315-970 São Paulo, SP, Brazil*
[4]*Departamento de Química Orgânica e Inorgânica, UFC, CEP: 60.451-970 Fortaleza, CE, Brazil*
[5]*Laboratório Imunopatologia Keizo Azami, UFPE, CEP: 50.670-901 Recife, PE, Brazil*
[6]*Departamento de Ciências Naturais, UFSJ, C. P.: 110, CEP: 36.300-000 São João del Rei, MG, Brazil*

Abstract. We present the photoluminescence spectra of α-glycine, L-alanine, and L-asparagine crystals. They are broad and structured, comprising green to ultraviolet emission in the 1.75-3.60 eV range. Absorption measurements show that the band gap energies of the crystals are of the order of 5.0 eV. *Ab initio* calculations of their electronic structures allow for the assignment of the observed peaks in the visible region to lattice-related processes of exciton nature associated with polaron levels. The very thin photoluminescence peaks in the ultraviolet region are assigned to intramolecular transitions, being a signature of the weakly interacting amino acid molecules in the crystals.

INTRODUCTION

Amino acid crystals are bio-organic materials with promising applications in UV and IR detectors and non-linear optics. As light-emitting materials, one should drive efforts for an improved description of their emission processes at the microscopic level. Due to their zwitterion states, a salt-bridge between the amino group NH_3^+ and the carboxylic group COO^- is formed, giving rise to interesting physical phenomena.

The photoluminescence and absorption spectra of α-glycine (Gly), L-alanine (Ala), and L-asparagine (Asn) crystals are presented and analyzed in this work. Their electronic structures and intramolecular transitions are calculated *ab initio* to allow for a better understanding of the luminescence processes. We demonstrate the existence of a molecular signature in the photoluminescence spectra of the amino acid crystals. We discuss technological applications for this molecular signature.

THE MOLECULAR SIGNATURE

Typical photoluminescence spectra of the solid amino acids, recorded at T = 300 K, are shown in Fig. 1. They are broad and structured, comprising green to ultraviolet emission in the 1.75-3.60 eV range. A strong peak occurs at 3.1 eV in all amino acid crystals, which does not shift when the temperature decreases [1]. Three other rather weaker peaks occur at 3.51, 2.75 and 2.35 eV, those in the visible region exhibiting the same temperature dependence. Only Asn shows an extra thin peak at 3.45 eV. The peaks around 3.51 eV remain thin(as well the 3.45 eV peak for the Asn), indicating that phonons are not involved in the UV emission process. We infer from our results that the peaks in the visible region can be assigned to lattice-related processes, while the peaks in the UV region could be due to the relaxation of excited molecular states. Consequently, they are a signature of the weakly interacting amino acid molecules forming each crystal.

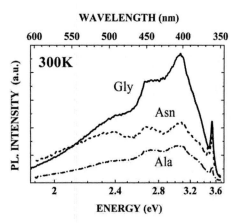

FIGURE 1. Photoluminescence of Gly (full line), Asn (dashed line), and Ala (dash-dot line) measured at 300 K.

We have evaluated the vertical and adiabatic transitions between ground(GS) and excited states of Gly, Ala, and Asn in the zwitterion form, within the framework of the Hartree-Fock(HF) method, together with single-excitation CI calculations and the 6-31G* basis set, as described in Ref. 1. It is interesting to note that, except for the $^1A \rightarrow$ GS transition in Asn, the calculated energies have the same values(see Table 1). Also, we have noted that the main contribution to the HOMO state comes from the COO^- ions, while to the LUMO state comes from the NH_3^+ ions. In amino acid molecular chains, these two molecular orbitals are at opposite extrema. So, in the interband transitions, the electron has to cross the whole chain, loosing energy to the vibrational modes of the crystal, contributing to the luminescence. Considering that both the salt-bridge and interchain interactions are not taken into account in our calculations, and the fact that HF method tends to overestimate the excitation energies, we can infer that the observed luminescence at 3.51 eV can be assigned to these two transitions, which also contribute to the the 3.1 eV emission.

TABLE 1. Calculated adiabatic transitions for Gly, Ala and Asn amino acids in their zwitterion modification.

Transition	Gly	Ala	Asn
$^3A \rightarrow$ GS	4.87 eV	4.83 eV	4.81 eV
$^1A \rightarrow$ GS	4.97 eV	5.01 eV	6.07 eV

In order to estimate the Gly band gap energy, we present in Fig. 2 its light absorption at 300 K As observed for Ala [1], the light absorption increases strongly around 5.1 eV, giving a direct gap of ~ 4.69 eV. By using the density-functional theory within the local density approximation, plane wave expansions and the pseudopotential method [2], we have calculated both the structural parameters and the band structure for Gly. The details of the calculations and the results for Ala are described in our previous work [1]. For Gly, our calculations show an 4.79 eV indirect gap very close to the X-point, and a 4.65 eV direct gap at the Γ-point. This suggests that both valleys should be effective in the luminescence processes. Also, at the X-point, the bands are very flat and asymmetric, indicating that polarons should contribute to the optical properties.

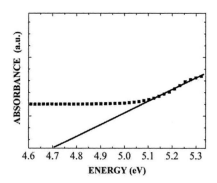

FIGURE 2. Light absorption of Gly measured at 300 K.

In summary, we presented both the photoluminescence and absorption spectra of Gly, Ala, and Asn crystals. We conclude that the amino acid crystals present a molecular signature in their luminescence, which is a very thin peak in the ultraviolet region. This molecular signature can be useful in the development of organic/semiconductor based optical biosensor devices. Optical detection of the vancomycin antibiotic to D-alanine is one of the possibilities.

ACKNOWLEDGMENTS

EWSC and MFZ, JRP would like to acknowledge the graduate fellowship at the Physics Department of the Universidade Federal do Ceará received from CAPES and CNPq, respectively, during the realization of this research. The authors would like to acknowledge the financial support received from the Science Funding Agency of the Ceará (FUNCAP), FINEP, and CNPq through the grant NanoSemiMat Project # 550.015/01-9.

REFERENCES

1. J. R. Pinheiro, E. W. S. Caetano, M. Z. S. Flores, V. N. Freire, G. A. Farias, J. R. L. Fernandez, J. R. Leite, M. C. F. de Oliveira, J. A. Pinheiro, B. S. Cavada, J. L. de Lima Filho and H. W. Leite Alves, to be published (2004).

2. X. Gonze, et al., *Comput. Mater. Sci.* **25**, 478(2002), and references therein.

Absorption and Emission of Excitons in thin PTCDA Films and PTCDA/Alq$_3$ Multilayers

H. P. Wagner, A. DeSilva, V. R. Gangilenka, T. U. Kampen*

Department of Physics, University of Cincinnati, Cincinnati, Ohio 45221-0011, U.S.A.
**Department of Molecular Physics, Fritz-Haber Institut der Max-Planck Gesellschaft, 14195 Berlin, Germany*

Abstract. We investigate the exciton absorption and emission in 3,4,9,10-perylene tetracarboxylic dianhydride (PTCDA) thin films and PTCDA/aluminium-tris-hydroxyqinoline (Alq$_3$) multilayers in the temperature range from 10 to 300 K. In the multilayers a strong recombination band is observed at 1.63 eV. This band is attributed to charge transfer transitions between stacked PTCDA molecules that are compressed by strain fields.

INTRODUCTION

Thin films of organic semiconductors have attracted growing interest because of various applications such as organic light emitting diodes [1,2]. One of the most intensively used molecules to study the influence of intermolecular interactions on the optical and electronic properties is 3,4,9,10,-perylene tetra-carboxylic dianhydride (PTCDA) [3,4]. Much less systematic investigations have been performed on crystalline PTCDA films or PTCDA/ aluminium-tris-hydroxyqinoline (Alq$_3$) multilayers. The microscopic knowledge of emission processes in such structures is important for improved device applications.

EXPERIMENTAL RESULTS

The investigated PTCDA films as well as PTCDA/Alq$_3$ multilayers were grown by organic molecular beam deposition (OMBD) at a base pressure of 10^{-8} mbar on (001) oriented Si and on amorphous Pyrex substrates. A detailed description of the growth process is given elsewhere [5]. X-ray diffraction measurements reveal a predominantly α-PTCDA crystal phase in thin films and in multilayers. A 50 Watt tungsten halogen lamp was used as light source for absorption measurements. In the photoluminescence (PL) measurements the samples were excited by frequency doubled 100 fs Ti-sapphire laser pulses at 2.84 eV (λ = 436 nm). To avoid any structural damage in the organic samples the average laser power was set below 0.1 mW. The absorption and PL spectra were analyzed by a monochromator and a GaAs photomultiplier. For variable temperature measurements between 10 and 300 K a closed-cycle He cryostat was used.

Room temperature absorption measurements on a 36 nm PTCDA film reveal a narrow 0-0 Frenkel exciton transition band at 2.22 eV followed by a broad and slightly structured band centered at 2.53 eV. The broad band is attributed to non resolved exciton transitions into higher vibronic subbands. The experimental result can be nicely reproduced by calculations that are based on a Frenkel exciton model [6]. Temperature dependent absorption measurements between 300 and 10 K reveal a red-shift of the 0-0 transition by 14 meV and an absorbance increase with decreasing temperature. In contrast the absorption bands in PTCDA/Alq$_3$ multilayers are nearly temperature independent but the 0-0 transition shows a blue-shift of 20 meV compared to the transition energy in PTCDA films at 300 K. The experimental observations are attributed to enhanced orbital overlap between PTCDA molecules and strain effects that are different in the pure films and in multilayers. Further theoretical investigations will elucidate the microscopic reasons for the observed energy shifts.

Temperature dependent PL studies of PTCDA films between 10 and 300 K show a very similar PL lineshape and temperature dependence as observed in single crystals [4]. The solid line in Fig. 1 displays the PL of a 36 nm thick PTCDA/Si film at 40 K. The emission reveals a high energy CT2-nr band at 1.95 eV that is assigned to a non-relaxed charge transfer exciton transition between stacked molecules and the dominating Frenkel exciton emission at 1.81 eV. With increasing temperature the weaker relaxed charge transfer transition CT1 between molecules within the same unit cell (at 1.82 eV) and relaxed CT2 between stacked molecules in different unit cells (at 1.68 eV) become the dominating emission bands. Also shown in Fig.1 are spectra of multilayers that comprise six alternating PTCDA and Alq_3 layers of 3 nm and 4 nm thickness (dash-dotted line) and twelve alternating layers of 1.5 nm and 2 nm thickness (dashed line), respectively. In addition to the PTCDA emission bands the spectra show the Alq_3 emission peaking at 2.37 eV. While the CT2-nr and Frenkel exciton emission lines are not energetically shifted and also have similar PL intensities as compared to the PTCDA film, the low energy emission band is strongly enhanced and red shifted with respect to the CT2 emission in pure films [5].

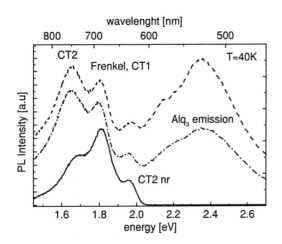

FIGURE 1. PL spectra at 40 K of a 36 nm PTCDA film (solid line) and of multilayers that comprise six alternating PTCDA and Alq_3 layers of 3 nm and 4 nm thickness (dash-dotted line) and twelve alternating layers of 1.5 nm and 2 nm thickness (dashed line), respectively. The samples are excited at at 2.84 eV (λ = 436 nm).

For a more specific explanation of this emission change we performed x-ray diffraction measurements on pure PTCDA layers and on multilayer samples. The multilayers reveal a broader full width at half maximum (FWHM) of 0.46° for the α-PTCDA (102) reflex compared to 0.37° in the pure films. The broader FWHM indicates inhomogeneous strain within the multilayer structure. We conclude that individual PTCDA crystallites in the multilayer are tilted with respect to the growth direction. This tilt causes compressive strain fields along the π-orbitals of stacked molecules that significantly affects the CT2 transition. The reduced distance between stacked molecules increases the charge transfer (CT) exciton binding energy and its formation probability causing a red-shift and a more pronounced emission line in the PL spectrum.

Strain dependent PL measurements on PTCDA films using a uniaxial piston pressure cell support this interpretation. The low energy CT2 emission shifts to lower energy and increases in intensity when uniaxial strain is applied to the PTCDA film [5].

In conclusion we studied the exciton absorption and emission in OMBD grown PTCDA films and PTCDA/Alq_3 multilayers. Temperature dependent absorption measurements reveal a red-shift and absorbance increase of the 0-0 Frenkel exciton emission that is attributed to enhanced π-orbital overlap and strain effects within the films and multilayers. Temperature dependent PL spectra reveal different recombination channels arising from Frenkel and charge transfer excitons. In PTCDA/Alq_3 multilayers the low energy emission band is strongly enhanced and red shifted with respect to the CT2 band in pure films. Strain dependent PL measurements on PTCDA films identify the dominating low energy band in multilayers as strain modified CT2 transition.

ACKNOWLEDGMENTS

The authors acknowledge Dr. P. Boolchand for performing x-ray diffraction measurements. Stimulating discussions with Dr. R. Scholz are kindly acknowledged.

REFERENCES

1. C. W. Tang and S. A. VanSlyke, *Appl. Phys. Lett.* **51**, 913 (1987).

2. M. A. Baldo, M. E. Thompson, S. R. Forrest, *Nature* **403**, 750 (2000).

3. A. Yu. Kobitski, R. Scholz, I Vragović, H. P. Wagner and D. R. T. Zahn, *Phys. Rev. B* **66**, 154204 (2002).

4. A. Yu. Kobitski, R. Scholz, D. R. T. Zahn and H. P. Wagner, *Phys. Rev. B* **68**, 150201 (2003)

5. H. P. Wagner, A. DeSilva, T.U. Kampen, *Phys. Rev. B,* submitted.

6. I. Vragović, R. Scholz, *Phys. Rev. B* **68**, 155202 (2003).

Optically Controlled Rotation Of PTCDA Crystals In Optical Tweezers

C. Starr, W. Dultz*, H. P. Wagner**, K. Dholakia***, H. Schmitzer

Department of Physics, Xavier University, Cincinnati, USA
**Physikalisches Insitut, Universität Frankfurt, Frankfurt a. Main, Germany*
***Department of Physics, University of Cincinnati, USA*
****School of Physics and Astronomy, University of St. Andrews, St. Andrews, Scotland*

Abstract. Small birefringent particles held in optical tweezers can rotate under the spin angular momentum transfer from polarized light. We tweezed and rotated organic semiconductor 3,4,9,10-perylenetetracarboxylic dianhydride (PTCDA) platelets with a HeNe Laser beam. The dynamics of this spin angular momentum transfer from photons to matter is discussed. For the first time the wobbling rotation is measured as a function of the ellipticity of the incident laser polarization.

INTRODUCTION

The dynamics of the spin transfer from photons to a birefringent plate depends on the thickness of the plate, its momentary orientation and the polarization of the light. We calculate the spin angular momentum transfer in a quantum optical picture instead of using Maxwell's equations [1] and discuss the subsequent motion of the rotating particle. We demonstrate the irregular rotation of small birefringent PTCDA crystals by measuring their angular velocity. We chose PTCDA because it has low weight, is therefore easy to hold in optical tweezers and can reach high terminal angular velocities. It is also highly birefringent and transparent from the infrared to the red spectrum [2]. The electro optical properties of PTCDA films have been studied extensively because of their various applications as an organic semiconductor.

THEORY

The Poincaré sphere is the adequate space to represent the polarisation state of the photon. In this representation an arbitrary polarisation P_0 is a superposition of the left and right circularly polarized basis vectors $|L\rangle$ and $|R\rangle$ and is given by [3]

$$|P_0\rangle = cos(\tfrac{1}{2}\alpha_0)exp(\tfrac{i}{2}\varepsilon_0)|L\rangle + sin(\tfrac{1}{2}\alpha_0)exp(-\tfrac{i}{2}\varepsilon_0)|R\rangle \quad (1)$$

$|L\rangle$ is the left circular eigenfunction of the spin angular momentum observable s_z with eigenvalue $+\hbar$ and $|R\rangle$ is the right circular eigenfunction of s_z with eigenvalue $-\hbar$. α_0 is the angle between P_0 and the north pole L on the Poincaré sphere and ε_0 is the azimuthal angle of P_0 on the sphere. The ellipticity ω_0 of the polarisation P_0 is given by $\omega_0 = (90° - \alpha_0)/2$. The expectation value of the spin angular momentum of the photon depends on the ellipticity ω_0 of the light's polarisation:

$$\langle s_z \rangle = \langle P_0|s_z|P_0\rangle = \hbar \cos\alpha_0 = \hbar \sin 2\omega_0. \quad (2)$$

We now derive the equation of motion of a birefringent plate with retardation Δ which absorbs spin angular momentum from photons of polarisation P_0. A photon of polarisation P_0 which enters the plate, has changed to polarisation P_1 upon exiting the plate. P_1 is found by rotating P_0 by the angle Δ around the axis F, which represents the fast axis of the plate and lies on the equator of the Poincaré sphere at an angle 2θ. The angle θ is the orientation of the fast axis of the disc shaped platelet in real space. Assuming that the friction is proportional to the angular velocity and using spherical geometry, we find the equation of motion for the platelet

$$\bar{I}\frac{d^2\theta}{dt^2} = N\hbar[\sin 2\omega_0 (1-\cos\Delta) - \cos 2\omega_0 \sin\Delta \sin(2\theta_0 - 2\theta)] - \eta\frac{d\theta}{dt} \quad (3)$$

$\bar{I}$ is the moment of inertia of the plate and N is the number of photons which pass through the plate per second. η is a form and friction factor. The torque (Eq. 3) is periodic with respect to a plate rotation of 180° and depends on the momentary orientation θ of the plate. Numerical integration shows that after a transient time this can lead to a stationary wobbling rotation of the plate.

EXPERIMENT

We demonstrated the varying torque (Eq. 3) on small PTCDA micro crystals. A HeNe-Laser with 40 mW at 633 nm was polarized with a linear polarizer and the beam was focused via a gold mirror, a quarter wave plate and a high aperture microscope objective onto the object slide with the sample suspension of the micro crystals in water.

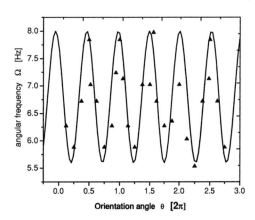

FIGURE 1. The angular frequency Ω of the PTCDA particle oscillates stationary under the influence of elliptically polarized light, $\omega_0 = \pm 25°$; the line is drawn to guide the eye.

The ellipticity ω_0 of the state of polarisation of the laser beam was tuned by rotating the $\lambda/4$ plate. The laser beam both held the platelet in the focus and transferred the spin angular momentum to it. A digital video camera observed the sample and the backscattered laser light through the gold mirror. Additional illumination was provided from a halogen lamp. At retardations $\Delta \neq \pi$ the crystals can wobble if the incident light is not circularly polarized. To demonstrate this experimentally we analyzed the frame sequences of the video camera and plotted the angular frequency Ω versus the orientation θ of the PTCDA platelet (Fig. 1).

DISCUSSION

The spin transfer from photons to a crystalline platelet consists of two terms. The second term depends on the orientation angle θ of the platelet. It can be interpreted as the action of the linear part of the electric light field on the induced electrical polarization of the dielectrically anisotropic platelet. The first term is the action of the circularly polarized part of the light; it propels the platelet, but does not depend on the orientation of the platelet. These competing effects generate the wobbling motion: The circular part of the light accelerates the particle's rotation until it is compensated by the friction (third term in Eq. 3), while the linear part generates an oscillatory motion around the potential minimum which is represented by the orientation of the slow axis (of high polarizability) of the crystal in the direction of the large axis of the elliptically polarized light wave. For small ellipticities the platelet rotates into an "off axis" position until it comes to rest due to damping. "Off axis" means that the slow axis of the particle does not align with the major axis of the elliptically polarized light unless $\omega_0 = 0°$.

ACKNOWLEDGEMENTS

This project was supported by the John Hauck Foundation. The technical advice and the assistance of D. Tierney and J. Brinkman are kindly acknowledged.

REFERENCES

1. Mootho, D. N., Arlt, J., Conroy, R. S., Akerboom, F., Voit, A., Dholakia, K., *Am. J. Phys.* **69**, 271-276 (2001).

2. Alonso, M. I., Garriga, M., Karl, N., Osso, J. O., Schreiber, F., *Organic Electronics* **3**, 23-31 (2002) and private communication

3. Hils, B., Dultz, W., Martienssen, W., *Phys. Rev. E* **60**, 2322-2329 (1999).

Ab-Initio Theory of Charge Transport in Organic Crystals

K. Hannewald and P.A. Bobbert

Group Polymer Physics, Eindhoven Polymer Laboratories, Technische Universiteit Eindhoven,
P.O. Box 513, 5600 MB Eindhoven, The Netherlands

Abstract. A theory of charge transport in organic crystals is presented. Using a Holstein-Peierls model, an explicit expression for the charge-carrier mobilities as a function of temperature is obtained. Calculating all material parameters from *ab initio* calculations, the theory is applied to oligo-acene crystals and a brief comparison to experiment is given.

INTRODUCTION

Organic semiconductors are promising materials for low-cost and easy-to-process (opto)electronic devices [1]. Besides π-conjugated polymers, an important material class are organic crystals. Due to their structural order, they are ideal candidates to study the *intrinsic* charge-transport phenomena in organic solids [2]. Organic molecular crystals exhibit weak intermolecular bonds, narrow electron bands, strong electron-lattice interaction, and hence polaron formation. In a pioneering work [3], Holstein developed a mobility theory using the polaron concept for *local* (on-site) electron-phonon coupling. The main prediction of his theory, i.e., the interplay between metallic (bandlike) conduction at low T and activated (hopping) transport at high T was, indeed, observed for the electrons in naphthalene but not for the holes [4,5,6]. This different behavior of electrons and holes as well as the microscopic origin of the crystallographic anisotropy in the mobility's T dependence and the influence of *nonlocal* couplings [7,8] have so far been poorly understood. Here, we address these questions by presenting an *ab-initio* theory of charge-carrier mobilities in organic crystals.

THEORY OF CHARGE TRANSPORT

We consider a Holstein-Peierls model described by the Hamiltonian $H = \sum_{mn} \varepsilon_{mn} a^+_m a_n + \sum_{Q=(q,\lambda)} \hbar\omega_Q (b^+_Q b_Q + \frac{1}{2}) + \sum_{mnQ} \hbar\omega_Q g_{Qmn} (b^+_Q + b_{-Q}) a^+_m a_n$. This model goes beyond the original Holstein model because the electron-lattice interaction described by g_{Qmn} contains both the *local* coupling to the on-site energies ε_{mm} (Holstein model) and the *nonlocal* coupling to the transfer integrals ε_{mn} ($m \neq n$, Peierls model).

The mobility μ_α in direction $\mathbf{e}_\alpha$ is obtained from the Kubo formula for electrical conductivity. This amounts to calculating a current-current correlation function where the current $\mathbf{j} = \mathbf{j}^{(I)} + \mathbf{j}^{(II)}$ is composed of a purely electronic term $\mathbf{j}^{(I)} = (e_0/i\hbar) \sum_{mn} (\mathbf{R}_m - \mathbf{R}_n) \varepsilon_{mn} a^+_m a_n$ and a phonon-assisted current $\mathbf{j}^{(II)} = (e_0/i\hbar) \sum_{mnQ} (\mathbf{R}_m - \mathbf{R}_n) \hbar\omega_Q g_{Qmn} (b^+_Q + b_{-Q}) a^+_m a_n$, with the latter one stemming solely from the *nonlocal* coupling. Using the method of canonical transformation, an approximate evaluation of the Kubo formula for the above Hamiltonian can be performed [9], with the results for dispersionless phonons ($\omega_Q \to \omega_\lambda$) as given below.

The mobility $\mu_\alpha^{(I)}$ due to $\mathbf{j}^{(I)}$ is found to be

$$\mu_\alpha^{(I)} = (e_0/2k_B T \hbar^2) \sum_{n \neq m} (R_{\alpha m} - R_{\alpha n})^2$$
$$\times \int_{-\infty}^{\infty} dt \left[\varepsilon_{mn} e^{-\sum_\lambda G_\lambda [1+2N_\lambda - \Phi_\lambda(t)]} \right]^2 e^{-\Gamma^2 t^2}, \quad (1)$$

where N_λ denotes the phonon occupation number and Γ is a phenomenological broadening parameter. Those terms in the exponent of Eq. (1) that contain $\Phi_\lambda(t) = (1+N_\lambda)e^{-i\omega_\lambda t} + N_\lambda e^{i\omega_\lambda t}$ describe *incoherent* scattering events involving actual changes in phonon numbers (hopping) whereas the remaining terms account for purely *coherent* scattering processes (bandwidth narrowing). Note that our result (1) is similar to the solution of the original Holstein model but, importantly, the T dependence is not only governed by the local coupling but by effective coupling constants $G_\lambda = (g_{\lambda mm})^2 + \frac{1}{2} \sum_{k \neq m} (\hbar \omega_\lambda g_{\lambda mk})^2$ that are composed of both the local and nonlocal ones.

The total mobilities μ_α are obtained by including both $\mathbf{j}^{(I)}$ and $\mathbf{j}^{(II)}$. As a result, we find the same expression as Eq. (1) but with the replacement [10]

$$(\varepsilon_{mn})^2 \rightarrow (\varepsilon_{mn} - \Delta_{mn})^2 + \frac{1}{2} \sum_\lambda (\hbar \omega_\lambda g_{\lambda mn})^2 \Phi_\lambda(t). \quad (2)$$

Here, in contrast to Eq. (1), the total mobilities μ_α can account for an anisotropic T dependence due to the extra term containing $\Phi_\lambda(t)$ in Eq. (2). This term describes an additional phonon-assisted hopping process, i.e., a novel transport-promoting effect of nonlocal electron-phonon coupling that is completely absent in any local-coupling theory.

As a key result, Eqs. (1) and (2) allow us to study the T dependence of the mobilities once the material parameters $\mathbf{R}_m$, ε_{mn}, $g_{\lambda mn}$, and ω_λ are determined. We obtain them using the total-energy program VASP [11] by means of a 3-step strategy described in Ref. [12].

NUMERICAL RESULTS

We proceed by applying our mobility theory to the oligo-acene crystals naphthalene ($C_{10}H_8$), anthracene ($C_{14}H_{10}$), and tetracene ($C_{18}H_{12}$). In Fig. 1, we present the calculated electron and hole mobilities $\mu_\alpha(T)$ for the a, b, and c' directions, with c' being perpendicular to the ab-plane of the molecular layers.

For naphthalene, we find three important agreements with the experiments of Karl et al. [5,6]. First, the calculated hole mobilities are generally larger than the corresponding electron mobilities, except for the in-plane mobilities at $T \approx 300K$. These lower electron mobilities are mainly due to the strong effective coupling between electrons and phonons [LUMO: $\Sigma_\lambda G_\lambda = (1.52)^2$, HOMO: $\Sigma_\lambda G_\lambda = (0.92)^2$]. Second, the *electron* mobilities exhibit a pronounced anisotropic T dependence. In particular, we find a metallic (bandlike) behavior within the ab-plane and a slightly activated (hopping) transport in the c' direction. These findings are in excellent agreement with the experiments of Ref. [6], which show a T-independent mobility around $T \approx 150K$ only in the c' direction. Third, our calculated *hole* mobilities obey at elevated T the power law $\mu_\alpha \sim T^{-2.5}$ that agrees well with experiment [6] where $\mu_\alpha \sim T^{-2.9}, T^{-2.5}, T^{-2.8}$. As explained in detail in Ref. [9], this power-law dependence is intimately related to the number of relevant phonon modes per molecule.

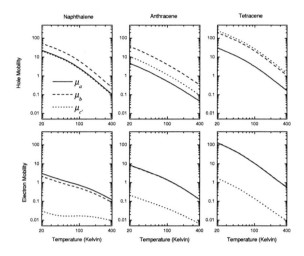

FIGURE 1. Charge-carrier mobilities $\mu_\alpha(T)$ in arbitrary units for naphthalene, anthracene, and tetracene crystals.

For the higher oligo-acenes, we predict that the mobilities generally increase with conjugation length (exception: holes in anthracene) and should exhibit a metallic (bandlike) behavior throughout the whole T range in experiments performed on ultrapure samples.

ACKNOWLEDGMENTS

Financial support by the Foundation for Fundamental Research on Matter (FOM) is acknowledged.

REFERENCES

[1] R.H. Friend et al., Nature **397**, 121 (1999).
[2] N. Karl, Synth. Met. **133 & 134**, 649 (2003).
[3] T. Holstein, Ann. Phys. (NY) **8**, 343 (1959).
[4] L.B. Schein et al., Phys. Rev. Lett. **40**, 197 (1978).
[5] W. Warta and N. Karl, Phys. Rev. B **32**, 1172 (1985).
[6] N. Karl, in *Landolt-Börnstein: Numerical Data and Functional Relationships in Science and Technology*, edited by K.-H. Hellwege and O. Madelung (Springer, Berlin, 1985), Group III, Vol. 17, Pt. i, p. 106.
[7] W.P. Su et al., Phys. Rev. Lett. **42**, 1698 (1979).
[8] R.W. Munn et al., J. Chem. Phys. **83**, 1854 (1985).
[9] K. Hannewald and P.A. Bobbert, Appl. Phys. Lett. **85**, xxx (2004); Phys. Rev. B **69**, 075212 (2004).
[10] The expression Δ_{mn} can be found in Ref. [9].
[11] G. Kresse et al., Phys. Rev. B **47**, 558 (1993); **49**, 14251 (1994); **54**, 11169 (1996).
[12] K. Hannewald et al., Phys. Rev. B **69**, 075211 (2004).

CHAPTER 4
MICROCAVITIES

Acoustic cavities

A. Fainstein*, P. Lacharmoise* and B. Jusserand[†]

*Instituto Balseiro and Centro Atómico Bariloche, CNEA, 8400 S. C. de Bariloche, Río Negro, Argentina.
[†]Laboratoire de Photonique et de Nanostructures, CNRS, Route de Nozay, 91460 Marcoussis, France.

Abstract. We describe semiconductor double resonators that confine light and sound and strongly enhance their interaction. We address the double optical resonant Raman interaction in these devices, and we describe in detail the main features of the Raman spectra that include the scattering with standing optical waves, the observation of the acoustic confined vibrations, the spectra of structures with one to three confined acoustic modes, finite-size effects, the cavity-mode parity and Raman selection rules.

MOTIVATION

Resonant cavities are the basis of numerous physical processes and devices. In the domain of optics, this includes, e.g., lasers, amplifiers, engineering of photon spectral and spatial densities, and control of light-matter interaction. [1] All these fundamental concepts can be extended to the domain of phonon physics. In fact, phonon amplification [2] and "lasing" [3] have been investigated for many years, and recently phonon engineering has been proposed has a means to control phonon lifetimes [4] and the electron-phonon interaction. [5] It is thus of interest to develop cavities for acoustic waves, and resonators where both light and sound are confined and their interaction enhanced.

A planar periodic stack of two materials with different refractive indices and optical thickness $\lambda/4$ reflects photons propagating normal to the layers within a stop-band around λ, and is termed a "Bragg reflector" (BR).[1] Similarly, a periodic stack of materials with contrasting acoustic impedances reflects sound. [6] A *phonon cavity* can be thus constructed by enclosing between two superlattices (SL's) a spacer of thickness $d_c = m\lambda_c/2$, where λ_c is the acoustical phonon wavelength at the center of the phonon minigap. [7, 8] The first $k = 0$ folded phonon mini-gap in a SL is maximum for a layer thickness (d) to velocity (v) ratio given by $d_1/v_1 = d_2/3v_2$. The relevant parameter in the design of this mirrors is the acoustic impedance mismatch, defined as $Z = v_1\rho_1/v_2\rho_2$. v_j and ρ_j are, respectively, the sound speed and density of material j. For the acoustic cavity mode to be centered at the phonon mini-gap, d_c must satisfy $d_c = 2v_c d_1/v_1 = 2v_c d_2/3v_2$ ($m = 1$). [7, 9]

The standing wave phonons described above can be generated and studied using standing wave photons through Raman scattering processes by placing the acoustic cavity inside an optical cavity. [10] Such geometry provides access to "forward scattering" (FS, $k = 0$) Raman processes which couple the light with the confined phonon modes, [7] and strongly enhances the interaction between light and sound. [10] In the experiment an incident photon is coupled to the optical cavity, a phonon is generated in the resonant acoustic cavity state, and a Raman photon is scattered out through an optical cavity mode. [7, 8, 10]

In this invited presentation we will briefly review and complete recent work related to the design and implementation of acoustic phonon cavities in the THz domain (i.e., nm wavelengths) using semiconductor multilayers. We will describe light-sound double resonators in which the interaction between photons and phonons is enhanced, and we will discuss in detail the important features of the observed acoustic phonon Raman spectra.

RESULTS AND DISCUSSION

One mode cavity

In Fig. 1 we present the calculated acoustical reflectivity and calculated and measured Stokes and anti-Stokes Raman spectra for a phonon cavity made with 11 period 74Å/38ÅGaAs/AlAs phonon mirrors and an $Al_{0.2}Ga_{0.8}As$ spacers of thickness $d_c = 50$ Å($\lambda/2$). The phonon cavity fits precisely as the spacer of a λ optical cavity defined by bottom(top) optical Bragg reflectors consisting of 20(16) $Al_{0.2}Ga_{0.8}As$/AlAs pairs. Calculations were performed for the nominal structures. The calculations are based on a photoelastic model [11] for the Raman efficiency. The confined phonon and photon modes used as input for these calculations are obtained using transfer matrix methods. Details of the calculations and experimental set-up can be found in Refs. [7] and [9].

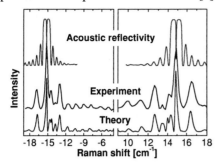

FIGURE 1. Calculated Reflectivity and calculated and measured Stokes and anti-Stokes spectra of a one-mode phonon cavity.

The spectra, which can be almost perfectly reproduced by the calculations, basically display the cavity-mode (coincident with the dip in the reflectivity stop band), and other peaks and oscillations. These and other features will be discussed in the following sections.

Multimode cavities

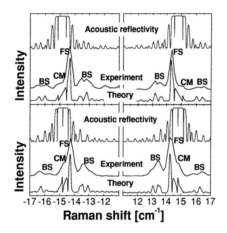

FIGURE 2. Calculated Reflectivity and calculated and measured Stokes and anti-Stokes spectra of two (top) and three-mode (bottom) phonon cavities. The labels are explained in the text.

Depending on d_c, different number of confined modes can be sustained by the cavity within a specific stop-band. This is demonstrated in Fig. 2 where we show the calculated acoustical reflectivity and measured and calculated Stokes and anti-Stokes Raman spectra for two phonon cavities made respectively with 19 and 16 period 74Å/38ÅGaAs/AlAs phonon mirrors, and $Al_{0.2}Ga_{0.8}As$ spacers of thickness $d_c = 80Å(16\lambda/2)$ and $d_c = 1500Å(30\lambda/2)$. The number of mirror periods are determined to make the structures fit precisely as 2λ spacers of optical cavities. The experimental spectra are limited by the spectrometer resolution (~ 0.3 cm^{-1}), but it is nevertheless clear that the main features are well reproduced by the calculations without any fitting parameter.

Double Optical Resonance

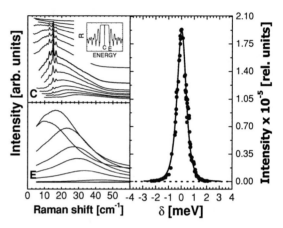

FIGURE 3. Left panel: Raman spectra obtained while detuning the laser energy away from DOR for the optical cavity (C) and edge (E) modes. C and E are indicated in the optical reflectivity shown in the inset. Right panel:Raman amplification scan as a function of cavity detuning δ. The full(dotted) curve corresponds to the photoelastic model calculations for the cavity(edge) modes, normalized to a similar phonon cavity without optical cavity.

The light-sound interaction is strongly enhanced in these double resonator devices. [9] To highlight this amplification, we show in Fig. 3 an experiment in which the double optical resonance (DOR) was detuned by shifting the laser energy δ away from the cavity mode. If the optical field in the cavity is expressed as $E = E_0 f(\delta)$, then the Raman intensity, which is proportional to the product of the squared incoming and scattered fields, depends on δ as $f^4(\delta)$. [10] In Fig. 3(left) we display Raman spectra obtained while detuning the laser energy out of DOR for the cavity mode, or through a reflectivity minimum on one of the optical stop-band edges (see the inset in Fig. 3). Note that no Raman peaks can be distinguished from luminescence and noise when coupling through the stop-band edge (E), while clear spectra can be extracted exploiting the cavity enhancement (C). The Raman amplification scan obtained from these spectra is displayed with full circles as a function of δ in Fig.3(right), and compared with theoretical curves obtained using the photoelastic model. The full(dotted) curve corresponds to the cavity(edge) mode. The curves are normalized with respect to a similar phonon cavity grown *without* an optical cavity. The experimental results, in turn, have been multiplied by a constant to fit the maximum intensity of the cavity mode amplification curve. Note that the light-sound interaction can be enhanced over five orders of magnitude in the described photon-phonon cavities. [9]

Raman scattering with confined photons

Due to the standing wave character of the optical mode in a cavity, both forward (FS) and back-scattering (BS) components contribute to the scattered light.[10, 12] This is demonstrated in Fig. 4 where we compare the one-mode cavity measured Stokes spectrum with calculated curves. The latter correspond to the FS (transferred $q = 0$), BS ($q = 2k_L$, with k_L the laser wavenumber) and total Raman intensities. From these spectra it is straightforward to identify the cavity-mode (CM), and the separate FS and BS contributions. The BS contribution is related to the standard folded-phonon spectra observed in Raman scattering in SL's [11], while the FS component corresponds to one of the two $q = 0$ mini-gap states which is Raman allowed. [11] Its position coincides with the phonon stop-band edge. Note, on the other hand, that the CM is only observed through the FS component of the total scattering. The corresponding assignments for the two and three mode cavities are given in Fig. 2.

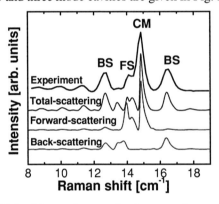

FIGURE 4. Measured one-mode phonon cavity spectra compared to the total, back-scattering (BS) and forward scattering (FS) calculated spectra.

Finite size effects

Besides the FS and BS related peaks, several other spectral features call the attention in Figs. 1 and 2. Namely, the low energy oscillations (particularly clear in Fig. 1), and the splitting of the BS and FS peaks (evident in the BS component of the two-mode cavity spectra in Fig. 2). The origin of these features can be traced down to finite-size effects originated in the existence of the cavity spacer on the otherwise periodic structure, and in the reduced number of periods that make the phonon mirrors. [13] This is demonstrated in Fig. 5, where calculated spectra corresponding to a mirror with a very large (300) number of periods is compared with the spectrum of the one-mode phonon cavity with ten period mirrors. In the latter, as for a diffraction pattern with few illuminated groves, the peaks are broadened and side oscillations become evident. In addition, the existence of the cavity spacer leads to interferences between the light scattered from the two mirrors that explains the appearance of dips in the spectra and, consequently, of apparent peak splittings. [13, 14]

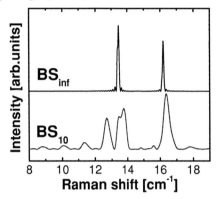

FIGURE 5. Calculated back-scattering Raman spectra for a 300 period phonon mirror (BS_{inf}) compared with that of the one-mode cavity with 10 period mirrors (BS_{10})

Phonon mode parity

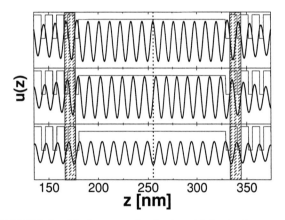

FIGURE 6. Displacement distribution $u(z)$ for the three confined modes of the three-mode phonon cavity. The thin lines identify the layers through their photoelastic constants. The vertical dashed line indicates the center of the phonon cavity. The shadowed regions highlight a mirror period.

Note in the calculated spectra in Fig. 2 that the confined phonons display alternate peaks and dips. This reflects that the modes alternate between allowed and forbidden, respectively. This can be understood analyzing the displacement distribution $u(z)$ associated with each of these modes, and shown for the three-mode cavity in Fig. 6. Note that while the first and third modes are even respect to the center of the cavity, the second (central) mode is odd. On the other hand, for the photoelastic mechanism it is the strain (du/dz) that is relevant. Having in mind that the parity is reversed for the strain, and that the Raman cross section is also proportional to the squared optical mode distribution [7, 11] that is even respect to the center of the cavity, it is straightforward to conclude that only modes with odd $u(z)$ are Raman allowed, and that subsequent standing phonon modes are allowed or forbidden according to their parity.

Cavity vs. mirror modes

Note that while in the single mode cavity (Fig. 1) the main feature corresponds to the confined acoustical mode, for the two and three-mode cavities the spectra are dominated by the FS contribution. The cavity and FS modes have essentially different character. In fact, while the acoustic cavity mode is confined within the cavity spacer and decays exponentially into the mirrors, the phonon mode that leads to the FS peak is concentrated in the SL's (see the phonon displacements depicted in Fig. 7). Since the SL's making the mirrors are considerably larger in the two and three-mode cavities, as compared to the one mode resonator, in the former the corresponding FS contributions are enhanced respect to the confined mode.

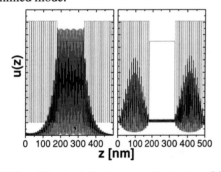

FIGURE 7. Calculated displacement distribution $u(z)$ for the cavity and forward-scattered modes of the three-mode phonon cavity. The thin lines identify the layers through their photoelastic constants.

CONCLUSION

We have reported on devices that have resonant cavities for acoustic phonons inside optical cavities. We presented experimental results on acoustic phonons generated by Raman processes in structures with one to three confined acoustical modes, and we described in detail the scattering with standing wave photons, finite size effects, the scattered phonon mode distribution, the phonon cavity-mode parity and Raman selection rules. The enhancement of the light-sound interaction was measured to be over five orders of magnitude in these devices.

While only a modest variation of the acoustic impedance mismatch can be obtained for e.g., Si/Ge and GaAs/AlAs ($Z = 0.73$ and 0.84, respectively), much larger contrasts exist when opening the scope to include other materials (for example oxides) that can also be grown into multilayers with epitaxy techniques. On the other hand, even for small variations of Z the impact on the cavity performance is enormous. Thus, research on other materials is of great actual interest. Concerning the physics, we believe that the technology is mature so that phonon devices with new characteristics and/or improved performances can be designed and fabricated using acoustical cavities. These include resonators for phonon "lasers" and amplifiers, and the engineering of phonon spectral and spatial distributions using cavities designed to enhance or reduce the coupling with electrons and optical phonons.

REFERENCES

1. Burstein, E., and Weisbuch, C., editors, Plenum, New York, 1995.
2. P. A. Fokker, J. I. Dijkhuis, and H. W. de Wijn, *Phys. Rev. B*, **55**, 2925 (1997).
3. I. Camps, S. S. Makler, H. M. Pastawski, and L. E. F. Foa Torres, *Phys. Rev. B*, **64**, 125311 (2001).
4. M. Canonico, C. Poweleit, J. Menéndez, A. Debernardi, S. R. Jhonson, and Y. -H. Zhang, *Phys. Rev. Lett.*, **88**, 215502 (2002).
5. J. Chen, J. B. Khurgin, and R. Merlin, *App. Phys. Lett.*, **80**, 2901 (2002).
6. V. Narayanamurti, H. L. Störmer, M. A. Chin, A. C. Gossard, and W. Wiegmann, *Phys. Rev. Lett*, **43**, 2012 (1979).
7. M. Trigo, A. Bruchhausen, A. Fainstein, B. Jusserand, and V. Thierry-Mieg, *Phys. Rev. Lett*, **89**, 227402 (2002).
8. J. M. Worlock, and M. L. Roukes, *Nature*, **421**, 802 (2003).
9. P. Lacharmoise, A. Fainstein, B. Jusserand, and V. Thierry-Mieg, *App. Phys. Lett*, **84**, 3274 (2004).
10. A. Fainstein, B. Jusserand, and V. Thierry-Mieg, *Phys. Rev. Lett*, **75**, 3764 (1995).
11. B. Jusserand, and M. Cardona, in *Light Scattering in Solids V*, edited by M. Cardona and G. Güntherodt, Springer, Heidelberg, 1989, p. 49.
12. A. Fainstein, M. Trigo, D. Oliva, B. Jusserand, T. Freixanet, and V. Thierry-Mieg, *Phys. Rev. Lett*, **86**, 3411 (2001).
13. M. Trigo, A. Fainstein, B. Jusserand, and V. Thierry-Mieg, *Phys. Rev. B*, **66** (2002).
14. M. Giehler, T. Ruf, M. Cardona, and K. Ploog, *Phys. Rev. B*, **55** (1997).

Manipulation of photons and electrons in photonic structures using surface acoustic waves

P. V. Santos*, M. M. de Lima, Jr.* and R. Hey*

Paul-Drude-Institut für Festkörperelektronik, Hausvogteiplatz 5–7, 10117 Berlin, Germany

Abstract.
The modulation of photonic cavities by surface acoustic waves opens new possibilities for the control of photons and electrons in semiconductor nanostructures. We show that the SAW modulation combined with the strong optical fields in the cavity leads to the formation of a dynamic optical superlattice with a folded dispersion and well-defined gaps. In addition, it allows for the optical detection of ambipolar transport of photogenerated carriers up to room temperature.

INTRODUCTION

Photonic crystals (PhCs) have received substantial attention during the last years due to their ability to tailor the density of electromagnetic modes in a medium by controlling the structural dimensions and the composition profile. They have been explored to modify the emission characteristics of light sources in the medium [1] as well as to confine light beams within dimensions of the order of the light wavelength [2]. These properties have made it possible to fabricate waveguides with sharp bends [3] and tailored transmission characteristics. As a result, PhCs have been envisaged as basic media for ultra small optical cavities as well as for integrated optics.

PhCs are usually passive structures. An additional degree of freedom for light control can be achieved if their dimensions or dielectric properties are modulated in real time. Different approaches have been proposed based on the modulation of the optical properties by a strain field, [4] or, in PhCs infiltrated by liquid crystals, via the application of electric or temperature fields. [5]

A powerful way of controlling the dielectric properties uses surface acoustic waves (SAWs), which are elastic vibrations propagating along a surface. In piezoelectric materials, SAWs can be electrically generated by interdigital transducers (IDTs). The strain and the piezoelectric fields carried by the SAW introduce a periodic modulation of the optical properties in the direction parallel to the surface through the acoustooptical and electrooptical effect, respectively, which constitute the basis of operation of a whole class of opto-electronic devices. [6]

In this paper, we review concepts for the realization of dynamic photonic devices via the modulation of semiconductor cavities by a SAW. These structures are one-dimensional PhCs with a sharp resonant state in the photonic band gap. The strong confinement of the optical fields in the cavity enables new approaches for the control of light beams by SAWs. Two of these approaches will be discussed here. The first one explores the modulation of the cavity resonance wavelength by the SAW strain field, which creates a lateral optical superlattice with a period dictated by the acoustic wavelength. [7] Photon manipulation is demonstrated by the huge Brillouin diffraction intensities in these structures, which reach values close to 50% of the incoming beam intensity. The second approach explores the strong interaction between the moving SAW piezoelectric potential (Φ_{SAW}) and electron-hole (e–h) pairs. [8, 9] Here, photon control is based on their conversion into e–h pairs within quantum wells (QWs) inside the cavity, the transport of the carriers by the moving Φ_{SAW}, and their subsequent recombination to deliver the output photons. The photonic cavity is employed in this case to control the carrier recombination lifetime, thus allowing for an efficient interconversion between photons and e–h pairs even at room temperature (RT).

EXPERIMENTAL DETAILS

The structures [cf. Fig. 1(a)] consist of a GaAs cavity layer (C) — with an optical thickness equal to a multiple of $\lambda/2$ — surrounded by GaAs/AlAs Bragg mirrors (BM_1 and BM_2) grown on (001) GaAs by molecular-beam epitaxy. The sample is modulated by Rayleigh SAWs generated by IDTs deposited on the surface. The IDTs are powered by a radio-frequency (rf) amplifier delivering continuous rf-power of up to $P_{rf} = 25$ dBm at the resonance frequency of $f_{SAW} = 540$ MHz. The linear power density (P_ℓ), defined as the acoustic power

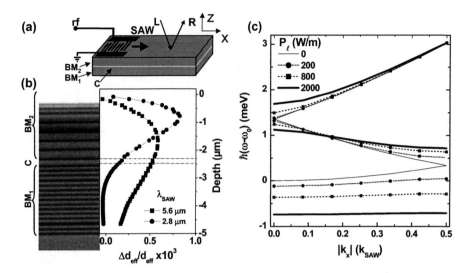

FIGURE 1. (a) Schematic diagram and (b) SEM micrograph displaying a sample cross-section. The light areas are GaAs layers, while the dark ones correspond to AlAs. The plot shows the modulation of the effective optical thickness ($\Delta d_{\text{eff}}/d_{\text{eff}}$) for a fixed phase as a function of depth, for different SAW frequencies, assuming an acoustic power density (P_ℓ) of 200 W/m. The dash-dotted lines mark the position of the cavity layer. (c) Calculated photon dispersion in an optical superlattice for $\lambda_{\text{SAW}} = 5.6\ \mu\text{m}$ for different P_ℓ. The thin solid curve indicates the dispersion for $P_\ell \to 0$. The energy scales are relative to the cavity resonance energy $\hbar\omega_0 = 1.3535$ eV.

flow per unit length perpendicular to the SAW beam, was determined from P_{rf} by taking into account the rf coupling losses measured using a spectrum analyzer.

The layer thicknesses were optimized to enhance the optic and acoustic fields in the cavity region for optic and acoustic wavelengths of $\lambda_0 = 920 - 950$ nm and $\lambda_{\text{SAW}} = 2.8 - 5.6\ \mu\text{m}$, respectively. Figure 1(b) displays a scanning-electron-microscopy (SEM) micrograph of the sample cross-section, where the dark areas correspond to the AlAs layers, while the light areas are GaAs. The lower Bragg mirror BM_1 is composed of 15 periods, each containing a $\lambda/4$ AlAs and a $\lambda/4$ GaAs layer. The first 5 periods of the upper mirror BM_2 (closest to the surface) are identical to those in BM_1, in the remaining 5 periods, which are located adjacent to a $\lambda/2$ cavity, the thickness of the GaAs layers was increased to $3\lambda/4$. While the larger GaAs thickness does not affect the optical properties, it reduces the effective acoustic velocity near the surface and helps to concentrate the SAW fields in this region.

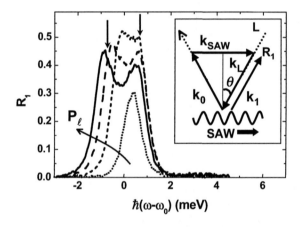

FIGURE 2. First-order diffraction intensity R_1 from a dynamic superlattice induced by SAWs with $\lambda_{\text{SAW}} = 5.6\ \mu\text{m}$ and $P_\ell = 24, 160, 310,$ and 500 W/m. The experimental configuration is shown in the insets, where L indicates the incoming light beam with wave vector k_L, while k_0 and k_1 denote the wave vectors of the reflected and diffracted modes, respectively. k_{SAW} represents the acoustic wave vector.

PHOTONIC CONTROL

The $\lambda/2$ cavities used display a sharp reflectivity minimum (quality factor $Q = 1200$) within the stop-band defined by the BMs, when the vertical (i.e., along z) component of the wave vector k_z of the incident light equals to π/d_{eff}. Here, $d_{\text{eff}} = n_c d_c$ denotes the optical thickness of the cavity layer with refractive index n_c and physical thickness d_c, respectively. For normal incidence, the resonance wavelength becomes $\lambda_0 = 2\pi/k_z$.

The SAW periodically changes the optical resonance wavelength λ_0 along its propagation direction (x-direction) according to $\Delta\lambda/\lambda_0 \sim \Delta d_{\text{eff}}/d_{\text{eff}} = \Delta n_c/n_c + s_{zz}$, where the last term denotes the mechanical modulation of the layer thicknesses through the vertical component $s_{zz} = \Delta d_c/d_c$ of the SAW strain field. [10, 11]

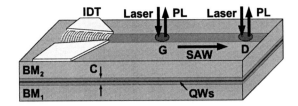

FIGURE 3. Cavity with resonant QWs (In,Ga)As for RT e–h transport. A focusing IDT generates a strong and collimated acoustic beam, thereby defining the transport path. Light spots at G and D are used to generate and detect the carriers.

The elasto-optic contribution $\Delta n_c/n_c$ depends on the light wavelength and polarization. Figure 1(c) shows depth profiles of $\Delta d_{\text{eff}}/d_{\text{eff}}$ induced by SAWs with a linear acoustic power density $P_\ell = 200$ W/m and different acoustic wavelengths λ_{SAW}. $\Delta d_{\text{eff}}/d_{\text{eff}}$ was determined from the spatial distribution of the SAW displacement field calculated with a numerical model. The position of the maximum of $\Delta d_{\text{eff}}/d_{\text{eff}}$ is close to the cavity layer for $\lambda_{\text{SAW}} = 5.6$ μm, but shifts towards the surface with decreasing SAW wavelengths. The calculations were performed for light polarized along y [cf. Fig. 1(c)]. For this polarization, the mechanical and elastooptical contributions have the same sign at the cavity position, with the first one approximately twice as large as the second.

The acoustic modulation creates a lateral optical superlattice characterized by a folded dispersion within the mini-Brillouin zone (MBZ) defined by $|k_x| \leq \pi/\lambda_{\text{SAW}}$, where k_x denotes the photon wave vector component along x. The thin solid curve in Fig. 1(c) shows this dispersion for a vanishing SAW intensity and corresponds to the folding into the MBZ of the quadratic photon dispersion $\hbar\omega = \hbar\omega_0\sqrt{1+[k_x/(n_c k_z)]^2}$ of a planar microcavity. Here, $\hbar\omega_0$ denotes the cavity resonance energy for normal incidence and coincides with the low-energy onset of the dispersion in the absence of a SAW. As the SAW acoustic power increases, gaps open up in the center ($k_x = 0$) and at the boundary ($|k_x| = k_{\text{SAW}}/2$) of the MBZ. Note that the lowest dispersion branch is essentially flat for high acoustic amplitudes, thus indicating that the resonance energy becomes omni-directional.

The modes defining the first stop-band at the edge of the MBZ were accessed via Brillouin scattering by illuminating the samples at the Bragg angle $\theta_B = \tan^{-1}[k_{\text{SAW}}/(2k_0)]$ using the configuration displayed in the inset Fig. 2(b) and measuring the intensity of the backscattered first-order diffraction beam (R_1). The spectra for $P_\ell < 100$ W/m are characterized by a single line, which only appears when the SAW is switched on. In this configuration, the structure behaves as an optical switch with a very high on/off contrast ratio. The stop-band width under these conditions is narrower than the characteristic width of the spectral lines, thus indicating a weak coupling between the zone-boundary photons. For $P_\ell > 100$ W/m, two lines appear with a splitting proportional to the square root of P_ℓ. These lines are attributed to the excitation of the two photon modes defining the stop-band at the edge of the MBZ. These results clearly demonstrate that dynamic superlattices can be created by combining SAWs with photonic cavities.

ELECTRONIC CONTROL

The structures for acoustically induced transport (AIT, Fig. 3) are essentially the same as in Fig. 1(a) except that (i) the cavity layer with an optical thickness λ now contains $In_{0.2}Ga_{0.8}As$ QWs with an emission energy matched with the cavity resonance at RT [12] and (ii) the SAWs are generated by focusing IDTs, which create collimated beams over several tens of μm. [13] The incoming photons are converted to e–h pairs at position G on the SAW path. Since the potential of e and h carriers decreases with the amplitude of the SAW field, the carriers remain inside the acoustic region during transport along the x direction.

The carrier transport is detected by mapping the photoluminescence intensity (PL) along the channel. In order to stop the AIT and induce carrier recombination at a remote position D, it becomes necessary to quench the piezoelectric field at D. This is achieved by generating a large carrier density at D using a second laser spot. These carriers locally screen the piezoelectric field at D and force the recombination of the carriers generated at G and transported by the SAW. The role of the cavity is to reduce the radiative recombination lifetime in order to retrieve most of the emitted photons at D.

The AIT at RT is demonstrated in Fig. 4. Figure 4(a) shows the PL profile in the absence of a SAW showing the emission excited by the laser spots at G and D. The PL intensity at G reduces with increasing SAW amplitude as shown in Fig. 4(b) and by the squares in Fig. 4(c). This reduction is attributed to the spatial separation and transport of the carriers by the SAW field. In contrast, the PL at the detection spot D increases with SAW power up to 22 dBm [cf. Fig. 4(c)]. This enhancement is assigned to the additional recombination of e–h pairs generated at G and transported to D by the SAW field. Finally, for high SAW amplitudes ($P_{\text{rf}} > 22$ dBm), the carrier density at D no longer sufficiently screens Φ_{SAW}, leading to a reduction of the PL intensity at D.

In general, the PL intensities at G and D depend on the light fluence as well as on the SAW intensity. The illumination conditions of Fig. 4(b) were chosen so that

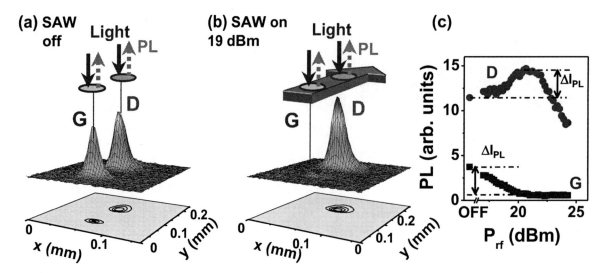

FIGURE 4. Spatially resolved PL displayed on a linear scale (a) without a SAW and (b) with a SAW with $P_{rf}=19.7$ dBm. (c) Dependence of the integrated intensities at spots G and D (cf. Fig. 3) on the SAW intensity expressed in terms of P_{rf}.

the PL at G becomes almost completely quenched for $P_{rf}=20$ dBm, while the carrier density at D is still sufficiently large to effectively screen Φ_{SAW}. Under these conditions, the PL enhancement ΔI_{PL} at D corresponds closely to the PL intensity at G in the absence of a SAW [cf. Fig. 4(c)]. The carriers are then transported with low loss over the approximately 100 μm-long path between the two spots.

The AIT in the structure of Fig. 3 provides a flexible mechanism for dynamic photon control by changing the intensity or position of the light beam focused at D. Another possibility involves the use of orthogonal SAW beams to control the transport direction of the carriers. [12, 14]

CONCLUSIONS

In summary, we have demonstrated that modulation of microcavities by SAWs opens new possibilities for the control of photons and electrons in semiconductor nanostructures. The strongly enhanced acousto-optic interaction in the cavities leads to the formation of a dynamic optical superlattice with a folded dispersion and well-defined gaps that can be used to fabricate active devices such as optical switches and modulators. Furthermore, the confinement of the optical field in cavity structures allows for the optical detection of ambipolar transport of photogenerated carriers up to room temperature. The latter can be explored for the design of electro-optic switches based on the AIT.

ACKNOWLEDGMENTS

We thank H. T. Grahn for comments and for a critical reading of the manuscript as well as S. Krauß and W. Seidel for the preparation of the samples. Support from the Deutsche Forschungsgemeinschaft is gratefully aknowledged.

REFERENCES

1. Yablonovitch, E., *Phys. Rev. Lett.*, **58**, 2059 (1987).
2. John, S., *Phys. Rev. Lett.*, **58**, 2486 (1987).
3. Mekis, A., Chen, J. C., Kurland, I., Fan, S., Villenwuve, P. R., and Joannopoulos, J. D., *Phys. Rev. Lett.*, **77**, 3787 (1996).
4. Kim, S. W., and Gopalan, V., *Appl. Phys. Lett.*, **78**, 3015 (2001).
5. Busch, K., and John, S., *Phys. Rev. Lett.*, **83**, 967 (1999).
6. Korpel, A., *Acousto-Optics*, Marcel Dekker, Inc., New York, 1997.
7. de Lima, Jr., M. M., Hey, R., Santos, P. V., and Cantarero, A., *Phys. Rev. Lett.*, **submitted** (June 2004).
8. Rocke, C., Zimmermann, S., Wixforth, A., Kotthaus, J. P., Böhm, G., and Weimann, G., *Phys. Rev. Lett.*, **78**, 4099 (1997).
9. Santos, P. V., Ramsteiner, M., and Jungnickel, F., *Appl. Phys. Lett.*, **72**, 2099 (1998).
10. de Lima, Jr., M. M., Hey, R., and Santos, P. V., *Appl. Phys. Lett.*, **83**, 2997 (2003).
11. de Lima, Jr., M. M., Santos, P. V., Hey, R., and Krishnamurthy, S., *Physica E*, **21**, 809 (2004).
12. de Lima, Jr., M. M., Hey, R., Stotz, J. A. H., and Santos, P. V., *Appl. Phys. Lett.*, **84**, 2569 (2004).
13. de Lima, Jr., M. M., Alsina, F., Seidel, W., and Santos, P. V., *J. Appl. Phys.*, **94**, 7848 (2003).
14. Alsina, F., Stotz, J. A. H., Hey, R., and Santos, P. V., *Solid State Commun.*, **129**, 453 (2003).

Polarization of Light Emission in Semiconductor Microcavities: Dispersion Mapping

Ł. Kłopotowski*, A. Amo*, M. D. Martín*, L. Viña,* and R. André[†]

Departamento Física Materiales, Universidad Autónoma de Madrid, Cantoblanco, E-28049 Madrid, Spain
[†]*Laboratoire de Spectrometrie Physique, Université Joseph Fourier, F-38402, Grenoble, France*

Abstract. We study the polarization of the light emission from polaritons in semiconductor microcavities. We focus on the non-linear regime and negative detunings. The behavior of the linear and circular polarization dynamics is shown to be determined by the polariton longitudinal-transverse splitting.

Strong coupling between cavity photons and quantum well (QW) excitons leads to appearance of new quasi-particles - exciton polaritons (for a review see [1]). Polaritons have a well defined spin, which makes them potential candidates for applications in many optoelectronic and spintronic devices. This perspective have resulted recently in a number of interesting theoretical and experimental reports. In particular, Aichmayr et al. observed that, at negative detunings and in the non-linear regime, the circular polarization of the $k = 0$ lower polariton (LP) emission exhibited temporal oscillations and that the time-integrated circular polarization was opposite to that of the excitation pulse [2, 3]. The oscillatory behavior was subsequently explained by Kavokin et al. within a pseudospin model taking into account the longitudinal-transverse splitting (Δ_{LT}) of the lower polariton [4].

In this work we address two issues: (i) The existence of a linear polarization which can be inferred from the model in Ref [4]. (ii) The existence of emission with positive circular polarization from certain k-states to compensate the negative value of the integrated circular polarization observed at $k = 0$ and therefore to assure conservation of the total angular momentum.

The sample is a typical wedge-shaped λ Cd$_{0.4}$Mg$_{0.6}$Te microcavity with four QWs placed at the antinodes of the confined electromagnetic field. The emission was excited by 2 ps pulses from a Al$_2$O$_3$:Ti laser pumped with an Ar^{2+} ion laser. The excitation energy was tuned to the first reflectivity minimum after the stop band. The signal was detected and time-resolved by a streak camera with an overall resolution of about 10 ps. We measured the signal at various angles with respect to the normal to the sample surface. The angular resolution was better than 1°, corresponding to a k-vector resolution of the about 10^3 cm^{-1}. Here, we present only the results obtained for negative detunings, where negative polarization and polarization oscillations were observed. The degree of circular/linear polarization (DCP/DLP) under circularly/linearly polarized excitation was evaluated as $P = \frac{I^{co} - I^{cross}}{I^{co} + I^{cross}}$, where I^{co} and I^{cross} are emission intensities of co- and cross-circular/linear signals. The linear polarization of the excitation pulse was horizontal.

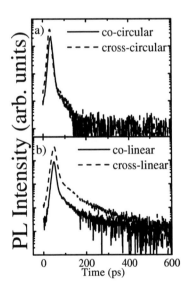

FIGURE 1. Time-resolved LP emission in two circular (a) and in two linear (b) polarizations under circularly σ^+ (a) and linearly (horizontal) (b) polarized excitation.

Figures 1a) and 1b) present temporal decays of LP emission at $k = 0$, in the nonlinear regime, for the circular and linear excitation, respectively. In the former case, in about 150 ps after the excitation, I^{co} and I^{cross} become equal. On the contrary, in the case of linear excitation, I^{cross} remains larger than and I^{co}, even 400 ps after excitation. These results are clearly evidenced in Fig 2,

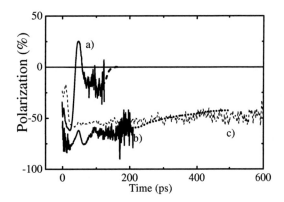

FIGURE 2. Temporal traces of DCP (a) and DLP (b) of LP emission. (c) DLP of LP emission in the linear regime. Dashed lines represent a mean value where the experimental data becomes very noisy.

where temporal traces of DCP (a) and DLP (b) are plotted. Oscillations of DCP, resulting in an ultrafast change from −60% to +25 % in just 27 ps are clearly resolved. According to Ref. [4], these oscillations occur due to beating between linear, longitudinal and transverse, polarization states separated by Δ_{LT}. This process is analogous to DCP oscillations of a bulk semiconductor photoluminescence in a magnetic field applied in Voigt geometry. On the other hand, a linear polarization, as high as

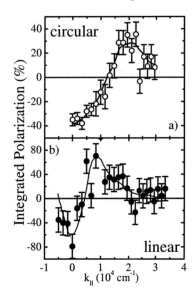

FIGURE 3. Time-integrated DCP (a,○) and DLP (b,●) polarization as a function of LP in-plane wave vector. Lines are guides to eye.

−50%, persists as long as there is an observable signal (Fig. 2). Negative DLP value indicates that the majority of polaritons occupies the lower energy state, as expected for a thermalized population. Since $\Delta_{LT} > 0$, the vertical polarization state has lower energy and thus we observe a negative polarization. This effect, in turn, is analogous to thermalization of linear excitons on the states split by a magnetic field in Voigt geometry. Moreover, even in DLP, we resolve an oscillatory behavior at short delays after the excitation. These oscillations are only present in the non-linear regime (absence of oscillations in Fig 2c). Oscillations in DLP suggest the that linear states are not the eigenstates of the system. The proper eigenstates therefore must have an admixture of circular ones, i.e. they are elliptical. In other words, the proper "bulk" analog is a magnetic field applied in an oblique direction. However, since we do not observe a significant, constant, equilibrium DCP, we conclude that the circular admixture (the deviance from Voigt configuration in the "bulk" analog") is small.

Figure 3 presents time-integrated DCP (a) and DLP (b) as a function of k. DCP increases from about −35% at $k = 0$ to +35% at $k = 2 \times 10^4 \text{cm}^{-1}$. With a further increase of k, DCP vanishes. This result proves the total angular momentum conservation in our system. Moreover, it points out that the stimulated scattering prefers the cross-circular polariton state. This effect can be related to a spin splitting of LP observed in emission excited non-resonantly with a circularly polarized laser pulse [2]. Due to a spin-dependent interaction between the excitonic parts of the polariton, the minority spin population state is shifted to lower energies [5]. Consequently, the stimulation to the cross-circular state occurs and a negative DCP at $k = 0$ is observed. Similar result is obtained for DLP. At $k = 0$ DLP is negative due to scattering to a lower, vertical linear polarization state. As k is increased, DLP becomes positive. In summary, we have shown that the polariton longitudinal-transverse splitting strongly affects the LP polarization dynamics. In particular, circular polarization exhibits oscillations and linear polarization reaches an equilibrium value. The dispersion mapping of the time-integrated polarization allowed us to demonstrate that the emission stimulation occurs in the spin subband of lowest energy.

This work was partially supported by the Spanish MCYT (MAT2002-00139), the CAM (07N/0042/2002) and the "Marie-Curie" MRTN-CT-2003-503677 "Clermont2" ("Physics of Microcavities").

REFERENCES

1. Semicond. Sci. Technol. Special Issue on Semiconductor Microcavities (ed. by J. J. Baumberg and L Vina) **18** (2003).
2. M. D. Martín *et al.*, Phys. Rev. Lett. **89**, 77402 (2002).
3. G. Aichmayr *et al.* Semi. Sci. Technol. **18**, 368 (2003).
4. K. V. Kavokin *et al.* Phys. Rev. Lett. **92**, 17401 (2004).
5. J. Fernández-Rossier *et al.* Phys. Rev. B **54**, 11582 (1996).

Optically pumped lasing from localized states in quantum-well and quantum-dot microdisks

T. Kipp*, K. Petter*, Ch. Heyn*, D. Heitmann* and C. Schüller*

*Institut für Angewandte Physik und Zentrum für Mikrostrukturforschung, Universität Hamburg, Germany

Abstract. We report on micro photoluminescence (PL) experiments on single semiconductor microdisks at low temperature T = 5 K. Two types of microdisks were investigated, containing (a) multiple quantum wells (QWs) of ternary AlGaAs alloys and (b) one layer of self-assembled InAs quantum dots (QDs). At low excitation strength we observe in both type of samples discrete peaks arising from the recombination of excitons localized, (a), at impurities within the ternary alloy QWs and, (b), in the InAs QDs. With increasing excitation strength, the emitted light is concentrated in several well-defined whispering gallery modes (WGMs) with typical widths of less than 1 meV distributed over an energy range broader than 200 meV. At even higher excitation strengths, we find a clear threshold behavior for the power emitted into the WGMs, suggesting the onset of lasing.

Semiconductor microdisks containing either quantum wells (QWs) or self-assembled quantum dots (QDs) have gained considerable interest in recent years, both for their possible application in new optoelectronic devices, like lasers [1, 2, 3, 4], and for experiments in fundamental cavity quantum electrodynamics [5, 6, 7]. Here, we compare excitation-power dependent micro-photoluminescence (PL) measurements at low temperature T = 5 K on single microdisks containing ternary alloy QWs with those on disks containing QDs. We find that localized excitons in the QW microdisks lead to spectra similar to the ones of QD microdisks.

Our microdisks are processed by laser-interference lithography, reactive-ion etching and selective wet-etching processes. Both types of disks have diameters of typically 1.6 μm, standing on pillars, 0.5 μm wide and 1 μm high. The QW microdisks consist of ten 10 nm-wide $Al_{0.2}Ga_{0.8}As$ QWs between $Al_{0.4}Ga_{0.6}As$ barriers on a GaAs pillar; the actual resonator disk has a height of $h = 240$ nm. Details on their fabrication process can be found in Ref. [8]. The resonator of the QD microdisks ($h = 248$ nm) is composed of one layer of self-assembled InAs QDs, surrounded by 100 nm GaAs, 20 nm $Al_{0.3}Ga_{0.7}As$ and 4 nm GaAs on both sides, standing on a $Al_{0.7}Ga_{0.3}As$ pillar. The preparation process of the QD microdisk is similar to the one described in Ref. [8], except that here a diluted HF solution is used as a selective etchant instead of the citric-acid based solution.

Micro-PL measurements on single microdisks were performed at T = 5 K in an optical continuous flow cryostat. The exciting laser was focussed on the sample by a microscope objective ($\times 80$, 0.75 NA). The emitted light

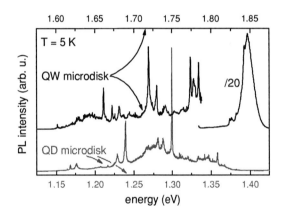

FIGURE 1. PL spectra of a QW microdisk (black) and of a QD microdisk (grey). Note the shifted energy ranges.

was collected by the same objective and then spectrally analyzed by a triple Raman spectrometer.

Figure 1 compares a spectrum of a QW microdisk with the one of a QD microdisk. Note, that both spectra belong to identically scaled but shifted energy axes. In both spectra, several WGMs are visible, distributed over a broad energy range of more than 200 meV, with typical widths of less than 1 meV. The QW microdisk spectrum at low temperatures resembles the room-temperature spectra published in Ref. [9]. The intense signal at about 1.85 eV originates in the recombination of free excitons of the QWs. The WGMs appear in a broad energy range below the free exciton emission, in which localized excitons of the alloy QWs emit. [9] In the QD microdisk spectrum, the WGMs enhance light arising from the re-

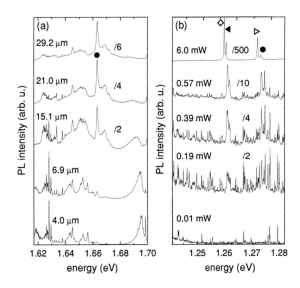

FIGURE 2. Parts of spectra from (a) a QW microdisk and (b) a QD microdisk. The spectra belong to different excitation powers as given in the figure.

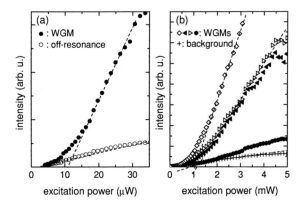

FIGURE 3. Intensities of the WGMs marked in Fig. 2 vs. excitation power. The broken lines point up the threshold behavior.

combination of excitons localized in the QDs. The overall slope of this spectrum is strongly affected by the sensitivity of our CCD camera: it is rapidly decreasing for energies lower than 1.25 eV. Macro-PL measurements with a Fourier transform spectrometer prove the ground-state transition of the QDs to be at 1.10 eV. In Fig. 1, we therefore observe WGMs only in the high energy range, in which excited states of the QDs emit.

Figure 2 shows parts of (a) the QW and (b) the QD microdisk spectra for different excitation powers. In both microdisk systems, sharp peaks of single localized excitons are resolved at low excitation power (lower spectra). These excitons are, for the QW microdisk, bound to impurity levels of the alloy QWs [Fig. 2(a)], and, for the QD microdisk, bound to excited levels of the self-assembled QDs [Fig. 2(b)]. With increasing excitation power, more of these peaks appear, finally summing up to a broadband background emission. Additionally, with increasing excitation power, WGMs develop (marked by symbols in Fig. 2), enhancing parts of the background emission. Note, that single localized excitons of the QW sample could not be observed in measurements at room temperature. [9]

Figure 3 depicts the intensity of the WGMs marked in Fig. 2 in dependence on the excitation power. Here, and also in Fig. 2, it becomes obvious that in the case of QD microdisks [Fig. 3(b)], WGMs develop at much higher excitation power as compared to the case of QW microdisks [Fig. 3(a)]. This is because the QDs have to be strongly pumped to accomplish emission out of excited states. The intensities of the WGM discussed in Fig. 3(a) (QW microdisk) and of at least three of the WGMs discussed in Fig. 3(b) (QD microdisk) show a clear nonlinearity, in contrast to the off-resonance intensities. This threshold behavior suggests the onset of lasing in the microdisks.

ACKNOWLEDGMENTS

We acknowledge financial support by the DFG via the SFB 508, the Graduiertenkolleg "Felder und lokalisierte Atome" and the Heisenberg grant SCHU1171/2.

REFERENCES

1. McCall, S. L., Levi, A. F. J., Slusher, R. E., Pearton, S. J., and Logan, R. A., *Appl. Phys. Lett.*, **60**, 289 (1992).
2. Baba, T., Fujita, M., Sakai, A., Kihara, M., and Watanabe, R., *IEEE Photonics Technol. Lett.*, **9**, 878 (1997).
3. Luo, K. J., Xu, J. Y., Cao, H., Ma, Y., Chang, S. H., Ho, S. T., and Solomon, G. S., *Appl. Phys. Lett.*, **77**, 2304 (2000).
4. Michler, P., Kiraz, A., Zhang, L., Becher, C., Hu, E., and Imamoglu, A., *Appl. Phys. Lett.*, **77**, 184 (2000).
5. Gayral, B., Gérard, J. M., Lemaître, A., Dupuis, C., Manin, L., and Pelouard, J. L., *Appl. Phys. Lett*, **75**, 1908 (1999).
6. Gérard, J.-M., and Gayral, B., *J. Lightwave Technol.*, **17**, 2089 (1999).
7. Michler, P., Kiraz, A., Becher, C., Schoenfeld, W. V., Petroff, P. M., Zhang, L., Hu, E., and Imamoglu, A., *Science*, **290**, 2282 (2000).
8. Petter, K., Kipp, T., Heyn, C., Heitmann, D., and Schüller, C., *Appl. Phys. Lett.*, **81**, 592 (2002).
9. Kipp, T., Petter, K., Heyn, C., Heitmann, D., and Schüller, C., *Appl. Phys. Lett.*, **84**, 1477 (2004).

Dephasing in cavity-polariton mediated resonant Raman scattering.

A. Bruchhausen*, A. Fainstein* and B. Jusserand[†]

*Instituto Balseiro and Centro Atómico Bariloche, CNEA, 8400 S. C. de Bariloche, Río Negro, Argentina.
[†]Laboratoire de Photonique et de Nanostructures, CNRS, Route de Nozay, 91460 Marcoussis, France.

Abstract. We present a phenomenological extension to the factorization model used to describe cavity polariton mediated Raman scattering, which takes polariton dephasing effects due to cavity-photon and exciton damping into account. The improved theory explains recent experimental results.

Recent experiments of cavity-polariton mediated resonant Raman scattering (RRS)[1–4], that cannot be accounted for by the used simplified theories, point into the relevance of including dephasing and damping effects in these descriptions [2, 4]. In particular, shifts of the Raman scan maxima have been observed in experiments in II-VI cavities that could be related to exciton lifetime effects.

These effects can be included phenomenologically using a factorization model as follows:

$$\sigma \propto T_i \, |\langle \mathbf{K}_s | H_{e-ph} | \mathbf{K}_i \rangle|^2 \, T_s \, \rho(k_\parallel^s) \, P(E_s) \big|_{E_i = E_s + \hbar\omega_{ph}} \quad (1)$$

The sequential processes described by (1), consists of three steps: (i) the transmission T_i of the incident photon into the cavity as an initial state ($|\mathbf{K}_i\rangle$). (ii) A phonon (ω_{ph}) induced inelastic scattering from $|\mathbf{K}_i\rangle$ to a final cavity polariton state $|\mathbf{K}_s\rangle$. (iii) The coupling of $|\mathbf{K}_s\rangle$ to the photon continuum outside the cavity (represented by the transmission factor T_s). Considering the outgoing RRS case, T_i is basically given by the Bragg mirror's residual transmittance, and can essentially be taken as constant[4]. T_s couples the final cavity internal state with the outside photon continuum, and is therefore proportional to the photon strength (S_p^s) of the final polariton state. Since the incoming channel is non-resonant, and the polaritons couple with phonons only through their exciton component, the squared matrix element $|\langle \mathbf{K}_s | H_{e-ph} | \mathbf{K}_i \rangle|^2$ is essentially proportional to the exciton strength of the final polariton state. For a three polariton branch system, i.e. a system with a double anticrossing between the cavity-photon and the 1s and 2s exciton states, the squared matrix element is proportional to $|a_{X_{1s}}^s + \alpha a_{X_{2s}}^s|^2 \propto S_{1s}^s + \alpha^2 S_{2s}^s + \alpha \beta S_{1s,2s}^s$[4]. Here $a_{X_n}^s$ stands for the nth exciton weight of the polariton final state, $S_n^s = |a_{X_n}^s|^2$ are the exciton strengths,

and $S_{1s,2s}^s = a_{X_{1s}}^{s*} a_{X_{2s}}^s + a_{X_{1s}}^s a_{X_{2s}}^{s*}$ is an interference term. $\alpha = \langle X_{2s}^s | H_{e-ph} | \mathbf{K}_i \rangle / \langle X_{1s}^s | H_{e-ph} | \mathbf{K}_i \rangle$, takes into account the difference of the scattering matrix elements between the common exciton initial state and the different exciton components of the final polariton state[5]. The coherence factor β has been introduced in the assumption of a partial coherence between the polarization associated with the two intervening exciton levels. The factor $\rho(k_\parallel^s)$, stands for the density of final states, which turns out to be detuning independent for experiments with a fixed $k_\parallel$ and a small collection cone, as those performed in Ref.[4]. In these experiments the detuning between the cavity-modes is changed by varying the spot position over the sample, grown with a cavity thickness gradient.

Finally, $P(E_s)$ describes, within Fermi's Golden rule[6], the *spectral density*, i.e. the distribution of probability, of the final polariton states. Non-inclusion of polariton lifetime effects implies that this distribution is a delta of energy centered at the final polariton energy, as has been done in Refs. [3, 4]. But, as experimental results have shown, polariton lifetime effects are not negligible. Cavity polaritons have a lifetime, which is indeed strongly detuning dependent [2–4, 8]. To take into account these effects the considered spectral density will be taken as a lorentzian probability function of the energy of line broadening Γ_s, centered at the final polariton energy [7]. Since the exact outgoing resonance case is analyzed here, due to the energy conservation condition $E_i = E_s + \hbar\omega_{ph}$ this distribution has to be specified at its maximum. Thus $P(E_s) = \Gamma_s^{-1}$. It can be shown, on the other hand, that the line broadening as a function of detuning, is in a good approximation $\Gamma_s = S_p^s \gamma_p + S_{1s}^s \gamma_{1s} + S_{2s}^s \gamma_{2s}$[9], where the γ_n correspond to the cavity-photon or exciton homogeneous line widths. From (1), the polariton mediated outgoing RRS efficiency for a three polariton branch cavity is then proportional to:

$$\sigma \propto \Gamma_s^{-1} S_P^s (S_{X_{1s}}^s + \alpha^2 S_{X_{2s}}^s + \alpha\beta S_{X_{1s,2s}}^s). \quad (2)$$

Using expression (2), we analyze the outgoing Raman profile for the medium polariton. Variations of the polariton modes as function of the spot position on the sample, as well as the Rabi-splitting between the cavity-photon (P) and exciton (X_{1s}, X_{2s}) modes ($\Omega_{1s} = 13$ meV and $\Omega_{2s} = 4$ meV respectively) have been chosen similar to those reported in Ref.[4]. The dispersion of the polariton lower, medium and upper (LP, MP and UP) modes (full lines), together with the "free" modes (dotted lines) are plotted in Fig.1(a). In Fig.1(b), the calculated exciton (X_{1s} and X_{2s}) strengths of the MP mode (dashed and dotted), as well as the interference term (dash-dotted), are shown. The thin full curve corresponds to the cavity-photon strength (S_P). The thick full curve corresponds to the factor the factor $\Gamma_{MP}^{-1} S_P^{MP}$ that takes the lifetime contribution into account. Γ_{MP} was normalized to unity in order to compare both curves. Figure 1(c) compares the calculated MP RRS profiles considering (full line) and *not* considering (dashed line) damping. Both curves have been normalized to unity. The parameters α and β have been chosen as 1 and 0 respectively, to better reproduce the experimental profile of Ref.[4], and $\gamma_P = 1$ meV and $\gamma_{X_{2s}} = \gamma_{X_{2s}} = 2$ meV.

As pointed out in previous works[3, 4], if damping is neglected, the RRS profile maxima are to be expected at the point where exciton and photon strengths are the same. This can be clearly observed in Fig.1 where the vertical doted lines indicate the maxima of the RRS profiles. Note also that the maxima coincide with the position where the distance between the modes LP-MP and MP-UP are smaller. If the damping factor is included, a notable shift of the maxima, towards a larger photonic weight of the MP polariton is observed. The shift of the first maximum is consistent with the experimentally observed shift of ~ 2 meV. If lifetime parameters are chosen in a way that $\gamma_P > \gamma_{X_n}$, the shift occurs towards the opposite direction, indicating that the exciton lifetime is mainly responsible for the experimentally observed shift.

Variations of the interference factor β have been discussed in Ref. [4]. It is observed that due to the interference effects, for increasing β the peaks are slightly moved apart towards opposite directions as the dip gets more pronounced. If the factor α is reduced, the second peak, due to the effect of the second anticrossing, becomes weaker until it disappears. The first maxima, on the other hand, is rather insensitive to α. For a less phenomenological inclusion of α, this matrix elements need to be evaluated rigorously.

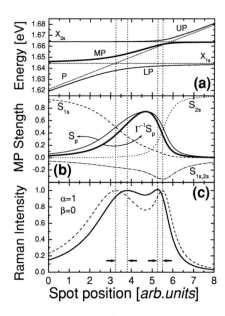

FIGURE 1. (a) Variation of the cavity-polariton modes (full lines) and non interacting modes (dashed lines) with spot position on the sample. (b) Medium polariton (MP) strengths (see text for details). (c) MP-RRS profile with (full line) and without (dashed line) dumping factor Γ_{MP}.

Summarizing, we briefly showed how the inclusion of polariton lifetime, in particular exciton damping, improves qualitatively the description of the RRS phenomena and is essential to explain resent experimental results.

REFERENCES

1. A. Fainstein, B. Jusserand, and V. Thierry-Mieg, *Phys. Rev. Lett*, **78**, 1576 (1997).
2. W. R. Tribe, D. Baxter, M. S. Skolnick, D. J. Mowbray, T. A. Fisher, and J. S. Roberts, *Phys. Rev. B*, **56**, 12429 (1997).
3. A. Fainstein, B. Jusserand, and R. André, *Phys. Rev. B*, **57**, R9439 (1998).
4. A. Bruchhausen, A. Fainstein, B. Jusserand, and R. André, *Phys. Rev. B*, **68**, 205326 (2003).
5. Note that in Ref.[4], α has been taken equal to unity.
6. See for example: Sakurai, J. J., *Modern Quantum Mechanics*, Benjamin/Cummings, Menlo Park, Claifornia, 1985, pp. 325–335.
7. A rigorous derivation of this can be achieved using Green's function formalism. See for example: M. L. Goldberger and K. M. Watson, *Collision Theory*, Wiley, NY, 1964.
8. P. Borri, J. R. Jensen, W. Langbein, and J. M. Hvam, *Phys. Rev. B*, **61**, R13377 (2000).
9. J. Wainstain, C. Delalande, D. Gendt, M. Voos, J. Bloch, V. Thierry-Mieg, and R. Planel, *Phys. Rev. B*, **58**, 7269 (98).

CHAPTER 5
ELECTRONIC AND OPTICAL PROPERTIES

- 5.1. Band Structures
- 5.2. Intersubband and Interband Transitions
- 5.3. Excitons and Condensates
- 5.4. High Frequency and Microwaves
- 5.5. Light Scattering and Nonlinear Optics

The Inverse Band Structure Approach: Find the Atomic Configuration that has Desired Electronic Properties

Alex Zunger, S.V. Dudiy, K. Kim, and W.B. Jones

National Renewable Energy Laboratory, Golden, CO 80401

Abstract. In conventional band theory, one specifies first the atomic configuration, and then solves for the electronic properties. We consider the inverse process where one specifies first some target electronic properties, and then searches for atomic configurations that have such properties. Application of the Genetic Algorithm, with additional advantages from its parallel computer implementation, significantly extends our capabilities in treating more demanding IBS problems, as illustrated by searching for the configuration of nitrogen impurity clusters in GaP that have given properties.

Material science often advances via accidental discoveries. After an interesting property is discovered -- such as the semiconductivity of silicon, the superconductivity of cuprates, or the magnetoresistivity of lanthanides -- an appropriately useful application is envisioned, and only then does basic research start. The shortcoming of this time-honored historical pattern is that the R&D period following accidental discovery is often very long, because one tries to figure out what it is that one actually found. The obvious question is whether one can reverse the process: Can one articulate a needed property, and then search within a large but specific family of materials which atomic configuration has this property? [1]. This Inverse Band Structure (IBS) approach (given the answer, find the question) makes sense for a number of reasons. First, "form controls function", *i. e.* atomic configuration does decide electronic properties, as can be gleaned *e.g.* by comparing the varying band gaps of different crystallographic ordering patterns of III-V alloys. Second, it is possible, within a certain range, to realize experimentally different atomic configurations of the same chemical system. Examples include the achievement of thermodynamically unstable [2], but extremely long-lived, III-V/III-V superlattices in vapor-phase growth [3], or the ability to position atoms via STM tip in unusual configurations [4]. Examples of the Inverse Questions could be: "Which configuration of AlAs/GaAs has the maximum band gap?"; or [5] "Which bulk arrangement of Si and Ge atoms fulfills the momentum conservation conditions needed to maximize impact ionization?", etc.

There are three obvious problems to solve:

(1) Quantum Chemistry and Solid State Theory often proceed by first stating the atomic configuration, then solving the Schrodinger equation, whereas the IBS approach requires deducing the configuration from the electronic structure. There is no intuitive path to do so. Clearly, we need to a method that learns the form vs. function relationship.

(2) There could be an astronomic number of possible configurations to explore. For example, an A_xB_{1-x} alloy described with a supercell of N lattice sites can have $N!/(xN)!/(N-xN)!$ possible configurations. Even for $x=0.25$ and $N=32$, this gives 10^{14} configurations. Which of these has, say, the largest band gap? Clearly, one needs a "fast learning" search algorithm that "visits" only a small fraction of all possible configurations, yet has a large probability of finding the right one.

(3) Predicting quantities such as band gaps or effective masses requires often an atomistic description, not just simplistic models such as "particle in a box". Indeed, the band gaps of $(AlAs)_n/(GaAs)_m$ for orientation $G=(001)$, (110), (111) have complex non monotonic dependencies on (n,m) [6], which can not be guessed by conventional "band gap engineering" based just on well-width. Clearly, one needs a fast way to solve the atomistic Schrodinger equation.

While previously we addressed these issues [1] in the context of electronic properties of semiconductors using the "Simulated Annealing" (SA) approach [7], we have recently used an alternative approach based on Genetic Algorithm (GA) [8] (with implementation of Refs. [9,10]). The electronic structure is evaluated as previously [1] (empirical pseudopotentials).

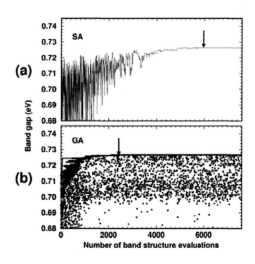
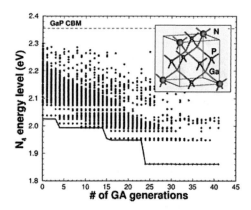

FIGURE 1. IBS search for maximal band gap of $In_{17}Ga_{15}As_{32}(001)$ superlattice using (a) Simulated Annealing (SA) and (b) Genetic Algorithm (GA). The plots show the calculated band gaps of tried structures vs. the number of band structure evaluation steps. The maximal band gap after 6000 steps in SA [arrow in (a)] is still smaller than the one reached by GA after 2400 steps [arrow in (b)].

In our calculation, we start with an initial set ("population") of configurations ("genomes"), represented as strings of integer numbers ("genes"), do band structure calculations to evaluate the "fitness" of each genome (the distance of the calculated property from the desired target). Then, we keep the fittest N_{pop}-N_{rep} genomes unchanged, whereas the least fit N_{rep} ones are replaced via "mating" (gene exchange between two "parent" strings) and "mutation" (swapping genes within a genome) of the fittest ones. The new set ("generation") of genomes is again subjected to evaluation of fitness. This cycle continues until one reaches the desired target, or exhausts the chosen budget of iterations.

Our extensive comparisons of GA and SA performance, using the IBS problem of band gap maximization for InGaAs/InP(001) supercells show that GA needs significantly less band structure calculations than SA to reach the same quality result, as illustrated by Fig. 1. The increase in efficiency of the IBS search algorithm due to replacement of SA with GA, combined with parallel implementation [9,10], significantly extends our capabilities of treating IBS problems to cases of larger configuration space and/or more computationally costly band structure evaluations. The characteristic example of the latter is our GA IBS study of nitrogen clusters in GaP, where N-N interactions go significantly beyond 1st nearest neighbors, and large simulation supercells are necessary [11]. We place N_n clusters (n=2-5) in a cubic 126 atom subcell within a large, 1728 atom, GaP supercell and search, via IBS, for those configurations with certain extremal properties, such as the deepest

FIGURE 2. (a) GA IBS search for an N_4 cluster configuration in GaP with the deepest level in the GaP gap. The horizontal dashed line shows the energy position of the GaP conduction band minimum (CBM). The insert shows the found optimal structure of N_4. Energy is with respect to the valence band maximum.

levels in the gap. An example of the GA history is presented in Fig. 2, where we efficiently identify the N_4 configuration with the deepest level in the GaP gap. The winning configuration is a tetrahedron with 4th nearest neighbor NN pairs as its edges. To find the answer, we had to calculate band structures of only ~1500 configuration out of 10^7 possible configurations within our supercell. We have applied this method [12] to identify the structure of N clusters in GaP which have given properties, such as the deepest or shallowest gap levels, the strongest absorption, or the minimal strain energy. The results are both interesting and unexpected [12].

This work is supported by DARPA and the U.S. Department of Energy, SC-BES-DMS Grant No. DEAC36-98-GO10337.

REFERENCES

1. A. Franceschetti and A. Zunger, *Nature* **402**, 60 (1999).
2. A. Zunger and S. Mahajan, in *Handbook of Semiconductors*, 2nd ed., edited by S. Mahajan (Elsevier, Amsterdam, 1994), Vol. **3**, p. 1399.
3. R. F. C. Farrow, *Molecular Beam Epitaxy: Application to Key Materials*, (Noyes Publications, New Jersey, 1995).
4. D. Eigler and E.K. Schweizer, *Nature* **344**, 524 (1990).
5. J.H. Werner *et al*, *Phys. Rev. Lett.* **72**, 3851 (1994).
6. K. Mader *et al*, *J. Appl. Phys.* **78**, 6639 (1995).
7. S. Kirkpatric *et al*, *Science* **220**, 67 (1983).
8. D.E. Goldberg, *Genetic Algorithms in Search, Optimization, and Machine Learning* (Addison-Wesley, Reading, MA, 1989).
9. D. Levine, Parallel Genetic Algorithm Library (1998).
10. K. Kim and W.B. Jones, IAGA code (2003).
11. P.R.C. Kent and A. Zunger, *Phys. Rev. B* **64**, 115208 (2004).
12. S. V. Dudiy and A.Zunger, to be published.

Energy-Band Structure Of Si, Ge And GaAs Over The Whole Brillouin Zone Via The k.p Method

Soline Richard, Frédéric Aniel and Guy Fishman

Institut d'Électronique Fondamentale, UMR 8622 CNRS, Université Paris Sud, 91405 Orsay Cedex, France

Abstract. We show that the **k.p** method taking into account spin-orbit coupling allows one to account for the band structure of Si, Ge and GaAs over the whole Brillouin zone on an extent of 5 eV above and 6 eV under the top of the valence band. The use of thirty bands provides effective masses in agreement with experimental data both for direct gap semiconductors (GaAs) and for indirect gap semiconductors (Si, Ge). The accuracy for the effective masses of the bottom of conduction band is of the order of one per cent as well for direct gap (GaAs) as for indirect gap (Si, Ge).

INTRODUCTION

The **k.p** method is usually used as a perturbation method to describe the details of the band structure around a given point of the Brillouin zone.[1] However this method may be used over a large range of the Brillouin zone if many bands are taken into account. More precisely Cardona and Pollak [2] showed that the use of fifteen bands leads to the description of the energy band structure of silicon and germanium, without spin, over the whole Brillouin zone. The aim of this paper is to show that it is possible to take into account the spin-orbit coupling with a thirty band **k.p** method and this leads to an accuracy of the same order or better than of comparable calculation by LCAO method.[3]

30-BAND k.p HAMILTONIAN AND BAND STRUCTURE

We start from the 15×15 Hamiltonian of Cardona-Pollak. For Si and Ge (O_h group) we keep the ten interband parameters introduced by Cardona-Pollak and we introduce three intraband spin-orbit splittings for the p valence band (Γ_{5V}^+), p conduction band (Γ_{4C}^-), the d conduction band (Γ_{5d}^+). This leads to a H_{30} 30×30 Hamiltonian. For GaAs (T_d group) we furthermore introduce one new interband parameter between Γ_{1C} and Γ_{5C}, this parameter being null in T_d group. *A priori* there are some other interband parameters but we have found no improvement if they are taken not equal to zero. The present calculation taking into account the real d levels in the H_{30} Hamiltonian makes unnecessary the use of Luttinger-like parameters so that eventually we use fewer parameters than in a sp^3s^* **k.p** Hamiltonian.[4]

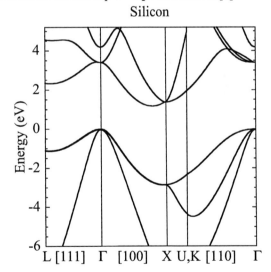

FIGURE 1. Band diagram of bulk Si at T = 0 K. Spin-orbit coupling is taken into account even if it does not appear on the diagram because of its scale. Δ_{so} = 44 meV in Si.

The continuity between U [1, ¼, ¼] and K [0, ¾, ¾], equivalent points of the Brillouin zone, is not obtained by construction as in pseudopotential or LCAO: on the contrary, it is the strongest numerical difficulty of this method. Figures 1, 2 and 3 show the band structures of Si, Ge, and GaAs obtained with our **k.p** model. The band structure is well reproduced on a width of about 11 eV: it describes correctly the valence band over a 6 eV scale and the lowest four conduction bands over a 5 eV scale.

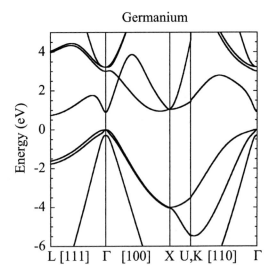

FIGURE 2. Band diagram of Ge at T = 0 K.

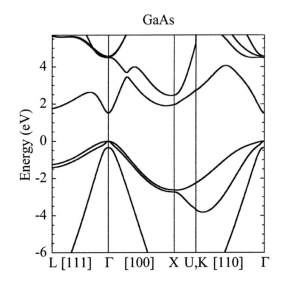

FIGURE 3. Band diagram of GaAs at T = 0 K.

Some numerical results are given Table 1. The accuracy is better than one per cent for all the effective masses of the bottom of conduction band. This is quite satisfactory especially for the indirect gap masses when considering that the H_{30} Hamiltonian is built with Bloch functions of the Brillouin zone center.

TABLE 1. Effective masses in the conduction band obtained with the 30 band k.p method, compared to experimental data (*in italic*).[5]

	$m(\Gamma)$	$m_t(X)$	$m_l(X)$	$m_t(L)$	$m_l(L)$
Si		0.1912 (*0.1905*)	0.9167 (*0.9163*)	0.128	1.65
Ge	0.038 (*0.038*)	0.195	0.93	0.0818 (*0.0815*)	1.593 (*1.59*)
GaAs	0.0676 (*0.067*)	0.23	1.16	0.108	1.67

CONCLUSION

We have shown that the 30 band **k.p** model including the spin-orbit coupling allows one to obtain a very accurate description of the band structure on the whole Brillouin zone with 10 (for O_h group) or 11 (for T_d group) adjustable parameters. Its accuracy is comparable to LCAO or pseudo-potential method. Compared to the 20-band **k.p** method, we have shown it is easier and more efficient to take into account real *d* states than to mimic them by Luttinger-like parameters and s* fictive bands. More particularly this 30-band model gives access to second conduction band which make possible to perform full band transport calculation including impact ionization or other high field effects.

REFERENCES

1. E. O. Kane, *Semiconductors and Semimetals*, edited by R. K. Willardson and A. C. Beer, Academic Press, New York (1966), Vol. 1, pp. 75-100.

2. M. Cardona and F. Pollak, *Phys. Rev.* **142**, 530 (1966).

3. J.M. Jancu, R. Scholz, F. Beltram, and F. Bassani, *Phys. Rev. B* **57**, 6493 (1998).

4. N. Cavassilas et al., *Phys. Rev. B* **64**, 115207 (2001).

5. *Semiconductors: Intrinsic Properties of Group IV Elements and III-V, II-VI and I-VII Compounds*, edited by O. Madelung, Landölt-Bornstein, New Series, Group III, Vol. 22, Part a (Springer, Berlin, 1987)

Screened Exchange Calculations of Semiconductor Band Structures

Michael C. Gibson*, Stewart J. Clark*, Stuart Brand* and Richard A. Abram*

*Department of Physics, University of Durham, DH13LE, United Kingdom

Abstract. We have calculated the band structures of several group IV and III-V semiconductors from first principles. The technique we have employed is the screened exchange method, which is an extension of the commonly used Kohn-Sham density functional theory (DFT). In standard DFT, exchange and correlation effects are usually treated with the local density (LDA), or generalised gradient (GGA), approximations. While good structural properties can be obtained for most materials with these approximations, the electronic band-gap is often severely underestimated. The screened exchange method employed here treats exchange non-locally, in a similar manner to the Hartree-Fock method. We find that the screened exchange method significantly improves results in most cases, although the computational cost is somewhat higher.

INTRODUCTION

Standard Kohn-Sham density functional theory (DFT) [1, 2] is a well established and commonly used method for calculating properties of condensed matter systems from first principles. DFT is, in principle, an exact theory of the electronic ground state, although in practice approximations must be made in the treatment of exchange and correlation effects. Computationally cheap exchange-correlation functionals such as the LDA [2] and GGA [3] provide very good descriptions of the structural properties of most materials. However, other properties, such as the electronic band gap, are not so well described. The reason for this may be in part due to the fact that DFT is only an exact theory of the ground state, and so should not necessarily be expected to predict exited state properties, but is mainly due to the approximations made in the LDA and GGA.

The screened exchange method [4] involves a generalisation of the Kohn-Sham formalism so that the exchange-correlation potential includes a non-local component in the form of an integral operator. This is essentially the Hatree-Fock method with the exchange interaction screened at long range. We have employed the screened exchange method to calculate the band structures of a number of group IV and III-V semiconductors.

In this article we briefly outline the screened exchange methodology and computational procedures. We then present and discuss the results of our screened exchange calculations in comparisson to the LDA.

THEORY

The standard Kohn-Sham scheme involves the mapping of the true many-electron wavefunction onto a fictitious non-interacting system with the same density profile, $\rho(\mathbf{r})$. The local Kohn-Sham potential for the non-interacting system includes a component associated with exchange and correlation effects. This exchange-correlation potential is defined as

$$v_{XC}(\mathbf{r}) = \frac{\delta}{\delta\rho(\mathbf{r})} E_{XC} \quad (1)$$

where E_{XC} is the exchange-correlation energy. It is this term that is approximated in the LDA and GGA. In the screened exchange scheme a non-local component is included in the exchange-correlation potential. It is given by

$$v_{SX}^{NL}(\mathbf{r},\mathbf{r}') = -\sum_i \frac{\phi_i^*(\mathbf{r}')\phi_i(\mathbf{r})}{|\mathbf{r}'-\mathbf{r}|} e^{-k_{TF}|\mathbf{r}'-\mathbf{r}|} \quad (2)$$

where the $\phi_i(\mathbf{r})$ are the single-particle Kohn-Sham orbitals. The exponential constant, k_{TF}, is the Thomas-Fermi screening length and is calculated from the average electron density of the system. Without the screening term, this simply reduces to the Hartree-Fock exchange potential. As well as this non-local operator, a local term, $v_{XC}(\mathbf{r})$, must also be added to take into account the exchange-correlation effects which are not included in v_{SX}^{NL}. This term is usually approximated in a similar way to the LDA.

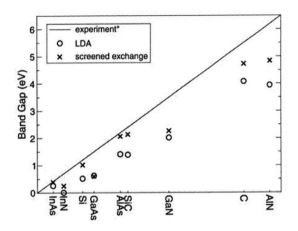

FIGURE 1. Band gaps of various semiconductors calculated with the LDA and screened exchange. * Experimental values from [6].

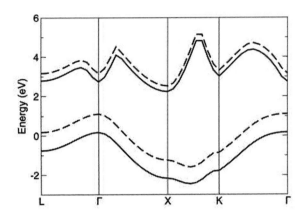

FIGURE 2. Band structure of highest valence and lowest conduction bands in AlAs calculated with the LDA (dashed line) and screened exchange (solid line). The Γ-X gap is larger using screened exchange.

COMPUTATIONAL PROCEDURE

We have implemented the screened exchange method within the CASTEP density functional code [5]. This code employs a plane-wave basis set to represent the Kohn-Sham orbitals, and pseudopotentials to describe the ionic cores. The computational cost of a screened exchange calculation scales as the number of basis functions cubed, and is significantly more expensive than the LDA or GGA. The code therefore makes extensive use of lookup-tables in order to shift as much effort as possible out of the innermost loops.

For band structure calculations, it is possible to use the LDA to perform the self-consistent electronic minimisation before using screened exchange for the band structure itself. This does not significantly alter the results and saves a great deal of computational time. Furthermore, a single evaluation of the screened exchange eigenvalues on a set of bands calculated with the LDA yields energies that differ by no more than a few percent from the self-consistent screened exchange results. This can also provide reductions in required computational effort.

In the calculations presented here we have taken a plane-wave cut-off energy of either 400eV or 700eV depending on the system under study. In all cases we have used a 4 × 4 × 4 Monkhosrt-Pack grid to sample the Brillouin zone. Norm-conserving Kleinman-Bylander pseudopotentials have been employed to describe the ionic cores. For both Ga and In compounds, we include the 3d electrons as valence rather than core, as this makes a significant difference to the obtained results.

RESULTS AND DISCUSSION

We have calculated the band gaps of Si, C (diamond), SiC, AlN, AlAs, GaN, GaAS, InN, and InAs using both the LDA and screened exchange, see Fig.1. We have also calculated the band structure of the highest conduction and lowest valence bands in AlAs. It is clear that, in most cases, the use of screened exchange increases the calculated band gap significantly, although the computational cost is typically an order of magnitude greater. In comparisson with experimentally determined values, the success of the method is somewhat mixed. Very good results are obtained in the case of Si and AlAs, however GaAs and GaN results differ only marginally from the LDA.

ACKNOWLEDGMENTS

We acknowledge the support of the EPSRC (GR/R25859/01).

REFERENCES

1. P. Hohenberg and W. Kohn, *Phys. Rev. B* **136**, 864 (1964).
2. W. Kohn and L. J. Sham, *Phys. Rev. A* **140**, 1133 (1965).
3. J. P. Perdew and Y. Wang, *Phys. Rev. B* **33**, 8800 (1986).
4. D. M. Bylander and L. Kleinman, *Phys. Rev. B* **41**, 7868 (1990).
5. CASTEP: M. D. Segall, P. J. Lindan, M. J. Probert, C. J. Pickard, P. J. Hasnip, S. J. Clark, M. C. Payne *J. Phys. : Cond. Matt.* **14**, 2717 (2002).
6. O. Madelung, *Semiconductors - Basic Data*, Publisher, Springer, 1996.

Material Design via Genetic Algorithms for Semiconductor Alloys and Superlattices

Kwiseon Kim*, Peter A. Graf* and Wesley B. Jones*

National Renewable Energy Laboratory, Golden, Colorado 80401

Abstract. We present an efficient and accurate method for designing materials for electronic applications. Our approach is to search an atomic configuration space by repeatedly applying a forward solver, guiding the search toward the optimal configuration using an evolutionary algorithm. We employ a hierarchical parallelism for the combined forward solver and the genetic algorithm. This enables the optimization process to run on many more processors than would otherwise be possible. We have optimized AlGaAs alloys for maximum bandgap and minimum bandgap for several given compositions. When combined with an efficient forward solver, this approach can be generalized to a wide range of applications in material design.

INTRODUCTION

The inverse problem for electronic properties in material design has been discussed previously [1, 2, 3]. Our approach to solving such inverse problems is to search the configuration space by repeatedly applying a forward solver, guiding the search toward the optimal configuration using an evolutionary algorithm. We call our PGAPack [4] based implementation of such a method IAGA. The configuration space here is the arrangement of different atoms in a supercell lattice. Optimization is performed using a genetic algorithm. The forward solver used to calculate the fitness value is the parallel folded spectrum electronic structure method with empirical pseudopotentials [5]. We employ a hierarchical parallelism for the combined forward solver and genetic algorithm. This enables us to run the optimization process on many more processors than would otherwise be possible. We have optimized AlGaAs alloys for maximum and minimum bandgaps. To date we have demonstrated that our IAGA method produces results that are consistent with the optimization work using simulated annealing [1]. However, due to parallelization and use of the genetic algorithm, the IAGA results are achieved in much less time. This is one step toward designing more realistic materials than has been possible to date.

OPTIMIZATION BY GENETIC ALGORITHM

Mathematically, the problem of finding the best atomic configuration with respect to any nontrivial property is a nonlinear combinatorial global optimization problem. Here we use a genetic algorithm (GA) - especially suited to such problems - to optimize the bandgap of semiconductor alloys.

The basic ingredients of a genetic algorithm are a "population" of potential solutions, a way of testing the "fitness" of any individual solution, and a mechanism for generating new solutions from the existing population based on this fitness. We have written a program, IAGA ("inverse method for alloys using genetic algorithm"), which calls the parallel genetic algorithm library known as PGApack [4], and the "forward solver" PESCAN [5] to perform these steps.

Here we summarize the operation of IAGA. First, parameters and options for the run are read in from a file. Then an initial set of atomic configurations are generated randomly. This set is called the "population", and its members are called "individuals." Now we test the "fitness" of each individual. This occurs in two stages. First, a "functional" value is generated (for example, by calling the forward solver to compute the bandgap of the material). From the functional, the fitness of the individual is determined depending on the "target" of the simulation, i.e. the maximum, minimum, or particular value of the functional. When each individual's fitness has been determined, we generate a new population by selecting from among the more fit individuals and performing "crossover" (combining two individuals into a single new one) and "mutation" (creating a new individual from a single existing one by randomly changing parts of it). These new individuals replace the least fit of the existing population, giving us the next "generation" of individuals. We repeat this process until a stopping criterion is reached, typically a specific number of generations.

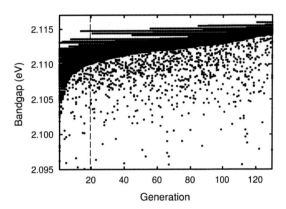

FIGURE 1. Bandgap vs generation

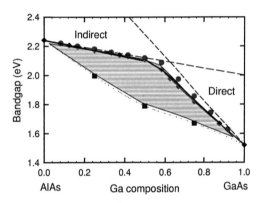

FIGURE 2. AlGaAs bandgaps vs composition. Circles: Maxima. Squares: Minima. Solid line: Bandgap of random alloy. The dashed lines connecting the the direct bandgaps of GaAs and AlAs at Γ and the indirect bandgaps of GaAs and AlAs at X indicate the direct to indirect bandgap transition.

ALGAAS BANDGAP OPTIMIZATION

Using IAGA we optimized the bandgap of AlGaAs alloys. An optimization by GA typically has two phases, an "exploratory" phase during the early generations followed by an "exploitation" phase as the system nears convergence.[6] During the exploratory phase, the population is diverse, and rapid improvement of the solution is typical. In the exploitation phase, the system approaches convergence on one individual or a set of similar individuals, and improvement of the solution slows. By changing the mutation and crossover rates, one can modify the length of these phases.[7] Figure 1 shows the typical progress toward convergence of an entire population. Figure 2 shows the bandgap of AlGaAs alloys with maximum and minimum bandgaps per composition. The bandgap of random AlGaAs alloys changes from direct to indirect below 0.55 Ga composition. The bending in

TABLE 1. Optimal $Al_xGa_{1-x}As$ configurations

x	E_g (eV)	SL direction	sequence
Maximum bandgap configuration			
0.25	1.86	[0 $\bar{1}$ 2]	$Al_1Ga_4Al_1Ga_2$
0.50	2.12	[0 $\bar{1}$ 2]	Al_2Ga_2
0.75	2.18	[0 $\bar{1}$ 2]	Al_3Ga_1
Minimum bandgap configuration			
0.25	1.67	"X" Cmmm structure	
0.50	1.79	[1 $\bar{1}$ 1]	Al_1Ga_1
0.75	2.00	"Luzonite" Cu_3AsS_4-type	

the maximum gaps (filled circles) indicates the direct to indirect gap transition. The shaded regions are the available bandgaps of AlGaAs alloy achievable by different atomic configurations. Table 1 shows some of the optimal configurations obtained by the IAGA method. Most of the optimal configurations are ordered structures with a small unit cell.

CONCLUSION

We have implemented an efficient and accurate method for solving inverse problems in the context of the electronic structure of semiconductor alloys and their superstructures. This incorporates an existing forward solving method with optimization techniques involving genetic algorithms. When combined with an efficient forward solver, the IAGA approach can be generalized to a wide range of applications in materials design.

ACKNOWLEDGMENTS

We thank L.-W. Wang for providing us with PESCAN codes and A. Zunger, A. Franceschetti, S. Dudiy, and G. L. W. Hart for fruitful discussions. The work was supported by U.S. DOE-SC-ASCR-MICS under Lab03-17 nanoscience initiative.

REFERENCES

1. A. Franceschetti and A. Zunger, Nature **402**, 60 (1999).
2. G. H. Jóhannesson, et al., Phys.Rev.Lett. **88**, 255506(2002).
3. T. Cwik and G. Klimeck, Proc. of 1st NASA/DoD Workshop on Evolvable Hardware, IEEE (1999).
4. D. Levine, PGAPack: Parallel Genetic Algorithm Library, 1998.
5. A. Canning, L.-W. Wang, A. Williamson, and A. Zunger, J. Comput. Phys., **160**, 29 (2000) and references therein.
6. Z. Michalewicz and D. B. Fogel, *How to Solve It: Modern Heuristics*, Berlin, Springer-Verlag, 2000
7. S. Dudiy, K. Kim, W. Jones, A. Zunger (unpublished).

Intra-impurity transitions in uniformly Iodine doped MBE CdTe/CdMgTe quantum well–with no energetic scaling

M.Szot[*], K.Karpierz[*], J.Kossut[#], M.Grynberg[*]

[*]*Institute of Experimental Physics, Warsaw, Poland*
[#]*Polish Academy of Sciences, Warsaw, Poland*

Abstract. Measurements of photoconductivity spectra for fixed transition energy between splitted by an external magnetic field levels of shallow donors, in the case of uniformly doped CdTe/$Cd_{0.8}Mg_{0.2}$Te quantum well structure, exhibit unexpected lack of sensitivity on photon energy causing the intra shallow impurity transition. One observes the line at the same position in magnetic field for different photon energies. The discussion of the shape of the obtained line based on the model of localization explaining lack of energetic scaling is presented.

INTRODUCTION

The well established understanding of the problem of shallow donors in 3D environment was achieved many years ago [1]. Within the era of possible getting the 2D structures, it seemed that the question of behavior of shallow donor was also both experimentally and theoretically answered [2]. The understanding is based on the Zeeman's splitting of the levels modified by an influence of external potentials, crystal lattice, dielectric properties etc. However there were reports showing certain unexpected discrepancies between theory and experiment [3]. The results led authors to the conclusion that the spatial distribution of potential fluctuations, present in the sample, is responsible for observed effects. In this paper we present further results of investigations of such behavior.

EXPERIMENT, SAMPLES

The change of sample conductivity due to intra-shallow donor transitions and absorption of the Far Infrared (FIR) light was measured. By splitting the levels in the increasing magnetic field one reaches the regime of resonant absorption of photons, what is associated with a change of conductivity. The photoconductivity (PC) was measured at 4.2K, using a Lock-in technique. During experiments the sample had to be additionally illuminated by visible light ($h\nu \approx 2$ eV) coming from the monochromator in order to have measurable sample resistance (~ 10 MΩ). All samples investigated in our experiments were MBE single CdTe/$Cd_{0.8}Mg_{0.2}$Te quantum well, deposited on the semiinsulating GaAs substrate with the deposited 8.0 µm buffer layer of an undoped CdTe. The role of such thick buffer layer is to separate the region of sample with the quantum well from the influence of GaAs - undoped CdTe interface. The 0.5 µm barrier of $Cd_{0.8}Mg_{0.2}$Te, 160 Å CdTe quantum well and 480 Å $Cd_{0.8}Mg_{0.2}$Te barrier were uniformly doped by Iodine at the level of ~10^{16} cm^{-3}. The quantum well was 200 meV deep in the conduction band. On the top of the sample the "home made" Indium contacts were carefully attached.

RESULTS AND INTERPRETATION

Figure 1 shows the PC spectra obtained for different FIR photon energies. One observes a reach structure of the spectrum. The arrows mark positions of the peaks due to intra shallow donor transitions in GaAs substrate. The most surprising is, however, the peak, which does not change its position with increasing transition energy (FIR photon energy) - marked by the dashed line on Fig.1.

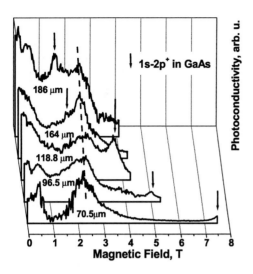

FIGURE 1. Photoconductivity spectra for different FIR wavelengths.

maximum of peak, one can estimate the size of fluctuation equal to ~350Å. Such a localized electron can occupy the existing donor state (within its Bohr radius). N_{D0} dependence on Fig.3 represents qualitatively the number of electrons localized on shallow donor ground state in that way.

For 2D system, energy of shallow donor ground state strongly depends on the position of centre in the quantum well [5]. The binding energy of shallow donor is the biggest for donors in the center of the well and decreases for donors located more away. For the case of the homogeneously doped 2D samples, there is a continuum of donor states due to theirs spatial localization in the well and barrier. Dashed line on Fig.3 marked as ρ shows density of available donor states. In that case, FIR photon of any energy from the range given by available density of states can be absorbed by electron. This process is possible only for $B > B^*$ - after having localized the electron.

For increasing B other factors (M_{1-2} - transition probability between ground and first excited donor state and $\sigma(B)$ - conductivity of electrons in the conduction band – see Fig.3) cause drop of the measured signal and the asymmetric shape of the observed line.

Its position is shown on Fig.2. For all transitions energies between 6.7 – 17.6 meV (wavelengths: 186 – 70.5 μm) the data points, within an experimental error, are placed along the horizontal line on Fig.2.

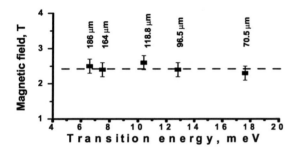

FIGURE 2. Position of the maximum of peaks marked on Fig.1. as a function of transition energy.

On the contrary to the GaAs, for mentioned peak we do not observe any energetic scaling due to the levels splitting in magnetic field. In order to understand such behavior, the mechanism of FIR absorption by shallow donor centers in the presence of local potential fluctuations has been proposed [3]. Its base is the presence of local potential minima in the conduction band in the quantum well originating mainly from ionized donors in the barriers and surface states on the top of the sample, concentration of which is estimated to be equal ~10^{11} cm^{-2} [4]. With increasing magnetic field B, the magnetic length associated with conducting electrons is decreasing. At certain field B^* it matches the size of fluctuations. This causes electron to be trapped in the fluctuation. Assuming that this process occurs at the magnetic field equal to the

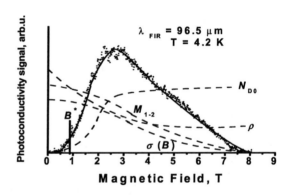

FIGURE 3. Comparison of the experimentally obtained shape of the peak (points) and qualitative fit (solid line) PC $\propto N_{D0} \times \rho \times M_{1-2} \times \sigma(B)$ – see text.

REFERENCES

1. Armistead C.J, Knowles P, Nadja S.P, Stradling R.A, J. Phys. C **17**, 6415 (1984).
2. Huant S, Grynberg M, Martinez G, Etienne B, Lambert B Regreny A, Solid State Commun. **65**, 1467-1472, (1988).
3. Szot M., Karpierz K., Kossut J., Grynberg M., Phys. Status Solidi C no. 2, 609-12, (2003).
4. Maslana W, Kossacki P, Bertolini M, Boukari H, Ferrand D, Tatarenko S, Cibert J, Gaj J.A, APL **82**, no.12, 1875-7, (2003).
5. G. Bastard, Phys. Rev. B **24**, 4714, (1981).

Radiative Processes In Layered Transition Metal Dichalcogenides

L. Kyuluk[1], E. Bucher[2], Luc Charron[3], D. Dumchenko[1], E. Fortin[3], C. Gherman[1]

[1]*Institute of Applied Physics, Academy of Sciences of Moldova, Chisinau, Republic of Moldova*
[2]*Department of Physics, University of Konstanz, Germany*
[3]*Department of Physics, University of Ottawa, Ottawa, Ontario*

Abstract. The near-infrared radiative properties of halogen-intercalated $2H$-MoS_2, $2H$-WS_2 and $2H$-WSe_2 are presented and described with a kinetic recombination model for an n-type material.

INTRODUCTION

The transition metal dichalcogenides compounds (TX_2) belong to a large class of the two-dimensional solids. Owing to their layered structure, the TX_2 compounds can be intercalated with foreign atoms and molecules. The intercalation leads to the changes of the electronic properties of these indirect band-gap semiconductors. Recently [1] we have reported the first observations of the bound exciton luminescence of synthetic molybdenite $2H$-MoS_2, intercalated by Cl_2 molecules. In this work we studied the radiative processes in $2H$-MoS_2:Cl_2, $2H$-WS_2:Br_2,I_2 and $2H$-WSe_2:I_2 layered crystals.

EXPERIMENT

The TX_2 single crystals were grown by chemical vapor method, using iodine, bromine and chlorine as a transport agents. The steady-state and time resolved photoluminescence (PL) measurements were carried out in the temperature range T=2-100K, excitation was provided by an Ar-ion laser (λ=514 nm) or a diode-pumped cw solid-state laser (λ=532 nm).

CHARACTERISTIC SPECTRA

Figure 1 presents the characteristic spectra of three samples. Two distinct regions were identified. The excitonic region, composed of several zero-phonon lines (ZPL) followed by their phonon sidebands (PS), is caused by radiative recombination of excitons bound to halogen impurities intercalated within the van der Waals gap of the crystals. The broad-band region is associated with radiative transitions between a deep donor center and the valence band in the conditions of a strong electron-phonon coupling [2].

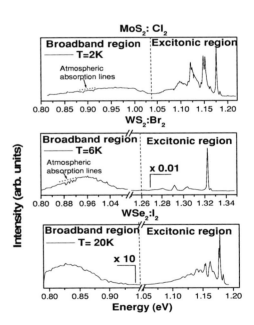

FIGURE 1. Characteristic spectra of every sample contain an excitonic and a broad-band region.

TEMPERATURE EVOLUTION

Figure 2 illustrates the excitonic ZPL PL temperature evolution. In each case, the redistribution of the PL intensity is played out through three major zero-phonon lines: A, B and C. At higher temperatures, on can observe a fast exponential thermal quenching and the excitonic emission is reduced.

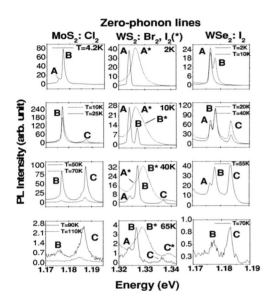

FIGURE 2. Temperature evolution of the ZPLs

KINETIC MODEL

The temperature evolution of the excitonic PL integral intensity measurements was described using a kintetic model similar to the one proposed for n-type nitrogen-doped GaP[3]:

$$\dot{N}_{T1}^{x} = p \cdot \upsilon_{th} \cdot \sigma_{px} \left(N_{T1}^{e} - N_{T1}^{x} \right) - N_{T1}^{x} \cdot \left(\frac{1}{\tau_{Rx}} + \frac{1}{\tau_{Nx}} + \frac{1}{\tau_{xp}} \right) \quad (1)$$

$$\dot{p} = G - p \cdot \upsilon_{th} \cdot \sigma_{px} \left(N_{T1}^{e} - N_{T1}^{x} \right) + \frac{N_{T1}^{x}}{\tau_{xp}} - p \cdot \frac{N_{T2}^{e}}{N_{T2}} \cdot \left(\frac{1}{\tau_{Rbb}} + \frac{1}{\tau_{Nbb}} \right) \quad (2)$$

where p is the minority carrier, $N_{T1,2}^{e}$ - concentrations of electron-occupied centers, N_{T1}^{x} - concentration of exciton-occupied centers, $\tau_{Rx(Nx)}$ - bound exciton radiative(nonradiative) lifetime, $\tau_{Rbb(Nbb)}$ - T_2 centers radiative(nonradiative) time, σ_{px} - hole capture cross-section by T_1 centers, υ_{th} - thermal velocity of p, τ_{xp} - isoelectronic trap hole exciton thermalization time, and G - excitation rate.

Solving these two equations under the assumption of thermal equilibrium yields the quantum efficiencies of the ZPL excitons $\eta_{Rx}(T)$ shown in Figure 3 as the solid lines.

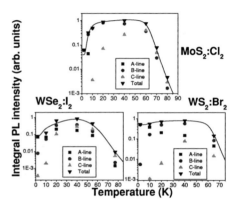

FIGURE 3. Integral ZPL PL intensity and theoretical fit.

CONCLUSION

Halogen molecules, used as transport agents in single crystal growth, can intercalate between the layers of transition metal dichalcogenides and can induce strong excitonic photoluminescence. This phenomenon has been observed in MoS_2, $2H-WS_2$, $2H-WSe_2$ and should likely be observed in other TX_2 compounds, leading to a new class of near-infra-red emitters.

REFERENCES

1. L. Kulyuk, L. Charron, E. Fortin, Phys. Rev. B, **68**, 075314 (2003).

2. C. H. Henry and D. V. Lang, Phys. Rev. B **15**, 989 (1977).

3. P. D. Dapkus, W. H. Hackett Jr., O. G. Lorimor, and R. Z. Bachrach, J. Appl. Phys. **45**, 4920 (1974).

How To Observe Charged Bosons in Quantum Wells?

L.D. Shvartsman*, D.A. Romanov**

*The Hebrew University of Jerusalem, Jerusalem 91904, Israel
** Temple University, Philadelphia, PA 19122, U.S.A.

Abstract. We discuss the prospects of experimental observation of biholes in quantum wells of various popular cubic semiconductors. Biholes are charged exciton-like complexes formed by a couple of holes belonging to different subbands and having negative reduced effective mass. The biholes should be manifested as extra peaks in inter-subband absorption. The necessity of both high binding energy and long lifetime reveals competing requirements to the parameters of the quantum well; thus, the optimal design of the heterostructure becomes the key ingredient for the success.

INTRODUCTION

We discuss the prospects of experimental observation of biholes in quantum wells of various popular cubic semiconductors. The bihole is charged compound boson consisting of two holes bound together by electrostatic repulsion. The concept of biholes as charged exciton-like complexes formed by a couple of holes belonging to different bands and having negative reduced effective mass appeared in connection with the observation of a reversed hydrogen-like absorption spectrum in bulk crystals of BiI_3 and ZnP_2 [1,2]. In our earlier papers we applied bihole concept to a couple of holes in quantum well (QW) of cubic semiconductors belonging to different subbands [3-5].

For the wide majority of popular cubic semiconductors the first excited hole subband has a W-shape dispersion, and the condition of negative reduced mass can be easily fulfilled when a hole in this subband couples with a hole in the ground subband. That is why the idea of excitation of this quasistationary charged boson with IR light looks feasible [3-5]. The target of this work is to bridge the theory and the experiment. We consider simultaneously the two key parameters of a bihole: the binding energy and the lifetime. The necessity of both high binding energy and long lifetime reveals competing requirements to the parameters of the quantum well; thus, the optimal design of the heterostructure becomes the key ingredient for the success.

BIHOLE BINDING ENERGY AND LIFETIME

Bihole ground state may be shown as a level located above the maximum of the first excited valence subband (Figure 1).

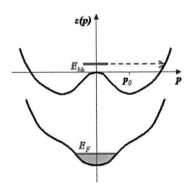

FIGURE 1. Two first hole subbands in QW of typical cubic semiconductor. Bihole state is shown and its isoenergetic decay is indicated by the dashed arrow.

To calculate the binding energy we reduce two-particle non-parabolic Shrödinger equation to the one-particle problem. This effective particle represents the relative motion of two holes when their center of mass is at rest. This particle has an effective W-dispersion being algebraic sum of the two subbands and moves in an effective repulsive potential that is partially 2D screened by the free holes of the ground subband.

The very existence of bound state in a repulsive potential requires negative value of reduced mass of the couple of holes, i.e. $|m_1|>|m_2|$, and $m_2<0$. As one can see in Table 1, these conditions are fulfilled in deep QWs of nearly all popular semiconductors, except Si, InAs, and InSb. Nevertheless, if the first inequality is too strong the bihole level becomes very shallow, as, for instance in a pure Ge QW. Appropriate design of solid solution lets adjust the value of masses to produce high enough values of the binding energy, however a lifetime constrain arising from the isoenergetic decay of bihole has to be minded as well.

Indeed, the bihole state coexists with isoenergetic free holes of the continuous spectrum [5]. In this situation, the stability of the bihole is determined by competition of the height of the negative mass section in the W-shaped reduced dispersion, Δ, and the height of the repulsive potential, U_0. If $U_0 > \Delta$, the bihole state of low orbital momentum cannot exist [5]. In more favorable situation, $U_0 < \Delta$, the bihole lifetime is limited by quantum decay, whose parameters are determined by interplay of W-dispersion and repulsive potential. Physically, this decay is a tunneling in the p-space from the bound state whose wavefunction comprises small-momentum components to the decoupled state corresponding to large momentum (see Figure 1). If the absolute value of the reduced mass is very high, the negative mass feature of the reduced dispersion is not well expressed. Although this condition favors high value of the binding energy, it also results in fast tunneling thus making the bihole unstable. To optimize both the binding energy and the lifetime one has to engineer W-dispersion with rather well expressed but not too sharp negative mass feature.

Another physical factor that has to be kept in mind is the total angular momentum conservation when bihole either forms or recombines in a one-photon process. (Here, the situation reminds that for ordinary QW excitons, though in the case of bihole these selection rules are significantly weakened by the substantial admixture of light (heavy) holes to any subband of heavy (light) for finite in-plane momenta.) One-photon processes are allowed if the absolute value of total momentum of the bihole is equal to one.

Otherwise "dark" biholes are formed. The ground subband for zero in-plane momentum always has heavy hole character (hh1) so that J=3/2. The first excited subband may be either "hh2" (J=3/2) or "lh1" (J=1/2) depending on the semiconductor material and QW width and depth. For instance, in Table 1 we noted by the asterisk materials having "lh1" character of first excited subband in deep QWs. In this case already for k=0 holes the total angular momentum conservation trivially favors one photon processes. For other compounds the design is less trivial. For instance, for pure GaAs infinite QW, as seen from Table 1, direct bihole excitation is problematic because of the total momentum selection rule and low binding energy. Our calculations show that for slight change in the system, as 100Å $In_{0.01}Ga_{0.99}As$ based QW surrounded by AlAs barriers both these factors can be improved substantially. In this case binding energy of bihole is around 3 meV while its intrinsic lifetime is in nanosecond range. Slightly lower (~2 meV) bihole binding energy and the same lifetime range one gets for 60 Å GaSb QW sandwitched between AlSb barriers. In this case first excited subband is already of "lh1" character.

	Si*	Ge	InAs	InSb	GaSb	ZnTe*	GaP*	GaAs
Ground subband (m_1/m_0)	0.3	0.072	0.044	0.33	0.08	0.217	0.245	0.125
Excited subband (m_2/m_0)	0.67	-0.009	-0.1	-0.684	-0.034	-0.186	-0.18	-0.017

TABLE 1. Hole effective masses of ground & first excited subbands of QWs of various cubic semiconductors

ACKNOWLEDGEMENT

We appreciate discussions with J.E.Golub in the early stages of this work.

REFERENCES

1. Gross, E.F., .Perel, V.I, and Shechmamet'ev, R.I., JETP Lett., **13**, 229-231, (1971).

2. Selkin, A.V., Stamov, I.G., Syrbu, N.U., Umenetz, A.G., JETP Lett., **35**, 57-59, (1982).

3. Chaplik, A.V., and Shvartsman, L.D., All Union School on Surface Physics: USSR Academy of Sciences, Chernogolovka edited by S.Dorozhkin 1983, p.123.

4. Shvartsman, L.D. and Golub, J.E., in Bose-Einstein Condensation, Cambridge University Press, 1995, edited by A.Griffin, D.W.Snoke, S.Stringary. pp.532-541.

5. Shvartsman, L.D., Romanov D.A., and Golub, J.E.,, Phys. Rev. A, **50**, R1969-1972 (1994).

Intersubband Absorption from 2-7 μm using Strain-compensated Double-barrier InGaAs Multiquantum Wells

K. T. Lai[1], R. Gupta[1], M. Missous[2] and S. K. Haywood[1]

[1]Dept. of Engineering, University of Hull, Hull, U. K.
[2]Dept. of Electrical Engineering and Electronics, UMIST, Manchester, U. K.

Abstract. We report observation of strong room temperature electron intersubband absorption in strain-compensated $In_{0.84}Ga_{0.16}As/AlAs/In_{0.52}Al_{0.48}As$ double barrier quantum wells grown on InP substrates. Multiple Γ-Γ intersubband transitions have been observed across a wide range of the mid infrared spectrum (2-7 μm) in three structures of differing InGaAs well width and therefore with a differing net strain. From the multiple Γ-Γ intersubband transitions observed in the 8 nm well, it is inferred that the electron effective masses and nonparabolicity parameters for the first two subbands differ significantly from each other. For the range $k \sim 5 \times 10^6 - 6 \times 10^6$ cm^{-1}, the difference in subband parameters results in a spread in transition energy of about twice the value calculated for the corresponding GaAs/AlGaAs quantum well.

INTRODUCTION

Quantum well infrared photodetectors (QWIPs) based on GaAs/AlGaAs have already been used to produce high performance, large uniform focal plane arrays for detection in the 6 to 18 μm range [1]. Detection below 6 μm requires a different material system due to the problems associated with using high aluminum content barriers [2]. The lattice-mismatched system, InGaAs/AlGaAs [2] is one option provided that the layer thicknesses do not exceed strain-dependent critical values. One way to avoid misfit dislocations is to choose well and barrier materials with opposing strain relative to the substrate thus maintaining the net strain in the system close to zero. In this paper, we present a study of $In_{0.84}Ga_{0.16}As/AlAs/In_{0.52}Al_{0.48}As$ double barriers QWs, (DBQWs) grown on InP substrates to produce differing strain compensation between the AlAs (tensile) and the InGaAs (compressive). This is achieved by changing the InGaAs well width in three samples, (L_W, = 3 nm, 4.5 and 8 nm) while other structural parameters are kept constant.

EXPERIMENT AND RESULTS

Figure 1 shows the 300K absorbance of the three QW samples. (Growth and detailed modeling of these samples are presented elsewhere [3].) Multiple electron intersubband (ISB) transitions, $E_i \rightarrow E_{i+1}$, were observed in the 300K absorption spectra (where E_i refers to the Γ subband energy level in the well, and i is the subband index).

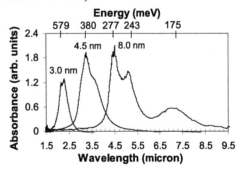

FIGURE 1. Absorption spectra of the samples, measured at 300K in 45⁰ waveguide geometry.

No peaks were observed in the parallel (forbidden) polarization. It can be seen that the peak detection wavelength for these structures can be tuned from 2-7 μm (620-177 meV) by simply varying the InGaAs well width. Modeling for the sample with the widest well (1554: L_W=8 nm) predicts four conduction Γ subbands within the well and the position of the Fermi level E_F indicates that three of these levels are occupied, leading to multiple ISB transitions. We have used these multiple transitions and the variation in the measured full-width half maximum (FWHM) to estimate the difference in the subband effective mass (m*) and non-parabolicity parameters (α). Figure 2 shows the 300K absorbance for this sample. Three absorption peaks are seen at energies between 177 meV and 280 meV. These are assigned to $E_1 \rightarrow E_2$, $E_2 \rightarrow E_3$ and $E_3 \rightarrow E_4$ Γ transitions in the InGaAs well [3]. FWHM, obtained from a Lorentzian fit of the spectrum, are: 31.6 meV, 32 meV and 23.6 meV, respectively.

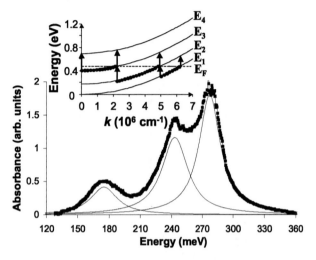

FIGURE 2. Measured absorption spectrum at 300 K (thick line) and Lorentzian curve fit (thin line) for sample 1554. The inset shows the energy dispersion of the quantized energy levels in the $k_{x,y}$ direction assuming parallel subbands where the arrows indicate the three ISB transitions.

From the position of E_F in the inset of Fig. 2, we can see that the transition $E_3 \rightarrow E_4$ samples a region of the Brillouin zone nearest to $k=0$ whereas $E_2 \rightarrow E_3$ and $E_1 \rightarrow E_2$ sample larger values of k. Taking ranges of k from the parabolic case (inset in Fig. 2) appropriate to each of the three transitions and using Eq. (1) in ref 4 below we calculate the spread in the $E_i \rightarrow E_{i+1}$ transition energy with k to be 21 meV, 21 meV and 3 meV for i=1, 2 and 3, respectively.

$$E_n(k) = \frac{E_G}{2}\left[1 + \frac{4E_n^{(0)}}{E_G} + \frac{2\hbar^2 k^2}{m_e E_G}\right]^{1/2} - \frac{E_G}{2} \quad (1)$$

where m_e is the Kane effective mass at the conduction band bottom and E_G is the semiconductor band gap and $E_n^{(0)} = \pi^2 \hbar^2 n^2 / (2m_e L_w^2)$ is the energy of the nth subband in the parabolic approximation. The calculated trend is in-line with the experimentally determined linewidth [3]. The spread in the $E_1 \rightarrow E_2$ transition energy is about twice the value calculated for the corresponding GaAs/AlGaAs QWs. Therefore the cancelling of the non-parabolicity contribution to linewidth due to depolarization effects [5], which is seen in some GaAs QWs, is not anticipated for high indium content InGaAs QWs. We have now used the subband dispersion obtained from Eq. (1), to extract values of m* and α for each subband by fitting the following equation [6]:

$$E_i(k_\perp) = E_{0i} + \frac{\hbar^2 k_\perp^2}{2m_i^*} + \alpha_i k_\perp^4 \quad (2)$$

We estimate the m* and α for the four subbands in the $In_{0.84}Ga_{0.16}As$ QWs to be $m_1^* \approx 0.05m_0$, $\alpha_1 \approx 4000$ eVÅ^4, and $m_2^* \approx 0.078m_0$, $\alpha_2 \approx 1800$ eVÅ^4, $m_3^* \approx 0.11m_0$, $\alpha_3 \approx 1100$ eVÅ^4, and $m_4^* \approx 0.14m_0$, $\alpha_4 \approx 800$ eVÅ^4. The change in m* and α for the higher subbands is due to a greater penetration of these wavefunctions into the barrier [6]. Note also that the electron effective mass of the lowest subband $m_1^* \approx 0.05m_0$ is higher than the linear interpolated value of $0.03m_0$ for the same reason.

CONCLUSION

In conclusion, we have grown three $In_{0.84}Ga_{0.16}As/AlAs/In_{0.52}Al_{0.48}As$ DBQWs, on InP substrates, with varying degrees of net strain due to different well widths. Together these three samples span the range from 2-7 μm. For the sample with the widest QW (L_W=8 nm), we have used these multiple intersubband transitions and the variations in the measured FWHM to estimate the difference in the subband parameters, m* and α.

REFERENCES

1. Gunapala S. D. *et al.*, *Infrared Phys. Technol.* **42**, 267-282 (2001) and references therein.

2. Choi, K. K., Bandara, S. V., Gunapala, S. D., Liu, W. K., and Fastenau, J. M., *J. Appl. Phys.* **91**, 551-564 (2002).

3. Gupta, R., Lai, K. T., Missous, M., and Haywood, S. K., *Phys. Rev. B* **69**, 033303(1)-033303(4) (2004).

4. Gelmont, B., Gorfinkel, V., and Luryi, S., *Appl. Phys. Letters* **68**, 2171-2173 (1996).

5. Zaluzny, M., *Phys. Rev. B* **43**, 4511-4514 (1991), and references therein.

6. Von Allmen, P., Berz, M., Reinhart, F. -K., and Harbeke, G., *Superlatt. Microstruct.* **5**, 259-263 (1989).

Effects of ZnSe Interlayer on Properties of (CdS/ZnSe)/BeTe Type-II Super-lattices Grown by Molecular Beam Epitaxy

B. S. Li, R. Akimoto, K. Akita and H. Hasama

National Institute of Advanced Industrial Science and Technology (AIST), Photonics Research Institute, Tsukuba, Ibaraki 305-8568, Japan

Abstract. An intersubband transition (ISB-T) down to 1.57 μm is realized in (CdS/ZnSe)/BeTe super-lattices for the first time. We studied the dependence of properties of super-lattices on the ZnSe interlayer by using in situ reflection of high energy electron diffraction, high-resolution X-ray diffraction, ISB-T, and high-resolution transmission electron microscopy. It is crucial to improve the growth mode, the structural and optical properties by inserting ZnSe interlayer between BeTe and CdS layers.

INTRODUCTION

The recent attempt to extend the ISB-T to near-infrared (NIR) ($\lambda < 3\mu m$), in particular to the 1.55μm fib-optic communication, has being developed exceedingly. The reports of ZnSe/BeTe and (CdS/ZnSe)/BeTe superlattices (SLs) has opened up applications in the NIR spectral region for II-VI semiconductor. [1, 2] We improved upon the previous results and obtain ISB-T down to 1.57 μm with a narrow full width at half maximum (FWHM) of 90 meV. So far, no reports covered the effects of ZnSe interlayer on the properties of (CdS/ZnSe)/BeTe SLs. In this paper, we studied the effects of ZnSe interlayer on the growth mode, structural and optical properties.

For studying the effects of ZnSe interlayer, we prepared 7 samples of (CdS/ZnSe)/BeTe SLs with different well width. Their parameters and experimental procedure will be introduced in detail elsewhere. The growth modes were detected using in situ RHEED system. The structural properties of SLs were studied by using a HRXRD (Philips, MRD). Absorption of samples was measured using a Fourier-transform spectrometer. HRTEM was conducted on structure parameters in the <110> direction via a Hitachi H-1500 high-voltage electron microscope, operated under 1.4 Å point-resolution at 800 kV.

RESULTS AND DISCUSSION

A streaky pattern of RHEED can be kept in the whole growth by inserting ZnSe interlayer, even if 0.5 ML, demonstrating a smooth, two-dimensional growth mode. As shown the top (with 2-MLs ZnSe interlayer) and middle curves (0.5-ML ZnSe) in Fig. 1, pronounced RHEED oscillations were observed for the growth start of S2 and S4. However, the growth mode degrades without ZnSe interlayer, as shown in the bottom curve in Fig. 1. A sporty pattern appears immediately when BeTe deposits on the smooth CdS surface. The RHEED intensity decreases permanently with no intensity modulation during the proceeding growth. In HRXRD spectra of samples S1-S6, sharp, intense satellite peaks exceeding to 14 orders further confirm the high structural quality of SLs with introducing ZnSe interlayer between BeTe and CdS layers.

Figure 2 show ISB absorption spectra of samples S1-S6. As shown in the top of Fig. 2, A resonant absorption at $\lambda = 1.57\mu m$ is observed with p-polarized light, disappears with s-polarization, indicating that ISB-T occurs. Shown at bottom in Fig. 2 are ISB-T spectra for samples S1-S6, which are deduced by taking the -log of the transmission ratio of Tp to Ts. Strong NIR ISB absorption with narrow full width at half maximum (FWHM) have been observed in samples S1, S2, and S3. The transition energy increases from 1.79 μm to 1.57 μm with the decrease of well width. However, there are no ISB absorption in the samples S4, S5, and S6 with 0.5 ML ZnSe interlayer. Although the (CdS/ZnSe)/BeTe SLs with 0.5 ML ZnSe interlayer also have a high structural quality, same as those with ZnSe interlayer ≥ 1 MLs identified by RHEED and HRXRD, There is an different performance of ISB-T in the SLs with 0.5 ML ZnSe interlayer from those with ZnSe interlayer ≥ 1 MLs.

Shown in Fig. 3 is the cross-sectional HRTEM micrographs of sample S3 along the [110] direction. A distinct interface is observed in the BeTe/ZnSe heterostructure, but the ZnSe/CdS/ZnSe well structure is indistinct

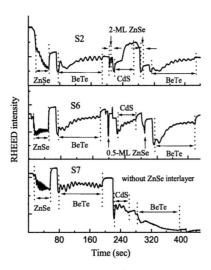
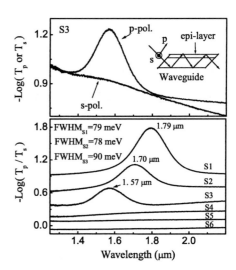

FIGURE 1. The in situ RHEED oscillations measured alone [100] azimuth during the growth start of S2, S4 and S7 SLs.

FIGURE 2. Top: absorption spectra of S3, calculated from transmission spectra Tp and Ts. Insert at upper right shows multipass waveguide geometry for intersubband absorption measurement. Bottom: absorption spectra for S1-S6 obtained by taking -log of the transmission ratio of p-polarized (Tp) to s-polarized (Ts) light. The ISB absorptions disappear in samples S4, S5 and S6 with 0.5 ML ZnSe interlayer.

and appears homogenous. As steps can not be observed in the HRTEM micrographs, the interface of the GaAs substrate with the epitaxy is confirmed to be flat on an atomic sacle for at least 40 nm. There are no misfit dislocations at the interface in the (CdS/ZnSe)/BeTe SLs, meaning a coherent 2-dimension growth. In distinct contrast to that of S3, A rough interface with more thickness is formed in SL structure with 0.5-ML ZnSe interlayer. The transition region between the barrier and the well broaden to more than 2 MLs. The thickness of this interface can be compared with the well width in this short period. A different interface is formed and results in the absence of ISB-T in those with 0.5 ML ZnSe interlayer.

CONCLUSIONS

An ISB-T of 1.57 μm was realized with a narrow FWHM in (CdS/ZnSe)/BeTe structure. The effects of ZnSe interlayer on the SLs properties were studied by in situ RHEED, HRXRD, ISB-T and HRTEM. The interface characteristic in (CdS/ZnSe)/BeTe short-period SLs plays an important role in the performance of ISB-T.

ACKNOWLEDGMENTS

This work was supported by the METI/AIST Research Project "Femtosecond Technolgy," TEM measurement was supported by the "Nanotechnology Support Project" of the Ministry of Education, Culture, Sports, Science and Technology (MEXT), Japan. We also thank the Japanese Society for the Promotion of Science (JSPS) for funding under fellowship grant No. P03537.

FIGURE 3. Cross-sectional high-resolution transmission electron microscopy images of sample S3. A sharp interface is formed between barrier and well layer in (CdS/ZnSe)/BeTe SLs in S3. The fluctuation in layer thickness is about 1 ML.

REFERENCES

1. R. Akimoto, Y. Kinpara, K. Akita, F. Sasaki, and S. Kobayashi, Appl. Phys. Lett. 78, 580 (2001).
2. R. Akimoto, K. Akita, F. Sasaki, and T. Hasama, Appl. Phys. Lett. 81, 2998(2002).

Transparency induced by coupling of intersubband plasmons in a quantum well[1]

Jianzhong Li* and C. Z. Ning*

Center for Nanotechnology, NASA Ames Research Center, Moffett Field, CA 94035, U.S.A.

Abstract. Coupling of intersubband plasmons associated with two cascading transitions in a quantum well has been studied. It is shown that the coupling can lead to the disappearance of the lower resonance amidst an anti-crossing behavior. Such spectral anomalies are of collective and resonant nature and provide the first example of Coulomb interaction induced transparency. The results from a microscopic theory are confirmed by an analytical model.

INTRODUCTION

It is well known that intersubband transitions in a quantum well (QW) are dominated by collective excitations such as intersubband excitons and intersubband plasmons (ISPs) [1, 2, 3, 4]. Such collective excitations and their coupling can dramatically change the spectral response of the quantum well. In this paper, we study the coupling of ISPs associated with two cascading transitions in a quantum well. We show that such coupling could lead to the striking disappearance of the lower resonance amidst an anti-crossing when the oscillator strengths of the two transitions exactly match.

METHOD AND RESULTS

We start from a Hamiltonian that includes quantum confinement by the QW structure, Coulomb interaction among charged particles (both electrons and donors), and light-electron dipole interaction. The equation of motion for polarization function p_k^{mn} (between subband m and n; k is the in-plane wavevector) is obtained as follows [5, 6]:

$$[\hbar(\omega + i\gamma_p^{mn}) - (\varepsilon_{nk} - \varepsilon_{mk})] p_k^{mn} \quad (1)$$
$$= (d_k^{mn} \cdot E_0 + \varepsilon_k^{mn})(f_{mk} - f_{nk}),$$

where ω (E_0) is the circular frequency (amplitude) of the incident light, γ_p^{mn} is the dephasing rate, d_k^{mn} is the dipole matrix element, and f_{mk} is the Fermi-Dirac distribution function. The single particle energy (ε_{mk}) and the *local field* correction (ε_k^{mn}) are, respectively, given by

$$\varepsilon_{mk} = E_{mk}^{(0)} - \Sigma_{lq} V_q^{mlml} f_{lk+q}, \quad (2)$$
$$\varepsilon_k^{mn} = -\Sigma_{j \neq l q} V_q^{njml} p_{k+q}^{jl} + \Sigma_{j \neq l q} V_0^{njlm} p_q^{jl}. \quad (3)$$

Subband dispersion $E_{mk}^{(0)}$ includes quantum confinement effect. Σ_{jq} sums over all subbands and q's. V_q^{ljmn}'s are Coulomb matrix elements (see [6]). The *local field* correction gives rise to excitons, via V_q^{njml} (Fock term), and ISPs, via V_0^{njlm} (Hartree term) [3, 4]. ISP coupling occurs thanks to the presence of other ISPs (p_q^{jl}'s) in the local field. Note that linearization in the polarization function has been invoked when deriving the above equations and a direct mixing term is found negligible and dropped.

Intersubband transitions are described by absorbance, given by $\omega W \mathscr{I}m[\chi(\omega)]/nc$, where W is the QW thickness, $\mathscr{I}m[\chi(\omega)]$ means the imaginary part of the optical susceptibility $\chi(\omega)$, n is the background refractive index, and c is the speed of light in vacuo. $\chi(\omega) \equiv P/\varepsilon_0 E_0$ and the material polarization P is defined as $2S/[(2\pi)^2 \mathscr{V}] \Sigma_{mn} \int dk d_k^{mn*} p_k^{mn}$. ε_0 is the electric constant and $\mathscr{V} = WS$ where S is a normalization area. The absorbance is obtained from Eq. (1) by matrix inversion.

We considered three subbands (1: ground; 2: second; 3: third) in a symmetric GaAs QW and thus two intersubband transitions: 1→2 and 2→3. V_q^{njml}'s are evaluated with the quantum box model. The subbands are simplified as parabolic ones with different masses. We assume $\gamma_p^{mn} = \gamma$ and $d_k^{jl} = d_{jl}$. The parameters and densities are given in Table I and Fig. 1. Figure 1 shows the intersubband spectra as the upper bandedge separation E_{32} is tuned while fixing the lower bandedge separation E_{21} to 50 meV (a 15 nm QW). The density is so large that ISPs dominate and excitons are negligible [4]. Two features

---

[1] The authors are with the University Affiliated Research Center at NASA Ames Research Center, managed by UC, Santa Cruz. This work was supported by AFOSR, and a joint NASA-NCI program.

TABLE 1. Separation-dependent subband populations

E_{32} (meV)	n_1	n_2 (10^{12} cm^{-2})	n_3
20	2.7410	1.3945	0.8645
30	2.8444	1.5045	0.6512
40	2.9478	1.6144	0.4378
50	3.0512	1.7244	0.2245
60	3.1476	1.8269	2.545×10^{-2}
70	3.1599	1.8401	3.603×10^{-6}
80	3.1599	1.8401	2.274×10^{-10}

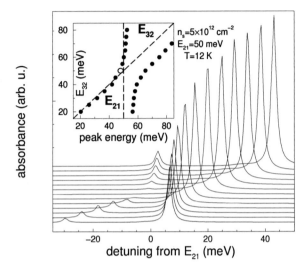

FIGURE 1. Intersubband spectra of a GaAs quantum well. Different curves, offset for clarity, correspond to E_{32} (upper bandedge separation) being varied from 20 to 80 meV with a 5 meV interval. Inset shows peak energies as a function of E_{32}. The open circle indicates where an induced transparency occurs at the lower resonance. Parameter values used: $m_1 = 0.068 m_e$, $m_2 = 0.073 m_e$, $m_3 = 0.080 m_e$; $d_{12} = -33.5$ e·Å and $d_{23} = -37.6$ e·Å; $\gamma = 1.0$ meV.

are conspicuous: an anti-crossing behavior and, most importantly, the vanishing of the lower resonance amidst anti-crossing—an induced transparency due to ISP coupling.

To understand the above results, we consider a case where an analytical solution is available: We neglect all the Fock terms. This is plausible because of an exact cancellation of all Fock terms for parabolic bulk bands [7, 8]. We use equal masses for all subbands for consistency. Then, Eq. (1) can be integrated and arranged as below:

$$\left(\hbar\omega - \tilde{E}_{21} + i\gamma\right) P_{12} - D_{1232} n_{12} P_{23} = -d_{12} n_{12} E_0 \quad (4)$$

$$-D_{1232} n_{23} P_{12} + \left(\hbar\omega - \tilde{E}_{32} + i\gamma\right) P_{23} = -d_{23} n_{23} E_0 \quad (5)$$

where $\tilde{E}_{ij} = E_{ij} + D_{jjii} n_{ji}$ is the depolarization-shifted transition energy, $D_{njml} \equiv V_0^{njml}$ is the depolarization factor, and $n_{jl} = \sum_k (f_{jk} - f_{lk})/S$. The equation for P_{13} is decoupled from these two equations. These two equations would be decoupled if without the depolarization term D_{1232}. Whereas the depolarization factors D_{iijj} ($i \neq j$) lead to the formation of ISPs, depolarization factor D_{1232} introduces coupling between such ISPs. The resonance frequencies ($\omega_\pm$) can be approximated by the poles in the total polarization $P = d_{12} P_{12} + d_{23} P_{23}$. They show an anti-crossing behavior and when $\gamma = 0$ the minimum separation is given as follows:

$$\hbar(\omega_+ - \omega_-)_{min} = 2 D_{1232} \sqrt{n_{12} n_{23}}, \quad (6)$$

at $\tilde{E}_{21} = \tilde{E}_{32}$. To explain the transparency, we calculate $\mathscr{I}m(P/E_0)$ at anti-crossing ($\tilde{E}_{21} = \tilde{E}_{32}$). It is given by

$$\frac{2 D_{1232}^2 n_{12} n_{23} \left[d_{12}\sqrt{n_{12}} \pm d_{23}\sqrt{n_{23}}\right]^2 + \gamma^2 \left(n_{12} d_{12}^2 + n_{23} d_{23}^2\right)}{\gamma(4 D_{1232}^2 n_{12} n_{23} + \gamma^2)}. \quad (7)$$

At high density, $2 D_{1232}^2 n_{12} n_{23}/\gamma^2$ is about 100 so that the lower resonance [with minus sign in Eq. (7)] is suppressed by the same order of magnitude (thus transparency) at anti-crossing when $d_{12}\sqrt{n_{12}} = d_{23}\sqrt{n_{23}}$. In sum, the analytical conditions for observing the induced transparency are as follows: (i) the two renormalized energies are degenerate ($\tilde{E}_{21} = \tilde{E}_{32}$); and (ii) the oscillator strengths ($\propto d_{ij}^2 n_{ij} \tilde{E}_{ji}$) match exactly ($d_{12}\sqrt{n_{12}} = d_{23}\sqrt{n_{23}}$). The second condition gives an electron density (n_s) for a 15 nm GaAs QW at 3.8×10^{12} cm^{-2}, in agreement with the value used in the simulation.

CONCLUSIONS

Coupling of intersubband plasmons associated with two cascading transitions in a quantum well is studied. We show that the coupling drives the lower absorption peak into transparency while the spectrum exhibits an otherwise typical anti-crossing behavior. Such a phenomenon opens a new way to achieve transparency in intersubband transitions, and a new horizon in applications.

REFERENCES

1. Ando, T., Fowler, A. B., and Stern, F., *Rev. Mod. Phys.*, **54**, 437–672 (1982).
2. Helm, M., *Intersubband Transitions in Quantum Wells: Physics and Device Application I*, Academic Press, San Diego, 2000, Chapt. 1.
3. Nikonov, D. E., et al., *Phys. Rev. Lett.*, **79**, 4633 (1997).
4. Li, J., and Ning, C. Z., *Phys. Rev. Lett.*, **91**, 097401 (2003).
5. Haug, H., and Koch, S. W., *Theory of the Electrical and Optical Properties of Semiconductors*, World Scientific, Singapore, 1994.
6. Kuhn, T., *Theory of Transport Properties of Semiconductor Nanostructures*, Chapman & Hall, London, 1998, Chapt. 6.
7. Waldmüller, I., et al., *Phys. Rev. B*, 205307 (2004).
8. Li, J., and Ning, C. Z., *Phys. Rev. B*, to be published (2004).

Nonlinear Intersubband Photoabsorption in Asymmetric Single Quantum Wells

H. O. Wijewardane* and C. A. Ullrich*

Department of Physics, University of Missouri-Rolla, Rolla, Missouri 65409, USA

Abstract. A density-matrix approach combined with time-dependent density-functional theory is used to calculate the intersubband photoabsorption in a strongly driven, DC-biased GaAs/AlGaAs single quantum well. For certain frequencies and intensities of the driving field, optical bistability is observed. Compared to a full time propagation of the density matrix, the commonly used two-level rotating wave approximation becomes less and less accurate for increasing asymmetry.

Intersubband (ISB) transitions in semiconductor quantum wells take place on a meV energy scale and are therefore attractive for THz device applications [1]. Nonlinear ISB dynamics has attracted particular attention, and many interesting effects have been studied: second- and third-harmonic generation [2], intensity-dependent saturation of photoabsorption [3, 4], directional control over photocurrents [5], generation of ultrashort THz pulses [6], plasma instability [7], or optical bistability [8, 9]. Inspired by the photoabsorption experiments by Craig et al. [4], we have recently performed a theoretical study of the optical bistability region in a strongly driven, modulation n-doped GaAs/Al$_{0.3}$Ga$_{0.7}$As quantum well [10]. We have demonstrated that ISB bistability can be manipulated on a picosecond time scale by short THz control pulses. This opens up new opportunities for experimental study of optical bistability, which in the long run may lead to new THz applications such as high-speed all-optical modulators and switches.

Most previous theoretical studies of nonlinear ISB dynamics were based on the semiconductor Bloch equations (SBE) in Hartree [11]–[13] or exchange-only [14]–[16] approximation. These studies showed that the collective ISB electron dynamics is strongly influenced by depolarization and exchange-correlation (xc) many-body effects. We account for these effects using time-dependent density-functional theory, which has the advantage of formal and computational simplicity.

The present study deals with a popular simplification of the ISB SBE: the 2-level rotating-wave approximation (RWA) [11]–[13], which was used by Załużny [11] to derive analytical expressions for nonlinear ISB photoabsorption. The 2-level RWA works well for symmetric quantum wells, but we will demonstrate numerically that it breaks down when the system becomes asymmetric under the influence of DC electric fields.

The conduction subbands are described in effective-mass approximation for GaAs, where $m^* = 0.067m$ and $e^* = e/\sqrt{\varepsilon}$, $\varepsilon = 13$, are the effective mass and charge. The ground state is characterized by single-particle states $\Psi^0_{j\mathbf{q}_\parallel}(\mathbf{r}) = A^{-1/2} e^{i\mathbf{q}_\parallel \cdot \mathbf{r}_\parallel} \psi^0_j(z)$, with $\mathbf{r}_\parallel$ and $\mathbf{q}_\parallel$ in the $x-y$ plane. The envelope function for the jth subband $\psi^0_j(z)$ follows self-consistently from a one-dimensional Kohn-Sham equation [17], with the ground-state density

$$n(z) = 2 \sum_{j,\mathbf{q}_\parallel} |\psi^0_j(z)|^2 \theta(\varepsilon_F - E_{j\mathbf{q}_\parallel}). \quad (1)$$

Here, $E_{j\mathbf{q}_\parallel} = \varepsilon_j + \hbar^2 q_\parallel^2/2m^*$, and ε_j and ε_F are the subband and Fermi energy levels. We consider electronic sheet densities N_s such that only the lowest subband is occupied, in which case $\varepsilon_F = \pi\hbar^2 N_s/m^* + \varepsilon_1$.

Under the influence of THz driving fields, linearly polarized along z, the time-dependent states have the form $\Psi_{j\mathbf{q}_\parallel}(\mathbf{r},t) = A^{-1/2} e^{i\mathbf{q}_\parallel \cdot \mathbf{r}_\parallel} \psi_j(z,t)$, with initial condition $\Psi_{j\mathbf{q}_\parallel}(\mathbf{r},t_0) = \Psi^0_{j\mathbf{q}_\parallel}(\mathbf{r})$. The time-dependent Hamiltonian

$$H(t) = -\frac{\hbar^2}{2m^*}\frac{\partial^2}{\partial z^2} + v_{\text{qw}}(z) + v_{\text{dr}}(z,t) + v_H(z,t) + v_{\text{xc}}(z,t) \quad (2)$$

features $v_{\text{dr}}(z,t) = eFzf(t)\sin(\omega t)$ describing the driving field, with electric field amplitude F, frequency ω, and envelope $f(t)$. $v_{\text{qw}}(z)$ is the bare quantum well potential, the Hartree potential v_H follows from Poisson's equation, and we use the time-dependent local-density approximation for v_{xc} [10]. The time-dependent density $n(z,t)$ follows by substituting $\psi_j(z,t)$ in Eq. (1).

To account for disorder or phonon scattering, we use a density-matrix approach. We expand the first conduction subband as $\psi_1(z,t) = \sum_{k=1}^{N_b} c_k(t)\psi^0_k(z)$. The associated $N_b \times N_b$ density matrix ρ has elements $\rho_{kl}(t) = c^*_k(t)c_l(t)$

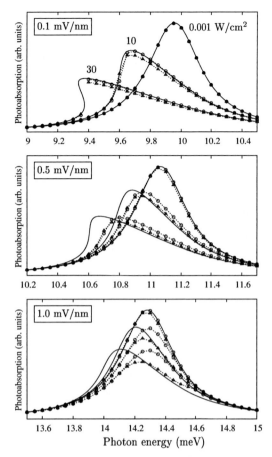

FIGURE 1. ISB photoabsorption for a 40 nm GaAs/AlGaAs quantum well with electron density 6.4×10^{10} cm^{-2} and DC bias 0.1, 0.5, and 1.0 mV/nm, driven by THz fields with intensities as indicated. Lines: 2-level RWA [11]. Symbols: full 2-level (circles) and 6-level (triangles) density-matrix propagation.

and initial condition $\rho_{kl}(t_0) = \delta_{kl}\delta_{1k}$. The time evolution of ρ follows from

$$i\hbar \frac{\partial \rho(t)}{\partial t} = [H(t), \rho(t)] - R, \qquad (3)$$

with the relaxation matrix $R_{kl} = \hbar[\rho_{kl}(t) - \rho_{kl}(t_0)]/T_{kl}$. For simplicity, $T_{kl} = T_1 \delta_{kl} + T_2(1 - \delta_{kl})$, where T_1 and T_2 are phenomenological relaxation and decoherence times.

We consider a 40 nm GaAs/Al$_{0.3}$Ga$_{0.7}$As square quantum well with $N_s = 6.4 \times 10^{10}$ cm^{-2} [4], with $\varepsilon_2 - \varepsilon_1 = 8.72$ meV and ISB plasmon frequency 9.91 meV at zero bias. We use $T_1 = 40$ ps and $T_2 = 3.1$ ps, consistent with recent values for similar systems [17]–[19]. In the following, we apply DC electric fields 0.1, 0.5, and 1.0 mV/nm, and we perform 2-level and 6-level density-matrix calculations ($N_b = 2, 6$). To describe ISB photoabsorption, we propagate Eq. (3) in the presence of THz driving fields, switched on at t_0 over a 5-cycle linear ramp and then kept at constant intensity for several hundred ps.

The photoabsorption cross section (the dissipated power) $\sigma(\omega)$ follows from the induced dipole moment [10].

Figure 1 shows results for the ISB photoabsorption, comparing Załużny's 2-level RWA [11] with our density matrix calculations, for THz intensities 0.001, 10 and 30 W/cm^2. At low intensities, $\sigma(\omega)$ has a Lorentzian shape, and the RWA and full calculations agree very well. The ISB plasmon peak Stark-shifts to higher frequencies under DC bias [17, 18]. At higher intensities, deviations from the Lorentzian lineshape are observed: population transfer into higher levels reduces the depolarization shift, predominantly at the peak position. At 0.1 mV/nm and 30 W/cm^2 this leads to bistability [10].

The 2-level RWA [11] and the full density-matrix calculations are close for small asymmetries, but discrepancies develop at increasing DC bias: the RWA tends to exaggerate the shift and change of shape of the absorption peak. The reason is that the RWA ignores all higher harmonics of ω in the induced density matrix, and thus the coupling to the diagonal matrix elements of the time-dependent potential, which are finite for asymmetric wells. We also observe more pronounced deviations between the 2-level and 6-level density-matrix calculations at larger asymmetries. A more complete analysis of the breakdown of the RWA for asymmetric systems will be presented in a forthcoming publication.

The authors acknowledge support from the donors of the Petroleum Research Fund, administered by the ACS. C.A.U. is a Cottrell Scholar of the Research Corporation.

REFERENCES

1. *Intersubband Transitions in Quantum Wells I*, edited by H. C. Liu and F. Capasso, Semiconductors and Semimetals Vol. 62 (Academic Press, San Diego, 2000).
2. J. N. Heyman et al., Phys. Rev. Lett. **72**, 2183 (1994).
3. F. H. Julien et al., Appl. Phys. Lett. **53**, 116 (1988).
4. K. Craig et al., Phys. Rev. Lett. **76**, 2382 (1996).
5. E. Dupont et al., Phys. Rev. Lett. **74**, 3596 (1995).
6. J. N. Heyman et al., Appl. Phys. Lett. **72**, 644 (1998).
7. P. Bakshi et al., Appl. Phys. Lett. **75**, 1685 (1999).
8. M. Seto and M. Helm, Appl. Phys. Lett. **60**, 859 (1992).
9. M. I. Stockman et al., Phys. Rev. B **48**, 10966 (1993).
10. H. O. Wijewardane and C. A. Ullrich, Appl. Phys. Lett. **84**, 3984 (2004).
11. M. Załużny, Phys. Rev. B **47**, 3995 (1993); J. Appl. Phys. **74**, 4716 (1993).
12. B. Galdrikian and B. Birnir, Phys. Rev. Lett. **76**, 3308 (1996).
13. A. A. Batista et al., Phys. Rev. B **66**, 195325 (2002).
14. D. E. Nikonov et al., Phys. Rev. Lett. **79**, 4633 (1997).
15. A. Olaya-Castro et al., Phys. Rev. B **68**, 155305 (2003).
16. J. Li and C. Z. Ning, Phys. Rev. Lett. **91**, 097401 (2003).
17. C. A. Ullrich and G. Vignale, Phys. Rev. B **58**, 15756 (1998); Phys. Rev. Lett. **87**, 037402 (2001).
18. J. B. Williams et al., Phys. Rev. Lett. **87**, 037401 (2001).
19. J. N. Heyman et al., Phys. Rev. Lett. **74**, 2682 (1995).

Intersubband Tunneling without Intrasubband Relaxation in Multi-quantum Wells

G. S. Vieira[1], J. M. Villas-Bôas[2], P. S. S. Guimarães[3], N. Studart[2], J. Kono[4], S. J. Allen[4], K. L. Campman[5] and A. C. Gossard[5]

[1]*Instituto de Estudos Avançados, Centro Técnico Aeroespacial, 12228-840 São José dos Campos, SP, Brazil*
[2]*Departamento de Física, Universidade Federal de São Carlos, 13565-905, São Carlos, SP, Brazil*
[3]*Departamento de Física, Universidade Federal de Minas Gerais, 30123-970, Belo Horizonte, MG, Brazil*
[4]*Center for Terahertz Science and Technology, University of California at Santa Barbara, CA, 93106, USA*
[5]*Materials Department, University of California at Santa Barbara, CA, 93106, USA*

Abstract. We present evidence of a new photon-assisted tunneling channel in multi-quantum wells in the presence of a magnetic field applied parallel to the quantum well layers. Electrons in one potential well that are photo-excited to anti-crossings of the dispersion curves of the conduction subbands tunnel resonantly to the neighboring well without intra-subband relaxation. This tunneling mechanism is to be distinguished from tunneling involving states located around the energy minimum of the conduction subbands. Measurements of the photocurrent under THz illumination show that both tunneling processes co-exist in a GaAs/AlGaAs multi-quantum well superlattice.

INTRODUCTION

In a multi-quantum well semiconductor heterostructure, a magnetic field applied parallel to its layers modifies the electron energy dependence on k_y, the wavevector component perpendicular to both the field and growth directions. At the anti-crossings of the k_y dispersion relations, the electron wave functions are extended over more than one well and therefore the states at the anti-crossings constitute tunneling channels [1,2]. If the electrons are at thermal equilibrium inside each subband, only those tunneling channels close to the bottom of the occupied subbands will be observed in transport measurements. In this process, when a higher energy subband is populated by photo-excitation, the photo-excited electrons will first relax to the lowest energy states of this subband before tunneling.

Here we propose a distinct resonant photon-assisted tunneling process, in which electrons are excited directly into extended states at the anti-crossings in the k_y dispersion relations, away from the bottom of any subband. Therefore, a significant fraction of these electrons can tunnel to a neighboring well before relaxing energy. We probed this tunneling channel in a GaAs/AlGaAs multi-quantum superlattice under a magnetic field parallel to the layers, with photocurrent measurements under THz radiation.

EXPERIMENT

The sample was grown by MBE over a semi-insulating GaAs substrate. It consists of ten layers of GaAs (33 nm) and eleven layers of $Al_{0.3}Ga_{0.7}As$ (4 nm), sandwiched between two layers of GaAs, each 50 nm thick. All these layers are uniformly doped with Si to 3×10^{15} cm^{-3}. This structure was again sandwiched between two 300 nm thick GaAs layers doped with Si at 2×10^{18} cm^{-3}. Small 2×4 μm^2 mesas were fabricated and integrated into a bow-tie antenna which includes a Schottky diode that provides a reference to the radiation power coupled to the device, as described elsewhere [3].

We performed a numerical non-perturbative calculation of the electronic energy subbands in the effective mass approximation. We find that, for our sample, at a magnetic field of 3.1 T there are anti-crossings at the bottom of the second subband of a well with the first subbands of the adjacent wells. Therefore, if the second subband is occupied by photo-excitation and a small fixed voltage is applied, a maximum in the photocurrent as a function of magnetic field should occur at this field value.

Figure 1 shows the calculated dispersion relations of three adjacent wells at a field of 5.1 T. The arrows indicate a possible photon-assisted tunneling channel in which electrons from the first subband of the wells are photo excited to states at the anti-crossing between the second subband of a given well and the first subband of the adjacent wells. If the dispersion relation of the first and second subbands were equal, as predicted by first order perturbation theory [4] (dotted lines in Fig. 1) this photon assisted tunneling channel would have the same resonance frequency as the photo-excitation between the bottoms of the first and second subbands. As shown in Fig. 1, the actual resonance frequency to excite electrons to the anti-crossings is significantly higher than the transition between the bottoms of the two subbands.

We performed photocurrent measurements as a function of applied magnetic field, at a small bias (15 mV), at radiation frequencies above 3.44 THz, the resonance between the bottoms of the two first subbands at 3.1 T. The results are shown in Figure 2.

maxima in the photocurrent arise as the result of the superposition of two peaks, one always at $B = 3.1$ T, where there is a maximum for the tunneling probability for electrons at the bottom of the second subband, and another at the resonance for photon-assisted tunneling without relaxation. The frequency of the latter resonance depends on the magnetic field. As f decreases, the two peaks get closer, leading to a narrowing of the resulting maximum in the photocurrent, as shown by the experimental results.

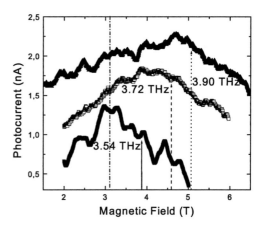

FIGURE 2. Photocurrent as a function of magnetic field, applied parallel to the MQW layers, for three radiation frequencies. The curves for $f = 3.72$ THz and $f = 3.90$ THz were moved up 0.4 and 0.8 nA, respectively. The resonances for photon-assisted tunneling are indicated.

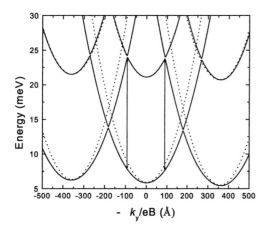

FIGURE 1. Dispersion relations of a three well structure at $B = 5.1$ T and a voltage drop of 0.4 mV per period, calculated using first order perturbation theory (dotted lines) and by a numerical non-perturbative calculation (solid line). Arrows show the excitation energies referred to in the text.

The photocurrent curve for $f = 3.90$ THz shows a broad asymmetric maximum near the expected field position for photon-assisted tunneling without relaxation at this frequency (arrows). As the radiation frequency goes down, the resulting maximum narrows and shifts to $B = 3.1$ T for $f = 3.54$ THz. The observed

ACKNOWLEDGMENTS

We thank the staff of the UCSB Center for Terahertz Science and Technology for their outstanding support and expertise and also M.A. Ruggieri Franco and V.A. Serrão for their help with the initial calculations. We acknowledge the financial support of CNPq, FAPESP, ONR and AFOSR.

REFERENCES

1. G. Platero, L. Brey and C. Tejedor, Physical Review B **40**, 8548-8551 (1989).
2. T. M. Fromhold, F. W. Sheard and G. A. Toombs, Surface Science **228**, 437-440 (1990).
3. G. S. Vieira, S. J. Allen Jr., P. S. S. Guimarães, K. L. Campman and A. C. Gossard, Physical Review B **58**, 7136-7140 (1998).
4. J. K. Maan, Festkörperprobleme **27**, 137-167 (1987).

Intersubband Lifetime Magnetophonon Oscillations in GaAs Quantum Cascade Lasers

O. Drachenko[1], D. Smirnov[2], J. Léotin[1], A. Vasanelli[3], C. Sirtori[3]

[1] *Laboratoire National des Champs Magnétiques Pulsés, 31432 Toulouse, France*
[2] *National High Magnetic Field Laboratory, Tallahassee, Florida, 32310, USA*
[2] *Laboratoire Matériaux et Phénomènes Quantiques, Université Paris VII, 75005 Paris, France*

Abstract. We studied the behavior of middle IR GaAs/AlGaAs quantum cascade lasers (QCL) subjected to a strong magnetic field perpendicular to the 2D planes. We observed that both laser threshold and differential quantum efficiency are modulated by a magnetic field. The intersubband lifetime vs magnetic field is derived from the field dependence of QCL threshold current. The magnetic field dependence of electron-LO phonon scattering rates derived from experiment shows good agreement with our calculations.

INTRODUCTION

The QCL is a unipolar semiconductor laser providing a cascade of radiative transitions between energy levels in MQW structure. In the MIR QCL, polar electron-LO phonon interaction is the dominant electron relaxation process limiting the upper laser level lifetime. A strong magnetic field applied perpendicular to the 2D planes breaks the continuum energy spectrum into a ladder of discrete Landau levels (LL). It has been shown recently that due to "intersubband magnetophonon resonance" effect, electron-LO phonon relaxation becomes strongly modulated by an external magnetic field, and causes giant oscillations of laser emission power [1,2]. In the present study, we show that not only laser emission power, but also both laser threshold I_{th} and differential quantum efficiency dP/dI are modulated by a magnetic field. We obtain the intersubband lifetime dependence on a magnetic field and compare to the calculated electron-LO phonon scattering rates.

EXPERIMENTS AND RESULTS

The devices investigated are MIR quantum cascade lasers based on a GaAs/AlGaAs heterostructures designed to emit near 11μm [3, 4]. The active region of each QCL`s period consists of three coupled quantum wells. Three subbands (n = 1, 2 and 3) are delocalised across this region and form the physical system that we investigated. The laser transition occurs between the n=3 and n=2 states and the lower level (n=1) helps a fast depopulation of the n=2 state by resonant optical phonon emission. At 4.2K, the laser transition 3 – 2 peaks at 111 meV (λ = 11.2 μm). The energy difference E2-E1 is about 37meV.

The magnetic field measurements were performed at T=5K with DC magnetic fields up to 45T (NHMFL) and with pulsed magnetic fields up to 60T (LNCMP). The QCL devices were mounted in the I//B geometry. Emitted light was detected with an external fast MCT detector. Current pulses 1-2 μs long with 1% repetition rate were used to avoid heating effects. In pulsed magnetic fields we measured the laser emission as a function of magnetic field L(B) for fixed values of current. In DC field studies, we measured light intensity vs current bias and magnetic field.

Laser emission intensity is strongly affected by the magnetic field and shows strong oscillations (Fig.1a). These oscillations are attributed to resonant electron-LO phonon scattering. In the frames of a simple 3-level model, the laser threshold current is inversely proportional to the lifetime of the upper level τ_3. Then, we can derive the magnetic field dependence of the intersubband scattering rate (Fig.1b). Within the same approximation, dP/dI does not depend directly on τ_3, and therefore it is not supposed to oscillate with a magnetic field. In our experiments we found a clear evidence of strong oscillations in both Ith (B) and dP/dI (B). These results suggest the field dependence of the injection rate assumed so far to be constant with a magnetic field, has to be taken into account.

CALCULATION OF ELECTRON – LO PHONON SCATTERING RATES

In our model [5] the lifetime of an electron is fundamentally controlled by the emission/absorption of an LO phonon. In order to describe the finite width of the transitions, intersubband transitions are envisioned as taking place in many micro-samples, which differ from one another by random values of their characteristic thicknesses. We assume a Gaussian distribution of these thicknesses and as a consequence, the delta-like peaks of the Landau levels (LL) density of states are smeared out in Gaussians of width σ.

We found that the population relaxation rate is an oscillating function of magnetic field. QCL gain depends on the scattering rate of electrons relaxing from the ground level of the Landau ladder 3 ($|3,0\rangle$) to lower Landau levels $|n,l\rangle$, $n<3$, and, therefore, it is also an oscillating function of the magnetic field. The resonances take place whenever the following condition is fulfilled: $E_{30} - E_{nl} = \hbar\omega_{LO}$. Calculations of the scattering rate $1/\tau_3$ ($1/\tau_3 = 1/\tau_{32} + 1/\tau_{31}$) have been performed by taking into account band nonparabolicity.

The calculated scattering rates are shown in Fig.1c. Peak positions are in good agreement with experimental results, except in the region between 15T and 20T. In this region, enhanced scattering is expected due to elastic relaxation processes from the LL $|3,0\rangle$ into excited LLs $|2,5\rangle$ and $|1,6\rangle$. The resonant elastic transitions made possible by scattering on ionized impurities, electron-electron scattering or acoustic phonon interaction, are not taken into account in our calculations.

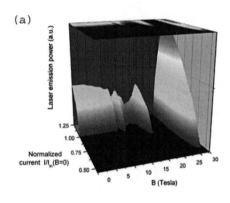
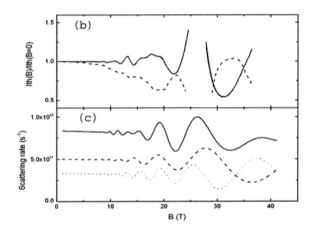

FIGURE 1. (a). Laser emission intensity as a function of current bias (normalized to the threshold current value) and magnetic field. T=5K. (b) Magnetic field dependence of the threshold current (solid line) and of the differential quantum efficiency dP/dI (dashed line). (c) Calculated magnetic field dependence of the total scattering rate $1/\tau_3$ (solid line) and contributions for each lower subband $1/\tau_2$ and $1/\tau_1$ for the LO-phonon emission.

ACKNOWLEDGMENTS

A portion of this work was performed at the NHMFL, which is supported by NSF Cooperative Agreement No. DMR-0084173, by the State of Florida, and by the DOE.

REFERENCES

1. D. Smirnov et al., Phys. Rev. B **66**, 125317 (2002).
2. D. Smirnov et al., Phys. Rev. B **66**, 121305 (R) (2002).
3. C. Sirtori *et al.*, Appl. Phys. Lett. **73**, 3486 (1998).
4. P. Kruck *et al.*, Appl. Phys. Lett. **76**, 3340 (2000).
5. C. Becker et al., Phys. Rev. B **69**, 115328 (2004).

Molecular Model for the Radiative Dipole Strengths and Lifetimes of the Fluorescent Levels of Mn^{2+} and Fe^{3+} in II-VI And III-V Compounds

R. Parrot * and D. Boulanger [†]

*Institut Universitaire de Formation des Maîtres et Institut d'Enseignement Supérieur de Guyane, BP 792, 97337 Cayenne Cedex, France

[†] Université de Paris-Sud, Laboratoire d'Informatique, Maîtrise de Sciences Physiques, Bâtiment 479, 91405 Orsay Cedex, France

Abstract. A molecular model is used to give an overall semi-phenomenological interpretation of the radiative transition probabilities (RTP) or radiative lifetimes (RL) of Mn^{2+} and Fe^{3+} in II-VI and III-V compounds. It is shown that the RTP's are primarily controlled by: (i) the mixing of the wavefunctions of the cation and of the ligands (ii) the molecular spin-orbit interaction which involves the spin-orbit coupling constants ζ_d of the d electrons of the cation and ζ_p of the p electrons of the ligands and (iii) the energies of the intermediate levels which appear in the perturbation model

INTRODUCTION

The aim of this paper is to present a semi-phenomenological covalent model for the RL's of d^5 ions in cubic II-VI and III-V compounds.

The model will be used to analyze the very scattered RL's of Mn^{2+} and Fe^{3+} in cubic symmetry. For Mn^{2+}, the RL's extend from 1.77 ms in cubic ZnS [1,2] to 0.22-0.24 ms in ZnSe [3,4] and 40-52 μs in ZnTe [5]. For Mn^{2+} in GaP, the RL's is of 1.5 ms. In the case of Fe^{3+}, the RL's are of 1.9 ms in GaAs [6,7] and 4.3 ms in cubic ZnS [8].

The molecular model for the RL's or RTP's is presented in Sec. II. The experimental results are interpreted in Sec. III.

MOLECULAR MODEL

The probability of spontaneous emission per unit time from an initial $^4T_1$ state to a final $^6A_1$ state is given by [9]:

$$P(^4T_1, {}^6A_1) = 1/\tau = (1/d_A)(1/3h) 64 e^2 \pi^4 \sigma^3 \chi S(^4T_1, {}^6A_1), \quad (1)$$

where τ is the radiative lifetime, d_A is the degeneracy of the initial state, σ is the energy of the emitted light, χ is a correcting factor which accounts for the effective electric field at the site of the impurity, and $S(^4T_1, {}^6A_1)$ is the total line strength of the transition.

In the proposed molecular model, $S(^4T_1 \rightarrow {}^6A_1)$ is given in terms of the matrix elements of the molecular spin-orbit interaction H_{SOmol} and of the matrix elements of the electric dipole moment $M = e\mathbf{r}$ by:

$$S(^4T_1, {}^6A_1) = 6f^2,$$

with,

$$f^2 = (1/6) \sum_{j, u, Ms, Ms'} |\sum_{q, v} \langle {}^6A_1 Ms'|H_{SOmol}|{}^4T_1^q v Ms\rangle$$
$$\langle {}^4T_1^q v Ms|r_j|{}^4T_1 u Ms\rangle [W({}^4T_1^q) - W({}^6A_1)]^{-1}$$
$$+ \sum_{q, v} \langle {}^6A_1 Ms'|r_j|{}^6T_2^q v Ms'\rangle$$
$$\langle {}^6T_2^q v Ms'|H_{SOmol}|{}^4T_1 u Ms\rangle [W({}^6T_2^q) - W({}^6A_1)]^{-1}|^2. \quad (2)$$

q represents the orbital triplets. u and v refer to the components of the orbital triplets.

The molecular spin-orbit interaction H_{SOmol} which describes the spin-orbit interaction in a covalent model can be written as [10]:

$$H_{SOmol} = \sum_i \zeta_C(r_{iC}) l_{iC} \cdot s_i + \sum_i \sum_L \zeta_L(r_{iL}) l_{iL} \cdot s_i, \quad (3)$$

where l_{iC} and l_{iL} are one electron orbital operators for the cation and the ligands respectively. ζ_C and ζ_L are the spin-orbit coupling constants of the electrons of the metal and of the ligands respectively, as defined by Blume and Watson [11]. In the following, the relevant spin-orbit coupling constants will be those of the electrons 3d of the metal and of the electrons np of ligands.

The matrix elements of H_{SOmol} must be calculated between the $|{}^6A_1\rangle$ and the $|{}^4T_1\rangle$ orbital triplets and between the $|{}^6T_2\rangle$ and $|{}^4T_1\rangle$ orbital triplets. We must emphasize the difficulty to calculate the matrix elements of the electric dipole moment which involves all molecular orbitals. The SO interaction which is localized around the cation and the ligands is decomposed into two independent terms [10].

PHENOMENOLOGICAL MODEL

A simplified phenomenological model is proposed whose main interest will be to determine the interactions controlling the RTP's.

First, we will assume that the multielectronic wave functions are almost identical for the clusters. This hypothesis is approximately justified from previous molecular calculations [12].

Second, we will also assume that the influence of the electric dipole moment is of the form $M = e\langle r\rangle$.

Finally, we will approximate the energy appearing in the denominators in equation (2) by the energy of the emission band.

Following these assumptions, the f's for the ${}^4T_1 \rightarrow {}^6A_1$ transitions are given by:

$$f({}^4T_1, {}^6A_1) = (\alpha\langle r\rangle \zeta_C + \beta\langle r\rangle \zeta_L)/\sigma, \qquad (4)$$

where σ is defined as the energy of the center of gravity of the fluorescent band and α and β are considered as constants for a common cation series.

Of course, a comparison with the experimental results will justify or not these assumptions.

Figure 1 represents $f\sigma$, as obtained from formula (1), in terms of ζ_L for Mn^{2+} in cubic ZnS, ZnSe, ZnTe, GaP, and for Fe^{3+} in GaAs and in cubic ZnS. It shows that for Mn^{2+} in the considered compounds, $f\sigma$ strongly increases in terms of ζ_L. In the case of Fe^{3+} we observe a lower increase of $f\sigma$ than for Mn^{2+}. This figure shows that for Mn^{2+}, α and β are constants not only for the considered II-VI compounds as predicted but also for GaP.

CONCLUSION

By considering the lifetimes of Mn^{2+} and Fe^{3+} in II-VI and III-V compounds of cubic symmetry, it has been shown that they almost linearly increase in terms of the spin-orbit constants of the p electrons of the ligands. Furthermore, a simplified covalent model has been elaborated which correctly accounts for the LF's of Mn^{2+} in several II-VI compounds and also in GaP. The model also accounts for the LF's of Fe^{3+} in ZnS and GaAs.

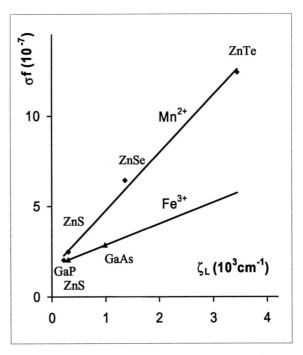

FIGURE 1. Representation of σf in terms of ζ_L.

REFERENCES

1. W. Busse, H.E. Gumlich, A. Geoffroy, and R. Parrot, *Phys. Stat. Solidi* **b 93**, 531 (1979).
2. W. Busse, H.E. Gumlich, W. Knack, and J. Schulze, *J. Phys. Soc. Japan* **49**, 581 (1980).
3. H.E. Gumlich, private communication.
4. T. C. Leslie and J.W. Allen, *Phys. Stat. Solidi* **a 65**, 545 (1981). S.G. Ayling and J.W. Allen, *J. Phys. C.* **20**, 4251 (1987).
5. M. Krause, A. Krost, and H. E. Gumlich, *Europhysics Conf. Abstracts* **9a** (1985).
6. K. Pressel, G. Bohnert, G. Rückert, A. Dörnen, and K. Thonke, *J. Appl. Phys.* **71**, 5703 (1992).
7. K. Pressel, G. Rückert, A. Dörnen, and K. Thonke *Phys. Rev.* **B 46**, 13171 (1992).
8. R. Heitz, P. Thurian, I. Loa, L. Eckey, A. Hoffmann, I. Broser, K. Pressel, B.K. Meyer, and E.N. Mokhov, *Appl. Phys. Letter* **67**, 2822 (1995).
9. T. Hoshina, *J. Phys. Soc. Japan* **48**, 1261 (1980).
10. M.C.G. Passegi and T. Buch, *J. Phys. C* **4**, 1207 (1971). A.A. Misetich and T. Buch, *J. Chem. Phys.* **42**, 2524 (1964).
11. M.Blume and R.E.Watson, *Proc. Roy. Soc. London* Ser. **A 271**, 565 (1963).
12. R. Parrot and D. Boulanger, *Phys. Rev.* **B 47**, 1849 (1993). D. Boulanger and R. Parrot, *J. Chem. Phys.* **87**, 1469 (1987).

On the origin of excitonic luminescence in quantum wells: direct measure of the exciton formation in quantum wells from time resolved interband luminescence

J. Szczytko, L. Kappei, J. Berney, F. Morier-Genoud,
M.T. Portella-Oberli, B. Deveaud

Institut de Photonique et Electronique Quantiques, Ecole Polytechnique Fédérale de Lausanne (EPFL) CH1015 Lausanne, Switzerland

Abstract. We present the results of a detailed time resolved luminescence study carried out on a very high quality InGaAs quantum well sample where the contributions at the energy of the exciton and at the band edge can be clearly separated. We perform this experiment with a spectral resolution and a sensitivity of the setup allowing to keep the observation of these two separate contributions over a broad range of times and densities. This allows us to directly evidence the exciton formation time, which depends on the density as expected from theory. We also denote the dominant contribution of excitons to the luminescence signal, and the lack of thermodynamical equilibrium at low densities.

The luminescence of free carriers in quantum wells has been an elusive and nevertheless very active matter over at least the last 15 years. In particular, time resolved luminescence is dominated, even at the shortest times, and under non resonant excitation, by an excitonic like luminescence line [1, 2]. A very interesting debate has been introduced by the group of Stefan Koch, aiming at understanding whether this excitonic like transition could be due to free carrier luminescence, without the need for bound pairs [3].

We present the results of a detailed time resolved photo-luminescence (TR-PL) study carried out on a very high quality InGaAs quantum well sample where the contributions at the energy of the exciton and at the band edge can be clearly separated as it is shown in Fig. 1. The sample used for this study is a very high quality single InGaAs 80 Å quantum well, with a low indium content of about 5%. Proper observation of the different luminescence transitions needs resonant excitation in the well and 4 orders of magnitude dynamical range. We perform this experiment with a spectral resolution and a sensitivity of the set-up allowing to keep the observation of these two separate contributions over a broad range of times and densities. More details about the sample and experimental setup can be found in [4].

The shape of the PL signal above the gap is the same within the two main theoretical descriptions (excitons plus free carriers [5, 6], or Coulomb correlated free carriers [3]). In both cases, we expect a Boltzmann like PL line provided electrons and holes are thermalized, which

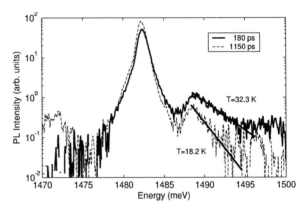

FIGURE 1. Spectral trace of the TR-PL $N_v = 2 \times 10^{10}$ photons/cm^2. The temperatures of the carriers estimated 180 ps and 1150 ps after the excitation are shown. Heavy-hole exciton $E_{1s} = 1.4823$ eV ($E_{2s} = 1.4882$ eV), $E_{fc} = 1.4888$ eV, light-hole exciton $E_{lh} = 1.4988$ eV.

is the case for times longer than 100 ps. In both models, the intensity at each energy is simply proportional to the product of the associated distribution functions f_e and f_h. Then, in the case of the low density regime we are studying, the integrated intensity of the free carrier PL I_{fc} is proportional to the concentration of the electrons n and holes p via the bimolecular recombination rate B [7], noting that $n = p$ intensity $I_{fc} \sim Bnp = Bn^2$. The bimolecular recombination rate B is inversely proportional to the carriers' temperature T at low temperatures in QW

[8], therefore:

$$n \propto \sqrt{I_{fc} T} \quad (1)$$

Thus knowing the time evolution of both quantities – the free carriers PL intensity $I_{fc}(t)$ and their temperature $T(t)$ we can deduce in a very simple and direct way the temporal evolution of the photoexcited free carrier density. This evolution is plotted in Fig. 2 (symbols). For I_{fc} we integrated the intensity of 1.487-1.495 eV. The relative values are real, and it is clear that carriers are missing at early times in the high density experiments. We will show in the following that this is a direct signature of fast exciton formation. We get two rate equations for the population of free carriers n and of excitons X:

$$\frac{dn}{dt} = -\gamma C n^2 + \gamma C N_{eq}^2 - \frac{n}{\tau_{nr}} - B n^2 \quad (2)$$

$$\frac{dX}{dt} = \gamma C n^2 - \gamma C N_{eq}^2 - \frac{X}{\tau_D} \quad (3)$$

where B is the electron-hole bimolecular recombination rate [8], τ_{nr} is the non-radiative decay time, τ_D is a thermalized exciton decay time which depends on the fraction of excitons in the radiative region only. In order to reduce the number of free fitting parameters, we took a reasonable constant value τ_D=700 ps. C is the rate calculated by Piermarocchi et al. [9]. It depends upon both carrier and lattice temperature through the interaction with optical and acoustic phonons. γ is a multiplication factor by which the measured formation rate in our InGaAs QW is changed compared to the theoretical value obtained for GaAs. $N_{eq}(n,X,T)$ is the equilibrium carrier concentration given by the Saha equation (for $n = p = N_{eq}$) [10]. The term $C N_{eq}^2$ describes exciton ionization. It ensures that the static solutions of Eq. 2 and 3 are the equilibrium values given by Saha equation. It corresponds to a first order approach of the rate equations valid in case the initial condition is a population of free carriers.

The presented model follows the description proposed by Piermarocchi [9]. Increasing the carrier density leads to a faster formation of excitons, because it gives rise to an increased probability of binding one electron and one hole through interaction with phonons. This theoretical prediction is indeed confirmed in our experiment. The formation time $\tau_f := (\gamma C n)^{-1}$, measured over two orders of magnitude in density 100 ps after the initial excitation, changes from less than 10 ps for highest density to 570 ps for the lowest one. This τ_f evolves as the plasma concentration and temperature change and, 1 ns after the excitation, binding of free carriers into excitons may be as long as 1100 ps. This description also brings along an initial drop of the carrier concentration because of rapid exciton formation. Figure 2 indeed shows that, within the first 5 ps, the carrier concentration drops by about 40 % for the highest densities. Such a rapid downfall is the consequence of the exciton formation through emission of LO phonons by the initially hot carriers [9]. This process is very efficient at high electron temperatures and densities. Our model includes this effect through an increase of C with temperature and the linear dependence of the formation rate on the carrier density ([9]). Thus a considerable fraction of an initially hot free carrier population can be transformed into excitons during the first picoseconds. Accounting for the rapid exciton formation at high densities, we get exciton concentration larger then a few percent in our system, as shown in Fig. 2 (dotted lines). Due to the large difference between the radiative rates of excitons and free carriers, the exciton contribution always dominate over that of free carriers.

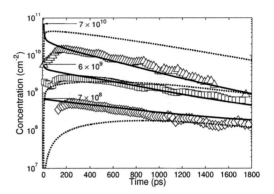

FIGURE 2. Concentration of free carriers as a function of time. The experimental data calculated according Eq. 1 for three different absorbed photon densities (given per pulse per cm^2) are shown together with the time decay fitting curves (solid lines, Eq. 2). The corresponding total (i.e. radiative and dark) exciton density X are marked as thick dotted lines.

ACKNOWLEDGMENTS

This work was partially supported by the Swiss National Research fund. We wish to thank T. Guillet, J.-D. Ganiere, S. Koch, V. Savona, D. Chemla and R. Zimmermann for fruitful discussions.

REFERENCES

1. B. Deveaud et al., Phys. Rev. Letters, **67**, 2355 (1991)
2. G.R. Hayes et al. Phys. Stat. Sol. a, **190**, 637 (2002)
3. M. Kira et al., Phys. Rev. Lett., **82**, 3544 (1999); M. Kira et al., Phys. Rev. Lett., **81**, 3263 (1998); S. Chatterjee et al., Phys. Rev. Lett., **92**, 067402
4. J. Szczytko et al., Phys. Stat. Sol. c **1**, 493 (2004)
5. B.K. Ridley, Phys. Rev. B **41**, 12190 (1990)
6. H.W. Yoon et al., Phys. Rev. B **54** 2763, 1996
7. G. Lasher, F. Stern, Phys. Rev. **133**, A553 (1964)
8. T. Matsusue et al., Appl. Phys. Lett., **50**, 1429 (1987)
9. C. Piermarocchi et al., Phys. Rev. B **55**, 1333 (1997)
10. R.T. Philips et al., Solid State Commun. **98** 287 (1996)

Why interacting excitons cannot be bosonized

M. Combescot and O. Betbeder-Matibet

CNRS, Université Pierre et Marie Curie, Paris

Abstract. We present a new many-body theory which allows to keep the composite nature of the excitons all over. From the obtained results on a few physical quantities, we conclude that interacting excitons cannot be bosonized, if we want to be sure not to miss terms, possibly dominant.

Excitons have been known to be composite bosons for a long time. However, up to now, they were considered as true bosons, even for their interactions, due to the lack of a many-body procedure adapted to composite bosons. Over the last few years, we have made an important break-through by developing such a theory [1], which is going to be of importance, not only for semiconductors, but also in many other fields, because essentially all bosons known in physics are in fact composite bosons.

This new many-body theory relies on four nicely simple commutators, namely

$$[B_m, B_i^\dagger] = \delta_{m,i} - D_{mi}, \quad (1)$$

$$[D_{mi}, B_j^\dagger] = 2\sum_n \lambda_{mnij} B_n^\dagger, \quad (2)$$

$$[H, B_i^\dagger] = E_i B_i^\dagger + V_i^\dagger, \quad (3)$$

$$[V_i^\dagger, B_j^\dagger] = \sum_{mn} \xi_{mnij}^{\text{dir}} B_m^\dagger B_n^\dagger, \quad (4)$$

where H is the system Hamiltonian, *i. e.*, here the semiconductor Hamiltonian composed of electron and hole kinetic energies plus Coulomb interactions between electrons, between holes and between electrons and holes. $B_i^\dagger$ is the creation operator of an exciton i, with center of mass momentum $\mathbf{Q}_i$ and relative motion index ν_i, *i. e.*, $(H - E_i) B_i^\dagger |v\rangle = 0$, with $E_i = \varepsilon_{\nu_i} + \hbar^2 \mathbf{Q}_i^2/(m_e + m_h)$. This generates two scatterings ξ_{mnij}^{dir} and λ_{mnij}. ξ_{mnij}^{dir} is a direct Coulomb scattering which corresponds to all possible Coulomb processes between the (i,j) excitons, the excitons m and i being made with the same electron-hole pair. λ_{mnij} corresponds to carrier exchanges between the (i,j) excitons, in the absence of any Coulomb process.

Using these four commutators, we can calculate any matrix elements we wish between N-exciton states. Indeed, to get

$$\langle v | B_{m_N} \ldots B_{m_1} f(H) B_{i_1}^\dagger \ldots B_{i_N}^\dagger | v \rangle, \quad (5)$$

we first push $f(H)$ to the right, using eqs. (3,4) which allow to obtain the commutator of $B_i^\dagger$ with any function of the Hamiltonian H. We then push the B_m's to the right according to eqs. (1,2); so that the matrix element (5) ultimately reads in terms of the products of various ξ_{mnij} by various λ_{pqkl}.

The major difficulty with interacting excitons is not so much the calculation of these matrix elements, but the determination of the appropriate ones for the problem at hand. Indeed, all our knowledge on interacting particles is based on perturbation, *i. e.*, on the existence of an interacting potential $V = H - H_0$. In the case of composite excitons, this V does not exist, so that we should find a way to write the quantity we wish in terms of H, not V. Although *a priori* possible, since H contains all the necessary informations on the possible interactions in the system, this is actually far from obvious in most problems.

Using this new many-body theory, we have first made the link with the usual effective bosonic Hamiltonian for excitons [2]. The exciton-exciton scattering it contains is just

$$\xi_{mnij}^{\text{dir}} - \sum_{rs} \xi_{mnrs}^{\text{dir}} \lambda_{rsij}, \quad (6)$$

while

$$\xi_{mnij}^{\text{dir}} - \frac{1}{2} \sum_{rs} \left(\xi_{mnrs}^{\text{dir}} \lambda_{rsij} + \lambda_{mnrs} \xi_{rsij}^{\text{dir}} \right), \quad (7)$$

would have allowed the effective Hamiltonian to be at least hermitian. Nevertheless, this is actually not enough to get correct answers in problems dealing with interacting excitons, because the scatterings ξ_{mnij}^{dir} and λ_{mnij} appear in much more complicated combinations than the ones of eqs. (6,7).

In order to invalidate the possible replacement of excitons by bosons, we have also shown that the link between the lifetime of an N-exciton state with all the excitons in the same 0 state and the scattering rates towards states in which two excitons are in $(i,j) \neq (0,0)$, is [3]

$$\frac{1}{\tau_0} = \alpha \sum_{ij \neq 00} \frac{1}{T_{ij}}, \qquad (8)$$

with $\alpha = 1$ if the excitons are bosonized and $\alpha = 1/2$ if their composite nature is kept. These sum rules come from closure relations which are different for elementary particles and composite particles, making the agreement of the corresponding quantities impossible, whatever are the scatterings we take in the effective bosonic Hamiltonian.

In view of this, it is necessary to reconsider all problems dealing with interactions between excitons, as well as all optical nonlinearities in semiconductors [4, 5], since these nonlinearities come from the interaction with the virtual excitons coupled to the photons.

REFERENCES

1. For a short review, see M. Combescot and O. Betbeder-Matibet, Cond-mat/0407068, and references therein.
2. M. Combescot and O. Betbeder-Matibet, Europhysics Lett. **58**, 87 (2002).
3. M. Combescot and O. Betbeder-Matibet, Phys. Rev. Lett. **93**, 016403 (2004).
4. M. Combescot, O. Betbeder-Matibet, K. Cho and H. Ajiki, Cond-mat/0311387.
5. M. Combescot and O. Betbeder-Matibet, Cond-mat/0404744, to appear in Solid State Com. (2004).

The trion as an exciton interacting with a carrier

M. Combescot and O. Betbeder-Matibet

CNRS, Université Pierre et Marie Curie, Paris

Abstract. Using a new many-body theory for excitons interacting with free carriers, in which the composite nature of the exciton is treated exactly, we can describe the trion as an exciton interacting with an electron, or a hole, through Coulomb and Pauli scatterings.

Even if the trion is often called charged exciton, theoretical works which have considered trions up to now, have seen it as two electrons plus one hole (or two holes plus one electron), not one exciton plus one carrier [1].

In order to reduce this three-body problem to a two-body problem, one exciton interacting with a carrier, it is in fact necessary to know how to handle the possible carrier exchanges properly. For that, we have generated a "commutation technique" for excitons interacting with free carriers [2, 3], similar to the one we have recently developed in our new many-body theory for excitons interacting with excitons [4, 5]. It relies on two scatterings: a direct Coulomb scattering between the carrier and the exciton, in which the exciton before and after scattering is made with the same electron-hole pair, and a "Pauli scattering", which is just a carrier exchange, without any Coulomb contribution.

Using this commutation technique, we can derive the trion oscillator strength in an easy way [6]. We have shown that it is one trion volume divided by one sample volume smaller than the exciton one, making the trion impossible to see in an infinite volume sample: Indeed, trions are essentially fermions, so that they cannot be piled up at the same energy as the boson-like excitons, or even as the excitons bound to a donor, which are in fact classical particles. This commutation technique also led us to generate new "exciton-electron diagrams" to represent the trion, which have to be contrasted with the usual electron-hole Feynman diagrams. We have actually shown that these standard Feynman diagrams are totally hopeless to represent a three-body object properly. The exciton-electron diagrams are going to be all the most useful to face the problem of a virtual exciton photocreated in a doped quantum well, commonly seen as a trion embedded in a Fermi sea. The state of the art on this quite difficult problem is quite unsatisfactory because, to get a trion binding energy much smaller than the exciton one, it is necessary to include the electron-electron repulsion on the same footing as the electron-hole attraction, not just as a screening of this attraction, as done up to now [7]. This makes the problem of a photocreated exciton in a Fermi sea far more complex than the standard Fermi edge singularities [8].

REFERENCES

1. M. Combescot, Eur. Phys. J. B **33**, 311 (2003).
2. M. Combescot and O. Betbeder-Matibet, Solid State Com. **126**, 687 (2003).
3. M. Combescot, O. Betbeder-Matibet and F. Dubin, Cond-mat/0402441.
4. M. Combescot and O. Betbeder-Matibet, Europhysics Lett. **58**, 87 (2002).
5. O. Betbeder-Matibet and M. Combescot, Eur. Phys. J. B **27**, 505 (2002).
6. M. Combescot and J. Tribollet, Solid State Com. **128**, 273 (2003).
7. P. Hawrylack, Phys. Rev. B **44**, 3821 (1991).
8. M. Combescot and P. Nozières, J. de Phys. (Paris) **32**, 913 (1971).

Resonant coupling of an excitonic state in a quantum disk with an exciton-polariton mode traveling through a nearby wave-guide

H. Takagi, H. Tanaka, M. Yamaguchi, N. Sawaki

*Department of Electronics, Nagoya University,
Chikusa-ku, Nagoya 464-8603, Japan*

Abstract. The propagation of an exciton-polariton in a coupled quantum disk-quantum well waveguide structure was investigated by analyzing the optical spectrum obtained at an end of the waveguide. On the basis of the experimental results, the mode pattern in the quantum disk area was analyzed numerically and it was found that the transmission probability of an exciton-polariton is modified around the resonance energy of the exciton defined within the disk.

INTRODUCTION

Recent developments in the nano-fabrication technology enable us to make structures involving the quantum dots/disks, which have a possibility to be basic components of a device utilizing the quantized nature of carriers. To realize a processing in such a device, it is important to extract information, i.e. probing the state of carrier, from a quantum dot and transfer it to the other keeping the information on the intensity/phase. If we have many dots with different quantum states, or if we have several states in a dot, the identification of a particular quantized state will be the issue in a real system. Thus a kind of scanning method is appropriate to do this.

The scanning near-field optical microscope (SNOM) provides us the map of the wavefunction of an exciton in a quantum disk[1]. In the SNOM, however, the focusing of the light/probe on a particular dot/disk is necessary to probe the excitonic state. This will be not practical in applying to the integrated device. In this paper, we will propose a novel method to detect the dot/disk states via the resonant coupling with the propagating light in a nearby wave-guide. In the analyses, we assume a quantum well (QW) wave-guide, and the propagating light is described by an exciton-polariton[2].

SAMPLE STRUCTURES AND METHOD OF EXPERIMENTS

The sample under study was made by molecular beam epitaxy (MBE) and selective etching method. The core of the waveguide was made of a 5 nm GaAs QW sandwiched by 500 nm superlattice (SL) layers (3nm GaAs and 4 nm $Al_{0.3}Ga_{0.7}As$). The core was grown on a 700 nm SL of 3 nm GaAs and 4nm $Al_{0.4}Ga_{0.6}As$ which serves as the lower clad. The upper clad was provided by the air. A 8 nm GaAs disk (QD) was put on the wave-guide as shown in Fig.1. The coupling of the excitonic states between the waveguide and the disk was investigated in detail. For the comparison use, we made a control sample (sample 2), which has no 8 nm QW in the disk area.

The He-Ne laser beam was focused on a position on the waveguide, and the PL spectrum was measured at an edge of the waveguide. During the propagation through the waveguide, the emission spectrum was modified by the optical spectrum of the waveguide. Thus the interaction with the disk was investigated as a function of energy/polarization. Since the interaction of the exciton-polariton with the exciton in the QD is not so strong, we investigated the difference of the spectrum from that of the control sample.

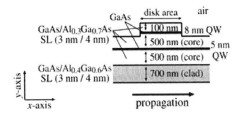

FIGURE 1. Schematic structure of sample1.

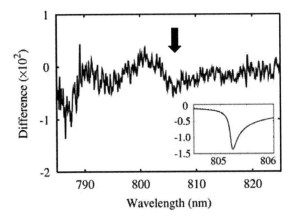

FIGURE 2. Difference intensity for TE mode measured at an edge of the waveguide. The inset shows the corresponding numerical result.

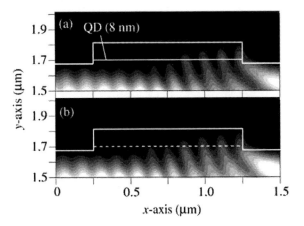

FIGURE 3. The intensity of electric field around the disk area at the energy of 1.5399 eV for (a) sample1 and (b) sample2, respectively, where the resonant energy of exciton in QD was 1.540 eV.

The numerical calculation was performed by solving the transfer matrix problem[3] for a general expression of two dimensional Maxwell's equation[4]. To clarify the effect of coupling, the length of disk (QD) was set at $1\,\mu$m in the numerical model, while it was $20\,\mu$m in the sample.

RESONANT COUPLING OF QD AND QW WAVE-GUIDE

Figure 2 shows the difference spectrum between the sample with QD and the control sample (sample 2). We could find a dip at 805 nm (1.54 eV) as indicated by an arrow, which is corresponding to that of the heavy-hole exciton in the 8 nm QD. The behavior is in agreement with the numerical results shown in the inset of Fig.2.

Numerically, when the wavelengths were longer than that of exciton resonance, the real part of the dielectric constant was enhanced. As the result, the individual peaks of confined modes of exciton-polariton became prominent (Fig.3(a)). This mode appearance enhanced the absorption in the QD, which in turn gives the decay of propagating wave beyond the QD. Thus we achieved the minimum of the transmission intensity at slightly lower energy of resonance.

SUMMARY

The resonant coupling of an exciton in a QD with the propagating exciton-polariton in a nearby QW wave-guide was studied by analyzing the optical spectrum. The results are in good agreement with the numerical results analyzing the propagating mode of the electric field.

ACKNOWLEDGMENTS

This work is partly supported by the 21st Century COE Program of the Ministry of Education, Culture, Sports, Science and Technology in Japan.

REFERENCES

1. K. Matsuda, T. Saiki, S. Nomura, M. Mihara, Y. Aoyagi, S. Nair, and T. Takagahara, *Phys. Rev. Lett.* **91**, 17740 (2003).
2. J. J. Hopfield, *Phys. Rev.* **112**, 176 (1958).
3. T. Usuki, M. Saito, M. Takatsu, R. A. Kiehl, and N. Yokoyama, *Phys. Rev.* **B 52**, 8244 (1995).
4. K. Cho, *J. Phys. Soc. Jpn.* **55**, 4113 (1986).

Excitonic Mott transition in spatially-separated electron-hole systems

V.V. Nikolaev[*,†], M.E. Portnoi[*] and A.V. Kavokin[**]

[*]*School of Physics, University of Exeter, Stocker Road, Exeter EX4 4QL, United Kingdom*
[†]*Present address: A.F. Ioffe Physico-Technical Institute, St Petersburg 194021, Russia*
[**]*LASMEA, UMR-6602, Université Blaise Pascal, 24, avenue des Landais, 63177 Aubière, France*

Abstract. We study an electron-hole system in double quantum wells theoretically using the Green's function technique. We demonstrate that there is a temperature interval over which an abrupt jump in the value of the ionization degree occurs with an increase of the carrier density or temperature. The opposite effect — the collapse of the ionized electron-hole plasma into an insulating exciton system — should occur at lower densities. In addition, we predict that under certain conditions there will be a sharp decrease of the ionization degree with increasing temperature — the anomalous Mott transition.

The excitonic Mott transition [1] is the avalanche ionisation of excitons as a result of screening and momentum-space filling effects. At high densities or temperatures the excitons are completely ionised and a free electron-hole plasma is formed. There is a long-standing question as to whether the transition between excitonic and plasma phases is smooth or abrupt, and despite significant effort there is still no firm theoretical understanding of this effect. A spatially-separated quasi-two-dimensional (2D) electron-hole system is very promising for experimental observation of the Mott transition, since the lifetime of excitons is relatively long and the intra-layer repulsion of like particles prevents the formation of electron-hole liquid droplets [2].

The main theoretical difficulty in investigation of the Mott transition is that it is inherently a medium-density effect, for which low-density and high-density approximations do not work. The correct approach would require careful handling of the Coulomb interaction depending on the thermodynamic state of the system.

Before studying the Mott transition, which is essentially a jump in the ionization degree of the electron-hole plasma (EHP), one should clarify the concept of ionization degree and exciton density at medium and high plasma densities. In the dilute case excitons can be considered as well-defined Bose particles, and the problem of finding the ionization degree can be solved simply by counting electrons and holes in bound states pairwise. This approach, however, runs into trouble when the density of carriers increases and the exciton wavefunctions start to overlap. Therefore, it appears more practical to formulate the theory in terms of primary quasi-particles, i.e. in terms of electrons and holes. Instead of considering excitons in the system we investigate the single-particle density at given temperature $T = 1/\beta$ and quasi-Fermi levels ξ_e and ξ_h. To this end we use many-body theory for a system of interacting quasi-particles within the ladder approximation. Starting with the self-energy, which is based on the electron-electron and electron-hole pair-interaction series, and performing a derivation similar to that of Zimmermann and Stolz for the 3D case [3], we obtain an expression for the electron density in the following form:

$$n_e = n_e^0(\xi_e) + n^{ee}(\xi_e) + n^{eh}(\xi_e + \xi_h), \quad (1)$$

where $n_e^0 = m_e/(\beta \pi \hbar^2) \ln(1 + \exp(\beta \xi_e))$ is the density of free quasi-particles, and n^{ee} (n^{eh}) originates from the electron-electron (electron-hole) interaction

$$n^{ab} = \frac{2}{\beta \pi \hbar^2} \sum_{m=-\infty}^{+\infty} \left[(1-\delta_{ab}) \sum_n M_{eh} L_{eh}(\varepsilon_{mn}) \right.$$
$$\left. + \frac{2}{\pi} M_{ab} \lambda_m^{ab} \int_0^\infty dk L_{ab}\left(\frac{\hbar^2 k^2}{2m_{ab}}\right) \sin^2 \delta_m^{ab}(k) \frac{d\delta_m^{ab}}{dk} \right], \quad (2)$$

$$L_{ab}(\varepsilon) = -\ln\left[1 - \exp(\beta(\xi_a + \xi_b - \varepsilon))\right]. \quad (3)$$

Here, m_e (m_h) is the electron (hole) effective mass, $M_{ab} = m_a + m_b$, $m_{ab} = m_a m_b/(m_a + m_b)$, $\lambda_m^{ab} = 1 - \delta_{ab}(-1)^m/2$, the quasi-Fermi levels are measured from the edges of the lowest quantized subbands, and ε_{mn} are the exciton energy levels. The quantities $\delta_m^{ab}(k)$ are the generalization of scattering phase shifts for the case of non-negligible k-space filling (see [4] for more details). The particle density expression, Eq. (1), can be divided into two parts according to the Fermi-level dependence. Indeed, the first two terms in Eq. (1) contain only the

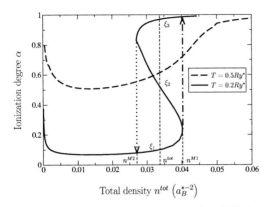

FIGURE 1. Ionization degree for an EHP in a DQW at two different temperatures. Arrows show possible phase transitions.

electron quasi-Fermi level, whereas the last, n^{eh}, depends on the sum $\xi = \xi_e + \xi_h$. We shall argue that the quantity n^{eh} can play the role of the excitonic density for the whole range of total concentrations. Firstly, the term n^{eh} becomes equal to the density of idealized bosons (excitons) in the limit of low densities. Secondly, a quantity which contains the quasi-Fermi levels in the form of a sum has "insulating" properties. This can be illustrated by the following consideration. Suppose there is an electric field inside the structure, which generates a perturbation to the local potential, δV. Then, the first two terms in the density expression, Eq. (1), "feel" the disturbance, and the local values of the density given by the sum of these two terms will be either increased or decreased, according to the sign of the perturbation potential and the type of particles considered. This represents metallic behavior. At the same time, the last term, n^{eh}, is not influenced by the local-potential perturbation, since $\delta \xi_e$ and $\delta \xi_h$ cancel each other. Thus, this part of the density is not disturbed by the external electric field and can be considered as insulating. Therefore, it is reasonable to introduce the ionization degree of the electron-hole plasma as

$$\alpha = 1 - n^{eh}/n_e = 1 - n^{eh}/n_h. \quad (4)$$

It can be shown [4] that in spatially separated systems screening by indirect excitons is of the same order as screening by the free-carrier plasma, and this is due to the non-zero dipole moment of the exciton. This questions the validity of applying previously developed theories of quasi-2D screening, which ignore screening by indirect excitons, to double quantum well (DQW) systems. Taking both types of screening into account we used a self-consistent procedure to calculate the parameters of the quasi-2D electron-hole plasma at a given temperature.

In Fig. 1 we plot the ionization degree, α, as a function of density for two different temperatures. One can see that for low enough temperatures there is a density region (confined between n^{min} and n^{max}) where α is a three-valued function. Physically, this situation means the following. Increasing the density of the carriers in the QW, one reaches a critical value n^{max}, at which a slight increase of the carrier concentration results in a large jump in the ionization degree. Effectively, this means a transition from a system consisting mainly of excitons to an almost completely ionized state. Clearly, this is what the Mott transition (avalanche ionization) is meant to be.

However, there is another transition mechanism. If the initial state of the system is a high-density ionized EHP and the density is slowly decreased, then the abrupt transition into an insulating state (ionized EHP collapse) may happen not at the Mott density n^{M1}, but at a lower density n^{M2} (the thick dotted arrow in Fig. 1). The reason for the difference between these two densities can be explained qualitatively as follows. When the system is in the insulating state, the screening due to excitons is comparatively weak, and accumulation of a substantial exciton density is required in order to trigger the avalanche ionization. After the avalanche ionization occurs, the screening in the system is mainly due to free carriers, i.e. it is much stronger, and one has to step back in density much further in order for the system to collapse into excitons.

We hope to encourage more experimental research which would lead to irrefutable observation of such a fundamental effect as excitonic Mott transition. Our theoretical results suggest the following method. The temperature should be set as low as possible and the pumping slowly increased to a high value, but so that the temperature kept constant. Then, as the system is placed in the stable insulating state below the normal Mott transition curve in Fig. 1, one should slowly increase the temperature. The Mott transition will manifest itself as the abrupt elimination of the excitonic peak from the luminescence spectra or an increase of the longitudinal photo-conductance. Once the ionized state is achieved, one can try to observe the anomalous enhancement of exciton luminescence with a decrease of the density (photo-excitation).

This work was supported by the British Council Alliance Programme.

REFERENCES

1. Mott N.F., *Philos. Mag.*, **6**, 287 (1961).
2. Rice T.M., "The Electron-Hole Liquid in Semiconductors: Theoretical Aspects," in: *Solid State Physics*, edited by H. Enrenreich, F. Seitz and D. Turnbull, Academic Press, New York, 1977, vol. 32, pp. 1-86.
3. Zimmermann R. and Stolz H., *Phys. Status Solidi B*, **131**, 151, 1985.
4. Nikolaev V.V., *Many-Particle Correlations in Quasi-Two-Dimensional Electron-Hole Systems*, Ph.D. thesis, University of Exeter, 2002.

Investigations of interface excitons at p-type GaAlAs/GaAs single heterojunctions in continues wave and time resolved magneto photoluminescence experiments

L. Bryja*, M. Kubisa*, K. Ryczko*, J. Misiewicz*, M. Kneip†, M. Bayer†, R. Stępniewski**, M. Byszewski**, M. Potemski**, D. Reuter‡ and A. Wieck‡

*Institute of Physics, Wrocław University of Technology, Wybrzeże Wyspiańskiego 27, 50-370 Wrocław, Poland
†Institute of Physics, Dortmund University, Otto-Hahn-Strasse 4, 44227 Dortmund, Germany
**Grenoble High Magnetic Field Laboratory, 25 av. Martrys, F-38042 Grenoble Cedex 9, France
‡Ruhr-Universität Bochum, Lehrstuhl für angewandte Festkörperphysik Universitätsstraβe 150, 44780 Bochum, Germany

Abstract. We report on investigations of polarisation resolved photoluminescence and photoluminescence excitiations from p-type $Ga_{1-x}Al_xAs$/GaAs single heterojunctions in magnetic field up to 23 T. The observed excitonic type transitions are attributed to neutral excitons rather then to the charged ones.

INTRODUCTION

Although optical properties of n- and p-type, two-dimensional semiconductor structures have been intensively studied in the past, the evolution of the optical response of these structures with carrier density is not clearly understood. The appearance of the Fermi Edge singularity [1] and formation of charged excitons [2] are two important effects to be mentioned in this context. A negatively charged exciton (also called trion) was predicted to be stable in bulk semiconductors but was not observed due to a very small binding energy. In quantum wells the binding energy of an extra carrier to a neutral exciton increase up to a few meV making negatively (X^-) and positively (X^+) charged excitons possible to detect in absorption and emission measurements.

In this paper we report on interface exciton studies using polarisation resolved low temperature photoluminescence (PL) and photoluminescence excitation (PLE) measurements from p-type $Ga_{1-x}Al_xAs$/GaAs single heterojunctions in continuous wave (cw) and time resolved conditions.

EXPERIMENTAL RESULTS

We have studied PL from five modulation doped p-type $Ga_{1-x}Al_xAs$/GaAs single heterojunctions with different 2D holes concentrations ranging from 2.0×10^{11} cm^{-2} up to 9.8×10^{11} cm^{-2} grown by MBE on (001) semi insulating GaAs substrates. All measurements were performed in temperature down to 2 K with magnetic field applied perpendicular to the growth direction in Faraday configuration. CW PL and PLE experiments were carried out using Bitter solenoids which provided magnetic fields up to 23 T. An optical fibber system was used. The spectra were analysed in the 1 m single grating monochromator with CCD nitrogen-cooled detector. Time resolved photoluminescence (TRPL) experiments were performed in optical split coil superconducting cryostat giving magnetic fields up to 7 T.

In our previous studies [3, 4] of PL from two samples with highest 2D holes densities we observed so called H-band lines, resulting from recombination of free GaAs conduction band electrons with holes bound in 2D Landau levels (LL). Detailed analysis of the positions of H-band lines allowed us to determine the dispersion of 2D valence band LL as shown in Fig. 1 for $p_1 = 7.6 \times 10^{11}$ cm^{-2}, in good agreeement with results of the theoretical calculations. In the energy range of H-band transitions we observed lines denoted as $E1$ and $E2$, which can not be attributed to free carrier transitions. They were of a main interested of our studies. They are observed in both samples. The line $E1$ appear in σ^- polarisation on the low energy wing of $H4$ transitions which in turn abruptly disappear from the PL spectra at filling factor $\nu = 6$. The similar behaviour is observed for lines $E2$ and $H3$ in σ^+ polarisation at $\nu = 3$. The origin of E-lines is not clear. As can be is seen in Fig. 1 they do not correspond to any 2D hole LL. The strong increase of the intensity of

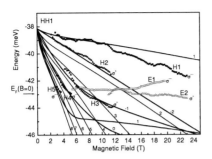

FIGURE 1. The 2DHG LL dispersion in magnetic field. Lines - theory, symbols - experimental data.

lines $E1$ and $E2$ with increase of magnetic field, their narrow widths and energy positions suggest that they are related to neutral excitons originating from 2D light hole subband. However, on the other hand, one has to consider the possibility of charged excitons creation with second hole originating from the heavy holes subband. Such charged exiton should be energetically more stable, because electron is attracted to interface by two positive charges insted of one. In all studied samples the intensity of lines attributed to GaAs acceptor bound excitons A^0X increase dramatically in magnetic fields above $\nu = 2$ and as the magnetic field is further increased they exceed the intensity of E lines. To omit this difficulty and in order to obtain more information about the nature of E-lines we performed PLE and TRPL measurements. In the PLE we recorded the whole spectra using CCD detector. The light source was tuneable Ti-sapphire laser. The spectra

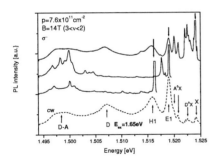

FIGURE 2. PL spectra for several excitations energies.

were collected with laser wavelength step equal 0.25 nm varied from the energy of $D-A$ up to the energy gap of GaAs. The PL spectra for the sample p_1 which is representative for PLE spectra of all samples, are presented in Fig. 2. In the lower part of the figure the PL spectrum for exciting energy higher than energy gap is shown. In the upper part a few PL spectra with different energies of laser excitation (marked by arrows) are presented. We see that all interface related lines (D, H and E) are detected in the spectra only when laser energy is equal or higher than the energy of free GaAs exciton. For lower laser energy we observe only the fine structure of resonant excitation of $D-A$. TRPL were performed using of 3 KHz Nd-YaG laser with 10 ns impulse. Time evolution of the PL spectra for the samples p_1 at $B = 7$ T is presented in Fig.3. The cw PL spectrum is also shown for

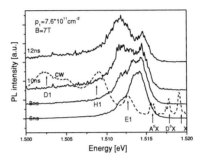

FIGURE 3. The evolution of TRPL spectra.

comparison. As it can be seen the GaAs related transitions disappear from the spectra recorded in the ns delay time and we observe only the interface transitions as in the case of cw measurements. We detected a very long time of photoluminescence of E-lines and H-band. These observations are an evidence of a low value of oscillator strengths of these transitions. This, in turns, indicates the considerably high space displacement of photo-excited electrons and 2D holes. According to our calculation this displacement observed in H-band transition in zero magnetic field for the sample p_1 is equal about 40nm. The magnetic fields applied in direction perpendicular to the growth direction brings electrons closer to the interface but spatial separation between holes and electron still remain larger that the mean distance between 2D holes. This shows that the attraction between the electron and the additional hole is smaller than the repulsion between holes and positively charged exciton is not be created.

REFERENCES

1. Skolnik M. S., Rorison J. M., Nash K. J., Mowbray D. J., Tapster P. R., Bass S. J., Pitt A. D., *Phys. Rev. Lett.*, **58**, 2130–2133 (1987).
2. Keng K., Cox R. T., Merle d'Aubigné Y., Bassani F., Saminadayar K., Tatarenko S *Phys. Rev. Lett.*, **71**, 1752–1755 (1993).
3. Bryja L., Kubisa M., Ryczko K., Misiewicz J., Larionov A., Bayer M., Forchel A., Sorensen C. B. *Solid State Commun.*, **122**, 379–384 (2002).
4. Kubisa M., Bryja L., Ryczko K., Misiewicz J., Bardott C., Potemski M., Ortner G., Bayer M., Forchel A., Sorensen C. B. *Phys. Rev.*, **B 67**, 035305(1-12) (2003).

Exciton line broadening due to plasma potential fluctuations

W. Bardyszewski*, P. Kossacki[†], J. Cibert** and S. Tatarenko**

*Institute of Theoretical Physics, Warsaw University, Hoża 69, 00-681 Warsaw, Poland
[†]Institute of Experimental Physics, Warsaw University, Hoża 69, 00-681 Warsaw, Poland
**Groupe " Nanophysique et Semiconducteurs ", CEA-CNRS-Université Joseph Fourier Grenoble, Laboratoire de Spectrométrie Physique, BP87, F-38402 Saint Martin d'Hères Cedex, France

Abstract. We report experimental and theoretical studies of the low temperature exciton absorption in a single quantum well in the presence of fully spin polarized free holes. Under these conditions only neutral exciton absorption line is observed. The evolution of this line with increasing density of holes is described by our theoretical model.

The interband absorption of light in semiconductors in the presence of free carriers is in principle a many body process which is difficult to describe theoretically in the general case. In the low carrier concentration regime, an electron and a hole created in the process of photon absorption form an exciton which interacts with a small number of particles in its vicinity. This interaction may lead to the formation of a charged exciton and in general introduces reduction of strength and broadening of the exciton spectral line due to scattering.

In this paper we present an experimental and theoretical study of the influence of free carriers on the exciton absorption in a modulation doped structure containing a single quantum well in the low and intermediate concentration regime. Our aim was to separate precisely the many body effects from the electronic structure modifications introduced by the quantum well confining potential.

In our theoretical approach we employ a concept of an exciton as a well defined composite particle scattered by the elementary excitations of the free carrier gas which are described by the non-local in time excitonic self-energy operator.

EXPERIMENT

The studied sample contained a single 8nm wide $Cd_{1-x}Mn_xTe$ quantum well embedded between (Cd,Zn,Mg)Te barriers grown pseudomorphically on a (100) $Cd_{0.88}Zn_{0.12}Te$ substrate [1]. It was designed in such a way that the combination of built-in strain and strong confinement resulted in the separation between the highest heavy hole and light hole levels as large as 30 meV. The free holes are provided by a layer doped with nitrogen located at the distance of $L_s = 20$ nm from the quantum well. The concentration of free holes estimated from the Moss-Burstein shift measurements was equal to 3×10^{11} cm^{-2}. The content of Mn atoms (on the order of 0.18%) in the quantum well region was high enough to provide substantial Zeeman splitting of the topmost heavy hole level in a weak magnetic field perpendicular to the QW plane without causing large exciton broadening. Due to the giant effective g-factors introduced by the exchange interaction with the Mn ions the complete spin polarization of free carriers in the conduction and valence bands could be achieved below $B = 0.5T$. The spatial part of the excitonic wave function is practically unaffected by the magnetic field in this range.

The concentration of holes in the QW is controlled by additional illumination by the halogen lamp in the spectral region above the energy gap in the barrier layer. Due to the diffusion of photo created electrons and rapid recombination, the steady state concentration of holes could be reduced by one order of magnitude.

The low temperature transmission was measured in Faraday configuration with magnetic field obtained from superconducting magnet. With the z axis directed along the magnetic field, we find that the upper heavy hole subband corresponds to the electron spin state $|\frac{3}{2}, -\frac{3}{2}>$ (hole spin up) so that only transitions to the conduction band $|-\frac{1}{2}>$ are allowed in the σ^+ polarization. Note that in the whole range of concentrations of holes only the hole spin up subband is occupied which corresponds to a full spin polarization. This configuration prevents formation of charged excitons (X$^+$) which would require two holes of opposite spins. Consequently only one absorption line was observed in the transmission spectra corresponding to the ground state of the neutral exciton for the whole range of hole gas concentrations. When increasing hole gas concentration we observed its blue shift, broadening and decrease of oscillator strength. These observations are consistent with the exciton bleaching under strong excitation by femtosecond pulse in a time resolved pump probe experiment.

THEORETICAL MODEL

According to the linear response theory the absorption coefficient is related to the Fourier transform of the retarded two-particle equilibrium Green's function which can be obtained from its time contour ordered counterpart defined by:

$$G^X(t,t') = \langle T_c[d^\dagger_{nk}(t) c_{mk}(t) c^\dagger_{m'k'}(t') d_{n'k'}(t')]\rangle, \quad (1)$$

where the angular brackets denote the average with respect to the equilibrium statistical ensemble with T_c denoting the chronological operator along the time contour. The evolution of the Green's function is governed by the Hamiltonian in which we include the interband interaction between the states in the conduction and valence band as well as the intraband interaction within the valence subband system.

The equation of motion for the Green's function (1) originates a set of equations for the infinite hierarchy of irreducible correlation functions which is usually truncated at some level in order to obtain a closed system. This general procedure [2] applied to the the excitonic Green's function yields:

$$\begin{aligned}i\partial_t G^X(t,t') &= iG_0^X + h^X G^X(t,t') \\ &+ \int_c dt'' M(t,t'') G^X(t'',t'),\end{aligned} \quad (2)$$

where G_0^X denotes the normalization factor. The exciton Hamiltonian represented by h^X describes the electron hole pair interaction while the non-local in time self-energy M is responsible for inelastic interaction with hole plasma. In order to close the equation (2) the M operator is approximately expressed in terms of products of the exciton Green's functions and plasma density correlation functions yielding a set of equations which have to be solved self-consistently. At the present level of approximation we neglect the self energy term and obtain the excitonic spectra by diagonalizing the Hamiltonian h^X. We assumed static screening of the interaction potential and included screened exchange corrections to the one particle energies. The quasi two-dimensional plasmon pole approximation was employed. In our multi-subband approach the exciton Hamiltonian is expressed in the basis set of cylindrical waves multiplying the quantum well envelope functions which are obtained by solving coupled Schrödinger and Poisson equations taking

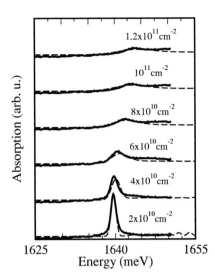

FIGURE 1. Measured (solid line) and theoretical (dashed line) heavy hole exciton absorption for the indicated concentrations of holes.

into account the exchange interaction with the Mn ions in the quantum well. Due to the large size of the resulting eigenvalue problem we have used the Lanczos reduction technique to obtain the lowest energy part of the excitonic spectrum. This approach was further verified by the full diagonalization. As a result of our calculation we obtained the energy and oscillator strength of the lowest energy exciton transition as a function of free hole concentration. In order to make comparison with the experimental absorption curves we have determined the experimental line widths by fitting the data with gaussians. Theoretical lines broadened using experimental line width parameters are then compared with experiment in Fig. 1. The magnitude of the theoretical curves was calibrated by by equating the magnitude of absorption only for the lowest concentration of holes.

In conclusion, using the single quantum well system with well controlled density of free holes we were able to follow the evolution of the exciton absorption line with increasing density of free holes. Our theoretical model correctly describes the position and intensity of the heavy-hole exciton.

REFERENCES

1. Kossacki, P., Cibert, J., Ferrand, D., d'Aubigné, Y. M., Arnoult, A., Wasiela, A., Tatarenko, S., and Gaj, J., *Physical Review B*, **60**, 16018–16026 (1999-I).
2. Tserkovnikov, Y. A., *Theor. Math. Phys.*, **49**, 993–1001 (1981).

Negatively charged excitons in a back-gated undoped heterostructure

S. Nomura*[†], M. Yamaguchi**[†], D. Sato*[†], T. Akazaki**[†], H. Tamura**[†], H. Takayanagi**[†], T. Saku[‡] and Y. Hirayama**[§]

*Institute of Physics, University of Tsukuba, 1-1-1 Tennoudai, Tsukuba 305-8571, Japan
[†]CREST, JST, 4-1-8 Honmachi, Kawaguchi, 331-0012, Japan
**NTT Basic Research Laboratories, NTT Corporation, 3-1 Morinosato-Wakamiya, Atsugi-shi, Kanagawa 243-0198, Japan
[‡]NTT Advanced Technology, 3-1 Morinosato-Wakamiya, Atsugi-shi, Kanagawa 243-0198, Japan
[§]SORST, JST, 4-1-8 Honmachi, Kawaguchi, 331-0012, Japan

Abstract. The observation of negatively charged excitons in a back-gated undoped GaAs/AlGaAs quantum well is reported in magnetic fields depending on the electron density between $1 \times 10^9 - 2 \times 10^{11}$ cm^{-2}. We find that a peak appears 0.3 meV below the singlet charged exciton state at 5 T, which develops to the lowest Landau-level with increase in the electron density.

INTRODUCTION

Photoluminescence (PL) measurements of a back-gated undoped heterostructure are reported in magnetic fields. Two-dimensional electron gas (2DEG) is induced by the back-gate bias in an undoped quantum well [1], unlike the conventional method of changing the electron density by depleting the electrons by a surface gate on Si modulation-doped quantum wells or single heterojunctions. This structure is free from the remote-impurity scattering by Si doping and is especially advantageous in investigation of the charged excitons [2, 3, 4] because of good electron density controllability in the extremely low electron density regime $1 \times 10^9 - 2 \times 10^{11}$ cm^{-2} and high electron mobility $2 \times 10^5 - 3 \times 10^6$ cm^2/Vs. High quality 2DEG in the low electron density regime have been investigated by measurements of transport properties. [1, 5, 6]

In this paper, we report measurements of PL in magnetic fields as a function of the electron density from the nominally undoped condition to 2×10^{11} cm^{-2} tuned continuously by a back gate structure.

EXPERIMENTAL

The sample layer consists of n-GaAs, superlattice barrier, 20-nm Al$_{0.33}$Ga$_{0.67}$As layer, 50-nm GaAs well and 250-nm Al$_{0.33}$Ga$_{0.67}$As layer. Ohmic contacts are formed on 1×1 mm^2 mesas from the surface side and on an n-GaAs substrate from the back side. A bias voltage (V_B) is applied between the surface and the back Ohmic contacts. The PL measurements were performed at 1.7 K with the incident laser excitation density of 25 μW/cm^2. At this optical excitation density, it has been verified by an independent transport measurement that the electron density is not affected by the continuous irradiation by possible effects of the vertical photoconduction through the barrier layer or the transfer of the carriers between 2DEG and the surface states

RESULTS AND DISCUSSIONS

Photoluminescence spectra at 0 T at V_B between -0.2 and 0.5 V are shown in Fig. 1 (a). The higher energy peak at around 1.5169 eV and the lower energy peak at 1.51618 eV at -0.20 V $\leq V_B \leq -0.09$ V are assigned to be a neutral exciton (X^0) and an exciton bound to a neutral donor (D^{0}X). Full width at half maximum (FWHM) of D^{0}X between $V_B = -0.20$ V and -0.09 V is 0.44 meV. Transition from D^{0}X to X$^-$ is accompanied by the shift of the transition energy and the narrowing of the spectral width. At $V_B = -0.08$ V, the peak position shifts 0.6 meV lower to 1.51612 eV, and FWHM narrows to 0.38 meV. The spectral width of X^0, on the other hand, shows no noticeable change and remains to be 0.24 meV. The peak energy of X^0 slightly shifts to the higher energy above -0.13 V as shown in Fig. 1(b). Because of the screening by the electron gas, X^0 peak disappears at 1 −

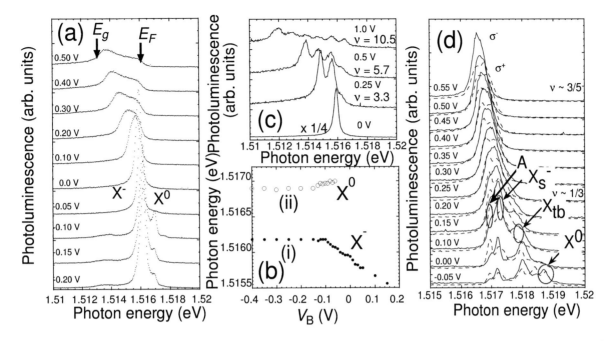

FIGURE 1. Photoluminescence spectra at (a) 0 T depending on V_B. The band-gap (E_g) and the Fermi-energy (E_F) are shown by the arrows. (b) Bias voltage dependence of the transition energies at 0 T for X^- (i) and X^0 (ii) excitons. (c) Photoluminescence spectra at 0.5 T depending on V_B. (d) Photoluminescence spectra for σ^- (solid line) and σ^+ (dashed line) polarizations at 5 T depending on V_B.

1.2×10^{10} cm^{-2}. The peak energy of X^- shifts to lower energy with increase in the bias voltage in the range $V_B \geq -0.09$ V, because of the band renormalization.

Figure 1 (c) shows well-developed equi-spaced 1 to 6 peaks due to the Landau-levels with a full width at half maximum of 0.38 meV are observed at 0.5 T depending on V_B, showing high controllability of the electron density.

Figure 1 (d) shows the evolution of the PL peaks with increase in the electron density from $\nu = 0.05$ to $\nu = 0.61$ at 5 T. Three peaks are observed in the σ^- polarization, which are assigned to be a singlet charged exciton (X_s^-), a bright triplet charged exciton (X_{tb}^-) and a X^0. With increase in V_B, first X^0, and then X_s^- loose their oscillator strengths due to the screening. Another peak (A) appears 0.3 meV below X_s^-, which grows with increase in V_B and is seen to be developed to the lowest Landau-level at 5 T. Peak A is polarized to the σ^+ polarization. The optical excitation power dependence of the intensity of peak A reveals that this peak saturate at lower optical excitation density than X_s^- and X_{tb}^- peaks. These results suggest that the peak A has small oscillator strength and may be classified as a "dark" symmetry forbidden state.

In summary, we have observed negatively charged excitons in a back-gated undoped quantum well in magnetic fields in the extreme low electron density region between $1 \times 10^9 - 2 \times 10^{11}$ cm^{-2}. A "dark" peak is found to appear 0.3 meV below the X_s^- exciton state at 5 T, which develops to the lowest Landau-level with increase in the electron density.

ACKNOWLEDGMENTS

This work was partly supported by Special Research Project on Nanoscience at University of Tsukuba.

REFERENCES

1. Hirayama, Y., Muraki, K., and Saku, T., *Appl. Phys. Lett.*, **72**, 1745 (1998).
2. Finkelstein, G., Shtrikman, H., and Bar-Joseph, I., *Phys. Rev. Lett.*, **74**, 976 (1995).
3. Yusa, G., Shtrikman, H., and Bar-Joseph, I., *Phys. Rev. Lett.*, **87**, 216402 (2001).
4. Wojs, A., Quinn, J., and Hawrylak, P., *Phys. Rev. B*, **62**, 4630 (2000).
5. Kawaharazuka, A., Saku, T., Kikuchi, C., Horikoshi, Y., and Hirayama, Y., *Phys. Rev. B*, **63**, 245309 (2001).
6. Lilly, M., Reno, J., Simmons, J., Spielman, I., Eisenstein, J., Pfeiffer, L., West, K., Hwang, E., and Sarma, S. D., *Phys. Rev. Lett.*, **90**, 056806 (2003).

Excitonic oscillator strengths in quantum wells containing a 2-D electron gas

R. T. Cox*, K. Kheng*, R. B. Miller[†], V. Huard*, C. Bourgognon*, K. Saminadayar* and S. Tatarenko*

*Equipe "Nanophysique et Semiconducteurs" of SP2M, CEA-Grenoble and Laboratoire de Spectrométrie Physique, Université J. Fourier, 38054 Grenoble-Cedex, France
[†]Department of Physics, La Trobe University, Victoria 3086, Australia

Abstract. Optical absorption spectra of modulation doped CdTe quantum wells with electron concentration n_e up to $\approx 0.1/\pi a_B^2$ (where a_B = excitonic Bohr radius) are interpreted in terms of few-body excitations : exciton, trion and quatron processes. The initial excitonic oscillator strength of the empty QW is shared with the trion and quatron processes as n_e increases in zero field and as v increases in magnetic field.

Adding a low concentration n_e of free electrons to a quantum well transforms its optical absorption spectrum as follows : (i) A resonance peak (T) corresponding to creation of the singlet trion X^- appears below the exciton resonance (X) and (ii) a broad wing develops on the high energy side of X, corresponding to exciton creation with simultaneous scattering of an electron accompanied by momentum-conserving exciton recoil.

Esser et al[1] predicted that, for $n_e \ll 1/\pi a_B^2$, the absorption intensity will remain nearly constant if summed over the two and three body processes : X, T, $e-X$ scattering. That is, the three processes share the initial excitonic oscillator strength of the empty QW. We have verified this [2], studying CdTe QWs where n_e could be varied up to $1 \times 10^{11} \text{cm}^{-2}$. This is $\approx 0.08/\pi a_B^2$ (for a quasi-2D $a_B \approx 5$ nm). The absorption integrated over the whole spectrum (i.e. T and broadened X) decreased by only $\approx 5\%$ on increasing n_e up to $1 \times 10^{11} \text{cm}^{-2}$.

We now present spectra for a 10 nm CdTe(0.3%Mn) single quantum well (sample S6) with higher equilibrium $n_e = 1.5 \times 10^{11} \text{cm}^{-2}$ and narrower absorption linewidths (the small Mn addition enhances circular polarization effects in magnetic field). Features additional to (i) and (ii) above become apparent, as seen in Fig.1.

First, increasing n_e induces a broadening to high energy of the T peak, not just the X peak. Also, the separation between the two peaks increases strongly with n_e. We label the broadened peaks ω_1 and ω_2, following Hawrylak's notation[3] but, with n_e still only $\approx 0.12/\pi a_B^2$, we propose to interpret them in a few-body rather than a many-body model. At ICPS 2002, Esser[4] outlined a theory for four-body or "quatron" processes.

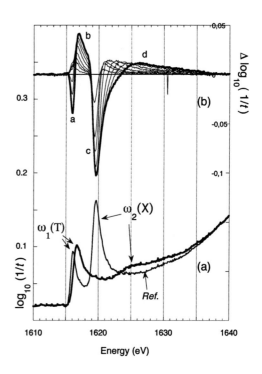

FIGURE 1. (a) Absorption $A = 1/\log(\text{transmission})$ of CdTe(0.3%Mn) QW sample S6 in zero field, at $n_e \approx 0.5 \times 10^{11} \text{cm}^{-2}$ (reference trace) and at $n_e = 1.5 \times 10^{11} \text{cm}^{-2}$ (dark trace). (b) Differential absorption $A - A_{(\text{Ref})}$ (not same scale) at various n_e (pale traces) up to $1.5 \times 10^{11} \text{cm}^{-2}$ (dark trace). Differences eliminate sloping baseline due to onset of substrate absorption.

Broadening of the trion resonance is one signature of such a process : one electron is bound into the trion and a second electron is scattered, with recoil of the trion. The very strongly broadened ω_2 peak could then logically be interpreted as exciton creation with scattering of *two* electrons. The ω_1-ω_2 splitting would contain the difference between the two electron and one electron scatterings, of the order of the Fermi energy E_F.

The differential absorption intensities in Fig 1(b) show that decreases in intensity a and c are balanced by intensity increases b (the broadened trion peak) and d (the exciton scattering wing extending well out in energy). Integrated from 1613 to 1638 meV in Fig.1, the total absorption intensity decreases by less than 15%. This implies there is a more general sum rule for the oscillator strength, that includes the four-body processes : The initial oscillator strength is maintained within the excitonic region of the spectrum and any effects of screening and phase space filling effects remain very small.

Resonance and scattering processes are clearly resolved under magnetic field when the free electron states are quantized. At high B, the scattering wing of X becomes discrete peaks (Z_i), corresponding to the "combined exciton and cyclotron-excitation" process ($X\&CR$) of Yakovlev et al[6]. That is, during exciton creation, a free electron is scattered from the lowest Landau level $n = 0$ to Landau level i, with momentum conserving exciton recoil. We have shown previously[2] how X shares its oscillator strength out to the $X\&CR$ and T processes with increasing n_e. Now we identify oscillator strength sharing with quatron magneto-processes.

Doing field sweeps for sample S6 at its maximum n_e of 1.5×10^{11}cm^{-2}, we see the broad ω_1 peak resolving into sharp peaks G_i ($i = \ldots, 3, 2, 1$). These move with field rather like "free carrier" transitions between valence band and conduction band Landau levels but, at our low n_e, we associate them with the "combined trion and cyclotron" scattering process ($T\&CR$) of Kochereshko et al[7]. At B= 3 and 6 T (filling factor $\nu = 2$ and 1), sharp trion and exciton resonances T and X emerge.

Fig.2 shows spectra at *fixed* field B =2.9 T for increasing n_e values. (Note : splittings do not correspond to $\hbar\omega_c$, due to interactions.) In σ^- polarization, resonance X shares intensity (mainly) to T and (less) to Z_1 (=$X\&CR_1$) as ν increases, until disappearing near $\nu = 1$. Resonance T reaches maximum intensity near $\nu = 1$ and then shares intensity out to peak G_1= $T\&CR_1$, until disappearing at $\nu = 2$. In polarization σ^+, the T resonance is never observed : X shares intensity out to Z_1 (=$X\&CR_1$) initially. From $\nu \approx 1$, X and Z_1 share to G_1= $T\&CR_1$, and also to peak F which we attribute to exciton creation with scattering of two electrons ($X\&2CR$). In field sweeps, it is peak F that merges into the ω_2 peak at zero field.

Despite the complex evolution of the individual intensities, the total absorption intensity, integrated over

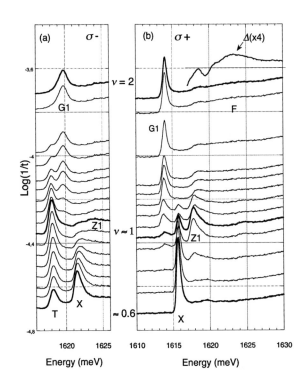

FIGURE 2. (a) Polarized absorption spectra of CdTe(0.3%Mn) QW sample S6 at B =2.9 T for various n_e values obtained by illuminating with 488 nm laser light. Spectra are shifted vertically proportionally to log(laser power); n_e varies rapidly in middle of spectrum set, slowly at bottom and top. Spectra in σ^- and σ^+ are split $\approx$ 5 meV by the giant Zeeman interaction with Mn (note that peaks are narrower in σ^-). Trace labelled $\Delta(\times4)$ at top is difference spectrum at $\nu = 2$, to show broad band F amplified $\times$ 4.

peaks X, T, Z, G and F in Fig.2, decreases by < 15% for $0 < \nu < 2$ (in either polarization), as in zero field. We are led to conclude that one need not invoke many-body screening and phase space filling to explain the disappearance of the X and T resonances with increasing ν in field, and their broadening with n_e in zero field. These effects can be interpreted in terms of oscillator strength sharing to three and four body scattering processes.

REFERENCES

1. A. Esser, R. Zimmermann, and E. Runge, Phys. Stat. Sol. (b) 227, 317 (2001)
2. R. T. Cox et al, Phys. Rev. B 69, 235303 (2004)
3. P. Hawrylak, Phys. Rev. B 44, 3821, 1991.
4. A. Esser, et al, Proc. ICPS26 Edinburgh 2002, page R 4.2.
5. V. Huard, et al, Phys. Rev. Lett 84, 187 (2000)
6. D. R. Yakovlev, et al, Phys. Rev. Lett. 79, 3974 (1997)
7. V. P. Kochereshko, et al Phys. Stat. Sol. (c) 0, 1463 (2003)

The Interaction of Quantum Well Excitons with Evanescent EM Waves and the Spectroscopy of Waveguide Polaritons

D.M. Beggs*, M.A. Kaliteevski*, S. Brand*, R.A. Abram*, and A.V. Kavokin[¶]

*Department of Physics, University of Durham, Durham, DH1 3LE, U.K.
[¶]LASMEA, Universite Blaise Pascal-Clermont-Ferrand II, Aubiere, 63177, France.

Abstract. A transfer matrix method has been used to calculate the dispersion relations of planar dielectric waveguides (WGs) containing quantum well (QW) excitons. It is found that WG polaritons are formed, whose dispersion displays the classic anti-crossing behaviour at the exciton resonance frequency. For a QW placed at the anti-node of the fundamental TE mode of the WG considered, the Rabi-splitting is calculated to be 6.55meV, with WG polaritons of lifetime ~ 7 ps, corresponding to a displacement along the WG axis of ~ 0.5 mm. A smaller splitting of the polariton modes (4.7 meV) was found for the case of the QW placed in the cladding of the WG.

INTRODUCTION

The aim of the present work is to investigate the interaction of a quantum well (QW) exciton with the electromagnetic field of an optical mode of a planar waveguide (WG). This study follows on from recent work, where it was shown that the excitonic modulation of the reflection spectrum from a QW behind a dielectric interface could be significantly increased if the angle of incidence was in the region of the critical angle for total internal reflection [1].

The optical mode of a planar WG has an electric field of the form $E_m(z)\exp(ik_xx - i\omega t)$, where ω is the angular frequency of the mode, k_x is the in-plane component of the wavevector, and $E_m(z)$ is the electric field profile of the mode across the thickness of the WG, and is illustrated in fig. 1 for the fundamental TE_0 mode.

If $\mathbf{M}(\omega)$ is the 2x2 transfer matrix [2] that propagates the fields across a layered dielectric medium, then the eigenfrequencies of the system are found from the matrix equation that couples the fields at the medium's left and right boundaries:-

$$A\begin{bmatrix}1\\-n_1\cos\theta_1\end{bmatrix} = \mathbf{M}(\omega)\begin{bmatrix}1\\n_1\cos\theta_1\end{bmatrix} \quad (1)$$

where n_1 is the refractive index of the cladding, θ_1 is the complex "angle of incidence" on the guiding layer of the WG, and A is the relative amplitude of the fields at the left and right boundaries. From the two equations represented by eq. (1), the constant A can be eliminated, and the resultant equation solved for the eigenfrequency. Hence, the dispersion relation $\omega(k_x)$ can be calculated.

INTERACTION OF AN EXCITON WITH A TE_0 MODE

For the case of the QW in the core of the WG, the transfer matrix (in TE polarisation) to propagate the fields across a QW with reflection coefficient r_{qw} is

$$\mathbf{M}_{qw} = \begin{bmatrix} 1 & 0 \\ 2n\cos\theta\dfrac{r_{qw}}{1+r_{qw}} & 1 \end{bmatrix} \quad (2)$$

where n is the background refractive index of the QW and θ is the angle of incidence, and r_{qw} is given by [3, 4]

$$r_{qw} = \dfrac{i\tilde{\Gamma}_0}{\tilde{\omega}_0 - \omega - i(\Gamma + \tilde{\Gamma}_0)} \quad (3)$$

$\tilde{\Gamma}_0 = \Gamma_0/\cos\theta$ is the exciton radiative broadening, $\tilde{\omega}_0$ is the renormalised resonance frequency (due to the polariton effect), and Γ is the non-radiative broadening of the exciton. To consider this case, the QW was modelled at the centre of the WG ($L = 0$), where it interacts with the anti-node of the TE_0 mode. A classic anti-crossing of the strongly coupled WG and exciton modes, with a Rabi splitting of 6.6 meV, can be seen in the dispersion curves of fig. 2a. This is the polariton effect, and the upper and lower polariton modes can be seen. The imaginary parts of the complex eigenfrequencies (or the mode-damping coefficient) for the case of $L = 0$ are shown in fig. 2b. At the anticrossing point, the two polariton modes have a lifetime of $\tau = 7$ ps, which corresponds to a propagation distance along the WG of $l = c\tau/n_{eff} = 0.5$ mm.

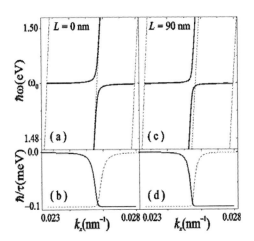

FIGURE 2. The dispersion relations for the TE_0 WG mode interacting with a QW exciton. The complex eigenfrequencies of the WG polaritons have been calculated, with the real parts shown in (a) and (c), and the imaginary parts shown in (b) and (d) (the solid and dashed lines for the lower and upper WG polariton modes respectively). The dashed lines in (a) and (c) show the dispersion of the empty WG. The QW exciton was modelled with the following parameters: $\hbar\omega_0 = 1.49$ eV, $\hbar\Gamma_0 = 0.03$ meV, $\hbar\Gamma = 0.1$ meV.

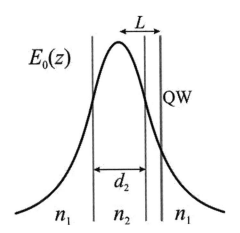

FIGURE 1. An illustration of the system under consideration. The QW is placed a distance L from the centre of the WG, as shown. The electric field profile of the fundamental TE_0 mode of the WG, $E_0(z)$, is shown. The parameters used in the numerical calculations were those appropriate for an AlAs:GaAs WG ($n_1 = 3.0$ and $n_2 = 3.7$) with a thickness of $d_2 = 150$ nm.

When placed in the cladding, the QW interacts with the evanescent EM wave present there, which is characterised by an imaginary wavevector $k_{1,z} = \pm i\chi$. The transfer matrix for the propagation of the fields through a QW interacting with an evanescent field can be found from eqs. (2) and (3) using a complex angle of incidence and imaginary wavevector. The case of $L = 90$ nm (see fig. 1) was considered. The dispersion curves are shown in fig. 2c, and the corresponding imaginary parts of the eigenfrequencies in fig. 2d. The Rabi splitting is 4.7 meV and is smaller than in the previous case, due to the electric field in the cladding being much smaller. As a whole, fig. 2 shows the polaritonic transition from WG-like modes, with an infinite lifetime and approximately linear dispersion, to exciton-like modes, with no dispersion and finite lifetimes.

CONCLUSIONS

To conclude, a transfer matrix method has been applied to study the properties of the polariton modes resulting from the interaction of a QW exciton with the TE_0 mode of a planar dielectric waveguide.

ACKNOWLEDGMENTS

This work has been supported by Marie-Currie Training Site contract HPMT-CT-2000-00191, and an ESPRC research grant.

REFERENCES

1. Beggs, D.M., Kaliteevski, M.A., Brand, S., Abram, R.A., Nikoleav, V.V., and Kavokin, A.V., *J. Phys.: Condens. Matter* **16**, 3401-3409 (2004).
2. Born, M., and Wolf, E., *Principles of Optics*, 6th ed., Cambridge University Press, (1980).
3. Ivchenko, E.L., and Kavokin, A.V., *Sov. Phys. Solid State* **34** (6), (1992).
4. Kavokin, A.V., and Malpuech, G., *Cavity Polaritons*, Amsterdam: Elseveir, ISBN: 0-125-33032-4 (2003).

Interplay of excitons, biexcitons, and charged excitons in pump-probe absorption experiments on a (Cd,Mn)Te quantum well

P. Płochocka[1], P. Kossacki[1], W. Maślana[1,2], C. Radzewicz[1], J. Cibert[2,3], S. Tatarenko[2], and J.A. Gaj[1]

[1]*Institute of Experimental Physics, Warsaw University, 69 Hoża, 00-681 Warszawa, Poland*
[2]*"Nanophysics and semiconductors" group, Laboratoire de Spectrométrie Physique, CNRS and Université Joseph Fourier Grenoble, B.P.87, 38402 Saint Martin d'Heres Cedex, France*
[3]*Laboratoire Louis Néel, CNRS, BP166, F-38042 Grenoble cedex, France.*

Abstract. We report time-resolved absorption studies of the influence of carriers on the neutral exciton (X) and charged exciton (X^+), in a modulation-doped (Cd,Mn)Te quantum well (0.2% Mn). We experimentally show that in p-doped II-VI quantum wells, phase space filling is less important than carrier-carrier interaction, and that the spin-dependent character of this interaction is a key to understand the oscillator strength of excitonic species. We relate the X, X^+ intensity to screening by the 2D hole gas. In particular, the X is screened mainly by one spin component of the hole gas.

INTRODUCTION

In a doped quantum well (QW) the presence of the charged excitons affects the character of the neutral exciton transition [1-3]. In such a structure three subsystems coexist and interact: charged excitons, neutral excitons and carriers. There are two possible ways to control the system. One method is changing the density of carriers, as was already done in CW measurements [2,3]. The second way is to change selectively the occupation of excitonic states, by using a strong optical excitation. This gives the opportunity to analyze the interaction between excitons, as was done for neutral excitons [4]. In this paper we present the experimental results of pump-probe experiments on a p-doped (Cd,Mn)Te QW. Using strong optical excitation, the charged or neutral excitonic states were selectively filled. We thus could analyze the interaction between the charged and neutral exciton populations and with the rest of the system [5].

EXPERIMENT

The time resolved study was carried out on a modulation doped structure consisting of a single QW of (Cd,Mn)Te, containing 0.2% Mn. Modulation p-type doping was assured by a nitrogen-doped layer placed at 200 Å from the QW. The density of the hole gas in the QW was controlled by additional illumination with photon energy above the gap of the barriers, as described in detail in [2]. The sample was mounted strain-free in a superconducting magnet and immersed in superfluid helium at 1.8K.

The light pulses were generated by a $Ti^{3+}:Al_2O_3$ laser tuned at 765 nm (1620 meV), at a repetition rate of 100 MHz. The spectral width of the 100 fs laser pulse was about 40 nm (80 meV), much broader than the splitting between the neutral exciton line (X) and the charged exciton line (X^+). The pump and probe pulses were focused on the sample to a common spot of diameter smaller than 100 μm, and the spectrum of the probe pulse transmitted through the sample was recorded as a function of the pump-probe delay. The power of both pulses was controlled independently, the pump-to-probe intensity ratio being at least 20:1. The pump pulse was shaped to a spectral width less than 1 nm and a duration of about 2 ps, and its average power was typically 300 μW, which results in the creation of a few times 10^{10} cm^{-2} excitons by each pulse. It was polarized circularly (by convention, σ^+, creating electrons with spin -1/2 in the conduction band, and holes of momentum +3/2 in the -3/2 spin-down valence subband). The probe beam was detected behind the sample in both circular polarizations, measuring the absorption associated with the creation of either X or X^+ having electron-hole pairs of the same spin as the pump, (σ^+, co-polarized), or opposite (σ^-, cross-polarized).

EXPERIMENTAL RESULTS

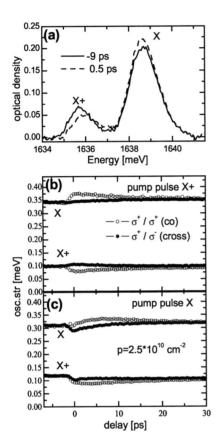

FIGURE 1. (a) Optical density for a QW with hole density 2.5×10^{10}cm^{-2}, at negative delay (-9 ps, solid line) and short positive delays (0.5 ps, dashed line), for a co-polarized pump pulse tuned to X+; (b,c) evolution of the X and X^+ oscillator strength with pump pulse tuned to X^+ (b) or tuned to X (c).

Figure1.a displays the measured optical density spectra for negative and short positive delays. Two absorption lines are observed, related to X and X^+. Under strong optical excitation, the X intensity increases and the X^+ intensity decreases. The intensities of both absorption lines, obtained by fitting Gaussian functions, are shown in Fig1b,c.

Let us start the discussion with the pump pulse tuned to resonance with X^+ (Fig.1b). In the co-polarized configuration, the intensity of X^+ decreases and that of X increases. In this experimental configuration the pump pulse creates directly charged excitons. The density of photocreated excitons is of the same order of magnitude as the density of preexisting holes. Therefore a sizable amount of holes within one spin subband becomes bound into charged excitons during the pump pulse. The co-polarized probe pulse cannot create charged excitons when there are no more holes available: The transition is partially blocked and its intensity is observed to decrease.

The concomitant increase of the neutral exciton is naturally explained by assuming that the neutral exciton is screened by the gas of free holes: This is in agreement with results obtained in CW experiments where the intensity of the neutral exciton line was observed to decrease when increasing the hole gas density. Then, a less effective screening by holes bound in charged excitons than by free holes explains the intensity recovery of the neutral exciton. [5].

When the pump pulse is tuned to resonance with neutral exciton, we observe similar effects, but for a slower rise of the neutral exciton intensity in co-polarization. This is related to the time required to form the charged exciton.

The time evolution of the intensities of both absorption lines is different for co- and cross-polarized pulses. From that fact we can conclude that the screening described above is spin-dependent. The neutral exciton is screened more efficiently by holes with spin opposite to that of its own hole [5]. Therefore the neutral exciton intensity change is much stronger in co-polarization. On the other hand, we expect a spin-independent screening of the trion, which contains two holes with opposite spins. This screening is visible in cross-polarization, when the pump pulse is tuned to resonance with trion, as a small increase of the charged exciton intensity.

When the pump pulse is tuned to resonance with the neutral exciton, in cross-polarization, we observe (not shown) the formation of biexcitons from neutral excitons created by the pump pulse, and a screening of the charged excitons by the photo created excitons.

This work has been partially supported by KBN grants: PBZ-KBN-044/P03/2001 and 2P03B 002 25, and Polonium program.

REFERENCES

1. K. Kheng et al., *Phys. Rev. Lett.*, **71**, 1752-1755 (1993).
2. P. Kossacki et al., *Phys. Rev. B,* **60**, 16018-16026 (1999).
3. R.T. Cox et al., *Phys. Rev. B,* **69**, 235303 (2004).
4. R. B. Miller et al., *J. Appl. Phys.*, **89**, 3734 (1999).
5. P. Płochocka et al., *Phys. Rev. Lett.*, **92**, 177402 (2004).

Free and Bound Exciton Dynamics in Bulk II-VI Semiconductors

M. D. Martín[1], A. Amo[1], L. Viña[1], G. Karczewski[2], J. Kossut[2]

[1] Dept. Física de Materiales, Universidad Autónoma de Madrid, Madrid, Spain
[2] Institute of Physics, Polish Academy of Sciences, Warsaw, Poland

Abstract. We have studied the recombination and spin dynamics of free and bound excitons in bulk CdTe. We have found that the energy relaxation dynamics is governed by exciton-LO phonon scattering, leading to a short lived emission. Additionally, the trapping of free excitons with center of mass momentum close to zero by impurities plays a major role in the relaxation process. The circular polarization degree of the emission decreases with increasing power. However, the spin flip time remains approximately constant for all the powers used, revealing that the free exciton spin dynamics is governed by exciton localization rather than carrier-carrier scattering.

INTRODUCTION

In the last years there has been an increasing interest on CdTe, not only because this material is a suitable candidate to produce gamma and X rays detectors, but also because of the development of a new solar cell technology based on CdTe [1]. It is therefore crucial to have a precise description of the fundamental properties of this material to determine which kind of elementary excitations are responsible for the absorption/transmission/conduction of those potential devices. In this paper we present a detailed study of the recombination dynamics of excitons in high quality CdTe epilayers, together with a description of the exciton spin dynamics.

EXPERIMENT

The two samples we have studied consist of 4.5 μm thick CdTe epilayers grown by Molecular Beam Epitaxy over a ZnTe buffer layer and a GaAs substrate. The optical characterization (by means of cw photoluminescence –PL– and photoluminescence excitation –PLE–, not shown) revealed a large contribution of bound states to the PL but yet a predominant peak due to free exciton (FX) recombination. The PLE spectra show a strong coupling of excitons (X) with LO phonons, as previously reported in the literature [2]. The low temperature (5 K) emission from these two samples has been time resolved by means of a spectrograph and a syncroscan streak camera, with an overall time resolution of 10 ps. To analyze the PL into its σ^+ and σ^- polarized components we have used a combination of λ/4 plates and linear polarizers.

RESULTS

Figure 1 shows the time integrated spectrum for the smallest excitation power used in our experiments (100 μW). A clear and dominant FX peak is observed, together with several bound exciton (BX) transitions. Increasing the excitation power leads to a red shift of the emission due to many body effects [3], and under these conditions the PL becomes completely dominated by the FX recombination.

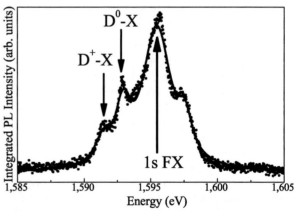

FIGURE 1. Time integrated PL spectrum for an excitation power of 100 μW.

The time evolution of the PL for both FX and BX recombination is displayed in Figure 2. It is clear that the time to reach the maximum (t_{max}) is slightly longer for the BX (dashed line) than for the FX (solid line). Furthermore, t_{max} for the FX (inset of Fig. 2, top) shows a non-monotonic dependence on excitation power. A similar behavior has been reported recently for bulk GaAs and attributed to the competition between the capture of FX with center of mass momentum close to zero by BX states and X-X scattering [4]. Increasing the excitation power turns into a larger FX population that progressively fills the BX states. When all these states are filled, t_{max} reaches its maximum value. A further increase of the power will revert in an enhancement of X-X scattering that will accelerate the dynamics again. However, in the present case the density of localization centers is much larger than in Ref. 4 and t_{max} saturates for very high powers, hindering the observation of a re-acceleration of the dynamics through X-X scattering.

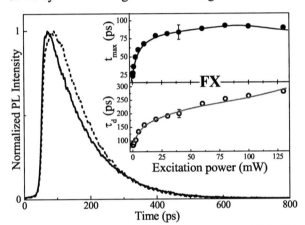

FIGURE 2. Time evolution of the PL from FX (solid line) and BX (dashed line). The inset shows t_{max} (above) and τ_d (below) for the FX emission as a function of excitation power. Lines are guides to the eye. The excitation power is 2.5 mW.

The FX PL's decay time (τ_d) (inset of Fig. 2, bottom) displays a monotonic increase with excitation power. This is again a consequence of the FX localization [5].

Figure 3 shows the maxima of the polarization degree under σ^+ excitation $[(I^+-I^-)/(I^++I^-)$, where $I^{+/-}$ denotes the $\sigma^{+/-}$ polarized emission] of the FX PL as a function of excitation power. A clear decrease with increasing power is observed. Simultaneously, the spin flip time (i.e. the decay time of the polarization degree) remains approximately constant with power (inset of Fig. 3). This fact evidences that increasing the carrier population has a negligible effect on the spin dynamics. The density of localization sites is large enough to trap most of the photocreated excitons [6] before any scattering event, through which spin relaxation will occur, can take place. Therefore, increasing the excitation power will gradually fill in more localization traps, leading to a reduction of the polarization degree of the PL, as we have experimentally observed.

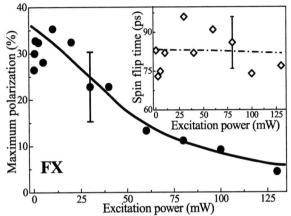

FIGURE 3. Maximum circular polarization degree of the FX PL versus excitation power. The inset displays the spin flip time extracted from a mono-exponential fit of the time evolution of the polarization. Lines are guides to the eye.

In conclusion, we have observed that the recombination and spin dynamics of excitons in high quality CdTe are governed by the localization of FX by BX states, even though the PL spectrum is dominated by FX recombination.

ACKNOWLEDGMENTS

This work was partially supported by the Spanish MCYT (MAT2002-00139), the CAM (07N/0042/2002) and the Network "Marie Curie: Physics of Microcavities" (MRTN-CT-2003-503677). A. A. acknowledges a fellowship of Spanish Secretaría de Estado de Educación y Universidades (MEC).

REFERENCES

1. D. Kraft, et al., *J. Appl. Phys.* **94**, 3589-3598 (2003); A. Romeo, et al., *Prog. Photovolt.: Res. Appl.* **12**, 93-111 (2004).
2. A. Nakamura, and C. Weisbuch, *Solid State Elect.* **21**, 1331-1336 (1978); M. Nawrocki, and A. Twardowski, *Phys. Stat. Sol. (b)* **97**, K61-K64 (1980); T. Schmidt, et al., *Phys. Rev. B* **45**, 8989-8994 (1992).
3. S. Schmitt-Rink, C. Ell, and H. Haug, *Phys. Rev. B* **33**, 1183-1189 (1986).
4. A. Amo, et al., *Phys. Stat. Sol. (c)* (in press) (2004).
5. J. F. Ryan, et al., *Phys. Rev. Lett.* **53**, 1841-1844 (1984).
6. G. Ghislotti, et al., *Solid State Commun.* **111**, 211-216 (1999).

Ultrafast dynamics of neutral and charged triplet magnetoexcitons in a high-mobility density-tunable GaAs-AlGaAs single quantum well

P. Schröter*, B. Su*, D. Heitmann*, C. Schüller*, M. Reinwald[†], H.-P. Tranitz[†] and W. Wegscheider[†]

Institut für Angewandte Physik und Zentrum für Mikrostrukturforschung, Universität Hamburg, 20355 Hamburg, Germany
[†]*Institut für Experimentelle und Angewandte Physik, Universität Regensburg, 93040 Regensburg, Germany*

Abstract. By means of spectrally resolved degenerate four-wave mixing spectroscopy, we have investigated a modulation doped AlGaAs/GaAs single quantum well. We obtain dephasing times of neutral and charged excitons, indicating that the charged excitons (trions) are not localized. A quantum beating between trion and exciton line shows that there is no spatial separation between the two.

INTRODUCTION

During the past decade there has been an intense discussion about the ground state properties of dilute two-dimensional electron systems (2DES) in optical experiments. It was found that the ground state is formed by negatively charged excitons, which can be singlet excitons or triplet excitons in a magnetic field.[1–4] There has been some controversy whether the charged excitons and the neutral excitons are spatially separated and/or whether or not they are localized due to potential fluctuations. Spatially resolved luminescence experiments under in-plane electric fields show a reasonable dependence of the sample quality on the trion mobility and thus on the localization of these quasiparticles.[5] Four-wave mixing experiments on p-doped CdTe quantum wells indicated that positively charged trions are localized.[6] There are also reports on exciton-charged exciton correlations [8] and the temperature dependence of their mobility [7] in n-doped II-VI quantum wells available.

In this paper we report on spectrally-resolved four-wave mixing (SR-FWM) experiments on a 25 nm GaAs/Al$_{0.3}$Ga$_{0.7}$As single quantum well, suggesting that charged excitons are not localized in high-quality samples and reside in the same spectral region as the neutral excitons.

EXPERIMENTAL DETAILS

Our sample is a modulation doped GaAs/Al$_{0.3}$Ga$_{0.7}$As quantum well, having a carrier density and mobility of of 1.5×10^{11} cm^{-2} and 2.7×10^6 cm^2/Vs, respectively. Via a metallic gate, the carrier density in the quantum well can be tuned in the SR-FWM experiments over one order of magnitude, down to about 1×10^{10} cm^{-2}. The experiments were carried out in an optical ^{3}He cryostat at temperatures between 0.7 K and 1.4 K. A standard degenerate FWM setup was employed, using two laser beams of equal intensity, supplied by a wavelength tunable Ti:sapphire laser with a pulsewidth of approximately 150 fs (FWHM) corresponding to a spectral width of 10 nm.

RESULTS AND DISCUSSION

Figure 1 shows SR-FWM spectra for a time delay $\Delta t = 0$ for different densities, starting from $n = 1.3 \times 10^{11}$ cm^{-2} (bottom spectrum) to the strongly diluted case. In the strongly diluted case at negative gate voltages, we observe coherent signals, which originate from neutral magnetoexcitons and, at lower energies, signals from negatively charged excitons, emerging from the 2DES signal at high electron density.

The evolution of these signals as a function of Δt for a fixed density ($n_s \sim 1 \times 10^{10}$ cm^{-2}) is shown in Fig. 2. Our main findings are: (i) We observe a quantum beating between the triplet exciton and a neutral magnetoexciton, and, (ii) the decay time of the SR-FWM signal of the triplet exciton (about 2.1 ps at B = 6 T) is shorter than the decay time of neutral magnetoexcitons at the same magnetic field (about 3.3 ps). Ongoing work, using spectrally narrow laser pulses to exite different levels selectively and thus suppressing the beating structure in the FWM signal, confirms this result.

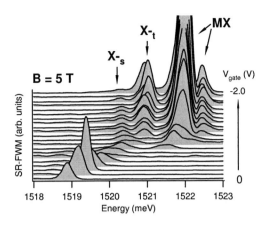

FIGURE 1. SR-FWM signal from a high mobility density-tunable single QW at $\Delta t = 0$ for different gate voltages corresponding to electron densities from $n_s = 1 \cdot 10^{11}$ cm^2/Vs (bottom trace) to $n_s = 1 \cdot 10^{10}$ cm^2/Vs (top trace).

From this we can draw two important conclusions: (i) In our high-mobility quantum well, the charged and neutral excitons are not spatially separated, otherwise there would be no quantum beating. (ii) The charged excitons are *not* localized. This can be deduced from the shorter decay time of the signal from the triplet exciton compared to the neutral exciton. If the charged exciton would be localized, scattering would be strongly reduced and its dephasing time should be longer than that of free neutral excitons.

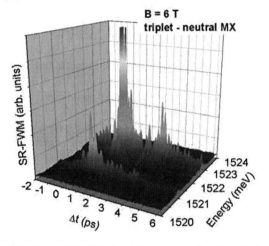

FIGURE 2. SR-FWM signal from charged and neutral excitons for σ^- circular polarization. The signal at 1521 meV is attributed to a charged triplet state while the higher energy peak is a signature of neutral excitons.

ACKNOWLEDGEMENTS

This work was supported by the Deutsche Forschungsgemeinschaft via projects SCHU1171/1 and SCHU1171/2.

REFERENCES

1. Kheng, K., Cox, R. T., d'Aubigné, Y. M., Bassani, F., Saminadayar, K., and Tatarenko, S., *Phys. Rev. Lett.*, **71**, 1752 (1993).
2. Yusa, G., Shtrikman, H., and Bar-Joseph, I., *Phys. Rev. Lett.*, **87**, 216402 (2001).
3. Shields, A. J., Osborne, J. L., Simmons, M. Y., Pepper, M., and Ritchie, D. A., *Phys. Rev. B*, **52**, R5523 (1995).
4. Schüller, C., Broocks, K.-B., Heyn, Ch., and Heitmann, D., *Phys. Rev. B*, **65**, 081301(R) (2002).
5. Pulizzi, F., Sanvitto, D., Christiansen, P. C. M., Shields, A. J., Holmes, S. N., Simmons, M. Y., Ritchie, D. A., Pepper, M., and Maan, J. C., *Phys. Rev. B*, **68**, 205304 (2003).
6. Brinkmann, D., Kudrna, J., Gilliot, P., Hönerlage, B., Arnoult, A., Cibert, J., and Tatarenko, S., *Phys. Rev. B*, **60**, 4474 (1999).
7. Portella-Oberli, M. T., Ciulin, V., Haacke, S., Garnière, J.-D., Kossacki, P., Wojtowicz, T., and Deveaud, B., *Phys. Rev. B*, **66**, 155305 (2002)
8. Portella-Oberli, M. T., Ciulin, V., Berney, H., Deveaud, B., Kutrowski, M., and Wojtowicz, T., *Phys. Rev. B*, **69**, 235311 (2004)

Radiative lifetime and dephasing of excitons studied by femtosecond time resolved intersubband spectroscopy

I. Marderfeld*, D. Gershoni*, E. Ehrenfreund* and A.C. Gossard[†]

*Physics Department, Technion-Israel Institute of Technology, Haifa 32000, Israel
[†]Materials Department, University of California, Santa Barbara, CA 93106, USA

Abstract. We used time resolved photoinduced intersubband absorption excitation spectroscopy in order to measure the dynamics of resonantly photoexcited excitons in multi-quantum well and superlattice samples. We show that they decay radiatively, much faster than thermalized excitons, and that their electron momentum is initially concentrated around the superlattice Brillouin zone center. Later, their electron momentum gradually spreads towards the zone edge, in which case the excitons cannot recombine radiatively.

In geometrically restricted semiconductor quantum structures, such as quantum wells (QWs) and superlattices (SLs), the translation symmetry is removed in one direction. Therefore, excitons with in-plane momentum, which is smaller than the momentum of light in the host crystal, can decay radiatively within their intrinsic radiative lifetime [1]. Accordingly, there is a marked difference between excitonic resonant excitation, in which excitons recombine radiatively immediately after pulse excitation, and non-resonant excitation, in which dephasing must occur before recombination kicks-in. Visible-infrared (IR) [2] and visible-TeraHertz [3] dual beam time-resolved spectroscopy were used recently, in order to study these processes directly. Our method is based on a 500 femtoseconds (fs) pulse excitation by tunable interband laser pulse, followed by 500 fs IR probe pulse, tuned to the intersubband (ISB) resonances of the system. The ISB probe pulse induces an optical transition of photoexcited electrons. Hence, it is very sensitive to the momentum of the carrier whose optical transition is being induced.[2] ISBA spectroscopy in SLs provides a sensitive tool for probing the electronic momentum component along the SL symmetry axis.

The $GaAs/Al_{0.33}Ga_{0.67}As$ heterostructures were grown by molecular beam epitaxy on a (100)-oriented GaAs substrate. They consisted of 33 periods of undoped GaAs quantum wells of thickness 6nm and $Al_{0.33}Ga_{0.67}As$ barriers of thickness 14.7 nm and 6.2 nm for the multi-QW (MQW) sample and for the SL sample, respectively. The thicknesses of the layers were measured by high resolution X-ray diffraction. Two edges of the sample were polished at 45° to the growth axis, in order to form a waveguide for the mid-IR radiation with an electric field component parallel to the growth direction. The measurements were done with the sample mounted on a cold finger cryostat at $\simeq 15$ K. We estimated an exciton density per period of $\simeq 2 \times 10^{10} cm^{-2}$ from the measured beam average power, repetition rate, and spot diameter on the sample, together with the calculated interband absorption coefficient [4]. Fig. 1 summarizes the measured experimental data from both samples. In Fig. 1a and Fig. 1c the cw measured photoinduced ISBA spectra of the MQW and of the SL sample, respectively, are shown. In Fig. 1b and Fig. 1d we present the temporally integrated PL and PL excitation (PLE) spectra together with the time resolved, photoinduced, intersubband absorption (ISBA) excitation (PIAE) spectra for both samples. The spectra are normalized at their highest energy edge ($\simeq 1.64$ eV).

We note that the PLE spectrum and the 3 ps PIAE spectrum of the MQW sample are almost identical, while the 16 ps PIAE spectrum is different. We clearly observe that the PIAE spectrum evolves with time in such a way that the relative intensity of the heavy hole exciton (HH_X) resonance is reduced. We attribute this reduction to radiative losses of low in-plane momentum excitons. The inset summarizes the temporal dependence of the HH_X resonance intensity on a logarithmic time scale.

The ISBA is a direct measure of the density of electrons in the first electronic subband of the QWs. Thus, the initial fast decay and the later slower decay of the PIAE excitonic resonance, demonstrate that resonantly excited excitons, "remember" their history, and they decay faster than non resonantly excited electron-hole (e-h) pairs. The SL sample offers further insight into this "excitonic memory". The spectral shape of the optical transition between the first electronic miniband, and the second miniband is mainly due to the dispersion of the

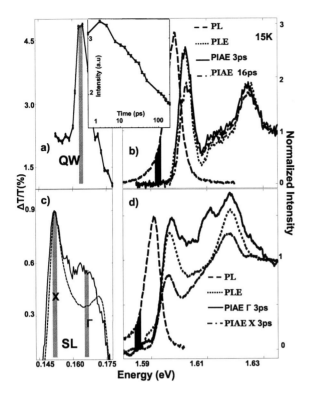

FIGURE 1. a),(c)) Temporally integrated photoinduced ISBA spectrum of the MQW (SL) sample. The dashed line in (c) represents thermal equilibrium calculations after Ref.[4]. b),(d)) Temporally integrated PL and PLE spectra together with time resolved ISBA excitation spectra for the MQW (SL) sample. The shadowed areas under the PL and under the ISBA spectra indicate the spectral ranges from which the PL and ISBA excitation spectra, respectively, were obtained. Inset: The exciton resonance intensity vs. time after the excitation pulse for the MQW sample.

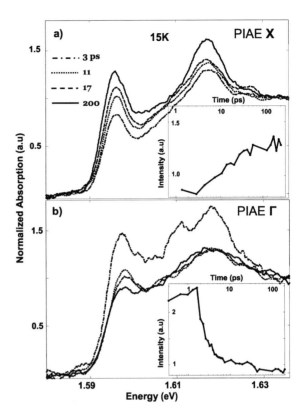

FIGURE 2. a),(b)) Time resolved PIAE spectra for various times after the excitation pulse for the SL sample probed at the X (Γ) point of the ISBA transition. Insets: The exciton resonance intensity vs. time after the excitation pulse.

second miniband along the growth axis. In particular, the SLs zone center (Γ) and zone edge (X) transitions are clearly identified spectroscopically (see Fig.1c), thus correlating the ISB transition energy and the momentum states of the probed photoexcited electron in the first miniband. By probing at different ISB energies, the population of resonant excitons with different momenta along the SL symmetry axis is measured. This is demonstrated in Fig. 1d, in which the PL and PLE spectra of the SL sample are quite similar to those of the MQW sample, but the 3 ps PIAE spectra are vastly different. In the PIAE spectrum probed at the X point, the excitonic resonance is much weaker than that observed in the PLE spectrum, while in the PIAE from the Γ point the excitonic resonances are stronger.

In Fig. 2a (2b) we present PIAE spectra probed at the X (Γ) ISBA energy for various times after the excitation pulse. The temporal dependence of the excitonic resonance is displayed in the insets. It is clearly seen that the excitonic strength increases with time when probed at the X point and drastically decreases with time when probed at the Γ point. Our observations can be qualitatively understood in terms of the electronic component of the excitonic momentum. For resonantly excited excitons, this component is initially concentrated around the superlattice Brillouin zone center as clearly shown by our measurements. This momentum redistributes in time, such that it gradually increases around the zone edge. At the zone edge the radiative decay of the excitons is prohibited, and therefore, at long times the strength of the resonance grows back to its original strength at the moment of excitation.

We acknowledge the Israel Science Foundation and the Technion Vice President Fund for the Promotion of Research.

REFERENCES

1. B. Deveaud et al., Phys. Rev. Letters **67**, 2355 (1991).
2. R. Duer et al., Phys. Rev. Letters **78**, 3919 (1997).
3. R.A. Kaindl et al., Nature **423**, 734 (2003).
4. D. Gershoni et al., IEEE J. QE, **29**, 2433 (1993).

Coherent Control of Bloch Oscillations by Means of Optical Pulse Shaping

R. Fanciulli*, A.M. Weiner*, M.M. Dignam[†], D. Meinhold** and K. Leo**

*Physics Department and School of Electrical and Computer Engineering, Purdue University, West Lafayette, Indiana, USA
[†]Physics department, Queen's University, Kingston, ON, Canada, K7L 3N6
**Institut für Angewandte Photophysik, Technische Universität Dresden, D-010609 Germany

Abstract. We create excitonic wavepackets in biased semiconductor superlattices (SLs) with spectrally-shaped ultrashort optical pulses. We tailor the shape and phase of the pulse spectrum in order to control the coherent dynamics of the excitonic wavepackets formed from a superposition of three excitonic states. The wavepacket evolution is monitored using spectrally-resolved four wave mixing. This ability to control the BOs provides a way to control the emitted THz radiation.

In recent years much work was done in the area of Bloch Oscillations (BO) in semiconductor superlattices (SL) [1]. Oscillating wavepackets have been optically created with fs laser pulses and observed both with non-linear optical techniques like degenerate four wave mixing (DFWM) or by directly observing the THz radiation derived from the electrons dynamics [1]. At the same time the introduction of laser pulse shaping techniques has allowed selective excitation of energy levels and has emerged as a key enabling technology for the new field of quantum coherent control [2]. By combining these two fields we have achieved full control over the quantum oscillatory dynamics of an electronic wavepacket displaying BO and the direct observation of the electronic motion.

We observe the BO in a GaAs/Al$_{0.3}$Ga$_{0.7}$As superlattice (SL) composed of 35 quantum wells (67Å/17Å). The structure is etched to be used in a transmission geometry and is cooled down to 10K. An external DC bias, F_{bias}, breaks the translational symmetry of the SL and breaks the minibands into equally spaced electronic levels, forming the so called Wannier-Stark ladder (WSL). The wavefunctions corresponding to these levels are localized and roughly centered in each well. A study of the interband oscillator strengths [3] allows us to consider only the transitions from the heavy hole to the lowest level in each well in the conduction band (1S excitonic levels). We can therefore label each transition unequivocally in terms of the WSL well number, n (see Fig. 1). In a simple non-interacting electron picture, the energies of the electronic transitions depend linearly on the field F_{bias} with a slope proportional to the well number n (0, ±1, ±2 ...) [1]. This property of the system is of

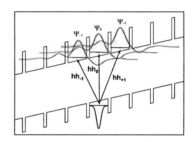

FIGURE 1. Representation of the superlattice. A schematic of the electronic wavefunctions and the transitions of interest in our experiment are reported.

great importance when monitoring the oscillatory motion of our excited electrons.

We observe the BO by measuring the spectrum of the degenerate four-wave mixing (DFWM) signal in the direction $2\vec{k}_2 - \vec{k}_1$ ($\vec{k}_1$ being the wavevector of the pump and $\vec{k}_2$ the one of the probe) as a function of the delay, τ, between pulse $\vec{k}_2$ and $\vec{k}_1$. Each transition covered by the laser gives a separate component in the DFWM spectrum that can be studied separately. Defining $\Psi_n(x)$ as the electronic wavefunction centered in the nth well, the pump pulse creates the superposition of three states given by:

$$\Psi(x) = A_{-1} e^{+i\omega_{BO}t} \Psi_{-1}(x) + A_0 \Psi_0(x) + \quad (1)$$
$$+ A_{+1} e^{-i\omega_{BO}t} \Psi_{+1}(x).$$

Here, A_n is a complex number proportional to the product of the laser complex spectral amplitude (resonant with the nth transition) and the interband dipole matrix element for that transition. Due to the alternating sign of the WSL wavefunction (see Fig. 1), M_{-1} and M_{+1} have opposite signs [3]. The expectation value for the position of the excited single electron is given by:

$$X_e(t) = \left(|A_{-1}|^2 x_{-1,-1} + |A_0|^2 x_{0,0} + |A_{+1}|^2 x_{+1,+1}\right) + \\ -2|A_{-1}^* A_0| \cdot |x_{-1,0}| \cos(\omega_{BO} t + \phi_{-1,0}) + \\ +2|A_{+1}^* A_0| \cdot |x_{+1,0}| \cos(\omega_{BO} t + \phi_{0,+1}) + \\ +2|A_{-1}^* A_{+1}| \cdot |x_{-1,+1}| \cos(2\omega_{BO} t + \phi_{-1,+1}), \quad (2)$$

where $\phi_{n,m}$ is the relative phase of the pump spectral components resonant with hh_n and hh_m, $x_{n,m} \equiv \int_{-\infty}^{+\infty} \Psi_n^*(x) x \Psi_m(x) dx$. Within a very good approximation $x_{-1,0}$ and $x_{+1,0}$ are equal [3], while the last term can be neglected due to the small ratio $|x_{-1,+1}|/|x_{-1,0}|$.

From eq. 2, one can see how, depending on the excitation conditions of the pump (A_n and $\phi_{n,m}$) the spatial oscillation of the electron can be symmetric (breathing mode (BM)) or not (non-breathing mode (NBM)).

The hole created by the pump in valence band is completely localized in its well of origin ($n = 0$) due to its larger effective mass. The spatial displacement between the center of mass of the electron (X_e) and the center of mass of the hole, creates an internal oscillating dipole field in the direction of the SL growth. The effect of this oscillating dipole, within a semi-static approximation, is to contribute to the bands bending in the SL alternatively screening or enhancing the effect of the external field F_{bias}. This effect along with the quasi-linear dependence of the transitions energy on the field, induce an oscillation in the transitions energy that can be monitored by measuring the spectrum of the DFWM for different delays τ of the probe pulse [4].

By amplitude-modulating the pump spectrum we leave only three components resonant with the three transitions $hh_{\pm 1}$, hh_0. Adjusting the relative amplitudes of these components and keeping their relative phases flat, we correct for the different oscillator strengths of the $hh_{\pm 1}$ and make the parameters $A_{\pm 1}$ to be the same. In this case the BM is excited and no internal AC dipole is expected. We observe that in Fig. 2(a) (excitation density of 5×10^9 e/cm^2 pulse well). We then apply different phases to the three spectral components in the pump spectrum. A [π,0,0] phase profile optimizes the magnitude of the oscillation of X_e. In this case we are exciting a NBM so that an internal dipole between the electron and the hole is created. This can be observed in Fig. 2(b). We are effectively controlling and observing the creation (NBM) and suppression (BM) of a THz train of pulses.

We can now keep the same shaping conditions (in amplitude and phase), but suppress the component of the pump resonant with the hh_0 transition. In this case the 4th term of eq. 2 becomes relevant (and the 2nd and 3rd disappear) and to overcome the limited overlap between the $\Psi_{\pm 1}$, we increase the excitation density by about 2.5 times. In

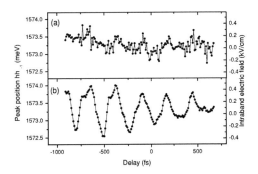

FIGURE 2. Peak position of the hh_{-1} in the case of (a) non-breathing mode (phase of the pump pulse [π,0,0])(b) breathing mode (phase of the pump pulse [0,0,0]). The right scale reports the estimated magnitude of the generated internal oscillatory THz field as estimated using the quasi-static model.

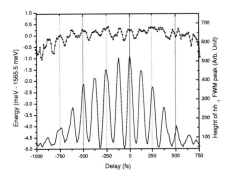

FIGURE 3. (bottom) Height of the hh_{-1} spectral component of the DFWM as a function of the delay of the probe. (top) Position of the hh_{-1} transition versus delay. The phase profile of the exciting pump spectrum is [0,π].

Fig. 3(a) we report the spectrally integrated DFWM as a function of the probe delay τ for this 2-color excitation scheme. From Fig. 3(b) we can see how the frequency has doubled and at the same time oscillations in the peaks position are still visible with an amplitude similar to the one observed in Fig. 2(b). A theoretical study that includes excitonic effects, the presence of a continuum of levels (unbound excitons) and a dynamical model is currently under development by one of our co-authors.

REFERENCES

1. Leo, K., *High-Field Transport in Semiconductor Superlattices*, Springer Verlag, Berlin Heidelberg New York, 2003.
2. Weiner, A., *Rev. of Scient. Instrum.*, **71**, 1929 (2000).
3. Dignam, M., Sipe, J., and Shah, J., *Phys. Rev. B*, **49**, 10502 (1994).
4. Lyssenko, V., Valusis, G., Löser, F., Hasche, T., Leo, K., Dignam, M., and Köhler, K., *Phys. Rev. Lett.*, **79**, 301 (1997).

Optical And Transport Properties Of Devices Utilizing Nanoscale Deep-Centers

Janet L. Pan

Yale University, P.O. Box 208284, New Haven, CT 06520-8284, USA

Abstract. We demonstrate how nanoscale deep-centers make possible novel devices for THz applications and 1.5um fiber-optic emitters on GaAs. We report the first GaAs light-emitting-diode (LED) emitting at 1.5υm fiber-optic wavelengths from arsenic-antisite deep-levels. This is an enabling technology for 1.5um fiber-optic components lattice-matched to GaAs ICs. We demonstrate experimental results for significant internal optical power (24mW), efficiency (0.6 percent), and speed (THz) from GaAs deep-level optical emitters. We demonstrate the first GaAs tunnel diodes utilizing arsenic-antisite deep-levels. At room temperature, our measured peak current density (16kA/cm2) is the largest ever in GaAs tunnel diodes. Our devices also show room-temperature peak-to-valley current ratios as high as 22. We ascertain the transport mechanisms which limit the peak and valley currents. Finally, we present the first fully-analytical multi-band model of the deep-center wave function. (e.g., symmetry and admixture of atomic orbitals). This wave function determines the general optical properties (optical selection-rules and transition strengths) of deep-centers in terms of simple materials parameters (bandgap energy, Kane dipole) and angular momenta quantum numbers. This model is applicable in a wide variety of materials, and is in agreement with experiment.

INTRODUCTION

Components for 1.55μm fiber optics fabricated on the same substrate as integrated circuits (ICs) are important for future high-speed communications. Consequently, industry has responded with a costly push to develop InP electronics. For fabrication simplicity, reliability, and cost, however, GaAs remains the established technology for integrated optoelectronics. Unfortunately, GaAs has a bandgap whose wavelength (0.85μm) is far too short for 1.5μm fiber-optics. Here we demonstrate the first GaAs light-emitting-diode [1] (LED) emitting at 1.5μm fiber-optic wavelengths from arsenic-antisite (As_{Ga}) deep-levels. This is an enabling technology for fiber-optic components lattice-matched to GaAs ICs. We demonstrate experimental results for significant internal optical power (24mW) and speed (THz) from GaAs deep-level optical emitters. Second, we present the first fully-analytical multi-band model [2,3] of the deep-center wave function. (e.g., symmetry and admixture of atomic orbitals). Our model determines the general optical properties of deep-centers in terms of simple materials parameters (bandgap energy, Kane dipole), and is thus applicable in a wide variety of materials. The strength of optical transitions from deep-centers ultimately limits the efficiency-bandwidth product of optical-emitters. Third, we demonstrate the first tunnel diodes utilizing deep-levels in low-temperature-grown (LTG) GaAs. At room temperature, our measured negative-conductance-per-area of 1/226Ω-μm^2 and peak current density of 16kA/cm^2 are the largest [4] ever in GaAs tunnel diodes. Our devices also show room temperature peak-to-valley current ratios as high as 22.

OPTICAL PROPERTIES

Recently, numerous groups [5,6] have reported optical emission from deep-states in silicon at both visible and infrared wavelengths. In the near future, much work will be devoted to understanding the physics of this optical emission. Much of this work will be numerical. Here, we go beyond a numerical approach to derive the general optical properties of semiconductor deep-levels. Our results are expressed in terms of simple materials parameters (e.g., bandgap energy), and are thus applicable in a wide variety of semiconductors. The solid curve in Fig. 1 shows the well-known energy dispersion of ideal GaAs. The new aspect of our model is the analytic continuation [2,3] of these energy bands into the band gap, which results in a dispersion curve for an imaginary wave vector, shown as the dashed curve in Fig. 1. This imaginary wave vector of our analytic continuation corresponds to both an evanescent penetration depth of an incident electron wave as well as to the deep-state radius. The details [2,3] of our model show that the deep state radius is twice the Kane dipole. Another new aspect of our model is that both the symmetry and admixture [2,3] of atomic orbitals in the deep-state wave function is easily obtained: the energy-eigenstates of a spherically symmetric deep-center are simultaneous eigenstates of total angular momentum (here, the spin plus the Bloch and envelope angular momenta). Our results predict that optical transitions are strong from midgap states to both the conduction- and valence-band. Our results are in good agreement [2,3] with scanning-tunneling-microscopy of the deep-state radius, and with measurements of optical transition strength and selection-rules.

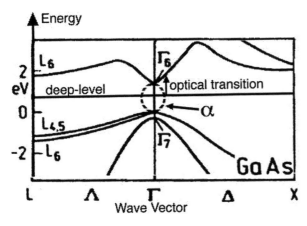

FIGURE 1. The dashed curve [2,3] shows the energy dispersion of an imaginary wave vector, resulting from our analytic continuation of the energy bands into the band gap. This imaginary wave vector corresponds both to the evanescent penetration depth of an incident electron wave, as well as to the deep state radius. The details of our model show the deep-state radius is twice the Kane dipole.

We have created a large number [7] (10^{20} cm^{-3}) of As$_{Ga}$ deep-levels by the molecular-beam-epitaxy (MBE) of low-temperature-grown (LTG-) GaAs. Figs. 2a and 2b denote as τ_R and τ_{NR}, respectively, the lifetimes for radiative and nonradiative transitions from an upper- to a lower state. The product of radiative efficiency η_R and bandwidth is $\eta_R \times (1/\tau_R + 1/\tau_{NR}) = 1/\tau_R$. This product is larger for strong optical transitions (fast τ_R). Absorption is one measure of the optical transition strength. Figure 3 shows that our measured absorption for LTG-GaAs at 1.1μm and 1.55μm is 10^4cm^{-1} and 1.6×10^3cm^{-1}, respectively. Thus, the measured absorption from deep-levels in LTG-GaAs is almost as large as for conventional conduction-to-valence band transitions.

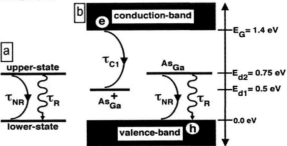

FIGURE 2. Energy levels and transition rates. (a) Radiative lifetimes (τ_R) and nonradiative lifetimes (τ_{NR}) associated with transitions from an upper-state to a lower-state. (b) Radiative lifetimes (τ_R), nonradiative lifetimes (τ_{NR}), and total transition lifetimes (τ_{C1}) associated with transitions[11] (to As$_{Ga}^+$ and from As$_{Ga}$) in LTG-GaAs.

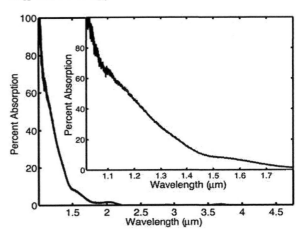

FIGURE 3. Absorption spectrum measured at room temperature of 1 μm of GaAs grown at 225°C. The inset shows the measured absorption between 1.0 μm and 1.8 μm.

TRANSPORT MECHANISMS AND TUNNEL DIODES

Our device consists of a P-layer, LTG-layer, and N-layer, all on GaAs. The contacts were not annealed. The n-contacts were weak Schottky diodes, which required 0.6v to reach mA currents. Surprisingly, our devices showed tunnel-diode behavior, with the largest [4] ever peak-current density of 16kA/cm^2 and the largest ever negative-conductance (NC) per-area of 1/226Ω-μm^2 at room temperature for a GaAs tunnel diode. See Fig. 4. These devices also show peak-to-valley current ratios as large as 22. Both the shape of

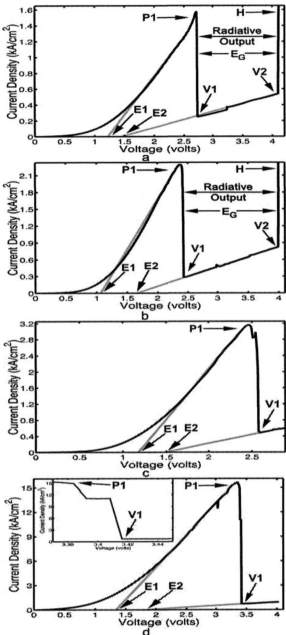

Figure 4. Measured current-voltage characteristics of four different devices on the same wafer. The device diameters, and measured room-temperature peak-to-valley current ratios, peak current densities, and negative resistances were, respectively, (a) 43μm, 6.3, 1.6kA/cm^2, 1.2kΩ-μm^2, (b) 34μm, 8.0, 2.3kA/cm^2, 844Ω-μm^2, (c) 34μm, 6.5, 3.2kA/cm^2, 795Ω-μm^2, (d) 28μm, 22, 16kA/cm^2, 226Ω-μm^2. The inset in Fig. 4d is an expanded view of the data in the proper part of Fig. 4d, where the points P1 and V1 denote the peak and valley currents. The NC was measured to be the slope of the straight line connecting the points P1 and V1. The points E1 and E2 are extrapolations of the measured I-V curve in straight-lines to the voltage-axis. The extrapolation of the measured curve through the points V1 and V2 in a straight-line reaches the voltage-axis at the point E2. E2 is approximately one E_G.

the current-voltage characteristic and the properties of the optical emission elucidate the transport mechanisms which limit the peak and valley current in our devices. Here, we show that the peak current in the measured tunnel-diode characteristic results from transport through the deep-levels within the bandgap of the LTG region. See Fig. 5a. We also show that the valley current in the measured tunnel-diode characteristic results from the transport of free holes and free electrons through the valence- and conduction-bands of the LTG region, with the deep-levels acting as recombination centers (thus reducing the valley current). See Fig. 5b and 5c. Thus, the large number of deep-levels contributes both to large peak currents as well as to small valley currents.

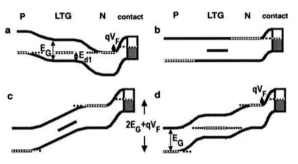

FIGURE 5. Energy-band diagrams in forward bias. (a) Band-diagram corresponding to the smaller voltages (between zero volts and the points P1) in Fig. 4. The band-diagram shows that transport occurs mainly through E_{d1}. No optical emission is observed at these voltages. V_F is the voltage drop at the contact. (b) Band-diagram corresponding to the points V1 in Fig. 4. (c) Band-diagram corresponding to the points V2 in Fig. 4. In (b) and (c), significant transport occurs through the conduction and the valence bands. At voltages between points V1 and V2 in Fig. 4, optical emission is observed. (d) Band-diagram corresponding to the points H in Fig. 4. At this voltage, significant transport occurs through E_{d1}. In (d), a voltage of E_G/q drops over both the P-LTG junction and the LTG-N junction, but almost no voltage drops over the LTG region.

Significantly, radiative emission is observed *only* after the sharp current drops (labeled P1) in Fig. 4. The absence of radiative emission at smaller voltages (between zero volts and the points P1 in Fig. 4) indicates a relative absence of free holes and free electrons. Thus, the peak current P1 is dominated by carrier transport through the deep-level E_{d1}, rather than through the conduction and valence bands (Fig 5a). The sharp drop in the total current between points P1 and V1 in Fig. 4 indicates a sharp change in the carrier transport mechanism. The relevant carrier transport mechanisms may be inferred from the shape of the measured characteristics in Fig. 4. The measured valley current (between points V1 and V2 in Fig. 4) has the characteristic of a large resistance in series with a diode of activation voltage 1.5v (see extrapolation to points E2). 1.5v is the size of the GaAs band gap E_G. Thus, the valley current of our devices can be modeled as a large resistance in series with a normal GaAs diode. (The large resistance measured between points V1 and V2 in Fig. 4 is a result of the small concentrations and mobilities of free carriers in the LT layer in Figs. 5b, 5c.) Thus, by analogy with normal GaAs diodes, we would expect the valley current of our devices to be dominated by the transport of free carriers through the conduction and valence bands. Moreover, our conclusion, that the valley current consists of free carrier transport through the conduction and valence bands, is consistent with our earlier observation of radiative emission (at both above E_G and below E_G) for the valley current (points V1,V2 in Fig. 4) but not the peak current (points P1).

LIGHT-EMITTING-DIODES

Figure 6a shows the first measurement [1] of room-temperature electroluminescence from As_{Ga} (using 200ns pulses and duty cycles less than 10^{-4}). Figures 3 and 6a show that both the absorption and electroluminescence of LTG-GaAs indicate transitions at 1.0μm and 1.55μm, which correspond to conduction-band-to-As_{Ga}^+ and As_{Ga}-to-valence-band transitions (Fig. 2b). The measured optical signal radiated by our LED into 2π steradians of free space was 560μW. It is well known that this 560μW represents only 2.4% of the internal optical power within the sample. This implies that the internal optical power within the GaAs sample was 24mW. At a lower injection current of 150mA, we measured an internal efficiency of 2×10^{-4}, which implies a nonradiative $\tau_{NR}=200fs$. Our measurement of $\tau_{NR}=200fs$ at low injection currents is consistent with an estimate of $\tau_{NR}=500fs$ based on previous work [7] on the capture of holes by As_{Ga}. Both the turn-off time and the turn-on time of optical emitters are determined by the below-threshold total relaxation rate $(1/\tau_R+1/\tau_{NR})$ of carriers from the upper- to the lower state of the optical transition. Our fast $\tau_{NR}=200fs$ in the total relaxation rate $(1/\tau_R+1/\tau_{NR})$ could make possible direct modulation of semiconductor lasers at THz speeds.

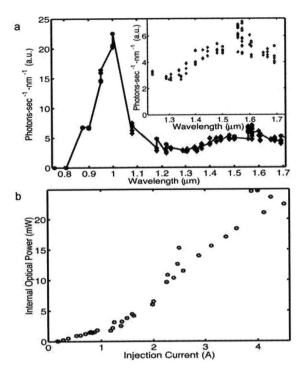

FIGURE 6. Room temperature electroluminescence. (a) Measured electroluminescence spectrum of our sample at room-temperature. The inset shows the measured electroluminescence between 1.2 μm and 1.7 μm. Both the absorption and the electroluminescence show transitions at 1.0 μm and 1.5 μm. (b) Room temperature electroluminescence measured as a function of injection current.

ACKNOWLEDGMENTS

JLP wishes to thank J. Woodall, NSF CAREER, and ONR YIP.

REFERENCES

1. Pan, J.L, McManis, J.E., Osadchy, T., Grober, L., Woodall, J.M., Kindlmann, P.J., *Nature Materials* **2**, 375-378, (2003).
2. Pan, J.L, *Opt. Exp.* **9**, 796-805, (2001).
3. Pan, J.L, *J. Appl. Phys.* **92**, 5991-6004, (2002).
4. Pan, J.L, McManis, J.E., Grober, L., Woodall, J.M., *Sol. St. Elect,* in press, (2004).
5. Ng, W., *Nature* **410**, 192, (2001).
6. Green, M., *Nature* **412**, 805, (2001).
7. Melloch MR, Nolte DD, Woodall JM, Chang JCP, Janes DB, Harmon ES, *Crit Rev in Sol St and Mater Sci* **21**,189-263 (1996).

Intraband Carrier Dynamics In Semiconductor Optical Amplifier-Based Switch

A. Gomez-Iglesias*, J. G. Fenn*, M. Mazilu*, A. Miller* and R. J. Manning[†]

*School of Physics and Astronomy, University of St Andrews, St Andrews, Fife KY16 9SS, UK
[†]Physics Department, University College Cork, 2200 Cork Airport Business Park, Cork, Ireland

Abstract. We present interferometric switching experiments using femtosecond pulses with an InGaAs semiconductor optical amplifier (SOA). Our measurements reveal significant refractive index changes occurring within a few picoseconds. The timescale and power dependence of this ultrafast feature, together with the predictions of a phenomenological model, strongly suggest carrier heating as responsible.

During the last decade, semiconductor optical amplifiers (SOA) have become promising candidates for all-optical switching applications. Interferometric switches exploiting the SOA nonlinearities, such as the Terahertz Optical Asymmetric Demultiplexer (TOAD) and the Mach-Zehnder interferometer [1, 2], have proven very successful. These systems are based on the refractive index dependence with carrier density in the conduction band but can still operate at faster rates than those imposed by the interband gain recovery time of the device. If ultrafast switching is to be achieved, femtosecond pulses must be used. However, despite the work carried out to date [3, 4], there is still much to be understood on the implications of the subpicosecond refractive index changes associated with the gain dynamics in switching operation. Previous gain pump-probe experiments and theoretical models [5] highlight the key role of ultrafast intraband phenomena, particularly carrier heating, in the SOA response on the shortest time scales. Here we report interferometric three-beam pump-probe experiments with an SOA, revealing the signature of carrier heating during the onset of switching, and consistent with the predictions of a phenomenological model.

Our experiments were carried out in a TOAD setup. The latter consists of a SOA placed asymmetrically within an optical loop mirror. We used an InGaAs amplifier and 700 fs pulses at 1.57 μm provided by an optical parametric oscillator. Every input pulse is split into two replicas which propagate around the loop in opposite directions, reaching the SOA at different times. In addition, a stream of strong control pulses is injected into the loop to deplete some of the device gain and therefore induce a change in the refractive index. This way, only when a control pulse arrives at the SOA between the two replicas, these will acquire a differential phase shift $\Delta\varphi$

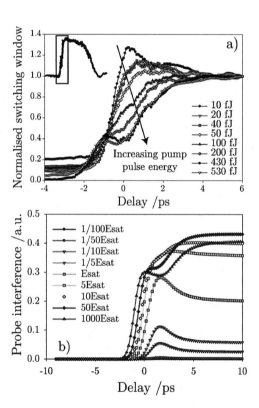

FIGURE 1. (a) Switching window edge for a range of pump pulse energies. These windows have been normalised so that their amplitudes equal unity after the ultrafast component has decayed. An example of a full switching window is shown in the upper left corner. (b) Numerical simulation of the onset of a TOAD switch for different pump pulse energies. E_{sat} is the pulse energy corresponding to a 3dB gain drop with respect to the small-signal value.

(ideally π) and give a transmitted pulse when interfering back at the input beam splitter. The switching window can thus be measured by varying the pump delay with respect to the probes (see inset in Fig. 1(a) for a typical example).

Interferometric switches based on gain depletion generally rely on the refractive index change associated with the interband resonant nonlinearity. This change may take up to a nanosecond to recover, depending on the current injection. The stimulated emission induced by the strong pump depletes the lowest energy levels in the conduction band. Simultaneously, the pump pulse increases the average energy of the electron population via two photon and free carrier absorption. These processes combined cause a rise of the carrier distribution temperature and a decrease in the available gain. Within a few picoseconds or less, carrier-phonon scattering balances the carrier and lattice temperatures. This finite thermalisation time of the carriers will also dictate the recovery of the gain/refractive index change induced by the pump. Therefore, the shape of the switching window should reveal the signature of this ultrafast component on the shortest timescales.

Our measurements focused on the onset of switching windows for different pump energies, as shown in Fig. 1(a). For weak pumps, gain depletion due to stimulated emission and the subsequent refractive index change are small resulting in a differential phase shift between probes well below π. The extra phase corresponding to the refractive index change induced by carrier heating therefore adds constructively giving a spike which decays in a few picoseconds. As the pump energy is increased, carrier density depletion alone gives a phase shift close to π, so the carrier heating contribution eventually overshoots and shows up as a dip in our measurements. So as to better display the evolution of the ultrafast component with pump energy, the switching windows in Fig. 1(a) have been normalised so that their amplitudes equal unity after the ultrafast component has decayed. As seen in Fig. 1(b), the window amplitude increases at larger pump energies until a maximum is reached when $\Delta\varphi = \pi$.

To better understand our results, a theoretical model based on rate equations has been developed. The SOA is split into thin slices and the temporal carrier dynamics induced by the incoming pump are solved for every slice. The subsequent time-dependent changes in gain and refractive index are then used to simulate the propagation of the two probes through the device. Apart from interband dynamics, this model also takes into account carrier heating effects via a gain compression factor. Such model structure facilitates the calculation of the evolution of pulses propagating in both directions, as required for the TOAD configuration. To simplify matters, it is assumed that the probe pulses are too weak to significantly modify the carrier population; being the only excitation due to the pump pulse. The switched signal is then given by the sum of the co- and counter-propagating probe pulse fields.

The model was used to simulate the evolution of the ultrafast component on the front edge of the window as the pump pulse energy is varied, see Fig. 1(b). The numerical results are consistent with experimental data, the carrier heating contribution starting as a spike and gradually becoming a dip as the pump energy is increased. However, the predicted ratio between peak transmission and the level reached after the ultrafast component has relaxed, is much larger than observed experimentally. This discrepancy will be the subject of further work. Note how the carrier heating contribution significantly reshapes the switching window edge. This may impose a limitation to the minimum window width attainable.

In conclusion, the SOA intraband dynamics become increasingly important for the device performance as subpicosecond pulses are used to boost the speed of all-optical processing applications. It is therefore crucial to understand their implications and opportunity for switching operation on the shortest timescales. The experimental signature of a refractive ultrafast component in the switching windows of a TOAD setup is reported. The timescale (~ 2 ps) and evolution of this feature with pump pulse energy point to carrier heating as the cause and was found to be consistent with the predictions of a numerical model. These results should also be relevant to other SOA-based switching schemes, such as Mach-Zehnder interferometers [4], etc.

REFERENCES

1. J. P. Sokoloff, P. R. Prucnal, I. Glesk, and M. Kane, *IEEE Photon. Technol. Lett.*, **5**, 787–790 (1993).
2. S. Nakamura, K. Tajima, and Y. Sugimoto, *App. Phys. Letters*, **65**, 283–285 (1993).
3. H. J. S. Dorren, X. Yang, D. Lenstra, G. D. Khoe, T. Simoyama, H. Ishikawa, H. Kawashima, and T. Hasama, *IEEE Photon. Technol. Lett.*, **15**, 792–794 (2003).
4. S. Nakamura, Y. Ueno, and K. Tajima, *App. Phys. Letters*, **78**, 3929–3931 (2001).
5. A. Mecozzi, and J. Mørk, *J. Opt. Soc. Am. B*, **14**, 761–770 (1997).

Terahertz Induced Photoconductivity of 2D Electron System in HEMT at Low Magnetic Field

Andrey Chebotarev and Galina Chebotareva

OpthUS, P.O. Box 20042, Stanford, CA 94309, USA

Abstract. A few results of our study of two-dimensional electron system (2DES) in low magnetic fields in GaAs/GaAlAs heterostructures by cyclotron resonance (CR) and photoconductivity techniques are presented. We have first discovered "CR-vanishing effect" in 2DES as well-defined crevasse on CR line in low magnetic fields, when Hall resistance is not quantized. "CR-vanishing effect" indicates vanishing longitudinal resistance & conductivity in these magnetic fields. Observed "CR-vanishing effect" demonstrates new correlated state of electrons in 2DES.

INTRODUCTION

Our study of 2D electron systems in low magnetic field in GaAs/AlGaAs HEMT heterostructures, having fundamental physics interest, is directed to development of new kind of THz devices. First results have been obtained by us since 1992 when we started to study Cyclotron Resonance (CR) of 2DES with high mobility at low magnetic fields. Interest of scientific society to behavior and properties of 2DES in GaAs/AlGaAs at low magnetic fields has been dramatically increased lately due to "microwaves induced vanishing resistance effect" discussed in literature [3].

RESULTS AND DISCUSSION

We have measured CR and Shubnikov-de-Haas (SdH) oscillations induced by THz radiation in samples of GaAs/GaAlAs heterostructures (MBE), with electron density $3-5 \times 10^{11} cm^{-2}$, mobility 10^5 and 10^6 $cm^2 V^{-1} s^{-1}$ at low magnetic fields around 0.3 T, at temperature $4,2K^0$ and frequencies 0.13 – 0.15 THz by photoconductivity technique (Fig.1-3). We have observed the vanishing of photocurrent & photovoltages as well-defined crevasse on CR line before CR maximum position in 2DES with mobility 10^6 $cm^2 V^{-1} s^{-1}$ (Fig1). We have observed that effect both in photovoltaic and photoconductivity measurements. The effect keeps value under attenuation THz power on 10 dB (Fig. 2), but does not appear in samples with lower mobility 10^5 $cm^2 V^{-1} s^{-1}$ (Fig.3). "CR-vanishing effect" indicates vanishing longitudinal resistance & conductivity (going to zero) in the limited regions of low magnetic fields. Hall resistance (R_H) has not been quantized at these low magnetic fields vs. R_H in experiments [4] at high magnetic fields. Observed "CR-vanishing effect" demonstrates new correlated state of electrons in 2DES.

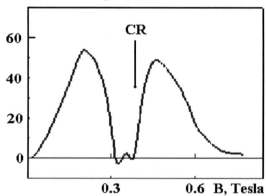

FIGURE 1. Magnetospectrum CR at 0.13 THz for 2DES with mobility 10^6 $cm^2 V^{-1} s^{-1}$

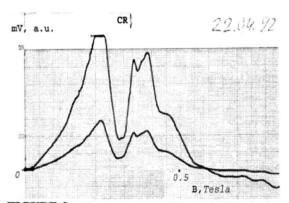

FIGURE 2. Magnetospectrum CR at 0.14 THz for 2DES with mobility 10^6 $cm^2 V^{-1} s^{-1}$. Lover curve was been measured under attenuation THz power on 10 dB. The both curves are similar in low fields including CR-vanishing region.

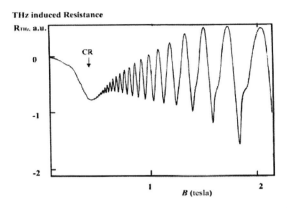

FIGURE 3. THz induced negative resistance in 2DES. CR and SdH oscillations caused by incident THz electromagnetic waves in 2DES with lower mobility (10^5 $cm^2V^{-1}s^{-1}$) at temperature $4,2K^0$ and F=0.13 THz. The magnetospectrum demonstrates very sharp line of CR without peculiarities except regular SdH oscillations that resolved up to maximum of CR-line.

Our research demonstrates that THz /Microwave radiation causes both photocurrent and photovoltages in 2DES of GaAs/AlGaAs heterostructures in low magnetic fields Therefore, "microwaves induced vanishing resistance effect" (discussed now in literature [3]) could be dependent on a value and sign of photovoltages induced by this radiation in the absence of applied driving current (Fig.4). Our results indicate that the photovoltaic response is connected with the sample surface area instead of edge effects mentioned in [3, 5]. We have observed CR-vanishing effect both in photovoltaic and photoconductivity measurements.

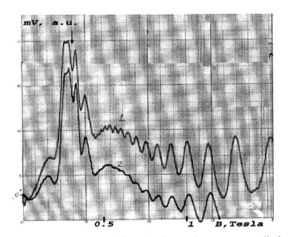

FIGURE 4. THz induced photovoltages at applied driving current (1) and photovoltages in the absence of applied driving current (2) in 2DES of GaAs/AlGaAs heterostructures with mobility 10^6 $cm^2V^{-1}s^{-1}$ at temperature $4,2K^0$ in low magnetic fields.

CONCLUSION

Our investigations of GaAs/AlGaAs heterostructures by cyclotron resonance and photoconductivity techniques have shown that the magnetospectra have the complex structure that depends on 2D-electrons' mobility & concentration, frequency and polarization of the incident radiation. We have demonstrated that THz/Microwave radiation induces both photocurrent and photovoltages in 2DES of GaAs/AlGaAs heterostructures. "CR-vanishing effect" (CRV) in 2DES at low magnetic fields, when the Hall resistance is not quantized, have been discovered by us. Observed "CR-vanishing effect" demonstrates new correlated state of electrons in 2DES. This CRV state is not depended on incident THz power.

ACKNOWLEDGMENTS

Authors are thankful to professors A.M.Dykhne, V.I Gavrilenko, E. Gornik, K. von Klitzing, R.A.Stradling, and V.I.Ryzhii for useful discussions and interest to our research in different years.

REFERENCES

1. A.P.Chebotarev " *Millimeter Spectroscopy of 2D-electron System in GaAS/GaAlAs Heterostructures*" Digest of 4th Intern. Conf. On Millimeter and Submillimeter Waves and Applications SPIE ,1998, p.112
2. A.P.Chebotarev, G.P.Chebotareva, A.P.Nikitin, E.Gornik. *"Deformation of cyclotron resonance spectrum of 2D-electron gas in GaAS/AlGaAs heterostructures at heating by pulsed electric field"* International Symposium "Nanostructures : Physics and Technology", Sankt-Petersburg, Russia, 1995, pp.38-41.
3. R. Fitzgerald " *Microwaves Induce Vanishing Resistance in Two-Dimensional Electron Systems*" Physics Today, April 2003, pp.24-27
4. D. Stein, G. Ebert, K. von Klitzing and G. Weimann *"Photoconductivity on GaAs/AlGaAs heterostructures "*. // Surface Science 142, 1984, pp. 406-411
5. C.T.Liu, B.E.Kane, D.S.Tsui. *"Far-infrared photovoltaic effect in a Landau level diode"*. // Appl. Phys. Lett. 55(2), 1989, pp.162-164

Observation of the Quantum Correction Term for the Microwave Magnetoresistance in Weakly Doped Ge

A.I.Veinger, A.G.Zabrodskii, T.V.Tisnek, and S.I.Goloshchapov

Ioffe Physico-Technical Institute, Russian Academy of Sciences, St.Petersburg, 194021, Russia

Abstract. The anomalous magnetoresistance (MR) of the interference origin was first observed in the weakly doped nongenerated n-Ge:Sb and p-Ge:Ga. Early proposed for the MR detection, the electron spin resonance technique was used for this investigation. It was found that the anomalous MR was observed in these materials in the temperature range of the classical (nonhopping) motion and in the magnetic fields not higher than some hundred Oersteds. Some peculiarities of the classical MR were observed and explained too.

INTRODUCTION

It is known that the resistance change in magnetic fields (magnetoresistance (MR)) can be caused by the different reasons. The classical MR resulted from the carrier trajectory change in the magnetic field is observed for the classical (nonhopping) electron transport in the weakly doped semiconductors. On the contrary, the so-called "anomalous" MR is observed usually when the classical MR is absent: in the metallic state and the hopping conductivity range. At the same time, the anomalous MR caused by the self-crossing quantum trajectory interference of the quantum correction theory [1,2] not needs necessarily in the high carrier density and could be detected in the weakly doped semiconductors. However, the low sensitivity and the low precision of the measurement technique cannot permit to detect this term of the MR.

We proposed the new technique for MR measurements based on the electron spin resonance (ESR) one [3]. This highly sensitive and precision enough un-contacted technique permits to measure the MR derivative with respect to magnetic field on the microwave band (~10 GHz). Early, it was used for anomalous MR investigation of highly doped semiconductors [3]. This report is devoted to the MR research in the weakly doped n- and p-Ge with $n=1.2\cdot10^{14}$ cm^{-3} and $p=3.5\cdot10^{14}$ cm^{-3} correspondingly in the range of classical (nonhopping) transport.

It is known from the electric networks theory that the microwave power absorption derivative dP/d$H \propto$ dρ/dH when $R_g>R$ and dP/d$H\propto$dσ/dH when $R>R_g$, (ρ,σ -resistivity and conductivity of the sample material, R_g, R–generator and sample resistance). In our case the relation $R>R_g$ remained always and so dP/d$H\propto$dσ/dH in the whole temperature range of the measurements. The investigations were made from $T=8$ K when the current carriers were appeared in the c- and v-bands up to $T=120$ K when all found low temperature MR peculiarities were decreased. The ESR spectrometer E-112 "Varian" with the flow cryostat ESR-9 "Oxford Instruments" was used for investigations.

EXPERIMENTAL RESULTS AND DISCUSSION

Figures 1, 2 show dP/d$H\propto$dσ/dH dependencies for the n- and p-Ge samples correspondingly.

Note, that It is known that from theory that dP/d$H\propto$dσ/dH=0 exactly in the point $H=0$ and dσ/d$H\propto-H$ in the weak fields. For more clear imagine, the curves are shifted along the y-axis. It is seen that these curves have some extremes at the low temperatures and varies with temperature. At the high temperature all curves consist one minimum corresponding to the classical MR.

One can write for the classical MR:

$$\mathrm{d}P/\mathrm{d}H \propto \mathrm{d}\sigma/\mathrm{d}H \propto \begin{cases} \propto -H,\ \mathrm{d}P/\mathrm{d}H<0\ \text{when}\ (\mu H)^2 \ll 1 \\ \to 0,\ \mathrm{d}P/\mathrm{d}H<0\ \text{when}\ (\mu H)^2 \gg 1. \end{cases} \quad (1)$$

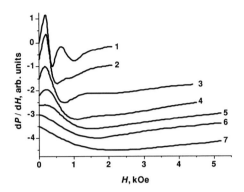

FIGURE 1. Field dependencies of microwave absorption derivative for n-Ge $n=1.2 \cdot 10^{14}$ cm^{-3}; T, K: 1–8, 2–10, 3–15, 4–20, 5–35, 6–50, 7–80.

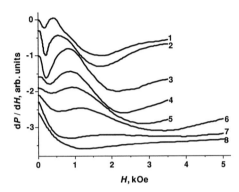

FIGURE 2. Field dependencies of microwave absorption derivative for p-Ge $p=3.5 \cdot 10^{14}$ cm^{-3}; T, K: 1–8, 2–10, 3–15, 4–20, 5–35, 6–50, 7–80, 8–120.

It means that the derivative curve has one minimum when

$$H_{ext} \approx 1/\mu = m/e\tau, \quad (2)$$

where μ - carrier mobility, m – carrier effective mass, τ - momentum relaxation time.

The quantum correction theory gives for the derivative for the MR interference term:

$$dP/dH \propto d\sigma(H)/dH = \begin{cases} (G_0/(D\tau_\varphi)^{1/2}) \, x/48; & x \to 0 \\ 0.151(G_0/(D\tau_\varphi)^{1/2}) \, x^{-1/2}; & x \to \infty, \end{cases} \quad (3)$$

where $G_0 = e^2/(2\pi\hbar) = 1{,}233 \cdot 10^{-5}$ Ом$^{-1}$, $x = 4eD\tau_\varphi H/\hbar c$, D–diffusion coefficient, τ_φ-dephasing time; the derivative sign is positive for the n-Ge and negative for p-Ge as a result of the more strong spin-orbit interaction in the last.

It is seen from (3) that this dependence has one extreme too at $x \approx 1$, or:

$$H_{ext} \approx \hbar c/4eD\tau_\varphi. \quad (4)$$

Thus, the field dependencies for n-Ge must reveal the maximum in the weak fields and minimum in the strong fields and for p-Ge – two minimums. Figures 1 and 2 confirm this conclusion at the low temperatures. When the temperature increases up to 35 K, only one minimum for the classical MR is observed for n-Ge. But for p-Ge, two minimums conserve until to temperature 80 K. This effect is connected with the light and heavy holes what are observed as the separated (no mixed) particles at so high frequencies until to such temperatures. In such case, the similar effect in n-Ge reveals until to 35 K and connects with the electrons in the different valleys of the conductivity band.

To distinguish experimentally the classical and the anomalous (quantum interference) MR terms, we studied the second derivative of the microwave absorption. For the classical MR:

$$d^2P/dH^2 \propto d^2\sigma(H)/dH^2 \propto \mu^2 \propto \tau^2 \propto -T^{-3}. \quad (5)$$

This relation fits well with the experiment at the high temperatures both for n-Ge and for p-Ge.

For anomalous MR under the dominant acoustical phonon scattering:

$$d^2P/dH^2 \propto (D\tau_\varphi)^{3/2} \propto T^{-9/4}. \quad (6)$$

This relation fits well with the experiment too at the low temperatures where this effect is observed.

Thus it was found that the anomalous interference MR was displayed together with the classical one in the lightly doped (nondegerate) semiconductor. These two different MR mechanisms gave the microwave absorption extremes in the different fields thus permitting to observe the both effects in the same experiment.

AKNOWLEGEMENTS

We thank the Russian Foundation for Basic Research (Grant 04-02-16587), RF President Foundation (Grant 223.2003.02), and Russian Academy of Sciences for support.

REFERENCIES

1. B.L.Altshuler and A.G.Aronov. In book: Electron – electron interactions in disordered systems, ed. by A.L.Efros and M.Pollak, North-Holland, 1985, p.1.
2. T.A.Polyanskaya, Ju.V.Shmartsev. Sov. Phys. *Semiconductors*, **23**, (1989), 3.
3. A.I.Veinger, A.G.Zabrodskii, T.V.Tisnek. Semicond. **36**, 772, (2002).

Microwave Time Resolved Cyclotron Resonance With Nanosecond Resolution

H. E. Porţeanu, O. Loginenko, and F. Koch

Physik-Department E16, TU München, Germany

Abstract. We developed a technique that allows to study the dynamics of free charge carriers in an intermediate time range namely between nanoseconds and milliseconds at microwave frequencies. The existing femtoseconds spectroscopy deals usually with optical relaxation and formation of excitons. In nanosecond range there are different processes visible like diffusion of carriers and interaction with acoustic phonons.

INTRODUCTION

An accurate determination of the electromagnetic field absorption can be done only in resonant conditions, where interferences are compensated. The existence of free or partially trapped carriers changes not only the conductivity, understood as real part, but also the dielectric constant i.e. the resonant conditions. Therefore the standard way to measure absorptions in microwave range is the cavity perturbation method. This works only in stationary regime of the resonating system (cavity +sample) and gives the changes of Q-factor and of the resonance frequency. S. Grabtchak [1] introduced an algorithm to determine quick changes of the resonance curve in a much shorter time than that of the slow scanning of frequency. One measures a reference resonance curve (Fig. 1, "dark") of the unperturbed resonator. Then instead of scanning, one measures the time dependent changes in the amplitude at different frequencies. The dynamic resonance curve is then computed.

ROOM TEMPERATURE INVESTIGATIONS

Two examples at 300 K show the strength of this method. Free photoexcited electrons in Ge contribute not only to an increase of the conductivity but also to the decrease of the dielectric constant (Fig. 2). Using the Drude formula for conductivity one gets $\tau = -\varepsilon_0 \Delta\varepsilon_r / \Delta\sigma \approx 2 \cdot 10^{-12}$ s. Porous TiO_2 shows on contrary an increase of the dielectric constant. The result is understood as localization of electrons near traps, having their free frequency higher than the driving microwave frequency. An analysis based on measurements in different ambient conditions gives a trap binding energy of the order of 5 meV. The sign of $\Delta\varepsilon_r$ is determined also by the size of nanoparticles, as we found in porous silicon [2].

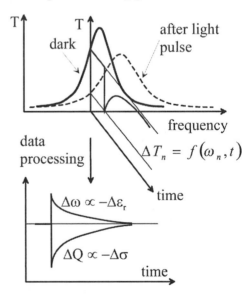

FIGURE 1. Principles of time resolved complex conductivity measurement. The time dependent changes in amplitude are used to determine the dynamic resonance curve. Fitting with Lorentz curve gives ΔQ and $\Delta\omega$.

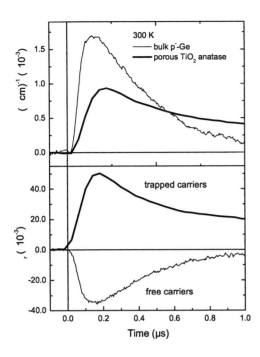

FIGURE 2. Room temperature time variation of σ and ε_r for free and trapped carriers.

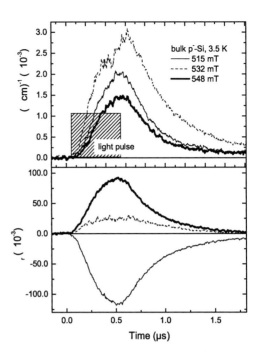

FIGURE 3. Time dependence of the real and imaginary ($\propto \Delta\varepsilon_r$) part of conductivity in bulk Si (100). $\Delta\varepsilon_r$ has an opposite sign left and right of CR.

TIME RESOLVED CR

The same procedure can be applied at low temperatures and in magnetic field. One can see in Fig. 3 that the localization in this case is given by the strength of the magnetic field. (positive or negative $\Delta\varepsilon_r$). We used intentionally a long light pulse (1.37 eV, 0.5 µJ, 500 ns) and very low microwave power (35 GHz, < 1µW) in order to evidence the relaxation of electrons after the optical excitation. The scattering on optical phonons takes usually place in femtoseconds time. However, the interaction with acoustic phonons for energies less than 50 meV takes place in nanoseconds range. This process was evidenced in Fig. 4. The spectra in Fig 4 were obtained from many transients presented in Fig. 3, measured at each indicated magnetic field. The variation of $\Delta\varepsilon_r$ is here not shown. If we associate the width of the CR curve with the scattering time, $2/\tau = e\Delta B/m_c$, we conclude that it increases within 500 ns from $1.3 \cdot 10^{-10}$ s to $2.2 \cdot 10^{-10}$ s. At higher microwave powers we observe a heating of electrons and therefore the opposite variation of τ.

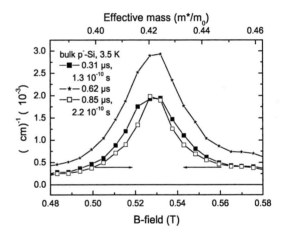

FIGURE 4. CR spectra of bulk Si (100). The arrows indicate the narrowing of the CR curve ("cooling of electrons").

REFERENCES

1. S. Y. Grabtchak and M. Cocivera, Phys. Rev. B **50**, 18219 (1994).

2. H. E. Porţeanu, et al., MRS Symp. Proc., **699** R4.8 (2002).

Generation and Remote Detection of Coherent Folded Acoustic Phonons

M. Trigo[*], T. A. Eckhause[*], J. K. Wahlstrand[*], R. Merlin[*],
M. Reason[†] and R. S. Goldman[†]

[*]*Department of Physics and* [†]*Department of Materials Science and Engineering,
University of Michigan, Ann Arbor MI 48109, USA*

Abstract. Coherent acoustic modes with frequencies in the THz range were generated in a superlattice using femtosecond optical pulses and detected, with little attenuation, after traversing 1.2 µm of bulk GaAs.

Folding of the Brillouin zone in superlattices allows one to excite monochromatic sound waves in the THz range using femtosecond laser pulses [1]. This is in contrast to sound created by the absorption of optical pulses in thin metal films, which is usually broadband and restricted to lower frequencies, typically a few hundred GHz [2]. The generation of THz modes holds promise for phonon imaging applications and studies of thermal transport in nanostructures [3] as well as a means to reduce the lifetime of longitudinal optical (LO) phonons [4]. The use of superlattices as a source of folded acoustic modes, however, has been limited by a lack of understanding of their behavior at interfaces, particularly, at the substrate-superlattice boundary, and by insufficient data available on the mean-free path of high-frequency sound. Here, we present experimental studies of the propagation of coherent longitudinal acoustic phonons with frequencies near 1 THz through bulk GaAs.

We consider two samples grown by molecular beam epitaxy. Sample A consists of a 25 period, [001]-oriented AlAs (34Å)/GaAs (12Å) superlattice, a sandwiched layer of 1.2-µm-thick GaAs, and an identical 25 period superlattice; see Fig. 1. Sample B is the same as sample A except that its sandwiched GaAs layer is 0.6-µm-thick. Our experiments were performed at 77 K. Spontaneous Raman spectra obtained with an Ar$^+$-ion laser at 514.5 nm reveal the first folded-phonon doublet at 0.96 THz and 1.06 THz and, in addition, the confined GaAs- and AlAs-like LO phonons. Photoluminescence spectra show a relatively broad feature centered at ~ 530 nm which is due to near gap recombination in the superlattice.

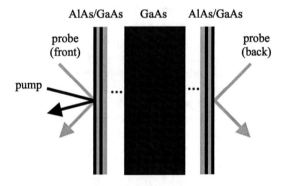

FIGURE 1. Schematic diagram of the sample. A thick GaAs layer is sandwiched between two superlattices. Folded phonons are generated in the superlattice at the front of the sample and detected, both, on the front and, after propagating through the bulk, on the backside.

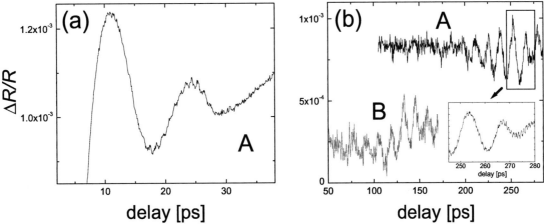

FIGURE 2. Time-domain data for samples A and B. Pump pulses generate sound waves on the front superlattice, which are later detected on the front (a) and the backside (b) of the sample. Trace (a) and the inset of (b), obtained on sample A, show folded-phonon oscillations superimposed on the larger, low-frequency oscillations associated with stimulated Brillouin scattering (see text).

To measure the propagation of the folded phonons, we performed time-resolved pump-probe experiments using an optical parametric amplifier which provided laser pulses of width ~ 60 fs. The central wavelength of the pulses, λ_C, was tuned to resonate with the bandgap of the superlattice at 530 nm. As shown schematically in Fig. 1, the pump pulses strike the front superlattice generating coherent sound, which we detect by measuring the pump-induced change in the reflectivity of the probe beam. Results are shown in Fig. 2. Figure 2(a) shows oscillations due to coherent phonons in the front superlattice soon after the arrival of the pump pulse. As shown in Fig. 2(b), after a transit time of ~230 ps and ~ 115 ps for samples A and B, respectively, we observe a transient change in reflectivity due to the arrival of sound at the backside superlattice. These values agree extremely well with the times it takes for sound to cross the bulk GaAs layer in the two samples. The power spectrum in Fig. 3, for the trace corresponding to sample A in Fig. 2(b), shows a dominant peak at ~ 67 GHz and the folded-phonon doublet at ~ 1 THz. Note that the relative change in reflectivity due to the arrival of folded phonons at the backside is comparable to that they produce on the same side where they were generated. This shows that ~ 1 THz acoustic phonons can propagate across 1.2 μm of GaAs with little attenuation. The low-frequency mode is ascribed to sound generated by light absorption, the spectrum of which is expected to be broadband [2]. The reason why we observe a single frequency is that light can only couple to sound of wavevector $q \sim 4\pi n/\lambda_C$ (n is the refractive index). The coupling mechanism is stimulated Brillouin scattering.

ACKNOWLEDGMENTS

This work was supported by the AFOSR under contract F49620-00-1-0328 through the MURI program.

REFERENCES

1. K. Mizoguchi, M. Hase, S. Nakashima, and M. Nakayama, *Phys. Rev. B* **60**, 8262 (1999).
2. H. T. Grahn, H. J. Maris, J. Tauc, *IEEE J. Quant. Electron.* **25**, 2562 (1989).
3. D. G. Cahill et al., *J. Appl. Phys.* **93**, 793 (2003).
4. J. Chen, J. B. Khurgin, R. Merlin, *Appl. Phys. Lett.* **80**, 2901 (2002).

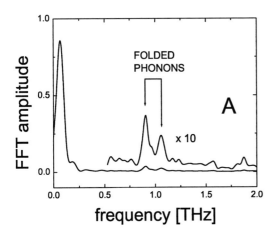

FIGURE 3. Fourier transform of the time-resolved trace for sample A in Fig. 2(b). The peak at ~ 67 GHz is due to coherent Brillouin scattering.

Direct Experimental Evidence of the Hole Capture by Resonant Levels in Boron Doped Silicon

S. T. Yen[1], V. Tulupenko[1,2], E. S. Cheng[1], A. Dalakyan[1,2], C. P. Lee[1], K. A. Chao[3], V. Belykh[2], A. Abramov[2] and V. Ryzhkov[2]

[1]*Department of Electronics Engineering, National Chiao Tung University, Hsinchu, 30050 Taiwan*
[2]*Donbass State Engineering Academy, Kramatorsk, 343913, Ukraine*
[3]*Depattment of Physics, Lund University, Sylvegutun 14A, S 23362 Lund, Sweden*

Abstract. The variation of hole population for the localized and resonant states of B-doped Si excited by sequences of short electric–field pulses has been investigated by the technique of time-resolved spectroscopy. The capture of holes by resonant levels is verified. A new spectral line earlier theoretically predicted was experimentally observed for the first time.

INTRODUCTION

There is a renewed interest in investigation of intra-centre optical transition for impurities in both bulk and QW structures. It is connected with discovering a THz radiation from uniaxially stressed p-Ge [1,2] and strained SiGe [3] QWs under a strong electric field. It is believed that the radiation arises as a result of inversion population of holes between resonant and localized impurity levels of shallow acceptors. The main point for the inversion to occur is a capture of holes by resonant levels while the localized levels are being emptied due to their impact ionization. Meanwhile there was not any direct experimental evidence for a resonance capture of the holes. Thus, the main task of the work was to get the evidence for capture of holes by resonant level(s) in a strong enough electric field.

EXPERIMENT

The idea of our experiment was to compare a change in the absorption between $p_{3/2}$ set of spectral lines in a gap and resonant $p_{1/2}$ set for boron-doped Si under applied electric fields by using the time-resolved spectroscopy. The optical phonon energy for Si (ε_0=63 meV) is larger than the spin-orbit splitting (Δ=44 meV), meaning that holes accelerated by the electric field can reach resonant levels attached to the split-off subband without being scattered by optical phonons. The capture of holes must result in decreasing absorption for corresponding spectral lines.

The samples were cut from Chochralsky grown along 100 crystallographic axis boron-doped Si wafer with the thickness 0.5 mm and with the room temperature resistivity 8 Ohm*cm. A continuous flow cryostat with the sample to which Ohmic contacts had been made was put into the IFS66v/s FTIR spectrometer. The experiments were made in a temperature range from 20 to 77 K. To avoid extremely high Joule overheating of the sample, the electric field was applied in a burst mode with pulse duration 0.2 µs and a duty cycle 0.01. There are 50 pulses in each burst. The repetition rate of the bursts was 15.6 Hz. Special experiments showed that overheating caused an increase in temperature no more than 4 K for the strongest electric field (2 kV/cm) at the initial temperature 27 K. We used a Si bolometer as a detector. Together with a set of filters it allows us to measure spectra of interest in the range of 240-685 cm^{-1} during one scan of the spectrometer. 100 coadditions were used to improve signal to noise ratio.

DISCUSSION

Delta absorbance for the first 10 (0-9) time slices with 40 µs between nearest ones taken at 27 K is shown in Fig.1. A similar, but not so prominent for $p_{1/2}$ set of lines, picture was also found for higher temperatures. It is seen that absorption for $p_{3/2}$ series is increased for the first 4 time slices and then it starts monotonously decreasing to approximately 250 µs after the end of the burst as it is shown in Fig.2 with dashed line. There are 2 reasons for such temporal dependencies of delta absorption. Firstly it might be connected with depletion of exited localized levels by electric field. And secondly, with still cooling down the sample for initial time slices due to the heat inertia (conditioned by heat capacitance consisting of the sample and copper pieces to which the sample was soldered with thick layer of In) and only after delta absorption passes the zero the temperature starts increasing. To check the last assumption we analyzed temporal dependence for delta absorption in the range of multiphonon absorption. This range (600-620 cm^{-1}) is particularly convenient as absorption here is only affected by temperature and not by the electric field. The results are presented in Fig.2. In this figure the time course of the delta absorption for 605 cm^{-1} at the same (27 K) temperature is shown. At this point the 1-st derivative $\partial A/\partial \nu$ has a local maximum for the multiphonon absorption and, consequently, the point is extremely sensitive to the change in temperature. Fig.2 clearly shows that the temperature is indeed being decreased for the beginning of a burst and only after about 250 µs it begins increasing. For convenience we also put in Fig.2 the time course for a line 4 belonging $p_{3/2}$ series as well.

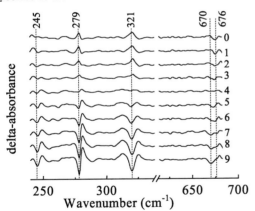

FIGURE 1. The delta electroabsorption spectra taken at different time slices for boron-doped silicon excited by sequential electric-field pulses. T=27 K, E=1850 V/cm.

Now let us consider lines near 670 cm^{-1}. At equilibrium (in the absence of electric fields) there is only one line at 670 cm^{-1}. However, one can see two lines for delta absorption spectra in Fig.1 We suppose that for the first time the line at 676 cm^{-1}, which was firstly theoretically predicted by Buzhko [4] was observed.

This line cannot be seen for equilibrium spectra as it corresponds to forbidden transition for the states with the same parity. With an electric field the selection rule is not longer valid and the line becomes observable. As to this peak being down for the first 4 slices when the temperature is still decreased, we can put forward the only reasonable explanation. Namely we believe that it is due to the fact that free holes accelerated by the electric field are captured by resonant levels. In such a way population of the levels is enhanced and that involves decreasing absorption. The further decreasing absorption (after 5-th) slice can be explained both by partial population of the resonant state and by decreased number of holes in the ground state with rising temperature.

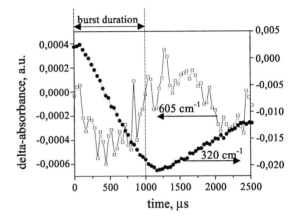

FIGURE 2. Time course of the lines: at 605 cm^{-1}-left scale and at 320 cm^{-1} – right scale. The temperature and electric field are the same as those in Fig.1.

CONCLUSION

For the first time it has been experimentally shown the capture of holes by resonant levels in p-type Si. Besides, a new spectral line earlier theoretically predicted was observed.

REFERENCES

1. I.V. Altukhov, E.G.Shirkova, M.S. Kagan, at al., Sov. Phys. JETP **74**, 404 (1992).
2. V.N. Bondar, A.T. Dalakyan, L.E. Vorob'ev, at al., JETP Letters, **70**, 265,(1999).
3. M.A. Odnoblyudov, I.N. Yassievch, M.S. Kagan, at al., Phys. Rev. Letters **83**, 644 (1999).
4. R. Buszko and F. Bassani, Phys. Rev. B **45**, 5838 (1992).

Two-phonon infrared processes in semiconductors

Hadley M. Lawler* and Eric L. Shirley[†]

*Department of Physics, University of Maryland, College Park, MD 20742
[†]Optical Technology Division, NIST, Gaithersburg, MD 20899-8441

Abstract. We report our detailed calculation of the infrared spectra of GaAs and GaP from first principles, and similar ongoing efforts for diamond-type spectra, which have recently been calculated.

LATTICE DYNAMICS

Over the last decade or so, density-functional perturbation theory (DFPT) and application of the $2n+1$ theorem have revolutionized the theoretical study of lattice dynamics [1]. Frozen-phonon algorithms have also been successful in calculating complete phonon dispersion relations [2]. These advances allow detailed knowledge of the vibrational continuum, to complement the realistic calculation of electronic band structure that has been possible for some time.

Beyond the harmonic approximation, with DFPT it has been possible to calculate anharmonic effects, such as widths, shapes, and shifts of Raman modes [3], and two-phonon absorption spectra in silicon and germanium [4]. Higher-order anharmonic effects can be treated by combining DFPT and frozen-phonon methods, and developing both methods may prove important to widening the scope of multiphonon interactions and their effects which can be realistically studied from theory.

Our focus has been on the frozen-phonon calculation of phonon-phonon and photon-phonon interactions. We have calculated the absorption spectra for GaAs and GaP from the far infrared through twice the frequency of the dispersion oscillator, or the zero-momentum transverse-optical phonon. The results demonstrate temperature dependence of the response in the far infrared, the asymmetric absorption profile of the dispersion oscillator and rich spectral structure caused by infrared light coupling to two-phonon states. The correspondence with experimental spectra are shown in Figure 1. We have also been able to relate prominent features in the two-phonon absorption spectra to very simple kinematical parameters of the primitive cell, the mass ratio of the basis ions in particular. This work combines calculations of Raman widths–as has been accomplished with DFPT [3]–with the modern theory of polarization [5] to calculate absorption spectra completely *ab initio*.

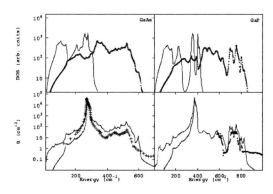

FIGURE 1. The one-phonon (line) and zero-momentum two-phonon (+) densities of states are plotted in top panels. In bottom panels, the calculated (line) and measured (+) absorption coefficients are plotted at 300 K for GaAs, and < 10 K for GaP. To point out the far-infrared temperature dependence, the theoretical spectra, in the far infrared, are plotted at low (GaAs) and room (GaP) temperatures (light dots). Plotted theoretical spectral features have been moved to 0.98 of their calculated position to facilitate comparison with the experimental spectra. The data are from Refs. [11, 12].

Currently we are working on a frozen two-phonon calculation of the weak infrared absorption in diamond-type materials. Calculations of this nature [6, 7], and a thorough DFPT calculation [4], have been reported. A point worth emphasizing is that, with our two-phonon calculation, the adiabatic evolution gives no change in the ionic moment with respect to the supercell origin; that is, the ionic contribution to the "current" is zero, because the the ionic displacements are given by standing waves.

The work here is within the frozen-phonon approach. In particular, the dynamical matrix and anharmonic matrix elements are calculated with wave-commensurate supercells, which map a phonon's periodicity to a supercell [8]. Born-von Karman force matrices can then be calculated as finite differences of Hellmann-Feynman forces. The technique has the advantage that force matrices are sampled in reciprocal space, and then interpolated, as is also the case in applications of DFPT. This allows implementation of the LO-TO splitting according to long-established formalisms [9], and avoids aliasing.

ZINCBLENDE CRYSTALS

Each of the above infrared-active processes is related to the interaction between zero-momentum two-phonon states and some macroscopic coordinate. In the case of GaAs and GaP, within the present theory, it is only the dispersion oscillator which is directly coupled to the macroscopic field, and the two-phonon states acquire oscillator strength through their anharmonic coupling to the dispersion oscillator. The relevant matrix element for this coupling is $M_{\alpha\beta\mathbf{k}} = \sum_{\tau\tau'ii'} \frac{\partial D_{ii'}^{\tau\tau'}(\mathbf{k})}{\partial u} \varepsilon_{\tau i}^{\mathbf{k}\alpha} \varepsilon_{\tau'i'}^{-\mathbf{k}\beta}$. The derivative of the dynamical matrix, $D_{ii'}^{\tau\tau'}(\mathbf{k})$, is taken with respect to the sublattice relative coordinate, and phonon polarizations are given on the right. Phonon branches are denoted α and β, Cartesian directions i and i', and basis ions τ and τ'. The matrix element above is actually the cubic term in the Born-Oppenheimer potential which gives interaction between three normal modes of wave vectors $\mathbf{k}, -\mathbf{k}$, and 0.

The two-phonon states, in addition to imprinting two-phonon character upon the spectra, also act as decay channels, giving the dispersion oscillator absorption feature finite linewidth and asymmetric shape. Hence, the anharmonic theory of infrared spectra in polar materials accounts for a broad class of infrared optical properties, from sensitive temperature dependence of the far-infrared absorption, to the profile of the reststrahlen and the two-phonon features. The dielectric function is [10]:

$$\varepsilon_{ii'}(\omega) - \varepsilon_{ii'}^{\infty} = \frac{4\pi}{\Omega} \sum_{\nu=TO} \sum_{\tau\tau'jj'} \frac{Z_{ij}^{\tau} Z_{i'j'}^{\tau'} \varepsilon_{\tau j}^{\nu} \varepsilon_{\tau'j'}^{\nu}}{2\sqrt{m_\tau m_{\tau'}} \omega_{TO}}$$
$$\left(\frac{1}{\omega + \omega_{TO} + \Sigma_{TO}(\omega)} - \frac{1}{\omega - \omega_{TO} - \Sigma_{TO}(\omega)} \right). \quad (1)$$

The numerator on the right side contains the Born effective charges, and the denominator contains the ionic masses. The frequency of the dispersion oscillator and its self-energy appear in parentheses. The self-energy is temperature-dependent, and is calculated via diagrams [10] and the matrix elements defined above. The results for GaAs and GaP are plotted in Figure 1, along with the experimental spectra.

DIAMOND-TYPE CRYSTALS

The Born effective charges express the integrated oscillator strength in Eq. 1, and are zero in a diamond-type material. Thus the absorption spectra follow from a different type of interaction. In fact, the electric field, E, at infrared frequencies is a macroscopic coordinate independent of the zero-momentum optical phonon coordinates, and the key quantity for the spectral calculation is, $\frac{\partial D_{ii'}^{\tau\tau'}(\mathbf{k})}{\partial E}$. Such calculations were recently performed with DFPT and methods for incorporating macroscopic fields into electronic calculations [4].

Another approach toward a calculation of the spectra is to evaluate the polarization change due to a two-phonon displacement: $\Delta P_i = \frac{\partial^2 P_i}{\partial u_{\mathbf{k}\alpha} \partial u_{-\mathbf{k}\beta}} u_{\mathbf{k}\alpha} u_{-\mathbf{k}\beta}$, with the modern theory of polarization. Initial work in this direction has been reported [6], and the absorption spectrum for amorphous silicon has been calculated with molecular dynamics and a time-domain analysis of the polarization [7]. Using wave-commensurate supercells, ionic displacements represent standing waves, and are appropriate to the momentum-conserving two-phonon geometry. For the diamond-type structure, overtones of the six branches make the desired quantity a third-rank tensor, which we are calculating in a Cartesian basis.

REFERENCES

1. Baroni, S., de Gironcoli, S., Dal Corso, A., and Giannozzi, P., *Rev. Mod. Phys.*, **73**, 515 (2001).
2. Kresse, G., Furtmüller, J., and Hafner, J., *Europhys . Lett.* **32**, 729 (1995).
3. Debernardi, A., Baroni, S., and Molinari, E., *Phys. Rev. Lett.* **75**, 1819 (1995).
4. Deinzer, G. and Strauch, D., *Phys. Rev. B* **69**, 045205 (2004).
5. King-Smith, R. D. and Vanderbilt, D., *Phys. Rev. B* **47**, 1651 (1993); Resta R., *Rev. Mod. Phys.* **66** , 899 (1994).
6. Strauch, D., Windl, W., Sterner, H.,Pavone, P., and Karch, K., *Physica B* **219-220**, 442 (1996).
7. Debernardi, A., Bernasconi, M., Cardona, M., and Parrineelo, M., *App. Phys. Lett* **71**, 2692 (1997).
8. Lawler, H. M., Chang, E. K., and Shirley, E. L., unpublished.
9. Gonze, X. and Lee, C., *Phys. Rev. B* **55**, 10 355 (1997).
10. Van Hove, L., Hugenholtz, N. M., and Howland, L. P., *Quantum Theory of Many-Particle Systems*, W.A. Benjamin, New York, 1961; Wallis, R. F. and Maradudin, A. A., *Phys. Rev.* **125**, 1277 (1962); Cowley, R. A., *Adv. Phys.* **12**, 421 (1963); Vinogradov, V. S., *Sov. Phys.–Solid State* **4**, 519 (1962).
11. Palik, E. D., *Handbook of Optical Constants of Solids*, edited by Palik, E. D., Academic, London, 1998.
12. Thomas, M. E., Blodgett, D., Hahn, D., and Kaplan, S., *SPIE Proceedings* **5078**, Windows and Domes Technologies VIII, April 22-23 (2003).

Time Resolved Cyclotron Resonance Studies In Semiconductors With Very Low And Very High Mobility

H. E. Porţeanu, O. Loginenko, and F. Koch

Physik-Department E16, TU München, Germany

Abstract. The aim of this work is to present two extreme examples where the dynamics of correlated carriers can be identified. Trapped electrons and holes in GaAs show a Fano-type shape of the CR, strongly dependent on density. Free electrons in Ge diffuses after optical excitation and redistribute their occupation on conduction band valleys.

INTRODUCTION

The study of transport with interacting charge carriers has a long history. The correlated motion, for example, in a high mobility 2D electron gas leads to the fractional quantum Hall effect. We used the time resolved cyclotron resonance technique [1, 2] in order to point out the dynamics of these correlations effects on nanoseconds till milliseconds scale.

EXPERIMENTAL SETUP AND RESULTS

The measurements have been done at 3 – 6 K and 35 GHz. The samples with a size of 2 ×2×0.05 mm are placed in the center of a cylindrical resonator and illuminated through an optical fiber with UV (3.67 eV, 5 ns, 10 µJ) or IR (1.37 eV, 150 ns, 50 nJ) pulses.

Coupled Cyclotron Resonance in bulk GaAs due to trapped centers

The GaAs sample is a Cr compensated semi-insulating substrate. We measured the real and imaginary part of conductivity successively with 2 light pulses above (UV) and below (IR) the E_g of GaAs. The time dependencies at two magnetic fields are plotted in Fig. 1. Especially for IR excitation, the shape of the decays are strongly dependent on magnetic field.

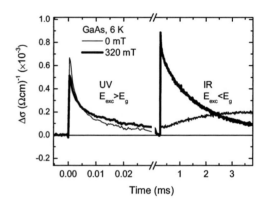

FIGURE 1. Time dependence of the real part of conductivity in bulk GaAs. The excitation energy strongly influences the dynamics.

The CR spectra are presented in Fig. 2. While for UV excitation the whole spectrum decays proportional with the carrier density, for IR excitation the shape changes substantially. An analysis of the real and imaginary part suggests a correlated motion of electrons and holes. At small densities remain predominantly trapped carriers. Their motion is influenced by the dipole - dipole coupling of electron-trap to hole - trap.

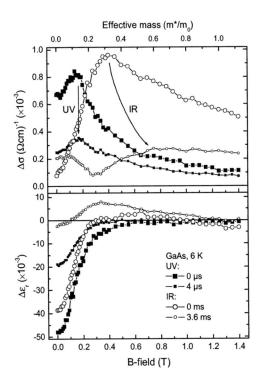

FIGURE 2. CR spectra for bulk GaAs for excitation above (UV) and below (IR) bandgap. The IR-excited CR changes not only in amplitude but also in shape with the decrease of the carrier density.

Electron Intervalley Scattering in Ge

The dynamics of electrons was investigated also in high purity Ge ($N_p < 10^{10}$ cm^{-3}). The time evolution of the complex conductivity is presented in Fig. 3 for 2 magnetic fields representing the 2 resonances ((111) orientation) for IR excitation.

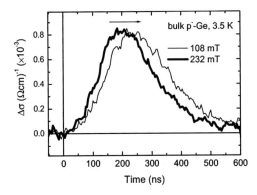

FIGURE 3. Time dependence of the real part of conductivity in bulk Ge (111). The first electronic resonance appears delayed compared to the second one.

The resulted CR spectra (Fig. 4) show an unchanged shape of the resonance, however, the relative intensities varies differently. The intervalley scattering was studied long time ago [3] in terms of redistribution from nonequilibrium to equilibrium (equal occupation). We observe directly how the diffusion process of photogenerated carriers in CR conditions leads to a nonequilibrium in the valley occupancy. The derived scattering time is of the order of 50 ns (see Fig. 3). This effect is slightly stronger for UV excitation (smaller penetration depth). In the same conditions there is no detectable redistribution between light and heavy holes.

Similar results were measured for silicon.

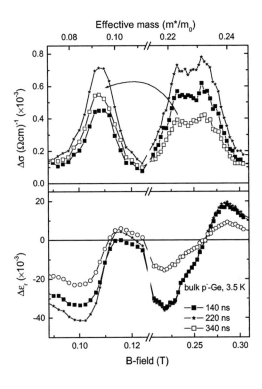

FIGURE 4. CR spectra of bulk Ge (111). The arrow indicates the transfer of electrons (intervalley scattering).

REFERENCES

1. H. E. Porţeanu, et al., MRS Symp. Proc., **699** R4.8 (2002).
2. H. E. Porţeanu, et al., ICPS 27 Symp. Proc., 2004.
3. A. A. Kapliansky, et al., Solid State Comm. **20**, 27 (2004).

THz Emitter Based On InAs-GaSb Coupled Quantum Wells: New Prospects For THz Photonics

Boris Laikhtman and Leonid D. Shvartsman

The Racah Institute of Physics, The Hebrew University, Jerusalem, 91904, Israel

Abstract. We suggest new types of THz emitter design that promise better performance than used so far.

THz radiation has very wide range of highly important applications in such fields of high importance as medical imaging, security, extra-sensitive biochemical spectroscopy, etc. Nevertheless, the actual progress in the field is hindered by the absence of low cost reliable sources similar to semiconductor lasers. Here we suggest new types of laser design that have some advantages as compared to those considered so far.

From the available experience and general consideration it is possible to conclude that an efficient lasing can be achieved in designs that meet at least one of two general requirements: 1. The structure has to be bipolar with the major gap lying in THz range. The radiating transition in such a structure takes place between the conduction and valence bands with a relatively large transition matrix element. 2. The radiative transitions have to take place between a number of pairs of equidistant subbands. It is easy to engineer structures with subband separation in THz range and an increase of the number of the transition rises the efficiency. Usually the first requirement is met in p-n junction lasers. But typical band gap in semiconductors is well beyond THz range. The second requirement is met in cascade lasers, but their efficiency is not large due to a small intersubband transition matrix element. We suggest designs based on InAs-GaSb coupled quantum wells, the only structure that is able to combine the advantages of both the p-n junction laser and the cascade laser. This combination is achieved by the engineering of a system of equidistant levels in InAs quantum well (QW) and resonant with them effective band gap between the conduction band of InAs QW and the valence band of GaSb QW. InAs-GaSb coupled QWs reveal a number of additional degrees of freedom in laser design that makes it indispensable in THz range. We demonstrate these features in two possible designs. One of them combines the interband and intersubband lasing in the same cascade (compare Ref.[1]) and while the other makes use of singularly high reduced density of states for interband optical transitions.

The most striking feature of InAs-GaSb based heterostructures is the 150 meV overlap between the conduction band of InAs and the valence band of GaSb. In wide enough coupled QWs the quantum size effect reduces this overlap but does not eliminate it. The resulting band crossing leads to the hybridization gap and W-dispersion[2] (Fig,1). The typical width of the hybridization gap is of the order of 10 meV, i.e., 3 THz and this gap can be used for lasing. Reduction of the QW width eliminates the band overlap and leads to usual quasi-parabolic V-dispersion. The band gap can be easily varied from zero up to a few hundred of meV. This gap also can be used for THz lasing. [3]

To illustrate the application of both the W- and the V-dispersion we consider two structures. The first one has a regular V-dispersion but to increase the gain in this laser we create another pair of levels in the InAs well, e2 and e3, with the same energy separation as between e1 and hh1. The small separation between e2 and e3 is created with the help of the InAlAs step in this well. To achieve population inversion between these levels the energy separation between e2 and e1 is made equal to the LO phonon energy. The second one has W-dispersion and the corresponding peculiarity in reduced density of states also results in better laser characteristics.

Below we study the gain of the lasing pairs of levels and compare these results with those for only intersubband or only interband lasing.

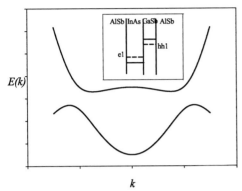

FIGURE 1. W dispersion in overlapping electrons and holes.

The material gain in this structure is the sum of the gains of the interband and intersubband transitions,

$$g = (4\pi^2 e^2 / m_0^2 \omega n_{eff} cL)$$
$$\times [v_{opt}(\omega)|M_{ib}|^2 (f_e + f_h - 1) + (2\tau/\pi)|M_{isb}|^2$$
$$\times (n_{e3} - n_{e2})],$$

where L is the total width of the QWs, n_{eff} is the refraction index of the composite material, ω is the radiation frequency, and M_{ib} and M_{isb} are transition matrix elements. The interband gain is proportional to the optical density of states $v_{opt}(\omega)$ that for the V-dispersion is m/π^2 where m is the reduced mass. The intersubband gain is proportional to the difference of the electron concentrations between the levels. This follows from the equidistance between the two dispersion curves, so that all electrons are able to participate in the radiative transitions and the transition probability is proportional to the electron relaxation time τ. The analysis of the gain for a system with a number of couples of lasing levels differs substantially from the ordinary case. The lasing may occur between levels 4-3 while the population inversion for another couple 2-1 is not reached yet. The meaning of the transparency threshold current is also different. One has to distinguish between integral transparency threshold current, J_{tr}, when the total gain is equal to zero, and transparency threshold currents for each couple of lasing levels. When the transparency threshold is surpassed for both pairs of levels the intersubband dominates at $J > J_{tr}$ and interband gain dominates at $J < J_{tr}$. According to our estimates J_{tr} is about 22 A/cm^2. A nonparabolicity makes J_{tr} larger, in the region about 100 A/cm^2 both contributions to the gain are close to each other. That is the threshold current is lower and the gain is higher than for each of the two level systems separately.

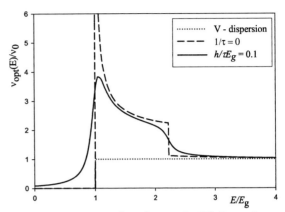

FIGURE 2. Density of states for W dispersion

The advantage of the design represented in Fig.2 is even more obvious. The density of states has a singularity, $V_{opt}(\omega) = 2m\omega/\pi \sqrt{(\omega)^2 - E_g^2}$, and at the threshold the photon emission probability is infinite. Scattering smears this singularity that gives $V_{opt} = (m/\pi^2)\sqrt{\omega\tau}$ at the threshold. If τ is 1ps then in comparison with the V-dispersion the probability of the photon emission increases by $\sqrt{\omega\tau} \approx 2.5$ times for 1 THz.

Our analysis illustrates the great prospects for THz laser structures based on InAs-GaSb coupled QWs. The variety of possible designs is rather high. The most straightforward development is the combination of both factors: interband -intersubband resonance and W-dispersion in the same structure. k-space cascade.

We acknowledge the support of the Israel Science Foundation founded by the Israel Academy of Sciences and Humanities

REFERENCES

1. N. Ulbrich, G. Scarpa, G. Böhm, G. Abstreiter, and M.-C. Amann, Appl. Physi. Lett. **80**, 4312 (2002).
2. B.Laikhtman, S. de-Leon, and L.D.Shvartsman, Solid State Commun. **104**, 257 (1997); Phys.Rev.B **60**, 1861 (1999).
3. J. D. Bruno, R. Q. Yang, J. L. Bradshaw, and J. T. Pham, D. E. Wortman, *Proc.SPIE* Vol. 4287 (2001)

Theory of Heterostructural Tunnel Emitters for Ballistic Transit-Time Terahertz-Range Oscillators

Zinovi S. Gribnikov*[¶] and George I. Haddad[¶]

Institute for Quantum Sciences, Michigan State University, East Lansing, Michigan 48824, USA,
[¶]*Department of EECS, University of Michigan, Ann Arbor, Michigan 4810, USA*

Abstract. We consider the high-frequency tunnel transparence of a rectangular heterostructural barrier perturbed by a small high-frequency potential. Is shown that this transparence experiences a very substantial growth for the intermediate amplitudes of the above-mentioned potential.

Transit-time oscillators with tunnel electron emission into a transit space (T-space), the so-called TUNNETT-oscillators, were proposed [1] almost half a century ago and repeatedly implemented [2-4]. A Zener tunnel $p^{++}n^+$-emitter as the electron source has been used in almost all of these implementations (excluding only one case [3]). Such an emitter requires a big bias voltage dropping mainly on a T-space (~7-9 V) [2, 4]. As a result, electron transportation across this space is always diffusive (with a drift velocity $\leq 2 \times 10^7$ cm/s) and does not allow one to reach oscillatory frequencies, $f = \omega/2\pi$, exceeding [4] $3.25 \times 10^{11} s^{-1}$. Recently, it has been suggested [5,6]:
1. To use thin semiconductor layers ($\leq 10^{-5}$ cm) with sufficiently high positions of noncentral valleys as T-spaces in order to provide ballistic or quasiballistic electron transport across T-spaces with saturated ballistic velocities $\geq (1-2) \times 10^8$ cm/s.
2. To use comparatively thin heterostructural tunnel barriers as tunnel emitters, that is, to change a $p^{++}n$-diode with a Zener tunnel emission through a bandgap by a unipolar n^+n-diode with a tunnel barrier adjoining an n^+-cathode.

The utilization of a thin T-space allows us to decrease electron transit times to $T_D = (0.25-1.0) \times 10^{-13}$ s and not only to reach the THz range but also to go deep into this range. Note the specificity of the ballistic transport across the T-space: in the case of sufficiently high voltages across this space, electrons overcome it having an invariable saturated ballistic velocity. Therefore, it is senseless to increase too much the voltage dropping across the T-space: such an increase does not result in increasing electron velocity but results only in an increase in the power dissipation. On the other hand, it is useful that the tunnel emitter injects electrons in the T-space with an initial velocity nearing the above-mentioned saturated ballistic velocity. From this point of view, heterostructural rectangular tunnel barriers with sharp barrier/T-space interfaces [6] are more preferable than quasi-triangular (Fowler-Nordheim) barriers with smooth boundaries.

With deepening into the THz-range, one more problem appears: a substantial frequency dependence of a tunnel emission for a vast area of the typical semiconductor tunnel-barrier parameters occurs just in the THz-range [7-9]. As a rule, in this case the subject of investigation is a direct tunnel current (DC) in the tunnel barrier exposed to an alternating (high-frequency) electric field. This exposure leads to an increase in a tunnel probability and, as a result, in the DC. Such an increase is usually interpreted as an absorption of the high-frequency photon, $\hbar\omega = hf$, by the tunneling electron during the time of tunneling. The absorption means a certain increase in the energy of the tunneling electron and in the barrier transparence. As a result, the DC increases. However, we are interested not only in the behavior of the DC but also and especially in the behavior of the alternating (high-frequency) tunnel current component induced by the same alternating (high-frequency) electric field [10-11]. In Ref.10 we have considered the model problems of a rectangular tunnel barrier ($0 \leq x \leq w$) with the invariable thickness, w, and with the oscillating height (see also [8]):

$$\varepsilon_0(t) = \varepsilon_0 + \varepsilon^{(1)} \cos\omega t. \quad (1)$$

or of the same stationary tunnel barrier containing inside the oscillating δ-barrier:

$$\varepsilon_0(x,t) = \varepsilon_0 + a\delta(x-x_1)\cos\omega t \quad (2)$$

where $0 \leq x_1 \leq w$. Finally, in Ref. 7 we have considered the same barrier with the added oscillating homogeneous electrical field:

$$\varepsilon_0(x,t) = \varepsilon_0 + eE_B x\cos\omega t. \quad (3)$$

We have assumed also that the energy of tunneling electron ε is noticeably smaller than the barrier height, ε_0, and $(\varepsilon_0-\varepsilon)$, $\varepsilon \gg hf$. The one-electron current can be presented for the above-mentioned three cases in the form:

$$j^{(1,2,3)} = j_0[1 - \Gamma^{(1,2,3)}\cos(\omega t - qx)] \quad (4)$$

Where j_0 is a stationary current component. The multipliers $\Gamma^{(1,2,3)}$ in Eq. (4) correspond to three versions of the oscillatory modulation of the tunnel barrier, described by Eqs. (1), (2) and (3) respectively and are equal to:

$$\Gamma^{(1)} = 4(\varepsilon^{(1)}/2\hbar\omega)\sinh Qw, \quad (5)$$

$$\Gamma^{(2)} = (a/a_0)\cosh Qw, \quad (6)$$

$$\Gamma^{(3)} = 8A\sinh^2(Qw/2)/Qw, \quad (7)$$

$$A = eE_B w/\hbar\omega, \quad Q = g\hbar\omega/2(\varepsilon_0-\varepsilon),$$

$$g = [2m(\varepsilon_0-\varepsilon)]^{1/2},$$

and m is the electron effective mass in the barrier. Equation (6) with $a_0 = a_0(\varepsilon)$ is written only for the case $x_1 = w$ when the oscillatory modulation is maximal. If $Qw \ll 1$ [that is $f \ll 2(\varepsilon_0-\varepsilon)/hgw$,] any frequency dependence in the formulae (5)-(7) disappears. They become quasistatic. The quasistatic derivation of Eq. (4) requires the obligatory small values of $\Gamma^{(1,2,3)}(f\rightarrow 0)$ $\ll 1$. In the opposite case we should add to the right sides of Eq. (4) higher harmonics with frequencies $n\omega$, $n>1$, in order to take into account the effect of nonlinearity of the static barrier transparence.

If $Qw \geq 1$ [that is $f \geq 2(\varepsilon_0-\varepsilon)/hgw$], we can observe the exponential increase in the coefficients $\Gamma^{(1,2,3)}$, which determine the high-frequency transparence of the tunnel barrier on the invariable static background. To obtain the growth of the static component of the tunnel current, we should take into account the components of the second order, which are proportional to $(\varepsilon^{(1)})^2$, a^2 or A^2. These components describe the above-mentioned tunneling with the absorption of the hf-photon taking place for an intensive high-frequency radiation. In the specific case of the current $j^{(3)}$ [see Eqs. (4) and (7)], consideration of the A^2-component leads to the formula:

$$j^{(3)}/j_0 = 1 + 4A^2 F_0(z) - 4AF_1(Z)\cos(\omega t-qx) +\ldots, \quad (8)$$

$F_0(Z) = (\cosh Z - 1) + \frac{1}{2} - \sinh Z / Z$, $F_1(Z) = (\cosh Z - 1)/Z$, $Z = Qw$. In Eq. (8) we omit the components corresponding to higher harmonics, which are nonexistent for our consideration. The absorption of the hf-photon predominates if

$$A\exp(Qw)/Qw \gg 1 \quad (9)$$

and leads to very substantial increase in the stationary current component. However we are interested in the intermediate situation when

$$A\exp(Qw)/Qw \cong 1. \quad (10)$$

In this case we have the maximal relation of the high-frequency tunnel current component to the stationary component for the given values of ω, ε, ε_0, w and other parameters of the barrier. This maximal relation nears 1: the high frequency current amplitude nears the stationary component. Such an effect is possible for the sufficiently thick barriers: $gw \gg 1$. This extreme of relative high-frequency transparence is very sharp because we deal with exponential dependencies.

This exponential high-frequency growth of the coefficients $\Gamma^{(1,2,3)}$ could be attributed to the virtual process: the electron absorbs the hf-photon during tunneling in the barrier and emits this photon to exit.

REFERENCES

1. J. Nishizawa and Y. Watanabe, Sci. Rep. Res. Inst. Tohoku Univ. A**10**, 91 (1958).
2. J. Nishizawa, K. Motoya, and Y. Okuno, IEEE Trans. Microwave Theory. Tech. **26**, 1029 (1978); H. Eisele and G. I. Haddad, IEEE Trans. Microwave Theory. Tech. **42**, 2498 (1994); ibid **43**, 210 (1995). ibid. **46**, 739 (1998); C. Kidner, H. Eisele, and G. I. Haddad, Electron. Lett. **28**, 511 (1992).
3. T. Bauer, M. Rosh, M. Claassen, and W. Harth, Electron. Lett. **30**, 1319 (1994).
4. P. Plotka, J. Nishizawa, T. Kurabayashi, and H. Makabe, IEEE Trans. Electron Devices, **50**, 867 (2003).
5. Z. S. Gribnikov, N. Z. Vagidov, V. V. Mitin, and G. I. Haddad, J. Appl. Phys. **93**, 5435 (2003); Physica E, **19**, 89 (2003).
6. Z. S. Gribnikov, N. Z. Vagidov, and G. I. Haddad, J. Appl. Phys. **95**, 1489 (2004).
7. R. Landauer and Th. Martin, Rev. Mod. Phys. **66**, 217 (1994).
8. M. Buttiker and R. Landauer, Phys. Rev. Lett. **49**, 1739 (1982).
9. B. I. Ivlev and V. I. Mel'nikov, Zh. Eks. Teor. Phys. **90**, 2208 (1986) {Sov. Phys. JETP, **63**, 1295 (1986)].
10. Z. S. Gribnikov and G. I. Haddad, J. Appl. Phys.**96** (accepted),(2004).
11. Z. S. Gribnikov and G. I. Haddad, (unpublished.) (2004).

Submillimeter Radiation - Induced Persistent Photoconductivity in $Pb_{1-x}Sn_xTe(In)$

Aleksander Kozhanov*, Dmitry Dolzhenko*, Ivan Ivanchik*, Dan Watson**, Dmitry Khokhlov*

Physics Department, Moscow State University, Moscow 119992, Russia
**Department of Physics and Astronomy, University of Rochester, 14627 New York, USA*

Abstract. Persistent photoconductivity in a $Pb_{0.75}Sn_{0.25}Te(In)$ alloy initiated by monochromatic submillimeter-range radiation at wavelengths of 176 and 241 µm was observed at helium temperatures. This photoconductivity is shown to be associated with optical excitation of metastable impurity states.

INTRODUCTION

Most of the presently used high-sensitivity non-thermal detectors intended for use in the far infrared range are based on doped silicon and germanium. The longest cutoff wavelength in such radiation detectors that has been reached in uniaxially strained Ge(Ga) was $\lambda_r = 220$ µm [1]. A viable alternative to the silicon- and germanium based radiation detectors is offered by lead telluride–based narrow-gap semiconductors. Doping lead telluride and its solid solutions by Group III elements gives rise to effects not characteristic of the starting material, such as Fermi level pinning and persistent photoconductivity [2]. In particular, the Fermi level in $Pb_{1-x}Sn_xTe(In)$ alloys with a tin content of $0.22 < x < 0.28$ is pinned within the band gap providing a semiinsulating state of the semiconductor at low temperatures. Because the effect of Fermi level pinning brings about homogenization of the electrophysical parameters of the material and because the characteristic energy parameters of the alloy, such as the band gap width and the impurity-state activation energy, are of the order of a few tens of millielectronvolts, the possibility of employing these semiconductors as radiation detectors in the far IR range appears extremely attractive. This possibility was realized [3], and it was found that the responsivity of a $Pb_{0.75}Sn_{0.25}Te(In)$-based IR radiometer substantially exceed those of their counterparts based on doped silicon and germanium.

However, the key question of the spectral response of the $Pb_{1-x}Sn_xTe(In)$-based radiation detector, in particular, of the red cutoff of the photoeffect in this material, has remained unexplored.

It was demonstrated in [4], that the radiation with wavelengths of 90 and 116 µm gives rise to strong persistent photoresponse in $Pb_{0.75}Sn_{0.25}Te(In)$ at the liquid helium temperature. An important point mentioned in [4] was that the radiation energy quantum corresponding to the used wavelengths is smaller than the thermal activation energy of the ground impurity state that pins the Fermi level. Therefore the persistent photoconductivity was defined in this case by the excitation of metastable, but not the ground local states.

However, the interpretation of the results mentioned above has remained open to question. Indeed, the thermal activation energy E_a was calculated from an analysis of the temperature dependence of the resistivity by using the relation $\rho \sim \exp(E_a/2kT)$ rather than $\rho \sim \exp(E_a/kT)$, as is usually accepted when of dealing with impurity states. The calculations made by using the first of the above relations were based on a study of the character of the pressure-induced motion of the impurity level [5]. However, this substantiation is of an indirect nature. If the activation energy of an impurity level is calculated using the second relation, the value of E_a will be one-half of that obtained from the first relation; i.e., it will correspond to a wavelength of 140 µm. Therefore, the conclusion that the metastable impurity states provide a major contribution to the photoresponse at wavelengths of 90 and 116 µm will be invalid.

EXPERIMENTAL RESULTS

In this paper, we report on the detection of a photoresponse in a $Pb_{0.75}Sn_{0.25}Te(In)$ film at wavelengths of 176 and 241 µm. $Pb_{0.75}Sn_{0.25}Te(In)$ films were grown through molecular-beam epitaxy on a BaF_2 substrate. The thermal activation energy of the impurity ground state calculated from the relation $\rho \sim \exp(E_a/2kT)$ was 20 meV.

Figure 1 plots the experimental data obtained for a sample voltage of 10 mV and a blackbody temperature of 300 K.

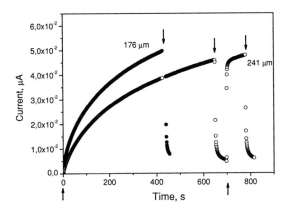

FIGURE 1. Kinetics of the rise and fall of the photocurrent plotted for a voltage of 10 mV across the sample and different radiation wavelengths. The arrows indicate the times the IR illumination is turned on and off.

A noticeable photoresponse was detected at both wavelengths of the radiation striking the sample. Because our sensitive measuring equipment made it possible to record currents of up to 0.25 µA, we could reliably measure the kinetics of the increase in the photocurrent only at low voltages across the sample, $U < 40$ mV. At higher voltages, the photocurrent grew so fast that the amplifier became overloaded in a time comparable to the time required to rotate the filter disk, i.e., within a few seconds.

DISCUSSION

The following features in the photoconductivity may be of interest here. First of all, the current rise observed after switching on the illumination follows a strongly nonlinear kinetics. Switching off the illumination triggers a fast decay of the photocurrent, with subsequent slow relaxation to the dark level. If, however, the illumination is switched on again a short time after its removal, the photocurrent rises very fast (in a time comparable to the "fast" relaxation time) to the value recorded just before the illumination removal, after which the previous, relatively slow dynamics of the increase in the photocurrent sets in again. The fast and the slow processes are apparently of essentially different natures. Another feature may also be noteworthy. The energies of photons corresponding to radiation wavelengths of 176 and 241 µm are substantially less than the thermal activation energy of the impurity ground state, even if this energy is calculated using the relation $\rho \sim \exp(E_a/kT)$. Thus, the results obtained in this study provide direct evidence for the persistent photoconductivity in $Pb_{1-x}Sn_xTe(In)$ originating from photoexcitation of metastable impurity states. The threshold energy for optical excitation of these states is very low. The wavelength of the corresponding photon is longer than at least 241 µm, which, as far as we know, is the largest for nonthermal radiation detectors. The photoconductivity cutoff of the materials studied here apparently lies at substantially longer wavelengths. One cannot rule out the possibility that the operating range of $Pb_{1-x}Sn_xTe(In)$-based photodetectors extends over the whole submillimeter range.

ACKNOWLEDGMENTS

This study was supported in part by the Russian Foundation for Basic Research (projects 04-02-16497, 02-02-17057, 02-02-08083), INTAS (project 2001-0184), and a NATO Collaborative Linkage Grant..

REFERENCES

1. .Haller, E. E., Hueschen, M. R., and Richards, P. L., *Appl. Phys. Lett.* **34**, 495 (1979).
2. Volkov, B. A., Ryabova, L. I., and Khokhlov, D. R., *Usp. Fiz. Nauk* **172**, 875 (2002) [*Phys. Uspekhi* **45**, 819 (2002)].
3. Chesnokov, S. N., Dolzhenko, D. E., Ivanchik, I. I., and Khokhlov, D. R., *Infrared Phys.* **35**, 23 (1994).
4. Khokhlov, D. R., Ivanchik, I. I., Raines, S. N., Pipher, J. L., and Watson, D. M., *Appl. Phys. Lett.* **76**, 2835 (2000).
5. Akimov, B. A., Zlomanov, V. P., Ryabova, L. I., Chudinov, S. M., and Yatsenko, O. B., *Fiz. Tekh.*

THz/subTHz Detection by Asymmetrically-Shaped Bow-Tie Diodes Containing 2DEG Layer

Dalius Seliuta*, Vincas Tamošiūnas*, Edmundas Širmulis*, Steponas Ašmontas*, Algirdas Sužiedėlis*, Jonas Gradauskas*, Gintaras Valušis*‡, Paul Steenson§, Wai-Heng Chow§, Paul Harrison§, Alvydas Lisauskas¶, Hartmut G. Roskos¶ and Klaus Köhler◊

*Semiconductor Physics Institute, A. Goštauto 11, 01108 Vilnius, Lithuania
§IMP, School of Electronic and Electrical Engineering, University of Leeds, Leeds LS2 9JT, United Kingdom
¶Physikalisches Institut, J.W. Goethe-Universität, Robert-Mayer Str. 2-4, 60054, Frankfurt/M, Germany
◊Fraunhofer-Institut für Angewandte Festkörperphysik, Tullastr. 72, 79108 Freiburg, Germany

Abstract. We present asymmetrically-shaped bow-tie diodes based on a modulation-doped GaAs/AlGaAs structure. One of the bow-tie leaves is metallized in order to concentrate the incident radiation into the apex of the other half which contains the 2DEG layer: Here the electrons are heated non-uniformly by incident radiation inducing a voltage signal over the ends of the device. The diode sensitivity at room temperature within 10 GHz – 0.8 THz is close to 0.3 V/W, while with an increase of frequency up to 2.52 THz it decreases due to weaker coupling. We consider options to improve the operation of the device.

INTRODUCTION

As a rule, two dimensional electron gas (2DEG) embedded in a GaAs/AlGaAs or GaN/AlGaN field-effect transistor can well be used to detect non-resonant THz/sub-THz radiation [1]. In this work, we present a novel sensor – asymmetric bow-tie diode based on a GaAs/AlGaAs modulation-doped structure – for this range of frequencies.

DESIGN OF BOW-TIE DIODES

The shape and the structure of the device are given in Fig. 1. One of the bow-tie leaves is metallized in order to concentrate the incident radiation into the apex of the active half which contains the 2DEG layer: Here the electrons are heated non-uniformly by incident radiation inducing a voltage signal over the ends of the device.

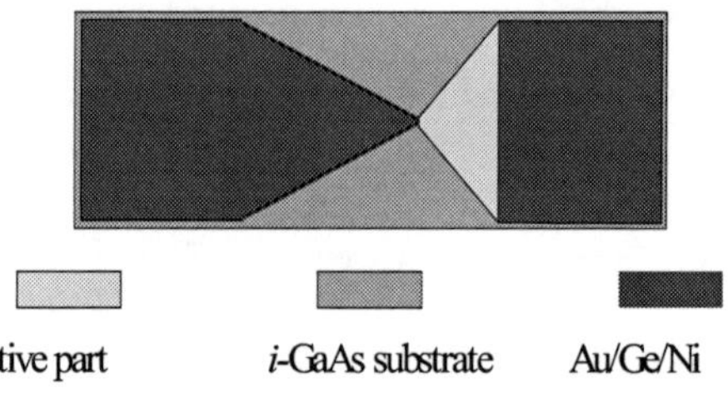

FIGURE 1. Device shape. The diode is processed into a 2 μm depth mesa and acts itself as a coupler of the radiation. The length of the device is 500 μm, the width is 100 μm whilst the size of the apex is 12 μm. The length of active part amounts of 50 μm. **Structure of the active part** (from the top) for 2DEG-A and 2DEG-B diodes, respectively: 20 nm *i*-GaAs cap layer; 80 nm Si-doped, 10^{18}cm^{-3}, layer of $Al_{0.25}Ga_{0.75}As$ and 60 nm Si doped, 2×10^{18}cm^{-3}, of $Al_{0.3}Ga_{0.7}As$; undoped spacers, 45 nm $Al_{0.25}Ga_{0.75}As$ and 10 nm $Al_{0.3}Ga_{0.7}As$; 1000 nm and 600 nm of *i*-GaAs; twenty and six periods of 9 nm AlGaAs/1.5 nm-GaAs layers; 0.5 μm and 0.6 μm of *i*-GaAs; semi-insulating substrate.

‡ author to whom correspondence should be addressed: valusis@pfi.lt

CP772, *Physics of Semiconductors: 27th International Conference on the Physics of Semiconductors*,
edited by José Menéndez and Chris G. Van de Walle

RESULTS AND DISCUSSION

The frequency dependence of the voltage sensitivity is presented in Fig. 2. One can see two characteristic ranges. The first one is from 10 GHz up to 0.8 THz, where the sensitivity is nearly independent of frequency and it is close to 0.3 V/W. The second range – above 0.8 THz – where the voltage sensitivity drops gradually. We explain this decrease by weaker coupling at higher frequencies as predicted by modeling using finite-difference time-domain method.

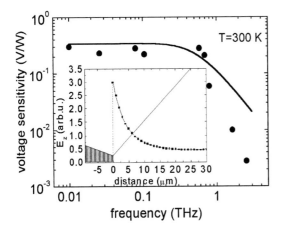

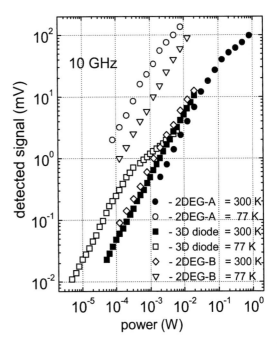

FIGURE 2. Voltage sensitivity as a function of frequency for the 2DEG-A diode. Dark circles denote experimental data, solid line shows fit using phenomenological approach [2]. **Inset**: distribution of the amplitude of electric field in the active part of the diode exposed to radiation of 0.75 THz frequency. Calculations are made using three-dimensional finite-difference time-domain method. Distance is taken from the narrowest part of the apex. The background of the insert is decorated schematically as the shape of the device.

Two options can be used to improve the voltage sensitivity. The first one relies on reducing the size of the diode apex. The experiments confirmed this prediction. For instance, when the apex size is about 800 nm, the voltage sensitivity at room temperature at 0.693 THz can reach the value of 2 V/W. The second way is to employ high mobilities of the 2DEG by cooling the diodes down to 77 K. This is evidenced by Fig. 3 where data for the sensors with different spacers are shown. One can see that at room temperature the data nearly overlap while at 77 K the signal increases proportionally to the increase of the electron mobility. The sensitivities reached are about 20 V/W for 2DEG-A diode and 5 V/W for 2DEG-B device.

The devices were studied using magnetron/klystron generators (10–37 GHz), backward wave oscillators (78–118GHz) and an optically-pumped molecular THz laser (0.584–2.52 THz). In all experiments the incident electric field E_Z was oriented along the diode.

FIGURE 3. Voltage-power characteristics of bow-tie diodes placed in 10 GHz frequency radiation at room and liquid nitrogen temperatures. Sheet electron densities and mobilities at 300 K and 77 K are, respectively: diode 2DEG-A – 5.5×10^{11}cm^{-2} and 4700 cm^2/Vs; 1.9×10^{11}cm^{-2} and 190600 cm^2/Vs; diode 2DEG-B – 5.6×10^{11}cm^{-2} and 8000 cm^2/Vs; 4.3×10^{11}cm^{-2} and 116000 cm^2/Vs. Data of a bulk n-GaAs-based device having the same design and dimensions [3] are given for comparison. Note that the signal increases linearly with incident power at both temperatures.

ACKNOWLEDGMENTS

The work was supported by NATO SfP project 978030, NATO CLG project PST.CLG.979121 and Lithuanian State Science and Studies Foundation under contracts V-04004 and V-04073.

REFERENCES

1. W. Knap, V. Kachorovskii, Y. Deng, S. Rumyantsev, J.-Q. Lü, R. Gaska, M. S. Shur, G. Simin, X. Hu, M. Asif Khan, C. A. Saylor and L. C. Brunel, *J. Appl. Phys.* **91**, 9346-9353 (2002).

2. S. Ašmontas and A. Sužiedėlis, *Int. J. Infrared Millim. Waves* **15**, 525-538 (1994).

3. A. Sužiedėlis, J. Gradauskas, S. Ašmontas, G. Valušis and H. G. Roskos, *J. Appl. Phys.* **93**, 3034-3038 (2003).

Terahertz emitters based on ion-implanted In$_{0.53}$Ga$_{0.47}$As

Masato Suzuki* and Masayoshi Tonouchi*

Institute of Laser Engineering, Osaka University, and CREST, Japan Science and Technology Corporation (JST), Osaka, Japan

Abstract. We have measured the terahertz radiated field from unimplanted and Fe-implanted dipole antenna based on In$_{0.53}$Ga$_{0.47}$As by 1.56 μm wavelength excitation of 5 and 7 mW. The maximum amplitude of the radiation waveforms from both the unimplanted and implanted emitters are found to be proportional to the applied electric field up to 5 and 12.5 kV/cm, respectively. The shape of the normalized THz radiation waveform from both the emitters is also independent of the magnitude of the applied field. Fe implantation induces a change of the THz radiated pulse width from 0.80 ps to 0.56 ps.

Among various methods as surface field generation and optical rectification effect, ultrafast switching of voltage biased photoconductive antenna by a femtosecond laser pulse is the most efficient method for generating and detecting electromagnetic field in the range of terahertz(THz) frequency .[1, 2, 3, 4, 5] The emission properties of the THz emitter are determined by the antenna structure and the material of the emitter. The frequency range of radiated field from a dipole antenna have higher components than that of radiation from a bow-tie antenna.[2] The emitter materials limit the switching time, breakdown biased field and the excitation laser wavelength for switching. low-temperature-grown GaAs (LT-GaAs) have very high breakdown field and can be switched by visible light.

Photoconductive antenna for converting infrared light including the optical communication wavelength of 1.55 μm to coherent THz radiation play an important role in new research fields of THz application and science. With the infrared emitter, fiber-optic transmission technologies including fiber-optic cable and less expensive and stable optical source is unitized for the THz technologies as time-domain spectroscopy and imaging. Baker et al. reported THz imaging system operated by the light of 1.06 μm wavelength with In$_{0.3}$Ga$_{0.7}$As-based bow-tie antenna for THz generation and detection.[6] Fe implantation to InGaAs also improve the required properties as carrier relaxation time for ultrafast devices.[7, 8, 9, 10]

In this letter, we report emission properties of lattice-matched In$_{0.53}$Ga$_{0.47}$As/InP switches excited by light of 1.56 μm wavelength, and a comparison of the properties of unimplanted and Fe-implanted InGaAs switches to develop potential infrared emitters.

A mode locked laser based on Er-doped fiber oscillator (Femtolite IMRA) was used as a optical pulse source of

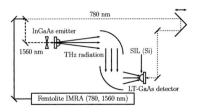

FIGURE 1. Schematic layout for the generation and detection of THz radiation.

both 1.55 μm and 780 nm wavelengths at the repetition rate of 48 MHz. The output of the fiber laser is split into two components by an optical dichroic filter. The fundamental 1.55 μm component with the pulse width of 200 fs is reflected by the filter and focused onto the gap of the dipole antenna on a THz InGaAs emitter. The other 780 nm component with the pulse width of 100 fs pass through the filter and gate a dipole antenna fabricated on a LT-GaAs detector. The emitter and the detector with the gap of 4 μm are mounted on the backs of silicon solid immersion lenses. Ultrafast switching of the biased emitter by the transmit light generates transient currents and produces a broadband THz radiation. The THz beam from the emitter is collimated and focused on the gap of the detector antenna by a pair of off-axis parabolic mirrors. At the gap of the detector, the electric field of the THz pulse acts as a transient bias field and induces transient current. Using a lock-in detection scheme referenced to the frequency of chopper for the fundamental pulses, we can observe the signal current. By delaying the timing of the gate pulse to the pump pulse, the waveform for the electromagnetic pulse was obtained.

Undoped 1.5-μm-thick In$_{0.53}$Ga$_{0.47}$As layers were

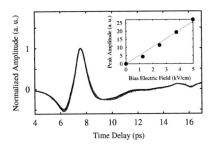

FIGURE 2. The amplitude of THz radiation from an unimplanted InGaAs emitter excited by 1.56 wavelength light of 5 mW in the applied field up to 5 kV/cm

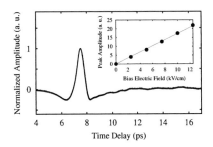

FIGURE 3. The amplitude of THz radiation from an Fe-implanted InGaAs emitter excited by 1.56 wavelength light of 7 mW in the applied field up to 12.5 kV/cm

grown by metalorganic chemical vapor deposition on semi-insulating InP:Fe substrates. Some of the layers were implanted with Fe ion of 340 keV at the dose of 1.0×10^{15} cm^{-2}. The 4 μm gap Au dipole antenna structure has been fabricated on the Fe-implanted and unimplanted layers by photolithographic technique.

Figure 2 and the inset show the normalized waveforms of the amplitude of THz radiation from the unimplanted emitter by 1.56 wavelength excitation of 5 mW at the applied fields of 1.25, 2.5, 3.75 and 5 kV/cm and linear applied field dependence of the maximum amplitude of the radiation, respectively. The pulse width of the waveforms in Fig. 2 is found to be 0.80 ps. As shown in Fig. 3, the normalized waveforms of THz radiated fields from the Fe-implanted emitter excited by the 1.56 μm light of 7 mW at the field of 2.5, 5, 7.5, 10, and 12.5 kV/cm also have no difference in the shape with the pulse width of 0.56 ps. The inset in Fig. 3 also shows that the maximum amplitude of the radiated field from the implanted emitter is also found to be proportional to the applied field. Therefore, the magnitude of the biased field in our experimental condition have no effect on the shape of the normalized waveforms of the radiated fields from both the emitters, and linear dependences of both the radiated peak amplitude. We have also found that the maximum amplitude of the radiated THz waveforms from the implanted emitter is of the similar order to that from a LT-GaAs dipole antenna excited by the light of 800 nm at the same condition. The pulse width of the radiated waveform from the implanted emitter by excitation power of 5 and 7 mW is observed as same value. Reduction in the pulse width by the implantation shows that carriers rapidly relax from conduction states to trapped states connected with the Fe ions and vacancies.

In conclusion, the THz radiation from the unimplanted and Fe-implanted dipole antenna on $In_{0.53}Ga_{0.47}As$/InP excited by 1.56 μm wavelength femtosecond laser pulse of 5 and 7 mW have been observed, respectively. With increase in the applied field up to 5 and 12.5 kV/cm, the shape of the normalized THz radiated waveform from the unimplanted and implanted emitters remain unchanged, respectively. The order of the maximum amplitude of radiation from the implanted emitter was same as that from a LT-GaAs switch excited by 800 nm wavelength under the same condition. Fe-implanted InGaAs photoswitch will be a potential candidate for application as an efficient THz emitter excited by the optical communication light of 1.55 μm wavelength.

This work was supported in part by the Strategic Information and Communications R&D Promotion Scheme Program from the Ministry of Public Management, Home Affairs, Posts and Telecommunications.

REFERENCES

1. Smith, F. W., Le, H. Q., Diadiuk, V., Hollis, M. A., Calawa, A. R., Gupta, S., Frankel, M., Dykaar, D. R., Mourou, G. A., and Hsiang, T. Y., *Appl. Phys. Lett.*, **54**, 890 (1989).
2. See, *for example, Sensing with Terahertz Radiation edited by D. Mittleman(Springer, Berlin, 2003)*.
3. Zhang, X. C., and Auston, D. H., *J. Appl. Phys.*, **71**, 326 (1992).
4. Chuang, S. L., Schmitt-Rink, S., Greene, B. I., Saeta, P. N., and Levi, A. F. J., *Phys. Rev. Lett.*, **68**, 102 (1992).
5. Suzuki, M., Kiwa, T., Tonouchi, M., Nakajima, Y., Sasa, S., and Inoue, M., *Physica E*, **22**, 574 (2004).
6. Baker, C., Gregory, I. S., Tribe, W. R., Bradley, I. V., Evans, M. J., Withers, M., Taday, P. F., Wallace, V. P., Linfield, E. H., Davies, A. G., and Missous, M., *Appl. Phys. Lett.*, **83**, 4113 (2003).
7. Carmody, C., Tan, H. H., Jagadish, C., Gaarder, A., and Marcinkevičius, S., *Appl. Phys. Lett.*, **82**, 3913 (2003).
8. Rao, M. V., Keshavarz-Nia, N. R., Simons, D. S., Amirtharaj, P. M., Thompson, P. E., Chang, T. Y., and Kuo, J. M., *J. Appl. Phys.*, **65**, 481 (1989).
9. Gulwadi, S. M., Rao, M. V., Berry, A. K., Simons, D. S., Chi, P. H., and Dietrich, H. B., *J. Appl. Phys.*, **69**, 4222 (1991).
10. Pearton, S. J., Abernathy, C. R., Panish, M. B., Hamm, R. A., and Lunardi, L. M., *J. Appl. Phys.*, **66**, 656 (1989).

Low frequency noise performance of quantum tunneling Sb-heterostructure millimeter wave diodes

Arttu Luukanen*[†], Erich N. Grossman*, Harris P. Moyer** and Joel N. Schulman**

*National Institute of Standards and Technology, 325 Broadway, Boulder, CO 80305, USA
[†]VTT Information Technology, Espoo, Finland.
**HRL Laboratories LLC 3011 Malibu Canyon Road, Malibu CA 90265-4797, USA

Abstract. Sb-heterostructure quantum tunneling diodes, fabricated from epitaxial layers of InAs and AlGaSb, are a recently proposed device for RF direct detection and mixing in the submillimeter wavelength range. These diodes exhibit especially high nonlinearity in the current-voltage characteristic that produces the rectification or mixing without bias. This is a highly desirable feature as the device does not suffer from large $1/f$ noise, a major shortcoming in other devices such as Schottky barrier diodes or resistive room temperature bolometers. In this paper we present the noise characteristics of the diode as a function of the bias voltage. At room temperature and zero bias, the device demonstrates a Johnson noise limited matched noise equivalent power of 1 pW/Hz$^{1/2}$.

INTRODUCTION

At present, millimeter wave imaging focal plane arrays are based on microwave monolithic integrated circuits (MMICs), and their pixel count is limited by the complexity and cost of the components and assembly. Moreover, the maximum operation frequency is limited by the availability of low noise amplifiers to <200 GHz. Bolometers can provide a low cost alternative and cover frequencies up to the infrared albeit with lower sensitivity [1]. Schottky diodes, the workhorse technology for high frequency mixers also can operate as direct detectors up to several THz, though typically displaying an RC-limited rolloff above a few hundred GHz. Their great drawback has been $1/f$ noise, brought about by the DC bias required for sufficient non-linearity of the I-V curve. In this paper we present the $1/f$ noise performance of a novel direct detector, the InAs/AlSb/GaSb heterostructure backward diode [2] The special feature of these diodes is that they exhibit highly non-linear current versus voltage characteristics near zero bias. In the ideal case, the absence of DC bias means that the devices do not suffer from $1/f$ noise. These properties are highly attractive for arrays of direct detectors for millimeter wave imaging.

DIODE $I(V)$ & NOISE MEASUREMENTS

Details on the fabrication can be found elsewhere [3]. Summarizing, the InAs/AlSb/GaSb tunnel diode layers are deposited using molecular beam epitaxy on semi-insulating GaAs substrates, using an InAs buffer layer process. The InAs layer is followed by a 2000 Å thick GaSb anode, 200 Å of undoped AlGaSb, a 32 Å thick undoped AlSb tunnel barrier, a 500 Å thick InAs cathode, and finally a n$^+$ InAs contact layer. Three diodes with a junction area of 2 μm × 2 μm were characterized by first measuring their DC $I(V)$ characteristics, while monitoring the differential conductance dI/dV and its derivative d^2I/dV^2 using lock-in technique. The measured $I(V)$ and dI/dV are shown in Fig 1. At $V = 0$, the three diodes had a mean junction resistance of $\overline{R}_j = (14.4 \pm 0.1 \text{ k}\Omega)$. The current responsivity (or curvature) of the diode is given by $\gamma = (d^2I/dV^2)/(dI/dV)$, and for the three measured diodes we obtained a mean curvature of 39.5 A/W at zero bias. A measured $\gamma(V)$ for one diode is shown in the inset of Fig. 1. The divergences at 43 mV and at 111 mV correspond to the points where dI/dV crosses zero.

Measurements of the diodes' noise can be used in combination with the $I(V)$ measurements to determine the true figure of merit for a direct detector, the noise equivalent power (NEP). The diode was again biased in series with the 10 kΩ current sensing resistor R_L using a programmable battery power supply. As the rectified RF power will produce a small positive voltage across the device, we concentrate on the noise on positive bias. The voltage noise spectral densities at two bias voltages are shown in Fig. 2. At $V = 0$ the noise corresponds to the Johnson noise of the junction resistance R_j. At finite values of V the $1/f$ and shot noise appears. To quantify the $1/f$ noise, each spectrum was fit with a voltage noise

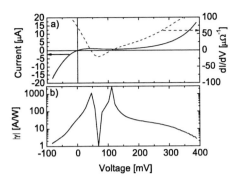

FIGURE 1. a) The $I(V)$ characteristics of the diode with the dI/dV curve as a function of the bias voltage. b) The absolute value of the current responsivity $\gamma(V)$.

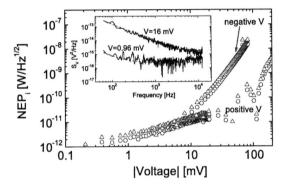

FIGURE 2. The matched (circles) and unmatched (triangles) NEP of a diode as a function of the absolute value of dc bias. The minimum NEP is obtained at $V = 0$ due to the rapid increase in $1/f$ noise which overcomes the increase in the responsivity at small positive bias. The inset shows noise spectra recorded at V=0.96 mV and V=16 mV. The dashed lines indicate the fits to the spectra.

power spectral density given by $S_{\text{fit}} = \alpha V^m / f^r + v_n^2$, where the white noise term v_n includes contributions from the Johnson noise and the shot noise. For the three diodes, the mean fitted values yielded $\alpha = 3.18 \cdot 10^{-6}$ $V^{2-m}Hz^{r-1}$, $m = 2.58$ and $r = 1.33$. The fact that the scaling with voltage deviates from that of the usual V^2 scaling is not surprising, as this strictly holds for uniform conductors at thermal equilibrium [4]. Even though the $1/f$ noise does not directly effect the NEP at zero bias, it should be noted that it does have an effect on the dynamic range of the device. The origin of the $1/f$ noise in our diodes is likely due to a similar scenario as that described in Ref. [5], according to which thermally excited traps states modulate the barrier height and thus its transmission coefficient. A noteworthy property of the tunnel diodes (with a series resistance $R_s = 17.6\ \Omega$ in series with the parallel combination of R_j and $C_j = 13.6$ fF) is that C_j fF is only weakly dependent of voltage [6]. In the case of zero bandwidth, when operated at zero bias, and at $\omega_c/2\pi \gg (R_j R_s C_j^2)^{-1/2} \approx 23$ GHz, the voltage responsivity yields with the given junction parameters $\mathscr{R}_{V0} \approx 16$ V/mW, with the noise given by the Johnson noise of R_j yielding $\text{NEP}_m|_{\Delta\omega=0} \approx \sqrt{16 k_B T R_j} \omega_c^2 C_j^2 R_s / \gamma \approx 1$ pW/$\sqrt{\text{Hz}}$ at $\omega_c/2\pi = 95$ GHz. Taking into account reflection loss only, the unmatched NEP is given by $\text{NEP}_u = \text{NEP}_m/(1-|\Gamma|^2)$ where Γ is the reflection coefficient. For example, when connected to a 100 Ω load, $\text{NEP}_u = 3.7$ pW/$\sqrt{\text{Hz}}$. As a comparison, room temperature antenna coupled microbolometers can reach a $\text{NEP}_m \approx 10$ pW/$\sqrt{\text{Hz}}$, but only at a modulation frequency of 50 kHz [7]. It should be noted that due to their purely real impedance, bolometers can be easily matched and have an matched NEP that is independent of operating frequency.

CONCLUSIONS

The measured the noise properties of the zero bias quantum tunneling diodes show that these non-optimized devices have already reached a sensitivity that challenges other existing room temperature detectors in this wavelength range. The high sensitivity, low cost and reproducibility of the diodes make them an attractive newcomer to the millimeter and submillimeter wave detector arena.

REFERENCES

1. Hwang, T.-L., Schwarz, S., and Rutledge, D., *Appl. Phys. Lett*, **34**, 773–776 (1979).
2. Schulman, J., and Chow, D., *IEEE Electron Device Letters*, **21**, 353–355 (2000).
3. Meyers, R. G., Fay, P., Schulman, J. N., S. Thomas, III, Chow, D. H., Boegeman, Y. K., Zinck, J., and Deelman, P., *IEEE Electron Device Letters*, **25**, 4–6 (2004).
4. Kogan, S., *Electronic noise and fluctuations in solids*, Cambridge University Press, Cambridge CB2 1RP, 1996, chap. 8, pp. 219–222.
5. Ng, S., and Surya, C., *J. Appl. Phys.*, **73**, 7504–7508 (1993).
6. Fay, P., Schulman, J., Thomas III, S., Chow, D., Boegeman, Y., and Holabird, K., "Performance and modeling of antimonide-based heterostructure backward diodes for millimeter-wave detection," in *IEEE Lester Eastman Conference on High Performance Devices*, IEEE, 2002, pp. 334–342.
7. Luukanen, A., *High performance microbolometers and microcalorimeters: from 300 K to 100 mK*, Ph.D. thesis, University of Jyväskylä, Department of Physics, P.O. Box 35 (YFL), FIN-40014 University of Jyväskylä, Finland (2003), http://www.phys.jyu.fi/theses/arttu_luukanen.pdf.

DC Field Response Of Hot Carriers Under Circularly Polarized Intense Microwave Fields In Semiconductors

Norihisa Ishida

Saitama Gakuen University
1510 Kizoro, Kawaguchi-Shi, Saitama 333-0831, Japan, e-mail: n.ishida@saigaku.ac.jp

Abstract. Hot carrier dynamics under intense microwave fields is investigated theoretically for the case that the dominant scattering process is optical phonon emission. When the microwave amplitude is appropriately large, an accumulated distribution of carriers in momentum space appears. The system of this motion is found to cause various peculiar dc fields response, e.g. strong non-linearity, negative differential conductivity, and negative response, under realistic physical conditions. In the proper strength of microwave and DC electric fields, especially in the case of circularly polarized microwave fields, the carrier motions are converged to some trajectories in momentum space. Resultantly a new type of accumulated distribution of carriers in momentum space appears. This situation is a sort of population inversion of carriers and causes various phenomena including a negative differential conductivity and/or a negative response appeared in the drift velocity vs. dc field relation not found in the case of linear polarized microwave fields.

INTRODUCTION

In many semiconductors, the carrier strongly interacts with the optical phonon. In rather pure crystals at low temperatures, the collision time of the carrier is long when the energy is less than the optical phonon energy (ε_{op}), while it is very short for the energy larger than ε_{op}, because of the spontaneous optical phonon emission. It has long been known that a system of such carriers exhibits various characteristic behaviors under intense electric fields [1].

Previously [2], a possibility of an accumulated distribution of such carriers in momentum space has been anticipated under appropriately intense microwave electric fields. The system would show various peculiar responses to DC fields applied parallel to the microwave fields. The only relevant experimental observation has been made by Komiyama et al.

CARRIER ACCUMULATION

Under the microwave fields, carriers act sinusoidal motion in momentum space. When the microwave amplitude E_{ac} is appropriately large near $\omega p_{op}/e$ (where p_{op} is a momentum corresponding to $\varepsilon_{op} = p_{op}^2/(2m)$, ω is a microwave frequency), carriers on most trajectories are accelerated above ε_{op} then scattered into low energy region within a period of the microwave field, while some carriers belonging to a region in momentum space do not arrive at ε_{op} and continue the free motion for their collision time. As a consequence, the carriers are accumulated in this region and move periodically in momentum space. This accumulated distribution is a dynamic population inversion so that it causes various peculiar effects. In additional DC fields applied with the microwave field, an accumulated distribution would be going to be missing as the DC field is increased. At very weak DC, the distribution still remain deforming or chipping their forms away, but in the large DC field strength, no specific forms in distribution could be found. Peculiarities in DC response of such carriers are restricted around the weak DC field strength in such a normal sense [2].

Remarkably, intense non-linearity including negative response would be found in DC response at large DC fields with circularly polarized microwave fields. This case has several complex properties discussed in next section. Limit cycle of the carrier motion is a key for a new type of accumulated distribution.

DC RESPONSE

The response to DC electric fields (E_{dc}) applied in the same plane to the existing circularly polarized microwave field exhibits many features. Figure 1 shows the computational result of the response (drift momentum p_x) to DC, where A_{dc} is expressed in dimensionless values instead of electric fields E_{dc} ($A_{dc} = eE_{dc}/\omega p_{op}$) using two dimensional simple band models considered in quantum wells.

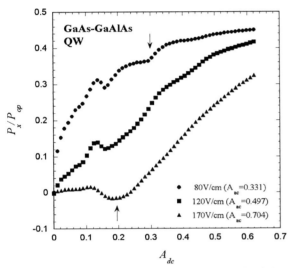

FIGURE 1. Drift velocity vs. DC electric field relation calculated for carriers in GaAs-GaAlAs quantum wells with 150 Angstrom, at 4.2K, 10GHz.

Three kinds of symbols denote different amplitudes of microwave (A_{ac}) scaled as for DC fields.

The change of carrier motions is simply determined by external field strength. Roughly four reasons would be exist in it: (1) Carrier accumulation at low DC fields, (2) Discontinuities of the collision time of distributed carriers as external fields change, (3) Synchronizations with microwave frequency, and (4) Trajectory convergence. The last one is only shown in the circularly polarized microwave fields.

(1) is caused by the distributed carriers, and the details reported in [2]. Discontinuities noted in (2) occur when carriers are lost or got its different period of collision time. Collision time could be understood by the equation of motion restricted by inelastic phonon emission. Conditions are expressed as,

$$\{-A_{ac}(\cos\theta_{i+1} - \cos\theta_i) + A_{dc}(\theta_{i+1} - \theta_i) + p_0/p_{op}\}^2$$
$$+ \{A_{ac}(\sin\theta_{i+1} - \sin\theta_i) + p_0/p_{op}\}^2 = 1$$

$\theta = \omega t$, where ω is a microwave frequency, t is the time. Suffix i+1 and i denote next and previous collision.

Equation above is assumed that a carrier-phonon interaction includes completely inelastic scattering. When the applied fields increased or decreased over a certain threshold, some part of carriers loses or gets a long period of time. Generally, the carrier with a long period has mobility much greater than a short one, so that the conductivity changed drastically around there. Corresponding field strength, A_{dc}=0.3, 0.16, 0.08, at A_{ac}=0.331 in Fig. 1 for example, all "dip" like structures are recognized.

Synchronizations at (3) is occurred in the condition:

$$A_{dc} = \frac{1}{2n\pi} \quad (n=1,2,\ )$$

A_{dc}=0.16 (1/2π), 0.08 (1/4π), 0.04 (1/8π), and so on are also recognized as a response in Fig. 1. In this case, all carriers are on the period of microwave and move simultaneously in momentum space. At this time, almost all carriers are in the same trajectory, and accumulated in a certain region.

Trajectory convergence cited as (4) is the only advent under circularly polarized electric fields. This is considered as a limit cycle of motion. Conditions to converge are still too complicated to determine.

For Fig.1, 8.0×10^{12} s^{-1} adopted to the collision rate for the polar optical phonon, much larger than microwave fields' 10^{10} s^{-1}. To avoid the collision except inelastic collision, higher frequency has advantage. Much higher rate of collision is obtained using much smaller wells, or inter-subband transitions, and so on.

REFERENCES

1. For Example, S. Komiyama, T. Kurosawa, and T. Masumi, Chapter 6, in Hot Electron Transport in Semiconductors, edited by L. Reggiani (Springer-Verlag, Berlin 1985).
2. N. Ishida, T. Kurosawa, Journal of Physical Society of Japan 64 (8) 1995, pp. 2994-3006.

Acoustoelectric Effects in Ge/Si Nanosystems with Ge Quantum Dots

I. L. Drichko*, A. M. Diakonov*, I. Yu. Smirnov*, Y. M. Galperin*†,
A. I. Yakimov** and A. I. Nikiforov**

*A. F. Ioffe Physico-Technical Institute of Russian Academy of Sciences, 194021 St. Petersburg, Russia
†Department of Physics, University of Oslo, PO Box 1048 Blindern, 0316 Oslo, Norway
**Institute of Semiconductor Physics, Siberian division of Russian Academy of Sciences, Novosibirsk, Russia

Abstract. Dense ($n = 3 \times 10^{11}$ cm^{-2}) arrays of Ge quantum dots (QDs) in Si host were studied using surface acoustic wave (SAW) propagating along the surface of a piezoelectric crystal located near the sample. The SAW magneto-attenuation coefficient, $\Delta\Gamma = \Gamma(\omega,H) - \Gamma(\omega,0)$, and change of the SAW velocity, $\Delta V/V = [V(H) - V(0)]/V(0)$, were measured in the temperature interval $T = 2\text{-}4.2$ K as functions of magnetic field H up to 6 T for the waves in the frequency range $f = 30 - 300$ MHz. Basing on the dependences of $\Delta\Gamma$ on H, T and ω, as well as on its sign, we believe that the AC conduction mechanism is hopping between the states localized in different QDs. The measured magnetic field dependence of the SAW attenuation is discussed basing on existing theoretical concepts.

INTRODUCTION

Acoustic methods allow one to study mechanisms of AC conductance and provide independent methods to determine material's parameters. Here we apply these methods to Si samples with high-density ($n = 3 \times 10^{11}$ cm^{-2}) arrays of Ge quantum dots (QD). At low temperatures, DC conductance of the samples is determined by inter-dot variable range hopping (VRH) strongly influenced by the Coulomb gap in the hole density of states [1]. The crucially important parameter in such situation is the hole localization length, ξ. One of the aims of the present work is to determine this quantity from AC conductance.

A surface acoustic wave (SAW) in the frequency range 30-300 MHz propagating along the surface of a piezoelectric substrate creates a wave of electric field. This wave acts on a the QD array "embedded" into the sample close to its surface causing an additional SAW attenuation, Γ, and change in its velocity, V. Such "hybrid" geometry allows finding electrical AC conductance, $\sigma^{(AC)}$, of a non-piezoelectric material by acoustic methods. In fact, we have measured the differences $\Delta\Gamma = \Gamma(\omega,H) - \Gamma(\omega,0)$ and $\Delta V/V = [V(H) - V(0)]/V(0)$ as functions of an external magnetic field, $H \leq 6$ T, perpendicular to the sample's surface, in the temperature interval $T = 2 - 4.2$ K.

EXPERIMENTAL SETUP AND RESULTS

The sketch of the experimental setup is shown in Fig. 1.

FIGURE 1. Sketch of the acoustoelectric device.

FIGURE 2. Scheme of the quantum dot array.

The samples were grown by MBE method on the (001) Si-substrate, on which an intrinsic Si buffer layer was placed followed by a B delta-doped silicon layer. Two samples with $N_B = 6.8 \times 10^{11}$ cm^{-2} (2.3 holes per QD) and $N_B = 8.2 \times 10^{11}$ cm^{-2} (2.7 holes per QD) were studied. Then a 10 nm undoped Si layer was grown, on top of which 8 Ge monolayers were placed. This structure was covered by a 200 nm i-Si layer. The self-organized Ge QDs had pyramidal shape with height of 15 Å and square 150×150 Å^2 base (Fig. 2).

The measured magneto-attenuation $\Delta\Gamma$ as a function of H for different SAW frequencies is shown in Fig. 3. One can see that the attenuation *decreases* with magnetic field, $\Delta\Gamma < 0$. We believe that under the experimental conditions $\Delta\Gamma$ is determined by AC hopping conductance between different QDs. For $\sigma < 10^{-7} \Omega^{-1}$ [2]:

$$\Delta\Gamma = \frac{\Delta\sigma^{(AC)} A(q) q e^{-2q(a+d)} \frac{4\pi}{\varepsilon_s V}}{2[(\varepsilon_0 + \varepsilon_1)(\varepsilon_s + \varepsilon_0) - (\varepsilon_1 - \varepsilon_0)(\varepsilon_s - \varepsilon_0)e^{-2qa}]^2},$$

$$A(q) = 4K^2 (\varepsilon_0 + \varepsilon_1) \varepsilon_0^2 \varepsilon_s, \qquad (1)$$

where q is the SAW wave vector, K^2 is the electromechanical coupling constant of the substrate (LiNbO$_3$), $\varepsilon_1 = 51$, $\varepsilon_0 = 1$, $\varepsilon_s = 12$ are the dielectric constants of LiNbO$_3$, vacuum and the sample, respectively. V is the sound velocity, a is the sample-niobate clearance, $d = 200$ nm is the sample cap layer. Thus, $\Delta\Gamma \propto \Delta\sigma^{(AC)}$, which decreases with magnetic field due to shrinkage of the hole wave functions. As follows from analysis

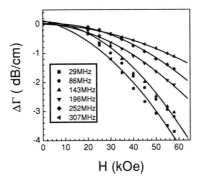

FIGURE 3. Experimental dependence of $\Delta\Gamma$ on H at different frequences for the sample with 2.7 holes per QD, $T = 4.2$K.

of Fig. 3, at small magnetic fields $\Delta\Gamma \propto H^2$ and in strong magnetic fields $\Delta\Gamma \propto H^{-2}$. This fact allows us to extract the zero-field attenuation, $\Gamma_0 \equiv \Gamma(H = 0)$. This quantity turns out to be *frequency independent* within the experimental error of $\approx 10\%$. At low temperatures $\Gamma_0(T) \propto T^4$ and then crosses over to T^1 at $T \approx 4$ K. We find this behavior compatible with the theoretical predictions, see Ref. [3], for AC hopping conductance in the regime $\omega\tau \gg 1$. Here $\tau(T)$ is the typical relaxation time for the population of relevant pairs of QDs.

To estimate the localization length we employ the fact that the crossover from low- to high-field behavior takes place at $\xi \approx (cH/e\hbar)^{1/2}$, see Fig. 4. As a result, we obtain $\xi \approx (1.2 - 1.4) \times 10^{-6}$ cm.

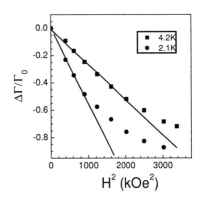

FIGURE 4. Illustrative dependence of $\Delta\Gamma$ on H^2 for the sample with 2.7 holes per QD. $f = 30$ MHz.

CONCLUSIONS

For the first time, AC magnetoconductance in a Ge-in-Si QD array is measured using the SAW technique. Mechanism of AC conductivity is hopping between the hole states localized in different QDs at $\omega\tau \gg 1$. The localization length of these states is determined. Large value of the relaxation time for the hole states is compatible with the energy relaxation time determined from the experiment on hole heating [4]. To verify this regime one should experimentally observe a crossover to the regime $\omega\tau \ll 1$. Unfortunately, the employed acoustic method does not allow us to observe this crossover since the acoustic response at low frequencies turns out to be very small. We plan to conduct electrical AC measurements to study the low-frequency regime.

ACKNOWLEDGMENTS

The work is supported by RFBR (Grants No. 04-02-16246 and 03-03-16536), MinNauki, Russia-Ukraine Program *Nanophysics and Nanoelectronics*, NATO-CLG.979355 grants.

REFERENCES

1. A. I. Yakimov et al, J. Phys.: Condens. Matter **11** (1999) 9715; Phys. Rev. B 61 (2000) 10868.
2. I.L. Drichko, A.M. Diakonov, I.Yu. Smirnov, Y.M. Galperin, and A. I. Toropov, *Phys.Rev.B* **62**, 7470 (2000).
3. Y. M. Galperin, V. L. Gurevich and D. A. Parshin, in "Hopping Transport in Solids", ed. By M. Pollak and B. I. Shklovskii, (North-Holland, 1991) p. 81.
4. G. O. Andrianov, I. L. Drichko, A. M. Diakonov, I. Yu. Smirnov, O. A. Mironov, M. Mironov, T. E. Whall, D. R. Leadley, Proc. of 12th Int. Symp. "Nanostructure: physics and technology", St. Petersburg, 2004, p.154

Cyclotron Resonance Study of Doped and Undoped InAs/AlSb QW Heterostructures

A. V. Ikonnikov*, V. Ya. Aleshkin*, V. I. Gavrilenko*, Yu. G. Sadofyev[†],
J. P. Bird[†], S. R. Johnson[†], and Y.-H. Zhang[†]

Institute for Physics of Microstructures, 603950, Nizhny Novgorod, GSP-105, Russia
[†]*Arizona State University, Tempe, AZ 85287, USA*

Abstract. We discuss the cyclotron resonance study of doped and undoped InAs/AlSb quantum wells with a two-dimensional electron gas density $n_s = 3 \cdot 10^{11} - 8 \cdot 10^{12}$ cm^{-2}. A remarkable increase of the cyclotron mass from $0.03 m_0$ to $0.06 m_0$ with the electron concentration is observed and shown to be a result of the conduction band nonparabolicity.

INTRODUCTION

In recent years there has been considerable interest in the InAs/AlSb quantum well (QW) system, which is characterized by a large electron confinement energy of 1.3 eV and motilities up to $9 \cdot 10^5$ cm^2/V·s (see, for example, [1-4]). Cyclotron resonance (CR) studies are known to be a powerful means to reveal the conduction band nonparabolicity and spin splitting of Landau levels in undoped InAs QWs with two-dimensional (2D) electron gas concentrations up to $1.4 \cdot 10^{12}$ cm^{-2} [5,6]. This paper reports CR characterization of high mobility AlSb/InAs/AlSb QWs with a 2D electron gas ranging from $2.7 \cdot 10^{11}$ to $8 \cdot 10^{12}$ cm^{-2}.

RESULTS AND DISCUSSION

The InAs/AlSb QW heterostructures under study were grown by molecular-beam epitaxy on semi-insulating GaAs(100) substrates. The structure starts with a 2.4 µm thick metamorphic AlSb or GaSb buffer layer followed by a ten-period GaSb/AlSb superlattice, a lower AlSb barrier, a 15 nm InAs QW, an upper AlSb barrier, and a 6nm GaSb cap [7]. Samples ##1-4 are not intentionally doped while samples ##5-8 are δ-doped with Te in the AlSb barriers on either side of the InAs QW. The Hall effect and Shubnikov-de Haas (SdH) measurements are carried out at T = 4.2 K. In the CR study, backward wave tubes (BWT) covering the spectral range from 160 to 710 GHz were used. The transmitted radiation is detected using an n-InSb crystal. Oscillations in the microwave photoconductivity (PC) (which are analogous to SdH oscillations) determine the 2D electron concentration.

The sample parameters are listed in Table 1. The 2D electron concentrations in the 1st, 2nd and 3rd electron subbands are obtained from Fourier analysis of the SdH oscillations (n_s^{SdH}) and from the microwave PC oscillations (n_s^{PC}). In the intentionally doped samples ## 5-8 the mobility decreases with electron concentration due to scattering by ionized donors in the δ-doped barriers.

Typical traces of the CR absorption spectra are given in Fig. 1. It is worth mentioning that electron mobility, determined from the CR absorption line width μ_c, is significantly less than that obtained from the DC measurements µ (see Table 1). This is due to saturation of the CR absorption caused by high electron mobility and/or concentration in the samples studied. The cyclotron mass m_c obtained from the resonant magnetic fields are summarized in Table 1. A remarkable increase in mass with the electron concentration is clearly seen. To interpret the experimental results we have calculated the electron energy spectrum and the CR masses using a simplified Kane model [8,9].

TABLE 1. Transport and CR properties of undoped (##1-4) and selectively doped (##5-8) samples.

Sample	n_s^{Hall}, 10^{12} cm^{-2}	n_s^{1SdH}, 10^{12} cm^{-2}	n_s^{2SdH}, 10^{12} cm^{-2}	n_s^{3SdH}, 10^{12} cm^{-2}	n_s^{IPC}, 10^{12} cm^{-2}	μ, 10^5 cm^2/V·s	μ_c, 10^5 cm^2/V·s	m_c/m_0
#1	-	-	-	-	0.27	-	0.45	0.029-0.031
#2	0.65	0.64	-	-	0.63	3.9	0.6-1.6	0.032-0.036
#3	0.68	0.66	-	-	0.67	2.5	0.4-1.6	0.033-0.037
#4	0.95	0.83	-	-	0.82	4.4	0.5-1.6	0.034-0.036
#5	2.4	1.8	0.6	-	-	1.0	0.4	0.042-0.045
#6	3.2	2.2	1.0	-	-	0.63	0.4	0.042-0.044
#7	4.3	2.8	1.5	-	-	0.53	0.4	0.044-0.048
#8	8.3	4.3	3.4	0.6	-	0.4	0.2	0.054-0.060

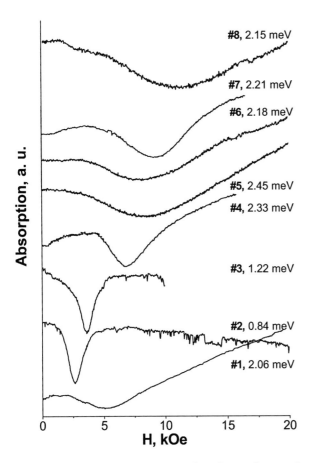

FIGURE 1. Typical CR traces. Sample numbers and radiation energy quanta are given at the curves.

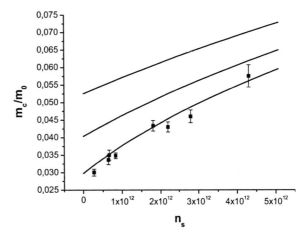

FIGURE 2. Cyclotron mass versus 2D electron concentration. Solid curves are calculated dependences for 1st, 2nd and 3rd electric subbands in InAs QW 20.5 nm wide.

The calculation results show that differences in CR masses at the Fermi energy in the 1st, 2nd and 3rd electric subbands are within the accuracy of the mass measurement in the absorption spectra (Fig. 2). The measured CR mass values agree fairly well with the calculated results if the InAs QW width is assumed to be 20.5 nm (Fig. 2) rather than 15.0 nm as determined from the growth parameters. To reveal the origin of this discrepancy independent measurements of QW width as well as calculations using 8x8 Kane Hamiltonian are needed.

REFERENCES

1. Tuttle G. et al., *J. Appl. Phys.* **65**, 5239 (1989).
2. Nguyen C. et al., *J. Electron. Mater.* **22**, 255 (1993).
3. Gauer Ch. et al, *Semicond. Sci. Technol.* **8**, S137 (1993).
4. Yang M. J. et al., *Appl. Phys. Lett.* **80**, 1201 (2002).
5. Yang M. J. et al., *Semicond. Sci. Technol.* **8**, S129 (1993).
6. Gauer C. et al., *Semicond. Sci. Technol.* **9**, 1580 (1994).
7. Sadofyev Yu. G. et al., *Appl. Phys. Lett.* **81**, 1833 (2002).
8. Bastard G. *Wave mechanics applied to semiconductor heterostructures.* Halsted Press, NY, 1988, pp.31-61.
9. Aleshkin V. Ya. et al., *Sov. Phys. Semiconductors*, **26**, 291 (1992).

Time Domain Terahertz Spectroscopy of the Magnetic Field Induced Metal-Insulator Transition in n:InSb

J. Y. Sohn, X. P. A. Gao, S. A. Crooker

National High Magnetic Field Laboratory, Los Alamos, New Mexico, 87545, USA

Abstract. Temperature (T) and frequency (ω) dependent conductivity measurements are reported for n-type indium antimonide (InSb) around the magnetic field induced metal-insulator transition (MIT). For the sample with electron density n= 2.15×10^{14} cm^{-3}, the critical field is observed at ~0.7 T in dc transport measurements. The frequency dependent conductivity $\sigma(\omega)$ measured via terahertz time domain spectroscopy indicates a higher critical field ~1.2 T. Both $\sigma_{dc}(T)$ and $\sigma_1(\omega)$ at low temperatures show power law dependence with exponents of α=1.2.

INTRODUCTION

It is well known that n-type indium antimonide (InSb) with a very low carrier concentration (~10^{14} cm^{-3}) undergoes a magnetic field induced MIT [1]. Although dc transport measurements in magnetic field suggest the existence of a field-induced MIT in InSb, the dynamics of the transition remain largely unexplored. To understand the dynamics of the field induced MIT in semiconductors in detail, frequency dependent measurements are necessary in the terahertz (THz) range. We performed high frequency (100 GHz ~ 1.5 THz) time domain spectroscopy on InSb under high magnetic fields.

For the purpose of these THz measurements, we employ a method [2] for THz time-domain spectroscopy directly in the cryogenic bore of high field magnets. Miniature, fiber-coupled THz emitters and receivers are constructed for working down to 1.5 K and up to 17 T. Under applied magnetic fields, we measured both the temperature dependent dc conductivity by conventional transport methods and the frequency dependent complex conductivity by time-domain spectroscopy. For the dc conductivity measurements, a sample with Hall bar configuration was measured in a Quantum Design physical properties measurement system (PPMS). The Hall measurement shows that the carrier concentration is 2.15×10^{14} cm^{-3} at 2 K near zero magnetic field.

For MIT in three dimensions (3D), it is established that the temperature dependence of the conductivity usually follows the power law [3]

$$\sigma_{dc}(T) = AT^\alpha + C, \qquad (1)$$

across the transition point, with $C>0$ for the metallic state and $C<0$ for the insulating state. The critical point of the transition corresponds to parameter C equal zero.

Scaling arguments can be used to describe the frequency dependence of the real conductivity at low temperature [4]. At low temperatures ($T\rightarrow 0$) and at fields close to the critical field, the conductivity should follow the power law

$$\sigma_1(\omega) = A'\omega^\alpha + C', \qquad (2)$$

also with a positive value of C' appropriate for the metallic state and a negative value of C' for the insulating state.

RESULTS

Figure 1 shows dc conductivity from 2 K to 12 K while varying the magnetic field from 0.3 T to 1.0 T in 0.1 T steps. Fitting the $\sigma_{dc}(T)$ data in different fields to Eq. (1), we found that α=1.2 best describes the data over all the temperature range in contrast to α=0.5 for the density tuned MIT in NbSi [5]. The temperature dependent dc conductivity clearly follows equation (1). The critical field, at which C=0, is found to be 0.7 T

Using the fiber coupled THz antennas, time-domain THz measurements were performed in magnetic fields in the Faraday geometry. Complex transmission can be obtained by Fourier transform of detected time-domain THz signal. From the complex transmission of THz pulses and reference THz pulses, we calculate the complex dielectric constant of InSb by iteratively solving Eq. (3) below [6].

$$\sqrt{\varepsilon(\omega)} - 1 = \frac{c}{i\omega d} \ln\left(\frac{E_t(\omega)}{E_0(\omega)} \frac{(1+n)^2}{4n}\right), \quad (3)$$

where $E_0(\omega), E_t(\omega)$ are the incident and transmitted THz fields, ε is the complex dielectric constant, n is the complex index, d is the sample thickness, and c is the speed of light. Complex conductivity can then be easily obtained from the complex dielectric constant.

$$\sigma(\omega) = i\varepsilon_0 \omega (\varepsilon(\omega) - 1), \quad (4)$$

where ε_0 is permittivity of free space.

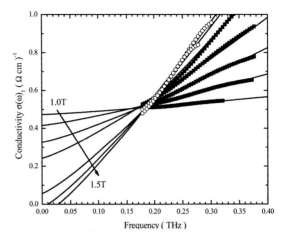

FIGURE 2. Real conductivity $\sigma_1(\omega)$ versus frequency as the magnetic field changes from 1.0 T to 1.5 T (from top to bottom) at T= 2 K. The lines are fitted curves to Eq. (2) with α=1.2.

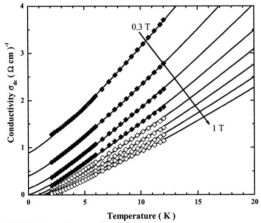

FIGURE 1. dc conductivity versus temperature as the magnetic field changes from 0.3 T to 1.0 T (from top to bottom). The critical field is observed around 0.7 T. The lines are fitted curves to Eq. (1) with α=1.2.

The frequency dependent real conductivity ($\sigma_1(\omega)$) is plotted in Fig. 2 with fits to Eq. (2). We only fit data in the frequency range $\hbar\omega \geq 3.5kT$, such that the quantum limit is satisfied, and we can also extract the dynamic exponent from $\sigma_1(\omega)$. The lines in Fig. 2 are power law fits to Eq. (2) with α=1.2, the same exponent from DC transport measurements. We see that α=1.2 also describes $\sigma_1(\omega)$ at T=2 K. These data suggests a critical point of 1.2 T where $\sigma_1(\omega)$ intercepts the origin.

In conclusion, we have studied the magnetic field induced MIT in InSb (n = 2.15×10^{14} cm^{-3}) by both temperature dependent dc conductivity and frequency dependent conductivity. Both $\sigma_{dc}(T)$ and $\sigma_1(\omega)$ can be fit by a power law dependence with exponents of α=1.2. However the temperature dependent dc conductivity gives a critical field ~0.7 T and the frequency dependent conductivity gives ~1.2 T.

ACKNOWLEDGEMENTS

Authors thank P. Littlewood, D. Smith, and S. Kos for valuable discussions.

REFERENCES

1. Shayegan, M., Goldman, V. J., and Drew, H. D., *Phys. Rev.* **B** 38, 5585 (1988).
2. Crooker, S. A., *Rev. Sci. Instrum.* **73**, 3258 (2002).
3. Lee, P. A. and Ramakrishnan, T. V., *Rev. Mod. Phys.* **57**, 287 (1985).
4. Lee, H. –L., Carini, J. P., Baxter, D. V., Henderson, W., and Grüner, G., *Science* **287**, 5453 (2000).
5. Lee, H. –L., Carini, J. P., Baxter, D. V., and Grüner, G., *Phys. Rev. Lett.* **80**, 4261 (1998).
6. Nuss, M. C. and Orenstein, J., "Terahertz Time-Domain Spectroscopy," in *Millimeter and Submillimeter Wave Spectroscopy of Solids*, edited by Grüner, G., Heidelberg: Springer-Verlag, 1998, pp. 7-50.

Cyclotron resonance revisited: the effect of carrier heating

H. Malissa*, Z. Wilamowski[†] and W. Jantsch*

Institut für Halbleiter- und Festkörperphysik, JKU Linz, Austria
[†]*Institute of Physics, Polish Academy of Sciences, Warsaw, Poland*

Abstract. We observe and investigate the cyclotron resonance (CR) of high mobility photo excited carriers in high resistivity (> 1000 Ωcm) bulk FZ-Si in a standard X-band electron paramagnetic resonance setup. At 2.5 K and low microwave (MW) power a series of very narrow CR lines is seen allowing for a very precise evaluation of the cyclotron masses. In addition, we make use of the narrow line-width to investigate electron heating caused by MW absorption. The minimum CR linewidth of 1.2 mT implies an electron mobility of more than $7 \cdot 10^6$ cm^2/Vs. Modelling the line-width we find that for a lattice temperature of 2 K, the electron temperature reaches 10 K at a MW power of 10 mW, and the temperature increases by 25 mK at 1 μW in spite of the fact that the sample is located in the MW cavity at the minimum of the electric field. At lower lattice temperature the electron heating is even bigger.

Fifty years after the famous paper by Dresselhaus, Kip, and Kittel [1] about cyclotron resonance (CR) in Si, we repeat such measurements on undoped bulk Si samples. Under band gap illumination we observe CR lines with a line-width of the order of 1 mT implying a mobility of photoelectrons of about $7 \cdot 10^6$ cm^2/Vs. This narrow line-width allows not only for a precise evaluation of the effective masses, but also for a detailed investigation of the effect of electron heating under CR conditions. This problem is of current interest because of the discovery of a new type of dc magneto-oscillations in an ultra high mobility electron gas under microwave (MW) radiation [2, 3] which are probably related to a specific type of electron heating [4].

Our experiments were done in a standard X-band electron paramagnetic resonance spectrometer operating at a MW frequency of approximately 9.48 GHz. The sample is located in the center of a rectangular TE$_{102}$ cavity, where the electric component of the MW field is minimum and the magnetic component is maximum. The sample is placed inside a liquid He continuous flow cryostat with a base temperature of $T = 2$ K. When the sample is illuminated with bandgap light, a series of CR lines appears in the spectra even for the lowest illumination power of 5 mW. The magnetic field is modulated with an amplitude of 0.5 mT for lock-in detection. The detected signal is thus the first derivative of CR absorption with respect to magnetic field.

In Fig. 1 an example of a CR spectra is shown. The most prominent features are the low-field CR line at $H \approx 0.065$ T and the high-field line at $H \approx 0.14$ T. These lines correspond to the extremal cross sections perpendicular to the external magnetic field of the six constant energy

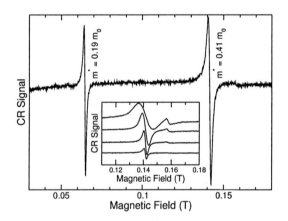

FIGURE 1. Differential CR spectra of Si with the magnetic field oriented parallel to the [001] direction. Inset: CR spectra measured at $T = 2$ K with different MW powers applied (from down to up: $P = 2$ μW, 20 μW, 200 μW, 2 mW).

ellipsoids in Si. For a magnetic field in [001] direction, the low-field line corresponds to the transverse mass $m_t = 0.19 m_0$ and the high-field line to a combination of transverse and longitudinal masses $(m_t m_l)^{1/2} = 0.41 m_0$. Both CR lines shift reproducibly as the sample is rotated inside the cavity.

Increasing the sample temperature leads to a significant broadening of the CR lines. In Fig. 2 (a) we plot the momentum relaxation rate (corresponding to the CR line-width; $\Delta \omega = e \Delta B / m^*$) as a function of sample temperature for minimum MW- and laser power. In the whole temperature range investigated ($T = 2 \ldots 20$ K) the momentum scattering rate of ultra high mobility electrons monotonically increases with $T^{3/2}$, indicating that,

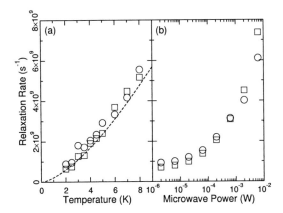
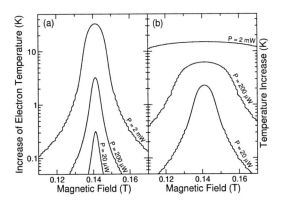

FIGURE 2. (a) Momentum relaxation rate, $\Delta\omega = e\Delta B/m^*$, evaluated from CR line-width, ΔB. The circles were evaluated from the low-field line ($m^* = 0.19 m_0$) and the squares from the high-field line ($m^* = 0.41 m_0$). These data were measured at low MW power $P = 2$ μW. The dashed line corresponds to the momentum relaxation originating from scattering by acoustic phonons. (b) The dependence of the CR line-width on MW power measured at $T = 2$ K. The circles were evaluated from the low-field line ($m^* = 0.19 m_0$) and the squares from the high-field line ($m^* = 0.41 m_0$).

FIGURE 3. The calculated temperature increase for different MW power caused by resonance absorption as a function of the magnetic field (a) when only the increase of electron temperature is assumed (b) when the electron and lattice heating is assumed.

down to the lowest temperature, momentum scattering is dominated by acoustic phonon scattering, with the electron velocity scaling with $T^{1/2}$ and the phonon density with T. This allows us to treat the CR resonance line-width as an effective thermometer for the electron temperature.

We investigate the change in electron temperature caused by MW absorption, which is a very efficient heating mechanism. In Fig. 1 (inset) the evolution of CR spectra with the MW power applied is shown. Even at a MW power of a few μW, a broadening of the CR lines can be observed already. The increase in CR line-width as a function of MW power is shown in Fig. 2 (b). A comparison of Figs. 2 (a) and (b) demonstrates how efficient heating by MW irradiation really is: the peak-to-peak line-width for a sample temperature of 2 K and an applied MW power of 1 mW corresponds to the line-width observed at a sample temperature of $T \approx 6$ K for 2 μW. This shows that CR absorption effectively increases the temperature of ultra high mobility electrons. The evaluation of the electron temperature is complicated by the fact that the electron temperature varies with applied magnetic field, according to the variation of MW absorption. The absorption spectra are thus a self consistent function of the actual temperature. Moreover, the energy relaxation of hot electrons, although ruled by acoustic phonon scattering, is modified by heating of the phonon distribution.

In Fig. 3 the estimated temperature increase caused by CR absorption is plotted as a function of magnetic field. The results shown in Fig. 3 (a) were obtained as the best fit to the experimentally observed line-width, assuming that the lattice temperature, and the phonon density correspond to the cryostat temperature, $T = 2$ K. Consequently, the momentum- and the energy relaxation rates were assumed to scale with $T^{1/2}$. The results in Fig. 3 (b) correspond to the assumption that electrons and lattice are in the thermal equilibrium. The momentum and the energy relaxation rates were assumed to scale with $T^{3/2}$. Comparison of the model and experimental line shapes and line amplitudes indicates that the assumption about the cold lattice and hot electrons, i.e., results presented in Fig. 3 (a), fit the observed spectra much better.

In conclusion, we have shown that electron heating caused by MW absorption is a highly efficient mechanism, suggesting that they should be taken into account in the interpretation of Mani's oscillations [2].

Work supported by KBN grants: PBZ 044/P03/2001 and 2 P03B 054 23 in Poland and in Austria by the FWF and ÖAD, both Vienna.

REFERENCES

1. G. Dresselhaus, A. F. Kip, and C. Kittel, *Phys. Rev.* **98** 368 (1955)
2. R. G. Mani, J. H. Smet, K. von Klitzing, V. Narayanamurti, W. B. Johnson, and V. Umanski, *Nature (London)* **420** 646 (2002)
3. M. A. Zudov, R. R. Du, J. A. Simmons, and J. L. Reno, *Phys. Rev. B* **64** 201311 (2001)
4. A. V. Andreev, I. L. Aleiner, and A. J. Millis, *Phys. Rev. Lett.* **91** 56803 (2003)

The nipnip-THz-emitter: photomixing based on ballistic transport

F. H. Renner*, O. Klar*, S. Malzer*, D. Driscoll[†], M. Hanson[†], A. C. Gossard[†], G. Loata**, T. Löffler**, H. Roskos** and G. H. Döhler*

*Institute for Technical Physics I, University Erlangen-Nürnberg, 91058 Erlangen, Germany
[†]Materials Departement, University of California, Santa Barbara, CA 93116, U.S.A.
**Physik. Institut, University of Frankfurt, D-60325 Frankfurt, Germany

Abstract. We report on a novel concept for THz-photomixers based on quasi-ballistic transport in an asymmetric nipnip-doping-superlattice. Due to tansport-optimized i-layers the emitted powers are not transit-time-limited up to $1\,THz$. Furthermore the capacitance and hence the RC-roll-off is minimized by increasing the number of pin-periods. The frequency-dependence of the nipnip-emitter proofs to be superior to corresponding pin-photomixers.

INTRODUCTION

THz-radiation is electromagnetic radiation between $0.1\,THz$ and $10\,THz$. Despite various applications in the fields of medical imaging, sensing, spectroscopy and telecommunication, there is still a lack of efficient and compact solid-state THz-emitters. One way of generating THz-radiation is photomixing. Photomixing can be described as the generation of high-frequency photocurrent by optical heterodyning of two slightly detuned laser beams in a fast photodetector, which is then radiated as THz-radiation by an attached planar antenna. State-of-the-art-photomixers are based on low temperature-grown GaAs (LT-GaAs) [1], yielding output powers on the order of $1\,\mu W$ at $1\,THz$. The major disadvantage of this approach was the very low conversion efficiency (10^{-5}) due to a low photoconductive gain and, therefore, low maximum emitted power. Furthermore, it can not be applied for the use in the $1.55\,\mu m$-wavelength regime due to the GaAs bandgap of approx. $870\,nm$. Using a pin-diode as the photoconductive element is another promising approach. In this case one has to consider the trade-off between *short carrier transit times*, favored by a *thin* intrinsic layer (yielding a high transit-time roll-off frequency) and *low capacitance* of the device favored by a *thick* intrinsic layer (resulting in a high RC-time roll-off frequency). In spite of a roll-off with the 4-th power of the THz-frequency, pin-emitter have proven the highest THz-powers, reported for photomixers so far [2]. But still this photomixer is limited by the two roll-off-terms, mentioned above, which both cannot be minimized at the same time. Our nipnip-THz-

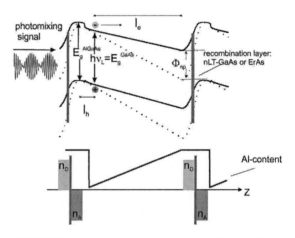

FIGURE 1. One period (p-i-n-p) of the nipnip-emitter.

emitter concept can minimize both roll-off-limitations, leading to a very efficient THz-photomixer.

THE NIPNIP-THZ-EMITTER-CONCEPT

In our concept, the transit-time roll-off constraint is overcome by the subtle use of the high-field transport-properties of III/V-compound semiconductors like GaAs. Electrons which are accelerated by an electric field are scattered from the Γ-valley into the L- or X-side-valleys as soon as they gain the corresponding threshold kinetic energy for this process. After the scattering, the carrier-velocity drops to the stationary saturation drift-velocity due to the larger effective mass and increased scattering rates. Before intervalley scattering, the electrons travel quasi-ballistically. Monte Carlo-simulation-data [3] indicates that for fields of $10..20\,kV/cm$ the electrons can travel a distance of $150\,nm$ within $0.5\,ps$, allowing them to follow the THz-photomixing excitation, before being scattered into side-valleys. The 3dB-transit-time frequency is therefore shifted towards $1\,THz$ and can be increased further, drastically reducing the transit-time-roll-off.

The RC-roll-off is minimized by using a periodic sequence of transport optimized pin-diodes on top of each other, connected by np-recombination-diodes. This leads to a substantial reduction of the capacitance of the photomixer.

In addition to minimizing the high-frequency losses, the overall power can be increased by embedding the photomixer into an asymmetric Fabry-Perot-cavity in order to enhance the absorption and, therefore, the responsivity.

One period of the nipnip-superlattice is depicted in fig. 1. In order to ensure that only the (faster) electrons perform the transport, the carriers are generated within a narrow region close to the p-contact due to a varying Al-content. The generated electron-hole-pairs are separated in the i-layer-field, creating a dipole-field which is screening the built-in electric field to an optimum value for ballistic transport. The electrons, gathered near the n-contact have to recombine with holes from the next period in order to prevent a complete flattening of the field and suppressing the transport. Therefore, the np-diodes are recombination-enhanced, allowing for the required current-densities of $1..10\,kA/cm^2$. This is achieved by introducing either ErAs or LT-GaAs inside the junction [4].

EXPERIMENTAL RESULTS

We have successfully fabricated GaAs nipnip-emitters with either ErAs and LT-GaAs recombination layers. The experimental results of the emitted THz-radiation vs. frequency are depicted in fig. 2 for a four-period-, a seven-period- and the corresponding pin-photomixer. Non-resonant log. periodic-antennas were attached to the photomixers. The pin-photomixer is clearly transit-time- *and* RC-limited, leading to a strong v^{-4}-decrease of the emitted THz-power. Since the optimum field inside the pin-nano-diodes can be set by photogenerated field-screening (amd a fine-tuning by a small external voltage), the transit-time roll-off can be overcome according to the carrier-transport-predictions and the emitted power is solely RC-limited (v^{-2}) with a significantly smaller capacitance compared to the pin-diode. This results in an increased THz-output of nearly 2 orders of magnitude

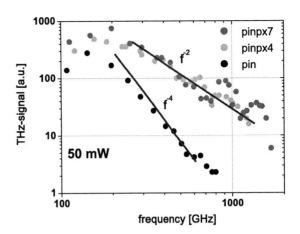

FIGURE 2. Comparison of the emitted THz-signal vs. the mixing-frequency for the four- and seven-periods-nipnip-emitter and the corresponding pin-emitter.

at $1\,THz$. Furthermore, the bias-dependent THz-power emitted by the nipnip-photomixer exhibits the expected external-voltage dependence.

CONCLUSION AND OUTLOOK

We have successfully demonstrated a novel THz-emitter concept, which allows to overcome the power-limiting roll-offs, namely the RC-roll-off and the transit-time-roll-off. As can be seen from comparison with conventional pin-photomixers, the AlGaAs-based nipnip-emitter proves to be highly efficient. Since the transport-properties of the (Al)InGaAs-material system allows for an even better performance of the emitter, the concept will also be adapted and realized in this material-system, creating an efficient THz-photomixer operating around $1.55\,\mu m$.

ACKNOWLEDGMENTS

This work has been supported by the Deutsche Forschungsgesellschaft (DFG) and the European Space Agency (under the reference number ESA-Photonic-THz-II/002).

REFERENCES

1. Brown, E. R., *Appl. Phys. Lett.*, **66**, 4903–4905 (1995).
2. Hirata, A., *Electr. Lett.*, **38**, 798–799 (2002).
3. Eckardt, M., *Physica E*, **17**, 629–630 (2003).
4. Pohl, P., *Appl. Phys. Lett.*, **83**, 4035–4037 (2003).

Short decay times of the THz photoresponse in quantum Hall Corbino detectors with spectral tunability

A. Hirsch[*], C. Stellmach[*], N.G. Kalugin[*,†], G. Hein[**], Yu. Vasilyev[‡] and G. Nachtwei[*]

[*]*Institut f. Technische Physik, TU-Braunschweig, Mendelssohnstr. 2, 38106 Braunschweig, Germany*
[†]*Department of Physics, Texas A&M University, College Station, TX 77843-4242, USA*
[**]*Physikalisch-Technische Bundesanstalt, Bundesallee 100, 38116 Braunschweig, Germany*
[‡]*A.F. Ioffe Physical Technical Institute, Polytekhnicheskaya 26, 194021 St. Petersburg, Russia*

Abstract. The THz photoconductivity of quantum Hall systems (QHS) around filling factor $\nu=2$ has been investigated. As radiation source a pulsed p-Ge Laser tunable in the frequency range form 1.7 THz to 2.5 THz is used. Via variation of the magnetic field, the carrier concentration (tuned by a gate) and the source drain voltage the spectral resolution of QHS as THz detector is determined. The spectral resolution is found best at voltages close to the breakdown value. In addition we report time resolved measurements with relaxations times from 20 ns to 150 ns.

INTRODUCTION

Far infrared (FIR) photoconductivity in quantum-Hall-(QH)-systems [1] is interesting both with respect to the applications for FIR detection and also with respect to the basic understanding of the conductivity mechanisms. QHS are promising for high sensitive and spectral tunable FIR detectors [2]. In these systems the transition from the QH in the dissipative state is used. Even so QH-detectors have been under research for over 20 years there are still open questions about the fundamental photoconductivity mechanisms.

In this work we present spectrally and time resolved photoresponse (PR) measurements of QH-systems in Corbino geometry. The FIR-source is a pulsed light-hole p-Ge Laser [3], which can be tuned in the frequency range from 1.7 THz to 2.5 THz (180 μm to 120 μm). The monochromatic radiation and the short switching times allow spectral and time resolved measurements.

EXPERIMENTAL SETUP

The Laser system (FIR source) and the QH-sample (FIR detector) are mounted in a He bath cryostat ($T \approx 4$ K). The magnetic fields (≤ 4 T for the Laser, ≤ 10 T for the detector) are generated by superconducting coils.

MBE grown GaAs/AlGaAs heterojunctions are used to realize a two dimensional electron system (2DES). Mobility and carrier concentration are $n_s = 2.1 \cdot 10^{15}\,\text{m}^{-2}$ and $\mu = 50\,\text{m}^2/(\text{Vs})$. By photolithography the sample is structured in circular Corbino geometry (inner/outer radius: 500 μm/1500 μm).

In the time resolved measurements an impedance matched detector circuit is used (time constant approx. 6 ns; see Figure 2A) and the Laser is driven by a special FET-based high power pulse generator, which provides very fast switching times (approx. 20 ns).

SPECTRAL RESOLUTION

The response of QH-samples to FIR radiation is due to two mechanisms: The bolometer effect, which is pinned at the QH minimum and cyclotron-resonance (CR), appearing at $E_\text{Laser} = \hbar\omega_\text{c} = \hbar eB/m^*$ (with the GaAs-effective mass $m^* = 0.067 m_\text{e}$) [4]. Variation of the magnetic field B enables spectral tunability.

The following method is used to qualify the spectral resolution: The Laser is adjusted at a certain wavelength. The PR is measured versus B around the filling factor $\nu=2$. This is done for different carrier concentrations, tuned by a gate voltage. At the maximum of the PR the relative position of the Fermi energy in the density of states of the 2DES is constant and the wavelength dependent effect on the spectral resolution is avoided.

Via variation of both the magnetic field and the carrier concentration a determination of the spectral resolution is possible. This is done for different source-drain voltages V_SD, as seen in Figure 1.

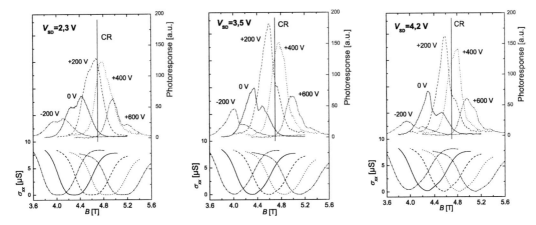

FIGURE 1. In the lower part transport curves around $\nu=2$ are shown. The upper part shows PR versus B for different gate voltages (marked at the curves). It is displayed for three different V_{SD}. The CR position corresponds to $E_{Laser} = 8.13$ meV.

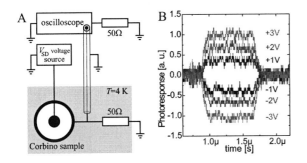

FIGURE 2. Time resolved PR measurements: A: Circuit with impedance matched signal line. B: PR versus time at filling factor $\nu = 2$, the photonenergy is E_{Laser}=8.28 meV, V_{SD} is marked at the curves.

TIME RESOLVED PHOTORESPONSE

With the pulsed Laser radiation time resolved PR is measured. Figure 2B shows the PR over the time for different V_{SD}. The start of the radiation causes a fast increase of the PR signal. During the illumination the PR stays constant. Finally the stop of the radiation causes fast relaxation of the PR. The magnitude of the PR depends on V_{SD}, as expected.

RESULTS AND DISCUSSION

Our investigations show, that the spectral resolution of the Corbino shaped FIR detectors is a function of the source-drain voltage V_{SD}. At low V_{SD} voltages the spectral resolution is relatively low. It reaches the maximum around the QH breakdown voltage V_c=3.5V. Further increasing of V_{SD} beyond the breakdown voltage causes a decreasing of the spectral resolution after all. Our data show the highest spectral resolution of $\Delta E/E \approx 0,07$ at a $V_{SD} = 3.5$ V. An earlier investigated sample (meander shaped) shows less spectral resolution, which is reached at currents above the breakdown value [5, 6].

In the time resolved measurements we see fast relaxation times from 20ns to 150ns. These time scales are comparable to those reported in [7] (estimated by an indirect method), but several orders of magnitude shorter in comparison to data published at meander type detectors [2, 4]. We find the relaxation time changing with variation of V_{SD}.

In conclusion, our Corbino devices are suitable for spectrally tunable FIR detectors with short response times.

ACKNOWLEDGMENTS

This work was supported by the Deutsche Forschungsgemeinschaft (DFG-Schwerpunktprogramm "Quanten-Hall-Systeme", project No. NA235/10-2). The authors thank J.M. Guldbakke, B. Coughlan and J. Wohlgemuth for supporting some of the experiments.

REFERENCES

1. K. von Klitzing et al., Phys. Rev. Lett. **45**, 494 (1980)
2. Y. Kawano et al., J. Appl. Phys. **89**, 4037 (2001)
3. E. Gornik, A. A. Andronov (Ed.), Opt. Quantum Electron., **23** Special Issue, 111 (1991)
4. N. G. Kalugin et al., Phys. Rev. B **66**, 085308 (2002)
5. A. Hirsch, Diploma thesis, TU-Braunschweig (2003)
6. C. Stellmach et al., Semicond. Sci. Technol. **19**, 454 (2004).
7. K. V. Smirnov et al., JETP Letters **71**, 31 (2000)

Acceleration Dynamics of Bloch Oscillating Electrons in Semiconductor Superlattices Investigated by Terahertz Electro-optic Sampling Method

Norihiko Sekine and Kaz Hirakawa

Institute of Industrial Science, University of Tokyo, 4-6-1 Komaba, Meguro-ku, Tokyo 153-8505, Japan
Phone: +81-3-5452-6261, Fax: +81-3-5452-6261, e-mail: nsekine@iis.u-tokyo.ac.jp

Abstract. We have investigated Bloch oscillations in semiconductor superlattices by terahertz (THz) electro-optic sampling method. The waveform of the emitted THz radiation reflects the motion of the photoexcited carriers in the miniband dispersion and is proportional to the acceleration of the Bloch oscillating electrons. Asymmetric positive and slightly larger negative amplitudes imply dynamical negative conductivities. Furthermore, the transient velocity was obtained by integrating the measured THz electric fields. By comparing the transient velocity with theory, the energy relaxation time of Bloch oscillating electrons has been estimated to be of the order of a few ps.

Recently, we have presented a strong experimental support for the terahertz (THz) gain of Bloch oscillating electrons in semiconductor superlattices (SLs) [1]. Since the acceleration/deceleration dynamics of electrons in SL minibands govern the Bloch gain, it is crucial to clarify how electrons traverse the miniband in the presence of scattering. In this work, we have systematically investigated the acceleration dynamics of Bloch oscillating electrons by the time-domain THz electro-optic (EO) sampling method.

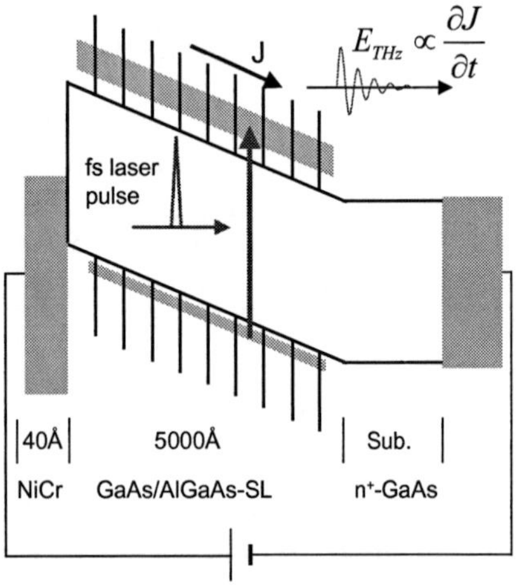

FIGURE 1. A schematic band diagram of the GaAs/AlGaAs (6.5 nm/ 2.5 nm) m-i-n diode. The miniband width was 30 meV.

The samples used in this work were undoped GaAs/$Al_{0.3}Ga_{0.7}As$ superlattice m-i-n diodes (Fig. 1). Electrons were photoexcited at the bottom of the ground miniband by femtosecond laser pulses. Transient THz electric fields that originate from subsequent acceleration of photoexcited carriers in biased minibands were detected by the THz EO sampling technique. The measurements were performed at 10 K for various bias electric fields.

It is found that the THz transients measured at low bias fields (~a few kV/cm) exhibit a positive signal right after the photoexcitation, which is, then, followed by a negative amplitude (Fig. 2). As the bias electric field is increased up to ~10 kV/cm, the temporal evolution of the THz signal becomes faster and exhibits an oscillatory behavior. As seen in the figure, the negative amplitudes are slightly deeper than the positive ones. Since the measured THz transients are proportional to the acceleration of electrons in the ground miniband, asymmetric positive and negative amplitudes in the THz transients indicate that the ensemble of photoexcited electrons do not traverse the

CP772, *Physics of Semiconductors: 27th International Conference on the Physics of Semiconductors,*
edited by José Menéndez and Chris G. Van de Walle

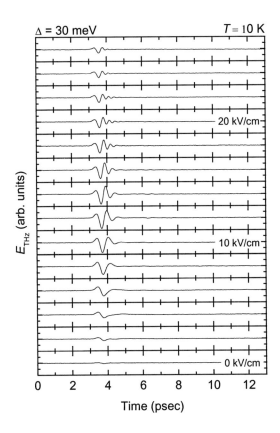

FIGURE 2. Dependence of the emitted THz waveforms on the bias electric fields measured at 10 K.

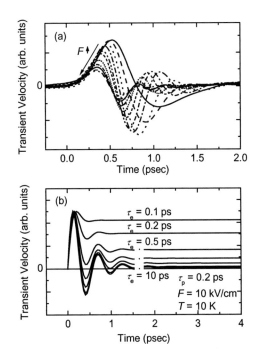

FIGURE 3. (a) Transient velocity obtained by integrating the measured THz electric field. (b) Transient velocity calculated for various energy relaxation times (τ_e). The calculated velocity was obtained by solving the Boltzmann equation [4].

Brillouin zone at a constant rate, suggesting the importance of scattering [2,3]. In addition, larger negative amplitudes imply dynamical negative conductivities in the THz range [1,4].

Furthermore, since the measured THz transient reflects the acceleration of the carriers as mentioned above, the transient velocity is obtained by integrating the THz waveform. Figure 3(a) shows the measured transient velocity. It is found that the steady-state velocity is much smaller (less than 10 %) than the transient peak velocity at all bias electric fields. This fact indicates that the energy relaxation time, τ_e, is rather long in the SL. Figure 3(b) shows transient velocities calculated for various τ_e by using a theory developed by Ktitrov et al. [4]. By comparing the measured velocity waveforms with the calculated ones, τ_e is found to be of the order of a few ps. This estimation is also supported by our previous work [Ref. 1], in which the relaxation time was obtained by fitting the square of the conductivity spectrum to that of the measured THz emission spectrum.

REFERENCES

1. Y. Shimada, K. Hirakawa, M. Odnoblioudov, and K. A. Chao, Phys. Rev. Lett. **90**, 46806 (2003).
2. H. Kroemer, cond-mat/0007428; E. Schomburg, N. V. Demarina, and K. F. Renk, Phys. Rev. B **67**, 155302 (2003).
3. E. Schomburg, N. V. Demarina, and K. F. Renk, Phys. Rev. B **67**, 155302 (2003).
4. S. A. Ktitorov, G. S. Simin, and V. Ya. Sindalovskii, Sov. Phys. Solid State **13**, 1872 (1971).

Carrier and Coherent Lattice Dynamics of Si Probed with Ultrafast Spectroscopy

D. M. Riffe* and A. J. Sabbah*

Physics Department, Utah State University, Logan, UT 84322-4415

Abstract. We have studied the ultrafast near-IR reflectivity of native-oxide covered Si(100) using 28 fs pulses from an 800 nm Ti:sapphire oscillator. Initial carrier excitation density is $(5.5 \pm 0.3) \times 10^{18}$ cm^{-3}. The reflectivity variations reveal both carrier and phonon dynamics on times scales from a few 10's of fs to several hundred ps. Carrier-dynamics contributions to the reflectivity variations include polarization-grating, anisotropic-distribution, state-filling, and free-carrier terms. From these contributions we obtain a momentum relaxation time of 32 ± 5 fs, an energy relaxation time of 260 ± 30 fs, and a surface-recombination velocity of $(3 \pm 1) \times 10^4$ cm sec^{-1}. The measurement also excites and detects coherent response of the Si zone-center phonon. The phase of the oscillations indicates impulsive stimulated Raman scattering (ISRS) as the excitation mechanism of the coherent vibrations. The 64.2 ± 0.1 fs period of the oscillation is in excellent agreement with standard Raman scattering measurements. The coherent phonon decay time of 2.9 ± 0.1 ps is significantly shorter than 3.4 ± 0.2 ps deduced from Raman-scattering data from the same samples, but is in agreement with data from heavily doped p-type samples.

The development of nanoscale semiconductor devices has emphasized the need to understand carrier-carrier and carrier-phonon interactions on increasingly shorter time scales. These dynamics can be probed in real time with fs laser pump-probe experiments. Here we present a study of Si dynamics using fs pump-probe reflectivity from native-oxide covered Si(001).

In these experiments electrons from the valence band are excited into the conduction band by 1.55 eV pump pulses that are linearly polarized along the (110) direction. Considering both one and two-photon absorption, the hot holes and electrons each have on average 0.3 eV of excess energy above their respective band edges. This excitation geometry also coherently excites the Si zone-center optical phonon.

The excitation and subsequent relaxation of the hot carriers and phonons is probed with orthogonally polarized probe pulses. Time resolved reflectivity of the probe pulses is shown in Fig. 1. The data (shown as dots) exhibit an initial rise (A) followed by a fast decrease (B), a slower decrease (C), and then a recovery (D) which, on a time scale of approximately 1 ns, eventually reaches the initial baseline. The coherent phonon oscillations are also evident.

A quantitative model of the real-time carrier dynamics, which we outline here, is used to least-squares fit the data. Part A of the data is dominated by two degenerate-four-wave-mixing coherent transients. To describe them we follow Wherrett et al.[1] The first transient is due to scattering of the pump pulse from the polarization grating formed by the pump and probe pulses; it is essentially an autocorrelation signal and contains no carrier dynamical information. The second transient arises from the anisotropic (in k space) part of the carrier distribution; its decay is governed by momentum relaxation of the carriers. The remainder of the reflectivity signal (B, C, and D) can be described by free-carrier (Drude) and state-filling responses of the excited carriers.[2] Most importantly, the slow decay exhibited in C is dominated by the decrease in optical effective mass (due to band nonparabolicity) through the free-carrier response as the hot electrons and holes relax to their respective band edges.[3] This decay thus reveals the carrier energy relaxation time.

A quantitative fit with these four model components (as well as a model for the phonon oscillations, see below) is also shown in Fig. 1 as the curve through the data points. Each individual contribution is also illustrated, offset to the right of the data. The analysis reveals the following: a momentum relaxation time of 32 ± 5 fs and an energy relaxation time of 260 ± 30 fs. From analysis of data obtained on a longer time scale of 120 ps (not shown) we also extract a surface recombination velocity of $(3 \pm 1) \times 10^4$ cm sec^{-1} for our native-oxide covered samples. Further details of the analysis can be found elsewhere.[4][5]

Figure 2 shows the background-subtracted coherent phonon oscillations. This part of the reflectivity signal is modeled as as an underdamped decaying oscillator with the equation $\exp(-\Delta t/\tau_{coh})\sin(2\pi\Delta t/T + \phi)$, where Δt

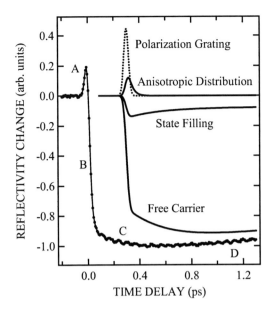

FIGURE 1. Reflectivity variations of Si(001) on a 1.3 ps time scale. Data are the dots; the fitted curve is the line through the data; components of the fit are offset to the right.

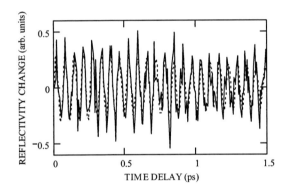

FIGURE 2. Coherent phonon variations in reflectivity. Data are the solid line; the fit is the dotted line.

is the pump-probe delay. A fit of the data with this equation is also shown in Fig. 2. (Note, the phonon data were actually fit over a 3 ps time scale.)

The phase ϕ can be used to identify the mechanism that excites the coherent phonon oscillations.[6] For a displacive type of excitation, such as is observed in the partially ionic semiconductor GaAs, $\phi = 90$ degrees. For purely covalent Si we expect an impulsive excitation. However, the impulse could come from either the pump-pulse photon field via impulsive stimulated Raman scattering (ISRS) or from the anisotropic part of the excited carrier distribution.[7] Because the ISRS driving term comes directly from the symmetric pump pulse, it will produce oscillations with $\phi = 0$ degrees. The peak of the anisotropic carrier distribution, however, is delayed by 13 fs and is somewhat asymmetric; based on the above results (shown in Fig. 1) we calculate that coherent phonons driven by the anisotropic distribution will have a phase of 70 degrees. From our fit of the phonon oscillations we extract a fitted phase of $\phi = 0 \pm 10$ degrees, thus indicating ISRS as the dominant excitation mechanism.

From our analysis we also extract $T = 64.2 \pm 0.1$ fs and $\tau_{coh} = 2.9 \pm 0.1$ ps. For comparison we have made standard Raman scattering measurements of our samples. The period T is in excellent agreement with the Raman determined value of $1/\nu = 64.12 \pm 0.03$ fs. The Raman Lorentzian line width indicates a coherence time of 3.4 ± 0.2 ps in that measurement. However, Raman scattering measurements of heavily doped p-type samples show a decrease in coherence time with increased doping.[8] Our measurement of 2.9 ± 0.1 ps for a carrier concentration of $(5.5 \pm 0.3) \times 10^{18}$ cm^{-3} falls directly in line with those measurements. Our extra width is thus explained by the same mechanism as in heavily doped p-type samples, namely the excitation of holes from the heavy-hole band to lighter-hole bands by the zone-center phonon.

In summary, our real-time measurements carrier and phonon dynamics has provided parameters for carrier-carrier, carrier-phonon, and phonon-carrier interactions in bulk Si.

ACKNOWLEDGMENTS

We thank Mark Holtz for making the Raman measurements on our samples.

REFERENCES

1. Wherrett, B. S., Smirl, A. L., and Boggess, T. F., *IEEE J. Quantum Electron.*, **19**, 680–690 (1983).
2. Bennett, B. R., Soref, R. A., and del Alamo, J. A., *IEEE J. Quantum Electron.*, **26**, 113–122 (1990).
3. Riffe, D. M., *J. Opt. Soc. Am. B*, **19**, 1092–1100 (2002).
4. Sabbah, A. J., and Riffe, D. M., *J. Appl. Phys.*, **88**, 6954–6956 (2000).
5. Sabbah, A. J., and Riffe, D. M., *Phys. Rev. B*, **66**, 165217 (2002).
6. Kütt, W. A., Albrecht, W., and Kurz, H., *IEEE J. Quantum Electron.*, **28**, 2434–2444 (1983).
7. Scholz, R., Pfeifer, T., and Kurz, H., *Phys. Rev. B*, **47**, 16229–16236 (1993).
8. Cerdeira, F., Fjeldy, T. A., and Cardona, M., *Phys. Rev. B*, **8**, 4734–4745 (1973).

Anharmonic Interactions in ZnO Probed with Impulsive Stimulated Raman Scattering

C. Aku-Leh[*], Jimin Zhao[*], R. Merlin[*], J. Menéndez[†] and M. Cardona[¶]

[*] *FOCUS Center and Department of Physics, University of Michigan, Ann Arbor, MI 48109-1120, USA*
[†] *Department of Physics and Astronomy, Arizona State University, Tempe, AZ 85287, USA*
[¶] *Max-Planck-Institut für Festkörperforschung, Heisenbergstraße 1, 70569 Stuttgart, Germany*

Abstract. Impulsive stimulated Raman scattering studies of ZnO show a dramatic increase in the lifetime of the low-energy E_2 optical phonon from 34 ps at 292 K to 211 ps at 5 K. The frequency of this mode at 5 K is (2.9789 ± 0.0002) THz, giving a quality factor $Q \sim 2000$. The temperature-dependent lifetime is dominated by two-phonon up-conversion processes and, at low temperatures, the lifetime is limited by isotopic disorder.

INTRODUCTION

Spontaneous Raman scattering is a powerful tool for the study of anharmonic interactions in solids [1]. In this paper, we use coherent impulsive stimulated Raman scattering (ISRS) [2] to study the temperature dependence of the lifetime and frequency of the low-frequency (LF) E_2 phonon in ZnO, referred to in the following as E_2(LF). Previous ISRS experiments at room temperature reported a lifetime of 29.2 ps [3]. Here, we find that the lifetime of this mode is significantly longer at low temperatures. We also show that the frequency of such long-lived modes can be measured with a high degree of accuracy using ISRS. Long-lived optical phonons hold promise for applications in semiconductor characterization, metrology and quantum information science.

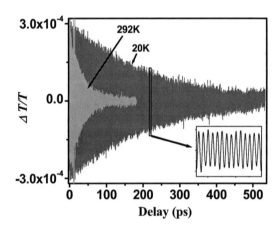

FIGURE 1. Differential transmission data showing E_2(LF) oscillations. The rectangle indicates the range of the scan shown in the inset.

EXPERIMENT

ISRS measurements were performed on a single crystal of ZnO. We used a standard pump-probe setup in the transmission geometry. The light source was a regenerative Ti:sapphire amplifier which provided pulses centered at 800 nm with width ~ 70 fs.

Figure 1 shows the pump-induced change in the transmission of the probe beam at 292 K and 20 K. Problems associated with the determination of time zero and multiple reflections of the beams cause the Fourier transform of the time-domain data to become a mixture of the real and imaginary parts of the phonon self-energy. To avoid these problems, we developed a phase-correction algorithm which gives a spectrum directly related to the imaginary part of the phonon self-energy and, hence, to the spontaneous Raman intensity. The so-obtained Fourier spectrum is shown in Fig. 2 for various temperatures. The frequency of the E_2(LF) mode at 5 K is $\Omega = (2.9789 \pm 0.0002)$ THz and the lifetime is (211 ± 7) ps. Thus, the frequency is obtained with five significant figures. The largest source of error is the uncertainty in the length of the delay line, followed by corrections arising from fluctuations

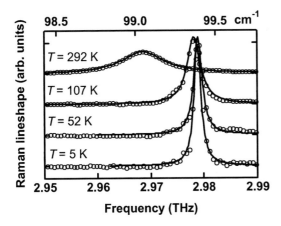

FIGURE 2. E_2(LF) Raman lineshapes obtained from the coherent phonon oscillations (circles). The solid lines are Lorentzian fits to the data.

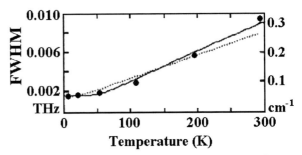

FIGURE 3. Full width at half maximum (FWHM) from the fits shown in Fig. 2. Dotted and solid lines correspond, respectively, to the down- and up-conversion fits; see text.

in the refractive index of air. Since there are relatively simple ways to reduce the experimental uncertainty, the ultimate precision of the ISRS technique has not been reached so far.

At temperatures that are low compared with the Debye temperature, the decay of optical phonons is determined by third-order anharmonic contributions [1]. The temperature-dependent linewidth Γ can be expressed as

$$\Gamma = \Gamma_0 + \Gamma_{DN}[1 + n(\omega_1) + n(\omega_2)] + \Gamma_{UP}[n(\omega_3) - n(\omega_4)]$$

where n denotes the Bose factor and $\omega_1 + \omega_2 = \omega_3 - \omega_4 = \Omega$. Here Γ_0, Γ_{DN} and Γ_{UP} are temperature-independent parameters associated with, respectively, decay caused by defects or disorder, and down- and up-conversion processes [4]. In ZnO, the E_2(LF) frequency is much lower than the maximum optical-phonon frequency. Therefore, we expect phonon down-conversion processes to make a small contribution to the phonon lifetime. Since the up-conversion term vanishes at zero temperature, however, the low-temperature lifetime in defect-free materials should still be dominated by down-conversion processes. The curves in Fig. 3 were obtained by considering only down or up-conversion contributions, in addition to the constant term Γ_0. It is apparent from a comparison of the two curves that the up-conversion fit is in better agreement with the experimental data. This strongly indicates that phonon up-conversion processes play a dominant role in the anharmonic decay of the E_2(LF) mode in ZnO. Theoretical estimates of the isotopic contribution to the linewidth of the E_2(LF) phonon provide additional support for the dominance of up-conversion processes. The corresponding phonon eigenvector consists mainly of Zn-displacements, so it is the isotopic disorder in the Zn sublattice which affects the phonon lifetime. Our natural ZnO crystals contain 48.6% ^{64}Zn, 27.9% ^{66}Zn, 4.1% ^{67}Zn, 18.75% ^{68}Zn, and 0.6% ^{70}Zn [6]. Therefore, the isotopic contribution to Γ should be sizable. A recent calculation by Serrano and co-workers, using *ab initio* phonon density of states, yields $\Gamma_0/\pi = 0.0018$ THz [5]. As shown in Fig. 3, this is very close to the experimental value at low temperatures, suggesting that the intrinsic anharmonic linewidth at these temperatures is significantly smaller, as expected for an up-conversion-dominated decay mechanism. This suggests that in isotopically pure samples the lifetime of the E_2(LF) phonon could be an order of magnitude longer (and perhaps even longer) than in our natural crystals. Lifetimes in the nanosecond range have not yet been reported for optical phonons in semiconductors.

ACKNOWLEDGMENTS

This work was supported by the AFOSR under contract F49620-00-1-0328 through the MURI program and by the NSF Focus Physics Frontier Center.

REFERENCES

1. See, e.g., Debernardi A., Baroni S. and Molinari E., *Phys.Rev.Lett.*, **75**, 1819-1822 (1995).
2. Merlin R., Solid State Commun. **102**, 207 (1997).
3. Lee I.H., Yee K.J, Lee K.G., Oh E. and Kim D.S., *J. Appl. Phys.*, **93**, 4939 (2003).
4. Wallis R.F. and Balkanski M., *Many Body Aspects of Solid State Spectroscopy* (North-Holland, Amsterdam, 1986).
5. Serrano J., Manjon F. J., Romero A. H., Lauck R. and Cardona M., *Phys.Rev.Lett.* **90**, 55510 (2003).
6. Rosman K. J. R. and Taylor P. D. P., Pure Appl. Chem. **70**, 217 (1998).

Non-resonant Raman efficiency in semiconductors under high pressure

C. Trallero-Giner*, K. Kunc[†,**] and K. Syassen[†]

*Dept. of Theoretical Physics, Havana University, Vedado 10400, Havana, Cuba
[†]Max Planck Institut für Festkörperforschung, Heisenbergstraße 1, D-70569 Stuttgart, Germany
[**]CNRS and Université Pierre and Marie Curie, T13-C80, 4 pl. Jussieu, F-75252 Paris-Cedex 05, France

Abstract. Intensities of the first- and second-order Raman scattering in zincblende-type semiconductors, under high hydrostatic pressure (P) and with laser energy below the fundamental gap, are analyzed. Theoretical description of the scattering efficiency and Raman cross section is carried out for optical and acoustic phonons in terms of electron-one-phonon and electron-two-phonon re-normalized interaction Hamiltonians. The calculations show a good agreement with measurements performed on ZnTe, and explain the observed intensity decrease (increase) with pressure for LO and TO optical modes (2TA acoustic phonon). The theory presented contains the essential ingredients of the Raman scattering response, which allows, in combination with experimental data, to extract information on electron-phonon interaction under hydrostatic pressure.

The effect of high pressure (P) on vibrational properties of II-VI and III-V semiconductors can be studied, among other methods, by Raman spectroscopy. The spectra corresponding to the first- and second-order processes, far from resonance, present nevertheless peculiar behavior under applied pressure [1]. It was shown that the integrated first-order scattering efficiency decreases as the pressure increases, while the intensity of the acoustic overtones increases with increasing P. Variation of the relative intensities of LO, TA, and LA combinations with applied pressure was found to be more complex. In the present work we study effect of pressure on the light scattering efficiency in bulk semiconductors with zincblende structure. We start from microscopic description of the scattering efficiency in terms of electron-one-phonon and electron-two-phonon re-normalized interaction Hamiltonians. Our theoretical representation is then tested against the available data on the first- and second-order Raman intensities in ZnTe. First-order Raman scattering on the TO-phonon takes place via electron deformation-potential (DP) while for the LO phonon the DP and interband Fröhlich interactions contribute to the light scattering. The DP interaction couples the heavy hole (HH) and light hole (LH) valence bands, while the electro-optical contribution (EO) couples the two upper conduction bands [2]. Assuming two independent HH and LH valence bands and scattering by a phonon with frequency ω_0 and wave vector $\mathbf{q}$, the Raman scattering efficiency can be written as

$$\frac{dS^{TO}}{d\Omega_S} = S_0(N+1)\left[d_0\frac{\omega_S}{\omega_L}\sqrt{\frac{\eta_S}{\eta_L}\frac{a_0}{\hbar\omega_0}}f(z,\omega_S)\right]^2, \quad (1)$$

where d_0 is the optical deformation potential, $z = \mu_{hh}/\mu_{lh}$, $\mu_{lh}(\mu_{hh})$ is electron-light- (heavy-) hole reduced mass, a_0 is the lattice constant, N_0 is the Bose-Einstein statistical factor, S_0 is a parameter independent on P, and f is given in Ref. [2]. For the LO-phonon we have to add the electro-optical contribution (EO) which couples the Γ_{1c} and Γ_{15c} conduction bands with the energy difference $\delta = E_{15c} - E_{1c}$. The electron-lattice Hamiltonian for a second order process contains contributions of the electron-two-phonon process to a first order and of the electron-one-phonon to second order. Far from resonance these interactions lead to a renormalized electron-two-phonon Hamiltonian, $\overline{H}^{(2)}_{E-L}$, containing both contributions with a renormalized electron-phonon deformation coupling constant $\overline{S}^{(2)}_{E-L}(\mathbf{q}_1,\mathbf{q}_2)$ [3]. Due to symmetry considerations the phonon scattering in the Γ_6 conduction band is forbidden and the Raman scattering efficiency for a pair of phonon branches ω_1 and ω_2 can be written as:

$$\frac{dS}{d\Omega_S} = \frac{3a_0^2}{(2\pi)^3}\frac{\overline{D}_1^2\eta_S}{2^3\eta_L}S_0$$
$$\int_{\omega=\omega_1+\omega_2}d\omega\frac{\mathcal{N}_A(\omega)}{\omega_1\omega_2}\left(\frac{\omega_L-\omega}{\omega_L}\right)^2$$
$$(N_1+1)(N_2+1)|f(z,\omega_L-\omega)|^2. \quad (2)$$

Here, $\mathcal{N}_A(\omega)$ is the two phonon density of states and $\overline{D}_1$ represents an average electron-two-phonon DP. In Ref. [1] the Raman intensifies in ZnTe were measured under hydrostatic pressure up to 9.5 GPa. To compare with the experimental data we took into consideration the pressure dependence of all parameters taking place in the scattering processes (see Ref. [2]). The energy difference between the Γ_{1c} and Γ_{15c} gap energies has been obtained by employing the pseudopotential plane-wave method within the Local Density Approximation (LDA) to the Density Functional Theory (DFT) [4]. The variation of $\delta = E_{15c} - E_{1c}$ with pressure is given in Ref. [2]. We found that variation of the deformation potential d_0 with pressure can be neglected. Figure 1 shows the measured integrated TO(Γ) and LO(Γ) intensities in ZnTe against pressure. The corresponding variation for the 2TA(X) is given in Fig. 2. The calculated Raman efficiencies (without any adjustable parameters) are shown by solid lines.

Analysis of the terms in Eqs. (1) allows to identify two main factors responsible for the observed decrease of the first-order integrated Raman intensity with pressure: variation of the energies of electronic transitions, as well as variation of the phonon frequency. The intensity decreases since both E_g and the optical phonon energy increase with the applied pressure. In case of the LO phonon also the effective electro-optical constant grows as the pressure rises. In the case of second-order scattering it is the combination of the phonon frequencies ω_1 and ω_2, the corresponding statistical factors N_1 and N_2, and the variation of E_g with pressure that are the main factors explaining the peculiar intensity behavior appearing in Fig. 2. For example, ω_{TA} decreases with the applied pressure and, following to Eq. (2), the scattering intensity should increase. Consequently, the scattering efficiency for overtones and combinations of optical and acoustic phonon modes may present different trends with the hydrostatic pressure (see the inset of Fig. 2). In conclusion, we have calculated, starting from first principles, Raman intensities of the first- and second-order Raman scattering on phonons, in zincblende semiconductors and far from resonance. The theory presented reproduces correctly the pressure variation of the integrated intensities of the TO(Γ), LO(Γ), 2TA(X), 2TA$_2$(K) and LO(X)+TA(X) peaks in ZnTe. Analysis of different contributions to the expression for the scattering cross-section identified the main mechanisms responsible for the pressure dependence of the intensities: variation with pressure of the phonon frequencies, statistical factors, and of the one-electron eigenvalues $E_n(\mathbf{k})$.

One of us (C. T-G.) would like to thank M. Cardona for helpful discussions and A. M. Alcalde for carefully reading of the manuscript. C. T-G. is also grateful to the A. von H. Foundation for financial support and to the MPI für FKF for hospitality.

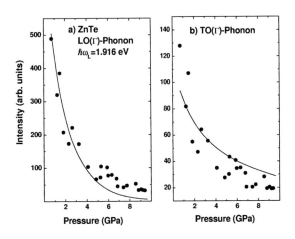

FIGURE 1. a) Integrated Raman intensity of the a) TO and b) LO phonon peaks in ZnTe as a function of pressure.

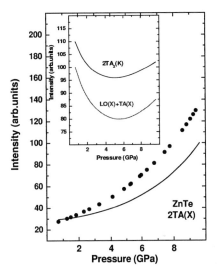

FIGURE 2. Integrated second order Raman intensity of the 2TA overtone in ZnTe as a function of pressure. The inset shows the calculated second order scattering efficiencies for the LO+TA combination and the 2TA$_2$ overtone phonon modes.

REFERENCES

1. Camacho, J., Loa, I., Cantarero, A., and Syassen, K., *J. Phys.: Condens. Matter*, **14**, 739 (2000).
2. Trallero-Giner, C., Kunc, K., and Syassen, K. (2004), to be published.
3. Alexandrou, A., Trallero-Giner, C., Kanellis, G., and Cardona, M., *Phys. Rev. B*, **40**, 1603 (1989).
4. Chang, K. J., Froyen, S., and Cohen, M. L., *Solid State Commun.*, **50**, 105 (1984).

Transient four-wave mixing of single exciton states: Exciton-exciton interaction and Rabi oscillations

B. Patton*, W. Langbein* and U. Woggon*

*Experimentelle Physik IIb, Universität Dortmund, 44221 Dortmund, Germany

Abstract. We present a novel heterodyne-detected four-wave mixing technique, which determines the third-order nonlinear polarization in phase and amplitude by spectral interferometry. It allows us to measure the resonant transient four-wave mixing of single localized excitons, both frequency and time-resolved. In the signal, the strength of the coupling between two probed exciton transitions is directly observable. Applying a strong optical pump pulse, we observe optical Rabi oscillations of the exciton polarization. Observing decay of the third-order polarization after the pump pulse, we can exclude the presence of significant excitation-induced dephasing.

INTRODUCTION

The coherent optical spectroscopy and manipulation of single exciton states is a prerequisite for their application in quantum information processing. Up to now, only few experimental results on this topic have been reported, such as non-degenerate four-wave-mixing [1], or transient pump-probe experiments on localized excitons [2, 3]. Rabi-oscillations of excitons in single InGaAs quantum dots were monitored instead by an indirect detection of the remaining exciton density via photoluminescence [4] or photocurrent [5]. We have implemented a novel heterodyne-detected four-wave mixing technique, which determines the third-order nonlinear polarization in phase and amplitude by spectral interferometry. It allows us to measure the resonant transient nonlinearity of single localized excitons, both frequency or time-resolved. In the third-order nonlinearity, two-exciton states contribute in the same order to the signal as the one-exciton states, so that the strength of the coupling between two probed exciton transitions is directly observable.

EXPERIMENTAL TECHNIQUE

A scheme of the experimental setup is shown in Fig. 1. We excite the transient four-wave mixing (FWM) of individual localized exciton transitions by two optical pulses 1,2 of a pulse duration tunable from 100 fs to 5 ps, which have a temporal separation τ controlled by a propagation delay line. In conventional four-wave mixing experiments on planar samples [6], the excitation pulses are send to the sample in different directions $\mathbf{k}_1$ and $\mathbf{k}_2$, so

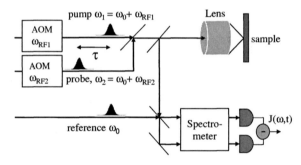

FIGURE 1. Schematical setup of the experiment. AOM: Acousto-optical modulators. Lens: High numerical aperture (0.85) microscope objective. Spectrometer: Imaging spectrometer of 15 μeV resolution

that the FWM signal can be collected in the background-free direction $2\mathbf{k}_2 - \mathbf{k}_1$. This method relies on a translational invariance of the sample, which is absent for single exciton states, which have a typical extension well below the wavelength of the exciting light. In such a case, the emission of the exciton state is emitted nearly isotropically in direction, and no directional selection can be employed. Additionally, in order to collect the signal, a lens of high numerical aperture is required. The selection of the FWM signal can be made using its phase coherence to the exciting pulses. By applying different phase shifts $\varphi_{1,2}(t) = \omega_{1,2}t$ to the excitation pulses 1,2 (using acousto-optical modulators), the FWM signal acquires the phase shift $\varphi_{\mathrm{FWM}}(t) = (2\omega_2 - \omega_1)t$. The phase shifts so slow that it can be considered constant within the pulse duration, but changes during the repetition period of the experiment (~ 13 ns). We interfere the signal field E_{sig} with a reference pulse field E_{ref}, resulting an

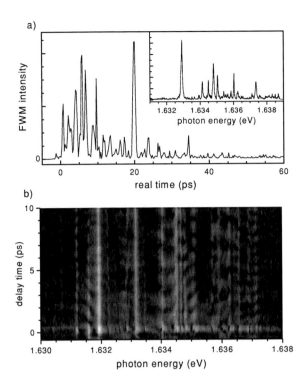

FIGURE 2. FWM intensity from a group of exciton states localized within an $(0.5\,\mu m)^2$ area. a) Fixed delay $\tau = 20\,\mathrm{ps}$. Time-resolved and spectrally resolved data. The time zero is the arrival time of pulse 2. b) Spectrally resolved versus delay time in a logarithmic grey scale over 3 orders of magnitude. The beating versus delay time evidences coherent coupling between exciton states.

FOUR-WAVE MIXING OF SINGLE EXCITON STATES

Using the described technique, we have investigated excitons localized by interface roughness in GaAs/AlAs single quantum wells. By a growth interruption at both interfaces, monolayer islands form, and excitons localized in these islands can be observed [9, 10]. A sample FWM spectrum is shown in the inset of Fig. 2a). Several emission lines are observable, with linewidths of 20-50 μeV. They arise from individual excitonic transitions. In the time-resolved FWM, the emission of these different transitions interferes, giving rise to an oscillating intensity. Interestingly, one can observe a precursor of the photon echo formation as a constructive interference at $t = \tau$ showing up as a peak (superradiance). In the delay-time dependent spectrally resolved intensity (Fig. 2b), the observed quantum beats are evidence of coherent coupling between exciton states. A quantitative analysis of the coupling strengths is readily possible.

ACKNOWLEDGMENTS

The sample was grown at the Research Center COM and the Niels Bohr Institute, Copenhagen University. The work was supported by the grant DFG WO477/14.

REFERENCES

1. Guest, J., Stievater, T. H., Chen, G., Tabak, E. A., Orr, B. G., Steel, D. G., Gammon, D., and Katzer, D. S., *Science*, **293**, 2224 (2001).
2. Guenther, T., Lienau, C., Elsaesser, T., Glanemann, M., Axt, V., Kuhn, T., Eshlaghi, S., and Wieck, A., *Phys. Rev. Lett.*, **89**, 57401 (2002).
3. Stievater, T. H., Li, X., Steel, D. G., Gammon, D., Katzer, D. S., and Park, D., *Phys. Rev. B*, **65**, 205319 (2002).
4. Htoon, H., Tagakahara, T., Kulik, D., Baklenov, O., Holmes, A. L., and Shih, C. K., *Phys. Rev. Lett.*, **88**, 087401 (2002).
5. Zrenner, A., Beham, E., Stufer, S., Findeis, F., Bichler, M., and Abstreiter, G., *Nature*, **418**, 612 (2002).
6. Shah, J., *Ultrafast Spectroscopy of Semiconductors and Semiconductor Nanostructures*, Springer, Berlin, 1996, chap. 2.
7. Langbein, W., Borri, P., Woggon, U., Stavarache, V., Reuter, D., and Wieck, A. D., *Phys. Rev. B*, **69**, 161301R (2004).
8. Lepetit, L., Chériaux, G., and Joffre, M., *J. Opt. Soc. Am. B*, **12**, 2467 (1995).
9. Leosson, K., Jensen, J. R., Langbein, W., and Hvam, J. M., *Phys. Rev. B*, **61**, 10322 (2000).
10. Savona, V., Langbein, W., and Kocherscheidt, G., *phys. stat. sol. (c)*, **1**, 501 (2004).

interference intensity $J(t) = 2\Re(E_{\mathrm{sig}}^* E_{\mathrm{ref}} e^{(2\omega_2 - \omega_1)t})$ that is oscillating with the frequency $2\omega_2 - \omega_1$. This property is then used in a Lock-In technique to isolate the interference of the FWM field. Additionally we note that the FWM of a single excitonic transition is a free polarization decay (FPD) rather than the photon echo (PE) observed for state ensembles. Due to the large temporal duration of the FPD relative to the reference pulse width, it is inefficient to use a direct balanced detection of the interference as we do in our FWM experiments on ensembles [7]. To overcome this limitation, we spectrally resolve the interference, and detect all spectral components of $J(\omega)$ simultaneously with a multichannel detector. Adjusting the reference delay to provide E_{ref} leading E_{sig}, we use spectral interferometry [8] to retrieve $E_{\mathrm{sig}}(\omega)$ in amplitude and phase. From this result, we can calculate both the time-resolved and the spectrally resolved FWM intensity.

Exchange Effects on Electronic States in QWs with e-h Plasma in an Electric Field

I. A. Fedorov[*], K. W. Kim[*], V. N. Sokolov[*,†], and J. M. Zavada[¶]

[*]*Department of Electrical and Computer Engineering, North Carolina State University, Raleigh, NC, USA*
[†]*V.E. Lashkar'ov Institute for Semiconductor Physics, Kiev, Ukraine*
[¶]*U.S. Army Research Office, Research Triangle Park, NC, USA*

Abstract. We study effects of electron-hole (e-h) plasma density N and a uniform electric field F on the ground and first excited eigenstates, energy levels and electron and hole wave functions, resulting from many-particle (Hartree and exchange) Coulomb interactions in a 2D e-h plasma. The coupled Schrödinger equations for electrons and holes are solved self-consistently in the Hartree-Fock approximation together with the Poisson equation. The solutions are analyzed treating N and F as independent parameters for quantum wells (QWs) with different width, d_{qw}. The calculations demonstrate that with decreasing d_{qw} and increasing N, the charge separation within the QW induced by the field becomes less effective and the relative contribution of the Hartree interactions to the energy level shifts is decreased. The results are applied to study possible bistable behavior of the QW electroabsorption under strong photoexcitation near the exciton resonance.

INTRODUCTION

In low-dimensional semiconductor heterostructures and quantum wells with free carriers (electrons or/and holes), the role played by the Hartree and exchange-correlation contributions to the total effective interparticle interaction may be important [1]. This leads to a substantial renormalization of electronic states: electron and hole band edges as well as excited subband energy levels, intersubband transitions [2], exciton properties [3, 4], etc. This fundamental phenomenon can be crucial for different nonlinear effects [5, 6] and various device applications utilizing near band-edge as well as intersubband transitions or resonant tunneling.

In this report, we investigate effects of the e-h plasma density N and uniform electric field F on the ground and first excited energy levels and corresponding wave functions resulting from many-body (exchange) interactions in a 2D e-h plasma. The coupled Schrödinger equations for electrons and holes are solved self-consistently in the Hartree-Fock approximation. Then, the solutions are analyzed treating N and F as independent parameters for different QW widths. In contrast to the Hartree potential, the exchange interaction terms depend on the electron/hole momentum $\mathbf{k}$. This leads to additional dependence of the electron and hole energies in excess of that corresponding to lateral movement in the QW [2]. Depending on the width d_{qw} and electric field F, our calculations demonstrate a strong competition between the exchange and Hartree interactions, which results in a nonmonotonous behavior of the electron and hole subband energies on the plasma density at a given F. The results are applied to study bistable regimes in the QW electroabsorption under strong photoexcitation near the exciton resonance.

GROUND-STATE ENERGIES AND WAVE FUNCTIONS

For numerical calculations, we choose the following parameters typical for GaN/AlGaN QWs, characterized by strong built-in electric field, similar to those used in [4]: $F < 1$ MV/cm, $N < 10^{13}$ cm^{-2}, where the electron and hole effective masses are $m_e = 0.3\ m_0$,

$m_h = 1\, m_0$, m_0 is the free electron mass, and the dielectric constant $\varepsilon = 7.3$.

The results obtained for the model of infinitely deep QW and zero temperature are shown in Fig. 1, where we depict the ground-state ($n = 0$, $k = 0$) electron energy E_e as a function of N for several electric field values. For weak fields, the charge neutrality

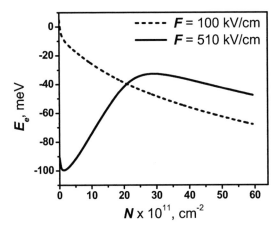

FIGURE 1. Electron ground-state energy.

minimizes direct Coulomb (Hartree) interactions and exchange contribution completely determines the energy shifts with increasing N. We found that with decreasing d_{qw}, the charge separation within the QW induced by the field becomes less effective and the relative contribution of the Hartree interactions to the energy shifts is also decreased. However, with increasing the QW width, the Hartree contribution becomes more considerable. Our calculations reveal a strong competition between the exchange and Hartree contributions resulting in a nonmonotonous dependence of the electron and hole (not depicted) energies on the plasma density at a given F.

In Fig. 2, we show the behavior of the normalized envelope wave functions $|\psi_{e,h}|^2$ corresponding to the ground-state energies calculated for two values of the plasma density, $N = 0$ and $N = 2\times10^{12}$ cm^{-2}, at the field $F = 510$ kV/cm. The results clearly demonstrate the charge separation within the QW induced by the field and its screening with increasing N. In contrast to the Hartree contribution, we reveal only a weak influence of the exchange interaction on the spatial shape of the wave functions.

ELECTRO-OPTICAL BISTABILITY

We apply the above results to the analysis of bistable regimes in the QW electroabsorbtion under photoexcitation at photon energies near the exciton resonance. These regimes can be described by the steady-state rate equation including the functions of electron-hole generation and recombination and a model of light absorption. For the absorption factor we assume a Lorentz shape as a function of the photon energy with the resonance (exciton) energy following the position of the electron and hole subbands. We found that the intrinsic bistability can be realized for both (positive and negative) values of detuning of the

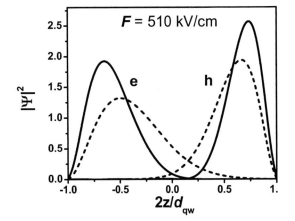

FIGURE 2. Electron (e) and hole (h) wave-functions.

photon energy relatively to the resonance. Interestingly, that the exchange interaction can lead to suppression (above the resonance) or initiation (under the resonance) of the revealed bistable regimes.

ACKNOWLEDGMENTS

The work was supported in part by the U.S. Army Research Office.

REFERENCES

1. Betbeder-Matibet O. and Combescot M., *Phys. Rev. B* **54**, 11375-11385 (1996).

2. Choe J.-W., Byungsung O, Bandara K.M.S.V., Coon D.D., *Appl. Phys. Lett.* **56**, 1679-1681 (1990).

3. Traetta G., Cingolani R., Di Carlo A., Della Sala F., and Lugli P., *Appl. Phys. Lett.* **76**, 1042-11044 (2000).

4. Nikolaev V.V. and Portnoi M.E., *Phys. Stat. Sol.* (a) **190** 113-119 (2002).

5. Di Carlo A., Reale A., Lugli P. *et al.*, *Phys. Rev. B* **63**, 235305-1-4 (2001).

6. Butov L.V. *et al.*, *Phys. Rev. Lett.*, **92** 117404-1-4 (2004); Rapoport R. *et al.*, *Phys. Rev. Lett.*, **92** 117405-1-4 (2004).

Rabi Oscillations of Ultrashort Pulses in 1.55-μm InGaAs/InGaAsP Quantum-well Optical Amplifiers

J.-Z. Zhang* and I. Galbraith*

Physics, School of Engineering and Physical Sciences, Heriot-Watt University, Edinburgh, EH14 4AS, UK.

Abstract. Using the Foreman effective mass Hamiltonian for $In_xGa_{1-x}As/In_yGa_{1-y}As_zP_{1-z}$ quantum wells, the propagation of a 150 fs pulse in an optical amplifier was calculated taking into account the multi-subband carrier dynamics and heating as well as the polarization dynamics. The intensity of the propagated pulse as well as its imposed frequency chirp, the carrier density and temperature show strong Rabi oscillations.

INTRODUCTION

In recent years, pulse propagation in semiconductor optical amplifiers (SOAs) has attracted considerable attention because of its prospective applications in optical communications. The gain dynamics during pulse amplification is governed by two nonlinear processes, namely spectral hole burning (SHB) and carrier heating (CH). The former is frequency-dependent and thus reflects the *local* carrier dynamics, as only the carriers energetically inside the pulse spectrum are involved. The mechanism for the latter is that the low energy carriers are taken away by stimulated emission, or the high energy carriers are injected via photon absorption, and thus the mean kinetic energy of the entire carrier system is increased, producing an elevated effective temperature. Thus CH governs a *global* carrier dynamics. A recent experiment [1] has demonstrated that, using CH, one can achieve ultrafast switching in semiconductor lasers.

There has been some experimental and theoretical work on pulse propagation in SOAs, however, most theoretical studies focused on bulk SOAs. In this paper, we theoretically study how the carrier dynamics, the SHB and the CH from stimulated emission, affect the pulse propagation in a realistic 1.55-μm quantum well (QW) SOA, including the bandstructure effects. As the SHB and CH occur simultaneously, for instance, leading to the gain saturation when the propagating pulse is or grows intense, we treat them on the same footing.

THEORY

The interband polarization and carrier population dynamics in QWs can be described by

$$\frac{\partial p_\mathbf{k}^{i_e i_h}}{\partial t} = -i(\omega_\mathbf{k}^{i_e i_h} - \omega_0 - i\gamma_p)\, p_\mathbf{k}^{i_e i_h} - \frac{i}{\hbar}\mu_\mathbf{k}^{i_e i_h} E(n_\mathbf{k}^{i_e} + n_\mathbf{k}^{i_h} - 1),$$

$$\frac{\partial n_\mathbf{k}^{i_e}}{\partial t} = \frac{i}{\hbar}\sum_{i_h}(\mu_\mathbf{k}^{i_e i_h} p_\mathbf{k}^{i_e i_h *} - \mu_\mathbf{k}^{i_e i_h *} p_\mathbf{k}^{i_e i_h})E - \gamma_e(n_\mathbf{k}^{i_e} - f_\mathbf{k}^{i_e}),$$

where i_e (i_h) is the conduction (valence) subband index, $f_\mathbf{k}^{i_e}$ is the Fermi-Dirac distribution function for electrons, E is optical electric field envelope, ω_0 is the center frequency of the pulse, γ_p is the polarization dephasing rate, and γ_e the electron-relaxation rate. $\hbar\omega_\mathbf{k}^{i_e i_h}$ and $\mu_\mathbf{k}^{i_e i_h}$ are the transition energies and dipole matrix elements. The population equation for the holes is similar. In above equations, the dependence on z (for propagation) and t were omitted for clarity.

A reduced wave equation is employed to describe the evolution of the electric field envelope,

$$\frac{\partial E(z,t)}{\partial z} + \frac{n}{c}\frac{\partial E(z,t)}{\partial t} = \frac{i\omega_0 \Gamma}{n\varepsilon_0 cV}\sum_{i_e i_h}\sum_\mathbf{k}\mu_\mathbf{k}^{i_e i_h}\, p_\mathbf{k}^{i_e i_h}(z,t),$$

which is coupled to the polarization equations via the polarization source term. n is the background refractive index, Γ the waveguide confinement factor, and V the QW volume. Numerically solving the coupled equations gives us the intensity and the phase distributions as the pulse travels through the amplifier.

Note that the functions $f_\mathbf{k}^i$ and thus the electron and hole chemical potentials as well as the carrier temperatures are both time and position dependent. We self-consistently computed them at each time step, using energy and carrier density conservation. Band structure calculations were performed within a multi-band $\mathbf{k}\cdot\mathbf{p}$ method and the Foreman effective mass Hamiltonian[2]

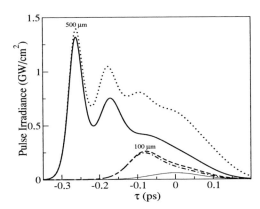

FIGURE 1. The pulse shape after 100 μm (dashed: without CH; dash-dotted: including CH) and 500 μm (dotted: without CH; solid: including CH) propagation distances. The thin solid line is for the shape of input pulse.

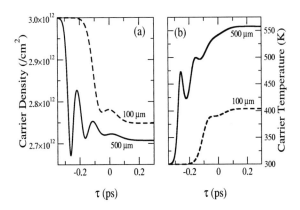

FIGURE 2. (a) The carrier density and (b) the carrier temperature in the amplifier after the pulse propagates 100 μm (dashed) and 500 μm (solid).

was used for the valence band. One conduction subband and three valence subbands were used to obtain converged chemical potentials.

RESULTS AND DISCUSSIONS

The investigated SOA structure is for strained $In_xGa_{1-x}As/In_yGa_{1-y}As_zP_{1-z}$ QWs on an InP substrate [3]. The well and barrier thicknesses are both 50 Å. The wells have 1.3% compressive strain. Γ is taken as 0.1. The initial carrier density and temperature are chosen to be $3 \times 10^{12}/cm^2$ and 300 K. The polarization and carrier-relaxation times, γ_p^{-1} and γ_e^{-1} (γ_h^{-1}), are taken to be 80 fs and 160fs, respectively. To follow pulse propagation, we choose a Gaussian input pulse of 150 fs FWHM irradiance, centered at 1.55 μm (i.e., 0.8 eV) and with a peak intensity of 60 MW/cm^2. Only transverse electric polarization is considered here.

Figure 1 shows the calculated intensity of the amplified pulse after it propagates 100 μm and 500 μm both with and without the CH effects. From Fig. 1 we can see that, for the 100 μm propagation, the pulse slightly re-shapes and advances. As the propagation distance increases (500 μm), however, the pulse re-shapes significantly, and its intensity exhibits strong oscillations. This is due to the Rabi oscillation [4], as we found that the population inversion in the amplifier oscillates across the traveling time of pulse, and the number of oscillations is determined by the pulse area $\hbar^{-1} \int dt \mu E(t)$. The pulse spectrum (not shown) is considerably distorted: the central peak is red-shifted with respect to the input spectrum, and with a low energy side band and a smaller sideband on the high energy tail of the main peak. The side bands tend toward the major peak with propagation, reflecting the carrier redistribution about the spectral hole. However, the asymmetry in the side bands is due to the gain dispersion.

Clearly, as can be seen in Fig. 1, the pulse intensities are strongly reduced by the CH for the longer-distance propagation. The pulse energy has reduced by 36%. During stimulated emission, the carrier density in the amplifier is depleted but the temperature is raised to 550 K, as shown in Fig. 2.

CONCLUSION

We studied the pulse propagation in strained $In_xGa_{1-x}As/In_yGa_{1-y}As_zP_{1-z}$ QW amplifiers, based on a band structure calculation and the use of microscopic interband polarization equations as well as multi-subband carrier population equations. We found that, due to the Rabi oscillation, the intensity of a propagated pulse as well as the carrier density and temperature in the amplifier show strong oscillations.

ACKNOWLEDGMENTS

We acknowledge the support from the EPSRC project 'The Ultrafast Photonics Collaboration'.

REFERENCES

1. Elsässer, M., Hense, S. G., and Wegener, M., *Appl. Phys. Letters*, **70**, 853–855 (1997).
2. Foreman, B. A., *Phys. Rev. B*, **48**, 4964–4967 (1993).
3. Dorren, H. J. S. et al., *IEEE Photon. Technol. Lett.*, **15**, 792–794 (2003).
4. Allen, L., and Eberly, J. H., *Optical Resonance And Two-Level Atoms*, John Wiley and Sons, New York, 1975, pp. 52–77.

Temperature Dependence of the Dephasing of Excitonic and Biexcitonic Polarization in a ZnSe Single Quantum Well

T. Voss*, L. Wischmeier*, H. G. Breunig*[†], I. Rückmann* and J. Gutowski*

*Institute for Solid State Physics, University of Bremen, D-28334 Bremen, Germany
[†]Institute for Physical Chemistry, University of Marburg, D-35032 Marburg, Germany

Abstract. The polarization decay of the exciton and bound-biexciton states in a ZnSe single quantum well was studied as a function of the temperature in real-time-resolved pulse-transmission and spectrally resolved four-wave-mixing experiments. Optical and acoustic-phonon scattering coefficients for both resonances were determined and found to be comparable. Additionally, the impact of the dephasing of the coherent excitonic polarization on coherent-control experiments was analyzed as a function of the sample temperature.

The dephasing time T_2 of coherent excitonic polarization is an important property of any semiconductor heterostructure [1, 2, 3, 4]. It is a measure of the time after which the coherence between the induced microscopic excitonic oscillators inside the sample is destroyed by various processes like Coulomb scattering or coupling to acoustic and optical phonons. If the radiative decay is much slower than the dephasing T_2 can be easily extracted from the linewidths of resonances or decay times of their transients in four-wave-mixing (FWM) experiments in which two ultrashort laser pulses with wave vectors k_1 and k_2 induce a transient excitonic polarization grating in the sample. This leads to the emission of signals in the diffraction directions $2k_2 - k_1$ and $2k_1 - k_2$.

By measuring T_2 one can therefore get a deeper insight into the fundamental processes which govern the ultrafast coherent dynamics of the excitonic polarization. Additionally, the dephasing processes also limit the time scale on which the coherent polarization can be actively manipulated by so-called coherent-control techniques. Since coherent control has become an important experimental method for quantum information processing [5] it is important to analyze how dephasing processes due to the coupling to phonons limit the control over the coherent excitations. Coherent-control experiments make use of ultrashort laser pulses which possess a spectral width of at least several nanometers. If applied to the exciton energy regime of a semiconductor not only the exciton transition (XT) but also the associated exciton-biexciton transition (EBT) is commonly excited.

To study the dephasing of the excitonic and biexcitonic polarization FWM experiments were carried out in transmission geometry. The experiments were performed at temperatures between 4 and 100 K with exclusive res-

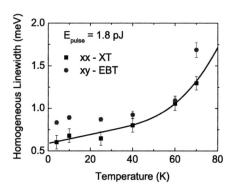

FIGURE 1. Linewidths of the XT and EBT determined in spectrally resolved FWM experiments. Solid line: numerical fit with $E_{LO} = 31$ meV to the XT data with the formula given in the text. Extracted linewidth without phonon broadening $\gamma_0 = (0.6 \pm 0.05)$ meV.

onant excitation of the heavy-hole exciton-biexciton system of a ZnSe single quantum well by 120 fs pulses.

From the temperature dependence of the decay of the nonlinear FWM polarization at the exciton and exciton-biexciton transition the acoustic- and optical-phonon-scattering coefficients were obtained for each of the two transitions according to the formula [1]

$$\Gamma^{\text{hom}}(T) = \gamma_0 + a \cdot T + b/[\exp(E_{LO}/k_B T) - 1]$$

with $\Gamma^{\text{hom}} = 2T_2/\hbar$. The temperature dependence of the homogeneous linewidths of the XT (measured with co-linearly polarized pulses) and EBT (measured with cross-linearly polarized pulses) is shown in Figure 1. Within the accuracy of the measurements the linewidth of the EBT was about 0.2 meV larger than that of the

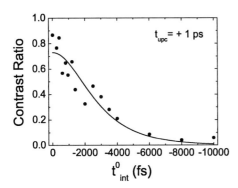

FIGURE 2. Contrast ratio of the coherent-control signal as a function of t_{int}^0 at 4 K. Circles: measured values, solid line: numerical fit yielding $T_2 = 1.9$ ps.

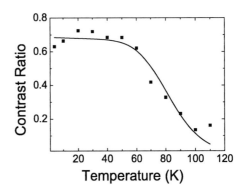

FIGURE 3. Contrast ratio of the coherent-control signal as a function of the temperature measured at $t_{upc} = +1.0$ ps. Squares: measured values, solid line: simulation.

XT for all measured temperatures. For both resonances the temperature dependence can be well described by the coupling to acoustic (coefficient $a \approx 2-9\,\mu\text{eV}$) and optical phonons (coefficient $b \approx 80-120\,\text{meV}$). The solid line is a fit with the formula given above.

In a second step, the impact of the dephasing of the coherent polarization on coherent-control experiments was analyzed. These experiments were conducted in a pulse-transmission configuration using a pair of phase-locked laser pulses that were analyzed after transmission with respect to their real-time shape by use of an up-conversion technique with a temporal resolution of $\Delta t_{upc} < 50$ fs. Each one of the two pulses induced a polarization in the sample. These two polarizations interfered constructively or destructively depending on the temporal separation $t_{int} = t_{int}^0 + \Delta t_{int}$ of the phase-locked pulses which is linked with their relative phase. t_{int}^0 denotes the basic temporal separation (in the fs-ps range) and Δt_{int} its fine tuning on an attosecond time scale. The pulse pairs were generated by use of an actively stabilized Michelson interferometer. Low excitation intensities were used for the pulse-transmission experiments so that only excitonic polarization without biexcitonic contributions was excited in the sample.

For these low excitation intensities the resulting contrast ratio $CR = \frac{I_{max}-I_{min}}{I_{max}+I_{min}}$ of the coherent switching signal of excitonic polarization was studied first as a function of t_{int} for a fixed temperature (Figure 2). The signal from the resulting coherent polarization was detected at $t_{upc} = 1$ ps, i.e., 1 ps after the arrival of the second phase-locked pulse. The measured dependence of the contrast ratio on t_{int} can be well explained by a linear interference of the two polarizations induced by each one of the phase-locked pulses. This leads to a change of the contrast ratio as a function of the basic temporal separation t_{int}^0 according to $CR(t_{int}^0, T_2) = CR_0 \times (\cosh(t_{int}^0/T_2))^{-1}$. The solid line in Figure 2 shows a numerical fit with this function which excellently reproduces the experimentally found dependence and yields $T_2 = 1.9$ ps. This value of T_2 is in agreement with the one determined from the signal decay in FWM experiments. The constant CR_0 accounts for the finite accuracy of the interference of the two phase-locked pulses for $t_{int}^0 = 0$ fs.

Figure 3 shows the contrast ratio as a function of the temperature measured for a fixed $t_{int}^0 = -800$ fs. If the decrease of CR with increasing T is entirely caused by a faster dephasing due to the enhanced interaction with phonons at elevated temperatures then the dependence should be completely determined by the parameters extracted from Figures 1 and 2. Indeed, the solid line which shows the simulated (not fitted!) dependence of $CR(T)$ is in an excellent agreement with the measured values (squares). This proves the linear interference of the induced polarizations in the coherent-control experiment at low excitation densities and shows that the dephasing of the coherent polarization significantly limits the achievable contrast ratio at temperatures above 60 K.

The authors thank W. Faschinger (deceased 2001), Würzburg University, for providing the sample. This work has been supported by the Deutsche Forschungsgemeinschaft (Grants Nos. Gu 252/12 and Ne 525/8).

REFERENCES

1. Langbein, W., and Hvam, J. M., *Phys. Rev. B*, **61**, 1692 (2000).
2. Breunig, H. G., Voss, T., Kudyk, I., Rückmann, I., and Gutowski, J., *phys. stat. sol. (c)*, **1**, 843 (2004).
3. Erland, J., Lyssenko, V. G., and Hvam, J. M., *Phys. Rev. B*, **63**, 155317 (2001).
4. Wang, H., Ferrio, K., Steel, D. G., Hu, Y. Z., Binder, R., and Koch, S. W., *Phys. Rev. Lett.*, **71**, 1261 (1993).
5. Li, X., Wu, Y., Steel, D., Gammon, D., Stievater, T. H., Katzer, D. S., Park, D., Piermarocchi, C., and Sham, L. J., *Science*, **301**, 809 (2003).

Microscopic theory for nonlinear polariton propagation

S. Schumacher*, G. Czycholl* and F. Jahnke*

Institute for Theoretical Physics, University of Bremen, 28334 Bremen, Germany

Abstract. The interplay of polariton propagation effects and optical nonlinearities in thin semiconductor layers is investigated within a microscopic theory. A density-matrix approach for the description of the excited semiconductor is used to fully include coherent nonlinearities up to third order in the optical field. The propagating light field is determined directly from Maxwell's equations coupled to the semiconductor medium. In contrast to the linear regime, Coulomb interaction turns out to strongly couple energetically well-separated polaritonic states in nonlinear optics.

Recently a microscopic theory for linear polariton propagation in semiconductor layers has been successfully applied to the description of linear optical transmission spectra [1, 2, 3]. A direct solution of the two-particle Schrödinger equation for the electron-hole-motion within the sample boundaries together with Maxwell's equations avoids ambiguities due to additional boundary conditions (ABCs). We present an extension to nonlinear polariton propagation that combines propagational effects under the influence of sample boundaries with excitonic and biexcitonic nonlinearities treated within the dynamics-controlled truncation (DCT) approach [4, 5].

We consider a two-band model with spin-degenerate conduction and valence bands to describe the energetically lowest interband transitions for a semiconductor layer with strain-split light- and heavy-hole valence bands. The spatially inhomogeneous polarization in the semiconductor,

$$\mathbf{P}(z,t) = \sum_{\text{eh}\mathbf{k}} \mathbf{d}_{\text{eh}}^* p_{\mathbf{k}}^{\text{eh}}(z,z,t) = \sum_{\text{eh}\mathbf{k}} \mathbf{e}_{\text{eh}} d_{\text{eh}}^* p_{\mathbf{k}}^{\text{eh}}(z,z,t), \quad (1)$$

can be given in terms of the non-local excitonic transition amplitude $p_{\mathbf{k}}^{\text{eh}}(z_e,z_h,t) = \langle \psi_{\mathbf{k}}^{h}(z_h) \psi_{\mathbf{k}}^{e}(z_e) \rangle$ with the electron and hole field operators $\psi_{\mathbf{k}}^{e}(z_e)$ and $\psi_{\mathbf{k}}^{h}(z_h)$, respectively. $\mathbf{k}$ is the in-plane electron momentum and the indices e,h simultaneously denote the band index and the z-component of the corresponding total angular momenta. d_{eh} is the amplitude of the dipole matrix elements with polarization $\mathbf{e}_{\text{eh}}$.

The dynamics of the excitonic transition amplitude can be given in terms of the Heisenberg equation of motion. To obtain a closed set of equations that truncates the hierarchy of higher-order correlation functions, the DCT approach is used. We consistently consider excitonic and biexcitonic coherent nonlinearities up to third order in the optical field. So far, this scheme has been successfully applied to quasi two-dimensional quantum-well (QW) systems [6], where propagation effects lead to radiative broadening and radiative coupling between the QWs. Here we extend this treatment to situations where the extension of the semiconductor is no longer small compared to the light wavelength such that true polariton effects become important. For the evaluation of the relevant equations of motion we use excitonic eigenstates $\phi_m(\mathbf{k}, z_e, z_h)$ calculated with respect to the microscopic boundary conditions [3]. This approach properly accounts for the finite extension of excitonic states within the spatially inhomogeneous system. In terms of these eigenstates and with time dependent coefficients $p_m^{\text{eh}}(t)$ the excitonic transition amplitude takes the form:

$$p_{\mathbf{k}}^{\text{eh}}(z_e, z_h, t) = \sum_m p_m^{\text{eh}}(t) \phi_m(\mathbf{k}, z_e, z_h). \quad (2)$$

Third order contributions beyond Hartree-Fock are given by the biexcitonic correlation function [7]. The expansion of its electronic singlet ($\lambda = -1$) and triplet ($\lambda = +1$) contribution in terms of excitonic eigenstates yields:

$$b_{\text{eh}}^{e'h'\lambda \,(\mathbf{k}_2,z_2,\mathbf{k}_1,z_1)}_{(\mathbf{k}_4,z_4,\mathbf{k}_3,z_3)}(t) =$$

$$\sum_{nm} \Big[\phi_n(\alpha\mathbf{k}_4 + \beta\mathbf{k}_3, z_4, z_3) \phi_m(\alpha\mathbf{k}_2 + \beta\mathbf{k}_1, z_2, z_1)$$

$$\times b_{nm}^{\text{ehe'h'}\lambda}(\mathbf{k}_4 - \mathbf{k}_3, t)$$

$$- \lambda \phi_n(\alpha\mathbf{k}_2 + \beta\mathbf{k}_3, z_2, z_3) \phi_m(\alpha\mathbf{k}_4 + \beta\mathbf{k}_1, z_4, z_1)$$

$$\times b_{nm}^{\text{ehe'h'}\lambda}(\mathbf{k}_2 - \mathbf{k}_3, t) \Big]. \quad (3)$$

The equations of motion for the coefficients $p_m^{\text{eh}}(t)$ and $b_{nm}^{\text{ehe'h'}\lambda}(\mathbf{q},t)$ of the excitonic and biexcitonic correlation functions, respectively, are similar to those obtained for the two-dimensional QW limit in [8]. However, in our case the matrix elements contain the additional space dependence for the inhomogeneous system. Note, that the

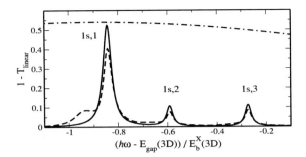
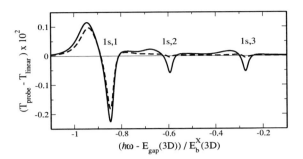

FIGURE 1. Calculated linear transmission spectrum for a $5a_0^X$-GaAs layer (solid line), spectral shape of the 120 fs laser pulses (dashed-dotted line) and nonlinear probe pulse transmission for opposite circular excitation (dashed line).

FIGURE 2. Calculated differential probe pulse transmission for e_+e_--configuration. Solid line: All Coulomb terms taken into account. Dashed line: Coulomb interaction between different excitonic states turned off.

discrete structure of the low-energy-part of the exciton spectrum is well-described with a finite number of states in the expansions (2), (3).

The dynamics of the optical field is given by Maxwell's equations, used in their one-dimensional form for propagation in z-direction perpendicular to the semiconductor layer [2]. Independent equations for the circular polarized components of the incident light fields are obtained which couple to the interband transitions according to the dipole selection rules.

The solid and dashed lines in Fig. 1 show the linear and nonlinear transmission for a GaAs layer of 5 exciton Bohr radii (a_0^X) thickness, see [2] for used material parameters. The energy is given relative to the bulk material band-gap $E_{gap}(3D)$ and normalized to the corresponding exciton binding energy $E_b^X(3D)$. For the chosen layer thickness, the confinement of carriers in the z-direction yields three polaritonic resonances in the displayed part of the spectrum which exhibit in-plane 1s symmetry.

In Fig. 2 the differential probe pulse transmission is displayed for a typical pump and probe setup with opposite circular (e_+e_-) polarization, no time-delay between the 120 fs pulses and a pump Rabi energy of $d_{eh}E_{pump} = 0.01 E_b^X(3D)$. For comparison Fig. 1 contains the nonlinear probe pulse transmission (dashed line) for the same situation but $d_{eh}E_{pump} = 0.07 E_b^X(3D)$. For the pulse shapes and their spectral position see the dashed-dotted line in Fig. 1. In the case where all Coulomb terms are taken into account we observe a bleaching of the excitonic resonances induced by the pump pulse. Additionally, on the low-energy side of the lowest polariton resonance the bound biexciton resonance appears and on the higher energy side a broad background in the probe pulse transmission due to excitation of the biexcitonic continuum (exciton-exciton scattering states) is observed. The nonlinearities around the higher polariton resonances are similar to that around the lowest one but with a decreased amplitude. With Coulomb interaction between excitonic states with different spatial distribution artificially turned off (dashed line in Fig. 2), we encounter only a slight change of the nonlinearities around the lowest polariton resonance whereas for higher peaks the influence of the pump pulse almost vanishes. Therefore, Coulomb interaction between different excitonic states is the main source for transmission changes around the higher polariton resonances. Furthermore, higher polariton states become more important with increasing layer thickness due to their decreasing energetical level-spacing.

In conclusion, we have presented a theory for nonlinear polariton propagation in spatially inhomogeneous semiconductors in the coherent limit. It incorporates both, propagational effects and excitonic and biexcitonic nonlinearities on a microscopic level. Calculated transmission spectra in a pump-probe geometry have been analyzed for a GaAs layer with finite thickness beyond the quasi two-dimensional QW limit. Coulomb interaction between different polariton states directly influences the nonlinear optical transmission.

REFERENCES

1. Tignon, J., Hasche, T., Chemla, D. S., Schneider, H. C., Jahnke, F., and Koch, S. W., *Phys. Rev. Lett.*, **84**, 3382 (2000).
2. Schneider, H. C., Jahnke, F., Koch, S. W., Tignon, J., Hasche, T., and Chemla, D. S., *Phys. Rev. B*, **63**, 045202 (2001).
3. Schumacher, S., Czycholl, G., and Jahnke, F., *Phys. Stat. Sol. B*, **234**, 172 (2002).
4. Axt, V. M., and Stahl, A., *Z. Phys. B*, **93**, 195 (1994).
5. Lindberg, M., Hu, Y. Z., Binder, R., and Koch, S. W., *Phys. Rev. B*, **50**, 18060 (1994).
6. Sieh, C., Meier, T., Jahnke, F., Knorr, A., Koch, S. W., Brick, P., Hübner, M., Ell, C., Prineas, J., Khitrova, G., and Gibbs, H. M., *Phys. Rev. Lett.*, **82**, 3112 (1999).
7. Schäfer, W., and Wegener, M., *Semiconductor Optics and Transport Phenomena*, Springer Berlin, 2002.
8. Takayama, R., Kwong, N. H., Rumyantsev, I., Kuwata-Gonokami, M., and Binder, R., *Eur. Phys. J. B*, **25**, 445 (2002).

Coherent control of the exciton-biexciton system demonstrated in four-wave-mixing experiments

T. Voss*, H. G. Breunig*[†], I. Rückmann*, J. Gutowski*, V. M. Axt** and T. Kuhn**

Institute for Solid State Physics, University of Bremen, D-28334 Bremen, Germany
[†]*Institute for Physical Chemistry, University of Marburg, D-35032 Marburg, Germany*
**Institute for Solid State Theory, University of Münster, D-48149 Münster, Germany*

Abstract. Systematic measurements and microscopic simulations were performed to study the optical coherent control of the exciton-biexciton systems in a ZnSe quantum well. Our results show that the spectral components in the four-wave mixing signal at the exciton and the exciton-biexciton transition, respectively, can be separately controlled in the coherent regime. This separate control also results in different beat structures on the real-time transients of the nonlinear wave-mixing polarization. With respect to all important features experiment and simulation are in excellent agreement.

The optical coherent control of elementary excitations in semiconductor nanostructures has become an important field of intense research in the last years [1, 2]. Among other topics the control of photocurrent, electronic states in quantum dots, spin orientation of excited carriers, as well as excitonic polarizations and populations has been proven experimentally and thoroughly investigated. The fascination about coherent control comes from the fact that this technique allows for ultrafast control and manipulation of different quantum mechanical excitations with regard to both their amplitudes and relative phases in an almost arbitrary way. In this work we focus on the interesting case of the simultaneous coherent manipulations of nonlinear polarizations of two coupled resonances, i.e., the exciton and exciton-biexciton resonances [3] in a ZnSe single quantum well which serves as a high-quality model system.

In order to study the coherent control of the nonlinear excitonic and biexcitonic wave-mixing polarization we have performed systematic transient four-wave-mixing experiments in transmission geometry on a free-standing ZnSe single quantum well at a temperature of 4 K. The sample was resonantly excited at both the heavy-hole ground-state-exciton (GET) and exciton-biexciton transition (EBT) with three 120 fs pulses. A phenomenological five-level energy diagram of the exciton-biexciton system is shown in Figure 1.

Phase-locked pulse pairs were generated by use of an actively stabilized Michelson interferometer (time resolution $\Delta t_{int} \approx 40$ as). The pulse pairs excited the sample from a direction k_1. An additional third pulse from a different direction k_2 was applied to the sample to generate a coherent four-wave-mixing (FWM) polarization.

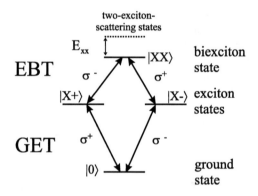

FIGURE 1. Energy diagram of the exciton-biexciton system. E_{xx}: biexciton binding energy, $\sigma^{+/-}$: circular polarization states of the photons.

The phase-locked pulse pairs in direction k_1 were used to achieve optical coherent control of the induced polarizations by selectively tuning and stabilizing the temporal separation $t_{int} = t_{int}^0 + \Delta t_{int}$ of each two locked pulses. In this notation t_{int}^0 denotes the basic temporal separation on a 10 fs time scale whereas Δt_{int} stands for the fine tuning of the delay with a sub-fs accuracy.

All experiments were performed in a configuration in which the single k_2 pulse reached the sample before the two pulses of the pair, i.e., for the measured direction $2k_2 - k_1$ a negative delay $t_{del} < 0$ was chosen for the generation of the coherent FWM polarization. Additionally, the linear polarization states of the k_1 and k_2 pulses were set to a cross-linear configuration which maximized the biexcitonic effects in the FWM signal.

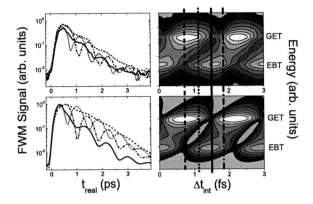

FIGURE 2. Experimental results (top row) and numerical simulations (bottom row) of the FWM signal for $t_{del} = -0.1\,\text{ps}$ and $t_{int}^0 = T_{beat}/2$. The left column shows the real-time transients of the FWM polarization, in the right column the spectrally resolved FWM intensity is depicted as contour diagrams on logarithmic scales spanning three orders of magnitude (white denotes maximum intensity). The vertical lines show the values of Δt_{int} which correspond to the transients with the same line shape.

The microscopic simulations of the coherent control of the GET and EBT were performed by use of a density-matrix formalism based on the dynamics-controlled truncation scheme (DCT) in the so-called coherent limit [4]. On this level of the theory which is exact within third order of the electric field the dynamics of the coherent polarization is described by the time evolution of the single-pair transition amplitude Y and the correlated two-pair transition amplitude $\bar{B}$. The coupled equations of motion of these two variables have been solved numerically for the excitation conditions used in the experiments.

The experimental results of the real-time and spectrally resolved FWM experiments are shown in Figure 2 together with the results of the microscopic simulations. For the experiments and corresponding simulations the single k_2 pulse reached the sample at $t_{del} = -0.1\,\text{ps}$ and the basic separation between the two phase-locked pulses was set to match one half of the beat period of the exciton-biexciton beats ($t_{int}^0 = T_{beat}/2$) which are clearly visible on the real-time transients (left column in Figure 2).

A background signal of $10^{-3} \cdot I_{max}$ has been added to the simulated real-time transients to account for the finite signal-to-noise ratio in the experiment. Experiment and simulation are in excellent qualitative agreement with respect to all features in the real-time resolved as well as in the spectrally resolved FWM signals.

In the spectrally resolved measurements and simulations a phase shift of π between the coherent switching signals at the GET and EBT occurs for the chosen excitation conditions. This behavior of the FWM signal clearly shows that the GET and EBT can be separately addressed by the phase-locked pulse pair in the coherent control experiments. In the real-time transients the separate control manifests itself by a change of the amplitude and phase of the exciton-biexciton beats which are superimposed to the transients. The cases exemplarily shown in the left column of Figure 2 correspond to an FWM signal with maximum intensity at the GET / minimum at the EBT (dashed line) and minimum intensity at the GET / maximum at the EBT (solid line). For these two transients the beats are strongly suppressed. The other two transients (dotted and dot-dashed) correspond to intermediate cases where in the spectrally resolved measurements and simulations significant contributions at both resonances are found. Although the corresponding spectra are identical the two transients strongly differ from each other since their beat structures are clearly out of phase. This is another direct proof that the GET and EBT are separately addressed by the phase-locked pulses.

Comparing the four real-time transients which are measured for different values of Δt_{int} it can be seen that in all cases the FWM signal is initially generated at $t_{real} \approx 0\,\text{ps}$ after the arrival of the single k_2 pulse and the first pulse from the interferometer. The signals start to deviate from each other upon the arrival of the second phase-locked pulse at a time $t_{real} = T_{beat}/2 \approx 430\,\text{fs}$. In a simple picture this second pulse also generates a coherent FWM polarization together with the remaining polarization previously induced by the k_2 pulse. The two parts of the FWM polarization then interfere which finally results in the coherent manipulation of the FWM polarizations at the GET and EBT. Additional calculations based on the microscopic theory which have been performed according to this model strongly support the given interpretation. The real-time resolved experiments and simulations therefore directly prove that the non-linear FWM polarization can be coherently manipulated after its initial creation.

The authors thank W. Faschinger (deceased 2001), Würzburg University, for providing the sample. This work has been supported by the Deutsche Forschungsgemeinschaft (Grants Nos. Gu 252/12 and Ne 525/8).

REFERENCES

1. Heberle, A. P., Baumberg, J. J., and Köhler, K., *Phys. Rev. Lett.*, **75**, 2598 (1995).
2. Bonadeo, N. H., Erland, J., Gammon, D., Park, D., Katzer, D. S., and Steel, D. G., *Science*, **1473**, 282 (1998).
3. Voss, T., Breunig, H. G., Rückmann, I., and Gutowski, J., *Opt. Comm.*, **218**, 415 (2003).
4. Axt, V. M., Victor, K., and Kuhn, T., *phys. stat. sol. (b)*, **206**, 189 (1998).

Exciton Associated Photorefractive Effect in ZnMgSe/ZnSe Quantum Wells

H. P. Wagner, H.-P. Tranitz, S. Tripathy

Department of Physics, University of Cincinnati, Cincinnati, Ohio 45221-0011, U.S.A.

Abstract. We observe a phase coherent all-optical photorefractive effect in a ZnSe single quantum well in four-wave-mixing configuration using ultrashort pulses. The microscopic origin of the observed effect is attributed to the coherent formation of an electron grating by interfering excitons.

INTRODUCTION

Since the recent fabrication of single photon emitters [1] and the observation of ultraslow and stopped light in solids [2] there has been substantial interest in its potential applications like optical data storage and optical computation. Further important building blocks towards the realization of these goals are photorefractive (PR) devices that are capable to store and process coherent optical information. Present multiple quantum well (QW) PR devices reveal high sensitivity and fast response [3] but do not use the coherence of exciton to conserve the phase and polarization information of incident light fields. All-optical PR QW structures using quantum coherent effects are therefore favored.

EXPERIMENTAL RESULTS

The investigated PR QW structure was pseudo-morphically grown on (001) oriented GaAs substrate by molecular beam epitaxy (MBE). It consists of a 10 nm wide ZnSe single QW sandwiched between two 30 nm $Zn_{0.9}Mg_{0.1}Se$ barriers, with a 20 nm thick ZnSe buffer layer between the barrier and the substrate [4]. A frequency doubled mode locked Ti-sapphire laser providing 90 fs pulses at a repetition rate of 80 MHz was used for excitation. The temporal width of the frequency doubled pulses was determined by two-photon-absorption in a SiC photodiode. Three-beam four-wave mixing (FWM) experiments involving circular polarized pulses k_1, k_2, and k_3, respectively, have been performed in a back scattering geometry with the sample mounted in a Helium flow cryostat at a temperature of 55 K. The focus diameter of the laser pulses on the sample was 100 μm. The time integrated and spectrally resolved four wave mixing signal was detected in direction $k_3+k_2-k_1$ by a combination of a spectrometer and an optical multichannel analyzer. We use notation $(\sigma^+\sigma^+\sigma^+)$, $(\sigma^+\sigma^-\sigma^+)$, $(\sigma^-\sigma^-\sigma^+)$ where the symbols indicate the circular polarization of pulse k_1, k_2, and k_3, respectively.

Figure 1 displays the $k_3+k_2-k_1$ signal response at the energetic position of the X_h transition for three different circular polarization configurations recorded at very low pulse intensities (I_1=48, I_2=32 and I_3=70 Wcm^{-2}). The delay time between k_2 and k_3 was kept fixed at $\tau_{23} = t_3-t_2 = 13$ ps while delay $\tau_{12} = t_2-t_1$ was varied. In $(\sigma^-\sigma^-\sigma^+)$ configuration we observe an almost symmetrical signal with exponential decay for positive and negative τ_{12} and a pronounced signal dip at $\tau_{12} = 0$. In $(\sigma^+\sigma^-\sigma^+)$ configuration we observe no signal around $\tau_{12} \approx 0$ but a strong symmetrical response with exponential decay around $\tau_{12} \approx -13$ ps with clear signal dip at $\tau_{12} = -13$ ps. In $(\sigma^+\sigma^+\sigma^+)$ configuration two symmetrical responses appear, one around $\tau_{12} \approx -13$ ps and a weaker one at $\tau_{12} \approx 0$. In $(\sigma^-\sigma^+\sigma^+)$ configuration (not shown) no diffracted signal was detected.

The experimental results unambiguously demonstrate that the diffracted $k_3+k_2-k_1$ signal at low intensities is not generated by a $\chi^{(3)}$ FWM process but is due to a resonant PR effect (PRE) caused by a laser induced grating. For example in $(\sigma^-\sigma^-\sigma^+)$ configuration the grating is written by pulses k_1 and k_2, and read by the following pulse k_3.

From earlier investigations we know that these samples allow to capture electrons in the QW which are optically excited in the GaAs substrate [5]. Accordingly the observed resonant PRE is attributed to the formation of a electron grating that is "written" into the QW in a multi step process: First an exciton grating is generated by the interference of coherent exciton polarizations created by pulses k_1 and k_2. Subsequently the optically excited GaAs electrons that are captured in the QW are redistributed by the exciton grating due to Colomb interaction and Pauli blocking. Finally, after about 100 ps the excitons recombine whereas the electron grating that is stabilized by localized positive charges at the ZnSe/GaAs interface remains in the QW. The created electron grating and longitudinal space charge fields within the QW and the GaAs/ZnSe interface cause a periodic modulation of the optical constants. This leads to a diffraction of exciton polarisation P_{k3} into direction $k_3+k_2-k_1$. The grating contrast depends on the coherence between the k_1 and k_2 excitons and decreases exponentially with the delay τ_{12} according to the exciton dephasing time. The efficiency of the PRE is proportional to the equilibrium electron density n_0 within the QW. The signal dip at $\tau_{12} = 0$ is attributed to increasing space charge fields between captured electrons and positive carriers within the bright areas of the light interference pattern that lead to a reduced QW electron density n_0.

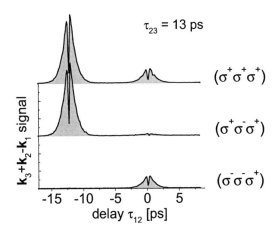

FIGURE 1. Traces of the diffracted $k_3+k_2-k_1$ PR signal at the X_h exciton transition for polarization configurations $(\sigma^+\sigma^+\sigma^+)$, $(\sigma^+\sigma^-\sigma^+)$ and $(\sigma^-\sigma^-\sigma^+)$ recorded at 55 K. The delay time between k_2 and k_3 was kept fixed at $\tau_{23} = 13$ ps while delay τ_{12} was varied.

In $(\sigma^+\sigma^-\sigma^+)$ configuration an electron grating can only be formed between pulses k_1 and k_3 since fields k_1 and k_2 are orthogonal. Thus polarization P_{k2} (which is generated in the following laser pulse cycle) is diffracted when τ_{12} approaches -13 ps where coherent k_1 and k_3 excitons are able to induce a phase coherent electron grating. The higher PR efficiency in this configuration is attributed to a larger grating period (the angle between pulses k_1 and k_3 is less than half the angle between k_1 and k_2) leading to a better defined grating contrast within the QW.

In $(\sigma^+\sigma^+\sigma^+)$ configuration an electron grating can be formed between pulses k_1 and k_2 and between pulses k_1 and k_2 causing a diffracted P_{k3} signal around $\tau_{12} \approx 0$ and a diffracted P_{k2} signal around $\tau_{12} \approx -13$ ps. In $(\sigma^-\sigma^-\sigma^+)$ no induced electron grating can be formed explaining the lack of any diffracted signal in this configuration.

In conclusion we observed a novel phase coherent PRE in ZnSe single QWs revealing a signal dip at pulse overlap. The observed effect is attributed to a stepwise formation of an electron grating that is induced by the interference of coherent excitons. The efficiency of the effect being clearly detectable at an average power level of a few μW makes it attractive for all optical applications. Of particular interest is the possibility to preserve phase and polarization of incident pulses and to erase the electron grating at pulse overlap. Additional temperature dependent measurements as well as excitation energy dependent measurements will further elucidate this interesting effect [6].

ACKNOWLEDGMENTS

Stimulating discussions with W. Langbein are kindly acknowledged. This work is supported by the National Science Foundation (DMR 0305076).

REFERENCES

1. S. Stauf, P. Michler, K. Klude, D. Hommel, G. Bacher, A. Forchel, *Phys. Rev. Lett.* **89**, 177403 (2002).
2. A. V. Turukin, V. S. Sudarshanam, M. S. Shahriar, J. A. Musser, B.S. Ham, P. R. Hemmer, *Phys. Rev. Lett.* **88**, 23602 (2002).
3. D. D. Nolte, *J. Appl. Phys.* **85**, 6259 (1999).
4. M Wörz, E. Griebl, Th. Reisinger, B. Flierl, D. Haserer, T. Semmler, T. Frey, W. Gebhardt, *Phys. Stat. Sol. B* **202**, 805 (1997).
5. H. P. Wagner, H.-P. Tranitz, R. Schuster, *Phys. Rev. B* **60**, 15542 (1999).
6. H. P. Wagner, S. Tripathy, to be published.

Coherent Photon Trapping In Doped Photonic Crystals And Dispersive Semiconductor Materials

I. Haque and Mahi R. Singh

Department of Physics and Astronomy
The University of Western Ontario, London, Ontario, Canada N6A 3K7

Abstract. A theory of photon trapping has been developed in photonic and dispersive polaritonic band-gap materials doped with an ensemble of five-level atoms. The atoms are prepared as coherent superpositions of the two lower states and interact with a reservoir and two photon fields. They also interact with each other by dipole-dipole interaction. It is found that, in a photonic crystal, when the resonance energies lie away from the band edges and within the lower or upper bands, trapping is observed. The requisite conditions for the trapping vary depending on the strength of the dipole-dipole interaction between the atoms. Also, if the photon fields are held constant, the population densities of the excited states of the atoms increase with increasing dipole-dipole interaction.

INTRODUCTION

Investigations of quantum coherence and correlation in quantum optics and radiation physics have led to many interesting and unexpected consequences such as the Hanle effect, lasing without inversion, coherent Raman beats, photon echo and self-induced transparency [1]. One particular phenomenon that has come under considerable scrutiny in the context of quantum coherence is coherent population trapping (CPT) in multi-level atomic systems. Recently, Singh and Haque [2] have studied CPT in photonic band-gap (PBG) and dispersive polaritonic band-gap (DPBG) materials, where the effect of the dipole-dipole interaction between atoms has been neglected. PBG and DPBG materials have gaps in their photon energy spectra [3]. In this paper, we have developed a theory of photon trapping in the presence of dipole-dipole interaction in the mean field approximation when the doped atoms are prepared as coherent linear combinations of the two lower states. The number density of the top state has been calculated by using the Schrödinger equation and the Laplace transform method. It is found that the dipole-dipole interaction between the atoms plays a very important role in determining the conditions required for CPT. It is also observed that this interaction directly influences the population densities of the excited states of the atoms.

THEORY

The energy levels of an atom are denoted by $|a>$, $|b>$, $|c>$, $|d>$ and $|e>$. The atoms are prepared as the coherent superposition $|\lambda, 0> = B(0)|b> + C(0)e^{i\psi}|c>$ where $B(0)$ and $C(0)$ are the initial amplitudes of $|b>$ and $|c>$, respectively. $|b>$ and $|a>$ are coupled by a Rabi frequency Ω_1 and $|c>$ and $|a>$ are coupled by a Rabi frequency Ω_2. Due to the interaction between the atom and the reservoir, $|b>$ and $|c>$ both decay spontaneously to $|e>$, and $|a>$ decays spontaneously to $|d>$ with rates Γ_b, Γ_c and Γ_a, respectively [4]. The decay rates of $|b>$ and $|c>$ are identical. In the dipole and rotating-wave approximations, the Hamiltonian for the system is written in energy space as $H_S = H_0 + H_{dd}$ where H_0 is taken as in reference [2]. The transitions from $|b>$ to $|a>$ and $|c>$ to $|a>$ induce a net dipole moment P in the atoms. The dipole interaction between the atoms is expressed as $H_{dd} = \sum_{i<j} J_{ij}(P_i.P_j)$. J_{ij} denotes the coupling between the atoms. Using the Schrödinger equation and the Laplace transform method, we calculate the expressions of the amplitudes in the mean field approximation. For lack of room, here we only present the expression for the upper state

$$A(t) = i \frac{\sin(\eta t)}{e^{(\Gamma_a + \Gamma_b)t/4}\eta} \left[\bar{\Omega}_1 Z_{ab} B(0) + \bar{\Omega}_2 Z_{ac} C(0) e^{i(\phi+\psi)} \right] \quad (1)$$

where, $\eta = \sqrt{\bar{\Omega}_1^2 Z_{ab}^2 + \bar{\Omega}_2^2 Z_{ac}^2 - (\Gamma_a - \Gamma_b)^2/4}$.

$\bar{\Omega}_1 = \Omega_1 + \Omega_m$ and $\Omega_m = 2N\lambda P_{ab}^2 (\rho_{ab} + \rho_{ac})/\hbar$ with $\rho_{ab} + \rho_{ac} = A(t)(B^*(t) + C^*(t))$.

N is the number of atoms and λ is the mean-field parameter, related to the coupling parameter J_{ij}. Z_{ij} are the atomic form factors for PBG or DPBG materials which are obtained as in reference [4]. Each amplitude expression has $A(t)(B^*(t) + C^*(t))$ on the right hand side. Therefore, in order to calculate the number density of state $\rho_{aa} = A^*(t)A(t)$, one has to solve the three expressions self-consistently.

RESULTS AND DISCUSSION

Numerical calculations for ρ_{aa} are shown in Figs. 1 and 2 for a PBG material. $B(0) = \cos(\theta/2)$ and $C(0) = \sin(\theta/2)$, respectively, where $0 \le \theta \le \pi$. We also let $\phi + \psi = \pi$. The remaining parameters used are as in reference [2]. Similar simulations have also been performed for DPBG materials, but are not presented here for lack of space. In Fig. 1, ρ_{aa} is plotted against the relative Rabi frequency (Ω_r) and N. It is found that, as N increases, the value of Ω_r at which the trapping occurs decreases. This is because a large value of N implies strong dipole-dipole interaction, thus requiring only a small relative frequency for destructive interference between the two transitions. Results are similar irrespective of whether the resonance energies lie within the lower or upper band. In Fig. 2, we observe that ρ_{aa} becomes zero as the resonance energy (ε_{ad}) approaches either of the edges of the band gap. This is due to the effect of the form factor – which becomes very large at the band edges – and can not be interpreted as CPT.

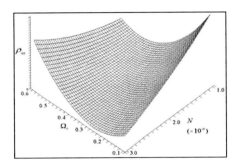

FIGURE 1. Three-dimensional plot of ρ_{aa} against Ω_r (relative Rabi frequency) and N (number of doped atoms).

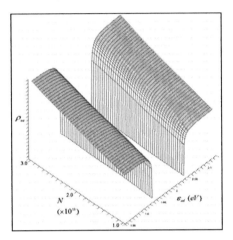

FIGURE 2. Three-dimensional plot of ρ_{aa} against ε_{ad} (energy) and N (number of doped atoms).

Figure 2 also shows that as the number of doped atoms increases, the number density of state $|a>$ increases when the photon fields are held constant. This is a direct consequence of the dipole-dipole interaction between the atoms.

ACKNOWLEDGMENTS

MRS is thankful to NSERC of Canada for financial support in the form of a research grant.

REFERENCES

1. Scully, M. O., and Zubairy, M. S., *Quantum Optics*, Cambridge: Cambridge University Press, 1997.
2. Singh, M., and Haque, I., *J. Mod. Optics* (communicated 2004).
3. John, S., *Phys. Rev. Letters* **58**, 2486-2489 (1987).
4. Rupasov, V. I., and Singh, M., *Phys. Rev. Letters* **77**, 338-341 (1996).

CHAPTER 6
TRANSPORT

    6.1. Carrier Dynamics and Magnetotransport

    6.2. Spin Dynamics for Spintronics

InAs-based Micromechanical Two-dimensional Electron Systems

H. Yamaguchi,[a] S. Miyashita,[b] Y. Tokura,[a] and Y. Hirayama[a,c]

[a] *NTT Basic Research Laboratories, NTT Corporation, Atsugi, Kanagawa 243-0198, Japan*
[b] *NTT Advanced Technologies, Atsugi, Kanagawa 243-0198, Japan*
[c] *SORST-JST, Kawaguchi, Saitama 332-0012, Japan*

Abstract. We review our recent experimental studies of InAs-based micromechanical two-dimensional electron systems. Integrating semiconductor low-dimensional structures in micromechanical cantilevers provides a stage for studying novel physics and developing new functional devices. In InAs-based systems, the Fermi level is pinned in the conduction band, which makes it easy to fabricate conductive freestanding structures, such as ultrathin freestanding Hall devices and novel displacement/force sensors based on quantum mechanical transport.

INTRODUCTION

The coupling of electronic/optical properties of semiconductor low-dimensional structures to mechanical ones in micro/nanomechanical beams and cantilevers has received increasing attention not only for practical device applications [1] but also for fundamental science, such as quantum Hall magnetometry [2,3], single spin sensing [4], and the detection of the quantized mechanical motion of nanoscale cantilevers [5,6]. In such measurements, the detection sensitivity of the mechanical systems plays a significant role and reducing the size of the mechanical objects to the nanometer scale can drastically improve the sensitivity [7,8].

We are studying InAs-based micro/nanomechanical systems because they allow us to reduce the size of semiconductor-device-based mechanical systems. The surface Fermi level of InAs is pinned in the conduction band, whereas in other semiconducting materials, it is pinned in the band gap. This property provides the possibility of fabricating very thin conductive mechanical objects [9-11]. A preliminary study has shown that the InAs thickness can be reduced to 10 nm with sufficient electric conduction [12]. This advantage is expected to lead to improved displacement/force sensitivity in ultra-small/thin mechanical systems [7,8]. It is also possible to fabricate curved two-dimensional electron systems using InAs-based heterostructures [13].

We have recently studied the transport properties of InAs-based micromechanical two-dimensional electron systems (2DES). Freestanding InAs and InAs/AlGaSb membranes, even with a thickness as small as 50 nm, were shown to have high conductivity, which is promising for the realization of highly sensitive piezoresistive displacement/force sensors or flexible semiconductor devices [9]. At low temperatures, the quantum transport in InAs-based low-dimensional systems strongly influences the piezoresistivity and an appropriate interference/resonance condition improves the sensitivity [14,15]. This sensing scheme was used to detect the magnetization of 2DES, where de Haas-van Alphen effect was clearly observed. In the following sections, we review these recent results on InAs-based mechanical systems.

FABRICATION

The first InAs-based freestanding membranes were fabricated by selectively etching an AlSb layer in InAs/AlSb heterostructures grown on GaAs substrates [16]. This is because, on conventional GaAs (001) substrates, a high-quality InAs layer can be grown only on a thick AlSb buffer layer [17], which includes a high density of dislocations and should be selectively etched as a sacrificial layer to obtain high-crystalline-quality membranes. We used InAs/ Al(Ga)Sb heterostructures grown on GaAs (111)A substrates. The advantage of using this substrate is that high quality heterostructures can be grown without a thick Al(Ga)Sb buffer layer [18]. Therefore, the substrate can be etched as a sacrificial layer and this material system allows us to fabricate ultrathin high-quality InAs-based freestanding heterostructures. NH$_4$OH solution, which etches GaAs more than 10^3 times faster than it does InAs/Al(Ga)Sb, was employed for the final selective etching. A supercritical drying technique was used to avoid the cantilever sticking caused by the surface tension of evaporating liquid [19].

EXPERIMENTAL RESULTS

To study the basic transport properties of InAs-based mechanical structures, suspended Hall bar structures were fabricated and their electrical properties were characterized [Fig.1(a)]. This study is particularly important in terms of the effects of phonon confinement on the electron transport. For thin membranes, the sacrificial layer etching results in an improvement of conductivity, where both the carrier concentration and mobility are increased.[9] This increase is induced by the relaxation of residual strain and the removal of the dislocation network at the highly mismatched heterointerfaces. The 50-nm thick freestanding InAs/Al$_{0.5}$Ga$_{0.5}$Sb QW showed a room temperature mobility of 10,000 cm^2/Vs and the carrier concentration of 1.4×10^{12} cm^{-2}. This highly conductive ultra-thin membrane may be applicable to highly sensitive force sensors and flexible compliant semiconductor devices.

The mobility showed little temperature dependence, indicating that it is limited by roughness scattering even at room temperature. We believe that it is not limited by surface scattering but by interface scattering due to unoptimized growth condition. This is because as-grown InAs/Al$_{0.5}$Ga$_{0.5}$Sb QW with a thicker Al$_{0.5}$Ga$_{0.5}$Sb buffer layer and a Al$_{0.5}$Ga$_{0.5}$Sb cap layer of similar thickness showed phonon-scattering-limited temperature dependence. We therefore expect that the effect of phonon confinement on the transport properties can be studied by improving the growth condition.

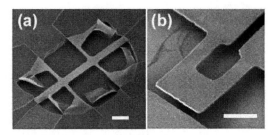

FIGURE 1. SEM viewgraphs of a fabricated (a) 50-nm-thick freestanding Hall bar and a (b) 300 nm-thick piezoresistive cantilever. The solid bars indicate the length of 10μm.

One of the most important properties caused by the couplings of mechanical and electrical ones in semiconductors is piezoresistivity. The mechanical deformation of a cantilever induces a strain, and the deformation potential, as well as the piezoelectric effect, modifies the energy band structure. This leads to a change in the mobility and carrier concentration, which changes the conductivity of the structures. We have fabricated piezoresistive cantilevers made from InAs/Al$_{0.5}$Ga$_{0.5}$Sb single heterostructures [Fig. 1(b)] and the piezoresistance, $\delta R/\varepsilon R$, where $\delta R/R$ is the fractional change in the device resistance and ε the induced strain, was measured as a function of InAs thickness. We found that the piezoresistance is sensitive to the InAs thickness when the thickness becomes comparable with the electron wavelength [11]. This can be explained by the finite size effects in the InAs film, where the small strain-induced change in energy band structure can cause a large change in the conductance when the Fermi level becomes closer to the quantum level [11].

The most significant quantum effect can be confirmed at a liquid helium temperature under a magnetic field. Figure 2(a) shows the measured piezoresistance of InAs/Al$_{0.5}$Ga$_{0.5}$Sb single heterostructures as a function of applied magnetic field, i.e. the magneto-piezoresistance R_{piezo}, which exhibited aperiodic oscillation. This aperiodic oscillation was reproducibly obtained in repeated measurements, but showed a different oscillation pattern when the sample was heated to 150 K and then cooled again to 2.5 K [14,15]. This "magnetofingerprints" behavior is similar to that observed in the magnetoresistance (MR) of Q1D systems, and is known as a universal conductance fluctuation (UCF). In fact, the MR of the same sample shows a similar, but not identical, aperiodic oscillation pattern as shown in Fig. 2(b).

Two major mechanisms cause the influence of conductance fluctuation on the piezoresistance. One is the modulation of the Fermi energy by the induced strain. The induced piezoresistance is, thus, not caused by the phase modulation in the local interference loops but by the wavelength modulation in them, although both cause aperiodic conductance oscillation as a function of magnetic field. The other is a rather "apparent" effect caused by the small cantilever misorientation. When the cantilever is misoriented even by a very small angle (say 2 or 3 degrees), its mechanical vibration modulates the perpendicular component of the magnetic field. As a result, the phase of electron interference is modulated, leading to a change in device conductance. In contrast to the former mechanism, where the oscillation amplitude has small magnetic field dependence, the latter causes a linear magnetic field dependence of the oscillation amplitude. A detailed analysis of our data has revealed that the strain-induced Fermi energy modulation is dominant in the low-magetic-field region (0-2T) and that the "apparent" piezoresistance is dominant in the high-magnetic-field region (2-7T) [15].

In terms of detection sensitivity, the sample with higher mobility has an advantage. Compared with the electron interference in Q1D systems, a much higher bias current can be used without losing the quantum effects. Higher bias current improves the detection sensitivity when the noise level is limited by the electronics. Figure 3 shows the magneto-piezoresistance and magnetoresistance obtained for the $Al_{0.5}Ga_{0.5}Sb/InAs/Al_{0.5}Ga_{0.5}Sb$ QW sample, which has an electron mobility of 63,000 cm^2/Vs at liquid-helium temperature. The magneto-piezoresistance curve shows the feature of Schbnikov de-Haas (SdH) oscillations as well as the conductance fluctuation. This SdH feature is also a result of the Fermi energy modulation and perpendicular magnetic field modulation. The maximum piezoresistance was obtained at 7.695 T [indicated by a solid arrow in Fig. 3(a)] and, with the bias current of 40 μA, we obtained an enhanced displacement sensitivity of about 0.01 $nm/Hz^{0.5}$, which corresponds to the force sensitivity of $10^{-12} N/Hz^{0.5}$.

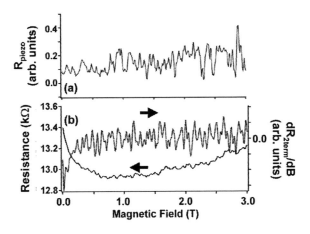

FIGURE 2. (a) Piezoresistance as a function of magnetic field measured at 2.5 K. (b) The two-terminal magnetoresistance of the same sample obtained after measuring (a). The dotted line (dR_{2term}/dB) is the derivative with respect to the magnetic field.

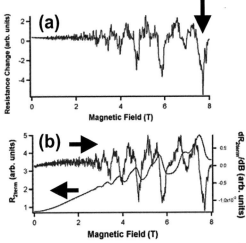

FIGURE 3. (a) Piezoresistance of the InAs/AlGaSb QW sample as a function of magnetic field measured at 2.5 K. (b) The two-terminal magnetoresistance and its derivative with respect to the magnetic field of the same sample.

This interpretation is also supported by the comparison between differential magnetoresistance (dR_{2term}/dB) and the piezoresistance. Our tight-binding conductance calculation suggests that the Fermi energy modulation results in much smaller similarity between these two curves than electron phase modulation. This shows a good agreement with our experimental results, where the crosscorrelation function has a larger peak in the high magnetic field region than in the low region [15]. This "apparent" effect is, however, more important than the "real" piezoresistance for improving the mechanical detection sensitivity because applying a high magnetic field enhances the piezoresistance.

This sensitive force-detection device was used to detect the magnetization of 2DES. It has been experimentally observed by some groups [2,3,20] that the de Haas-van Alphen effects in 2DES induce a periodic torque on mechanical cantilevers. We could detect the torque in our InAs/AlGaSb 2DES as a shift in the resonance frequency. Figure 4 shows the piezoresistance and the shift in the resonance frequency calculated from the measured mechanical resonance peaks in the piezoresistance. The shift in the resonance frequency corresponding to the Landau levels could be observed only with the cantilever with the 2DES mesa, indicating that it was caused by the magnetization of the 2DES. This InAs-based 2DES magnetometer has the advantage of size reduction and might be applicable for the study of magnetization of semiconductor nanostructures.

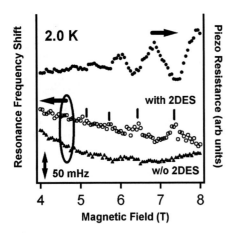

FIGURE 4. Resonance frequency shift and piezoresistance of a InAs/AlGaSb QW piezoresistive cantilver with and without (w/o) an additional 50 μm x 90 μm 2DES mesa on the cantilever.

CONCLUSION

We have reviewed our recent experimental studies of the InAs-based micromechanical low-dimensional systems. Freestanding membranes with a thickness as small as 50 nm show high conductivity, suggesting their applicability for highly sensitive force/displacement detection. By appropriately choosing the inteference/resonance conduction, the use of quantum effects in low-dimensional structures can also improve the sensitivity. By combining these two novel techniques, the properties of semiconductor low-dimensional structure is expected to be further investigated.

ACKNOWLEDGMENTS

We thank Dr. H. Takayanagi for his continuous encouragement in this study. Our work was partly supported by the NEDO International Joint Research Program; *"Nanoelatsticity"*, and a Grant-in-Aid for Scientific Research from the Japan Society for the Promotion of Science.

REFERENCES

1. Craighead, M. P., *Science* **290**, 1532-1535 (2000).
2. Eisenstein, J.P. et al., *Phys. Rev. Lett.* **55**, 875-878 (1985).
3. Harris, J. G. E. et al., *Phys. Rev. Lett.* **86**, 4644-4647 (2001).
4. Stipe, B. C. et al., *Phys. Rev. Lett.* **87**, 277602 (2001).
5. Knobel, R. G., and Cleland, A. N., *Nature* **424**, 291 (2003).
6. LaHaye, M.D. Buu, O., Camarota, B. and Schwab, K.C., *Science* 304, 74-77 (2004).
7. Mamin, H. J., and Rugar, D., *Appl. Phys. Lett.* 79, 3358-3360 (2001).
8. Roukes, M.L. "Nanoelectromechanical Systems" in *Technical Digest of the 2000 Solid-State Sensor and Actuator Workshop*, 2000, pp. 1-10.
9. Yamaguchi, H., Dreyfus, R., Hirayama, Y., and Miyashita, S., *Appl. Phys. Lett.* 78, 2372-2374 (2001).
10. Yamaguchi, H., Hirayama, Y., *Appl. Phys. Lett.* 80, 4428-4430 (2002).
11. Yamaguchi, H., Miyashita, S., and Hirayama, Y., *Appl. Phys. Lett.* 82, 394-396 (2003).
12. Yamaguchi, H., Homma, Y., Kanisawa, K., and Hirayama, Y., *Jpn. J. Appl. Phys.* 38, 635-644 (1999).
13. Voro'bev, A., Prinz, V., Preobrazhenskii, V. P., and Semyagin, B., *Jpn. J. Appl. Phys.* 42, L7-L9 (2003).
14. Yamaguchi, H., Miyashita, S., and Hirayama, Y., *Physica* E 21, 1053-1056 (2004).
15. Yamaguchi, H., Tokura, Y., Miyashita, S., and Hirayama, Y., *Phys. Rev. Lett.* to be published.
16. Yoh, K. et al., *Semicond. Sci. Technol.* 9, 961-965 (1994).
17. Tuttle, G., Kroemer, H., and English, J. H., *J. Appl. Phys.* 65, 5239-5242 (1989).
18. Yamaguchi, H., Fahy, M.R., and Joyce, B.A., *Appl. Phys. Lett.* 69, 776-778 (1996).
19. Namatsu, H., Yamazaki, K., and Kurihara, K., *Microelectron. Eng.* 46, 129-131 (1999).
20. Meinel, I., Hengstmann, T., Gundler, D., and Heitmann, D., *Phys. Rev. Lett.* 82, 819-821 (1999).

Shot noise as a probe of electron dynamics in hopping and resonant tunnelling

S.S. Safonov*, A.K. Savchenko*, S.H. Roshko*, D.A. Bagrets[†], O.N. Jouravlev[†], Y.V. Nazarov[†], E.H. Linfield** and D.A. Ritchie**

*School of Physics, University of Exeter, Stocker Road, Exeter, EX4 4QL, U.K.
[†]Department of Applied Physics, Delft University of Technology, Lorentzweg 1, 2628 CJ Delft, The Netherlands
**Cavendish Laboratory, University of Cambridge, Madingley Road, Cambridge, CB3 0HE, U.K.

Abstract. We show that shot noise can be used for studies of electron dynamics in hopping and resonant tunnelling between localised electron states. In this work an enhancement of shot noise is observed in resonant tunnelling through a single impurity due to Coulomb interaction between two resonant tunnelling impurities. In hopping Coulomb interaction between electrons as well as the distribution of the hopping barriers between the states is detected by shot noise.

In short barriers, where transport is dominated by resonant tunnelling through a single localised state, a suppression of shot noise is expected [1]. In this study of shot noise in resonant tunnelling we have observed not only the suppression of noise, but also its significant enhancement [2]. We show that this effect is caused by Coulomb interaction between two parallel tunnelling channels conducting the current in a correlated way.

With increasing the sample length and temperature, we study shot noise in the regime of phonon-assisted hopping. In 2D variable-range hopping a suppression of shot noise was detected in [3] and it was suggested that the Fano factor $F = S_I/2eI$ is close to L/L_c, where L_c is the typical distance between the dominant hops in the hopping network [4] and L is the sample length. In this work by studying mesoscopic structures with much shorter gates we are able to realise 1D hopping through localised states with uniform and non-uniform distribution of the barriers between them. By changing the ratio of the channel length and the distance to the gate we can see the effect of the Coulomb interaction between hopping electrons on shot noise.

The experiment was carried out on a n-GaAs MESFET consisting of a GaAs layer of 0.15 μm thickness (donor concentration $N_d = 10^{17}$ cm^{-3}). On the top of the GaAs layer Au gates were deposited with dimensions $L = 0.2 - 1$ μm (in the direction of the current) and $W = 4 - 20$ μm. By applying a negative gate voltage, V_g, a lateral potential barrier is formed between the source and drain with impurity states in it.

In the case of resonant tunnelling through a single

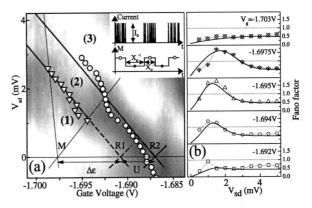

FIGURE 1. (a) Grey-scale plot of the conductance at $T = 1.85$ K (darker regions correspond to higher conductance). Lines show the positions of the conductance peaks of R and M. Inset: Schematic representation of the modulation of the current through impurity R by changing the occupancy of modulator M. (b) Fano factor at different V_g. Solid lines show the results of the numerical calculation.

impurity shot noise has been studied on a sample with $L = 0.2$ μm and $W = 20$ μm at $T = 1.85$ and 4.2 K. By varying the gate voltage V_g and the source-drain bias V_{sd} and measuring the differential conductance we have been able to identify the V_g-V_{sd} region where the current is carried by two interacting impurities, R and M, Fig.1a (region 2), [2]. If impurity M (modulator) gets charged with leak rates $X_{e,c}$, level R is shifted upwards by the Coulomb energy U. If V_{sd} is small enough and the upper state of R is above the Fermi level in the source, electrons can only be transferred via R with the rates $\Gamma_{L,R}$

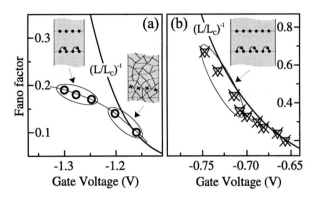

FIGURE 2. Fano factor as a function of gate voltage for hopping conduction along chains of states in different samples: (a) $L = 0.4$ μm - A-2D (circles) and (b) $L = 1$ μm (triangles), with a schematic representation of hopping paths in each case. Crosses show dominant hops. Dotted lines are guides to the eye.

when M is empty. As a result, the Fano factor in such a modulation regime is $F \simeq \frac{\Gamma_L^2 + \Gamma_R^2}{(\Gamma_L + \Gamma_R)^2} + 2 \frac{\Gamma_L \Gamma_R}{\Gamma_L + \Gamma_R} \frac{X_c}{(X_e + X_c)^2}$ [2]. The enhancement of F (second term) comes from the bunching of current pulses, Fig.1a(inset), which can be treated as an increase of the effective charge in electron transfers.

As a function of V_{sd} this enhancement has been observed as a pronounced peak in noise with an unexpectedly large Fano factor, $F > 1$, Fig1b. The increase of shot noise appears only in a specific range of V_g-V_{sd} corresponding to region of the modulated current.

We have also performed numerical calculations of the current, shot noise and the Fano factor using the generalisation of the above simple model for any relation between X and Γ [2]. The dependences $I(V_{sd}, V_g)$ and $S_I(V_{sd}, V_g)$ have been successfully fitted to the experimental data, Fig1b.

With decreasing the temperature down to $T = 70$ mK we have observed a further increase of the Fano factor up to $F \sim 5$. This regime corresponds to the quantum case $\hbar\Gamma \gg k_B T$ where the master equation approach is not applicable. We can assume that this increase is also related to the interaction between different resonant tunnelling channels, although more theoretical input is required to understand this effect.

To study shot noise behaviour in 1D hopping with uniform and non-uniform barriers we have used a longer sample ($L = 0.4$ μm and $W = 4$ μm) with a split in the gate: $\omega = 0.3$ μm. This sample shows different behaviour from one cooldown to another: two-dimensional (referred to as A-2D) and one-dimensional (A-1D). This difference is caused by randomly trapped charge near the 1D constriction [5].

In 1D case, the Fano factor increases from 0.07 to 0.15 when the negative V_g is increased. Assuming that $F = 1/N$ [6], where N is the number of barriers in a uniform chain of states, the number of hops in this range changes from 14 to 7. With a further increase of negative gate voltage the Fano factor rapidly increases to 0.8. As the position of the channel is defined by the split, the distribution of the resistances of the random hops is exponentially broad, so that only a single hop dominates the whole conductance and F is close to 1.

One can see in Fig.2a that in the 2D case the Fano factor slowly increases from 0.1 to 0.2 in the range V_g from -1.18 to -1.34 V, Fig.2a. This increase agrees with $N = L/L_c$ found from the T-dependence of the conductance and presented in Fig.2a as a solid line [7, 3]. With applying more negative V_g, the Fano factor shows a saturation around $F \sim 0.2$. This behaviour can be explained by the fact that at these gate voltages hopping is dominated by a set of most conductive hopping chains with the close and even position of localised states [8]. A relatively large number of states, $N \sim 5$, gives a smaller Fano factor.

To study the effect of Coulomb interaction on shot noise suppression in hopping we have measured the 2D sample with larger length of the gate, $L = 1 \mu m$, but with the same separation between the channel and the gate. One can see in Fig.2b that the Fano factor for this sample changes from 0.2 to 0.7 with increasing negative gate voltage. We explain this behaviour to be due to the suppression of Coulomb interaction in longer hopping chains due to its more efficient screening by the longer metallic gate. As a result, the $1/N$ suppression model for the uniform hopping chains is not applicable and a large F is expected [9].

In summary, we have demonstrated that shot noise is a valuable tool for studying the details of electron transport in the insulating regime of conduction.

REFERENCES

1. Y.V. Nazarov and J.J.R. Struben, Phys. Rev. B **53**, 15466, (1996).
2. S.S. Safonov *et.al.*, Phys. Rev. Lett., **91**, 136801 (2003).
3. V.V. Kuznetsov, E.E. Mendez, G.L. Snider, E.T. Croke, Phys. Rev. Lett. **85**, 397, (2000).
4. B.I. Shklovskii and A.L. Efros, *Electronic Properties of Doped Semiconductors* (Springer, Berlin, 1984).
5. S.H. Roshko *et. al.*, Physica E **12**, 861, (2001).
6. R. Landauer, Physica B **227**, 156 (1996).
7. V.A. Sverdlov *et. al.*, Phys. Rev. B **63**, 081302 (2001).
8. M.E. Raikh and A.N. Ruzin, in *Mesoscopic Phenomena in Solids*, eds B.L. Altshuler, P.A. Lee, and R.A. Webb (Elsevier Science, Amsterdam, 1991).
9. A.N. Korotkov and K.K. Likharev, Phys. Rev. B **61**, 15975 (2000).

Preserved Symmetries in Nonlinear Electric Conduction

C.A. Marlow[a], A. Löfgren[b], I. Shorubalko[b], R.P. Taylor[a], P. Omling[b], L. Samuelson[b], and H. Linke[a]

[a] *Material Science Institute, Physics Department, University of Oregon, Eugene OR 97403-1274*
[b] *Solid State Physics and The Nanometer Consortium, Lund University, Box 118, S-22100 Lund, Sweden*

Abstract. For the general case of a mesoscopic, two-terminal device with no geometrical symmetry, conductance symmetries break down in the non-linear regime. Using basic symmetry arguments, we predict and experimentally confirm a set of symmetry relations that are preserved for electric conductors with respect to bias voltage, V, and B in the non-linear regime.

INTRODUCTION

Electron transport in the linear response regime is characterized by a high degree of symmetry with respect to both the sign of the voltage and the direction of an applied magnetic field, B. As described by the reciprocity relation [1], the electric conductance of mesoscopic, two-terminal devices is symmetric with respect to the direction of an external magnetic field and the sign of the voltage in the linear response regime. For two-terminal devices the reciprocity theorem reduces to the relation $G_1(B) = G_1(-B)$. The conductance of mesoscopic, two-terminal devices is symmetric with respect to magnetic field with a current flowing from contact 1 to 2. Also in linear response the orientation of the measurement leads and the sign of the source-drain bias voltage are of no consequence, $G_1(B) = G_2(B)$.

The non-linear regime is fundamental to our understanding of mesoscopic devices since linear response is limited to very small voltages [2-5]. Beyond linear response the reciprocity theorem is not valid. In the general case, where the conductor has no geometrical symmetry $G_{12}(V) \neq G_{12}(-V)$, regardless of magnetic field [2-5]. Symmetry in conductance with respect to B is also not expected if the conductor has no symmetry, $G_{12}(V, B) \neq G_{12}(V, -B)$ [6].

The reciprocity relation breaks down in the non-linear transport regime [2, 3], recently the possibility of surviving symmetries in this regime has begun to receive attention [6]. Symmetry of non-linear transport requires geometrical symmetry of the conductor – an experimental challenge in terms of fabrication and material quality. We consider mesoscopic phase-coherent devices with geometrical symmetry and without significant disorder. Based on basic symmetry arguments, we propose symmetries of the electric conductance with respect to V and B in the non-linear regime [7].

ESTABLISHING SYMMETRIES

We consider triangular conductors, designed with and without symmetry axes parallel and perpendicular to the current. To further simplify we only consider "rigid" devices, by rigid it is meant that any circuit-induced asymmetry [7] (also referred to as "self-gating" [8]) is not significant. When the symmetry axis is parallel to the current, we call a device up-down (UD) symmetric and left-right (LR) symmetric when it the symmetry axis is perpendicular to the current direction. We measure the billiard conductance for a given V as a function of perpendicular B under reversal of V and B.

In the non-linear regime, regardless of the sign of B, an absence of LR-symmetry implies $G_{12}(V) \neq G_{12}(-V)$ [3, 6, 8]. In the presence of UD-symmetry, a reversal of B perpendicular to the device for a given V should be of no consequence for electron transport.

$$G_1(V, B) = G_1(V, -B) \quad \text{(UD)} \quad (1)$$

In the case of a "rigid" LR-symmetric device we expect, independent of UD-Symmetry,

$$G_1(V, B) = G_1(-V, -B) \quad \text{(LR, rigid)} \quad (2)$$

SAMPLES/ EXPERIMENTAL DETAILS

These symmetries are observed in GaInAs/InP billiards designed with and without symmetry axes parallel and perpendicular to the current (for fabrication details see Ref. 7). Equilateral triangles were created with contact openings positioned to form an UD-symmetric or LR-symmetric billiard such that the electrons' elastic mean free path and the phase-coherence length were greater than the device dimensions. Transport was phase-coherent through the devices and conductance fluctuations (CF) due to quantum interference effects were observed. The CF serve as a "magneto-fingerprint" of the electrostatic potential experienced by the electrons and the details of different traces can be compared to ascertain any symmetry present in the device. A small ac signal was added to a tunable dc bias voltage and two-terminal magneto-conductance measurements were taken.

RESULTS/ DISCUSSION

Relationship 1 predicts CF should be unaltered with the direction of B in the presence of perfect UD symmetry. The differential conductance, g_{ij}, for the UD device is shown as a function of B in Fig. 3 for an applied bias of $|V| = 2$ mV. The four individual traces shown were taken in the four possible configurations of the sign of V and direction of B. By comparing the pairs of traces in the individual panels in Fig. 1 (e.g. A and B, or C and D), one can see the CF are almost perfectly symmetric under a reversal of B as predicted. Any deviations from perfect symmetry in B could be attributed to imperfections of the device geometry.

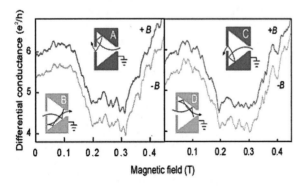

FIGURE 1. CF for UD billiard measured in 4 different possible configurations of the sign of the bias voltage, direction of magnetic field and lead configuration. The lower trace in each panel has been offset by 0.5 e^2/h, (a) shows $g_{12}(V = +2$ mV, $\pm B)$, (b) $g_{12}(V = -2$ mV, $\pm B)$.

Fig. 2 shows the CF for the LR-symmetric device at a bias of $|V| = 1$ mV. The top trace is for $V = 0$ (linear regime) and as expected the conductance is symmetric in B. By comparing the marked features on the lower two traces taken at finite V, one can see striking similarities, as predicted by Eq. 2. Also the traces are not symmetric in B as expected in the absence of UD symmetry.

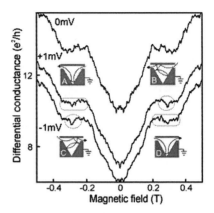

FIGURE 2. CF for "rigid" LR billiard at $V = 0$, $V = +1$ mV and $V = -1$ mV. Data offset for clarity. Note that $g_{12}(V, B) \neq g_{12}(V, -B)$, while $g_{12}(V, B) \approx g_{12}(-V, -B)$.

We have demonstrated and confirmed a set of proposed symmetry relations for electric conductance in the non-linear regime and in the presence of a magnetic field using a sensitive experimental test for mesoscopic, phase-coherent devices. [9].

ACKNOWLEDGMENTS

Financially supported by an NSF IGERT fellowship (C.A.M.), an NSF CAREER award (H.L.), a Cottrell scholarship (R.P.T).

REFERENCES

1. Büttiker, M., Phys. Rev. Lett., 1986. **57**(14): p. 1761.
2. Al'tshuler, B.L. and D.E. Khmel'nitskii,. JETP Lett., 1985. **42**(7): p. 359.
3. Landauer, R., *Nonlinearity: Historical and Technological View*, in *Nonlinearity in Condensed Matter*, A.R. Bishop, et al., Editors. 1987, Springer Verlag: Berlin.
4. Büttiker, M., *Capacitance, admittance and rectification properties of small conductors*. J. Phys.: Condens. Matter, 1993. **5**: p. 9361-9378.
5. Christen, T. and M. Büttiker, Europhys. Lett., 1996. **35**(7): p. 523-528.
6. Linke, H., et al., Phys. Rev. B, 2000. **61**(23): p. 15914-15926.
7. Löfgren, A., et al.,. Phys. Rev. Lett., 2004. **92**(4): p. 046803.
8. Kristensen, A., et al., Phys. Rev. B, 2000. **62**: p. 10950-10957.
9. Linke, H., et al., Europhys. Lett., 1998. **44**: p. 341-347.

Effects of Scale-Free Disorder on the Metal-Insulator Transition

Macleans L. Ndawana*, Rudolf A. Römer* and Michael Schreiber[†]

*Department of Physics and Centre for Scientific Computing,
University of Warwick, Coventry CV4 7AL, United Kingdom
[†]Institut für Physik, Technische Universität, 09107 Chemnitz, Germany

Abstract. We investigate the effects of scale-free disorder on the metal-insulator transition. Scale-free disorder is characterized by an autocorrelation function decaying asymptotically as $r^{-\alpha}$. We study the dependence of the localization-length exponent ν on the correlation-strength exponent α. We find that for fixed disorder W, there is a critical α_c, such that for $\alpha < \alpha_c$ we have $\nu = 2/\alpha$ while for $\alpha > \alpha_c$, ν retains the value of the uncorrelated system in accordance with the extended Harris criterion. At the band center, ν is independent of α but equal to that of the uncorrelated system.

INTRODUCTION

We study the effects of long-range power-law correlated disorder — so called scale-free disorder — on the metal-insulator transition (MIT). The effects of scale-free disorder on the critical properties of physical systems have recently received much renewed attention [1, 2, 3]. For the Anderson model of localization previous investigations of scale-free disorder have concentrated on the one- (1D) [4, 5] and two-dimensional (2D) [6, 7] cases. For the 1D Anderson model, it has been shown that for energies close to the band edge the presence of scale-free diagonal disorder causes states to be strongly localized [8], while at the band center the states tend to have localization lengths ξ comparable to the system size [4]. In the 2D Anderson model with scale-free disorder an MIT of the Kosterlitz-Thouless transition type [6, 7] has been observed.

Our calculation is based upon the Anderson tight-binding Hamiltonian in site representation $\mathcal{H} = \sum_{\langle i,j \rangle} |i\rangle\langle j| + \sum_i \varepsilon_i |i\rangle\langle i|$, where $\langle i,j \rangle$ denotes a sum over nearest neighbors and $|i\rangle$ is an atomic-like orbital at site i. The random on-site energies ε_i are drawn from a Gaussian distribution with zero mean. The autocorrelation function of the disorder decays asymptotically as $\langle \varepsilon_i \varepsilon_{i+r} \rangle \sim r^{-\alpha}$. The average is done over spatial positions and disorder realizations. Large α corresponds to the nearly uncorrelated case which shall serve as our point of reference, while small α is the strongly correlated case. We generate scale-free disorder by use of the modified Fourier-filtering method as outlined in [9]. The Hamiltonian is evaluated via a variant of the transfer-matrix method (TMM) as detailed

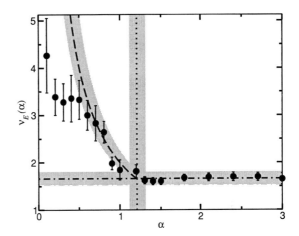

FIGURE 1. The critical exponent $\nu_E(\alpha)$ as a function of the correlation-strength exponent α at $W = 12$. Error bars reflect one standard deviation. The horizontal line indicates the uncorrelated value $\nu_0 = 1.66 \pm 0.06$ and the vertical dotted line $\alpha_c = 1.21$. The dashed line for $\alpha < \alpha_c$ gives the extended Harris criterion (1). The grey areas denote error bounds of one confidence interval arising from the error in ν_0. Deviations from the extended Harris criterion for small $\alpha < 0.5$ are due to finite-size effects. The system sizes used are $M = 5, 7, 9, 11$ and 13.

in [10] to obtain the localization length λ for quasi-1D bars of cross-section $M \times M$. This modified TMM has previously been used for two interacting particles [11].

The critical behavior can be determined by numerically establishing for the reduced localization lengths $\Lambda = \lambda/M$ that the one-parameter scaling hypothesis holds [12, 13] in the vicinity of the MIT. And accordingly, the scaling parameter ξ diverges when x, which

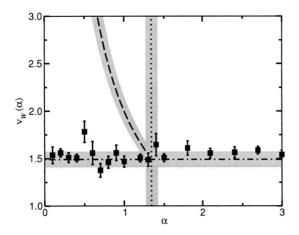

FIGURE 2. Same as Fig. 1 but for $\nu_W(\alpha)$ at the band center, $E = 0$. The horizontal line indicates the uncorrelated value $\nu_0 = 1.49 \pm 0.03$ and the vertical dotted line $\alpha_c = 1.34$. The system sizes used are $M = 5, 7, 9, 11, 13$ and 15.

can be the disorder strength W or the energy E, is tuned to its critical value x_c as $\xi(x) \sim |x - x_c|^{-\nu}$. The value of ν is estimated via finite-size scaling [12, 13].

THE EXTENDED HARRIS CRITERION

The extended Harris criterion (EHC) can be stated as;

$$\nu = \begin{cases} 2/\alpha & \text{if } \alpha < \alpha_c \\ \nu_0 & \text{if } \alpha > \alpha_c \end{cases}, \quad (1)$$

where ν_0 is the critical exponent for a system without correlations in the disorder. Eq. (1) implies that there is a well-defined critical value $\alpha_c = 2/\nu_0$, below which correlations are relevant and above which correlations are irrelevant for a second-order phase transition characterized by ν.

We define ν_E and ν_W as critical exponents of the localization length $\xi \propto |E - E_c|^{-\nu_E}$ at fixed disorder strength W and $\xi \propto |W - W_c|^{-\nu_W}$ at fixed energy E, respectively. Our results show that ν_E obeys the EHC such as demonstrated in Fig. 1. On the other hand at the band center ν_W is independent of α as can be seen in Fig. 2. This means that scale-free disorder increases the critical exponent ν_E for small α while leaving ν_W unchanged.

SUMMARY

We have demonstrated that the MIT obeys the EHC when the phase boundary is crossed along the energy axis. More specifically, the critical exponent ν_E satisfies the EHC, Eq. (1). However, we showed that the EHC fails at the band center and we trace [10] this failure to the different mechanisms governing the MIT in the vicinity of the band center and outside. We emphasize that such different scaling behavior at the band center and at the band edges has indeed been speculatively discussed for a long time [14, 15], although numerical studies of ν_E and ν_W for the uncorrelated system suggest a common value.

ACKNOWLEDGMENTS

This work is supported by the Deutsche Forschungsgemeinschaft via the SFB 393 and in part via the priority research program "Quanten-Hall-Systeme".

REFERENCES

1. Bouchaud, J.-P., and Georges, A., *Phys. Rep.*, **195**, 127 (1990).
2. Cain, P., Römer, R. A., Schreiber, M., and Raikh, M. E., *Phys. Rev. B*, **64**, 235326 (2001).
3. Sandler, N., Maei, H., and Kondev, J., *Phys. Rev. B*, **68**, 205315 (2003).
4. Russ, S., Kantelhardt, J. W., Bunde, A., Havlin, S., and Webman, I., *Physica A*, **266**, 492 (1999).
5. Russ, S., Kantelhardt, J. W., Bunde, A., and Havlin, S., *Phys. Rev. B*, **64**, 134209 (2001).
6. Liu, W., Chen, T., and Xiong, S., *J. Phys. A: Math. Gen.*, **11**, 6883 (1999).
7. Liu, W.-S., Liu, S. Y., and Lei, X. L., *Eur. Phys. J. B*, **33**, 293 (2003).
8. Russ, S., Havlin, S., and Webman, I., *Phil. Mag. B*, **77**, 1449 (1998).
9. Makse, H. A., Havlin, S., Schwartz, M., and Stanley, H. E., *Chaos, Solitons and Fractals*, **6**, 295 (1995).
10. Ndawana, M. L., Römer, R. A., and Schreiber, M. (2004), ArXiv: cond-mat/0402018.
11. Römer, R. A., and Schreiber, M., *Phys. Rev. Lett.*, **78**, 4890 (1997).
12. MacKinnon, A., and Kramer, B., *Z. Phys. B*, **53**, 1 (1983).
13. Slevin, K., and Ohtsuki, T., *Phys. Rev. Lett.*, **82**, 382 (1999).
14. Bulka, B., Kramer, B., and MacKinnon, A., *Z. Phys. B*, **60**, 13 (1985).
15. Bulka, B., Schreiber, M., and Kramer, B., *Z. Phys. B*, **66**, 21 (1987).

Non-perturbative Scattering of Electrons by Charged Dislocations

L. B. Hovakimian

Institute of Radiophysics and Electronics, Armenian Academy of Sciences, Ashtarak-2, 378410, Armenia

Abstract. In this study the eikonal-scattering formalism is utilized to obtain information on the impact of charged dislocations on the Hall mobility of conduction electrons in wide-bandgap semiconductors.

INTRODUCTION

The lowest-order Born approximation has been extensively used in the development of the theory of electron scattering by charged dislocation lines in various semiconductor systems [1-5]. In this work, we use a non-perturbative eikonal approach [6] for calculating the dislocation-scattering Hall mobility in cases where the interaction of impinging electrons with charged dislocations is sufficiently strong and/or the temperature T is not high.

MODEL AND THEORY

According to theory of electrically active edge dislocations in non-degenerate n-type semiconductors [2,7], at temperatures where shallow donors with volume concentration n_d are completely ionized, the properties of the dislocation scattering potential $U(r)$ depend crucially on the value of $\Gamma = U_0/k_B T$. In the high-T limit, $\Gamma \ll 1$, this potential is given by [1-3]

$$U(r) = 2U_0 K_0(r/r_D), \quad U_0 = fe^2/\chi a, \quad (1)$$

where r is the radial distance from the dislocation axis, K_0 is the modified Bessel function, f denotes the fraction of filled acceptor-like traps along the charged line, and the Debye screening radius is given by $r_D = (\chi k_B T/4\pi e^2 n_d)^{1/2}$, in which χ is the dielectric constant, and e is the electron charge.

In the opposite limit, $\Gamma \gg 1$, the Debye mechanism loses its significance due to reduced kinetic energies of the screening electrons, and the scattering potential takes the following form [7],

$$U(r) = U_0\{2\ln(R/r) - [1-(r/R)^2]\}, \quad r \leq R, \quad (2)$$

where $R = (f/\pi a n_d)^{1/2} \sim \Gamma^{1/2} r_D$ is the Read radius.

For the case where the current flows perpendicular [1-5] to a network of parallel dislocations, the electron Hall mobility μ_H can be evaluated from

$$\mu_H = e\langle \tau^2 \rangle / m^* \langle \tau \rangle, \quad (3)$$

$$\tau^{-1} = N_D \mathrm{v} \int_0^{2\pi} d\theta (1-\cos\theta)|A(\theta)|^2, \quad (4)$$

where $\langle \tau \rangle$ denotes an average [1] of the relaxation time, $\mathrm{v} = \hbar k/m^*$ is the component of electrons velocity perpendicular [1-5] to dislocations with density N_D, and $A(\theta)$ is the scattering amplitude. Under the condition $k^{-1} \ll (r_D, R)$ the structure of $A(\theta)$ can be studied via eikonal formula [6]

$$A(\theta) = (k/2\pi i)^{1/2} \int_{-\infty}^{\infty} dx e^{-iq(\theta)x}[e^{i\Delta(x)} - 1], \quad (5)$$

in which $q(\theta) = 2k\sin(\theta/2)$ is the momentum transfer upon scattering, and the eikonal phase is given by

$$\Delta(x) = -(\hbar \mathrm{v})^{-1}\int_{-\infty}^{\infty} dy U(r), \quad r = (x^2+y^2)^{1/2}.$$

In the limit $\Gamma \ll 1$ we employ Eq. (1) to compute the phase as $\Delta(x) = -\Delta \exp[-|x|/r_D]$, $\Delta = 2\pi v$, where $v = U_0 r_D / \hbar v = k_0^2 r_D / k$ is the Born parameter [6], and it is assumed in what follows that U_0 is such that $r_D^{-1} \ll k_0 = (m^* U_0 / \hbar^2)^{1/2}$. Due to the above form of $\Delta(x)$, (5) becomes $A(\theta) = r_D (k/2\pi i)^{1/2} \times \{(-i\Delta)^{iqr_D} \gamma(iqr_D, -i\Delta) + (-i\Delta)^{-iqr_D} \gamma(-iqr_D, -i\Delta)\}$, in which $\gamma(\alpha, z)$ is the incomplete gamma-function [8]. Performing the angular integration in Eq. (4) with the aid of this eikonal $A(\theta)$, one finds in the strong coupling situation ($v \gg 1 \gg k_0/k$) the result $\tau(k) = (k/k_0)^3 \tau(k_0)$, where $\tau(k_0) = \tau_0 /(2\pi)^2 (k_0 r_D)$, and $\tau_0 = m^*/\hbar N_D$ [$\mu_0 = e\tau_0/m^* = e/\hbar N_D$].

In the case $\Gamma \gg 1$ the scattering problem at hand can be analysed by arguments identical to those for the case $\Gamma \ll 1$. Utilizing Eq. (2) and the condition $k_0 R \gg 1$, one obtains $\tau(k) = \tau_0 / 2kR$ ($R^{-1} \ll k \ll k_0$).

Finally, using in Eq. (3) the above results for $\tau(k)$, we arrive at the following expression:

$$\mu_H / \mu_0 \sim \begin{cases} \Gamma^{-3/2} / k_0 r_D, & 1 \gg \Gamma \gg (k_0 r_D)^{-2}, \quad (6a) \\ \Gamma^{1/2} / k_0 R, & \Gamma \gg 1. \quad (6b) \end{cases}$$

DISCUSSION

We will consider the salient features of μ_H in the wide-bandgap limit, where the influence of T on f can be neglected [9] if $k_B T / E_D \ll 1$ (E_0 is the ionization energy of deep dislocation traps). Under this condition Eqs. (6a)-(6b) show that $\mu_H(T)$ passes at a temperature $T_m \sim (f/\chi)(e^2/k_B a)$ through a minimum described by $\mu_m = \mu_H(T_m) \sim (a/fN_D)(\chi n_d/m^*)^{1/2}$.

Physically, this minimum in mobility is seen to occur when $R/r_D(T_m) \sim \Gamma^{1/2} \sim 1$, i.e., when the kinetic energies of Boltzmann electrons become comparable to the characteristic Coulomb energy of the scattering field. If we use for order-of-magnitude estimates the values $a \approx 5A$ and $\chi \approx 10$, we can conclude that for typical values of $f \approx 0.1$ in Si [7] the expected magnitude of T_m is roughly $T_m \sim 10^2 K$. One should note that in plastically deformed n-Si crystals a minimum in mobility due to charged dislocation scattering has been experimentally observed at $T \sim 100 \div 150 K$ [10].

Next we consider the problem of threading dislocation scattering in n-GaN epilayers [3-5]. For this purpose we pay careful attention to the fact that according to theory [4] and experiment [11] the values of f for highly charged threading dislocations in GaN are almost ~ 1. In view of this we are led to conclude that T_m in this material should be high (roughly, $T_m \sim 10^3 K$). As a consequence, an interesting picture arises, where the essentially non-perturbative scattering regime $\Gamma > 1$ remains operative even at elevated $T \sim 300 K$. From pertinent Eq. (6b) we then can see that due to fundamental relation $R \propto n_d^{-1/2}$ the room temperature $\mu_H(n_d)$ increases with n_d according to $\mu_H \propto n_d^{1/2}/N_D$. Such a scaling behavior for n-GaN has been observed in transport experiment [12].

ACKNOWLEDGMENTS

This work was supported by the NFSAT under grant no. G-2004/04.

REFERENCES

1. B. Podor, *Acta Phys. Hungaricae* **23**, 393-405 (1967).
2. K. Seeger, *Semiconductor Physics*, Vienna: Springer, 1997, pp. 221-225.
3. N. G. Weimann, L. F. Eastman, D. Doppalapudi, H. M. Ng, and T. D. Moustakas, *J. Appl. Phys.* **83**, 3656-3659 (1998).
4. D. C. Look and J. R. Sizelove, *Phys. Rev. Letters* **82**, 1237-1240 (1999); D. C. Look et al., *Solid State Commun.* **117**, 571-575 (2001).
5. M. N. Gurusinghe and T. G. Andersson, *Phys. Rev. B* **67**, 235208 (2003).
6. L. D. Landau and E. M. Lifshitz, *Quantum Mechanics*, Oxford: Butterworth-Heinemann, 1997, pp. 611-619.
7. V. B. Shikin and Yu. V. Shikina, *Usp. Fiz. Nauk* **165**, 887-917 (1995).
8. I. S. Gradshteyn, and I. M. Ryzhik, *Tables of Integrals, Series and Products*, New York: Academic, 1994.
9. Yu. V. Kornyushin, *Fiz. Tekh. Poluprovodn.* **16**, 1679-1680 (1982).
10. L. S. Milevsky and A. A. Zolotukhin, *Pis'ma Zh. Eksp. Teor. Fiz.* [*JETP Letters*] **19**, 478-481 (1974).
11. D. Cherns, C. G. Jiao, H. Mokhtari, J. Cai, and F. A. Ponce, *Phys. Stat. Solidi* (b) **234**, 924-930 (2002).
12. H. W. Choi, J. Zhang, and S. J. Chua, *Mater. Sci. Semicond. Processing* **4**, 567-570 (2001).

Picosecond Raman Studies of Electron and Hole Velocity Overshoots In a GaAs-based p-i-n Semiconductor Nanostructure

W. Liang,[1] K. T. Tsen,[1] C. Poweleit,[1] J. M. Barker,[2] D.K. Ferry,[2] H. Morkoc[3]

[1] *Department of Physics and Astronomy, Arizona State University, Tempe, Arizona, 85287*
[2] *Department of Electrical Engineering, Arizona State University, Tempe, AZ 85287*
[3] *Department of Electrical Engineering, Virginia Commonwealth University, Richmond, VA 23284*

Abstract. Picosecond Raman spectroscopy has been employed to study electron and hole transport in a GaAs-based p-i-n nanostructure. Electron as well hole velocity overshoots are observed. It has been demonstrated that due to the relatively long laser pulse used in the experiments the extent of overshoot is about the same in both cases.

INTRODUCTION

Recent progress in microelectronic fabrication processing has made the size of electronic devices down to the order of 0.1 μm or smaller. This means that a very strong electric field intensity exists in an electronic device if a typical device operation voltage (which is of the order of 1 V) is applied. Under very high electric field intensity, the carrier transient transport phenomenon, which normally lasts for1 picosecond or shorter, has been demonstrated to exhibit drastically different behavior from that of the steady state.[1]

EXPERIMENTAL TECHNIQUE AND RESULTS

The GaAs-based p-i-n nanostructure used in this work has been described elsewhere.[2] The excitation laser having photon energy larger than the bandgap of GaAs is incident onto the intrinsic region of GaAs to generate electron-hole pairs. These hot electron and holes, which are drifted in opposite directions due to the application of an electric field, are then probed by the trailing portion of the same laser pulse by Raman spectroscopy.

The laser source is a DCM dye laser, which is synchronously pumped by the second harmonic of a cw mode-locked YAG laser.[3] The laser has a repetition rate of 76 MHz; and its pulse width is tuned to either 2ps or 20ps, depending on the experimental needs. The laser photon energy is about 2.0 eV.

$Z(X,Y)\bar{Z}$ scattering configuration is used, where $X=(100)$, $Y=(010)$, $Z=(001)$, so that single-particle scattering (SPS) associated with spin-density fluctuations can be detected.[4] All the experiment results were obtained at $T=300K$.

Figure 1(a) shows a typical SPS spectrum for the GaAs-based p-i-n nanostructure, taken with an excitation laser having a pulse width of 2ps. The injected electron-hole pair density is estimated to be $n \sim 5\times10^{17}$ cm^{-3}. The effective applied electric field intensity is about E = 15 kV/cm. The SPS spectrum has been found to sit on a smooth luminescence background coming from the hot electron-hole recombination. This background has been found to be very well fit by an exponential curve within the range of frequency shift of our interest, i.e., from -800 cm^{-1} to 1000 cm^{-1}.[5] After the subtraction of the luminescence background, the SPS spectrum is obtained. The spectrum is then transformed into the electron velocity distribution,[1] which is shown in Fig 1(b). It is clear from Fig. 1(b) that the electron distribution is shifted toward the opposite direction of the applied electric field, as expected. The electron drift velocity in this case is deduced to be $(2.5\pm0.3)\times10^{7}$ cm/sec.

We note that hole distribution along the wavevector transfer direction (in this case, z-direction) can be deduced once electron distribution along the wavevector transfer direction and the luminescence intensity of electron-hole pair recombination are determined.[6] The corresponding E_0 bandgap luminescence spectra are shown in Fig. 2(a) for the applied electric field intensity of E = 15 kV/cm. If we assume that the hole distributions under our experimental conditions can be described by a shifted Fermi-Dirac function determined completely by two parameters: the hole temperature T_h and the hole drift velocity V_{dh}, the hole distribution can be indirectly deduced from a knowledge of electron distribution and electron-hole pair luminescence.

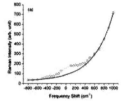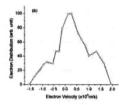

FIGURE 1. (a) SPS spectrum for a GaAs p-i-n nanostructure. (b) The electron velocity distribution after the subtraction and nonparabolicity modification.

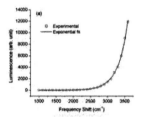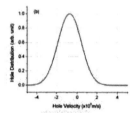

FIGURE 2 (a) electron-hole pair luminescence spectrum (b) the deduced hole velocity distribution.

Figure 2(a) shows the electron-hole pair luminescence spectrum and its fit for the laser pulse width of 2ps. The parameter set best fit our data is: T_h= (315± 30)K, V_{dh}=(7.2 ± 0.7) ×10^6 cm/sec. The deduced hole distribution along the wavevector direction is shown in Fig. 2(b).

We have also carried out similar experiments on the sample except that the laser pulse width is ~20ps. The electron drift velocity has been found to be (1.5±0.2)×10^7 cm/sec. The parameter set best fit the electron-hole pair luminescence is: T_h = (315 ± 30)K, V_{dh}=(4.0 ± 0.4) ×10^6 cm/sec. All the experimental results are summarized in Table 1. Since experimental results for a laser pulse width as long as 20 ps represents very closely to the steady-state value, based upon Table 1, we conclude that electron and hole velocity overshoots have been observed and demonstrated in our experiments.

DISCUSSIONS

The different response of the electrons and the holes under applied electric field can be understood by the nature of the scattering process different carrier undergoes. For the electrons, they are rapidly accelerated in the field to higher energies, subsequently scattered from Γ valley to the L and X valleys of the conduction band. This process is very efficient and within short time period (~1.0ps), those carriers at higher energy position in Γ valley can be monitored. Experimentally, we observe the velocity overshoot. For a longer period of time (about the order of 20ps), the electrons in the Γ valley are mostly those returned from the satellite valleys. Those electrons have been randomized efficiently due to the intra-valley scattering. Hence more low energy electrons, compared with the energetic ones, weigh in the electron distribution, and we expect a much lower drift velocity on this time scale. The data in Table 1 indicate it is exactly the case. The electron drift velocity overshoots its steady-state value by a factor of about 1.7. We attribute this small electron overshoot factor to the relatively long laser pulse width (~2ps) used in our experiments. On the other hand, the dominant scattering process for holes is intra- Γ-valley TO modes. This process is ultrafast and as a result, the hole drift velocity overshoots its steady-state value by a factor of about 1.8.

TABLE 1 Drift velocities of electron and hole as excited by laser pulse with different durations

Drift velocity	2ps	20ps
V_{de} (cm/sec)	(2.5±0.3)×10^7	(1.5±0.2)×10^7
V_{dh} (cm/sec)	(7.2±0.7)×10^6	(4.0±0.4)×10^6

In summary, electron and hole overshoot in a GaAs-based p-i-n nanostructure are directly observed by picosecond Raman spectroscopy. The experimental results are discussed and explained in terms of electron and hole scattering processes.

REFERENCES

1. See for example, K.T. Tsen in "Ultrafast Phenomena in Semiconductors", edited by K.T. Tsen (Springer-Verlag, New York, 2001), p. 191.
2. E.D. Grann, S.J. Sheih, K.T. Tsen, O. F. Sankey, S.E. Gunser, D.K. Ferry, A. Salvador, A. Botcharev and H. Morkoc. Phys. Rev. B51, 1631 (1995).
3. K.T. Tsen, Keith R. Wald, Tobias Ruf, P.Y. Yu and H. Morkoc. Phys. Rev. Lett. 67, 2557-2560 (1991).
4. C. Chia, O. F. Sankey and K. T. Tsen. Modern Phys. Letts. B, Vol. 7, No. 6, 331(1993).
5. D.S. Kim, and P.Y. Yu, Phys. Rev. B43, 4158 (1991).
6. K.T. Tsen in "Ultrafast Dynamical Processes in Semiconductors", edited by K.T. Tsen ,Volume #92 in the series – Topics in Applied Physics (Springer-Verlag, Heidelberg, 2004), p. 193

Disorder-induced non-Ohmic steady-state flow of hopping carriers

S. A. Baily[*] and David Emin[†]

[*]Space Vehicles Directorate, Air Force Research Laboratory, KAFB, NM USA
[†]Department of Physics and Astronomy, University of New Mexico, Albuquerque, NM USA

Abstract. We review the formal theory of multi-phonon hopping conductivity in a disordered medium beyond the linear response regime. Attention is focused on the relationship between disorder and non-Ohmic current flow. Some numerical studies of physically meaningful one- and two-dimensional models are reported and discussed.

The application of a large electric field induces many non-crystalline semiconductors to switch from semiconducting to conducting states [1]. The electrical conductivities of such semiconductors rise with electric-field strength before they switch. Transport measurements on a number of these materials imply that the charge carriers move with low mobility via thermally activated hopping. This study addresses the electric-field dependencies of the electrical conductivity, non-Ohmic behavior, associated with charge carriers that move between energetically disparate sites by phonon-assisted hopping under the influence of large electric fields.

The net current flow between any pair of sites i and j, I_{ij} is described by the master equation:

$$I_{ij} = q[P_i(1-P_j)R_{ij} - P_j(1-P_i)R_{ji}], \quad (1)$$

where P_i is the probability that site i is occupied, and R_{ij} is the rate with which a carrier of charge q can hop from a site i to site j. The probability of a site i with energy E_i being occupied during steady-state current flow is expressed in terms of a local quasi-electro-chemical potential (QECP), μ_i, that is defined via the relation $P_i \equiv 1/\{\exp[(E_i-\mu_i)/(k_BT)]+1\}$ [2]. Use of Fermi statistics ensures only single occupancy of each site. This restriction is equivalent to modeling the Coulomb repulsion between carriers with strong on-site (Hubbard-type) repulsion.

In equilibrium the QECP of all sites equals the electro-chemical potential. The carrier density is determined by the difference between the electro-chemical potential and the electronic energy levels. In the steady state with an applied emf the QECP varies monotonically. With disorder the variation of the QECP becomes non-uniform albeit remaining monotonic. Concomitantly, steady-state flow in a disordered system produces an electric-field induced redistribution of charge amongst sites. In addition, the current garners a non-linear dependence on the emf. In these non-Ohmic regimes the conductivity also depends on the ordering of difficult hops.

We restrict our attention to high enough temperatures for the atomic motions associated with phonon-assisted hopping to be treated as classical (typically $k_BT > h\nu/3$, where ν is the characteristic vibrational frequency of atomic vibrations with which carriers interact). We also assume that the electronic transfer energies that link sites between which carriers hop are great enough for electronic carriers to adiabatically adjust to atoms' vibratory motion. Concomitantly, we restict our attention to hops between nearest-neighbor sites. The jump rate governing such semiclassical adiabatic hops is:

$$R_{ij} = \nu \exp[-(4\varepsilon_a + E_j - E_i)^2/(16\varepsilon_a k_BT)], \quad (2)$$

where ε_a is the small-polaron hopping activation energy, and E_i and E_j refer to the electronic energy levels at sites i and j, respectively. [3].

Current flow is studied by imposing a difference of the electro-chemical potential across an array of sites. The resulting inter-site motion is described by Eqs. (1) and (2). Steady-state behavior is ensured by requiring the vanishing of the net current from any site.

The electrical current is calculated for different values of the site energies, small-polaron hopping activation energy, temperature, and applied electric field. The computer calculations utilize the simultaneous over-relaxation method with Chebyshev acceleration. When disorder of site energies was taken as Gaussian, psuedo-random numbers were generated with the R250 method [4]. The steady-state condition is concidered fulfilled when each iteration yields only small changes of the QECP on every site. The sufficiency of this criterion for steady-state flow is verified in 1-D calculations by verifying the constancy of the current flowing between each

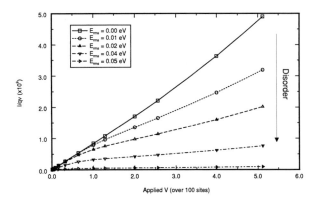

FIGURE 1. The room-temperature steady-state current along a 100-site chain vs. applied emf. The site energies are distributed (with Gaussian distribution widths shown in the legend) about a mean value that is 0.3 eV away from the chemical potential. The small-polaron activation energy $\varepsilon_a = 0.2$ eV.

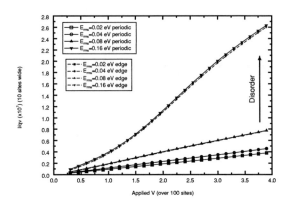

FIGURE 2. The steady-state current across a 100-site long and 10-site wide array of sites vs. applied emf. The parameters not shown are the same as in Fig. 1. These results are insensitive to whether the first and tenth sites represent a bounding edge or are linked to one another with periodic boundary conditions.

pair of adjacent sites. An analogous technique is used for 2-D calculations; it is verified that the currents flowing across different lines traversing the current flow essentially equal one another. Furthermore, the results of our computer calculations agree with those obtained for several 1-D models that were solved analytically [2].

In the absence of disorder, the current associated with the hopping of a constant (electric-field independent) density of carriers is proportional to $\sinh(qEa/(2k_BT))$, where E is the electric-field strength and a is the inter-site separation; $Ea \equiv V/N$ for an emf V applied across N sites. This factor dominates the electric-field dependence of the hopping conductivity except for exceptionally strong electric fields, $qEa > 4\varepsilon_a$ [3]. As a result, the hopping conductivity increases with applied voltage as illustrated by the top curve of Fig. 1. The imposition of Gaussian energetic disorder affects current flow. In particular, the current between a pair of sites is proportional to $\exp[-(E_i + E_j)/(2k_BT)]$, where E_i and E_j are the local electronic energies at the two sites. As a result, increasing energetic disorder reduces steady-state current flow along a linear chain as carriers are forced to negotiate energy barriers associated with high-energy sites. This feature is evident in Fig. 1. These I-versus-V curves manifest significant non-Ohmic (non-linear) behavior. A qualitatively different situation prevails when comparable energetic disorder is imposed on a two-dimensional array of sites. Then, as illustrated in Fig. 2, increasing Gaussian energetic disorder increases current flow. In these instances energetic disorder opens up low-energy current paths that avoid high-energy sites. That is, the low-energy sites facilitate current flow rather than acting as "traps." In contrast to the curves of Fig. 1, these I-versus-V curves of Fig. 2 only manifest non-Ohmic behavior when the disorder energy is extremely large.

Variations of the electronic transfer energy linking sites also affects hopping transport. For example, with very small electronic transfer energies the rates for hops between sites are greatly reduced from their adiabatic values. With a significant reduction (a factor of 10^9) of a sizable fraction (1/3) of the transfer energies, percolation effects become important. With highly constrained current flow, non-Ohmic behavior similar to that of a one-dimensional chain is obtained with modest energetic disorder ($E_{rms} = 0.04$ eV).

Moderate energetic disorder produces non-Ohmic transport on a chain. However, extreme energetic disorder is required to produce non-Ohmic behavior for hopping in a 2-D array. Thus, it seems unlikely that simple energetic disorder can account for the non-Ohmic behavior of chalcogenide glasses. The non-Ohmic behavior may be associated with variations of electronic transfer energies. Indeed, bond-angle variations produce large variations of the transfer energies between states with directed orbitals [5]. Space-charge limited current may also contribute to the observed non-Ohmic behavior.

This research was performed while S. A. Baily held a National Research Council Research Associateship Award at the Space Vehicles Directorate, Air Force Research Laboratory, Kirtland Air Force Base, NM.

REFERENCES

1. Adler, D., *CRC Critical Reviews in Solid State Sciences*, CRC Press, New York, 1971, vol. 2, chap. Amorphous Semiconductors, p. 317.
2. Emin, D., *phys. stat. sol. (b)*, **205**, 69–71 (1998).
3. Emin, D., *SPIE*, **2850**, 159–170 (1996).
4. Kirkpatrick, S., and Stoll, E., *J. Comp. Phys.*, **40**, 517–526 (1981).
5. Emin, D., and Bussac, M.-N., *Phys. Rev. B*, **49**, 14290–14300 (1994).

Alkali Metals Transport at High Temperatures in the Presence of an Electric Field

I. Rapoport[*], P. Taylor[*], V. Mart[*], J. Kearns[*], and F. Kirscht[¶]

[*]*Sumitomo Mitsubishi Silicon Group, OR 97302, USA*
[¶]*Institute for Crystal Growth, 12489 Berlin, Germany*

Abstract. The transport of Sodium from silicon wafers in the presence of an electric field was studied. A dual polarity silicon boat was designed to apply the electric potential to silicon wafers during annealing. Secondary ion mass spectrometry (SIMS) and atomic absorption spectroscopy (AAS) were used to measure sodium contamination levels. Results indicate that alkali metals desorb from the silicon oxide surface, diffuse through gas boundary layer, and are transported into the gas flow. This process is strongly depends on process temperature, ambient gas pressure and electric field conditions. A DC-electric field was found to be an effective means to manage the desorbed ionized Na^+ species diffusion through the near-surface stagnant layer and transporting it into the gas phase.

INTRODUCTION

Migration of Na^+ ions is known to be responsible for causing silicon devices instabilities and failure [1]. The sodium contamination level in silicon wafers after thermal treatment depends on purity of furnace hardware and incoming materials. The utilization of electric field at elevated temperatures was proposed to reduce metal contamination levels during high temperature processing as alternative to the traditional chlorine based cleaning process [2].

After metal contaminants out-diffuse to silicon or silicon oxide surfaces, desorption is ruled by bonding enthalpy and vapor pressure [3]. Na vapor pressure at 400°C is already at over 10^{-1} Torr.

The high viscosity gas boundary layer is always present over a surface [4]. Most of desorbed species do not have enough energy to diffuse through gas boundary layer and re-adsorb back to the surface. Therefore metal removal from silicon surfaces during thermal treatment is limited by transport in gas phase.

DC-electric field was implemented to accelerate or suppress (depending on applied field direction) the desorbed Na^+ diffusion through the stagnant gas layer.

A schematic of the experiment set-up is presented on fig. 1. The samples are 6" silicon wafers with a 900Å thermal oxide. Some of silicon samples were Na contaminated by ion implantation (Dose = $1 \times 10^{13} cm^{-2}$, E = 50keV).

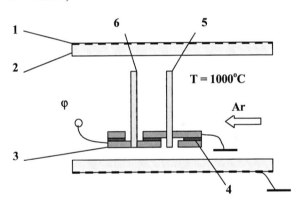

FIGURE 1. The IR heating chamber schematic: 1 – external metal electrode (grounded); 2 – quartz tube; 3 – dual polarity silicon boat; 4 – quartz insulators; 5 – grounded silicon wafer; 6 – sodium implanted silicon wafer connected to potential φ.

A Sodium implanted silicon wafer was electrically connected to the dual polarity boat so that either a positive or a negative bias φ could be applied. A second Si wafer is inserted to the grounded portion of the bipolar boat. Gas pressure conditions varied from 760 Torr to 0.4 mTorr.

Sodium contamination was measured applying SIMS (in-depth profiles) and AAS (in surface oxide films).

RESULTS AND DISCUSSION

Figures 2 and 3 summarize the Sodium (Na) SIMS data for samples after argon anneal at 1000°C with a DC-electric field applied.

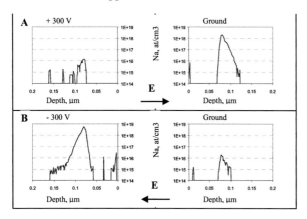

FIGURE 2. Sodium SIMS profiles after argon anneal (1000°C, 760 Torr) with direct (A) and inverse (B) electric field applied. Potential +/- 300V was applied to sodium implanted wafer (left). Second wafer was grounded.

At atmospheric pressure (fig. 2) we could observe effective transport of Na from the ion-implanted sample towards the grounded wafer.

At low pressure conditions (fig. 3) Na is transported from implanted sample by high velocity gas stream and does not adsorb on the grounded wafer.

In case of reverse field polarity, the Na transport from the contaminated sample is suppressed at both atmospheric and low pressure conditions.

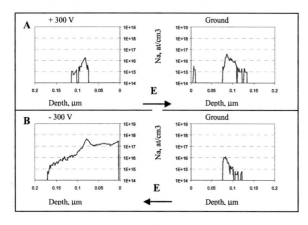

FIGURE 3. Sodium SIMS profiles after argon anneal (1000°C, 0.4 mTorr) with direct (A) and inverse (B) electric field applied. Potential +/- 300V was applied to sodium implanted wafer (left). Second wafer was grounded.

At low pressure, the gas stream velocity is high and boundary layer is thin which explains the effective Na transport. In case of reverse polarity, the negatively biased electrode can also adsorb the Na contamination, such as from the quartz tube.

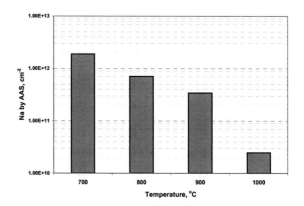

FIGURE 4. Oxide layer Na contamination measured by AAS after the Na implanted samples thermal treatment at different temperatures with an applied electric field (+600V/cm).

The oxide contamination levels measured by AAS (see fig.4) indicate the dramatic Na reduction after the argon anneal at 1000°C with DC-electric field applied. At lower temperatures Na transport is limited by desorption from the wafer surface.

CONCLUSIONS

Na transport from the intentionally contaminated wafers at 1000°C in presence of electric field is limited by the desorption of Na^+. Applied DC-electric field is effective in accelerating the Na^+ species diffusion through near-surface gas stagnant layer, preventing re-adsorption. Sodium transport in gas phase is determined by gas stream velocity, which is much higher at low pressure conditions.

REFERENCES

1. Snow, E. H., Grove, A. S., Deal, B. E., and Sah, C. T., *J. Appl. Phys.* **36**, 1664 (1965).
2. Mahoney, J. F., and Perel, J., U.S. Pat. 4,462,806 (1984).
3. Helms, C. R., *J. Electrochem. Soc.* **142**, L125-L128 (1995).
4. Zangwill, A., *Physics at Surfaces*, Cambridge: Cambridge University Press, 1988, pp. 454-472.

A theory of low-field, high-carrier-density breakdown in semiconductors

K. Kambour*[†], Harold P. Hjalmarson[†] and Charles W. Myles*

*Texas Tech University, Lubbock TX USA [1]
[†]Sandia National Laboratories, Albuquerque NM USA [2]

Abstract. Collective impact ionization has been used to explain lock-on, an optically-triggered electrical breakdown occurring in some photoconductive semiconductor switches (PCSS's). Lock-on is observed in GaAs and InP but not in Si or GaP. Here, a rate equation implementation of collective impact ionization is discussed, and it leads to new insights both about intrinsic electrical breakdown in insulating materials in general and about lock-on specifically. In this approach, lock-on and electrical breakdown are steady state processes controlled by competition between carrier generation and recombination. This leads to theoretical definitions for both the lock-on field and the breakdown field. Our results show that lock-on is a carrier-density dependent form of electrical breakdown which exists in principle in all semiconductors. Results for GaAs, InP, Si, and GaP are discussed.

INTRODUCTION

A photoconductive semiconductor switch (PCSS) is fabricated by attaching electrical contacts to a bulk semiconductor. It is activated into a conductive state when the surface is illuminated. Such switches exhibit conventional photoconductivity whereby they become conductive while illuminated [1]. In this case, each absorbed photon generates at most a single electron-hole pair, and it is necessary to continually replenish the carriers by photogeneration in order to maintain the photoconductivity.

Some PCSS's, such as those made from semi-insulating GaAs or InP, can also be triggered into a self-sustaining "on" state, called "lock-on" [1, 2, 3]. In the lock-on mode, the field across the switch is "locked-on" to an almost constant field [1] and the current flows in filaments, visible in the infrared [1]. The lock-on mode of operation has the advantage that once the initial trigger is applied, the current continues without the need for optical replenishment.

Collective impact ionization (CII) theory [4] explains the main characteristics of lock-on. The physical mechanism of CII theory is that, for high carrier densities, the heating of high kinetic energy carriers becomes more effective because carrier-carrier (cc-) scattering redistributes the energy of the carriers.

THEORY

Our approach is to write an expression for the time evolution of carrier density after the trigger is discontinued. We include impact ionization, Auger recombination, and recombination at defects, with corresponding quantum mechanical rates of r_{ii}, r_{Auger} and $r_{defects}$. We obtain

$$\frac{dn}{dt} = \int f_k (r_{ii} - r_{Auger} - r_{defects}) d^3k \quad (1)$$

in which the integral spans the first Brillouin zone and f_k is the carrier distribution function.

Qualitative Solution

The approximate carrier density dependence for each term in Eq. (1) [5, 6] is known. Substituting these approximate dependences into Eq. (1) gives:

$$\frac{dn}{dt} = C(F,n)n - an^3 - rn = R(F,n)n, \quad (2)$$

where F is the electric field and $C(F,n)$, a, and r are the impact ionization, Auger, and defect recombination rates. We seek the non-trivial steady state solutions which correspond to the cases when $R(F,n) = 0$.

In the case of no cc-scattering, the impact ionization rate $C(F,n)$ is independent of the carrier density

---

[1] CWM supported in part by a grant from the AFOSR MURI program.
[2] Sandia is a multiprogram laboratory operated by Sandia Corporation, a Lockheed Martin company, for the United States Department of Energy under contract DE-AC04-94AL85000.

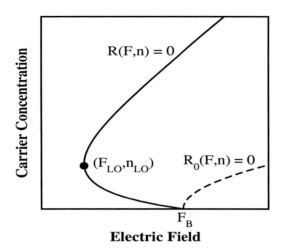

FIGURE 1. Schematic plot of the steady state carrier density n as a function of electric field F. The dashed curve is the low carrier density result and the solid curve is the result for high carrier density.

$(C(F,n) = C(F))$. Assuming $C(F) = \alpha e^{-\beta/F}$ [7, 8], the steady state carrier density n can be obtained as a function of field, and it is plotted schematically as the dashed curve in Fig. 1 with the breakdown field, F_B, defined as the minimum field for which a non-trivial solution exists.

In the case of significant cc-scattering (e.g., at high carrier density), the carrier density dependence of the impact ionization rate must also be included. For this qualitative discussion, we assume a linear dependence of the form $C(F,n) = C(F)(1 + n/n_0)$, where n_0 is a constant. The resulting $n(F)$ is shown schematically as the solid curve in Fig. 1. Clearly, including the effects of cc-scattering changes the solution for $n(F)$ both qualitatively and quantitatively. At the breakdown field, F_B, there is now a sudden jump in the carrier density conforming to our expectations about a catastrophic breakdown event. Furthermore, the solutions can be characterized by a new minimum field for which a non-zero steady state carrier density is possible. We define this field to be the lock-on field, F_{LO} (see Fig. 1).

Quantitative Calculations

Three types of calculations were undertaken. First, collective impact ionization (CII) theory [4] was implemented by using an ensemble Monte Carlo (EMC) method to solve the Boltzmann equation. Second, because the effects of cc-scattering are minimal at low carrier densities, and for this case we use an EMC simulation without including the effects of cc-scattering. Third, for high carrier densities, CII theory [4] predicts that a quasi-equilibrium steady state will be achieved. In these calculations, the carrier distribution function is approximated as a Maxwellian distribution function.

RESULTS

We did extensive calculations for GaAs and InP, two materials in which lock-on has been observed. The predicted intrinsic breakdown fields are 177 kV/cm and 173 kV/cm, respectively, and the predicted lock-on fields are approximately 90 and 40 kV/cm, both of which are larger then the experimental lock-on fields of $\approx$ 5 and 14.4 kV/cm.

We did similar calculations for Si and GaP, two materials in which lock-on has not been observed. Our predicted breakdown fields are 104 and 192 kV/cm, and our predicted lock-on fields are 77 and 176 kV/cm. For these materials, the predicted breakdown and lock-on fields are close to each other. This means that an observation of lock-on would require that the switch be triggered near the breakdown field, and we suggest this makes lock-on difficult to distinguish from breakdown.

SUMMARY/CONCLUSIONS

We have used CII theory to develop a both a theory of lock-on in PCSS's and a new approach to the classic problem of steady state electrical breakdown in insulators. We suggest that our theory predicts the correct qualitative lock-on behavior. We also suggest that the lock-on effect is common to all semiconductors.

REFERENCES

1. Rosen, A., and Zutavern, F., editors, *High-Power Optically Activated Solid State Switches*, Artech House, Boston, 1994.
2. Loubriel, G. M., O'Malley, M. W., and Zutavern, F. J., , in *Proc. 6th IEEE Pulsed Power Conf., Arlington, VA*, IEEE, New York, 1987, p. 145.
3. Loubriel, G., Zutavern, F., Hjalmarson, H., and O'Malley, M., , in *Proc. 7th IEEE Pulsed Power Conf., Arlington, VA*, IEEE, New York, 1989, p. 45.
4. Hjalmarson, H., Loubriel, G., Zutavern, F., Wake, D., Kang, S., Kambour, K., and Myles, C., , in *Proc. 12th IEEE Pulsed Power Conference, Monterey, CA*, IEEE, New York, 1999, p. 299.
5. O'Dwyer, J., *The Theory of Dielectric Breakdown of Solids*, Oxford University Press, London, 1964.
6. Beattie, A. R., and Landsberg, P. T., *Proc. Phys. Soc. A*, **249**, 16–29 (1959).
7. Wolff, P. A., *Phys. Rev.*, **95**, 1415–1420 (1954).
8. Baraff, G. A., *Phys. Rev.*, **128**, 2507–2517 (1962).

Quantum Hall Ferromagnetism in Magnetic Heterostuctures and Wires

J. Jaroszyński[*,†], T. Andrearczyk[†,*], E. A. Stringer[*,‡], G. Karczewski[†], T. Wojtowicz[†], J. Wróbel[†], Dragana Popović[*] and T. Dietl[†]

[*]National High Magnetic Field Laboratory, Florida State University, Tallahassee FL 32310
[†]Institute of Physics, Polish Academy of Sciences, Warszawa, PL 02688
[‡]The University of the South, Sewanee, TN 37383

Abstract. Magnetotransport measurements in modulation doped n-(Cd,Mn)Te quantum wells reveal an expected square root dependence of the electronic exchange energy on magnetic field B. Tilted field experiments show that this exchange contribution depends on the tilt angle (θ). At the same time a parallel field suppresses the Ising easy axis anisotropy of the quantum Hall ferromagnet (QHFM). The QHFM spikes disappear in high electron density samples, however, they are clearly seen in a 2μm wide wire.

In a (Cd,Mn)Te diluted magnetic semiconductor, Mn ions are electrically neutral. However, in external B they contribute to a macroscopic magnetization M described by modified Brillouin function [1]. In turn, strong s-d exchange coupling between band and localized d-electrons leads to a giant spin splitting Δ^{s-d} which is proportional to M. As a consequence a number of striking spin-dependent transport phenomena were observed in 3D [2, 3], 2D [4, 5], and 1D [6, 7] systems. In particular, in 2D electron systems of interest here, Δ^{s-d} can substantially exceed the cyclotron energy E_c, especially in the low-B range. Indeed, since Δ^{s-d} is proportional to M, its dependence on B is strongly nonlinear. This brings Landau levels (LL) with opposite real spin into coincidence at selected magnetic fields B_c, without any component of $B_\parallel$ parallel to the sample plane involved.

This offers a worthwhile opportunity to examine the quantum Hall ferromagnetism (QHFM) at crossings of real-spin subbands in the perpendicular configuration, and moreover, to study how the QHFM evolves under $B_\parallel$ applied. Indeed, our previous study [5] evidences Ising QHFM formation with the critical temperature as high as $T_C \approx 2$ K in a n-(Cd,Mn)Te/(Cd,Mg)Te:I heterostructure.

Here we report on new findings in the 2D and 1D samples with a broad range of electron densities 1.6×10^{11} cm$^{-2} < n_s < 6 \times 10^{11}$ cm^{-2}, and peak mobilities $\mu \approx 2.5$ m^2/Vs. High T_C and low B involved made it possible to collect magnetoresistance (MR) data over a wide range of T, B, and θ. The QHFM in our devices is manifested by metastable (i.e. hysteretic, noisy, and slowly relaxing) MR spikes around integer fillings ν. The spikes are accompanied by the usual Shubnikov-de Haas

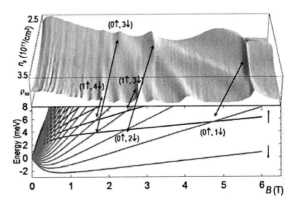

FIGURE 1. ρ_{xx} in the $B - n_s$ plane and energy level diagram calculated within the independent electron model. Arrows show selected LL's crossings and corresponding QHFM spikes.

(SdH) maxima, strongly modified when the LLs overlap.

Figure 1a shows several QHFM spikes at different LL's (m, n) crossings, where m, n are subband indices.

It is clearly seen that the experimental positions B_c of the spikes are substantially shifted towards higher B with respect to these predicted by the one electron model. This makes it possible to determine the strength of interaction. The observed shift stems mainly from the exchange interactions with frozen LL's↓ lying well below Fermi energy. This contribution reflects the Pauli exclusion principle by lowering LL↓ at the crossing when there are fully occupied and spin down polarized LL↓ below. We deal here with an unique situation when a number of frozen LL↓ increases as B decreases.

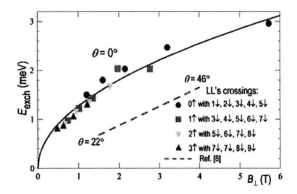

FIGURE 2. Interaction energy inferred from the experimental positions of QHFM spikes at several different LL crossings (points). The solid and the dashed lines are fits to the data of the present work and Ref. [8], respectively.

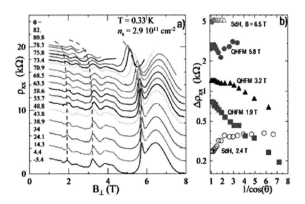

FIGURE 3. a) ρ_{xx} as a function of $B_\perp$ for different θ. The dashed lines shows trajectories of LL crossings calculated within noninteracting model. b) Amplitudes of different QHFM spikes and SdH maxima as a function of $1/\cos(\theta)$.

The QHFM forms at B_c where the effective Zeeman coupling $b = g^*\mu_B B_c + \Delta^{s-d}(x_{\text{eff}}, B_c) - (m-n)E_c(B_c) - E_{\text{exch}}(B_c)f(m,n)$ is zero [9, 10]. Here the first and the second terms are the band and the s-d exchange real Zeeman couplings, respectively, the third term is the cyclotron energy, and the last term is the electronic exchange with the low lying LL's↓; $g^* = -1.67$ for CdTe, x_{eff} is the effective concentration of Mn ions, and $f(m,n)$ is a coefficient specific for each LL's crossing [11]. Taking x_{eff} and E_{exch} as fitting parameters we were able to collapse experimental data onto one curve, shown in Fig. 2. We found $E_{\text{exch}} = 1.2\sqrt{B_\perp[T]}$ meV. The same calculations yield $J = 0.22$ meV, where J is an effective Ising anisotropy parameter, which agrees with the observed critical temperature T_C as well as with the theoretical predictions [10].

As it is seen in Fig. 3, measurements in tilted fields reveal that exchange contribution increases with θ up to

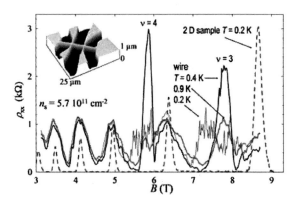

FIGURE 4. ρ_{xx} vs. B in 2D and 1D samples.

$\theta \approx 70^0$ and then start to decrease. It appears that this could be a reason why previous B-tilt studies show (see, e.g. [8, 12]) linear instead of expected $\sqrt{B_\perp}$ dependence of interaction energy on $B_\perp$, since in these experiments results for higher $B_\perp$ were taken at higher tilt angles θ.

At the same time an increase of the in-plane $B_\parallel$ suppresses the amplitude of the QHFM spikes. This strongly suggests a transition from Ising QHFM to a different magnetic phase with easy-plane anisotropy at high θ [9].

The QHFM does not show up in the high-density ($n_s \geq 4 \times 10^{11}$ cm^{-2}) 2D heterostructures [13]. However, as Fig. 4 shows, anomalous spikes with critical temperature $T \approx 0.4$ K are clearly seen in a submicron wire ($n_s = 5.7 \times 10^{11}$ cm^{-2}) at $\nu = 3$ and 4.

The authors are grateful to A. Piotrowska, E. Kamińska, and E. Papis from the Institute of Electron Technology in Warsaw for the fabrication of samples' gates. This work was supported by NSF Grant No. DMR- 0071668, NHMFL through NSF Cooperative Agreement No. DMR-0084173, and EC IST-2001–33334 SPINOSA.

REFERENCES

1. Gaj, J. A., et al., *Phys. Rev. B*, **50**, 5512 (1994).
2. Wojtowicz, T., et al., *Phys. Rev. Lett.*, **56**, 2419 (1986).
3. Sawicki, M., et al., *Phys. Rev. Lett.*, **56**, 508 (1986).
4. Smorchkova, I. P., et al., *Phys. Rev. Lett.*, **78**, 3571 (1997).
5. Jaroszyński, J., et al., *Phys. Rev. Lett.*, **89**, 266802 (2002).
6. Jaroszyński, J., et al., *Phys. Rev. Lett.*, **75**, 3170 (1995).
7. Jaroszyński, J., et al., *Phys. Rev. Lett.*, **80**, 5635 (1998).
8. De Poortere, E. P., et al., *Science*, **290**, 1546 (2000).
9. Jungwirth, T., et al., *Phys. Rev. Lett.*, **81**, 2328 (1998).
10. Jungwirth, T., and MacDonald, A. H., *Phys. Rev. Lett.*, **87**, 216801 (2001).
11. Glasser, M., and Horing, N., *Phys. Rev. B*, **31**, 4603 (1985).
12. Desrat, W., et al., *Phys. Rev. B*, **69**, 245324 (2004).
13. De Poortere, E. P., et al., *Phys. Rev. Lett.*, **91**, 216802 (2003).

Imaging Transport: Monitoring the Motion of Charge through the Detection of Light

Nancy M. Haegel, Vu D. Hoang, and Will Freeman

Physics Department
Naval Postgraduate School
833 Dyer Rd.
Monterey, CA 93943 USA

Abstract. We present a technique that allows for direct monitoring of charge transport in semiconductors and other luminescent materials via the spatial imaging of recombination luminescence. Drift behavior has been imaged in high purity epitaxial GaAs. We demonstrate the role of sample geometry in the measurement of luminescent spot size, with the goal of developing a contact-free method to determine local diffusion lengths.

INTRODUCTION

The ability to directly image charge carrier motion provides new insight into transport behavior, particularly with regard to localized phenomena and spatial variation. We have developed a system to image the two-dimensional transport of carriers in luminescent materials [1]. By generating charge at a fixed point, it is possible to observe diffusion and drift by imaging the associated recombination luminescence. The technique differs from conventional cathodoluminescence (CL) because spatial information from the recombination is maintained. In scanning CL, one assigns the total photon signal to the generation point. Transport imaging takes advantage of the breakdown of that assumption, in cases where the drift or diffusion of charge produces luminescence at positions removed from the point of generation.

The approach is similar to the spatially resolved photoluminescence (PL) by Höpfel et. al. [2] to illustrate carrier drag. Benefits of e-beam excitation, however, include application to wide bandgap materials, as well as easy control, on the sub-micron scale, of the point of excitation. Our work uses a JEOL 840A scanning electron microscope. An optical microscope is inserted into the chamber and connects externally to a thermoelectrically cooled Apogee CCD camera. The electron beam is incident through a small hole in a mirror assembly that then directs light to a lens system for reimaging at the CCD focal plane.

IMAGING DRIFT BEHAVIOR

The technique has been used to image the drift motion of charge in epitaxial layers of n-type GaAs, grown by liquid phase epitaxy. Figure 1 shows an image of drift behavior for a field of 550 V/cm (60 V applied across planar contacts 1.1 mm apart). The image was created by taking an image with bias applied and then subtracting a 0 V bias image to isolate the drift behavior.

FIGURE 1. Image of charge transport and recombination with E = 550 V/cm. Image size is 180 x 60 μm.

A model has been developed for minority carrier motion in 2D including diffusion and a one-dimensional applied field [1]. This results in a minority carrier distribution $\varpi(r)$ of

$$\varpi(x,y) = \frac{1}{2\pi} e^{\frac{Sx}{2L^2}} K_0\left(\frac{\sqrt{S^2 + 4L^2}}{2L^2} r\right) \quad [1]$$

where K_o is a zeroth order modified Bessel function of the second kind, S is the drift length ($\mu\tau E$), L is the diffusion length ($\sqrt{\mu\tau kT/e}$) and $r = (x^2 + y^2)^{1/2}$. Figure 2 presents an example of modeling for transport at 550 V/cm, with L = 4.3 µm. The final image is produced by subtraction of images with and without applied bias, as with the experimental data.

FIGURE 2. Simulation of transport for E = 550 V/cm and L = 4.3 µm. Image size is 180 x 60 µm.

IMAGING DIFFUSION BEHAVIOR

The technique may also be used as a contact-free way to measure minority carrier diffusion lengths from the luminescent spot size. In this, it resembles a Haynes-Shockley experiment [3], but one that is highly spatially resolved, while requiring no contacts or external field. For e-beam generation in bulk material, the excess minority carrier density should decrease as (1/r)exp(-r/L), where r is the distance from the generation point. The distribution is not strongly dependent on L. Donolato [4] has shown that, as a consequence of this weak dependence, the diffusion length in bulk samples does not limit the resolution in conventional CL. In transport imaging, this suggests that the luminescent spot size for bulk samples will be determined primarily by the incident electron beam energy, i.e., by the generation volume and not the material's diffusion length.

In two dimensions, however, Eqn. [1] reduces to Ko(r/L) when S = 0, with a stronger dependence on L. This suggests that while the area or diameter of the luminescent region resulting from point generation in a bulk sample will be determined primarily by the generation volume, for 2D (or better yet, 1D) samples it should be possible to extract diffusion lengths from the dimension of the luminescent spot. This assumes that the substrate material does not produce a significant luminescent signal, or that the epilayer signal can be optically isolated.

To test the role of sample geometry, we have measured luminescent spot diameter as a function of probe current for a bulk (3D) sample of n type GaAs and a ~ 1 µm layer of AlGaAs. Figure 3 shows the comparison of the measured spot diameters over a probe current range from 10^{-10} to 10^{-7} A. Increasing probe current should increase the e-beam spot size at the surface, but these diameters remain small compared to the generation diameter. Hence, the bulk case shows spot size independent of probe current (it varies, as expected, with e-beam energy). In 2D, however, an increase in luminescent area with increasing probe current is observed.

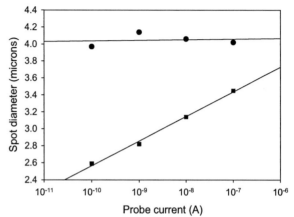

FIGURE 3. Variation in luminescent spot diameter with probe current for a 2D epilayer (■) compared to a bulk (3D) sample (●). E-beam energy is fixed at 10 kV.

The current system is limited to a resolution of ~ 0.4 µm/pixel, making it hard to measure diffusion lengths in materials with L ~ 1 µm or less. However, longer diffusion lengths can be extracted by comparing the luminescent distribution to the modeling. Preliminary measurements have shown increasing diffusion length in GaAs epilayers with increasing growth temperature. We intend to increase magnification of the system for use in higher bandgap materials (GaN, diamond, etc.) where diffraction limits will be less severe.

ACKNOWLEDGEMENTS

We thank B. Cardozo and E. E. Haller of LBNL for the GaAs samples. We acknowledge the contributions of J. D. Fabbri and M. C. Coleman of Fairfield University in the modeling. This work is supported by NSF Grant DMR2003397 and was initiated with the support of Research Corporation Award CC 4411.

REFERENCES

1. N. M. Haegel, J. D. Fabbri and M. C. Coleman, *Appl. Phys. Letters* **84**, 1329-1331 (2004)..
2. R. A. Höpfel, J. Shah, P. A. Wolff and A. C. Gossard, *Phys. Rev. Letters* **56**, 2736 (1986).
3. R. A. Haynes and W. Shockley, *Phys. Rev.* **81**, 835 (1951).
4. C. Donolato, *Optik* **52**, 19-36, 1978.

Calibrated Scanning Capacitance Microscopy for Two-Dimensional Carrier Mapping of n-type Implants in p-doped Si–Wafers

W. Brezna[a], B. Basnar[a], S. Golka[a], H. Enichlmair[b], J. Smoliner[a]

[a] Institut für Festkörperelektronik, TU-Wien, Floragasse 7, A-1040 Wien, Austria
[b] Austria Mikro Systeme International AG, Schloss Premstätten, A-8141 Unterpremstätten, Austria

Abstract. In our previous work we showed that a reproducible monotonic relation between Scanning Capacitance Microscopy (SCM) signal and the sample doping is obtained under appropriate SCM operating conditions. In this work, we quantitatively determine the two-dimensional carrier distribution of an n-type ion-implanted collector region of an industrial transistor sample. As the structures were large enough so that tip geometry effects could be neglected, the SCM data were converted into a two-dimensional doping distribution via a simple lookup table, which was generated from a one-dimensional doping profile measured on a reference wafer. In case the depth distribution of carriers is unknown, we show that the lookup table can also be calculated and a calibration can be carried out using e.g. the substrate doping as reference point.

INTRODUCTION

The determination of two-dimensional doping profiles with high lateral resolution is of crucial importance for the future development of semiconductor structures. Due to this, many different analytical tools[1] have been employed to the task of determining the carrier profile in various semiconductors. Among these methods, scanning probe microscopes[2] (SPM) and especially the scanning capacitance microscope[3] (SCM) have proven to be very successful candidates for achieving this goal[4].

In order to achieve reproducible and quantifiable results, the most important experimental parameters are the quality of the insulating layer, in most cases SiO$_2$, and the performance and longevity of the tips. Many investigations have been performed to investigate the former criterion, leading to several different procedures like oxidation in an oven[5,6], irradiation with UV-light[7], polishing with silica slurry[8,9], or any combination of these[10,11]. The latter parameter, the tip quality, has long been the major drawback for all SPM methods. However, the introduction of a new generation of cantilevers coated with highly doped conducting diamond has brought a probe on the market exhibiting high resistance against mechanical damage (e.g. abrasion) and, therefore, prolonged lifetime, making comparative measurements with a single cantilever possible.

So far, several authors have approached the task of determining the doping profiles[4,9,12] for different structures such as staircases or implants. Most of the work related to samples containing pn-junctions was carried out in regards to junction delineation. All of these experiments, however, showed a certain lack of reproducibility due to instable tip conditions and sample preparation problems. In addition, rather sophisticated calibration schemes had to be applied, which usually required intensive simulation[13]. Especially on small structures, a quantitative determination of the 2D doping distribution is obviously impossible without inverse modeling[14] or intensive simulations using empirical databases[15].

In this work we report calibrated SCM measurements on n-type implanted domains in p-doped silicon. Using the newly developed diamond coated cantilevers, a reproducible and quantitative correlation of the SCM signal with spreading resistance profiling (SRP) results is obtained. As the devices are large enough so that geometry effects can be neglected, the SCM data can be calibrated in a simple way and the two dimensional doping distribution is determined quantitatively. This method is a fast and useful tool for device characterization and failure analysis.

EXPERIMENTAL

For all SCM measurements, a Dimension-3100 atomic force microscope (AFM) with integrated SCM module (Digital Instruments, Santa Barbara, USA) was used. As tips we employed silicon cantilevers coated with conducting diamond (p-doped, $N_A=10^{20}$ cm^{-3}, Nanosensors, Germany). Due to their high resistance against abrasion these tips were found to be superior to metal coated tips and stable image contrast was obtained with one and the same tip for many hours of operation even on different samples. The samples we used were p-doped Si(100) wafers taken out of a transistor fabrication process at AMS (Austria Mikro Systeme International AG, Austria). On the wafers we investigated As-implanted areas (size 7μm x 7μm) which represent the collector regions of the transistors. In order to overcome the usual problems related to polishing procedures, the cross-sectional samples were prepared by cleaving. The insulating oxide was obtained by UV/ozone treatment under ambient conditions. The final oxide thickness obtained in this way is about 1.5 nm and was confirmed by ellipsometry. More details about our sample preparation processes and the performance of the diamond tips can be found in reference [16].

As we have shown in a previous publication[17], a stable and monotonic dependence of the SCM signal on the doping level is only achieved if the tip bias is adjusted in a way that the sample underneath the tip is in "accumulation". Then, a decreasing SCM signal is obtained with increasing donor concentration in n-type areas at sufficiently high positive bias, whereas in p-type areas, the SCM signal is negligibly small. At sufficiently high negative bias, the situation is reversed. As a consequence, the tip bias can only be adjusted in a way that monotonic contrast is achieved either in n-type or in p-type areas of the sample, but not in both areas simultaneously, Further, the SCM contrast behavior in the space charge region of pn-junctions was found to be ambiguous and could not be treated in simple models. We therefore restrict our considerations to the n-type ion implanted areas of our sample and stay away from the space charge region of the pn-junction. Choosing an appropriate tip bias for monotonic contrast (V_{tip} =+0.5 V on our samples), the SCM signal in the n-type regions can then be converted into carrier concentrations by simple lookup tables. The actual value of the appropriate tip bias depends on various parameters such as oxide quality and oxide thickness. If V_{tip} is chosen too small, the contrast can become non monotonic, if V_{tip} is too large, the SCM signal becomes unnecessarily small. Thus, it is very helpful to know the doping profile at least qualitatively.

RESULTS AND DISCUSSION

Figure 1(a) shows an SCM image of the n-type collector region, Figure 1(b) a section through this image along line L. The image size is 10 μm x 10 μm. For convenience, the phase setting of the internal lockin-amplifier was adjusted in a way that a negative SCM signal is obtained in n-type regions and positive signal in p-type areas. In the bottom graph the detailed features of the implant region. Due to the chosen positive tip bias, the SCM signal is negligibly small in the p-type substrate region (A). In region (B) the SCM signal becomes more negative. This region represents the space charge region of the pn-junction between the low doped substrate and the ion implanted region. The width of this region is in reasonable agreement with the results of a simple calculation. The position of the

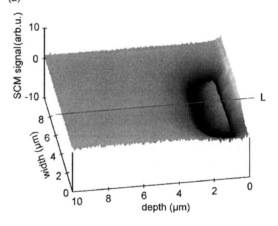

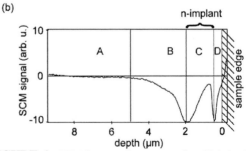

FIGURE 1. SCM-image of an n-type ion (As) implanted region in p-doped silicon recorded at V_{tip}=+0.5 V. (a) shows a 3-dimensional representation of the SCM signal, (b) depicts the cross-section through the implant along line L. The behavior of the SCM signal in regions (A)-(D) is discussed in the text.

minimum at a depth of 2μm can be identified with the beginning of the entirely n-type region (C). As the carrier concentration is small in the lower tail of the implant profile, the SCM signal is strongly negative. Towards the center of the implanted area, the doping level increases. Therefore, the SCM signal becomes smaller towards the position of the highest doping level approximately 0.7 μm below the surface. Then, the SCM signal becomes more negative again, because the implant concentration decreases towards the sample surface. In region D, the SCM signal goes to zero for purely geometrical reasons because a part of the tip starts to hang over the edge of the sample and therefore the active area of the measurement is reduced in this regime.

Under normal circumstances, industrial ion implantation processes are highly reproducible so that one can assume that the implanted profiles are identical at least over a batch of wafers. For process control, such a batch of wafers normally contains a reference wafer, which is used to monitor the depth distribution of implanted carriers using secondary ion mass spectroscopy (SIMS) or SRP measurements. Figure 2 shows the depth profile of the doping concentration obtained by SRP measurements on a reference wafer, which was processed in the same batch like our transistor sample. The known depth distribution of implanted carriers can now be used to calibrate the SCM, so that the lateral or in general the two-dimensional distribution of dopants can be determined on the transistor samples. As long as the structures are large enough so that tip-geometry effects can be neglected, there are two simple possibilities to calibrate the SCM data. First and most simple, one can create a lookup table from the SRP data to calibrate the SCM. Assuming, that the general sample properties (e.g. oxide thickness, stray capacitance) will not change significantly across the sample surface, the lookup table is valid for all datapoints of the sample, so that a quantitative representation of the 2-dimensional carrier distribution is obtained.

As second possibility, one can calculate a so called "calibration curve", a relation between dC/dV and doping at a given bias. On relatively large structures like our sample, conventional one-dimensional conventional Metal-Oxide-Semiconductor (MOS) theory[18,19] can be used for this purpose and yields surprisingly good results. On small structures more advanced models including tip geometry effects have to be applied. As next step, the instrument sensitivity has to be included. The instrument sensitivity is generally unknown and depends on a number of SCM operating parameters, some of which are quantitatively unknown or cannot be controlled. However, the instrument sensitivity can be determined using the substrate doping, which is always known, as reference point. This procedure has the advantage, that it also works when the depth profile of the implant is unknown. As third advantage, the calibration on the sample substrate eliminates the errors related to the preparation of additional reference samples. Even higher accuracy is achieved, if two concentrations. i.e. the highest and lowest doping level are known on the sample.

Figure 2 shows the results of such a two-point calibration procedure using conventional one dimensional MOS theory. The solid line represents the SRP data, the dashed curve shows the depth profile of the doping obtained from the SCM data. As one can see, the SCM and SRP data are in good agreement over the whole range of concentrations. Both the position of the maximum as well as the shape of the carrier concentration profile are well reproduced when compared to the SRP results. The maximum deviation is in the order of a factor of 2 above 10^{16} cm^{-3} and

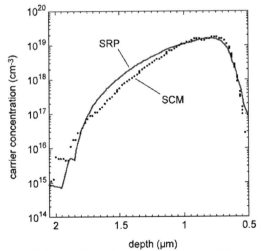

FIGURE 2. one-dimensional carrier profile of the n-type implant determined on a reference wafer by SRP (solid line) and the converted SCM data (dashed line) measured on the transistor sample. The SRP data were provided by AMS.

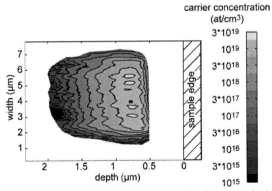

FIGURE 3. 2-dimensional dopant distribution calculated from SCM data. The diagram shows the converted SCM signal for the n-implanted region only.

increases to about 5 below 10^{16} cm^{-3}. However, there is a general tendency for lower carrier concentrations being derived from SCM data. This is especially obvious on the left side of the implant depicted in Figure 2. An explanation for this behavior can probably be found in geometry effects. More detailed investigations on this behavior are currently in progress. The most interesting consequence of this good agreement between the SRP and SCM data is the possibility to determine the dopant distribution in lateral direction directly beneath the surface. Figure 3 shows a contour-plot of the 2-dimensional carrier profile after conversion of the SCM data. As mentioned above, information is only obtained from the entirely n-type regions, the SCM signal from the space charge region of the pn-junction and the p-type areas cannot be analyzed by our simple method and is therefore omitted in this plot. Nevertheless the results are good enough to study e.g. lateral diffusion processes, which is a parameter of great importance in process control and process simulation.

Finally it should be noticed that another advantage of this approach lies in the short time that it takes to obtain such a result. The sample preparation (cleaving and oxidation) can be performed within 30 minutes, the measurement itself takes about 15-30 minutes and the evaluation of the data as well as the conversion can be carried out in a few seconds when the mathematical procedures are being automated. Thus, a complete analysis and data evaluation can be achieved in less than an hour, making this a promising tool for a routine two-dimensional carrier distribution analysis.

SUMMARY

In summary we have determined the two-dimensional carrier distribution in an n-type ion-implanted collector region of an industrial transistor sample using SCM and a simple straightforward calibration scheme. The two most important results of this investigation are the applicability of calibration tables for 2D carrier profiling on relatively large structures and the low time consumption for data evaluation, both of which are important parameters for practical applications in failure analysis.

ACKNOWLEDGMENTS

This work was sponsored by "Fonds für innovative Projekte an der TU Wien", "Gesellschaft für Mikroelektronik (GMe)" "Fonds zur Förderung der wissenschaftlichen Forschung (FWF)" project No P16337, and the European Community ("NESTOR"-project).

REFERENCES

1. P.DeWolf, R.Stephenson, T.Trenkler, T.Clarysse, T.Hantschel, W.Vandervorst, J. Vac. Sci. Technol. **B18**, 361(2000)
2. G. Friedbacher, H. Fuchs, Pure Appl. Chem. **71**, 1337 (1999)
3. D.D. Bugg, P.J. King, J. Phys. **E21** 147, (1988)
4. V.V. Zavyalov, J.S. McMurray, C.C. Williams, Rev. Sci. Instr. **7**, 158 (1999)
5. G.H.Buh, H.J.Chung, C.K.Kim, J.H. Yi, I.T.Yoon, Y.Kuk, Appl. Phys. Lett. **77**, 106, (2000)
6. J. Schmidt, D.H. Paoport, G. Behme, H.J.Fröhlich, J. Appl. Phys. **86**, 7094 (1999)
7. J.J. Kopanski, J.F. Marchiando, B.G. Rennex, J. Vac. Sci. Technol. **B18.** 409 (2000)
8. F. Giannazzo, F. Prioli, V. Raineri, V. Privitera, Appl. Phys. Lett. **76**, 2565 (2000
9. J.S. McMurray, J. Kim, C.C. Williams, J. Vac. Sci. Technol. **B15**, 1011 (1997)
10. V.A. Ukraintsev, F.R. Potts, R.M. Wallace, L.K. Magel, H. Edwards, M.-C. Chang, AIP Conf. Proc. **449** 736 (1998)
11. V.V. Zavyalov, H.S.McMurray, S.D. Stirling, C.C.Williams, H. Smith, J. Vac. Sci. Technol. **B18**, 549 (2000)
12. T. Clarysse, M.Caymax, P. DeWolf, T.Trenkler, W.Vandervorst, J.S.McMurray, J.Kim, C.C.Williams, J.G.Clark, G.Neubauer, J. Vac. Sci. Technol. **B16**, 394 (1998)
13. J.F.Marchiando, J.J.Kopanski, J.Albers, J. Vac. Sci. Technol. **B18**, 414 (2000)
14. Y.Huang, C.C.Williams, J.Slinkman, Appl. Phys. Lett. **66**, 344 (1995)
15. J.F.Marchiando, J.J.Kopanski, J.R.Lowney, J. Vac. Sci. Technol. **B16**, 463 (1998)
16. B.Basnar, S.Golka, E.Gornik, S.Harasek, E.Bertagnolli, M.Schatzmayr, J.Smoliner, J. Vac. Sci. Technol. **B19**, 1808 (2001)
17. J. Smoliner, B. Basnar, S. Golka, E. Gornik, B. Löffler, M. Schatzmayer, H. Enichlmair, Appl. Phys. Lett., to be published (October 2001), preprint available via ftp://macmisz.fke.tuwien.ac.at/pub/preprints/SCM_contrast.pdf
18. S.M. Sze, Physics of Semiconductor Devices, John Wiley & Sons Ltd, New York, 1981
19. E.H. Nicollian, J.R. Brews, MOS (metal oxide semiconductor) physics and technology, John Wiley & Sons Ltd, New York 1982

Evaporative Cooling of Electrons in Semiconductor Devices

Thushari Jayasekera, Kieran Mullen and Michael A. Morrison

Department of Physics and Astronomy, University of Oklahoma, Norman, Oklahoma, 73019, USA.

Abstract. We discuss the theory of cooling of electrons in solid-state devices by evaporative emission. Our model is based on filtering electron subbands in a quantum-wire device. When the higher-subband electrons scatter out of the initial electron distribution, the system equilibrates at a different chemical potential and temperature than those of the input system. Our calculation shows that such filtering can give considerable cooling. We discuss the effect of device geometry on cooling.

INTRODUCTION

As electronic devices become smaller, they must be described by quantum physics. Many classical quantities (e.g. resistance) must be reinterpreted when they are examined on a mesoscopic level. One such classical concept is that of the refrigerator - a device that uses an external source of work to cool a gas. It is interesting to ask if this classical concept can be applied to an electron gas so that by applying a voltage we cool the system.

We investigate the theory of electron cooling in solid-state devices via evaporative cooling. Evaporative cooling is the preferential removal of the higher energy particles from a system and the subsequent relaxation of the remaining system to a temperature lower than the temperature of the initial system. A different kind of electron cooling mechanism in semiconductor devices is presented by G.Rego et.al[1] based on quasi-static expansion of a two dimensional electron gas. Our system differs from their design as we do not use an additional external field to achieve cooling. Moreover our approach has simple analogs in classical refrigeration[2].

Below we discuss the theory of evaporative cooling of electrons via subband-filtering in quantum wires. We use the Landauer formula[3] to analyze the cooling characteristics of the system. To calculate the scattering amplitudes that the Landauer formula requires, we use an extension of R-matrix theory[5]. Using these techniques we calculate the ratio of input to output temperature for our device. We then investigate how the cooling changes as we change the device geometry.

ANALYSIS

The proposed theory has a classical analog in the working principle of the Hilsch vortex tube[2]. Hilsch vortex tube separates high-pressure air into high temperature and low temperature flows using a T shaped assembly of pipe.

We apply the same idea to cooling electrons in a quantum mechanical system. A multi-lead quantum-wire configuration is used as a subband filter that eliminates the higher-subband electrons from the system.

Electrons are injected into the system from the input lead where the electrons are in thermal equilibrium at temperature T_i and chemical potential μ_i. All filtering occurs in the scattering region. Electron will re-equilibrate at a different temperature, T_f, and a different chemical potential, μ_f, in the output lead. We use R-matrix theory (RMT) to calculate the transmission coefficients of the electrons in different leads in a single-particle picture (ignoring electron-electron scattering). RMT was originally developed to solve scattering problems in nuclear physics[5] and is now commonly used in atomic and molecular physics[6]. Recently RMT has been used in device physics[7]. We have developed a new formulation of RMT that can analyze scattering in a multi-lead system. We discuss this new formulation elsewhere[4].

Using the calculated transmission coefficients, and conservation of particle and energy, we calculate T_f and μ_f. Our calculation shows that for some device geometries the final equilibrium temperature T_f is less than the initial temperature T_i, which is the desired cooling effect. We characterize this cooling property by a dimensionless parameter $\eta = T_f/T_i$, where $\eta < 1$ means cooling.

We calculate the cooling parameter η using the calculated transmission coefficients for our model. We can control scattering in the system by changing the geometry. Note for a fermionic system, losing high-energy electrons helps cooling, but losing low-energy electrons heats the system. We can avoid this heating by changing the geometry of the system. We propose a device geometry (Fig.1) that scatters off the higher-energy electrons giv-

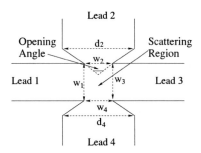

FIGURE 1. A schematic of a quantum-wire configuration that cools electrons. The electrons are injected from the lead 1. Electrons scatter out of the system in the scattering region. Those that remain in lead 3 will come to a new equilibrium at T_f and μ_f resulting in cooling.

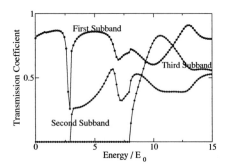

FIGURE 2. Transmission coefficients of electrons in different subbands in lead 3 as a function of the incident energy for the device parameters $w_3 = 1.0, w_2 = w_4 = 0.5, d_2 = d_4 = 0.75$ and $\theta = 170^o$. The lengths are measured in units of the width of the input wire, w_1 and the energy is measured in units of the first subband energy E_0 of the input wire, $E_0 = \hbar^2/2mw_1^2$.

ing the desired cooling effect.

Fig.2 shows transmission coefficients vs energy for the device parameters given in the figure caption.

Using the transmission coefficients, we calculate the cooling parameter for different geometries. We change the opening angle of the wedge in our device geometry and calculate the cooling parameter η. The result is shown in Fig.3a. Fig.3b shows the cooling parameter for different values of the width of the opening, d_2.

CONCLUSION

We have presented a prototype of a device to cool electrons in a single particle picture. Such a device could be used to cool the photo-detection electron population. We obtained more than 15% cooling ($T_f = 0.85 T_i$, Fig.3) using this device. Greater cooling could be obtained by cascading these devices.

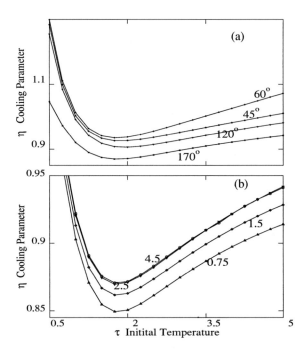

FIGURE 3. The cooling parameter η as a function of the input temperature τ. The cooling parameter η is the ratio of the input to output electron temperature. The temperature τ is measured in units of E_0/k_B where E_0 is the first subband energy of the input wire, $E_0 = \hbar^2/2mw_1^2$. Fig. 3a shows η for different angles for $w_3 = 1.0, w_2 = w_4 = 0.5$ and $d_2 = d_4 = 2.5$ where lengths are measured in units of the width of the input wire w_1. The greatest cooling was for $\theta = 170^o$. Fig. 3b shows η for different values of d_2 ($= d_4$) for $w_3 = 1.0, w_2 = w_4 = 0.5$ and $\theta = 170^o$. The greatest cooling was for $d_2 = d_4 = 0.75$. Note that the result for the case $d_2 = 2.5$ and $d_2 = 4.5$ are similar because the scattering effects are not important when we go further away from the scattering center. In all these cases the initial chemical potential μ_i is kept at the first subband energy of the input wire.

ACKNOWLEDGMENT

This work was supported by NSF MRSEC DMR-0080054, NSF EPS-9720651 and PHY-0071031.

REFERENCES

1. G.Rego and G.Kirzenow. Appl. Phys. Lett. **75**, 2262 (1999).
2. V. Deemter, Applied Scientific Research **3**, 174 (1952).
3. M. Buttikker, Phy.Rev.Lett **14**, 1761(1986).
4. Thushari Jayasekera, Kieran Mullen and Michael A. Morrison, to be published.
5. E.Wigner and L. Eisenbud, Phy. Rev. **72**, 29(1941).
6. R.K.Nesbet, Stephane Mazevet and Michael A. Morrison, Phy. Rev.A, **64**, 034702 (2001).
7. Y.Avishai and Y.B.Band, Phy. Rev. A **60**, 8992 (1999).

Mesoscopic phonon-electric effect

D. W. Horsell*, A. K. Savchenko*, Y. M. Galperin[†], V. I. Kozub** and V. M. Vinokur[‡]

*School of Physics, University of Exeter, Stocker Road, Exeter, EX4 4QL, U.K.
[†]Department of Physics, University of Oslo, PO Box 1048 Blindern, 0316 Oslo, Norway
**A. F. Ioffe Physico-Technical Institute of Russian Academy of Sciences, 194021 St. Petersburg, Russia
[‡]Argonne National Laboratory, 9700 S. Cass av., Argonne, IL 60439, U.S.A.

Abstract. We propose a new mechanism for an observed dc current generation in a nano-scale transistor structure where no driving voltage is applied. This current is shown not to be an artefact of the measurement. The model considers the non-equilibrium electrons which are excited by acoustic phonons created by the gate leakage current. The asymmetry of the mesoscopic barrier is an essential feature in the generation of this dc current.

The mesoscopic fluctuations found to exist in nano-scale metallic and insulating structures have been the subject of considerable interest over the past few decades. The reason for this interest is that these fluctuations reveal deep physical processes otherwise hidden in macroscopic monotonic behaviour. They have been observed in the diffusive [1] and insulating [2] regimes of conduction.

In an insulating, nano-scale system, where strong fluctuations are seen in the conductance due to hopping and resonant tunnelling processes, we observe a dc current in response to no applied driving voltage. We attempt to account for its origin, which we determine to be due to a novel, non-equilibrium process mediated by electron–phonon interaction.

The sample we study is a GaAs MESFET (metal–semiconductor field-effect transistor). A $0.15\,\mu\text{m}$ thick $1 \cdot 10^{17}\,\text{cm}^{-3}$ silicon doped layer is grown on an undoped GaAs substrate. A (Au) gate of length $0.15\,\mu\text{m}$ (along the current direction) and width $9\,\mu\text{m}$ defines a small region along a standard Hall bar geometry sample. The conductance measured at 30 mK by a standard lock-in technique is shown in Fig. 1, which is inset with a schematic of the sample and circuit. The gate voltage V_g depletes the region of the channel directly beneath the gate and dominates the conductance of the device below $-1.8\,\text{V}$.

In this circuit we shunt the source–drain voltage source V and measure an unexpected dc current i_0. This current, shown in Fig. 2(a), has several striking features. The *first* is that it is observed only in the region of strong fluctuations of the conductance, between the diffusive and strongly insulating regimes of conduction. The *sec-

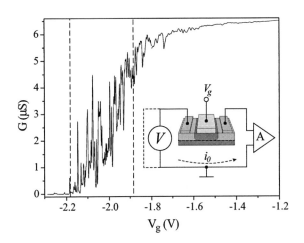

FIGURE 1. Differential conductance measured at $T = 30\,\text{mK}$. The dashed region delimits range over which i_0 is resolved. The inset shows the schematic of the setup and sample.

ond is that this current has a strong reproducible 'fingerprint' as a function of the gate voltage which is only changed in character by thermally cycling the sample to room temperature or changing the charged impurity distribution by breakdown at large gate voltages. The *third* is that the direction of this current around the source–drain circuit is determined by the value of the gate voltage.

To understand the origin of this current we eliminate from the measurement all possible known contributions to such a current. These are: a dc voltage offset produced by the pre-amplifier, i_d; rectification of stray interference coupled to the circuit, i_r; and the leakage current from

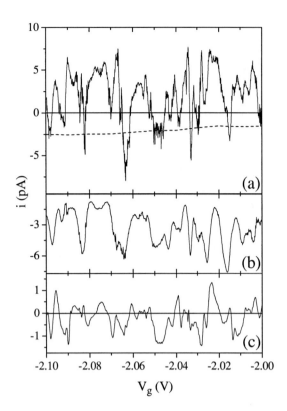

FIGURE 2. (a) i_0 (solid line), i_g (dashed line), (b) i_d, and (c) i_r as a function of the gate voltage.

the gate, i_g.

We account for i_d by first realising the condition $i_d = 0$ at $V_g = 0$ (where i_g is not present and i_r is negligible). Over the course of an experiment ($\sim$ 2 hours) there is natural drift of the dc voltage. The resulting current should mirror the measured conductance and in Fig. 2(b) we show the result of multiplying this drift by the measured conductance. We see that this current is monopolar, is small compared to, and has little correlation with the current i_0 (after first removing the contribution i_d from the measured current).

To substantially reduce the effects of interference the experiment was conducted in a steel enclosure and all connections to the sample were carefully filtered. We measure explicitly the second harmonic response to an ac signal (proportional to the second derivative) which correlates with the rectified current i_r. Using a Taylor expansion we find $i_r(\omega) = (1/4)(d^2i/dV^2)V_\omega^2$, where V_ω is the magnitude of the interference voltage at frequency ω, so the total rectified current i_r is an integral over all such frequencies. Using a spectrum analyser we convert the spectrum of noise measured at the output of the preamplifier to the integral voltage noise drop across the sample (by using the full complex transfer function of the pre-amplifier). The resultant i_r is shown in Fig. 2(c). Although bipolar, this current is smaller by almost an order of magnitude and does not show correlation with i_0.

The gate leakage current measured in the gate–channel circuit is shown in Fig. 2(a). Only a fraction of this current passes down the branch of the circuit containing the pre-amplifier. No fluctuations are seen in this monopolar, monotonic current and along with its much smaller magnitude it makes only a negligible contribution to the measured current.

We propose the following explanation of the observed effect. It bears similarity to the the photo-galvanic effect [3, 4] where, in a diffusive system containing non-equilibrium electrons excited by microwave radiation, a dc current is generated by virtue of the asymmetry of the potential landscape.

The source of power for our system comes from the gate voltage and we find $i_g V_g \gg i_0^2 R$ (R is the total resistance of the source–drain circuit) by several orders of magnitude. Electrons from the gate are injected into the channel with excess energy of 2 eV. These hot electrons quickly emit a cascade of optical phonons. Each phonon decays into two acoustic phonons of characteristic frequency ω_f, such that the number of acoustic phonons in the channel is $eV_g/\hbar\omega_f \sim 100$. Each phonon then excites a localised electron in the channel which effectively amplifies the original gate leakage current.

The electrons excited in this way move isotropically towards the contacts. However, the asymmetry in the recharging of the ionised impurities by the electrons in the contacts results in a potential drop between the contacts. This potential difference drives the current i_0 measured in the external circuit. The degree of asymmetry and therefore the direction of the current i_0 is controlled in the experiment by the gate voltage. This gives a unique fingerprint which is uncorrelated with both the linear and nonlinear conductance of the system.

REFERENCES

1. Umbach, C. P., Washburn, S., Laibowitz, R. B., and Webb, R. A., *Phys. Rev. B*, **30**, 4048 (1984).
2. Orlov, A. O., and Savchenko, A. K., *JETP Lett.*, **44**, 41 (1986).
3. Fal'ko, V. I., and Khmel'nitskii, D. E., *Sov. Phys. JETP*, **68**, 186 (1989).
4. Bykov, A. A., Gusev, G. M., Kvon, Z. D., Lubyshev, D. I., and Migal, V. P., *JETP Lett.*, **49**, 13 (1989).

Theory of electric-field-induced quantum diffusion in semiconductors

P. Kleinert[*] and V.V. Bryksin[†]

[*]*Paul-Drude-Institut für Festkörperelektronik, Hausvogteiplatz 5–7, 10117 Berlin, Germany*
[†]*Physical-Technical-Institute, Politekhnicheskaya 26, 194021 St. Petersburg, Russia*

Abstract. We derive a general quantum-mechanical approach to electric-field driven carrier diffusion in semiconductors, which is applicable both for the band and hopping transport regime. This theory of quantum diffusion is extended to multi-band systems. Applications refer to the electric-field-domain formation in superlattices subject to quantizing magnetic fields, to enhanced space charge waves, and to the diffusion in multi-band heterostuctures.

INTRODUCTION

Based on semiclassical Boltzmann or balance equations as well as on the nonequilibrium Green's function technique, the quantum theory of high-field carrier transport in semiconductors has been developed to a high level of sophistication. However, except for complete numerical studies, there is, to our knowledge, no systematic treatment of diffusion phenomena in strongly biased semiconductors. The distinction between the field-dependent drift velocity and the diffusion coefficient has a clear physical origin. Whereas at least an initial inhomogeneity of the carrier ensemble is necessary for diffusion processes to occur, one assumes a homogeneous carrier distribution far away from the contacts for the description of the stationary carrier transport. Consequently, in contrast to the drift velocity, the diffusion coefficient turns out to be characterized by a quantity, which does not have the meaning of the carrier distribution function and does not solve the Boltzmann equation or it quantum-mechanical extensions. Therefore, it is necessary to develop a specific quantum-mechanical approach that allows the calculation of the field-dependent diffusion tensor. In this paper, we propose such a general microscopic theory, which allows the treatment of quantum diffusion in strongly biased semiconductors.

GENERAL APPROACH

A physically transparent, unified approach for the drift velocity and the diffusion coefficient can be derived within the hopping picture on the basis of the Wannier-Stark representation. In this picture, both quantities are expressed by the effective field-dependent lateral distribution function and the scattering rates in a similar manner. In the hopping regime, the longitudinal diffusion coefficient is expressed by [1]

$$D_{zz} = \frac{1}{2} \sum_{\mathbf{k}_\perp, \mathbf{k}'_\perp} \sum_{m=-\infty}^{\infty} n(\mathbf{k}'_\perp)(md)^2 \widetilde{W}_{0,m}^{0,m}(\mathbf{k}'_\perp, \mathbf{k}_\perp), \quad (1)$$

where both the effective scattering probability $\widetilde{W}$ and the lateral carrier distribution function n are solutions of kinetic equations. In Eq. (1), the hopping length md is given by the lattice constant d. The vector of the lateral quasi-momentum is denoted by $\mathbf{k}_\perp$. Based on the principle of detailed balance, an asymptotic relationship between the drift velocity v_z and the diffusion coefficient D_{zz} can be derived that is applicable in the ultra-quantum limit $\Omega\tau \gg 1$ (where $\Omega = eEd/\hbar$ denotes the Bloch frequency, E the electric field, and τ an effective scattering time). We obtain

$$D_{zz} = \frac{v_z d}{2} \coth\left(\frac{\hbar\Omega}{2k_BT}\right), \quad (2)$$

which reproduces the Einstein relation in the limit of sufficiently high temperatures T ($\hbar\Omega < 2k_BT$). From the hopping picture of quantum diffusion, one can switch to the equivalent band model in an exact manner. [1, 2] The generalization of this approach to quantum diffusion in multi-band systems is straightforward. [3]

APPLICATION

We shall discuss various applications of the general approach to quantum diffusion. From the general approach, one identifies an interesting quantum effect, namely electro-phonon resonances, not only in the drift velocity, but also in the diffusion coefficient. [1] These resonances are due to the electric-field dependence of the effective scattering probability. Another application refers to the quasi-elastic scattering regime, for which both the drift velocity and the diffusion coefficient can be analytically evaluated. [4] Both quantities, which are expressed by an electron temperature, are used to treat the onset of electric-field-domain formation in superlattices on the basis of a linear stability analysis. Depending on parameters such as the carrier density or lattice temperature, the carrier diffusion may strongly influence the onset of domain formation by stabilizing a homogeneous field distribution. The microscopic analysis of domain formation becomes more complicated, when a quantizing magnetic field is applied perpendicular to the layers. [5] Both in the drift velocity and in the diffusion coefficient, combined cyclotron-Stark-phonon resonances occur. Due to these resonances, new branches with a fan-like structure appear in the electric- and magnetic-field diagram, for which both the density profile and the electric-field distribution are uniform.

As another application, the general approach to quantum diffusion can be used to treat space-charge waves in semiconductors with an N-shaped current-voltage characteristic. [6] Two eigenmodes are predicted to appear, which refer to oscillations of the free electron gas and to trap-recharging waves. In the region of negative differential conductivity, the Maxwellian relaxation time becomes negative. Consequently, the resonances are enhanced because the effective scattering rate decreases in this electric-field regime. The space-charge waves may be excited by various methods such as, for instance, by an oscillating grating or an ac electric field.

Other interesting applications of the microscopic approach to quantum diffusion refer to multi-band (multi-subband) semiconductors. For a simple two-band model, we encounter an additional contribution to the diffusion coefficient, which is proportional to the elements of the subband velocity tensor. [7] For a superlattice exhibiting negative differential conductivity, this contribution gives rise to a non-monotonous electric-field dependence. This excess diffusion, which results from different drift velocities in the subbands, has a simple physical origin. Let the carrier ensemble be composed of particles in both subbands with an initial δ-function like character. If intrinsic diffusion and intersubband transitions are absent, the peak splits into two parts moving with different subband velocities by maintaining their δ-function like shapes. Spreading of the carrier packets results from intersubband transitions, which mix carriers moving with different velocities. Consequently, an additional contribution to the diffusion coefficient appears. Due to this specific contribution, maxima and minima are expected to appear in the electric-field dependence of the diffusion coefficient, which can be determined on the basis of the Price relation by measuring the noise temperature. Applying high-frequency noise measurements, these peculiarities have been observed in experimental studies both on GaAs/(Ga,Al)As quantum wells and superlattices.

Another interesting application of the general approach is the treatment of quantum diffusion associated with electric-field-induced interband (intersubband) tunneling. [3] From the general microscopic theory, we obtain that both the drift velocity and the longitudinal diffusion coefficient decompose into a scattering- and tunneling-induced contribution. In the tunneling contribution of the diffusion coefficient, which is most relevant at the Zener tunneling resonance, a term appears that is proportional to the difference of the subband velocities. Depending on the superlattice parameters (such as, for example, the barrier width, the scattering times, and the miniband widths), this term may dominate giving rise to a Zener resonance or antiresonance in the quantum diffusion. This result has to be contrasted with the behavior of the drift velocity that always exhibits a Zener resonance, but never an antiresonance. This discrepancy underlines the different physical origin of both quantities. As a consequence, the Einstein relation between the drift velocity and the diffusion coefficient is not valid for arbitrary electric-field strengths. We conclude that in general the calculation of the diffusion coefficient requires the application of a specific approach, which differs from the kinetic theory used to determine the drift velocity or the current density.

ACKNOWLEDGMENTS

Partial support by the Deutsches Zentrum für Luft- und Raumfahrt is greatefully acknowledged.

REFERENCES

1. Kleinert, P., and Bryksin, V., *Physics Letters A*, **317**, 315–323 (2003).
2. Bryksin, V. V., and Kleinert, P., *J. Phys.: Condens. Matter*, **15**, 1415–1425 (2003).
3. Kleinert, P., and Bryksin, V., *J. Phys.: Condens. Matter*, **16**, 4441–4454 (2004).
4. Bryksin, V. V., and Kleinert, P., *Physics Letters A*, **308**, 202–207 (2003).
5. Kleinert, P., and Bryksin, V., *Physica B*, **334**, 413–424 (2003).
6. Bryksin, V. V., Kleinert, P., and Petrov, M. P., *Physics of the Solid State*, **45**, 2044–2052 (2003).
7. Kleinert, P., and Bryksin, V., *Appl. Phys. Lett.*, **submitted** (2004).

Quantum Electron Transport in Finite-size Flat-band Kagome Lattice Systems

Hiroyuki Ishii and Takashi Nakayama

*Department of Physics, Faculty of Science, Chiba University,
1-33 Yayoi, Inage, Chiba 263-8522, Japan*

Abstract. Electron transport properties are investigated for the Kagome-lattice chain systems. It is found that the electric current flowing along the chain suddenly increases the magnitude when a very small electric field is applied to the system. This large current originates from the symmetry breaking of the large energy degeneracy of flat-band states in the Kagome lattice and is nearly independent of the size of the Kagome-lattice chain.

INTRODUCTION

Recent advances in nanotechnology have enabled us to fabricate and arrange semiconductor quantum wires/dots periodically on semiconductor substrates. These periodic systems are called artificial lattices. The artificial lattices have two fascinating natures. One is the flexibility to design various shapes of lattices. The other is the controllability of the electron numbers in the lattice, which is realized by applying the gate voltage. Among various lattices, the Kagome lattice is located at a special position because it has the complete flat electronic band structure. This flat band appears reflecting the specific geometry of the lattice and the interference of electron wavefunction. One can choose eigenstates of a flat band slightly overlapping each other but completely localized around one plaquette. Since these eigenstates have the same eigenenergies, a sum of them produces the large energy degeneracy and forms the flat band.

The large energy degeneracy of flat-band states induces interesting physical properties for the two-dimensional Kagome lattice systems; Mielke *et al.* showed that the ferromagnetism occurs when the flat band is half filled with electrons [1]. On the other hand, we found that the existence of flat bands remarkably increases the binding energy of exciton, which is much larger than that of one-dimensional lattice system [2]. With respect to the conductive properties, flat-band states do not contribute to the current because of their large mass. Kimura *et al.* calculated the Drude weights and demonstrated that the Kagome-lattice system becomes conductive when a magnetic field is applied [3]. However, since the magnetic field they used to break the flat-band feature is so strong, it is not clear what relation exists between the current and the large energy degeneracy of flat-band states. The purpose of this work is to clarify the effects of large energy degeneracy of flat-band states on conductive properties. To realize this purpose, we used a finite-size Kagome-lattice system, applied a small electric or magnetic field, and investigated the current in the system because such a small field just a little removes the degeneracy.

METHODS, RESULTS, AND DISCUSSIONS

To simplify the analysis, we used the Kagome-lattice chain with a finite number of plaquettes, which is shown in Fig.1. Left and right edges of this system are connected to two one-dimensional electric leads. We investigated the conductive properties of this joint system under a small external field, by employing the non-equilibrium Green function method and the tight-binding approximation, where we consider no interactions between electrons and use the electron transfer energies shown in Fig.1 with t as an energy unit. The current channel opens when the energy levels of the Kagome-lattice chain are changed by applying the gate voltage to the chain to be located between Fermi energies of right and left electric leads. In this paper, we concentrate on the electric-current vs. gate-voltage characteristics.

In the case of no external fields, flat-band states have no contributions to the current that flows from the left to the right of the Kagome-lattice chain. This is because most flat-band states are localized around plaquettes. However, this situation suddenly changes even when a small electric field, which is directed from the bottom to the top of Fig.1 and has the magnitude of $0.05t/a$, is applied to the system. Here, a is the width of the Kagome lattice chain. Figure 2 shows the calculated electric-current vs. gate-voltage characteristics.

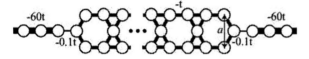

FIGURE 1. Schematic picture of the Kagome-lattice chain connected to two one-dimensional electric leads. The electrons can transfer between the nearest-neighboring sites along the solid lines.

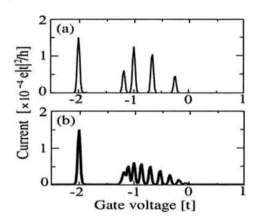

FIGURE 2. The calculated electric-current vs. gate-voltage characteristics of the Kagome-lattice chain with (a) four and (b) ten plaquettes.

The current peaks around $-1.3t$ to 0 correspond to the channels due to the normal-band electronic states of the Kagome-lattice chain and exist independent of external field. On the other hand, the large peak at $-2t$ originates from the flat-band states, which channel is closed in the case of no external field. The external field weakly breaks the symmetry in the Kagome lattice and promotes the hybridization of localized flat-band states to produce the extended states with almost the same energies. This change of the character of degenerate flat-band states opens the current channels around $-2t$.

Then we consider the system-size dependence of electric current. As seen in Figs.2(a) and 2(b), the current channels due to the normal-band states broaden and decrease the magnitude as the size of the Kagome-lattice chain increases, which is the normal situation expected for the dispersive-band states in the finite systems. On the other hand, the flat-band-state channel has almost the same magnitude independent of the system size. This is because the current magnitude of a single flat-band state decreases while the number of flat-band states increases as the system size increases. As a result, as long as the energy difference between these states is small, the sum of the current, which is roughly obtained by multiplying such magnitude and number, has almost the same magnitude. The fact that the applied small field remarkably changes the character of flat-band states by merely introducing a small energy variation is the general tendency observed in the degenerate perturbation theory. In this sense, one can say that the conductive properties explained in this work originate from the large energy degeneracy of flat-band states in the Kagome lattice.

CONCLUSIONS

The electron transport properties of the Kagome-lattice chain system were studied, by employing a non-equilibrium Green function method and the tight-binding model. We found that when a very small external field is applied to the system a large electric current flows originating from the large energy degeneracy of flat-band states and the magnitude of this current is almost independent of the system size.

ACKNOWLEDGMENTS

This work was supported by the CREST program of JST, Japan, the Futaba Memorial Foundations, and the 21COE program of Chiba University. We thank the Super Computer Centers, ISSP, University of Tokyo and Chiba University for the use of facilities.

REFERENCES

1. A. Mielke and H. Tasaki, *Commun. Math. Phys.* **158**, 341 (1993).

2. H. Ishii, T. Nakayama, and J. Inoue, *Phys. Rev. B* **69**, 085325 (2004).

3. T. Kimura, H. Tamura, K. Shiraishi, and H. Takayanagi, *Phys. Rev. B* **65**, 081307 (2002).

Spin-Orbit Interaction in InSb Thin Films Grown on GaAs (100) Substrates by MBE: Effect of Hetero-Interface

S. Ishida[*], K. Takeda, A. Okamoto[A] and I. Shibasaki[A]

Faculty of Science & Engineering, Tokyo University of Science Yamaguchi, Onoda, Yamaguchi 756-0884, JAPAN
[A] *Asahi Kasei Corporation, Samejima 2-1, Fuji, Shizuoka 416-8501, JAPAN*

Abstract. Magnetoresistance (MR) effects caused by quantum interference have been investigated in order to search into the spin-orbit interaction (SOI) in the InSb films grown on GaAs(100) substrates by MBE. The positive MR in the accumulation layer at the InSb/GaAs interface arises from the two-dimensional (2D) weak anti-localization (WAL) and is explained by taking account of the spin-Zeeman effect on the SOI caused by the asymmetric potential at the hetero interface (Rashba term with SO scattering rate $\tau_{so}^{-1} \sim 1.7 \times 10^{12}$ s^{-1}). The negative MR in extremely weak magnetic fields found for Sn-doped films dramatically crossovers to the positive MR with decreasing the film thickness. The results have been analyzed using a two-layer model for the films; the SO scattering rate in the intrinsic InSb film due to the bulk inversion asymmetry (Dresselhaus term) has been found to be as small as $\tau_{so}^{-1} \leq 2 \times 10^9$ s^{-1} and when the interface is approached in the film the WL crossovers to the WAL with the increase of SOI in the layers caused by the increased influence of the Rashba electric field.

INTRODUCTION

Among the narrow-gap III-V semiconductors, InSb has the highest mobility with the smallest electron effective mass (m* ~ 0.014m$_0$) and the largest negative effective g-factor (|g*| = 51 ~ 35 according to the carrier concentration n = 10^{14} ~ 10^{17} cm^{-3} [1]). Its thin layers are often grown on semi-insulating GaAs substrates using MBE for device applications. However, a large lattice mismatch of 14.6 % between InSb and GaAs induces high-density misfit dislocations at the interface resulting in an extraordinarily large carrier accumulation [2]. The effective thickness of the accumulation layer d_a is roughly estimated as ~ 20 nm from the in-plane transverse magnetoresistance (MR) [3]. The positive MR with anisotropy between parallel and perpendicular field orientation arising from the two-dimensional (2D) weak anti-localization (WAL) in the accumulation layer found for a 0.1 μm thick undoped film was explained by taking account of the spin-Zeeman effect on the SOI caused by the asymmetric potential at the hetero interface (Rashba term) [3, 4]; the zero-field spin splitting energy of $\Delta_0 \sim 13$ meV with SO scattering rate $\tau_{so}^{-1} \sim 1.7 \times 10^{12}$ s^{-1} was inferred from the fits of MR with the 2D WL theory. In this paper, the crossover from WL to WAL found in extremely weak magnetic fields for Sn-doped films with decreasing the film thickness from 1 to 0.1 μm have been analyzed assuming a two-layer model for the film in order to search into the spin-orbit interaction (SOI) inside the InSb films.

RESULTS AND DISCUSSIONS

InSb thin films were grown directly on the semi-insulting GaAs(100) substrates, ignoring the large lattice mismatch between InSb and GaAs. The films studied in this work are 7×10^{16} cm^{-3} Sn doped ones with the thickness ranging $d = 0.1 \sim 1.0$ μm prepared under the same growth conditions. The film parameters at 1.4 K are given in Table 1. The MR measurements have been carried out in both perpendicular and parallel magnetic fields to the film plane at liquid He temperatures. In these films with $d \geq 0.5$ μm show the clear Shubnikov-de Haas (SdH)

TABLE 1. Film parameters at 1.4 K.

d (mm)	n_s (10^{12}cm^{-2})	R_s(kΩ)	μ (10^4cm^2/Vs)	l_e (nm)	$k_F l_e$
0.1	0.84	7.56	0.0978	8.74	1.2
0.2	1.16	1.08	0.495	39.1	4.7
0.3	1.75	0.273	1.13	87.3	10
0.5	3.35	0.0960	2.05	166	20
1.0	8.07	0.0215	3.64	321	43

oscillations corresponding to the high Landau quantum numbers ($N \geq 4$) below 1.5 T at low temperatures, but for $d \leq 0.3$ µm the oscillations are not observed. The WL effect on the MR manifests itself more pronounced when $k_F l_e > 1$ (k_F: Fermi wave number, l_e: mean free path) and $B < B_{tr} = \hbar/2el_e^2$ (transport magnetic field) for perpendicular field because it assumes the diffusion process where $l_B = (h/eB)^{1/2} > 2^{1/2} l_e$. The MR due to WL observed in extremely weak magnetic fields (where $\sigma_s(B) \sim \sigma_{xx}$ since $\mu B \ll 1$) far before the appearance of the SdH oscillations has been transfigured from the negative one with weak SOI into the positive one with strong SOI with decreasing the film thickness from $d = 1$ µm to 0.1 µm. These data for magnetoconductance (MC) $\Delta\sigma_s(B)$ have been analyzed for both perpendicular and parallel orientation of magnetic field using a two-layer model in which the composition of upper layer under the surface and lower one adjacent to the InSb/GaAs interface of the film yielding parallel conduction channels is assumed; the sheet conductance for each decomposed layer has been inferred as $\sigma(0.5\sim0.3) = \sigma(0.5) - \sigma(0.3)$, for example, where $\sigma(0.5\sim0.3)$ means the sheet conductance σ_s for the upper part 0.2 µm thick of the film with $d = 0.5$ µm and $\sigma(0.5)$ for the entire film. The inelastic diffusion length l_e is roughly estimated to be larger than 4, 4, 1 and 0.6 µm for the decomposed layer d(1.0~0.5), d(0.5~0.3), d(0.3~0.2) and d(0.2~0.1), respectively, so that $l_e > t$ (the thickness for each layer). We find that B_{tr} for the layer d(1.0~0.5), d(0.5~0.3), d(0.3~0.2) and d(0.2~0.1) is smaller than 3, 12, 46 and 220 mT, respectively. Thus, both parallel and perpendicular MC for each layer can be fitted in the extremely weak magnetic fields.

Fig. 1(a), (b) and (c) show the variations in perpendicular and parallel ($B\perp I$ & $B//I$) MC for the decomposed layers, respectively, where the best fits to the 2D WL theory obtained are depicted. In the fitting routine, the value of $|g^*| \sim 37$ for bulk n-InSb with the same carrier concentration [1] was used. The values of τ_{so} inferred from the fits of parallel and perpendicular MC for the decomposed layers are listed in Table 2. One can see that τ_{so} of the layer decreases abruptly with approaching the interface from the distance of $\sim$ 0.5 µm. From the above result, the variation of MC in the extremely weak magnetic fields for each

TABLE 2. SO scattering time τ_{so} inferred from the fits of perpendicular and parallel ($B\perp I$ & $B//I$) MC for the decomposed layers.

	d(1.0~0.5)	d(0.5~0.3)	d(0.3~0.2)	d(0.2~0.1)
$\tau_{so\perp}$ (ps)	950	430	46	6.9
$\tau_{so//}$ (ps)($B\perp I$)	490	190	47	7.1
$\tau_{so//}$ (ps)($B//I$)	500	200	48	9.9

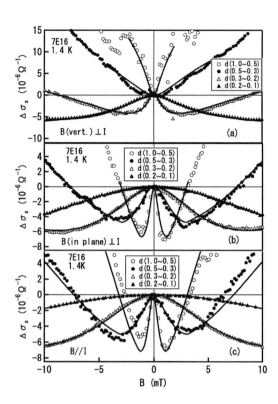

FIGURE 1. Variation of transverse MC in extremely weak magnetic fields for four decomposed layers $\sigma(1.0\sim0.5)$, $\sigma(0.5\sim0.3)$, $\sigma(0.3\sim0.2)$ and $\sigma(0.2\sim0.1)$ at 1.4 K are fitted: (a) perpendicular MC, (b) parallel MC ($B\perp I$) and (c) parallel MC ($B//I$).

decomposed layer has been explained as the crossover from the 2D WL to WAL with the increase of SOI in the layers, being caused by the increased influence of the asymmetric potential at the hetero interface (Rashba term) [5]. Probably, our result of $\tau_{so}^{-1} \sim 10^9$ s^{-1} for the bulk-like decomposed layer d(1.0~0.5) barely measures the SOI induced by bulk inversion asymmetry (Dresselhaus term) [5] in the intrinsic InSb film.

REFERENCES

1. Gluzman, N. G., Sabirzayanova, L. D. and Tsidil'kovskii, I., Sov. Phys. Semicond. **13**, 275-281 (1979).
2. Tanaka, T., Washima, M., and Sakaguchi, H., Jpn. J. Appl. Phys. **38**, 1107-1110 (1999).
3. Ishida, S., Takeda, K., Okamoto, A., and Shibasaki, I., Physica E **20**, 255-259 (2004).
4. Maekawa, S., and Fukuyama, H., J. Phys. Soc. Jpn. **50**, 2516-2524 (1981).
5. See, e.g., Knap, W. et. al, Phys. Rev. B **53**, 3912-(1996) and references therein.

Spin-dependent Non-equilibrium Transport in Mesoscopic 2D Electron Systems

A. Ghosh, M. H. Wright, K. DasGupta, M. Pepper, H. E. Beere and D. A. Ritchie

Semiconductor Physics Group, Cavendish Laboratory, University of Cambridge, UK.

Abstract. We have investigated the non-equilibrium differential conductance of a large number of ultra-clean 2D electron systems (2DES) at mesoscopic dimensions. At zero magnetic field and low carrier densities a strong enhancement of conductance was observed as the bias across the sample was reduced, followed by a dip at the Fermi energy. This behavior was found to be universal with sample-specific energy-scales depending on the background disorder, carrier density and thermal history of the sample. From the magnetic field and temperature dependence of conductance, we propose that this non-equilibrium structure is a manifestation of a spontaneous spin-polarization of the electrons in the 2DES at low-density regime.

INTRODUCTION

Ever since the observation of Kondo screening of a single spin-1/2 electron within an artificial quantum dot[1], non-equilibrium transport measurements have been used to identify the spin-structure of various low-dimensional systems. Experiments in 1D ballistic point contacts, which are not separated from the leads through an intentional tunnel barrier, have also revealed a strong enhancement in the zero-bias conductance that was attributed to a Kondo-like effect[2]. Theoretical investigations [3] have indicated a possible ferromagnetic spin-ordering within the 1D region with an effective spin magnetic moment of 1/2. Motivated by these theoretical and experimental developments, we have measured the non-equilibrium transport of a large number of quasi-ballistic 2D electron systems at mesoscopic length scales.

DEVICE AND MEASUREMENT

We have used high mobility GaAs/AlGaAs heterostructures, lithographically patterned to length scales of mesoscopic dimensions (typically, 5μm × 5μm). Details of the device structure and measurement technique are given elsewhere [4]. In this work we have used both undoped and δ-doped heterostructures to ensure generality of the observed non-equilibrium effects. In the latter, the spacer thickness (δ_{sp}) was varied to modulate the background disorder. At zero gate-voltage ($V_g = 0$), the elastic scattering length (l) was ~ 6 – 8 μm, implying a quasi-ballistic transport within the active device area. By varying V_g, the carrier density (n) could be lowered down to ~ 5×10^9 cm^{-2}, which corresponds to a strongly interacting regime (r_s ~ $1/a_B\sqrt{(\pi n)}$ ~ 7 -8). All measurements were performed in dilution fridges (base temperature ≈ 40 mK), with magnetic field (B) applied parallel to the plane of the 2DES in order to avoid any orbital effect.

RESULTS

In Fig.1 we show the differential conductance (dI/dV_{SD}) in two of the devices as a function of the source-drain field (V_{SD}). Traces with increasing zero-bias conductance (G) correspond to increasing value of V_g. The enhancement in dI/dV_{SD}, followed by the dip at $V_{SD} = 0$, can be observed in both samples, even though the details of the profiles are different. In most samples the separation of the local peaks in dI/dV_{SD} were found to be ~ 0.10 – 0.15 meV. At higher V_g, the peak separation (Δ) was found to increase weakly.

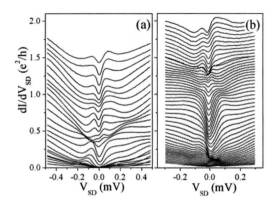

FIGURE 1. Non-equilibrium conductance as function of source-drain bias (a) doped system (δ_{sp} = 60 nm) and (b) undoped sample. Different traces correspond to different V_g.

Fig.2 illustrates the dependence of dI/dV_{SD} and Δ on a B-field applied parallel to direction of current in the doped sample. At a given V_g and small B, the peaks initially close in, accompanied by an enhancement in the zero-bias conductance G (Fig.2a). Further increase in B results in a strong suppression of G as the peaks move away from each other. Fig.2b shows the variation of Δ with B at various values of V_g. Note that the qualitative behavior is similar, even though the critical B at which the upturn in Δ occurs depends on V_g. The near-linear dependence of Δ at large B is common for all values of V_g and indicates a Zeeman-type splitting with effective g-factor $g^* \approx 0.6$ (for bulk GaAs, $|g^*| = 0.44$).

DISCUSSIONS

The split-peaks in dI/dV_{SD}, and its behavior in parallel B, indicate an unusual property of the mesoscopic low-density 2DES. If we assume the system behaves as a large, nearly open quantum dot with the two upper-most energy-levels forming a triplet at $B=0$, then the observations indicate a two-stage Kondo screening [5]. Alternatively, transport in a high-spin ($S \geq 1$) ground state has also been addressed in the same model [6], which fits the B and T-dependence of G in our system very well (not shown). Other theoretical investigations on tunneling density of states of a spontaneously spin-polarized 2D region have also indicated additional zero-bias anomalies at energies $\approx \pm\sqrt{(\Delta_0^2+\Delta_Z^2)}$, where Δ_0 and Δ_Z are B-dependent energy scales of spontaneous spin-splitting and Zeeman splitting respectively [7]. In both cases, the large-B dynamics will be determined by Zeeman splitting. Hence, while the precise mechanism giving rise to the observed non-equilibrium is still unclear, we have shown that the explanation will possibly involve a high spin state of the 2D region. In view of the low-density nature of the system, the most likely origin of the spin-polarization is the strong exchange interactions between the electrons.

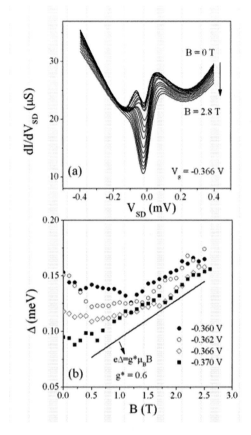

FIGURE 2. (a) Bias-dependence of dI/dV_{SD} at various B. (b) B-dependence of Δ at different values of V_g.

REFERENCES

1. Goldhaber-Gordon D. et al., Nature **391**, 156 (1998); Cronenwett S. M. et al., Science **281**, 540 (1998).
2. Cronenwett S. M. et al., Phys. Rev. Lett. **88**, 226805 (2002).
3. Berggren K. –F., and Yakimenko I. I., Phys. Rev. B **66**, 085323 (2002); Hirose K. et al., Phys. Rev. Lett. **90**, 026804 (2003).
4. Ghosh A. et al., Phys. Rev. Lett. **92**, 116601 (2004).
5. van der Weil et al., Phys. Rev. Lett. **88**, 126803 (2002); Hofstetter W., and Schoeller H, Phys. Rev. Lett. **88**, 016803 (2002).
6. Glazman L. I., and Pustilnik M., Phys. Rev. Lett. **87**, 216601 (2001).
7. Apalkov V. M., and Raikh M. E., Phys. Rev. Lett. **89**, 096805 (2002).

Ferromagnetic/DMS hybrid structures: one- and zero-dimensional magnetic traps for quasiparticles

P. Redliński*, T. Wojtowicz[†*], T. G. Rappoport*, A. Libal*,
J. K. Furdyna* and B. Jankó*

*University of Notre Dame, Notre Dame, IN 46556, USA
[†]Institute of Physics, Polish Academy of Sciences, Warsaw, Poland

Abstract. We investigated possibility of using local magnetic field originating from ferromagnetic island deposited on the top of semiconductor quantum well to produce zero- and one-dimensional traps for quasi-particles with spin. In particular we considered two shapes of experimentally made magnets - cylindrical and rectangular. In the case of ferromagnetic micro-disk the trap can localize spin in three dimensions, contrary to the rectangular micro-magnet which creates a trap that allows free propagation in one direction. We present in detail prediction for absorption spectrum around the main absorption edge in both type of micro-magnets.

Recently there is an increasing interest in using the spin of particles, instead of their charge, as a basis for the operation of a new type of electronic devices. In this work we show via theoretical calculations that spin degrees of freedom can be utilized for achieving spatial localization of both charged quasi-particles (electrons, holes, trions) as well as of neutral complexes (excitons). Such localized states are of interest from spintronic application point of view.

The hybrid structure we consider is build of CdMnTe/CdMgTe quantum well (QW) buried at nanometers distances (d) below two types of experimentally important magnetic Fe islands: with rectangular [1] and cylindrical [2] shape. In order to make localization effects sizeable we used diluted magnetic CdMnTe semiconductor (DMS) QW instead of classical semiconductor because of a giant Zeeman effect that exist in DMS (for a review see ref. [3]). Theoretically, DMS can be described by a Zeeman term with a very large g-factor. Dietl [4] et al. reported electron g-factor $|g_e|=500$ in sub-Kelvin experiment, which assuming known ratio of exchange constant in CdMnTe: $\beta/\alpha = 4$, gives hole g-factor $|g_h|=2000$. These giant, but realistic values of g-factors were used in our calculations.

The magnetic field produced by the magnetic islands was calculated by solving magneto-static equations with magnetization distribution determined from micromagnetic simulations in the case of micro-disk, and by direct integration of Maxwell's magneto-static equations in the case of the rectangular magnet, where homogenous magnetization was assumed. Then, the energy spectrum of conduction band as well as the spectrum of a valence

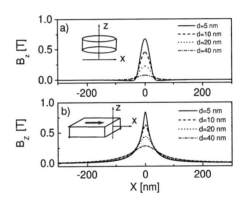

FIGURE 1. Upper panel: magnetic field B_z at a distance d below the center of magnetic micro-disk. Lower panel: B_z at distance d below one of the two ends of the rectangular micromagnet. Both magnets have the same thickness $D_z=5$ nm. With increasing d maximum of B_z is decreasing.

band were calculated by approximating the Schrödinger differential eigen-equation with a finite difference algebraic equation. In the case of μ-disk, valence band electrons were modeled by Luttinger Hamiltonian. Inclusion of mixing between heavy and light holes is important because of anisotropic spin-splitting [5] of valence energy levels in DMS QW in the presence of an external magnetic field with an arbitrary direction. Using Fermi's Golden Rule, together with previously calculated energies and wave functions, we could calculate absorption coefficient $\alpha(E)$. Each transition line was broadened with a Gaussian function in order to simulate the realistic absorption data.

We first consider a Fe micro-disk [2] (diameter $R = 1\,\mu$m and thickness D_z=50 nm, $\mu_0 M_s$ =2.2 T) in the so called vortex state. In this state the magnetization lies mainly in plane of the disk but in its center it is forced out of plane. The diameter R_c of the core, where magnetization is pointing out of plane, extends only over 60 nm. It is important to mention that only M_z, the z-component of the total magnetization $\vec{M}$, produces magnetic field $\vec{B}$ which couples to the quasi-particle's spin. The calculated profiles of the B_z component, that gives largest confinement effect, are presented in Fig. 1 a. At a distance of d=10 nm below the Fe magnet, the maximum magnetic field is $|B|_{max}$=0.46 T and its spatial extension is over a distance 80 nm. In Fig. 2 we present $\alpha(E)$ for three distances d between micro-magnetic disk and QW: d=15 nm, d=10 nm, and d=5 nm. Energy is measured relative to the energy of the main absorption edge in QW without deposition of magnetic island. Vertical bar height represent oscillator strengths of the optical transitions. As expected, low-energy peaks appear at lower energies with decreasing d because the magnitude of $\vec{B}$ increases. Separation between peaks is approximately the same (3 meV) for all three distances d.

Next, we analyze rectangular μ-magnet [1] (size: D_x=6 μm, D_y=2 μm, and D_z=0.15 μm, $\mu_0 M_s$=2.2 T) in a single domain state, with magnetization pointing in x-direction. Please note that the thickness of this magnet is 3 times larger than cylindrical μ-disk discussed above. The corresponding field profile is plotted in Fig. 1 b. The magnetic field is constant along the y edge of the magnet and therefore, the field produces one dimensional confining potential. In Fig. 3 we present the absorption spectra for three distances: d=10 nm, d=30 nm, and d=60 nm. In panel for d=10 nm, the numbers (nm) indicate that corresponding line is associated with transition from m-hole state to n-electron state. The interesting property of field-induced confining potential is that non-diagonal transitions ($n \neq m$) are very strong, and can even be stronger than diagonal ones ($n = m$). For example, transition (24) is stronger than transition (33), which is not even visible in this scale.

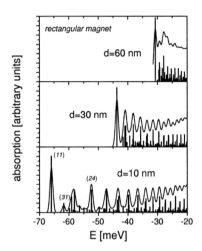

FIGURE 3. Absorption spectra (full lines) in the case of rectangular ferromagnetic island at three distances d between island and QW. For d=10 nm, numbers (nm) indicate that given line corresponds to the transition from m-hole state to n-electron state.

Additionally, we have performed calculations for various magnet thickness. We found that although the spatial extent of the field increases for thicker micromagnet the effect of increasing amplitude of the field is dominating and therefore the energetic distance between (11) transition and main absorption peak of QW, as well as distance between peaks in the spectrum increase with the increasing micromagnet thickness. Therefore it is more promising to make hybrid structures with relatively thick micromagnets.

This work was supported by the NSF-NIRT grant DMR02-01519.

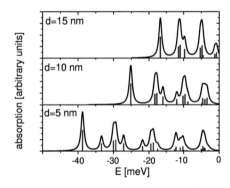

FIGURE 2. Absorption spectra (full lines) of a hybrid structures composed of a Fe magnetic island deposited on the surface of DMS QW structure at distance d=15 nm, d=10 nm, and d=5 nm from QW. In our calculations each transition line (bar) was broaden by Gaussian function with line width of 1 meV.

REFERENCES

1. Kossut, J., et al., *App. Phys. Lett.*, **79**, 1789 (2001).
2. Berciu, M., and Jankó, B., *Phys. Rev. Lett.*, **90**, 246804 (2003).
3. Furdyna, J. K., and Kossut, J., *Diluted Magnetic Semiconductor*, Academic, San Diego, 1988.
4. Dietl, T., et al., *Phys. Rev. B*, **43**, 3154 (1991).
5. Kuhn-Heinrich, B., et al., *Solid State Commun.*, **91**, 413 (1994).

Controlling hole spin relaxation in charge tunable InAs/GaAs quantum dots

S. Laurent*, B. Eble*, O. Krebs*, A. Lemaître*, B. Urbaszek†, X. Marie†, T. Amand† and P. Voisin*

Laboratoire de Photonique et Nanostructures-CNRS, Route de Nozay, 91460 Marcoussis, France
†*Laboratoire de Nanophysique, Magnétisme et Optoélectronique-INSA, avenue de Rangueil, Toulouse, France*

Abstract. We report on optical orientation of excitons and trions (singly charged excitons) in individual charge-tunable self-assembled InAs/GaAs quantum dots (QD). When the number of electrons varies from 0 to 2, the trion photoluminescence under quasi-resonant excitation gets progressively polarized from zero to ∼100%. We discuss this behavior as the quenching of the anisotropy induced exciton spin quantum beats due to the trion formation. This indicates a long hole spin relaxation time larger than the radiative lifetime, confirmed by time-resolved photoluminescence measurements carried out on a QD ensemble.

Schottky diodes containing a layer of self-assembled quantum dots (QDs) close to the n^+-electrode have been recently used for tuning the number of resident electrons in QDs by applying a gate voltage[1, 2, 3]. In the context of using electrons trapped in QDs for quantum information processing, such systems appear very promising. As, in contrast to chemical doping, the QD charge can be precisely tuned, they offer a unique way to investigate the spin relaxation of carriers trapped in QDs and their mutual spin-dependent interaction. For example, in a negative (positive) trion complex, as the exchange interaction vanishes in the ground state, the circular polarization of the photoluminescence (PL) should reflect the spin state of the hole (electron) as already reported in [3], giving access to intrinsic spin relaxation mechanisms.

Here, we report on optical orientation experiments of (individual and ensemble) charge-tunable QDs. We focus on the negative trion (X^-) luminescence and show that its circular polarization depends on the QD equilibrium charge (1 or 2) before the excitation. Time-resolved measurements directly demonstrate the hole spin relaxation quenching in a trion complex.

We studied two samples grown using molecular-beam epitaxy on a [001]-oriented semi-insulating GaAs substrate. The field-effect structure consists of a 200nm-thick n^+-GaAs layer, followed by 25nm of intrinsic GaAs, ∼ 2 monolayers of strained InAs, 30nm of GaAs, 100nm of $Al_{0.3}Ga_{0.7}As$ and, finally, a 20nm GaAs cap layer. The density of the QDs grown in the Stranski-Krastanov mode is estimated to be less than $10^9 cm^{-2}$ in sample A designed for single QD spectroscopy and a few $10^{10} cm^{-2}$ in sample B. To control the QD charge

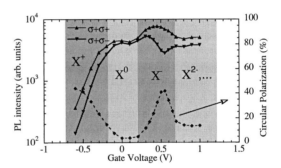

FIGURE 1. PL and polarization of a QD ensemble (sample B) detected at 1.27eV under excitation at 1.34 eV against gate voltage. Gray-shaded areas indicate roughly the stability domains of (charged) excitons.

a bias is applied between a semi-transparent Schottky contact (50Å-$Ni_{0.5}Au_{0.5}$) and a Ni/Ge/Au ohmic contact diffused downto the n^+-layer. For individual QD spectroscopy, 1µm-wide apertures were made through an optical mask deposited on the sample surface.

From micro-photoluminescence and capacitance-voltage measurements carried out respectively on samples A and B, we could determine the voltage regime of QD charging. The bias range of X^- stability, which basically corresponds to the QD filling with $N = 1$ or $N = 2$ electrons, depends for a particular QD, on the precise energy position of the quantized levels. On average it spreads over ≃ 500mV, from ∼ 200mV to ∼ 700mV, as schematically shown in Fig.1 for a QD ensemble. These values agree well with the conduction band profile of the structure characterized by a lever-arm of ∼7. We report

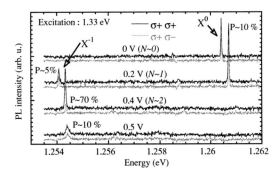

FIGURE 2. Polarization-resolved PL spectra of a single QD for different bias. The trion feature appearing 6.7 meV below the exciton line appears with a weak polarization and gets strongly polarized at 0.4 V.

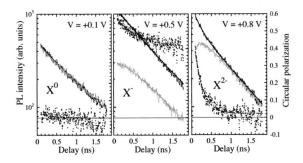

FIGURE 3. Time-resolved PL intensity (solid lines) in $\sigma^+\sigma^+$ (black) and $\sigma^+\sigma^-$ (gray) configurations, and circular polarization (black dots) of sample B excited at 1.34 eV.

here results obtained at low temperature ($\sim 15\,\mathrm{K}$), under quasi-resonant (or *intra-dot*) excitation, the laser energy being fixed 70 meV above the detection. The circular polarization of the Ti:Sapphire laser was set to σ^+, and the PL polarization σ^+ or σ^- was analyzed by a quarter-wave plate placed in front of a fixed Glan-Thomson linear polarizer. The circular polarization is defined by $(I_{\sigma^+\sigma^+} - I_{\sigma^+\sigma^-})/(I_{\sigma^+\sigma^+} + I_{\sigma^+\sigma^-})$.

Figure 1 shows the PL intensity and polarization of a QD ensemble as a function of the gate voltage. For negative biases, the PL intensity rapidly decreases due to the increasing internal field which sweeps out the photo-created carriers. In the following, we mainly discuss what happens for a positive bias: below 300 mV the PL polarization is very weak, then increases strongly with the bias up to 42% at 550 mV, when the dots are filled with 2 electrons. This behavior is more clearly observed when studying individual QDs as shown in Fig.2. The X^0 neutral exciton is weakly polarized like the trion feature X^- which appears 6.7 meV below at 0.2 V. The latter gets however rapidly polarized up to 70% when increasing the bias, and then above 0.4 V broadens while its polarization collapses. Trion polarization above 95% could be measured for single QDs studied without optical mask.

The precise *intra-dot* excitation process is not clearly identified but did not turn out to be critical in order to observe this effect. Therefore we believe that the excitation mainly involves transitions between a confined hole state and the conduction quasi-2D continuum. The hole relaxes rapidly with an oriented spin while the photo-electron tunnels towards the n^+-electrode in a few ps. The trion thermalization time τ_{X^-} is then governed by the bias-dependent coupling with this reservoir of electrons. It can be tuned from a few ps when the dot already contains 2 electrons ($N=2$) up to a few ns when the trion and neutral exciton coexist ($N=0 \rightarrow 1$). During this thermalization, the hole spin can relax under the anisotropic exchange interaction (AEI) with electrons. The latter is responsible for the splitting δ_1 of the bright excitons with angular quantum number $j = \pm 1$ into linearly polarized eigenstates $|x\rangle, |y\rangle$[4], which amounts to a few tens μeV for the InAs/GaAs system. This prevents observation of a sizeable PL circular polarization of neutral excitons in our time-integrated experiments. For a trion in its ground state, the electron-hole exchange interaction vanishes because of the 2-electron singlet configuration. Therefore, as soon as the trion is thermalized the hole spin precession in the anisotropic exchange field is quenched. In practice, if τ_{X^-} is greater than some effective precession period τ^*_{AEI} the hole spin polarization is averaged to zero, but it can get very large when $\tau_{X^-} \ll \tau^*_{AEI}$.

To support this interpretation and in particular that the trion spin orientation is frozen once thermalized, we performed time-resolved spectroscopy of sample B. Figure 3 gives a rapid glimpse of our results. At +0.1 V the PL of excitons is basically unpolarized during all the recombination (for delays above 100 ps), while the trion PL at +0.5 V shows a high degree of polarization with a component which does not relax within our temporal window ($\tau_{spin} > 20$ ns). At higher bias, the initial polarization remains high, as a signature of the photo-created hole spin, but damps rapidly in ~ 300 ps probably because of nonzero AEI in muti-charged excitons.

In conclusion, we have performed optical orientation of trions in (single) charge-controlled InAs QDs, demonstrating the possible quenching of hole spin relaxation by reducing with an applied bias the trion thermalization time.

REFERENCES

1. Finley, J. J., et al. *Phys. Rev. B*, **63**, 161305 (2001).
2. Urbaszek, B., et al. *Phys. Rev. Lett.*, **90**, 247403 (2003).
3. Flissikowski, T., et al., *Phys. Rev. B*, **68**, 161309 (2003).
4. Bayer, M., et al., *Phys. Rev. B*, **65**, 195315 (2002).

Whole Spectrum Of The Spin Polarized Two Dimensional Electron Gas

Florent Perez[♦], Bernard Jusserand[♦], Grzegorz Karczewski[♣]

[♦]CNRS/LPN, Route de Nozay, 91460 Marcoussis, France
[♣]Institute of Physics, Polish Academy of Sciences, Aleja Lotnikow 32/46, Warszawa, Poland

Abstract. Collective and individual spin-flip and spin-conserving excitations of a two-dimensional spin-polarized electron-gas in a semimagnetic $Cd_{1-x}Mn_xTe$ quantum well are observed by resonant Raman scattering. Application of a magnetic field splits the spin subbands and a static spin polarization is induced. Wavevector and field experimental dispersions are very well reproduced by calculations one when both exchange and correlation corrections are taken into account. A maximum polarization rate of 40% is deduced from measurements.

Semimagnetic quantum wells based on CdMnTe, recently developed [1], provide the reverse situation to that of GaAs [2] : under application of moderate magnetic fields, spin quantization dominates over orbital quantization. Indeed, a giant Zeeman splitting is induced in these materials by exchange interaction between the conduction electrons and those localized on Mn ions [3]. Understanding the low-energy spectrum of the spin-polarized electron gas is of particular interest, both because of the new information it provides about exchange and correlations corrections between electrons and also because these low-energy excitations are responsible for the spin-transport in spin-based electronics.

We have investigated by Resonant Raman Scattering (RRS) the low temperature (1.5K) spectrum of a two dimensional electron gas confined in a high mobility $Cd_{1-x}Mn_xTe$ quantum well under magnetic field B ranging from 0 to 4 Tesla, parallel to the plane of the gas in quasi-Voigt configuration with different angles of incidence to vary the transferred wavevector q from 0 to $16\mu m^{-1}$ (for further details see ref. 5). Compared with previous work [2,4], we claim this is the first time the spectrum of all known excitations of the spin polarized system is measured as a function of both B and q.

In the RRS process, when the polarizations of the two electromagnetic fields involved are crossed in exact Voigt configuration, the process acts like a weak magnetic field rotating around the spin quantization axis. The resulting "Crossed spectrum" is dominated by Spin Flip wave (SF), and spin flip Single Particle Excitations (SPE). Due to Larmor's theorem, the SF energy at vanishing wavevectors equals the bare Zeeman splitting $Z(B)$ which follows the thermal distribution of Mn-spins (Brillouin function), as verified in Figure 1b. Exchange and correlation corrections shift (see Fig. 1a) the SF energy below the $SPE_{\downarrow\uparrow}$ continuum (excitation of one majority spin down electron to spin up free state).

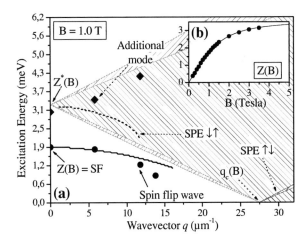

FIGURE 1. (a) q-dispersion of spin flip excitations. Symbols are experimental values. Curves are calculated dispersions with exchange and correlations corrections. The domain is the non-corrected SPE continuum whose renormalized maximum follows the dotted curve. $Z^*(B)$ is the renormalized Zeeman splitting. (b) Bare Zeeman splitting measured from spin flip wave energy at $q=0$ and fitted with the Brillouin function : $x_{Mn}=0.75\%$, $T=2.7K$.

For finite q, the SF disperses negatively and disappears when its wavevector reaches the critical value $q_c(B)=k_{f\downarrow}-k_{f\uparrow}$, difference of the Fermi wavevector associated with each spin subbands, where the reverse SPE$_{\uparrow\downarrow}$ continuum starts. Figure 2a illustrates this typical behavior for $q=11.8\mu m^{-1}$. When B increases from 0, q is above the critical value and the SPE splits into two lines : the increasing SPE$_{\downarrow\uparrow}$ and the decreasing SPE$_{\uparrow\downarrow}$ which reaches zero for B such that $q=q_c(B)$. For larger B, q is below $q_c(B)$ and the spin flip wave emerges and screens the SPE$_{\downarrow\uparrow}$ line whose maximum position slightly disperses with q. An additional mode at higher energies, presently not understood, emerges from the SPE continuum. This mode exhibits a positive q-dispersion (see Fig. 1a) in contrast to the spin flip wave.

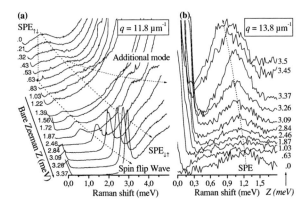

FIGURE 2. (a) Crossed Raman spectra measured at $q=11.8\mu m^{-1}$ for varying magnetic field (labeled by the corresponding Zeeman energy). The high Raman shift slope is due to luminescence (b) Parallel Raman spectra at $q=13.8\mu m^{-1}$.

We calculated the spectrum and the dispersions with linear response theory in the Local Spin Density Approximation (LSDA) completed with very recent estimations of exchange and correlation energies in a spin-polarized two-dimensional system [5]. We neglected dynamic coupling with Manganese spins. Figure 3 compares experimental and theoretical typical magnetic dispersions of the spin flip excitations. Apart from the additional mode not given by the model, very good agreement is obtained without fitting parameters for the density $n=4.4\ 10^{11}cm^{-2}$ determined independently (see Ref. 4). This result clearly demonstrates the validity of the model and the importance of correlation effects.

When polarizations are parallel, the RRS process acts like a scalar potential and plasmons and spin conserving SPE, are measured as shown on Figure 2b. The plasmon is not considered as it doesn't provide any information about spin properties. Surprisingly, the spectrum reduces to only one line associated mainly to spin conserving SPE from the minority spin up subband. Indeed, SPE from the spin up subband are screened by the spin-conserving spin wave (damped) dispersing between the two SPE continua (paramagnon renormalization). The calculated spectrum (not shown) reproduces exactly this observation. Its maximum position follows $k_{f\uparrow}$ and varies with the spin-polarization rate $\zeta=(n_\uparrow-n_\downarrow)/(n_\uparrow+n_\downarrow)$ like $(1+\zeta)^{0.5}$. From this behavior we deduced from experimental spectra that a maximum polarization rate of 40% is reached in this sample.

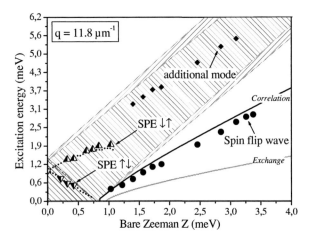

FIGURE 3. Z-dispersion of spin flip excitations. Symbols are experimental values, for SPE : the maximum position is shown. Curves are calculated with indicated corrections. The domains are non-corrected SPE continuum, dotted curves are the corresponding renormalized maximum position.

In conclusion, we have measured and understood q-dispersion and B-dispersion of all known excitations of the spin polarized system : plasmon, spin conserving SPE, spin flip wave and spin flip SPE. Calculated spectra and dispersions within LSDA reproduce very well the experimental data and demonstrate the importance of correlation effects. We deduced a maximum polarization rate reached in the sample of 40% which will be improved in further developments in order to achieve full polarization.

REFERENCES

1. G. Karczewski, J. Jaroszynski, A. Barcz, M. Kutrowski, T. Wojtowicz, J. Cryst. Growth **184/185**, 814 (1998).
2. A. Pinczuk, B. S. Dennis, D. Heiman, C. Kallin, L. Brey, C. Tejedor, Phys. Rev. Lett. **68**, 3623 (1992).
3. J. A. Gaj, R. Planel, and G. Fishman, Solid State Comm. **29**, 435 (1979).
4. B. Jusserand, F. Perez, D. R. Richards, G. Karczewski, T. Wojtowicz, Phys. Rev. Lett. **91**, 086802 (2003)
5. D. Marinescu, J. Quinn, Phys. Rev. B **56**, 1114 (1997). C. Attacalite, Phys. Rev. Lett. **88**, 256601 (2002).

Side-Gate Control of Rashba Spin Splitting in a $In_{0.75}Ga_{0.25}As/In_{0.75}Al_{0.25}As$ Heterojunction Narrow Channel: Toward Spin-Transistor Based Qubits

Tomoyasu Kakegawa[1], Masashi Akabori and Syoji Yamada

Center for Nano Materials and Technology, JAIST
1-1, Asahidai, Tatsunokuchi, Ishikawa 923-1292 Japan

Abstract. Side-gate (SG) control of Rashba spin-orbit interaction (SOI) in a diffusive wire made at InGaAs/InAlAs narrow-gap heterojunction has been studied. The wires have four-terminal structure and side-gates are prepared between the voltage probes on both side of the wire via small air-gap. By applying negative voltages in single (one gate biased and another grounded) and dual (two gates equally or unequally biased) bias conditions, spin-orbit coupling constant, α, estimated from low-field Schubnikov de-Haas oscillation was found to show drastic increase followed by saturation or reduction. This behavior could be attributed to the asymmetric lateral electric field, E_{asym}, created also by SG bias, since it can additionally contribute to the total SOI via the new effective magnetic field, $B_{eff} \propto v$ (electron velocity) $\times E_{asym}$.

INTRODUCTION

As an effort for realizing a spin-field effect transistor (FET) [1], top-gate (TG) control of SOI has been reported in narrow-gap two-dimensional electron gas systems [2-5]. However, independently controlled two SOI configurations against different axes of effective magnetic field, $B_{eff} \propto v \times E_{asym}$ (asymmetric electric field), are necessary to create spin FET based qubit devices [6,7]. This is realized, for example, by making a bend narrow wire spin-FET or by making a narrow wire spin-FET with both TG and SG electrodes. However, the SG control of SOI has so far not succeeded nor reported.

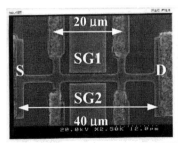

FIGURE 1. Long wire sample with side-gates fabricated by self-align method. The structure width of the wire is typically 2 μm. The air-gap between the gate and the side surface of the wire is 400 nm.

In this work, we report quantum transport experiments in narrow wires with gate electrodes on both the side walls. By applying a negative voltage, V_{SG}, to the SGs, one can *not only* decrease the width as well as the electron density (n_s) of the wire *but also* produce E_{asym} in the lateral direction with in the wire. We estimated SO coupling constant, α, from magneto-resistance oscillations taken for a variety of pairs of V_{SG}s.

SAMPLE AND EXPERIMENTAL

Plan view SEM photograph of the long wire sample with side gates is shown in Fig. 1. Base wafer is a modulation-doped high In-content (~80%) InGaAs/InAlAs heterojunction grown via InAlAs step-graded buffer on GaAs substrates. Maximum electron mobility, μ_e, and α so far obtained in this type of heterojunctions were ~5×10^4 cm^2/Vsec and ~ 40×10^{-12} eVm, respectively, at 1.5 K. These values correspond to the electron mean free-path of 6 μm and effective spin-splitting ($2\alpha k_f$) of over 10 meV. Four-terminal wire with structure width of 2 μm and length between the voltage probes of 20 μm was fabricated. Ti/Au SGs were fabricated on the etched surface via thin SiO_2 insulator by self-align method. The length and air-gap from the wire side surface were typically 15 and 0.4 μm, respectively. Low-field (< 5 Tesla) magnetoresistances were measured at 1.5 K against a lot of negative SG bias voltage pairs, V_{SG1} and V_{SG2}, applied to the electrodes, SG1 and SG2,

---

[1] Corresponding author: t-kakega@jaist.ac.jp

respectively. α was deduced by analyzing beating oscillations appeared in the resistances by fast Fourier transformation (FFT).

RESULTS AND DISCUSSIONS

Typical result when only one side-gate is negatively biased and another side-gate is grounded was a drastic enhancement of α. For example, it increased from ~ 9.8×10^{-12} eVm at $V_{SG} = 0$ V to the maximum value of 46.7×10^{-12} at -12V followed by a decrease to 22.4×10^{-12} at -17 V. Similar results were confirmed in almost all samples, although the value of α at $V_{SG} = 0$ V is different among them. Indeed, the sample with $\alpha \sim 0$ at $V_{SG} = 0$ V often revealed the finite enhanced value for $V_{SG} < 0$ V. The reason for the decrease of α at deep negative V_{SG}s is not clear at present. As an origin of this interesting result, we consider the effect of lateral asymmetric electric field, $E_{asym, side}$, created by the $V_{SG} < 0$. This field is newly added to the original vertical built-in field, $E_{asym, vericalt}$. In other words, $E_{asym, total} = E_{asym, side} + E_{asym, vertical}$ determines the total SOI in the wire. Decrease of α at deep $V_{SG} < 0$ is probably due to the decrease of n_s at those conditions.

Second experimental results are shown in Figs. 2 and 3, which are 2D plots of $n_s = n_{s, up\ spin} + n_{s, down\ spin}$, and α against the gate voltages, V_{SG1} and V_{SG2}, applied to the side-gate electrodes attached to the both side of the wire. If the assumption of $E_{asym, side}$ is true, the effect should be symmetric for the exchange of bias voltages to the side-gates. Moreover, if both the side-gates were equally biased, additional spin-orbit interaction due to the $E_{asym, side}$ should disappear due to the cancellation of electric fields in opposite directions. When we focus on the upper-right triangle in Fig. 3 (low bias region), it is found that the effect of both the gate biases is almost symmetric and α decreased in the center region where $V_{SG1} \sim V_{SG2}$. Those results can almost be expected from our $E_{asym, side}$ assumption. The result obtained in the lower-left triangle in Fig. 3 is somewhat different from that in the former triangle. In this region, as seen in Fig. 2,

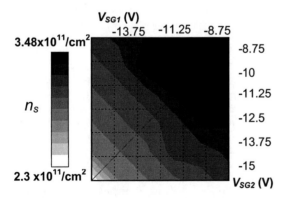

FIGURE 2. 2D plot of total sheet electron density, $n_s = n_{s, up\ spin} + n_{s, down\ spin}$, against voltages, V_{SG1} (V) and V_{SG2} (V) applied to the side-gate electrodes.

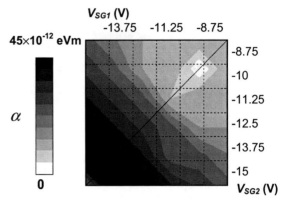

FIGURE 3. 2D plot of spin-orbit coupling constant (α) against voltages, V_{SG1} (V) and V_{SG2} (V) applied to the side-gate electrodes.

the decrease of total sheet electron density (n_s) became pronounced and there occurred no α decrease in the center region. This result suggests the appearance of implicit effect of the SG bias, which is probably raising the bottom of the confining potential in the wire. If this is the case, α could be enhanced further as the results observed in the experiments [3,4] adopting the inverted modulation-doped samples. The results obtained here thus clearly demonstrate the SG control of SOI in the narrow wire structures and hence gives the first and promising experimental basis for studies toward spin-FET quantum computing devices.

ACKNOWLEDGMENTS

This work is partially supported by a Grant-in-Aid for Scientific Research in Priority Areas "Semiconductor Nano Spintronics" (No. 140762139) of The Ministry of Education, Culture, Sports, Science and Technology, Japan and by Mitsubishi and SCAT Foundations for Science and Technology.

REFERENCES

1. S. Datta et al, *Appl. Phys. Lett.*, **56**, 665, (1990).
2. M. Schultz et al, *Semicond. Sci. Technol.* **11**, 1168 (1996).
3. J. Nitta et al, *Phys. Rev. Lett.*, **78**, 1335 (1997).
4. T. Schäpers et al, *J. Appl. Phys.*, **83**, 4324 (1998).
5. Y. Sato et al, *J .Appl. Phys.*, **89**, 8017 (2001).
6. A. E. Popescu et al, cond-mat/0306401 v1.
7. S. Yamada, *Science and Technology of Advanced Materials*, **5**, 301 (2004).

Microphotoluminescence study of disorder in a ferromagnetic (Cd,Mn)Te quantum well

W. Maślana[1,2], P. Kossacki[1,2], P. Płochocka[1], A. Golnik[1], J.A. Gaj[1], D. Ferrand[2], M. Bertolini[2], S. Tatarenko[2], J. Cibert[2,3]

[1] *Institute of Experimental Physics, Warsaw University, 69 Hoza, 00-681 Warszawa, Poland*
[2] *"Nanophysics and semiconductors" group, Laboratoire de Spectrométrie Physique, CNRS and Université Joseph Fourier Grenoble, B.P.87, 38402 Saint Martin d'Heres Cedex, France*
[3] *Laboratoire Louis Néel, CNRS, B.P.166, 38042 Grenoble Cedex 9, France*

Abstract. We present a study of static fluctuations in p-doped (Cd,Mn)Te quantum wells, exploiting the high sensitivity of photoluminescence spectra to local magnetization and carrier density. We find that carrier density fluctuations have a much stronger influence on the magnetically ordered phase than fluctuations of the Mn concentration.

INTRODUCTION

It has been proposed theoretically [1], and shown experimentally [2], that the presence of a 2D carrier gas in (Cd,Mn)Te quantum wells (QWs) induces a ferromagnetic interaction between localized Mn spins. The most interesting configuration is that of p-type modulation doped structures. The first results have been obtained in (Cd,Mn)Te QWs [2,3], and satisfactorily described in terms of a mean field model. The critical temperature was found to be proportional to the density of free Mn spins x_{eff}, which increases with the Mn concentration x for small values of x. The spontaneous magnetization was found to increase linearly with x_{eff} and with the hole density [4]. However, measurements with a Mn content above 5% show that increasing alloy disorder results in deviations from the mean field model predictions [5].

In this paper we present a study of static fluctuations in p-doped (Cd,Mn)Te QWs, with submicron resolution. The heterostructure chosen for this study might be used as a model system for studies of magnetic properties of carrier induced ferromagnetism and gives us an insight into the influence of disorder. The excellent optical quality of the samples gives us the opportunity to use photoluminescence (PL) spectra as a probe of the local magnetization.

EXPERIMENTAL

The sample chosen for micro-PL experiments contained a single, 80Å wide, (Cd,Mn)Te QW embedded between (Cd,Mg)Te barriers (27% Mg). It was grown by Molecular Beam Epitaxy on a (001) oriented 4% (Cd,Zn)Te substrate. The QW contained 4.65% of Mn, as determined by reflectivity experiments in magnetic fields up to 4.5 T. The QW was covered by a cap layer thin enough (250 Å) that a hole gas of density about 2×10^{11} cm^{-2} be generated by acceptor surface states [5].

Micro-PL was measured with a reflection microscope lens immersed in the helium bath in the cryostat of a superconducting magnet [6]. Pumping the He vapor allowed us to control the temperature from 4.2 K down to 1.5 K. PL mapping of square areas up to 32×32 micrometer in size was accomplished by rotating a 10 mm thick plane-parallel quartz plate around two perpendicular axes. A semiconductor diode laser (680 nm) was used for excitation. The laser beam was focalized to a spot of diameter smaller than 0.5 micrometer. The PL was collected by the same scanning optical system and then analyzed by a spectrograph equipped with a CCD camera.

RESULTS AND DISCUSSION

The most straightforward way to observe the magnetic transition is an experiment at zero magnetic field. The PL line corresponding to the energy gap in the QW region shows a splitting and a shift of its spectral position when cooling down below a Curie temperature T_C [2,3]. The splitting is due to Zeeman effect caused by the spontaneous magnetization. In our previous work[7], it was shown that even in a micro-PL experiment with resolution better than 1 µm the PL is averaged over different domains. In spite of that, the zero field splitting corresponds to the local magnetization within the domains. Figure 1 shows a typical map of the zero-field PL splitting.

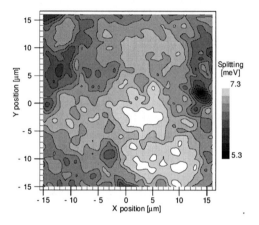

FIGURE 1. Micro-PL map of zero field splitting (in meV).

The second parameter determined in our micro-PL studies is the local density of the hole gas. We used the particular shape of the PL line in σ^+ polarization [8]: under conditions of complete spin polarization of the holes, the PL in σ^+ assumes the form of a double line. The two components of this double line are interpreted as transitions involving hole states of k vector close to zero, and close to k_F, respectively. The splitting between the two components of the double line is therefore a measure of the hole density.

The third parameter needed to characterize the ferromagnetic phase, is the local concentration of Mn ions. Its value is obtained from the giant Zeeman effect. We used the shift of the σ^+ polarized line between 1 T and 3 T as a measure of the Zeeman effect and thus of the local Mn concentration.

Thus, by measuring the micro-PL spectra in the Faraday configuration in σ^+ circular polarization, in magnetic fields of 0 T, 1 T and 3 T, we were able to determine maps of all three parameters needed for further analysis. All maps exhibit significant irregular fluctuations (see Fig. 1). The characteristic size of them is a few micrometers. We interpret these inhomogeneities as originating from fluctuations of the local magnetization, hole density and Mn content, respectively, for maps of the zero-field splitting, splitting at $B=1$ T, and Zeeman shift from 1 to 3 T.

The aim of our study was to measure the fluctuations of the spontaneous magnetization and to find their origin. We thus analyzed the correlations between the fluctuations of the local magnetization and those of the Mn content and hole density. Fig. 2 shows the correlation of the zero field splitting with the splitting of the double line in σ^+ polarization in magnetic field. A significant correlation is observed. No such correlation was observed between the zero-field splitting and the Zeeman shift (which is our measure of the Mn content).

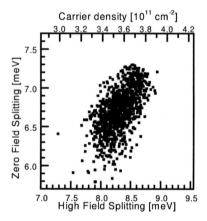

FIGURE 2. Plot of the zero field splitting as a function of the splitting measured at high field. The top scale is deduced from the measured data using the calibration from ref 8.

The main conclusion is that the dominant mechanism responsible for the fluctuations of magnetization in a p-doped (Cd,Mn)Te QW is the presence of fluctuations of the hole density.

This work was partially supported by Polish State Committee for Scientific Research (KBN) grants 2P03B 002 25 and PBZ-KBN-044/P03/2001.

REFERENCES

1. T. Dietl et al., Phys. Rev. B 55, 3347, (1997).
2. A. Haury et al., Phys. Rev. Lett. 79, 511, (1997).
3. P. Kossacki et al., Physica E 6, 709, (2000).
4. H.Boukari et al., Phys. Rev. Lett. 88 207204, (2002).
5. W. Maślana et al., Appl. Phys. Lett. 82, 1875 (2003).
6. J.A. Gaj, et al., Phys. Status Solidi (c) 1, 831 (2004).
7. P. Kossacki, et al., *Physica A*, 12(1-4):344 - 350, (2002).
8. P. Kossacki, et al., cond-mat\0404490, (2004)

Spin dynamics of Mn-ion system in diluted-magnetic-semiconductor heterostructures based on ZnMnSe

D.R.Yakovlev[1,3], M.Kneip[1], M.Bayer[1], A.A.Maksimov[2], I.I.Tartakovskii[2], A.V.Scherbakov[3], A.V.Akimov[3], D.Keller[4], W.Ossau[4], L.W.Molenkamp[4], and A.Waag[5]

[1] *Experimental Physics 2, University of Dortmund, D-44227 Dortmund, Germany*
[2] *Institute of Solid State Physics, Russian Academy of Sciences, 142432 Chernogolovka, Russia*
[3] *Ioffe Physico-Technical Institute, Russian Academy of Sciences, 194021 St. Petersburg, Russia*
[4] *Physikalisches Institut der Universität Würzburg, D-97074 Würzburg, Germany*
[5] *Institute of Semiconductor Technology, Braunschweig Technical University, D-38106 Braunschweig, Germany*

Abstract. A comprehensive study of a spin-lattice relaxation dynamics of magnetic Mn-ions in DMS quantum wells based on ZnMnSe is reported. We exploit an internal thermometer of the Mn spin temperature, which is provided by the high sensitivity of the giant Zeeman splitting of excitons (band states) to the polarization of the magnetic Mn-ions. The spin-lattice relaxation time varies by five orders of magnitude from 10^{-3}s down to 10^{-8}s, when the Mn content increases from 0.4% up to 11%. We establish an important role of free carriers in the *heating* of the Mn system, by its interaction with photoexcited carriers with excess kinetic energy, and in the *cooling* of the Mn system in the presence of cold background carriers provided by modulation doping. For $Zn_{0.89}Mn_{0.11}Se$ quantum well structures, where the spin-lattice relaxation process is considerably faster the characteristic lifetimes of nonequilibrium phonons, the phonon dynamics and it contribution to the heating of the Mn system is investigated.

INTRODUCTION

In the recent years much attention has been attracted to the fast developing new research area called "spintronics". Diluted magnetic semiconductors (DMS) hold promise for closing the gap between information storage and information processing in microelectronics. The unique possibility to tailor electronic and magnetic properties independently in these materials has triggered a variety of visionary device concepts, based on controlling the interaction of carrier spins and the spins of magnetic ions. The prerequisite of spintronics implementations is the detailed knowledge about spin-lattice relaxation (SLR) dynamics of magnetic Mn-ions. It has been shown by us that in II-VI DMS like ZnMnSe the SLR time τ_{SLR} is varied by five orders of magnitude from 10^{-3}s down to 10^{-8}s by increasing the Mn contents from 0.4% to 11% [1]. We present here a sequel of these studies.

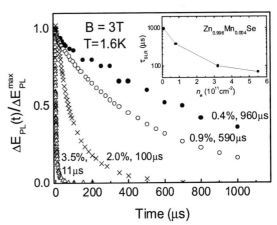

FIGURE 1. Temporal evolution of PL line spectral shift ΔE_{PL} after a laser pulse in $Zn_{1-x}Mn_xSe/Zn_{0.94}Be_{0.06}Se$ QWs with different Mn content x. Spin-lattice relaxation time τ_{SLR} decreases from 960 to 11µs with x increase from 0.4 to 3.5%. *In the inset:* dependence of τ_{SLR} on the concentration of free electrons n_e in the samples with x=0.4%.

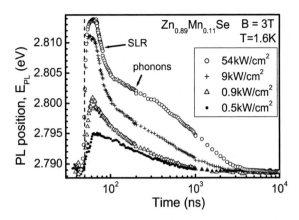

FIGURE 2. The PL line energy E_{PL} of a $Zn_{0.89}Mn_{0.11}Se/Zn_{0.94}Be_{0.06}Se$ QW in a very wide interval of times during and after laser pulse for various excitation power densities P. A dashed line indicates the maximum of laser pulse at 50ns. The fast blue shift of E_{PL} (during first ~10ns) is due to Mn heating by means of photoexciting carriers. The following in time red shift of E_{PL} shows *two stages*: the fast decay due to spin-lattice relaxation and the slower decay determined by the nonequilibrium phonons kinetics.

EXPERIMENTAL RESULTS AND DISCUSSION

We have studied $Zn_{1-x}Mn_xSe/Zn_{0.94}Be_{0.06}Se$ multiple-quantum well structures with different Mn contents from $x=0.004$ to 0.11. Details on samples and an experimental technique for optical detection of the SLR dynamics of the Mn-system excited by means of interaction with photogenerated carriers are given in Ref.[1]. Optical measurements were performed at a bath temperature $T=1.6K$ and in a magnetic field $B=3T$ applied in the Faraday geometry. A pulsed laser has a wavelength of 355nm, pulse duration of 8ns, maximum peak intensity ~1kW, and repetition rate of 3 kHz. Gated CCD with time resolution of 2 ns was used for detection the photoluminescence (PL). Additionally a *cw* HeCd laser at 325nm was used to excite a PL at delays exceeding the pulse duration. The changes in spectral position of PL line E_{PL} in magnetic field are proportional to the magnetization of Mn ions. This allows us measure and evaluate the spin-lattice relaxation times τ_{SLR} in samples with Mn content $x<3.5\%$ (Fig.1). The obtained values of τ_{SLR} follow the trend known for CdMnTe and support the idea that the relaxation process is provided by anisotropic spin interactions of magnetic ions and is rather insensitive to the host material [1]. Also the strong dependence of τ_{SLR} in $Zn_{0.996}Mn_{0.004}Se$ samples on the free carrier concentration provided by modulation doping was observed (see inset of Fig.1). In this case strong exchange interaction of Mn-spins with free electrons bypass the intrinsic SLR mechanism and accelerate the SLR dynamics [2].

Photocarriers generated by laser pulses heat directly the Mn system and generate nonequilibrium phonons which decay within ~1μs. They can be detected for samples with $x=0.11$, where τ_{SLR} are shorter than the phonon times. An example is given in Fig.2, where the initial fast decay is due to SLR and at the longer delays the Mn-system temperature follow the decay of nonequilibrium phonons. Both contributions depend on the laser excitation density (Fig.3). τ_{SLR} increases from ~20ns to ~70ns with decreasing P that could be related to strong temperature dependence of SLR [3]. It is natural that nonequilibrium phonons kinetics shows an opposite trend.

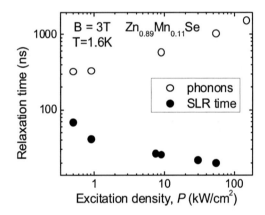

FIGURE 3. Dependences of spin-lattice and nonequilibrium phonons relaxation times on excitation density.

ACKNOWLEDGMENTS

This work has been supported by the Deutsche Forschungsgemeinschaft for the research stay of AAM in Dortmund (DFG 436 RUS 17/81/04), and by the RFBR via the grants 02-02-16873 and 04-02-16852.

REFERENCES

1. D.R. Yakovlev et al., *phys. stat. sol. (c)* **1**, 989 (2004).

2. A.V. Scherbakov et al., *Phys. Rev. B* **64**, 155205 (2001).

3. D. Scalbert, *phys. stat. sol. (b)* **193**, 189 (1996).

Charge Redistribution Spectroscopy as a Probe of Spin Phenomena in Quantum Dots

A.S.Sachrajda[1], M.Pioro-Ladrière[1,2], M.Ciorga[1], S.Studenikin[1], P.Zawadzki[1], P.Hawrylak[1], J.Lapointe[1], Z.R.Wasilewski[1] and J.A.Gupta[1]

[1] *Institute for Microstructural Science, NRC, Ottawa, Canada K1A 0R6*
[2] *Centre de Recherche sur les Propriétés électroniques de Matériaux Avancés, Université de Sherbrooke, Sherbrooke, Canada, J1K 2R1*

Abstract. Two transport techniques are described for performing charge redistribution spectroscopy at a fixed electron number. In the first method we simply apply the well established 'Cambridge' QPC charge detector technique to a dot with no leads to external reservoirs. In the second technique we use two dots in series in the regime where the electrochemical potential of only one of the dots is aligned with the external leads. In this case the transport involves co-tunneling processes through the second dot. Charge rearrangements occurring in one dot are picked up through shifts in the electrochemical potential of the other. We perform proof of concept demonstrations using well characterised transitions in the vicinity of filling factor 2 to confirm that both techniques can be used for charge redistribution spectroscopy. We discuss why such spectroscopy can provide new information about complex many-body spin phenomena in quantum dots.

INTRODUCTION

Few electron single and coupled quantum dots have been studied intensively for over a decade using transport techniques. The spectroscopy of these 'artificial atoms and molecules' has been performed using capacitance[1], Coulomb blockade[2] and spin blockade spectroscopy[3]. The first two techniques measure the addition spectrum of the quantum dot. The latter, applicable when the device is connected to spin selective leads, uses the amplitude of Coulomb blockade peaks to determine whether two neighboring ground states differ by spin +1/2 or spin -1/2. These techniques have resulted in the discovery of a variety of spin related many body effects in quantum dots which modify the addition spectrum[4,5]. However, there have been very few reports related to the presence of electron 'correlations' manifesting themselves in the properties of quantum dots weakly coupled to the reservoirs. This is in contrast to the case of dots more strongly coupled to the leads where the Kondo effect has been the subject of much interest[6].

We have confirmed in our spectroscopic measurements on few electron dots that correlation effects do indeed play important roles in specific regimes of the spectrum. Specifically we observe strong evidence for the existence of sequences of spin textures that accompany the well known spin flip events[7,8]. These spin texture states, however, often involve little overlap with the ground states for an electron droplet containing one less or more electron, leading to just an amplitude suppression in Coulomb blockade spectroscopy (nb. Coulomb blockade peaks involve *two* neighbouring electron numbers). Similarly a complex behaviour in the non-linear regime is observed whose interpretation requires a detailed knowledge of the spin properties of the excited states of both electron numbers involved. It is clear, therefore, that spectroscopy at a fixed electron number would provide a very valuable tool for studying these complex spin phenomena in more detail. In this paper we suggest two techniques to achieve this. They both make use of the charge reconfigurations that accompany spin transitions. We refer to this idea as 'charge redistribution spectroscopy' since one is directly detecting the charge redistributions associated with ground state transitions. One technique is the rather obvious adaptation of the Quantum point contact (QPC) non-invasive charge detector technique developed by Field et al., [9]. The second method involves co-tunnelling through two dots in series and has the added advantage that simultaneous spin blockade information can be derived. We now describe the two techniques in more detail and present experimental proof of concept demonstrations for both.

QPC'S AS CHARGE RECONFIGURATION DETECTORS

Non-invasive charge detection has become an important tool for transport investigations of quantum dots. A QPC is set up to be as sensitive as possible to changes in the local charge environment. When positioned in the vicinity of a quantum dot it can be used to detect the addition or subtraction of electrons to the quantum dot. Recently it was shown, in the Kondo coupling regime of a few electron quantum dot, that such a QPC charge detector could also detect changes in the charge configuration within the quantum dot[10]. In Fig. 1 we illustrate that this is also true in the weak coupling regime.

A SEM of a device similar to the one used is shown in Fig. 1. A rather large dot (0.7 micron diameter) is defined electrostatically in a high mobility 2DEG by the application of appropriate voltages to two gates. One gate is used to outline the circumference of the dot and also form one side of the QPC detector. The second narrow gate is a plunger gate used for changing the number of confined electrons in the dot. A third gate is used to define the remaining side of the QPC. In this device there exist no tunneling barriers to connect the dot to the 2DEG reservoirs. The properties of the dot are measured exclusively by means of the QPC charge detector. It was found experimentally that the absence of tunneling barriers made changing the number of confined electrons more complicated than for more standard devices. Here we just note that we are able to confine a fixed number of electrons (~100) in the dot at equilibrium with the 2DEG. To confirm whether we are able to detect spin transitions through charge reconfiguration changes we place the device between filling factor 2 and 1 by applying a suitable perpendicular magnetic field. As a function of increasing magnetic field we expect in such a large dot, to see the electrons transferred one by one from the second spin resolved Landau level at the center of the dot to the lowest spin resolved Landau level at the edge of the dot[11]. These are the well known 'spin-flip' events which are required to spin polarize the dot at high magnetic fields. They are indeed observed by our charge detector. Fig. 1 shows traces over a narrow field range illustrating a few spin-flips events taken at different sweep rates and in both sweep directions. The spin-flip is observed as a sharp dip in QPC conductance as the electron is transferred to the edge within the dot. It is followed by a gradual increase in conductance as the magnetic field gradually contracts the confined charge which triggers the next spin flip event and so on.

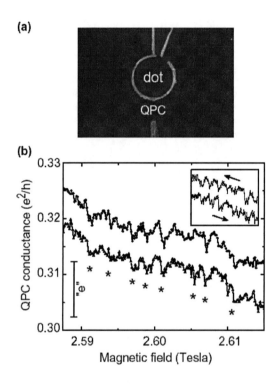

FIGURE 1. (a) A SEM of a device similar to the one used. (b) Conductance through the QPC with the dot in the spin flip regime (1<v<2). The asterisks indicate individual spin flip events, detected by the QPC due to the resulting charge redistribution. The two offset curves represent two different sweep rates 0.01T/min. (top) and 0.02T/min. (bottom). The solid line marked "e" represents the change in conductance on adding or removing an electron. The inset shows offset traces taken with sweeps in opposite directions.

It is interesting to note that in Coulomb blockade spectroscopy spin flip features have a larger width which is related to the span in magnetic field over which the spin flip occurs for the two consecutive electron numbers associated with the Coulomb blockade peak[12]. In Fig. 1 it can be seen that charge reconfiguration spectroscopy reveals a very sharp event since this spectroscopy probes one ground state transition only. It is also interesting to observe the spacing between spin flips, even in such a large dot is clearly non-uniform. The relative amplitude of the charge configuration event can be compared directly to that which occurs on adding an electron to the dot (solid line in the figure) measured in experiments using the plunger gate. Subtle differences between up and down field sweeps as well as the non-uniform spacing will require further both experimental and theoretical investigation.

QUANTUM DOTS AS CHARGE REDISTRIBUTION DETECTORS

A second technique for detecting charge redistributions is illustrated in Fig. 2. The inset shows a SEM of our device for two few electron dots in series. To achieve our spectroscopy the device is set up such that the electro-chemical potential of only one dot (aligned) is in resonance with the 2DEG leads (see schematic in Fig. 2). A weak current can still occur, however, because transport through the second dot (unaligned) can take place through higher order elastic co-tunneling processes. We note that this is the regime away from the triple points in the stability diagrams of two quantum dots in series. In a simple picture we would expect the Coulomb blockade peak position to simply measure the addition spectrum of the aligned dot.

However, we show below that charge reconfigurations associated with spin transitions of the unaligned dot are also picked up in the peak position. The aligned dot acts as a charge detector for the unaligned dot whose electron number remains constant. We have demonstrated in previous papers that the amplitude of Coulomb blockade peaks in our devices contains both spatial and spin information. The former originates from the lateral nature of the electron injection process and the latter from a spin selectivity that is a consequence of the formation of spin resolved magnetic edge states in the 2DEG. In the present configuration the amplitude modulation is still related to these two processes. We find, in fact, that the amplitude modulation can be used to identify the dominant co-tunneling process. Here, however, we focus only on the information obtainable from the peak position. Our device geometry allows us to directly compare data from the two dots in series with the individual component dots simply by applying appropriate voltages to specific gates.

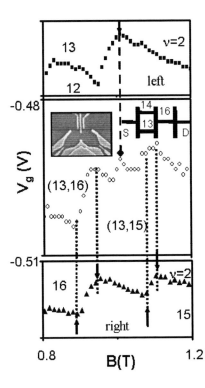

FIGURE 2. The left inset shows a SEM of our device. The right inset is a schematic of the level alignment configuration during the double dot trace. The top (bottom) curve is taken from left (top) single dot measurements and plots the 12 to 13 (15 to 16) Coulomb blockade peak position. The center panel is the peak position from the double dot results for the (13,15) to (13,16) transition. The single dot curves are aligned with the double dot traces at the transition to the ν=2 regime (see text for details) which is at a slightly different field position for the left and right dots. The black diamond marks the charge redistribution detection in the right dot of the transition in the left dot described in the text. The dotted lines reveal the direct connection between the single dot and double dot traces (see text for details.)

In Fig. 2 we plot two curves showing the peak positions versus magnetic field from single dots for peaks 15 → 16 and 12 → 13 for the right and left dots respectively (m-1 → m refers to adding the mth electron to a dot containing m-1 electrons) and (13, 15) → (13,16) for the double dots ((a, b) refers to the number of electrons in left dot, a, and right dot, b). For comparison purposes magnetic field traces for single and double dots are accurately aligned at the very marked transition to the well understood ν =2 regime [12,13]. This is achieved through small shifts to the single dot traces, ~50 mT (i.e. the 15 → 16 peak from the right dot is aligned to the double dot peak (13, 15) → (13,16) peak and the 12 → 13 peak from the left dot is aligned with the double dot peak (12,15) → (13,15) which is not shown). Based on the above arguments we expect the (13, 15) → (13, 16) peak to contains the same ground state transitions as the 15 → 16 peak. This can be seen to be the case in Fig. 2 (vertical dotted lines). However, over this field range an extra transition marked with a diamond can also be observed. A comparison with the 12 → 13 trace immediately identifies this with the transition to the ν =2 regime for 13 electrons in the left dot (vertical dashed line). At this transition, as the magnetic field is lowered, an electron is transferred from an orbital at the edge of the dot of the lowest Landau level to an orbital in the center of the dot belonging to second Landau level. This charge reconfiguration in the left dot at constant electron number is detected by the right dot through a change in the electrochemical potential and a resulting shift in the position of CB peak. It is found and will be presented in more detail elsewhere that this is a very general result i.e. that other expected charge rearrangements in the quantum dot spectrum can be detected in this way including those that involve spin transitions.

In conclusion, we have shown proof of concept demonstrations of two techniques for charge redistribution spectroscopy. It is hoped that these techniques which provide information on ground state transitions at a fixed electron number will be useful for probing correlation effects associated with spin textures in quantum dots. Experiments are currently under the way to verify this.

ACKNOWLEDGMENTS

A.S.S. S.S. and P.H. acknowledge support from CIAR. We also acknowledge valuable discussions with Guy Austing and Marek Korkusinski. M.P.-L. acknowledges financial support from NSERC and FCAR.

REFERENCES

1. Ashoori, R. C., Stormer, H. L., Weiner, J. S., Pfeiffer, L. N., Pearton, S. J., Baldwin, K. W. and West, K. W. *Physical Review Letters* **68** (1992) 3088-3091

2. Tarucha, S., Austing, D. G., Honda, T., van der Hage, R. J. and Kouwenhoven, L. P., *Physical Review Letters* **77** (1996) 3613-3616

3. Ciorga, M., Sachrajda, A. S., Hawrylak, P., Gould, C., Zawadzki, P., Jullian, S., Feng, Y. and Wasilewski, Z. *Physical Review B* **61** (2000) R16315-R16318

4. Tarucha, S., Austing, D. G., Tokura, Y., van der Wiel, W. G. and Kouwenhoven, L. P. *Physical Review Letters* **84** (2000) 2485-2488

5. Ciorga, M., Wensauer, A., Pioro-Ladriere, M., Korkusinski, M., Kyriakidis, J., Sachrajda, A. S. and Hawrylak, P. *Physical Review Letters* **88** (2002) 256804-256808

6. Goldhaber-Gordon, D., Gores, J., Kastner, M. A., Shtrikman, H., Mahalu, D. and Meirav, U. *Physical Review Letters* **81** (1998) 5225-5228

7. Gould, C., Hawrylak, P., Sachrajda, A., Feng, Y., Zawadzki, P. and Wasilewski, Z. *Physica E: Low-dimensional Systems and Nanostructures* **6** (2000) 461-465

8. Sachrajda, A. S., Korkusinski, M., Hawrylak, P., Ciorga, M., Pioro-Ladriere, M. and Zawadzki, P. *Journal of Magnetism and Magnetic Materials* **272-276** (2004) E1273-E1274

9. Field, M., Smith, C. G., Pepper, M., Ritchie, D. A., Frost, J. E. F., Jones, G. A. C. and Hasko, D. G. *Physical Review Letters* **70** (1993) 1311-1314

10. Sprinzak, D., Ji, Y., Heiblum, M., Mahalu, D. and Shtrikman, H. *Physical Review Letters* **88** (2002) 176805-9

11. McEuen, P. L., Foxman, E. B., Kinaret, J., Meirav, U., Kastner, M. A., Wingreen, N. S. and Wind, S. J. *Physical Review B (Condensed Matter)* **45** (1992) 11419-11422

12. Sachrajda A.S., Hawrylak, P.H. and Ciorga M., "Nanospintronics with Lateral Quantum Dots," in *Electron Transport in Quantum Dots,* edited by J.P.Bird, Boston, Kluwer Academic Publishers, 2003, pp. 87-123

13. Pioro-Ladriere, M., Ciorga, M., Lapointe, J., Zawadzki, P., Korkusinski, M., Hawrylak, P. and Sachrajda, A. S. *Physical Review Letters* **91** (2003) 026803-7

Bulk Inversion Asymmetry Spin-splitting in L-valley GaSb Quantum Wells

J.-M. Jancu[1], R. Scholz[2], G. C. La Rocca[1], E. A. de Andrada e Silva[3] and P. Voisin[4]

[1]*Scuola Normale Superiore and INFM, Piazza dei Cavalieri 7, I-56126 Pisa, Italy*
[2]*Institut fur Physik, Technische Universitat, D-09107 Chemnitz, Germany*
[3]*Instituto Nacional de Pesquisas Espaciais, C.P. 515, 12201-970 São José dos Campos - SP, Brazil*
[4]*Laboratoire de Photonique et de Nanostructure, CNRS, route de Nozay, F91000, Marcoussis, France*

Abstract. Very large spin-orbit, or zero-field, spin splittings are predicted for thin GaSb/AlSb symmetric quantum wells (QWs) with the absolute conduction minimum deriving from the L point of the bulk Brillouin zone. The electronic structure calculations performed with an improved tight-binding model and reproduced by a 4x4 kp effective Hamiltonian, including valley mixing and k-linear spin splittings derived from the GaSb bulk, are briefly described and specific results for [100] QWs with large splittings and valley mixing are shown which provide direct insight into L-valley III-V nanostructures.

INTRODUCTION

A promising route to spintronics is based on non-magnetic semiconductor nanostructures and spin-dependent properties originating from the spin-orbit interaction [1]. Zero-field, or spin-orbit splittings play the crucial role in phenomena like the precession of the electron spin in an applied electric field, spin resonance spectroscopy, antilocalization, and oscillatory magneto-transport [2].

Most studies however have been focused on states near the center of the Brillouin zone. Close to the Γ_{6c} conduction band minimum, the spin splitting induced by bulk inversion asymmetry is of third order in the Cartesian components of the wave vector k [3]. In heterostructures, additional k-linear contributions may originate from mesoscopic inversion asymmetry [4] and from microscopic interface asymmetry [5].

At the L point on the other hand, the spin splitting is forbidden along the [111] direction of the valley axis, while k-linear splittings are expected, on the basis of double-group symmetry considerations, for wave vectors transverse to [111]. The effective bulk Hamiltonian of the conduction band near the L minimum can in fact be written as

$$H = \frac{\hbar^2 k_l^2}{2m_l} + \frac{\hbar^2 k_t^2}{2m_t} + \alpha(\vec{k}\times\vec{\sigma})\cdot\hat{n} \quad (1)$$

where k_l and k_t are the longitudinal and transverse wave vectors with respect to the L-valley, m_l and m_t the band masses along these directions, σ the vector of the Pauli matrices, n // [111] and α a material dependent parameter for the k-linear term. Note that this term is of the Rashba form but refer to a bulk inversion asymmetry contribution as the Dresselhaus one.

In [100] GaSb/AlSb quantum wells with GaSb layers with less than 15 monolayers (~4.5 nm), the lowest confined conduction subbands derive from the GaSb L-valleys, which have $m_t = 0.087$, $m_l = 1.3$ and α = 0.84 eVA. The four-fold L-valley degeneracy of the bulk is broken whenever the valley axes are orientated differently with respect to the growth direction. On top, the broken translation symmetry along the growth direction also leads to valley mixing. For [111] QWs, for example, the longitudinal valley (along [111]) is split from the other three; while for [100] QWs there is a strong valley mixing within each of the two pairs of coupled valleys.

We have performed tight-binding (TB) calculations within an $sp^3d^5s^*$ nearest neighbor model including the spin-orbit coupling [6]. It is a 40-band model which adequately reproduces the measured band edge effective masses, interband transition energies and the cubic Dresselhaus spin-orbit splitting around Γ. The existence of large spin splittings in L-valley GaSb nanostructures is demonstrated.

RESULTS

Figure 1 shows, for example, the obtained spin-orbit split lowest L-subbands in [100] GaSb QWs with 9 and 10 monolayers. Energies are measured from the GaSb bulk L_{6c} band edge and wave-vectors from the 2D L-point. The dashed lines give the kp modeling which includes the anisotropy due to the misalignment between the growth and valley axis, treats the spin-orbit k-linear term in first-order perturbation theory and includes the valley mixing with two phenomenological parameters chosen to best fit the TB results. The details of this minimal effective Hamiltonian compatible with the QW symmetry will be published elsewhere with further results.

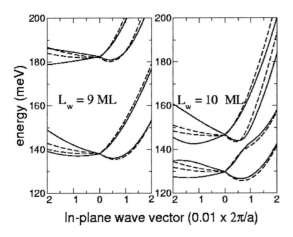

FIGURE 1. In-plane dispersion of the lowest L-conduction subbands, for $(GaSb)_9$ / $(AlSb)_{20}$ (left panel) and $(GaSb)_{10}$ / $(AlSb)_{20}$ (right panel) as a function of k along the directions [1-10] (to the right) and [110] (to the left). Solid lines give the TB calculation and dashed ones the kp modeling, a is the lattice parameter.

Besides the large splittings obtained, well exceeding 10 meV, in Fig. 1, we also note that the L-valley mixing in these QW structures is large and highly sensitive on whether it is a QW with even or odd number of monolayers. The spin splittings instead do not depend much on the well width. The deviations of the kp model seen for k along [110] represent the maximum deviation, which is limited however to a narrow range of directions. For all the other directions the agreement obtained between the TB and kp calculations is very good.

CONCLUSION

In conclusion, we have demonstrated that large spin splittings can be realized in symmetric GaSb/AlSb L-valley QWs. By a careful combination with asymmetries of the confining potential or by external electric field gating, additional freedom for the control of the spin properties of the electronic bands can be achieved. The large zero-field spin splittings of L valley heterostructures point to a new direction for future work on spin electronics.

ACKNOWLEDGMENTS

The authors thank P. Kruger for unpublished LDA calculations and F. Bassani for clarifying discussions. Financial support from FAPESP-Brazil and Scuola Normale Superiore, Italy, is also acknowledged.

REFERENCES

1. S. Datta and B. Das, Appl. Phys. Lett. **56**, 665 (1990). D. D. Awschalom, D. Loss, and N. Samarth (ed.), "Semiconductor Spintronics and Quantum Computation", Nanoscience and Technology Series, series ed. K. von Klitzing, H. Sakaki and R. Wiesendanger (Springer, Berlin 2002). E.A. de Andrada e Silva and G.C. La Rocca, Phys. Rev. B **67**, 165318 (2003).

2. B. Jusserand et al., Phys. Rev. B **51**, 4707 (1995). P.D. Dresselhaus et al., Phys. Rev. Lett. **68**, 106 (1992). J. B. Miller et al., Phys. Rev. Lett. **90**, 076807 (2003). J. Luo et al. Phys. Rev. B **41**, 7685 (1990). C.-M. Hu et al., Phys. Rev. B **60**, 7736 (1999).

3. G. Dresselhaus, Phys. Rev. **100**, 580 (1955).

4. Yu.A. Bychkov and E.I. Rashba, Sov.-Phys.- JETP Lett. **39**, 78 (1984).

5. O. Krebs and P. Voisin, Phys. Rev. Lett. **77**, 1829 (1996); L. Vervoort, R. Ferreira and P. Voisin, Semic. Sci. Techn. **14**, 227 (1999).

6. J.-M. Jancu et al.,Phys. Rev. B **57**, 6493 (1998).

Experimental separation of Rashba and Dresselhaus spin-splittings

S.D. Ganichev*, Petra Schneider*, S. Giglberger*, W. Wegscheider*, D. Weiss*, W. Prettl*, V.V. Bel'kov[†], L.E. Golub[†] and E.L. Ivchenko[†]

*University of Regensburg, 93040 Regensburg, Germany
[†]A.F. Ioffe Physico-Technical Institute, 194021 St. Petersburg, Russia

Abstract. The Rashba effect, whose experimental access is usually masked by the Dresselhaus effect, allows manipulation of spins in semiconductor spintronics. Based on the far-infrared radiation induced spin-galvanic effect, we present a unique way to separate both types of spin-orbit coupling.

INTRODUCTION

The manipulation of the spin of charge carriers in semiconductors is one of the key problems in the field of spintronics. In the paradigmatic spin transistor, e.g. proposed by Datta and Das [1], the electron spins injected from a ferromagnetic contact into a two-dimensional electron system are controllably rotated during their passage from source to drain by means of the Rashba spin-orbit coupling [2]. The coefficient α, which describes the strength of the Rashba spin-orbit coupling, and hence the degree of rotation, can be tuned by gate voltages. This coupling stems from the inversion asymmetry of the confining potential of two-dimensional electron (or hole) systems. In addition to the Rashba coupling, caused by structure inversion asymmetry (SIA), also a Dresselhaus type of coupling of strength β contributes to the spin-orbit interaction. The latter is due to bulk inversion asymmetry (BIA) including phenomenologically inseparable interface inversion asymmetry (IIA) [3]. Both, Rashba and Dresselhaus couplings result in spin-splitting of subbands in k-space (Fig. 1) and give rise to a variety of spin dependent phenomena.

However, usually it is impossible to extract the relative contributions of Rashba and Dresselhaus terms to the spin-orbit coupling. To obtain the Rashba coefficient α, the Dresselhaus contribution is normally neglected. Here we show that angular dependent measurements of the spin-galvanic photocurrent in the far-infrared [4] allow to separate contributions due to Dresselhaus and Rashba terms. We make use of the fact that these terms contribute differently for particular crystallographic directions.

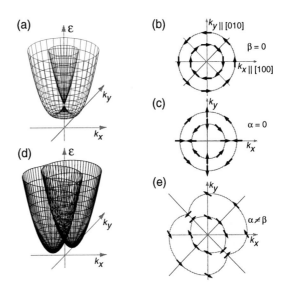

FIGURE 1. Schematic 2D band structure with k-linear terms for C_{2v} symmetry for different relative strengths of SIA and BIA and the distribution of spin orientations at the 2D Fermi energy: (a) shows the case of only Rashba or Dresselhaus spin-orbit coupling and (d) represents the case of simultaneous presence of both contributions. Arrows indicate the orientation of spins.

SPIN-ORBIT INTERACTION

We consider quantum wells (QWs) of zinc-blende structure grown in [001] direction. For the corresponding C_{2v} symmetry the spin-orbit part $\hat{H}_{SO}$ of the Hamiltonian $\hat{H} = \hbar^2 k^2/2m^* + \hat{H}_{SO}$ contains the Rashba term as well as the Dresselhaus term according to

$$\hat{H}_{SO} = \alpha(\sigma_x k_y - \sigma_y k_x) + \beta(\sigma_x k_x - \sigma_y k_y) \quad (1)$$

where k is the electron wavevector and σ is the vector of the Pauli matrices. The x-axis is aligned along the [100]-direction, y along [010], and z is the growth direction.

The resulting energy dispersion $\varepsilon(k)$ and contours of constant energy in the k_x,k_y plane for different α and β are illustrated in Fig. 1. For $\alpha \neq 0$, $\beta = 0$ and $\alpha = 0$, $\beta \neq 0$ the dispersion has the same shape and consists of two parabolas shifted in all directions (Fig. 1a). However, Rashba and Dresselhaus terms result in a different pattern of the eigenstate's spin orientation. The distribution of this spin orientation is obtained by diagonalizing $\hat{H}_{SO}$ in Eq. (1). In the presence of both Rashba and Dresselhaus spin-orbit couplings, relevant for C_{2v} symmetry, the $[1\bar{1}0]$ and the $[110]$ axes become strongly non-equivalent yielding an anisotropic dispersion sketched in Fig. 1 d,e.

EXPERIMENTAL

The experiments were carried out at room temperature on (001)-oriented n-type InAs/Al$_{0.3}$Ga$_{0.7}$Sb single QWs of 15 nm width. The free carrier densities and mobilities at room temperature were about $1.3 \cdot 10^{12}$ cm^{-2} and $\approx 2 \cdot 10^4$ cm^2/(Vs), respectively. Eight pairs of contacts on each sample allow to probe the photocurrent in different directions (see Fig. 2b). For optical spin orientation we use pulsed NH$_3$ laser at $\lambda = 148\,\mu$m.

The photocurrent j_{SGE} is measured in unbiased structures in a closed circuit configuration [5]. It is detected for right (σ_+) and left (σ_-) handed circularly polarized radiation. The spin-galvanic current j_{SGE}, studied here, is extracted after eliminating current contributions which are helicity independent: $j_{SGE} = (j_{\sigma_+} - j_{\sigma_-})/2$.

RESULTS AND DISCUSSION

We employ the spin-galvanic effect to extract the ratio of the Rashba and the Dresselhaus contributions. The spin-galvanic current is driven by the electron in-plane average spin $S_\parallel$ according to [4, 5]:

$$j_{SGE} \propto \begin{pmatrix} \beta & -\alpha \\ \alpha & -\beta \end{pmatrix} S_\parallel \qquad (2)$$

Therefore, j_{SGE} for a certain direction of $S_\parallel$ consists of Rashba and Dresselhaus coupling induced currents, j_R and j_D (see Fig. 2a). Their magnitudes are $j_R \propto \alpha |S_\parallel|$, $j_D \propto \beta |S_\parallel|$ and their ratio is $j_R/j_D = \alpha/\beta$.

The non-equilibrium in-plane spin polarization $S_\parallel$ is prepared as described recently [4]: Circularly polarized light at normal incidence on the QW plane polarizes the electrons resulting in a monopolar spin orientation in the z-direction (Fig. 2b). An in-plane magnetic field ($B = 1$ T) rotates the spin around the magnetic field axis

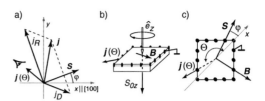

FIGURE 2. Angular dependence of the spin-galvanic current (a) and geometry of the experiment (b and c).

and results in a spin polarization $S_\parallel \propto \omega_L \tau_s$, where ω_L is the Larmor frequency and τ_s is the spin relaxation time. In the range of the applied magnetic field strength j_{SGE} in the present samples at room temperature rises linearly with B indicating $\omega_L \tau_s < 1$ and, thus, the Hanle effect is not present. The angle between the magnetic field and $S_\parallel$ in general depends on details of the spin relaxation process. In the InAs QW investigated here, the isotropic Elliott-Yafet spin relaxation mechanism dominates. Thus the in-plane spin polarization $S_\parallel$ is always perpendicular to B and can be varied by rotating B around z as illustrated in Fig. 2c. This excess spin polarization $S_\parallel$ leads to an increase of the population of the corresponding spin-polarized states. Due to asymmetric spin relaxation an electric current results [4].

To obtain the Rashba- and Dresselhaus contributions the spin-galvanic effect is measured for a fixed orientation of $S_\parallel$ for all accessible directions Θ (see Fig. 2c). According to Eq. (2) the current j_R always flows perpendicularly to the spin polarization $S_\parallel$, and j_D encloses an angle -2φ with $S_\parallel$. Here φ is the angle between $S_\parallel$ and the x-axis. Then, the current component along any direction given by angle Θ can be written as a sum of the projections of j_R and j_D on this direction

$$j_{SGE}(\Theta) = j_D \cos(\Theta + \varphi) + j_R \sin(\Theta - \varphi). \qquad (3)$$

Evaluating the measurements using this equation yields the ratio between Rashba and Dresselhaus terms. The best value obtained here is $j_R/j_D = \alpha/\beta = 2.1$ in a good agreement to theoretical results [6] which predict a dominating Rashba spin-orbit coupling for InAs QWs.

REFERENCES

1. S. Datta, and B. Das, Appl. Phys. Lett. **56**, 665 (1990).
2. Y.A. Bychkov, and E.I. Rashba, Pis'ma ZhETF **39**, 66 (1984) [Sov. JETP Lett. **39**, 78 (1984)].
3. U. Rössler, and J. Kainz, Sol. St. Com. **121**, 313 (2002).
4. S.D. Ganichev et al., Nature (London) **417**, 153 (2002).
5. S.D. Ganichev, and W. Prettl, J. Phys.: Condens. Matter **15**, R935 (2003), and references cited therein.
6. G. Lommer, F. Malcher, and U. Rössler, Phys. Rev. Lett. **60**, 728 (1988).

Spectroscopy and Characteristic Energies of a Spin-Polarized Hole Gas

H. Boukari[a], P. Kossacki[b], D. Ferrand[a], J. Cibert[c],
M. Bertolini[a], S. Tatarenko[a], J.A. Gaj[b]

(a) Groupe « Nanophysique et Semiconducteurs », CEA-CNRS-Université Joseph Fourier Grenoble, Laboratoire de Spectrométrie Physique, BP87, F-38402 Saint Martin d'Hères cedex, France.
(b) Institute of Experimental Physics, Warsaw University, Hoza 69, PL-00-681 Warszawa, Poland.
(c) Laboratoire Louis Néel, CNRS, BP166, F-38042 Grenoble cedex, France.

Abstract. The giant Zeeman splitting present in a quantum well made of a diluted magnetic semiconductor allows us not only to completely polarize the hole gas it contains, but also to destabilize the charged exciton singlet state (which emits at small spin splitting) in favor of an electron-hole state involving a majority-spin hole. At low carrier density, this is the neutral exciton, and at higher carrier density, a weakly correlated electron-hole pair. By comparing spectra recorder under different conditions, we access different characteristic energies such as the true binding energy of the charged exciton, the spin splitting necessary to fully polarize the hole gas, and the energy of excited states of the hole gas which are involved in optical transitions.

INTRODUCTION

Quantum wells (QW) incorporating a diluted magnetic semiconductor and modulation doped p-type, offer the unique opportunity to fully polarize the hole gas at very low values of the applied field, and to make the giant spin splitting larger than characteristic values of the system. We have studied a large series of $Cd_{1-x}Mn_xTe$ QWs, with x below 0.01, with a hole density up to 5×10^{11} cm^{-2}. In some samples, the carrier density was controlled optically, or through a bias in a *pin* diode [1]. Particular care was taken to assess the carrier density and the position of the states involved in the optical transitions. Our main results [2] bear on the nature of the photoluminescence (PL) when the spin splitting is large enough to destabilize singlet states, and on the dependence on carrier density of some characteristic energies.

PHOTOLUMINESCENCE SPECTRA

In all samples, at low values of the hole spin splitting, we observe [2] PL in both σ^+ and σ^- polarizations, with a continuous evolution from the lowest non-zero values of the carrier density, up to the highest one achieved in our samples (5×10^{11} cm^{-2}). The splitting between the two lines in σ^+ and σ^- is governed by the giant Zeeman splitting. When the hole gas is fully polarized, the σ^- transition takes place between the ground state and the bare charged-exciton singlet state, both in PL and in excitation (or transmission, or reflection) spectroscopy (range b in Fig.1). The σ^+ transition involves also excitations of the hole gas, which gives rise to a shift between PL and transmission, similar to the Moss-Burstein shift known for band-to-band transitions. This is also the case in σ^- polarization when the hole gas is not fully polarized (range a in Fig.1), so that a clearcut determination of the spin splitting needed to fully polarize the hole gas is obtained.

When the hole spin splitting is larger than the binding energy of the charged exciton (range c in Fig.1), a double line is observed in σ^+. The high energy component of the double line exhibits a clear phonon replica; its intensity increases and it approaches the neutral exciton when the carrier density decreases. The low-energy component is direct in k-space, it dominates the spectra and tends to the vertical band-to-band transition at large carrier density.

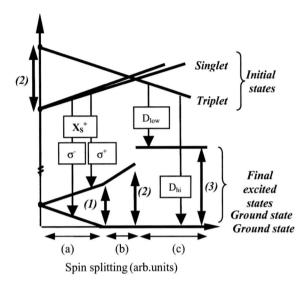

FIGURE 1. Scheme of PL transitions in a QW containing a spin-polarized hole gas. Thick lines denote the initial states (singlet state of the charged exciton and "triplet" state favored at large spin splitting) and final states (successively at increasing spin splitting: (a) hole gas with one electron at $k=0$; (b) ground state and one spin-flipped hole; (c) ground state and excited hole gas without spin flip). Single arrows mark the optical transitions (charged exciton X^+ and double line D_{hi} / D_{low}). Double arrows indicate the characteristic energies deduced from PL: (*1*) spin splitting needed to polarize the hole gas, (*2*) or to destabilize the singlet charged exciton, which also gives the binding energy of the charged exciton, (*3*) splitting of the double line.

CHARACTERISTIC ENERGIES

The analysis of the nature of the different PL lines, and of the initial and final states involved, allows us to study quantitatively the dependence on carrier density of several energies characterizing the hole gas in the presence of an interband excitation, or after it recombined. We thus measure (Fig.2):

(*1*) The spin splitting needed to fully polarize the hole gas, which demonstrates a strong enhancement of the spin susceptibility of the hole gas as expected from carrier-carrier interactions.

(*2*) The spin splitting needed to destabilize the singlet state of the charged exciton, which shows that the binding energy of the charged exciton hardly depends on the carrier density. This is in contrast with the splitting measured between the neutral and the charged exciton transitions, which is observed to increase with the carrier density in the same samples, an effect attributed to the formation of excitations of the carrier gas [3].

(*3*) The excitations of the hole gas (with and without a spin flip), created by the recombination of the charged exciton in its singlet state; they coincide with expectations for the excitation of a hole from the bottom of the spin subband, to the Fermi level.

(*4*) The energy of the excitations of the hole gas in the presence of a charged exciton, which are created by the absorption of a photon: this process, combined with the previous one (3), results in the observation of an "excitonic Moss-Burstein shift" (not shown).

(*5*) The excitations of the hole gas (without spin flip) created by the recombination of the triplet state at large spin splitting. It is only at large values of the carrier density, that the triplet state tends to a weakly correlated electron-hole state, so that the splitting of the double line tends to the Fermi energy.

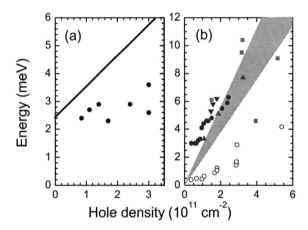

FIGURE 2. (a) Binding energy of the charged exciton singlet state (symbols), compared to the splitting between the neutral and charged exciton lines (solid line), as a function of the carrier density; (b) full symbols: splitting between the two components of the double line observed at large values of the spin splitting; open symbols: spin splitting needed to fully polarize the hole gas; grey array: Fermi energy for two extreme assumptions of the transverse heavy hole mass.

REFERENCES

1. H. Boukari, P. Kossacki, M. Bertolini, D. Ferrand, J. Cibert, S. Tatarenko, A. Wasiela, J. A. Gaj, T Dietl, Phys. Rev. Lett. **88**, 207204 (2002).
2. P. Kossacki, H. Boukari, M. Bertolini, D. Ferrand, J. Cibert, S. Tatarenko, J.A. Gaj, B.Deveaud, V.Ciulin, M. Potemski, Phys. Rev. B in press., cond-mat 0404490.
3. A.Esser, R. Zimmermann, E. Runge, Phys. Stat. Sol. B **227**, 317 (2001).

Observation of coherent hybrid intersubband-cyclotron modes in a quantum well

J. K. Wahlstrand*, J. M. Bao*†, D. M. Wang*, P. Jacobs*, R. Merlin*, K. W. West** and L. N. Pfeiffer**

*FOCUS Center and Department of Physics, The University of Michigan, Ann Arbor, MI 48109, USA
†Present address: Division of Engineering and Applied Sciences, Harvard University, Cambridge, MA 02138, USA
**Bell Laboratories, Lucent Technologies, Murray Hill, NJ 07974 USA

Abstract. We report on the coherent excitation of hybrid intersubband-cyclotron modes in a high-mobility semiconductor quantum well by ultrafast optical pulses. Pump-probe experiments were performed in the reflection geometry at 7 K in tilted magnetic fields, which mix motions in-plane and along the growth direction and lead to hybrid modes. The modes observed in the pump-probe experiments are all consistent in frequency and selection rules with spontaneous Raman scattering data on the same samples.

Raman spectroscopy is a useful tool for probing electronic excitations in semiconductor quantum wells (QW's) [1]. It compliments infrared spectroscopy because it is sensitive to excitations of different symmetry. Impulsive stimulated Raman scattering (ISRS) is an ultrafast technique in which an optical pulse drives a system impulsively, leading to coherent oscillations. These oscillations are sampled by a probe pulse which follows the initial pump pulse after a time delay τ. The coherent excitation modifies the optical constants of the system and leads to a change in the probe pulse intensity as a function of τ. By shaping the pulse, it is possible to excite one particular mode of the system preferentially; this is known as coherent control. Recently, coherent charge- and spin-density excitations generated this way were reported for a GaAs QW [2]. Here we present ISRS studies of a doped QW in the presence of a tilted magnetic field.

A high-mobility modulation-doped single QW $Al_xGa_{1-x}As$/GaAs sample was used, 400 Å thick and doped on one side; see Fig. 1. From photoluminescence spectra, we estimate that the Fermi level is 7 meV above the bottom of the conduction band, placing it between the lowest two subbands. The areal electron density is 1.6×10^{11} cm^{-2}. The spontaneous Raman spectra (Fig. 1b) taken with a cw Ti:sapphire laser at zero magnetic field, show charge-density (CD), spin-density (SD), and single particle (SP) excitations [3]. The bare intersubband transition energies are $E_{01} = 3.0$ THz and $E_{12} = 3.8$ THz; see [2] for details.

Pump-probe experiments were performed in a similar setup to that described in [2]; see Fig. 1c. A modelocked

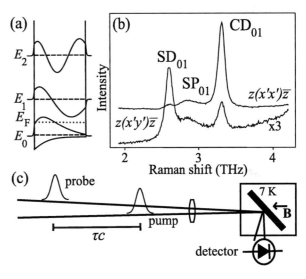

FIGURE 1. (a) QW energy diagram and lowest-energy wavefunctions. (b) Backscattering spontaneous Raman spectra at zero magnetic field in two configurations [2]. (c) Pump-probe experimental setup.

Ti:sapphire laser produced 60 fs pulses at a repetition rate of 80 MHz with central wavelength 803 nm, which overlaps with a sharp exciton resonance as observed in spontaneous Raman scattering. The pulses were linearly polarized. The samples were cooled down to 7 K in a cryostat and subjected to magnetic fields of up to 6.8 T, produced by a split-coil superconducting magnet. The magnetic field was oriented at 45° with respect to the

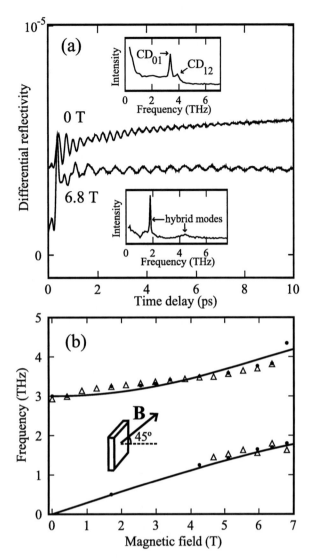

FIGURE 2. Experimental results. (a) Pump-probe data. The insets show the Fourier spectra. (b) Frequencies as a function of magnetic field. Triangles are Raman peaks, dots are peaks in the Fourier transform of pump-probe data. For clarity, a subset of the data points is shown. The solid curve is the magnetic field dependence of the hybrid modes for a parabolic well [5].

growth direction. The pump-probe data was analyzed using discrete Fourier transform and singular value decomposition techniques.[4]

Results are shown in Fig. 2. The pump-probe signal, examples of which are shown in Fig. 2a, consists of a non-oscillating background, mostly due to excited carriers, and an oscillating signal. At 0 T, the data consists of charge-density oscillations, associated with E_{01} and E_{12} [2]. At 6.8 T, the data is dominated by a lower-frequency, long-lived oscillation and a second one that is much shorter-lived and of higher frequency.

Pump-probe and Raman results match well. Our Raman results as a function of magnetic field are similar to those seen previously in doped QW's. Raman scattering by Landau levels is normally forbidden by selection rules but, in a magnetic field with a component in the quantum well plane, Landau levels couple to allowed inter-subband transitions, leading to hybrid modes which are Raman-active [6]. For a parabolic well, the dispersion of these modes can be calculated analytically [5]. Selected peaks in the Fourier transform and in spontaneous Raman spectra as a function of magnetic field are shown in Fig. 2b. The theoretical curve in Fig. 2b, calculated in the parabolic well approximation, matches our data reasonably well. In addition to the hybrid modes, other excitations were observed (not shown) in which intersubband and inter-Landau level transitions occur simultaneously [7, 8]. These excitations emerge from the single-particle excitation frequency and show avoided crossings near the CD and SD modes due to the in-plane magnetic field component.

As discussed recently [2], coherent charge-density oscillations represent movement of the charge back and forth along the growth direction in the quantum well. The tilted magnetic field in the experiments described here couples the oscillatory motion of the charge-density wave with collective cyclotron motion; the hybrid modes thus combine the two motions.

ACKNOWLEDGMENTS

This work was supported by the NSF Focus Physics Frontier Center.

REFERENCES

1. G. Abstreiter, R. Merlin, and A. Pinczuk, *IEEE J. Quantum. Electron.*, **QE-22**, 1771–1784 (1986).
2. J. M. Bao, R. Merlin, K. W. West, and L. N. Pfeiffer, *Phys. Rev. Lett.*, **92**, 236601 (2004).
3. A. Pinczuk, S. Schmitt-Rink, G. Danan, J. P. Valladares, L. N. Pfeiffer, and K. W. West, *Phys. Rev. Lett.*, **63**, 1633–1636 (1989).
4. H. Barkhuijsen, R. de Beer, W. M. M. J. Bovee, and D. van Ormondt, *Journal of Magnetic Resonance*, **61**, 465–481 (1985).
5. R. Merlin, *Solid State Commun.*, **64**, 99–101 (1987).
6. R. Borroff, R. Merlin, J. Pamulapati, P. K. Bhattacharya, and C. Tejedor, *Phys. Rev. B*, **43**, 2081–2087 (1991).
7. G. Brozak, B. V. Shanabrook, D. Gammon, and D. S. Katzer, *Phys. Rev. B*, **47**, 9981–9984 (1993).
8. V. E. Kirpichev, L. V. Kulik, I. V. Kukushkin, K. von Klitzing, K. Eberl, and W. Wegsheider, *Phys. Rev. B*, **59**, R12751–R12754 (1999).

Room temperature spin dependent current modulation in an InGaAs-based spin transistor with ferromagnetic contact

Kanji Yoh,[1,2] Marhoun Ferhat[2], Alexandru Riposan[3], Joanna M. Millunchick[3]

[1]Research Center for Integrated Quantum Electronics, Hokkaido University, Sapporo, 060-8628 Japan
[2]Japan Science and Technology Agency (JST)
[3]Materials Science and Engineering Dept., University of Michigan, Ann Arbor, MI 48109 U.S.A.

Abstract. We have observed clear current oscillations when the electrodes were magnetized along the channel current at room temperature. The drain current oscillation dependence on gate voltage agreed with the estimation. We believe that this is the first observation of spin transistor operation.

INTRODUCTION

Spin transistor structure based on semiconductor and ferromagnetic metal electrodes has been proposed [1] as a candidate for solid-state version of quantum information control device. We have been demonstrating that high efficiency spin injection is possible without tunneling barriers between ferromagnetic metal electrode and InAs[2][3][4]. We have fabricated a spin transistor structure based on strained InAlAs/InGaAs heterostructure with ferromagnetic metal (Fe) electrodes and high-indium content $In_{0.81}Ga_{0.19}As$ channel. Here we show clear current oscillations as a function of gate voltage. The experimental results are compared with simulations.

SAMPLE STRUCTURE

In order to obtain spin splitting in the channel, low band gap materials are preferred in general. Pure InAs is known to have a large Rashba parameter α, which is an indicator of strong spin-orbit interaction, as do large indium content InGaAs alloys. Experimentally, $\alpha \approx 30 \times 10^{-12}$ eV/m has been obtained[5] in $In_xGa_{1-x}As$ (x>0.75), which corresponds to a spin split energy ranging from 12 meV ($n_s=6\times10^{11}cm^{-2}$ in case of reference 11) to 22 meV ($n_s=2\times10^{12}cm^{-2}$ in the present case) at the Fermi level depending on the sheet carrier concentration. α is expressed as

$$\alpha = \Delta_F / 2k_F \qquad (1)$$

where Δ_F is the energy split of two spin bands, and k_F represents the wave number of the electrons at Fermi surface. We have chosen Fe to act as "a spin injector" and "a spin detector" for the source and the drain electrodes, because high efficient spin injection efficiency has already been verified in Fe/InAs spin LED system by optical measurements[2][3][4]. A high indium content $In_{0.81}Ga_{0.19}As$ channel was used instead of pure InAs in order to reduce the lattice mismatch on these strained layers, with InGaAs/InAlAs cladding layers nominally lattice matched to the InP(001) substrate. The schematic device structure is illustrated in Fig.1.

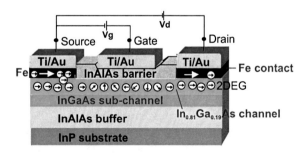

FIGURE 1. Schematic diagram of a spin transistor structure.

The heterostructure InGaAs/AlInAs was grown by MBE on InP substrate with strained $In_{0.81}Ga_{0.19}As$ channel layer inserted between InGaAs subchannel and AlInAs barrier layer. The growth of relatively thick $In_{0.81}Ga_{0.19}As$ channel layer of 80 Å was made possible by increasing critical thickness of strained $In_{0.81}Ga_{0.19}As$ channel layer by appropriate growth condition[6]. The room temperature mobility was approximately 9,350 cm2/Vs. The crystal quality bcc Fe thin film[7] was grown on $In_{0.81}Ga_{0.19}As$ channel layer in a ultra-high-vacuum with e-beam evaporator. Selectively etched source/drain region were formed down to the top surface of the 80 Å of $In_{0.81}Ga_{0.19}As$ channel layer prior to the Fe growth. The non-ferromagnetic metals were deposited on top of Fe and lifted-off.

RESULTS AND DISCUSSIONS

Decent Ohmic and pinch-off characteristics were obtained in the FET performance. The overall performance was 20 mS/mm in 2-µm gate device. When the Fe electrodes were magnetized in parallel to the current, peculiar current voltage characteristics were obtained. A clear current oscillation was observed as a function of gate voltage. After magnetization of the ferromagnetic electrodes in the "parallel condition", a steep increase in the current was observed (Fig.2) at high drain voltages. This peak is followed by oscillations in the drain current with increasing gate voltage. When the electrodes are magnetized in the "perpendicular condition", the anomalous current behavior is greatly reduced. This strongly suggests that these abnormal current modes are caused by Rashba effect, because it is most effective when electrodes are magnetized in the "parallel condition" (and the "out-of-plane condition"), but no effect is expected in "perpendicular condition". The spin dependent current modulation observed in Fig. 2 is determined by the degree of spin precession along the channel, which may be calculated with the formula

$$\Delta\theta = 2m^*\alpha L / \hbar^2 \quad (2)$$

where m* is an effective mass of an electron in the channel, α is the Rashba parameter expressed in (1), and L is the electron propagation distance. α is along the channel, so that the final change of the spin orientation by the time the injected electron reaches to the drain end of the channel must be calculated by integrating the incremental difference $\alpha(x)dx$ from source to drain.

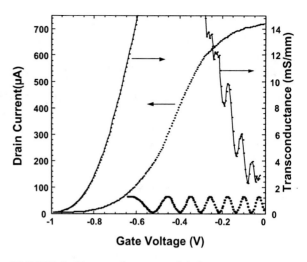

FIGURE 2. Transconductance and drain current versus gate voltage with Vds=0.6 V. Current oscillation is clearly seen. Inset curve at the bottom is the estimated current oscillation based on the separate measurement of Rashba parameter in similar heterostructures.

The boundary condition at zero gate voltage was chosen to coincide with the measured value[7] of $\alpha=30\times10^{-12}$ eV/m. The spin precession is estimated by assuming α is proportional to the Fermi wave-number and the vertical field. The result of this is that the drain current varies sinusoidally, as shown in Fig. 2 (solid black line). The period of the predicted oscillation decreases with increasing gate voltage, in reasonable agreement with the observed data. Monte Carlo simulation supports our room temperature result[8]. We believe that this is the first demonstration of spin transistor operation with the gate voltage control of spin-orbit interaction of electrons in the channel.

ACKNOWLEDGMENTS

The authors would like to acknowledge Hideo Ohno, E.Mendez, and K.Ensslin, for valuable comments.

REFERENCES

1. S.Datta and B.Das; Appl.Phys.Lett. **56** (1990) 665.
2. H.Ohno, Kanji Yoh et al, Jpn.J.Appl.Phys. Express Lett. **42** (2003) pp.L1-L3.
3. Kanji Yoh et al, Semi.Sci.Technol. **19** (2004) S386-S389
4. Kanji Yoh et al, J.Vac.Sci.& Technol. B **22** (2004) 1432
5. J.M.Millunchick,J.Crystal Growth **236** (2002) 563.
6. Hiroshi Ohno, Kanji Yoh et al, J.Vac.Sci.& Technol. B **19** (2000) 228
7. Y.Sato et al, J.Appl. Phys. **89** (2001) 8017-8021.
8. T.Tsuchiya and K.Yoh, unpublished.

Dynamic Nuclear Polarization In GaAs : Hot Electron And Spin Injection Mechanisms

M.J.R. Hoch, J. Lu, P.L. Kuhns and W.G. Moulton

National High Magnetic Field Laboratory, Florida State University, Tallahassee, FL 32310, USA

Abstract. The detailed understanding of electron-nucleus interactions and spin manipulation processes in semiconductors is of crucial importance in the development of new technology based on spin degrees of freedom. We have shown that dynamic polarization of nuclei may be induced by charge transport in just–insulating GaAs at low temperatures. The dynamic nuclear polarization (DNP) effect is highly sensitive to the current density. The results have been interpreted using a model involving the hyperfine interaction of electrons in localized donor states with nearby nuclei within a Bohr radius of these sites. A similar mechanism for DNP in polarized spin injection processes is likely and an experiment has been designed to test this suggestion.

INTRODUCTION

The development of semiconductor devices using spin degrees of freedom is currently receiving much attention. Dynamic nuclear polarization (DNP) effects in GaAs, and $Al_xGa_{1-x}As$ quantum well heterostructures due to polarized spin injection have been detected using optical methods [1,2]. Little work using conventional pulsed NMR techniques has been carried out on these systems. In the present experiments DNP effects produced by charge transport in GaAs have been investigated using bulk NMR. Previous experiments by Clark and Feher [3] on InSb have shown that DNP can be induced by charge transport. This Feher effect has been ascribed to out of equilibrium hot electrons coupling to the nuclei. A different model for the DNP effect in GaAs involving hyperfine coupling of localized electrons with nuclei is discussed.

EXPERIMENTAL

Two samples were used in the experiments. Sample I was cleaved from an n-type just-insulating GaAs substrate (001) with donor concentration $5.9 \times 10^{15} cm^{-3}$. Sample II, prepared in a similar way, was just-metallic with a donor concentration of $7 \times 10^{16} cm^{-3}$. Both platelet and bar samples were used with annealed indium ohmic contacts. The DNP measurements, using ^{71}Ga as the nuclear probe, were made using a pulsed NMR spectrometer with sample temperatures in the range 2 to 10 K and magnetic fields of up to 5T.

RESULTS AND DISCUSSION

The DNP enhancement factor, $^{71}\eta$, measuring the increase in nuclear polarization, is shown in Fig. 1(a) for sample I in a magnetic field of 1.53 T as a function of current density for three T values. Figure 1(b) plots the current density, j, versus applied electric field, E, determined during the DNP experiment. η increases in magnitude with increasing j, reaches a maximum value, corresponding to a negative enhancement, and then decreases to a roughly constant value at high j values. Comparison of Figs. 1(a) and 1(b) reveals that the peak in η occurs close to the j-value at which there is a marked change in the slope of the j -E curve. Results obtained at other temperatures in the range 2 to 10 K, and at other magnetic fields, show similar characteristics to Fig.1. For the just-metallic sample II no enhancement of the NMR signal was detected for j values in the same range as used for sample I.

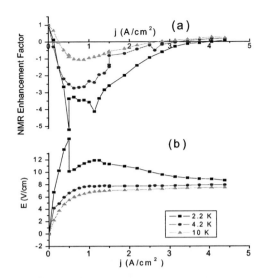

FIGURE 1. (a) NMR enhancement factor η as a function of current density j for just-insulating sample I. (b) j versus applied E field corresponding to (a).

Optically detected DNP effects in GaAs have been interpreted using a model involving itinerant electrons in the conduction band and localized electrons in donor states [4]. Exchange coupling between the two electron reservoirs results in a common spin temperature and equal electron polarizations. The hyperfine coupling of the localized electrons with surrounding nuclei is dominant in determining DNP properties. We have adapted this model to the present charge transport case and obtain the following expression for η under steady state conditions neglecting any competing nuclear spin lattice relaxation processes

$$\eta = 1 - [(\gamma_e/\gamma_I)\frac{4I(I+1)}{3S(S+1)}](\frac{\Delta T}{T_R}) \quad (1)$$

with $\Delta T = T_R - T_L \succ 0$ where T_R is the electron spin temperature and T_L is the lattice temperature. Equation (1) shows that large negative enhancements may in principle be achieved. In the present experiments η is comparatively small reaching a peak value of –5.

Applying Equation (1) to the results shown in Fig.1 (a) suggests that for sample I ΔT initially increases with j until the highly non-linear region of the j-E curve is reached. At higher current densities impact ionization effects are important with excitation of carriers into the conduction band. The hyperfine coupling between itinerant electrons and nuclei is weak [4] reducing the efficiency of polarization transfer between electron and nuclear reservoirs. Competing spin lattice relaxation processes can reduce the DNP effect when polarization times become long.

Electrical conductivity measurements on sample I show that for low j values hopping conduction is the primary transport mechanism at the temperatures of interest. Our Hall effect results are consistent with the itinerant carrier density increasing significantly with j in the highly non-linear region. This observation provides support for the proposed mechanism for the reduction of DNP effects in the highly non-linear region of the j-E curve.

The mechanism responsible for the observed DNP effects in the present charge transport experiments is applicable to other situations such as polarized spin injection in GaAs. Spin injection experiments are planned to confirm this suggestion.

CONCLUSION

DNP effects produced by charge transport have been observed in a just –insulating sample of GaAs using conventional NMR techniques at temperatures of 10 K and below. The DNP effects are small and depend sensitively on the current density. A maximum DNP enhancement is found for current densities just below the value at which highly non-linear transport behavior is observed with a steady decrease to much smaller values at higher current densities. The form of the enhancement curve is found to be almost T and B-field independent. For a just-metallic sample no DNP effects due to charge transport were found for current densities similar to those used in the just-insulating sample. The small hyperfine interaction between itinerant electrons and nuclei in GaAs is important in explaining the observations. Localized states with large hyperfine couplings to nuclei within a Bohr radius of the donor atom play an important role in the DNP process.

ACKNOWLEDGMENTS

Support from DARPA is gratefully acknowledged.

REFERENCES

1. Ohno,Y.,Young,D.K.,Beschoten,B.,Matsukura,F., Ohno,H., and Awschalom, D.D., Nature **402**,790 (1999).
2. Strand,J.,Schultz,B.D..,Isaković,A.F.,Palmstròm,C.J.,and Crowell,P.A., Phys.Rev. Lett. **91**. 036602 (2003).
3. Clark,W.G.,and Feher,G.,Phys,.Rev. Lett.**10**, 134 (1963).
4. Paget, D., Phys. Rev. B **25**, 4444 (1982).

Investigation of a GaMnN/GaN/InGaN structure for spin LED

F.V. Kyrychenko*, C.J. Stanton*, C.R. Abernathy[†], S.J. Pearton[†], F. Ren**,
G. Thaler[†], R. Frazier[†], I. Buyanova[‡], J.P. Bergman[‡] and W.M. Chen[‡]

*Department of Physics, University of Florida, P.O. Box 118440, Gainesville, FL 32611
[†]Department of Materials Science and Engineering, University of Florida, Gainesville, FL 32611
**Department of Chemical Engineering, University of Florida, Gainesville, FL 32611
[‡]Department of Physics and Measurement Technology, Linköping University, Linköping S-58183, Sweden

Abstract. Theoretical and experimental studies of GaMnN/GaN/InGaN structure for a spin LED device were performed. Strong electron spin relaxation was experimentally observed in a InGaN/GaN quantum well. It is shown that the strong spin relaxation might result from the built-in piezoelectric field in strained wurzite heterostructures. A five level $\mathbf{k}\cdot\mathbf{p}$ model was used for microscopic calculations of the structure inversion asymmetry induced spin-orbit interaction. The magnitude of this interaction is shown to be comparable with that in InGaAs/GaAs quantum structures.

GaMnN appears to be an important material for possible spintronic applications for two reasons. First, it was predicted that in GaMnN diluted magnetic semiconductor (DMS) the ferromagnetic ordering might occur at or above room temperature [1], thus making this material an ideal source of spin polarized electrons. Second, GaN has a weak spin-orbit interaction, $\Delta_{so} \approx 15$ meV. This, in combination with a wide energy gap (~ 3.5 eV) should lead to a long spin relaxation time. Indeed, recent calculations [2] showed that the spin life time in *bulk* GaN is about three orders of magnitude longer than that in *bulk* GaAs. This is an important advantage of GaN since spintronic devices usually require a long spin coherence time. Therefore, GaN appears to be a natural candidate for spintronic devices.

In this paper we report our results on creating a GaN based spin light emitting diode (LED). The schematic view of the structure is presented in Fig. 1. Spin polarized electrons are created optically within the ferromagnetic GaMnN layer, transferred through a GaN spacer to a nonmagnetic InGaN quantum well and then recombine. The degree of spin polarization was detected through the polarization of photoluminescence (PL) signal.

It should be noted that the optical detection (and excitation) of spin-polarized electrons in the nitrides is more difficult than in GaAs. Due to the small spin-orbit interaction, the optical matrix elements of the heavy hole and light hole transitions involved the same conduction electron spin state are nearly the same. Since these transitions occur in different circular polarizations, for accurate detection of electron spin it is important to distinguish the A

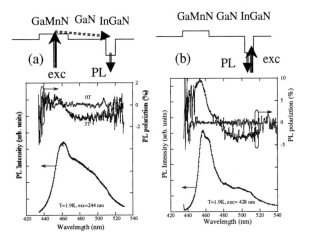

FIGURE 1. Schematic diagram of spin LED and optical spectra for (a) optical excitation within the magnetic layer and (b) optical excitation within the nonmagnetic well. In both cases no appreciable polarization was detected at zero magnetic field, while small polarization appears at 3 T.

and B exciton optical transitions. These transitions were not resolved in our experiment.

No polarization was observed without an applied magnetic field. When a magnetic field of 3 Tesla was applied, however, a small but detectable polarization signal appeared, see Fig. 1a. The magnetic field might affect the resutls in two ways. First, it induces magnetic ordering in the DMS layer, which might otherwise have no spontaneous magnetization. This would increase the spin polarization of injected electrons. And second, the magnetic field can suppress the spin relaxation mechanisms. To determine what process dominates in our sample, we measured the polarization of PL signal after exciting electrons directly within nonmagnetic quantum well, Fig. 1b. Again we obtained no polarization without the magnetic field and a small polarization in the magnetic field. We might therefore conclude that there is a strong spin relaxation mechanism in InGaN quantum well, that could be suppressed by the magnetic field.

We have performed also the series of time-resolved measurements without the magnetic field and observe no induced PL polarization in both long time (with time resolution 1 ns) and short time (with time resolution 20 ps) windows. The spin relaxation thus seems to be very fast, with the relaxation time shorter than 20 ps. However, as mentioned above, the difficulty with optical creation and detection of spin polarized electrons in InGaN might affect this estimation.

We attribute the strong spin-relaxation in GaN/InGaN heterostructures to the structure inversion asymmetry (SIA) induced spin-orbit interaction, also known as the Rashba spin-orbit interaction [3]. The Rashba term in the conduction band Hamiltonian

$$\hat{H}_{SIA} = \alpha [\sigma \times \mathbf{k}]_z \quad (1)$$

arises from the inversion asymmetry of the confining potential and leads to spin splitting of states with the same wave vector.

In symmetrical InGaAs/GaAs/InGaAs quantum wells, the confining potential has reflection symmetry and the Rashba spin-orbit interaction is therefore relatively weak. At a first glance, one might expect the same to hold for symmetrical InGaN/GaN quantum wells. The latter system, however, has a wurzite crystal structure where the biaxial strain give rise to large piezoelectric fields (up to several MeV/cm) directed along the growth axis. This piezofield breaks the reflection symmetry of confining potential (inset in Fig. 2) leading to a strong Rashba spin-orbit interaction.

To microscopically calculate the Rashba coefficient α in wurzite crystals we used a 5 level $\mathbf{k} \cdot \mathbf{p}$ approach [4]. Details of our calculations will be published elsewhere. The valence band contribution to α has the form

$$\alpha = P_{xy} P_z \frac{\partial}{\partial z} \left(\frac{\Delta_3}{(E_{lh} - \varepsilon)(E_{ch} - \varepsilon) - 2\Delta_3^2} \right), \quad (2)$$

where P_{xy} and P_z are Kane's momentum matrix elements, ε is conduction electron eigenenergy, Δ_3 is spin-orbit interaction and E_{lh} and E_{ch} are diagonal elements of $\mathbf{k} \cdot \mathbf{p}$ matrix for light and crystal field split holes that includes the strain contribution and the piezoelectric field. In evaluating the Eq. (2) one should carefully take into account jumps of E_{lh} and E_{ch} at interfaces. The contribution from the upper conduction band has the same form.

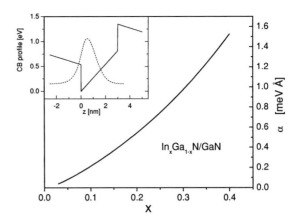

FIGURE 2. Rashba coefficient α as a function of In content in $In_xGa_{1-x}N/GaN$ strained heterostructures. The inset shows the conduction band profile and the square of the electron wave function for $In_{0.2}Ga_{0.8}N/GaN$ structure.

In Fig. 2 we plot the Rashba coefficients α in $In_xGa_{1-x}N/GaN$ heterostructures as a function of In content x. In spite of the weak spin-orbit interaction in InGaN, the Rashba coefficient in this structure can be of the same order of magnitude (about 1-2 meV A) as in GaAs based quantum structures, see e.g., Ref. [5].

The Rashba spin-orbit interaction lifts the two fold degeneracy within the conduction band. The conduction band electrons then are subject to an effective spin-relaxation through the D'yakonov-Perel' (DP) mechanism [6], which can be understood as spin precession between scatterings around the random field. The spin relaxation time for DP mechanism depends on the particular type of momentum scattering. However, regardless of the scattering mechanism, the magnetic field suppresses the DP spin relaxation by aligning the directions of the random fields, the result observed in our experiment.

ACKNOWLEDGMENTS

This work was supported by the National Science Foundation through grant DMR 032547.

REFERENCES

1. T. Dietl, H. Ohno, and F. Matsukura, Phys. Rev. B **63**, 195205 (2001)
2. S. Krishnamurthy, M. van Schilfgaarde and N. Newman, Appl. Phys. Lett. **83**, 1761 (2003)
3. Yu.A. Bychkov, E.I. Rashba, J. Phys. C **17**, 6039 (1984)
4. U. Rossler, Solid State Commun. **49**, 943 (1984); P. Pfeffer and W. Zawadzki, Phys. Rev. B **41**, 1561 (1990)
5. F.G. Pikus, G.E. Pikus, Phys. Rev. B **51**, 16928 (1995)
6. M.I. Dyakonov and V.I. Perel, Sov. Phys. JETP **33**, 1053 (1971)

Ultrafast Exciton Spin Dynamics in $Cd_{1-x}Mn_xTe$ Quantum Wells

K. Seo, K. Nishibayashi, Z.H. Chen, K. Kayanuma, A. Murayama, and Y. Oka

IMRAM, Tohoku University, Katahira 2-1-1, Sendai 980-8577, Japan

Abstract. Dynamics of spin-polarized excitons has been studied in diluted magnetic semiconductor quantum wells by femto-second pump-probe spectroscopy in magnetic fields. Transient processes of injection and spin relaxation of excitons are measured in the quantum wells. The injection time and the spin-flip time of excitons are obtained as 5 ps and 41 ps, respectively.

INTRODUCTION

Injection and transport processes of carrier spins in semiconductor quantum well structures have extensively been studied recently due to the possibility of applications for spin-electronic devices. To clarify the detailed time-dependent processes, we need to measure fast carrier dynamics in the quantum wells (QWs). Transient absorption spectroscopy by pump-and-probe technique using ultrashort laser pulses is one of the powerful tools to detect such ultrafast carrier dynamics. We have investigated the exciton spin dynamics in diluted magnetic semiconductor (DMS) QWs by femto-second pump-probe spectroscopy [1, 2]. The result resolves fast dynamics of excitons and their spins in the DMS-QWs.

RESULTS AND DISCUSSION

A sample studied is multi-QWs composed of $Cd_{1-x}Mn_xTe$ (x=0.05) wells and $Cd_{1-y}Mg_yTe$ (y=0.25) barriers. The thickness of the wells and the barriers is 4.7 and 10.2 nm, respectively. Figure 1 (a) shows the absorption and photoluminescence (PL) spectra of the QWs. The lowest levels of the heavy hole (hh) and light hole (lh) excitons in the DMS QWs locate at 1.78 and 1.82 eV. The exciton energy in the barrier layers is at 2.08 eV. The hh exciton PL peak appears at 1.76 eV, which is shifted by 20 meV from the hh exciton absorption peak energy. The Stokes shift energy is caused by the localization effect of the excitons in the DMS QWs, where the localization takes place due to the fluctuations of the well width and the composition in the DMS QWs.

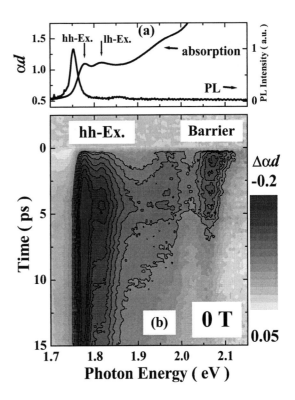

FIGURE 1(a). Absorption and PL spectra of the excitons in the $Cd_{1-x}Mn_xTe$ QWs. **(b).** Transient differential absorption spectra of the barrier and QW excitons.

Pump-probe measurement using intense optical pumping pulses (2.4 eV, 100 fs, 1 μJ/pulse) derives the transient absorption spectra around the exciton region of 1.7-2.2 eV in the QWs. Figure 1 (b) displays the transient differential absorption spectra with a contour map. At 2.08 eV, the absorption saturation for the barrier exciton arises with a fast rising time less than 1 ps and decays within 8 ps. The absorption saturation of the hh-exciton in the magnetic QWs appears at 1.78 eV with much slower decay. The lh-exciton at 1.82 eV also shows the absorption saturation in the time range of 0-12 ps.

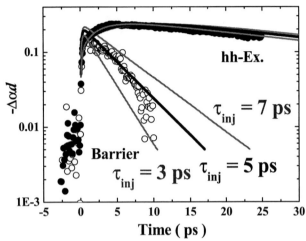

FIGURE 2. Time variation of the excess population of the barrier excitons and the hh excitons in the QWs. Solid lines are calculated results.

Figure 2 shows the time variation of the saturated absorption intensities of the barrier excitons (open circles) and the excitons in the magnetic wells (solid circles). The variation shows the transient excess population of excitons in the barriers and QWs. The barrier exciton population shows rapid rise within 1 ps and decreases with a decay time constant of 5 ps. The hh exciton population in the QWs increases with a fast rising time of 1-5 ps and decays with a decay time of 40 ps. The population of the hh excitons in the magnetic QWs takes place by the direct relaxation of optically pumped carriers in the upper energy levels of the DMS QWs and also by the injection of the carriers or excitons (generated by the pump photons) from the barriers. The transient exciton populations calculated with rate equations in the present QWs are shown by the solid and dark lines in Fig. 2. Good agreement is obtained in the calculation using the injection time of 5 ps for the barrier exciton and the lifetime of 40 ps for the exciton in the DMS QWs.

Figure 3 shows the time variation of the integrated excess population of the excitons in the barriers and the magnetic QWs at a magnetic field of 5 T. The barrier exciton population (open circles) shows decrease with the time constant of 5 ps, which is similar decay behavior as in the case at zero-magnetic field. Since spin levels of the exciton in the magnetic QWs at 5 T split by the giant Zeeman effect, the spin relaxation of excitons arises from the upper lying spin-up level to the lower lying spin-down level. The excess population of the hh exciton in the spin-up level (shown by triangles) decays with a time constant of 32 ps, while the lower lying spin-down hh exciton population (solid circles) decays with a slower time constant of 147 ps. The latter time constant almost agrees with the exciton lifetime observed in the photoluminescence.

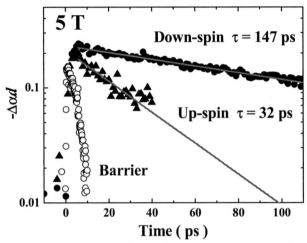

FIGURE 3. Time variation of the excess populations of the spin split excitons in the QWs and the barrier excitons.

The large difference in the decay time of the up and down exciton-spin population shows the dominant exciton-spin relaxation from the up-spin state to the down-spin state. From the difference of the decay time, the spin-flip relaxation time is calculated as 41 ps. The spin splitting of the hh exciton by the giant Zeeman effect is 80 meV at 5 T. Since the LO phonon energy is 21 meV, the fast spin relaxation time of 41 ps is determined by the multi-LO phonon relaxation.

In summary, exciton spin dynamics in the DMS QWs of $Cd_{1-x}Mn_xTe$ with $Cd_{1-y}Mg_yTe$ barriers is studied by the pump-probe transient absorption spectroscopy. The result shows the fast injection process of excitons from the barrier to the QWs with the time constant of 5 ps. In magnetic fields, the spin-flip time of hh excitons in the DMS QWs is determined as 41 ps. The obtained results for the exciton spin dynamics are useful to design efficient spin injection processes in the coupled QWs of DMSs and nonmagnetic semiconductors. This work is supported by NEDO Nanotechnology Program and the Ministry of Education, Science and Culture, Japan.

REFERENCES

1. H. Sakurai, K. Seo, Z.H. Chen, K. Kayanuma, T. Tomita, A. Murayama, and Y. Oka, *Phys. Stat. Sol. (c)* **1**, 981-984 (2004).
2. Z.H. Chen, H. Sakurai, T. Tomita, K. Kayanuma, A. Murayama, and Y. Oka, *Physica E* **21**, 1022-1026 (2004).

Control of Excitonic Motion in Modulation-Doped (Cd,Mn)Te QW by Magnetic and Electric Fields

Fumiyoshi Takano[*], Takashi Tokizaki[*], Hiro Akinaga[*], Shinji Kuroda[†], and Kôki Takita[†]

[*]*Nanotechnology Research Institute (NRI) and Research Consortium for Synthetic Nano-Function Materials Project (SYNAF), National Institute of Advanced Industrial Science and Technology (AIST), 1-1-1 Umezono, Tsukuba, Ibaraki 305-8568, Japan*
[†]*Institute of Materials Science, University of Tsukuba, 1-1-1 Tennodai, Tsukuba, Ibaraki 305-8573, Japan*

Abstract. We successfully observed the time development of an electric-field-induced drift of a *negatively-charged* exciton X^- formed in an *n*-type modulation-doped (Cd,Mn)Te quantum well. The drift of X^- was found to be promoted by magnetic fields. This could be resulting from the suppression of an exciton-magnetic-polaron formation.

INTRODUCTION

In diluted magnetic semiconductor (DMS), the *sp-d* exchange interaction originates novel magnetic phenomena, such as the giant *g*-factor, the magnetic-polaron formation and so on. On the other hand, it was demonstrated that the in-plane motion of the *negatively-charged* exciton (X^-) in a modulation-doped GaAs quantum well (QW) can be controlled by electric fields [1]. As the results of the coexistence of the *sp-d* exchange interaction and the excitonic properties in modulation-doped DMS QW, one may expect to control the excitonic motion not only by an electric field but also by a magnetic field. In this work, we demonstrate the unique behaviors reflecting the feature of DMS regarding the excitons' motion manipulated by magnetic and electric fields.

EXPERIMENTAL

Our sample was grown using a molecular beam epitaxy (MBE) method on a GaAs substrate. A $Cd_{0.95}Mn_{0.05}Te$ well layer (~100Å) was sandwiched between $Cd_{0.8}Mg_{0.2}Te$ barrier layers, and an iodine(I)-doped layer (~80Å) was separated by a spacer layer (~80Å) from the well layer. The 2D electron density and Hall mobility were estimated at $2\times10^{11} cm^{-2}$ and $5,000 cm^2 \cdot V^{-1} \cdot s^{-1}$, respectively.

Time- and spatial-resolved circular-polarized photoluminescence (PL) measurement was performed on this sample using a fluorescence microscope at 4K under magnetic fields (up to 0.5T) and/or electric fields. Laser pulses with the wavelength of 659nm and duration of 62ps were focused on the sample by a 100× microscope objective. The emitted light was collected by the same objective and detected by a charge-coupled device (CCD) detector. The spatial and time resolutions were 1μm and 300ps, respectively.

RESULTS AND DISCUSSION

Figure 1 shows the CCD images of the laser spot and the spatial-resolved PL in the respective magnetic fields. The PL spatial extent exceeded the laser spot in both the left circular polarization (LCP) and the right one (RCP). In LCP, the PL extent showed a tendency to contract with an increase in the magnetic fields, although that of RCP increased with the magnetic fields. These results are understood by considering the mobile *neutral* exciton X, the localized X^- (*spin-singlet*) by the Lorentz force and the suppression of the exciton-magnetic-polaron (EMP) formation [2]. The observed difference between LCP and RCP originates in the difficulty of X^- formation from X, which is resulting from the redistribution of conduction electrons between the two spin subbands. That is, X^- of LCP is formed at a very early time after the photoexcitation, though X^- of RCP is formed at a later time. This difference is more enhanced by the giant Zeeman splitting due to the *sp-d* exchange interaction.

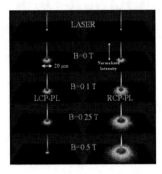

FIGURE 1. CCD images of the laser spot and the spatial-resolved PL in LCP and RCP under the respective magnetic fields at a time of 0.1ns (See Fig. 2).

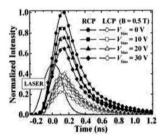

FIGURE 2. Time dependence of the spatial- and energy-integrated PL intensities at 0.5T under the respective bias voltages within an 8μm radius from the excitation center.

Figure 2 shows the time dependence of the energy- and spatial-integrated PL intensity of RCP and LCP at their respective bias voltages under a 0.5T magnetic field. The spectra for a zero bias voltage were almost polarized in RCP due to the giant Zeeman splitting. As is shown in Fig. 2, we found that the spin-polarization degree showed a consistent tendency to decrease with the increase in the bias voltage, that is, the PL intensity of RCP (LCP) decreased (increased) with the bias voltage. The hot electron effect may be suggested as a proper reason for this phenomenon. At the present stage, however, the effect has not been fully investigated yet.

We show in Fig. 3 the cross-sectional profiles of the spatial-resolved PL at the respective bias voltages. In the absence of a magnetic field, no position shift in the profile was observed under the bias voltage of V_{bias}=10V (Fig. 3a). In the 0.5T magnetic field, however, the position shift of the profile was vividly observed even for V_{bias}=10V in both LCP (Fig. 3b) and RCP (Fig. 3c). These results remind us of the effect of the EMP formation, that is, the localized excitons (X and X^-) due to the EMP formation at zero magnetic fields were released by magnetic fields. Furthermore, the time when the shift in RCP-PL profile starts became earlier with the increase in the bias voltage (Fig. 3c-3e). This result has sufficient consistency with the aforementioned model. In the case when a higher bias voltage is applied, the degree of spin-polarization is more decreased as shown in Fig. 2. As a result, the formation of RCP-X^- should become easier and begin at the earlier time.

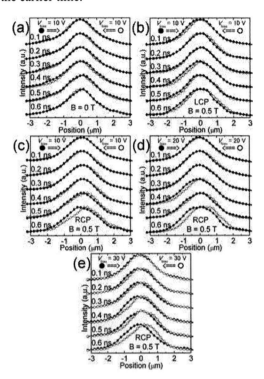

FIGURE 3. Time development of the PL profile shift under the bias voltage. The bias voltage was applied in the transverse direction of Fig. 1. In order to obtain a sufficient S/N ratio for analysis, the PL profiles within a 2μm space including the luminescence center were averaged.

ACKNOWLEDGMENTS

This work was partly supported by NEDO under the Nanotechnology Program and Grant-in-Aids for Scientific Research from the Ministry of Education, Science, Sports and Culture.

REFERENCES

1. Sanvitto, D., Pulizzi, F., Shields, A.J., Christianen, P.C.M., Holmes, S.N., Simmons, M.Y., Ritchie, D.A., Maan, J.C., and Pepper, M., Science **294**, 837 (2001).
2. Takano, F., Akinaga, H., Tokizaki, T., Kuroda, S., and Takita, K., Appl. Phys. Lett., **83**, 2853 (2003).

Intrinsic & phonon-induced spin relaxation in quantum dots

C. F. Destefani[*], Sergio E. Ulloa[*] and G. E. Marques[†]

[*]Department of Physics and Astronomy, Ohio University, Athens, Ohio 45701-2979
[†]Departamento de Física, Universidade Federal de São Carlos, 13565-905, São Carlos, São Paulo, Brazil

Abstract. Spin relaxation rates and effective g-factor of GaAs quantum dots are studied. Both piezoelectric and deformation potentials for acoustic phonons, as well as the intrinsic mixing induced by both Rashba and Dresselhaus spin-orbit couplings, are considered. Influence of different quantum dot parameters is analyzed, and good agreement with experiments is found.

One of the main causes of both spin relaxation and g-factor variation in quantum dots (QDs) is related to the spin-orbit (SO) interaction. In 2D zincblende QDs there are two possible forms of SO coupling, namely, the Rashba [1] and Dresselhaus [2] interactions; the former is due to the surface inversion asymmetry (SIA) induced by the 2D confinement, while the latter is caused by the bulk inversion asymmetry (BIA) present in those structures. The SO coupling itself is able to mix spins with different alignment in the Zeeman sublevels at finite magnetic fields, giving origin to an intrinsic (no-phonon induced) spin relaxation rate.

In this work, in addition to such intrinsic SO mixing, we study spin relaxation rates caused by acoustic phonons in the QD. We consider both piezoelectric and deformation potentials, and analyze how different parameters influence relaxation rates and the effective QD g-factor. Our focus is on GaAs in an in-plane field **B**.

The QD is defined by an in-plane parabolic confinement, $V(\rho) = m\omega_0^2\rho^2/2$, where m ($\omega_0 = E_0/\hbar$) is the electronic effective mass (confinement frequency). The perpendicular confinement $V(z)$ is strong enough so that only the state in the first quantum well subband is relevant. The QD Hamiltonian H used is the Fock-Darwin model [3] H_0 with the addition of terms that decribe SO interactions: $H_{SO} = H_{SIA} + H_{BIA}$ [4]. The SIA term, given by $H_{SIA} = \alpha\sigma \cdot \nabla V(\rho,z) \times \mathbf{k}$, can be separated as $H_{SIA} = H_R + H_{LAT}$, where $H_R = \alpha/l_0(-dV/dz)[\sigma_+L_-A_- + \sigma_-L_+A_+]$ is the Rashba term caused by the interfacial electric field $-1/edV/dz$, and $H_{LAT} = \alpha/l_0^2 E_0\sigma_z L_z$ is the contribution due to the lateral confinement E_0 [4]. The BIA term can be separated as $H_{BIA} = H_D^L + H_D^C$, where $H_D^L = -i\gamma\langle k_z^2\rangle/l_0[\sigma_+L_+A_+ - \sigma_-L_-A_-]$ and $H_D^C = i\gamma/l_0^3[\sigma_-L_+^3A_1 + \sigma_+L_-^3A_2 + \sigma_-L_-A_3 + \sigma_+L_+A_4]$ are, respectively, the linear and cubic Dresselhaus contributions [4]. The z-confinement is given by an infinite well of width z_0, so that $\varphi(z) = \sqrt{2/z_0}\sin(\pi z/z_0)$ and $\langle k_z^2\rangle = (\pi/z_0)^2$.

The phonon-induced spin relaxation rate between the two lowest Zeeman sublevels is $\Gamma_{fi} = 2\pi/\hbar \sum_{j,\mathbf{Q}}|\gamma_{fi}(\mathbf{q})|^2|Z(q_z)|^2|M_j(\mathbf{Q})|^2(n_Q+1)\delta(\Delta E + \hbar c_j Q)$, where the sum is over the emitted phonon modes j ($j = LA, TA1, TA2$) with momentum $\mathbf{Q} = (\mathbf{q}, q_z)$. Also, $Z(q_z) = \langle\varphi_z|e^{iq_z z}|\varphi_z\rangle$ ($\gamma_{fi}(\mathbf{q}) = \langle f|e^{i\mathbf{q}\cdot\mathbf{r}}|i\rangle$) is the form factor perpendicular (parallel) to the 2D-plane, while n_Q is the Bose-Einstein phonon distribution with energy $\hbar c_j Q$; energies $\Delta E = \varepsilon_f - \varepsilon_i$ and states $|i\rangle$, $|f\rangle$ are obtained after diagonalization of the full $H = H_0 + H_{SO}$, so that the SO mixing is fully taken into account and not only perturbatively. Also, $M_j(\mathbf{Q}) = \Lambda_j(\mathbf{Q}) + i\Xi_j(\mathbf{Q})$ includes both piezoelectric Λ_j and deformation Ξ_j potentials; in zincblende structures [5], they become $\Xi_{LA}(\mathbf{Q}) = \Xi_0 A_{LA}\sqrt{Q}$ (only LA is present), $\Lambda_{LA}(\mathbf{Q}) = 3/2\Lambda_0 A_{LA}\sin(2\theta)q^2q_z/Q^{7/2}$, $\Lambda_{TA1}(\mathbf{Q}) = \Lambda_0 A_{TA}\cos(2\theta)qq_z/Q^{5/2}$, and $\Lambda_{TA2}(\mathbf{Q}) = 1/2\Lambda_0 A_{TA}\sin(2\theta)q^3/Q^{7/2}(2q_z^2/q^2 - 1)$ (both TA1 and TA2 modes are shown together in the results below), where $A_j = \sqrt{\hbar(2N_0Vc_j)^{-1}}$ and $\Lambda_0 = 4\pi eh_{14}/\kappa$. The values for GaAs are: $\Xi_0 = 7$ eV and $eh_{14} = 0.14$ $eV/\text{Å}$ for the phonon coupling constants, density $N_0 = 5.32 \times 10^{-27}$ $Kg/\text{Å}^3$, velocities $c_{LA} = 4.73 \times 10^{13}$ $\text{Å}/s$ and $c_{TA} = 3.35 \times 10^{13}$ $\text{Å}/s$, dielectric constant $\kappa = 12.4$, $m = 0.067 m_0$, bulk g-factor $g_0 = -0.44$, and SO constants $\alpha = 4.4$ Å^2 and $\gamma = 26$ $eV\text{Å}^3$. When no other numbers are specified in the graphs, we consider $z_0 = 40$ Å, $dV/dz = -0.5$ $meV/\text{Å}$ and $T = 0$ K. The effective g-factor is obtained via an energy definition ($=\Delta E/\mu_B B$) or by the spin mean-value of Zeeman sublevels ($=\Delta S/2$).

In the left panel of Fig. 1 we show the phonon-induced spin relaxation rates as a function of B for different confinements (curves with symbols). It is clear that larger

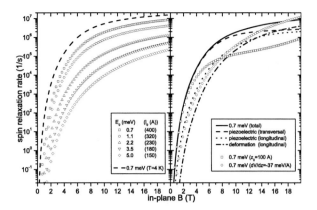

FIGURE 1. Left panel shows phonon-induced spin relaxation rates as function of in-plane magnetic field for different confining energies (curves with symbols) at $T=0$; dashed line refers to the first case at $T=4\,K$. Right panel shows the influence of the different couplings in forming the total phonon rate; also shown are the cases for larger well (squares) and stronger electric field (circles).

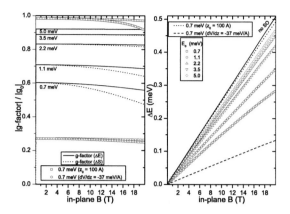

FIGURE 2. Left panel shows effective g-factor as function of in-plane field for the confinements of Fig. 1; solid (dotted) lines are obtained from Zeeman sublevel energy (spin) difference. Right panel shows the corresponding energy differences (curves with symbols) as compared to the pure $g_0\mu_B B$ value (diagonal line). Larger well ($z_0 = 100$ Å) and electric field ($dV/dz = -37$ meV/Å) are also considered in both panels.

energies or smaller QDs present smaller rates (larger relaxation times): as the confinement increases, the orbital levels become more separated and the SO coupling becomes less important. Notice the strong dependence of the phonon-induced rates with B and E_0; for example, at $B = 2\,T$, one goes from $2.5 \times 10^2\,s^{-1}$ at $E_0 = 0.7\,meV$ to $2.5 \times 10^{-1}\,s^{-1}$ at $E_0 = 5.0\,meV$, while at $B = 15\,T$ one goes from $5.2 \times 10^6\,s^{-1}$ to $6.9 \times 10^4\,s^{-1}$ at the same confinement values. Finite temperature (dashed line) enhances the phonon rates. The intrinsic rates obtained from the possible SO spin mixing are not relevant in GaAs, where the in-plane fields induce no anticrossings in the range explored. The situation is different in narrow gap materials such as InSb, as will be shown elsewhere.

The right panel of Fig. 1 shows the isolated contributions from the three kinds of phonons (curves without symbols) for the $0.7\,meV$ QD. The transverse piezoelectric coupling dominates at $B < 16\,T$; at larger fields, the deformation potential takes over; the situation is the same at larger E_0. Larger wells (squares) present lower rates because the linear BIA coupling gets smaller; larger electric fields (circles) make the Rashba coupling comparable to the BIA one [4], and produce also smaller rates.

The right panel of Fig. 2 shows how the SO coupling alters the splitting of the Zeeman sublevels at distinct confinements (curves with symbols). ΔE significantly deviates from the result without SO coupling (diagonal solid line) for smaller E_0 because then the SO influence becomes stronger. The effective g-factors for the same confinements are shown in the left panel of Fig. 2, where solid (dotted) lines refer to data extracted from the ΔE (ΔS) calculation. Both forms give the same value of g at stronger confinements, however, one sees that for QDs with 0.7 and 1.1 meV they differ at higher B, since the SO-induced spin mixing becomes stronger. Notice that g decreases significantly as E_0 decreases. Larger wells ($z_0 = 100$ Å) make both ΔE and g values essentially equal to the case without SO. Larger Rashba fields ($dV/dz = -37\,meV/Å$) produce strong deviations from the case without SO; ΔE is strongly suppressed, and so is g. However, one should notice that here, even at $0.7\,meV$, g obtained from ΔS and ΔE is basically the same; this confirms that by increasing the Rashba coupling the spin mixing is not necessarily enhanced, which is at the origin of the corresponding rate discussed in Fig. 1.

We conclude with experimental comparisons. Ref. [6] reported a lower bound of 50 μs for the relaxation time of a GaAs QD with $1.1\,meV$ at $7.5\,T$; from Fig. 1 we get $4.2\,\mu s$. They also report energy splittings, where for example $\Delta E \simeq 200\,\mu eV$ at $10\,T$; Fig. 2 yields $\Delta E \simeq 180\,\mu eV$. The experiment yields $|g| = 0.29 \pm 0.01$ at $B = 10\,T$, and Fig. 2 gives $|g| = 0.31$ ($|g| = 0.30$) from the ΔE (ΔS) data, both in excellent agreement.

We thank support from FAPESP-Brazil, NSF-NIRT, and CMSS at OU.

REFERENCES

1. Y. A. Bychkov, E. I. Rashba, J. Phys. C **17**, 6039 (1984).
2. G. Dresselhaus, Phys. Rev. **100**, 580 (1955).
3. L. Jacak, A. Wojs, P. Hawrylak, *Quantum Dots* (Springer, Berlin, 1998).
4. C. F. Destefani, S. E. Ulloa, G. E. Marques, Phys. Rev. B **69**, 125302 (2004).
5. V. F. Gantmakher, Y. B. Levinson, *Carrier Scattering in Metals and Semiconductors* (North-Holland, Amsterdam, 1987).
6. R. Hanson, B. Witkamp, L. M. K. Vandersypen, L. H. W. van Beveren, J. M. Elzerman, L. P. Kouwenhoven, Phys. Rev. Lett. **91**, 196802 (2003).

Optically Induced Zero-Field Magnetization Of CdMnTe Quantum Dots

S. Mackowski[1], T. Gurung[1], H.E. Jackson[1], L.M. Smith[1], G. Karczewski[2], and J. Kossut[2],

[1] *Department of Physics, University of Cincinnati, 45221-0011 Cincinnati OH, USA*
[2] *Institute of Physics Polish Academy of Sciences, 02-660 Warszawa, POLAND*

Abstract. We study the influence of size and/or Mn concentration of CdMnTe quantum dots on the magnitude of the optically induced magnetization at zero magnetic field. We find that the polarization of the emission decreases with increasing Mn content due to stronger anti-ferromagnetic interaction between neighboring Mn ions.

INTRODUCTION

Recently it has been demonstrated [1] that one can optically align magnetic impurities confined to CdMnTe quantum dots (QDs) by suitably polarized resonant excitation. This ferromagnetic alignment is attributed to formation of exciton magnetic polarons (EMPs) [2] along the direction determined by the polarization of photo-created excitons. Moreover, in contrast to CdMnTe quantum wells [3] and bulk samples [4], the spin polarization of CdMnTe QDs persists up to 160K. This remarkable spin stability is attributed to confinement-induced increase of the EMP binding energy.

In this work we study the dependence of this spin polarization on the QD size and/or the average Mn concentration by measuring resonantly excited photoluminescence (PL) of QD ensembles. In the case of as-grown CdMnTe QDs samples we find larger polarization for the sample with smaller average Mn concentration (~2%). This behavior is ascribed to Mn-Mn interaction, which forces anti-ferromagnetic alignment between nearest neighbors. On the other hand, we show that after rapid thermal annealing of the sample with high Mn content (~5%), the polarization of the emission increases considerably (by 10%). Since annealing has been shown to increase the average dot size and may also change the Mn content in CdMnTe QDs [5], both these processes could contribute to the observed increase of polarization.

SAMPLE AND EXPERIMENT

The samples containing CdMnTe QDs were grown by molecular beam epitaxy on GaAs substrates. Magnetic ions (Mn) were incorporated by opening the Mn shutter for several second prior to the growth of the CdTe QD layer. In this way the magnetic ions located within the exciton Bohr radius may, through the exchange interaction, strongly influence the magnetic properties of the structure. Indeed, previous results of single dot PL spectroscopy and magnetic field dependent measurements demonstrate this strong spin-spin interaction between the excitons and the Mn ions [5]. In order to increase the average QD size and/or change the Mn concentration in QDs the high Mn content sample was annealed at 420C for 15 seconds in argon atmosphere [5].

The PL measurements were carried out at T=6K with the samples mounted in a continuous-flow helium cryostat. In order to create spin-polarized excitons directly into the ground states of CdMnTe QDs we use LO phonon-assisted absorption. In this approach we suppress any possible scattering of the exciton spin that might occur during the capture time and/or energy relaxation from the ZnTe barrier. For resonant excitation a dye laser with Rhodamine 6G is used. Polarization of the excitation and emission were determined by a set of Glan-Thompson polarizers and Babinet-Soleil compensators. The emission was dispersed by a DILOR spectrometer (subtractive double mode) and detected by a liquid nitrogen cooled CCD camera.

RESULTS AND DISCUSSION

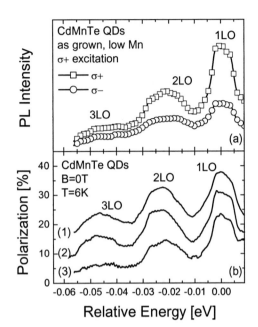

FIGURE 1. (a) Low temperature (T=6K) resonantly excited PL spectra obtained for the as-grown CdMnTe QDs with low Mn concentration. The excitation was σ+ - polarized. (b) Polarization of the PL emission obtained for all three samples studied. For the sake of comparison the energy was shifted to the energy of the 1LO replica.

Figure 1a displays the resonantly excited PL spectrum of the as grown QD sample with low Mn content obtained at B=0T. All spectra in Fig. 1 are shifted so that the energy of the first LO replicas for all samples coincide. We observe strong σ+-polarized PL when exciting with σ+-polarized excitation. This strong polarization of the emission demonstrates that by using suitably polarized laser one could at B=0T control the spin alignment of the EMPs confined to CdMnTe QDs. The transfer of exciton spin to the Mn ions in QDs is very effective since the strong QD confinement of excitons provides an instantaneous point of localization for formation of the EMPs. Moreover, this QD localization also suppresses exciton spin scattering. This allows efficient formation of polarized EMPs in these magnetic dots.

To study this effect in greater detail, we calculate the polarization of the emission at each photon energy using the formula: $P=(I^+-I^-)/(I^++I^-)$, where I^+ and I^- are the intensities of σ^+ and σ^- emissions, respectively. The resulting spectra for each of the three samples are shown in Fig. 1b. Spectra (3) and (1) correspond to the as-grown samples with high and low Mn concentration, respectively. Spectrum (2) was obtained for the sample with high Mn content after annealing. It is important to note that all of the spectra were obtained with different excitation energies, as the energy distribution of QDs is different for all three samples. However, in each case the laser energy is tuned to approximately 1LO phonon above the maximum of the non-resonant PL emission. As one can see for the as-grown samples, the overall polarization is significantly higher for QDs with lower Mn content. We attribute this effect to weaker contribution from the antiferromagnetic Mn-Mn nearest neighbor interaction in lower density dots. In addition, a substantial increase of the polarization is observed for the sample with high Mn content after rapid thermal annealing. As shown previously, annealing increases the average size of the QDs but also may induce interdiffusion of Mn out of QDs [5]. Since, presumably, both these effects would have similar qualitative impact on the Mn-Mn interaction, we cannot at present distinguish which of the two plays dominant role. We believe that experiments on single CdMnTe QDs should allow separation of these two contributions.

CONCLUSIONS

We find that the optically induced magnetization of magnetic polarons in CdMnTe QDs increases for the samples with lower Mn concentration and/or larger average QD size.

ACKNOWLEDGMENTS

We acknowledge the support of NSF Grant 0071797 (US) and PBZ-KBN-044/P03/2001 (Poland).

REFERENCES

1. Mackowski, S., Gurung, T., Nguyen, T.A., Jackson, H.E., Smith, L.M., Karczewski G., and Kossut, J., *Applied Physics Letters* **84**, 3337 (2004).

2. Mackh G., Ossau, W., Yakovlev, D.R., Waag, A., Landwehr, G., Hellmann, R., and Gobel, E.O., *Phys. Rev.* **B 49**, 10248 (1994).

3. Awschalom D.D., Warnock, J., and von Molnar, S., *Phys. Rev. Letters* **58**, 812 (1987).

4. Krenn H., Zawadzki, W., and Bauer, G., *Phys. Rev. Letters* **55**, 1510 (1985).

5. Mackowski S., Jackson, H.E., Smith, L.M., Heiss, W., Kossut, J., and Karczewski, G., *Applied Physics Letters* **83**, 3575 (2003).

Control of Electron-Spin Precession in Quantum Well Through the E Field Influence on the Interface Asymmetry

Y. G. Semenov[1], and S. M. Ryabchenko[2]

[1]*Institute of Semiconductor Physics NAS of Ukraine, Kiev, Ukraine*
[2]*Institute of Physics NAS of Ukraine, Kiev, Ukraine*

Abstract. Symmetrical [001] $A^2_{1-x}A'_xB^6$ quantum well (QW) in $A^2_{1-y}A'_yB^6$-compounds, creates the two interface C_{2V} potentials, which compensate each other. The electric field E normal to QW plane makes sum of these potentials t unequal to zero that strongly influence the optical polarization (OP) of electron-heavy hole transition and its dependence on rotation sample around normal to QW plane in a magnetic field [so called OP Anisotropy (OPA)]. We study theoretically E-field influence on OPA that allows to find the dependence $t=t(E)$ and predict the control of heavy hole transversal g-factor by an electric field.

INTRODUCTION

The linear polarization ρ of photo-luminescence (PL) in quantum wells (QWs) is sensitive to relatively weak low symmetry interactions V, which mix the light hole (LH) and heavy hole (HH) states [1-5]. One could expect due to this virtue the polarization magnitude about $\varepsilon = |V|/\Delta_{HL} \ll 1$ (Δ_{HL} is HH - LH energy splitting). In contrast with such estimation of ε a strong polarization of PL from [001]-oriented QW $Cd_{1-x}Mn_xTe$ /CdTe/ $Cd_{1-x}Mn_xTe$ and its anisotropy (i.e. dependence ρ on sample rotation about QW normal) has been observed in Refs. [2,4]. The polarization increased sharply with an increase of the in-plane magnetic field B and reaches a few tens of percents.

In this paper, we analyze the OPA caused by different low-symmetry interactions of a hole in [001] - oriented QW under the influence of electric field E normal to QW plane. These interactions are shown to reveal the various OPA dependencies on in-plane magnetic field B rotation. We analyze the C_{2v} perturbation that leads to in-plane g-factor and OPA of HH and show that this g-factor can be strongly affected by applied E-field.

THEORETICAL BACKGROUND

The 1e - 1HH PL-spectrum involves four optical transitions from two ($k=1,2$) electron spin sub-levels to two ($j=1,2$) HH ones. The electron (or HH) spin splitting $\omega = \omega_e$ (or $\omega = \omega_h$) is assumed to be described by the matrix Hamiltonian in certain basis $|n>$, n = 1, 2. The probabilities of optical transitions, $W^\alpha_{k,j} = |M^\alpha_{k,j}|^2 = |\langle \psi^k_c | \hat{V}_\alpha | \psi^j_v \rangle|^2$, between electron states ψ^k_c, $k = \pm 1$, and HH ones ψ^j_v, $j = \pm 1$ are

$$W^\alpha_{k,j} \propto \begin{cases} \sin^2(3\varphi/2 + \alpha - \Theta/2), & k = j; \\ \cos^2(3\varphi/2 + \alpha - \Theta/2), & k \neq j; \end{cases}$$

where φ is an angle between x-axis and magnetic field, α the angle of rotation of optical polarization plane, Θ the phase of hole spin ψ-function. Taking into account the sub-level of electron and HH populations with electron T_e and HH T_h spin temperatures, one can found the OPA in the form

$$\rho_\alpha = -P_{eh} \cos(3\varphi + 2\alpha - \Theta); \quad (1)$$
$$P_{eh} = \tanh(\omega_e/2T_e)\tanh(\omega_h/2T_h). \quad (2)$$

THE HH INTERACTIONS

We take into account the interactions, which mix the HH and light hole (LH) states that influence the effective HH g-factor and OPA. In terms of Pauli matrices σ_x and σ_y determined on the basic functions of HH states, they take the following form in lowest order of perturbations:

1. Zeeman interaction

$$V^{(3)}_Z = \frac{3}{4}\Delta_{HL}h^3(\sigma_x \cos 3\varphi + \sigma_y \sin 3\varphi); \quad (3)$$

2. Non-Zeeman interaction with a magnetic field

$$V_q^{(1)} = \frac{3}{4}\Delta_{HL}q_1 h(\sigma_x \cos\varphi - \sigma_y \sin\varphi); \quad (4)$$

3. Interface potential of C_{2V} symmetry

$$V_{ht}^{(2)} = -\frac{3}{2}\Delta_{HL} h t(\sigma_x \sin\varphi - \sigma_y \cos\varphi). \quad (5)$$

Here Δ_{HL} is energy splitting between lowest HH and LH confinement energies, h and t are Zeeman energy and interface potential in units of Δ_{HL}. Each of interactions [Eqs (3)-(5)] (if considered separately) results to isotropic HH spin splitting and some specific OPA. Their interference leads to non-trivial effects of HH splitting anisotropy and OPA. [5]. As an example we consider HH spin splitting and relevant OPA that corresponds to experimental situation described in the Ref. [2] for the structure $Cd_{1-x}Mn_xTe/CdTe/Cd_{1-x}Mn_xTe$. One can see, that our theory can very precisely describe the experimental data (Fig.1) if we assume t=0.001 and q_1=−0.0006.

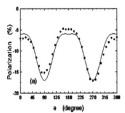
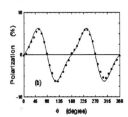

FIGURE 1. Comparison of the OPA calculated in terms of Eqs (3)-(5) (solid lines) with experimental data [2] recorded parallel (α=0°) to the magnetic field polarization plane (a) and plane with α=45° rotated relative to α=0° by 45° (b). θ=φ-45°.

One can see that dominant period of OPA is π which stems from interface anisotropy [Eq.(5)]; small contribution with $\pi/2$ period arises due to interaction (4).

Two correspondent potentials of the left (t_L) and right (t_R) interfaces contribute to coefficient t= t_L+t_R in Eq.(5):

$$t_{L,R} = \pm \frac{rx}{\Delta E_{HL}} \frac{\hbar^2}{2m_0 a} |\psi(z_{L,R})|^2. \quad (6)$$

Here r is dimensionless parameter, m_0 the electron mass, a the lattice constant, $\psi(z_{L,R})$ the envelope function at the left (right) interface. Thus, the difference of $\psi(z_L)$ and $\psi(z_R)$ results in finite value of t. Note that applied electric field in Z-direction (perpendicular to QW plane) can control the difference $D=|\psi(z_L)|^2 - |\psi(z_R)|^2$. Quantitatively, this effect can be considered in terms of variational calculations performed in the Ref.[6]. Introducing the envelope function $\psi_0(z)$ for zero electric field and corresponded difference D_0, one can find

$$D = D_0 - c_1 \frac{\Delta U m L_w^2}{\hbar^2}(|\psi_0(z_L)|^2 + |\psi_0(z_R)|^2), \quad (7)$$

where c_1=1/6−1/π^2 (see [6]), ΔU is an electric potential difference between two interface locations, m the effective mass of HHs, L_w the QW width.

The numerical value of coefficient r=0.091 for CdTe/Cd$_{1-x}$Mn$_x$Te interface was found in [3]. Using this value as well as our estimation t=0.001 and other known parameters for the structure studded in Ref. [2] for QW width L_w=60Å, we found that very small potential difference $|\Delta U|$=3.5meV mediated by the electric field can fully suppress the initial asymmetry of confinement potential and suppress OPA or doubles OPA depending on sign ΔU. Above magnitude ΔU corresponds to electric field 5.8 10^3 V/cm, which can be achieved by applying the voltage of several Volt to the structure under investigation. These evaluations demonstrate also the high electric field sensitivity of spin manipulations in valence band with respect to that in the conductivity band.

ACKNOWLEDGMENTS

This paper was partly supported by grant INTAS 03-51-5266.

REFERENCES

1. Krebs, O., and Voisin, P., *Phys. Rev. Lett.* **77**, 1829-1832 (1996).

2. Kusrayev, Yu. G. et al., *Phys. Rev. Lett.* **82**, 3176-3179 (1999).

3. A. Kudelski, A. et al., *Phys. Rev. B* **64**, 045312-1-6 (2001).

4. Ryabchenko, S. M. et al., in *Optical Properties of 2D Systems with Interacting Electrons*, Ed. by W.Ossau and R.Suris, Kluwer Acad., Dordrecht, 2003, pp. 217-222.

5. Semenov, Yu. G., and Ryabchenko, S. M., *Phys. Rev. B* **68**, 045322-1-8 (2003).

6. Bastard, A. G. et al., *Phys. Rev. B* **28**, 3241-3245 (1983).

Theory of Electronic Structure and Magnetic Interactions in (Ga,Mn)N and (Ga,Mn)As

P. Bogusławski[1,2] and J. Bernholc[2]

1. Institute of Physics, Polish Academy of Sciences, 02668 Warszawa, Poland
2. Department of Physics, North Carolina State University, Raleigh, NC 27695

Abstract. Using density functional theory we investigate the electronic structure of Mn ions and the magnetic Mn-Mn coupling in (Ga,Mn)N and (Ga,Mn)As. Our calculations show that the magnetic interactions critically depend on the position of the Fermi level and thus not only on Mn concentration but also on the presence of dopants.

INTRODUCTION

A considerable effort was recently devoted to search of semiconductors with ferromagnetism controlled by band carriers, and Curie temperatures T_C exceeding the room temperature for spintronic applications [1], with a particular focus on (Ga,Mn)N [2]. While a consensus on the electronic structure of an isolated Mn ion in GaN is emerging [3,4], a consistent picture of magnetism of (Ga,Mn)N is lacking. In particular, the observed Mn-Mn coupling is either antiferromagnetic (AF) [5,6], or ferromagnetic (FM), with T_C exceeding 300 K, and observed in n-type [7] and p-type samples [8]. Consequently, it is necessary to investigate properties of (Ga,Mn)N as a function of the Fermi energy E_F, which, as we show, affects the positions of the Mn-induced levels, the sign of the Mn-Mn coupling, etc. While we provide a consistent interpretation of many of the experimental data, our results do not explain ferromagnetism observed at high temperatures. In contrast, (Ga,Mn)As is by now rather well understood. The agreement between theory and experiment is also very good. For example, we explained the observed increase of T_C with low-temperature annealing, obtaining a *quantitative* agreement with the experimental data [9].

The calculations were performed within the Local Spin Density Approximation, using a plane wave code [10], and ultrasoft pseudopotentials. We consider the wurtzite phase of GaN, and use a large unit cell with 72 atoms. The results for (Ga,Mn)As were obtained for a unit cell with 64 atoms. The positions of all atoms in the unit cell were allowed to relax.

ENERGY LEVELS OF MN IN GAN

We find the spin-up d(Mn)-derived states in the lower half of the band gap, see Fig. 1. The d shell is split into a doublet e and a triplet t_2, separated by about 1.3 eV. With decreasing E_F, Mn ion assumes charge states ranging from Mn^{2+} to Mn^{5+} (i.e., from d^5 to d^2, respectively). The configuration of the neutral Mn is d^4. The energies of the t_2 and e states strongly depend on the charge state. For Mn^{3+}, they are 1.7 and 0.3 eV above the top of the valence band, respectively, which agrees with previous results [11]. For the Mn^{2+} ions, the d-derived bands are higher by about 0.6 eV than for Mn^{3+}. For Mn^{4+} and Mn^{5+}, the e level is a resonance degenerate with the valence band. Spin-up and spin-down states are separated by the exchange energy, which decreases from 3.3 eV for d^5 to 1.4 eV for the d^2 configuration.

For the sequence $Mn^{2+} \rightarrow Mn^{5+}$, the calculated Mn-N bond lengths are 2.01, 1.94, 1.86, and 1.79 Å, respectively, and the calculated Ga-N bond length is 1.93 Å. The result for Mn^{2+} explains the observed increase by ~4% of the bong lengths in n-type samples [12,7]. The impact of atomic relaxations on the d(Mn) levels can be neglected only for the neutral Mn^{3+}. For the d^5 (d^3) configurations the relaxation decreases (increases) the energy of t_2 by about 0.5 eV. The relaxation energies for d^5 and d^3 are ~0.5 eV, which agrees well with the experiment [4]. The calculated energy levels agree well with the available experimental data [3,4].

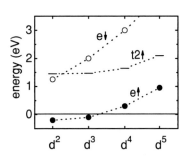

FIGURE 1. Energy levels of $e\uparrow$ and $t_2\uparrow$ and $e\downarrow$ levels relative to the top of the valence bands.

range part of the interaction, which is stronger in the latter material.

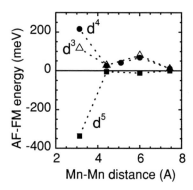

FIGURE 2. The energy difference per unit cell ΔE^{A-F} as a function of the Mn-Mn distance for the case when all Mn ions are in (i) d^5 (squares), (ii) d^4 (circles), and (iii) d^3 (triangles) configurations, and x=0.056.

MAGNETIC INTERACTIONS

Magnetic interactions are investigated by placing two Mn ions in the unit cell, and comparing the energies of the ferromagnetic (FM) and antiferromagnetic (AF) alignment of their spins. The calculated energy differences ΔE^{A-F} are given in Fig 2. The position of E_F determines the sign, the magnitude, and the range of the coupling. In intrinsic (Ga,Mn)N the interaction is FM, but its dependence on the distance is strongly non-monotonic. A similar quasi-oscillatory behavior in MnGe has been ascribed to a RKKY-like coupling [13]. When E_F is lowered by co-doping with acceptors and Mn ions are in the d^3 state, the coupling is close to that in the intrinsic case: the main change is a weakening of the NN interaction. Co-doping with donors leads to Mn ions in the d^5 state; this changes the type of interaction from FM to AF and the coupling becomes short range. This may explain the AF coupling seen in Refs. 5 and 6, but not the FM order in weakly n-type samples.

The results for (Ga,Mn)As are qualitatively similar to those for (Ga,Mn)N. In particular, the influence of p-type and n-type doping on the coupling is the same, and the strength of the NN interaction is very close in both crystals. The difference between intrinsic (Ga,Mn)N and (Ga,Mn)As mainly concerns the long-

ACKNOWLEDGMENTS

This work is supported by FENIKS project (EC:G5RD-CT-2001-00535), grant PBZ-KBN-044/P03/2001, and grants from US ONR and DoE.

REFERENCES

1. S. Koshihara *et al.*, *Phys. Rev. Lett.* **78**, 4617 (1997); Y. Ohno *et al.*, *Nature* **408**, 944 (2000).
2. T. Dietl *et al.*, *Science* **287**, 1019 (2000). Further references are given in the contribution by T. Dietl at this Conference.
3. T. Graf *et al.*, *Phys. Rev.* B **67**, 165215 (2003).
4. A. Wołoś *et al.*, *Phys. Rev.* B **69**, 115210 (2004).
5. M. Zajac *at al.*, *Appl. Phys. Lett.* **79**, 2432 (2001).
6. S. Dhar *at al.*, *Phys. Rev.* B **67**, 165205 (2003).
7. R. Giraud *et al.*, *Europhys. Lett.* **65**, 553 (2004).
8. K. H. Kim *at al. Appl. Phys. Lett.* **82**, 4755 (2003); S. V. Novikov *et al.*, *Semicond. Sci. Technology* **19**, L13 (2004).
9. K. Edmonds *et al.*, *Phys. Rev. Lett.* **92**, 37201 (2004).
10. S. Baroni *et al.*, http://www.pwscf.org.
11. P. Mahadevan and A. Zunger, *Phys. Rev.* B **69**, 115211 (2004); L. M. Sandratskii *et al.*, *ibid.* **69**, 195203 (2004), and the references therein.
12. R. Bacewicz *at al.*, *J. Phys. Chem. Solids* **64**, 1469 (2003).
13. Y. J. Zhao *at al.*, *Phys. Rev. Lett.* **90**, 47204 (2003).

Electrical properties of Ni/GaAs and Au/GaAs Schottky contacts in high magnetic fields

Holger v. Wenckstern[1,3], Rainer Pickenhain[1], Swen Weinhold[1], Michael Ziese[2], Pablo Esquinazi[2], and Marius Grundmann[1]

[1] *Abteilung Halbleiterphysik*, [2] *Abteilung Supraleitung und Magnetismus*
Universität Leipzig, Fakultät für Physik und Geowissenschaften, Institut für Experimentelle Physik II,
Linnéstrasse 5, 04103 Leipzig, Germany,
[1] *corresponding author e-mail: wenckst@physik.uni-leipzig.de*

Abstract. Rectifying metal-semiconductor contacts are possible candidates for measuring the spin asymmetry of a ferromagnetic contact. We examined this by the investigation of the Schottky barrier height of a Ni/GaAs diode and a Au/GaAs reference diode in dependence on an applied longitudinal magnetic field varied between 0 and 9 T. The measurements were carried out for temperatures ranging from 5 to 293 K. The observed behavior is similar for Ni/GaAs and Au/GaAs and reveals effects that depend stronger on the magnetic field than the effect predicted due to the spin asymmetry of Ni on GaAs.

INTRODUCTION

In recent years the realization of devices based on the spin of charge carriers attracts more and more attention. The detection of the spin as well as the determination of spin asymmetries P in ferromagnetic components play an essential role for progress in this field. A method outlined in [1,2] describes the determination of P by measuring the current across a Schottky diode (SD) made with a ferromagnetic contact metal in dependence on an applied magnetic field B. We examined this dependence for a Ni/GaAs SD and a Au/GaAs reference SD for magnetic fields ranging from 0 to 9 T and for temperatures between 5 and 293 K. For our SDs the dominant current transport process changes from tunneling to thermionic emission as the temperature exceeds 50 K, so we are able to investigate differences in the dependence on B for the different transport mechanisms.

SAMPLES

The samples were grown by metal organic vapor phase epitaxy on a n-GaAs(001) substrate with a misorientation of 2° in the (110) direction. The Si doping concentration of the substrate is 2×10^{18} cm^{-3}, that of the 2 μm thick epitaxial GaAs layer 4×10^{16} cm^{-3}. The ohmic back contacts are an eutectic mixture of Au and Ge. The Schottky contacts were realized by the ferromagnet Ni and the non-ferromagnetic Au as reference. No ferromagnetic behavior is seen in magnetization curves of the Au/GaAs SD whereas the Ni/GaAs SD shows a clear ferromagnetic response as confirmed by SQUID measurements. The Schottky diodes were characterized by current-voltage (I-V) and capacitance-voltage (C-V) measurements.

EXPERIMENTS

The I-V measurements at 293 K without applied magnetic field showed the high quality of the used diodes having an ideality factor of 1.03 for the Ni/GaAs and 1.05 for the Au/GaAs SD, respectively. The current density at a reverse bias of 0.5 V is about 5×10^{-8} Acm^{-2} for both diodes. C-V measurements confirmed the nominal doping concentration. The change of the I-V characteristic in presence of a magnetic field has been predicted by [1,2] to be

$$I = A^*T^2 \exp\left(-\frac{e\Phi_B + \Delta\Phi}{k_B T}\right)\left(e^{eU/k_B T} - 1\right), \quad (1)$$

where Φ_B is the barrier height without magnetic field, $\Delta\Phi = \mu_B PB$ is the change of the barrier height due to an applied magnetic field B, U the applied voltage, and A^* is the effective Richardson constant. We calculated $\Delta\Phi$ for the Ni/GaAs SD for $B = 9$ T to be 0.21 meV ($P = 0.4$). Equ. 1 considers only thermionic emission but it was also used to estimate relative changes of the saturation current density and with that of the barrier height for T < 50 K.

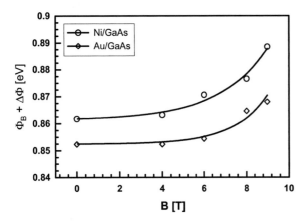

FIGURE 1. Dependence of the barrier height of Ni/GaAs and Au/GaAs SDs on the applied magnetic field at room temperature. The lines are exponential fits to the data.

C-V measurements at 293 K reveal a decrease of the capacitance of the SD's space charge region for increasing B for both Ni/GaAs and Au/GaAs. Similar behavior of the two SDs is also observed in the I-V characteristics. The saturation current density decreases with increasing magnetic field which means that $\Delta\Phi$ increases. The dependence of $\Delta\Phi$ on B is not linear as expected from (1). We found that the dependence on B is fitted best by exponential growth $a+\exp(c(B-b))$ (see Fig. 1) or by $xB+yB^2$ which is similar to the first terms of the expansion of $\exp(cB)$. At room temperature $\Delta\Phi$ amounts to 17 meV for Au/GaAs and to 27 meV for Ni/GaAs. These values are about a factor of hundred larger than the calculated value for Ni/GaAs. The behavior at 10 K is similar to that at room temperature. The, in this case, apparent barrier heights determined using Equ. (1) are with about 80 meV much smaller than those at 293 K and the ideality factors are of the order of 10. The product of the ideality factor and these apparent barrier heights yields values that are similar to Φ_B obtained at 293 K. This behavior is known and explained in, e. g., [3,4]. The dependence of the barrier heights determined at 10 K is again similar for Au/GaAs and Ni/GaAs. The fitting parameter b is similar for all measurements whereas c is about a factor of 2 smaller at higher temperatures suggesting a stronger dependence on B for low T. This is confirmed by the investigation of the forward current measured at a constant voltage in dependence on the applied magnetic field. For that we chose a voltage for which the influence of the series resistance is negligible. We observed different behavior for tunneling or thermionic emission being the dominant current transport process, respectively (Fig. 2). The relative decrease of the forward current for low B is stronger at low temperatures. Again, the Ni/GaAs and the Au/GaAs SDs behave similarly.

These findings suggest that the observed changes of the I-V characteristics in presence of a magnetic field are originating from the semiconductor part of the SDs. The influence of the spin asymmetry of a ferromagnetic contact metal described by Equ. (1) is small compared to the observed effects. In conclusion, we believe that Schottky contacts made of a ferromagnetic metal are not suitable for investigations of the spin asymmetry P of this metal. The increase of the barrier height with increasing B requires further experimental but also theoretical investigations.

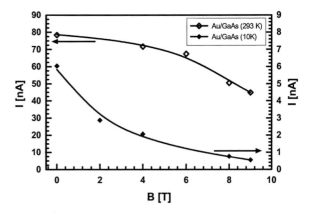

FIGURE 2. Dependence of the forward current at a constant voltage for Au/GaAs at 10 K and 293 K. The lines are guides to the eye.

REFERENCES

1. J. F. Gregg, „Introduction to Spin Electronics" in *Spin Electronics* edited by M. Ziese and M. J. Thornton, Heidelberg: Springer Verlag, 2001, p. 24
2. J. F. Gregg, I. Petej, E. Jouguelet and C. Dennis, *J. Phys. D* **35**, R121–R155 (2002).
3. Mönch, W., *Semiconductor Sufaces and Interfaces*, Berlin Heidelberg: Springer, 2001, pp. 392-399.
4. J. H. Werner, H. H. Güttler, U. Rau, *Mat. Rec. Soc. Symp. Proc.* **260**, 311-316 (1992).

Square-wave conductance through a chain of rings due to spin-orbit interaction

B. Molnár[*], P. Vasilopoulos[†] and F. M. Peeters[*]

[*]*Departement Fysica, Universiteit Antwerpen, (Campus Drie Eiken), B-2610 Antwerpen, Belgium*
[†]*Department of Physics, Concordia University, 1455 de Maisonneuve Quest, Montréal, Canada H3G 1M8*

Abstract. We study ballistic electron transport through a finite chain of quantum circular rings in the presence of spin-orbit interaction (SOI) of strength α. The transmission and reflection coefficients for a single ring, obtained analytically, lead to the conductance for a chain of rings as a function of α and of the wave vector k of the incident electron. Due to destructive spin interferences the chain can be totaly opaque for certain ranges of k the width of which depends on the value of α. A periodic modulation of α widens up the gaps considerably and produces a nearly binary conductance output.

Recently the study of spintronics devices, which utilize the spin rather than the charge of an electron, has been intensified mainly because they are expected to operate at much higher speeds than the conventional ones and have potential applications in quantum computing. One such device is a single ring in the presence of the Rashba SOI coupling [1] which results from asymmetric confinement of semiconductor nanostructures. It is important in materials with a small band gap such as In-GaAs. An important feature of electron transport through a ring is that, even in the absence of an external magnetic field, the difference in the Aharonov-Casher phases of carriers travelling clockwise and counterclockwise produces spin-sensitive interference effects.[2], [3] Here we build on this fact by studying electron transport through a chain of identical rings in the presence of SOI and the influence on the conductance of periodic modulations of α and of finite temperatures. Here are the results.

Single-ring transport. In the presence of SOI the Hamiltonian operator for a one-dimensional ring reads

$$H = -\hbar\Omega(\partial^2/\partial\varphi^2) - i\hbar\omega_{so}(\cos\varphi\sigma_x + \sin\varphi\sigma_y)(\partial/\partial\varphi) - i\hbar\omega_{so}(\cos\varphi\sigma_y - \sin\varphi\sigma_x)/2, \quad (1)$$

with $\hbar \equiv h/2\pi$, $\Omega = \hbar/2m^*a^2$, $\omega_{so} = \alpha/\hbar a$, σ_x, σ_y, and σ_z the Pauli matrices. The parameter α represents the average electric field along the z direction. For an InGaAs-based system α can be controlled by a gate voltage with values in the range $(0.5-2.0) \times 10^{-11}$eVm. [4]

The eigenvalues $E_n^{(\mu)}$ and unnormalized eigenstates $\Psi_n^{(\mu)}$ pertaining to Eq. (1) are given by [5]

$$E_n^{(\mu)} = \hbar\Omega(n - \frac{\Phi_{AC}^{(\mu)}}{2\pi})^2 - \frac{\hbar\omega_{so}^2}{4\Omega}, \quad \Psi_n^{(\mu)}(\varphi) = e^{in\varphi}\chi^{(\mu)}(\varphi) \quad (2)$$

the orthogonal spinors $\chi^{(\mu)}(\varphi)$ can be expressed in terms of the eigenvectors $[1,0]^T$, $[0,1]^T$ of the matrix σ_z as

$$\chi^{(1)} = [\cos\theta, e^{i\varphi}\sin\theta]^T, \chi^{(2)} = [\sin\theta, -e^{i\varphi}\cos\theta]^T \quad (3)$$

with $\chi^{(\mu)} \equiv \chi^{(\mu)}(\varphi)$, T denoting the transpose of the row vectors, and $\theta = \arctan[\Omega/\omega_{so} - (\Omega^2/\omega_{so}^2 + 1)^{1/2}]$. The spin-dependent term $\Phi_{AC}^{(\mu)}$ is the Aharonov-Casher phase $\Phi_{AC}^{(\mu)} = -\pi[1 + (-1)^\mu(\omega_{so}^2 + \Omega^2)^{1/2}/\Omega]$.

Reflection and transmission coefficients. The ring connected to two leads is shown in the inset to Fig. 1. At the intersections the wave function and the spin current density [5] must be continuous. This connects [3] the expansion coefficients $f^{(\mu)}, r^{(\mu)}$ of the wave function in lead I to those, $g^{(\mu)}, t^{(\mu)}$ in lead II by a 2×2 transfer matrix L^μ

$$L^\mu[r^{(\mu)}, f^{(\mu)}]^T = [g^{(\mu)}, t^{(\mu)}]^T. \quad (4)$$

The matrix L^μ depends on $T^{(\mu)}$ and $R^{(\mu)}$ where

$$T^{(\mu)} = i\cos(\Phi_{AC}^{(\mu)}/2)\sin(ka\pi)/D^{(\mu)} \quad (5)$$

$$R^{(\mu)} = [1 + 3\cos(2ka\pi) - 4\cos(\Phi_{AC}^{(\mu)})]/8D^{(\mu)} \quad (6)$$

$$D^{(\mu)} = \cos^2(\Phi_{AC}^{(\mu)}/2) - [\cos(ka\pi) - i\sin(ka\pi)/2]^2 \quad (7)$$

Assuming $f^{(\mu)} = 1$ and $g^{(\mu)} = 0$, i. e., that there is no incident electron current from the right of the ring, the transmission $t^{(\mu)} = -L_{12}^\mu L_{21}^\mu/L_{11}^\mu + L_{22}^\mu$ and reflection coefficients $r^{(\mu)} = -L_{12}^\mu/L_{11}^\mu$ are exactly $T^{(\mu)}$ and $R^{(\mu)}$, respectively, obeying the standard relation $|T^{(\mu)}|^2 + |R^{(\mu)}|^2 = 1$.

Multi-ring conductance. For a chain N rings the single-ring result can be easily generalized if the rings only touch each other, cf. Fig. 2(c1). First, one has to calculate the joint transfer matrix $\tilde{L}^\mu$

$$\begin{bmatrix} g_N^{(\mu)} \\ t_N^{(\mu)} \end{bmatrix} = \tilde{L}^\mu \begin{bmatrix} r_1^{(\mu)} \\ f_1^{(\mu)} \end{bmatrix} = L_N^\mu ... L_1^\mu \begin{bmatrix} r_1^{(\mu)} \\ f_1^{(\mu)} \end{bmatrix}, \quad (8)$$

then apply the boundary condition $g_N = 0$, that is, after the last ring there is only outgoing wave function. Then the reflection ($\tilde{R}^{(\mu)}$) and transmission ($\tilde{T}^{(\mu)}$) coefficients are written in terms of $\tilde{L}^\mu$ and the conductance G reads

$$G = \frac{e^2}{h} \sum_{\mu=1}^{2} |\tilde{T}^{(\mu)}|^2 = \frac{e^2}{h} \sum_{\mu=1}^{2} |\tilde{L}_{12}^\mu \tilde{L}_{21}^\mu / \tilde{L}_{11}^\mu - \tilde{L}_{22}^\mu|^2 \quad (9)$$

Numerical results. In Fig. 1(a) the conductance $G(ka)$, through a chain of $N = 101$ rings, is shown as a function of the incident wave vector ka for various values of α. Because $G(ka)$ is an even and periodic function of ka with period 1 we show it only for $5 \leq ka \leq 6$. In the absence of SOI G oscillates with high values but it never drops to zero. That is, a chain without SOI is *never* totally reflexive. But for finite values of α, denoted by $\alpha_n \equiv (\tilde{h}/2m^*a)[4(n+1)^2-1]^{1/2}$, n integer, the difference between the phases $\Phi_{AC}^{(1)} = (2n+1)\pi$ and $\Phi_{AC}^{(2)} = -2\pi - \Phi_{AC}^{(1)} = -(2n+3)\pi$, for electron spins travelling clockwise and counter-clockwise, respectively, renders the spin interference destructive and leads to the widest gap as each ring is non-transparent for all values of ka. With $m^* = 0.023m_0$, for InAs, and a ring radius $a = 0.25\mu m$ the smallest value α_0 that produces a total reflection is $\alpha_0 = 1.147 \times 10^{-11}$eVm. The rapid oscillations and the square profile of $G(k)$ stem from the fact the chain contains many identical rings. In general, the low transmission values for a single ring become almost zero for a chain with many rings while the values near the maximum 1 remain nearly unchanged. However, this is not true if ka is a half integer. For such a ka, G depends on the parity of N: if N is odd G equals the single ring conductance; if N is even G is a discontinuous function and vanishes only for $\alpha = \alpha_n$. Otherwise $G = 2e^2/h$.

The influence of the temperature T on G is shown in Fig. 2(b) where G is plotted as a function α. As shown, increasing T smoothens the $T = 0$ profile of the conductance but does not alter its periodic character.

Fig. 2(a) is a grey-scale contour plot tof G versus ka and α with α the same for all rings. The darkest regions correspond to ranges of ka and α for which the chain is opaque ($G = 0$) and the white ones to those for which the maximum G is $2e^2/h$ and the chain is transparent.

A similar behavior of G is shown in Fig. 2(b) if α is changed from ring to ring as shown in panel (c1). Many more gaps appear now in a "regular" manner along ka. The results are similar to those of a single waveguide with periodically varying α.[6] Relative to Fig. 2(a) we see more gaps for constant α and variable ka; they can

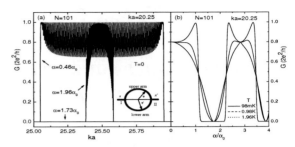

FIGURE 1. (a) Conductance vs ka for some values of α at zero temperature. The inset shows the ring and the leads. (b). Conductance vs α for different temperatures.

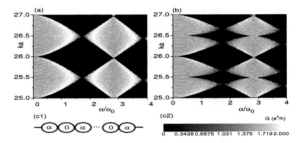

FIGURE 2. (a) Contour plot of the conductance vs ka and α). (b) As in (a) but with α varied periodically as shown in (c1). (c2) Grey color meter.

be understood with the help of Eqs. (2) and (10): upon varying α, ω_{so} and the energy levels change from ring to ring thus creating the usual superlattice barriers or wells and the resulting gaps or bands in the transmission.

The results presented here, valid for chains of strictly one-dimensional rings, can be extended to rings of finite width w for $w \ll a$, if, holds and, e.g., a square-well confinement is assumed along the radial direction. Then the radial and angular motion are decoupled and the energy levels $E_n^{(\mu)}$ are shifted by $\tilde{h}l^2/2m^*w^2$, l integer. The results presented above pertain to the $l = 1$ mode.

The full gaps in the conductance shown in Fig. 2 occur whether the incident electrons are not spin polarized, as we assumed, or spin polarized. Accordingly, the results are pertinent to the development of the spin transistor [7] where a spin-dependent and binary conductance output is necessary with as good a control as possible.

REFERENCES

1. E. I. Rashba, Sov. Phys. Solid State **2**, 1109 (1960).
2. Y. Aharonov and A. Casher, Phys. Rev. Lett. **53**, 319 (1984); S. L. Zhu and Z. D. Wang, *ibid* **85**, 1076 (2000).
3. B. Molnar, F. M. Peeters, and P. Vasilopoulos, Phys. Rev. B **69**, 155335 (2004).
4. D. Grundler, Phys. Rev. Lett. **84**, 6074 (2000).
5. J. B. Xia, Phys. Rev. B **45**, 3593 (1992).
6. X. F. Wang and P. Vasilopoulos, Appl. Phys. Lett. **82**, 940 (2003).
7. S. Datta and B. Das, Appl. Phys. Lett. **56**, 665 (1990).

Magnetic circular dichroism in ZnSe/Ga$_{1-x}$Mn$_x$As hybrid structures with Be and Si co-doping

R. Chakarvorty[1], K. J. Yee[1], X. Liu[1], P. Redlinski[1], M. Kutrowski[1,2], L.V. Titova[1], T. Wojtowicz[1,2], J.K. Furdyna[1], B. Janko, M. Dobrowolska[1]

[1]*Department of Physics, University of Notre Dame, Notre Dame, IN 46556, USA*
[2]*Institute of Physics, Polish Academy of Sciences, 02-668 Warsaw, Poland*

Abstract. Magnetic circular dichroism (MCD) experiments were performed on Ga$_{1-x}$Mn$_x$As samples with a wide range of Mn and free carrier concentrations. We studied two types of samples: a series consisting of undoped Ga$_{1-x}$Mn$_x$As with x ranging from 0.008 up to 0.06; and a series of Ga$_{1-x}$Mn$_x$As with a fixed Mn content ($x = 0.01$) and different levels of Be or Si co-doping. The observed dependence of MCD on photon energy, magnetic field, and temperature could not be described by the $k \cdot p$ model and the Moss-Burstein effect. The results suggest that a more general model involving band renormalization may be required to understand MCD in magnetic Ga$_{1-x}$Mn$_x$As alloys.

Although many aspects of ferromagnetic (FM) semiconductors, such as Ga$_{1-x}$Mn$_x$As, have been extensively studied, our understanding of magneto-optical effects in these materials is very incomplete. For example, the exchange parameter β – which is of key importance in magneto-optical effects – is still not fully understood [1-5]. In this article we present a systematic study of magnetic circular dichroism (MCD) on ZnSe/Ga$_{1-x}$Mn$_x$As hybrid structures with a wide range of Mn and carrier concentrations, in the hope of providing additional input toward the understanding of magneto-optical phenomena in this group of materials.

Two series of Ga$_{1-x}$Mn$_x$As samples were grown by low temperature molecular beam epitaxy (MBE) on semi-insulating (001) GaAs substrates followed by a 0.4 μm thick ZnSe buffer layer. One series consisted of thin (~ 0.3 μm thick) Ga$_{0.99}$Mn$_{0.01}$As films with different levels of Be or Si co-doping. The second series consisted of undoped Ga$_{1-x}$Mn$_x$As thin films, 1.0 to 2.0 μm thick, with x ranging between 0.008 and 0.06. Carrier concentration of the samples was estimated from room temperature Hall measurements.

To allow transmission measurements, the GaAs substrate was removed by mechanical polishing and chemical etching. Here the ZnSe layer plays a double role: it acts as an excellent etch stop; and it allows transmission experiments up to 2.8 eV. The dependence of MCD on photon energy, magnetic field and temperature was measured on all samples. A white light source and a monochromator were used, together with a photo-elastic modulator which switches between left and right circular polarizations. Magnetic field was applied parallel to the direction of propagation (i.e., perpendicular to the sample surface). The MCD is given by:

$$\text{MCD} = \frac{I(\sigma+) - I(\sigma-)}{I(\sigma+) + I(\sigma-)}, \quad (1)$$

where $I(\sigma+)$ and $I(\sigma-)$ are, respectively, transmission intensities of right- and left-circularly polarized light. Here one should note that for transitions at the center of the Brillouin zone ($k = 0$), the sign of MCD will be determined directly by the sign of β. When MCD is associated with transitions away from the zone center, however, the relation between MCD and β is far from understood [6]. In particular, curvatures of the participating bands may then have to be taken into account in describing MCD.

With the exception of Ga$_{1-x}$Mn$_x$As with $x = 0.008$ and of samples co-doped with Si, all of our samples were ferromagnetic (FM). As expected, the Curie temperature T_C and the magnetization M of the FM samples increased with increasing x. However, in the sample series with different doping we observe a non-monotonic dependence of M on the hole concentration p: M first increases with p, but begins to drop for higher doping levels. The likely reason for this is that increasing Be doping results in an increased concentration of Mn interstitials [7].

Figure 1 shows the MCD signal as a function of photon energy for a Be co-doped sample with $p = 3.4 \times 10^{19}$ cm^{-3}. One can see a broad MCD spectrum, peaking at ~1.76 eV. We observe very similar spectra for *all* our FM samples, with maxima at the same energy, independent of the carrier concentration. Very similar MCD spectra for $Ga_{1-x}Mn_xAs$ have been reported by Beschoten *et al.* [3], who inferred from their data that $\beta < 0$ at top of the valence band ($k = 0$).

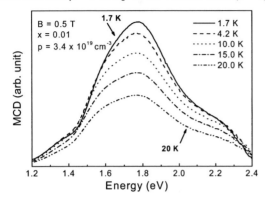

FIGURE 1. Energy dependent MCD of the *p*-type $ZnSe/Ga_{0.99}Mn_{0.01}As$:Si sample at different temperatures.

Figure 2 shows MCD spectra obtained on the Si-doped n-type sample (which is paramagnetic) for several values of magnetic field. Here we observe a *sign change* of the MCD signal at the energy 1.57 eV. This feature is also observed in *p*-type sample with low Mn concentration ($x = 0.008$). Surprisingly, the positions of the zero (at 1.57 eV) and of the negative and positive peaks do not depend on temperature, magnetic field, or carrier concentration.

We attempted to interpret these data using the $k \cdot p$ 8-band model, including disorder and the Moss-Burstein effect, as proposed by Szczytko *et al.* [4]. As seen in Fig. 3, this model clearly predicts that the positions of the crossing point of the MCD and of its

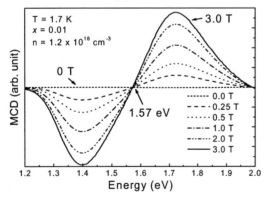

FIGURE 2. Energy dependent MCD of the n-type $ZnSe/Ga_{0.99}Mn_{0.01}As$:Si sample at different magnetic fields.

two extrema depend strongly on the magnetization, in stark disagreement with the experimental observation shown in Fig. 2, as well as with similar behavior seen in the p-type paramagnetic sample with $x = 0.008$ shown in Fig. 3. The similarity of the MCD data for both n- and p-type paramagnetic samples (Figs. 2 and 3) provides a strong indication that – in addition to the Moss-Burstein effect – some other effect must be responsible for the observed energy dependence. Furthermore, the fact that MCD data observed in all our FM samples (i.e., materials with widely varying values of *M* and *p*) show essentially the same behavior, characterized by a peak around 1.76 eV that does not vary with temperature or magnetic field (see Fig. 1) also cannot be explained exclusively by Moss-Burstein effect.

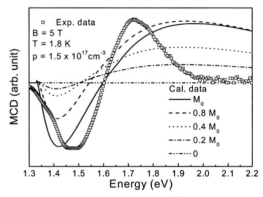

FIGURE 3. Comparison between experimental results and the $k \cdot p$ calculations for the *p*-type $ZnSe/Ga_{0.992}Mn_{0.008}As$ sample.

Since *all* our attempts to describe the data by the $k \cdot p$ model have failed, we believe that it is necessary to go beyond this model. A likely reason for the failure of the $k \cdot p$ approach is that – due to the very large number of impurities in the samples – one must, additionally, include band renormalization in the calculations. Such renormalization would result in different band shapes of the two spin sub-bands, and consequently in different spectral shapes of the absorption coefficients for the σ^+ and σ^- polarizations. It is our hope that the results reported here will lead to further theoretical work aimed at addressing these fundamental but as yet unresolved issues.

Work supported by NSF Grant DMR02-45227.

REFERENCES

1. K. Ando *et al., J. Appl. Phys.* **83**, 6548-6550 (1998).
2. J. Szczytko *et al., Phys. Rev. B* **59**, 12935-12939 (1999).
3. B. Beschoten *et al., Phys. Rev. Lett.* **83**, 3073 (1999).
4. J. Szczytko *et al., Phys. Rev. B* **64**, 075306 (2001).
5. W. Heimbrodt *et al., Physica E* **10**, 175-180 (2001).
6. K. Ando, "Magneto-Optics of Diluted Magnetic Semiconductors: New Materials and Applications", in *Magneto-Optics,* edited by S. Sugano and N. Kojima, Berlin: Springer, 2000, pp. 211-244.
7. K. M. Yu *et al., Phys. Rev. B* **68**, 041308(R) (2003).

Extension of the Semiconductor Bloch Equations for Spin Dynamics

C. Lechner* and U. Rössler*

Institut für Theoretische Physik, Universität Regensburg, D-93040 Regensburg, Germany

Abstract. We present a theoretical approach to spin dynamics in semiconductors and semiconductor heterostructures based on the density matrix formalism. It consists in an extension of the coherent Semiconductor Bloch equations for the two-level system to a six-level system which accounts for the spin and polarization degrees-of-freedom and includes carrier-carrier scattering beyond the Hartree-Fock truncation. This extension provides the concept to formulate spin dynamics including spin relaxation and dephasing times on a microscopic level.

INTRODUCTION

Recently spin dynamics has received a lot of attention with respect to visionary applications in the field of electronic devices operating with the spin of the carriers [1] and in the field of quantum computation [2]. Key quantities in both areas are the spin relaxation and dephasing times. Experimentally, spin dynamics is accessable by time-resolved spectroscopy with circularly polarized light [2], leading to optical spin orientation [3]. In a single-particle picture the dynamics of spin-orientation is described by equations of motion (EOM) for the spin-density matrix which include spin-orbit coupling and single-particle scattering [4]. On the other hand carrier dynamics in a driven two-level system including carrier-carrier interaction is described by the Semiconductor Bloch equations (SBE) [5]. Unfortunately the SBE do not account for the spin and polarization degrees-of-freedom. To overcome this shortcoming the two-level system has been extended to a six-level system [6]. This extension of the equations of motion (EOM) of carriers in semiconductors or semiconductor heterostructures has been formulated taking into account spin-split subbands due to spin-orbit interaction, the interaction with a light field of arbitrary polarization in dipole approximation and the carrier-carrier interaction. Due to the many-particle terms, the EOM end up in an infinite hierarchy, which needs to be properly truncated. In Ref. [6] the established *Hartree-Fock truncation* [5] was used leading to a closed set of EOM for the *coherent* dynamics. Here we go beyond the Hartree-Fock truncation by including the first order corrections to the coherent EOM caused by the carrier-carrier interaction. These corrections are essential for the description of scattering and dephasing in semiconductors and semiconductor heterostructures.

SPIN-DEPENDENT SCATTERING EXTENSIONS

The Hamiltonian of the system in second quantization consists of the following parts:

$$\mathcal{H} = \mathcal{H}_0 + \mathcal{H}_{light} + \mathcal{H}_{carrier}, \quad (1)$$

where

$$\mathcal{H}_0 = \sum_{\mathbf{k}'m'_c} \varepsilon_{m'_c}(\mathbf{k}') c^\dagger_{m'_c}(\mathbf{k}') c_{m'_c}(\mathbf{k}') + \sum_{\mathbf{k}'m'_v} \varepsilon_{m'_v}(\mathbf{k}') v_{m'_v}(\mathbf{k}') v^\dagger_{m'_v}(\mathbf{k}') \quad (2)$$

is the single-particle term of the Hamiltonian. $\varepsilon_{m_i}(\mathbf{k})$ is the energy dispersion in the subband m_i, with $m_c = \pm\frac{1}{2}$ for the electron subband and $m_v = \pm\frac{3}{2}$ ($\pm\frac{1}{2}$) for the heavy (light) hole subband. The energy eigenvalues include non-parabolic corrections due to higher order $\mathbf{k} \cdot \mathbf{p}$ coupling [7], in particular spin-orbit coupling, which leads to the removal of the spin degeneracy. The annihilation (creation) operator for electrons (holes) with spin m_i is given by $c^{(\dagger)}_{m_c}$ ($v^{(\dagger)}_{m_v}$). $\mathcal{H}_{light}$ describes the interaction with the light field in dipole approximation. The carrier-carrier interaction can be split up into different parts, namely the electron-electron interaction, the hole-hole interaction and the electron-hole contribution each due to both direct and exchange Coulomb interaction. For brevity reasons we refer to Ref. [6] for their explicit structure. In this communication we demonstrate our concept for the electron subsystem, given by a 2×2 spin density matrix, and the direct electron-electron interaction. It can be generalized to the 6×6 density matrix. Furthermore we restrict ourselves to the case of the diagonal entry of the density matrix. As starting point we take the coherent EOM of $\rho_{m_c m_c}(\mathbf{k})$ including only the first direct electron-electron interaction as given in Ref. [6]:

$$i\hbar \dot{\rho}_{m_c m_c}(\mathbf{k}) = -\sum_{\mathbf{k'q}} \sum_{\substack{m'_c m_{c_1} \\ m_{c_2}}} \{ \mathscr{V}^{ee}_{m_{c_1} m_c m'_c m_{c_2}}(\mathbf{k'}, \mathbf{k}, \mathbf{q}) \times \\ \langle c^{\dagger}_{m_{c_1}}(\mathbf{k'}+\mathbf{q}) c^{\dagger}_{m_c}(\mathbf{k}) c_{m'_c}(\mathbf{k}+\mathbf{q}) c_{m_{c_2}}(\mathbf{k'}) \rangle \}, \quad (3)$$

where $\mathscr{V}^{ee}_{m_{c_1} m_c m'_c m_{c_2}}(\mathbf{k'}, \mathbf{k}, \mathbf{q})$ is the interaction matrix element. Due to the choice of the *energy eigenstates* as basis, it is labeled with spin indices. To achieve the scattering contributions we formulate the EOM of the *reduced density matrix* [5], defined as the difference between the four-point density matrix in Eq. (3) and the corresponding term:

$$\rho_{m_{c_1} m_{c_1}}(\mathbf{k}+\mathbf{q}) \rho_{m_c m_c}(\mathbf{k}) \quad (4)$$

in the Hartree-Fock truncation. We confine the further treatment of the reduced density matrix to the *Boltzmann limit*, which is equivalent to taking into account only the leading order of the electron-electron interaction. After applying the adiabatic and Markov approximation [8] and inserting the result in Eq. (3) we get:

$$i\hbar \dot{\rho}_{m_c m_c}(\mathbf{k}) = -\Gamma^{out}_{ee} \rho_{m_c m_c}(\mathbf{k}) + \Gamma^{in}_{ee} (1 - \rho_{m_c m_c}(\mathbf{k})), \quad (5)$$

with the in- and out-scattering terms Γ^{in}_{ee} and Γ^{out}_{ee}:

$$\Gamma^{out}_{ee} = \frac{\pi}{\hbar} \sum_{\mathbf{k'q}} \sum_{\substack{m_{c_1} m_{c_2} \\ m'_c}} |\mathscr{V}^{ee}_{m_{c_1} m_c m'_c m_{c_2}}(\mathbf{k'}, \mathbf{k}, \mathbf{q})|^2 \times \\ \delta(\varepsilon_{m_{c_1}}(\mathbf{k'}+\mathbf{q}) + \varepsilon_{m_{c'}}(\mathbf{k}-\mathbf{q}) - \varepsilon_{m_c}(\mathbf{k}) - \varepsilon_{m_{c_2}}(\mathbf{k'})) \times \\ (1 - \rho_{m_{c'} m_{c'}}(\mathbf{k}-\mathbf{q}))(1 - \rho_{m_{c_1} m_{c_1}}(\mathbf{k'}+\mathbf{q})) \rho_{m_{c_2} m_{c_2}}(\mathbf{k'}) \quad (6)$$

and

$$\Gamma^{in}_{ee} = \frac{\pi}{\hbar} \sum_{\mathbf{k'q}} \sum_{\substack{m_{c_1} m_{c_2} \\ m'_c}} |\mathscr{V}^{ee}_{m_{c_1} m_c m'_c m_{c_2}}(\mathbf{k'}, \mathbf{k}, \mathbf{q})|^2 \times \\ \delta(\varepsilon_{m_{c_1}}(\mathbf{k'}+\mathbf{q}) + \varepsilon_{m_{c'}}(\mathbf{k}-\mathbf{q}) - \varepsilon_{m_c}(\mathbf{k}) - \varepsilon_{m_{c_2}}(\mathbf{k'})) \times \\ \rho_{m_{c'} m_{c'}}(\mathbf{k}-\mathbf{q}) \rho_{m_{c_1} m_{c_1}}(\mathbf{k'}+\mathbf{q})(1 - \rho_{m_{c_2} m_{c_2}}(\mathbf{k'})). \quad (7)$$

It is obvious that we have recovered a Boltzmann-like structure for the scattering distribution, which explicitely includes the spin of the electrons. A corresponding equation can be derived for the off-diagonal entry of the density matrix $\rho_{m_c - m_c}(\mathbf{k})$:

$$i\hbar \dot{\rho}_{m_c - m_c}(\mathbf{k}) = \Sigma^{ee}_{m_c - m_c}(\mathbf{k}) \rho_{m_c - m_c}(\mathbf{k}), \quad (8)$$

with the *self-energy* $\Sigma^{e-e}_{m_c - m_c}(\mathbf{k})$, which is proportional to the in- and out-scattering terms of Eqs. (6) and (7). It is an extension of the corresponding result for the two-level SBE [9].

For comparison with measured relaxation and dephasing times one has to transform the EOM into the standard basis, using spin-up and spin-down states with respect to a fixed direction (e.g. the growth direction of the quantum structure), for which spin-orbit terms are not diagonal. This can be achieved by an unitary transformation, which depends on the type of spin-orbit interaction to be considered. For a more detailed description of the transformation we refer to Ref. [10].

SUMMARY

We have derived the incoherent part of the spin-dependent SBE to leading order in the carrier-carrier interaction. Consequently scattering and dephasing contributions were given in the basis of energy-eigenstates where non-parabolic corrections like structural and bulk inversion asymmetry are taken into account. Further we made the connection with the basis states used in experiments. For a more detailed description of the relevant steps for the derivation we refer to Ref. [10], where the corresponding derivation was made for the electron-phonon interaction.

ACKNOWLEDGMENTS

We thankfully acknowledge financial support from the DFG via Forschergruppe 370/2-1 "Ferromagnet-Halbleiter-Nanostrukturen".

REFERENCES

1. Žutić, I., Fabian, J., and Das Darma, S., *Rev. Mod. Phys.*, **76**, 323 (2004).
2. Awschalom, D., *Semiconductor spintronics and quantum computation*, Springer-Verlag, Berlin, 2002.
3. Meier, F., and Zakharchenya, B. P., editors, *Optical Orientation*, North-Holland, New York, 1984.
4. Ivchenko, E. L., and Pikus, G. E., *Superlattices and other heterostructures*, vol. 110 of *Springer series in solid state sciences*, Springer, Berlin, 1997.
5. Haug, H., and Koch, S. W., *Quantum theory of the optical and electronic properties of semiconductors*, World Scientific, Singapore, 1993.
6. Rössler, U., *phys. stat. sol. (b)*, **234**, 385 (2002).
7. Rössler, U., *Solid State Commun.*, **49**, 943 (1984).
8. Kuhn, T., and Rossi, F., *Phys. Rev. B*, **46**, 7496 (1992).
9. Lindberg, M., and Koch, S. W., *Phys. Rev. B*, **38**, 3342 (1988).
10. Lechner, C., and Rössler, U., *to be published* (2004).

Self-Sustaining Resistance Oscillations by Electron-Nuclear Spin Coupling in Mesoscopic Quantum Hall Systems

G. Yusa[1], K. Hashimoto[2], K. Muraki[1], T. Saku[3], and Y. Hirayama[1,2]

[1.] *NTT Basic Research Laboratories., NTT Corporation, Atsugi JAPAN*
[2.] *SORST-JST, Kawaguchi JAPAN*
[3.] *NTT-AT, Atsugi JAPAN*

Abstract. We study electron-nuclear spin coupling implemented in mesoscopic fractional quantum Hall (FQH) devices. We find that longitudinal resistance in such systems oscillates with a period of several hundreds of seconds driven by a constant voltage instead of a constant current. The anomalous behavior suggests that an average nuclear spin polarization self-sustainingly oscillates between randomized and polarized states, which reveal nonlinear nature of the mesoscopic electron-nuclear spin coupled systems.

Nuclear spin is considered to have many applications for quantum information processing and computation [1]. In effort to control nuclear spin flip processes by conducting electrons, electron-nuclear spin coupled systems have been investigated in two-dimensional electron systems. In such systems transport coefficients, such as longitudinal resistance R_{xx}, shows hysteresis or monotonic development on a large time scale, reflecting the slow nature of spin-lattice relaxation of nuclei [2-4]. In this work, we studied microscopic electron-nuclear spin coupling and observed that R_{xx} oscillated with a period of several hundreds of seconds.

In order to examine microscopic coupling, we studied point contact structures defined by the width, $W=5$, and the length, $L=1$ μm in mesoscopic fractional quantum Hall (FQH) devices [5] and measured the time dependence of R_{xx} by constant-current or -voltage mode.

At a temperature of 200-300 mK the R_{xx} oscillated for more than fifteen hours only in the vicinity of spin-transition point at Landau level filling factor, $\nu=2/3$. The condition for the perpetual oscillations is limited to within ±1 % in ν and slightly away from this region R_{xx} oscillations are damped and have finite lifetime.

The period and the amplitude of such self-sustaining oscillations are resonantly modulated by radio frequency radiation (RF), which corresponds to

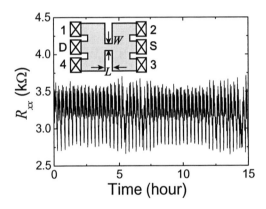

FIGURE 1. Time dependence of the longitudinal resistance, R_{xx}, of a point contact device. The measurement was performed at the spin-transition point of $\nu=2/3$ at $B=7.5$ T, and at $T=250$ mK. (Inset) Schematic illustration of our device.

nuclear magnetic resonance (NMR) of ^{75}As nuclei. The R_{xx} oscillates only in the constant-voltage mode; in the constant-current mode, it increases monotonically and saturates.

These results suggest that the nuclear spin polarization oscillates between randomized and polarized states. The resistance depends on the current, which reveals nonlinear nature of the electron-nuclear

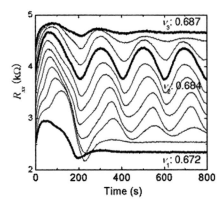

FIGURE 2. The resistance oscillations for ν between 0.672 and 0.687 at 7.5 T and at 200 mK. We confirmed that the transition point between the spin-polarized and -unpolarized states was in the vicinity of $\nu \sim 0.69$ at this magnetic field by obtaining the dependence of R_{xx} on B and ν

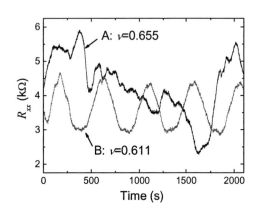

FIGURE 3. Time dependence of the resistance at 50 mK. B=6.25 T. The R_{xx} oscillates periodically at ν=0.611, while it shows non-sinusoidal behavior at ν=0.655.

spin coupling implemented in mesoscopic FQH devices.

When the temperature is decreased to 50 mK, the sinusoidal oscillations observed at 200 mK become non-periodic behavior depending on the filling factor. Figure 3 and 4 show the time dependence of R_{xx} measured at 50 mK and 6.25T. The R_{xx} oscillates periodically at ν=0.611, while it shows non-sinusoidal behavior at ν=0.655, as shown in Fig. 3. Such complicated time dependence of R_{xx} is clearly observed at ν above 0.63 at the experimental conditions. It suggests chaotic characteristics in the system at lower temperatures.

ACKNOWLEDGMENTS

The authors are grateful to T. Inoshita, K. Ono, S. Tarucha, Y. Tokura, K. Takashina, and N. Kumada for fruitful discussions.

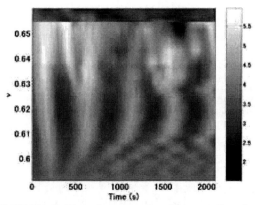

FIGURE 4. Time dependence of R_{xx} as a function of ν. The R_{xx} is gray scale coded in the unit of kΩ. The measurement was performed at 50 mK and 6.25 T. The R_{xx} oscillates periodically at low ν, however, the periodicity decays above 0.63.

REFERENCES

1. V. Privman, I. D. Vagner, and G. Kventsel, Phys. Lett. A **239**, 141 (1998).
2. S. Kronmüller, et. al., Phys. Rev. Lett. **81**, 2526 (1998).
3. J. H. Smet, et. al., Phys. Rev. Lett. **86**, 2412 (2001)
4. K. Hashimoto et al., Phys. Rev. Lett. **88**, 176601 (2002).
5. G. Yusa et al., Phys. Rev. B **69**, 161302 (2004).

Electron Spin Decoherence due to Hyperfine Coupling in Quantum Dots

L. M. Woods[1] and T. L. Reinecke[2]

[1]*Department of Physics, University of South Florida, Tampa, FL 33620,*
[2]*Naval Research Laboratory, Washington DC 20375*

The decoherence of an electron spin due to the hyperfine interaction in a realistic quantum dot with a large number of nuclei and with varying degrees of initial nuclear polarization is obtained. A nonperturbative method for calculating the spin decay time is given. The spin decay is described by the powers of a single function $I(t)$ that arises from the inhomogeneity of the hyperfine coupling in the quantum dot. The behavior of an ensemble of dots is decidedly different from that of a single dot.

I. INTRODUCTION

The spin degree of freedom of electrons is of current interest in applications ranging from 'spintronics' to implementations for quantum computing. The spin decoherence time becomes a key issue in these contexts.

Recently the effects of phonon scattering on spin decoherence rates in QD's have been shown to give relatively long decoherence times, on the order of 10^{-4}- 10^{-5} s (Ref. 1, 2) A second mechanism for decoherence, which arises from the hyperfine coupling also has been identified[3,4], and it is now believed to provide the dominant contribution to their decoherence rates. A solution for the fully polarized nuclear system has been given[3] and numerical simulations have been presented for dots with very small numbers ($N = 19$) of nuclei[4]. Realistic dots, however, have large numbers of nuclei ($\sim 10^5$) and have varying initial polarizations, which makes the problem a formadable task.

Here we treat the time evolution of an electron spin in a QD interacting with a large number of nuclear spins and with varying nuclear polarizations in an external magnetic field B. We show that the electron spin decay is described in terms of a decay function $I(t)$ originating from the inhomogeneous hyperfine coupling. We find that larger initial nuclear polarizations give longer decay times. We also solve for the dephasing of an ensemble of dots.

II. ELECTRON SPIN DYNAMICS

Consider the Hamiltonian of an electron coupled to a nuclear system in a QD with an applied magnetic field **B** along the z-direction:

$$H = (\epsilon + \gamma_z)s_z + \gamma_- s_+ + \gamma_+ s_- \quad (1)$$

$$\gamma_z = \sum_i A_i I_{i,z}, \quad \gamma_{-(+)} = \sum_i \frac{A_i}{2} I_{i,-(+)}, \quad \epsilon = \mu_B g B \quad (2)$$

where μ_B is the Bohr magneton, g is the Zeeman g-factor, $\mathbf{I}_i$ is the nuclear spin. The hyperfine coupling is $A_i = av_0|\psi(\mathbf{r}_i)|^2$, where a is the strength of the hyperfine interaction, $v_0 = 1/n_0$ is the inverse density of the nuclei, and $\psi(\mathbf{r}_i)$ is the electron wave function at the nuclear position $\mathbf{r}_i$. The shape of the dot is taken to be a sphere, and $|\psi(\mathbf{r}_i)|^2 \sim \frac{1}{(R/2)^{3/2}}e^{-4r_i^2/R^2}$ for the electron wavefunction. For a typical QD with radius $R \sim 20$ nm, $N \sim 10^5$. We take the nuclei to have spin 1/2, and values for the hyperfine coupling for GaAs. Dipole-dipole coupling becomes important only at times longer than 10^{-4} s and is not included in the Hamiltonian in Eq. (1). The nuclear Zeeman interaction is much smaller than that for the electron and is not included.

The time evolution of the wave function describing the electron-nuclear system is given by $|\psi(t)> = e^{-iHt/\hbar}|\psi(0)>$ with H in Eq. 1 and with $|\psi(0)> = |\downarrow;\mathcal{N}>$ the wave function at t=0. The arrow stands for the electron spin (chosen down for definiteness), and $\mathcal{N}$ is the nuclear spin configuration in the dot. We calculate the following function that describes the electron spin dynamics:

$$<s_z(t)> = <\psi(0)|e^{iHt/\hbar}s_z(0)e^{-iHt/\hbar}|\psi(0)> \quad (3)$$

Fully Polarized Nuclei: $|\mathcal{N}> = |\Uparrow\Uparrow ... \Uparrow>$ For this case the hyperfine interaction couples the initial state to a state with one nuclear spin flipped down and the electron spin flipped up. The electron spin dynamics is described by

$$<s_z(t)> \sim -\frac{1}{2}[1 - 2\frac{\bar{\omega}^2}{(\epsilon+A)^2} - \frac{\bar{\omega}^2}{\Omega^2}\sin^2\frac{\Omega t}{\hbar}$$
$$+ 4\frac{\bar{\omega}^2}{(\epsilon+A)^2}F(t)\sin\frac{\Omega t}{\hbar}] \quad (4)$$

$$F(t) = Im(e^{-i(\epsilon+A)t/\hbar}I(t)) \quad (5)$$

where $I(t) = \sum_{i=1}^{N}(\frac{A_i}{2\bar{\omega}})^2 e^{iA_i t/2\hbar}$. We use $A = \sum_i^N \frac{A_i}{2}$, $\bar{\omega}^2 = \sum_i^N (\frac{A_i}{2})^2$, and $\Omega^2 = (\epsilon+A)^2/4 + \bar{\omega}^2$. This result is obtained using the approximation $\bar{\omega}^2 - (A_i^2/2)^2 \sim \bar{\omega}^2$, since there are many nuclei. The evolution of the system is that of N two-level states that oscillate with a single Rabi frequency Ω. The first three terms describe the dynamics of a single two level system composed of the electron and the colection of N nuclei in a magnetic field. The last term describes the effective decay of the spin and it is controlled by the function $I(t)$. This last term is smaller than the first two terms by order 1/N because $\bar{\omega}^2 \sim a^2/N$.

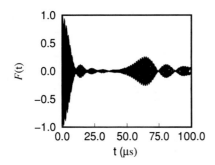

FIG. 1: $F(t)$ for the hyperfine coupling a=90 μeV for GaAs

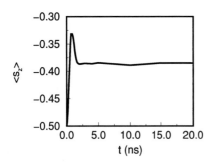

FIG. 2: Electron spin time decay for an ensemble of dots.

The function $F(t)$ from Eq. (5) is shown in Fig. 1. It has fast oscillations arising from the factor $e^{-i(\epsilon+\bar{A})t/\hbar}$. It also decays on a slower time scale, which is on the order of 0.1 μs. This decay gives decoherence to the electron spin. $F(t)$ also exhibits 'revivals' at longer times. If the coupling constant were uniform, $A(\mathbf{r}_i) \to \bar{a}$ for all nuclei N, there is no decoherence in the system because $I(t) = \sum_{i=1}^{N}(\frac{A_i}{2\bar{\omega}})^2 e^{iA_i t/\hbar} = e^{i\bar{a}t/\hbar}$ is simply a phase factor.

Partially Polarized Nuclei: We can derive a general expression for $<s_z>$ for m nuclei initially down and M up such that $m + M = N$ and $m << M$. Then, $<s_z>$ has the form

$$<s_z> \sim -\frac{1}{2}[OF_0 + OF_1 \times Im(I(t)) + ... + OF_m \times Im(I^m(t))] \quad (6)$$

where OF stands for an oscillating function. The result is found using similar approximation as in the fully polarized case.

When larger numbers of nuclei are initially down, more phase space for the electron-nuclear system is available, which gives rise to the many terms in $<s_z>$. The term $OF_{n+1} \sim \Pi_1^n (A_n/2)^2 \sim 1/N^{2n}$, and therefore terms with increasing n give decreasing magnitudes. Here Π_1^n stands for a product from 1 to n. The characteristic decay time of $F(t)^m$ for 85% initial nuclear polarization is ~ 0.05 μs, which is several orders of magnitude faster than that of $F(t)$.

III. ENSEMBLE OF DOTS

In an ensemble of dots the electron will experience a different nuclear configuration in each dot, and an averaging over nuclear configurations[5] is needed to describe the dephasing of the ensemble of spins. This can be done by treating the nuclear field as a classical variable for which the spin function can be written exactly using Eq. (3)

$$\bar{s}_z = -\frac{1}{2}[1 - \frac{\gamma_\perp^2}{\Omega^2}(1 - \cos\frac{\Omega t}{\hbar})] \quad (7)$$

where $\gamma_\perp^2 = \gamma_x^2 + \gamma_y^2$ and $\gamma^2 = \gamma_z^2 + \gamma_\perp^2$ are no longer operators. This is a new result that describes the entire dynamics of an ensemble of dots in an external magnetic field of arbitrary magnitude. In Fig. 2 we give the time decay of $\bar{s}_z$ for a magnetic field on the order of the hyperfine field with $\epsilon = 90$ μeV. The time dependence is characterized by oscillations that are non-exponentially damped with a characteristic time on the order of several ns.

IV. SUMMARY

We have presented a non-perturbative result describing the full time evolution of the electron spin in realistic systems with large numbers of nuclei and with varying initial nuclear polarizations for a first time. The characteristic time of the effective spin decoherence is on the order of 10^{-5} for fully polarized nuclei, and it decreases to the order of 0.05×10^{-6} for a nuclear polarization of 85%. The time decay for an electron spin in an ensemble dots is also given and it is on the order of several ns.

This work was supported by the US Office of Naval Research and by the DARPA QuIST program. One of us (LMW) acknowledges an NRC/NRL Associateship.

---

[1] A. V. Khaetskii and Y. V. Nazarov, Phys. Rev. B **61**, 12639 (2000); A. V. Khaetskii and Y. V. Nazarov, Phys. Rev. B **64**, 125316 (2001).
[2] L. M. Woods, T. L. Reinecke, and Y. Lyanda-Geller, Phys. Rev. B **66**, 161318(R) (2002).
[3] A. V. Khaetskii, D. Loss, and L. Glazman, Phys. Rev. Lett. **88**, 186802 (2002).
[4] J. Schliemann, A. V. Khaetskii, and D. Loss, Phys. Rev. B **66**, 245303 (2002).
[5] I. A. Merkulov, Al. L. Efros, and M. Rosen, Phys. Rev. B **65**, 205309 (2002).

Spin-Dependent Tunneling In III-V Semiconductors

Soline Richard[1], Henri-Jean Drouhin[2], Guy Fishman[1], Nicolas Rougemaille[2]

[1] *Institut d'Electronique Fondamentale, UMR 8622 CNRS, Université Paris Sud, 91405 Orsay Cedex, France*
[2] *Laboratoire de Physique de la Matière Condensée, UMR 7643-CNRS, Ecole Polytechnique, 91128 Palaiseau Cedex, France.*

Abstract. We calculate evanescent waves in GaAs-like III-V semiconductors throughout the forbidden band gap taking into account both the absence of inversion center and the spin-orbit coupling. We find that the evanescent energy bands are spin-split and that the evanescent wave functions only exist in limited energy and wave-vector domains. Such tunnel barriers can be used as solid-state spin injectors.

INTRODUCTION

Evanescent states in semiconductors have been studied for a long time. For instance Chang [1] considered semiconductors oriented in the [100], [111] and [110] directions, with inversion center (O_h group) or without inversion center (T_d group), but without taking into account the spin-orbit coupling. Chang and Schulman [2] performed a detailed calculation of the band structure of silicon, which belongs to the O_h group. Schuurmans and t'Hooft [3] studied semiconductors belonging to the T_d group but explicitly discarded terms which lead to odd k terms so that practically they studied GaAs and AlAs as if they were belonging to the O_h group. Briefly speaking, calculations about evanescent waves taking into account both the T_d group symmetry and the spin-orbit coupling have not been carried out. Here we calculate the evanescent band structure in the fundamental gap of GaAs-like III-V semiconductors including both the spin-orbit coupling and the lack of inversion symmetry, within a 30×30 **k.p** Hamiltonian framework (H_{30}). These properties enable the design of original solid state spin injectors.

CALCULATION IN H_{30} FORMALISM

We have calculated the H_{30} eigenvalues for a GaAs semiconducting barrier. The energy origin is set at the top of the valence band. The most important is that our parameters yield the γ value of 24 eV.Å³,[4] a parameter which characterizes the strength of the D'yakonov-Perel' field, responsible for the real conduction band spin splitting. The result of the calculation in the **k** = [iK, Q, 0] direction, where K and Q are real, is plotted in Fig. 1 for Q/K = 3/7. These directions are indexed by the angle θ they form with respect to the imaginary K axis (tan θ = Q / K)

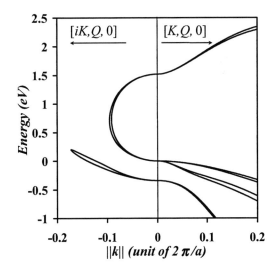

FIGURE 1. Evanescent states connecting the spin-split conduction band to the upper valence bands throughout the fundamental gap. Q/K = 3/7 and a is the cubic lattice constant.

Figure 2 presents calculations performed at different θ. When θ lies between 45° and 90°, no evanescent state exists in the gap. When decreasing θ below 45°, a small branch first appears in the band gap, with a small energy splitting. At smaller angles, this branch extends into the gap with an increasing energy splitting. For θ = 23°, there is an energy region with no allowed states. At θ ≈ 23°, the evanescent bands connect the conduction band with the heavy- and light-hole valence bands. A further decrease of θ does not change this situation but the energy splitting decreases to become equal to zero when k becomes aligned with the imaginary Ox axis. Such band diagrams, restricted to limited wave-vector and energy domains have no equivalent in real band structures.

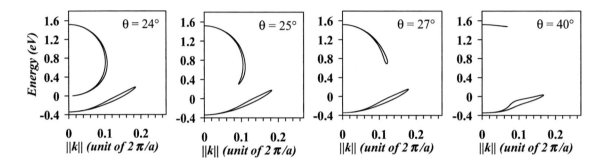

FIGURE 2. GaAs band structure in the gap along the **k** = [iK, Q, 0] direction for different values of θ ($tan\ \theta = Q/K$).

The evanescent-state spin splitting leads to large effects in tunneling phenomena. Consider a [100] GaAs barrier of height V and thickness b and an incident electron with the longitudinal energy ε_l and the transverse energy ε_t, associated with the wave vector Q. We assume that $\varepsilon_l \ll V$ so that tunneling arises at the energy V-ε_t below the bottom of the conduction band of the barrier material. Let be $iK_\pm$ the up-spin (down-spin) electron wave vector inside the barrier and $T_\pm$ the tunnel transmission and define $\Delta K = |K_+ - K_-|$ and $K = (K_+ + K_-)/2$. The polarization efficiency of the barrier is $\Delta T = |T_+ - T_-| / (T_+ + T_-)$. When $bK \ll 1 \ll b\Delta K$, $\Delta T/T \approx (bK)(\Delta K/K)$. Taking [$iK_\pm$, Q, 0] directions, with Q =0.03 Å^{-1}, which corresponds to $\varepsilon_t \approx$ 50 meV, we calculate $\Delta K/K$ values between 2% and 3% for V-ε_t between 0.4 eV and 1 eV. For bK=3, $\Delta T/T$ ranges from 6% up to 9%. This constitutes an extension of the calculation of [4], performed in the effective mass approximation. For the same bK value, a spin filter effect of only 3% in GaAs was estimated and in GaSb, a semiconductor where γ = 187 eV Å^3, a 12% effect was expected.

CONCLUSION

The spin-split evanescent states in the band gap of GaAs have been accurately calculated. Specific features, consisting of closed-loop branches, are observed. The spin-injector structure we consider is built according to the proposal given in [4]. It can be viewed as a HEMT-type structure, with an ultra-thin n-type GaAs gate, an undoped GaAs insulating barrier and a smaller band gap quantum well channel. An in-plane gate current results in nonzero average wave vector Q and thus into a net spin polarization of tunneling carriers. The electron polarization is controlled by the current magnitude.

ACKNOWLEDGMENT

H.-J. D. thanks the Délégation Générale pour l'Armement for support.

REFERENCES

1. Chang, Y. C., *Phys. Rev. B* **25**, 605 (1982).

2. Chang, Y. C. and Schulman, J. N., *Phys. Rev. B* **25**, 3975 (1982).

3. Schuurmans, M. and 't Hooft, G., *Phys. Rev. B* **31**, 8041 (1985).

4. Perel', V. et al., *Phys. Rev. B* **67**, R201304 (2003).

Interplay between in-plane magnetic fields and spin-orbit coupling in InGaAs/InP

P. T. Coleridge*, S. A. Studenikin*, G. Yu* and P. J. Poole*

*Institute for Microstructural Sciences, National Research Council, Ottawa, Ontario, Canada K1A 0R6

Abstract. The coupling between the Zeeman splitting and the Rashba spin-orbit interaction, with tilted magnetic fields, has been investigated in a gated (InGa)As/InP quantum well structure. It is demonstrated that the "spin zeroes" in the Shubnikov-de Haas oscillations can be enhanced by tilt and why they are not always present. A simple expression is given that allows the Rashba parameter to be extracted from the position of the spin zeroes in tilted fields. Values of the parameter obtained in this way are in good agreement with those estimated from the width of the weak anti-localisation peak.

In quantum well samples the strength of the Rashba spin-orbit interaction, which can be tuned using the gate voltage, can be obtained from the position of the "spin-zeroes" seen in the Shubnikov-de Haas (SdH) oscillations. The zeroes occur when the spin splitting δ_s is equal to half the cyclotron spacing $\hbar\omega_c$. If they appear at too low a field to be visible they can sometimes be moved to higher fields and made visible by tilting the magnetic field and increasing the contribution of the Zeeman term. Investigation of this behaviour in a gated (InGa)As/InP quantum well structure [1] is presented.

Measurements were made, at 0.3 K, by rotating the sample in a split-coil, tranverse access solenoid. At higher densities ($\sim 4\times 10^{15} \mathrm{m}^{-2}$) well defined spin zeroes could be observed (see Fig.1) associated with a Rashba energy splitting [2, 3] $\Delta_R = 2\alpha k_F$ [1]. As expected the spin zeroes move to higher fields with tilt.

The Rashba term mixes spin states on different Landau levels. With parallel fields it is necessary, in principle, to treat the mixing between multiple levels numerically [3] but an approximate analytical expression can be derived using the four Landau levels closest to the Fermi energy

$$(\delta_s/\hbar\omega_c)^2 = P - 2Q^{1/2},$$
$$P = 2(1-\beta)^2 + 2(\beta^2 + \beta_x^2) + (\Delta_R/\hbar\omega_c)^2,$$
$$Q = [(1-\beta)^2 - (\beta^2 + \beta_x^2)]^2 + (\Delta_R/\hbar\omega_c)^2(1+\beta_x^2) \quad (1)$$

where $\beta = g\mu_B B/2\hbar\omega_c$ with g the Zeeman g-factor, $\beta_x = \beta\tan\theta$ with θ the tilt angle and ω_c involves only the perpendicular component of the field. As expected this expression reduces to the usual Zeeman term when $\Delta_R = 0$

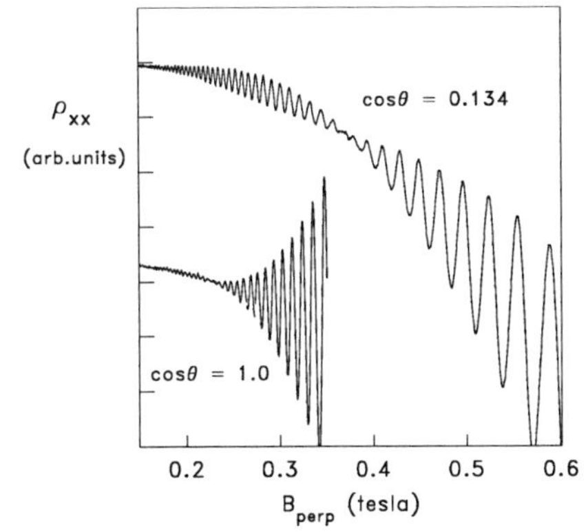

---

[1] Note, in the literature, Δ_R is sometimes defined as half this quantity, ie αk_F where k_F is the Fermi wavevector.

FIGURE 1. SdH oscillations at a density of $4.7\times 10^{15} \mathrm{m}^{-2}$ in perpendicular and tilted magnetic fields.

and to the standard perpendicular field expression [2, 3] $(|\delta_s| = [(1-2\beta)^2(\hbar\omega_c)^2 + \Delta_R^2]^{1/2} - \hbar\omega_c)$ when $\theta=0$.

Some results obtained using this expression, for a range of parameters appropriate to the sample measured here, are shown in Fig.2. For the values shown the errors associated with truncating the number of Landau levels are less than 1%. This plot shows that while sharp and well defined spin zeroes are expected at small tilt angles the curves eventually run approximately parallel to the $\delta_s/\hbar\omega_c = 1/2$ line and careful manipulation of the tilt is needed if the zeroes are to be detected.

Experimentally, it was found empirically (see Fig.3) that the spin-splitting as a function of tilt is given by an expression of the form $\delta_s = [c_0 + c_2 B_{tot}^2]^{1/2}$ where c_0 and c_2 are constants and B_{tot} is the total field. Fitting this data to eqn. 1 (using the measured value $0.044 m_e$ for the cyclotron mass) gives a somewhat poorer fit with $\Delta_R = 0.64$ meV and $\beta = -0.032$, ie a g-factor of -2.9. The fact

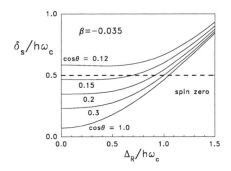

FIGURE 2. Calculated values of δ_s from eqn.1.

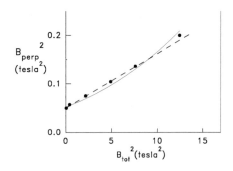

FIGURE 3. Values of perpendicular and total magnet fields at the positions of the spin-zero, as a function of tilt. Dotted line is empirical linear fit, solid line fit using eqn. 1.

that the empirical quadratic expression actually provides a better fit probably reflects a weak tilt dependence of the g-factor. The value of Δ_R is given predominantly by the intercept at zero tilt so despite the imprecision of the fit it provides an accurate means of extrapolating tilted field measurements to zero tilt. Values of Δ_R determined in this way (expressed in terms of α) are shown in Fig.4 as a function of density.

An alternative means of determining the spin-orbit coupling constant when spin zeroes are not visible is to analyse the weak antilocalisation feature seen at low

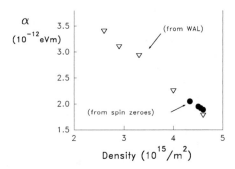

FIGURE 4. Values of the Rashba coefficient obtained from positions of the spin zeroes (solid points) and from the weak antilocalisation feature (open triangles).

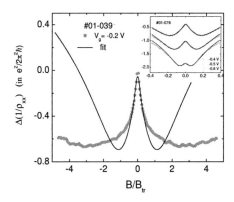

FIGURE 5. Typical weak antilocalisation curve with an attempt to fit the data in the range $B/B_{tr} \leq 0.5$. Inset shows good fits obtained in another sample, over the same range of B/B_{tr}, when the WAL features appear at lower fields.

magnetic fields [1, 4]. In this sample (see Fig.5) the feature is characterised by the unusually large values of field needed to induce the turn-over from negative to positive magnetoconductance; several times larger than the characteristic field $B_{tr} = \hbar/4eD\tau_{tr}$ where D is the diffusion constant. We know of no theory able to describe the data under these conditions. In other samples (see inset), where the turnover occurs at much lower values of B/B_{tr}, the data can be fitted well by the theory of Iordanski *et al* [5] (developed in terms of the Dyakonov-Perel mechanism) . It is found then that B_{so} ($= 2\Delta_R^2 \tau_{tr}/4eD\hbar$) is equal to the value of B at the turn-over. Although this theory is not expected to be valid for $B > B_{tr}$, and indeed (see figure 5) does not fit the data, we make the *empirical* assumption that the "turn-over" still occurs at B_{so}. Corresponding values of α, as a function of density, are plotted in figure 3.

The good agreement between the two sets of values obtained in different ways gives us confidence that either technique can be used to make reliable experimental estimates of α , even when good spin zeroes are not apparent in the perpendicular field SdH oscillations.

REFERENCES

1. Studenikin, S. A., Coleridge, P. T., and Poole, P., *Phys. Rev.B*, **68**, 035317 (2003).
2. Bychkov, Y. A., and Rashba, E. I., *J.Phys.C*, **17**, 6039 (1984).
3. Das, B., Datta, S., and Reifenberger, R., *Phys. Rev.B*, **41**, 8278 (1990).
4. Koga, T., Nitta, J., Akazaki, T., and Takayanagi, H., *Phys. Rev. Lett.*, **89**, 046801 (2002).
5. Iordanskii, S. V., Lyanda-Geller, Y. B., and Pikus, G. E., *JETP Lett.*, **60**, 206 (1994).

Zeeman Measurement of Quantum Wire Array in High Magnetic Field Exhibiting Abrupt Change at Elliptical Landau Orbit Formation

I.T. Jeong, T.S. Kim, S. Ahn, B.C. Lee, D.H. Kim, M.G. Sung and J.C. Woo

School of Physics, Seoul National University, Seoul 151-747, Korea

Abstract. The anomalous change in the Zeeman separation obtained from the ground state excitonic recombination in GaAs/AlGaAs quantum wire array is reported. At the applied magnetic field, where the magnetic length is comparable with the quantum wire width, an abrupt change in g-value is observed. Also observed is the oscillatory field dependence at high field region. The measurement is performed in polarization dependent magneto-luminescence at the field region of 0~32 T in LHe temperature.

INTRODUCTION

Recently, the electronic spin state in low-dimensional (low-D) confinement attracts interests due to its binary states conjunction with spin memory[1] and spin quantum computation.[2] Quantum dot exhibited single electron trapping.[3] In 1-D structure, the localization effect related to the Landau quantization has been reported.[4] In this work, the spin state in 1-D confinement transiting to pseudo-0D is studied by measuring Zeeman effect in quantum wire superlattice (QWR) under strong magnetic field. Zeeman shift is measured by polarization dependent magneto-photoluminescence (PL) with spectral resolution of 0.04 meV in the field up to 32 T at LHe temperature, and the GaAs/Al$_{0.5}$Ga$_{0.5}$As QWR sample is grown with fractional monolayer deposition by Petroff method using molecular beam epitaxy (MBE).[5]

RESULT AND DISCUSSION

Magneto-PL is shown that the PL peak positions are tends to be blue shift, and its line width more sharpen while its intensity increases, as magnetic field increases. The energy dispersion of each subband can be expressed,

$$E_{\lambda\pm}(k, B_a) = \hbar\omega_0 \alpha(B_a)\left(\lambda + \frac{1}{2}\right) + \frac{\hbar^2 k^2}{2m^*}\frac{1}{\alpha(B_a)} \pm \frac{1}{2} g_{eff} \mu_B B_a \quad (1)$$

where m^* is the effective mass, $\omega = \sqrt{k/m^*}$ the frequency related to the harmonic potential, $\omega_0 = eB_a/m^*c$ the cyclotron frequency, $\alpha(B_a) = \sqrt{1+\omega^2/\omega_0^2}$ and g_{eff} is the effective g-value of the exciton. This line width sharpening is due to the delocalization of excitons, and for GaAs/AlGaAs QW under a magnetic field of Faraday configuration was reported.[6] And the integrated intensity of magneto-PL increases because the overlap of the electron and hole wave function will increase and the diameter of the exciton will decrease, as magnetic field increases. The photon energy shifts that it is best fit to linear to B_a in high field region and quadratic to B_a in low magnetic field are displayed in Fig. 1.

Abrupt changes of slopes in Zeeman splittings can be seen in Fig. 2, near the magnetic fields where magnetic length $l = \sqrt{\hbar/eB_a}$ becomes comparable to wire width l_W. The effective g-factor obtained from this slopes changed at B_{c1} and $B_{c2} \approx 3B_{c1}$, where $l \approx l_w$ and $l \approx l_w/2$, respectively. The observed values of ε_l are equivalent to effective internal fields of B_{c1}=6.5 to 7 T

TABLE 1. The g-value and zero-field separation obtained from the slope of Zeeman splittings.

Sample (nm)	$\omega'(ps^{-1})$	$\alpha(B_{c1})$	g_0^*	g_1^*	ε_1 (meV)	g_2^*	ε_2 (meV)
(9+9)×9	3.5	1.1	3.4	1.7	0.7	0.7	1.9
(9+9)×18	5.3	1.2	1.7	1.0	0.4	-	-

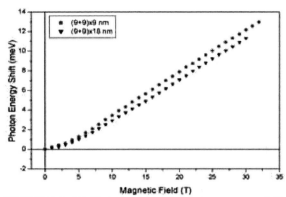

FIGURE 1. Field-dependent PL photon energy shifts of QWR samples obtained from the average value of σ^+ and σ^- spectra

for 9 nm wires. Introducing a quantized elliptical orbit model,[7] this zero-field splitting is explained similarly with spin-orbit coupling. The fluctuation of g^* by samples and spots is closely related to the effective cyclotron mass, which is greatly influenced by anisotropic lateral confinement. And, the quantization of magnetic flux as following,

$$\Phi_B = B_{eff} S_{eff} = n \frac{h}{e} S_{eff} \qquad (2)$$

where n is an integer and $S_{eff} = \pi \, \alpha(B_a) \, l_w^2$ is the effective area of exciton elliptical orbit. We can say that the internal magnetic field may be formed to the cyclotron motion of electrons. And the linear blue shift of magneto-PL to B_a for $B_{c1} < B_a < B_{c2}$ and its slope change around B_{c2} are supportive of the formation of Landau orbit. Since this spin-Landau orbit coupling, zero-field Zeeman splitting, ε_1 and ε_2, appears. Also, the oscillatory behavior, shown as Fig. 2, may be caused by the array periodicity. A downward bending of Zeeman splitting, similar with our work, has been reported for a strained QWR of 40 and 70 nm.[8] The effective g-factors and the zero-field separation of QWR obtained from the slope of Zeeman splitting are summarized in Table 1.

In summary, we performed the circular polarization dependent magneto-PL for a couple of QWR samples. The field-dependent Zeeman separation obtained from the photon energy difference of σ^+ and σ^- transition shows that anomalous changes of the effective g-value occur at B_{c1} where the magnetic length is comparable with the QWR width. The presence of zero-field splitting whose magnitude is equivalent to B_{c1}=6.5 to 7 T is observed at $B_a > B_{c1}$. This value of internal field is unusually large and its origin cannot be clearly understood. However, the elliptical cyclotron orbit introduced by a 1-D harmonic potential can be one of the possibilities.

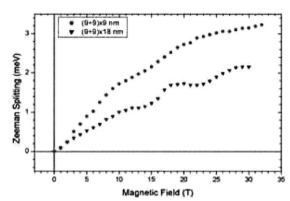

FIGURE 2. Zeeman splitting of the QWR depends on the applied magnetic field.

ACKNOWLEDGMENTS

This Work is supported in part by KRF 2003-015-C00185 and MOE of Korea. The magneto-PL experiment is performed using the facility of NHMFL, Tallahassee, FL, USA.

REFERENCES

1. P. Recher, E.V. Sukhorukov and D. Loss, Phys. Rev. Lett. **85**, 1962 (2000)
2. L.M. Woods, T.L. Reinecke, and Y. Lyanda-Gellar, Phys. Rev. **B 66**, 161318 (2002)
3. D. Loss, B.P. DiVincenzo, Phys. Rev. **A 57**, 120 (1998)
4. S.J. Bending, K. von Klitzing and K. Ploog, Phys. Rev. Lett. **65**, 1060 (1990)
5. P.M. Petroff, A.C. Gossard and W. Wiegmann, Appl. Phys. Lett. **45**, 620 (1984)
6. H. Sakaki, Y. Arakawa, M. Nishioka, J. Yoshino, H. Okamoto and N. Miura, Appl. Phys. Lett. **46**, 83 (1985)
7. A. Sommerfeld, Ann. Phys. **51**, 1 (1916).
8. M. Notomi, J. Hammersberg, J. Zeman, H. Weman, M. Potemski, H. Sugiura and T. Tamamura, Physica **B 249-251**, 171 (1998)

Spin relaxation in CdTe/ZnTe quantum dots

Ye Chen, Tsuyoshi Okuno, Yasuaki Masumoto

Institute of Physics and Venture Business Laboratory, University of Tsukuba, Tsukuba, Ibaraki 305-8571, Japan

Yoshikazu Terai, Shinji Kuroda, Koki Takita

Institute of Material Science, University of Tsukuba, Tsukuba, Ibaraki 305-8573, Japan

Abstract. We have measured photoluminescence (PL) spectra and time-resolved PL in CdTe quantum dots (QDs) under the longitudinal magnetic field up to 10T. Circular polarization of PL increases with increasing magnetic field, while its linear polarization is absent under linearly polarized excitation. Time-resolved PL measurements clarified that this behavior is caused by the suppression of spin relaxation induced by the longitudinal magnetic field. We believe that this behavior is related to the hyperfine interaction of electron spin with magnetic momenta of lattice nuclei.

INTRODUCTION

The spin state of electron in QDs is considered as one of the most promising candidates of spintronic and quantum information techologies. Contrary to bulk and quantum well, the main spin relaxation mechanisms connected with the spin-orbit interaction of carriers and inelastic processes are strongly suppressed due to the absence of the carrier's space motion in QDs. Recently Merkulov et al.[1] pointed out that the hyperfine interacting with nuclei may become the dominant mechanism of electron spin relaxation in QDs. However, no direct experimental evidence proved the importance of hyperfine interaction with nuclei responsible for spin relaxation in QDs till now. Here we investigated the spin relaxation mechanisms of carriers in self-assembled CdTe QDs. Our results indicate that the spin relaxation in QDs is mainly caused by the nuclear hyperfine field.

EXPERIMENTS AND DISCUSSIONS

The CdTe QD was grown by MBE on a thick (0.6-1.0μm) ZnTe buffer layer formed on a GaAs (100) substrate. The sample was undoped intentionally. The average size of CdTe QDs are ~20nm in diameter and ~2.7nm in height. Optical measurements were performed in magnetic fields up to 10T and at 10K in Faraday configuration. An optical parametric amplifier of a Ti:sapphire laser was used as an excitation source with a photon energy of 2.217eV corresponding to a quasi-excitation condition.

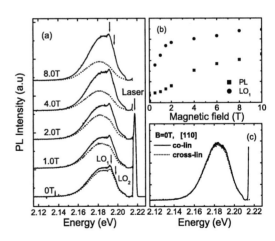

FIGURE 1. (a) Circularly polarized PL spectra in co- and cross-circular geometries at various magnetic fields. (b) Magnetic field dependence of circular polarization of QDs PL peak and LO_1 phonon. (c) Linearly polarized PL spectra in co- and cross-linear geometries along [110] direction.

The circularly polarized PL spectra are shown in Fig. 1(a). Two sharp features, LO_1 and LO_2, are related to LO phonons of ZnTe and CdTe. At zero field, the PL band shows very weak circular polarization (7%). With increasing magnetic field the circular polarization of PL band increases. The phonon replica is not polarized at zero field, while it becomes strongly polarized in co-circular geometry with applying magnetic field. The magnetic field dependence of circular polarization of the emission peak and LO_1 phonon replica are shown in Fig. 1(b).

This behavior is quite similar to the conversion from optical orientation to alignment induced by the anisotropic exchange interaction of excitons,[2] which has been observed in elongated QDs. However, we did not observe any linear polarization at zero field under linearly polarized excitation along [110] and [1$\bar{1}$0] directions as shown in Fig.1(c). These experimental results indicate that anisotropic exciton fine structure is not responsible for our results.

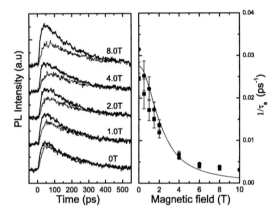

FIGURE 2. (a) Time dependence of circularly polarized PL for co-circular (solid lines) and cross-circular (dotted lines) geometries at various magnetic fields. (b) Magnetic field dependence of the spin relaxation rate. The experimental values from time-resolved PL (squares) and the estimated values from PL spectra (circles) are shown together. The solid line is the fitting according to Eq. (1).

The temporal profiles of circularly polarized PL under two geometries are shown in Fig. 2(a). We can see that the spin relaxation time becomes longer with increasing magnetic field. The time-resolved PL measurements proved that the spin relaxation in QDs was suppressed by the external magnetic field. We can obtain the spin relaxation time by fitting the decay curves of the polarization $P(t) = (I^+ - I^-)/(I^+ + I^-)$. The obtained spin relaxation rates, $1/\tau_s$, are shown with the estimated one from PL spectra together in Fig. 2(b). We can see that they are in agreement very well. Correspondingly the polarization of phonon structures can be explained by the assumption that spin flip time becomes longer than the time for the relaxation mediated by the LO phonon plus acoustic phonons with the increase of the magnetic field.

Such suppression of spin relaxation by external longitudinal magnetic field has been reported for the case of spin relaxation of electrons localized at donors in bulk crystal.[3,4] The spin relaxation of electrons localized in potential wells is mainly caused by hyperfine interaction of electron spins with nuclei. In strong magnetic field, the nuclear hyperfine fields only perturb the precession frequency of the electron spin about the external magnetic field. As a result, the spin component parallel to B is conserved, while the transverse spin component precesses with Larmor frequency. This mechanism should also work for localized electrons in QDs. Following this model we can fit the dependence of the electron spin relaxation time τ_s on magnetic field well by the expression:

$$\frac{1}{\tau_S} = \frac{1}{\tau_S(0)(1+\Omega^2 \tau_C^2)}. \quad (1)$$

Here $\tau_S(0)$ is the spin relaxation time under the zero magnetic field, Ω is the precession frequency of the electron spin in the field B, and τ_C is a characteristic time of the order of correlation time of the fluctuating magnetic field responsible for the spin relaxation.

Merkulov et al. studied theoretically electron spin relaxation in InAs and CdSe QDs via interaction with nuclear spins.[1] In CdTe QDs only a fraction of the nuclei (25% of the Cd ions) have magnetic moment ($I=1/2$). As a result, the electron spin interaction with the nuclei in these QDs is weak. The hyperfine interaction constants in CdTe are not experimentally determined. Reliable quantitative calculation is difficult to be obtained. Further theoretical and experimental works are needed to understand the effect of hyperfine field on spin relaxation in QDs.

REFERENCES

1. Merkulov, I. A., Efros, A. L. and Rosen, M., Phys. Rev. B 65, 205309 (2002).
2. Paillard, M., Marie, X., Renucci, P., Amand, T., Jbeli, A. and Gérard, J.M., Phys. Rev. Lett. 86, 1634 (2001).
3. Berkovits, V. L., Ekimov, A. I. and Safarov, V. I., Sov. Phys. JETP 38, 169 (1974) [Zh. Eksp. Teor. Fiz. 65, 346 (1973)].
4. Parsons, R. R., Can. J. Phys. 49, 1850 (1971).

Hole-spin reorientation in $(CdTe)_{0.5}(Cd_{0.75}Mn_{0.25}Te)_{0.5}$ tilted superlattices grown on $Cd_{0.74}Mg_{0.26}Te(001)$ vicinal surface

T. Kita[1], S. Nagahara[1], Y. Harada[1], O. Wada[1], L. Marsal[2], and H. Mariette[2]

[1]*Department of Electrical and Electronics Engineering, Faculty of Engineering, Kobe University, Rokkodai 1-1, Nada, Kobe 657-8501, Japan*
[2]*Laboratoire de Spectrometrie Physique, Universite J. Fourier, Grenoble I, CNRS (UMR 5588), Boite Postal 87, F-38402 Saint Martin d'Heres Cedex, France*

Abstract. We have studied anisotropic exchange interaction in $(CdTe)_{0.5}(Cd_{0.75}Mn_{0.25}Te)_{0.5}$ tilted superlattices fabricated by fractional monolayer growth onto a 1° off vicinal surface. The Zeeman shift in the Voigt configuration depends on the direction of the external magnetic field in the (001) plane. When applying the external magnetic field parallel to the wire direction, linear polarization of the magneto-PL reveals reorientation of hole spins into the external field direction. Moreover, in the parallel magnetic field the valence-band mixing occurs for fields larger than ~3.5 T, which results in anisotropic Zeeman shifts.

INTRODUCTION

Low-dimensional diluted magnetic semiconductor (DMS) heterostructures have attracted much interest in realizing enhanced magneto-optical phenomena caused by the quantum confined excitonic states. Recently, significant magneto-optical phenomena have been demonstrated in quantum wires (QWRs) and quantum dots.[1] On the other hand, spin-induced optical properties of excitons separated spatially from magnetic ions are expected to be adjusted by controlling the wave function.[2] In this paper, we have studied magneto-photoluminescence (PL) from $(CdTe)_{0.5}(Cd_{0.75}Mn_{0.25}Te)_{0.5}$ tilted superlattices (TSLs) grown on a $Cd_{0.74}Mg_{0.26}Te(001)$ vicinal surface. We demonstrate hole-spin reorientation in the CdTe-rich wire structure of the TSLs and discuss origin of anisotropic magneto-optical effects [3] found in the Voigt configuration.

TSL GROWTH

CdTe/CdMnTe wire structures have been fabricated by a self-regulated method for the growth of TSLs using molecular beam epitaxy.[3] $Cd_{0.96}Zn_{0.04}Te$ (001) substrate misoriented 1° toward the [110] direction was used in this study. This substrate has steps aligned along the [1-10] direction. After growing a $Cd_{0.74}Mg_{0.26}Te$ buffer layer, $(CdTe)_{0.5}(Cd_{0.75}Mn_{0.25}Te)_{0.5}$ TSLs were fabricated by repeated deposition of 1/2 ML of CdTe followed by 1/2 ML of $Cd_{0.75}Mn_{0.25}Te$. The repetition was 30 times. As a consequence, wires with almost square cross-section of 9.3x9.7 nm^2 can be formed. This size is comparable with the bulk exciton Bohr radius (~6.5 nm) of CdTe.

ONE-DIMENSIONAL OPTICAL PROPERTIES OF TSL

Figure 1 shows PL and PL excitation (PLE) spectra of the $(CdTe)_{0.5}(Cd_{0.75}Mn_{0.25}Te)_{0.5}$ TSLs. The sample temperature was 3.8K. The solid line indicates the PL spectrum. The PL from the CdTe-rich wires and $Cd_{0.74}Mg_{0.26}Te$ barrier are clearly observed at 1.81 eV and 2.03 eV, respectively. Linearly polarized PL intensities for the CdTe-rich wires and $Cd_{0.74}Mg_{0.26}Te$ barrier are shown in Fig. 2 as a function of the polarization angle defined from the [110] direction. Open and solid circles indicate results for the CdTe-rich wires and the $Cd_{0.74}Mg_{0.26}Te$ barrier, respectively. The PL intensity of the $Cd_{0.74}Mg_{0.26}Te$ barrier is independent of the polarization angle. Nevertheless, a clear sinuous dependence is observed for the PL intensity of the CdTe-rich wires. This result demonstrates quantum confinement with the quantization direction perpendicular to the wire. A dashed line in Fig. 1 shows the PLE spectrum detected at 1.810 eV. The Stokes shift is about 18 meV, which is due to the exciton-magnetic polaron formation and a non-magnetic localization on alloy potential fluctuation and wire-size fluctuation.

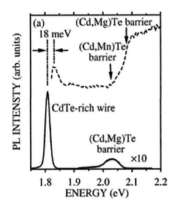

FIGURE 1. PL (solid line) and PLE (dashed line) spectra of a $(CdTe)_{0.5}(Cd_{0.75}Mn_{0.25}Te)_{0.5}$ TSL.

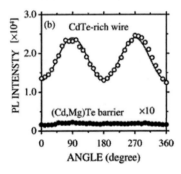

FIGURE 2. PL Polarization angle dependence of PL intensities for CdTe-rich wire and $Cd_{0.74}Mg_{0.26}Te$ barrier.

HOLE-SPIN REORIENTATION

In the Voigt configurations, the Zeeman shift of PL from the CdTe-rich wires in the magnetic field perpendicular to the wire direction is larger than that in the parallel field.[3] Such anisotropic behavior has never been observed in bulk materials and quantum wells. The anisotropic Zeeman shift results from the anisotropic exchange interactions. The CdTe-rich wire has an initial quantization direction of the hole spin perpendicular to the wire. When applying the external magnetic field parallel to the wire direction, the hole spin is expected to be reoriented into the direction of an external magnetic field. Figure 3 shows magnetic field dependence of PL-linear polarization of the CdTe-rich wires. The polarization P is defined as $(I_{E//wire} - I_{E\perp wire})/(I_{E//wire} + I_{E\perp wire})$. In the perpendicular magnetic field, the polarization increases with the magnetic field up to ~100%. On the other hand, the polarization in the parallel magnetic field decreases in the magnitude, then the polarization converges to a constant negative level, which reveals that hole spins are reoriented in the CdTe-rich wire into the external field direction. The non-zero constant negative level reflects the valence-band mixing. Since the valence-band mixing reduces the Zeeman splitting, the observed anisotropic Zeeman shift can be attributed to this effect.

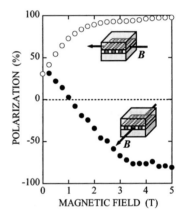

FIGURE 3. Magnetic field dependence of PL polarization as a function of magnetic field in the Voigt configurations. Open and closed circles indicate PL polarizations measured with the magnetic field applied either perpendicular or parallel to the wire direction, respectively.

REFERENCES

1. Takahashi, N., Takabayashi, K., Souma, I., Shen, J., and Oka, Y., *J. Appl. Phys.* **87**, 2000, pp. 6469-6471.

2. Takeyama, S., "Spin-induced optical phenomena in diluted magnetic semiconductors" in *OPTICAL PROPERTIES OF LOW-DIMENSIONAL MATERIALS*, edited by Ogawa, T. and Kanemitsu, Y., Singapore, World Scientific, 1998, pp. 179-230.

3. Nagahara, S., Matsuura, Y., Naganuma, R., Kita, T., Wada, O., Marsal, L., and Mariette, H., *Physica E* **21**, 2004, pp. 345-348.

Towards Using Multiferroism in Optoelectronics and Spintronics: Tunneling, Confinement and Optical Properties of Si/BiMnO$_3$ Systems

L. M. Rebelo*, F. F. Maia Jr.*, E. W. S. Caetano*, V. N. Freire*, G. A. Farias*, J. A. P. da Costa**, E. F. da Silva Jr.***

*Departamento de Física, Universidade Federal do Ceará, Caixa Postal 6030, 60455-900 Fortaleza, Ceará, Brazil
**Departamento Física Teórica e Experimental, Universidade Federal do Rio Grande Norte, Caixa Postal 1641, 59072-970 Natal, Rio Grande do Norte, Brazil
***Departamento de Física, Universidade Federal de Pernambuco, 50670-901 Recife, Pernambuco, Brazil

Abstract. Confinement and optical properties in Si/BiMnO$_3$/Si quantum wells are addressed. The characteristics of the confined excitons are determined varying the band offset, which was not measured or calculated yet. The ground state e-hh exciton binding energy is shown to be in the 3-4 meV range, and exciton based 1600-1700 nm optical emission is demonstrated to be feasible by varying the quantum well width from 3 nm to 10 nm. The characteristics of electron tunneling across BiMnO$_3$/Si/BiMnO$_3$ barriers are also presented.

INTRODUCTION

Bismuth manganite (BiMnO$_3$) is a multiferroic material, *i.e.* presents simultaneous ferroelectric and ferromagnetic ordering in the same phase. This class of materials is promising for applications in ferroelectric random access memories and future non-volatile memory devices [1,2]. Advances on integration of ferroelectric and ferromagnetic oxides with silicon by MBE allow for the possibility of using multiferroism in optoelectronics [3]. Structural, magnetic and electric properties of multiferroic BiMnO$_3$ with a highly distorted perovskite structure was studied [4], allowing to estimate as ~100/ϵ_o the BiMnO$_3$ dielectric constant at room temperature, and a less than 1% dielectric constant change induced by magnetization. Although bulk BiMnO$_3$ is not stable at 1 atm pressure, recently epitaxial thin films of BiMnO$_3$ were deposited on SrTiO$_3$ [5], pointing to its integration with silicon layers in a near future. This was already accomplished with YMnO$_3$ [6], and others RMnO$_3$ manganites are also of great interest [7].

In this work, tunneling, confinement and optical properties in Si/BiMnO$_3$ systems are addressed. *Ab initio* calculations were performed within the Density Functional Theory (DFT) to calculate the BiMnO$_3$ band structure, from which carrier's effective masses in high symmetry directions were obtained. The feasibility and characteristics of exciton confinement and tunneling properties of Si/BiMnO$_3$ quantum wells and single barriers, respectively, are determined varying the conduction band offset.

RESULTS AND DISCUSSION

The electron-heavy hole (e-h) confined excitons and tunneling properties of BiMnO$_3$/Si systems were studied by solving Schrödinger-like equations [8, 9]. The electron and heavy hole effective masses, which were calculated using first principles methods, are presented in Table 1.

Table 1. Si and BiMnO$_3$ parameters used in the numerical calculations. m_o is the free space electron mass. The crystal structures are cubic.

	BiMnO$_3$	Si
m_e/m_0	3.2097	0.173
m_h/m_0	1.8237	0.567
E_g (eV)	0.7225	1.052
ϵ/ϵ_o	100	11.7

Figure 1 displays the ground state exciton binding and total energies, E_b and E_T, for well widths in the 3-10 nm range. Note that the barriers are Si layers, and the wells are BiMnO$_3$ layers. Three values of the conduction band offset, Q_e=0.3, 0.5, and 0.7 were considered. They have apparently no effect on the confinement properties because the effective mass of the carriers in the BiMnO$_3$ are very large (see Table 1). Typical heavy hole (electron) confinement energies are in the 2-19 meV range (1-12 meV) as the well width decreases from 10 to 3 nm. For well widths larger than 7 nm, the exciton binding energy is of the same order of magnitude of the confinement energies. The ground state electron-heavy hole recombination energy ranges within the infrared (1600–1700 nm) due to the small BiMnO$_3$ band gap. Our calculations show that the carrier confinement energy is higher for holes for a given well width due to the larger effective mass of electrons in BiMnO$_3$.

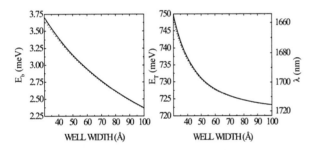

FIGURE 1. Exciton binding (left) and total (right) energies in a BiMnO$_3$/Si abrupt quantum well for three different band-offsets. Q_e =0.3(solid), 0.5(dashed), 0.9 (dotted).

Since the electron effective mass in Si is small, the conduction band offset influences strongly the electron tunneling across BiMnO$_3$/Si/BiMnO$_3$ barriers. This is shown in Fig. 2 for 50 Å wide Si barriers. Shifts up to 175 meV of resonance peaks are due to 0.4 band offset variations.

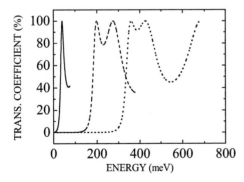

FIGURE 2. Transmission coefficient of electrons through BiMnO$_3$/Si/ BiMnO$_3$ single barriers considering conduction band offsets Q_e = 0.1 (solid), 0.5 (dashed), and 0.9 (dotted).

Recent growth of metastable BiMnO$_3$ films allows for the possibility of silicon dots in this material by ion implantation. The small quantum well depth and the high dielectric constant point to application in silicon-based flash memories. Related perovskite oxides like BaTiO$_3$, YMnO$_3$ etc, and their integration with silicon offer new research possibilities. Finally, work is on progress to consider the spin contribution to the optoelectronic and spintronic properties of silicon integrated ferroic systems.

In conclusion, our calculations suggest the possibility of optoelectronic applications for Si/BiMnO$_3$ systems in the infrared. Tuning of the BiMnO$_3$ dielectric constant by an applied bias could open the possibility of using these systems in Si-based spintronics. Due the large dielectric constant of BiMnO$_3$, charge image effects must be considered for an improved description of the confinement and tunneling properties of Si/BiMnO$_3$ systems.

ACKNOWLEDGMENTS

EWSC would like to acknowledge the graduate fellowship at the Physics Department of the Universidade Federal do Ceará received from CAPES during the realization of this research. VNF, GAF and EFSJr. would like to acknowledge the financial support received from the Science Funding Agency of the Ceará (FUNCAP) and of the Pernambuco (FACEPE) states in Brazil, and the Brazilian National Research Council (CNPq) through the grant NanoSemiMat Project #550.015/01-9.

REFERENCES

1. W. C. Yi, J. S. Choe, C. R. Moon, S. I. Kwun, J. G. Yoon, Appl. Phys. Lett. **73**, 903 (1998).
2. D. Ito, N. Fujimura, T. Yoshimura, and T. Ito, J. Appl. Phys. **93**, 5563 (2003).
3. K. J. Hubbard and D. G. Schlom, J. Mat. Res. **11**, 2757 (1996).
4. T. Kimura, S. Kawamoto, I. Yamada, M. Azuma, M. Takano, and Y. Tokura, Phys. Rev. B **67**, 180401 (2003).
5. A. Moreira dos Santos, A. K. Cheetham, W. Tian, X. Pan, Y. Jia, N. J. Murphy, J. Lettieri, D. G. Schlom, Appl. Phys. Lett. **84**, 91 (2004).
6. D. C. Yoo, J. Y. Lee, I. S. Kim, and Y. T. Kim, Thin Solid Films **416**, 62 (2002).
7. T. Kimura, T. Goto, H. Shintani, K. Ishizaka, T. Arima, and Y. Tokura, Nature **426**, 55 (2003).
8. Y. Ando and T. Itoh, J. Appl. Phys. **61**, 1497 (1987).
9. Ji-Wei Wu, Solid State Communic. **69**, 1057 (1989).

Exciton Spin Relaxation In Symmetric Self-Assembled Quantum Dots

S. Mackowski[1], T. Gurung[1], H.E. Jackson[1], L.M. Smith[1], G. Karczewski[2], J. Kossut[2], M. Dobrowolska[3], and J.K. Furdyna[3]

[1] *Department of Physics, University of Cincinnati, 45221-0011 Cincinnati OH, USA*
[2] *Institute of Physics Polish Academy of Sciences, 02-660 Warszawa, POLAND*
[3] *Department of Physics, University of Notre Dame, 46556-5670 Notre Dame IN, USA*

Abstract. We show that by using circularly polarized quasi-resonant spectroscopy, one can probe the exciton spin processes in symmetric self-assembled quantum dots, where at B=0T the excitonic levels are degenerate. While in the case of CdTe dots we observe no net polarization of the emission, a strongly circularly polarized luminescence is detected for CdSe quantum dots. We ascribe this difference to longer spin relaxation time in CdSe quantum dots.

INTRODUCTION

Semiconductor quantum dots (QDs) are believed to be a promising structure for using the spin degree of freedom in future technology. Achieving this goal requires a complete understanding of spin relaxation processes in these structures. It has been recently found for InAs QDs that when spin polarized excitons are created by a resonant polarized excitation, the emission at B=0T is unpolarized [1]. This has been ascribed to a distribution of the exchange splitting within an ensemble. Similarly, unpolarized emission has also been observed for the ensemble of CdTe QDs where approximately 20% of all the QDs were symmetric [2]. In this case, however, the unpolarized emission is the result of an extremely rapid exciton spin relaxation in symmetric QDs [2].

In this work, we compare the results of circularly polarized resonant spectroscopy of CdTe QDs and CdSe QDs. In CdSe QDs we find that for circularly polarized excitation the emission at B=0T is predominantly polarized as the excitation. This is a qualitatively different result than observed for CdTe QDs, where under the same excitation conditions, the emission is completely unpolarized. Using a very simple model we estimate the spin relaxation times to be 50ps and 10ps in CdSe QDs and CdTe QDs, respectively. To our knowledge, this is the first observation of optical orientation of excitons in symmetric QDs. We note that due to degeneracy of the exciton levels in symmetric QDs, the spin relaxation is significantly shorter than for QDs where this degeneracy is lifted wither by asymmetry or external magnetic field [1].

Both structures containing self-assembled QDs were grown by molecular beam epitaxy on GaAs substrates. Details of the growth can be found elsewhere [3,4]. The measurements were carried out at T=6K with the sample mounted in a continuous-flow helium cryostat. Spin polarized excitons were photo-excited directly into the ground states of QDs by using LO phonon-assisted absorption [5] by a Pyrromethane and Rhodamine 6G dye laser for CdSe QDs and CdTe QDs, respectively. Polarization of the excitation was controlled by a combination of Glan Thompson polarizer and Babinet-Soleil compensator. Similar optics was used to analyze the QD emission. The photoluminescence (PL) signal was dispersed by a DILOR spectrometer (subtractive mode) and detected by a liquid nitrogen cooled CCD camera.

RESULTS AND DISCUSSION

Figure 1 shows low temperature (T=6K) resonantly excited PL spectra of (a) CdTe QDs and (b) CdSe QDs obtained for σ+ polarized excitation. Both circularly polarized PL components are displayed. In addition, resonantly excited PL spectra are compared to the emission measured for non-resonant excitation (shaded areas). All the spectra were shifted in such a way that the energy corresponding to the first LO phonon replica below the laser energy becomes the origin. In the case of CdTe QDs, we observe identical intensity and shape of both $\sigma+$ (squares) and $\sigma-$ (circles) polarized PL [2]. In contrast, CdSe QDs exhibit strong predominant polarization of the PL, which is identical to that of the excitation. This difference in polarization behavior is clearly visible in Fig. 2, where we plot the PL polarization of both QD systems calculated using the formula: $P=(I^+-I^-)/(I^++I^-)$, where I^+ and I^- are the intensities of σ^+ and σ^- emissions, respectively. As can be seen, a polarization of the CdSe QDs emission is as high as 8%, whereas in the case of CdTe QDs no signal is detected within the experimental resolution (~3%).

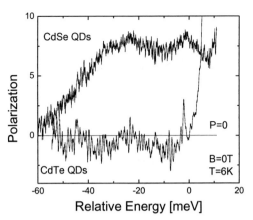

FIGURE 2. Polarization of the QD PL emission obtained for both studied QD systems.

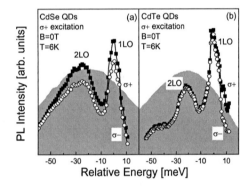

FIGURE 1. Resonantly excited PL spectra of (a) CdSe QDs and (b) CdTe. The excitation is $\sigma+$ polarized while $\sigma+$ (squares) and $\sigma-$ (circles) emissions are measured. Shaded areas correspond to non-resonantly excited PL.

With circularly polarized excitation at B=0T, we probe only the exciton spin dynamics in symmetric QDs, where the electronic levels are degenerate. These results demonstrate qualitatively different regimes of the exciton dynamics in these two QDs systems. To a first approximation, the PL polarization is determined by the relation between the exciton spin relaxation (τ_S) and exciton recombination (τ_R) times by the relation [6]: $P=1/(1+\tau_R/\tau_S)$. Here we assume that the polarization of the photo-created excitons is complete. The values of τ_R are equal to 300ps and 500ps for CdTe QDs and CdSe QDs, respectively. Therefore, we estimate the respective values of the exciton spin relaxation time in these two similar systems to be 10ps and 50ps. We believe that experiments on single QDs should allow us to obtain a more comprehensive description of the exciton spin dynamics in symmetric QDs.

CONCLUSIONS

The results of resonantly excited polarized PL of symmetric QDs show that the exciton spin relaxation time at B=0T is significantly shorter than the exciton recombination time. We estimate the characteristic times of spin scattering for degenerate excitonic levels to be 10ps and 50ps for CdTe QDs and CdSe QDs, respectively.

ACKNOWLEDGMENTS

We acknowledge the support of NSF through Grants DMR 0071797, 0216374 and 0245227 and DARPA SpinS Program (United States) and PBZ-KBN-044/P03/2001 (Poland).

REFERENCES

1. Paillard M., Marie, X., Renucci, P., Amand, T., Jbeli, A., and Gerard, J.M., *Phys. Rev. Letters* **86**, 1634 (2001).

2. Mackowski S., Nguyen, T.A., Jackson, H.E., Smith, L.M., Kossut, J., and Karczewski, G., *Applied Physics Letters* **83**, 5524 (2003).

3. Kim, C.S., Kim, M., Lee, S., Furdyna, J.K., Dobrowolska, M., Rho, H., Smith, L.M., Jackson, H.E., James, E.M., Xin, Y., and Browning, N.D., *Phys. Rev. Letters* **85**, 1124 (2000).

4. Mackowski S., *Thin Solid Films* **412**, 96 (2002).

5. Nguyen T.A., Mackowski, S., Jackson, H.E., Smith, L.M., Karczewski, G., Kossut, J., Dobrowolska, M., Furdyna, J.K, and Heiss, W., *Phys. Rev. B*, accepted.

6. Meier F., and Zakharchenya B., *Optical Orientation*, Modern Problems in Condensed Matter Sciences Vol. 8, North-Holland, Amsterdam, 1984.

A graphite-diamond hybrid structure as a half-metallic nano wire

Koichi Kusakabe and Naoshi Suzuki

Graduate School of Engineering Science, Osaka University, Toyonaka, 560-8531, Japan

Abstract. Design of a nanometer-scale half-metallic wire made of a graphene structure on a diamond (100) surface was performed theoretically. The wire consists of a nano-graphene ribbon covalently bonded to the diamond surface. Due to spin polarization of edge states, the graphene structure shows a finite magnetic moment. Stability of the structure and existance of spin moment ($1\mu_B$ per an edge carbon atom) are confirmed by the first-principles calculation with the local-spin-density approximation. The half-metallicity appears when a shift ($\sim$ 120meV) in the Fermi level is given.

Carbon and related materials have unexpected flexibility to fabricate new device structures. This would be true for the spin-electronics. Actually, several authors have suggested spin-polarization in carbon nanotubes,[1, 2, 3, 4] which may be relevant for new spin-polarized electron source. As another candidate, we propose to consider a wire structure of hydro-carbon on a diamond surface. To design high-density spintronics devices, magnetic wires in the atomic scale would be demanded. We demonstrate that a graphene structure may be an answer.

The purpose of the present study is twofold. We will show that creation of graphite-diamond hybrid structure automatically results in a π network with magnetism. The designed structure is an extension of Balaban-Klein-Folden's diamond-graphite hybrids.[5] However, our structure is shown to be stable and to have a finite moment similar to our magnetic nanographite.[1] The second purpose is to show that this graphitic structure covalently bonded to the diamond surface may be utilized for a highly spin-polarized electrode. We will discuss the band-structure, from which a way to obtain half-metallic states is proposed.

In the design of magnetic nanographite, we utilized two types of zigzag edges of graphene. One is an ordinal zigzag edge studied by Fujita *et al.* for magnetism[6] and the other is a modified edge by Klein.[7] The latter edge can be realized by di-hydrogenation of edge carbon.[1] Combination of these edges creates a nearly flat band, which is completely spin polarized, and the system becomes a half-semiconductor.

Formation of Klein's edge is expected in the diamond-graphite hybrid.[5] If we make sp^2-sp^3 interface, topology of π network is modified and the structure may satisfy our demand. However, too densely created graphitic layers on a diamond surface make the system unstable. Our calculation suggests that sp^3 bonds of original diamond-graphite hybrids are dissolved even in a calculation with a fixed unit cell.

Thus, we consider hydrogenated nanographite structures created perpendicular to a diamond (100) surface as shown in Figure 1. The π network of the attached graphene becomes a bipartite network with different numbers of sublattice sites. Therefore a high-spin state can appear in this one-dimensional structure. In addition, by separating graphene ribbons from each other, the structure is stabilized as shown below.

We now confirm stability and magnetism of our graphite-diamond hybrid structure shown in Fig. 1 using the first-principles calculations with the local-spin-density approximation.[8] Between graphene ribbons, we have a diamond surface. To consider possible formation of asymmetric dimers on this (100) surface, we have chosen a 3×2 cell as a unit cell for computation. We consider a slab model with 8 carbon layers. The back surface is hydrogen terminated. The size of the cell along the surface is the same as that of ideal diamond. The vaccume layer has a thickness of 11.4Å. Using the plane-wave expansion with the energy cut-off of 25Ry and the ultrasoft pseutopotential, we optimized the structure and determined the spin S of the ground state. Each component of inter-atomic forces becomes less than 1×10^{-4} H/a.u. Thus the structure is optimized. The optimized structure has a dimer row. We do not find a clear sign of asymmetric dimer. Thus the structure is a 3×1 structure on a diamond (100) surface.

The total spin is $S = 1$ per a unit cell. Since the cell has two edge carbon atoms in the graphene ribbon, a spin moment of $1\mu_B$ per an edge carbon atom is concluded assuming that $g = 2$. We have checked convergence of the result against calculational conditions by changing thick-

FIGURE 1. Spin density of a graphite-diamond hybrid structure on a diamond (100) surface. Carbon atoms and hydrogen atoms are represented by the ball-and-stick model. Dark isosurfaces on the graphene ribbon structure represent the spin density of up electrons. Isosurfaces of down electrons are too small to be seen.

ness of the diamond layers from 4 layers to 8 layers and by increasing the sampling k-points from one (at the Γ point) to 6×6 in plane. We always have the $S = 1$ ground state and the band structure is essentially the same. We show the spin density in Fig. 1. The distribution shows ferrimagnetic spin structure, which is just the expected one in the magnetic nanographite. The moment is only seen on the graphene structure.

The bandstructure of our graphite-diamond hybrid is shown in Figure 2. We see strong spin splitting in bands just around the Fermi level E_F. These branches, *i.e.* two spin-up branches just below E_F and two spin-down branches just above E_F, come from a single band, which is folded in the doubled unit cell and is strongly spin-splitted. This surface band is the expected nearly flat band of the edge states, which appears in the gap of the diamond. We see that, if E_F is slightly shifted by $\sim$ 120meV by applied bias voltage or by electron (or hole) doping to the wire, the system becomes half-metallic.

We have desinged an atomic scale wire of graphene on a diamond (100) surface, which is stable. The structure can show half-metallicity, when the Fermi level is shifted by $\sim$ 120 meV. Existence of graphitic structure is specific to carbon. Thus, our graphite-diamond hybrid structure may be special for the diamond. The present finding may open a new application of diamond surfaces for spin-electronics devices.

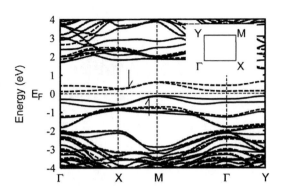

FIGURE 2. The band structure of the graphite-diamond hybrid. The x-direction is along the graphene structure and the y-direction is perpendicular to it. The solid lines and the dashed lines represent spin-up branches and spin-down branches, respectively Two branches indicated by the arrows are originated from the surface band of edge states.

ACKNOWLEDGMENTS

This work was partly supported by the Ministry of Education, Culture, Sports, Science and Technology of Japan (MEXT) under MEXT Special Coordination Funds for Promoting Science and Technology (Nanospintronics Design and Realization), a Grant-in-Aid for Scientific Research (No.15GS0213 and No.15310086), the 21st Century COE Program supported by Japan Society for the Promotion of Science, New Energy and Industrial Technology Development Organiation under the Materials and Nanogechnology Program and Japan Science and Technology Corporation under Research and Development Applying Advanced Computational Science and Technology program. We utilized the Tokyo *ab-initio* program package deveploped by Tsukada's group in University of Tokyo.[9] The calculation was partly done using the computer facility of ISSP, University of Tokyo.

REFERENCES

1. Kusakabe, K., and Maruyama, M., *Phys. Rev. B*, **67**, 092406 (2003).
2. Okada, S., *J. Phys. Soc. Jpn.*, **72**, 1510 (2003).
3. Kim, C., Seo, K., Kim, B., Park, N., Choi, Y., Park, K., and Lee, Y., *Phys. Rev. B*, **68**, 115403 (2003).
4. Kim, Y.-H., Choi, J., Chang, K., and Tománek, D., *Phys. Rev. B*, **68**, 125420 (2003).
5. Balaban, A., Klein, D., and Folden, C., *Chem. Phys. Lett.*, **217**, 266 (1994).
6. Fujita, M., Wakabayashi, K., Nakada, K., and Kusakabe, K., *J. Phys. Soc. Jpn.*, **65**, 1920 (1996).
7. Klein, D., *Chem. Phys. Lett.*, **217**, 261 (1994).
8. Perdew, J., and Wang, W., *Phys. Rev. B*, **45**, 13244 (1992).
9. Yamauchi, J., Tsukada, M., Watanabe, S., and Sugino, O., *Phys. Rev. B*, **54**, 5586 (1996).

Exciton Spin Manipulation In InAs/GaAs Quantum Dots: Exchange Interaction And Magnetic Field Effects

M. Sénès*, B. Urbaszek*, X. Marie*, T. Amand*, J. Tribollet[†], F. Bernardot[†], C. Testelin[†], M. Chamarro[†] and J.M. Gérard**

*Laboratoire de Nanophysique, Magnétisme et Optoélectronique, INSA 135 avenue de Rangueil, 31077 Toulouse Cedex 4, France
[†]Groupe de Physique des Solides-Universités Paris-VI et Paris-VII, CNRS UMR 7588, 2 place Jussieu, 75251 Paris Cedex 05, France
**CEA/DRFMC/SP2M, Nanophysics and Semiconductors Laboratory, 17, rue des Martyrs, 38054 Grenoble Cedex, France

Abstract. We report quantum beat phenomena between different polarisation states in InAs quantum dots measured in time resolved photoluminescence and pump probe experiments. The polarisation of the emitted light can be changed from linear to circular (and vice versa) by applying a small magnetic field. These experiments enable us to measure the anisotropic exchange energy ΔE_{XY} and the exciton g-factor. We find $\Delta E_{XY} \simeq 30 \mu eV$ and $g \simeq 2.5$ for the quantum dot excitons.

Long optical coherence [1] and spin relaxation times [2] support the numerous proposals for future applications of quantum dots (QDs) in spin dependent devices. In particular, for applications of QDs as single photon emitters, the anisotropic exchange interaction between electrons and holes (AEI) is crucial as it can determine both the polarisation and entanglement of the emitted photons [3]. For perfectly symmetrical dots the heavy-hole exciton states $|+1\rangle$ and $|-1\rangle$ can be excited with circularly polarised light. However, the reduced symmetry of real quantum dots leads to mixing of the exciton spin doublet via AEI, resulting in two linearly polarised transitions which are aligned along the orthogonal in-plane axes of the dot structure, separated by an energy ΔE_{XY} of up to $150 \mu eV$, as measured in single dot spectroscopy [4]. Here we report spin quantum beat phenomena for InAs/GaAs QDs in degenerate pump-probe experiments and in time resolved (TR) photoluminescence (PL) experiments after quasi-resonant excitation that enable us to extract the energy splittings between different polarisation eigenstates and to estimate the relaxation time of these spin states. For a magnetic field of B=0 we observe quantum beats in the circular polarisation, for B=0.4T we observe beats of the linear polarisation and for an intermediate field of B=0.21T we observe beats in both circular and linear polarisation. We explain this change in quantum beat polarisation in terms of an interplay between AEI and the applied magnetic field.

The sample consists of 40 planes of InAs self assembled QDs, separated by 15 nm thick GaAs spacer layers. The average QDs density is 3.10^{11} cm^{-2} with a typical dot height of 4 monolayers and a diameter of 15nm. At T=10K the ensemble PL is centred around 1.33 eV with a full width at half maximum (FWHM) of 60 meV. The sample is excited by 1.5 ps pulses from a Ti-sapphir laser. The TR PL is recorded with an upconversion technique (temporal resolution of 1.5ps) or a streak camera system (8ps). We denote linearly and circularly polarised photons as Π^X, Π^Y and σ^-, σ^+, respectively. The linear and the circular polarisation degrees of the luminescence are defined as $P_{lin} = (I^X - I^Y)/(I^X + I^Y)$ and $P_C = (I^+ - I^-)/(I^+ + I^-)$, respectively.

For a resonantly photo-generated electron-hole pair on the QD groundstate, neither the electron nor the hole spin relax during radiative lifetime of the exciton[2], allowing in principle the observation of spin quantum beats. In practice, the PL signal is initially obscured by backscattered laser light following resonant excitation. This can be avoided by exciting quasi-resonantly [5]: The excitation laser energy is tuned to an energy of one GaAs LO phonon (36meV) above the ground state.

Following strictly resonant excitation of the sample with linearly polarised light we observe a degree of linear polarisation $P_{lin} = 70\%$ that remains constant during the exciton lifetime. Also in the case of quasi-resonant excitation we observe a constant P_{lin}, and therefore no spin relaxation during the radiative lifetime, although the absolute value of P_{lin} dropped down to 40%. Following σ^+ excitation, the two eigenstates $|X\rangle$ and $|Y\rangle$ of the exciton are populated as the laser linewidth is larger than the

AEI splitting, resulting in an oscillating PL signal shown in figure 1a, with a period reflecting the energy splitting between $|X\rangle$ and $|Y\rangle$.

The energy difference between $|X\rangle$ and $|Y\rangle$ varies from dot to dot and due to this inhomogeneous distribution the observed oscillations are damped, as we average the signal over many dots. As the oscillations become damped we find a remaining circular polarisation rate of $P_C = 10\%$, which reflects the presence of unintentionally n-doped QDs in our sample [6]. Finally the PL signal decays on a time scale of about 1 ns which is due to the exciton recombination.

The oscillation period T and decay time τ were deduced from a simple equation which reproduced the measured oscillations accurately: $P_{c/Lin} = P_{c/Lin}(0)e^{-t/\tau}cos(2\pi t/T + \psi)$. We obtain an oscillation period of 135 ps, corresponding to an average splitting of $\Delta E_{XY} \simeq 30 \mu eV$, and a decay time of 30ps. The ΔE_{XY} measured is in agreement with values reported by other groups on similar samples from single dot PL [4] and differential transmission experiments on ensembles [7].

We have performed resonant transient dichroism measurements on the same sample using a pump-probe arrangement, similar to the set-up described in [8]. The pump polarisation was either σ^+ or σ^-. After transmission through the sample the electric field of the Π^X probe is separated into two orthogonal components Π^{X+Y} and Π^{X-Y} and analysed with an optical bridge. The difference between the intensities of the two components reveals a linear dichroism sensitive to the coherent superposition of the X and Y states induced by the pump resulting in oscillations of the linear dichroism with T=130ps and τ=26 ps. This result, obtained in a resonant configuration at zero magnetic field, confirms the values obtained in the time and polarisation resolved PL measurements after quasi-resonant excitation. This agreement is a strong indication that the quantum beats observed in PL are due to the creation of a coherent superposition $|X\rangle + i|Y\rangle$ and not due to optical interference effects of from dots in either state $|X\rangle$ or $|Y\rangle$.

A magnetic field applied along the growth axis of the sample (Faraday configuration) was used to investigate the behaviour of the excitonic fine structure states split by the magnetic field. For an intermediate field of B=0.21T we observe a beating of the *circular* polarisation following *circularly* polarised excitation, as at B=0, but with a shorter period of 90ps. In addition, we observe oscillations of the *linear* polarisation following *linearly* polarised excitation. At B=0.4T we observe only this second type of oscillations, shown in the TR co-polarised and cross-polarised PL in figure 1b with an even shorter period of 65ps, demonstrating the tunability of the spin quantum beat period in magnetic fields. By increasing the magnetic field, the Zeeman splitting between the

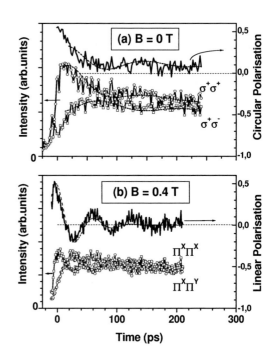

FIGURE 1. TR PL, co- and cross-polarised after quasi-resonant excitation. (a) for B=0 oscillations in the *circular* polarisation with a period of 135ps following σ^+ excitation and (b) for B=0.4T oscillations in the *linear* polarisation with a period of 65ps following Π^X excitation are observed.

$|+1\rangle$ and $|-1\rangle$ states starts to dominate the AEI splitting and $|+1\rangle$ and $|-1\rangle$ become the optically active eigenstates [9]. To fit the dependence of the energies deduced from the oscillation periods as a function of magnetic field, we use the function $\Omega_z = \sqrt{\omega_{exch}^2 + (g_{ex}\mu_B B/\hbar)^2}$ which takes the exchange energy $\Delta E_{XY} = \hbar\omega_{exch}$ and the Zeeman splitting $\hbar\Omega_z = g_{ex}\mu_B B$ into account [4]. This enables us to deduce an effective exciton g-factor of $g \simeq 2.5$ for the dots investigated here.

REFERENCES

1. Borri, P., and et al, *Phys. Rev. Lett*, **87**, 157401 (2001).
2. Paillard, M., and et al, *Phys. Rev. Lett*, **86**, 1634 (2001).
3. Santori, C., and et al, *Phys. Rev. B*, **66**, 045308 (2002).
4. Bayer, M., and et al, *Phys. Rev. Lett.*, **82**, 1748 (1999).
5. Flissikowski, T., and et al, *Phys. Rev. Lett*, **86**, 3172 (2001).
6. Cortez, S., and et al, *Phys. Rev. Lett*, **89**, 207401 (2002).
7. Lenihan, A. S., and et al, *Phys. Rev. Lett*, **88**, 223601 (2002).
8. Tribollet, J., and et al, *Phys. Stat. Sol. C*, **1**, 585 (2004).
9. Dzhioev, R. I., and et al, *Phys. Rev. B*, **56**, 13405 (1997).

Photo-induced ferromagnetism in bulk-$Cd_{0.95}Mn_{0.05}Te$ via exciton magnetic polarons

Y. Hashimoto, H. Mino, T. Yamamuro, D. Kanbara, T. Matsusue,[1] S. Takeyama,[2] G. Karczewski,[3] T. Wojtowicz,[3] J. Kossut[3]

Graduate School of Science and Technology, Chiba University, Chiba, Japan
[1]*Faculty of Engineering, Chiba University, Chiba, Japan*
[2]*The Institute for Solid State Physics, University of Tokyo, Chiba, Japan*
[3]*Institute of Physics, Polish Academy of Sciences, Warsaw, 02-668, Poland*

Abstract. Free exciton magnetic polarons in bulk CdMnTe with low Mn concentration have been investigated through the time-resolved and spectral-resolved photo-induced Faraday rotation measurements. The time-resolved photo-induced Faraday rotation measurements have revealed the spin relaxation dynamics and show a finite rotation angle lasting longer than 13 ns. This previously unreported long decay is attributed to dark exciton magnetic polarons.

INTRODUCTION

In diluted magnetic semiconductors (DMS), there have been many reports on "magnetic polarons" (MP), i.e., the exciton wearing ferromagnetically aligned Mn spin clouds via the sp-d exchange interaction. After the first report of free exciton magnetic polaron (FEMP) by Gornik [1], much attention has been directed to the DMS due to the possibility of the optical manipulation of magnetism. A systematic study of localization energy of the FEMP as a function of Mn concentration has been performed on Cd1-xMnxTe with $0.1 < x < 0.35$ [2], and the maximum value of the energy is expected to occur at a Mn concentration of 5 to 10%. The lack of investigations on Cd1-xMnxTe in this low Mn concentration regime arises from the difficulties in growing high quality samples with a reduced amount of bound exciton photoluminescence (PL) lines and showing dominant free exciton luminescence. However, recent development in molecular beam epitaxy (MBE) has allowed the growth of high quality bulk crystals in this low Mn concentration regime. In this study, the spin relaxation dynamics and the magnetic polaron formation process in bulk CdMnTe with low Mn concentration have been investigated through the time-resolved- (TR-) and spectral-resolved- photo-induced Faraday rotation (SR-PIFR) measurements.

EXPERIMENTAL SETUP

The samples of bulk-$Cd_{1-x}Mn_xTe$ were grown on (100)-GaAs substrates with $Cd_{1-y}Mg_yTe$ buffer layers by molecular beam epitaxy (MBE). The Mn molar fraction of the samples was 5%. All the measurements were performed at 1.4 K. In Figs. 1(a) and 1(b), the experimental set-up for the spectral-resolved and the time-resolved photo-induced Faraday rotation (SR-

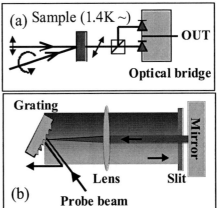

FIGURE 1. Schematics of (a) experimental set-up of PIFR measurement and (b) Fourier transform spectrum filter.

and TR-PIFR) measurements are shown schematically. The excitation source was a mode-locked Ti:sapphire laser producing 200 fs pulses with 80MHz. The pump pulse was circularly polarized and the probe pulse was linearly polarized. They were focused on the same position of the sample surface with a spot size of 200 μm in diameter. The excitation density was estimated to be 0.1 μJ/cm^2. In order to spectrally resolve the PIFR signal, a Fourier transform spectrum filter (FTSF) shown in Fig. 1(b) was used. The FTSF narrowed the spectral half width of the probe pulse down to 0.7 nm and varies the energy of the probe pulse by varying the slit position. The rotation of the linearly polarized probe light was measured by means of an optical bridge with a precision better than a milli-degree [Fig. 1(a)].

RESULTS AND DISCUSSIONS

In Fig. 2(a), the temporal profiles of the PIFR at 1.4 K are shown. The solid and dotted curves were obtained by right-handed and left-handed circularly polarized pumping, respectively. The photon energy of the pump and probe pulse was set at the band edge exciton (EX) resonance. The rise time of the signal is limited by the laser pulse width, and the decay contains three contributions of very different relaxation time. The first decay time, less than 1 ps, is attributed to hole spin relaxation. The second decay of 8 ps transient corresponds to exciton spin relaxation. The third process has a long decay time and appears even in negative delay region. This indicates that the signal persisted more than 13 ns, which is the period of the mode-locked laser used for the pumping. We attribute this long decay to the dark exciton magnetic polarons (DEMP) that remain for a long time as the radiative recombination is forbidden. In the following, we will briefly discuss the formation process of the DEMP. Assuming the right-handed circularly polarized excitation, the pump pulse excites bright exciton with +1 angular momentum. Individual spin relaxation of the hole and electron spins results in the formation of the light hole dark excitons with 0 angular momentum and the heavy hole exciton with −2 angular momentum. In the diluted magnetic semiconductor, the heavy hole dark exciton will form the DEMP via the p-d exchange interaction. The spin-polarized DEMP will act as a spin reservoir that feeds the bright exciton states via electron and hole spin flip. In order to clarify the nature of the long decay signal in PIFR, we spectrum resolve the PIFR signal at negative delay region. The spectral profile of the PIFR signal in the negative delay region is shown in Fig. 2(b). One can see that the PIFR spectrum shows the maximum value at the band edge exciton resonance. Theoretical analysis of the PIFR spectrum [3] indicates that the observed PIFR spectrum implies the energy splitting between the exciton states with +1 and −1 angular momentum. This means the ferromagnetically aligned Mn spins via the DEMP formation cause the Zeeman splitting of the bright exciton states. The ferromagnetic orientation of the Mn spins via the DEMP formation can be interpreted as the ferromagnetism mediated by the dark exciton.

SUMMARY

Spin-dependent transient absorption and spectrum- and time-resolved photo-induced Faraday rotation measurements were performed. The temporal profile of the PIFR shows a long decay, longer than 13 ns. The PIFR spectrum at negative delay region shows the maximum value at the band edge exciton resonance and attributed to the Zeeman splitting caused by the ferromagnetically aligned Mn spins via the dark exciton magnetic polaron.

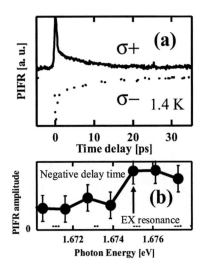

FIGURE 2. (a) Photo-induced Faraday rotation for right- and left- circularly polarized pump as a function of the delay time between pump and probe pulse. (b) Photo-induced Faraday rotation spectrum at negative delay region

REFERENCES

1. A. Golnik et al., J. Phys. C **16**, 6073 (1983).
2. S. Takeyama et al., Phys. Rev. B **51**, 4858 (1995); M. Umehara, Phys Rev. B **68**, 193202 (2003).
3. W. Maslana et al., Phys. Rev. B **63**, 165318 (2001).

Ac conductivity and magneto-optical effects in the metallic (III,Mn)V ferromagnetic semiconductors from the infrared to visible range

E. M. Hankiewicz[*], T. Jungwirth[†,**], T. Dietl[‡,§], C. Timm[¶] and Jairo Sinova[*]

[*]*Department of Physics, Texas A&M University, College Station, TX 77843-4242*
[†]*Institute of Physics, ASCR, Cukrovarnická 10, 162 53 Praha 6, Czech Republic*
[**]*Department of Physics, University of Texas, Austin, TX 78712-0264*
[‡]*Institute of Physics, Polish Academy of Sciences, al. Lotników 32/46, PL-02-668 Warszawa, Poland*
[§]*ERATO Semiconductor Spintronics Project, Japan Science and Technology Agency, al. Lotników 32/46, PL-02-668 Warszawa, Poland*
[¶]*Institut für Theoretische Physik, Freie Universität Berlin, Arnimallee 14, D-14195 Berlin, Germany*

Abstract. We study the ac conductivity as well as magneto-optical effects in a wide range of spectrum from infrared to visible parts. The calculations are based on the kinetic exchange model, which treats the disorder effects perturbatively within the Born approximation. We present the results for the ten-band Kohn-Luttinger (KL) model with six valence, two conduction, and two dispersionless bands. The latter simulate upper-mid-gap states induced by As antisites or Mn interstitials. We report on new features in the magneto-optical effects. We also analyze an influence of lifetime broadening on ac conductivity. The comparison of the calculated and experimental data yields quasiparticle lifetime broadening of the order of 100meV.

An attractive opportunity to apply the ferromagnetic semiconductors in the spintronics and the fabrication of the magnetorecoding devices have been triggered intensive theoretical efforts to understand the electronic structure of these materials. However, one of the main obstacles for the industrial application of (III-Mn)V semiconductors is the formation of defects such as Mn interstitials and As antisites. While substutional Mn ions introduce localized spins as well as mobile holes to III-V compounds, mentioned defects act as the double donors decreasing the hole concentration p and consequently the critical temperature T_c for the ferromagnetic order. Moreover, a strong anomalous Hall effect in these materials makes the estimation of the compensation by defects very cumbersome. This makes the deeper understanding of the influence of these defects on transport and optical properties even more crucial. Here we report on an influence of compensation effects on the ac conductivity and magnetic circular dichroism (MCD). We show that the localized states induced by defects should be accounted in order to explain recent experiments [1]. We also propose a new method of the estimation of the carrier concentration based on analysis of ac conductivity and magneto-optical effects.

Our theoretical approach starts from the coupling of the valence-band electrons with $S = 5/2$ Mn local moments via a semi-phenomenological local exchange interaction: $H = H_{KL} + J_{pd}\sum_{I,i} \mathbf{S}_I \cdot \mathbf{s}_{i,\text{holes}} \delta(\mathbf{r}_i - \mathbf{R}_I) + J_{sd}\sum_{I,i} \mathbf{S}_I \cdot \mathbf{s}_{i,\text{elec}} \delta(\mathbf{r}_i - \mathbf{R}_I)$. Here H_{KL} is the 10 band Kohn-Luttinger Hamiltonian which describes the band structure of GaAs, J_{pd} and J_{sd} are the exchange integrals between localized d-electrons and delocalized p-holes and s-electrons, respectively, while $\mathbf{S}_I$ and $\mathbf{s}_i$ denote localized d-spin and carrier spin [2]. The form of H_{KL}, values of exchange integrals as well as details of calculations can be found elsewhere [3, 4]. The disorder effects are introduced through valence-band quasiparticle lifetime broadening Γ as well as by introduction of two dispersionless bands simulating the upper-mid-gap defects states. Figure 1 shows calculated and experimental [1] real part of the diagonal conductivity as a function of the photon energy for different Γ. Exact diagonalization studies [5] as well as magnetotransport experiments [6] show that for "weakly" metallic samples used in the experiment by Singley et al. [1] with conductivities around $\sim 100\ \Omega^{-1}\text{cm}^{-1}$ the Drude peak is strongly suppressed because of localization effects. Hence, we subtracted the intraband contribution from the numerical simulations. In Fig.1 one can see that 10 band KL model with $\Gamma \approx 100$ meV describes the experimental data well. Moreover, the peak becomes dramatically sharper with decreasing Γ while it smears out for larger Γ, as expected. The agreement of experimental and calculated data for $\Gamma = 100$ meV is in accordance with estimation

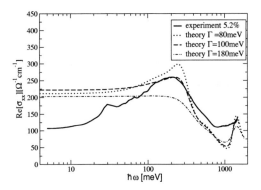

FIGURE 1. The comparison of experimental data of real part of the diagonal conductivity, $\text{Re}[\sigma_{xx}(\omega)]$ as a function of frequency for $x = 5.2\%$ Mn with theoretical simulations for different quasiparticle lifetime broadening Γ's.

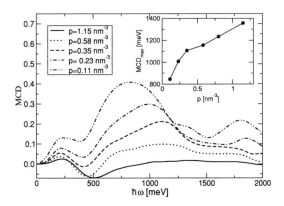

FIGURE 2. The MCD spectra as a function of the photon energy for different hole concentrations within ten band KL model, $\Gamma = 100$ meV. The inset: the position of the main MCD maximum as a function of hole concentration

of Γ within six band model where screening and compensation effects are included [4]. Importantly, our calculations made it possible to estimate hole concentration $p \approx 0.35$ nm^{-3}. Moreover, a systematic fitting to experimental data by Singley et al. [1] yields compensation level $\sim 70\%$–80% in samples studied. This level of compensation was confirmed by Hall measurements in as grown samples [7].

The magnetic circular dichroism (MCD) signal in the thin-film geometry is defined as follows: MCD $\equiv \frac{\alpha^- - \alpha^+}{\alpha^+ + \alpha^-}$ where α^- is the absorption for the left and α^+ for the right-circularly-polarized light, respectively. Figure 2 shows MCD as a function of the photon energy for different hole concentrations. The maximum at small energies, around 220 meV corresponds to heavy-hole $\rightarrow$ light-hole transitions. The main maximum at higher frequencies around 0.8-1.4 eV is even more interesting. It is attributed to the inter-valence transitions, transitions between valence and conduction bands, as well as between valence and defect bands. As one can see, the maximum shifts as a function of the compensation making its evaluation possible by means of MCD. For $p = 1.15$ nm^{-3}, when holes from all the Mn acceptors are electrically active (zero compensation) the position of the maximum corresponds to the valence $\rightarrow$ conduction band transition. When compensation increases, the magnitude of the main maximum increases. At the same time it shifts towards lower energies where the valence to localized states transitions can dominate. Indeed, the MCD spectra of LT-GaAs [8] show that the MCD for antisites lies around 1 eV in agreement with our theoretical predictions. However the intensity of the MCD is very weak in undoped GaAs, of the order of $5 \cdot 10^{-3}$. Our calculations show that the MCD is strongly enhanced in the ferromagnetic semiconductors. We also found a huge enhancement of the Kerr effect in the (III,Mn)V which gives a potential application in the magnetorecording devices.

For $Ga_{0.95}Mn_{0.05}As$ and $p = 0.11$ nm^{-3} the Kerr angle is around 3 deg. The inset to Fig. 2 shows the positions of the main maximum of MCD as a function of p. The sensitivity of the amplitudes and the spectral positions of MCD maxima to the compensation offers a worthwhile opportunity to estimate carrier concentration from optical measurements. The comparison of the theoretical predictions of MCD spectra performed within eight [3] and ten (Fig. 2) band KL models seems to indicate that dominant transitions observed in MCD experiments [7, 9] are connected with valence -conduction bands transitions. However more experiments on metallic (III,Mn)V is necessary to confirm this point.

ACKNOWLEDGMENTS

We thank W. A. Atkinson, A. H. MacDonald, H. Ohno, K. S. Burch, D. N. Basov and J. Mašek for useful discussions. The work by TD was partly supported by FENIKS project (EC:G5RD-CT-2001-00535). This work was further supported by the Welch Foundation, the DOE under grant DE-FG03-02ER45958, and the Grant Agency of the Czech Republic under grant 202/02/0912.

REFERENCES

1. Singley, E. J., et al., *Phys. Rev. B*, **68**, 165204 (2003).
2. Dietl, T., et al., *Science*, **287**, 1019 (2000).
3. Hankiewicz, E. M., et al., *cond-mat/0402661* (2004).
4. Sinova, J., et al., *Phys. Rev. B*, **67**, 235203 (2003).
5. Yang, S.-R., et al., *Phys. Rev. B*, **67**, 045205 (2003).
6. Matsukura, F., et al., *Physica E*, **21**, 1032 (2004).
7. Beschoten, B., et al., *Phys. Rev. Lett.*, **83**, 3073 (1999).
8. Meyer, B. K., et al., *Phys. Rev. Lett.*, **52**, 851 (1984).
9. Ando, K., et al., *J. Appl. Phys.*, **83**, 6548 (1998).

Carrier Concentration Dependencies of Magnetization & Transport in $Ga_{1-x}Mn_xAs_{1-y}Te_y$

M.A. Scarpulla[1,2], K.M. Yu[2], W. Walukiewicz[2], O.D. Dubon[1,2]

[1]*Department of Materials Science & Engineering, University of California Berkeley, Berkeley, CA 94720*
[2] *Lawrence Berkeley National Laboratory, Berkeley, CA 94720*

Abstract. We have investigated the transport and magnetization characteristics of $Ga_{1-x}Mn_xAs$ intentionally compensated with shallow Te donors. Using ion implantation followed by pulsed-laser melting, we vary the Te compensation and drive the system through a metal-insulator transition (MIT). This MIT is associated with enhanced low-temperature magnetization and an evolution from concave to convex temperature-dependent magnetization.

INTRODUCTION

$Ga_{1-x}Mn_xAs$ has been established as the prototypical III-V system displaying ferromagnetism mediated by (fairly) delocalized carriers. Despite recent advances in film processing that have allowed the T_C to reach 150 K [1], its dependence on the hole concentration is still poorly understood. Compensation by Mn interstitials, which are both electrically and magnetically active, has complicated magnetic and transport characterization of this system - thus obscuring the underlying physics. Previously, Satoh *et al.* investigated Sn compensation of $Ga_{0.987}Mn_{0.013}As$ films grown by LT-MBE [2]. In this work, we have endeavored to further probe the dependencies of magnetization and charge transport on hole concentration by introducing compensating shallow Te donors in films having much higher Mn concentrations.

EXPERIMENTAL

Semi-insulating GaAs (001) wafers were implanted with 160 keV Te^+ to varying doses followed by a dose of 1.84×10^{16} /cm² 80 keV Mn^+. Mn was implanted first to reduce sputtering effects and water cooling was employed to ensure full amorphization. Samples were irradiated in air with single pulses of varying fluence from a KrF excimer laser as we have described previously [3,4]. SIMS measurements reveal that both the Mn and Te concentrations vary with depth and that the peak Mn concentration is roughly x=0.07 for the samples discussed herein. The total amounts of Te and Mn remaining in the films after processing were measured using RBS and PIXE, and we define the overall Te:Mn ratio as γ [4]. We emphasize that no post-irradiation annealing is necessary to achieve high T_Cs using our process because our films are virtually free of Mn interstitials due to the liquid-phase epitaxial regrowth. All samples were etched in HCl for 15 minutes to remove metallic droplets and Mn oxides from the surface. Magnetization measurements along {110} in-plane directions were carried out using a DC SQUID magnetometer and were shown to be unchanged by the HCl etching. Resistivity measurements were conducted using pressed In contacts in the van der Pauw geometry.

RESULTS & DISCUSSION

Figure 1 presents the sheet resistivity of representative $Ga_{1-x}Mn_xAs_{1-y}Te_y$ samples spanning the compensation-induced MIT. For clarification, we emphasize that this MIT is distinct from the MIT commonly observed at low temperatures in $Ga_{1-x}Mn_xAs$. This MIT should be compared to the Anderson localization transition in doped semiconductors, keeping in mind that magnetic interactions may also be significant. The MIT appears to occur in the range of γ~2/3-3/4 for different processing conditions. The γ=0 and 0.64 samples both

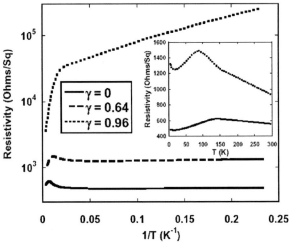

FIGURE 1. [main] Resistivity vs. 1/T for $Ga_{1-x}Mn_xAs_{1-y}Te_y$ samples irradiated at 0.2 J/cm^2 and having Te:Mn ratios $\gamma=0$, $\gamma=0.64$, and $\gamma=0.96$. [inset] Sheet resistivity vs. temperature for the $\gamma=0$ (solid) and $\gamma=0.64$ (dashed) samples.

display metallic behavior typical of $Ga_{1-x}Mn_xAs$ with $x>\sim0.03$. Both samples exhibit a strong resistivity peak near their T_Cs of 130 and 99 K which is commonly attributed to critical scattering. No distinct peak is present for the insulating $\gamma=0.96$ sample, which displays thermally activated transport between 4.2 and 300 K. However, a drastic change in activation energy occurs near its ferromagnetic T_C of ~68 K.

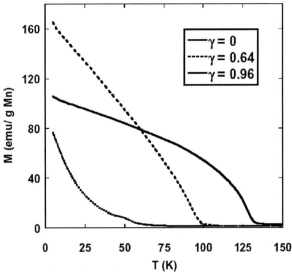

FIGURE 2. Magnetization vs. temperature for the set of $Ga_{1-x}Mn_xAs_{1-y}Te_y$ samples presented in Fig. 1.

Figure 2 presents low-field (50 Oe) measurements of magnetization vs. temperature for the same set of films. The uncompensated $\gamma=0$ sample exhibits a concave dependence, the $\gamma=0.64$ sample near the MIT exhibits a nearly linear dependence, and the strongly-compensated $\gamma=0.96$ sample displays a convex dependence. These changes in magnetization curvature with compensation are qualitatively consistent with recent mean field models [5]. An as-yet unexplained enhancement in low- and high- field magnetizations is present at low temperatures in samples near the MIT.

CONCLUSIONS

The magnetization and resistivity of Te compensated $Ga_{1-x}Mn_xAs$ films produced using ion implantation and pulsed laser melting were investigated as functions of temperature. A metal insulator transition was identified which delineates metallic films showing concave magnetization curves from insulating films exhibiting thermally activated conduction at all temperatures and convex magnetization curves.

ACKNOWLEDGMENTS

This work was supported by the Director, Office of Science, Office of Basic Energy Sciences, Division of Materials Sciences and Engineering, of the US Department of Energy under Contract No. DE-AC03-76SF00098. MAS acknowledges support from an NSF Graduate Research Fellowship. ODD acknowledges support from the Hellman Family Fund.

REFERENCES

1. Ku, K.C., et al., *Appl. Phys. Letters* **82**, 2302-2304 (2003).

2. Satoh, Y., et al., *Physica E* **10**, 196-200 (2001).

3. Scarpulla, M.A., et al., *Appl. Phys. Letters* **82**, 1251-1253 (2003).

4. Scarpulla, M.A., et al., *Physica B* **340-342**, 908-912 (2003).

5. Das Sarma, S., et al., *Phys. Rev. B* **67**, 155201 (2003).

Ferromagnetism and carrier polarization of Mn-doped II-IV-V$_2$ chalcopyrites

P. R. C. Kent* and T. C. Schulthess[†]

University of Tennessee, Knoxville, TN 37996
[†]*Oak Ridge National Laboratory, Oak Ridge, TN 37831*

Abstract. Density functional calculations are used to investigate the magnetic properties of the Mn-doped II−IV−V$_2$ chalcopyrites CdGeP$_2$, ZnGeP$_2$, ZnGeAs$_2$, ZnSnP$_2$, and ZnSnAs$_2$. We find the Zn-based compounds have significantly larger ferromagnetic interactions than CdGeP$_2$. Calculated valence band spin-splittings indicate the possibility of high carrier polarization in these systems. We do not find a simple relationship between lattice constant and exchange coupling, as suggested by simple models. Our results suggest the II−IV−V$_2$ chalcopyrites are promising alternate high temperature dilute magnetic semiconductors.

INTRODUCTION

Since the discovery of dilute magnetic semiconductors (DMS) with Curie Temperatures (T_c) above liquid nitrogen temperatures[1], the search has focused on potential room temperature DMS materials. Primarily focused on III-V materials, e.g. GaAsMn, progress has been difficult due to the intrinsic low Mn solubility. Model calculations[2] suggested that wide band-gap/small lattice constant III-V and II-VI materials, e.g. GaN and ZnO, would have high T_c, but success to date has been limited. As an alternative to these conventional materials, we have investigated the II−IV−V$_2$ chalcopyrite (pnictide) alloys. These "virtual III-V" materials, consisting of two interposing zinc-blende lattices, are compatible with conventional III-V materials while permitting high degrees of Mn incorporation: Mn^{2+} ions may easily substitute on the group II sites[3]. Although experimental investigations have been made for CdGeP$_2$[3, 4], ZnGeP$_2$[5, 4], ZnSnAs$_2$[6], and ZnGeSiN$_2$[7], no systematic trends in T_c have yet emerged.

Ferromagnetism in the chalcopyrites is more complex than in the III-Vs and II-VIs, because Mn may substitute on different sites, depending on growth conditions[8]. We have performed a systematic *ab-initio* investigation of the magnetic properties of Mn-doped CdGeP$_2$, ZnGeP$_2$, ZnGeAs$_2$, ZnSnP$_2$, and ZnSnAs$_2$. We investigated dilute Mn alloys and Mn-related defects, where ferromagnetism may be promoted through the simultaneous creation of spins *and* carriers[3, 8, 12], to uncover any trends in the magnetic properties. We also calculated the degree of splitting at the valence band edges to determine any trends in attainable carrier polarization.

METHODS

We utilized density functional theory for our calculations, using the projector-augmented wave approach, VASP code[9, 10], and Perdew-Wang 91 GGA functional[11]. For Brillouin zone integrations of bulk systems, we used an 8x8x8 Monkhorst-Pack k-point mesh, corresponding to 56 k-points in the irreducible wedge. For 64 atom supercell calculations, we utilized a coarser 2x2x2 k-point mesh.

We calculated the difference in energy (2J, the Heisenberg spin coupling) between ferromagnetic and anti-ferromagnetic spin arrangements of "near neighbor" Mn atoms, and use the strength of the ferromagnetic interaction as an approximate measure of the ferromagnetic T_c. We examined (i) Mn$_{II}$ − Mn$_{II}$ pairs, e.g. where Mn substitutes neighboring Zn sites in ZnGeAs$_2$, (ii) Mn$_{II}$ − Mn$_{IV}$, e.g. where Mn substitutes a Zn and adjacent Ge site, and (iii) Mn$_{IV}$ − Mn$_{IV}$. In the latter two cases, both holes and spins are introduced, enabling carrier induced ferromagnetism. These pair substitutions have been shown to be energetically favorable under certain growth conditions[8], e.g. group IV poor.

RESULTS

The calculated energy differences between ferromagnetic and anti-ferromagnetic spin configurations of different Mn configurations are shown in Fig. 1. We observe, for Mn$_{II}$ − Mn$_{II}$ pairs (Fig. 1a), a strong preference for anti-ferromagnetic interaction, in agreement with expectations[3]. For other pairings, the interactions

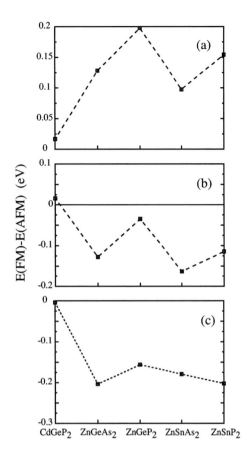

FIGURE 1. Relative energy of ferromagnetic vs. antiferromagnetic spin configurations of Mn-pairs, for 64 atom supercells computed with GGA density functional theory. Mn pairs on neighboring (a) group II sites, (b) II-IV sites, (c) IV sites (see text). Negative energies indicate ferromagnetism is preferred. Lines drawn are a guide to the eye.

are predominantly ferromagnetic (Fig. 1b,c).

Overall, the weakest exchange couplings were calculated in $CdGeP_2$. The *magnitude* of the couplings for all the Zn compounds is considerably larger than for $CdGeP_2$. Strong ferromagnetic couplings are found for pairs where at least one Mn resides on a non-group II site, indicating a potential route to large-scale ferromagnetism. In the III-V compounds large T_c are predicted for small lattice constant materials owing to the assumed RKKY interaction[2]. However, in the chalcopyrites we find no consistent trend between host lattice constant and calculated interactions (Fig. 1), e.g. $ZnGeP_2$ has the smallest lattice constant of the materials considered here, but does not differ dramatically in calculated exchange constants.

We also calculated the valence band spin splittings for each of the chalcopyrite systems. The spin-splittings, measured as the energy difference between the highest valence band up-spin and down-spin states were 0.12 eV ($Zn_{0.94}Mn_{0.06}GeAs_2$), 0.21 eV ($Cd_{0.94}Mn_{0.06}GeP_2$), 0.21 eV ($Zn_{0.94}Mn_{0.06}SnAs_2$), 0.26 eV ($Zn_{0.94}Mn_{0.06}GeP_2$), and 0.26 eV ($Zn_{0.94}Mn_{0.06}SnP_2$). These calculated values compare favorably to those of 0.39 eV in GaAsMn[13] and 0.06 eV in GaNMn[13]. Whereas small splittings are found for GaNMn (due to the low valence band maximum in GaN), higher spin-splittings are obtained for the chalcopyrite compounds, indicating the possibility of larger carrier polarization.

CONCLUSIONS

Zn-based $II-IV-V_2$ chalcopyrites (IV=Ge or Sn, V=As or P) have significantly larger calculated ferromagnetic exchange interactions than $CdGeP_2$, and therefore likely higher ferromagnetic transition temperatures, T_c. Significant valence band spin splittings are found for all examined chalcopyrites. Work supported by the Defense Advanced Research Project Agency and by the Division of Materials Science and Engineering, U.S. Department of Energy, under Contract DE-AC05-00OR22725 with UT-Battelle LLC.

REFERENCES

1. Ohno, H., Shen, F., A. anf Matuskura, Oiwa, A., Endo, A., Katsumoto, S., and Iye, Y., *Appl. Phys. Lett.* (1996).
2. Dietl, T., Ohno, H., Matsukara, F., Cibert, J., and Ferrand, D., *Science*, **408**, 944 (2000).
3. Medvedkin, G. A., Ishibashi, T., Nishi, T., Hayata, K., Hasegawa, Y., and Sato, K., *Jpn. J. Appl. Phys.*, **39**, L949 (2000).
4. Sato, K., Medvedkin, G., and Ishibashi, T., *J. Cryst. Growth*, **237-239**, 1363 (2002).
5. Medvedkin, G., Hirose, K., Ishibshi, T., Nishi, T., Voevodin, V. G., and Sato, K., *J. Cryst. Growth*, **236**, 609 (2002).
6. Choi, S., Cha, G.-B., Hong, S. C., Cho, S., Kim, Y., Ketterson, J. B., Jeong, S.-Y., and Yi, G., *Solid State Commun.*, **122**, 165 (2002).
7. Pearton, S. J., Overberg, M. E., Abernathy, C. R., Theodoropoulou, N. A., Hebard, A. F., Ghu, S. N. G., Osinsky, A., Fuflyigin, V., Zhu, L. D., Polyakov, A. Y., and Wilson, R. G., *J. Appl. Phys.*, **92**, 2047 (2002).
8. Mahadevan, P., and Zunger, A., *Phys. Rev. Lett.*, **88**, 047205 (2002).
9. Kresse, G., and Furthmuller, J., *Phys. Rev. B*, **54**, 11169 (1996).
10. Kresse, G., and Joubert, J., *Phys. Rev. B*, **59**, 1758 (1999).
11. Perdew, J. P., and Wang, Y., *Phys. Rev. B*, **45**, 13244 (1992).
12. Zhao, Y.-J., Picozzi, S., Continenza, A., Geng, W. T., and Freeman, A. J., *Phys. Rev. B*, **65**, 094415 (2002).
13. Mahadevan, P., and Zunger, A., *Phys. Rev. B*, **69**, 115211 (2004).

Search For Hole Mediated Ferromagnetism In Cubic (Ga,Mn)N

M. Sawicki[1], T. Dietl[1], C.T. Foxon[2], S.V. Novikov[2], R.P. Campion[2], K.W. Edmonds[2], K.Y. Wang[2], A.D. Giddings[2], B.L. Gallagher[2]

[1] *Institute of Physics, Polish Academy of Sciences, al. Lotników 32/46, Warszawa, Poland.*
[2] *School of Physics and Astronomy, University of Nottingham, Nottingham, United Kingdom*

Abstract. Results of magnetisation measurements on p-type zincblende-(Ga,Mn)N are reported. In addition to a small high temperature ferromagnetic signal, we detect ferromagnetic correlation among the remaining Mn ions, which we assign to the onset of hole-mediated ferromagnetism in $Ga_{1-x}Mn_xN$.

INTRODUCTION

For widespread technological usage of ferromagnetic semiconductors [1], a Curie temperature T_C significantly above 300 K is necessary. In this context, the mean-field p-d Zener model prediction of the room temperature hole-mediated ferromagnetism in (Ga,Mn)N [2] stimulated much interest. However, contrary to (Ga,Mn)As, Mn does not seem to be an efficient acceptor in GaN. Nevertheless, there are numerous observations of room temperature ferromagnetism in (Ga,Mn)N, even in samples that appear to be n-type [3]. The origin of the ferromagnetism in these cases is unresolved but there is little evidence that the ferromagnetic-like behaviour is brought about by carrier mediated coupling. Hole-induced ferromagnetism appears more probable in cubic (Ga,Mn)N, as Mn incorporation is favoured in this phase and since large polar effects, which hinder effective p-type doping, are absent. In fact, we recently demonstrated that this material can be highly *p*-type, with carrier concentrations approaching 10^{18} cm^{-3} at room temperature for desirable Mn contents [4]. Here we report on magnetic studies of such a material, which point to the presence of the onset of the hole-mediated ferromagnetism.

SAMPLES AND EXPERIMENTAL

Cubic (Ga,Mn)N layers, of typical thickness 300 nm, were grown on semi-insulating GaAs (001) substrates by plasma-assisted molecular beam epitaxy using arsenic as a surfactant to initiate the growth of cubic phase material [5]. Films were grown under N-rich conditions at growth temperatures from 450 to 680°C. The Mn concentration in the films was set using the *in-situ* beam monitoring ion gauge, and calibrated *ex situ* by secondary ion mass spectrometry (SIMS). Arsenic concentration in the layers is below 0.5 %, which is known to reduce GaN band gap less than 50 meV [6]. To ensure electrical isolation as well as rule out the possibility of Mn diffusion into the GaAs layer being responsible for the observed electrical and magnetic properties, an undoped cubic GaN (~150 nm thick) followed by a cubic AlN buffer layers (50-150 nm thick) were introduced between the GaAs and cubic (Ga,Mn)N layers in some samples. Hall effect measurements unambiguously reveal that the layers are *p*-type. The nature of relevant acceptor is unknown at present, but we note that layers grown under the same conditions but without Mn are *n*-type. The hole density p_{Hall} generally increases with increasing Mn up to ~4 %, reaching a plateau or decreasing slightly above this value [4]. Since p_{Hall} at 300 K is found to be in the range 10^{17} to 10^{18} cm^{-3}, the doping level is below the Mott critical concentration, and the samples show localisation of the carriers at low temperatures.

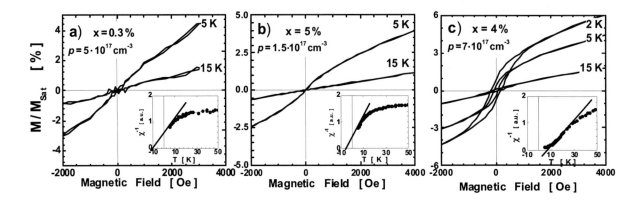

FIGURE 1. Hystersis loops at selected temperatures for zb-(Ga,Mn)N with various hole and Mn concentrations (as established by SIMS). The data are plotted relative to the expected saturation for the stated Mn contents.

Magnetic measurements are performed in a SQUID magnetometer, with magnetic fields of up to 4 kOe applied in the plane of the sample. Room temperature measurements evidence the presence of a weak, on average below 0.1 μ_B/Mn, ferromagnetic signal. Similarly to earlier reports, it persists to above 400 K, so MnAs formation due to Mn diffusion to GaAs substrate can be excluded. The origin of this ferromagnetic signal is not presently known, however there are several Mn_xN_y phases with ferromagnetic transition temperature above room temperature which could account for the observed behaviour [7]. Here we concentrate on investigating the majority of Mn that is not involved in the high temperature coupling. The measured magnetisation of the room temperature ferromagnetic phase is nearly temperature-independent below 200 K. We therefore subtract magnetisation curves measured at 50 K from those measured at lower temperatures. Starting with samples having low concentration of either Mn ions (Fig. 1a) or holes (Fig. 1b) we find the relative magnetisation, $\Delta M(H,T) = M(H,T) - M(H,50 K)$, to be nearly linear with the external magnetic field, characteristic of a paramagnetic phase. A considerably modified behaviour is observed for samples having higher hole densities (as shown on Fig. 1c). There is a much stronger low field curvature and a clear hysteresis is developing on lowering temperature. This is a clear indication of an additional, low temperature magnetic coupling between the Mn ions in this sample. This conjecture is further supported by low temperature $\chi^{-1}(T)$ (plotted in the insets), which changes from a negative value for the first two samples to clearly positive value for the most heavily doped sample. We ascribe this to the presence of holes, which favours a *ferromagnetic* alignment of substitutional Mn at low T. We also note that the magnetisation curves of Fig. 1c are reminiscent of those observed for p-type (Zn,Mn)Te single crystals [8]. Even though those samples exhibited insulating behaviour at low T, recent inelastic neutron measurements indicated local ferromagnetic ordering mediated by weakly localised holes [9]. However, other mechanisms may also give rise to ferromagnetic ordering at low temperatures in p-type materials, including percolation of bound magnetic polarons [10]. Clearly, more work is required to resolve this issue.

This work was supported by EU FENIKS project (EC:G5RD-CT-2001-00535), UK EPSRC (Gr/46465 and GR/S81407), and Polish KBN grant PBZ-KBN-044/P03/2001.

REFERENCES

1. T. Dietl and H. Ohno, MRS Bulletin, October 2003, p. , and references therein.
2. T. Dietl *et al.*, Science **287**, 1019 (2000).
3. See, S.J. Pearton *et al*, J. Appl. Phys. **93**, 1 (2003).
4. S.V. Novikov *et al.*, Semicond. Sci. Technol. **19**, L13 (2004), and to be published.
5. T.S. Cheng et al., Appl. Phys. Lett. **66** 1509 (1995).
6. C.T. Foxon et al.,J. Phys: Condens. Matter., **14**, 3383 (2002).
7. M. Zając *et al.*, J. Appl. Phys. **93**, 4715 (2003).
8. M. Sawicki *et al.*, Phys. Stat. Sol. B **229**, 717 (2002).
9. H. Kępa et al, Phys. Rev. Lett. **91**, 087205 (2003).
10. A. Kaminski and S. Das Sarma, Phys. Rev. Lett. **88**, 247202 (2002); A. C. Durst, R. N. Bhatt, and P. A. Wolff, Phys. Rev. B **65**, 235205 (2002).

Clear Spin Valve Signals in Conventional NiFe/In$_{0.75}$Ga$_{0.25}$As-2DEG Hybrid Two-Terminal Structures

Masashi Akabori, Katsushige Suzuki and Syoji Yamada

Center for Nano Materials and Technology (CNMT), Japan Advanced Institute of Science and Technology (JAIST), 1-1 Asahidai, Tatsunokuchi, Ishikawa 923-1292, Japan

Abstract. We investigated the transport properties of ferromagnetic/semiconductor hybrid two-terminal structures utilizing an In$_{0.75}$Ga$_{0.25}$As/In$_{0.75}$Al$_{0.25}$As 2DEG formed on a GaAs (001) substrate with In$_x$Al$_{1-x}$As step-graded buffer layers. We used NiFe as ferromagnetic electrodes for injection/detection of spin-polarized electrons, which were formed on side walls of the semiconductor mesa to contact electron channel directly. We measured spin valve properties at low temperatures, and successfully found clear spin valve signals as well as clear channel width dependence. The results with such clear dependence suggest successful spin transport in the present samples.

INTRODUCTION

Spin field effect transistors (spin-FETs) proposed by Datta and Das [1] has been paid much attentions, because their operation is based on the combination of spin filter by ferromagnetic (FM) electrodes and spin precession by spin-orbit (SO) interaction in the channel. In order to realize spin-FETs, we have studied spin transport in original In$_{0.75}$Ga$_{0.25}$As two-dimensional electron gas (2DEG) [2], and demonstrated spin injection signals from a FM in the non-local configuration [3]. However, the previous samples were affected by the local Hall effect (LHE) via fringe field of FM contacts, therefore we could not obtain spin valve signals in the conventional configuration. In this paper, we report on the fabrication of simple two-terminal samples, and demonstrate clear spin valve signals without LHE.

EXPERIMENTAL PROCEDURE

Our 2DEG structures having large SO interaction were a metamorphic In$_{0.75}$Ga$_{0.25}$As/In$_{0.75}$Al$_{0.25}$As modulation doped heterostructure grown by molecular beam epitaxy. Its electron concentration and mobility at 1.6 K are $N_S = 5.1 \times 10^{11}$ cm^{-2} and $\mu = 5.4 \times 10^4$ cm^2/Vs, respectively. Figure 1 shows top view and schematic cross-section of a fabricated hybrid structure. We fabricated three kind of samples with different channel width, W = 1.6 µm, 2.5µm, and 6.2 µm. The channel length, L, was fixed to 2.2 µm which is about 3.7 times as long as the estimated mean free path l_e = 0.6 µm. We used electron-beam lithography and wet etching with H$_2$SO$_4$-base solution for mesa fabrication with 400-nm-depth. The FM electrodes consisted of 50-nm-thick Ni$_{40}$Fe$_{60}$ were fabricated on side walls of the mesa structures by radio-frequency sputtering process. We note that Ar-ion etching during 2 minutes in the same chamber was carried out before the formation of FM electrodes to remove native

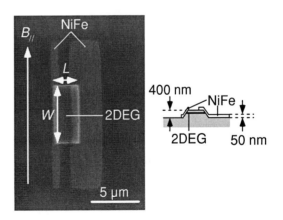

FIGURE 1. Top view and schematic cross-section of a NiFe/2DEG hybrid two-terminal structure.

oxides on FM/2DEG contact area.

We measured transport properties by AC lock-in technique in a conventional liquid He cryostat with a super conducting magnet. Typical measurement temperature was 1.5 K.

RESULTS AND DISCUSSIONS

Figure 2 shows spin valve properties of the samples at 1.5K. In the present samples, symmetrical spin valve signals were clearly observed without hysterisis loops. The results suggest the present design indeed reduces the LHE. Additionally, we note that the results were obtained in longer channels with $L = 2.2$ µm compared to the mean free path, $l_e = 0.6$ µm. The fact indicates the spin relaxation should not be induced by the elastic scattering.

Furthermore, we found some unique features of spin valve peaks. Figure 3 shows the plots of peak positions and full-width half-maximum (FWHM). From the plots, as the channel width became narrower, the peak shape became sharpened and the positions were shifted close to zero field. We think such peak behaviors originate from the coercive force of FM contacts. Thus, it unfortunately depended on the aspect ratio of effective contact regions corresponding to mesa side-walls and/or excess FM parts on mesa-top. However, the results also suggest successful spin transport in the present samples due to these channel width dependence.

To confirm magnetic structures in FM contacts, we

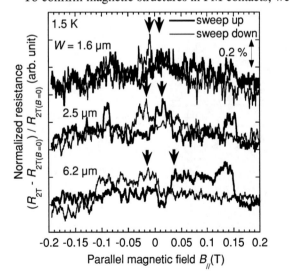

FIGURE 2. Spin valve properties of NiFe/2DEG hybrid two-terminal structures.

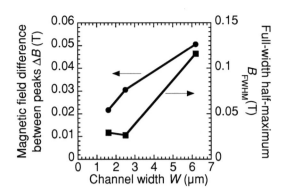

FIGURE 3. Channel width dependences of spin valve peaks.

measured magnetic force microscopy (MFM) of the hybrid structures, although the images are not shown here. We found clear domain walls at the edges of mesa-top as well as many domains in both the FM contacts on isolation area. However, we could not observe magnetic dipoles at mesa side-walls due to the limit of MFM resolution. At the present, we conclude that the deep mesa structures resulted in complex domain structures in FM contacts as well as channel width dependence of spin valve properties.

SUMMARY

We investigated spin valve properties of NiFe/ $In_{0.75}Ga_{0.25}As$-2DEG hybrid two-terminal structures utilizing a metamorphic modulation doped heterostructure. We successfully found clear spin valve signals as well as clear channel width dependence at 1.5 K. With such clear dependence and MFM measurements, we think that the results suggest successful spin transport in the present structures.

ACKNOWLEDGEMENTS

This work is partially supported by a Grant-in-Aid for Scientific Research in Priority Areas "Semiconductor Nanospintronics" (No.14076213) of The Ministry of Education, Culture, Sports, Science and Technology, Japan, and by Mitsubishi, SCAT, and Nakajima Foundations for Science and Technology.

1. S. Datta and B. Das, Appl. Phys. Lett. 56, (1990) 665.

2. S. Gozu et al., Jpn. J. Appl. Phys. 37 (1998) L1501.

3. Y. Sato et al., Jpn. J. Appl. Phys. 40 (2001) L1093.

Tunneling magnetoresistance: The relevance of disorder at the interface

Michael Wimmer and Klaus Richter

Institut für Theoretische Physik, Universität Regensburg, D-93040 Regensburg, Germany

Abstract. The effect of disorder on the tunneling magnetoresistance (TMR) of a semiconductor tunnel barrier is investigated using a single-band model including elastic scattering. We find that disorder can decrease the TMR ratio significantly. Furthermore, we show that impurities close to the barrier interface are most effective in reducing the TMR ratio.

INTRODUCTION

Tunneling magnetoresistance (TMR) has attracted a lot of interest from both fundamental and applied research. Most of the TMR experiments have been performed on amorphous oxide barriers, e.g. aluminum oxide [1], but theoretical predictions, based on *ab initio* calculations, are usually limited to epitaxial systems.

Band structure considerations predicted a very large TMR ratio ($\approx 100\%$) for epitaxially grown semiconductor barriers (such as ZnSe or GaAs) on Fe [2]. However, experiments on single-crystalline GaAs barriers only yielded a very low value of the TMR ratio ($\approx 1\%$) [3]. Some of the experimental evidence hints at the importance of disorder at the Fe/GaAs-interface and spin-flip scattering [3, 4].

It has already been shown, that impurities inside the tunneling barrier can have a great influence on spin-dependent tunneling [5]. In this paper we extend this analysis and especially consider the importance of disorder at the barrier interface.

MODEL AND TECHNIQUE

The model of the magnetic tunnel junction is based on a single-band description [6, 7], using a position-dependent effective mass $m^*(\mathbf{x})$ [8] and exchange-split bands for the ferromagnet. The Hamiltonian then reads

$$H = -\frac{\hbar^2}{2}\nabla \frac{1}{m^*(\mathbf{x})}\nabla + \frac{\Delta}{2}\mathbf{M}\cdot\boldsymbol{\sigma} + V(\mathbf{x}), \quad (1)$$

where $\mathbf{M}$ is the direction of the magnetization in the ferromagnetic leads, Δ the exchange splitting and $V(\mathbf{x})$ the (barrier) potential. The material parameters for a Fe/GaAs junction are given in Table 1. We now introduce disorder into the system through an impurity potential

$$V_{\text{imp}}(\mathbf{x}) = \sum_i \bar{U}_i \delta(\mathbf{x}-\mathbf{x}_i) \quad (2)$$

with random impurity positions $\mathbf{x}_i$ and $\bar{U}_i$ a random variable taken from the uniform distribution $[-\bar{U}_0, \bar{U}_0]$.

The conductance of the heterostructure is calculated using Landauer-Büttiker theory [9]. To solve the problem numerically, we discretize the Hamiltonian (1) using a grid with lattice spacing a [10] yielding an effective tight-binding description of the problem. The impurity potential then reduces to the Anderson model [11] where the impurity energies are taken from a random distribution $[-U_0, U_0]$ ($\bar{U}_0 = U_0 a^3$). The transmission coefficient is then calculated using a recursive Green's function algorithm [12] and the Fisher-Lee relation [13].

RESULTS AND DISCUSSION

We now calculate the effect of disorder on the TMR ratio. The TMR ratio is defined as

$$TMR = \frac{\langle G_P\rangle - \langle G_{AP}\rangle}{\langle G_{AP}\rangle}, \quad (3)$$

where $\langle...\rangle$ denotes an average over impurity configurations, $G_{P(AP)}$ is the conductance for parallel (antiparallel) alignment of the magnetizations in the ferromagnets.

In our numerical calculations we choose the lattice constant $a=1\,\text{Å}$. This gives results within 10% of the continuum limit $a \to 0$. Since this model is only expected

TABLE 1. Material parameters: $k_{F,\uparrow/\downarrow}$: Fermi wave vector of Fe, V_0: GaAs barrier height (from [14]).

$k_{F,\uparrow}$	$k_{F,\downarrow}$	m_{Fe}	V_0	m_{GaAs}
0.86 Å^{-1}	0.29 Å^{-1}	1.0 m_e	0.75 eV	0.067 m_e

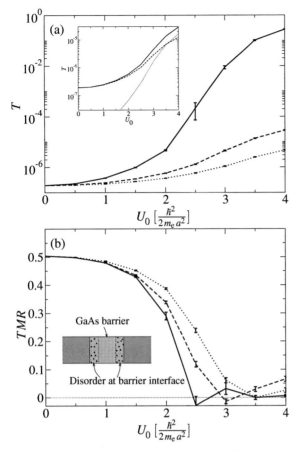

FIGURE 1. Transmission probability (for parallel magnetizations) (a) and TMR ratio (b), Eq. (3), as a function of the disorder strength for a (A) fully disordered barrier (solid line), disorder within (B) 1 nm (dashed line) and (C) 0.5 nm (dotted line) of the barrier interface. The inset in (a) shows the specular (dashed line) and scattered part (dotted line) of the total transmission of B (solid line)

to yield *trends* rather than a quantitative description of experiment, the agreement is reasonable. For computational reasons we only consider a two-dimensional system with a width of 100 grid points. Given that tunneling is a quasi one-dimensional process for a low tunneling barrier such as GaAs (decaying wave vector $\kappa \ll k_{F,\uparrow/\downarrow}$) [6, 7], the results of a three-dimensional simulation are not expected to differ significantly from two-dimensional calculations. Simulations for different systems have confirmed this assumption.

The calculations have been performed for a 7 nm GaAs barrier and different distributions of the impurities: disorder everywhere in the barrier (sample A) and disorder located within 1 nm (B) and 0.5 nm (C) of the barrier interface (see inset in Fig. 1 (b)). We find a strong increase in the transmission probability due to resonant, impurity-assisted tunneling. The maximum current is considerably less for disorder localized at the interfaces, since then the transmission is bounded by the remaining 5 nm (B) and 6 nm (C) of clean barrier.

With increasing disorder strength we also find a strong decrease in the TMR ratio. Surprisingly, the decrease for samples B and C is only slightly weaker than for sample A, although the disorder only represents a small fraction of the total barrier in these cases. Since there is still a sizeable, clean barrier left, the reduction of the TMR ratio cannot be due to resonant coupling of the leads as suggested in [5].

The effect of disorder on the transmission is twofold: First, disorder changes the electronic structure and thus the density of states of the system. Second, it breaks the transversal translational symmetry of the tunnel junction leading to the mixing of conduction channels with different transversal momentum $k_{\parallel}$. An analysis of the transmission probability for disorder within 1 nm of the interface (B) however shows (inset in Fig. 1 (a)), that the specular ($k_{\parallel}$-conserving) part dominates over the scattered part of the transmission for $U_0 < 3.5$, when the TMR effect has already vanished almost completely.

We can therefore conclude that indeed the change in the density of states is responsible for the reduction of the TMR effect. Disorder close the interface proves to be most effective in reducing the TMR ratio in this regard.

ACKNOWLEDGMENTS

We acknowledge support from the Deutsche Forschungsgemeinschaft within GRK 638.

REFERENCES

1. Moodera, J. S., Kinder, L. R., Wong, T. M., and Meservey, R., *Phys. Rev. Lett.*, **74**, 3273 (1994).
2. Mavropoulos, P., Papanikolaou, N., and Dederichs, P. H., *Phys. Rev. Lett.*, **85**, 1088 (2000).
3. Zenger, M., Moser, J., Wegscheider, W., Weiss, D., and Dietl, T. (2004), cond-mat/0404351.
4. Kreuzer, S., Moser, J., Wegscheider, W., and Weiss, D., *Appl. Phys. Lett.*, **80**, 4582 (2002).
5. Tsymbal, E. Y., and Pettifor, D. G., *Phys. Rev. B*, **58**, 432 (1998).
6. Slonczewski, J. C., *Phys. Rev. B*, **39**, 6995 (1989).
7. Bratkovsky, A. M., *Phys. Rev. B*, **56**, 2344 (1997).
8. BenDaniel, D., and Duke, C., *Phys. Rev.*, **152**, 683 (1966).
9. Büttiker, M., Imry, Y., Landauer, R., and Pinhas, S., *Phys. Rev. B*, **31**, 6207 (1985).
10. Mains, R. K., Mehdi, I., and Haddad, G. I., *Appl. Phys. Lett.*, **55**, 2631 (1989).
11. Anderson, P. W., *Phys. Rev.*, **109**, 1492 (1958).
12. MacKinnon, A., *Z. Phys. B*, **59**, 385 (1985).
13. Fisher, D. S., and Lee, P. A., *Phys. Rev. B*, **23**, 6851 (1981).
14. Kreuzer, S., Ph.D. thesis, Universität Regensburg (2001).

Comparison of spin injection and transport in organic and inorganic semiconductors

P. P. Ruden*, D. L. Smith[†], and J. D. Albrecht[¶]

University of Minnesota, Minneapolis, MN 55455
[†] *Los Alamos National Laboratory, Los Alamos, New Mexico 87545*
[¶] *Air Force Research Laboratory, Wright-Patterson Air Force Base, Ohio 45433*

Abstract. We present a theoretical model to describe spin transport in a structure consisting of a ferromagnetic metal injector, a thin (usually undoped) semiconductor layer, and a ferromagnetic metal collector. In thermal equilibrium the magnetic contacts are spin-polarized whereas the semiconductor is unpolarized. Due to the large ratio of the metal to semiconductor conductivities, the semiconductor needs to be driven far out of local thermal equilibrium to achieve efficient injection of spin-polarized electrons. This requires a barrier to injection that may be due either to a large Schottky barrier or to an insulating tunnel barrier. Since carrier mobilities (and other relevant parameters) in inorganic and organic semiconductors differ by orders of magnitude, the conditions for achieving a state far from equilibrium at the injecting contact are quite different for the two types of materials.

INTRODUCTION

Conjugated organic semiconductors are promising candidate materials for spintronic devices because long spin relaxation times are expected for carriers in the relevant conduction and valence bands that originate from carbon π and π* orbitals. Organic semiconductors are usually undoped, and given the nature of the relevant deposition process, control over doping profiles is limited. However, charge carriers may be injected from metallic contacts and organic electronic devices can be fabricated from disordered thin films. Electrical conduction results from carrier hopping between localized sites, and carrier mobilities are therefore quite low. At metal contacts to organic semiconductors, the Schottky energy barrier usually scales directly with the contact metal workfunction. To date, the experimental investigation of spin injection and transport has focused on inorganic semiconductors, with a few notable exceptions [1,2].

MODEL DESCRIPTION

We model electrical spin injection, transport, and detection in semiconductor device structures consisting of a thin semiconductor film (either a polymer or silicon) sandwiched between two ferromagnetic metal contacts. A device model is generalized to include spin injection and transport [3]. The coupled Poisson and drift-diffusion equations are solved and boundary conditions for the drift-diffusion equation are determined by interfacial carrier injection. We include the possibility of a spin-selective interfacial tunnel barrier due to a thin insulating layer between the metal and the semiconductor.

For each spin type, the injected current is the sum of a thermionic injection current, an interface recombination current, and a tunneling current, which enables electron transport through the potential barrier of the Schottky contact. Tunneling is spin-selective due to its dependence of the contact metal wavefunctions for spin up and spin down electrons at the Fermi level. In the numerical calculations we take spin-up electrons to have twice the transmission probability of spin-down electrons. Image charge induced Schottky barrier lowering, which is comparatively strong in the organic semiconductors due to their small dielectric constants, is included. If an additional thin insulating interface layer is present, it is treated as a spin-dependent series resistance. The current boundary condition at the electron collecting contact is formulated in direct analogy to that for the injecting contact. In the semiconductor the steady state spin current density and the spin polarized electron density are connected by spin relaxation.

RESULTS AND DISCUSSION

Considering the case of no interfacial barrier in addition to a Schottky contact, it is readily seen that the low mobility typical of organic semiconductors implies that the thermionic injection is nearly completely canceled by interface recombination (its time reversed process) for cases of low Schottky barrier and/or low applied bias. Only for large Schottky barriers and high applied bias is tunneling through the Schottky barrier the dominant injection process, and the junction may be driven far out of equilibrium (Fig. 1). In that case, the strong spin dependence of tunneling can yield spin polarization of 0.333 for the injected current [3].

In inorganic semiconductors the situation is quite different. Due to the much higher mobility, thermionic injection is not balanced by interface recombination, and the junction may be driven well out of equilibrium even with small bias. However, the weak spin dependence of the thermionic process prevents injection of currents that are significantly spin polarized. In principle, large applied bias enables tunneling in the case of inorganic semiconductors as well, but for undoped structures the electric field required is so high that breakdown due to impact ionization is likely to occur (Fig. 2). A possible way to enable tunneling through the Schottky barrier is to augment the electric field at the interface by incorporating a suitable doping profile [4].

The injection of a spin-polarized current is improved substantially by a thin insulating interface layer between the semiconductor and the ferromagnetic electrode. Strongly spin-dependent tunneling through that layer enables large spin-polarization of the injected current. Organic semiconductors may present excellent opportunities for the fabrication of such (organic) insulating layers through self-assembly techniques.

The spin current may be detected through the dependence of the device's electrical resistance on the relative orientation of the magnetization of the injecting and collecting contacts (spin valve). For a given current nearly all of the voltage difference across the device originates from the collecting contact. Because of the field direction, Schottky barrier tunneling does not contribute to the current at that contact in undoped devices. However, a tunneling barrier due to a thin insulating layer can be very effective in providing the spin-dependent mechanism necessary for inducing a measurable spin valve effect.

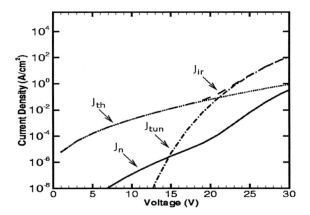

FIGURE 1. Magnitudes of thermionic injection, interface recombination, tunneling, and total injected electron current for a 0.8eV Schottky barrier on a polymer.

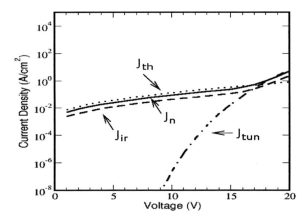

FIGURE 2. Results equivalent to Fig. 1 for 0.6eV barrier on silicon device.

ACKNOWLEDGMENTS

We thank J.W. Balk for his assistance. This work is supported in part by NSF-ECS, by an LANL LDRD program, and by AFOSR.

REFERENCES

1. V. Dediu, M. Murgia, F.C. Matacotta, C. Taliani, and S. Barbanera, Solid State Commun. **122**, 181 (2002).

2. Z.H. Xiong, D. Wu, Z. V. Vardeny, and J. Shi, Nature **427**, 821 (2004).

3. P.P. Ruden and D.L. Smith, J. Appl. Phys. **95**, 4898 (2004).

4. J.D. Albrecht and D.L. Smith, Phys. Rev. B **68**, 035340 (2003).

Spin-polarized And Ballistic Transport In InSb/InAlSb Heterostructures

Hong Chen[1], J. A. Peters[1], A. O. Govorov[1], J. J. Heremans[1],
N. Goel[2], S. J. Chung[2], M. B. Santos[2]

[1]*Ohio University, Dept. of Physics & Astronomy, & The Nanoscale & Quantum Phenomena Institute, Athens, OH 45701 USA*
[2]*University of Oklahoma, Dept. of Physics & Astronomy, & Center for Semiconductor Physics in Nanostructures, Norman, OK 73019 USA*

Abstract. We describe and experimentally demonstrate a method to create spin-polarized ballistic electrons through spin-orbit coupling in a two-dimensional electron system in an InSb/InAlSb heterostructure. Reflection of a spin-unpolarized injected beam from a lithographic barrier creates two fully spin-polarized side beams, in addition to an unpolarized specularly reflected beam. Antidot lattices were also fabricated in the InSb/InAlSb heterostructure. The temperature dependence of the antidot magnetoresistance from 0.4 K to 50 K was studied to characterize ballistic electron transport in InSb/InAlSb.

INTRODUCTION

The spin of electrons and holes in semiconductors has attracted much interest [1]. In heterostructures, spin can manifest itself through spin-orbit interaction (SOI) terms [2]. Semiconductor heterostructures with long carrier mean free path make it possible to fabricate mesoscopic devices in which SOI can be exploited for spin manipulation, and for the preparation of spin-polarized carrier states. InSb/InAlSb heterostructures in this respect combine a high mobility, and a strong SOI [3]. Our samples are fabricated from *n*-type InSb/InAlSb heterostructures grown on GaAs substrates by molecular beam epitaxy [3]. Electrons reside in a 20 nm wide InSb well, flanked by $In_{0.91}Al_{0.09}Sb$ barrier layers, with Si δ-doped layers on both sides. The material as grown yields a mobility $\mu = 1.6 \times 10^5$ cm^2/Vs, and two-dimensional density $N_S = 2.3 \times 10^{11}$ cm^{-2}, at a temperature $T = 0.4$ K, resulting in a mobility mean-free-path of ~1.3 µm. Mesoscopic structures are patterned on top of Hall bars by standard electron beam lithography and wet etching.

EXPERIMENT

We present a method to create spin-polarized beams of ballistic electrons by utilizing elastic scattering off a barrier in an open geometry. As illustrated in Fig. 1a and 1b, a beam of electrons in a two-dimensional system is injected towards a barrier. Both energy and the momentum parallel to the barrier are conserved during the scattering event off the barrier. However, in the presence of SOI, scattering off the barrier leads to spin-flip events, resulting in two fully spin-polarized side beams, in addition to an unpolarized specularly reflected beam. The spin-polarized reflected beams can then be captured through suitably positioned apertures (Fig. 1a). The multi-beam reflection process can be utilized to create spin-polarized electron populations without the use of ferromagnetic contacts. Figure 1c shows the sample geometry. Equilateral triangles of inside dimensions 3.0 µm feature ~0.4 µm wide apertures on two sides, while the left side forms the scattering barrier. Several triangles are measured in parallel. Carriers enter the geometry from the top, travel ballistically to the left barrier, reflect off the latter, and exit through the bottom aperture. The total distance, including the reflection, between the apertures, amounts to 2.6 µm. The resistance over the triangles is measured as a function of the magnetic field *B* applied perpendicular to the plane of the pattern. In the semi-classical limit, *B* serves to slightly deflect the carriers from linear trajectories, and thus to sweep the trajectories over the exit aperture. The interaction with the barrier gives rise to three reflection angles, and the exit (lower) aperture

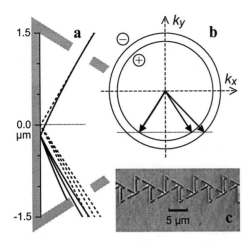

FIGURE 1. (a) Schematic of the geometry. Electrons are injected at the upper aperture, scattered from the left barrier and collected at the lower aperture. A perpendicular magnetic field B allows the trajectories to sweep the lower exit aperture. Trajectories are indicated for $B=0$ (dotted line) and $B=7$ mT (solid line). (b) Geometrical interpretation of the spin-polarized scattering event, with incident and reflected wave vectors at the Fermi surface (for clarity only scattering of incident + spin states is depicted). Energy and the momentum parallel to the barrier are conserved. (c) Image of the samples.

is sufficiently wide to accommodate the three exiting beams. Varying B in either direction causes the beams to be sequentially cut off. Each cutoff leads to a stepwise rise in the resistance. Three cutoffs on each side of the sweep, or six steps in total, are expected. Figure 2 contains experimental data for two separate samples, plotted as the four-contact resistance R versus applied B. The top curve shows 6 minima at low B, as a result from the stepwise increase in resistance added to the negative magnetoresistance weak-localization background. The uncertainty in the dimensions of the structures allows for the non-centralization of the 6 minima around $B=0$ and for the observation of less than 6 minima (lower curve in Fig. 2). The average

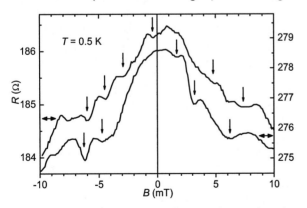

FIGURE 2. The four-contact resistance of two triangular structures versus the applied magnetic field B. The arrows indicate the values of B where beam cutoffs occur.

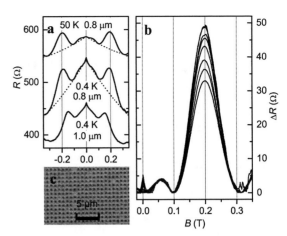

FIGURE 3. (a) Longitudinal magnetoresistance R of the 0.8 μm antidot lattice at 0.4 K and 50 K, with their background fits shown as dotted lines, as well as R of the 1.0 μm lattice at 0.4 K. (b) Ballistic transport peaks, after subtracting background, at different temperatures — from top to bottom: 0.4, 2.0, 5.0, 10, 20, 32, 40 and 50 K. (c) Image of the 0.8 μm antidot lattice.

difference between the minima is ~3 mT, close to the theoretically calculated result of the average distance between the cutoff magnetic fields, 3.4 mT. We also note that spin transport is coherent in the structures, since a calculation of the spin coherence length based on the Elliott-Yafet mechanism [4], yields ~ 20 μm.

To demonstrate ballistic transport in the InSb/InAlSb heterostructure, we fabricated and measured antidot lattices (Fig. 3c). Characteristic features of ballistic transport were observed at temperatures up to 50 K (Fig. 3a and 3b). The temperature dependencies of the ballistic peaks reveal a scattering time, particular to anti-dot lattices, with a linear dependence on temperature in the range 0.4 K to 50 K [5].

ACKNOWLEDGMENTS

The authors acknowledge support from the National Science Foundation under Grant No. DMR-0094055, DMR-0080054 and DMR-0209371.

REFERENCES

1. G. A. Prinz, *Science* **282**, 1660 (1998); Y. Ohno et al., *Nature* **402**, 790 (1999); D. Loss et al., *Phys. Rev. A* **57**, 120 (1998).
2. Y. A. Bychkov et al., *J. Phys. C* **17**, 6039 (1984); J. Luo et al., *Phys. Rev. B* **38**, 10142 (1988); J. Nitta et al., *Phys. Rev. Lett.* **78**, 1335 (1997).
3. K. J. Goldammer et al., *J. Cryst. Growth* **201/202**, 753 (1999); S. J. Chung et al., *Physica E* **7**, 809 (2000).
4. P. Murzyn et al., *Phys. Rev. B* **67**, 235202 (2003); J.-N. Chazalviel, *Phys. Rev. B* **11**, 1555 (1975).
5. Hong Chen et al., *Appl. Phys. Lett.* **84**, 5380 (2004).

High Spin Filtering Under the Influence of In-plane Magneto-electric Field and Spin Orbit Coupling

SG Tan[1], Mansoor BA Jalil[2], KL Teo[2], Thomas Liew[1], TC Chong[1]

1) Data Storage Institute, Singapore
2) ISML, ECE Department, National University of Singapore

Abstract. Periodic delta magnetic barriers are applied in the plane of a III-V semiconductor 2DEG across the electron conduction path of a HEMT device. Because of structural inversion asymmetry in the 2DEG, electron spin is coupled to both the crystal field and the in-plane magnetic field, producing highly spin-polarized current in the device. To maximize spin polarization |P|, an array of repeating zero-gauge type of magneto-electric barriers is used. Our calculation for electron transmission T for 10 repeating barrier units shows resonant-like peaks for different electron spin, resulting in ideal |P| of 100% over a wide energy range of 0.5-0.9 E_F. It was also found that a perpendicular E field can also modulate |P| in the 2DEG channel.

INTRODUCTION

Recent works [1,2,3] have indicated that magneto-electric fields (δB_z-E_z) applied perpendicular to the 2DEG plane of a HEMT device can induce the spin polarization |P| of electrons in the 2DEG. However, because of structural inversion asymmetry [4] in the 2DEG, electron spin is also coupled to the electric field within the zinc-blende type crystal lattice of narrow gap type of III-V semiconductors. This Rashba spin-orbit coupling (RSOC) effect constitutes a form of spin relaxation mechanism that resembles the D'yakonov Perel' [5] relaxation effect of wide bandgap semiconductor. Previous calculations [1,2,3] of ballistic electron transport influenced by external field have neglected this effect. The possibility of using an in-plane B field to counter-act spin-dephasing due to the D'yakonov-Perel' effect has also not been discussed.

DEVICE

In this article, we proposed the use of the in-plane delta magnetic field barriers instead of perpendicular fields used in ref. [1,2,3] to induce spin-polarized current in the 2DEG. The combined effect of in-plane magnetic and out-of-plane electric barriers (δB_y-E_z) and SOC effect on |P| is studied. Figure 1 shows the device model, which is based on the spin-FET structure first proposed by Datta and Das [4], in which the RSOC is solely responsible for the splitting of the conduction band dispersion curve. The δB_y-E_z are applied through a series of ferromagnetic gates deposited on top of the heterostructure. To maximize |P|, multiple gates are used in order to produce a periodic system of zero-gauge [6] type of magneto-electric barriers. Each repeating zero-gauge unit (Fig. 1) consists of four magnetic and two electrostatic barriers, whose height and orientation are such that an electron passing through requires zero net kinetic energy in the transverse direction. This condition is crucial as it allows the enhancement of |P| by having multiple barriers, without reducing electron T at energies close to the Fermi level E_F. Details of such transport are explained in ref. [6]. The HEMT device enables near-ballistic electron transport in the 2DEG especially at low temperatures. Thus, a ballistic model was used in computing the transmission and spin polarization of current. The minimal coupling type of free-electron model was assumed.

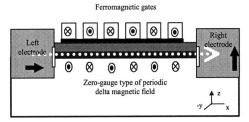

FIGURE 1. Schematic diagram of device considered with ferromagnetic gates magnetized alternately along ±y.

Rashba Spin Orbit Coupling

The Hamiltonian that describes electron transport in a 2DEG in the presence of RSOC (without δB_y-E_z fields) is:

$$H = H_0 + H_R = \frac{p_{//}^2}{2m} + \eta_R [k_y \sigma_x - k_x \sigma_y] \quad (1)$$

where H_0 and H_R are terms corresponding to the kinetic energy of the electron and the Rashba effect, η_R is the Rashba constant in eVm. Considering $k_y=0$ only, we use the Ansatz $\psi^+ = e^{i(k_z + eA_z)z}(A^+ e^{ik_R x} \chi_y^+ + B^+ e^{ik_L x} \chi_y^-)$ and $\psi^- = e^{i(k_z + eA_z)z}(A^- e^{iq_R x} \chi_y^- + B^- e^{iq_L x} \chi_y^+)$ to represent wavefunction that contains right-moving spin up and right-moving spin down electron, respectively. A, B are amplitudes of right-, left-moving wavefunctions, respectively; χ^+, χ^- are eigenspinors of σ_y; k_R, k_L, q_R, q_L are the degenerate wavevectors of the spin-split conduction band electron, $|k_R|=|k_L|$, $|q_R|=|q_L|$.

In-plane Magneto-electric Field

In the presence of δB_y-E_z, with the Landau gauge $A=(0, 0, -A_z)$, the Hamiltonian becomes:

$$H = \frac{-\hbar^2}{2m^*}\frac{\partial^2}{\partial x^2} + \frac{-\hbar^2}{2m^*}\frac{\partial^2}{\partial z^2} + U + \frac{e\hbar g^*}{4m_0}\sigma_y B_y + \sigma_y \eta_R \frac{i\partial}{\partial x} \quad (2)$$

Solution of (2) yields the wavevector equation (3) with k_x denoting any of the four degenerate wavevectors k_R, k_L, q_R, q_L,

$$k_x^2 = \left(-\sigma\eta \frac{m}{\hbar^2} \pm \sqrt{\left(\eta \frac{m}{\hbar^2}\right)^2 + \frac{2m}{\hbar^2}E - \left(0 + \frac{e}{\hbar}A_z\right)^2}\right)^2 \quad (3)$$

NUMERICAL RESULTS

Figure 2 shows that reasonably large |P| can be achieved in the presence of RSOC and electrical barriers but without δB_y. Figure 3 shows that |P| is magnified to close to 100% from 0.2-1.0 E_F when a zero-gauge, periodic δB_y is applied in the plane of the 2DEG. In addition, an electron spin coupled to the total in-plane field (δB_y and RSOC) would be resistant to the spin dephasing effects of the Dy'akonov-Perel'-like mechanism. This mechanism has been neglected in previous calculations in devices with perpendicular B_z fields. The extra contribution of RSOC also enables high |P| to be achieved with fewer magneto-electric barriers. The inset of Fig. 3 shows that appreciable value of |P|=50% can be achieved with only one double-pair barriers ($n=1$). By contrast, previous calculations based on a perpendicular field and neglecting RSOC show negligible |P| for $n=1$, and appreciable |P| only for $n \sim 30$. The reduced number of barriers required means that the active device length can be greatly reduced, thus making it easier to achieve the ballistic condition assumed in the calculations.

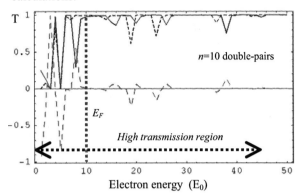

FIGURE 2. T and |P| curves show spin transport under the influence of RSOC.

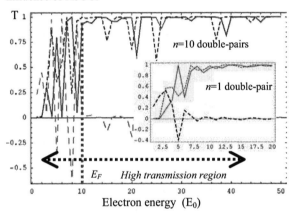

FIGURE 3. T and |P| curves show spin transport under the influence of RSOC and external in-plane fields.

REFERENCES

[1] G. Papp, F.M. Peeters, *Appl. Phys. Letters* **78** 2184 (2001).
[2] Yong Guo, Binglin Gu, and Wenhui Duan, Yu Zhang, *Phys. Rev. B* **55** 9314 (1997).
[3] Y. Jiang, Jalil M.B.A., and T.S. Low, *Appl. Phys. Letters* **80** 1673 (2002).
[4] Supriyo Datta, and Biswajit Das, *Appl. Phys. Letters* **56** (7), 665 (1989).
[5] M.I. D'yakonov, and V.I. Perel', *Zh. Eksp. Teor. Fiz.* **60** 1954 (1971) [*Sov. Phys. JETP* **33**, 1053 (1971)].
[6] M.B.A. Jalil, S.G. Tan, T. Liew, K.L. Teo, and T.C. Chong, *J. Appl. Phys.* **95** 7321 (2004).

Spin Precession In A Model Structure For Spintronics

M. Ghali[1] J. Kossut[1] E. Janik[1] F. Teppe[2], M. Vladimirova[2] and D. Scalbert[2]

[1]*Institute of Physics, Polish Academy of Sciences, Warsaw, Poland.*
[2]*Groupe d'Etudes des semiconducteurs, Université Montpellier 2, Montpellier, France.*

Abstract. Transverse electron and hole spin relaxation times were measured in both parts of a generic spintronic structure device consisting of two thin layers: diluted magnetic $Cd_{0.96}Mn_{0.04}Te$ (spin aligner) and a layer non-magnetic CdTe (spin accumulator). By using time-resolved magnetoptical Kerr rotation as a fast time-resolved magnetization probe, we find that the carriers in $Cd_{0.96}Mn_{0.04}Te$ lose their initial spin coherence within ~1ps. The undoped spin accumulator is characterized by relatively long electron spin relaxation time that of the order of several hundreds ps.

INTRODUCTION

A generic spintronic device combines two fundamental elements: a spin aligner (SA) and a spin accumulator (SAC). Optimization of their properties requires, respectively, that SA can polarize spins of carriers fast while the spin relaxation in SAC is suppressed. Before such optimization is attempted one has to study spin relaxation times in that characterize materials making both layers.

EXPERIMENT AND RESULTS

As a model structure we used in this study an MBE-grown heterostructure consisting of layers of $Cd_{0.96}Mn_{0.04}Te$ (0.38μm wide, E_g=1.65eV) as SA and CdTe (1.3μm wide, E_g= 1.59eV) as SAC. We used time-resolved magneto-Kerr rotation (TRKR) technique to study the carrier spin dynamics in both layers. The one color pump-and-probe experiment was performed utilizing pulses from a 100fs mode-locked tunable Ti: sapphire laser with 82MHz repetitions [1]. Energy of the light was tuned to the lowest hh exciton in the SA material (i.e., to 1.65eV in the absence of an external magnetic field). A mechanical delay line generates time delays between pump and probe pulses of up to 2.5ns. At a delay time, Δt, the linearly polarized probe pulse measures the spin magnetization S_x along the direction of light propagation. The experiments were done with the sample immersed in helium bath at about 2K.

Figure 1 shows the TRKR oscillations of electron spins in the SA. The oscillations frequency reaches several THz (~13THz at 5T) indicating a very fast electron spin precession. This is due to the strong *s-d* exchange interaction between the electron spin and that of the Mn ion which leads to a giant Zeeman spin splitting. Beyond ~14ps the oscillations in Fig.1 disappear. This means that the spin coherence time (T_2) is below 14ps. The design of our sample, under the conditions of pumping with energy close to the CdMnTe gap, facilitates injection of the polarized electronic spins into the entire structure. This is because the CdTe conduction band edge is only by 20meV below that in CdMnTe SAC, which is comparable to the 100fs pulse spectral width.

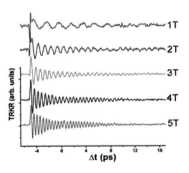

FIGURE 1. TRKR oscillations of the linearly polarized probe pulse at the fundamental transition of CdMnTe in different magnetic fields at 2K.

However, it is the probe energy, which determines which spins are detected. We focus now only on probing spin dynamics in the SA region of the structure, leaving the spin dynamics in SAC for a discussion below. We carried out our measurements in external magnetic field (up to 5T) directed in the plane of the sample. For such orientation the Larmor precession of the optically injected spins can be detected as oscillation of the Kerr rotation angle of the linearly polarized probe pulse [1]. We were able to fit the oscillations with three components

$$\theta_k = A_1 e^{-\Delta t/T_2^{C_1}} \cos(\omega_1 t + \Phi_1) + A_2 e^{-\Delta t/T_2^{C_2}} \cos(\omega_2 t + \Phi_2) + A_3 e^{-\Delta t/T_2^{C_3}}$$

where θ_k is the Kerr rotation angle, A_i is the amplitude of each component i, $T_2^{C_i}$ (the transverse spin lifetime) and Φ_i being the fit parameters ($i=3$). The relaxation times for the three components of the TRKR signal are obtained from the fit to the curves shown in Fig. 1. The resulting times are summarized in Fig 2. The two first terms in the above expression describe the electron spins precessions with very similar frequencies. The corresponding relaxation times are of the same order of magnitude (below 10ps) and decrease with magnetic field. This behavior can be ascribed to the spin resonance broadening due to g-factor distribution [3].

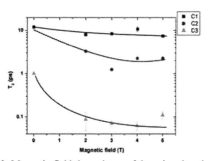

FIGURE 2. Magnetic field dependence of the spin relaxation times of the conduction electron (closed squares and circles) and the heavy hole spin relaxation time (closed triangles). The line is to guide the eyes.

A very short (around 0.1ps at B>2T) decay component is related to the hole spin relaxation. It is so fast because of (1) mixing of the valence band spin states in our 3D bulk-like epilayer, (2) strong exchange p-d interaction. At high fields, however, where T_2^h ~0.1ps, the precision of our measurements is limited because the decay time becomes comparable with the laser pulse duration. The values of T_2^h determined by us are consistent with previous results from Ref. [3] obtained on ZnSe-based structures. In contrast, by bringing the pump energy to resonance with CdTe, we observe slow oscillations corresponding to a long relaxation time (~500ps) in SAC. The large difference in the spin lifetimes in SA and SAC suggests that the presence of Mn^{2+} in the former is the main origin of the spin decoherence. By fitting these slow oscillations with the above formula we find again the presence of two terms with different frequencies ω_1 and ω_2, and a third component, which decays exponentially.

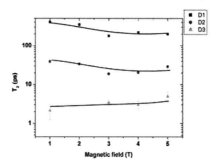

FIGURE 3. Magnetic field dependence of the spin relaxation times of the oscillatory components (D1 closed squares, and D2 closed circles) and the heavy hole spin relaxation time (D3 closed up-triangles). The line is to guide the eyes.

The two frequencies increase roughly *linearly* with the field as is typical for non-magnetic semiconductors. From $g = \hbar\omega / \mu_B B$ we obtain the g-factors corresponding to each frequency: $g_1=1.66$ and $g_2=1.69$, close to known value for electrons in bulk CdTe. Figure 3 shows the spin relaxation time of the three components. The exponential decay shows T_2 that grows with the field, which is different from the behavior in SA layer. We ascribe this component to holes in SAC. The oscillating components (with T_2 450 and 40 ps at 1T, respectively) both decrease slightly in amplitude with the magnetic field.

ACKNOWLEDGMENTS

Work supported by the Foundation for Polish Science (subsidy 8/2003) and by the Polonium project.

REFERENCES

1. C. Camilleri et al. Phys. Rev. B **64**, 085331 (2001).

2. A. Abragam, The *Principles of Nuclear Magnetism*, Oxford, Clarendon Press, 1961.

3. S. Crooker et al., Phys. Rev B **56**, 7574-7588 (1997).

Observation of spin diffusion, drift and precession in bulk n-GaAs using a spatially resolved steady-state technique

M. Beck*, C. Metzner*, S. Malzer* and G. H. Döhler*

*Erwin-Rommel-Str. 1, 91058 Erlangen, Germany

Abstract. We report on the spin transport of photoexcited spin polarized electrons in n-type GaAs under steady state conditions. Spin-diffusion, -drift and -relaxation are measured via spatially resolved Faraday rotation. If strain is applied to the sample, spin precession is observed for strain is along [110], due to the k-linear spin splitting induced by the off-diagonal strain components. No significant effect is seen for strain along [100]. From the results, we obtain a spin splitting constant of $C_3 = 7.6$ eVÅ. We believe that the described method for the determination of this constant is more direct than previously reported approaches. Spin transport is described theoretically in a drift-diffusion model taking into account the D'yakonov-Perel' and Elliot-Yafet spin relaxation mechanisms and the heating of the electron gas.

INTRODUCTION

Spin-dephasing, -transport and -manipulation are key issues of potential future spintronics devices. Whereas spin lifetimes of electrons in n-GaAs at low temperatures are well-known at thermodynamic equilibrium (e.g. [1, 2]), not much attention has been paid to the spin relaxation of electrons drifting in an electric field. Kikkawa and Awschalom [1] reported that "applying a voltage does not introduce any severe decoherence". Recently, Kato et al. [3] reported on spatiotemporally resolved measurements of spin transport in n-GaAs and n-InGaAs films, associated with spin precession at zero magnetic field. Although they could ascribe the observed effect to the influence of strain, the quantitative influence of strain could not be extracted from their data since the relaxed InGaAs layers were strained in an uncontrolled manner.

In the following, we describe the experimental method and the obtained experimental results together with the theoretical predictions. The theory itself will be discussed elsewhere. It should be noted here that a good agreement between theory and experiment is only obtained by correctly taking into account Fermi statistics in the intermediate regime between strong degeneracy and non-degeneracy and by avoiding the small angle approximation used by D'yakonov and Perel' for the scattering cross-section of ionized impurity scattering. Here, the cross-section introduced by Ridley [4] is used, which reduces the calculated spin lifetimes by a factor 2...3 in the regime of interest ($N_D = 10^{16\pm1}$ cm^{-3}, $T_{el} = 4...100$ K).

EXPERIMENTAL TECHNIQUE

Spin polarized electron hole pairs are created in a film of Si-doped GaAs (lifted from the substrate by epitaxial lift-off and attached to glass by van-der-Waals bonding; $N_D = 1.4 \times 10^{16}$ cm^{-3}; different thicknesses between 750 μm and 3 μm have no significant effect on the results) by circularly polarized light from a cw laser diode (pump; resonant with the exciton energy, $P \approx 2$ mW). The Faraday rotation of linearly polarized light from a second cw laser diode (probe; about 3 meV below the exciton energy, $P \approx 0.2$ mW) is used to measure the electron spin polarization. A stepper motor varies the pump-probe distance along the direction of the electric field. In unstrained samples, the spin polarization decays exponentially along the drift direction. Knowing the drift velocity, we obtain the spin relaxation time. Strain is applied to the sample by bending the glass substrate. We determine the magnitude of strain by measuring the focal length of light reflected from the glass.

SPIN TRANSPORT

In an electric field F, the steady state distribution of spin polarization is given by $n^\uparrow - n^\downarrow = (n^\uparrow - n^\downarrow)_0 e^{\mp \Delta x/\Lambda_{d,u}}$ (upper sign and index d for negative $\Delta x F$) with the downstream and upstream diffusion lengths

$$\Lambda_{d,u} = \Lambda_s \left(\sqrt{\left(\frac{F}{2F_c}\right)^2 + 1} \mp \frac{F}{2F_c} \right)^{-1} \quad (1)$$

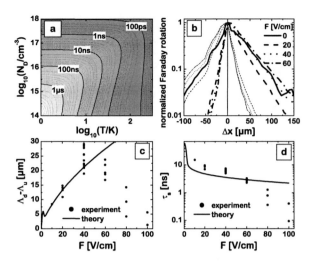

FIGURE 1. a: Calculated spin relaxation times vs. temperature and donor density (D'yakonov-Perel' and Elliot-Yafet mechanism; localization of carriers is neglected); b: example of measured Faraday rotation, thin lines: opposite field direction; c: calculated and measured spin drift lengths ($\Lambda_d - \Lambda_u$); d: calculated and measured spin relaxation times (all shown experiments at T_L=5 K).

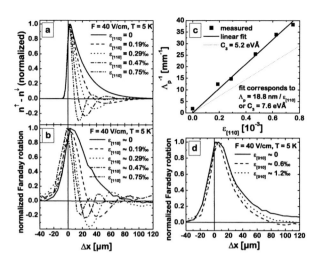

FIGURE 2. Spin transport at $F = 40 V/cm$ in strained samples. a: expected spin polarization for the measurements shown in b; b: Faraday rotation for drift along $[1\bar{1}0]$ and strain along $[110]$; c: determined spin precession length and linear fit corresponding to $C_3 = 7.6$ eVÅ; d: Faraday rotation for drift and strain along $[100]$

$$\text{with} \quad \Lambda_s = \sqrt{D_e \tau_s} = \sqrt{-\frac{\mu}{e} \frac{\int g(E) f(E) dE}{\int g(E) \frac{\partial f(E)}{\partial E} dE} \tau_s},$$

where $g(E)$ is the density of states, $f(E)$ the Fermi function, μ the electron mobility, and τ_s the spin relaxation time. F_c defines a critical field $F_c = \frac{D_e}{\mu_e \Lambda_s}$. For zero field, $\Lambda_{u,d} = \Lambda_s$. In a finite electric field, $\Lambda_d - \Lambda_u = v_{dr} \tau_s$, with the drift velocity v_{dr}. The electron temperature is given by $T_{el} = T_L + \frac{2e}{3k_B} \mu_e F^2 \frac{\langle E \rangle}{\langle E/\tau_E(E) \rangle}$, the angular brackets denoting the averaging over the distribution function, T_L being the lattice temperature and $\tau_E(E)$ the energy relaxation rate. Some theoretical and experimental results are shown in Fig. 1. Deviations between theory and experiment for $F > 50$ V/cm are probably due to an increase of the lattice temperature caused by the electric current.

STRAIN-INDUCED SPIN PRECESSION

Strain induces spin precession according to [5]

$$\Omega_z = \frac{C_3}{\hbar}(\varepsilon_{xz} k_x - \varepsilon_{yz} k_y) \quad + \text{cyclic permutation}. \quad (2)$$

This enhances spin relaxation, and the average spin precession of drifting electrons will generally be different from zero. Since the precession is proportional to k, a complete precession of 2π will take place on the field-independent precession length $\Lambda_p = \frac{2\pi v_{dr}}{\Omega_\perp}$, $\Omega_\perp$ being the component of Ω perpendicular to the field. For strain and drift along $[110]$ and equivalent directions, one obtains $\Lambda_p = \frac{4\pi \hbar^2}{m \varepsilon C_3} = 27.5$ nm/ε (using $C_3 = 5.2$ eVÅ) [6]. We obtain $C_3 = 7.6$ eVÅ (c.f. Fig 2c) with an estimated error of about 30%, the main uncertainty being the quantification of strain. As theoretically predicted, we observe no significant influence of strain along $[100]$.

CONCLUSIONS

We have demonstrated, that spin relaxation times at finite electric field can be determined from steady state measurements without magnetic field and that the strain-induced spin splitting can be directly determined from the spin precession length.

REFERENCES

1. Kikkawa, J. M., and Awschalom, D. D., *Nature*, **397**, 139–41 (1999).
2. Dzhioev, R. I., Kavokin, K. V., Koronev, V. L., Lazarev, M. V., Meltser, B. Y., Stepanova, M. N., Zakharchenya, B., Gammon, D., and Katzer, D., *Phys. Rev. B*, **66**, 245205–1–7 (2002).
3. Kato, Y., Myers, R. C., Gossard, A. C., and Awschalom, D. D., *Nature*, **427**, 50–3 (2004).
4. Ridley, B. K., *J. Phys. C*, **10**, 1589–93 (1977).
5. Dresselhaus, G., *Phys. Rev*, **100**, 580–6 (1955).
6. D'yakonov, M. I., Marushchak, V. A., Perel', V. I., and Titkov, A. N., *Sov. Phys. JETP*, **63**, 655 (1986).

Optical Study of Spin Injection Dynamics in Double Quantum Wells of II-VI Diluted Magnetic Semiconductors

K. Kayanuma[1], T. Tomita[1], A. Murayama[1], Y. Oka[1], A.A. Toropov[2], S.V. Ivanov[2]
I.A. Buyanova[3], W.M. Chen[3]

[1]*Institute of Multidisciplinary Research for Advanced Materials, Tohoku University,*
2-1-1 Katahira, Sendai 980-8577, Japan
[2]*A.F. Ioffe Physico-Technical Institute, Russian Academy of Sciences,*
St. Petersburg, 194021, Russia
[3]*Department of Physics and Measurement Technology, Linköping University,*
58183 Linköping, Sweden

Abstract. The spin injection process is studied in double quantum wells of ZnMnSe and ZnCdSe with a tunneling barrier by circularly-polarized transient photoluminescence spectroscopy. The result shows two types of spin injections in the time ranges of 30 and 1000 ps. The observed spin injection processes are quantitatively interpreted by the individual tunneling of electrons and holes based on the rate equation analysis.

INTRODUCTION

Dynamical spin properties of carriers in semiconductors have extensively been studied recently owing to possible applications for spin-electronic devices. The spin injection of carriers has been reported in the double quantum wells (DQWs) system with a magnetic quantum well (MW) and a non-magnetic well (NW) [1-3]. However, the dynamical process in the spin injection is not sufficiently clarified until now, although the knowledge of such detailed dynamics is crucial to development of ultrafast spin devices. The spin injection efficiency is determined by the spin injection rate from the MW to the NW relative to the spin relaxation rate in both wells. The spin injection efficiency is decreased if the spin injection to the NW occurs faster than the spin polarization in the MW followed by the spin relaxations in the well. Therefore the insertion of a proper tunneling barrier between the MW and NW can increase the spin injection efficiency. The objective of the present study is to increase the spin injection efficiency from the MW to the NW by designing the structure of the DQWs.

RESULTS AND DISCUSSION

The DQWs structure studied is shown in FIGURE 1. (a), which consists of the NW of $Zn_{0.76}Cd_{0.24}Se$ and the superlattice MW of $Zn_{0.96}Mn_{0.04}Se/CdSe$. These NW and MW are separated by the barrier of ZnSe. The well widths of NW and MW were fixed to 7 nm and 40 nm, respectively. The barrier width (L_B) was varied from 4 to 8 nm to study the variation of the spin injection dynamics depending on the carrier tunneling rate. Transient and circularly polarized photoluminescence from the DQWs was measured by a spectrometer and streak camera system. Magnetic field up to 6 T was applied by a split coil superconducting magnet in the Faraday configuration.

The PL spectrum of the DQW (L_B = 8 nm) is shown in FIGURE 1. (b). The exciton emissions from NW and MW appear at 2.55 and 2.74 eV at B = 0 T. Although the MW with the well width of 40 nm is sufficiently wide to give rise to PL in the MW, the

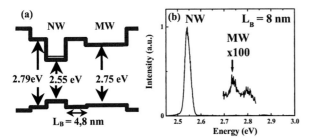

FIGURE 1. (a) Energy diagram of the DQW sample. (b) PL spectra of the DQWs (L_B = 8 nm).

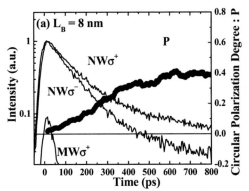
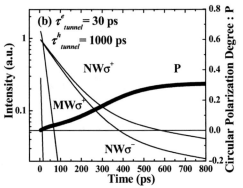

FIGURE 2. (a) Transient PL intensity and circular polarization degree P in DQWs of L_B = 8 nm at B = 3 T. (b) Calculated results of transient PL intensity and P by the individual spin tunneling model with τ^e_{tunnel} = 30 ps and τ^h_{tunnel} = 1000 ps.

exciton PL in the MW is weak. This suggests the dominant tunneling of excitons or carriers from MW to NW in the DQWs. When a magnetic field is applied, the PL peak of the MW exciton with left-circular (σ^+) polarization shows a giant Zeeman shift to the lower energy side, which becomes 43 meV at 6 T. In the NW, the exciton PL peak energy remains unchanged over the range of the magnetic field. For the optical excitation in magnetic field, carriers generated in the MW are populated mostly to the lower lying spin state (the spin down state) due to the fast spin relaxations. Therefore the carrier spins are highly polarized to the spin down configuration. Then the polarized spin-down carriers tunnel into the NW. The circularly polarized exciton PL arises in the NW, if the spin relaxation is not significant in the NW. We measured the time resolved PL by exciting with the linearly polarized light of 2.774 eV at B = 3 T and T = 2 K. The circular polarization degree is defined as P = $(I(\sigma^+)-I(\sigma^-))/(I(\sigma^+)+I(\sigma^-))$, where $I(\sigma^\pm)$ is the circularly polarized PL intensity (with σ^+ and σ^- components). In the DQW sample with L_B=4 nm, the PL lifetimes of NW was 115 ps and the circular polarization degree is 0.20 at the time just after the excitation. On the other hand, in the sample with L_B=8 nm (shown in FIGURE 2. (a)), the PL lifetime of the NW was 138 ps and the circular polarization degree increases from 0.03 to 0.40 with a rising time of 400 ps. The increase of the circular polarization degree in the NW with the time constant of 400 ps can be considered as a result of the exciton tunneling from MW to NW with the long time constant. However, in the present case, the PL lifetime of excitons in the MW is much short (30 ps). Therefore, a model of slow exciton-tunneling cannot derive the short lifetime of the MW exciton. Here we propose a model of individual tunneling of electrons and holes to interpret quantitatively the experimental results. We calculate the injection and recombination rates of electrons and holes by the individual tunneling model in the DQWs with the aid of rate equations. The calculated results for the time developments of the MW and NW exciton PL intensity are shown in FIGURE 2. (b), where the electron spin tunneling time is 30 ps and the hole spin tunneling time is 1000 ps, respectively. The calculated results agree well with the observed transient PL of the MW and NW in the DQWs with L_B= 8nm. The weaker experimental PL intensity of the MW as compared with the calculated value. PL efficiency of the MW is lower than the NW, which is not taken into account in the present calculation. Therefore the individual electron and hole tunneling model reproduces well the observation, where the PL lifetime of the MW is short (30 ps), while the PL lifetime of the NW increases in magnetic field and the circular polarization degree builds up gradually due to the slower hole tunneling.

In summary, we have observed the high spin injection process in the DQW of L_B = 8 nm. The observed time dependence of polarization is interpreted quantitatively by the individual tunneling of electrons and holes through the barrier in the DQW system. Present results demonstrate that the spin injection process can be controlled by the design of the tunneling barrier in the DQWs and the highly polarized spin injection of carriers and excitons can be realized by the suitable barrier width.

One of the authors (K. K.) is indebted to Japan Society for the Promotion of Science for the financial support on the present research. This work is supported by the NEDO nanotechnology program.

REFERENCES

1. I. A. Buyanova et al., Appl. Phys. Lett. **81**, 2196-2198 (2002).
2. K. Kayanuma et al., Physica E **10**, 295-299 (2001).
3. K. Kayanuma et al., Physica B **340-342**, 882–885 (2003).

Excitonic Spin Dynamics in Coupled Quantum Dots of Diluted Magnetic Semiconductors

A. Murayama, A. Uetake, I. Souma, K. Kayanuma, T. Asahina, K. Hyomi, T. Tomita, and Y. Oka

Institute of Multidisciplinary Research for Advanced Materials, Tohoku University, 2-1-1 Katahira, Sendai 980-8577, Japan

Abstract. Spin-polarized exciton dynamics is studied in coupled quantum dots (QDs) of II-VI diluted magnetic semiconductors. In double QDs fabricated by electron-beam lithography, an efficient spin-injection process is demonstrated via spin-polarized exciton tunneling with the tunneling time of 500 ps. In high-density self-organized CdSe QDs, the exciton lifetime is shorter in smaller dots with higher energies, indicating the energy transfer and the correlation of wavefunctions. Circular polarization of excitonic photoluminescence is observed at 0 T with an opposite sign to that of the excited light and with the rise time of 50 ps.

INTRODUCTION

Quantum dots (QDs) of semiconductors are promising nanomaterials for efficient optical devices and potential applications to spintronics and quantum computing. In addition, remarkable magneto-optical properties, such as, the giant Zeeman effect, appear in diluted magnetic semiconductor (DMS) QDs.[1] Therefore, a spin aligner and efficient magneto-optical devices can be designed using the DMS QDs. We have fabricated two types of coupled QDs based on exciton tunneling or energy transfer, and have studied the spin dynamics. One was a double QD with the diameter of 30 nm, which was fabricated from a double quantum well (QW) of DMS $Zn_{1-x-y}Cd_xMn_ySe$ and non-magnetic $Zn_{1-x}Cd_xSe$.[2] Here, spin-polarized excitons can be injected from the DMS QD into the $Zn_{1-x}Cd_xSe$ QD. Another was self-organized QDs of CdSe and $Cd_{1-x}Mn_xSe$. Also, coupled quantum structures of CdSe QDs with a $Zn_{1-x}Mn_xSe$ layer were fabricated with a ZnSe spacer layer.

EXPERIMENTAL PROCEDURE

QWs and self-organized QD layers were grown by molecular beam epitaxy on GaAs substrates with proper buffer layers. After the growth, those layers were fabricated into nano-pillars by electron beam lithography. A micro-photoluminescence (PL) system using a cw-GaN laser (λ = 408 nm) was used at 4.2 K for the QD structures in magnetic fields with the Faraday geometry. The area of the self-organized QD layer was restricted by lithography so that we resolved PL spectra from single dots. The spin dynamics was also measured for plane samples in magnetic fields by time-resolved PL using a ML-Ti:S laser (λ = 390 nm).

RESULTS AND DISCUSSION

Coupled QD Fabricated by Lithography

Double QDs of $Zn_{1-x-y}Cd_xMn_ySe$ DMS were fabricated with the lateral diameter d ranging from 10 to 100 nm by lithography, where a laterally quantum confinement effect on the exciton energy was observed for the QDs with $d \leq 30$ nm as well as the giant Zeeman effect.[2] Therefore, we apply the DMS QD to a coupled QD system with non-magnetic $Zn_{1-x}Cd_xSe$. A typical SEM image for our QD structure is shown in Fig. 1 (a). The normalized PL intensity from the DMS QD is 5 times higher than that in the DMS QW prior to the dot fabrication, when the DMS QD is placed below the $Zn_{1-x}Cd_xSe$. It means that the photo-excited carriers or excitons in the buffer layer flow into the DMS QD. The down spin state of the $Zn_{0.70}Cd_{0.22}Mn_{0.08}Se$ exciton is designed to be 100 meV higher than that of the

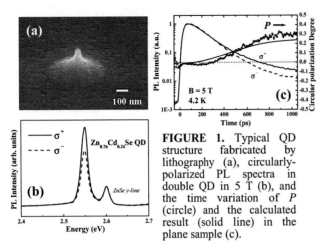
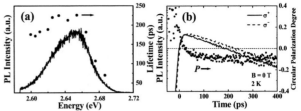

FIGURE 1. Typical QD structure fabricated by lithography (a), circularly-polarized PL spectra in double QD in 5 T (b), and the time variation of P (circle) and the calculated result (solid line) in the plane sample (c).

FIGURE 2. PL intensity and lifetime (circle) as a function of energy in self-organized CdSe QDs grown on ZnSe (a), and the PL intensities with σ^+ and σ^- polarizations and the resultant P as a function of time (b), where the excitation was made by σ^+ light and thus the negative P means opposite polarization (σ^-) to the excitation one.

$Zn_{0.76}Cd_{0.24}Se$ exciton state in 5 T, with a 3 nm-thick ZnSe barrier. Therefore, spin-polarized excitons can be migrated from the DMS to the $Zn_{0.76}Cd_{0.24}Se$ QD in the laterally confined nanostructure. Moreover, the number of exciton injected from the DMS as a spin aligner will increase in the $Zn_{0.76}Cd_{0.24}Se$, because of the increased carriers captured from the buffer layer. The circularly polarized PL from the $Zn_{0.76}Cd_{0.24}Se$ QD shows enhanced circular polarization degree P of 0.16 ± 0.05 in the QD structure with $d = 30$ nm, while P is 0.10 ± 0.01 in the QW (Fig. 1 (b)). Therefore, the ratio of the number of injected spin-polarized exciton to that of unpolarized exciton is concluded to be enhanced in the $Zn_{0.76}Cd_{0.24}Se$ spin detector. The P rises at 300 ps and increases gradually up to 0.30 from direct observation of the time variation of P in the plane QW, as shown in Fig. 1 (c). It can be quantitatively explained by slow tunneling of excitons (500 ps) due to relatively low tunneling probability of a heavy-hole from the DMS (a solid line in the figure).

Self-organized CdSe QDs

In self-organized CdSe QDs, the dot diameter and the density were determined typically as 3 nm and $5000/\mu m^2$, respectively, from the energy and the number of individual sharp PL peak with the width less than 0.5 meV in micro-PL measurements. Lifetimes of the excitonic PL in the higher energy side are significantly shorter (~70 ps) than those in the lower energy side (~220 ps), as shown in Fig. 2 (a). The micro-PL shows that the broad PL band originates from superposition of individual sharp PL peaks, in which the peak intensity at the high energy side is lower and corresponds to the lifetime. Therefore, the energy transfer of excitons from smaller dots to larger ones occurs, since the dot density is rather high. The exciton tunneling is also suggested due to overlapping of QD wavefunctions, although the density of state is discrete. In addition, we observe significant transient change in P at 0 T, as shown in Fig. 2 (b), with the excitation using a circularly polarized light. At the initial stage less than 30 ps, the fast relaxation of P with the same polarization as the excitation polarization is observed. However, after the relaxation, the sign of P changes to opposite. The P value increases gradually and then saturates in this time region. It directly indicates spin accumulation in the non-magnetic dots, although the physical origin of the opposite sign of P is unclear at the moment. The giant Zeeman effects of $Cd_{1-x}Mn_xSe$ QDs are also investigated by micro-PL under magnetic fields. We have prepared several types of dot structures including coupled quantum structures of CdSe QDs with a $Zn_{1-x}Mn_xSe$ layer. In such structures, the giant Zeeman shift of excitonic PL appears with the shift energy up to 20 meV in 5 T. The field dependence is well expressed by a Brillouin function and is explained by overlapping of the QD wavefunctions with the DMS layer. The lifetime of exciton is as 40 ps and is markedly shorter than that of the CdSe dots without Mn ions, which is understood by the energy transfer process from the exciton in the dot into the internal d-d transitions of the Mn ion in the DMS layer.

ACKNOWLEDGMENTS

This work is supported in part by the Ministry of Education, Science, and Culture, Japan, and also NEDO Nanotechnology Materials Program. The authors acknowledge support of the Murata Science Foundation.

REFERENCES

1. Oka, Y., Kayanuma, K., Shirotori, S., Murayama, A., Souma, I., and Chen, Z.H., *J. Lumin.* **100**, 175-190 (2002).
2. Uetake, A., et al., *Phys. Stat. Sol.* (c), **1**, 941-944 (2004).

Spin Relaxation Dynamics in Highly Uniform InAs Quantum Dots

A. Tackeuchi[1,4*], Y. Suzuki[2], M. Murayama[1], T. Kitamura[1], T. Kuroda[1], T. Takagahara[3] and K. Yamaguchi[2]

[1]*Department of Applied Physics, Waseda University, Tokyo 169-8555, Japan*
[2]*Department of Electronic Engineering, University of Electro-Communications, Chofu, Tokyo 182-8585, Japan*
[3]*Department of Electronics and Information Science, Kyoto Institute of Technology, Kyoto 606-8585, Japan*
[4]*PRESTO, Japan Science and Technology Corporation*
*E-mail: atacke@waseda.jp

Abstract. We have investigated carrier spin dynamics in highly uniform self-assembled InAs quantum dots. The highly uniform quantum dots allowed us to observe the spin dynamics in the ground state and that in the first excited state separately, without the disturbance of inhomogeneous broadening. The spin relaxation times in the ground state and the first excited state were measured to be 1.0 ns and 0.6 ns, respectively. Our measurements reveal the absence of the carrier density dependence of the spin relaxation time. The measured spin relaxation time decreases rapidly from 1.1 ns at 10 K to 200 ps at 130 K. This large change in the spin relaxation time is well explained in terms of the mechanism of acoustic phonon emission.

INTRODUCTION

Recently, interesting features of electron spin in III-V quantum dots (QDs) have been revealed. For example, carriers in quantum dots have a long spin relaxation time of approximately 1 ns or longer. Also, an antiferromagnetic order is found to be present between QDs by inter-dot exchange interaction.[1] The interesting carrier spin features may be applicable to the quantum computation. The spin related phenomena in III-V quantum confined structures became observable from the 1990s using time resolved measurements.[2] However, the inhomogeneous broadening of QDs has made it difficult to understand spin-related phenomena because the ground-state energies of some QDs are equal to the first excited state energies of the other QDs. To overcome this problem, we have adopted highly uniform QDs as samples, which typically have the narrowest PL line width of 18.6 meV.[3]

EXPERIMENT

The highly uniform QD sample was grown on semi-insulating GaAs substrates by molecular beam epitaxy.[3] The InAs dots were grown at 500 °C under an As pressure of 3×10^{-7} Torr and low growth rate of 0.035 monolayer/s. The InAs coverage was 2.65 monolayers. Enhanced surface migration due to the low arsenic pressure of 6×10^{-7} Torr during the GaAs capping growth leads to a narrow size distribution.

The transient spin polarization was time-resolved by spin-dependent photoluminescence (PL) measurement. The optical source was a Ti:sapphire laser which generated 100 fs pulses. To generate and measure the spin polarization, we used the optical transition selection rule between carrier spin and circularly polarized light. The excitation laser wavelength was tuned to near the band gap of GaAs. The PL was detected with a synchroscan streak camera with a time resolution of 15 ps.

RESULTS AND DISCUSSION

The PL spectra time-integrated during 0.9 ns after photo-excitation at 10 K are shown in Fig. 1. The PL peaks of the ground state and the first excited state are clearly separated. The spin polarization, $(I_+ - I_-)/(I_+ + I_-)$, of 16 % in the first excited state is greater than that of 5 % in the ground state. This is due to spin Pauli blocking.[4] The spin relaxation time is equal to twice the decay time of the spin polarization. The evaluated spin relaxation times in the ground state and the first excited state are 1.0 ns and 0.6 ns, respectively.

In an ideal QD structure, the D'yakonov-Perel' process, which is the most significant spin relaxation mechanisms for III-V quantum wells should be suppressed. The other candidates are the Bir-Aronov-Pikus (BAP) process and Elliott-Yafet (EY) process. The BAP process arises from the exchange interaction between electrons and holes. However, the measured spin relaxation time does not depend on the excitation power density. The absence of the carrier density dependence indicates that the BAP process is not significant.

The EY process depends on temperature, because some parameters, such as the carrier scattering time, depend on temperature. The temperature dependence of the spin relaxation time is plotted in Fig. 2. The spin relaxation time decreases rapidly from 1.1 ns at 10 K to 200 ps at 130 K. The large change suggests that scattering related mechanisms become significant at high temperature. We have assumed that the spin relaxation rate is proportional to the acoustic phonon emission rate, as follows.

$$\frac{1}{\tau_s} \propto 1 + \frac{1}{\exp(E/kT) - 1} \qquad (1)$$

Here, E is the acoustic phonon energy. The experimental result is well fitted by the curve with the energy, E, of 2.7 meV as shown in Fig. 2. This energy value is reasonable in reference to recent experimental and theoretical results on the exciton fine structures in a single QD.[5,6] Thus, at temperatures from 10 to 130 K, the acoustic phonon process between the exciton ground state and the excited exciton state is dominant spin relaxation mechanism.

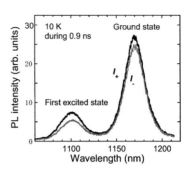

FIGURE 1. Time-integrated PL spectra during 0.9 ns after photoexcitation. The black and gray curves indicate the PL intensity of the same (I_+) and opposite (I_-) circular polarizations from the pump laser, respectively. The difference between black and gray curves corresponds to the spin polarization.

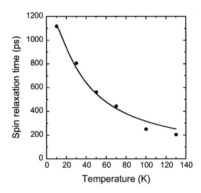

FIGURE 2. Temperature dependence of the spin relaxation time (closed circle) of the ground state.

REFERENCES

1. A. Tackeuchi et al., Jpn. J. Appl. Phys. **42**, 4278 (2003).

2. A. Tackeuchi et al., Appl. Phys. Lett. **56**, 2213 (1990).

3. K. Yamaguchi et al., Jpn. J. Appl. Phys. **39**, L1245 (2000).

4. V. K. Kalevich et al., Phys. Rev **64**, 045309 (2001).

5. D. Gammon et al., Science **273**, 87 (1996).

6. T. Takagahara, Phys. Rev. B **62**, 16840 (2000).

Spin Transport And Spin Relaxation In Ge/Si Quantum Dots

A.F. Zinovieva, A.V. Nenashev, A.V. Dvurechenskii

Institute of Semiconductor Physics, Siberian Branch of Russian Academy of Science,
630090 Novosibirsk, Russia

Abstract. The spin relaxation for localized hole state in single Ge quantum dot and for hole tunneling between two coupled quantum dots were investigated theoretically. The time of spin relaxation due to acoustic phonon coupling (direct one-phonon processes) is $\tau \sim 10^{-5}$s at temperature T=4K. A new mechanism of spin dephasing due to the structure-inversion asymmetry (SIA) of Ge quantum dot is offered. At the tunneling the angular momentum of hole is turned at the small angle and this provokes the spin flip after several events of tunneling. Simple estimations of spin relaxation caused by SIA mechanism give $\tau \sim 10^{-5}$s.

INTRODUCTION

A problem of spin relaxation in semiconductor structures caused a great interest due to recent proposals of quantum computing schemes and spin devices basing on the manipulation with spin of carriers. Here we investigate theoretically the spin relaxation in array of Ge/Si quantum dots (QD). A Ge/Si system is one of more attractive systems for exploiting a spin degree of freedom, because spin-flip mechanisms caused by bulk-inversion-asymmetry are absent in this system. The Ge QDs are usually fabricated by molecular beam epitaxy and accumulate in two-dimensional layer. Large valence band offset in Ge/Si system leads to effective confinement of holes in Ge regions. We consider the following situation: one hole is created in spin *up* state in any QD and then we trace its spin orientation when the hole moves across the array of QDs. The spin flip event can occur at the following stages of transport: when hole is located inside Ge QD and at the tunneling between coupled quantum dots.

SPIN RELAXATION IN THE SINGLE GE QD

The first step of our investigation of spin transport in Ge/Si quantum dot array is calculation of rate of spin relaxation in single isolated Ge QD. Our studies are based on the calculation of the energy spectra of holes and its wave functions in tight-binding approach [1]. The typical, pyramidal shaped, Ge QDs (height h=1.5 nm, base size l=15 nm) were investigated. As analog of electron spin we consider the projection of angular momentum J_z for hole states. We take into account only interaction with acoustic phonons in the presence of the magnetic field at temperatures T=4K. In magnetic field the Kramers degeneracy of hole level in Ge QD is lifted. The time reversal invariance can not apply in an interaction involving a magnetic field. Consequently the matrix element of interaction with phonons for transition between Zeeman sublevels is not zero. We consider the first order of perturbation theory and the change of wave functions is linear in magnetic field. We use standard form of electron-phonon interaction H_ϵ for valence band states in the basis sp^3. We taken into account only long-wavelength transverse phonons and obtain finally:

$$\Gamma \sim \frac{1}{2} \langle |M'|^2 \rangle \frac{E_{Zeeman}^3}{\rho \, s_t^5 \hbar^4},$$

where M' is reduced matrix element of electron-phonon interaction. The rate dependent on magnetic field as $\sim H^5$, three orders come from E_{zeeman} and two orders come from wave functions dependence. Then we obtain the time of spin relaxation $\tau_{ph} \sim 10^{-5}$s for ground state in the magnetic field H=1T. For first excited state our calculations give $\tau_{ph} \sim 10^{-6}$s.

SPIN RELAXATION AT THE TUNNELING BETWEEN GE QDS

Here we calculate the probability of hole spin flip for resonant tunneling through discrete energy levels in QD. For calculation of tunneling probability we consider an infinite well-ordered array of coupled quantum dot. Our studies are based on the calculation of overlap integrals between neighboring QDs. Our results show that for ground state the tunneling occurs with conservation of spin mainly: only one spin flip falls at hundred tunneling events. For excited states the probability of spin flip is higher. At the average one spin flip event falls at 5-10 tunneling events. The probability of tunneling with and no spin flip as dependent upon the size and the shape of quantum dot were calculated (Fig.1). It was found that the probability of spin flip depends on the aspect ratio h/l. For lens shaped quantum dot the probability is defined by effective curvature of upper side of quantum dot. The bigger the curvature the higher the probability of spin flip. The origin of spin flip is the structure-inversion-asymmetry. This leads to appearance of effective magnetic field (SIA-field), which turns the angular momentum of holes. This field can be considered as analog of Rashba field for two-dimensional systems. The action of the SIA field on the angular momentum is presented in the Fig.2. Based on these results one can conclude that every tunnelling event is accompanied by the turning of angular momentum on small angle. And this provokes the spin flip after several events of tunneling. Finally we give simple estimations of spin flip rate at the tunneling using the properties of real Ge QD array. We consider the recent experiments where the two-dimensional hopping conductance in Ge quantum dot array was investigated [2]. The SIA-field affects on the spin of hole only when carrier located inside quantum dot. This time is reciprocal of rate of hopping of carriers between centers i and j in QD array, $\tau_{hopping} \approx 1/\Gamma_{ij}$. The conductance between sites i and j is defined $G_{ij} \approx e^2 \Gamma_{ij}/kT$. The conductance of typical Ge/Si heterostructure with array of Ge quantum dots was obtained in experiments at helium temperatures and it is near $G \approx 10^{-4}(e^2/h)$ (T≈4K). Using this value we obtain $\tau_{hopping} \approx 10^{-7} s$. Since the spin flip probability is 100 times smaller than tunneling probability we can estimate the time of spin relaxation as $\tau_{SIA} \approx 10^{-5} s$.

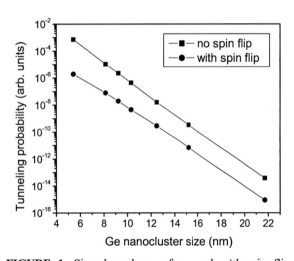

FIGURE 1. Size dependence of *no* and *with* spin flip tunneling probability for ground hole state in Ge quantum dot. The height of Ge nanocluster *h*=1.5 nm, lateral size *l* is changed.

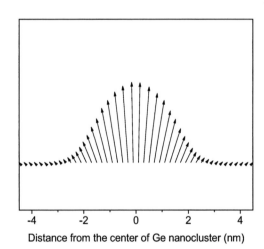

FIGURE 2. The distribution of angular momentum for ground state across the QD (*h*=1.5 nm, *l*=15 nm). For better representation we multiply the turning angle by factor of 5.

ACKNOWLEDGMENTS

This work has been supported by Russian Foundation of Basic Research (Grant 02-02-16020), INTAS-2001-0615 and Programs of RAS and SB RAS (No.27) for young scientists.

REFERENCES

1. Dvurechenskii, A.V., Nenashev, A.V., Yakimov, A.I., Nanotechnology **13**, 75 (2002).
2. Yakimov, A.I., Dvurechenskii, A.V., Nikiforov, A.I., and Bloshkin, A.A., JETP Letters **77**, 376 (2003).

Quantum antidot as a controllable spin injector and spin filter

I. V. Zozoulenko* and M. Evaldsson*

Department of Science and Technology (ITN), Linköping University, 601 74 Norrköping, Sweden

Abstract. We propose a device based on an antidot embedded in a narrow quantum wire in the edge state regime, that can be used to inject and/or to control spin polarized current. The operational principle of the device is based on the effect of resonant backscattering from one edge state into another through a localized quasi-bound states, combined with the effect of Zeeman splitting of the quasibound states in sufficiently high magnetic field. We outline the device geometry, present detailed quantum-mechanical transport calculation and suggest a possible scheme to test the device performance and functionality.

The most ambitious and challenging long-term aim of semiconductor-based spintronics is the practical implementation of quantum information processing based on the spin properties of the electron. One of the most promising practical realization areas of quantum logic in solid state physics are low dimensional semiconductor structures like quantum dots, antidots and related systems [1]. In the present paper report detailed numerical simulations of a controllable spin injector and spin filter based on the antidot system. We propose a new method to control the spin freedom of an electron by magnetic field in the edge state regime by making use of an antidot embedded in a narrow quantum wire.

The geometry of the device is presented in Fig. 1. An antidot is defined in one arm of a quantum wire in a three-terminal geometry. Leads 2 and 3 are kept at the same voltage (e.g. grounded) such that there is no current flow between them. We assume that magnetic field is sufficiently high allowing only two spin-resolved channels. The operational principle of the device relies on the Zeeman splitting of the energy of the quasi-bound state circulating around the antidot. As the magnetic field B or the Fermi energy of incoming electrons varies, the resonant backscattering occurs when the energy of the incoming electrons matches the resonant energy of the quasibound state [2, 3, 4]. The transmission through such system thus exhibits a series of split peaks, where each split peak in the split pair corresponds to a condition of resonant backscattering for spin-up or spin-down electrons.

In order to test feasibility of the proposed device we perform full quantum mechanical transport calculations for the case of an antidot system defined in GaAs heterostructure [5]. The system is described by Hamiltonian

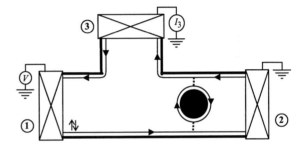

FIGURE 1. Schematic geometry of the device. An antidot is defined in one arm of a quantum wire in a three terminal geometry. The wire supports two spin-resolved channels (i.e. the lowest edge state with spin-up and spin-down electrons).

$H = \sum_\sigma H_\sigma$, that in the Landau gauge $\mathbf{A} = (-By, 0)$ reads

$$H_\sigma = \frac{\hbar^2}{2m^*}\left[\left(\frac{\partial}{\partial x} - \frac{ieBy}{\hbar}\right)^2 + \frac{\partial^2}{\partial y^2}\right] + V(x,y) + V_Z \quad (1)$$

where $m^* = 0.067 m_e$ is the effective electron mass for GaAs, $V(x,y)$ is the external confining potential. The last term in Eq. (1), $V_Z = g\mu_B \sigma B$, accounts for Zeeman energy where $\mu_B = e\hbar/2m_e$ is the Bohr magneton, $\sigma = \pm\frac{1}{2}$ describes spin-up and spin-down states, $\uparrow, \downarrow$, and the g factor of GaAs is $g = -0.44$. For the case of phase-coherent electron dynamics, the transport through the device is described within the Landauer-Büttiker formalism that relates the conductance of the multi-terminal device to its scattering properties [6]. In a sufficiently high magnetic field there is no overlap between the edge states running along opposite sides of the channel, such that $T_{23} = T_{12} = T_{11} = 0, T_{13} = 1$, where T_{ji} defines the transmission probability from lead i to lead j. Define the transmission coefficients of the system as a fraction of electrons that are injected from lead 1 and after being backscattered by the antidot are transmitted into lead 3, $T \equiv T_{31}$ (note $T_{31} + T_{21} = N$, where the number of spin-resolved channels $N = 2$). Using the Landauer-Büttiker formula we obtain that the current flowing out of the lead 3 is simply determined by the transmission coefficient T, $I_3 = GV, G = (e^2/h) T$.

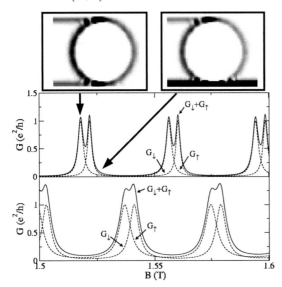

FIGURE 2. The conductance of the device $G = I_3/V$ as a function of a perpendicular magnetic field B. The middle and lower panels correspond to the cases of weak and strong coupling where the antidot diameter is $d = 318$ nm and $d = 320$ nm respectively. The width of the wire $w = 400$nm, and the sheet electron density $n_s = 10^{15} \text{m}^{-2}$. The upper panel shows the current density for spin-up electrons for the resonant and off-resonant transmission.

Figure 2 shows the three-terminal conductance of the device $G = G^\uparrow + G^\downarrow$ as a function of magnetic field, where $G^\uparrow$ and $G^\downarrow$ represent the conductance of the individual spins channels. The conductance $G = G(B)$ exhibits periodic oscillations, where the periodicity (i.e. the distance between two pairs of the split peaks) is related to the addition of one flux quantum $\phi_0 = h/e$ to the total flux $\Phi = BS$ (S being the area of the antidot). By varying the magnetic field the conductance of the device can be tuned within each of the split peaks such that the current in lead 3 will be due exclusively to spin-up or spin-down electrons. Thus, the proposed device can operate as a controllable spin injector. If the incoming state in the lead 1 is already spin-polarized (say, only spin-ups are injected into the lead 1), tuning the energy of the quasi-bound state to the one corresponding to spin-down electrons would suppress the current completely, i.e. the device would operate as a spin filter or switcher (for details see [7]).

We would like to stress that in our calculations we used a one-electron Hamiltonian where electron-electron interaction is effectively included in form of an external self-consistent potential. This one-electron description was successfully used to explain various features observed in a number of experiments on antidot structures [4, 8, 9]. It has been argued however that charging and interaction effects may be important for understanding of such features in the conductance as double-frequency ($h/2e$) Aharonov-Bohm oscillations [10, 11, 12]. It was suggested that a gradual transition from conventional Zeeman splitting to the interacting picture may take place as B increases [11]. Thus the detailed nature of $h/2e$ oscillations still remains an open question and its full understanding requires accounting for spin and charging effect in a self-consisted way. We currently undertake full quantum-mechanical transport calculation combined with the local-spin-density approximation in an attempt to answer this question.

To conclude, based on the proposed device, a spin population of quantum dots or antidots can be achieved in a controlled way, and the read-out spin information can be converted into transport properties. On the basis of our simulations, we provide recommendations on the device design and suggest possible schemes to test a device performance.

We thank Andy Sachrajda for stimulating discussions. Financial support from Swedish Research Council is greatly acknowledged. M.E. acknowledges a support from National Graduate School in Scientific Computing.

REFERENCES

1. D. Loss and D. P. DiVincenzo, Phys. Rev. A **57**, 120 (1998). P. Recher, E. V. Sukhorukov, and D. Loss, Phys. Rev. Lett. **85**, 1962 (2000); R. M. Potok *et al.*, Phys. Rev. Lett. **89**, 266602 (2002).
2. J. K. Jain and S. A. Kivelson, Phys. Rev. Lett. **60**, 1542 (1988).
3. Y. Feng *et al.*, Appl. Phys. Lett. **63**, 1666 (1993); C. Gould *et al.*, Phys. Rev. Lett. **26**, 5272 (1996)
4. G. Kirczenow *et al.*, Phys. Rev. Lett. **72**, 2069 (1994); G. Kirczenow *et al.*, Phys. Rev. B **56**, 7503 (1997).
5. I. V. Zozoulenko, F. A. Maaø, and E. H. Hauge, Phys. Rev. B **53**, 7975 (1996); **53**, 7985 (1996); **56**, 4710 (1997).
6. M. Büttiker, Phys. Rev. Lett. **57**, 1761 (1986).
7. I. V. Zozoulenko and M. Evaldsson, arXiv:cond-mat/0404431 (unpublished).
8. Y. Takagaki and D. K. Ferry, Phys. Rev. B **48**, 8152 (1993); G. Kirczenow, Phys. Rev. B **50**, 1649 (1994); Y. Takagaki, Phys. Rev. B **55**, R16021 (1997); C. C. Wan, T. De Jesus, and Hong Guo, Phys. Rev. B **57**, 11907 (1998).
9. D. R. Mace *et al.*, Phys. Rev. B **52**, R8672 (1995).
10. C. J. B. Ford *et al.*, Phys. Rev. B **49**, 17456 (1994)
11. M. Kataoka *et al.*, Phys. Rev. B **62**, R4817 (2000).
12. M. Kataoka *et al.*, Phys. Rev. Lett **83**, 160 (1999) (2000); M. Kataoka, *et al.*, Phys. Rev. B **68** 153305 (2003);

Magnetic field induced polarized optical absorption in europium chalcogenides

A.B.Henriques*, L.K.Hanamoto*, E.Abramof[†], A.Y.Ueta[†] and P.H.O.Rappl[†]

*Instituto de Física, Universidade de São Paulo, Caixa Postal 66318, 05315-970 São Paulo, Brazil
[†] LAS-INPE, Av. Astronautas, 1758, 12227-010 S. J. dos Campos, Brazil

Abstract. The optical absorption spectrum of EuTe epitaxial layers grown by molecular beam epitaxy was studied at low temperatures and high magnetic fields, using circularly polarized light. At high magnetic fields a narrow line (width 41 meV) emerges at the low energy side of the absorption threshold. The line shows a huge displacement with magnetic field and is attributed to a magnetic exciton. The emergence of this line has a strong effect on the Faraday rotation, which becomes strongly non-linear.

Europium chalcogenides have interesting optical properties due to the d-f exchange interaction between band-edge electrons and localized Eu^{2+} spins [1,2]. Early investigations showed that the band-edge absorption spectrum in EuTe is associated with a $4f^7 \rightarrow 4f^6 5d$ electronic transition, characterized by a very broad band seen at ~2.3 eV (full width at half maximum (FWHM) of almost 1 eV). It also presented a broad photoluminescence (PL) (FWHM 150 meV) with a large Stokes shift (>0.6 eV). Recently, molecular beam epitaxy (MBE) was used to produce EuTe, which has lead to the observation of a much narrower photoluminescence line (FWHM 10 meV), and at a much photon higher energy (1.92 eV) than in previous investigations [3].

In this work we studied the $4f^7 \rightarrow 4f^6 5d$ absorption of epitaxial layers of EuTe. The layers (thickness in the range 0.18 – 2 μm) were grown by molecular beam epitaxy on (111) BaF_2 substrates. The x-ray rocking curve of the (222) Bragg reflection of the thicker layers showed a peak corresponding to a bulk lattice parameter of 0.6600 nm, with a full width at half maximum of only 400 arcsec. PL measurements at T=2K showed a sharp luminescence (FWHM 10 meV) at 1.922~eV. The optical absorption spectra were measured at 1.8K, using left and right circularly polarized light, in magnetic fields of intensity up to 17 Tesla. Measurements were done in the Faraday configuration, using optical fibers coupled to *in situ* focusing optics and circular polarizers. At zero magnetic field, the optical absorption spectrum is described by a threshold around 2.3~eV (Fig. 1). When the intensity of the applied magnetic field is increased above 6 T, a very sharp absorption line (full width at half maximum of 41 meV) appears at the low energy side of the absorption onset. With increasing field this line shows a huge red shift (~36 meV/T).

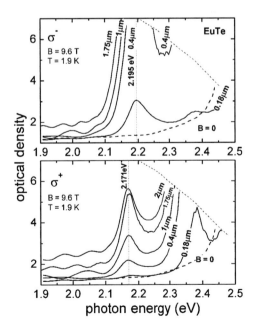

FIGURE 1. Optical absorption in EuTe layers.

At high fields the absorption intensity is much greater in the σ^- polarization than in the σ^+ one. The optical density shows a linear dependence on the thickness of the layers, demonstrating that it is a bulk effect (Fig. 2).

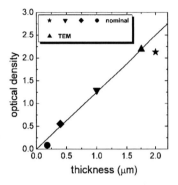

FIGURE 2. Thickness dependence of the σ^+ absorption.

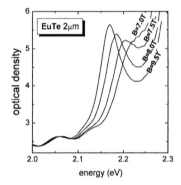

FIGURE 3. σ^+ absorption as a function of magnetic field.

The position of the σ^+ absorption line was studied as a function of the magnetic field intensity (Fig. 3, 4). Solid line in figure 4 is the theoretical result, assuming the formation of a magnetic exciton whose exchange energy is 0.15 eV and a bandgap of 2.321 eV [4].

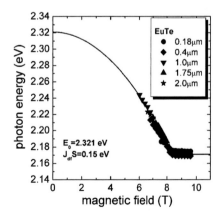

FIGURE 4. Position of the σ^+ absorption peak.

The Faraday rotation (FR) in the transparency gap was also studied at T=2K (Fig. 5). FR is linear at low fields. However, at fields when the new absorption line emerges, Faraday rotation becomes strongly non-linear. This strong FR non-linearity is in sharp contrast to earlier EuTe investigations [5]. The non-linear FR can be qualitatively explained by the emergence of the optical absorption line at high fields. The theoretical line shown in Fig. 5 is given by [6]

$$\Theta_F = \left[\frac{A}{E_0^2 - (h\nu)^2} + B \frac{f_X}{E_X^2 - (h\nu)^2} \right] M(B) \quad (1)$$

where E_0=2.321 eV is energy gap, E_X is the position of the magnetic exciton line, f_X is the oscillator strength (taken to be proportional to the E_X absorption intensity), $M(B)$ is the magnetization, and A and B are adjustable constants.

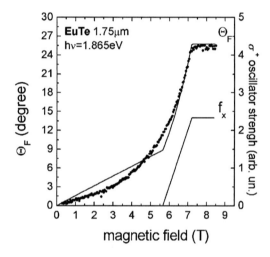

FIGURE 5. Faraday rotation for $h\nu$=1.865 eV.

CNPq and FAPESP – Brazilian agencies, supported this work.

REFERENCES

1. T. Kasuya and A. Yanase, Rev. Mod. Phys. **40**, 684 (1968).
2. A. Mauger and C. Godart, Phys. Rep. **141**, 51 (1986).
3. W. Heiss *et al*, Appl. Phys. Lett. **8**, 484 2001).
4. L. K. Hanamoto *et al*, J. Phys. C **16**, 5597 (2004).
5. H. Hori et al, Physica B **201**, 438 (1994).
6. M. Bloembergen *et al*, Phys. Rev. **120**, 2014 (1960).

Spin injection and spin loss in GaMnN/InGaN Light-Emitting Diodes

I.A. Buyanova[*], M. Izadifard[*], W.M. Chen[*], J. Kim[†], F. Ren[†], G. Thaler[‡], C.R. Abernathy[‡], S.J. Pearton[‡], C.-C. Pan[*‡], G.-T. Chen[*‡], J.-I. Chyi[*‡], and J. M. Zavada[**]

[*]*Dept of Physics and Measurement Technology, Linköping University, S-581 83 Linköping, Sweden*
[†]*Dept of Chemical Engineering, University of Florida, Gainesville, FL32611, USA*
[‡]*Dept of Materials Science and Engineering, University of Florida, Gainesville, FL32611, USA*
[*‡]*Dept of Electrical Engineering, National Central University, Chung-Li 32054, Taiwan, ROC*
[**]*US Army Research Office, Research Triangle Park, NC 27709, USA*

Abstract. Electrical and optical spin injection efficiency of GaMnN/InGaN spin LEDs is evaluated. At room temperature, the spin LEDs are shown to exhibit negligible optical and electrical spin injection efficiency despite that the n-type GaMnN spin injector employed is ferromagnetic. On the other hand, carrier supply from GaMnN at low temperatures is accompanied by a reduction (by 1-5%) in optical polarization of the InGaN spin detector (SD) from its intrinsic values measured without carrier supply from GaMnN. This observation seems to indicate some degrees of spin injection from GaMnN with the spin orientation opposite to that of the lowest spin state of the SD. The very low degree of polarization, however, implies efficient spin loss during the spin injection process. From cw- and transient resonant optical orientation studies, the spin loss is attributed to fast spin relaxation within the InGaN spin detector.

The novel concept of spintronics requires spin injection from a magnetic layer, which provides a desired spin orientation and is also compatible with the rest of device structures from non-magnetic semiconductors [1-3]. Among natural choices for spin injectors are dilute magnetic semiconductors (DMS) that are ferromagnetic at or above room temperature (RT), such as GaMnN alloys [4]. However, progress in realizing GaMnN-based spintronic devices is so far rather limited. In this work we evaluate electrical and optical spin injection efficiency in GaMnN/InGaN spin light emitting devices (LEDs) and aim to determine possible origin of spin loss.

The spin-LED structures studied in this work were grown on sapphire substrates and contain: (1) a 2-μm-thick layer of Mg-doped p-type GaN for electrical injection of holes into the spin detector; (2) a non-magnetic spin detector consisting of 5 periods of $In_{0.40}Ga_{0.6}N$ (3 nm)/GaN:Si (10 nm) multiple quantum wells (MQW); (3) a 20nm-thick non-magnetic GaN:Si spacer; (4) a spin injector of a 100nm-thick $Ga_{0.97}Mn_{0.03}N$ layer; and finally a top layer of n-type Si-doped GaN (100nm). As a reference, identical LED structures but without the top GaMnN layer were also studied. The detailed description of the structures can be found in Ref.5. Magneto-optical experiments were performed in the Faraday configuration in magnetic fields 0-5T. Photoluminescence (PL) was excited either with the 244 nm of a frequency-doubled Ar$^+$ laser (when studying spin injection from the GaMnN layer) or a tunable dye-laser (420-435 nm) when evaluating intrinsic polarization of the InGaN MQW. The polarization degree was defined in percentage by $100(\sigma^+ - \sigma^-)/(\sigma^+ + \sigma^-)$, where $\sigma^+(\sigma^-)$ is the PL intensities measured at the right (left) circular polarization.

Figure 1a shows the hysteresis behavior of in-plane magnetization measured at RT from the GaMnN layer, indicative of ferromagnetic ordering. Unfortunately, no measurable circular polarization of the light emission from the InGaN MQW was detected during electroluminescence (EL) and PL measurements performed in magnetic fields up to 5T - Fig.1b. A

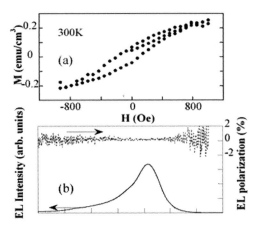

FIGURE 1. (a) Magnetization M of the GaMnN spin injector as a function of applied magnetic fields H measured in in-plane geometry at RT. (b) The EL spectrum measured at RT under forward bias of 16 V from the spin LED, as well as its polarization at 5T (the upper part of the figure).

possible reason could be cancellation of the emission polarization due to a spectral overlap of the radiative transitions involving heavy-holes (hh) and light holes (lh) with opposite polarization.

To avoid this complication, the measurements were repeated at low temperatures. A gradual increase of the PL polarization from the InGaN MQW was observed in an applied magnetic field. Interestingly, a consistent reduction in a PL polarization degree was detected with carrier supply from the magnetic GaMnN layer as compared to the intrinsic InGaN polarization measured either from the reference non-magnetic LED or under resonant excitation of the MQW- Figure 2. This observation seems to indicate some degrees of spin injection from GaMnN with the spin orientation opposite to that of the lowest spin state of the spin detector. However, the observed field dependence of the PL polarization seems to be rather different from that of the DMS magnetization - Fig.1a which makes questionable the origin of the observed PL polarization. Moreover, the very low degree of polarization implies efficient spin loss during the spin injection process.

In order to shed light on the possible origin for the weak spin polarization of the InGaN MQW, optical orientation experiments were performed. In this case, a chosen spin orientation of electron-hole pairs/excitons in the InGaN MQW was generated by the corresponding circularly polarized excitation light resonantly pumping the InGaN MQW. No sizable spin polarization, however, was observed at 0T neither during cw- nor during time-resolved PL measurements with time delays exceeding about 20 ps after an excitation pulse. The absence of the PL polarization

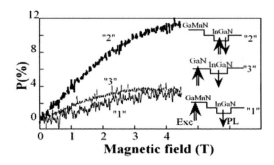

FIGURE 2. Field dependence of the PL polarization measured at 2K in the Faraday geometry under excitation conditions as specified in the Figure.

during optical orientation can be attributed to a very short spin relaxation time in the InGaN MQW (<20 ps that is the resolution of our measurement system). This by itself will destroy any spin polarization of carriers created either due to injection from the DMS layer or due to optical pumping.

According to our most recent theoretical calculations, the short spin-relaxation time in GaN/InGaN heterostructures can result from the Rashba effect [6]. A large Rashba term in the conduction band Hamiltonian has been suggested to arise from the built-in piezo-electric field which breaks the reflection symmetry of confining potential. In highly strained structures, such as the InGaN MQW studied is this work, this confinement induced spin-orbit interaction is expected to be the main source of spin mixing facilitating fast spin relaxation via Elliott-Yafet and/or D'yakonov-Perel' mechanisms.

REFERENCES

1. T. Dietl, *Semicond. Sci. Technol* **17**, 377-392 (2002).
2. D. Ferrand, A. Wasiela, S. Tatarenko, J. Cibert, G. Richter, P. Grabs, G. Schmidt, L. W. Molenkamp, and T. Dietl, *Sol. State Commun.* **119**, 237-241 (2001)
3. T. Dietl, H. Ohno, F. Matsukura, J. Cibert, and D. Ferrand, *Science* **287**, 1019-1022 (2000).
4. For a recent review, see e.g. S. J. Pearton, C. R. Abernathy, G. T. Thaler, R.M. Frazier, D. P. Norton, F. Ren, Y. D. Park, J. M. Zavada, I. A. Buyanova, W. M. Chen, and A. F. Hebard, *J. Phys. Condens. Matter* **16**, R209–R245 (2004)
5. I. A. Buyanova, M. Izadifard, W. M. Chen, J. Kim, F. Ren, G. Thaler, C.R. Abernathy, S.J. Pearton, C. Pan, G. Chen, J. Chyi and J.M. Zavada, *Appl. Phys. Lett.* **84**, 2599-2601 (2004)
6. F.V. Kyrychenko, C.J. Stanton, C.R. Abernathy, S.J. Pearton, F. Ren, G. Thaler, R. Frazier, I. A. Buyanova, W. M. Chen, to be presented at the 27th Int. Conf. Physics of Semiconductors, Flagstaff, AZ, July 2004.

Transient Faraday Rotation and Circular Dichroism Induced by Circularly-Polarized Light in InGaN

T. Matsusue[1], D. Kanbara[1], H. Mino[2], M. Arita[3], and Y. Arakawa[3]

[1]Deptartment of Materials Technology, Chiba University, 1-33 Yayoi-cho, Inage-ku, Chiba 263-8522, Japan
[2]Deptartment of Physics, Chiba University, 1-33 Yayoi-cho, Inage-ku, Chiba 263-8522, Japan
[3]Institute of Industrial Science, University of Tokyo, 4-6-1 Komaba, Meguro-ku, Tokyo 153-8505, Japan

Abstract. Spin dynamics of InGaN is successfully examined with an optimized sample design and experimental conditions. Relaxation of spin initially created by a circularly-polarized pump pulse is monitored by Faraday rotation and circular dichroism of a probe pulse at various wavelengths and temperatures. It is found that temporal decay consists of a fast and a slow components. Amplitude and decay time of the slow component have the second maximum against wavelength, which is related to band structures of InGaN. The fast and slow decay times are 1.1ps and 140ps at 10K, and decrease with temperature.

Spin dynamics of InGaN is an interesting subject. It could exhibit peculiar and useful properties compared with InGaAs since InGaN has distinct band structures and in practice is strongly affected by built-in electric field, defects, and fluctuation of chemical content. However, spin-related phenomena of InGaN[1] remain almost veiled mainly due to experimental difficulties in selective excitation and monitor of carrier spin.[2] That is because inverse spins are excited with nearly degenerated hole bands by a circularly-polarized light, in contrast to the case of InGaAs. In this paper, we report the first observation of spin dynamics in InGaN with an optimized sample structure and experimental conditions. We have clarified basic properties of spin dynamics with wavelength and temperature, and found novel behaviors.

We grew a sample of 120-nm thick undoped $In_{0.016}Ga_{0.986}N$ capped by 10-nm GaN following GaN buffer layer on a c-axis sapphire substrate by metal-organic chemical vapor deposition. We used a bulk sample eliminate structural irregularity, designed the In content to be as small as possible to suppress inhomogeneous broadening due to fluctuation of In content but enough to observe absorption peak separately from that of GaN buffer layer. In order to examine spin dynamics, spin was excited by circularly-polarized pump pulse, and its relaxation process was monitored by delayed probe pulse using Faraday rotation for linearly-polarized light and dichroism for circularly-polarized light, which provide complimentary information. Second harmonic light of fs mode-locked Ti:sapphire laser was split to pump and probe pulses. Probe pulse passed through the sample was detected by lock-in technique with chopping pump and probe beams at different frequencies. In order to increase detectivity for Faraday rotation, probe beam after the sample was divided into two at mutually perpendicular polarization with an equal intensity at no pump, using polarized beam splitter. The intensity difference induced by pump was detected by a balanced amplifier. The spectral width of the beam was adjusted as narrow as 10meV and the temporal width was 100fs.

In absorption spectrum, an exciton peak of InGaN was observed separated from absorption tail of GaN. Transient optically-induced circular dichroism has been observed near the absorption peak, as shown in Fig. 1, where spin excited by a pump pulse relaxes to a reverse direction and the initial spin polarization is nearly 60%. Figure 2 shows Faraday rotation, where transient response consists of a fast and a slow decay. We ascribe fast decay to relaxation of hole spin, and partially of orbital moment and coherence, while slow decay to that of electron spin. A dip in the decay curve around 3ps appears in detailed measurement for early period, whose origin is not clear at present. Transient response depends remarkably on wavelength, where the slow decay component and its decay time become maximal at 362 and 360.5nm. In

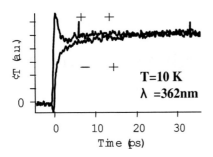

FIGURE 1. Transient differential transmittance for co- and anti-circular polarization of pump and probe.

valence band of InGaN, there exist A- and B- bands with an energy difference of near 10meV. The A-band is configured with orbital angular momentum and spin of hole in parallel directions, whereas the B-band in anti-parallel directions. Assume photons are circularly polarized in + direction. As wavelength decreases from a wavelength well above the exciton absorption peak, it becomes that excitons are efficiently created, where A-excitons consisting of electrons and holes with down spins are predominant. Therefore, relaxation process of polarized electrons by a slow decay is appreciable. As wavelength decreases further, B-excitons consisting of electrons and holes with up spins become to be created in mixture to A-excitons. Hence, there exists electrons of opposite spins associated with A- and B-excitons. This suppresses component of electron spin relaxation process. Moreover, relaxation of B-exciton to A-exciton without spin-flip changes polarization of interacting photons, which decreases Faraday rotation, equivalently to spin relaxation. As wavelength decreases further more, it happens that B-excitons consisting of electrons and holes with up spins are predominantly created. The mechanism explains the dependence of the result on wavelength. Note that the spectral width of laser is comparable to the separation energy of A- and B-excitons, incompletely selective excitation of A- or B-excitons is feasible.

Figure 3 shows decay time of fast and slow components as a function of temperature at a wavelength associated with A-exciton. At 10K, the fast decay is as fast as 1.1ps. The slow component has a decay time of 140ps, which corresponds to 280ps for lifetime of electron spin. Both decay times decrease with temperature. Previous works for GaAs and other semiconductors have clarified that D-P and E-Y mechanisms are important for spin relaxation, which change spin relaxation time proportionally to T^{-3} and T^{-2}, respectively, for electrons.[3] Our results depend differently on T. InGaN has small spin-orbit splitting and large band gap, which may lead the above mechanisms to be suppressed and other mechanisms to

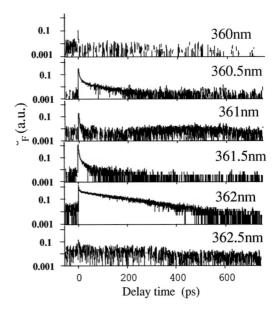

FIGURE 2. Optically induced Faraday rotation at various wavelengths.

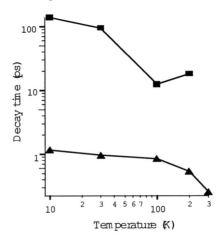

FIGURE 3. Decay time of fast and slow component in Faraday rotation as a function of temperature.

dominate. Effects of exciton, localization and other peculiar properties such as internal electric field may affects. Further investigation is necessary for clarify the mechanism of spin relaxation in InGaN.

We thank Profs. S. Takeyama, K. Muro and K.Oto for support and useful discussions, and S. Harada for assistance.

REFERENCES

1. B. Beschoten et al., *Phys.Rev.* **B63,** 121202 (2001).
2. A. Tackeuchu et al., *Appl. Physca* **E65**, 1011 4 (2000).
3. P.H. Song et al., *Phys.Rev.* **B66,** 35207 (2002).

Field dependence of spin lifetimes in nitride heterostructures

J. A. Majewski, P. Vogl

Walter Schottky Institute, Technische Universität München, Am Coulombwall 3, 85748 Garching, Germany

Abstract. We present first-principles calculations of the zero field spin splitting of energy bands in nitride heterostructures. Our calculations reveal that the huge electric fields originating from strong piezo- and pyroelecric character of nitrides do not increase the spin splitting of band in nitride heterostructures. This implicates long spin lifetimes in quantum structures based on these materials.

INTRODUCTION

Recently, the room temperature ferromagnetism in diluted magnetic semiconductor GaN:Mn has been predicted theoretically and later confirmed in numerous experiments [1]. This opens new perspectives for spin devices entirely based on nitrides. The possible design of such spintronic devices requires the understanding of the spin dynamics and spin relaxation processes, which are, in turn, determined by the zero field spin splitting of the bands. Unfortunately, the spin splitting of the bands, induced by the spin orbit interaction, is mostly unknown in nitride quantum structures. In standard semiconductors, the electric field dependence of the spin splitting of energy bands (Rashba effect) has a marked influence on the spin dynamics and spin relaxation processes. This issue becomes much more relevant in nitride heterostructures. In view of their huge (of order few MV/cm) electric fields that originate in their strong piezo- and pyroelectric character, one may anticipate very significant spin splittings.

In this paper, we present a systematic study of the influence of these built-in electric fields on the spin splitting in nitride heterostructures. These studies are based on first-principles, relativistic local density calculations that provide very accurate description of the spin splitting in nitride bulks [2]. Here we concentrate on the lowest order in the wave vector k spin splitting, i.e., terms linear in k.

RESULTS AND DISCUSSION

For wurtzite bulks and superlattices grown along the *c*-axis, the effective 2x2 Hamiltonian describing k-linear terms can be written in the form $H = \sigma \cdot B_{eff}$, with $B_{eff}(k_x, k_y) = \alpha_R(k_y, -k_x, 0)$, where σ is the vector of Pauli matrices, k_x and k_y are wave vector components in the plane perpendicular to the hexagonal axis, and α_R is a band dependent constant that determines the isotropic spin splitting energy $\Delta E_{spin} = 2 \alpha_R (k_x^2 + k_y^2)^{1/2}$ of a certain band. In the wurtzite structure, the combination of crystal-field and spin-orbit interactions leads to a three-edge structure of the top of the valence band in the Γ point, known as A, B, and C edges. Two of these three edges are of Γ_7 and one of Γ_9 symmetry, while the lowest conduction state has Γ_7 symmetry. The valence states are, in order of decreasing energy, $\Gamma_9, \Gamma_7, \Gamma_7$ for GaN, and $\Gamma_7, \Gamma_9, \Gamma_7$ for AlN (i.e., like in ZnO). It is known that the Γ_9 states of wurtzite are not linearly split. In Fig. 1 we depict calculated values of

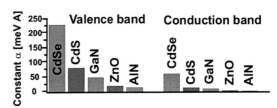

FIGURE 1. Linear-k spin splitting (in eVÅ) of the conduction band and the higher Γ_7 valence band in few wurtzite semiconductors.

constant α_R for the highest Γ_7 valence and conduction band states in AlN and GaN and compare with values for other semiconductors. As can be seen in Fig. 1, in all wurtzite semiconductors, the linear-k spin splitting is considerably smaller in the conduction band than in the valence band.

In further discussion, we focus on the spin splitting of the conduction band in AlN/GaN superlattices. We have performed calculations for AlN/(GaN)$_n$ superlattices with various width of the GaN quantum well, measured by the number of GaN layers n. The divergence of the large electric polarization (spontaneous and piezoelectric) causes a negative and positive bound sheet charges located at the interfaces, which leads to huge electric fields in the structure of the order of few MV/cm. A schematic band lineup of (AlN)$_4$/(GaN)$_8$ superlattice together with the calculated wave functions of the top of the valence band and bottom of the conduction band are shown in Fig. 2. The wave function of the valence band top is strongly localized at one interface, just giving rise to the negative polarization charge. In the (AlN)$_4$/(GaN)$_n$ superlattices on GaN substrate, the calculated values of the constant α_R are equal to 0.4, 0.9, 0.6, and 0.8 meVÅ for n = 4, 8, 12, 16, respectively. They are therefore an order of magnitude smaller than the corresponding value of the α_R constant for bulk GaN (α_R = 9 meVÅ). It means that the macroscopic electric field in the well that originates from interfaces charges counteracts the effect of spontaneous polarization of the bulk material and leads to the strong reduction of the conduction band spin splitting. This reduction is only weakly dependent on the width of the well.

To investigate the role of the macroscopic electric field further, we calculated also the spin splitting in the AlN/(GaN)$_n$ superlattices grown on AlN. In such structures GaN quantum wells are subject to the strong compressive biaxial strain that results from the lattice mismatch (-2.5%). This strain causes the considerable increase of the conduction band spin splitting, giving the value of α_R equal to 20 meVÅ. For the (AlN)$_4$/(GaN)$_n$ superlattices on AlN, the calculations yield the values of α_R equal to 8, 11, and 12 meVÅ for n = 4, 8, 16, respectively. The reduction of α_R constant caused by the polarization induced macroscopic electric field is of the same order as in the case of AlN/GaN superlattices grown on GaN.

We would like to stress that the linear in k spin splitting terms dominate the spin splitting around the Γ point. Even for wave vectors of the length 0.25 1/Å (roughly ¼ of the Brillouin Zone) the linear term constitutes approximately 75% of the total spin splitting. The strong dependence of the constant α_R on the growth induced strain (mostly unknown) hinders direct comparison with experiment. On the other hand, the present calculations shed light on the physical origin of contradictory values of the spin splitting of the conduction band electrons as extracted from various experiments [3].

Altogether, the conduction band spin splitting of the AlN/GaN heterostructures remains small in spite of the huge electric fields, indicating long spin relaxation times in the nitride heterostructures. For comparison, the α_R for [111] AlAs/GaAs heterostructure is by factor 10 larger (equal to 140 meVÅ) than the highest value of α_R observed in AlN/GaN structures. On the other hand, the spin splitting of the conduction band can be only marginally influenced by the external electric field. The calculations show that the external bias of 250 kV/cm changes the constant α_R by approximately 1%.

The presented calculations strongly suggest that GaN heterostructures are a promising candidate for some spintronic applications, where long spin lifetime is needed. GaN quantum wells are not particularly suitable for spin transistor (rather weak steering possibility) but they can be very good for static Qbits.

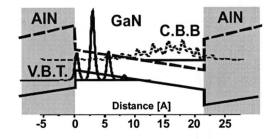

FIGURE 2. Band lineup and wave functions of the valence (VBT) and conduction band (CBB) edges. For both bands, the laterally averaged first-principles wave-functions are shown.

REFERENCES

1. Dietl T., Ohno H., Matsukara F., Cibert J., Ferrand D., *Science* **287**, 139 (1998); for a review of experiments see Graf T., Goennenwein T.B., Brandt M.S., *phys. stat. sol. (b)* **239**, 277 (2003).
2. Majewski J.A., Städele M. and Vogl P., *Mat. Res. Soc. Symp. Proc.* **449**, 887 (1997).
3. Tsubaki K., Maeda N., Saitoh T. and Kobayashi N., *Appl. Phys. Letters* **80**, 3126 (2002).

Current-Induced Spin Polarization at a Single Heterojunction

A. Yu. Silov*, P. A. Blajnov*, J. H. Wolter*, R. Hey[†], K. H. Ploog[†] and N. S. Averkiev**

COBRA Inter-University Research Institute, Eindhoven University of Technology, PO Box 513, NL-5600 MB Eindhoven, The Netherlands
[†]*Paul-Drude-Institut für Festkörperelektronik, Hausvogteiplatz 5-7, 10117 Berlin, Germany*
**Ioffe Physico-Technical Institute, Polytekhnicheskaya 26, 194021 St. Petersburg, Russia*

Abstract. We have experimentally achieved spin-polarization by a lateral current in a single non-magnetic semiconductor heterojunction. The effect does not require an applied magnetic field or ferromagnetic contacts. The current-induced spin orientation can be seen as the inverse of the circular photo-galvanic effect (also often referred to as spin-photocurrents): the nonequilibrium spin changes its sign as the current reverses.

The current-induced spin orientation in semiconductors had been theoretically discussed by Dyakonov and Perel [1] already in the 1970s, but, up to now, was not realized experimentally. Such spin polarization by an electric current can be seen as the inverse of the circular photo-galvanic effect and differs radically from the methods employed so far [2, 3]. The effect is allowed in quantum wells of III-V compounds due to the removal of the spin degeneracy [4].

In low-dimensional systems (quantum wells and single heterojunctions), the theory of spin polarization in an electric field was developed in [5–7]. In systems lacking inversion symmetry there exists an additional term of the spin-orbit interaction that is linear in the wave vector, k (see, for instance, Ref. [8]). This interaction removes the spin degeneracy away from the zone centre. Previously, the current-induced spin polarization was theoretically considered only for the conduction electrons. However, the k-linear terms are not unique for the conduction band only, and, in zinc-blende semiconductors, are significantly larger in the valence band.

The sample was MBE-grown on a (001) GaAs substrate. The layer sequence starts with a nominally undoped GaAs (50 nm) buffer layer followed by a 25-period GaAs/AlAs (5 by 5 nm, respectively) superlattice. Subsequently, the active layer (20 nm) of undoped GaAs was deposited. On top of the active layer, a Be-doped (1.33×10^{18} cm^{-3}) $Al_{0.33}Ga_{0.67}As$ with a 20-nm spacer was grown. Finally, the $Al_{0.33}Ga_{0.67}As$ layer was capped with a 8-nm-thick layer of GaAs. Annealed Ni/ZnAu contacts were made to the 2DHG confined by the potential notch at the $Al_{0.33}Ga_{0.67}As$/GaAs heterojunction. The hole concentration (N_{Hall}) was 4.15×10^{11} cm^{-2} with the mobility (μ_{Hall}) of 46500 cm^2/V s at 5.8 K.

The angle between the mean spin and the electric current is governed by the interplay between the Dresselhaus and Bychkov-Rashba effects. It can be shown, for instance, that in the (311)-grown quantum wells the mean spin density will have a component along the growth direction. For the quantum wells on (001)-surface, however, the spin density has only planar components [7].

To detect the spin-polarization, we have measured the degree of circular polarization, P, of the 2DHG photoluminescence, a method originally proposed for detecting the electrical spin injection through a ferromagnetic contact [9]. This experimental procedure has become a proven method for probing spin polarization [2]. The sample was cleaved into bars of 1×2.5 mm^2 with the current flowing along the long side, which is in the [1$\bar{1}$0] direction. The photoluminescence (PL) was excited with 633-nm line from a helium-neon laser. The PL was collected from the cleaved (110) facet of the sample. The baseline of the circularly polarized PL was carefully checked by recording a difference between PL$_{\sigma-}$ and PL$_{\sigma+}$ spectra when no current was delivered through the sample.

Figure 1 shows the low-temperature PL spectra both with a current flow turned on or switched off. The PL contains a strong radiative band centred at about 1.52 eV. This band originates from recombination of the two-dimensional holes with electrons photoexcited into the

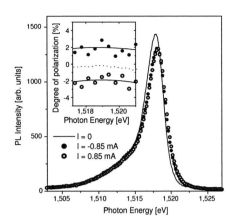

FIGURE 1. Photoluminescence spectra at 5.1 K in absence of a current (solid line) and with current of ±0.85 mA in opposite directions (empty and filled circles, respectively). Spectral resolution is 1.8 meV, excitation density is 1.2 mW/cm². Inset shows a degree of circular polarization within the full width at half maximum. The dotted line gives the polarization baseline. The solid lines are are a guide for the eye.

GaAs conduction band [10]. The PL intensity is decreased by passing the current through the sample, while the emission wavelength is blueshifted. This spectral shift towards the higher energy side is explained by the extra energy acquired by the 2DHG in the lateral electric field.

The inset in Fig. 1 shows the observed P with the dotted line representing a baseline of the circular polarization. The observed degree of polarization yields a maximum of 2.5 %. It is the important feature of the data that the degree of polarization inverses its sign as the electric current reverses. This is a direct prove of the current-induced spin orientation, in complete agreement with the results on the spin photocurrents in quantum wells [11].

The spin orientation, S, can be estimated following calculations for 2D electrons from Ref. [6]:

$$S = 6Q\frac{\gamma F e \tau_p}{\hbar E_F}, \quad (1)$$

where Q is a coefficient of approximately unity depending of the scattering mechanism, E_F is the DHG Fermi energy, F is an electric field, e is elementary charge, τ_p is the momentum relaxation time, and γ describes the k-linear spin splitting. In a field of 6.5 V/cm, with $\gamma = 7 \times 10^{-2}$ eV Å (from linear interpolation of data [12]), $\tau_p = 10^{-11}$ s and $E_F = 2$ meV from our transport measurements, the spin polarization is estimated to be 10%. The experimentally observed P of 2.5% yields only a low-limit estimate of the nonequilibrium spin-polarization. A strong admixture of the light-hole character to the ground (heavy-hole) state of the 2DHG reduces the value of the circular polarization degree. Nonetheless, even $S = 2.5\%$ exceeds the spin polarization that has been initially achieved by the spin injection from ferromagnetic metals into GaAs [13].

Note added. After submission of the abstract, we learned of the independent experiments on the current-induced spin polarization in strained bulk material [14] and in p-doped quantum wells [15].

ACKNOWLEDGMENTS

This work was financially supported by the Dutch Organization for Advancement of Research (NWO). The research at the Ioffe Institute was also supported by INTAS, RFBR, scientific programs of RAS and Russian Ministry of Industry, Science and Technology.

REFERENCES

1. Dyakonov, M. I., and Perel, V. I., *Phys. Lett. A*, **35**, 459 (1971).
2. Awschalom, D., Loss, D., and Samarth, N., editors, *Semiconductor spintronics and quantum computation*, Springer, Berlin, 2002.
3. Meier, F., and Zakharchenya, B., editors, *Optical orientation*, North–Holland Physics, Amsterdam, 1984.
4. Ganichev, S. D., and Prettl, W., *J. Phys.: Condens. Matter*, **15**, R935 (2003).
5. Edelstein, V. M., *Solid State Communications*, **73**, 233 (1990).
6. Aronov, A. G., Yu. B. Lyanda-Geller, and Pikus, G. E., *Sov. Phys. JETP*, **73**, 537 (1991).
7. Chaplik, A. V., Entin, M. V., and Magarill, L. I., *Physica E*, **13**, 744 (2002).
8. Rashba, E. I., and Sheka, V. I., "Electric-Dipole Spin Resonances," in *Landau Level Spectroscopy*, edited by G. Landwehr and E. I. Rashba, North–Holland, Amsterdam, 1991, vol. 1, p. 178.
9. Pikus, G. E., and Aronov, A. G., *Sov. Phys. Semicond.*, **10**, 698 (1976).
10. Silov, A. Yu., Haverkort, J. E. M., Averkiev, N. S., Koenraad, P. M., and Wolter, J. H., *Phys. Rev. B*, **50**, 4509 (1994).
11. Ganichev, S. D., Ivchenko, E. L., Danilov, S. N., Eroms, J., Wegscheider, W., Weiss, D., and Prettl, W., *Phys. Rev. Lett.*, **86**, 4358 (2001).
12. Störmer, H. L., Schlesinger, Z., Chang, A., Tsui, D. C., Gossard, A. A., and Wiegmann, W., *Phys. Rev. Lett.*, **51**, 126 (1983).
13. Jonker, B. T., *Proc. IEEE*, **91**, 727 (2003).
14. Kato, Y., Myers, R. C., Gossard, A. C., and Awschalom, D. D. (2004), cond-mat/0403407 (unpublished).
15. Ganichev, S., Danilov, S., Schneider, P., Bel'kov, V., Golub, L., Wegscheider, W., Weiss, D., and Prettl, W. (2004), cond-mat/0403641 (unpublished).

Optical Studies of Spin Coherence in Organic Semiconductors

C. Yang, C. Liu, Z.V. Vardeny

Physics Department, University of Utah, 115 S 1400 E, Rm201, Salt Lake City, UT84112

Abstract. Spin coherence in organic oligomer and polymer semiconductors is a crucial property in determining the performance of organic spin devices. Using the techniques of photo-induced absorption (PA) detected magnetic resonance (PADMR) and photoluminescence (PL) detected magnetic resonance (PLDMR), we have studied the influence of temperature, conjugation length, excitation energy, magnetic nano-particles mixing, and in-chain heavy metal atoms on the spin coherence in organic semiconductors. From the PADMR measurements in addition, we could infer the formation ratio of singlet/triplet excitons from electron-hole pairs in these materials, and found that the formation of singlet excitons is preferred in polymers. We also found that spin randomization induced by magnetic nanoparticles or heavy atoms lead to faster polaron recombination dynamics and lack of optically detected magnetic resonance signal.

INTRODUCTION

Organic semiconductors have emerged as the active materials in organic light emitting diodes, thin film transistors, photovoltaic cells and spin valve devices in the past decade [1-2]. By virtue of having large spin coherence, spin dynamics of charged carriers (polarons) have played an important role for the applications of these materials in spintronic devices [2]. The optically detected magnetic resonance (ODMR) technique has been extensively used in organic semiconductor thin films and devices for studying spin-dependent phenomena, such as excited states spin characterization, spin-dependent polaron recombination, triplet-triplet annihilation, and photoluminescence quenching processes [3-5]. Spin coherence in organic semiconductors has been always implied to be large, but has not been directly measured as yet.

In this work we propose to use the ODMR technique as a tool to study spin coherence of photogenerated charged polarons at different conditions that include variation of the temperature, excitation photon energy, and also by mixing the polymer chains with magnetic nanoparticles. We found that the processes of magnetic nanoparticles mixing, raising the temperature, and binding in-chain heavy atoms enhance the spin randomization rate. In our experiments the change of spin coherence has been monitored by PADMR and PLDMR measurements.

EXPERIMENT

The materials and suppliers used for this study were: regiorandom (RRa-) poly(3-hexylthiophene-2,5-diyl) (P3HT) from Sigma-Aldrich Co, and Fe_3O_4 nanoparticles from MACH II Inc. The samples were drop-cast films on glass substrates with peak optical density OD ≈ 2. The blends were mixed in a supersonic bath for 30 minutes. The PL, PA and ODMR measurements were described elsewhere [4].

RESULTS AND DISCUSSION

The absorption, PL, PA, PADMR and PLDMR spectroscopies have been used to characterize the polymers, as shown in Figs. 1 and 2 for RRa-P3HT. PADMR, which measures the ratio δPA/PA, is obtained in the polymer film due to the following process: the photogenerated polaron pairs, with a characteristic PA band in the mid-ir spectral range (P_1 in Fig. 1), are spin polarized to some extent due to spin-dependent recombination rates, where pairs with parallel spins recombine slower compared to pairs with anti-parallel spins. Microwave absorption in resonance conditions with the Zeeman-splitted spin ½ polaron sublevels under the influence of an external magnetic field induces spin flips that redistribute the polaron spins causing a change in the steady state population and hence a change δPA is obtained. Same is also true for spin triplet sublevels of photogenerated triplet excitons (band T in Fig. 1) with half-field and full-field powder patterns (Fig. 2). Thus the PADMR signal depends on the steady state spin polarization process, which, in turn depends on spin coherence in the material. A faster spin decoherence process would decrease δPA. The PLDMR signal is directly related to the PADMR signal [3], but is easier to measure.

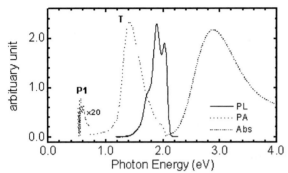

FIGURE 1. Absorption, PL, and PA spectra of RRa-P3HT film. The PA bands P_1 and T have been assigned to polarons and triplet excitons, respectively, by PADMR (Fig. 2).

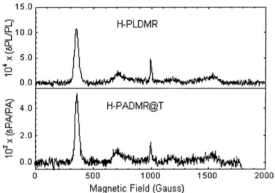

FIGURE 2. H-scan ODMR spectra for RRa-P3HT showing spin-1/2 polarons at ca. 1010 Gauss, and half-field and full-field powder patterns of triplet excitons at ca. 350 and 700 (1500) gauss, respectively.

In Fig. 3 we show the effect on ODMR of mixing the polymer chains with magnetic nanoparticles (Fe_3O_4). It is seen that adding nanoparticles actually kills the spin ½ ODMR signal. The reason for this ODMR decrease is the induced enhancement of spin 1/2 randomization rate. Lifetime measurement of the polarons verified that these randomization mechanisms indeed increase the recombination rate. In addition, raising temperature, adding charge transfer channels, increasing excitation energy and adding heavy atoms to the chain all result in smaller PADMR signal, indicating a decrease of spin ½ coherence in the sample [6].

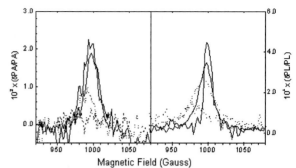

FIGURE 3. The dependence of the ODMR signal on the concentration of Fe_3O_4 magnetic nanoparticles in RRa-P3HT films. The thick line, black line, dotted line and dashed line correspond to nanoparticle molar concentration of 0%, 1%, 2% and 5%, respectively.

CONCLUSION

ODMR serves as a very useful tool not only for studying spin related recombination phenomena, but also for measuring spin coherence. Enhanced spin randomization rates by mixing the polymer chains with magnetic nanoparticles decrease the spin coherence that results in a decrease of the ODMR signal. The same effect was achieved by increasing temperature and adding heavy atoms to the chain.

ACKNOWLEDGMENTS

We thank Drs. Viner and Delong for lab assistance and useful discussions. This work was supported in part by the DOE Grant No. FG 02-04ER46109.

REFERENCES

1. Friend, R. et al, *Nature* **397**, 121-128 (1999).
2. Xiong, Z. H. et al, *Nature* **427**, 821-824 (2004).
3. Vardeny Z. V. and Wei X., *handbook of conducting polymers*, Marcel Dekker, Inc., 1998, pp. 639-666.
4. Wohlgennant, M. and Vardeny, Z. V., *J. Phys.: Condens. Matter* **15**, R83-R107 (2003).
5. Swanson, L. et al, *Phys. Rev. B* **46**, 15072-15077 (1992).
6. Yang, C. et al, in preparation.

Twisted exchange interaction between localized spins in presence of Rashba spin-orbit coupling

Hiroshi imamura*, Patrick Bruno[†] and Yasuhiro Utsumi[†]

Graduate School of Information Sciences, Tohoku University, Sendai 980-8579, Japan
[†]*Max Planck Institute of Microstructure Physics, Weinberg 2, D-06120 Halle, Germany*

Abstract. We theoretically study the RKKY interaction between localized spins embedded in a 1D- or 2DEG with Rashba spin-orbit coupling. We show that rotation of spin of conduction electrons due to the Rashba spin-orbit coupling causes a twisted RKKY interaction between localized spins which consists of three different terms: Heisenberg, Dzyaloshinsky-Moriya, and Ising interactions. We also study the spin-configuration of linear- and ring-shaped artificial molecules consisting localized spins coupled via the twisted RKKY interaction. We find that the square-norm of the total spin of an artificial molecule oscillates with the twist angle θ and has maxima at θ = an odd integer times π. The square-norm of the total spin of the ring-shaped artificial molecules change drastically at certain values of θ where the lowest two energy levels cross each other.

There has been a great deal of interest in the field of spintronics where spin degrees of freedom of electrons are manipulated to produce a desirable outcome. Eminent examples are given by the giant magnetoresistance (GMR) effect and the interlayer exchange coupling in magnetic multilayers. The interlayer exchange coupling is explained in the context of Ruderman-Kittel-Kasuya-Yosida (RKKY) interaction[1]. Recently, much attention has been focused on the effect of the Rashba spin-orbit (RSO) coupling in two-dimensional electron gases (2DEG)[2, 3, 4] . It has been established the RSO coupling can be controlled by means of a gate voltage. The RSO coupling has also been observed in 2DEG formed from surface states at metal surfaces such as Au(111)[5]. It has also been found that confinement of the surface state due to atomic steps on vicinal surfaces leads to quasi one-dimensional surface states, which also exhibit the RSO coupling[6].

We consider the system consisting of localized spins embedded in a 1D or 2DEG with RSO coupling. The Hamiltonian of two localized spins, S_1 and S_2, located at positions R_1 and R_2 is given by

$$H = -\frac{\hbar^2}{2m}\nabla^2 + \alpha(-i\hbar\nabla\times\hat{z})\cdot\sigma + J\sum_{i=1,2}\delta(r-R_i)S_i\cdot\sigma, \quad (1)$$

where α represents the strength of the spin-orbit coupling, $\hat{z}$ is a unit vector along the z-axis, σ is the vector of Pauli spin matrices, and J represents the strength of the s-d interaction. We assume that the conduction electrons are confined in a wire along the x-axis (one-dimensional system) or in the $x-y$ plane (two-dimensional system). The direction of the effective electric field of spin-orbit coupling is taken to be along the z-axis for both one- and two-dimensional systems.

The RKKY interaction between S_1 and S_2 is calculated from the second order perturbation theory as

$$H_{1,2}^{RKKY} = F_d(R)\left[\cos(2k_R R)S_1\cdot S_2 \right.$$
$$\left. +\sin(2k_R R)(S_1\times S_2)_y + \{1-\cos(2k_R R)\}S_1^y S_2^y\right], \quad (2)$$

where $k_R \equiv m\alpha/\hbar^2$, $R \equiv |R_1-R_2|$, and $F_d(R)$ is the range function for d-dimensional system[7]. We take the coordinate system so that the vector R is aligned with the x-axis, i. e., $R_1-R_2 = R\hat{x}$. The resulting RKKY interaction of Eq.(2) consists of three physically quite different interactions: Heisenberg, Dzyaloshinsky-Moriya (DM) and, Ising interactions. The Heisenberg and Ising interactions favor a collinear alignment of localized spins. On the contrary, the DM interaction favors a non-collinear alignment of localized spins.

This peculiar twisted coupling of localized spins can be easily understood by introducing the twisted spin space where the spin quantization axis of the second localized spin S_2 is rotated by an angle $\theta \equiv 2k_R R$ around the y-axis. One can easily show that the RKKY interaction of Eq. (2) can be expressed as

$$H_{1,2}^{RKKY} = F_d(R)S_1\cdot S_2(\theta). \quad (3)$$

Eq.(3) implies that the twisted RKKY interaction results in a collinear coupling of localized spins in the θ-twisted

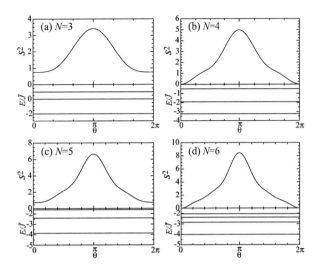

FIGURE 1. The square-norm of the total spin S^2 of the ground states and energy levels of linear-shaped AMs with $N = 3, 4, 5$, and 6 are plotted against the twist angle θ in panels (a), (b), (c) and (d), respectively. Energy is normalized by the coupling constant of the RKKY interaction, $\tilde{J}$, and the twist angle takes the same value θ for all bonds.

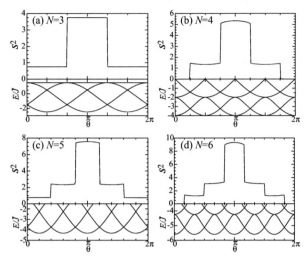

FIGURE 2. The same plot as Fig. 1 but for ring-shaped AMs.

spin space. The energy eigenvalues do not depend on θ and the total spin in the θ-twisted spin space is a conserved quantity.

In order to capture the basic physics of the twisted-RKKY interaction, we set hereafter $F_d(R) = \tilde{J} > 0$. The wavefunction of the ground state of two localized spins is given by $\psi \propto \cos\frac{\theta}{2}(|\uparrow,\downarrow\rangle - |\downarrow,\uparrow\rangle) + \sin\frac{\theta}{2}(|\uparrow,\uparrow\rangle + |\downarrow,\downarrow\rangle)$. Then the square norm of the total spin of the ground state, $S^2 = 2\sin^2\frac{\theta}{2}$, is an oscillating function of θ and takes the maxima at $\theta = \pi$. One should note that the ground state in the θ-twisted spin space is always the singlet. However, the singlet in the π-twisted spin space is in fact the triplet with $S^2 = 2$ ($S = 1$) in the real spin space. The wavefunction in the real spin space changes from singlet → triplet → singlet as we increase the twist angle θ[8]. For semiconductor artificial molecules (AMs), quantum dots connected via spin field effect transistors, the twist angle θ can be controlled by a gate voltage[3].

The above discussions can be generalized to the linear shaped AMs with $N(\geq 3)$ localized spins, since the Hamiltonian can be mapped to the usual RKKY Hamiltonian by the rotation of the spin quantization axis. In Figs. 1 (a)-(d), we show the square-norm of the ground state and energy levels for the linear-shaped AM consisting of $N = 3, 4, 5$, and 6. The energy levels do not depend on θ. Without RSO coupling, $\theta = 0$, the ground state has the antiferromagnetic spin configuration with $S = 0$ or $1/2$. The square-norm S^2 is again an oscillating function of θ with a period of 2π.

On the contrary to the linear-shaped AM, the square-norm S^2 of the ring-shaped AM is not a smooth function of the twist angle as shown in Figs. 2 (a)-(d). Due to the boundary condition we cannot twist the spin-wavefunction by an arbitrary angle θ. Therefore, the square-norm S^2 jumps at a certain value of angle θ where the lowest two energy levels cross each other as shown in Figs. 2 (a)-(b).

ACKNOWLEDGMENTS

This work was supported by MEXT.KAKENHI(Grant No. 14076204 and 16710061), NAREGI Nanoscience Project, and BMBF (Grant No. 01BM924).

REFERENCES

1. Kittel, C., *Solid State Physics*, Academic, New York, 1968, vol. 22, p. 1.
2. Rashba, É. I., *Fizika Tverdogo Tela*, **2**, 1224 (1960), [Sov. Phys. Solid State **2**, 1109 (1960)].
3. Nitta, J., Akazaki, T., Takayanagi, H., and Enoki, T., *Phys. Rev. Lett.*, **78**, 1335 (1997).
4. Engels, G., Lange, J., Schäpers, T., and Lüth, H., *Phys. Rev. B*, **55**, R1958 (1997).
5. LaShell, S., McDougall, B. A., and Jensen, E., *Phys. Rev. Lett.*, **77**, 3419 (1996).
6. Mugarza, A., et al., *Phys. Rev. B*, **66**, 245419 (2002).
7. Imamura, H., Bruno, P., and Utsumi, Y., *Phys. Rev. B*, **69**, 1231303(R) (2004).
8. Imamura, H., and Ebisawa, H., *J. Magn. Magn. Mater.*, **272-276**, 1909 (2004).

AC-Driven Double Quantum Dots as Spin Pumps

E. Cota*, R. Aguado** and G. Platero**

*Centro de Ciencias de la Materia Condensada, UNAM, Ensenada, Mexico.
**Instituto de Ciencia de Materiales (CSIC), Cantoblanco, E-28049 Madrid, Spain.

Abstract. We investigate the current through an ac-driven double quantum dot in the pumping configuration, in the presence of an external magnetic field. By means of the density matrix formalism in the Markov approximation, we calculate the current through the system, for both, spin up and spin down polarizations. We show that the current through the double quantum dot is spin polarized and the ac signal allows to control the direction and the degree of its spin polarization. In particular, increasing the ac frequency the current polarization can be fully reversed from fully spin down to spin up.

In recent years, there has been great interest in artificial atom devices, the so called quantum dots. The interest stems from a variety of fields such as nanoelectronics or quantum computation. The application of an ac signal to a double quantum dot (DQD) has shown to produce dramatic effects in the electron dynamics such as coherent destruction of tunnelling[1]. Recently, Cota et al. investigated pumping of spin polarized electrons[2] through a DQD attached to non ferromagnetic leads in the presence of an uniform magnetic field B by applying an ac gate voltage V_{ac} between the dots. Interestingly, the application of an V_{ac} allows to control the degree of spin polarization of the current flowing through the DQD even in the case of contact leads which are not spin polarized. This is of importance, for understanding the behavior of spins in nanostructures which has become a subject of intense investigation due to its relevance to quantum information processing and spintronics.

In this paper, we have extended the work by Cota et al., to include doubly occupied excited states, a triplet intra-dot state, in the transport window. As we will see below, it will allow not only to obtain fully spin polarized current I, either spin up or down, but it will allow as well to invert the spin polarization by tuning the frequency of the V_{ac}. We have analyzed, by means of the reduced density matrix (DM) formalism, within the Markov approximation, the time evolution of the current, and its spin polarization as a function of the ac frequency ω_{ac}, in the pumping configuration.

Our system consists of a DQD connected to two non ferromagnetic leads which are in equilibrium with reservoirs kept at the chemical potentials μ_α, $\alpha = L, R$. Using a standard tunneling Hamiltonian approach, we write for the full Hamiltonian:

$H_l + H_{DQD} + H_T$, where $H_l = \sum_\alpha \sum_{k_\alpha,\sigma} \varepsilon_{k_\alpha} c^\dagger_{k_\alpha\sigma} c_{k_\alpha\sigma}$ describes the leads, H_T describes the tunnelling between leads and each QD, and $H_{DQD} = H^L_{QD} + H^R_{QD} + H_{L \Leftrightarrow R}$ describes the DQD. H^α_{QD} describe each dot and include the charging energies of the dot electrons. Here, $H_{L \Leftrightarrow R}$ is the hopping between the dots. It is assumed that only one orbital in the left dot participates in the spin-polarized pumping process whereas two orbitals in the right dot are considered. The isolated left dot is thus modelled as a one-level Anderson impurity: $H^L_{QD} = \sum_\sigma \varepsilon_L d^\dagger_{L\sigma} d_{L\sigma} + U_L n_{L\uparrow} n_{L\downarrow}$, whereas the isolated right dot is modelled as: $H^R_{QD} = \sum_{i\sigma} \varepsilon_{Ri} d^\dagger_{Ri\sigma} d_{Ri\sigma} + U_R(\sum_i n_{Ri\uparrow} n_{Ri\downarrow} + \sum_{\sigma,\sigma'} n_{R1\sigma} n_{R2\sigma'}) + J\mathbf{S}_1 \mathbf{S}_2$. The index $i = 1, 2$ denotes the two levels and $\mathbf{S}_i = (1/2)\sum_{\sigma\sigma'} d^\dagger_{Ri\sigma} \sigma_{\sigma\sigma'} d_{Ri\sigma'}$ are the spins of the two levels. U_L, U_R are the charging energies for each dot and J the exchange coupling in the right dot. We have considered the DQD parameters such that $E_{T_-} > E_{T_0} > E_{T_+}$, (where $|T_+\rangle = d^\dagger_{R1\uparrow} d^\dagger_{R2\uparrow}|0\rangle$, $|T_0\rangle = (1/\sqrt{2})(d^\dagger_{R1\uparrow} d^\dagger_{R2\downarrow} + d^\dagger_{R1\downarrow} d^\dagger_{R2\uparrow})|0\rangle$ and $|T_-\rangle = d^\dagger_{R1\downarrow} d^\dagger_{R2\downarrow}|0\rangle$). Also the triplet $|T_+\rangle$ is higher in energy than the singlet $|S_0\rangle = (1/\sqrt{2})(d^\dagger_{R1\uparrow} d^\dagger_{R1\downarrow} - d^\dagger_{R1\downarrow} d^\dagger_{R1\uparrow})|0\rangle$[2]. We include an V_{ac} acting on the dots such that the single particle energy levels become: $\varepsilon_{L(R)}(t) = \varepsilon_{L(R)} \pm \frac{eV_{AC}}{2} \cos \omega t$.

If one drives the system, initially prepared in a state with $n = n_L + n_R = 3$ electrons: $|L = \downarrow\uparrow, R = \uparrow 0\rangle$, at a ω corresponding to the energy difference between the singlets in both dots, namely $\hbar\omega_\downarrow = E^S_R - E^S_L$, pumping of $\downarrow$ spin is obtained in the regime where the chemical potential for taking $\downarrow$ electrons out of the right dot fulfils $E^S_R > \mu_R$ while the chemical potential for taking $\uparrow$ electrons out of the right dot fulfils $E^S_R - \Delta_z < \mu_R$ where $\Delta_z = g\mu_B B$ is the Zeeman splitting due to B $\vec{B} = (0,0,B)$ where g is the effective g-factor. Then, a spin-polarized

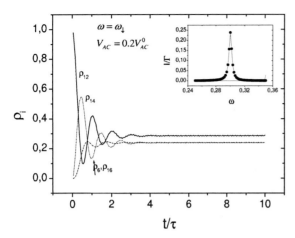
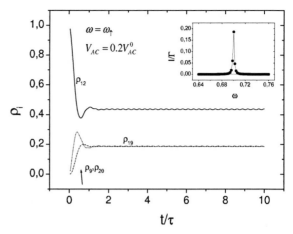

FIGURE 1. Diagonal DM elements as a function of time in units of τ, the period of Rabi oscillations, for $\hbar\omega_\downarrow = E_R^S - E_L^S$. The inset shows $I_\downarrow/\omega$ Parameters are (all in units of meV): $t = .005, \Gamma_L = \Gamma_R = .001, U_L = 1, U_R = 1.3, \Delta_1 = \Delta_2 = .026, J = .2, \mu_L = \mu_R = 1.31, eV_{AC}^0 = .2\hbar\omega_\uparrow$

FIGURE 2. Diagonal DM elements as a function of t in units of τ, for $\hbar\omega_\uparrow = E_{T_+} - E_L^S$. The inset shows $I_\uparrow/\omega$.

pump is realized through the sequence: $(\downarrow\uparrow,\uparrow 0) \overset{AC}{\Leftrightarrow} (\uparrow,\downarrow\uparrow 0) \overset{\Gamma_R}{\Rightarrow} (\uparrow,\uparrow 0) \overset{\Gamma_L}{\Rightarrow} (\downarrow\uparrow,\uparrow 0)$ or $(\downarrow\uparrow,\uparrow 0) \overset{AC}{\Leftrightarrow} (\uparrow,\downarrow\uparrow 0) \overset{\Gamma_L}{\Rightarrow} (\downarrow\uparrow,\downarrow\uparrow 0) \overset{\Gamma_R}{\Rightarrow} (\downarrow\uparrow,\uparrow 0)$ If now one drives the system at a ω corresponding to the energy difference between the singlet in the left dot and the spin up triplet in the right one, i.e., $\hbar\omega_\uparrow = E_{T_+} - E_L^S$, spin up current flows through the sequence: $(\downarrow\uparrow,\uparrow 0) \overset{AC}{\Leftrightarrow} (\downarrow,\uparrow\uparrow) \overset{\Gamma_R}{\Rightarrow} (\downarrow,\uparrow 0) \overset{\Gamma_L}{\Rightarrow} (\downarrow\uparrow,\uparrow 0)$ or $(\downarrow\uparrow,\uparrow 0) \overset{AC}{\Leftrightarrow} (\downarrow,\uparrow\uparrow) \overset{\Gamma_L}{\Rightarrow} (\downarrow\uparrow,\uparrow\uparrow) \overset{\Gamma_R}{\Rightarrow} (\downarrow\uparrow,\uparrow 0)$.

Finally, we analyze the dynamics by means of the reduced DM which is formulated in terms of the eigenstates and eigenenergies of each isolated QD. In Fig. 1 we plot the time evolution of the diagonal elements of the DM, i.e., the occupation probabilities for the states involved in the process, for photon energy $\hbar\omega_\downarrow$. We observe that mainly four states participate in the dynamics: $|12\rangle = |\uparrow\downarrow,\uparrow 0\rangle$ which is the initial state, and states $|14\rangle = |\uparrow,\uparrow\downarrow 0\rangle, |6\rangle = |\uparrow,\uparrow 0\rangle$ and $|16\rangle = |\uparrow\downarrow,\uparrow\downarrow 0\rangle$, in accordance with the sequence for producing spin down polarized current discussed before. For higher field intensity, these four diagonal elements of the DM converge in the stationary limit. Here, the occupations of states other than the initial state $|12\rangle$ are slightly lower. In the inset, we plot I/ω, centered at $\hbar\omega_\downarrow$ and which corresponds to fully spin down polarized current: one electron in the left dot absorbs one photon and tunnels to the right one and further to the collector.

Increasing ω_{ac} up to $\omega_\uparrow$, we obtain fully spin up polarized current by one photon absorption processes. The diagonal elements of the DM which participate in the dynamics are plotted in Fig. 2: $|12\rangle = |\uparrow\downarrow,\uparrow 0\rangle$, $|19\rangle = |\downarrow,\uparrow\uparrow\rangle, |9\rangle = |\downarrow,\uparrow 0\rangle, |20\rangle = |\uparrow\downarrow,\uparrow\uparrow\rangle$, where the last three occupations converge to their stationary values while the difference with respect to the initial state $|12\rangle$ is larger than in the previous case. Here, the ratio V_{AC}/ω, which determines the frequency of Rabi oscillations [3] is smaller, which results in a decrease in the population of accessible states other than the initial one. In the inset, $I(\omega)$ is plotted. At $\omega_\uparrow$, the current presents a peak which corresponds to the electron tunnelling from the left dot through the triplet state by one photon absorption and which is fully spin up polarized.

In summary, using the Markovian master equation approach, we have studied the dynamics of an ac driven DQD in the pumping configuration, in the presence of B. We have shown that, with unpolarized leads, the DQD can produce fully spin polarized current, whose polarization can be modified and even inverted by tuning ω_{ac}. The effect of spin relaxation processes will be explored in a future work.

Work supported by Programa de Cooperación Bilateral CSIC-CONACYT (E.C. and G.P.), the MCYT of Spain, grant MAT2002-02465 (R.A. and G.P.), the "Ramón y Cajal" program (R.A.) and by the EU Human Potential Programme: HPRN-CT-2000-00144 (G. P.).

REFERENCES

1. C.E.Creffield and G. Platero, Phys. Rev. B **65**, 113304 (2002); G. Platero and R.Aguado, Phys. Rep. **395**,1 (2004).
2. E. Cota, R. Aguado, C.E. Creffield and G. Platero, Nanotechnology **14**, 152-156 (2003); E. Cota, R. Aguado and G. Platero, to be submitted.
3. C.A. Stafford and N.S. Wingreen, Phys. Rev. Lett. **76**, 1916 (1996)

Spin splitting in open quantum dots

M. Evaldsson*, I. V. Zozoulenko*, M. Ciorga[†], P. Zawadzki[†] and A. S. Sachrajda[†]

*Department of Science and Technology (ITN), Linköping University, 601 74 Norrköping, Sweden
[†]Institute for Microstructuaral Science, National Research Council, K1A 0R6, Ottawa, Canada

Abstract. We demonstrate that the magnetoconductance of small lateral quantum dots in the strongly-coupled regime (*i.e.* when the leads can support one or more propagating modes) shows a pronounced splitting of the conductance peaks and dips which persists over a wide range of magnetic fields (from zero field to the edge-state regime) and is virtually independent of the magnetic field strength. Our numerical analysis of the conductance based on the Hubbard Hamiltonian demonstrates that this is essentially a many-body/spin effect that can be traced to a splitting of degenerate levels in the corresponding closed dot. The above effect in open dots can be regarded as a counterpart of the Coulomb blockade effect in weakly coupled dots, with the difference, however, that the splitting of the peaks originates from the interaction between the electrons of opposite spin.

Due to possible applications in the emerging fields of spintronics and quantum information there has been a surge of interest in the spin properties of semiconductor quantum dots. Typically Coulomb blockade (CB) experiments are used to probe the nature of the spin states [1, 2, 3, 4, 5, 6, 7, 8]. The CB regime corresponds to a weak coupling between the dot and the leads, so that the number of electrons in the dot is integer and each peak signals an addition/removal of one electron to/from the dot. In strong contrast to the Coulomb blockade regime in the *open dot* dot regime electrons can freely enter and exit the dot via leads that support one or more propagating modes. In this case the charge quantization no longer holds and one may expect that the conductance is mediated by two independent channels of opposite spin resulting in a total spin $S = 0$ in the dot. In the present paper we present experimental evidence supported by theoretical calculations that in small open dots two spin channels are correlated and therefore *the spin degeneracy can be lifted*.

A gate device layout scheme has been recently developed which enables the number of electrons confined within an electrostatically defined quantum dot to be controllably reduced to zero [5]. These few electron devices were used to study the spin properties of quantum dots in the CB regime using Coulomb and spin blockade spectroscopic techniques [6, 7]. The measurements in this paper are on these same devices but in the strongly coupled regime using a resistance bridge with ∼1nA current with an estimated number of electrons in the dot from 25 to 90. Details of the two device designs and the AlGaAs/GaAS wafer used for the measurements are given elsewhere [6, 7]. For the open dot experiments the following experimental procedure was used. All the gates defining the quantum dot were swept simultaneously. Altogether measurements were made on four different quantum dots. Figure 1 illustrates typical experimental results. As can be seen clearly in the data there exists a remarkable splitting of all of the conductance peaks/dips. The features discussed in this paper were present in all four dots and on several cooldowns. The amount of splitting varies from doublet to doublet with a typical value of 0.2meV. We stress that the observed splitting is practically independent on the external perpendicular field B, extends to zero field $B = 0$ and is almost two orders of magnitude larger than Zeeman splitting (which is 0.005meV at B=0.5T).

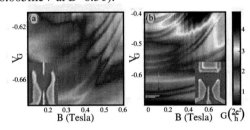

FIGURE 1. Experimental conductance as a function of magnetic field B and gate voltage V_G obtained from different dots. The corresponding device layouts are shown in the insets. The lithographic size of the dots is ∼ 450nm.

The magnetoconductance through the dot is modeled by a tight-binding Hubbard Hamiltonian in the mean-field approximation [10, 11, 12, 13, 2, 14], $H = H_\uparrow + H_\downarrow$,

$$H_\sigma = -\sum_{\mathbf{r},\delta} t_{\mathbf{r},\delta} a^\dagger_{\mathbf{r}\sigma} a_{\mathbf{r}+\delta\sigma} + U \sum_{\mathbf{r}} \langle a^\dagger_{\mathbf{r}\sigma'} a_{\mathbf{r}\sigma'} \rangle a^\dagger_{\mathbf{r}\sigma} a_{\mathbf{r}\sigma}, \quad (1)$$

where σ, σ' describe the two opposite spin states, $a^\dagger_{\mathbf{r}\sigma}$ and $a_{\mathbf{r}\sigma}$ are the creation/annihilation operators at the lat-

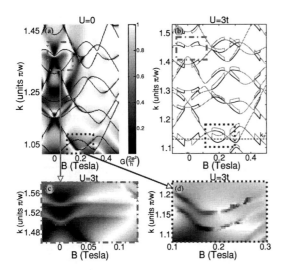

FIGURE 2. (a) The conductance $G = G(k_F, B)$ for $U = 0$. Solid lines depict the eigenspectrum of the corresponding closed dot (k_F is the fermi wave vector). (b) The eigenspectrum of the corresponding closed dot for the same strength of $U = 3t$. Dotted line indicates the k_F. (c),(d) The conductance $G = G(B, k_F)$ for $U = 3t$. Note that because of significant computational time, the splitting of the peaks are shown as a 2D grey scale plots for two selected regions only.

tice site $\mathbf{r}$, δ corresponds to nearest neighbors, and $t_{\mathbf{r},\delta}$ is a hopping matrix between the neighboring sites where the magnetic field B is incorporated as the phase factor in the form of Peierls' substitution [11]. U is the Hubbard constant describing the onsite Coulomb interaction between electrons of different spin. On the basis of the experimental findings we expect that the observed effect of the spin-splitting is generic to small quantum dots and thus rather insensitive to a detailed shape of the potential. We thus use the hard-wall confinement for the dot of a rectangular shape with the size $0.21 \times 0.36 \mu m$ that is connected to infinite ideal leads with width $w = 80nm$. The conductance of the dot is given by the Landauer formula $G = G_\uparrow + G_\downarrow = \frac{e^2}{h}(T_\uparrow + T_\downarrow)$, where T_σ is the transmission coefficient for different spin channels and is computed from retarded Green function $\mathscr{G}_\sigma = (E - H_\sigma + i\varepsilon)$ using the standard recursive Green function technique [11, 15]. The expectation value for the electron number at site $\mathbf{r}$ for the spin σ is given by [15]

$$\langle N_{\mathbf{r}\sigma}\rangle = \langle a^\dagger_{\mathbf{r}\sigma} a_{\mathbf{r}\sigma}\rangle = -\frac{1}{\pi}\int_{-\infty}^{E_F} \text{Im}[\mathscr{G}_\sigma(\mathbf{r},\mathbf{r},E)]dE, \quad (2)$$

where E_F is the Fermi energy, and $\mathscr{G}_\sigma(\mathbf{r},\mathbf{r},E)$ is the Green function in the real space representation. Equations (1), (2) are solved self-consistently. We neglect the Zeeman term as its effect on the dot conductance in the chosen field interval is negligible.

Figure 2a shows a linear conductance vs magnetic field B and Fermi wave vector k_F which includes no spin or Coulomb effects ($U = 0$). For comparison the single-particle spectrum of the corresponding closed dot is superimposed onto the conductance plot in order to underline the relationships between them (see [9] for a detailed discussion of the relationships between single-particle spectrum and conductance). The eigenspectrum of the dot for the case of $U = 3|t|$ is shown in Fig. 2b for a representative k_F in the dot. A major difference in comparison with the single-particle spectrum in figure 2a is that for certain regions of magnetic field the spin degeneracy is lifted and thus spin-up and spin-down eigenenergies are split. This comes about as either a spin-up or spin-down state close to the fermi level becomes occupied and through the Hubbard interaction increases the eigenenergy for the opposite spinstate. Because the bound states in the closed dot can, in a non-trivial way, be related to the quasi-bond states in the open dot (see [9]), the spin splitting in the eigenspectrum leads to a splitting of the conductance peaks/dips as illustrated in Fig. 2(c),(d).

In conclusion, we demonstrate experimentally and theoretically that conductance peaks and dips are split in small few electron open quantum dots. The numerical analysis of the conductance and the dot eigenspectrum shows that this effect is related to a spin splitting in the corresponding closed dot when the interactions between the electrons with opposite spins is taken into account.

REFERENCES

1. Tarucha S. et al. *Phys. Rev. Lett.*, **77**, 3613 (1996).
2. Sivan U. et al. *Phys. Rev. Lett.*, **77**, 1123 (1996).
3. Tarucha S. et al. *Phys. Rev. Lett.*, **84**, 2485 (2000).
4. Lüscher S. et al. *Phys. Rev. Lett.*, **86**, 2118 (2001).
5. Ciorga M. et al. *Phys. Rev. B*, **61**, R16315 (2000).
6. Hawrylak P. et al. *Phys. Rev. B*, **59**, 2801 (1999).
7. Ciorga M. et al. *Phys. Rev. Lett.*, **88**, 256804 (2002).
8. Fuhrer A. et al. *Phys. Rev. Lett.*, **91**, 206802 (2003).
9. Zozoulenko I. V. et al. *Phys. Rev. Lett.*, **83**, 1838 (1999).
10. Anderson P. W. *Phys. Rev.*, **124**, 41 (1961).
11. Lee P. A. and Fisher D. S. *Phys. Rev. Lett.*, **47**, 882 (1981).
12. Henrickson L. E. et al. *Phys. Rev. B*, **50**, 4482 (1994).
13. Berkovits R. *Phys. Rev. Lett.*, **81**, 2128 (1998).
14. Nonoyama S. et al. *Phys. Rev. B*, **50**, 2667 (1994); Takagaki Y., Tokura Y., S. Tarucha *Phys. Rev. B*, **53**, 15 462 (1996); H. Chen et al. *Phys. Rev. B*, **55**, 1578 (1997).
15. Datta S. *Electronic Transport in Mesoscopic Systems*, Cambridge University Press, Cambridge, 1995
16. Wojs A. and Hawrylak P. *Phys. Rev. B*, **53**, 10841 (1996); Koskinen M., Manninen M. and Reimann S. M. *Phys. Rev. Lett.*, **79**, 1389 (1997); Yakimenko I. I., Bychkov A. M. and Berggren K.-F. *Phys. Rev. B*, **63**, 165309 (2001).

Square-wave, spin-dependent transmission through periodically stubbed electron waveguides

X. F. Wang and P. Vasilopoulos

Department of Physics, Concordia University, 1455 de Maisonneuve Ouest, Montréal, Canada H3G 1M8

Abstract. We study spin-dependent electron transmission T through waveguides (WGs) with *periodically* varied width c and strength α of the spin-orbit interaction. c is varied by attaching stubs to the WG, and α by applying gates. T can exhibit a nearly **square-wave** behavior as a function of the stub dimensions, of α, of the length of the gated subunit, and of the incident energy if only one mode propagates in the WG. If mode mixing is allowed the results become more complex but remain qualitatively the same. The transmission through a superlattice, with alternating segments of lengths l_1, l_2, and strengths α_1, α_2, is a periodic function of α_j and l_j, $j = 1, 2$. By simultaneously varying c and α one can very effectively control the gaps in the transmission, make them square and wide, block either spin state, and select the outgoing spin state.

INTRODUCTION

Recently the study of spintronics devices, which utilize the spin rather than the charge of an electron, has been intensified mainly because they are expected to operate at much higher speeds than the conventional ones and have potential applications in quantum computing. In nanostructures with asymmetric confinement the dominant term of the spin-orbit interaction (SOI), which depends on the electron spin, is the Rashba term [1]. It is important in materials with a small band gap such as In-GaAs and has been used to describe the spin-dependent transmission in a WG of constant [2] or periodically varied [3] width c or one with constant c and periodically varied [4] SOI strength α. Here we build upon this work by considering WGs with periodically varying α and c. The study is pertinent to the development of the spin transistor.[2] Here are the context and the results.

FORMALISM

In a quasi one-dimensional electron gas, at zero magnetic field, the spin degeneracy of the energy bands at $\mathbf{k} \neq 0$ is lifted by the coupling of the electron spin with its orbital motion [2, 3, 5]. The relevant Hamiltonian is

$$H_{so} = \alpha(\vec{\sigma} \times \vec{p})_z/\hbar = i\alpha[\sigma_y \partial/\partial x - \sigma_x \partial/\partial y]. \quad (1)$$

Here the waveguide is along the y axis and the confinement along the x axis. The parameter α measures the strength of the coupling; $\vec{\sigma} = (\sigma_x, \sigma_y, \sigma_z)$ denotes the spin Pauli matrices, and $\vec{p}$ is the momentum operator.

We treat H_{so} as a perturbation. With $\Psi = |n, k_y, \sigma\rangle = e^{ik_y y}\phi_n(x)|\sigma\rangle$ the eigenstate in each region the unperturbed states satisfy $H^0|n, k_y, \sigma\rangle = E_n^0|n, k_y, \sigma\rangle$ with $E_n^0 = E_n + \lambda k_y^2$, $\lambda = \hbar^2/2m^*$, and $\phi_n(x)$ obeys $[-\lambda d^2/dx^2 + V(x)]\phi_n(x) = E_n\phi_n(x)$, where $V(x)$ is the confining potential assumed to be square-type and high enough that $\phi_n(x)$ vanishes at the boundaries. The eigenfunction for $H_{so} \neq 0$ is written as $\sum_{n,\sigma} A_n^\sigma \phi_n(x)|\sigma\rangle$. H_{so} is a 2×2 matrix. Combining it with the 2×2 diagonal matrix H^0, $H\Psi = (H^0 + H_{so})\Psi = E\Psi$, and setting $\bar{E}_n = E_n^0 - E$ gives

$$\begin{bmatrix} \bar{E}_n & \alpha k_y \\ \alpha k_y & \bar{E}_n \end{bmatrix} \begin{pmatrix} A_n^+ \\ A_n^- \end{pmatrix} + \alpha \sum_m J_{nm} \begin{pmatrix} A_m^- \\ -A_m^+ \end{pmatrix} = 0, \quad (2)$$

with n labeling the discrete subbands resulting from the confining potential $V(x)$. For negligible subband mixing we can take $J \approx 0$ and Eq. (2) leads to the eigenvalues

$$E^\pm(k_y) = E_n + \lambda k_y^2 \pm \alpha k_y. \quad (3)$$

The eigenvectors corresponding to E^+, E^- satisfy $A_n^\pm = \pm A_n^\mp$. Accordingly, the spin eigenfunctions are taken as $|\pm\rangle = \begin{pmatrix} 1 \\ \pm 1 \end{pmatrix}/\sqrt{2}$. For the same energy the difference in wave vectors k_y^+ and k_y^- for the two spin orientations is $k_y^- - k_y^+ = 2m^*\alpha/\hbar^2 = \delta$. For the same energy E there are four k_y values and a phase shift δ between the positive or negative k_y^+ and k_y^- values of the branches E^+ and E^-.

The procedure outlined above applies to all waveguide segments. When the electron energy is low and only one subband is occupied, we omit the subband label n and use the segment label i instead. Then the eigenfunction ϕ_i of energy E in segment i reads

$$\phi_i(x, y) = \sum_\pm \left[c_i^\pm e^{ik_y^\pm y} |\pm\rangle + \bar{c}_i^\pm e^{-ik_y^\mp y} |\pm\rangle \right] \sin\frac{\pi x}{c}. \quad (4)$$

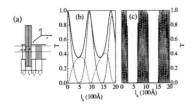

FIGURE 1. (a) Geometry of the unit. The SOI strength α is finite (zero) in the shaded (unshaded) regions. (b) Transmission through one stubless unit vs l_h with $l'_h = 0$ and $h = c$. (c) Transmission through five stubless units at temperature 1 K, with $l'_h = 0$, $\alpha = 5 \times 10^{-11}$ eVm, $m^* = 0.05m_0$, $c = h = 500$Å, and energy $E = 3.1 meV$. The dashed (dotted) curves in (b) and (c) show T^+ (T^-). The solid curves show the total transmission.

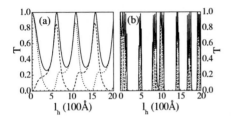

FIGURE 2. (a) Transmission through *one* composite unit with $l'_h = l_h$ and $h = 2c$ as a function of the stub width l_h. (b) Transmission through *five* composite units with the parameters of (a) and Fig.1. The curves are marked as in Fig. 1.

In the stub regions, with $h > c$, more subbands are occupied and the wavefunction takes the form of Eq. (4) with coefficients $u_n^\pm$ and $\bar{v}_n^\pm$ instead of $c_i^\pm$ and $\bar{c}_i^\pm$

At the interfaces between different segment of the unit, cf. Fig. 1, we match, in line with Ref. [5], the wave function and its flux. The velocity operator is given by

$$\hat{v}_y = \frac{\partial H}{\partial p_y} = (1/\hbar) \begin{bmatrix} -2i\lambda \partial/\partial y & \alpha \\ \alpha & -2i\lambda \partial/\partial y \end{bmatrix}. \quad (5)$$

The continuity of the wave function at the interface $y = y_{i,i+1}$, between the i and $i+1$ segments, gives $\phi_{i+1}(x, y_{i,i+1}) = \phi_i(x, y_{i,i+1})$ and that of the flux $\hat{v}_y \phi_{i+1}(x,y)|_{y_{i,i+1}} = \hat{v}_y \phi_i(x,y)|_{y_{i,i+1}}$. This connects the coefficients in the input region, $\hat{L}^T = (c_1^+, \bar{c}_1^+, c_1^-, \bar{c}_1^-)$, in the stub region $\hat{U}_n^T = (u_n^+, \bar{u}_n^+, u_n^-, \bar{u}_n^-)$, and the output region $\hat{R}^T = (c_2^+, \bar{c}_2^+, c_2^-, \bar{c}_2^-)$, where T denotes the transpose matrix. This gives $\hat{L} = \hat{T}\hat{R}$ with

$$\hat{L} = (\sum_n \hat{N}_n \hat{M}_n^{-1} \hat{P}_n - \hat{\eta})^{-1}(\hat{\beta} - \sum_n \hat{N}_n \hat{M}_n^{-1} \hat{Q}_n)\hat{R}, \quad (6)$$

where $\hat{N}_n, \hat{M}_n, \hat{P}_n, \hat{\eta}, \hat{\beta}$ are 4×4 matrices. If there are more than one units, we denote the transfer matrix of the i-th stub by $\hat{T}_i$, that of the i-th waveguide segment by $\hat{P}_i$, and obtain the total transfer matrix as $\hat{T} = \prod_i \hat{T}_i \hat{P}_i$.

Assuming we input electrons from the left of the device and measure the transmission at its right, the reflection coefficient at its right should be zero. The transmission $\hat{M}_t$ and reflection $\hat{M}_r$ matrices are given by

$$\hat{M}_t^{-1} = \begin{bmatrix} T_{11} & T_{13} \\ T_{31} & T_{33} \end{bmatrix}, \quad \hat{M}_r \hat{M}_t^{-1} = \begin{bmatrix} T_{21} & T_{23} \\ T_{41} & T_{43} \end{bmatrix}, \quad (7)$$

where T_{ij} are the matrix elements of $\hat{T}$. The transmission process is then embodied in the matrix $\hat{M}_t$ in the manner $(c_1^-, c_1^+)^T = \hat{M}_t (c_2^-, c_2^+)^T$. Then T^+ (T^-) is given by

$$T^\pm = |c_2^+ \pm c_2^-|^2 / 2[|c_1^+|^2 + |c_1^-|^2]. \quad (8)$$

NUMERICAL RESULTS

We calculate the spin transmission along a waveguide composed of identical composite units as shown in Fig. 1(a). Each unit has four segments of width l_c, l_h, l_c, and l'_h, stub height h and asymmetry d, d being the distance between the centerlines of the waveguide and the stub. The SOI strength α is finite only in the grey areas. First we calculate the total transmission T vs l_h for a stubless unit, i.e., for $h = c$ and $l'_h = 0$ in Fig. 1(a). T is a periodic function of l_h. The spin-up (T^+) and spin-down (T^-) transmissions are given by $T^+ = T\cos(\delta l_h/2)$ and $T^- = T\sin(\delta l_h/2)$. In Fig. 1(c), we plot T for a waveguide of five identical stubless units, with $h = c$ and $l'_h = 0$. Here T has two gaps and acquires a nearly square-wave form.

In Fig. 2(a), we consider one composite unit with $h = 2c$, $d = 0$, and $l'_h = l_h$. Four oscillations appear and the patterns of T^+ and T^- change. For five such composite units, the transmission dips of Fig. 2(a) develop into a series of transmission gaps as shown in Fig. 2(b). This behavior of transmission can be used to establish a spin transistor where spin flips and spin currents can be controlled by changing the size of the stubs and the strength α. Comparing the results of Fig. 1, where no stubs exist, with those of Fig. 2, we see that more gaps develop. A further control of the transmission is obtained by changing the asymmetry parameter d [3].

SUMMARY

We studied spin transmission through a waveguide with periodically modulated width and SOI strength. In this way, we can further control the spin transmission and improve the behavior of the spin transistor studied previously [3, 4]. Full details will be given elsewhere.

REFERENCES

1. E. I. Rashba, Sov. Phys. Solid State **2**, 1109 (1960).
2. S. Datta and B. Das, Appl. Phys. Lett. **56**, 665 (1990).
3. X. F. Wang, P. Vasilopoulos, and F. M. Peeters, Phys. Rev. B **65**, 165217 (2002).
4. X. F. Wang and P. Vasilopoulos, Appl. Phys. Lett. **82**, 940 (2003).
5. L. W. Molenkamp, G. Schmidt, and G. E. W. Bauer, Phys. Rev. B **64**, 121202 (2001).

Optical Orientation Of Electron And Nuclear Spins In Negatively Charged InP QDs

S. Yu. Verbin[1,2], I. Ya. Gerlovin[3], I. V. Ignatev[1,2], and Y. Masumoto[4]

[1] *Institute of Physics and Venture Business Laboratory, University of Tsukuba, Tsukuba, 305-8571, Japan*
[2] *V.A.Fock Institute of Physics, St-Petersburg State University, 198504 St-Petersburg, Russia*
[3] *Vavilov State Optical Institute, St-Petersburg, Russia*
[4] *Institute of Physics, University of Tsukuba, Tsukuba, Japan*

Abstract. Light-induced spin orientation in negatively charged InP quantum dots is shown experimentally to be conserved for about 1 ms.

INTRODUCTION

The long-term orientation of spin systems in semiconductor structures attract in recent years considerable attention as a promising way of recording and storage of optical information [1]. The most interesting, from this point of view, are the structures with quantum dots (QD), where, due to confinement of the carrier motion, the main spin-relaxation processes appear to be suppressed. According to theoretical estimates [2], the spin lifetime in such structures may reach units of ms and more. The so long lifetimes of the optically oriented electron spins in QDs have not been observed thus far. The longest lifetime of spin orientation (15 ns) was found experimentally in [3] in the studies of n-doped InAs QDs. In this communication, we present the results of experiments demonstrating conservation of the spin orientation for much longer time intervals.

EXPERIMENT AND DISCUSSION

We studied kinetics of circularly polarized photoluminescence (PL) of the negatively charged InP QDs. It was found that the degree of circular polarization of the PL shows a slow component with a decay time of about 1 ns. The amplitude of the slow component was the greatest when each QD possessed, on average, a single resident electron. For the Stokes shift $\Delta E_{st} > 25$ meV, the degree of polarization was negative.

In conformity with conclusions of [3], the negatively polarized PL is emitted by QDs in which spins of the photoexcited and resident electrons are parallel. When the resident electron spin is preferentially oriented along that of the photoexcited electron, the amplitude of the degree of the negative circular polarization (NCP) should increase. Using the amplitude of the NCP as a measure for the degree of orientation of the electron spin, we have carried out a series of experiments to determine the spin orientation lifetime.

In the experiments, the PL was excited by two laser pulse trains obtained by splitting the beam of a mode-locked Ti:sapphire laser. One of these trains was shifted in time using an optical delay line. The time delay was about 1 ns. Polarizations of the beams could be varied independently.

Figure 1a shows the kinetics of the degree of polarization of the PL excited by the second (probe) beam for two different polarizations of the first beam considered as a pump. As is seen, passing from the co-polarized to cross-polarized pump results in more than two-fold decrease of the NCP created by the probe pulse. This means that the spin orientation created by the first beam is held till the moment of arrival of the second beam, i.e., for about 1 ns.

To strongly increase the time interval between the two beams, we used a mechanical chopper, which alternatively blocked the pump and the probe beams with a frequency of 120 Hz. A typical result of the experiment is presented in Fig. 1b. It is seen that the NCP of PL after probe pulse still reveals a well reproducible change, $\Delta\rho \approx 5\%$, caused by the change of polarization of the pump pulse. This result demonstrates in a straightforward way conservation of the spin orientation during the modulation period of units of ms.

So long conservation of the spin orientation is impossible without a stabilizing factor that suppresses the electron spin precession in random laboratory magnetic fields. In this case, the role of this factor is likely to be played by the fluctuating exchange field produced by nuclear spins [4]. To evaluate the fluctuations of the nuclear field, we have studied the behavior of the NCP at co- and cross-polarized pumping (with no chopper) in small transverse magnetic fields (Voigt configuration).

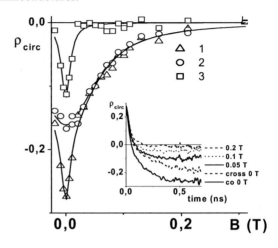

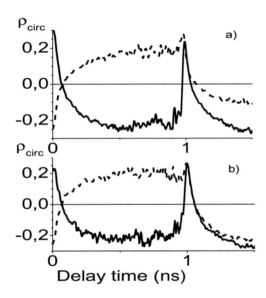

FIGURE 1. Kinetics of circular polarization degree at the two-pulse excitation: (a) without chopper, (b) with chopper blocking the beams alternatively. Solid (dashed) curves are for co- (cross-) polarized pump and probe beams.

FIGURE 2. Dependences of NCP on trans-verse magnetic field: 1 is for co- and 2 for cross-polarized pump and probe, 3 is the difference between 1 and 2. Dots are experimental data, solid curves are fits by Lorentzians. Inset: kinetics of NCP at various magnetic fields.

The results of these studies are shown in Fig. 2. As one can see, dependences of the NCP on magnetic field, in the Voigt configuration, for the co- and cross-polarized pumping are strongly different. The difference between the values of the NCP for co- and cross-polarized pumping characterizes the spin orientation. Dependence of the degree of orientation on magnetic field is seen to be well described by a Lorentzian with a half-width of 0.01 T (curve 3 in Fig.2). According to conclusions of [4], this value corresponds to the mean value of the exchange field produced by fluctuations of the nuclear spins in the QD.

It follows from the above experimental data that the light-induced spin orientation in the negatively charged InP QDs is held during the time of about 10^{-3} s. This value exceeds by approximately 4 orders of magnitude the value of 150 ns, given in [5], which was considered until now a record value for semiconductor nanostructures.

ACKNOWLEDGMENTS

Authors are thankful to Dr. K. Kavokin for fruitful discussions. The work is supported by Grant-in-Aid for Scientific Research on the Priority Area "Nanospintronics" (No. 16031203) from MEXT of Japan, by INTAS (1B 2167) and by RFBR (03-02-16858).

REFERENCES

1. D.D.Awschalom J.M.Kikkawa, Phys.Today, **52**, 33 (1999).
2. A. V. Khaetskii, Yu. V. Nazarov, Phys. Rev. B, **61**, 12639 (2000)
3. S. Cortez, O. Krebs, S. Laurent, M. Senes, X. Marie, P. Voisin, R. Ferreira, G. Bastard, J-M. Gerard, T. Amand, Phys. Rev. Lett., **89**, 207401 (2002).
4. A. Merkulov, Al. L. Efros, M. Rosen, Phys. Rev. B, **65**, 205309 (2002)
5. R.I. Dzhioev, V.L. Korenev, B.P. Zakharchenya, D. Gammon, A.S. Bracker, J.G. Tischler, and D.S. Katzer, Phys. Rev. B **66**, 153409 (2002)

Ferromagnetism in partially spin-polarized GaInAs quantum wells

Felix Vogt[*,†], Georg Nachtwei[*], Günter Hein[**] and Harald Künzel[‡]

[*]*Institut für Technische Physik, Technische Universität Braunschweig D-38106, Germany*
[†]*Hochmagnetfeldanlage der Technischen Universität Braunschweig D-38106, Germany*
[**]*Physikalisch Technische Bundesanstalt, Braunschweig D-38116, Germany*
[‡]*Heinrich-Hertz-Institut für Nachrichtentechnik, Berlin D-10587, Germany*

Abstract. We performed measurements of the Shubnikov-de Haas effect (SdH) and of the Quantum-Hall effect (QHE) at GaInAs quantum wells confined by AlInAs barriers. At odd filling factors (1 and 3), the two-dimensional electron system (2DES) of the quantum well is partially spin-polarized, leading to an enhanced effective Landé factor g^*. In this case, asymmetric and hysteretic SdH-oscillations were predicted [1]. At filling factor 1 (higher spin polarization) the asymmetry could be confirmed experimentally. At filling factor 3, the predicted asymmetry could be observed at a sample of higher electron density in tilted magnetic fields above 12 T. A hysteresis was not resolvable due to a reduced spin polarization.

INTRODUCTION

In the region of filling factor $\nu = 1$ the behaviour of the spin-polarized 2DES in the GaInAs quantum well appears to be distinct from a quantum Hall ferromagnet (QHFM) [2]: (a) It seems unlikely that the spin polarisation in GaInAs could be influenced by Skyrmions (correlated spin states, see Ref. [3, 4]), since $\Delta E_s/E_c$ is about 10 times larger than in GaAs. (b) In contrast to a QHFM a bistable, hysteretic behaviour of the longitudinal (R_{xx}) and transversal resistance (R_{xy}), accompanied by a ferromagnetic memory of the spin polarization, was observed near $\nu = 1$ [1]. Therefore this ferromagnetism is explained as a feedback effect between the current-dependent population of different spin levels and the exchange-influenced spin gap [1]. Hartree-Fock calculations confirm this model not only for filling factor one, but also for partially spin-polarized 2DES and predict a similar behaviour of R_{xx} and R_{xy} near $\nu = 3$.

In this article we present recent experimental results, which can prove a more general validity of the model for this type of ferromagnetism in Quantum Hall systems.

EXPERIMENTAL DETAILS

The $Ga_{0.47}In_{0.53}As/Al_{0.48}In_{0.52}As$ quantum-well structures studied were characterized by Hall mobilities of $\mu_H = 2.6 \times 10^4$ cm^2/Vs (sample A), $\mu_H = 5.6 \times 10^4$ cm^2/Vs (sample B) and electron densities of $n_s = 2.3 \times 10^{11}$ cm^{-2} (sample A), $n_s = 4.1 \times 10^{11}$ cm^{-2} (sample B). Hall bars of 300 μm (sample A) and 500 μm (sample B) width and length between the potential probes were used. Magnetotransport measurements were performed at temperatures of 1.7 K$\leq T \leq$ 4.2 K, magnetic fields up to 17.5 T, and DC currents of 1 μA$\leq I \leq$ 110 μA.

RESULTS AND DISCUSSION

The measurements of the longitudinal (R_{xx}) and transversal resistance (R_{xy}) in the region of $\nu = 3$ were performed at sample B. By tilting the sample with respect to the magnetic field, the spin splitting ΔE_s in relation to the cyclotron energy $\hbar\omega_c$ is increased. The effectively larger spin gap shifts the feedback effect between the tunneling rate and the exchange-enhanced spin splitting to higher currents. This shows Figure 1 (a) for $\nu = 1$ (sample A): the 0 ↓-SdH peak at $I = 30$ μA and $\phi = 54°$ looks similar as the peak at $I = 20$ μA and $\phi = 0°$.

Figure 1 (b) shows the current dependent evolution of the 1 ↓-SdH peak, performed with a tilting angle of 63°. At low currents ($I \approx 10$ μA) a rudimental plateau is observable in the region of $\nu = 3$. With increasing current the 1 ↓-SdH peak shifts towards higher magnetic fields and becomes more and more asymmetric. The asymmetry is the result of the feedback effect: Starting from a dissipative state of the system ($R_{xx} \gg 0$, $\nu > 3$) the tunneling barrier between localized states near the Fermi level E_F and extended 1 ↓-states will be increased by increasing the magnetic field. Consequently the occupation of ex-

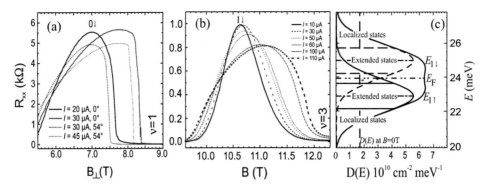

FIGURE 1. Effect of tilted magnetic fields at $\nu = 1$. The feedback effect shifts towards higher currents (a). The current dependent evolution of the $1\downarrow$-SdH peak (b). The estimated density of states $D(E)$ at $\nu = 3$ (c).

tended $1\downarrow$-states decreases and the spin polarisation of the system grows, which leads via exchange interaction to a wider spin gap and therefore the tunneling barrier increases further. Finally this self-amplifying process leads to a tunneling blockade at a certain critical magnetic field B_j which results in a drastical reduction in the occupation of extended $1\downarrow$-states and R_{xx} drops close to zero. Also the shift towards higher magnetic fields, if the current is increased, corresponds with the feedback model: At higher currents, or Hall electric fields the tunneling blockade to the extended $1\downarrow$-states requires higher magnetic fields, because a higher energy difference between E_F and the $E_{1\downarrow}$-spin band is necessary to reach the critical tunneling barrier. As a result, the SdH-maximum and B_j shift to higher fields.

In accordance with the model the change between dissipative and quantum Hall conduction should be bistable and hysteretic[1]. We only observed an increasing asymmetry (as mentioned above) and an apparently steeper transition when increasing the current, but no bistable, hysteretic switching. To clarify the reason, we performed temperature-dependent measurements of the SdH curves, from which we could estimate the density of states $D(E)$ and other parameters at $\nu = 3$ (see table 1). Figure 1 (c) shows $D(E)$, interpreted as overlap of two spin levels (Gaussian profile) seperated from each other by ΔE_s. The density of states is above the two-dimensional $D(E)$ at $B = 0$ and above the value at $\nu = 1$. This indicates a considerable overlap of the spin levels, leading to a lower spin polarisation as compared to the value of $1/3$[2]. In accordance with these results, we observed a smaller spin gap in comparison to $\nu = 1$. Also the number of localized states within the spin gap is clearly reduced, relative to the situation at $\nu = 1$. Therefore only a small rudi-

TABLE 1. Density of states $D(E)$, spin polarization P_s, g^*-factor, level width Γ, spin splitting ΔE_s, and localization L at filling factor 3 in comparison to filling factor 1. The parameters at $\nu = 1$ come from previous measurements [1] with observable hysteresis.

Parameter	Recent meas.	Previous meas.
ν	3	1
$D(E)[\text{cm}^{-2}\text{meV}^{-1}]$	7.2×10^{10}	6.3×10^{10}
$P_s[\%]$	24	60
$|g^*|$	3.6	6.2
$\Gamma[\text{meV}]$	2.3	5.1
$\Delta E_s[\text{meV}]$	2.09	3.55
$L[\%]$	44.5	87

mental plateau is at $\nu = 3$ observable. In previous measurements [1] the hysteresis could be observed at $\nu = 1$, therefore the low spin polarization at $\nu = 3$ may explain the absence of bistability and hysteresis.

The current dependent asymmetry of the $1\downarrow$ SdH peak shows however a more general validity of the model:

- the feedback effect is observable at another odd filling factor and
- the model is valid for a sample with a clearly higher electron density (in comparison to previous measurements).

REFERENCES

1. G. Nachtwei, A. Manolescu, N. Nestle, and H. Künzel, Phys. Rev. B **63**, 045306 (2000).
2. A. H. MacDonald, Solid State Commun. **102**, 143 (1997).
3. T. H. R. Skyrme, Proc. R. Soc. London, Ser. A **262**, 233 (1961).
4. S. L. Sondhi, A. Karlhede, S. A. Kivelson, and E. H. Rezayi, Phys. Rev. B **47**, 16419 (1993).

---

[1] If the process starts from the higher spin gap, by decreasing the magnetic field, the critical tunneling barrier will be reached at a lower magnetic field (reason for the hysteresis).
[2] Value of P_s at $\nu = 3$, without overlap of the spin levels.

Electrically Controlled Spin Device Concepts

D. Z.-Y. Ting[1], X. Cartoixà[2],*, and Y.-C. Chang[2]

[1] *Jet Propulsion Laboratory, California Institute of Technology, Pasadena, CA 91109 USA*
[2] *Dept. of Physics, U. Illinois at Urbana-Champaign, Urbana, IL 61801 USA*

Abstract. We discuss spin device concepts that exploit spin-orbit coupling in nonmagnetic semiconductor heterostructures, including the resonant interband tunneling spin filter, the bi-directional spin pump, the bulk inversion asymmetry enhanced spin filter, and resonant spin lifetime devices.

INTRODUCTION

Spin degeneracy in non-magnetic semiconductor heterostructures results from time reversal and spatial inversion symmetries, and can be lifted by the presence of structural inversion asymmetry (SIA) or bulk inversion asymmetry (BIA). SIA- and BIA-induced spin splittings can be described by the Rashba effect [1] and the Dresselhaus term [2], respectively. They can be exploited to build spin devices using only non-magnetic semiconductor heterostructures, without external magnetic fields or optical excitation. In this paper we briefly describe two types of non-magnetic spin devices. The first uses resonant tunneling to filter spins, and can be used to create a source of spin polarized current. The second exploits the interplay between BIA and SIA to control spin lifetimes for device applications. The proposed devices are electrically controllable, potentially amenable to high-speed modulation and integration in optoelectronic devices for added functionality.

RESONANT TUNNELING DEVICES

The resonant tunneling spin filter idea originated with Voskoboynikov *et al.* [3]. It uses an asymmetric resonant tunneling structure in which quantized conduction subband states are spin split by the Rashba effect due to SIA, and relies on resonant tunneling for spin selection. Because Rashba spin split states in a given subband with opposite in-plane momenta have opposite spins, thus yielding no net spin upon subband averaging, a mechanism for lateral momentum selection such as lateral E-field in the emitter [3], one-sided collector [4], and side gating [5] is needed. Rashba spin splittings in the conduction subbands are typically small, and vanish at the zone center. Typical predicted spin filtering efficiencies in the original conduction band based designs are low, although a spin-blockade scheme could be used for improvement [4]. Below we describe additional device concepts.

Resonant Interband Tunneling Spin Filter

The asymmetric resonant interband tunneling diode (aRITD) [6,7] exploits large valence band spin-orbit interaction to provide strong spin selectivity while minimizing the effects of faster hole relaxation. Interband tunneling through the heavy-hole 1 (hh1) subband not only exhibits strong spin-dependent tunneling, but also eliminates tunneling through states near the zone center, where spin splitting vanishes. The spin-filtering efficiency of aRITD has been predicted to be significantly higher than in conduction subband resonant tunneling diodes [6]. A prototype side gated aRITD has been fabricated [5]. An RITD grown on [110] to take advantage of BIA spin splitting has also been proposed [8].

---

* Current address: *Computational Research Division, Lawrence Berkeley National Laboratory, Berkeley, CA 94720 USA*

Bi-Directional Spin Pump

A bi-directional resonant tunneling spin pump [9] is essentially a resonant tunneling spin filter operated under zero bias along the growth direction, but with a small lateral E-field in the emitter region. This results in a forward (emitter to collector) current with one spin polarization, and a backward current with the opposite spin polarization, due to the special properties of the spin filter. The bi-directional spin pump induces the simultaneous flow of oppositely spin-polarized current components in opposite directions through spin-dependent resonant tunneling, and can thus generate significant levels of spin current with very little net electrical current across the tunnel structure, a condition characterized by a greater-than-unity current spin polarization.

BIA Enhanced Spin Filter

The resonant tunneling spin filter and spin pump concepts were developed to exploit the Rashba effect [1], which is a consequence of spin–orbit interaction and the presence of SIA. The effect on spins due to the presence of BIA in zincblende semiconductors can also be exploited in spin devices. It can be shown that the efficiency of nonmagnetic resonant tunneling spin devices can be improved significantly when SIA and BIA effects are combined properly [10]. Best of all, the design changes required to take advantage of this improvement are minimal: we only need to be specific in selecting the direction of one-sided collectors or the lateral E-field [10].

RESONANT SPIN LIFETIME DEVICES

An interesting property arising from the interplay between SIA and BIA effects is that the D'yakonov-Perel' mechanism of spin relaxation becomes suppressed when $\alpha_{SIA}=\alpha_{BIA}$ [11], where α_{SIA} and α_{BIA} are coefficients describing the strengths of SIA and BIA effects respectively. This led us to the concept of a variant of the Datta-Das spin transistor [12] based on [001] grown structures called the resonant spin lifetime transistor (RSLT) [13]. In the RSLT, electrons are emitted and collected at ferromagnetic contacts. The switching action is accomplished by electrically controlling the spin lifetimes of electrons in the channel, where the size of α_{SIA} can be tuned by gate biasing while α_{BIA} remains essentially fixed. The RSLT is similar in concept to the non-ballistic spin-field-effect transistor [14], and can be used in non-volatile memories and magnetic readout heads [13]. The RSLT may also be implemented in [110] grown structures, where BIA effects play a dominant role [8]. But perhaps the best way to implement the RSLT is to use [111] grown heterostructures where, under the right conditions, the lowest-order-in-k component of the spin relaxation tensor can be made to vanish for *all* three spin components at the same time, in contrast to [001] or [110] devices where spin relaxation can be suppressed for one of the three spin components [15].

ACKNOWLEDGMENTS

We thank O. Voskoboynikov, A. T. Hunter, D. L. Smith, D. H. Chow, J. S. Moon, T. C. McGill, T. F. Boggess, J. N. Schulman, P. Vogl, and T. Koga for helpful discussions. This work was sponsored by the DARPA SpinS Program through HRL Laboratories. A part of this work was carried out at the Jet Propulsion Laboratory, California Institute of Technology, through an agreement with NASA.

REFERENCES

1. Y. A. Bychkov and E. I. Rashba, *J. Phys. C - Solid State Phys.* **17**, 6039 (1984).
2. G. Dresselhaus, *Phys. Rev.* **100**, 580-6 (1955).
3. A. Voskoboynikov, S. S. Lin, C. P. Lee, and O. Tretyak, *J. Appl. Phys.* **87**, 387 (2000).
4. T. Koga, J. Nitta J, H. Takayanagi, and S. Datta, *Phys. Rev. Lett.* **88** 12661 (2002).
5. J. S. Moon et al., to appear in *Appl. Phys. Lett.* (2004).
6. D. Z-Y. Ting and X. Cartoixà, *Appl. Phys. Lett.* **81**, 4198 (2002).
7. D. Z-Y. Ting, X. Cartoixà, D. H. Chow, J. S. Moon, D. L. Smith, T. C. McGill and J. N. Schulman, *IEEE Proceedings* **91** 741 (2003).
8. K. C. Hall, W. H. Lau, K. Gundogdu, M. E. Flatte and T. F. Boggess, *Appl. Phys. Lett.* **83**, 2937 (2003).
9. D. Z-Y. Ting and X. Cartoixà, *Appl. Phys. Lett.* **83**, 1391 (2003).
10. D. Z.-Y. Ting and X. Cartoixà, *Phys. Rev. B* **68**, 235320 (2003).
11. N. S. Averkiev and L. E. Golub, *Phys. Rev. B* **60** 15582 (1999).
12. S. Datta and B. Das, *Appl. Phys. Lett.* **56,** 665 (1990).
13. X. Cartoixà, D. Z.-Y. Ting and Y.-C. Chang, *Appl. Phys. Lett.* **83**, 1462 (2003).
14. J Schliemann, J C Egues and D Loss, *Phys. Rev. Lett.* **90** 146801 (2003).
15. X. Cartoixà, D. Z.-Y. Ting and Y.-C. Chang, *cond-mat*/0402237 (2004).

Strong Enhancement of Rashba Effect in Strained p-type Quantum Wells

D. M. Gvozdić and U. Ekenberg

Department of Microelectronics and Information Technology
Royal Institute of Technology, Electrum 229, SE-16440 Kista, Sweden

Abstract. One of the most studied spintronic devices is the spin transistor proposed by Datta and Das. The mechanism behind this transistor is the Rashba effect: The inversion asymmetry caused by the gate voltage gives rise to a spin splitting. We show that the relevant spin splitting in k-space is typically two orders of magnitude larger in unstrained p-type quantum wells compared to n-type quantum wells. We also show that further order-of-magnitude improvement can be obtained by utilizing the frequently ignored lattice-mismatch between GaAs and AlGaAs.

INTRODUCTION

A new active research area is called spintronics where the spins of the carriers (electrons or holes) are utilized rather than their charge [1]. We here consider possibilities to improve the performance of the spin transistor proposed by Datta and Das [2]. The fundamental mechanism behind this transistor is called the Rashba effect: The gate voltage gives rise to a spin splitting even without an applied magnetic field [3]. We here demonstrate ways to optimize the Rashba effect, i.e. to create a large spin splitting with a small gate voltage. Using an 8×8 **k·p** Hamiltonian which simultaneously gives the spin splitting of electrons and holes we have calculated subband structures for 20 nm quantum wells grown along the [001] direction.

RASHBA EFFECT FOR ELECTRONS

For the lowest electron subband the spin splitting is well described by the so called Rashba term [3]. Then the spin splitting in k-space relevant for a spin transistor becomes $\Delta k = (2 m^* e\, \alpha_{so}\, F)/\hbar^2$, where α_{so} gives the strength of the spin-orbit interaction and F is the electric field [2]. To be able to compare the strength of the Rashba effect between different structures we define an electric-field-normalized figure of merit $G \equiv \partial(\Delta k)/\partial F$. The common way to get a strong Rashba effect is to choose materials with heavy atoms having strong spin-orbit coupling. We obtain $G = 0.13$ V^{-1} for a GaAs and $G = 1.2$ V^{-1} for an In$_{0.74}$Ga$_{0.26}$Sb quantum well. These results are almost independent of energy, electric field and strain.

IMPROVEMENT USING HOLES

The situation for hole subbands is much more complex because of the strong interaction between heavy and light holes in low-dimensional structures. The subbands become strongly nonparabolic and the spin splitting usually has a maximum a few meV below the top of the uppermost hole subband, where the Fermi energy is situated for common hole concentrations. Because of the stronger spin-orbit interaction in the valence band the energy spin splitting ΔE is much larger for holes than for electrons and the wave vector spin splitting Δk is enhanced even more. We have evaluated G at the optimal energies and obtained $G = 94$ V^{-1} for GaAs and $G = 1080$ V^{-1} for In$_{0.74}$Ga$_{0.26}$Sb, about three orders of magnitude larger than for electrons. For the performance of a spin transistor instead of taking the derivative it is more appropriate to consider the increase of Δk divided by the finite electric field required to switch the transistor, but even then the result for holes becomes several hundred times those for electrons.

IMPROVEMENT USING STRAIN

With small biaxial tension the HH_1 - LH_1 separation decreases and the increased interaction can cause the HH_1 spin subbands to have almost flat energy dispersions for a range in k. This is clearly and conveniently implemented in GaAs quantum wells with $Al_{0.27}Ga_{0.73}As$ barriers. Fig. 1a shows how strain modifies the subband dispersion and Fig. 1b shows how ΔE gradually increases with F. But because of the flat dispersion one can get a large Δk even with a small ΔE. The dependence of Δk on F is shown in Fig. 2 for a number of energies for the unstrained and strained cases. It is remarkable that the small lattice-mismatch between GaAs and AlGaAs, that is usually ignored, is sufficient to get these strong effects. (The lattice constant of AlAs is 0.2 % larger than that of GaAs). For a strained GaAs quantum well grown on an $Al_{0.27}Ga_{0.73}As$ buffer layer determining the in-plane lattice constant we obtain $G = 2230$ V^{-1} which is more than an order of magnitude larger than for unstrained GaAs and twice as large as for unstrained $In_{0.74}Ga_{0.26}Sb$. Compared to n-type GaAs we get an enhancement by a factor 17000 which remains tremendous even if a finite change of F is considered.

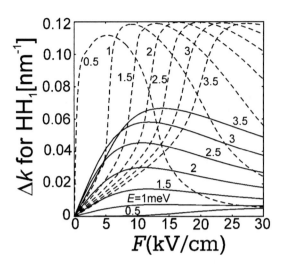

FIGURE 2. Wave vector splitting Δk vs. F for an unstrained (solid lines) and a strained (dashed lines) GaAs quantum well. The energies at which Δk is evaluated are shown.

DISCUSSION AND CONCLUSION

The implications for spin transistors are evaluated elsewhere. The switch energy E_{sw} is a common figure of merit for transistors. For n-type $In_{0.74}Ga_{0.26}Sb$ we get $E_{sw} = 40$ aJ and for strained p-type GaAs $E_{sw} = 0.002$ aJ although a closer analysis is required in the latter case. $E_{sw} = 2$ aJ has been obtained for Si MOS transistors [4]. While n-type spin transistors hardly will become competitive, p-type spin transistors are much more promising. For future work it could be worthwhile investigating more complex quantum well structures, quantum wires, other growth directions etc.

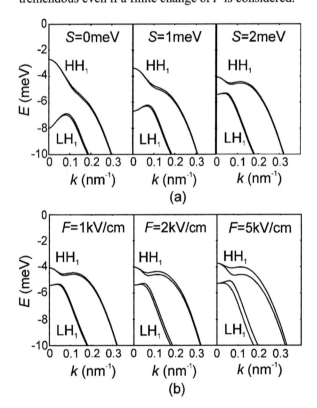

FIGURE 1. Hole subbands in a GaAs quantum well for (a) constant $F = 1$ kV/cm and increasing strain and (b) constant $S = 2$ meV and increasing F. The heavy-hole band edge is lowered by S compared to the unstrained case and the light-hole band edge raised by about the same amount.

ACKNOWLEDGMENTS

We thank the Swedish Research Council for financial support and L. Thylén and P.-E. Hellström for valuable discussions.

REFERENCES

1. For a recent review, see e.g. Žutić, I., Fabian, J. and Das Sarma, S., *Rev. Mod. Phys.* **76**, 323 (2004).
2. Datta, S. and Das, B., *Appl. Phys. Lett.* **56**, 665 (1990).
3. Bychkov, Y.A. and Rashba, E.I., *J. Phys. C* **17**, 6039 (1984).
4. *International Technology Roadmap for Semiconductors (ITRS) (2003)*
http://public.itrs.net/Files/2003ITRS/Home2003.htm

CHAPTER 7
QUANTUM INFORMATION

Strongly Tunable Coupling Between Quantum Dots

Y. B. Lyanda-Geller[1,*], G. Bacher[2], T. L. Reinecke[1], M. K. Welsch[3], A. Forchel[3], C. R. Becker[3] and L. Molenkamp[3]

[1]*Naval Research Laboratory, Washington DC 20375 USA*
[2]*Universität Duisburg-Essen, Duisburg, 47057 Germany*
[3]*Iniversität Würzburg, Am Hubland, Würzburg, 97074 Germany*

Abstract. The incorporation of Mn ions in one of the quantum dots (QD) results in markedly different spin splittings for states of single QD molecules fabricated from coupled quantum well structures of (Cd,Mn,Mg)Te alloy materials. By comparing the observed magnetic field dependence of the photoluminescence polarization with detailed calculations, we show that the coupling between the dots is coherent and is tunable with magnetic field. We show that spin splittings and g-factors of these coupled dots can be manipulated by orders of magnitude. We also develop an approach for the fast control of spin splittings in a QD with an electric field in these structures.

The search for physical implementations for quantum bits (qu-bits) and quantum gates is attracting much research interest. Spin splittings in a magnetic field provide a natural representation for two level qu-bits. A quantum dot (QD) spin has a long lifetime [1]. A major challenge is to achieve also control of the single QD spin for qu-bit operations and control of couplings between QD's for quantum gates. Local magnetic fields can be used to rotate a spin, but the speed with which magnetic fields can be changed is slow compared to the requirements for quantum computation. Manipulation of spins by an electric field can offer greater speed. To date, however, no such control of single spins has been achieved.

In the present joint experimental and theoretical work, we (i) develop a new coupled quantum dot system based on Mn doped II-VI quantum dots, (ii) demonstrate coherent coupling between the two dots, and (iii) develop a method for electric field control of the spin splitting. Our coupled QDs have large lateral confinement energies ~ 10 meV and strong and tunable interdot couplings. Selective interdiffusion of coupled quantum wells leads to QD pairs with well-defined interdot separations. Mn ions selectively doped into one of the QD's give two QD's with widely different spin splittings in a magnetic field B. This allows us to control the interdot coupling by changing B. We show that the QD coupling is coherent by comparing data on circular polarization of photoluminescence with calculations. In addition, we show that the spin splitting in a non-magnetic QD of the coupled QD pair can be tuned by applying an external electric field. This creates the opportunity to use Mn-doped QDs as auxiliary QD's for quantum computing. Static magnetic fields can then be used to bring the spin states of the auxiliary QD into coupling with those of the non-magnetic (target) dot. Then an electric field is used to modulate the target dot spin splittings by changing the relative energies of the two dots. We find that the spin splitting of the target dot can be enhanced by orders of magnitude.

We show in the Fig.1 the measured photo-luminescence (PL) energies vs. B for QD pairs with barrier thickness d_{SP} 9 and 3 nm. In both cases, the PL signals from the CdMnTe and the CdTe quantum dots of the pair approach one another with increasing B, but the spectral lines do not come close in either case. The PL polarizations are shown in the bottom panel of Fig.1. In both cases the signal originating from the CdMnTe QD is completely σ^+-polarized. In contrast, the polarization of the signals from the CdTe QDs depends on the interdot separation d_{SP}. For $d_{SP} = 9$ nm, the CdTe signal is unpolarized at B=0 and becomes σ^-- polarized with increasing B. The situation is quite different in the case of $d_{SP} = 3$ nm. For it, the PL

---

*Permanent address: Department of Physics, Purdue University, West Lafayette IN 47907 USA

exhibits σ^--polarization for small B, but the polarization changes sign with increasing B.

We have also calculated the exciton properties of these QD pairs. We find that the tunnel coupling (resulting from the wave function overlap) for dots separated by d_{SP} = 9 nm is negligible, whereas for d_{SP} = 3nm it is non-negligible. The qualitative behavior of the exciton energies is sketched in Fig.2. The upper pair of exciton states arises from the CdMnTe dot, is composed of exciton states with angular momentum $L = \pm 1$ and has a large splitting with B. The lower pair of $L = \pm 1$ exciton states arises from the CdTe dot, has a small splitting and oppositely ordered states due to its negative g-factor. The total interaction between excitons arises from the tunnel coupling of electrons and holes and from the Coulomb interactions

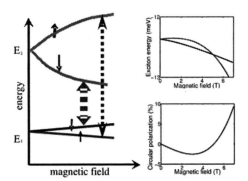

FIGURE 2. Left: Scheme of energy levels of the QDs. Right, top: crossing of energy levels in the ground state of the pair. Right, bottom: degree of circular polarization of emission from the ground state in magnetic field

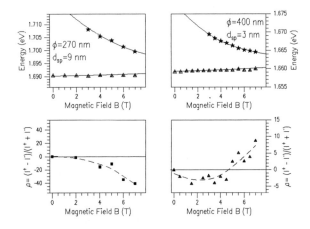

FIGURE 1. Top: Energy shift of the σ^+-polarized photoluminescence signal with magnetic field. Bottom: polarization degree of the emission from the CdTe QD.

between them. Tunneling of excitons occurs mostly between states with the same L. For for d_{SP} = 9nm for which it is small, the lower state of the doublet from the CdTe dot becomes increasingly thermally occupied for increasing B giving a σ^--polarization in agreement with the experiment. In the case of significant tunnel coupling for d_{SP} = 3nm, the L = 1 state of the CdMnTe dot approaches the L = -1 state of the CdTe dot. These two states interact most strongly (thick vertical arrow) and repel one another. As a result the L=-1 state of the CdTe crosses below the L=1 state. This is shown in detail in the top right panel in Fig.2. By taking the two lowest exciton states to be populated thermally we obtain the polarization of the lowest state shown in the bottom left panel. These results account well for the experimental data in Fig.1. This effect would not occur in the absence of quantum mechanical tunnel coupling between the quantum dots. Thus we conclude that the CdMnTe and CdTe dots are coupled coherently.

An electric field can change the relative exciton energies of the two isolated dots. At a given B, it brings the L = 1 state of the upper exciton doublet of states closer to the L = -1 state of the lower doublet in Fig.2 and affects exciton Zeemann splitting. The two electron spin levels are also of interest for quantum computing. The electric field can control the coupling of the electron spins originating from two dots. In this way the electric field can modulate electron spin splittings as shown in Fig.3. Such a modulation can be used for fast single qu-bit operations.

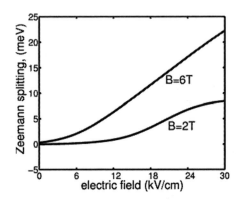

FIGURE 2. Zeemann splitting of electrons in CdTe/CdMnTe pair of QDs with d_{SP} =5nm vs. electric field.

REFERENCES

1. J. A. Gupta, *et al.*, Phys. Rev. B **59**, 10421 (1999)

2. M. Bayer, et al., Phys. Rev. Lett. 90, 086404 (2003).

Optically-Generated Many Spin Entanglement in a Quantum Well

J. Bao[*,¶], A. V. Bragas[*,†], J. K. Furdyna[**] and R. Merlin[*]

[*] *FOCUS Center and Department of Physics, University of Michigan, Ann Arbor, MI 48109-1120, USA*
[¶] *Present address: Division of Engineering and Applied Sciences, Harvard University, Cambridge, MA 02138, USA*
[†] *Present address: FCEyN, Universidad de Buenos Aires, Ciudad Universitaria, 1428 Buenos Aires, Argentina.*
[**] *Department of Physics, University of Notre Dame, Notre Dame, IN 46556, USA*

Abstract. A many-particle system may exhibit non-local behavior in that measurements performed on one of the particles can affect a second particle that is far apart. These so-called entangled states are essential for the implementation of quantum information protocols and gates for quantum computation. Here, we use ultrafast optical pulses and coherent techniques to create spin entangled states of non-interacting electrons bound to donors in a CdTe quantum well.

INTRODUCTION

The problem of quantum entanglement has attracted much attention since the early days of quantum mechanics. Following the proposal by Deutsch for a quantum computer in 1985, the building of a quantum cryptography machine by Bennett et al. in 1989, and the discoveries by Shor in 1994 and by Grover in 1996 of quantum algorithms, the question of entanglement has now acquired practical significance [1]. Here, we propose and demonstrate experimentally a novel method for many-particle entanglement involving spin states of *non-interacting* paramagnetic impurities in a quantum-well (QW) [2].

QUBITS AND ENTANGLEMENT

Our qubits are embodied by Zeeman-split spin states of donor-bound electrons in a $Cd_{1-x}Mn_xTe$ QW [2]. The N-electron entangled states relevant to our work are coherent superpositions of the form

$$\psi = \sum_{k=0}^{2S} C_k e^{-ik\Omega_o t} |S-k\rangle \qquad (1)$$

where $|S-k\rangle$ is an eigenstate of the Zeeman energy operator with $S = N/2$, $\Omega_0 = g\mu_B B/\hbar$ (g is the gyromagnetic factor and B is the external magnetic field), and C_k are parameters that can be controlled by optical means [2]. Our entanglement method relies on the exchange coupling between the hole of the photoexcited exciton and the donor electrons. This interaction can be described in terms of an effective magnetic field $\mathbf{B}_e = \kappa \mathbf{J}$ where $\mathbf{J}$ is the spin of the heavy-hole and κ is an exchange constant. Due to a combination of spin-orbit-coupling and quantum-confinement effects, the heavy-hole spin and, consequently, $\mathbf{B}_e$ are parallel to the QW growth axis (the z-axis) at small external fields [3]. Thus, $J_z = \pm 3/2$ and, provided the external field is not parallel to the same axis, the quantization axes in the absence and in the presence of the exciton are different and, since the exciton does not modify S, the corresponding states are not orthogonal to each other. It follows that, for an electric-dipole allowed exciton, Raman coherences between arbitrary states can be established by using two laser fields tuned to resonate with the exciton energy, E_e, so that their frequency difference is $|m-l|\Omega_0$.

As it is well known for generic Raman coherences [4], we expect in the limit $\Omega_0 \ll E_e$ that the effect of the coherent spin states of Eq. (1) on the optical properties be that of a modulation involving the harmonics of Ω_0. This modulation can be probed in the time domain using standard ultrafast pump-probe methods [2]. We notice that overtones of Ω_0 should not be observed for non-interacting electrons, and that, because the electron spin is $s = ½$, the signature for an entanglement of N electrons is the observation of the Nth harmonic.

EXPERIMENT

Our sample consists of 100 periods of 58-Å-thick $Cd_{1-x}Mn_xTe$ ($x = 0.0039$) with 19-Å-thick MnTe barriers grown by molecular beam epitaxy on a CdTe substrate along the [001] z-axis. The sample is nominally undoped. However, spin flip Raman experiments reveal the presence of isolated donors in the wells (possibly indium) with a concentration of $\sim 5 \times 10^{16}$ cm^{-3}.

In our time-domain experiments, we used circularly-polarized pump pulses to generate the Raman coherences which we probed by measuring the pump-induced shift of the polarization angle of the reflected probe field, $\Delta\theta$ (magnetic-Kerr effect) [2]. Our mode-locked Ti-sapphire laser provided ~ 130 fs pulses of central energy ω_C in the range 1.47-1.72 eV (720–840 nm) at a repetition rate of 82 MHz.

Time-domain data for ω_C slightly above the QW gap is shown in Fig. 1. At short times, the signal is dominated by electron Zeeman quantum beats, i. e., oscillations showing a field- and temperature-behavior consistent with that of spin-flips of electrons [2]. The oscillations that persist above ~ 15 ps are associated with the paramagnetic resonance (PR) of the Mn^{2+} ions with $g \approx 2$. The inset shows the Fourier decomposition spectrum with parameters gained from linear prediction fits (the PR is not shown) [2]. In addition to the peak labeled 1SF, associated with single-electron spin-flips, the data reveal features 2SF and 3SF, at nearly twice and three times the frequency of 1SF, which are assigned to multiple spin-flips of donors. The relatively large spectral width of these lines is ascribed to inhomogeneous broadening arising from fluctuations in the manganese concentration [2]. The observation of two spin-flip overtones indicates the establishment of Raman coherences and, hence, entanglement involving at least three donor impurities. As discussed in [2], the spin-flip harmonics as well as the Mn^{2+}-PR exhibit a significant enhancement when ω_C is tuned to resonate with localized exciton states below the gap. Excitation at energies close to but above the QW gap, as in Fig. 1, leads to an enhancement of the signal of the overtones with respect to the fundamental. These results clearly indicate that the bound-electron entanglement benefits from the mediation of states in the continuum. However, the process by which the entan-

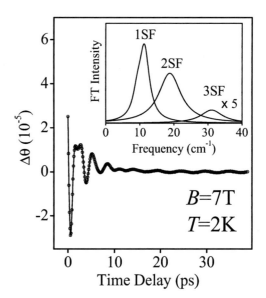

FIGURE 1. Differential magnetic Kerr data at ω_C = 1.71 eV. Inset: Fourier decomposition spectrum.

glement is attained is not well understood. We note that a mechanism involving the RKKY interaction between localized electrons and delocalized excitons has been previously proposed [5].

ACKNOWLEDGMENTS

Work supported by the NSF under Grants No. PHY 0114336 and No. DMR 0245227, and by the DARPA-SpinS program. Acknowledgment is made to the donors of The Petroleum Research Fund, administered by the ACS, for partial support of this research.

REFERENCES

1. See: M. A. Nielsen and I. L. Chuang, *Quantum Computation and Quantum Information* (University Press, Cambridge, 2000); and references therein.
2. J. M. Bao, A. V. Bragas, J. K. Furdyna, and R. Merlin, Nature Mater. **2**, 175 (2003); Solid State Commun. **127**, 771 (2003).
3. R. W. Martin, R. J. Nicholas, G. J. Rees, S. K. Haywood, N. J. Mason, and P. J. Walker, Phys. Rev. B **42**, 9237 (1990).
4. R. W. Boyd, *Nonlinear Optics*. (Academic Press, San Diego, 1992) Ch. 9, pp. 365-397.
5. C. Piermarocchi, P. Chen, L. J. Sham and D. G. Steel, Phys. Rev. Lett. **89**, 167402 (2002).

Indirect spin coupling between quantum dots

G. Ramon[*], Y. Lyanda-Geller[*], T. L. Reinecke[*] and L. J. Sham[†]

[*]*Naval Research Laboratory, Washington DC 20375-5320*
[†]*Department of Physics, University of California San Diego, La Jolla, California 92093-0319*

Abstract. An indirect exchange interaction between spins in two quantum dots is investigated theoretically. The coupling is mediated by virtual delocalized excitons driven by an off-resonance laser. We present a formulation of the exchange interaction between the localized and itinerant electrons that includes the effects of hybridization of continuum and dot states and the double occupancy in the dot. These are shown to produce a contribution comparable to conventional exchange. Among the studied geometries, the largest couplings are found for vertically stacked dots in a quasi one-dimensional host.

Coherent coupling between spins in quantum dots is much sought after for implementation of gates between qubits for quantum computation. Several coupling schemes have been proposed including direct wavefunction overlap [1], exchange of a cavity mode between dots [2] and indirect coupling [3]. Here we focus on the last case where the interaction is mediated by virtual delocalized carrier excitations in the host material. This is analogous to the RKKY interaction between two magnetic impurities. The virtual excitations are driven by an interband off-resonance laser. The proposed scheme has the advantages of ultrafast optical control and long spin coherence times that are maintained due to the virtual nature of the excitations.

The effective spin-spin coupling is calculated by considering the self-energy correction in the continuum electron propagator due to its Coulomb interaction with each of the localized spins and is given by [3]

$$J_{12}(R) = \frac{|\Omega|^2}{16} \int \frac{d^d k d^d k'}{(2\pi)^{2d}} \frac{|J(k,k')|^2 e^{-i(\mathbf{k}-\mathbf{k}')\cdot \mathbf{R}}}{\left(\delta + \frac{k^2}{2\mu}\right)^2 \left(\delta + \frac{k^2}{2m_h} + \frac{k'^2}{2m_e}\right)} \quad (1)$$

where R is inter-dot distance, δ is the detuning of the laser with respect to the electron-hole continuum, Ω is the Rabi energy for the electron-hole pair and μ is it's reduced mass. $J(k,k')$ is the exchange interaction between the localized and itinerant electrons and constitutes a major part in evaluating the dot-dot spin coupling.

We calculate $J(k,k')$ by taking into account the effects of hybridization of continuum and dot states and double occupancy in the dot. The Hamiltonian is constructed by considering the kinetic energy, the dot potential relative to the host material and the electron-electron Coulomb interactions and is given by

$$\mathcal{H} = \mathcal{H}_0 + \mathcal{H}_M + \mathcal{H}_1 \quad (2)$$

where

$$\mathcal{H}_0 = \sum_\sigma E_d n_\sigma + \sum_{k,\sigma} E_k c^\dagger_{k\sigma} c_{k\sigma} + U n_\uparrow n_\downarrow \quad (3)$$

$$\mathcal{H}_M = \sum_{k,\sigma} \left[V_k c^\dagger_{k\sigma} c_{d\sigma} + T_k c^\dagger_{k\sigma} c_{d\sigma} n_{-\sigma} + h.c. \right] \quad (4)$$

Here, $c^\dagger_{d\sigma}$ ($c^\dagger_{k\sigma}$) is the creation operator for a localized (itinerant) electron and $n_\sigma = c^\dagger_{d\sigma} c_{d\sigma}$. $V_k = \int d\mathbf{r} \varphi^*_k(r) V_d(r) \varphi_d(r)$ is the hybridization potential where $\varphi_k(r)$ [$\varphi_d(r)$] is the itinerant (localized) electron wave function. $\mathcal{H}_1$ contains all the two-electron processes that are not included in $\mathcal{H}_0$ and $\mathcal{H}_M$. In particular it contains the conventional (Heitler-London) spin exchange term $\sum_{kk',\sigma} C^{ex}_{kk'} c^\dagger_{k\sigma} c_{d\sigma} c^\dagger_{d-\sigma} c_{k'-\sigma}$. The Coulomb interaction is also responsible for a population dependent hybridization of the dot and continuum states that is provided by the second term in (4). An additional contribution to the spin exchange interaction is found by applying a canonical transformation on $\mathcal{H}' = \mathcal{H}_0 + \mathcal{H}_M$

$$\bar{\mathcal{H}}' = e^S \mathcal{H}' e^{-S}. \quad (5)$$

The unitary operator S is constructed to eliminate $\mathcal{H}_M$ to first order by requiring $\mathcal{H}_M = -[S, \mathcal{H}_0]$, and is given by

$$S = \sum_{k\sigma} [\beta_k + (\alpha_k - \beta_k) n_{-\sigma}] c^\dagger_{d\sigma} c_{k\sigma} - h.c. \quad (6)$$

where

$$\alpha_k = \frac{V_k + T_k}{U + E_d - E_k} \; ; \; \beta_k = \frac{V_k}{E_d - E_k}. \quad (7)$$

This is a generalized form of the Schrieffer-Wolff transformation [4], and it produces a contribution to the exchange interaction given to first order by

$$J^{(1)}_{kk'} = \frac{1}{2} [\beta_k V^*_{k'} - \alpha_k (V_{k'} + T_{k'})^*] + [k \leftrightarrow k']^*. \quad (8)$$

This contribution vanishes when correlation effects are neglected ($U, T_k \to 0$). We find that the first order result, given in (8), is inadequate as it requires that $\mathcal{H}_M$ would be a small perturbation to $\mathcal{H}_0$, which is not the case here. It is therefore necessary to sum up the infinite series in the transformed Hamiltonian

$$\bar{\mathcal{H}}' = \mathcal{H}_0 + \sum_{n=1}^{\infty} \left(\frac{1}{n!} - \frac{1}{(n+1)!} \right) [S, \mathcal{H}_M]_n, \quad (9)$$

where $[S, \mathcal{H}_M]_n = [S, [S, ..., [S, \mathcal{H}_M]...]]$. The first term in this series is calculated as

$$[S, \mathcal{H}_M]_1 = \sum_{\mathbf{k}\mathbf{k}',\sigma} \left[J_1(k,k') \left(c^{\dagger}_{\mathbf{k}\sigma} c_{d\sigma} c^{\dagger}_{d-\sigma} c_{\mathbf{k}'-\sigma} + \right. \right.$$
$$\left. n_{-\sigma} c^{\dagger}_{\mathbf{k}\sigma} c_{\mathbf{k}'\sigma} \right) + P_1(k,k') \left(c^{\dagger}_{\mathbf{k}\sigma} c_{d\sigma} c^{\dagger}_{\mathbf{k}'-\sigma} c_{d-\sigma} + h.c. \right)$$
$$\left. - G^c_1(k,k') c^{\dagger}_{\mathbf{k}\sigma} c_{\mathbf{k}'\sigma} \right] + \sum_{\sigma} \left[G^d_1 n_{\sigma} + I_1 n_{\sigma} n_{-\sigma} \right] \quad (10)$$

where the functions $J_1(k,k'), P_1(k,k'), G^c_1(k,k'), G^d_1, I_1$ in Eq. (10) are found in terms of α_k, β_k and V_k, T_k and represent effective interactions that renormalize the terms in the original Hamiltonian. In higher orders one can bring the even orders to the same form of $\mathcal{H}_M$ and the odd orders to the form of Eq. (10). We do this by neglecting off-diagonal contributions to higher order correlations. With this approximation we then sum the series by solving a set of recursion relations [5]. The exchange contribution is collected from the odd orders of the series. Since the result of the series summation deviates appreciably from the first order result, the summation on the even orders entails an appreciable residual hybridization. We are thus led to perform a second canonical transformation which eliminates the next order in the hybridization terms and further corrects the resulting exchange contribution. This process is reiterated until one eliminates the hybridization part of the Hamiltonian to infinite order.

In figures 1-2 we show the results for the dot-dot coupling [Eq. (1)], incorporating all the exchange contributions. Figure 1 shows the spin coupling for the case of lateral dots in a two-dimensional quantum well. The results for vertically stacked quantum dots in a quasi one-dimensional wire are given in figure 2. Here we used $m_e = 0.07m$, $m_h = 0.5m$ and $\Omega = 0.1 meV$. It is seen that the spin coupling is more than an order of magnitude larger for the one-dimensional case as compared with the two-dimensional case. The dot-dot coupling is enhanced by two orders of magnitude due to the Coulomb interactions between the intermediate virtual electrons and holes (see the right axes of figures 1-2).

We have shown that the effect of hybridization of continuum and dot states produces a sizable contribution to the exchange coupling between localized and itinerant electrons. We point out that a canonical transformation

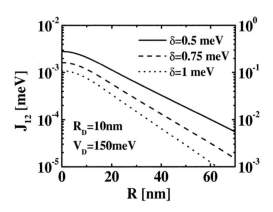

FIGURE 1. Optically induced exchange coupling in a 2D host vs. the distance between the dots for dot radius $R_D = 10nm$, potential height $V_e = 150meV$ and several values of detunings. Right axis shows that coupling values after excitonic corrections.

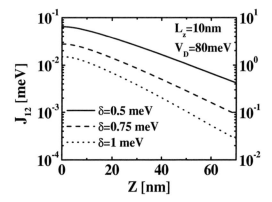

FIGURE 2. Same as figure 1 for a quasi 1D host and cylindrical dots with $L_z = 10nm$, $R_D = 5nm$ and potential height $V_e = 80meV$.

with partial summation over higher order terms provides a useful tool to evaluate the spin exchange coupling.

This work was supported by ONR, DARPA and ARDA/ARO. G. R. wishes to thank Dr. A. K. Rajagopal for helpful discussions.

REFERENCES

1. D. Loss and D. P. Di Vincenzo, Phys. Rev. A **57**, 120 (1998).
2. A. Imamoglu et al., Phys. Rev. Lett. **83**, 4204 (1999).
3. C. Piermarocchi, Pochung Chen, L. J. Sham and D. G. Steel, Phys. Rev. Lett. **89**, 167402 (2002).
4. J. R. Schrieffer and P. A. Wolff, Phys. Rev. **149**, 2 (1966).
5. R. Chan and M. Gulacsi, cond-mat 0308405 (2003).

Ruderman-Kittel-Kasuya-Yosida Interaction in Quantum Dot Arrays

Hiroyuki Tamura[*†], Kenji Shiraishi[**] and Hideaki Takayanagi[*†]

[*]*NTT Basic Research Laboratories, NTT Corporation, Atsugi, Kanagawa 243-0198, Japan*
[†]*CREST-JST, 4-1-8 Honmachi, Kawaguchi, 331-0012, Japan*
[**]*Institute of Physics, University of Tsukuba, 1-1-1, Tennodai, Tsukuba 305-8571, Japan*

Abstract. We discuss theoretically the Ruderman-Kittel-Kasuya-Yosida (RKKY) interaction in self-organized quantum-dot (QD) arrays. The RKKY interaction between QDs is modulated by changing the Fermi energy, and the magnitude or even the sign of the exchange interaction can be tuned. We propose a tunable magnetic device based on self-organized quantum-dot arrays.

INTRODUCTION

The Ruderman-Kittel-Kasuya-Yosida (RKKY) interaction is known as an indirect exchange interaction between local magnetic impurities mediated by the Fermi sea of metals. The RKKY interaction in semiconductor QDs has several interesting features [1]: The Fermi wavelength λ_F is typically several tens of nanometers, which is very long, and it is possible to form neighboring dots that are within λ_F, where the RKKY interaction is ferromagnetic and its magnitude is very large. Moreover, the RKKY interaction between QDs can be controlled by changing the electron density with gate electrodes, which leads to a tunable magnetic transition in QD devices.

In this paper, we propose a magnetic device based on self-organized quantum-dot arrays and estimate the magnitude of the RKKY interaction in the QD device.

THEORETICAL MODEL

We assume that QDs with a local spin $\mathbf{S}_n$ ($S = 1/2$, $n = 1, 2, \cdots$) are coupled to the Fermi sea, where the interaction can be described by the Kondo Hamiltonian with s-d exchange interaction H_{ex} given by $H_{\text{ex}} = \sum_{nkk'\sigma\sigma'} J_n(\mathbf{k},\mathbf{k}') c^\dagger_{k'\sigma'} \sigma_{\sigma'\sigma} c_{k\sigma} \cdot \mathbf{S}_n$, where $J_n(\mathbf{k},\mathbf{k}') = -V_n^*(\mathbf{k}')V_n(\mathbf{k})U_n/\{(U_n + \varepsilon_n)\varepsilon_n\}$, $V_n(\mathbf{k})$ is the coupling strength between conduction electrons and a quantum dot n having a single-level energy ε_n and a charging energy U_n.

The indirect RKKY exchange interaction between two local spins is obtained from the second order perturbation for the ground state of conduction electrons at zero temperature as

$$H_{\text{RKKY}} = \sum_\alpha \frac{|\langle 0|H_{\text{ex}}|\alpha\rangle|^2}{E_0 - E_\alpha} = -J_{\text{RKKY}} \mathbf{S}_1 \cdot \mathbf{S}_2, \quad (1)$$

$$J_{\text{RKKY}} = \frac{8mJ_1J_2}{\hbar^2\Omega^2} \sum_{\substack{|\mathbf{k}_1|<k_F \\ |\mathbf{k}_2|>k_F}} \frac{\cos\{(\mathbf{k}_1 - \mathbf{k}_2)\cdot\mathbf{R}\}}{k_2^2 - k_1^2}, \quad (2)$$

where $|\alpha\rangle = c^\dagger_{\mathbf{k}_2\sigma} c_{\mathbf{k}_1\sigma}|0\rangle$ and $J_n = -|V_n|^2 U_n/\{(U_n + \varepsilon_n)\varepsilon_n\}$, m is the effective mass of an electron in the Fermi sea, $\mathbf{R} = \mathbf{r}_2 - \mathbf{r}_1$, and Ω is the volume of the conduction electron system. Then, the RKKY interaction is obtained for three ($d=3$) [2], two ($d=2$) [3, 4], and one ($d=1$) [5, 6] dimensions as

$$J_{\text{RKKY}}(R) = 16\pi E_F \tilde{J}_1 \tilde{J}_2 F_d(2k_F R), \quad (3)$$

where E_F is the Fermi energy of the conduction electrons. The dimensionless Kondo parameter $\tilde{J}_n = -\Gamma_n U_n/[4\pi(U_n + \varepsilon_n)\varepsilon_n]$ ($\Gamma_n = \pi\rho|V_n|^2$ is the elastic broadening of the energy level in dot n due to tunneling and ρ is the density of states at E_F), and the range functions are given by $F_3(x) = (-x\cos x + \sin x)/4x^4$, $F_2(x) = -[J_0(x/2)N_0(x/2) + J_1(x/2)N_1(x/2)]$, $F_1(x) = -\text{si}(x) = \int_x^\infty dt (\sin t/t)$, where J_n and N_n are the n-th Bessel functions of the first and second kind, and $\text{si}(x)$ is the sine-integral function.

DISCUSSIONS

Figure 1 illustrates one example of quantum dot arrays coupled with conduction electrons. Self-organized

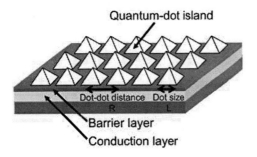

FIGURE 1. Schematic quantum-dot island.

TABLE 1. Maximum value of the RKKY interaction constant $J_{RKKY}(R)$ in three-dimensional conduction layer for different dot diameters.

	Maximum $J_{RKKY}(R)$ for $10^{17} < n_{3D} < 10^{18}$ cm^{-3}		
	$R = 3$ nm	10 nm	30 nm
GaAs ($m^* = 0.067$)	470 K	42 K	1.2 K
Si ($m^* = 0.2$)	157 K	14 K	0.4 K

quantum-dot islands are grown on the three-dimensional conduction layer separated by a thin barrier. To estimate the magnitude of the RKKY interaction, we assume that all dots have the same Kondo parameter $\tilde{J}_n = \tilde{J} = \Gamma/\pi U = 0.15$, estimated from a typical Kondo device [7]. In Table 1, we estimate the maximum magnitude of the RKKY interaction in Eq. (3) assuming GaAs and Si in the three-dimensional conduction region of electron density of $10^{17} < n_{3D} < 10^{18}$ cm^{-3} for different dot diameters. Quantum dots of GaAs [8], Si [9, 10] and Ge [11] with diameters of 3-10 nm have already been fabricated. Dot arrays with a dot density of 10^{13} cm^{-2} (or inter-dot distance of 3 nm), grown on a thin barrier layer on top of a three-dimensional quantum well of Si or GaAs, have $J_{RKKY} \sim 470$ K for GaAs and 157 K for Si quantum well. Therefore, using current technology, it is already possible to measure the effect of the RKKY interaction at room temperature.

In summary, we presented a theoretical discussion of the RKKY interaction in self-organized quantum dots. We estimated the magnitude of the RKKY interaction in the QD devices.

ACKNOWLEDGMENTS

We are grateful to L. I. Glazman and W. Izumida for valuable discussions. This work was supported by the NAREGI Nanoscience Project, Ministry of Education, Culture, Sports, Science and Technology, Japan.

REFERENCES

1. H. Tamura, K. Shiraishi, and H. Takayanagi, Jpn. J. Appl. Phys. **43**, L691 (2004).
2. C. Kittel, in *Solid State Physics*, edited by F. Seitz, D. Turnbull, and H. Ehreinreich (Academic, New York, 1968), Vol. 22, p. 1.
3. B. Fischer and M. W. Klein, Phys. Rev. B **11** (1975) 2025.
4. M. T. Beal-Monod, Phys. Rev. B **36** (1987) 8835.
5. Y. Yafet, Phys. Rev. B **36** (1987) 3948.
6. V. I. Litvinov, Phys. Rev. B **58** (1998) 3584.
7. W. G. van der Wiel, S. De Franceschi, T. Fujisawa, J. M. Elzerman, S. Tarucha, and L. P. Kouwenhoven, Science **289** (2000) 2105.
8. K. Ueno, K. Saiki, and A. Koma, Jpn. J. Appl. Phys. **40** (2001) 1888.
9. S. Oda and K. Nishiguchi, J. Phys. IV (France) **11** (2001) 1065.
10. A. A. Shklyaev and M. Ichikawa, Phys. Rev. B **65** (2001) 045307.
11. A. A. Shklyaev and M. Ichikawa, Appl. Phys. Lett. **80** (2002) 1432.

Tunable Electronic-Nuclear Dynamics and Current Instability in Double Quantum Dots

Takeshi Inoshita and Seigo Tarucha

ERATO, JST and Dept. Appl. Phys., Univ. Tokyo

Abstract. Nonlinear dissipative dynamics of the coupled electronic-nuclear system in double quantum dots in the spin-blockade regime is studied. When the singlet and triplet double dot states are made to cross by a magnetic field, enhanced nonlinearity brings about multiple stability and self-oscillation of the nuclear moment, which is observable through current. Nuclear polarization is widely tunable by means of the magnetic field and bias voltages.

INTRODUCTION

Nuclear magnetic fields felt by electrons in solids can reach several teslas if the nuclear spins are fully polarized. Usually this effect is negligible due to the smallness of thermal nuclear polarization. The recent discovery of dramatic nuclear effects in nanostructures triggered renewed interest in this field [1,2]. In this paper, we explore the nonlinear electronic-nuclear dynamics in double quantum dots (DQDs) [3].

OUTLINE OF THEORY

We consider a vertical DQD coupled via weak tunneling to two leads and biased in the spin-blockade regime [Fig.1(a)]. A magnetic field $\boldsymbol{B}=(0,0,B)$ is applied along the dot plane. The DQD is strongly-coupled (artificial molecule) and modeled by a Hubbard-like Hamiltonian H_{dot}. Among the eigenstates of H_{dot}, the most important for our discussion are the singlet (S) and triplet (T) two-electron states. As B is varied, either the $S_z=+1$ or the -1 branch of T [T(+) or T(−)] crosses S as shown in Fig. 1(b).

The total Hamiltonian is $H=H_{dot}+H_n+H_l+H_{ld}+H_{en}$, where H_n is the nuclear Zeeman energy, H_l the two leads, H_{ld} the dot-lead tunnel coupling, and H_{en} the hyperfine interaction between the electrons and nuclei which we decompose as $H_{en}=H_{n1}+H_{n2}$, $H_{n1}=(S_+<J_-> + S_-<J_+>)/2 + S_z<J_z>$, $H_{n2}= [S_+ (J_- -<J_->)+ S_- (J_+-<J_+>)]/2 +S_z (J_z-<J_z>)$, $J_\mp=J_x\mp iJ_y$ and $S_\mp=S_x\mp iS_y$. Here, $J_{x,y,z} = N^{-1}\sum_n (I_{x,y,z})_n$ is the average of the nuclear spins I_n ($n=1,2,\cdots,N$; $N \cong 10^6$), S is the electron spin, and the angular brackets denote expectation values. Note that H_{n1} represents the interaction between an electron and the classical nuclear polarization, whereas the fluctuation H_{n2} is quantum and smaller by a factor of $N^{-1/2}$. We then use $H_{dot}^0 = H_{dot} + H_{n1}$ as the new dot Hamiltonian which describes electrons in an effective (external + nuclear) field.

The dissipative dynamics of our system can be described by the density matrix equation of motion (EOM) $id\rho/dt=[H,\rho]$. To make the problem tractable, we first obtain the stationary *electron* density matrix, $idR/dt=[(H_{dot}^0+H_l+H_{ld}),R]=0$, by tracing out H_l in the Born-Markov approximation. Writing ρ as $\rho=R\,\rho_n$ (adiabatic approximation) and using this in the original equation, one gets an EOM for the nuclear density matrix ρ_n, from which follows the EOM for $<J>$.

This scenario leads to six equations for the six variables $<J_{Li}>$ and $<J_{Ri}>$ ($i=x,y,z$) where $J_{R,L}$ denotes the average nuclear spin in the right (left) dot. However, if the two dots are identical, the equations separate and one merely has to consider the two variables $<J_z>=(<J_{Lz}>+<J_{Rz}>)/2$ and $K=((<J_{Rx}>-<J_{Lx}>)^2+(<J_{Ry}>-<J_{Ly}>)^2)^{1/2}/2$. They satisfy the EOM:

$$\dot{K} = GJ_z - (1+2J_z)F_1 - (1-2J_z)F_2 \\ - F_3 K - K/T_2, \quad (1)$$

$$\dot{J}_z = -GK - (1+2J_z - 2K^2)F_1 \\ -(-1+2J_z+2K^2)F_2 - (J_z - J_{z0})/T_1. \quad (2)$$

(For simplicity, $\langle J_z \rangle$ was written as J_z.) Here $G(K, Jz)$ originates from H_{n1} and represents a classical torque. The terms with $F_i(K, Jz)$ represent nuclear relaxation through tunnel-assisted electron-nuleus mutual spin flip (H_{n2} plus H_{ld}). The F_1 term tends to align the nuclear spins in the -z direction ($J_z=-1/2, K=0$) while the F_2 term aligns them to +z: Thus they compete with each other. The last terms of Eqs. (1) and (2) take account of other relaxation mechanisms, where J_{z0} is the thermal nuclear polarization.

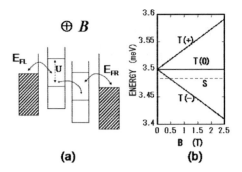

RESULTS

The solutions of Eqs. (1) depend critically on whether S crosses T(-) or (T+), which, in turn, depends on the bias condition. Figures 2(a) and (b) present typical stationary solutions (SSs) of Eqs. 1 versus B for the former case. For small B, the only SS is a stable node (thick line) but in the S-T crossover region, multiple SSs appear which can be classified into nodes (stable, unstable), saddles (unstable) and spirals (stable, unstable). These solutions are oriented in the direction of B ($K=0$) with the exception of the spiral which has $K>0$ [Fig. 2(b)]. This spiral is stable near $B=1.1$ T (open circles) where two stable fixed points coexist (bistability). This spiral fixed point becomes unstable for $B>1.2$ T (filled circles). For $B>2$ T, no stable fixed point exists. The integration of eqs. (1) reveals that the only stationary state in this case is an oscillation (limit cycle), leading to a spiky oscillation of the current [Fig.2(c)]. We believe that this explains the observed oscillation of current [2].

Similar result for S-T(+) crossover is plotted in Fig. 2(c). The system is more stable in this case with all the SSs aligned along B. No solution with $K>0$ was found. A salient feature is the existence of two stable SSs with large polarization near $B=0$. One can easily select one of these SSs by a proper preparation of the initial state. Optimizing the system design, it should be possible to achieve nearly perfect polarization at low B. This large polarization would be useful for the application of DQDs as nuclear memories [4].

ACKNOWLEDGMENTS

T. I. is grateful to Sheila M. P. Paradero for valuable assistance and to Keiji Ono, Vladimir Kalevich, Akira Shimizu, and Mikio Eto for illuminating discussions.

FIGURE 1. (a): Potential profile of the DQD system we consider. (b): B dependence of the energies of the singlet (S) and triplet (T) two-electron states. (The +,- and 0 denote the z-component of the total spin.) The parameters were chosen to simulate the sample used in [2].

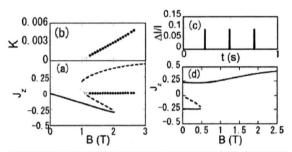

FIGURE 2. (a) is the stationary solutions (SSs) of Eqs. (1) for S-T(-) crossing. These SSs come in various stablility classes: Stable node (thick line), unstable node (crosses), (unstable) saddle (dashed line), unstable spiral (filled circles) and stable spiral (open circle). Multiple SSs arise when $B\approx 1.1$ T due to the enhanced hyperfine coupling and nonlinearity. Only the spiral SS has $K>0$ as shown in (b). For B>2 T, no stable fixed point exists, and the only stationary solution is oscillatory, leading to oscillatory current (c). Similar result for S-T(+) crossing is shown in (d). There are no SSs with $K>0$ or oscillatory solutions.

REFERENCES

1. S. Kronmuller et al., Phys. Rev. Lett. **82** (1999) 4070; K. Hashimoto et al., Phys. Rev. Lett. **88** (2002) 176601; T. Machida et al., Appl. Phys. Lett **82** (2003) 409.

2. K. Ono et al., Phys. Rev. Lett. **92** (2004) 256803.

3. T. Inoshita et al., J. Phys. Soc. Jpn. **72** (2003) Suppl. A 183; T. Inoshita et al., Physica E **22** (2004) 422.

4. J. M. Taylor et al., Phys. Rev. Lett. **90** (2003) 206803.

Controlling and measuring a single donor electron in silicon

K. R. Brown, L. Sun, B. Bryce, and B. E. Kane

Laboratory for Physical Sciences, University of Maryland, 8050 Greenmead Drive, College Park, Maryland 20740

Abstract. We present measurements of single electron transistors, on various substrates, and in the presence of strong electric fields, showing clear differences between doped and undoped samples. We also present evidence of charge motion between resonant charge states. These measurements are relevant to recently proposed charge-based spin measurements in particular, and to gaining a better understanding of single electron effects in general.

INTRODUCTION

Individual electronic or nuclear spins associated with impurities in Si are attractive qubits for a future quantum computer due to their long decoherence times, identical nature, and potential for scaling.[1] Nevertheless, measurement of a single electronic spin remains an open problem. There have been several proposals to measure individual spin states indirectly, by taking advantage of spin-dependent charge transfer.[2, 3, 4] A single electron transistor (SET), with demonstrated charge sensitivity better than 10^{-5} $e/\sqrt{Hz}$,[5] is sensitive enough to measure such charge motion, provided that it can be fabricated close enough to the donor. For example, it should be possible to observe the spin-dependent ionization of single donors in fields ≈ 100 kV/cm.[2, 6, 7] SET-based measurement of donors in Si faces formidable experimental obstacles in practice due to unwanted mobile charges in the substrate and due to the nonlinear nature of the measuring device itself. We have begun measurements of metallic SETs fabricated on various Si substrates in an attempt to sort out these difficulties.

SAMPLE FABRICATION

Figure 1 shows a schematic of our experimental configuration. Sample fabrication starts with a float-zone Si wafer, resistivity $\rho > 10,000$ Ω-cm. We grow 200 Å thermal oxide at 950° C in O_2 for 30 min. and in N_2 for 30 min. An optional ion-implantation step, for example P^+ at 35 keV, places impurities at an average depth of 50 nm with 20 nm straggle. We have tried implant densities between $10^{10}/cm^2$ and $10^{12}/cm^2$. This is followed by annealing at 850° C in N_2 for 30 min. to activate the donors and at 450° C in forming gas to passivate the Si/SiO_2 in-

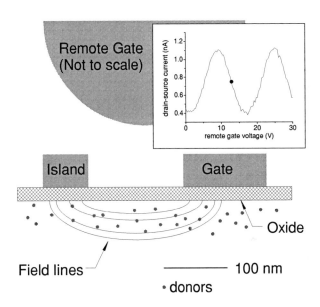

FIGURE 1. Schematic of SET-based donor measurements. The SET has two gates: one lithographically defined local gate and a wire floating above the entire sample (remote gate.) Voltages applied to the local gate create an electric field near the donors. The remote gate fixes the operating point of the SET. The inset shows a plot of drain-source current as a function of gate voltage, with a black dot marking the bias point for optimum charge sensitivity.

terface. Ti/Pt/Au bond pads are defined with photolithography and metallization in an electron-beam evaporator, and the Au is etched away near the center of the chip, exposing the thin (120 Å) Pt layer, to which a subsequent Al layer can make contact. The Al SET and (local) gate are defined with standard electron-beam lithography and double-angle thermal evaporation. A wire suspended above the sample forms another (remote) gate that allows us to set the bias point of the SET independently of the

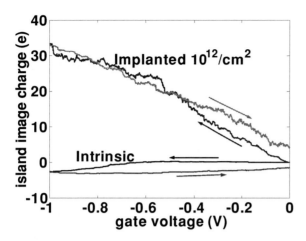

FIGURE 2. Remote gate feedback voltage (in units of SET island charge) as a function of local gate voltage, for a P ion-implanted sample ($10^{12}/cm^2$) and for an intrinsic sample, with the local gate swept from 0 V to -1 V and back. Each trace has had a linear slope subtracted, corresponding to the response of the feedback circuit near zero gate voltage. The intrinsic sample shows almost flat behavior, corresponding to little charge motion, whereas the implanted sample shows an induced charge of 30 e over this voltage range.

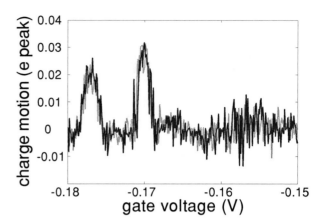

FIGURE 3. 200 Hz component of the SET drain-source current, in phase with an applied 200 Hz electric field, as a function of local gate voltage. Away from resonance the response is zero within the $\sim$ 3 me noise level. As the dc electric field from the local gate brings two states (defects?) into resonance, a charge tunnels between them in phase with the applied 200 Hz field. This yields distinct peaks in the SET response, peaks whose width is determined by the widths of the applied ac potentials.

local gate. In contrast with the local gate, we expect the fields from this remote gate to be largely screened near the donors by the Al metallization and by the Si dielectric.

EXPERIMENTS AND DISCUSSION

We have made two different but related measurements on these samples. The first, a quasi-dc measurement, consists of ramping the voltage on the local gate from zero to a few volts, either negative or positive, over about an hour. This yields electric fields in the substrate of order 10 - 100 kV/cm. A feedback circuit controls the voltage on the remote gate to maintain a constant current through the SET, and this voltage is recorded. Any deviations from linearity in the feedback voltage as a function of the local gate voltage represent additional charges moving near the SET. As shown in Fig. 2, we see differences from linearity in P ion-implanted samples compared with non-implanted samples, deviations that we interpret as donor electrons being driven toward the SET island.

The second type of measurement was motivated by a desire to look for controlled switching of two-level systems in these samples. We begin by applying 200 Hz sine waves, one to each gate, and 180° out of phase with each other. We monitor the in-phase component of the SET drain-source current at 200 Hz with a lock-in amplifier and adjust the applied ac amplitudes so that there is no measureable response. This means that the applied 200 Hz signals have no effect on the local potential of the SET island, yet they should not cancel in other areas, for example near the implanted donors. The voltage on the local gate is ramped in the same way as in the dc measurement, using feedback on the remote gate to keep a constant SET current. The feedback loop bandwidth is much smaller than 200 Hz, so that the feedback is unaffected by the ac potentials even in the absence of complete cancellation. As the dc voltage on the local gate is increased we see reproduceable peaks in the measured 200 Hz response as shown in Fig. 3. We interpret these peaks as charge moving in phase with the 200 Hz potentials as two states are brought into resonance by the applied electric field. However, these features appear in both implanted and intrinsic samples, so we do not believe that most of them are coming from donors.

This work was supported by ARDA and the NSA.

REFERENCES

1. Kane, B. E., *Nature*, **393**, 133 (1998).
2. Kane, B. E., et al., *Phys. Rev. B*, **61**, 2961 (2000).
3. Friesen, M., et al., *Phys. Rev. Lett.*, **92**, 037901 (2004).
4. Hollenberg, L. C. L., et al., *Phys. Rev. B*, **69**, 233301 (2004).
5. Schoelkopf, R. J., et al., *Science*, **280**, 1238 (1998).
6. Martins, A. S., Capaz, R. B., and Koiller, B., *Phys. Rev. B*, **69**, 85320 (2004).
7. Smit, G. D. J., et al., *Phys. Rev. B*, **68**, 193302 (2003).

Manipulation of Charge States of an Isolated Silicon Double Quantum Dot through Microwave Irradiation

Emir Emiroglu*, David Hasko*, David Williams[†]

*Microelectronics Research Centre, Cavendish Laboratory, University of Cambridge, Madingley Road, Cambridge, CB3 0HE United Kingdom.
[†]Hitachi Cambridge Laboratory, Hitachi Europe Ltd., Cavendish Laboratory, Madingley Road, Cambridge CB3 0HE, United Kingdom.

Abstract. We report results from low-temperature electron transport measurements conducted on a single quantum dot single-electron transistor that is capacitively coupled to an isolated double quantum dot and in-plane electrical gates for the application of DC and microwave fields. We observe photon assisted tunneling of an electron within the double dot at several frequencies. These resonant frequencies are demonstrated to be tunable by the static electric fields, and have high quality factors, indicating long lifetimes in the excited charge states.

INTRODUCTION

An isolated double quantum dot (IDQD) is a system which is completely decoupled from external elements. It is expected that this decoupling from the environment will result in long lifetimes of excited electronic states for confined electrons that occupy the eigenstates of the double well system. An indirect measurement of electron excited state lifetimes in IDQD systems in silicon is through spectroscopy. In this work, we present experimental evidence for photon assisted tunneling within the IDQD, measured in the DC current of a capacitively coupled single-electron transistor (SET).

DEVICE LAYOUT AND RESULTS

The layout of the device used in experiments is similar to a design already demonstrated [1], and is schematically illustrated in Fig. 1. All device elements are patterned on chip using high resolution electron-beam lithography and reactive-ion etching. This results in the so called trench-isolated profile. The ~25 nm wide constrictions that connect the quantum dots act as tunnel barriers for the conduction electrons due to surface depletion. The quantum dot connected to leads is used as a single-electron transistor and tuned to a high-sensitivity region. The SET current is directly influenced by the charge state of the IDQD. This results in a noninvasive read-out mechanism of the IDQD charge state. The electrochemical potentials of all quantum dots can be influenced by the DC control gates (gates 1 to 4), while continuous wave high-frequency signals are carried down through the RF gate.

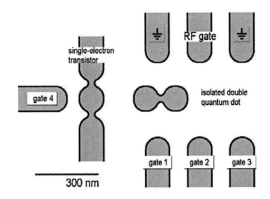

FIGURE 1. Device schematic. The active (Si) region of each quantum dot in practice is around 30 nm^3, although their planar physical diameter is around 70 nm because of the oxidation step in processing which grows additional oxide while reducing the Si region dimensions.

The RF gate is patterned as a coplanar waveguide up to the room temperature electronics. All measurements are taken with the sample immersed in liquid helium at an ambient temperature of 4.2K, so that the device is operating in the Coulomb blockade regime. The SET is operated in the linear transport regime with a source-drain potential of $V_{sd} = 150$ μV. The potential on gate 4 is adjusted so that the SET maintains a constant DC current near half the maximum conductance, where dI/dV_g is maximized. The sample is then irradiated with a continuous microwave field at a constant input power of -20 dBm, although the amplitude of the signal at the IDQD is likely to be less due to attenuation and imperfect matching along the high frequency network.

Figure 2 shows a measurement where the applied frequency f to the RF gate is swept between 1.878 GHz and 1.886 GHz. We observed three highly repeatable absorption features within this range all of which resulted in a sharp change in the SET current. The voltage on gate 3 was tuned at $V_{g3} \sim 1.90$ V, so that an electric field component along the IDQD axis was already present. Through the absorption of a photon of the appropriate energy, an electron that is initially localized on one of the quantum dots of the IDQD may be excited to an occupation level where the probability of tunneling to the other quantum dot is greatly increased. When the electron tunnels to the other quantum dot, the SET electrostatic energy is modified by an approximate effective gating term e/C_d, where C_d is the capacitance between the SET and the IDQD. This results in a change in the SET current. The energies needed to overcome tunneling correspond to ~ 7 μeV.

This result, considering the thermal energy of 362 μeV, is indicative of a strong decoupling between electrons in this system and the relaxation processes that arise from the surrounding environment. Additional evidence of the origin of the observed peaks is obtained when the experiment is repeated at slightly different gate voltages, which should change the energy difference between the electrochemical potentials of the two dots of the IDQD. As expected, we observed a shift in the center frequency of the peaks at different voltages on gate 3. The SET current was also shifted along its Coulomb oscillation due to the nonzero direct coupling between the SET and gate 3. We also verified that changing the voltage on gate 4 hardly influenced the peak positions (data not shown). This strongly suggests that the observed features are indeed a result of single electron polarization of the IDQD through photon assisted tunneling.

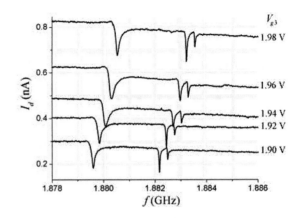

FIGURE 2. Peaks in the current through the single-electron transistor as a function of the irradiation frequency and the voltage applied to gate 3. Each trace is taken at a slightly different voltage on gate 3. The high quality factor of $\sim 10^5$ for some of the observed peaks indicates a lifetime of ~ 100 μs. The peaks are shifted in energy as the voltage on gate 3 is incremented.

CONCLUSION

We have presented experimental evidence for photon assisted electron tunneling in a silicon isolated double quantum dot structure through read-out of the DC current of a capacitively-coupled single-electron transistor. We have found that the photon energy required to stimulate the tunneling process is tunable by the applied gate electric fields.

ACKNOWLEDGMENTS

The authors would like to acknowledge partial support from the LINK project *Nanoelectronics at the Quantum Edge*.

REFERENCES

1. E. G. Emiroglu, D. G. Hasko, and D. A. Williams, Appl. Phys. Lett. 83, 3942 (2003).

Silicon-based spin quantum computation and the shallow donor exchange gate

Belita Koiller*, R. B. Capaz*, Xuedong Hu† and S. Das Sarma**

*Instituto de Física, Universidade Federal do Rio de Janeiro, Cx. P. 68528, 21941-972 Rio de Janeiro, Brazil
†Department of Physics, University at Buffalo, the State University of New York, Buffalo, NY 14260-1500
**Condensed Matter Theory Center, Department of Physics, University of Maryland, MD 20742-4111

Abstract. Proposed silicon-based quantum-computer architectures have attracted attention because of their promise for scalability and their potential for synergetically utilizing the available resources associated with the existing Si technology infrastructure. Electronic and nuclear spins of shallow donors (e.g. phosphorus) in Si are ideally suited candidates for qubits in such proposals, where shallow donor exchange gates are frequently invoked to perform two-qubit operations. An important potential problem in this context is that intervalley interference originating from the degeneracy in the Si conduction-band edge causes fast oscillations in donor exchange coupling, which imposes significant constraints on the Si quantum-computer architecture. We discuss the theoretical origin of such oscillations. Considering two substitutional donors in Si, we present a systematic statistical study of the correlation between relative position distributions and the resulting exchange distributions.

INTRODUCTION

As semiconductor devices decrease in size, their physical properties tend to become increasingly sensitive to the actual configuration of dopant substitutional impurities [1]. A striking example is the proposal of donor-based silicon quantum computer (QC) by Kane [2], in which the monovalent ^{31}P impurities in Si are the fundamental quantum bits (qubits). This intriguing proposal has created considerable recent interest in revisiting all aspects of the donor impurity problem in silicon, particularly in the Si:^{31}P system.

Two-qubit operations for the donor-based Si QC architecture, which are required for a universal QC, involve precise control over electron-electron exchange[2, 3, 4] and electron-nucleus hyperfine interactions (for nuclear spin qubits). Such control can presumably be achieved by fabrication of donor arrays with accurate positioning and surface gates whose potential can be precisely controlled [5, 6, 7, 8]. However, electron exchange in bulk silicon has spatial oscillations[9] on the atomic scale due to the valley interference arising from the particular six-fold degeneracy of the bulk Si conduction band. These oscillations place heavy burdens on device fabrication and coherent control [10], because of the very high accuracy requirement for placing each donor inside the Si unit cell, and/or for controlling the external gate voltages.

The potentially severe consequences of these problems for exchange-based Si QC architecture motivated us and other researchers to perform further theoretical studies, going beyond some of the simplifying approximations in the formalism adopted in Ref. [10], and incorporating perturbation effects due to applied strain[11] or gate fields [12]. These studies, performed within the standard Heitler-London (HL) formalism [13], essentially reconfirm the originally reported difficulties regarding the sensitivity of the electron exchange coupling to donor positioning, indicating that these may not be completely overcome by applying strain or electric fields. The sensitivity of the calculated exchange coupling to donor relative position originates from interference between the plane-wave parts of the six degenerate Bloch states associated with the Si conduction-band minima. More recently [14] we have assessed the robustness of the HL approximation for the two-electron donor-pair states by relaxing the phase pinning at donor sites, which could in principle eliminate the oscillatory exchange behavior. Within this more general theoretical scheme, the *floating-phase* HL approach, our main conclusion is that, for all practical purposes, the previously adopted HL wavefunctions are robust, and the oscillatory behavior obtained in Refs. [10, 11, 12] cannot be taken as an artifact.

SINGLE DONOR

We describe the single donor electron ground state using effective mass theory. The bound donor electron Hamiltonian for an impurity at site $\mathbf{R}_0$ is written as

$$H_0 = H_{SV} + H_{VO}. \qquad (1)$$

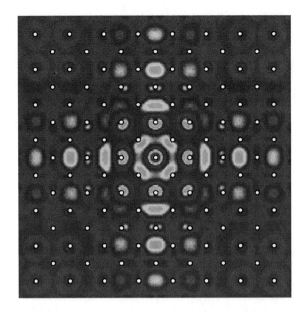

FIGURE 1. Electron probability density on the (001) plane of bulk Si for the ground state of a donor in Si within the envelope function approximation. The white dots give the in-plane atomic sites. The grey-scale scheme runs from light (high density) to dark (low density). The highest density is at the central site, corresponding to the impurity site. A figure in color is given in Ref. [16]

The first term, H_{SV}, is the single-valley Kohn-Luttinger Hamiltonian [15], which includes the single particle kinetic energy, the Si periodic potential, and the impurity screened Coulomb perturbation potential

$$V(\mathbf{r}) = -\frac{e^2}{\varepsilon|\mathbf{r} - \mathbf{R}_0|}. \quad (2)$$

For shallow donors in Si, we use the static dielectric constant $\varepsilon = 12.1$. The second term of Eq. (1), H_{VO}, represents the inter-valley scattering effects due to the presence of the impurity which breaks the bulk translational symmetry.

The electron eigenfunctions are written on the basis of the six unperturbed Si band edge Bloch states $\phi_\mu = u_\mu(\mathbf{r})e^{i\mathbf{k}_\mu \cdot \mathbf{r}}$ [recall that the conduction band of bulk Si has six degenerate minima ($\mu = 1,\ldots,6$), located along the Γ–X axes of the Brillouin zone at $|\mathbf{k}_\mu| \sim 0.85(2\pi/a)$ from the Γ point]:

$$\psi_{\mathbf{R}_0}(\mathbf{r}) = \frac{1}{\sqrt{6}} \sum_{\mu=1}^{6} F_\mu(\mathbf{r} - \mathbf{R}_0) u_\mu(\mathbf{r}) e^{i\mathbf{k}_\mu \cdot (\mathbf{r} - \mathbf{R}_0)}. \quad (3)$$

The phases of the plane-wave part of all band edge Bloch states are naturally chosen to be pinned at $\mathbf{R}_0$: In this way the charge density at the donor site [where the donor perturbation potential Eq. (2) is more attractive] is maximum, thus minimizing the energy for $\psi_{\mathbf{R}_0}(\mathbf{r})$.

In Eq. (3), $F_\mu(\mathbf{r} - \mathbf{R}_0)$ are envelope functions centered at $\mathbf{R}_0$, for which we adopt the anisotropic Kohn-Luttinger form, e.g., for $\mu = z$, $F_z(\mathbf{r}) = \exp\{-[(x^2 + y^2)/a^2 + z^2/b^2]^{1/2}\}/\sqrt{\pi a^2 b}$. The effective Bohr radii a and b are variational parameters chosen to minimize $E_{SV} = \langle \psi_{\mathbf{R}_0} | H_{SV} | \psi_{\mathbf{R}_0} \rangle$, leading to $a = 25$ Å, $b = 14$ Å and $E_{SV} \sim -30$ meV when recently measured effective mass values are used in the minimization [10]. The periodic part of each Bloch function is pinned to the lattice, independent of the donor site.

The H_{SV} ground state is six-fold degenerate. This degeneracy is lifted by the valley-orbit interactions [17], which are included here in H_{VO}, leading to the nondegenerate (A_1-symmetry) ground state in (3). Fig. 1 gives the charge density $|\psi_{\mathbf{R}_0}(\mathbf{r})|^2$ for this state, where the periodic part of the conduction band edge Bloch functions were obtained from *ab-initio* calculations, as described in Ref. [14]. The impurity site $\mathbf{R}_0$, corresponding to the higher charge density, is at the center of the frame. It is interesting that, except for this central site, regions of high charge concentration and atomic sites do not necessarily coincide, because the charge distribution periodicity imposed by the plane-wave part of the Bloch functions is $2\pi/k_\mu$, incommensurate with the lattice period.

The oscillatory behavior of the single donor wave functions in Si, illustrated in Fig. 1, is well established experimentally [18] and theoretically [14, 19]. This behavior does not bring significant consequences for conventional applications in Si-based devices (n-doped Si). A recent study of the single-qubit operations (A-gate) in the Kane QC shows that the A-gate operations do not present additional complications due to the Si band structure interference effects [20].

DONOR PAIR EXCHANGE COUPLING

The HL approximation is a reliable scheme for the well-separated donor pair problem (interdonor distance much larger than the donor Bohr radii) [13]. Within HL, the lowest energy singlet and triplet wavefunctions for two electrons bound to a donor pair at sites $\mathbf{R}_A$ and $\mathbf{R}_B$, are written as properly symmetrized combinations of $\psi_{\mathbf{R}_A}$ and $\psi_{\mathbf{R}_B}$ [as defined in Eq.(3)]

$$\begin{aligned}\Psi_t^s(\mathbf{r}_1, \mathbf{r}_2) &= \frac{1}{\sqrt{2(1 \pm S^2)}} [\psi_{\mathbf{R}_A}(\mathbf{r}_1)\psi_{\mathbf{R}_B}(\mathbf{r}_2) \\ &\pm \psi_{\mathbf{R}_B}(\mathbf{r}_1)\psi_{\mathbf{R}_A}(\mathbf{r}_2)],\end{aligned} \quad (4)$$

where S is the overlap integral and the upper (lower) sign corresponds to the singlet (triplet) state. The energy expectation values for these states, $E_t^s = \langle \Psi_t^s | H | \Psi_t^s \rangle$, gives the exchange splitting through their difference, $J = E_t - E_s$. We have previously derived the expression for

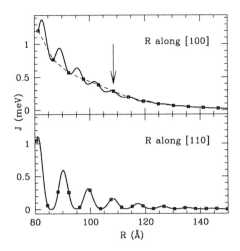

FIGURE 2. Exchange coupling between two phosphorus donors in Si along the indicated directions in the diamond structure. Values appropriate for impurities at substitutional sites are given by the squares. The dashed line in the $R\|[100]$ frame is a guide to the eye, indicating that the oscillatory behavior may be ignored for donors positioned exactly along this axis.

the donor electron exchange splitting [11, 14], which we reproduce here:

$$J(\mathbf{R}) = \frac{1}{36} \sum_{\mu,\nu} j_{\mu\nu}(\mathbf{R}) \cos(\mathbf{k}_\mu - \mathbf{k}_\nu) \cdot \mathbf{R}, \quad (5)$$

where $\mathbf{R} = \mathbf{R}_A - \mathbf{R}_B$ is the interdonor position vector and $j_{\mu\nu}(\mathbf{R})$ are kernels determined by the envelopes and are slowly varying [10, 11]. Note that equation (5) does not involve any oscillatory contribution from $u_\mu(\mathbf{r})$, the periodic part of the Bloch functions [12, 14]. The physical reason for that is clear from (3): While the plane-wave phases of the Bloch functions are pinned to the donor sites, leading to the cosine factors in (5), the periodic functions u_μ are pinned to the lattice, regardless of the donor location.

The exchange energy calculated from Eq. (5) for a pair of donors as a function of their relative position along the [100] and [110] crystal axis is given in Fig. 2. This figure vividly illustrates both the anisotropic and the oscillatory behavior of $J(\mathbf{R})$, which is well established from previous studies [9, 10, 11, 12]. It is interesting to note that for substitutional donors with interdonor position vectors exactly aligned with the [100] crystal axis, the oscillatory behavior may be ignored in practice, as indicated by the dashed line in the figure. This behavior is qualitatively similar to the exchange versus donor separation dependence assumed in Kane's proposal [2], where the Herring and Flicker expression [21], originally derived for H atoms, was adapted for donors in Si. Therefore one might expect that reliable exchange gates operation would be possible if donors are exactly aligned along the [100] crystal axis.

NANOFABRICATION ASPECTS

Aiming at the fabrication of a P donor array accurately positioned along the [100] axis, and given the current degree of control in substitutional P positioning in Si of a few nm [5, 6, 7, 8], we investigate the consequences of interdonor positioning uncertainties in the values of the corresponding pairwise exchange coupling. We define the *target* interdonor position $\mathbf{R}_t$ along [100], with an arbitrarily chosen length of 20 lattice parameters (~ 108.6 Å) indicated by the arrow in Fig. 2. The distributions for the interdonor distances $R = |\mathbf{R}_A - \mathbf{R}_B|$ when $\mathbf{R}_A$ is fixed and $\mathbf{R}_B$ "visits" all of the diamond lattice sites within a sphere centered at the *target* position are given in Fig. 3. Different frames give results for different uncertainty radii, and, as expected, increasing the uncertainty radius results in a broader distribution around the *target* distance. Note that the geometry of the lattice implies that the distribution is always centered and peaked around R_t, as indicated by the arrows. The additional peaks in the distribution reveal the discrete nature of the Si lattice.

The respective distributions of exchange coupling between the same donor pairs in each *ensemble* is presented in Fig. 4, where the arrows give the exchange value at the *target* relative position: $J(\mathbf{R}_t) \sim 0.29$ meV. The results here are qualitatively different from the distance distributions in Fig. 3, since they are neither centered nor peaked at the *target* exchange value. Even for the smallest uncertainty radius of 1 nm in (a), the exchange distribution is peaked around $J \sim 0$, bearing no semblance to the inter-donor distance distributions. Increasing the uncertainty radius leads to a wider range of exchange values, with a more pronounced peak around the lowest J values.

From the perspective of current QC fabrication efforts, ~ 1 nm accuracy in single P atom positioning has been recently demonstrated [6], representing a major step towards the goal of obtaining a regular donor array embedded in single crystal Si. Distances and exchange coupling distributions consistent with such accuracy are presented in Figs.3(a) and 4(a) respectively. The present calculations indicate that such deviations in the relative position of donor pairs with respect to perfectly aligned substitutional sites along [100] lead to order-of-magnitude changes in the exchange coupling, favoring $J \sim 0$ values. Severe limitations in controlling J would come from "hops" into different substitutional lattice sites. Therefore, precisely controlling exchange gates in Si

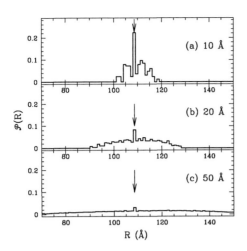

FIGURE 3. Interdonor distance distributions for a *target* relative position of 20 lattice parameters along [100] (see arrows). The first donor is fixed and the second one "visits" all of the Si substitutional lattice sites within a sphere centered at the *target* position, with uncertainty radii (a)10 Å, (b)20 Å and (c)50 Å.

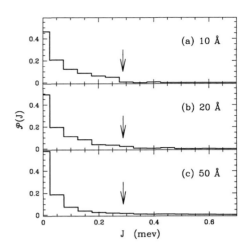

FIGURE 4. Exchange distributions for the same relative position *ensembles* in Fig. 3. The arrow indicates the *target* situation. Contrary to the distance distributions, the exchange distributions are not centered or peaked around the target value.

remains an open challenge.

ACKNOWLEDGMENTS

This work was partially supported by CNPq and by Instituto do Milênio de Nanociências in Brazil, by ARDA and LPS at the University of Maryland and by ARDA and ARO at the University at Buffalo.

REFERENCES

1. Voyles, P., Muller, D., Grazui, J., Citrin, P., and Grossmann, H.-J., *Nature*, **416**, 827 (2002).
2. Kane, B., *Nature*, **393**, 133 (1998).
3. Vrijen, R., Yablonovitch, E., Wang, K., Jiang, H., Balandin, A., Roychowdhury, V., Mor, T., and DiVincenzo, D., *Phys. Rev. A*, **62**, 012306 (2000).
4. Hu, X., and Das Sarma, S., *Phys. Rev. A*, **61**, 062301 (2000).
5. O'Brien, J. L., Schofield, S. R., Simmons, M. Y., Clark, R. G., Dzurak, A. S., Curson, N. J., Kane, B. E., McAlpine, N. S., Hawley, M. E., and Brown, G. W., *Phys. Rev. B*, **64**, 161401 (2001).
6. Schofield, S. R., Curson, N. J., Simmons, M. Y., Rueß, F. J., Hallam, T., Oberbeck, L., and Clark, R. G., *Phys. Rev. Lett.*, **91**, 136104 (2003).
7. Buehler, T. M., McKinnon, R. P., Lumpkin, N. E., Brenner, R., Reilly, D. J., Macks, L. D., Hamilton, A. R., Dzurak, A. S., and Clark, R. G., *Nanotechnology*, **13**, 686 (2002).
8. Schenkel, T., Persaud, A., Park, S. J., Nilsson, J., Bokor, J., Liddle, J. A., Keller, R., Schneider, D. H., Cheng, D. W., and Humphries, D. E., *J. Appl. Phys.*, **94**, 7017 (2003).
9. Andres, K., Bhatt, R. N., Goalwin, P., Rice, T., and Walstedt, R., *Phys. Rev. B*, **24**, 244 (1981).
10. Koiller, B., Hu, X., and Das Sarma, S., *Phys. Rev. Lett.*, **88**, 027903 (2002).
11. Koiller, B., Hu, X., and Das Sarma, S., *Phys. Rev. B*, **66**, 115201 (2002).
12. Wellard, C. J., Hollenberg, L. C. L., Parisoli, F., Kettle, L., Goan, H.-S., McIntosh, J. A. L., and Jamieson, D. N., *Phys. Rev. B*, **68**, 195209 (2003).
13. Slater, J. C., *Quantum Theory of Molecules and Solids*, vol. 1, McGraw-Hill, New York, 1963.
14. Koiller, B., Capaz, R. B., Hu, X., and Das Sarma, S., *arXiv:cond-mat/0402266* (2004), to appear in Phys. Rev. B.
15. Kohn, W., *Solid State Physics Series*, vol. 5, Academic Press, 1957, edited by F. Seitz and D. Turnbull, p.257, and references therein.
16. Koiller, B., Capaz, R. B., Hu, X., and Das Sarma, S., *arXiv:cond-mat/0407183* (2004).
17. Pantelides, S. T., *Rev. Mod. Phys.*, **50** (1978).
18. Feher, G., *Phys. Rev. B*, **114**, 1219 (1959).
19. Overhof, H., and Gerstmann, U., *Phys. Rev. Lett.*, **92**, 087602 (2004).
20. Martins, A. S., Capaz, R. B., and Koiller, B., *Phys. Rev. B*, **69**, 085320 (2004).
21. Herring, C., and Flicker, M., *Phys. Rev.*, **134**, A362 (1964).

Coherent control of tunneling in a quantum dot array

J. M. Villas-Bôas[*,†], Sergio E. Ulloa[*] and Nelson Studart[†]

[*]*Department of Physics and Astronomy, and Nanoscale and Quantum Phenomena Institute, Ohio University, Athens, Ohio 45701-2979, USA*
[†]*Departamento de Física, Universidade Federal de São Carlos, 13565-905, São Carlos, São Paulo, Brazil*

Abstract. In this theoretical work we demonstrate that it is possible to use an ac electric field to coherently control the position of an electron in a finite-size quantum-dot array or stack. By tuning the amplitude (or frequency) of the external ac electric field we can selectively suppress the tunneling between dots, trapping the particle in a chosen region of the array. The dynamics is obtained non-perturbatively using Floquet theory and employing a one-band nearest-neighbor tight-biding approximation (NNTB) within the dipole interaction for the external field.

Semiconductor quantum dots (QDs) have been shown to be excellent systems to implement basic ideas of quantum system control. A good example of such control is the well known coherent destruction of tunneling (CDT) of a driven two-level system [1, 2, 3], in which the tunneling of one particle in a symmetric double well potential is suppressed for special values of frequency and field intensity where the quasienergy spectrum presents crossings. This was predicted to happen in quantum wells at high frequency ac electric field (high compared to the interwell tunneling splitting). After this initial prediction numerous other works have been published to support and/or extend this idea (see for example [3] and references therein). We have recently shown that the CDT can also occur at lower frequency but with a lower degree of localization [4], and that this result is highly dependent on the phase of the driving field [5]. We have also shown that a new kind of CDT is expected in a finite QD array at level anticrossings [6].

In this work we make use of CDT in conjunction with our latest predictions, in order to effectively *manipulate* the electron *position* in a QD array. These two distinct phenomena occur at different field amplitudes and produce CDT in different barriers, resulting in a system where one can selectively suppress the tunneling between individual quantum dots by tuning the external ac electric field. By suitable variation of frequency and applied ac field amplitude, one can *precise the location* of one electron in the multidot array/stack. The dynamics is analyzed using Floquet theory and direct integration of the time-dependent Schrödinger equation, using a one-band nearest-neighbor tight-biding approximation (NNTB), and dipole coupling with the external field.

The Hamiltonian for an electron in an array of identical QDs under ac field within NNTB is written as

$$H = \sum_j T_e(a_{j+1}^\dagger a_j + h.c.) + \sum_j eFdj a_j^\dagger a_j \sin(\omega t), \quad (1)$$

where T_e is the hopping matrix element, $a_j^\dagger$ (a_j) is the electron creation (annihilation) operator in dot j, e the electronic charge, F the field intensity, d the separation between dots, and ω the field frequency.

Since H is periodic in time [$H(t) = H(t+\tau)$, where $\tau = 2\pi/\omega$ is the period] we can make use of the standard Floquet theory [3, 4] and write the solutions of the time-dependent Schrödinger equation as $\psi(t) = \exp(-i\varepsilon t/\hbar)u(t)$, where $u(t)$, the so-called Floquet state, is also periodic in time with the same period τ, and ε is the Floquet characteristic exponent or quasienergy, which can be obtained from the eigenvalue equation

$$\left(H - i\hbar\frac{\partial}{\partial t}\right)u(t) = \varepsilon u(t). \quad (2)$$

Figure 1 shows the collapse region of the quasienergy spectrum as function of the field intensity for the ratio $T_e/\hbar\omega = 0.2$. We can see a strong repulsion between levels one and three and two and four, which results in localized eigenstates, as one can see in the lower inset. There we show the probability density as a function of the dot position for all eigenstates, ordered from bottom to top, for the corresponding field intensity indicated by the arrow pointing down. These localized eigenstates exhibit a "trapped" electron either in the middle or in the borders of the QD array for such field. To prove this statement in Fig. 2(a) we show the occupation probability of each dot and its time evolution, assuming that we start the system with the electron in dot 2. We can see that the

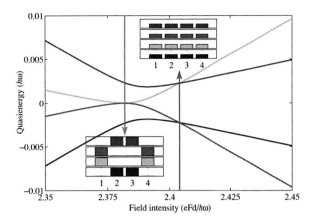

FIGURE 1. Quasienergy collapse region in the first Brillouin zone (in units of $\hbar\omega$) as function of ac field intensity ($eFd/\hbar\omega$), for $T_e/\hbar\omega = 0.2$. Lower inset shows spatial component of the probability density of Floquet eigenstates for the field intensity given by arrow pointing down (anticrossing). Notice that eigenstates are strongly localized either in the middle or in the outer dots. Upper inset is for field intensity given by arrow pointing up (crossing) and shows that eigenstates are equally distributed.

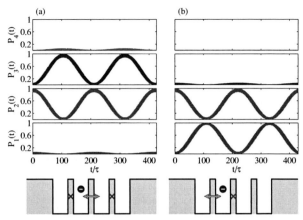

FIGURE 2. (a) Time evolution of probability to find one particle in each one of the QDs in a four-dot array for $eFd/\hbar\omega \simeq 2.38$ (anticrossing), and starting the system with particle in dot 2. Lower panel is schematic representation of dynamics of the system, as seen in the time evolution of the occupation probability of each dot. The electron represents the position of the particle at time zero, crosses indicate the suppression of tunneling through that barrier and arrows indicate that tunneling is possible. (b) Same analysis for $eFd/\hbar\omega \simeq 2.4$ (crossing).

electron is trapped in dots 2 and 3, meaning that the outer barriers are closed for tunneling. The bottom cartoon is a schematic representation of the electron dynamics for this initial condition, where a cross in the barrier means that the tunneling is not allowed, and an arrow means that the electron can tunnel back and forth through this barrier.

Figure 1 also shows a level crossing at the field intensity given by the arrow pointing up. The probability density for the eigenstates at this field present completely different behavior, as we can see in the upper inset. The eigenstates are all equally distributed in the dot array. A simple analysis of this situation, assuming the initial condition to be the electron in dot 1, gives $P_1(t) = \cos^2 \frac{1}{2}\Delta t$, $P_2(t) = \sin^2 \frac{1}{2}\Delta t$, and $P_3(t) = P_4(t) = 0$ where $\Delta = (\varepsilon_1 + \varepsilon_2) = -(\varepsilon_3 + \varepsilon_4)$. Similar results can be obtained assuming dot 4 as the initial location, so that the electron is localized in either pair of dots, 1 and 2 or 3 and 4, such that the middle barrier is closed for tunneling. The occupation probability for this choice of field can be seen in Fig. 2(b) and its respective dynamics representation is shown in the lower panel.

These examples represent an exciting result since by simply tuning the field intensity from the anticrossing point ($eFd/\hbar\omega \simeq 2.38$) to the crossing point ($eFd/\hbar\omega \simeq 2.4$), one can choose which barrier is "open" or "blocked" for tunneling. This allows us to manipulate the electron position in an array of dots. For example, suppose we start (or manipulate the system to start) with the electron in dot 4, then, tuning the field to $eFd/\hbar\omega \simeq 2.4$ will make the electron tunnel back and forth between dots 3 and 4. When the electron is in dot 3 (after a time $t \simeq 100\tau$), we switch the field to $eFd/\hbar\omega \simeq 2.38$, which opens the middle and closes the outer barrier. The electron is then trapped in the middle of the array, oscillating between dots 2 and 3 [as in Fig. 2(a)]. Switching back the field to $eFd/\hbar\omega \simeq 2.4$ when the electron is in dot 2 would allow it to oscillate between dot 1 and 2 [as in Fig. 2(b)]. When the electron is in dot 1 we can again tune the field to $eFd/\hbar\omega \simeq 2.38$ and trap the electron there. This sequential application of ac field amplitudes after suitable switching times would allow for a controlled "electron-shuttle" in the QD array.

In conclusion, we have shown that one can selectively suppress the tunneling between quantum dots by tuning the intensity of the applied ac field. With this tool one could manipulate the position of the particle in a quantum-dot array or stack, and assist in the initialization and control of a possible charge qubit.

We thank support of FAPESP-Brazil, US DOE grant no. DE–FG02–91ER45334, and the Indiana 21st Century Fund.

REFERENCES

1. Grossmann, F., Dittrich, T., Jung, P., and Hänggi, P., *Phys. Rev. Lett.*, **67**, 516 (1991).
2. Bavli, R., and Metiu, H., *Phys. Rev. A*, **47**, 3299 (1993).
3. Grifoni, M., and Hänggi, P., *Phys. Rep.*, **304**, 229 (1998).
4. Villas-Boas, J. M., Zhang, W., Ulloa, S. E., Rivera, P. H., and Studart, N., *Phys. Rev. B*, **66**, 085325 (2002).
5. Villas-Boas, J. M., Ulloa, S. E., and Studart, N., *unpublished*.
6. Villas-Boas, J. M., Ulloa, S. E., and Studart, N., *Phys. Rev. B*, **70**, 041302 (2004).

Fano resonance in quantum dots with electron-phonon interaction

Akiko Ueda and Mikio Eto

Faculty of Science and Technology, Keio University, 3-14-1 Hiyoshi, Kohoku-ku, Yokohama 223-8522, Japan

Abstract. We theoretically study the nonequilibrium transport properties in an Aharonov-Bohm ring with an embedded quantum dot, in the presence of electron-phonon interaction. The differential conductance is calculated under finite bias voltages, using Keldysh Green function method. By the perturbation of the electron-phonon interaction, we find that the Fano resonance is more suppressed with increasing bias voltage. This bias-voltage dependence of the dephasing effect cannot be obtained using the canonical transformation method of electron-phonon interaction.

INTRODUCTION

Fano resonance has been observed recently in an Aharonov-Bohm (AB) ring with an embedded quantum dot by the transport measurement [1]. The Fano resonance shows an asymmetric line shape with both peak and dip, which is caused by the interference between discrete states (in a quantum dot) and continuum of states (in the ring). The asymmetric line shape disappears under large bias voltages, which implies that the observation of the resonance requires high coherence kept in the whole system. To elucidate the dephasing effect on the Fano resonance, we theoretically study the electron-phonon (e-ph) interaction which is a major dephasing mechanism under a finite bias. The Keldysh Green function method is adopted to examine the nonequilibrium transport properties.

MODEL

The model which we consider is illustrated in the inset to Fig. 2. In a quantum dot, an energy level (ε_0) can be tuned using a gate voltage. There are two paths between external leads, one path connects the leads directly by $We^{-i\varphi}$ and the other path connects the leads through a quantum dot by t_L and t_R. We fix $t_L = t_R$ in this paper. Using the density of states in the leads, v, the coupling strength by the former path is characterized by the transmission probability, $T_b = [2x/(1+x^2)]^2$ with $x = \pi v W$, whereas that by the latter path is characterized by the line broadening of the dot level, $\Gamma = 2\pi v(t_L^2 + t_R^2)$. The conductance is expressed by an extended Fano form with a complex parameter q in the presence of a magnetic flux enclosed in the ring ($\varphi \neq 0$) [2], without e-ph interaction.

We show the conductance at $\varphi = 0$, as a function of the dot level ε_0, by dotted line in Fig. 2.

We consider the e-ph interaction in the quantum dot. The Hamiltonians of the interaction and phonons are written as $H_{e-ph} = \sum_q M_q(a_q + a_{-q}^\dagger)d^\dagger d$, $H_{ph} = \sum_q \omega_q a_q^\dagger a_q$, where $a_q^\dagger$ and a_q ($d^\dagger$ and d) are the creation and annihilation operators of phonon q (an electron in the dot). In quantum dots of size L, phonons with wave number $|q| < 2\pi/L$ effectively couple to electrons.

We adopt the Keldysh Green function method to examine transport properties under finite bias voltages, $eV = \mu_L - \mu_R$ [μ_L (μ_R) is the chemical potential in lead L (R)]. The current is written in terms of the retarded Green function for the quantum dot, $G^r_{dd}(t,t') = -i\theta(t-t')\langle\{d(t),d^\dagger(t')\}\rangle$ [3]. The self-energy in $G^r_{dd}(\omega)$ is calculated to the second order in H_{e-ph}: $\Sigma^r = \Sigma^t + \Sigma^<$ with

$$\Sigma^t(\omega) = \frac{i}{2\pi}\int d\omega' G^{t(0)}_{dd}(\omega - \omega')D^t(\omega'), \quad (1)$$

$$\Sigma^<(\omega) = -\frac{i}{2\pi}\int d\omega' G^{<(0)}_{dd}(\omega - \omega')D^<(\omega'), \quad (2)$$

where $G^{t(0)}_{dd}(\omega)$ and $G^{<(0)}_{dd}(\omega)$ [$D^t(\omega)$ and $D^<(\omega)$] are the time-ordered and lesser Green functions for the dot electrons in the unperturbed system [for the phonons].

CALCULATED RESULTS

We consider longitudinal optical (LO) phonons ($\omega_q = \omega_0$) to observe the effect of phonons clearly. The coupling strength of the e-ph interaction is characterized by $\lambda = \sum M_q^2/\omega_0^2$. The temperature is set to be zero.

First, we examine a resonant tunneling through a quantum dot, neglecting the direct tunnel coupling be-

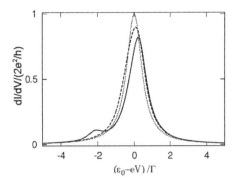

FIGURE 1. Differential conductance as a function of the dot level ε_0, in the case of resonant tunneling through a quantum dot ($T_b = 0$), with LO phonons. The phonon energy is $\omega_0 = 2\Gamma$, whereas the strength of e-ph interaction is $\lambda = 0.8\Gamma$. The bias voltage is $eV = 3.0\Gamma$ (solid line) and $eV = 0.5\Gamma$ (broken line). The conductance in the absence of e-ph interaction is indicated by dotted line.

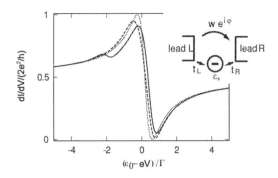

FIGURE 2. Differential conductance as a function of the dot level ε_0, in an AB ring with an embedded quantum dot ($T_b = 0.5$, $\varphi = 0$), with LO phonons. The phonon energy is $\omega_0 = 2\Gamma$, whereas the strength of e-ph interaction is $\lambda = 0.8\Gamma$. The bias voltage is $eV = 3.0\Gamma$ (solid line) and $eV = 0.5\Gamma$ (broken line). The conductance in the absence of e-ph interaction is indicated by dotted line. Inset: Model that we consider for an AB ring with an embedded quantum dot.

tween the leads, $T_b = 0$. In Fig. 1, the differential conductance dI/dV is shown as a function of the dot level ε_0. In the absence of e-ph interaction, the conductance shows a Briet-Wigner resonance (dotted line). The peak height is $2e^2/h$, whereas the peak width is given by Γ. The peak height is suppressed by the e-ph interaction. The suppression is visible even when the bias voltage is smaller than the phonon energy ($eV < \omega_0$; broken line). Although a real process of phonon emission is not possible, electrons can emit and absorb a phonon in a virtual way, which disturbs a coherent transport through the quantum dot ("pure dephasing"). With increasing bias voltage, the suppression of the peak height is more prominent. When $eV > \omega_0$ (solid line), a subpeak appears which corresponds to the phonon emission.

We compare the calculated results by the second-order perturbation with those by the canonical transformation method of the e-ph interaction [4, 5]. The latter method enables us to treat the e-ph interaction exactly (with an approximation for the tunnel couplings). We obtain sublevels of polarons (coherent states consisting of an electron and phonons) in the quantum dot, before considering the tunnel couplings between the dot and leads. Although we can explain the suppression of main peak and the appearance of subpeaks by this method, we cannot obtain the bias-dependence of the conductance. Therefore, the canonical transformation method is not suitable to examine the dephasing effect in the nonequilibrium transport.

In Fig. 2, we present the calculated results for an AB ring with a quantum dot ($T_b = 0.5$, $\varphi = 0$). The e-ph interaction diminishes the amplitude of the Fano resonance. The amplitude is more suppressed under larger bias voltages. When $eV > \omega_0$, a subpeak appears due to the emission of a phonon.

We perform similar calculations for longitudinal acoustic (LA) phonons. We find that the amplitudes of the resonant tunneling and Fano resonance are more suppressed with increasing bias voltage. A subpeak structure does not appear in this case.

CONCLUSIONS

We have examined the nonequilibrium transport properties in an AB ring with an embedded quantum dot, using Keldysh Green function method. The electron-phonon interaction has been taken into account by the second-order perturbation. The suppression of the Fano resonance is enhanced with increasing bias voltage, which cannot be explained by the calculations using the canonical transformation.

This work was partially supported by a Grant-in-Aid for Scientific Research in Priority Areas "Semiconductor Nanospintronics" (Nos. 14076105, 14076216) of the Ministry of Education, Culture, Sports, Science and Technology, Japan.

REFERENCES

1. K. Kobayashi, H. Aikawa, S. Katsumoto, and Y. Iye, Phys. Rev. Lett. **88**, 256806 (2002).
2. A. Ueda, I. Baba, K. Suzuki, and M. Eto, J. Phys. Soc. Jpn. **72**, Suppl. A, 157 (2003).
3. W. Hofstetter, J. König, and H. Schoeller, Phys. Rev. Lett. **87**, 156803 (2001).
4. G. D. Mahan, *Many-Particle Physics* (Plenum Press, New York, 1990).
5. J.-Xin. Zu and A. V. Balatsky, Phys. Rev. B **67**, 165326 (2003).

Electric Field Induced Charge Noise in Doped Silicon: Ionisation of Phosphorus Dopants

A. J. Ferguson, V. Chan, A. R. Hamilton & R. G. Clark

Centre for Quantum Computer Technology, University of New South Wales, Sydney, NSW 2052

Abstract. We present measurements of aluminum single electron transistors fabricated on two silicon substrates of different doping densities. On the lower doped substrate ($n=10^{14}$cm^{-3}) the usual coulomb blockade behaviour is seen. However in the case of the more highly doped material ($n=10^{17}$cm^{-3}) a change in the coulomb oscillation period and a supplementary charge noise are noticed. These effects are attributed to the electric field induced ionisation of phosphorus dopants.

INTRODUCTION

The charge sensitivity of the single electron transistor (SET) allows electrical phenomena to be probed in a wide variety of materials [1,2,3]. We make use of a gate nearby to the SET (~300nm) to apply electric fields in the range of $E \sim$ kVcm^{-1} to a small volume of silicon. The motivation for this work is quantum information processing with individual dopants in semiconductors [4,5]. Coupling between dopants is expected to be controlled by electrostatic gates [6], hence it is important experimentally to understand the effect of electric fields on the system. Two silicon substrates bulk doped with phosphorus are measured. One is deep in the insulating regime ($n=10^{14}$cm^{-3}), with the other more highly doped ($n=10^{17}$cm^{-3}), but beneath the metal-insulator transition ($n_{MIT}=3.45 \times 10^{18}$ cm^{-3}).

FABRICATION AND MEASUREMENT

In order to make a direct comparison both materials undergo identical processing. First, a thermal oxidization process grows 5nm of oxide on the silicon. The Ti/Au metallic gates are then defined with electron beam lithography (EBL). Finally a separate EBL step patterns the SETs in standard bilayer resist. The SETs are then evaporated by double angle evaporation, with the tunnel junctions defined by a 5 min oxidization at an O$_2$ pressure of 30 mTorr.

The measurements are performed in a dilution refrigerator (T~100 mK). A 1T magnetic field is applied in order to quench superconductivity. The electrical measurements are made using an SR830 lock-in amplifier, the oscillator frequency is set to 1 kHz with an amplitude of Vac=20 μV rms.

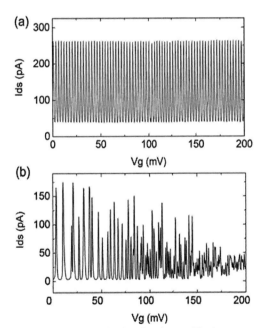

FIGURE 1. (a) Coulomb oscillations on the $n=10^{14}$ cm^{-3} material. (b) The same measurement on the $n=10^{17}$cm^{-3} material showing a much increased noise level and decreased oscillation visibility.

COULOMB OSCILLATIONS

A difference between the two substrates is seen in a gate voltage sweep. The current though the SET is measured while the electrostatic gate's potential is ramped. Figure 1a shows the response to this gate sweep on the $n=10^{14} cm^{-3}$ substrate. As expected for a metal SET, there is a periodic modulation of the current. However for the $n=10^{17} cm^{-3}$ substrate, shown in fig 1b, there is a departure from this behaviour. As two samples were measured on each substrate, and the SETs were made in a well-defined, repeatable process the difference can be attributed to the substrate rather than a variation in the SET itself.

Two main effects are seen for the doped substrate.
  i. There is a period contraction of the oscillations. This is evident in the voltage range up to $V_g=100 mV$.
  ii. Above $V_g=100$ mV a supplementary charge noise may be observed which causes a loss in visibility of the oscillations and, by $V_g=200$ mV, their disappearance.

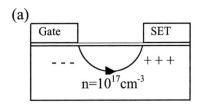

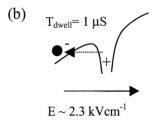

FIGURE 2. (a) Schematic of the device. (b) In a simple WKB approximation an $E= 2.3$ kVcm^{-1} causes ionisation in 1 µS.

The difference between the substrates is explained by the presence of dopants. For the $n=10^{14}$ cm^{-3} substrate, the average dopant spacing is 220 nm and for the $n=10^{17}$ cm^{-3} material it is 22 nm. Hence in a hemisphere diameter 300 nm (roughly the volume of significant electric field between the gate and SET) there are N~6 and N~6000 dopants respectively. Hence, if there is ionisation of Si:P caused by the electric field, there would be just a few events visible for the lower doped material. However for the more highly doped material there will be a significant redistribution of charge as measured by the SET.

Ionisation of the P dopants explains the behaviour of the highly doped substrate [7]. As the gate voltage is made more positive, the electric field experienced by some of the dopants becomes sufficiently high (E~2 kVcm^{-1}) to cause ionisation of the bound electron (see fig 2b for a schematic). These electrons tunnel from the dopant towards the gate where they form a layer of charge. The remaining positive charges are more strongly capacitively coupled to the SET island than the free electrons under the gate due to their location. Hence on ionisation, the SET measures a net positive charge.

CONCLUSIONS

A significant difference between the electric field responses of two silicon substrates with different doping densities has been observed. In the higher doped substrate ($n=10^{17}$ cm^{-3}) an electric field dependent charge noise is observed, consistent with ionisation of the phosphorus dopants.

ACKNOWLEDGMENTS

The authors thank Lloyd Hollenberg and Keith Schwab for helpful discussions. This work was supported by the Australian Research Council, the Australian Government, the U.S. National Security Agency (NSA), the Advanced Research and Development Activity (ARDA) and the Army Research Office (ARO) under contract number DAAD19-01-1-0653.

REFERENCES

1. Ilani, and Yakoby, A., *Science* **292**, 1354 (2001)
2. Lu, W., Ji, Z., Pfeiffer, L., West, K. W., and Rimberg, A. J. *Nature* **423**, 422 (2003)
3. Zimmerli, G., Eiles, T. M., Kautz, R. L., and Martinis, J. M., *Appl. Phys. Lett.* **61**, 237 (1992)
4. Kane, B. E, *Nature* **393**, 133 (1998)
5. Vrijen, R., Yablonovitch, E., Wang, K. , Jiang, H. W., Balandin, A., Roychowdhury, V., Mor, T., and D. DiVincenzo *Phys. Rev. A* **62**, 012306
6. Martins, A. S., Capaz, R. B., and Koiller, B., *Phys. Rev. B* **69**, 085320 (2004)
7. Smit, G. D. J., Rogge, S., Caro, J., and Klapwijk, T. M. *Phys. Rev. B* **68**, 193302 (2003)

Exciton Coherence Times and Linewidths in InGaAs Quantum Dots

S. Rudin[1] and T. L. Reinecke[2]

[1] *U.S. Army Research Laboratory, Adelphi, Maryland 20783, USA*

[2] *Naval Research Laboratory, Washington, D.C. 20375, USA*

Abstract. We have evaluated the contributions of several exciton-acoustic-phonon scattering mechanisms to the homogeneous linewidths of excitons in self assembled InGaAs quantum dots. These contributions determine the temperature dependence of the linewidths. We have considered both inelastic phonon scatterings and also lattice relaxation effects. We have expressed the exciton linewidth in terms of the phonon linewidths and have evaluated the effects of all phonon lifetime mechanisms, including anharmonic phonon effects, impurity scattering, alloy fluctuations, and boundary and interface phonon scattering in the mesa structures. We have found that only the boundary phonon scattering on the mesa produces linewidth of the magnitude needed to account for experiment. Our results are consistent with the mesa size dependence of the linewidths observed in recent experiments.

Exciton decoherence in quantum dots (QD's) is a key limitation in their use in implementations for quantum computing. The interaction of acoustic phonons with excitons in QD's have been found to control the exciton decoherence at low temperatures [1,2,3]. There is also recent evidence that the corresponding homogeneous exciton linewidths in single QD's depend on the size of mesas used to isolate the dots [4]. To date, however, there is no quantitative understanding of exciton decoherence in QD's.

We consider dots with cylindrical symmetry, and the Coulomb correlation in the electron-hole pair is included through variational wave-functions for their ground and excited states. First, we have evaluated in detail the inelastic acoustic phonon scattering contribution to the homogeneous exciton linewidth, and we find that it is orders of magnitude too small to explain the experimental observations [1-4]. Second, we have considered the effects of finite phonon lifetimes on exciton linewidth. This was done using an independent boson model of the interaction of localized electron-hole states with acoustic phonons [5,6].

In the subspace of the electron-hole pairs the interaction can be written in terms of phonon and confined exciton operators a and B. To the linear order in phonon operators,

$$H_{int} = \sum_{n,m} \hat{B}_n^+ \hat{B}_m \sum_{q,\sigma} M_{nm}(q,\sigma)(\hat{a}_{q,\sigma} + \hat{a}^+_{-q,\sigma}), \quad (1)$$

M_{nm} is the interaction matrix element. The summations are over the bound states of the interacting electron-hole pair in the quantum dot and the wavevector and polarization of bulk phonons. If only the diagonal part of the coupling is retained, one obtains an exactly solvable model. We will treat the off-diagonal part as a perturbation. Following reference 6, we will also take into an account the finite lifetime of the phonons and explicitly consider various phonon lifetime limiting processes.

The absorption spectrum can be obtained as a Fourier transformation of a one-particle time-dependent exciton Green's function, which gives the interband polarization. Using the method of linked cluster expansion, one can easily generalize the treatment in reference 5 to include finite phonon lifetime in the derivation of the electronic correlation functions. At lower temperatures we consider the interaction of electrons and holes with acoustic phonons only. We included the band deformation potential interaction, the piezoelectric interaction, and the electron-phonon interaction produced by the phonon induced deformation of the confining potential. At low temperature the spectrum has a form of a "zero-phonon" line plus a broad background from many-phonon transitions. The zero-phonon line is an optical analogue of the Mössbauer line in the impurity nuclei transitions [6].

In the absence of the off-diagonal exciton terms in the coupling, the linewidth of the central peak, defined as full width at half maximum, is given by

$$\Gamma = 2\sum_{q,\sigma} \frac{|M_{00}(q,\sigma)|^2 \gamma_q}{\hbar^2 \omega_{q\sigma}^2 + \gamma_q^2}(2n_q + 1). \quad (2)$$

where $n_q=[\exp(\hbar\omega_q/k_BT)-1]^{-1}$ and phonon lifetime is $\tau_q=\hbar/\gamma_q$. In the evaluation of the exciton-phonon interaction matrix elements we consider a cylindrically shaped $In_xGa_{1-x}As$ quantum dot (a disk) of height L and base radius R. The coulomb correlation in the electron-hole pair is included through variational wave functions for the ground and excited pair states. For the case of $In_{0.6}Ga_{0.4}As/GaAs$ dot the band gap discontinuity is 0.74 eV and we took a 40%-60% split between valence and conduction band edges in evaluation of confining potentials. The electron mass in the dot is 0.0404 in units of free-electron mass, and 0.0665 in the barrier material. The hole mass parameters in the dot are $m_z=0.34$, $m_{xy}=0.09$.

Before evaluating the effect of the finite phonon lifetime in the Eq.(2) we consider effects of the off diagonal exciton terms in Eq.(1). Such terms were included in reference 7 in evaluation of exciton dephasing in larger disk-like dots. The order of magnitude of the corresponding contribution to the exciton linewidth, $\Gamma(0\rightarrow 1)$, can be obtained from the ground to excited confined exciton state with one-phonon absorption. For the R=100 A dot we obtain that at T=100 K $\Gamma(0\rightarrow 1)\sim O(10^{-4}$ meV). We conclude then that the contribution of the off diagonal terms in Eq.(1) to the experimentally observed linewidth for smaller dots is negligible.

When the phonon linewidth γ is included in the model, the resulting width Γ of the zero-phonon peak in the absorption spectrum of the dot is given in Eq. (2). We consider several sources of the finite lifetime of acoustic phonons: anharmonicity induced three-phonon processes, scattering of phonons by impurities and alloy fluctuations, and boundary phonon scattering [8]. We find that for temperatures well bellow the room temperature, the acoustic phonon lifetimes are limited primarily by the scattering at the side surfaces of the mesa for smaller mesas and the top surface of the structure for large mesas.

We obtain the phonon lifetime from the steady state Boltzmann equation for phonons. In the evaluation of Eq. (2) we assume the following linear dimensions: base radius of the dot R=10 nm, dot hight L=6 nm, distance to the top GaAs layer $\Lambda_z = 6$ nm. Then we obtain the width of the zero phonon line Γ as a function of temperature T and the lateral mesa size Λ. At temperatures above 5 K we find that the width is a nearly linear function of T,

$$\Gamma(T) = \Gamma(0) + \alpha T, \quad (3)$$

and we evaluated the linear coefficient α as a function of the mesa size Λ. The results shown in the Fig. 1 are in quantitative agreement with the mesa size dependence of the linewidths observed in recent experiments [4].

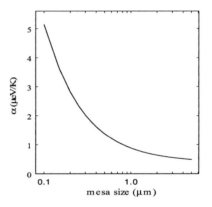

FIGURE 1. The linear coefficient of the linewidth temperature dependence is shown as a function of the lateral mesa size.

REFERENCES

1. M. Bayer and F. Forchel, Phys. Rev. B **65**, 04130(R), (2002).
2. I. Favero *et. al.*, Phys. Rev. B **68**, 233301 (2003).
3. P. Borri *et al*, Phys. Rev. Lett. **89**, 187401 (2002).
4. G. Ortner *et al.*, unpublished.
5. C. B. Duke and G. D. Mahan, Phys. Rev. B **139**, A1965 (1965).
6. M. A. Krivoglaz, Sov. Phys.- Solid State **6**, 1340 (1964).
7. T. Takagahara, Phys. Rev. B **60**, 2638 (1999).
8. J. M. Ziman, *Electrons and Phonons*, (Clarendon Press, Oxford, 1960).

Anisotropic g-factor Dependence of Dynamic Nuclear Polarization in n-GaAs/AlGaAs (110) Quantum Wells

S. Matsuzaka[1], H. Sanada[1], K. Morita[2,1], C. Y. Hu[1,3], Y. Ohno[1,3], and H. Ohno[1,2]

[1] *Laboratory for Nanoelectronics and Spintronics, Research Institute of Electrical Communication, Tohoku University, Sendai 980-8577, Japan*
[2] *Ohno Semiconductor Spintronics Project, ERATO, Japan Science and Technology Agency*
[3] *CREST, Japan Science and Technology Agency*
Phone: +81-22-217-5555 Fax: +81-22-217-5555 E-mail: shun@riec.tohoku.ac.jp

Abstract. In this work, we investigated the well width dependence of dynamic nuclear polarization in n-GaAs/Al$_{0.3}$Ga$_{0.7}$As (110) quantum wells. We prepared a set of samples with different well width and thus different g-factor, and measured the degree of dynamic nuclear polarization in these samples by time-resolved Faraday rotation measurements. Our calculation, considering simply the projection of time-averaged electron spins along the external magnetic field, reproduced overall shapes of the nuclear fields as a function of anisotropy of g-factor.

INTRODUCTION

From past studies, anisotropy of electron g-factor in semiconductor quantum structures is one of the key parameters for dynamic nuclear polarization (DNP) via hyperfine interaction [1], which has attracted growing interest from the viewpoint of implementation of solid-state quantum computers. Since it can be modified by changing structural parameters or applying external field, one can control DNP by clarifying the dependence of DNP on the anisotropy of the electron g-factor. In this work, we investigated the well width dependence of DNP in n-GaAs/Al$_{0.3}$Ga$_{0.7}$As (110) quantum wells (QWs). n-type QWs grown on (110) substrates provide a 2D electron gas with long spin lifetime [2], which is necessary to polarize nuclear spins efficiently. Here we prepared a set of samples with different well width and thus different g-factor [3], and measured the degree of DNP in these samples by time-resolved Faraday rotation (TRFR) measurement.

EXPERIMENTAL

The sample structure used here is n-GaAs/Al$_{0.3}$Ga$_{0.7}$As single QW grown on GaAs (110) substrate by molecular beam epitaxy, and the doping concentration of QW is 4×10^{11} cm^{-2}. A set of samples with different well width was prepared by cleaving the QW films including well width distribution. The sample is placed in the cryostat with a superconducting magnet and optical windows. Magnetic field ***B*** was applied perpendicular to the pump and probe beams, while the QW plane is tilted α against ***B*** in order to enhance DNP. We used a mode-locked Ti:Sapphire laser which generates ~100 fs pulses at 76 MHz tuned at the lowest heavy hole absorption edge of the QW. A right ($\sigma+$) or left ($\sigma-$) circularly polarized pump beam excites spin-polarized electrons sample normal, and the FR signal for a time-delayed linear-polarized probe beam is measured by using a polarized beam splitter and balanced receiver.

RESULTS AND DISCUSSION

First we evaluated the g-factor in- ($g_{x\|[001]}$) and perpendicular ($g_{z\|[110]}$) to the QW plane of each sample from the angle dependence of the precession frequencies measured with low excitation intensities where nuclear field caused by DNP is reduced. The experimental results of g-factor with respect to absorption edge of each sample are shown in Fig.1 (A). The anisotropy of g-factor is varied with different well width. Then DNP at low magnetic field and high excitation intensities was evaluated for each sample. From the precession frequencies affected by polarized nuclear spins and anisotropic g-factor obtained above, we evaluated actual nuclear field. Fig. 1(B) shows well width dependence of nuclear field. The nuclear field changes from 5 to -0.5 GHz for the samples with the well width of 5~10 nm.

In Fig.1, the g-factors change linearly with the absorption edge, whereas the nuclear fields do not behave in the same way. In order to explain this, the contribution of time-averaged electron spins to nuclear orientation is considered. When **B** is applied, electron spins **S** precess around not **B** but the effective field $\Omega(=\hat{g}\mu B/\hbar)$. Then the time-averaged electron spin <**S**> appears along Ω the direction of which is governed by anisotropic g-tensor $\hat{g}$. In DNP geometry, the nuclear field Ω_N directs in the direction parallel to **B** [1]. Thus a component of <**S**> parallel to **B** contributes Ω_N. Based on the above model, magnitude of DNP should be proportional to the projection of <**S**> along Ω. We calculated the contribution of **S** to Ω_N as a function of anisotropic parameter g_z/g_x of the g-factor, which is determined from linear fitting of experimental g. Fig. 2 shows experimental Ω_N and calculated values. The calculation suggests that Ω_N has a peak with respect to

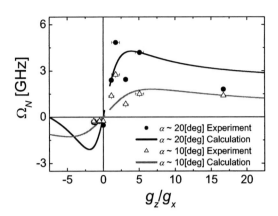

FIGURE 1. The nuclear field Ω_N is plotted with the calculation line as a function of anisotropic parameter g_z/g_x with different angle.

g_z/g_x, which reproduces the overall shape of the experimental $\Omega_N(g_z/g_x)$, indicating that Ω_N can be tuned by tuning g_z/g_x in semiconductor quantum structures.

CONCLUSION

The well width dependence of DNP was investigated in n-GaAs/AlGaAs (110) QW by TRFR measurement. The results are analyzed by taking into account the effect of anisotropic electron g-factor obtained from the experiments. DNP is closely related to <**S**> which can be tuned by the anisotropy of g-factor.

ACKNOWLEDGMENTS

This work is partly supported by Ministry of Education, Culture, Sports, Science and Technology (MEXT), Japan Science and Technology Agency (JST), Japan Society for the Promotion of Science (JSPS) and the 21st Century COE at Tohoku University.

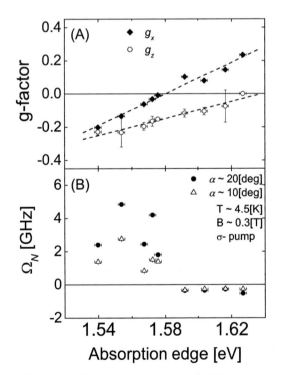

FIGURE 2. (A)The electron g-factor and (B)the nuclear field Ω_N vs absorption edge of the set of samples.

REFERENCES

1. H. Sanada et al., Phys. Rev. B **68**, 241303 (2003).
2. Y. Ohno, et al., Phys. Rev. Lett. **83**, 4196 (1999).
3. E.V. Ivchenko, et. al., Sov. Phys. Semicond. **26**, 827 (1992).

Two-Dimensional Nuclear Magnetic Resonance in Optically Pumped Semiconductors

Anil Patel and Clifford R. Bowers

University of Florida Chemistry Department and NHMFL, Gainesville, FL 32611 USA

Abstract. We demonstrate that 2DNMR can be used to detect optical pumping enhancement of dipolar order in bulk GaAs and InP. Zero and/or double quantum nuclear spin coherences were stimulated by a single non-selective RF pulse. This would ordinarily require at least two non-selective pulses separated by a preparation delay. A density matrix calculation indicates that these transitions could stem from either dipolar or Zeeman spin order at low spin temperature, but the optical pumping time-dependence of the double quantum amplitude supports the conclusion that it is the latter that is the source of the observed zero and double quantum coherences.

INTRODUCTION

Optically pumped NMR (OPNMR) in bulk semiconductors such as GaAs and InP involves optical excitation at energies above the band gap, generating a non-equilibrium conduction electron spin polarization which, upon subsequent electron-nuclear cross-relaxation, can produce significantly enhanced nuclear spin polarization that may be observed by standard NMR techniques. In the widely accepted description of optical pumping (OP) in bulk semiconductors, the trapping of photoexcited electrons at shallow donor sites leads to an enhanced electron-nuclear cross-relaxation rate for nuclei in close proximity to the defect. Since the OPNMR signal is inverted with respect to the thermal equilibrium signal in both InP and GaAs, it has be suggested that the dominant hyperfine cross-relaxation mechanism is dipolar in InP and scalar in GaAs [1], It has also been suggested that the appearance of OP enhanced dipolar order among the ^{115}In nuclei in bulk InP could involve the dipolar hyperfine relaxation process [2].

Here we report the use of 2DNMR to detect the presence of optically pumped dipolar nuclear spin order in both GaAs and InP. It will be shown that both zero and double quantum coherences (0QC and 2QC) can be stimulated from a state of dipolar order among the nuclear spins by a single non-selective pulse. In contrast, a spin system described by a density operator of pure Zeeman ($\hat{I}_z$) symmetry can yield only single quantum coherence (1QC) when stimulated by a single non-selective RF pulse [3]. Since the detection of 0QC and 2QC requires conversion to observable 1QC by a subsequent RF pulse, the OPNMR experiments presented here are necessarily two-dimensional.

EXPERIMENTAL

OPNMR was performed on undoped bulk GaAs and InP samples illuminated with 795 nm laser light at an optical intensity of 500-700 mW/cm^2. The laser system consists of a fiber array packaged solid-state laser diode array (Coherent). The output of the laser was brought to the sample via an 0.6mm core optical fiber which terminated about 1 cm above the sample. Experiments were conducted in an Oxford Instruments CF1200 cryostat inserted into a high homogeneity Oxford Instruments superconducting magnet. The 7×8×1 mm InP sample (Showa Denko, Lot No. 3161) was of n-type conductivity with a $100 \pm 0.1°$ surface orientation, and a (300 K) mobility of 3800 cm^2/V·s. The 5×7×1mm GaAs sample is a semi-insulating (nondoped) substrate single crystal (ingot no. S8870) obtained from Crystal Specialities, Intl. with a $\langle 100 \rangle 2.0° \langle 110 \rangle$ surface orientation and a (300 K) mobility of 4500 cm^2/V·s. The NMR was excited and detected using a flat 6-7 turn copper wire (AWG 36) coil wrapped directly around the sample. The ^{31}P spectra were recorded using a $\pi/2$ RF pulse duration of $0.8-1.0\mu s$, while the ^{71}Ga spectra were recorded using a $\pi/2$ RF pulse duration of $5.0\mu s$.

RESULTS AND DISCUSSION

The observation of zero and double quantum coherence can be explained using a three-spin density operator calculation. The thermal equilibrium density operator is given by:

$$\hat{\rho}_{eq} = \exp(-\hat{H}/k_B T)/Tr\{\exp(-\hat{H}/k_B T)\}$$

where $\hat{H} = \hat{H}_{ZI} + \hat{H}_{II} + \hat{H}_{IS}$ contains the Zeeman (ZI), and dipolar (II,IS) interactions. Under OP, it is postulated that the density operator becomes [2]:

$$\hat{\rho}_{op} \propto \exp\{-\hbar\hat{H}_{zI}/k_B T_z\}\exp\{-\hbar(\hat{H}_{II} + \hat{H}_{IS})/k_B T_D\}$$

where the Zeeman and Dipolar spin temperatures T_z and T_D, respectively, are possibly unequal and spatially varying. The density operator following a resonant rf pulse yields the homonuclear two spin coherence amplitudes, as follows:

$$0QC : \frac{1}{8}\cos\kappa\left[-3+\cos(2\kappa)-2\cos^2(\kappa)e^{\hbar d_{II}/2k_B T_D}\right.$$
$$\left.+4e^{-\hbar d_{II}/2k_B T_D}\cosh(\hbar\omega_0/k_B T_z)\right]$$

$$2QC : \frac{1}{8}\left[1+\cos(2\kappa)-e^{\hbar d_{II}/2k_B T_D}\left[\cos(2\kappa)-3\right]\right.$$
$$\left.-4e^{-\hbar d_{II}/2k_B T_D}\cosh(\hbar\omega_0/k_B T_D)\right]$$

where $2\tan\kappa = d_{IS}/d_{II}$ and d_{IS}, d_{II} are the hetero- and homonuclear dipolar coupling constants. The calculation indicates that 0QC and 2QC could stem from either dipolar or Zeeman spin order at low spin temperature. However, the optical pumping time-dependence of the double quantum coherence amplitude exhibits an exponential build-up with a time constant of 24 s, while the build-up of Zeeman order occurs over a much longer timescale. This additional data supports the conclusion that it is the enhanced dipolar order that is responsible for the observed zero and double quantum coherences.

Several mechanisms to account for OP enhancement of dipolar order have been proposed. The dipolar order may be created by nuclear spin diffusion in a field gradient, as described originally by Bloembergen for crystals containing paramagnetic dopants. Such a hyperfine field gradient is present in the vicinity of the shallow donors when they are occupied.

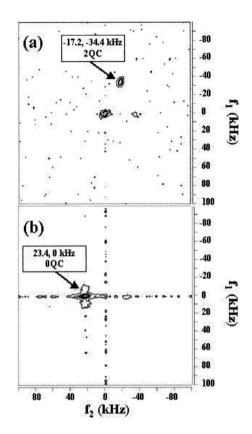

FIGURE 1. 2D OPNMR spectra of ^{31}P in undoped InP. Spectrum (a) was obtained at 51.920 MHz using a 2QC filter (rotating frame offset of –17.2 kHz) while (b) was recorded at 51.880 MHz using a 0QC filter (rotating frame offset of 23.4 kHz). Both spectra exhibit cross-peaks corresponding to 2QC and 0QC coherence, respectively. No signal was obtained when the experiments were repeated without OP.

In conclusion, the results demonstrate the feasibility of using 2DNMR to study optically pumped spin order in semiconductors. This approach may also be applicable to other bulk or low dimensional semiconductors in which optical pumping enhancement of NMR has been observed.

REFERENCES

1. A. Patel, O. Pasquet, J. Bharatam, E. Hughes, and C. R. Bowers, Phys. Rev. B **60**, R5105 (1999).
2. C.A. Michal, R. Tycko, Phys. Rev. Lett. **81**, 3988 (1998).
3. S. Wimperis, G. Bodenhausen, Chem. Phys. Lett. **140**, 41 (1987).

Clean Thermal Processing at Elevated Temperatures

I. Rapoport*, P. Taylor*, B. Orschel*, J. Kearns*

Sumitomo Mitsubishi Silicon Group, OR 97302, USA

Abstract. Clean thermal processing (CTP) was developed to reduce silicon wafer metal contamination. Silicon carbide (SiC) and quartz boats were investigated to find and eliminate Fe contamination sources at 1050 and 1250°C. Contamination originates from (a) metal diffusion in direct wafer to boat contact area (MDD – Metals Direct Diffusion) and (b) as adsorption of metal species from gas phase (MAG – Metals Adsorption from Gas Phase). To reduce both, MDD and MAG contamination, additional cleaning procedures were implemented using Trans-LC treatment at 1250°C, followed by steam oxidation. Metal contamination levels were evaluated using surface photo-voltage (SPV), photo-conductance decay (PCD) and photoluminescence (PL) high-resolution mapping. SPV iron, PCD lifetime and PL intensity maps indicate cleaning procedures are effective for both Quartz and SiC boats.

INTRODUCTION

Fe contamination is known to be responsible for quality degradation during device manufacturing [1]. HCl Oxidation is widely used to reduce Fe contamination during thermal treatment [2]. SiC and Quartz furnace components have been shown to be major metal contamination sources at elevated temperatures [3, 4].

MDD and MAG contamination from Quartz and SiC boats was studied after 1050 and 1250°C Oxidation in diluted oxygen for 1 hour. Trans-LC treatment at 1250°C was implemented to reduce metal contamination. P-type wafers (10 Ωcm) were loaded to Quartz and SiC boats horizontally (touching the top two rods) to monitor Fe contamination. SPV, PCD and PLM high-resolution mapping is effective to clarify the contamination patterns origin.

RESULTS AND DISCUSSION

A dramatic reduction of Fe contamination at 1050 and 1250°C was found after the standard (Trans-LC Oxidation plus Steam Oxidation) and after the ultimate (Trans-LC Oxidation I plus Steam Oxidation plus Trans-LC Oxidation II) cleaning at 1250°C (see fig.1).

Trans-LC cleaning at 1250°C helped to reduce MDD Fe contamination for SiC boats by creating a Fe diffusion barrier on top of SiC CVD layer.

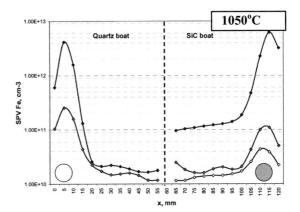

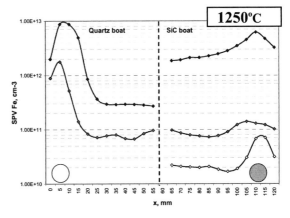

FIGURE 1. SPV Fe after Oxidation: ◆ - initial, ◈ - standard cleaning, ◇ – ultimate cleaning.

Ultimate Trans-LC cleaning for SiC works to reduce the impact of SiC CVD layer defects and to suppress MAG contamination at 1250°C by dramatic Fe out-diffusion reduction from SiC cast.

MDD Fe contamination reduction for Quartz boat is not as effective because Trans-LC cleaning only impacts the Quartz surface but does not create a barrier for metals out-diffusion. MAG Fe contamination from the Quartz boat is suppressed after Trans-LC cleaning. Finally, the Quartz boat loses its shape after the long thermal treatment at 1250°C.

PCD lifetime and PLM mapping results indicate the similar metal contamination reduction after Trans-LC treatment (see figs. 2 and 3).

PCD technique is highly sensitive to monitor Fe contamination. PLM could be used for Fe contamination monitoring at concentrations exceeding 10^{10} cm^{-3}. In addition, it was found that PLM signal is highly sensitive to wafer surface conditions.

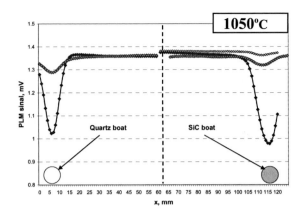

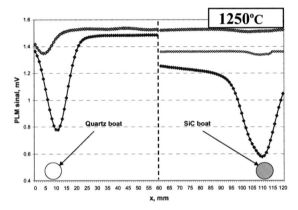

FIGURE 3. PLM signal after Oxidation: ◆ - initial, ◈ - standard cleaning, ◇ – ultimate cleaning.

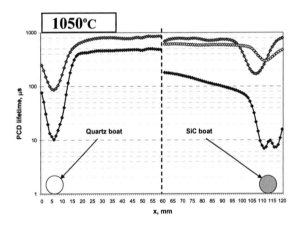

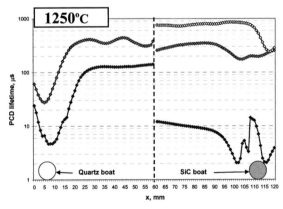

FIGURE 2. PCD lifetime after Oxidation: ◆ - initial, ◈ - standard cleaning, ◇ – ultimate cleaning.

CONCLUSIONS

Trans-LC cleaning is effective to suppress MDD and MAG metal contamination from SiC boats. For Quartz boats MAG contamination was reduced. Trans-LC cleaning was not as effective to reduce MDD metal contamination from Quartz boats.

REFERENCES

1. Istratov, A.A., Hieslmair, H., and Weber, E.R., *Appl. Phys.* **A 70**, 489-534 (2000).
2. Itsumi, M., Akiya, H., Tomita, M., Ueki, T., and Yamawaki, M., *J. Appl. Phys.* **80**, 6661-6665 (1996).
3. Itsumi, M., *J. Electrochem. Soc.* **141**, 1304-1308 (1994).
4. Hellmann, D., Falter, T., Berger, R., and Burte, E., Microelectronic Manufacturing' 94, SPIE Proceedings Vol. 2334 (1994).

Decoherence Control Of Excitons By A Sequence Of Pulses

A. Hasegawa[A], T. Kishimoto[B], Y. Mitsumori[C],
M. Sasaki[A] and F. Minami[B]

[A]*Communications Research Laboratory
[B]Department of Physics, Tokyo Institute of Technology
[C]Research Institute of Electrical Communication, Tohoku University
*National Institute of Information and Communications Technology

Abstract. The control of the dephasing of excitons was performed in GaSe by using successive three femtosecond pulses, i.e., six-wave-mixing. By changing the pulse interval conditions, we observed for the fist time the suppression of the dephasing of the excitons by π pulse irradiation.

INTRODUCTION

Phase relaxation (dephasing) is one of the most important relaxation processes which gives information of the interaction between the electron system and the thermal reservoir. In the ultrashort time region comparable to the correlation time τ_c of the thermal motion of the reservoir, the phase relaxation in matter exhibits the characteristic time variation inherent in the non-Markovian nature of the matter system. This behavior cannot be explained by the conventional phenomenological theory based on the relaxation time T_2.

The time evolution of non-Markovian dynamics can be reversed by using optical π pulses [1-3]. Due to this time reversibility in the non-Markov regime, the dephasing time can be extended up to the energy relaxation time T_1 by irradiating a sequence of π pulses. Here we present the first observation, to our knowledge, of the suppression of exciton dephasing in a semiconductor by optical pulse irradiation. The experiment was performed for the excitons in GaSe by using three successive short pulses, i.e., six-wave-mixing (SWM) configuration. We compared the signal profile between the SWM and four-wave-mixing (FWM) signals, and confirmed that the exciton dephasing is suppressed by an additional π pulse.

EXPERIMENTS

The FWM and SWM experiments were performed for the 1S exciton (2.11 eV) in layered semiconductor GaSe by using an optical parametric oscillator (OPO), with a repetition rate of 76 MHz and a pulse duration of ~200 fs, pumped synchronously by a mode-locked Ti:sapphire laser.

The excitation light was divided into three beams, and sent into the sample from three different directions with wavevectors $\mathbf{k}_1$, $\mathbf{k}_2$, and $\mathbf{k}_3$. In this article, the pulse with the $\mathbf{k}_i$ vector will be denoted as #i, and the relative delay time between #i and #j is written as T_{ij}. In the case of SWM, the echo signal appears $2T_{23}$ away from the arriving time of the #1 pulse, because of inhomogeneous broadening in our sample [4]. In the SWM experiments, T_{23} was scanned with a fixed value of T_{12} and the signal in the $\mathbf{k}_1-2\mathbf{k}_2+2\mathbf{k}_3$ direction was observed. In this phase-matching direction, we can detect only the signals generated by the process subjected to an additional π-pulse irradiation, as compared to the FWM case.

For the FWM, the signal intensity in the $-\mathbf{k}_1+2\mathbf{k}_3$ direction was measured as a function of T_{13}. It is possible to select the SWM and FWM processes only by choosing the observation direction without changing any other conditions. Therefore, the pump intensity and measured temperature (5K) are the same in both experiments.

RESULTS AND DISCUSSION

Figure 1 shows the signal profiles of the SWM as a function of T_{23} obtained for several values of T_{12}. It can be seen from the figure that the decay profile shifts toward longer delays with the delay of the incident timing of the #2 pulse. On the other hand, the FWM profile was unaffected by the temporal position of the #2 pulse. The decay profile of the SWM for $T_{12}=0$ ps is almost the same as that of the FWM.

The dephasing of excitons arises from the random motion of the thermal reservoir, which rapidly modulates the exciton transition frequency and therefore causes degradation of the phase coherence. When the observation time scale is much shorter than the reservoir correlation time τ_c (non-Markovian limit), however, the modulation by the reservoir cannot be regarded as a fully random process. In this case, the slow frequency modulation due to the reservoir plays the role similar to that of the inhomogeneous broadening [5]. In the non-Markovian regime, therefore, the #2 pulse causes some of excitons to precess back towards their initial state when the #1 pulse was encountered. This non-Markovian behavior explains perfectly what we have observed [6]. To our knowledge, this is the first observation of the dephasing-suppression of the excitonic state. This shows that the coherence time can be lengthened to the theoretical lifetime limit of the exciton by irradiating a sequence of π pulses.

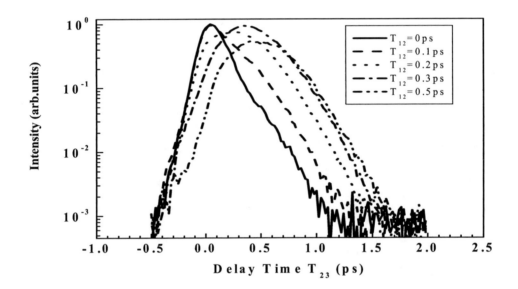

FIGURE 1. The T_{12} dependence of the T_{23} scanned SWM profile

ACKNOWLEDGMENTS

This work was supported by a 21st Century COE Program at Tokyo Tech "Nanometer-Scale Quantum Physics" by the Ministry of Education, Culture, Sports, Science and Technology, by the Grant-in-Aid for Scientific Research from the Ministry of Science, Education and Culture of Japan, and by the Strategic Information and Communications R&D Promotion Scheme.

REFERENCES

1. M. Ban, J. Mod. Opt. **45**, 2315 (1998).
2. L. Viola and S. Lloyd, Phys. Rev. **A58**, 2733 (1998).
3. C. Uchiyama and M. Aihara, Phys. Rev. **A66**, 032313 (2002).
4. A. Hasegawa, T. Kishimoto, Y. Mitsumori, M. Sasaki, and F. Minami, J. Lumin. **108**, 211 (2004).
5. M. Aihara, Phys. Rev. **B25**, 53 (1982).
6. M. Sasaki, A. Hasegawa, Y. Mitsumori and F. Minami, J. Lumin. **108**, 215 (2004).

Imaging Electron Interferometer

A.C. Bleszynski[1], K.E. Aidala[2], B.J. LeRoy[1,4], R.M. Westervelt[1,2], E.J. Heller[1,3], K.D. Maranowski[5] and A.C. Gossard[5]

[1]*Dept of Physics,* [2]*Division of Engineering and Applied Sciences, and*
[3]*Division of Chemistry and Chemical Biology, Harvard University, Cambridge, MA 02138 USA*
[4]*Present Address: Delft University of Technology, Delft, The Netherlands*
[5]*Materials Department, University of California Santa Barbara, Santa Barbara, CA 91306 USA*

Abstract. An imaging electron interferometer was created in a two-dimensional electron gas (2DEG) using a liquid-He cooled scanning probe microscope (SPM). Electron waves emitted from a quantum point contact (QPC) return to the QPC along two paths: reflection from a concave mirror formed by a gate, and backscattering from the depleted disc underneath the charged SPM tip. Interference of these waves when they return to the QPC produces strong interference fringes in images of electron flow. A quantum phase shifter is formed by moving the mirror via its gate voltage - the fringes move a corresponding amount. The coherent fringes are robust to thermal averaging when the lengths of the two paths are within $\ell_T = \hbar v_F / \pi k_B T$ of each other.

INTRODUCTION

Devices that rely on the coherence of electrons promise new ways of computing and sensing. However electron coherence in microscale devices is delicate and is easily lost through interactions with the environment. It is therefore imperative to gain a good understanding of the coherence properties of electrons in these devices. A scanning probe microscope (SPM) is a powerful tool to probe the properties of electrons in a two-dimensional electron gas (2DEG) [1] and to image the flow of electron waves [1-5].

Thermal averages over many electrons can smear the effects of coherence even though the coherence of individual electrons is not lost. In the samples discussed below, the phase coherence length ℓ_φ is longer than the ballistic thermal length $\ell_T = \hbar v_F / \pi k_B T$, where v_F is the Fermi velocity. Therefore one would expect thermal averaging over the thermal distribution to mask interference effects before they are truly lost. We demonstrate an imaging mechanism that detects coherence and is robust to thermal averaging.

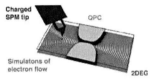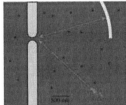

FIGURE 1 (Left) Schematic diagram of the imaging mechanism. A negatively charged tip backscatters electrons to the QPC; the change in QPC conductance images electron flow. (Right) The interferometer includes a circular reflector gate; electron waves passing along two paths - QPC to mirror and QPC to SPM tip - interfere, changing the QPC conductance. The imaged area is shown in blue above.

IMAGING ELECTRON INTERFEROMETER

An imaging electron interferometer can be created in a 2DEG by using a liquid-He cooled SPM [4-5]. The flow of electrons can be imaged [1-3] using the technique shown in Fig. 1. A conducting SPM tip is scanned above the surface of a GaAs/AlGaAs heterostructure. The negatively charged tip depletes a disc in the 2DEG below, and backscatters electron waves arriving from the QPC, thus reducing the QPC conductance. By recording the conductance as the tip is scanned over the sample, an image of the flow of electron waves can be obtained [1-3]. An imaging electron interferometer is made by adding a circular gate that forms an electron mirror when it is energized, as shown in Fig. 1. Electron waves leaving the QPC return along two paths: reflection from the mirror and backscattering from the depleted disc under the SPM tip. Interference of these waves at the QPC changes its conductance and produces fringes in the recorded images. The 2DEG had density 4.2×10^{11} cm^{-2} and mobility 1.0×10^6 cm^2/Vsec; all images were recorded at 4.2K.

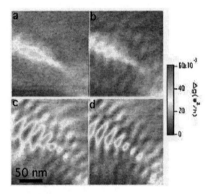

FIGURE 2. Images of the flow of electron waves in the blue box in Fig 1 located $\cong 2$ μm from the QPC for four reflector gate voltages: (a) 0.0 V, (b) – 0.5 V, (c) – 1.0 V and (d) – 2.0 V. Energizing the reflector gate strongly enhances the interference fringes.

Figures 2(a)-(d) show how the imaging electron interferometer operates. Interference fringes emerge in the images of electron flow as the reflector gate is energized. The interference fringes clearly demonstrate the coherence of electron waves. These fringes are observed when the roundtrip path lengths between the QPC and the SPM tip and between the QPC and the mirror are the same, within the thermal length $\ell_T \cong 170$ nm at 4.2K for this sample.

The interferometer acts as a quantum phase shifter - the location of interference fringes in an image can be moved by moving the mirror, by changing its gate voltage. Figures 3(a) and (b) show how the position of the fringes moves with reflector gate voltage. Remarkably, the fringes move twice as quickly with gate voltage when the imaged area is twice as far away from the QPC as the reflecting gate. This occurs because the roundtrip path lengths must be the same length, within ℓ_T, for the interference fringes to survive thermal averaging. A double bounce between the QPC and the mirror is required to match the length of the QPC to tip roundtrip as shown in Fig. 3(d).

Higher temperatures make it easier to resolve events in the time domain, because the collection of paths that interfere is limited to those with comparable lengths. The spatial resolution of our SPM, allows us to make interferometric measurements corresponding to frequencies up to $E_F/h \sim 3$THz and times as short as 50 fsec, where E_F is the Fermi energy.

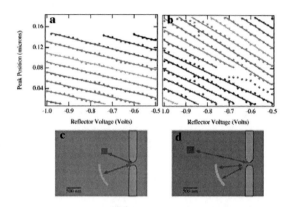

FIGURE 3. (a) Movement of peaks for interference fringes vs. reflector gate voltage for images recorded at the same distance away from the QPC as the mirror, as shown in (c). (b) Peak movement vs. reflector voltage for images recorded twice as far away, as shown in (d). For (b) the fringes move twice as quickly with gate voltage because the interfering path is a double bounce between the QPC and mirror.

ACKNOWLEDGEMENTS

This work was supported by the Nanoscale Science and Engineering Center (NSEC) based at Harvard under NSF Grant No. PHY-0117795.

REFERENCES

1. M.A. Topinka, R.M. Westervelt and E.J. Heller, Physics Today 56, (12) 47 (2003), and references therein.
2. M.A. Topinka, B. LeRoy, S. Shaw, E. Heller, R.M. Westervelt, K.D. Maranowski and A.C. Gossard, Science 289, 2323 (2000).
3. M.A. Topinka, B.J. LeRoy, R.M. Westervelt, S.E.J. Shaw, R. Fleischmann, E.J. Heller, K.D. Maranowski and A.C. Gossard, Nature 410, 183 (2001).
4. R.M. Westervelt, M.A. Topinka, B.J. LeRoy, A.C. Bleszynski, K. Aidala, S.E.J. Shaw, E.J. Heller, K.D. Maranowski and A.C. Gossard, Proc. FSNS2003, NTT Atsugi, Physica E, in press (2004).
5. B.J. LeRoy, A.C. Bleszynski, K.E. Aidala, R.M. Westervelt, A. Kalben, E. J. Heller, S. E. J. Shaw, K.D. Maranowski and A.C. Gossard, submitted for publication (2004).

A New Concept On A Quantum Computer Based On Schockley-Read-Hall Recombination Statistics In Microelectronic Devices.

K. Theodoropoulos[1], D. Ntalaperas[1], I. Petras[1], A. Tsakalidis[1,2], N. Konofaos[1,2]

[1]Computer Engineering and Informatics Dept., University of Patras, Patras, Greece
[2]Research Academic Computer Technology Institute, Riga Feraiou 61, 26110, Patras, Greece.

Abstract. In this paper a quantum computer based on the recombination processes happening in semiconductor devices is presented. A "data element" and a "computational element" are derived based on Schokley-Read-Hall statistics and they can later be used in order to manifest a simple and known quantum algorithm. Such a paradigm is shown by the application of the proposed technology onto the Shor's period-finding algorithm.

INTRODUCTION

Quantum computers [1] based semiconductor properties [2] have recently been examined [3]. The theory of recombination of electrons and holes as proposed by Shockley, Read and Hall [2] describes the behavior of the recombination centers during the transition of an electron from the conduction to the valence band. The situation of these centers forms the main part of this newly proposed computational mechanism. The "datum" will be the ambiguous state in which each recombination center lies. The statistical nature of the transition provides a basis for quantum computation by determining the statistical transition using an external signal applied to the semiconductor device and evaluation of the computation will be provided by measuring electrical quantities of the device [2]. This mechanism is formulated in the paragraphs below and an application considering the solution of a well known quantum algorithm [4] using the above mechanism is also presented.

THE COMPUTER AND AN APPLICATION

The "datum" will be the state of each recombination center. Two bases for computation are considered. The first is the "occupied" state and the second is the "unoccupied" state of the trap. The Fermi-Dirac distribution provides the probability of occupation during equilibrium [2]:

$$f(E) = \frac{1}{1+e^{(E-E_F)/kT}} \quad (1)$$

where E is the energy of the trap, E_F is the Fermi level and T is the temperature. Therefore, the state of the trap can be written by a sum of two states that are the two possible results of a measurement i.e. "occupied" and "unoccupied":

$$|S\rangle = \sqrt{f(E)}|occupied\rangle + \sqrt{1-f(E)}|unoccupied\rangle \quad (2)$$

$$= \sqrt{f(E)}|0\rangle + \sqrt{1-f(E)}|1\rangle$$

This formula defines the "qubit". The selection of the state amplitudes is straightforward from quantum mechanics since the probability of getting the desired outcome is the square of the projection of S to one of the basis states.

In order to describe a quantum computer, implementation of a universal quantum computing machine is needed [1,4]. The gate that can be provided is the CNOT gate which is the quantum analog of the XOR gate [1]. The CNOT has a control qubit and a target qubit which changes its state depending on the state of the control qubit. The physical analog based on our assumption of the qubit will be the following: Initially, we allow the trap to be in one certain state characterized by its energy level and position in the semiconductor gap. Another parameter characterizing the trap is its time constant, or traps lifetime, τ. Then, we consider the trap with the immediately lower energy as the control qubit. Then a signal of frequency equal to the characteristic frequency (1/τ) of the lower trap and of energy enough to emit and then discharge the lower trap is applied across the device. If the lower trap was occupied by an electron (control state = |1>) the higher trap (which constitutes the "target") will change its situation. If the above trap was occupied by an electron then the electron of the lower trap cannot occupy the target bit (Pauli's principle) and it will be forced to discharge itself to a lower trap. The discharge of the electron will cause a hole emission that will make the occupied target trap, unoccupied. When the above trap is not occupied then the electron of the lower trap will be captured, thus changing the state to "occupied". After that, the lower trap is restored to its previous condition by replacing the electron. This is illustrated in figure 1. The coherence criteria are obeyed since only trap-to-trap procedures are involved, without any interaction with reservoir of carriers (i.e. a metal or a valence band).

One way to determine the mechanism is to measure a macroscopic property of a device (i.e. a MOS transistor or a Schottky diode) that is directly related to the traps situation. Such a quantity is the MOS conductance due to interface traps which can be measured using an LCR meter or a DLTS signal applied on well known trapping systems such as the DX center in a GaAs diode. The signal should have the proper frequency in order to achieve the resonance and above all, provide the information of the traps situation.

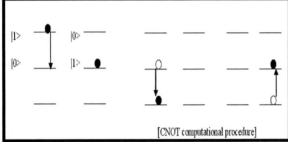

FIGURE 1. The computation mechanism for the CNOT procedure.

An application is considered by applying Shor's quantum algorithm in order to find the period of a string of symbols all chosen from a finite alphabet. The series of symbols that constitute the string in mind are: S_1, S_2, S_3 ... S_n constituting the aplhabet. Then each symbol S_i is specified by:

$$Q(S_i) = <S_i>$$

Then the string will be constituted by a series of symbols selected from the alphabet with repetitions allowed. For each position of the string its value is the value of the symbol that bears. Therefore we seek the period of the resulting number theoretical function F that operates on the positions of the string and that period will indicate the period of the string. This algorithm is a mixture of a classical and a quantum program. It consists of a classical preparation of the system before the quantum program is executed. The quantum algorithm is based on the Quantum Fourier Transform, proposed by Shor and Simon [1]. The circuit in figure 2 demonstrates the QFT, where H stands for the Hadamard transformation and R_k is the k-th fraction of the full rotation of the qubit's phase.

Schematically $R_k = \begin{bmatrix} 1 & 0 \\ 0 & e^{\frac{2\pi i}{2^k}} \end{bmatrix}$.

Using the above circuit and after translating all the gates into our technology the quantum program for finding the period of the specified string is defined. Hence the function in question, the CNOT gate, the value of the qubits and the two quantum registers are given. The first register has as much qubits as they are needed to store a value of the function and the second has as much qubits needed to store the period. The circuit implementing the algorithm is then applied, and a solution procedure is followed. The measurement will be made as mentioned before, using an electrical technique capable to derive the traps "occupied"-"unoccupied" state. Then the result will provide the value of the register and by applying the inverse QFT we directly determine the continued fraction expansion of the measurement thus getting the desired period. Simulations run on this algorithm using the above assumptions provided successful solutions for various paradigms.

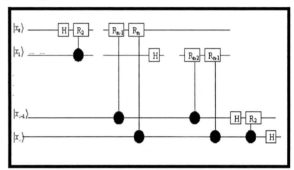

FIGURE 2. The Quantum Fourier Transform.

CONCLUSIONS

A new concept on a quantum computer based on recombination procedures has been demonstrated by presenting the qubit, the computation mechanism, a measurement scheme and an application. Future work involves exploring problems such as the effect of the temperature stability and the role the number of the recombination centers play on the "average decoherence time" which this method requires. Such a work is currently on.

REFERENCES

1. Nielsen M.A., Chuang I.L., *Quantum Computation and Quantum Information*, Cambridge University Press, 2000.
2. Grove A.S., *Physics and Technology of Semiconductor Devices,* New York, John Wiley & Sons, 1967 pp. 126-140.
3. Hollenberg, L.C.L., Dzurak, A.S, Wellard, C., Hamilton, A.R., Reilly, D.J., Millburn, G.J., Clark, R.G., Physical Review B 69 113301 (2004).
4. Barenco, C.H,. Bennett, R,. Cleve, D.P., diVincenzo, N. Margolus, N., Shor, P.W., Sleator, T., Smolin, J., Weinfurter, H., *Phys. Review A* **52** 3457 – 3467 (1995).

Quantum information processing using Coulomb-coupled quantum dots

J. Danckwerts*, J. Förstner* and A. Knorr*

Institut für Theoretische Physik, Technische Universität Berlin, 10623 Berlin, Germany

Abstract. A system of two quantum dots coupled by dipole-dipole interaction is investigated within a density matrix approach. We compute the temporal evolution of the system in the linear and nonlinear optical regime and discuss the possibility of performing basic quantum information gates. The influence of the Förster energy transfer on Rabi oscillations is discussed.

Quantum dots are widely discussed as basic building units for quantum computers. The requirements for the realization of a quantum computer are quantum mechanical two-level systems that can be prepared, manipulated and measured with high fidelity and a coupling mechanism that allows the preparation of entangled states.

Electronic states in quantum dots have been proposed as representatives of qubits with the Coulomb interaction between these states as the main coupling mechanism [1]. The possibility to implement basic quantum information gates in such a system has been shown. These gates depend on system parameters like Coulomb interaction strength, quantum dot size, electron-phonon coupling and dipole moments. While the principal mechanisms have been investigated, a quantitative analysis of the gate fidelity depending on the system parameters is lacking. In this work, the influence of the Coulomb interaction is studied for an elementary single-qubit gate acting on a two-quantum dot system.

We describe strongly confined electrons in two quantum dots with an interdot distance that allows dipole-dipole interaction between the dots, analogous to the model systems considered in Refs. [1, 2].

The model Hamiltonian for the considered system consists of the free electron part H_e, the part describing coherent interaction with an electromagnetic field H_{el} and a part describing the dipole-dipole Coulomb interaction H_{ee}, i.e. $H = H_e + H_{el} + H_{ee}$. In second quantization picture the individual terms take the form

$$H_e = \sum_{\lambda,n} \varepsilon_{\lambda,n} a^+_{\lambda,n} a_{\lambda,n} \quad (1)$$

$$H_{el} = \sum_{a,b,c,d} V_{abcd} a^+_{\lambda_a n_a} a^+_{\lambda_b n_b} a_{\lambda_c n_c} a_{\lambda_d n_d} \quad (2)$$

$$H_{ee} = -\mathbf{E}(t) \cdot \mathbf{d}_{cv} \sum_i (a^+_{c,n_i} a_{v,n_i} X_{cv,n} + a^+_{v,n_i} a_{c,n_i} X_{vc,n}) \quad (3)$$

where $a^+_{\lambda,n}$ ($a_{\lambda,n}$) is the creation (destruction) operator of an electron in level λ in the n-th quantum dot, V_{abcd} the Coulomb matrix element, $\mathbf{E}(t)$ the classical light field, $\mathbf{d}_{cv}$ the dipole matrix element for the transition between valence and conduction level.

As shown in [2] the dipole-dipole interaction part of the Coulomb Hamiltonian in this case consists of diagonal matrix elements and one non-diagonal matrix element. Therefore the two main effects of the Coulomb interaction are (a) the biexcitonic shift $\Delta \varepsilon$ that depends on the diagonal matrix elements of the Hamiltonian and (b) the Förster energy transfer that depends on the nondiagonal element, named Förster energy V_F.

Qubits in this system are represented by the two states of each quantum dot corresponding to an excited or ground state electron. The two quantum dots then constitute a system of two qubits with the basis states $|00\rangle, |01\rangle, |10\rangle, |11\rangle$, where $|0\rangle$ represents an electron in the lower (valence) level and $|1\rangle$ an electron in the upper (conduction) level. To gain insight into the energy levels of the interacting electron system one can diagonalize the Hamiltonian $H_e + H_{ee}$, leading to new basis states $|\Psi_i\rangle$ and new energies $\hbar \omega_i$. The states with one excited electron $|01\rangle$ and $|10\rangle$ superimpose to form the new basis states $|\Psi_2\rangle = c_1|01\rangle + c_2|10\rangle$ and $|\Psi_3\rangle = -c_2|01\rangle + c_1|10\rangle$ where the coefficients c_1, c_2 depend on the Coulomb matrix elements and the difference Δ of the excitation energies [2]. As a consequence the states $|01\rangle$ and $|10\rangle$ are not stationary states anymore.

We compute the system's density matrix elements by numerically solving the Heisenberg equation of motion. The electromagnetic field is taken as an external quantity which is not changed by the quantum dot polarization. Processes representing different quantum gates can be realized by varying the strength and number of pulses.

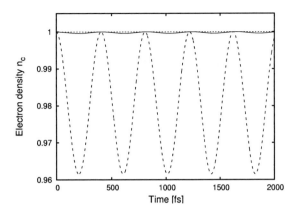
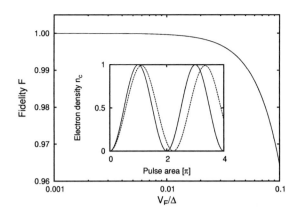

FIGURE 1. Time dependent electron density n_c for different values of V_F at constant Δ (dotted line: $V_F/\Delta = 10^{-3}$, solid line: $V_F/\Delta = 10^{-2}$, dashed line: $V_F/\Delta = 10^{-1}$)

FIGURE 2. Fidelity of a π-rotation on the first qubit for different values of V_F at constant Δ, inset: Rabi-oscillation of the electron density for $V_F/\Delta = 10^{-1}$ (dashed line) and $V_F/\Delta = 0$ (solid line)

For our numerical analysis we consider a pair of quantum dots with a difference of the excitation energies of $\Delta = 10$ meV. We refer to the quantum dot with the smaller excitation energy as the first dot and to the other one as the second dot. The temporal evolution of the electron density $n_c = \langle a_{c1}^+ a_{c1} \rangle$ in the conduction level of the first quantum dot is shown in Fig. 1 for different values of V_F. The initial state of the system is $\psi_0 = |10\rangle$, the value of the biexcitonic shift $\Delta\varepsilon$ is taken as 1 meV. The electron density n_c shows oscillations that get more pronounced with increasing Förster energy V_F. Not only the amplitude but also the frequency of these oscillations depends on the magnitude of the Förster energy. Therefore, when measuring the originally excited quantum dot, with a certain probability the dot may be found in the ground state. This means that the qubit state has changed without gating, which must not happen in a system used for quantum computing. The consequence for the applicability of this system for quantum information processing has to be investigated further.

π-pulse excitation of an uncoupled two-level electron system leads to a complete inversion of the electron state and can be regarded as the implementation of the single qubit gate that a performs an inversion of the qubit Bloch-vector. The performance of the gate in the presence of coupling can be quantified by the gate fidelity F [3]. The fidelity of this gate applied to the initial state $\Psi_0 = |00\rangle$ with a pulse resonant to the first dot is given by $F = \langle 10|\rho|10\rangle$ where ρ is the system's density matrix after the pulse. The fidelity for different values of the Förster matrix element V_F is shown in Fig. 2. The biexcitonic shift was taken as $\Delta\varepsilon = 1 meV$ for all calculations, since the dependence of the fidelity on $\Delta\varepsilon$ is much smaller than that on V_F. In the inset the final electron density n_c in the conduction band after pulse excitation is shown for varying pulse area, showing Rabi oscillations. One can see that with increasing Förster energy V_F the fidelity decreases. There are two reasons for this effect: first, the period of Rabi-oscillation increases (analogous to field renormalisation effects in the Hartree-Fock approximation). As a consequence, excitation with π-pulses does not yield the maximum value of conduction band density. Second, the maximum of the Rabi-oscillation decreases. So, even with a renormalized pulse area, one cannot reach complete inversion anymore.

For the applicability of a gate, the error δ, given by $\delta = 1 - F$, must not exceed a tolerance level of roughly 10^{-5} [4]. In realistic systems the Förster energy can be made small in relation to the biexcitonic shift and the difference of the excitation energies, for example by applying a static electric field [2]. But as our results show, even for small values of V_F the error for the considered single qubit gate lies in a range where it cannot be simply ignored but has to be taken into account. If a real quantum dot system is to be used for quantum information processing it must be possible to restrict the strength of the Förster energy to very small values. In consequence, for a given experimental setup the quantities mentioned above have to be determined carefully.

REFERENCES

1. Biolatti, E., Iotti, R., Zanardi, P., and Rossi, F., *Phys. Rev. Lett.*, **85**, 5647–5650 (2000).
2. Lovett, B. W., Reina, J., Nazir, A., and Briggs, A. D., *Phys. Rev. B*, **68**, 205319-1–205319-18 (2003).
3. Nielsen, M. A., and Chuang, I. L., *Quantum Computation and Quantum Information*, Cambridge University Press, Cambridge, 2000.
4. Knill, E., Laflamme, R., and Zurek, W. H., *Science*, **279**, 342–345 (1998).

Decoherence of Charge Qubit Systems

A. Weichselbaum and S. E. Ulloa

Dept. of Physics & Astronomy, Ohio University, Athens, OH 45701

Abstract. Quantum computation in its binary concept requires a set of two different quantum states, a quantum two-level system that physically realizes the quantum bit (qubit). Amongst the many other existing proposals, we discuss qubits based on the charge distribution of a few electrons. Especially the role of higher lying states is investigated with respect to their role as source of decoherence for the two-level system of the qubit constructed in the ground state.

Qubits are investigated widely in a large variety of physical systems. Natural two-level systems such as spin 1/2 appear appealing but they bear their own problems [1]. The venue of superconducting qubits [2] is also very interesting to follow, yet we emphasize our discussion on *electrical* qubits, i.e. on qubits encoded in the charge distribution of a few electrons [3, 4]. We further constrain ourselves to systems of tunnel-coupled quantum dots with *constant* tunneling in between the dots as for example motivated by constant oxide layers in metallic quantum dots. It has been shown by the author in earlier papers [5, 6] that for these type of capacitively coupled quantum networks with up to two operational electrons it is essential to have a dynamic external magnetic field. As for the 2×2 array discussed in [6], the magnetic field by adding a phase to the tunneling can suppress the tunneling and thus creates a tunable *effective* tunneling amplitude. With geometrical symmetry in the system, a ground-state two-level system can be formed well separated from the higher lying states, which also implies that the tunneling amplitude should be kept relatively weak.

The effect of the higher lying states is to perturb the system and eventually to dephase it. This is what is analyzed in more detail in this paper. The simplest system one can imagine is a third dot coupled to a two dot system with one operating electron as shown in the inset of Fig. 1. Only one state per quantum dot is considered. The system can also be thought of as modeling the previously mentioned 2×2 array yet with an effective tunneling amplitude that can be turned off completely. The level spectrum in Fig. 1 is plotted versus the potential applied to the gate acting on the third dot. As intuitively clear, applying a negative voltage decouples the two-dot system (dots 1 and 2 in Fig. 1) with the third dot lying far above in energy. The ground-state space is then determined by the symmetric/antisymmetric wavefunction (note that with a *static* external magnetic field the role of symmetric and

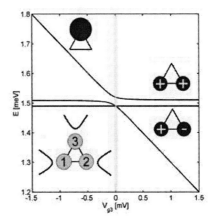

FIGURE 1. Two-dot system coupled to third dot – Level spectrum in dependence of the potential V_{g3} on the gate acting on the third dot. Inset lower left – schematic setup of the system. Remaining insets - state distribution on the array (ring) of three dots for the part of the energy spectrum where they are shown. The +/− signs indicate symmetric/antisymmetric states. Typical parameters chosen for the capacitive coupling are around 16 aF. The tunneling is chosen as 10 μeV. The yellow line through the origin shows the degeneracy of the antisymmetric state with the state on the third dot if V_{g3} where set to zero.

antisymmetric can be flipped such that the antisymmetric state can actually be made the ground-state; magnetic field is not considered otherwise in this paper).

The role of the higher lying (the third) state is analyzed as follows. The necessary two quantum gates [7] to realize full single qubit operation in the two-dimensional ground state state space are assumed to be given. For the triangle in Fig. 1 the qubit space consists of the symmetric/antisymmetric states on dots 1 and 2. With this, the two quantum gates are realized via an asymmetric voltage drag through the gates acting on these dots and via an effective handle on the tunneling amplitude in between the two dots. Decoherence is then studied with respect to both excitation to higher lying states as well as dephasing within the qubit space. This is done in dependence of the two parameters which play a direct role in coupling the higher lying state, namely the rise and fall time σ and the gate voltage V_{g3} which acts on the third dot and thus brings the third state closer to the qubit energy space. The rise and fall time σ of the step–like gate operation should in general be smooth in order for the transfer of probability to the third dot to be negligible.

For the numerical study of the decoherence, the qubit initialized in the ground state is driven with a random sequence of the two single qubit gates. The duration of each pulse is tailored to be a $\pi/2$ pulse which typically appear in qubit operation. Having decided on some random sequence of the two quantum gates, it is straightforward to extract the decoherence time τ related to the loss of probability to the higher lying state by fitting an exponential to the weak loss of probability within the qubit space. Phase coherence, on the other hand, can be checked by comparing the time evolution of the Bloch vector of two *close-by* trajectories in the sense that the parameters σ and V_{g3} are varied slightly and the effect on the time evolution is recorded. In the ideal case these trajectories would be the same because then the influence of the higher lying state is negligible.

The results of the numerical study are shown in Fig. 2. The data is averaged over 50 sequences of 15 randomly chosen quantum gates. The *quality factor Q* shown in the vertical measures how many full gate operations like full Bloch sphere rotations can be performed until the qubit is either lost to other states or its phase is out of control. Panel (a) shows the strong decay with the gate voltage V_{g3} with a rather weak dependence on the transition time σ. The lowest values ($Q \leq 5$) just reflect that the qubit space is already compromised to the higher lying state within a few gate operation. Small V_{g3}, namely approaching the central region with the anticrossing in Fig. 1, clearly encourages the transition to the third state. On the other hand if the state in the third dot is reasonably far elevated from the anticrossing, the state in the qubit space can live over many gate operations. However, small reversible virtual and adiabatic transitions to the third level bring the Bloch vector (the qubit state) out of control and this clearly limits the time window which allows for controlled qubit operation. This dephasing is shown in panel (b) relative to a 1% change of the parameters σ and V_{g3}. For the case of real dephasing, the environment plays the role of varying parameters like V_{g3} through charge fluctuations in the gate voltage. Also the pulse shape is never exactly the same which consequently gives an uncertainty width to the pulse transition time σ.

In conclusion, the simple system of the double dot

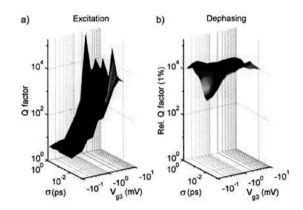

FIGURE 2. Decoherence as a function of the gate switching time σ and the proximity of a third level determined via the voltage gate V_{g3} in terms of the quality factor Q (see text). Panel a) Amplitude decay in the qubit space and thus excitation to higher lying states. Panel b) Phase decoherence determined from the recording of the evolution of configurations with σ and V_{g3} varied by 1%. The sharp spikes in both panels are statistical noise.

system coupled to a third dot provides an interesting model system to investigate decoherence effects in charge qubits. For the well–decoupled groundstate quantum two–level systems, the excitation to the higher lying state is negligible in a good approximation, yet the dephasing due to the presence of higher lying states clearly presents a strong mechanism of decoherence. In order to implement quantum error correction codes, a Q factor of $\sim 10^6$ is required [8] indicating that besides a well–decoupled two–level system in the ground state the control on the qubit parameters must be well below the 1% accuracy.

We gratefully acknowledge support from NSF Grant NIRT 0103034, and the Condensed Matter and Surface Sciences Program at Ohio University.

REFERENCES

1. Stepanenko, D., Bonesteel, N. E., DiVincenzo, D. P., and Burkard, G., *PRB*, p. 115306 (2000).
2. Yamamoto, T., Pashkin, Y. A., Astafiev, O., Nakamura, Y., and Tsai, J. S., *Nature*, p. 941 (2003).
3. Hayashi, T., Fujisawa, T., Cheong, H. D., Jeong, Y. H., and Hirayama, Y., *PRL*, **91**, 226804 (2003).
4. Fujisawa, T., Hayashi, T., Cheong, H. D., Jeong, Y. H., and Hirayama, Y., *Physica E*, **21**, 1046–1052 (2004).
5. Weichselbaum, A., and Ulloa, S. E. (2004), cond-mat/0401106 (submitted to PRA).
6. Weichselbaum, A., and Ulloa, S. E. (2004), condmat/0403120 (submitted to PRB).
7. Nielsen, M., and Chuang, I., *Quantum Computation and Quantum Information*, Cambridge University Press, 2000.
8. Knill, E., Laflamme, R., and Zurek, W. H., *Science*, **279**, 342 (1998).

Modelling of open quantum devices within the closed-system paradigm

Remo Proietti Zaccaria, Rita C. Iotti and Fausto Rossi

*Istituto Nazionale per la Fisica della Materia and Dipartimento di Fisica
Politecnico di Torino, Corso Duca degli Abruzzi 24, 10129 Torino, Italy*

Abstract. We present a new simulation strategy to study the non-equilibrium carrier dynamics in quantum devices with open boundaries. We propose a kinetic description of the system-reservoir thermalization process: the partial carrier thermalization induced by the device spatial boundaries is treated within the standard Boltzmann-transport approach via an effective scattering mechanism between the highly non-thermal device electrons and the thermal carrier distribution of the reservoir.

The overall behavior of state-of-the-art quantum optoelectronic devices often results from a complex interplay between phase coherence and dissipation, the latter being mainly due to the presence of spatial boundaries [1]. In this context, a generalization to open systems (that is, systems with open boundaries coupling them to external charge reservoirs), of the well-known semiconductor Bloch equations has been recently proposed [2, 3]. However, such microscopic treatments, which are essential for the basic understanding of the quantum phenomena involved, are often extremely computer-time consuming. Therefore they cannot be employed in standard device modelling, where, in contrast, partially phenomenological models are usually considered.

Within the typical approach employed in the simulation of open quantum systems, given that f_α is the carrier distribution over the electronic states α of the device, the equation governing hot-carrier transport/relaxation phenomena may be schematically written as [4]:

$$\frac{d}{dt}f_\alpha = \left.\frac{d}{dt}f_\alpha\right|_{scat} + \left.\frac{d}{dt}f_\alpha\right|_{res}. \quad (1)$$

Here, the first term describes scattering dynamics within the device active region and is usually treated at a kinetic level via a Boltzmann-like collision operator of the form:

$$\left.\frac{d}{dt}f_\alpha\right|_{scat} = \sum_{\alpha'}\left(P^s_{\alpha\alpha'}f_{\alpha'} - P^s_{\alpha'\alpha}f_\alpha\right), \quad (2)$$

where $P^s_{\alpha\alpha'}$ is the total scattering rate (i.e., summed over all relevant interaction mechanisms) from state α' to state α. The last term in Eq. (1) accounts for the open character of the system and describes injection/loss contributions from/to the (at least two) external carrier reservoirs. These processes are usually modelled by a relaxation-time-like term of the form [5]:

$$\left.\frac{d}{dt}f_\alpha\right|_{res} = -\gamma_\alpha(f_\alpha - f^\circ_\alpha) = G_\alpha - \gamma_\alpha f_\alpha, \quad (3)$$

here γ_α^{-1} may be interpreted as the device transit time for an electron in state α, while f°_α is the carrier distribution in the external reservoirs.

In this contribution we present an alternative simulation strategy. In particular, we propose to replace the conventional relaxation-time-like term in Eq. (3) with a Boltzmann-like operator of the form:

$$\left.\frac{d}{dt}f_\alpha\right|_{res} = \sum_{\alpha'}\left(P^r_{\alpha\alpha'}f_{\alpha'} - P^r_{\alpha'\alpha}f_\alpha\right), \quad (4)$$

which has the same structure of the scattering operator in Eq. (2); however, the new scattering rates $P^r_{\alpha\alpha'}$ describe electronic transitions within the simulated region induced by the coupling to the external reservoirs. We stress that, contrary to the conventional injection/loss term in Eq. (3), in this case there is no particle exchange between device active region and thermal reservoirs. The total number of simulated particles is therefore conserved.

Let us now discuss the explicit form of the rates $P^r_{\alpha\alpha'}$ entering Eq. (4). In the absence of scattering processes ($P^s_{\alpha\alpha'} = 0$), the steady-state solution of the conventional injection/loss model in Eq. (3) is $f_\alpha = f^\circ_\alpha$, i.e., the carrier distribution inside the device coincides with the distribution in the external carrier reservoirs. As a first requirement, we therefore impose the same steady-state solution ($f_\alpha = f^\circ_\alpha$) to the new collision operator in Eq. (4). This, in turn, will impose conditions on the explicit form of the scattering rates $P^r_{\alpha\alpha'}$. More specifically, from the detailed-balance principle we get:

$$\frac{P^r_{\alpha\alpha'}}{P^r_{\alpha'\alpha}} = \frac{f^\circ_\alpha}{f^\circ_{\alpha'}}. \quad (5)$$

It follows that our transition rates should be of the form:

$$P^r_{\alpha\alpha'} = \mathscr{P}_{\alpha\alpha'} f^\circ_\alpha, \quad (6)$$

where $\mathscr{P}$ can be any positive and symmetric transition matrix ($\mathscr{P}_{\alpha\alpha'} = \mathscr{P}_{\alpha'\alpha} > 0$). What is important in steady-state conditions is the ratio of the scattering rates in Eq. (5) and not their absolute values which are, in contrast, crucial in determining the transient nonequilibrium response of the system.

Since our aim is to replace the injection/loss term in Eq. (3) with the Boltzmann-like term in Eq. (4), as second requirement we ask that the relaxation dynamics induced by the new collision term corresponds to the phenomenological relaxation times in Eq. (3). This corresponds to imposing that the total out-scattering rate coincides with the relaxation rate γ_α:

$$\Gamma_\alpha \equiv \sum_{\alpha'} P^r_{\alpha'\alpha} = \gamma_\alpha. \quad (7)$$

By assuming $\mathscr{P}_{\alpha\alpha'} = p_\alpha p_{\alpha'}$, Eq. (7) reduces to the following system of equations for the unknown quantities p_α: $\sum_{\alpha'} p_{\alpha'} f^\circ_{\alpha'} = \frac{\gamma_\alpha}{p_\alpha}$. Since the sum on the left is α-independent, we get: $p_\alpha \propto \gamma_\alpha$. Starting from this result, we finally obtain:

$$\mathscr{P}_{\alpha\alpha'} = p_\alpha p_{\alpha'} = \frac{\gamma_\alpha \gamma_{\alpha'}}{\sum_{\alpha''} \gamma_{\alpha''} f^\circ_{\alpha'}}. \quad (8)$$

The explicit form of the desired system-reservoir scattering rates entering the Boltzmann-like collision term in Eq. (4) may be derived by combining Eqs. (6) and (8).

The proposed kinetic formulation in terms of Boltzmann-like collision operators only, is particularly suited for standard ensemble-Monte Carlo (MC) simulations, where one deals with a fixed number of particles. In this respect, contrary to the phenomenological model in Eq. (3), in the present closed-system formulation the total carrier density is not fixed by the external reservoirs and the resulting transport equation is homogeneous.

To test the proposed simulation strategy, we have developed a three-dimensional (3D) MC simulator, using as basis states α the product of scattering states along the field/growth direction, and two-dimensional plane waves accounting for the in-plane dynamics. To properly describe phonon-induced energy and momentum relaxation within the device active region, carrier-phonon scattering in a 3D fashion has been included, in addition to the new scattering-like thermalization mechanism in Eq. (4).

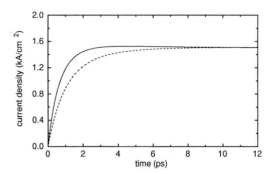

FIGURE 1. Charge current density as a function of time, corresponding to our simulated experiments with the conventional (dashed line) and the proposed (solid line) models.

In particular, we have considered an extremely simple transport problem: a GaAs mesoscopic bulk system of length $l = 200\,\text{nm}$ sandwiched between two reservoirs with different chemical potentials ($\mu_{left} - \mu_{right} = 50\,\text{meV}$). We have applied to this problem our alternative simulation strategy [see Eq. (4)] and have compared the results with those of the conventional simulation approach [see Eq. (3)]. The total carrier density, which is now a free parameter, has been set equal to the steady-state value of the conventional model.

Figure 1 shows the charge current density as a function of time corresponding to our simulated experiments with the conventional [see Eq. (3)] (dashed curve) and the alternative [see Eq. (4)] (solid curve) approaches. At time $t = 0$ the current is in both cases equal to zero; this is however ascribed to different reasons. In the former model, at $t = 0$ the carrier density is equal to zero while the mean velocity is different from zero; in the latter, the mean velocity is equal to zero while the carrier density is different from zero. In spite of a slightly different transient, both curves reach almost the same steady-state value, therefore confirming the validity of the proposed simulation strategy.

REFERENCES

1. See, e.g., W. Frensley, Rev. Mod. Phys. **62**, 3 (1990) and references therein.
2. F. Rossi, A. Di Carlo, and P. Lugli, Phys. Rev. Lett. **80**, 3348 (1998).
3. R. Proietti Zaccaria and F. Rossi, Phys. Rev. B **67**, 113311 (2003).
4. See, e.g., R.C. Iotti and F. Rossi, phys. stat. sol. (b) **238**, 462 (2003).
5. M.V. Fischetti, Phys. Rev. B **59**, 4901 (1999) and references therein.

Nuclear Spin Polarizer for Solid-State NMR Quantum Computers

A. Goto[*], T. Shimizu[*], K. Hashi[*], S. Ohki[¹], T. Iijima[*], S. Kato[*], H. Kitazawa[*], and G. Kido[*]

[*]*National Institute for Materials Science, Sakura, Tsukuba, Ibaraki 305-0003 JAPAN*
[¹]*CREST, Japan Science and Technology Agency, Honcho, Kawaguchi, Saitama 332-0012 JAPAN*

Abstract. We have been developing a nuclear spin polarizer for solid-state NMR quantum computers (QCs). The scheme utilizes the methods of optical pumping and polarization transfer, and enables initialization of the qubits in the QC materials where the optical pumping is not directly applicable. The current status of the development is described in both the instrumentation and the material aspects.

INTRODUCTION

Nuclear spin systems in solids have a great potential as a basis for scalable quantum computers (QC), but their extremely low polarization at thermal equilibrium states necessitates effective initialization scheme, i.e., a hyperpolarization method. One of the promising methods for this purpose is the optical pumping in semiconductors. It is so effective that nuclear polarization up to 50 % can be achieved inside a semiconductor [1]. The only drawback is that ideal QCs are not necessarily semiconductors. To relieve this constraint, we have proposed a scheme of the effective nuclear spin polarizer, i.e., an "optical pumping qubit initalizer (OPQI)" [2].

The OPQI is a composite system of semiconductors (polarizers) and QC devices, which are in close contact with each other on a nanometer scale. A circularly polarized light tuned to the band gap of the semiconductor hyperpolarizes nuclear spins in the semiconductors by the optical pumping effect. The hyperpolarization thus obtained is transferred (injected) to the QCs through the interfaces, and further diffuses into the inner part of the QCs by the spin diffusion process via homo-nuclear interactions. As a result, hyperpolarization is achieved even in the materials other than semiconductors.

We have been engaged in the development of the scheme for a few years. The development has two aspects; instrumentations for the optical pumping and the polarization transfer, and material fabrications for nano-scale contacts between the semiconductors and the QCs. In this paper, we describe the overview of the current status of the development in both the instrumentation and the material aspects.

INSTRUMENTATIONS

The instrumentations include an optical system, NMR magnet, cryostat and an optical pumping NMR probe. The optical system consists of a tunable laser, optical fibers and a fiber coupler, whose details are described in Ref. [3]. The characteristic feature of our system is a usage of the polarization maintaining fiber, which transmits linearly polarized light emitted from the tunable laser to the tip of the NMR probe in the cryostat installed in the 270 MHz (6.4 T) NMR magnet with conserving the linear polarization. The system has been successfully applied to the enhancement of ^{31}P nuclear polarization in the semi-insulating III-V semiconductor, InP:Fe, and ^{31}P nuclear polarization of more than 30 % has been achieved [4].

In order to transfer polarization from the semiconductors to the QCs, one needs an optical pumping NMR probe of "2 + 1 dimensions"; i.e., two radio frequencies and one infrared light. The configuration is schematically shown in Fig. 1. The two rf fields are provided by crossed coils. They are used to cross-polarize nuclei in the QCs using hyperpolarized nuclei in the semiconductors. A sample is mounted on the crossing region of the two coils and an infrared light led through the optical fiber is irradiated to the sample from the bottom. We have succeeded in constructing the probe, and a cross-polarization between ^{31}P and ^{115}In in InP:Fe has been performed with the probe developed.

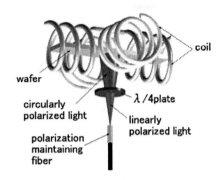

FIGURE 1. Schematic illustration of the tip of the 2+1 dimension NMR probe. Crossed coils provide two rf fields for the polarization transfer, while the optical pumping is rendered by polarized light irradiated through an aperture of the coils. The polarization is converted from linear to circular ones by a quarter-waveplate.

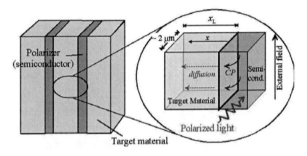

FIGURE 2. An example of the possible nano-structures, where target materials and semiconductors (polarizer) are multilayered. Circularly polarized infrared light hyperpolarizes the nuclear spins in the semiconductors, whose polarization is transferred to the target materials through the interfaces and further diffuses in the target materials. As a result, hyperpolarization is realized in the target materials.

NANO-STRUCTURES

The transfer of the polarization from the semiconductors to the QCs requires interfaces, and at the same time, one needs to secure a clean surface on which infrared light for optical pumping is irradiated. Some of the possible configurations that fulfill these requirements include multilayer and porous systems [5]. The former case is illustrated in Fig. 2. This configuration is realized with the gratings etched on the wafers of the III-V semi-insulating semiconductors such as non-doped GaAs and InP:Fe. The dimensions of the gratings are a few tens of nanometers in width and 1~2 μm in depth. These dimensions are determined by the nuclear spin diffusion length and the "penetration depth" of the infrared light in the semiconducting layers.

In the case of the porous samples, one can invoke a fabrication technique for porous aluminas [6]. A porous alumina can be used as a mold to fabricate porous samples with fillers of semiconductors (e.g., GaAs) and pores of the QCs. We have testing these two types of configurations for the present purpose.

CONCLUSION

In this paper, we have described the current status of the development of the nuclear spin polarizer for solid-state NMR QCs. Although the scheme still needs further refinements in both the instrumentation and the material aspects, its prospects has been shown to be quite good.

ACKNOWLEDGMENTS

This work has been supported by the Industrial Technology Research Grant Program in '04 from New Energy and Industrial Technology Development Organization (NEDO) of Japan.

REFERENCES

1. Meier, F., and Zakharchenya, B. P., eds., *Optical Orientation*, Amsterdam, Elsevier Science, 1984.

2. Goto, A., Shimizu, T., Hashi, K., Kitazawa, H., and Ohki, S., *Phys. Rev. A* **67**, 022312 (2003).

3. Goto, A., Miyabe, R., Hashi, K., Shimizu, T., Kido, G., Ohki, S., and Machida, S., *Jpn. J. Appl. Phys. Pt. 1* **42**, 2864 (2003) 2864.

4. Goto, A., Hashi, K., Shimizu, T., Miyabe, R., Wen, X., Ohki, S., Machida, S., Iijima, T., and Kido, G., *Phys. Rev. B* **69**, 075215 (2004).

5. Goto, A., Shimizu, T., Hashi, K., Ohki, S., Iijima, T., and Kido, G., *IEEE Trans. Appl. Supercond.*, **14**, 150 (2004).

6. Kato, S., Kitazawa, H., Kido, G., *J. Magn. Magn. Mat.*, **272-276**, 1666 (2004).

Small Metallic Contacts in the System Metal/ Barrier/ Semiconductor as the Single-Electron Qubits

Zinovi S. Gribnikov*[§] and George I. Haddad*

[§]*Institute for Quantum Sciences, Michigan State University, East Lansing, Michigan 48824,USA,*
*Department of EECS, University of Michigan, Ann Arbor, Michigan 48109,USA.

Abstract. We consider single-electron qubits induced in a semiconductor by positively charged microscopic metallic gates separated from this semiconductor by an isolating heterostructural barrier. Our primary attention is paid to double quantum dots formed by intersection of the filament-gate induced quantum wires.

An extensive plane positively charged metallic gate separated from a semiconductor (S) surface by an isolating barrier (B) induces a 2DEG layer in the semiconductor with electron states quantized in the normal direction to the B/S-interface and forms a self-consistent quasi-triangular quantum well. A small positively charged metallic gate whose sizes in the xy-plane parallel to the B/S-interface are restricted from all the sides also induces a certain electron gas in the semiconductor. But this gas contains a limited number of electrons and electron states are quantized in all the directions. It is the so-called 0DEG or the quantum dot (QD). Such QDs induced by charges of one or several small metallic gates could serve as the qubits of a quantum computer. Here, we are interested only in the single-electron QDs having the pronounced two-level electron states: the ground and the first excited states should be separated from all the energetically-higher states by the gap, which is much wider than the clearance between them. The latter should also be very small in the absolute sense to maximally limit a probability of interaction between the qubit electron and acoustic phonons. The QDs serving as the qubits should be easily controlled by external DC and AC potentials, interact (entangle) with each other, and so on [1]. Taking into account the above-mentioned requirements, we are interested first of all in double quantum dots (DQDs) with the very small symmetric-antisymmetric (S-AS) splitting of the ground state of the halves. The simple evaluations of the electronic spectrum for the induced quantum dots can be produced in the two specific cases considered below.

1). The specific case of the so-called size-quantized QD (SQQD) appears when the inducing metallic gate is a plane plate parallel to the B/S-interface and having a comparatively simple geometric form (for example, a round disk, a rectangular plate, etc.) Such a gate can be described by a small set of the sizes a_n: a diameter, side lengths, etc., and each of them is much larger than a distance z_0 between the gate and the B/S-interface as well as the size of the electron wave function in the normal direction to the B/S-interface. The latter is controlled by the value of gate potential V measured in relation to the remote earthed n^+-contact in the semiconductor. In this case, the positive charge is homogeneously distributed on the lower gate surface and we have the size electron quantization under the B/S-interface determined only by the boundary condition: $\Psi(\Sigma) = 0$ where Σ is the contour of the gate in the xy-plane. Such a boundary condition can be substantiated by the image force effect at the B/S-interface. We have assumed that the dielectric constant of the semiconductor, ε_{DS}, is noticeably larger than the dielectric constant of the barrier medium, ε_{DB}. For example, in Si $\varepsilon^0_{DS} = 11.7$ and in SiO_2 $\varepsilon^0_{DS} < 4$ (in Si_3N_4 $\varepsilon^0_{DS} < 7$.) Therefore the barrier pushes back electrons in the depth of the semiconductor from the B/S-interface outside the gate, but under the gate we have the attraction.

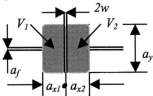

FIGURE 1

In Fig. 1 we demonstrate a gate in the form of a slit rectangular consisting of two halves with the same sizes: $a_{x1} = a_{x2}$. The slit width, $2w$, should provide the necessary tunnel connection between the induced SQQD electron states of the halves. Let us emphasize that such an inter-half-tunnel connection takes place only for the induced electron states but does not occur for electrons of the inducing gate halves, which are isolated from each other carefully and reliably by the $2w$-clearance. The potentials of the half-gates are approximately equal: $V_1 \cong V_2 \cong V$ but can be somewhat different: $\delta V = V_1 - V_2 \neq 0$. The value of the tunnel transparence depends substantially on the potential V. The small difference, δV, allows us to compensate a small possible inequality of the widths a_{x1} and a_{x2} and other technological errors. The gate contacting could be established by thin metallic strips (filaments) with transverse sizes a_f, which are substantially small in comparison to $a_{x1,2}$ and a_y (see Fig. 1). As a result of the assumed smallness, the electron states in the strips (filaments) are energetically much higher than the gate states and do not perturb them.

2). The second QD form available for comparatively simple evaluations can be induced as an intersection [2] of quantum wires (QWrs).

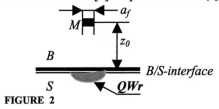

FIGURE 2

The necessary QWrs in the semiconductor can be induced by metallic filaments (strips) with the transverse sizes a_f, which are negligibly smaller in comparison to the distance z_0: $z_0 >> a_f$ (Fig. 2). Note that the effective transverse size of the QWr itself is not so small. But for the sufficiently large filament potentials V it can be also noticeably smaller than z_0. An intersection of the induced QWr by another induced QWr forms the QD with the completely localized electron states around the intersection point. The crossed metallic filaments inducing the above-mentioned QD can be positioned in different distances z_{01} and z_{02} from the B/S-interface and have different potentials V_1 and V_2. We need only the intersection of their projections on the interface. To obtain the DQD, we need to form the double QWr (DQWr) and to intersect it by the ordinary QWr. The DQWr is induced by the two parallel metallic filaments M_1 and M_2 placed in the same xy-plane relatively far from each other: the distance $2w$ [see Fig. 3, (A)] should exceed the distance z_0 with the sufficient reserve. To obtain the accurate resonance of the electron states in the QWrs forming the DQWr, we should establish the independent potentials V_1 and V_2 for the filaments M_1 and M_2. The distance $2w$ can be decreased (down to zero) and controllability of the system can be increased to add the third metallic filament M_3 placed in the equal distances from the M_1 and M_2 noticeably closer to the B/S-interface [$z_3 < z_1 = z_2$ with a certain reserve, see

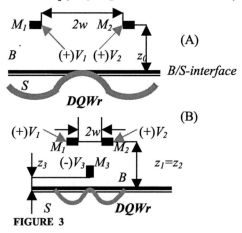

FIGURE 3

Fig. 3, (B)] and negatively charged. The relatively small potential V_3, unlike the positive potential $V_1 \cong V_2 \cong V$, should be negative. This negative potential just serves as the inductor of the tunnel barrier transforming the initial QWr (formed by the positive potentials $V_1 \cong V_2 \cong V$ of the close to each other filaments M_1 and M_2) to the DQWr. As a result of the intersection of the DQWr on the basis of two or three metallic filaments by the one more QWr induced by the crossing positively charged metallic filament M_{tr}, we obtain the desirable DQD (see Fig. 4). This DQD can be controlled by three or four independent potentials of the filaments M_1, M_2, M_3, and M_{tr} to adjust the balance of the half-dots and their tunnel connection.

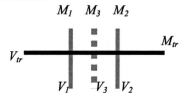

FIGURE 4

REFERENCES

1. See for example, B. Golding and M. I. Dykman, cond-mat/ 0309147; M. I. Dykman, P. M. Platzman, and P. Seddighrad, Phys. Rev. B 67, 155402 (2003).
2. R. L. Schult, D. G. Ravenhall, and H. W. Wyld, Phys. Rev. B 39, 5476 (1989); M. Grundman and D. Bimberg, Phys. Rev. B55, 4054 (1997); G. Schedelbeck, W. Wegscheider, M. Bichler, and G. Abstreiter, Science 278, 1792 (1997).

Tuning nanocrystal properties for quantum information processing

G. Medeiros-Ribeiro*, E. Ribeiro* and H. Westfahl Jr.*

Laboratório Nacional de Luz Síncrotron, PO Box 6192, 13084-971 Campinas - SP, Brazil

Abstract. The prescription for quantum information processing (QIP) devices requires a series of stringent requirements that must be fulfilled for successful operation. The capability of storing electrons, one by one in a predictable and controllable manner makes self-assembled quantum dots (QDs) a promising candidate for the use of the spin degree of freedom of trapped electrons as qu-bits. The embodiment of excitons as qu-bits can be also implemented in QDs, given the relatively long dephasing times. Here two relevant parameters are explored aimed at QIP device implementations: a) valence band line-up in QDs and b) g-factor engineering in QDs.

INTRODUCTION

The possibility of producing small structures will enable, besides higher device integration, the possibility of harnessing quantum degrees of freedom permitting novel functionalities in solid state devices. However, the task of implementation of quantum information processing (QIP) devices in a solid state setting poses tremendous challenges. The creation of nanostructured systems by self-assembly methods [1] could be a viable route. For instance, one can tune the sizes of island and wires exhibiting quantum effects by using growth kinetics and thermodynamics. Besides size, shape and density, other properties that may bear significance to QIP devices, most specifically quantum dots (QDs) used for the storage of either electrons or excitons, concern issues such as composition gradients, band line up and for the case of the spin degree of freedom, the electronic g-factor. In addition to the stringent requirements for QIP [2], any successful implementation prescribes some degree of control over these fundamental properties. There is a limited realm of possibilities for QDs, basically given by materials choice [3] as well as growth kinetics and thermodynamics [4]. In order to assess these properties and control them, novel structural and electronic characterization techniques have to be implemented for QDs. For instance, the composition profile in QDs can be evaluated by Synchrotron X-ray techniques [5], permitting a more realistic modelling of band structure. From dephasing times measurements [6], one can infer the prospects for QIP implementation in QDs. Finally, the observation of single photon emission [7] and Rabi oscillations in simple device structures [8] grant QDs a promising platform for further QIP implementations.

Here two issues will be dealt with: a) band-line up in type-I-type-II ternary alloy quantum dots, and b) g-factor engineering on QDs. By modifying the band line-up from type-I to type-II, one can increase the electron-hole recombination times, by spatially separating the corresponding wavefunctions. Thus, one could conceivably extend dephasing times. g-factor engineering on its own is key for spintronic applications using QDs. $k \cdot p$ theory can be effectively applied to describe g-factors in bulk semiconductors [9, 10]. QDs present a few characteristics that complicate g-factor calculations, since quantum effects, composition and strain change g-factors. Composition and strain profile may induce different g-factor profiles, which in turn is important for g-tensor modulation resonance [11]. Here the electronic g-factor in InAs QDs is controlled by straining the QDs with InGaAs stressor layers [12, 13].

EXPERIMENTAL ASPECTS

For the band line-up studies, $InAs_xP_{1-x}$ QDs were grown by Metal-Organic Chemical Vapor Deposition (MOCVD) on GaAs (001) substrates. The growth procedure and stoichiometry control is described elsewhere [14]. The investigated structures contained a QD layer embedded in an undoped GaAs matrix, for further photoluminescence (PL) and magneto PL experiments. For the g-factor engineering investigations, InAs QDs were grown by Molecular Beam Epitaxy (MBE) on semi-insulating GaAs (001). A reference sample (A) was prepared, consisting of an undoped 1 μm thick GaAs buffer layer grown at 600 ^{o}C, followed by a back contact, Si doped nominally to 1×10^{18} cm^{-3} and 80 nm thick. The temperature was then lowered to 530 °C, and an undoped 25 nm thick GaAs tunneling barrier layer thickness (t_b) was grown, above which the InAs QD layer was nucleated. The QD density was about $1-2 \times 10^{10}$ cm^{-2}. The structure was capped with a 150 nm thick undoped GaAs layer (t_i). The other samples (B,C) were grown with the same layer structure, except for inserting a stressor layer, also called strain reducing layer (SRL) [12]. Here, a $In_{0.2}Ga_{0.8}As$ 30 Å thick SRL was deposited over the QD layer after QD nucleation. Finally, an additional sample (D) was prepared with the same structure as sample A, for the sake or checking the reproducibility of g-factors in QDs.

The photoluminescence spectra of all samples were measured and exhibited narrow lines (~40 meV for the ground state transition) and shell structure up to 4 discrete levels of the trapped carriers for the InAs QDs. For the band line-up investigations, PL and magneto-PL experiments were performed at 2.2 K with the magnetic field parallel to the growth direction spanning the 0-12 T range. For the g-factor studies, Schottky diodes were processed over different portions of the wafers, using standard lithography with AuGeNi ohmic back contacts and 1 mm diameter Cr top Schottky gates. Capacitance experiments were performed at 2.2 K with lock-in amplifiers at frequencies ranging from 1 kHz to 10 kHz, and an ac amplitude of 4 mVrms superimposed on a varying dc bias. The magnetic field was oriented parallel to the growth direction [001] as well. The signal/noise ratio for these experiments was above 10^5.

BAND LINE-UP

The $InAs_xP_{1-x}$ QDs grown by MOCVD presented reasonably good uniformity, as can be inferred from their PL characteristics [3]. Controlling the stoichiometry posed a challenge, since for IIIV_xV_{1-x} ternary alloys the group V presents a volatility and incorporation kinetics that depend both on the temperature and strain in the system [14], as opposed to $III_xIII_{1-x}V$ material where only the In re-evaporation at high temperatures may lead departures on the composition. Evidences of type-I-type-II transition for the $InAs_xP_{1-x}$ on GaAs system can be inferred quite simply by comparing the electron-hole recombination energies of the QD material in bulk form, the surrounding matrix and the QD transition. Should the QD transition be lowest, one can infer a type-II alignment. A more in depth analysis can be carried out by micro-PL experiments as a function of the excitation power, and for the case of InP QDs [15], the type-II alignment was particulary evident. Thus, a control on the band line-up of a QD relative to the host material can be implemented. The spatial carrier separation induced by the

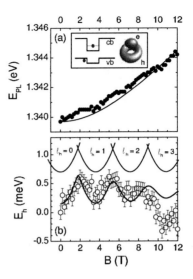

FIGURE 1. (a) Photoluminescence peak energy as a function of the magnetic field for InP QDs (solid circles). The curve represents the two first terms of eq. (1). The inset presents a sketch of the band alingment for a InP QD and a representation of the electron and hole configurations. (b) Hole energy as a function of the magnetic field (open circles), presenting an oscillatory behavior with period $\Phi_0 = 3.61$ T. The parabolas are fits to each oscillation period following the last term in (1). The solid curve is a calculation of the hole energy including the effect of QD size dispersion.

type-II alignment [(see inset in Fig. 1(a)] gives rise to a hole ring around the InP QD [16]. By placing this system in a magnetic field, the hole energy minimum should oscillate due to the Aharonov-Bohm effect [17], corresponding to different angular momenta states. Thus, the excitonic spectrum of the InP QD should also oscillate due to the hole contribution.

Fig. 1(a) shows the exciton photoluminescence energy, E_{PL}, as a function of the magnetic field (solid circles). Instead of a monotonic, featureless curve, some plateau-like structures are visible in E_{PL}. One can reason that E_{PL} is composed of three main components:

$$E_{PL}(B) = E_g + \sqrt{(\hbar\omega_0)^2 + \left(\frac{\hbar\omega_c}{2}\right)^2} + \frac{\hbar^2}{2m_h^*R^2}\left(\ell_h - \frac{\Phi}{\Phi_0}\right)^2, \quad (1)$$

where $\hbar\omega_0 = (5.6 \pm 0.9)$ meV (determined from capacitance measurements), m_h^* is the hole effective mass, ℓ_h is the hole angular momentum, $\Phi_0 = hc/e$ is the flux quantum and $\Phi = \pi R^2 B$ is the total magnetic flux through the ring of radius R. Since the diamagnetic shift of the electron confined into the QD is known, one can subtract both the constant and the electron contributions from the experimental data. The sum of the two first terms in (1) can be seen in Fig. 1(a) (solid curve). The hole energy is then the difference between the experimental data and this curve, which is plotted in Fig. 1(b) (open circles). The hole energy clearly oscillates as a function of the magnetic field and each period of the oscillations can be independently fitted by the last term in (1), using the appropriate value for ℓ_h. The results are the parabolas displayed in Fig. 1(b). From these fittings a ring radius of (19.1 ± 0.4) nm was inferred, in excellent agreement with the InP QD size determined by TEM, (16 ± 3) nm [3]. The ring radii should indeed be larger than the QD radius, since the hole ring is located outside the QD in the GaAs barriers.

It is worth noting that an ensemble of islands and not a single QD was probed. Nevertheless, the size dispersion did not suppress the AB oscillations. Its main effect on E_h is a progressive attenuation of the oscillation intensity, as one would expect, and it is consistent with the decrease in the amplitude of the data in Fig. 1(b).

This clear evidence of Aharonov-Bohm oscillations in the exciton spectrum of InP/GaAs quantum dot ensembles allows the following conclusions to be drawn: a) type-II band alignment, which may lead to longer dephasing times due to both an increase on electron-hole wavefunctions spatial separation, and the transition between levels of different angular momenta (for example, between an electron s-state level to a hole level with $l_h = 1$); b) despite the presence of a size dispersion, the AB oscillations were observable, indicative of a robust effect, and c) the observation of up to three flux quanta trapped inside the ring trajectory of the hole, for a reasonable span of magnetic fields, suggesting the phase coherence for holes moving around QDs for higher ℓ_h.

G-FACTOR ASSESSMENT

Knowledge of the g-factor is paramount for any application that requires the control of the spin degree of freedom of electrons and holes. Therefore, to successfully implement spintronics into semiconductor heterostructures, one has to achieve g-factor assessment, control, and engineering. Traditionally, g-factors are measured via electron-spin resonance (ESR) techniques [18], from which the g-factor as well as the dephasing and relaxation times can be precisely evaluated. The difficulty in applying ESR technique to ensembles of QDs lies on the smallest number of spins that can be measured, which hovers around 10^{12} spins for the lowest detection limit. In lithographically defined QDs [19] as well as in self-assembled quantum dots [20], and in metallic nanoparticles [21] one can assess the g-factor by using transport spectroscopy experiments.

Here the electron g-factor was evaluated using capacitance spectroscopy, which provides direct information on the QD density of states (DOS). In what follows we will

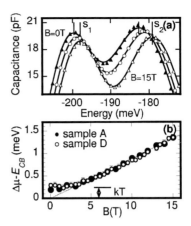

FIGURE 2. (a) s1 and s2 peak dependence on the magnetic field for $B = 0$, 7.5 and 15 T. (b) Dependence on the magnetic field of the s1-s2 difference excluding the Coulomb charging energy contribution. The different symbols refer to samples A and D, grown without the strain reducing layer (SRL).

focus on the s shell. It has been demonstrated that InAs QDs can be modeled quite accurately utilizing a lateral parabolic confinement approximation [22], for which the eigen-energies are known as a function of an external magnetic field [23]. Including the effect of the applied bias, the Coulomb charge interaction between carriers sequentially added to the QD ensemble, and the Zeeman splitting, the energies of states with one and two electrons in the s shell are [24]:

$$E_{s1}(B) = E_z + \hbar\sqrt{\omega_0^2 + \frac{\omega_c^2}{4}} - \frac{|g|\beta B}{2} - \mu_1 \quad (2)$$

$$E_{s2}(B) = 2E_z + 2\hbar\sqrt{\omega_0^2 + \frac{\omega_c^2}{4}} + E_{CB} - 2\mu_2, \quad (3)$$

where E_z is the vertical confining potential, $\hbar\omega_0$ the lateral confining potential characteristic energy, ω_c the cyclotron frequency, $|g|$ the g-factor modulus, β the Bohr's magneton, B the magnetic field intensity, E_{CB} the Coulomb charging energy for the two electrons in the QD ground state, and μ_1 (μ_2) the QD chemical potential at which the first (second) electron tunnel into the ground state. From equations (2) and (3) and band-bending corrections [13], by measuring the energy difference from one to two electrons, we have:

$$\Delta\mu = \mu_2 - \mu_1 = E_{CB} + |g|\beta B. \quad (4)$$

From the dependence of $\Delta\mu$ on the magnetic field, one can therefore obtain the g-factor modulus. Figure 2(a) shows the DOS for s1 and s2 for 0, 7.5 T and 15 T for sample A. The peak spacing is equal to E_{CB} for $B = 0$.

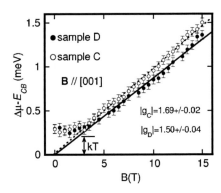

FIGURE 3. Spin-splitting of the s state reflecting g-factors differing in about 13% for samples C and D.

Figure 2(b) shows the spin splitting dependence on the magnetic field, obtained from $\Delta\mu - E_{CB}$ for samples A and D. From equation (4), the g-factor modulus can be found from the slopes of those curves and were $|g| = 1.53 \pm 0.03$ and 1.50 ± 0.04 respectively.

Samples A and D exhibited the same g-factor, within the experimental uncertainties. This is an extremely important result, evidencing the reproducibility of the procedure for the growth runs, sample processing and measurement. In contrast, the effect of the SRL on the electronic properties and on the g-factor of samples B (not shown) and C is quite significant [13], as shown in Fig. 3. From the data presented here, one can infer that the g-factor depends mostly on the strain configuration. The g factor obtained for sample C is larger than for samples A and D by 13%.

CONCLUSIONS

Stressing InAs:GaAs QDs with thin InGaAs capping layers can be quite effective, permitting a 13% change by using a 3nm thick, 10% In rich overlayer. Finally, the band line-up of QDs can be tuned to spatially separate electron and holes, a resource that can be used to modify recombination times as well as to enable effects such as Berry phases to be observed. These effects were observed in InP:GaAs QDs exhibiting an oscillatory ground state transition on the applied magnetic field. Thus, several parameters important for the implementation of QIP in QDs can be controlled within useful ranges, as these preliminary but promising results indicate.

ACKNOWLEDGMENTS

This work was funded by CNPq, FAPESP and HP Brazil. G.M.R wishes to acknowledge the use of the high-field magnet at the GPO group at the IFGW, UNICAMP, and the IFSC/USP for the access to the MBE apparatus.

REFERENCES

1. D. Bimberg, M. Grundmann, N. N. Ledentsov, Quantum Dot Heterostructures (John Wiley & Sons, New York, 1999).
2. D. P. DiVincenzo: *e-print quant-ph* 0002077.
3. E. Ribeiro, R. L. Maltez, W. Carvalho Jr., D. Ugarte, and G. Medeiros-Ribeiro, *Appl. Phys. Lett.* **81**, 2953 (2002).
4. R. E. Rudd, G. A. D. Briggs, A. P. Sutton, G. Medeiros-Ribeiro, and R. Stanley Williams, *Phys. Rev. Lett.* **90**, 146101 (2003).
5. A. Malachias, S. Kycia, G. Medeiros-Ribeiro, R. Magalhães-Paniago, T. I. Kamins, and R. Stanley Williams, *Phys. Rev. Lett.*, **91**, 176101 (2003).
6. P. Borri, W. Langbein, S. Schneider, U. Woggon, R. L. Sellin, D. Ouyang, and D. Bimberg *Phys. Rev. Lett.* **87**, 157401 (2001).
7. P. Michler, A. Kiraz, C. Becher, W. V. Shoenfeld, P. M. Petroff, L. Zhang, E. Hu, and A. Imamoglu, *Science* **290**, 2282 (2000).
8. A. Zrenner et al., *Nature (London)* **418**, 612 (2002).
9. H. Kosaka, A. A. Kiselev, F. A. Baron, K. W. Kim, E. Yablonovitch, *Electronics Letters* **37**, 464 (2001).
10. A. A. Kiselev, K. W. Kim, E. Yablonovitch, *Appl. Phys. Lett.* **80**, 2857 (2002).
11. Y. Kato, R. C. Myers, D. C. Driscoll, A. C. Gossard, J. Levy, D. D. Awschalom, *Science* **299**, 1201 (2003).
12. K. Nishi, H. Saito, S. Sugou, J.-S. Lee, *Appl. Phys. Lett.* **74**, 1111 (1999).
13. G. Medeiros-Ribeiro, E. Ribeiro and H. Westfahl Jr., *Appl. Phys. A* **80**, 4229 (2003).
14. R. L. Maltez, E. Ribeiro, W. Carvalho Jr., D. Ugarte, and G. Medeiros-Ribeiro, *J. Appl. Phys.* **94**, 3051 (2003).
15. M. K. K. Nakaema, F. Iikawa, M. J. S. P. Brasil, E. Ribeiro, G. Medeiros-Ribeiro, W. Carvalho Jr., M. Z. Maialle, and M. H. Degani, *Appl. Phys. Lett.* **81**, 2743 (2002).
16. E. Ribeiro, A. O. Govorov, W. Carvalho Jr., and G. Medeiros-Ribeiro, *Phys. Rev. Lett.* **92**, 126402 (2004).
17. Y. Aharonov and D. Bohm, *Phys. Rev.* **115**, 485 (1959).
18. A. Abragam, *Electron Paramagnetic Resonance of Transition Ions* (Clarendon Press, Oxford 1970).
19. S. Lindemann, T. Ihn, T. Heinzel, W. Zwerger, K. Ensslin, K. Maranowski, A. C. Gossard, *Phys. Rev. B* **66**, 195314 (2002).
20. G. Medeiros-Ribeiro, M. V. B. Pinheiro, V. L. Pimentel, E. Marega, *Appl. Phys. Lett.* **80**, 4229 (2002).
21. J. R. Petta, D. C. Ralph, *Phys. Rev. Lett.* **89**, 156802 (2002).
22. M. Fricke, A. Lorke, J. P. Kotthaus, G. Medeiros-Ribeiro, P. M. Petroff, *Europhys. Lett.* **36**, 197 (1996); R. J. Warburton, C. S. Durr, K. Karrai, J. P. Kotthaus, G. Medeiros-Ribeiro, P. M. Petroff, *Phys. Rev. Lett.* **79**, 5282 (1997); P. Hawrilak, G. A. Narvaez, M. Bayer, A. Forchel, *Phys. Rev. Lett.* **85**, 389 (2000).
23. V. Fock, *Z. Phys.* **47**, 446 (1928).
24. G. Medeiros-Ribeiro, F. G. Pikus, P. M. Petroff, A. L. Efros, *Phys. Rev. B* **55**, 1568 (1997).

CHAPTER 8
DEVICES

8.1. Electronic Devices
8.2. Photonic Devices
8.3. Frontiers in Device Physics

Remarkably Strong Image Potential Effects in SrTiO$_3$/Si and HfO$_2$/Si Tunneling Structures

T. A. S. Pereira*, J. A. K. Freire*, J. S. Sousa**, V. N. Freire*, G. A. Farias*,
L. M. R. Scolfaro**, J. R. Leite**, E. F. da Silva Jr. ***

Departamento de Física, Universidade Federal do Ceará, Caixa Postal 6030, 60455-900 Fortaleza, Ceará, Brazil
**Instituto de Física, Universidade de São Paulo, Caixa Postal 66318, 05315-970 São Paulo, São Paulo, Brazil*
Departamento de Física, Universidade Federal de Pernambuco, 50670-901 Recife, Pernambuco, Brazil

Abstract. The contribution of the charge image to the tunneling properties of electrons through Si/SrTiO3/Si and Si/HfO$_2$/Si single barriers is studied. It gives rise to a structure interfacing potential which blue shifts several meV the transmission resonance peaks.

INTRODUCTION

Nowadays research focusing high-dielectric constant (high-κ, $\epsilon \geq 20$) materials aims to extending Moore's law to equivalent gate oxide thickness below 10 Å through silicon dioxide replacement [1,2]. SrTiO$_3$ and HfO$_2$ are two important high-κ candidates, the former with dielectric $\epsilon_{SrTiO3} \cong 300$ and the latter $\epsilon_{HfO2} \cong 25$, while the conduction band offsets are $\Delta E_{c,SrTiO3} \cong 0.1$ eV and $\Delta E_{c,HfO2} \cong 1.5$ eV in respect to silicon. The dielectric mismatches Si-SrTiO$_3$ and Si-HfO$_2$ are much higher than in GaAs-AlAs and Si-SiO$_2$, indicating that charge image effects may be important in high-κ/Si-based heterostructures. Recently, Pereira et al. [3] have demonstrated that charge image effects in Si-SrTiO$_3$ and Si-HfO$_2$ single quantum wells give rise to structured interfacing potentials which can trap carriers. Charge image related upward shifts of the carrier recombination energy up to few hundred meVs were obtained by Pereira et al. [3].

In this work, calculations are performed to evaluate image potential effects in Si/SrTiO$_3$ and Si/HfO$_2$ tunneling structures. Due to the strong dielectric mismatch and barrier height, the charge image related displacements of the resonance peaks in these tunneling structures are remarkably (several meV) in comparison with that of GaAs/Al$_{0.3}$Ga$_{0.7}$As and Si/SiO$_2$ systems.

THE TUNNELING

The HfO$_2$ and SrTiO$_3$ carrier's effective masses (not yet measured!) were obtained through all-electronself-consistent linear augmented plane-wave *ab initio* calculations, within the framework of the local density approach and the generalized gradient approximation taking into account full--relativistic contributions [4,5]. Charge image effects on the tunneling properties were studied by solving Schrödinger-like equations [6,7] for the single Si/SrTiO$_3$/Si and Si/HfO$_2$/Si barriers $V(z)=V_o(z)+V_{im}(z)$, where V_o is due to the band offset, and V_{im} is the charge image contribution to the barrier, which was calculated resorting to the classical approach of Stern [8]. Table 1 presents the parameters used in the numerical calculations. The charge image gives rise to potential structures (sharp and deep) close to the barrier interfaces, which can trap carriers in Si/SrTiO$_3$ but not in Si/HfO$_2$ heterostructures (see Fig.1).

Table 1 Si, HfO$_2$ and SrTiO$_3$ parameters used in the numerical calculations.

	Si	HfO$_2$	SrTiO$_3$
$m_e^\perp/m_0$	0.173	0.836	0.676
E_g (eV)	1.1	5.6	3.3
ε	11.7	25	300

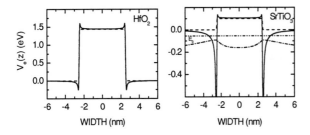

FIGURE 1. Left: Si/HfO$_2$/Si potential profile with (solid) and without (dashed) the charge image contribution; Right: Si/SrTiO$_3$/Si potential profile with (solid) and without (dashed) the charge image contribution.

The charge image contribution increases the mean width of the barriers, which produces a decrease of the intensity and an upward shift of the resonance peaks in the tunneling structures, as depicted in Fig. 2. The first transmission resonance peak in both systems is blue shifted by approximately 10 meV. The blue shift is shown to increase with the order of the resonance.

Recently, wave-mechanical calculations of leakage current through-stacked dielectrics for nanotransistor metal-oxide-semiconductor design were performed [9]. Approaches for determining the effective tunneling mass of electrons in HfO$_2$ and other high-κ alternative gate dielectrics for advanced CMOS devices were proposed [10,11]. However, these works have not considered the contribution of charge image effects, which according our results must be take into account if an improved description of data related with tunneling through high-κ dielectrics is sought.

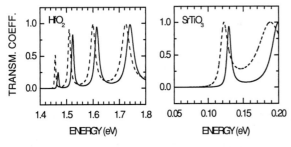

FIGURE 2. Transmission coefficient of electrons through single Si/HfO$_2$/Si (left) and Si/SrTiO$_3$/Si (right) barriers, calculated disregarding (dashed) and considering (solid) the charge image contribution.

A drawback of the present calculations is the sharp interface picture used to obtain the charge image contributions. Actual Si/high-κ dielectrics interfaces are complicated, with SiO$_2$ layers and/or smooth interfacial regions. Work is in progress for the calculation of the charge image contribution to the tunneling properties of graded Si/high-κ dielectrics/Si tunneling barriers.

In conclusion, the charge image contribution to the electron tunneling behavior cannot be disregarded in high-κ systems, being as important as graded interface effects. Surprisingly, it gives rise to a structured interfacing potential close to the barriers which can trap carriers in Si/SrTiO$_3$/Si but not in Si/HfO$_2$/Si tunneling structures. Despite developments on the growth of high-κ dielectrics in silicon, there is no report yet of Si/SrTiO$_3$/Si and Si/HfO$_2$/Si tunneling devices for a direct confirmation of the results presented here. It is the authors' hope that this work will stimulate efforts to examine the surprisingly charge image potential effects in these and similar high-κ dielectrics systems.

ACKNOWLEDGMENTS

T. A. S. Pereira. would like to acknowledge the graduate fellowship at the Physics Department of the Universidade Federal do Ceará received from FUNCAP during the realization of this research. The authors would like to acknowledge the financial support received from the Science Funding Agency of the Ceará (FUNCAP) and the Brazilian National Research Council (CNPq) through the grant NanoSemiMat Project #550.015/01-9.

REFERENCES

1. P. Packan, Science **285**, 2079 (1999).
2. G. D. Wilk, R. M. Wallace, and J. M. Anthony, J. Appl. Phys. **89**, 5243 (2001), and references therein.
3. T. A. S. Perreira, J. A. K. Freire, V. N. Freire, G. A. Farias, L. M. R. Scolfaro, J. R. Leite, E. F. da Silva Jr., Appl;. Phys. Lett., accepted for publication (2004).
4. M. Marques, L. K. Teles, V. Anjos, L. M. R. Scolfaro, J. R. Leite, V. N. Freire, G. A. Farias, E. F. da Silva Jr., Appl. Phys. Lett. **82**, 3074 (2003).
5. J. C. Garcia, L. M. R. Scolfaro, J. R. Leite, V. N. Freire, A. T. Lino, G. A. Farias, E. F. da Silva Jr., submitted to Appl. Phys. Lett. (2004).
6. Y. Ando and T. Itoh, J. Appl. Phys. **61**, 1497 (1987).
7. Ji-Wei Wu, Solid. State Communic. **69**, 1057 (1989).
8. F. Stern, Phys. Rev. B **17**, 5009 (1978).
9. M. Le Roy, E. Lheurette, O. Vanbesien, D. Lippens, J. Appl. Phys. **93**, 2966 (2003).
10. W. J. Zhu, T. P. Ma, T. Tamagawa, J. Kim, Y. Di, IEEE Elec. Dev. Lett. **23**, 97 (2002).
11. C. L. Hinkle, C. Fulton, R. J. Nemanich, G. Lucovsky, Microelec. Eng. **72**, 257 (2004).

Electrical Properties Of Modulation Doped Si/SiGe Heterostructures Grown On Silicon On Insulator Substrates

K. Alfaramawi, A. Sweyllam, L. Abulnasr, S. Abboudy, E. F. El-Wahidy

Physics Department, Faculty of Science, Alexandria University, Alexandria, Egypt

L. Di Gaspare, F. Evangelisti

Dipartimento di Fisica, universita degli Studi Roma Tre, Via della vasca Navale 84, 00146 Roma Italy.

Abstract. We report on the growth and characterization of high-mobility 2DEGs on tensile strained Si grown on cubic SiGe alloy on SOI substrates. The electrical properties were investigated in the 4.2-300 K temperature range as a function of growth conditions and sample structure. Hall mobilities as high as 82000 cm^2/V.s at T=4.2 K were obtained in the best 2DEGs on SOI. We found that the main mechanism limiting the mobility at low temperature is the ionized impurity scattering.

INTRODUCTION

Si/SiGe heterostructures are grown in such a way that the Si channel, epitaxially grown on a relaxed SiGe buffer, undergoes a tensile strain [1]. In order to produce the confinement of the carriers within the Si-channel, a tensile Si film must be grown on top of a relaxed Si$_{1-x}$Ge$_x$ layer [2]. The best results are obtained by using a compositionally graded SiGe buffer layer grown on a Si surface followed by a thick SiGe layer at a fixed composition [3,4]. In this way, it is possible to confine the dislocations within the graded layer and to obtain a dislocation density at the sample surface lower than 10^6 cm^{-2} [5].

EXPERIMENTAL

Silicon on Insulator (SOI) substrate was cleaned in situ by heating up to 1100 °C in H$_2$ atmosphere. The virtual substrate (VS) was grown at 700 °C on top of 3 μm thick Si buffer layer deposited at 850 °C. The graded part of the VS consisted of seven alloy layers with the Ge concentration increasing from 7 % up to 19 %. Each layer was 400 nm thick and the top layer at the final concentration was 2.4 μm thick. The two dimensional electron gases (2DEGs) were obtained by depositing the following layers on the VS sequentially: (i) a tensile Si channel (thickness 11 nm), (ii) a Si$_{0.81}$Ge$_{0.19}$ spacer layer (thickness: 5.7 to 11.4 nm) and (iii) a n-doped Si$_{0.81}$Ge$_{0.19}$ layer (thickness 5.7 to 11.4 nm). The structures were completed by a final 15 nm Si cap layer to protect the system from indepth oxidation.

RESULTS AND DISCUSSION

Figure 1 shows a TEM image of sample A. All the interfaces are very sharp and the thicknesses of the films are uniform. Figure 2 shows the carrier density and the mobility as a function of temperature (T) for sample A. The mobility increases by lowering down the temperature. The recorded mobility at 4.2 K is 82000 cm^2/V.s. Series of samples were grown with thicker doped layer, as 5-10 times as that of sample A. Nominal doping concentration was chosen in order to maintain the same total amount of dopant atoms.

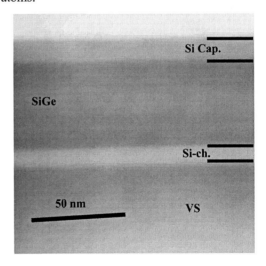

FIGURE 1. Cross-sectional TEM image of 2DEG on SOI

Other series of samples with thicker spacer layer, as two times as that of sample A, were also grown. Table 1 shows the types of samples and their characteristics. Figure 3 shows a plot of the mobility versus 1/T for samples of table 1.

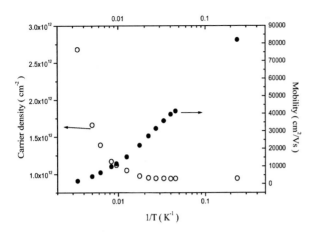

FIGURE 2. The carrier density and the mobility versus 1/T for SC sample.

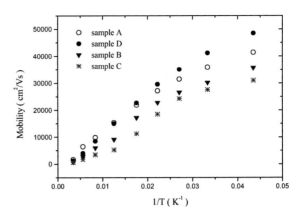

FIGURE 4. The mobility versus 1/T for different samples grown on Si(001) substrate.

TABLE 1. Sample codes and their characteristics.

Sample code	Characteristics
A	Standard
B	Thick doped layer (28.5 nm)
C	Thick doped layer (57 nm)
D	Thick spacer layer (11.4nm)

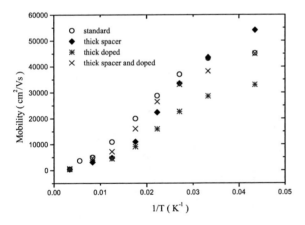

FIGURE 3. The mobility versus 1/T for the 2DEG samples grown on SOI substrate.

With the same system followed by using SOI substrate we have grown series of samples on Si(001) substrates. Figure 4 represents the mobility versus 1/T for such series.

CONCLUSIONS

In this study we have managed to grow structures with high mobility 2DEGs on SOI substrate using the same deposition procedure adopted for the growth on Si(001) substrates. The increase of the spacer layer thickness leads to an increase in the mobility, while the increase of the doped layer thickness causes a drop in the mobility at the same temperature. The increase of the spacer layer thickness is thought to minimize the electron-ionized impurity scattering leading to higher mobility.

ACKNOWLEDGMENTS

The authors acknowledge the support from the Italian CNR-MADESS II project. K. Alfaramawi acknowledges the support of ICTP, Trieste, Italy.

REFERENCES

1. Ismail, K., Mayerson, B. S. and Wang, P. J, *Appl. Phys. Lett.*, **58**, 2117(1991).
2. Di Gaspare, L.,. Alfaramawi, K.,. Evangelisti, F.,. Palange, E., Barucca , G.and Majni, G. ,*Appl. Phys. Lett.* ,**79** ,2031-2033(2001).
3. Di Gaspare, L., Scappucci, G., Palange, E., Alfaramawi, K., Evangelisti, F., ,Barucca, G. and Majni,G. , *Materials Science and Engineering* **B89**, 346-349(2002).
4. Schaffler, F., Tobben, D., Hertzog, H.J., Abstreiter,G. and Hollander, B., *Semicond. Sci. Technol.*, **7**, 260(1992).
5. Di Gaspare, L., Palange, E., Capellini, G. and Evangelisti, F. , *J. Appl. Phys.* ,**88**, 120-123(2000).

Transient Quantum Drift-Diffusion Modelling of Resonant Tunneling Heterostructure Nanodevices

Nenad Radulovic, Morten Willatzen, and Roderick V.N. Melnik

Mads Clausen Institute for Product Innovation, University of Southern Denmark, Grundtvigs Alle 150, DK-6400 Sonderborg, Denmark

Abstract. In the present work, double-barrier GaAs/AlGaAs resonant tunneling heterostructure nanodevices are investigated. Numerical results, obtained by in-house developed software, based on a transient quantum drift-diffusion model, are presented and discussed. In the model, quantum effects are incorporated via parameter dependencies on the carrier density gradients. Particular emphasis is given to the carrier densities and quasi-Fermi levels as a function of applied bias, and electrostatic potential profiles inside the resonant tunneling diodes and superlattices.

INTRODUCTION

In what follows, two types of GaAs/AlGaAs resonant tunneling heterostructures are analyzed. The double-barrier resonant tunneling diodes (RTD) with quantum well (QW) width of 2, 5, and 10 nm, barriers and space layers 5 nm each, and contact regions of 26.5, 25, and 22.5 nm, respectively; and the superlattices (SL) made of 2, 3, and 5 QWs, with QWs, barriers and space layers 5 nm each, and contact regions 25 nm. The active regions (QWs, barriers and space layers) are n-type (10^{21} m^{-3}), contact regions are n$^+$-type (10^{24} m^{-3}), and only the barriers are made of AlGaAs (65% of Al). The bias is applied on the right contact of the nanodevices, while the left contact is grounded. The devices operate at $T = 77$ K.

THEORY AND NUMERICAL SIMULATION

In the numerical simulation, a transient quantum drift-diffusion model, which is a set of macroscopic semiconductor equations, is employed. The model is originally proposed by Ancona et al. [1], and references therein, and is based on the idea to include density-gradient correction into a generalized transport equation. Hence, the model is powerful enough to account for quantum effects and, at the same time, computationally not very expensive. The model is presented in [2], and a detailed derivation of the model is given in the references therein. The scaled system of three coupled partial differential equations, where the dependent variables are electron density n, quasi-Fermi level F, and electrostatic potential V, is discretized in time and space. The system of nonlinear discretized equations is solved numerically, using matrix form of the Newton iteration procedure.

RESULTS

The maximums of the carrier density, inside the QW of the RTDs (Fig.1), are strongly nonlinear as a function of bias. For small bias, the increase is rapid, and afterwards the maximums reach "saturation" in three orders of magnitude higher region. The increase is monotonous, and only for the RTD with QW = 10 nm, and bias higher than 0.7 V, a light decrease of the maximum value, is observed. In the case of the SL (Fig.2), it is noticeable that the maximums are quite different in different QWs, for the same bias, except for the balance-bias (0.3 V), when the maximums in all QWs reach almost the same value. In the case of RTDs, the maximums of the carrier density for zero bias are strongly different, while in the case of SLs, the maximums are almost the same, for all QWs in all SLs under consideration. In the SLs, the maximums are definitely not monotonously increasing functions. The same behavior of the maximums is observed in the case of the SLs with 2 and 3 QWs where the balance-biases are 0.175 V and 0.225 V, respectively. It is common for all SLs that the maximums in the QW close to the grounded contact firstly show a decrease and afterwards reach the highest value for the bias above the balance-bias. The balance-bias has always a value less than the peak-voltage in the N-shape I-V characteristics.

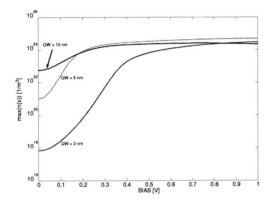

FIGURE 1. The maximums of the carrier density inside the QW of the RTDs as a function of the bias.

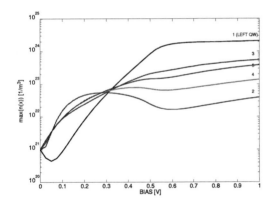

FIGURE 2. The maximums of the carrier density inside the QWs of the SL (with 5 QWs) as a function of the bias.

The quasi-Fermi levels inside the QW of the RTDs (Fig.3), are weakly nonlinear. The same behavior is observed in the case of the quasi-Fermi levels inside the QWs of the SL (Fig.4). In general, the nonlinearity is stronger in the QWs, which are closer to the grounded contact of the SLs. The quasi-Fermi levels in the SLs are not equidistant. The first derivative of the quasi-Fermi levels inside the QWs, as a function of the bias, can indicate the presence and position of the current peak in the N-shape I-V characteristics. Inside the RTDs and the SLs, the quasi-Fermi levels have constant values (in steady state) in the QWs, and in the right contact regions, while in the barriers and in the left contact regions they are definitely not constant.

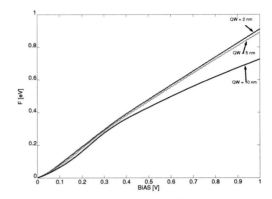

FIGURE 3. The quasi-Fermi levels inside the QW of the RTDs as a function of the bias.

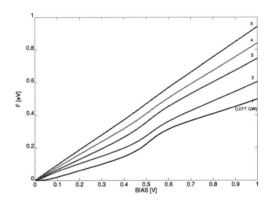

FIGURE 4. The quasi-Fermi levels inside the QWs of the SL (with 5 QWs) as a function of the bias.

The electrostatic potential profiles inside the RTDs and the SLs are nonlinear and the potential drop mainly occurs in the active region (two linear drops with different slopes), and the left contact region (nonlinear drop), while the potential is almost constant in the right contact region (negligible drop).

REFERENCES

1. Ancona, M.G., and Iafrate, G.J., *Phys. Rev. B* **39**, 9536-9540 (1989).
2. Radulovic, N., Willatzen, M., and Melnik, R.V.N., "Resonant Tunneling Heterostructure Devices – Dependencies on Thickness and Number of Quantum Wells", in *Lecture Notes in Computer Science* 3045, edited by A. Lagana et al., ICCSA 2004 Conference Proceedings, Berlin: Springer-Verlag, 2004, pp. 817-826.

The Role of Non-abrupt Interfaces in SiC MOS Devices: Quantum Mechanical Simulations and Experiments

[1]E. L. de Oliveira, [1]J. S. de Sousa, [1]V. N. Freire, [2]E. A. de Vasconcelos and [2]E. F. da Silva Jr.

[1]*Departamento de Física, Universidade Federal do Ceará, Campus do Pici, Caixa Postal 6030, 60455-970 Fortaleza, Ceará, Brazil*
[2]*Departamento de Física, Cidade Universitária 50670-901 Recife, Pernambuco, Brazil*

Abstract. A numerical analysis of SiC/SiO$_2$ MOS devices is performed focusing on the role of non-abrupt interfaces. We show that non-abrupt interfaces affects the shape of capacitance-voltage curves increasing accumulation and inversion capacitances in comparison to ideal abrupt models which may lead to underestimates of physical characteristics of devices.

INTRODUCTION

Silicon Carbide (SiC) is a promising material for high-power, high-speed and high-temperature electronics, but the successful realization of these applications highly depends on the quality of the gate insulator [1]. Long term operation of SiC MOS with thermal SiO$_2$ at temperatures higher than 250°C may be not practical [2] and new insulators and processes have been investigated to improve reliability [3]. Obtaining a high quality oxide has proven to be a challenge due to the presence of carbon at the SiC/SiO$_2$ interface and nonabrupt interfaces [4-5]. A number of studies reported the effects of interfacial sub-oxide (SiO$_x$) transition regions on Si-based devices [6]. However, much less work has been done regarding SiC-based devices. In this work, we performed theoretical calculations focusing on the role of non-abrupt interfaces and how it affects the interpretation of experimental results by means of ideal models with abrupt interfaces.

MODEL

We address SiC/SiO$_2$ MOS device with a metallic Al gate and 10^{17} cm^{-3} p-doped substrate, which are fully ionized. In the semiconductor/oxide interface, we consider an intermediate SiO$_x$ layer represented by a linear variation of both dielectric constant and band gap discontinuity in this region. Morever, we consider that there is no doping in this region. Electrons in the inversion layers are described both semi-classical and quantum mechanically [7]. Holes are described semi-classically. Our semi-classical (SC) model is two-fold. We consider that the interfacial layer has no thermal generated mobile charges (Boltzmann occupation statistics). This model is named as M1. The opposite situation is named as M2. We also disregard fixed oxide charges and interfaces traps.

RESULTS

Fig. 1 shows a low frequency capacitance-voltage (CV) curve of a 3C-SiC/SiO$_2$ MOS device with 10 nm wide oxide thickness (SiO$_2$ + SiO$_x$). SC models M1 (dash-doted line) and M2 (dashed line) present slightly different behaviors. They both have larger capacitances in accumulation (C_{AC}) and inversion (C_{INV}) regimes in comparison to the ideal abrupt case (solid line). C_{AC} is larger for M1, while C_{INV} are nearly equal for both M1 and M2. Ideal models state that $C_{INV} \cong C_{AC}$. Fig. 1 shows that this is not true for M2 where $C_{INV} > C_{AC}$. Moreover, oxide thicknesses (T_{OX}) are, in general, directly extracted from CV measurements by using the expression $C_{AC} = \varepsilon A/T_{OX}$, where T_{OX} is the oxide thickness and A is the gate area. Therefore, if samples are not appropriately characterized by complimentary measurements (eg. electronic microscopy), this expression can underestimate the oxide thickness and disregard the existence of non-abrupt interfaces.

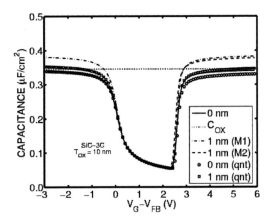

FIGURE 1. Low frequency CV curve of a 3C-SiC/SiO$_2$ MOS device with abrupt and non-abrupt interfaces (1 nm wide). The total oxide thickness T_{OX} includes both SiO$_2$ and sub-oxides.

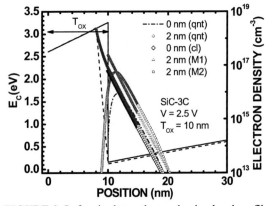

FIGURE 2. Left axis shows the conduction band profile of abrupt (solid) and non-abrupt (dashed) interfaces of SiC-based MOS device. The inversion charge densities for all models addressed in this work are shown in the right axis.

Concerning quantum calculations, we obtained that C_{INV} are, in general, smaller than SC ones. For abrupt interfaces (circles), it is known that the center of the inversion charge density is shifted away from the interface (see Fig. 2) because of the shape of the wavefunctions (WF's). This effectively creates an additional layer that separates semiconductor charges from the gate charges, which decreases the C_{INV} in comparison to SC models. This is more important in the limit of small oxide thicknesses [7]. When nonabrupt interfaces are considered (squares), C_{INV} increases in comparison to the abrupt case, and it becomes nearly equal to the abrupt SC result. Due to the non-abruptness of the potential barrier, WF's are allowed to penetrate more than in the abrupt case. This effect decreases the separation between gate charges and the center of the inversion charges and increases C_{INV}.

In conclusion, our numerical simulations SiC/SiO$_2$ MOS devices have shown that non-abrupt interfaces affects the shape of CV curves increasing C_{AC} and C_{INV} in comparison to ideal abrupt models. Therefore, if experimental results are interpreted by means of ideal models, physical characteristics of devices can be underestimated. Moreover, there are additional differences between quantum and SC models which become non-negligible for very thin oxide thicknesses. Within the authors knowledge, these effects have not been observed yet and we expect to motivate experimental works in this matter.

ACKNOWLEDGMENTS

E. L. de Oliveira would like to acknowledge the graduate fellowship at the Physics Department of the Universidade Federal do Ceará received from CAPES during this research. The authors would like to acknowledge the financial support received from the Brazilian National Research Council (CNPq) through the grant NanoSemiMat Project #550.015/01-9.

REFERENCES

1. B.J. Baliga, Microelectronic Engineering **28**, 177 (1995).

2. M. M. Maranowski and J. A. Cooper Jr., IEEE Trans. Electron Dev. **46**, 520 (1999).

3. X. Wang, Z. J. Luo and T. P. Ma, IEEE Trans. Electron Dev. **47**, 458 (2000).

4. R. Buczko, S. J. Pennycock and S. T. Pantelides, Phys. Rev. Lett. **84**, 943 (2000).

5. V. V. Afanas'ev, M. Bassler, G. Pensl, M. J. Schulz and E. S. von Kamienski, J. Appl. Phys. **79**, 3108 (1996).

6. H. Yang, H. Niimi, Y. Wu, G. Lucovsky, J. W. Keister and J. E. Rowe, Microelectronic Engineering **48**, 307 (1999).

7. D. Vasileska, Journal of Modeling and Simulation of Microsystems **1**, 49 (1999).

A Comprehensive Model for Low Frequency Noise in Poly-Si Thin-Film Transistors

I. K. Han[1], J. I. Lee[1*], M. B. Lee[2], S. K. Chang[3], and A. Chovet[4]

[1]*Korea Institute of Science and Technology, Seoul 136-791, Korea*
[2]*Kyungpook National University, Daegu 702-701, Korea*
[3]*Yonsei University, Seoul 120-749, Korea*
[4]*IMEP, ENSERG, CNRS/INPG, Grenoble 38016, France*

Abstract. This paper presents a comprehensive model for 1/f noise in poly-Si thin-film transistors in the linear regime where the current is controlled by the gate bias with a small fixed drain bias. For small gate biases the conduction and noise generation is dominated by thermionic emission at the grain boundary where as for large gate biases the conduction and noise generation is dominated by drift in the grain bulk region. The generation of the 1/f noise in the grain bulk can explained by the 'Unified Model' for crystalline Si-MOSFET's. However, the noise from grain boundary should be explained by number fluctuation models such as thermal activation model via barrier height modulation. An expression for the critical barrier height or gate bias, which severs these two regimes, is derived.

INTRODUCTION

Polycrystalline silicon (poly-Si) thin film transistors (TFT's) have attracted a lot of research efforts due to their important application such as active elements in driving circuits for liquid crystal displays. 1/f noise is an important characteristic for diagnosis and improvement of the device performance. The grain boundary plays an essential role in the conduction for poly-Si TFT's and can lead the generation of noise [1]. In this paper, we first consider the current conduction with fixed small drain bias and varying gate bias. The noise generation mechanisms are discussed in two separate current conduction regimes, thermal emission regime with high barrier height, and drift conduction regime with low barrier height. Also an analytic expression for the critical barrier height is derived, which differentiates the two regimes.

CURRENT CONDUCTION

Conduction in poly-Si films is believed to be due to the thermionic emission across the barriers formed at the grain boundary. However, in TFT's where the current is controlled by gate bias, the barrier height can so low that the drift conduction in grain bulk can be important at high gate bias. The conducting channel can be separated into two regions, the depleted grain boundary region and the neutral grain bulk region. The grain boundary region can be considered as a symmetric Schottly Barrier and the drain current of a TFT with the width W, can be expressed as follows,

$$I_d = W t_{ch} A^{**} e^{\frac{-qV_b}{kT}} \tanh\left[\frac{q(V_d - R_s I_d)}{2kTN_g}\right], \quad (1)$$

where, A^{**} is the effective Richardson constant and the barrier height V_b is related to the gate bias as follows [2],

$$V_b = \frac{t_{ch}(qN_{GB})^2}{8\varepsilon C_{ox}(V_g - V_0)}. \quad (2)$$

N_{GB} is the defect density in the grain boundary, ε the dielectric constant of poly Si, V_0, a characteristic constant of the grain boundary. If source/drain resistance is negligible, R_s is the resistance of the two-dimensional electron gas channel in the grain bulk region and can be expressed as,

$$R_s = \frac{L}{WC_{ox}(V_g - V_{th})\mu_{eff}}, \quad (3)$$

where L is the length of the channel, V_{th} the threshold voltage, and μ_{eff} the effective mobility of the electron. $N_g (=L/a)$ is the number of grain along the direction of conduction with a the grain size, assuming a uniform distribution of grains.

NOISE GENERATION

The total drain voltage noise of the TFT can be represented as follows, assuming the noise sources in the two regions are uncorrelated,

$$S_{V_d} = S_{Ig}R_{gdy}^2 + S_{Ib}R_s^2. \quad (4)$$

The current conduction and noise generation is dominated by grain boundary in this regime. One can utilize number fluctuation model for the Schottky barrier. For carrier number fluctuation, there are three mechanisms for current fluctuation by trapping and detrapping of carriers at the depletion region of the barrier and the interface, via barrier height modulation. The mechanisms are thermal activation, tunneling involving bulk traps, and random walk of carriers into the interface states [3]. At room temperature for low or moderately doped samples, thermal activation dominates the tunneling. The current noise spectral power density for multi grain resistor due to thermal activation is given as,

$$\frac{S_{Ig}}{I_d^2} = \frac{1}{f}\left(\frac{q}{4\varepsilon}\right)^2 \frac{qD_t}{Wt_{ch}N_g\pi N_d V_b}, \quad (5)$$

where D_t is the bulk trap states density in the grain boundary depletion region, N_d the doping concentration, ε the dielectric constant of Si. The model predicts basically quadratic current dependence of the noise density.

Low frequency noise in the grain bulk is essentially the same as the case of crystalline Si MOSFET's and the 'Unified Model' [4] can be utilized such as,

$$\frac{S_{I_d}}{I_d^2} = \left(1 + \alpha\mu_{eff}C_{ox}\frac{I_d}{g_m}\right)^2 \frac{g_m^2}{I_d^2} S_{V_{fb}}, \quad (6)$$

where g_m is the transconductance of the device and $S_{V_{fb}}$ is the flat band voltage noise intensity which can be expressed as,

$$S_{V_{fb}} = \frac{kT\lambda(qN_{ox})^2}{fWLC_{ox}^2}. \quad (7)$$

From eq. (4), one can see the current noise is weighted by the square of the (dynamic) resistances of these two regions. Then, one can define the critical barrier height, V_{bc}, which gives the portion of the applied drain biases at the grain boundary and at the grain bulk the same, Utilizing eqs. (1)-(3), with proper approximation, one can derive expressions for the critical barrier height that satisfies the condition defined above such as,

$$V_{bc} = \frac{kT}{q}\ln\left[\frac{qt_{ch}A^{**}T^2(l-2a)}{2kTC_{ox}(V_g - V_{th})\mu_{eff}}\right]. \quad (8)$$

Note that the critical barrier height for two-dimensional case does not directly depend on the doping level. Grain size and the gate oxide thickness are the major parameters to decide the value of the critical barrier height. Empirical relationship between 1/f noise and the grain boundary potential barrier height in poly-Si TFT's has been observed and the existence of the critical barrier height independent of doping density has been confirmed [5].

ACKNOWLEDGMENTS

This work was supported in part by the Ministry of S&T via Nano R&D Program and NRL (HIK) Program.

REFERENCES

1. Rigaud, D., Valenza M., and Rhayem, J., *IEE Proc.- Circuit Devices Syst.* **149**, 75-82 (2002).

2. Chen, H.-L. and Wu, C. W., *IEEE Trans. Electron Devices* **45**, 2245-2247 (1998).

3. Lee, J. I., Brini, J., Chovet, A., and Dimitriadis, C. A., *Solid-St. Electron.* **43**, 2185-2189 (1999).

4. Ghibaudo, G., Roux, R., Nguyen-Duc, Ch., Valestra, F., and Brini, J., *phys. stat. sol.* (a) **124**, 571-580 (1991).

5. Angelis, C. T. *et al.*, *Appl. Phys. Letters* **76**, 118-120 (2000).

Hydrogen Cleaving of Silicon at the Sub-100-nm Scale

O. Moutanabbir[*], B. Terreault, N. Desrosiers, A. Giguère and G. G. Ross

INRS-EMT, Université du Québec, 1650 Boul. Lionel-Boulet, Varennes, Québec J3X 1S2, Canada

M. Chicoine and F. Schiettekatte

Département de physique, Université de Montréal

Abstract. Silicon-on-insulator (SOI) fabrication is based on ion cleaving. Using low energy ions to produce such structures with sub-100 nm dimensions, two effects with implications in silicon physics and device engineering were found: i) H-ion blistering (or cleaving when bonded to a "handle" wafer) can only be achieved in a narrow fluence window, and ii) for D-ion blistering the window is shifted up by a surprising factor of 2 to 3. Here we briefly examine the underlying mechanisms.

Silicon-on-Insulator (SOI) is a most promising approach to achieve low voltage, low power integrated circuits. SOI is a thin layer of single crystal Si bonded to a dielectric layer on top of a wafer. The *Ion-Cut* process of layer splitting by ion implantation and thermal annealing is a versatile method of transferring a thin layer from one substrate onto another substrate.[1] It is based on the blistering effect, which can be induced in Si by H-ion implantation. Cleavage results from the lateral coalescence of pressurized sub-surface gas bubbles, which form *via* a thermal evolution in Si-H bonding ending in the formation of planar microcavities.[2,3] Using low-keV ions in order to obtain thinner ion-split layers has revealed new facts:[4] i) cleaving is only achieved in a narrow fluence window; and ii) H and D display a giant isotope effect. Here we briefly examine the microscopic mechanisms behind these observations. The temperature evolution of Si-H/D bonding was studied by Raman Scattering Spectroscopy (RSS) upon step-wise annealing with successive ramps of 20 ^{0}C/min, up to 550 °C.

With H-ions, maximum blistering occurs for 2×10^{16} H/cm^2 [Fig. 1c]. The Si-H stretch mode region of the Raman spectra [Fig. 2] shows discrete modes associated with isolated point defects superimposed on a broadband.[5] At 200 °C, the main features are the "low frequency" (LF) broadband ($k < 2050$ cm^{-1}) due to monohydride-terminated multivacancies (V_nH_m, $m \leq n$) in disordered regions, and a few peaks at 2025 cm^{-1} (divacancies V_2H or V_2H_2), ~2125 cm^{-1} (VH_2 or a coupled mode $Si(100){:}H$ involving H atoms adsorbed on (001) internal surfaces), and 2182 cm^{-1} (VH_3 and/or V_2H_6). At 300°C, we see a general attenuation of the LF broadband. At ~400 °C, the reduction is more pronounced for the V_nH_m, and accompanied by the emergence at higher frequency (HF, > 2050 cm^{-1}) of VH_n peaks. At 500 °C, the spectrum is dominated by the 2182 cm^{-1} peak (VH_3/V_2H_6) and the feature at ~2120 cm^{-1} [$Si(100){:}H$]. Previous work[2] attributes a crucial role to that feature. There is a loss of bound H and agglomeration of the remaining trapped H at existing vacancies. Blistering develops simultaneously.

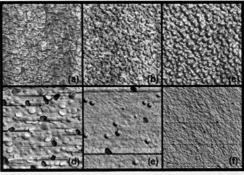

FIGURE 1. AFM images of Si(100) implanted with 5-keV H ions to fluences of: (a) 1×10^{16} H cm^{-2}, (b) 1.5×10^{16} H cm^{-2}, (c) 2×10^{16} H cm^{-2}, (d) 3.5×10^{16} H cm^{-2}, (e) 4.5×10^{16} H cm^{-2}, (f) 6×10^{16} H cm^{-2}, and then annealed up to 550 °C.

---

[*] Author to whom correspondance should be addressed:
Present address: Dept. of Applied Physics and Physico-Informatics
Keio University, 3-14-1, Hiyoshi, Yokohama 223-8522 Japan
electronic mail: moutanab@appi.keio.ac.jp

The interplay between VH_3/V_2H_6, VH_4 and Si-H on internal surfaces is compatible with theoretical scenarios for the nucleation of *H-platelets* in silicon.[6]

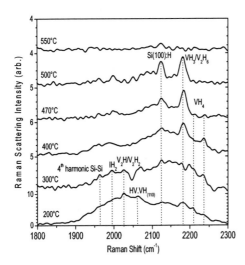

FIGURE 2. Temperature evolution of Raman spectra for a blistering fluence: 2×10^{16} H cm^{-2}.

Blistering appears at low fluence but surprisingly disappears at slightly higher fluence [Fig. 1]. Figure 3 (right) shows RSS spectra for high fluence (6×10^{16} H/cm^2 – no cleaving). Note the different ordinates for Fig. 2 vs. Fig. 3 (right), showing that H is more stably retained at higher fluence: for 6×10^{16} H/cm^2 the broadband is stable up to 300 °C, and the peak due to hydrided internal (100) surfaces remains remarkably strong at all temperatures. This indicates the formation of more and more H-decorated internal surfaces, believed to be the embryos of flat cavities and blisters. Yet, the abundance of these structures at high fluence fails to induce cleaving. Radiation-induced stresses and fracture toughening may play roles in inhibiting cleavage at high fluence. Rutherford Backscattering in channelling mode shows that the damage profile widens at higher fluence,[7] and the resulting stress configuration likely prevents crack propagation at a definite depth in the implanted region.

For D-implantation no blisters are observed below 3.5×10^{16} D/cm^2 and maximum blistering is only reached for $(6-7)\times10^{16}$ D/cm^2. Figure 3 (left) shows the evolution of the Si-D stretch modes for a subthresold fluence of 2×10^{16} D/cm^2. At 200 °C, D is mostly trapped in LF V_nD_m complexes ($k < 1490$ cm^{-1}), dominated by the 1403 plus 1472.5, and 1453 cm^{-1} peaks, associated with the V_2D/V_2D_2 and D^+_{BC}. The obvious difference between Si-D [Fig. 3 (left)] and Si-H [Fig. 2] is the strength of LF multivacancy peaks, the relative weakness of the LF broadband and the paucity of high frequency modes with D. After annealing to 300 °C, strong attenuation of the LF modes takes place, affecting both the broadband and the peaks. The modes at 1555, 1587 and 1621.6 cm^{-1}, associated with VD_2 (also to 3rd harmonic Si-Si), VD_3/V_2D_6 and VD_4 are enhanced relative to the as-implanted spectrum. The growth of VD_3/V_2D_6 and VD_4 may be explained by re-trapping into existing vacancies of D released from $VnDm$ centers. Nevertheless, the intensity of the HF monovacancy modes remains too weak to trigger blistering. Contrary to H [Fig. 2], which is completely detrapped after annealing at 550°C, some D is still present at this temperature, bound in good part to the self-interstitial as ID_2. The puzzling dependence on ion mass appears to be mainly connected with the nature of radiation damage.[3] H is more efficient in "preparing the ground" for blistering by nucleating platelets parallel to the surface, due to its ability to agglomerate into VH_n complexes that evolve into hydrogenated extended internal surfaces. By contrast, D is preferentially trapped in the surprisingly stable V_nD_m.

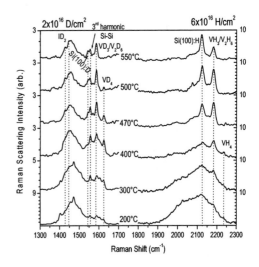

FIGURE 3. Temperature evolutions of Raman spectra for: (left) a subthreshold D fluence: 2×10^{16} D cm^{-2}; (right) an H overdose: 6×10^{16} H cm^{-2}.

1. M. Bruel, Electron. Lett. **31**, 1201 (1995).
2. Y. J. Chabal et al., Physica B **273-274**, 152 (1999); M. K. Weldon et al., J. Vac. Sci. Technol. B. **15**, 1065 (1997).
3. O. Moutanabbir et al., J. Chem. Phys. **121**, 7973 (2004).
4. O. Moutanabbir al., Appl. Phys. Lett. **84**, 3286 (2004).
5. M. Budde et al., Phys. Rev. B **57**, 4397 (1998).
6. F. A. Reboredo et al., Phys. Rev. Lett. **82**, 4870 (1999); S. B. Zhang et al., ibid. **87**, 105503 (2001).
7. O. Moutanabbir et al., accepted for publication in Applied Physics A, October 19, 2004.

Quantised Vortex Flows And Conductance Fluctuations In High Temperature Atomistic Silicon MOSFET Devices

John R Barker

Nanoelectronics Research Centre, Department of Electronics and Electrical Engineering,
University of Glasgow, Glasgow G12 8LT, Scotland UK
jbarker@elec.gla.ac.uk, Tel:+44(0)141 330 5221, Fax:+44(0) 141 330 4907

Abstract. Green function simulation of electron transport in a 25 nm channel Si MOSFET at 300K is used to predict that quantized vortices in the carrier velocity field make significant reductions in the channel transmission function and ultimately lead to statistical spread in the current-voltage characteristics for devices that differ only in the micro-spatial configuration of the dopants. High fields suppress the effect but interface roughness scattering induces further micro-vortices and subsequent conductance fluctuations. These processes are not predicted by semi-classical models.

INTRODUCTION

Recent demonstrations of advanced device architectures point to the continuation of silicon-based MOSFET technology from the imminent 65 nm node down to channel lengths ~ 5 nm. Conventional semi-classical device modelling fails in this regime because the transport is overtly quantum mechanical even for T>300K: inelastic coherence lengths exceed the channel length. Green function methods have been deployed to study MOSFETs quantum mechanically at 25 nm scales [1] although scattering processes were treated perturbatively. However, below 65 nm scales, the impurity distribution cannot be treated as continuous [2] and the small numbers of dopants in the device volume leads to fluctuations in conductance due to non-self-averaging of the interference processes [3-5]. Previous studies [3-6] have shown that the current flow in ballistic MOSFET channels comprises open meandering flows connecting source to drain interspersed with closed quantized vortex flows that essentially block the channel. In the present paper we report on the role of the quantized vortices using more realistic device modelling based on Green function methods [3-5] that incorporate a non-perturbative treatment of atomistic impurity scattering and interface roughness scattering in self-consistent electric fields. The simulated device is based on the 25 nm X 25 nm Si MOSFET reported in [1, 5].

DEVICE MODELLING

In principle, atomistic simulation requires a full self-consistent 3D model, but, following [5] we use an embedded 2D approximation in which the quantum transport of the carriers in the lateral channel 2DEG is computed for the self-consistent fields calculated for a vertical geometry 2D semi-classical device (Fig. 1). We neglect carrier heating in source and drain, so that the current I_{ds} is determined by integrating the difference between the source and drain supply functions over the energy-dependent channel transmission function $T(E, V_g, V_{ds})$, readily obtained from Green function simulation. Inhomogeneous transport is readily understood and visualized by computing the carrier charge density $n(\mathbf{r},E)$ and current density $\mathbf{j}(\mathbf{r},E)$, at energy E together with the velocity field $\mathbf{v}(\mathbf{r},E)= \mathbf{j}(\mathbf{r},E)/ n(\mathbf{r},E)$ and angular momentum field $\mathbf{L}=m^*\mathbf{r}\times\mathbf{v}$. In pure states

$$\frac{d\mathbf{L}}{dt} \equiv \frac{\partial \mathbf{L}}{\partial t} + \mathbf{v}.\nabla \mathbf{L} = -\mathbf{r} \times \nabla(V+V_Q) \qquad (1)$$

where V, V_Q are the total potential energy and quantum potential. $\mathbf{L}$ evidently varies throughout most of the inhomogeneous system. However, quantized vortices in the velocity field arise around the strong nodal points $\mathbf{r}_N$ of the charge density and there: $\mathbf{L}' = m^*(\mathbf{r} - \mathbf{r}_N) \times \mathbf{v}(\mathbf{r}) =$ constant. In 2D the vortices occur at nodal points; generally the strong nodes are open lines and loops giving vortices in 3D.

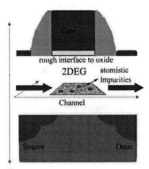

FIGURE 1. Simulation domain of atomistic MOSFET.

RESULTS AND DISCUSSION

Figure 2 shows a typical velocity field superposed on the charge density of the channel for an energy close to kT at 300K, computed non-perturbatively for an interface roughness potential that is exponentially random-distributed with $\Delta = 0.15$ nm, $\lambda = 3$ nm.

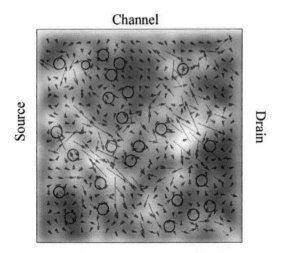

FIGURE 2. Current density and charge density fields in a 25 nm X 25 nm Si MOSFET. The quantized vortices are located by circles. Smaller charge densities are coded darker.

The channel contains three ionized impurities also treated non-perturbatively and it is subject to a weak source-drain electric field. Vortex flows clearly block the flow. Figure 3 shows the transmission function $T(E)$ for each configuration of three impurities located randomly at fixed relative distances but rotated through an angle between 0 and 2π about the channel centre. The fluctuations relate to changes in the number of occupied transverse states. A high driving field (4 kV/cm) improves the transmission by suppressing some interference processes, whereas the rough interface decreases the transmission by introducing additional vortices. Figure 4 exhibits the current-voltage characteristics for a full simulation with a fixed number of dopants in four different micro-configurations. A significant statistical spread occurs.

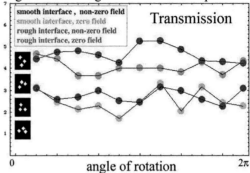

FIGURE 3. Transmission as a function of configuration rotation angle about centre of channel.

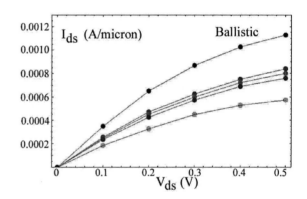

FIGURE 4. Current (I_{ds})-Voltage (V_{ds}) characteristics for four configurations of three impurities in a 25 nm channel. V_g=1 V. Ballistic continuum limit is shown for comparison.

REFERENCES

1. Svizhenko, A, Anatram, M.P, Govindan, T.R, Biegel, B and Venugopal, R., *J.Appl.Phys.*, **91** 2343 (2002).
2. Asenov, A, Saini, S., *IEEE Trans. Electron. Dev.* **47** 805 (2000).
3. Barker, J.R., *J. Computational Electronics* **2**, 153-161 (2003).
4. Barker, J.R.,*Semiconductor Science & Technology*, **19**S 56-60 (2004).
5. Barker, J.R., *Superlattices and Microstructures*, in the press (2004).
6. Fischetti, M.V, Laux, S, and Kumar, A., *Proc.IEDM* (2003).

Raman and XRD Strain Analysis of 3D Bonded And Thinned SOI Wafers

S. Pozder, M. Canonico, S. Zollner, R. Liu, K. Yu, J.-Q. Lu[*]

Freescale Semiconductor, Inc., Technology Solutions Organization, 3501 Ed Bluestein Blvd., Austin, TX 78721
[*]*Rensselaer Polytechnic Institute, Center for Integrated Electronics, 110 8th Street, Troy, NY 12180-3590*

Abstract. For development of three-dimensional integrated circuits, a device-quality silicon-on-insulator (SOI) wafer was bonded to a second Si wafer using an organic benzocyclobutene adhesive, followed by removal of the substrate of the SOI wafer through grinding, polishing, and wet-etching. This bonded structure was then bonded again to a third Si wafer, followed by removal of the substrate of the second wafer. The thicknesses of the layers in the resulting structure were measured with spectroscopic ellipsometry. The residual strain in the device-quality SOI layer, which affects transistor performance, was determined using UV Raman spectroscopy and triple-axis high-resolution x-ray diffraction.

INTRODUCTION

Three-dimensional integrated circuits (3D-ICs) are novel because devices are fabricated using processes optimized for a device or wafer type, then integrated vertically by bonding a wafer on top of another to build devices with multiple active layers [1]. Silicon-on-insulator (SOI) wafers are ideal for 3D-IC wafer bonded circuits because the buried oxide (BOX) acts as an etch stop during the critical step of wafer thinning after bonding. But removal of the substrate Si and bonding adhesive properties may affect the strain of a bonded and thinned SOI wafer. Typical bond types are Si-Si, SiO_2-SiO_2, Cu to Cu, and dielectric adhesives such as benzocyclobutene (BCB). Of these bond materials the Young's modulus of BCB (2 GPa) is less than Si making it the most compliant bonding material [2].

PROCEDURE AND RESULTS

This paper reports on strain analysis of two sets of bonded SOI wafers: a 70 nm/140 nm Si/BOX or a 110 nm/200 nm Si/BOX, which is bonded on a Si handler wafer using BCB as shown in Fig. 1. A normal SOI wafer is first bonded on a Si wafer (Si-I). The SOI Si substrate is removed by grinding, polishing and wet-etching. The bonded structure is then bonded again on another Si wafer (Si-II). Finally, the Si-I wafer is completely removed, and the BCB on the SOI layer is etched. The final structure is a TEOS/SOI Si layer/BOX/BCB/Si-II, that is similar to a 3D-IC interconnect stack [3]. The layer thickness and bond robustness were confirmed with scanning electron microscopy and spectroscopic ellipsometry, using BCB optical constants determined independently on a Si substrate.

As strain could affect the SOI Si by altering the carrier mobility or inducing defects, the pre and post wafer bond and thin strain was analyzed in our bonded SOI Si layers using UV micro-Raman spectroscopy [4] and high-resolution x-ray diffraction (HRXRD) utilizing Si (004) reciprocal space maps [5]. UV-Raman has advantages in microanalysis of stress and crystallinity in modern Si-based devices. First, the shorter wavelength improves the lateral spatial resolution, with a spot size of 0.4 μm at wavelength λ=325 nm. Second, the much shorter optical penetration depth in the UV (<10 nm at 325 nm in Si)

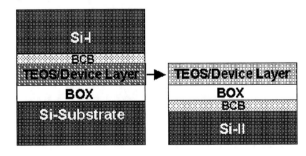

FIGURE 1. Schematic of the bonding process and final wafer structure.

allows probing ultra thin layers such as SOI in CMOS devices, which is impossible using a visible laser with a much larger penetration depth (~1 μm at λ=500 nm).

UV micro-Raman analysis of the SOI Si stress for prime SOI and two BCB bonded wafers is summarized in Table 1. After TEOS etch a compressive shift from 280 MPa to -337 MPa is observed for the 70 nm SOI and a compressive shift from 180 MPa to 43 MPa for the 110 nm SOI. The unprocessed SOI wafer is slightly tensile at 20 MPa using the lowest UV power, which induces minimal thermal expansion shifts in the Si LO phonon frequency, while the 70nm and 110nm BCB bonded wafers show a compressive stress of −64 MPa and −17 MPa, respectively. In contrast, HRXRD shows the 70 nm Si is essentially unstrained while the 110 nm Si is 66 MPa tensile. Potential contributions that could explain the differences include issues related to the sampling volumes, the inherent instrument resolutions, and an inhomogeneous strain field associated with the samples. Although the absolute stress magnitudes differ, both techniques indicate a compressive shift in stress for the thinner 70 nm BCB SOI Si.

Local laser induced heating (thermal expansion) can affect the Raman determination of stress, therefore the apparent or equivalent shift in stress as a function of the incident laser power density at the sample was measured. Figure 2 shows that both prime SOI and 110 nm BCB bonded SOI shift toward a tensile state, while the 70 nm BCB SOI becomes more compressive with increasing laser power density.

TABLE 1. UV-Raman and HRXRD Stress Measurements. Stress values +/- (tensile/ compressive) force. The resolution of the Raman and HRXRD instruments is 25 and 30 MPa, respectively.

SOI thickness	Post Bond and Thin (MPa)	Post TEOS Etch (MPa)
70nm (271 kW/cm²), BCB	280	-337
110nm (271 kW/cm²), BCB	180	43
70nm (9 kW/cm²), BCB		-64
110nm (9 kW/cm²), BCB		-17
HRXRD 70nm, BCB		0
HRXRD 110nm, BCB		66

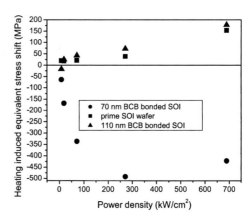

FIGURE 2. Equivalent stress shift from the Si LO phonon frequency as a function of the UV incident laser power density at the sample [6].

in stress for prime SOI Si and for 110 nm BCB bonded SOI Si and a compressive shift for 70 nm BCB bonded SOI Si with increasing laser power density. Although the most accurate stress evaluations are made under conditions of minimal local laser induced heating, the data reflects the sensitivity, in particular for thin BCB bonded SOI, under which the laser power density can significantly influence the apparent in-plane biaxial stress.

SUMMARY AND CONCLUSION

The Raman in-plane analysis of BCB bonded SOI Si demonstrates compressive shifts in stress after TEOS removal for 70 nm and 110 nm SOI Si. This trend towards a compressive regime for thinner SOI Si was confirmed by HRXRD. The thermal heating effect of UV-Raman results in an apparent tensile shift

ACKNOWLEDGMENTS

The Freescale authors acknowledge Ha Le and David Theodore. The work at Rensselaer is partially supported by DARPA, MARCO, and NYSTAR through the Interconnect Focus Center (IFC).

REFERENCES

1. J.-Q. Lu, Y. Kwon, J.J. McMahon, A. Jindal, B. Altemus, D. Cheng, E. Eisenbraun, T.S. Cale, and R.J. Gutmann, in *20th International VLSI Multilevel interconnection Conference*, pp. 227-236 (2003).
2. K. Rim et al., IEEE IEDM, p. 50, 2003.
3. S. Pozder et al., IITC, p.102, 2004.
4. R. Liu and M. Canonico, in *Characterization and Metrology for ULSI Technology,* edited by D.G. Seiler et al., AIP Conference Proceedings 683, Melville, New York, 2003, pp. 738-743.
5. D. K. Bowen and B. K. Tanner, *High Resolution X-ray Diffractometry and Topography*, Taylor & Francis, London, 1998.
6. Equivalent stress is calculated from the shift in the Si LO phonon frequency from unstrained Si and the Si LO phonon strain deformation coefficient in the UV.

The Impact Of Soft-Optical Phonon Scattering Due To High-κ Dielectrics On The Performance Of Sub-100nm Conventional And Strained Si n-MOSFETs

L. Yang[*], J. R. Watling, J. R. Barker and A. Asenov

*Device Modelling Group, Department of Electronics and Electrical Engineering,
University of Glasgow, Glasgow, UK*
[*]L.Yang@elec.gla.ac.uk, Tel: +44-(0)141 330 4792, Fax: +44-(0)141 330 4907

Abstract. Self-consistent Poisson/Ensemble Monte Carlo simulations are used to investigate performance degradations due to soft-optical phonon scattering introduced by high-κ dielectrics within sub-100nm conventional Si and strained Si MOSFETs. A degradation in the drive current of ~25% at V_G–V_T=1.0V and V_D=1.2V are observed for devices with a 2.2nm EOT HfO$_2$, as compared to devices with the same EOT SiO$_2$. However, the inherent mobility degradation associated with the high-κ gate stacks can be compensated for by the introduction of strained Si channels.

INTRODUCTION

High-κ dielectrics are expected to replace SiO$_2$ at or after the 65nm technology node in order to reduce the gate leakage and enable further scaling [1]. However, they introduce a number of technological problems as well as a fundamental drawback in the mobility degradation due to strong soft-optical (SO) phonon scattering. This scattering arises from the coupling between carriers in the inversion layer and surface longitudinal optical (LO) phonons at the silicon/high-κ interface due to increased polarizability of the high-κ dielectric [2]. In this paper, we use a self-consistent Ensemble Monte Carlo (EMC) device simulator that includes SO phonon scattering to study the impact of high-κ dielectrics on sub-100nm conventional and strained Si (SSi) n-type MOSFETs.

Our Monte Carlo simulator includes all relevant scattering mechanisms: optical intervalley phonon, inelastic acoustic phonon, ionized impurity scattering, interface roughness (IR) and SO phonon. The simulator has been carefully calibrated against transport behaviour in bulk strained Si. Our recently developed perturbative IR scattering model [3] used in these simulations is based on the semi-classical Fuchs-Boltzmann theory and is incorporated as a boundary condition for the Boltzmann Transport Equation. The model has been successfully validated and calibrated with respect to experimental universal mobility behaviour and device characteristics for both Si and strained Si devices [3]. The SO phonon scattering results from the strong ionic polarizability of the high-κ material, which also determines the large value of the dielectric constant. Electrons scatter from these phonons via a Fröhlich interaction [2].

DEVICE STRUCTURE

The simulated device structures are based on the Si and SSi MOSFETs with 67nm effective channel length fabricated by IBM [4]. The SSi MOSFET, illustrated in Figure 1(a), consists of a 10nm SSi channel; 2.2nm oxide and a relaxed SiGe buffer featuring a 15% Ge concentration. The Si device is assumed to have an identical structure and process conditions as the SSi MOSFET. The SSi MOSFET exhibits a 35% drive current enhancement compared to the Si MOSFET [4]. The Monte Carlo simulated I_D-V_G characteristics are plotted in Figure 1(b) and compared with the experimental data [4]. The calibrated IR parameters are an RMS height of 0.5nm and a correlation length

(CL) of 1.8nm for Si MOSFETs and an RMS of 0.5nm and a CL of 3.0nm for SSi MOSFETs. The 'smoother' interface required for the calibration of the SSi MOSFET indicates that reduced IR scattering contributes to the observed performance enhancement in SSi MOSFETs [5]. To investigate the impact of high-κ dielectrics on these devices, we have replaced the 2.2nm SiO_2, with the leading high-κ dielectric HfO_2, with the same equivalent oxide thickness (EOT). All other structural device parameters are assumed to remain the same in the simulations, although this may be an optimistic scenario given the immaturity of high-κ fabrication.

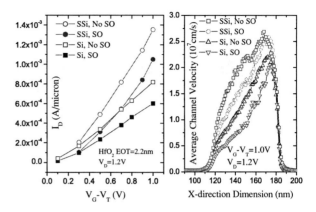

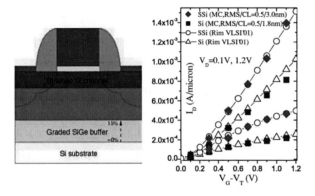

FIGURE 1. (a) Schematic of simulated strained Si MOSFET; (b) Monte Carlo calibrated I_D-V_G characteristics of the IBM 67nm strained Si MOSFET.

FIGURE 2. (a) I_D-V_G characteristics of the 67nm Si and SSi MOSFET with HfO_2 gate dielectrics; (b) Average channel velocities in the 67nm Si and SSi MOSFETs with HfO_2.

RESULTS AND DISCUSSIONS

The same EOT for different gate dielectrics in simulated Si and SSi devices leads to the same gate capacitances and enables identical electrostatic gate control. However, in the presence of HfO_2 carriers within the inversion layer are subjected to increased SO phonon scattering, leading to a reduction in the mobility of carriers and drive current.

Figure 2(a) illustrates the I_D-V_G characteristics for the 67nm Si and SSi devices with HfO_2. Simulations including SO phonon scattering due to the HfO_2 gate dielectric exhibit a ~27% reduction in drive current for the Si MOSFET and a ~23% reduction for the SSi MOSFET at $V_D = 1.2$ V and $V_G - V_T = 1.0$ V. For comparison such current reduction is less than 5% in the SiO_2 case. The differences in current degradations due to SO phonon scattering can be explained by the larger phonon energies and reduction in the difference between the static and optical permittivities, as we move from HfO_2 to SiO_2.

Figure 2(b) compares the average channel velocities with and without including SO phonon scattering at the same gate overdrive for both Si and SSi devices, demonstrating that the introduction of high mobility strained channels could be used to partially counteract the performance degradation due to SO phonon scattering. However, the infancy of high-κ gate fabrication techniques means that other performance degrading scattering mechanisms are likely to be present including a strong interface roughness contribution. Thus the overall performance degradations associated with high-κ gate dielectrics are expected to be worse than the predictions made in this paper.

ACKNOWLEDGMENTS

This work was supported by the UK EPSRC, under grant number: GR/N65677/01.

REFERENCES

1. ITRS 2003, http://public.itrs.net.

2. Fischetti M. V., et al., *J. Appl. Phys.* **90**, 4587-4608 (2001).

3. Watling J. R., et al., *Solid State Elec.* **48**, 1337-1346 (2004).

4. Rim K., et al., *Symp. on VLSI Tech. Dig. Tech. Papers*, 2001, pp. 59-61.

5. Fischetti M. V., et al., *J. Appl. Phys.* **92**, 7320-7324 (2002).

Quantization Conditions for Variable Operations of pHEMT

S. Mil'shtein, S. Gudimetta and C. Gil

Advanced Electronic Technology Center, ECE Department, U Mass, Lowell MA 01854, U.S.A.

Abstract. The performance of pHEMTs is defined by the shape of the quantum well along the channel, as well as by the set of available energy levels in 2DEG and position of Quasi Fermi Level (QFL). Existence of an excessive electric field at the drain end of the channel is a reason for many events that downgrade pHEMT performance. Following our model of tri-gate p-HEMT with 120Å, and 170Å wide channels we were able to control the electron density by changing the QFL position along the channel. As we changed operational voltages at the device terminals (the electrical field was tailored) in pHEMT the gate leakage currents in 3rd and 4th quadrants decreased by 28 – 50%.

INTRODUCTION

Studying pHEMT, various research groups observed negative peaks in the third and fourth quadrants of the gate diode characteristics. These were explained on the basis of impact ionization, tunneling out of 2DEG domain, etc [1-2]. Modeling of pHEMT with 120Å, and 170Å preceded experiments with multigate HEMTs. The wider channel allowed reduction of the quantization energy along the channel. Commercial software Silvaco^{TM} was used to predict the device performance and to compare the model with experimental results. Tri-gate GaAs/Al$_x$Ga$_{(1-x)}$As/In$_x$Ga$_{(1-x)}$As pHEMT was modeled for two biasing conditions: one when the gates are connected together, and another, when gates are biased separately. Properly selected ratio of gate voltages (higher at the first gate, lower at the second and even lower at the third) reduces the field in the channel towards the drain. It also reshapes the heterostructure potential. This method of tailoring of electrical field in MESFETs and MOSFETs described elsewhere produced higher and at the same time more linear gain, higher frequency of operation etc [3-4].

In the current study, we attempted to minimize the amplitudes of negative peaks observed in the leakage current of the gate diode by shaping the quantum well from source to drain and changing the electric field along the channel. That in turn, generated an almost uniform set of available electron energy levels. Effective control of the population of these energy sites was executed by changing the voltage applied to the three gates of a pHEMT in the manner $V_{G1} > V_{G2} > V_{G3}$, i.e. changing the QFL position [5]. As a result strength of electric field was reduced in the channel and more linear gain was observed.

EXPERIMENTAL RESULTS

Characteristics of p-HEMT, fabricated by M/A-COM's AlGaAs/InGaAs process, were measured. AuGe/Ni/Au ohmic contacts were formed on a heavily-doped GaAs cap layer. The 0.5μm long gates were made with a Ti/Pt/Au metal lift-off process in a double recessed channel. The entire die was covered with a Si$_3$N$_4$ passivation layer.

Our model provided set of energy levels E_1=31.56meV; E_2=123.87meV; E_3=265.9meV existing in the 120Å thick channel and E_1=17.5meV; E_2=69.4meV; E_3=153.78meV; E_4=264.25meV in the 170Å thick channel as well as QFL for different biases [6]. It can be seen that for the energy levels are dropping down for the wider channel pHEMT.

Figures 1 and 2 present the negative peaks in the third and fourth quadrants of I_g-V_g characteristics for a tri-gate pHEMT in non-tailored and tailored regimes of operation respectively. The negative resistance segments in the I_g-V_g curves increases with drain voltage changing from 3V to 11V. We observed that for the same drain voltage the fourth quadrant peak is reduced significantly for the pHEMT with tailored field.

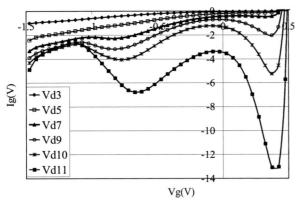

FIGURE 1. I_g-V_g characteristics of heterostructure gate junction of a tri-gate pHEMT: non-tailored condition.

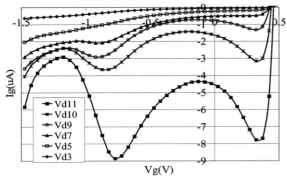

FIGURE 2. I_g-V_g characteristics of heterostructure gate junction of a tri-gate pHEMT: tailored condition.

CONCLUSIONS

The theoretical model we developed demonstrated that proper set of available energy levels in the channel and sufficient control of both, the shape of quantum well and QFL position allow to have pHEMT operation with reduced leakage current of the gate. The model was supported by measurements of Ig-Vg characteristics. The tilt of the well and shift of QFL was experimentally controlled by variation of biases at the terminals of tri-gate pHEMT. The large leakage current was measured when voltages $V_{G1}=V_{G2}=V_{G3}$ (non-tailored field) were applied. With gate biases $V_{G1}>V_{G2}>V_{G3}$ applied (tailored field) the leakage current in the 4th quadrant was reduced 28-50% for given drain voltage.

The observed improvement of HEMT performance can be attributed to increased distance between the top energy levels and edges of quantum well. In addition tailoring the field made the QFL more uniform along the channel and reduced the electron population of the top energy level. Obvious correlation between the quantization condition in the channel and improved performance of the pHEMT makes 170 Å thick channel a better device. Our model predicts the set of lower energy along the thicker channel and more efficient control of QFL. Next step of this study is to produce the pHEMT with 170 Å thick channels.

ACKNOWLEDGEMENT

This research is supported by a grant from MA/COM Co, Tyco Electronics Ltd. The authors appreciate discussions with of technical staff of Aerospace and Defense Division.

REFERENCES

1. R. Menozzi, P. Cova, C. Canali, and F. Fantini, "Breakdown Walkout in Pseudomorphic HEMTs", Trans. Electr. Dev., 43, 4, pp.543-546 (1996).
2. A.S.Wakita, H.Rohdin, C.Y.Su, N.Moll, A.Nagy, V.M.Robbins, "Drain resistance degradation under high fields in AlInAs/GaInAs MODFETs", Hewlett Packard Labs, pp. 376-379 (1997).
3. S. Mil'shtein, "New Phenomena in Transistor with Tailored Field" in a book, "Physics of Semiconductors", World Scientific, 2, 1665-1668 (1992).
4. S. Mil'shtein, S. Sui, "Study of 2DEG in MOSFET with Tailored Field", Proc. 25th Int. Conf. Phys. Semicond., 1755-1759 (2001).
5. S. Mil'shtein, P. Ersland, S. Somisetty, C. Gil, "p-HEMT with Tailored Field", Microelectronic J., 34, No.5-8, pp.359-361 (2003).
6. S. Gudimetta, S. Mil'shtein, "Improved Quantization of 2DEG of p-HEMT", International Semiconductor Device Research Symposium, pp. 524-525 (2003).

Design and Optimization of Vertical CEO-T-FETs with Atomically Precise Ultrashort Gates by Simulation with Quantum Transport Models

J. Höntschel[1], W. Klix[1], R. Stenzel[1], F. Ertl[2], G. Abstreiter[2]

[1] *University of Applied Science Dresden, Friedrich-List-Platz 1, D-01069 Dresden, Germany,*
Phone: +49-352-462-2548, Fax: +49-351-462-2193, Email: hoentsch@et.htw-dresden.de
[2] *Walter Schottky Institute, TU Munich, Am Coulombwall, D-85748 Garching, Germany*

Abstract. The cleaved-edge overgrowth (CEO) technique offers an innovative approach to designing novel quantum sized field-effect transistors (FETs) with a T-like gate-to-channel structure. Numerical simulations of vertical CEO-T-FETs have been carried out to optimize the structure and predict device performance. For the simulation the 2D/3D device simulator SIMBA is used, which is capable of dealing with complex device geometries as well as with several physical models represented by certain sets of partial differential equations.

INTRODUCTION

One intention of modern semiconductor technology is the reduction of device length and width. With the realization of these nanometer scale structures several physical effects appear such as short channel effects and overshoot behaviour, as well as quantum mechanical effects. With the CEO technique it is possible to produce vertical CEO-T-FETs based on AlGaAs/GaAs heterostructures with atomically precise ultrashort gates that eliminate short channel behavior [1]. The idea of a CEO-T-FET is to build up a planar epitaxial structure with source, barrier, atomically precise ultrashort gate, barrier and drain. In the second growth step, a conducting channel, a supply and a cap layer are grown perpendicular to the original planar layers. Numerical investigations of this CEO-T-FET show good device behaviour without short channel effects.

SIMULATION MODELS

The simulation models of our device simulator SIMBA are described. The equations are shown only for electrons. SIMBA calculated the equations for electron and holes. The Poisson equation

$$\nabla(\varepsilon_s \varepsilon_0 \nabla(\varphi)) = -q\,(p - n + N_D - N_A) \quad (1)$$

where N_D^+, N_A^- are the ionized donor and acceptor densities, the continuity equation

$$\nabla \cdot \mathbf{J}_n = q\left(R - G + \frac{\partial n}{\partial t}\right) \quad (2)$$

and the transport equation

$$\mathbf{J}_n = -qn\mu_n \nabla(\varphi + \lambda_n + \Theta_n) + D_n q \nabla(n) + k_B n \mu_n \nabla(T_n) \quad (3)$$

are solved self-consistent in the Gummel algorithm to get the device characteristics at different bias conditions. In the drift gradient of the transport equation an accessory expression is included. This additional equation describes a quantum potential for the carriers

$$\lambda_n = \frac{\gamma_n \hbar^2}{6\,m_n q} \frac{\nabla^2 \sqrt{n}}{\sqrt{n}} \quad (4)$$

where m_n is the constant effective mass. The problem of the anisotropic effective mass is handled by the fitting factor γ_n. Non-equilibrium device phenomena, like short-channel and overshoot behaviour are taken into account by the energy balance equation and the energy flux density, which are included as additional equations in the self-consistent Gummel algorithm. The energy flux density equation

$$\mathbf{S}_n = -\kappa_n \nabla(T_n) + \frac{5}{2}\frac{k_B}{q}T_n \mathbf{J}_n + \frac{3}{2}\lambda_n \mathbf{J}_n \quad (5)$$

and the energy balance equation

$$\nabla \cdot \mathbf{S}_n = \mathbf{J}_n \mathbf{E}^* - \frac{3}{2}k_B n\frac{(T_n - T_L)}{\tau_{wn}} - \frac{3}{2}k_B \frac{\partial}{\partial t}(nT_n) - \frac{3}{2}k_B T_n (R-G) + \frac{1}{2}q\lambda_n \left(\frac{n}{\tau_{wn}} - (G-R)\right) + \frac{1}{2}q\frac{\partial}{\partial t}(n\lambda_n) \quad (6)$$

are likewise extended to include quantum effects by the same quantum correction potential for the carriers. τ_{wn} is the energy relaxation time. With the quantum hydrodynamic model it is possible to consider quantum mechanical effects. The Poisson equation (1), the continuity equation (2) and the transport equation (3) without the temperature gradient are described together as quantum drift diffusion (QDD) model. The quantum potential does not account for the classical hydrodynamic (HD) and drift diffusion (DD) models.

RESULTS

Figure 1 shows the CEO-T-FET structure with a gate length of $L_G = 20$ nm that was used for simulation together with the doping densities as well as the layer thicknesses.

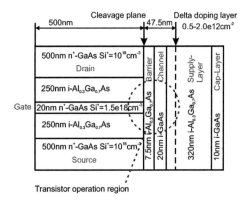

FIGURE 1. Structure of the 20nm CEO-T-FET.

The calculated output and transfer characteristics are represented in Fig. 2 and 3. They are simulated with two classical models and do not indicate short channel or overshoot effects. The transfer characteristics for different γ_n of the QDD model are shown in Fig. 4. The increase of the drain current at lower gate-source voltage results from tunneling and leakage currents through the gate and can be calibrated by γ_n in the QDD model. The lower drain current results from the smaller calculated electron density distribution by the QDD model.

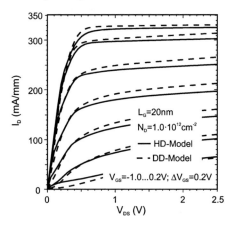

FIGURE 2. Output characteristics.

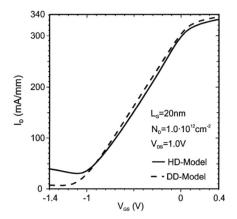

FIGURE 3. Transfer characteristics.

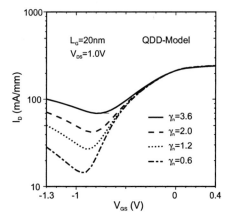

FIGURE 4. Transfer characteristics for different γ_n.

REFERENCES

1. Stormer, H. L., Baldwin, K. W., Pfeiffer, L. N. and West, K. W., *Appl. Phys. Letters* **59**, 1111-1113 (1991).

Double quantum dot with integrated charge readout fabricated by layered SFM-lithography

M. Sigrist*, A. Fuhrer*, T. Ihn*, K. Ensslin*, D. D. Driscoll[†] and A. C. Gossard[†]

*Solid State Physics, ETH Zürich, 8093 Zürich, Switzerland
[†]Materials Department, University of California, Santa Barbara, CA 93106, USA

Abstract. A double dot system with integrated charge readout is fabricated by combining direct oxidation of the GaAs surface of a Ga[Al]As-heterostructure with the local oxidation of a thin titanium film evaporated on top. The charge stability diagram is measured in direct transport through the dots as well as in the detector signal.

Patterning techniques based on scanning force microscopes (SFMs) have been developed increasing the possibilities to fabricate semiconductor nanostructures. A useful technique is to oxidize substrates locally by applying a negative voltage between the SFM tip and the substrate. On shallow Ga[Al]As-heterostructures, the two-dimensional electron gas is depleted below the oxide lines leading to mutually isolated regions of electron gas [1]. Similar oxidation techniques have been used to divide thin titanium films into areas separated by oxide lines [2]. One option for combining top gates and lateral gates is the local oxidation of the heterostructure surface followed by evaporating top gates defined by electron beam lithography [3]. In order to fabricate more complex and tunable nanostructures we have extended local oxidation to a two step lithography process. In our approach, the direct oxidation of the heterostructure surface is combined with the SFM patterning of a thin titanium film evaporated on top [4].

The sample has been fabricated based on an AlGaAs-GaAs heterostructure containing a two-dimensional electron gas (2DEG) 34 nm and a backgate 1.4 µm below the surface. Ohmic contacts and metallic top gate fingers for contacting the Ti film are provided by optical lithography. The nanostructure is defined in a first step by locally oxidizing the surface with a biased SFM tip (see Figure 1). Thereby, the electron gas is depleted below the oxide lines (bright). In a second step, a 6-7 nm thick titanium film is evaporated on top. This film is then in a third step locally oxidized by a biased SFM tip. The resulting Ti oxide lines (black dashed) divide the film into mutually isolated top gate segments. The completed double dot structure is connected via quantum point contacts (QPCs) to four distinct leads labelled A to D. The in-plane gates pg1 and pg2 formed by the 2DEG regions next to the dot structure act as plunger gates. Each of the

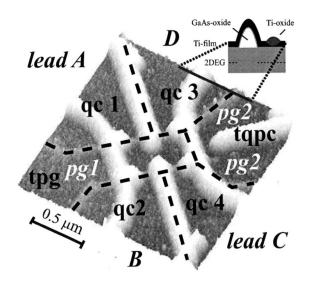

FIGURE 1. SFM-micrograph of the double-dot system. The inset shows a section of the sample schematically. The 2DEG is depleted below the oxide lines (bright) on the GaAs surface. This forms two coupled quantum dots connected to four leads (A..D) and a QPC at the side of the upper right dot. An evaporated thin Ti film is divided into mutually isolated top gate regions by oxide lines (black dashed). Each of the six QPCs can be controlled by a top gate region, the two isolated 2DEG regions act as plunger gates on the dots.

four QPCs connecting the dots to the leads can be controlled by a separate top gate (qc1..qc4). The coupling between the two dots can be tuned by the top plunger gate (tpg). The in-plane gate pg2 contains an additional QPC next to the dot system forming an integrated charge read-out. This QPC is tunable with another top gate region (tqpc).

Conductance is simultaneously measured for each dot

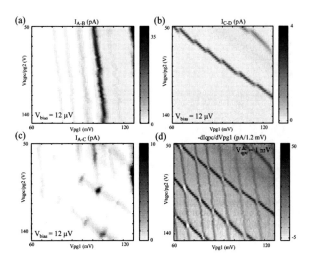

FIGURE 2. Conductance measurements of the double-dot system are shown as a function of the in-plane gates pg1 and pg2. The topgate tqpc is tuned in a linear relation to pg2 in order to keep the conductance of the detector QPC constant. (a) Conductance through the lower left dot. (b) Conductance through the upper right dot. (c) Current from lead A to lead C. (d) Detector signal from a transconductance measurement.

by using lock-in technique at different frequencies and passing the currents through the corresponding tunnel contacts. Coulomb oscillations are observed at an electronic temperature of about 200 mK. The coupling QPC between the two dots is tuned in the weak coupling regime by applying a negative voltage to the top plunger gate. Figure 2 (a) and (b) show the conductance of each dot as a function of both in-plane plunger gates pg1 and pg2 in a parallel setup. In addition, a negative voltage is applied to tqpc in a linear relation to pg2 in order to keep the conductance of the detector QPC constant. Applying a bias voltage to A and measuring current at C (B and D in the tunneling regime connected to ground), we observe the double dot charging in a series configuration see Figure 2 (c). By tuning the charge detector QPC into the tunneling regime we are able to detect charging of the individual dots with single electrons shown in Figure 2 (d). The characteristic hexagon pattern can be mapped out as a function of the three tuning gates. The charge detector is closer to the upper dot and shows therefore a stronger signal if this dot is charged [5]. This is observed in lines with a smaller slope in agreement with the relative location of this dot to the three tuning gates. The lines for the other dot show a weaker signal and a larger slope. All lines lie exactly on the conductance maxima measured in Figure 2 (a) and (b).

A double dot system with integrated charge read-out is fabricated by using a layer by layer oxidation with an biased SFM tip. Our transport measurements demonstrate individual control over nanostructured in-plane and top gate electrodes. This opens up a new way to realize innovative designs of semiconductor nanostructures with SFM-lithography.

Financial support from the Swiss National Science Foundation (Schweizerischer Nationalfonds) is gratefully acknowledged.

REFERENCES

1. A. Fuhrer, A. Dorn, S. Lüscher, T. Heinzel, K. Ensslin, W. Wegscheider, and M. Bichler, Superl. Microstr. **31**, 19 (2002).
2. R. Held, T. Heinzel, P. Studerus, K. Ensslin, and M. Holland, Appl. Phys. Lett. **71**, 2689 (1997).
3. M.C. Rogge, C. Fühner, U.F. Keyser, R.J. Haug, M. Bichler, G. Abstreiter, and W. Wegscheider, Appl. Phys. Lett. **83**, 1163 (2003).
4. M. Sigrist, A. Fuhrer, T. Ihn, K. Ensslin, and D. Driscoll, to be published
5. L. DiCarlo, H.J. Lynch, A.C. Johnson, L.I. Childress, K. Crockett, C.M. Marcus, M. P. Hanson, and A.C. Gossard, Phys. Rev. Lett. **92** , 226801 (2004).

Rad-Hard Silicon Detectors

Marco Giorgi

University and INFN of Perugia, via Pascoli 06123 Perugia Italy

Abstract. For the next generation of High Energy Physics (HEP) Experiments silicon microstrip detectors working in harsh radiation environments with excellent performances are necessary. The irradiation causes bulk and surface damages that modify the electrical properties of the detector. Solutions like AC coupled strips, overhanging metal contact, <100> crystal lattice orientation, low resistivity n-bulk and Oxygenated substrate are studied for rad-hard detectors. The paper presents an outlook of these technologies.

INTRODUCTION

Silicon microstrip detectors are widely used in HEP Experiments mainly thanks to a good spatial resolution (few microns) and fast signals readout (order of tens of nanoseconds).
One microstrip detector is basically an array of p+ junctions on n bulk, working in inverse polarization. The signal (due to an ionizing energy loss process) generated by the crossing particle is collected on the strips close to the impact point and elaborated by the readout electronic.
Only detectors with a certain design operate optimally in environmental conditions characterised by very high particle fluxes. The aim of this paper is to make an overview of the technological solutions investigated to obtain rad-hard silicon microstrip detectors.

RADIATION EFFECTS ON SILICON

Exposing detectors to a heavy radiation environment induces bulk and surface defects. Bulk defects are generated by the interaction of hadrons (order of 10^{14} n_{eq}/cm^2) with silicon atoms, causing their displacement from the reticular positions. Surface defects are mainly due to the interaction of charged particles and gammas with the isolation layer generating a deposit of trapped charges on the Silicon-oxide interface. The defects cause a change in the macroscopic electrical behavior of the detectors. The main effects of bulk damage are the increment of the leakage current, the variation of the depletion voltage and the decrease of the charge collection efficiency (CCE). Surface damage is responsible of the increment of the interstrip capacitance.

Leakage current and interstrip capacitance have a direct impact on the detector electronic noise. Increased depletion voltage involves increased bias voltage, resulting in higher breakdown risk. Decrease of CCE causes a loss of signal and means a deterioration of the S/N ratio.

RAD-HARD THECNIQUES

The detector lifetime in a radiation environment is mainly limited by the increase of the depletion voltage with fluence and consequently by the maximum voltage that can be applied. Over-depleting the sensor permits to improve the CCE, partially recovering the loss due to irradiation. The increase of the bias voltage is limited by breakdown and thermal run away effects. The occurrence of breakdown discharge phenomena is inherently localized at some device critical regions, in particular on the edge of p+ implants. The adoption of overhanging metal contacts (i.e. an extension of the metal layer over the oxide on the interstrip gap) change the electric field profile with an effect of "dilution" of field lines near the edge of the implants [1]. This has been demonstrated as an effective way to reduce breakdown risks.

The depletion voltage is proportional to the absolute value of the effective space charge density, N_{eff}. The N_{eff} variation with respect to the fluence and annealing time has been parameterized as:

$$N_{eff}^{\Phi} = N_{eff}^{0} - N_C - N_{ann} \quad (1)$$

where N_{eff}^{0} is the initial space charge density, N_{ann} represent the annealing component (both beneficial and reverse annealing) and N_C is the stable damage component. With high particle fluxes the substrate become effectively p-type (type inversion). As showed in Fig. 1, lower resistivity n-type material presents higher inversion fluence and, consequently, lower depletion voltage after heavy irradiation dose. The lower limit on the resistivity is determined by the initial depletion voltage, which is inversely related to the resistivity value.

The leakage current can be largely reduced by operating the detector at low temperature:

$$J \propto T^2 e^{-E_a/KT} \quad (2)$$

with J the current density and E_a the activation energy of the defects. The minimum operating temperature is set by the general experimental conditions and by the need not to freeze the beneficial annealing of N_{eff}.

AC coupled strips allow to block the DC component of leakage current and to reduce the electronic noise of device. The detector is biased by connecting all strips to a common bias line using polysilicon resistors with high resistance: with large currents the noise will increase of a small value since this is proportional to $\sqrt{\tau/R_{bias}}$, where τ is the sampling time of the readout electronics [2].

Growing crystals along different directions leads to a different number of dangling bonds, with a different trapped charge density at the Silicon-Oxide interface. The effect of this charge is the creation of an accumulation layer of electrons into the n substrate in the interstrip region and a consequent increase of the interstrip capacitance. In the <100> crystal orientation, the trap density at interface is about ten times smaller than in <111>. The lower density of trapped charge typical of <100> crystal orientation allows to reduce the effects of the surface radiation damage.

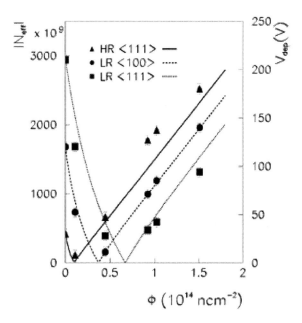

FIGURE 1. Measured doping concentration | N_{eff} | and depletion voltage V_{dep} vs fluence. The parameterized | N_{eff} | dependence is showed as line.

The use of Oxygen enriched detectors substrate permits a higher radiation tolerance [3]. A concentration of the order of 10^{17} cm^{-3} gives significant improvements in the variation of N_{eff} with respect to fluence and better overall detector performances, as demonstrated in beam test study [4].

Silicon detectors featuring ad-hoc characteristics can be operated in hostile radiation environment. An overview of these characteristics has been presented.

REFERENCES

1. D.Passeri et al., "Analysis and test of Overhanging-Metal Microstrip detector", *in proc. Of IEEE Nuclear Science Symposium 2000*, Lyon (France)

2. C.Bozzi, "Signal-to-Noise Evaluation for the CMS Silicon Microstrip Detectors", CERN *CMS internal note* 1997/026.

3. G.Lindsrom et al., *Nuclear Instruments and Methods in Physics Research* **A 466**, 2001, pp. 308-326.

4. V. Zhukhov et al., "Test of CMS silicon detector modules in a 350 Mev/c hadron beam", CERN *CMS internal note* 2001/049.

Modeling, Fabrication and Test Results of a MOS Controlled Thyristor – MCT - with high controllable current density.

Evgeny Chernyavskiy[a], Vladimir Popov[a] and Bert Vermeire[b]

[a] *Institute of Semiconductor Physics Siberian Branch Russian Academy of Sciences, Lavrentev pr. 13, Novosibirsk, Russia*
[b] *Ridgetop Group Inc., 6595 N. Oracle Road, Tucson AZ 85704, USA*

Abstract. A simple 2D static model for the evaluation of the maximum controllable current density in a MOS Controlled Thyristor (MCT) will be presented. It is shown that a 2D model of a P-I-N diode with a cathode PMOS transistor is appropriate for simulating the carrier distribution in an MCT. The maximum controllable current density can be modeled accurately using a static model of the free electron and acceptor concentrations. The simple physics rationale for this is discussed. MCT test results validate this simplified modeling approach. Punchthrough technology (PT) for high voltage application was used to manufacture an MCT with breakdown voltage of 2500 V and a maximum controllable current of 33 A with active area 0.33 cm^2. The total number of N-type cathode emitter cells is 144,042. For optimization, transient power dissipation electron irradiation with 2 MeV energy was used. This increased the maximum controllable current to 50 A. To our knowledge, our current density values of 100 A/cm^2 for non irradiated and 150 A/cm^2 for irradiated MCTs are the highest that have been reported for large area devices.

INTRODUCTION

The development of MOS-gated power thyristors is of interest for applications where power devices with high blocking voltage are needed. In the MOS-controlled thyristor (MCT) [1], the turn-off is obtained by shorting the emitter/base junction of one of two coupled transistors in the thyristor structure. The purpose of this paper is to report a new simplified approach to simulating maximum controllable current in MCTs.

DEVICE STRUCTURE

A two-dimensional cross section is shown in Fig. 1. The device contains a vertical thyristor region and has shorted cathode emitter cells. The thyristor can be turned on by the application of a positive bias to the gate. This results in the formation of an n-type inversion layer at the surface of the P-base region through which electrons can flow into the N-drift region. The on-state current conduction in the MCT is accompanied by the injection of a high concentration of both electrons and holes into the lower base or drift region.

When turn-off is desired, the PMOSFET is turned on to short the P base/N+ emitter junction of the upper NPN transistor in the thyristor path. All or part of the holes entering the P base region are diverted to the turn-off channel, and, as a consequence, the holding current level of the thyristor increases. The success in turning off the MCT relies on the fact that the latching condition can be broken. Therefore, the amount of current that an MCT can turn off is mainly decided by the conductivity of the inversion channel and the resistance of the upper P base region. As long as the thyristor's latched condition is broken (N+ emitter stops injection), the turn-off then proceeds like an open-base transistor with a rapid current fall followed by a current tail while excess carriers inside the N- drift region recombine. The fabricated N-MCT device [2] has a turn-off channel length of 1 µm and a turn-on channel length of 4 µm. The N-drift region has a doping concentration of 10^{13} cm^{-3} and a thickness of 540 µm. The N$^+$ emitter and the P base region have surface concentrations of 8 x10^{19} cm^{-3} and 9 x10^{16} cm^{-3}, and junction depths of 2 µm and 8 µm, respectively. Use of punchthrough technology (PT) has allowed us to achieve breakdown voltages as high as 2500 V.

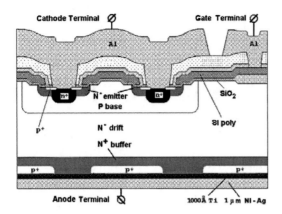

FIGURE 1. Cross-sectional view of the MCT structure.

DEVICE SIMULATION

Complete modeling of the regenerative feedback mechanism between the two coupled transistors within the thyristor requires transient simulation. However, this demands huge computer resources. Therefore, a simplified modeling approach is needed. It is well known that the thyristor on-state field and carrier distribution follows the 1D p-i-n diode model. We propose using a 2D p-i-n diode model with the N^+ emitter shunted by a PMOSFET. We believe that this model is physically equivalent to the MCT for realistic current conditions. The total simulated area is 16 x 540 µm.

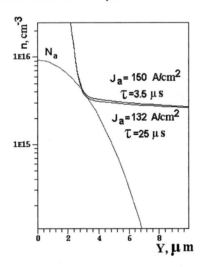

FIGURE 2. Acceptor and electron concentration comparison in 2D p-i-n diode model (U_g = -15 V, J_a is the anode current density before applying gate voltage).

The dependence of the anode current density on the gate voltage was simulated. As the initial current density increases, the current density modulation is reduced. We believe this to be caused by negative charge compensation by holes remaining under the N^+ emitter. Figure 2 shows the acceptor and electron concentrations in the 2D p-i-n diode model with carrier lifetimes of 25 and 3.5 µs. This allows us to formulate a criterion for evaluating the maximum controllable current density in MCTs: if the electron density is lower than the acceptor concentration ($n < N_a$) when a negative gate voltage is applied, then current density is controllable.

EXPERIMENTAL RESULTS

The maximum controllable current density obtained experimentally is 100 A/cm^2 with a carrier lifetime of 25 µs. After electron irradiation, its value is 150 A/cm^2 with a carrier lifetime 3.5 µs [3].

CONCLUSIONS

A high voltage MCT with buffer layer and anode short structure has been successfully developed and characterized. The active area of the MCT is 0.33 cm^2. Electron irradiation was used for carrier lifetime control. The simulation results of 2D p-i-n diode model shows simple evaluating criteria can be formulated for maximum controllable current density in MCT.

REFERENCES

1. F. Bauer et al, "Design Aspects of MOS-Controlled Thyristor Elements: Technology, Simulation, and Experimental Results," *IEEE Trans. Electron Devices*, vol. 38, No. 7, pp 1605-1611, 1991.

2. Chernyavskii, E.V., Popov, V.P., Pakhmutov, Yu.S., Krasnikov, Yu.I., and Safronov, L.N., "MOS Controlled Thyristor 2500 V", in Materials and Processes for Power Electronic Devices, Thesis 5th APAM Topical Seminar, 2001, p. 36.

3. Chernyavskii, E.V., Popov V.P., Pakhmutov Yu.S., Krasnikov Yu.I., and Safronov L.N., "Carrier Lifetime and turn-off current control by electron irradiation of MCT", Nuclear Instruments and Methods in Physic Research B, 2002, No. 2 ,v.186, pp. 157-160.

Triode with Heterostructure Filament

C. Gil, S. Mil'shtein

Advanced Electronic Technology Center, ECE Department, U Mass, Lowell MA 01854, U.S.A.

Abstract. A novel solid-state based triode, replicating vacuum tubes, was modeled, designed, fabricated and tested. The anode was made of a heterostructure junction, similar to the structure of HEMTs. Using planar technology, the gate (grid) was positioned between the anode and one of the ohmic contacts in the same layer as the anode. The filament made of 120Å thick InGaAs layer carries current in the range of 100-200mA/mm. The electron flow along the filament is deflected towards the anode by its positive potential. Variation of the potential at the gate (grid) allows control of the current through the filament. The I-V characteristics of the new device resemble that of a vacuum triode.

INTRODUCTION

The I-V characteristics of the semiconductor triode presented in this study replicate the performance of vacuum tubes [1]. The proposed device operates quite differently from known vertical structures such as permeable transistors, gated diodes or real space transfer (RST) transistors [2-5]. The motivation of this design was to achieve ballistic electron transport through the short distance between the filament and the anode while still utilizing photolithography with conventional resolution in the fabrication procedures. Small output capacitance, i.e. short RC delay time is important for high frequency operation of the novel device. We present in this study one of the first iterations of the design and fabrication of the triode with heterostructure filament.

EXPERIMENTAL RESULTS

The anode was made of a heterostructure junction similar to the methods of fabricating gates for HEMTs. The gate (grid) was positioned between the anode and one of the ohmic contacts (see Figure 1) in the same layer as the anode [6]. The device was fabricated by an AlGaAs/InGaAs process at M/A-COM. AuGe/Ni/Au ohmic contacts are formed on a heavily-doped GaAs cap layer to connect the filament. The 0.5μm long by 1mm wide anode was optically patterned and made with a Ti/Pt/Au metal lift-off process. The entire die was covered with a Si_3N_4 passivation layer. The filament was made of a 120Å thick InGaAs layer, which carries current in the range of 100-200mA/mm. The electron flow along the filament is deflected towards the anode by its positive potential resulting in anode current of 12- 15mA/mm. Figure 1 shows a very important part of the design, namely $V_{anode}=+V_{Fil}$. The same voltage drop across the filament and the anode minimizes the mutual influence of electrode potentials so common for vacuum tubes [1]. This interference of electrostatic potentials at different terminals is stronger with submicron distances in the semiconductor triode. Additional control is provided by the gate (grid) positioned close to the anode. The current-voltage characteristics of the new device resembled the output of a vacuum triode.

The discrete triode die was mounted on a Cascade probe station and probed directly using 150 pitch Ground-Signal-Ground probes. DC bias was applied using an HP4142b source-monitor. The RF Signal was applied using an 8510C vector network analyzer (VNA) and the RF frequency was swept from 500 MHz to 9 GHz. Figure 2 demonstrates that I-V characteristics of semiconductor triode resembled ones of vacuum tubes. One can learn from Figure 3 that the transconductance of the novel device varies very little for most of the midrange voltages applied to the grid. Figure 4 shows the decay of parameter S21, which could be attributed to still long distance between filament and anode.

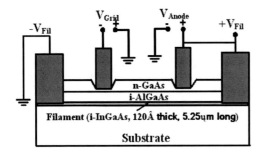

FIGURE 1. Sketch of semiconductor triode

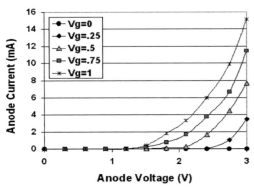

FIGURE 2. I-V characteristics of triode semiconductor triode with minimized voltage interference between anode voltage and filament bias.

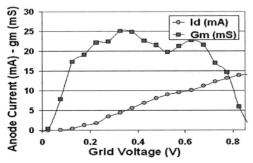

FIGURE 3. Transfer and transconductance characteristics of triode with minimized interference between anode voltage and filament bias.

CONCLUSIONS

From the viewpoint of its physical design and operation, this novel triode represents a vertical structure, which could be compared, among other, with vertical gated diodes [3-4]. The anode control in our triode and gate control in the RST devices is similar [5]. Unlike in vacuum tubes the grid of the semiconductor triode controls the filament current.

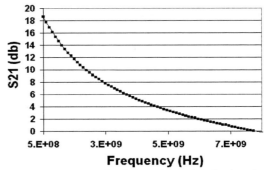

FIGURE 4. S21 of triode with minimized voltage interference between anode voltage and filament bias

As in any vertical structure, the fabrication of a permeable transistor does not depend on the resolution of photolithography, but from conceptual point of view, it operates as vertical JFET. Also the S21 of the very first semiconductor triode decays with increase of frequency. Good RF performance is expected as the design and control of the filament is improved.

ACKNOWLEDGMENT

This research is supported by a grant from M/A-COM Co. The authors appreciate discussions with technical staff of Aerospace and Defense Division. The access to the new design software, provided by Silvaco Co., is also appreciated.

REFERENCES

1. H. Raich, "Principle of Electron Tubes", AA Publishing house, (1995).
2. V. Camarchia, E. Bellotti, M. Goano, and G. Ghione, "Physics-Based Modeling of Submicron GaN Permeable Base Transistors", IEEE Electr. Dev. Lett., 23, 6, 303-305, (2002).
3. P. Speckbacher, J. Berger, A. Asenov, F. Koch, and W. Weber, "The Gated-Diode Configuration in MOSFETs, A Sensitive Tool for Characterizing Hot-carrier Degradation",IEEE Trans. Electr. Dev., 42, 1287-1296, (1995).
4. M. Dellow, H. Beton, M. Henini, P. Main, L. Eaves, S. Beaumont, C. Wilkinson,. "Gated Resonant Tunneling Devices" Electr. Lett. 27, 2, 134-136, (1991).
5. K. Hess "Negative Differential Resistance through Real-Space-Electron Transfer" Appl. Phys. Lett. 35, 469-471 (1979).
6. S. Mil'shtein., C. Gil. "Semiconductor Triode" Patent Applications (July 2004).

From Vacuum Tubes to a Semiconductor Triode

S. Mil'shtein

Advanced Electronic Technology Center, ECE Department, U Mass, Lowell MA 01854, U.S.A.

Abstract. Current study presents a brief review of an electronic technology evolution: from vacuum tubes, to transistors, to a novel, recently developed semiconductor triode, where electrons travel vertically about 600 angstroms from the filament to the anode. We plotted I-V and transfer curves for the semiconductor triodes. The very first prototypes proved to carry a maximum gain of about 15db and f_T= 8GHz. Filaments of variable length were produced to study mutual electrostatic interaction of the electrodes in the triode.

INTRODUCTION

The evolution of electronics from vacuum tubes to solid state devices was driven mainly by microminiaturization trends [1-2]. As solid state electronics established itself with conventional diodes and transistors, the era of quantum well devices began with the goal to further improve power, frequency response, and noise parameters of electronic devices [3]. Simultaneously, development of vacuum microtubes with field emitters had been started [4]. The fabrication of vacuum tubes on silicon wafers meant to combine the high electron group velocity in a vacuum with the submicron distances between the cathode and anode to shorten the transit time. However, low reliability of field emitters limited the progress of this development [4]. Ballistic transistors were also designed to achieve short transit times for electrons [5]. With the same goal in mind, a variety of vertical and V-groove structures were proposed. Among them were permeable transistors, real space transfer transistors, and gated diodes [6-8].

We discuss in current study the operation of the semiconductor triode, which is conceptually different from all known semiconductor devices including vacuum tubes despite the resemblance of their output characteristics. In vacuum tubes, the grid controls the transport of electrons, already freed from a cathode, toward a plate (anode). In our novel device, the grid controls the filament current itself, a portion of which is collected by the anode. The heterostructure anode requires a very small voltage for operation, and so does the bias of a filament. Low terminal voltages are required for "power hungry" applications such as, cellular phones and pagers etc. Also of great advantage are the small internal capacitances of the novel triode. The grid/filament capacitance is about the same as that of a gate/source in HEMT, $C_{gf} = C_{gs}$. However, the anode/filament capacitance is much smaller than that of the drain/gate: $C_{af} \ll C_{gd}$.

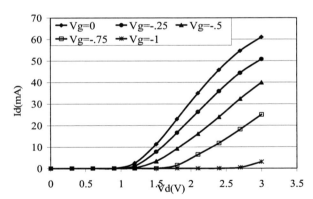

FIGURE 1. I-V characteristics for triode with non interfering anode and filament biases.

EXPERIMENTAL RESULTS

This new device consists of an InGaAs filament with the grid and the plate (anode) designed as small (0.5 µm by 1000 µm) heterojunction AlGaAs/InGaAs contacts. Both the grid and the anode are susceptible to a bias by the voltages applied at the filament terminals. We designed and fabricated two semiconductor triodes with the filaments of 7.5μm and 5.25μm in length. Variation of the filament dimension allows changing of the location of both the grid and the anode in reference with the terminals of the filament. The design with shorter filament is presented in current proceedings [9]. In current study the measurements are performed with a 7.5μm long filament. The procedures of DC and RF testing, as well as the design of the triode are described in detail elsewhere [9-10]. Figure 1 shows the I-V characteristics of the triode. The filament and the anode voltage were swept simultaneously from 0 to 3 volts, and the grid voltage was stepped from 0 to -1 volt. Figure 2 shows the transfer and transconductance characteristics of the triode biased with the filament metal and anode voltage the same. The anode voltage and filament metal voltage were set to a constant 3 volts, and the Anode was then swept from -1.5 to 1 volt. Figure 3 shows the S21 characteristics of the triode. The S-parameters were measured with -0.5 V gate voltage and +3 V anode/filament voltage.

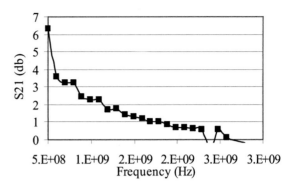

FIGURE 3. S21 for triode with non interfering anode and filament biases.

ACKNOWLEDGMENTS

This research is supported by a grant from M/A-COM Co. The authors appreciate discussions with technical staff of Aerospace and Defense Division. The access to the new design software, provided by Silvaco Co., is also appreciated.

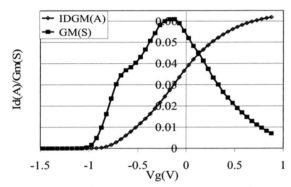

FIGURE 2. Transfer and transconductance characteristics of triode with non interfering anode and filament biases.

CONCLUSIONS

The very first prototypes of the novel triode demonstrated the ability to operate with a maximum gain of 15 db and f_T= 8GHz with the short filament, and with a 6db gain and f_T= 2.3GHz for the long filament. All parameters: transconductance, transfer characteristics, gain, and f_T, varied with longer filament and relative position of the grid and anode. For the 7.5μm long filament we recorded reduced filament grid interaction and the anode current increase by 20%. At the same time the maximum gain dropped to 6db and f_T reduced to 2.3GHz. Overall, the lower terminal voltages and smaller internal capacitances, in comparison with the best transistors, make the novel triode a promising device for systems with limited power.

REFERENCES

1. H. Raich, "Principle of Electron Tubes", AA Publ.,(1995).
2. T.Ohmi, S. Sugawa, K.Kotani, M. Hirayama and A. Morimoto, "New Paradigm of Silicon Technology", Proc. IEEE, 89, 3, 394-411, (2001).
3. H. Morkoc, S. Strite, G.Gao, M. Lin, B. Sverdlov, and M.Burns, "Large –band – gap SiC, III –V Nitrides and ZnS-based Semiconductor Device Technologies" J. Appl. Phys. 76 (3) 1363, (1994).
4. S. Mil'shtein, C. A. Paludi, Jr., P. Chau and J. Awrach, "Perspectives and Limitations of Vacuum Microtubes," Jour. Vac. Techn. A 11(6), 3126-3129 (1993).
5. M. Shur and L.Eastman, "Ballistic Transport in Semiconductors at Low Temperatures for Low Power High Speed logic", IEEE Electr. Dev. 26, 11, 1677-1683, (1979).
6. V. Camarchia, E. Bellotti, M. Goano, and G. Ghione, "Physics-Based Modeling of Submicron GaN Permeable Base Transistors", IEEE Electr. Dev. Lett., 23, 6, 303-305, (2002).
7. K. Hess "Negative Differential Resistance through Real-Space-Electron Transfer" Appl. Phys. Lett., 35, 469-471, (1979).
8. M. Dellow, P. Beton, M. Henini, P. Main, L. Eaves, S. Beaumont, C. Wilkinson, "Gated Resonant Tunneling Devices" Electr. Lett. 27, 2, 134-136, (1991).
9. C. Gil, S. Mil'shtein, "Semiconductor Triode with Heterostructured Filament" Proc. ICPS-27, (2004).
10. S. Mil'shtein, C. Gil. "Semiconductor Triode" Patent Applications (July 2004).

Steering of quantum waves: Demonstration of Y-junction transistors using InAs quantum wires

Gregory M. Jones[*], Jie Qin[*], Chia-Hung Yang[*], and Ming-Jey Yang[†]

[*]Dept. of Electrical and Computer Engineering, University of Maryland, College Park, MD 20742
[†]Naval Research Laboratory, Washington DC 20375

Abstract. In this paper we demonstrate using an InAs quantum wire Y-branch switch that the electron wave can be switched to exit from the two drains by a lateral gate bias. The gating modifies the electron wave functions as well as their interference pattern, causing the anti-correlated, oscillatory transconductances. Our result suggests a new transistor function in a multiple-lead ballistic quantum wire system.

INTRODUCTION

The original proposal of the Y-channel transistor, or Y branch switch (YBS) [1] came from an electron wave analogy to the fiber optic coupler. The semiconductor version of YBS has a narrow electron waveguide patterned into a "Y" configuration with one source and two drain terminals. A lateral electric field along the plane of the YBS device steers the injected electron wave into either of the two outputs. There are advantages of YBS as a fast switch. Based on electron wave steering, the ultimate size can be downscaled to nanometers, leading to the switching speed of devices in the THz range. In addition, most interestingly, the switching can be accomplished by a voltage of the order of $\hbar/(e\tau_t)$, where τ_t is the transit time of electrons. Then, the switching voltage for a YBS can become smaller than the thermal voltage, $k_B T/e$, as opposed to 40-80 times of $k_B T/e$ needed for the current transistors. Here k_B is the Boltzmann constant and T is the absolute temperature. That would make such devices less noisy and consuming less power.

InAs QUANTUM WIRE SYSTEM

We demonstrate a Y branch switch in the quantum regime using InAs electron wave guides. All transport characteristics reported here are acquired with an excitation voltage less than the thermal energy. Thus the system is kept near equilibrium. Our YBS shows significant deviation from classical transport in their characteristics. These novel characteristics arise, to a great extent, because we have used InAs quantum wires in our YBS's.

Our transistors are built on InAs single quantum wells grown by molecular beam epitaxy. A typical structure has a 2 μm undoped GaSb buffer, GaSb/AlSb smoothing superlattice, a 100 nm AlSb bottom barrier and a 17 nm bare InAs quantum well (QW). The two-dimensional (2D) electron gas in InAs QWs can have a long mean free path and long coherence length. In addition, the InAs/AlSb system has a number of properties that are advantageous for nanofabrication and for studying low-dimensional physics. First, the surface Fermi level pinning position in InAs is above its conduction band minimum, allowing for InAs conducting wires with widths as small as nanometers. [2] Second, the small InAs electron effective mass (m_e^* = 0.023 m_0) leads to a large quantization energy. Magnetotransport studies are first used to calibrate the 2D electron concentration (n_{2D}) and mobility (μ_{2D}) of the as-grown sample. Quantum Hall plateaus and Shubnikov de Haas (SdH) oscillations, both characterizing 2D electrons, are clearly observed on photo-lithographically patterned Hall bars. We obtained an n_{2D} of 3.09×10^{12} cm^{-2}, 2.08×10^{12} and 1.08×10^{12} cm^{-2}, and a μ_{2D} of 1.06×10^4 cm^2/Vs, 1.67×10^4 cm^2/Vs and 1.54×10^4 cm^2/Vs, at 300 K, 77 K and 4K, respectively. The corresponding Fermi wavelength (λ_F), Fermi energy, and l_e, are calculated to be 14 nm, 140 meV, and 307 nm at 300K, 17nm, 114 meV, and 397 nm at 77K, and 24nm, 85 meV, and 264 nm at 4K, respectively.

EXPERIMENTAL RESULT

The YBS devices are fabricated by electron-beam lithography and wet etching. Fig. 1 shows an atomic force micrograph of a YBS device. The Y junction is designed in such a way that the source (Source) and the two drains (Drain1 and Drain2) are all tapered toward the immediate junction area in order to avoid significant backscattering. The two side-gates (gate1 and gate2) are shaped to be with a sharp angle, aiming at localizing the lateral electric field in the junction region, thereby facilitating the observation of the mode switching effect.

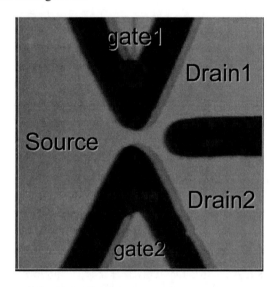

FIGURE 1: Atomic force micrograph of a finished YBS. The narrowest neck is 76nm in width. The dark region is etched away. The terminals in characterization are labeled.

Because of the ballistic, coherent transport, the characteristics of the gated YBS are expected to be drastically different from their classical counterparts, such as differential pair amplifiers. Figure 2 shows the transfer characteristics of our YBS with 76nm wide junction. There is no measurable gate leakage current (<< 1pA) in our measurements. The transconductances through Drain1 and, separately, Drain2 are shown as a function of the sweeping differential gate voltage: $-0.83V < V_{gate1} < 0.83V$, with $V_{gate1} = -V_{gate2}$. When V_{gate1} ($= -V_{gate2}$) is swept, the electric field in the lateral direction steers the wave functions and the interference pattern of the injected electrons. Under such differential gating, the conductances through Drain1 and Drain2 show peaks and valleys, and these oscillatory features are anti-correlated. These anti-correlated, oscillatory trans-conductances provide strong evidence of quantum wave steering in the YBS. For $|V_{gate1}| < 0.25V$, there is little gating effect, and we attribute this lack of response to electrostatic screening by the InAs surface states. That is, the Fermi level pinning position must be shifted to allow for the gating effect to occur, and the nonlinear Poisson equation applied to a three dimensional structure can account for the observed nonlinear response.

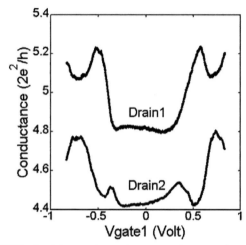

FIGURE 2: Measured conductances from Drain1 and Drain2 are shown as a function of the differential gate voltage.

The reported results here not only have verfied the quantum steering of electron wave functions in a semiconductor wave guide, but also open up possibilities for further studies of quantum switches in multiple-terminal, nanometer-scale structure for information processing.

ACKNOWLEDGMENTS

This work is supported in part by LPS and ONR. CHY acknowledges discussion with Prof. P.T. Ho at UMCP, and Prof. Thylén and Dr. Forsberg at KTH.

REFERENCES

1. T. Palm and L. Thylén, App. Phys. Lett. **60**, 237 (1992).
2. K. A. Cheng, C. H. Yang, and M. J. Yang, Appl. Phys. Lett. **77**, 2861 (2000)

InGaN-Based Nanorod Array Light Emitting Diodes

Hwa-Mok Kim*,[1], Yong Hoon Cho[†], Deuk Young Kim*, Tae Won Kang*, Kwan Soo Chung[¶]

*Quantum-functional Semiconductor Research Center, Dongguk University, Seoul 100-715, Korea
[†]Department of Physics, Chungbuk National University, Cheongju 361-763, Korea
[¶]Department of Electronic Engineering, Kyunghee University, Yongin 449-701, Korea

Abstract. We demonstrate the realization of the high-brightness and high-efficiency light emitting diodes (LEDs) using dislocation-free indium gallium nitride (InGaN)/gallium nitride (GaN) multi-quantum-well (MQW) nanorod (NR) arrays by metal organic-hydride vapor phase epitaxy (MO-HVPE). MQW NR arrays (NRAs) on sapphire substrate are buried in silicon dioxide (SiO_2) to isolating individual NRs and to bring p-type NRs in contact with p-type electrodes. The MQW NRA LEDs have similar electrical characteristics to conventional broad area (BA) LEDs. However, due to the lack of dislocations and the large surface areas provided by the sidewalls of NRs, both internal and extraction efficiencies are significantly enhanced. At 20 mA dc current, the MQW NRA LEDs emit about 4.3 times more light than the conventional BA LEDs, even though overall active volume of the MQW NRA LEDs is much smaller than conventional LEDs.

INTRODUCTION

Group III-nitride semiconductors have been attracted much attention especially for their light-emitting device applications, such as highly efficient light emitting diodes (LEDs) and cw laser diodes (LDs). [1] Recently, studies have shown that micron-size array LEDs fabricated by dry etching offer higher light output efficiencies than conventional BA LEDs. [2-3] Unfortunately, as in the conventional BA LEDs, many threading dislocations are produced in these micro-size LEDs. On the other hand, threading dislocations do not occur during the growth of nanorod (NR), since the growth mechanism of NR is completely different. Therefore, the formation of *dislocation-free* InGaN/GaN multi-quantum-wells (MQWs) in the NR have the potential for negligible non-radiative recombination loss, and thus the efficiency of light output power is much higher than in InGaN/GaN MQW layers. Moreover, the sidewalls of the NR play an important role in extracting light efficiently. A larger surface area is desirable, since it provides more pathways by which generated photons can escape. In spite of these great advantages, LEDs and LDs composed of NR (or nanowire (NW)) arrays (NRAs or NWAs) have not yet been fabricated and studied. Only a few researchers and we have reported on the single-wire (rod) scale devices. [4-6] Here, we demonstrate the realization of high-brightness and high-efficiency LEDs using dislocation-free InGaN/GaN MQW NRAs by metalorganic-hydride vapor phase epitaxy (MO-HVPE). The benefits of the MQW NRA LEDs are examined in this study, and their characteristics are compared to those of conventional BA LEDs.

RUSULTS AND DISCUSSION

Figure 1 shows a cross-sectional schematic diagram of the fabricated InGaN/GaN MQW NRA LED structure. The fabrication processes are almost identical to those of conventional BA LEDs, except for the SiO_2 coating process. Conventional BA LEDs with the same area were also fabricated by conventional MOCVD for comparison studies.

---

[1] To whom correspondence should be addressed. E-mail : khmmkjs@passmail.to (H.M.K.), Tel:+82-2-2260-3952, Fax:+82-2-2260-3945

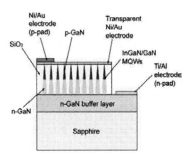

FIGURE 1. Schematic diagram of cross-sectional MQW NRA LED structures.

Carrier transportation in the NRA LEDs is physically confined to the vertical direction due to the geometry of the one-dimensional structure. In turn, this structure of carrier movement ensures efficient use of the injected carriers. The turn-on voltages of MQW NRA LEDs are slightly higher than that of conventional BA LEDs, as seen from the $I-V$ characteristics shown in Fig. 2 (A). The higher turn-on voltages can be attributed to the effective contact area of the p-type GaN NR in the devices. The MQW NRA LEDs are essentially an interconnected array of individual nano-sized LEDs in the SiO_2, each with their own ohmic contact, so that a larger resistive drop may be expected across the contact. The slightly higher threshold voltages in MQW NRA LEDs can also be attributed to the effective contact area of the p-type GaN NR in the devices. Similar factors apply in case of the micro-ring LEDs, [3] which have larger contact areas than do the MQW NRA LEDs.

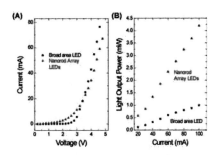

FIGURE 2. Electrical properties of InGaN/GaN MQW NRA LEDs (▲) compared to conventional BA LEDs (■).

Figure 2 (B) shows the light output power versus forward current ($L-I$) for the InGaN/GaN MQW NRA LEDs and for the conventional BA LEDs measured from the top of the samples. An important result, shown in Fig. 2 (B), is that the light power output of MQW NRA LEDs is significantly higher than the output of conventional BA LEDs. At an injection dc current of 20 mA, the MQW NRA LEDs emit about 4.3 times more light than the conventional BA LED, providing solid evidence that the sidewalls facilitate light extraction.

CONCLUSION

We have fabricated and characterized high-brightness and high-efficiency LEDs, by using dislocation-free InGaN/GaN MQW NRAs. Device characteristics have been measured and compared with those of conventional BA LEDs. The extraction efficiency was found to be much higher in MQW NRA LEDs than the conventional BA LED. Light output measurements indicated that light output from MQW NRA LEDs was about 4.3 times that of conventional BA LEDs. Enhanced light emission efficiency makes MQW NRA LEDs clearly superior in many applications, most notably super-bright LEDs for lighting, full color displays, etc.

ACKNOWLEDGMENTS

This work was supported by the KOSEF through the QSRC at Dongguk Univ. in 2004.

REFERENCES

1. Nakamura, S., Fasol, G., *The Blue Laser Diode*, New York, Springer, 1997.

2. Jin, S. X., Li, J., Li, Z., Lin, J. Y., Jiang, H. X., *Appl. Phys. Letters* **76**, 631-633 (2000).

3. Choi, H. W., Dawson, M. D., Edwards, P. R., Martin, R. W., *Appl. Phys. Letters* **83**, 4483-4485 (2003).

4. Kim, H. –M., Kang, T. W., Chung, K. S., *Adv. Materials* **15**, 567-569 (2003).

5. Gudiksen, M. S., Lauhon, L. J., Wang, J., Smith, D. C., Lieber, C. M., *Nature* **415**, 617–620 (2002).

6. Kim, H. –M., Cho, Y. H., Kang, T. W., *Adv. Materials* **15**, 232-235 (2003).

Spiral-shaped microcavity laser: a new class of semiconductor laser

Michael Kneissl[1], Mark Teepe[1], Naoko Miyashita[1], Grace D. Chern[2], Richard K. Chang[2], and Noble M. Johnson[1]

[1]*Palo Alto Research Center Inc. (PARC), Palo Alto, CA 94304, U.S.A.*
[2]*Department of Applied Physics, Yale University, New Haven, CT 06520, U.S.A.*

Abstract. A spiral-shaped microcavity structure is used to achieve uni-direcitonal laser emission from a microresonator. This was previously demonstrated with an InGaN MQW microdisk laser emitting near 400 nm and is demonstrated here with room-temperature laser operation of an AlGaAs heterostructure with pulsed threshold current densities as low as 400 A/cm^2, emission wavelength near 730 nm, and output power of more than 4 mW. Continuous-wave operation of an AlGaAs microdisk laser is also demonstrated for a disk radius of 50 μm.

INTRODUCTION

Laser devices based on microresonator structures such as microdisks, micro-cylinders, and microspheres have been the focus of intense research for more than a decade because of the extremely high Q-factors and very low threshold power densities that can be achieved with such configurations [1-9]. Microresonator lasers emit from so-called whispering gallery modes (WGM), which circulate around the perimeter of the microcavity and are confined by total internal reflection at the dielectric interface of the resonator sidewalls. However, due to the high degree of symmetry inherent in the simple geometric shapes the laser emission tends to be spatially distributed and is therefore difficult to efficiently extract and collect for practical purposes. In this paper we describe a novel resonator design, which confines the emission of a microcavity laser into a narrowly defined direction.

EXPERIMENTAL RESULTS

The basic design of our microdisk lasers is illustrated in Fig. 1(a) with a spiral-shaped cross-section is shown. Previously we have used this design to demonstrate uni-directional emissions from InGaN multiple-quantum-well (MQW) lasers emitting in the

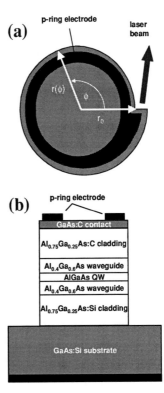

FIGURE 1. (a) Top view showing the cross-section of a spiral-shaped microcavity disk laser diode. (b) Schematic of an AlGaAs single quantum well laser heterostructure.

vicinity of 400 nm [10,11]. The structure consisted of a microcavity disk laser diode with a spiral cross-section, which is defined by

$$r(\phi) = r_0 \cdot \left(1 + \frac{\varepsilon \cdot \phi}{2\pi}\right),$$

where ε is the deformation parameter and r_0 is the radius at $\Phi = 0$. The varying cross-sectional radius creates an indentation at the perimeter of the spiral microdisk at $\Phi = 2\pi$, which enables the efficient out-coupling of the light from the laser resonator through diffraction of the whispering gallery modes at the indentation [11]. We have now applied the concept to a different materials system and demonstrate here room-temperature laser operation of AlGaAs spiral microdisk laser diodes.

A schematic diagram of an AlGaAs single-quantum-well (SQW) laser heterostructure is shown in Fig. 1(b). The AlGaAs heterostructure was grown by metalorganic chemical vapor deposition (MOCVD) on Si-doped (100) GaAs substrates. The active region of the device is comprised of a single 80Å-thick $Al_{0.2}Ga_{0.8}As$ quantum well surrounded by 100nm thick un-doped $Al_{0.4}Ga_{0.6}As$ waveguide layers. The active region is surrounded by Si- or C-doped 1000 nm thick $Al_{0.75}Ga_{0.25}As$ cladding layers providing transverse mode optical confinement. The heterostructure is capped by a 50nm thick C-doped GaAs contact layer. After MOCVD growth the laser heterostructures were etched into spiral cross-sections by chemically assisted ion beam etching with disk radii ranging between 50 μm and 350 μm. A subsequent wet-etch step was used to smooth the microdisk sidewalls. Figure 2 shows an optical microscope image of a spiral-shaped AlGaAs microdisk laser operating under forward bias. The AlGaAs SQW region was designed to emit in the red spectral region at a wavelength near 730nm. The "notch" of the spiral disk is clearly visible on the right. Most of the spiral disk is covered by the opaque top metal-electrode (and the probe tip) so that only light close the perimeter of the spiral disk can be seen.

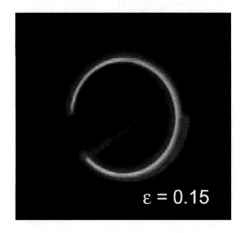

FIGURE 2. Optical microscope image of a spiral-shaped (ε=0.15, r_0=150 μm) AlGaAs SQW microdisk laser operated under forward bias.

Figure 3 (a) shows a room-temperature emission spectrum, measured above threshold, of a spiral microdisk laser with a disk radius of 50 μm. The laser emission peaks near 730 nm with multiple WGMs observable in the spectrum. The mode spectrum changes with drive current due to the excitation of different WGM in the spiral microdisk. Figure 3 (b)

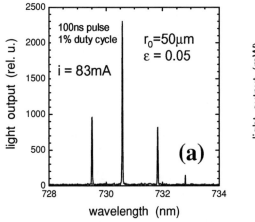 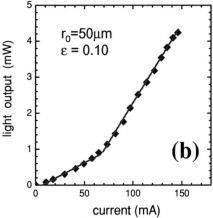

FIGURE 3. (a) Room-temperature emission spectrum of an AlGaAs SQW spiral microdisk laser (disk radius 50 μm, deformation parameter ε=0.05). The laser diode was operated under pulsed current-injection condition with a pulse width of 100ns and a duty cycle of 1%. (b) Light-output vs. current (L-I) characteristic of an AlGaAs microdisk laser diode.

shows the light-output vs. current (L-I) characteristic of a spiral microdisk laser with the same disk radius of 50 μm but with a larger deformation parameter of $\varepsilon=0.1$. A kink in the L-I curve is evident at ~70mA injection current, which corresponds to the onset of laser operation. An output power of more than 4 mW was obtained for laser emission wavelengths near 730 nm. The threshold current density is dependent on the disk radius as well as the deformation parameter as can be seen in Fig. 4. Room-temperature pulsed (100ns pulse width, 1% duty cycle) laser operation was achieved with threshold current densities as low as 400 A/cm^2, which is comparable to the threshold current-density obtained in conventional Fabry-Perot type laser devices fabricated from the same material. As can be seen in Fig. 4, the threshold current density increases with decreasing disk radius due to the higher out-coupling loss and reduced gain per roundtrip. We also observe an increase of threshold current density with increasing deformation parameter. This can be explained by the higher out-coupling efficiency and greater loss for larger deformation parameters ε. The disk radius and ε can be treated as design parameters, such that a low threshold current density can be traded off for a higher out-coupling efficiency.

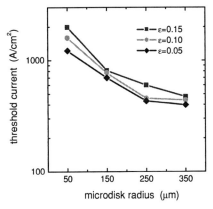

FIGURE 4. Threshold current densities for AlGaAs SQW microdisk laser diodes with different disk radii r_0 and deformation parameters ε.

Figure 5 shows the emission spectrum of an AlGaAs spiral microdisk laser diode operating continuous-wave. The threshold current for the 50 μm radius microdisk was ~60mA (dc) with the emission wavelength near 730 nm. This is the first demonstration of cw operation of a spiral microdisk laser diode. It is anticipated that by further reducing the disk radius and by eliminating unwanted optical loss through scattering of light due to sidewall roughness, the threshold current density can be further reduced.

The concept underlying a spiral microdisk laser is, of course, applicable to other material systems emitting at different wavelengths. For example, microcavity disk lasers can be made from InGaAlAs or InGaAsP compounds that emit in the near infrared, a spectral range of great importance in telecommunication applications. It might also be possible to combine different microresonator elements, to enable new functionalities such as wavelength filters or master-oscillator-power-amplifier configurations.

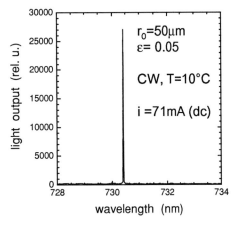

FIGURE 5. CW laser emission spectrum from an AlGaAs spiral microdisk, measured above threshold.

ACKNOWLEDGMENTS

The work was partially supported by the Defense Advanced Research Projects Agency SUVOS program monitored by LTC Dr. J. Carrano.

REFERENCES

1. S.-X. Qian, J.B. Snow, H.-M. Tzeng, R.K. Chang, Science 231, pp. 486-488 (1986).
2. S.L. McCall, A.F.J. Levi, R.E. Slusher, S.J. Pearton, R.A. Logan, Appl. Phys. Lett. 60, pp. 289-291 (1991).
3. A.F.J. Levi, R.E. Slusher, S.L. McCall, S.J. Pearton, W.S. Hobson, Appl. Phys. Lett. 62, pp. 2021-2023 1993).
4. Y. Yamamoto, E. Slusher, Physics Today 46, pp. 66-73 (1993).
5. D.K. Armani, T.J. Kippenberg, S.M. Spillane, K.J. Vahala, Nature 421, pp. 925-928 (2003).
6. S.M. Spillane, T.J. Kippenberg, K.J. Vahala, Nature 415, pp. 621-623 (2002).
7. R.K. Chang and A.J. Campillo, Optical Processes in Microcavities, Chap. 10, (World Scientific, Singapore, 1996).

8. J.U. Nöckel, A.D. Stone, Nature 385, pp. 45-49 (1997).
9. C.Gmachl, F. Capasso, E.E. Narimanov, J.U. Nöckel, A.D. Stone, J. Faist, D.L. Sivco, A.Y. Cho, Science 280, pp. 1556-1569 (1998).
10. G.D. Chern, H.E. Tureci, A.D. Stone, R.K. Chang, M. Kneissl, N.M. Johnson, Appl. Phys. Lett. 83, pp. 1710-1712 (2003).
11. M. Kneissl, M. Teepe, N. Miyahsita, N.M. Johnson, G.D. Chern, R.K. Chang, Appl. Phys. Lett. 84, pp. 2485-2487 (2004).

Room-temperature operation of a green monolithic II-VI vertical-cavity surface-emitting laser

C. Kruse, K. Sebald, H. Lohmeyer, B. Brendemühl, R. Kröger, J. Gutowski and D. Hommel

Institut für Festkörperphysik, Universität Bremen, Otto-Hahn-Allee NW1, 28359 Bremen, Germany

Abstract. We report on the realization of an optically pumped monolithic vertical-cavity surface-emitting laser operating at a wavelength of 511 nm. The structure consists of a ZnSSe λ-cavity containing three ZnCdSSe quantum wells surrounded by two Bragg reflectors using ZnSSe for the high-index material and MgS/ZnCdSe superlattices for the low-index material. The microresonator has a quality factor of 3200 while the threshold excitation power density for the onset of lasing is 22 kW/cm^2 at room temperature. Micropillars of different diameter fabricated out of this structure show discrete optical modes due to the three dimensional optical confinement of the optical wave.

Up to now, only II-VI-based laser diodes are able to provide laser emission in the blue-green spectral region [1]. The realization of an electrically pumped vertical-cavity surface-emitting laser (VCSEL) is of high interest concerning data transmission using plastic optical fibers (POFs), since they have the damping minimum in the green. Furthermore, CdSe quantum dots (QDs) show single photon emission up to 200 K [2]. The integration of such QDs into a microresonator with a high quality factor (Q) could lead to an efficient single photon source with highly directed emission.

The growth of the VCSEL was performed by molecular beam epitaxy at a substrate temperature of 280°C. The two distributed Bragg reflectors (DBRs) of the VCSEL (18 periods bottom DBR, 15 periods top DBR) consist of 48 nm thick ZnS$_{0.06}$Se$_{0.94}$ for the high refractive index layers and a MgS(1.9 nm)/ZnCdSe(0.6 nm) superlattice (SL) of 24.5 periods for the low index layers. The ZnS$_{0.06}$Se$_{0.94}$ λ-cavity contains three ZnCdSSe quantum wells (QWs) at the antinode position of the optical wave (details in Refs. [3,4]). The cylindrically shaped micropillars were prepared out of the planar VCSEL structure using a focussed gallium-ion beam (FIB). The II-VI material around the micropillars has been removed in a radius of 15 μm down to the GaAs substrate in order to make sure that only emission of the pillar is detected by the micro-photoluminescence (μ-PL) setup. For the μ-PL measurements a cw-excitation at a wavelength of 458 nm (Ar$^+$ laser) was used (spot diameter ~1.5 μm). The optical pumping of the planar VCSEL structure was performed with an excimer pumped dye laser (455 nm, 4 ns FWHM pulses, 25 Hz repitition rate) in a conventional PL setup (spot diameter ~500 μm).

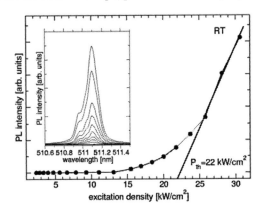

FIGURE 1. PL excitation series of the planar VCSEL structure (spot diameter ~500 μm) at room temperature (RT). Above threshold of 22 kW/cm^2 the spectrum is dominated by a narrow emission line. The Q-factor of the structure is 3200.

In Fig. 1 the development of the PL intensity for the planar VCSEL structure under increasing excitation density is shown at RT (corresponding spectra in the inset). Above the threshold of 22 kW/cm^2 stimulated emission sets in and a single emission line becomes dominant. From the FWHM of the spectrum below threshold a Q-factor of 3200 can be estimated. The FWHM of the PL emission measured for a reference structure without microresonator containing the same active region as the VCSEL is about 8 nm at RT.

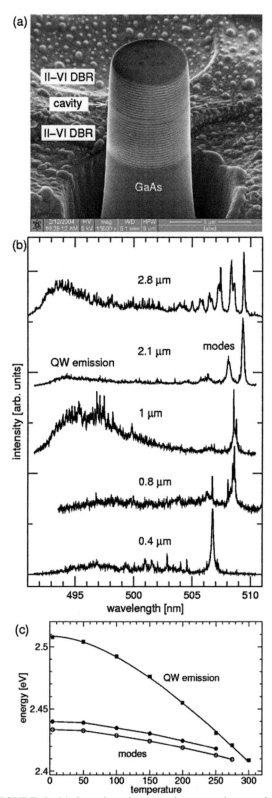

FIGURE 2. (a) Scanning electron microscopy image of a micropillar. (b) Micro-PL spectra for micropillars with different diameters (4 K). (c) Temperature dependence of QW emission and the first two modes of the 2.1 μm pillar.

In Fig. 2(a) a scanning electron microscopy image of a micropillar with a diameter of 2.8 μm is depicted. Sidewalls of the pillar prepared by the FIB are smooth and Bragg pairs of bottom and top DBR are clearly visible as well as the cavity with the active region in the center. Micro-PL spectra of the pillars with different diameter at a temperature of 4 K are shown in Fig. 2(b). Discrete modes are formed due to the additional confinement of the optical wave in lateral direction by the refractive index step between semiconductor and air. With decreasing diameter the first (fundamental) modes shift torwards shorter wavelengths, i.e. higher energies because of the increasing optical confinement. Furthermore, an increased spectral distance between the modes is observed for decreasing diameter of the micropillar, which is consistent with theoretical expectations. The Q-factor of the fundamental mode is 3600 (diameter 2.8 μm) and 2600 (d=2.1 μm) as determined from the FWHM. The origin of the sharp lines superpimposed to some modes in the spectra is unclear at the moment and still under investigation. They might be due to localized states in the low energy tail of the QW emission.

In order clearify the origin of the broad emission band on the short wavelength side of the spectra, temperature dependent PL measurements on the d=2.1 μm micropillar have been performed. The energy shift of the broad emission compared to the first two modes from 4 to 300 K is depicted in Fig. 2(c). The broad emission shifts about 100 meV an can be fitted with the empirical Varshni equation using typical parameters for the ZnCdSSe bandgap. The shift of the modes is only about 25 meV, due to the lower variation of the refractive index with temperature [lines in Fig. 2(c) guide to the eye].

In conclusion, a monolithic II-VI-based VCSEL shows laser emission at a wavelength of 511 nm under optical excitation at RT (threshold excitation density 22 kW/cm^2). For micropillars of different diameter etched out of this structure discrete modes with Q-factors up to 3600 are observed.

We like to thank P. Michler and F. Jahnke for helpful discussions. This work was approved by the Volkswagen-Stiftung (I/76 142).

REFERENCES

1. M. Klude, G. Alexe, C. Kruse, T. Passow, H. Heinke, D. Hommel, phys. stat. sol. (b) **229**, 497 (2002).
2. K. Sebald, P. Michler, T. Passow, D. Hommel, G. Bacher, A. Forchel, Appl. Phys. Lett. **81**, 2920 (2002).
3. C. Kruse, G. Alexe, M. Klude, H. Heinke, D. Hommel, phys. stat. sol. (b) **229**, 111 (2002).
4. C. Kruse, S. M. Ulrich, G. Alexe, R. Kröger, P. Michler, J. Gutowski, D. Hommel, phys. sat. sol. (b) **241**, 731 (2004).

Terahertz Emission and Detection by Plasma Waves in Nanoscale Transistors

F. Teppe[1,2], J. Łusakowski[1], N. Dyakonova[1], Y. M. Meziani[1], W. Knap[1,2], T. Parenty[4], S. Bollaert[4], A. Cappy[4], V. Popov[5], F. Boeuf[6], T. Skotnicki[6], D. Maude[7], S. Rumyantsev[2], and M.S. Shur[2]

[1] *GES CNRS-Universite Montpellier2 UMR 5650 34900 Montpellier, France*
[2] *Center for Broadband Data Transport and ECSE, Rensselaer Polytechnic Institute, Troy, New York 12180*
[4] *IEMN-DHS UMR CNRS 8520, Avenue Poincaré, 59652 Villeneuve d'Acsq France*
[5] *Institute of Radioengineering and Electronics, Russian Academy of Sciences 410019 Saratov, Russia*
[6] *ST Microelectronics, BP 16, 38921 Crolles, France*
[7] *Grenoble High Magnetic Field Laboratory, CNRS-MPI 38450 Grenoble, France*

Abstract. We report on the detection of the sub-THz and THz radiation by silicon FETs and on the voltage tunable emission of terahertz radiation from InGaAs/AlInAs HEMTs with nanoscale gate lengths. The observed photo-response is in agreement with the predictions of the plasma wave response theory. The spectrum of the emitted signal has two peaks. The lower peak is interpreted as resulting from the Dyakonov – Shur instability of the gated two dimensional electron fluid. The emission measurements in a magnetic field show that the threshold voltage remains close to the transistor saturation voltage that increases with magnetic field due to geometric magnetoresistance.

INTRODUCTION

Terahertz emission and detection using plasmon resonances in two-dimensional electron gas field effect transistors was predicted in the early 90-s [1,2]. However, experiments demonstrating resonant detection [3-5] and emission [6] have been reported only recently. The plasma wave THz emission was observed from 60-nm gate GaInAs high electron mobility transistors [6]. Non-resonant detection of sub-terahertz radiation was recently observed in Si FETs at room temperature [7] In this paper, we report on the new results on the non-resonant detection of the THz radiation by silicon FETs and on the resonant (voltage-tunable) emission of terahertz radiation from a gated 2D electron gas in InGaAs/AlInAs lattice-matched HEMTs.

THZ DETECTION BY Si MOS

The photo-response of *n*-channel Si MOSFETs to 120 GHz and 3 THz radiations was measured for several transistor lengths, including ultra-short transistors with the 30 nm gate length (see Fig.) The solid lines in Fig.1 are fits of the plasma wave theory (Ref. [4]) to the experimental results. The theory reproduces relatively well the position and the shape of the observed maximum.

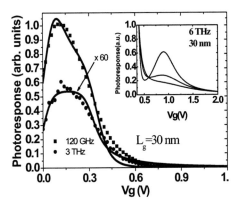

FIGURE 1. Drain-source signal as a function of the gate voltage for *f*=120 GHz (squares) and *f*=3THz (circles)) for the 30nm gate length Si transistor at 300K. The inset shows the calculated photo-response for mobilities of 200cm^2/V.s, 300cm^2/V.s, and 500cm^2/V.s.

The calculations based on the extracted parameters using the model described in Ref. [2] predict that a resonant detection at THz frequencies might be possible by investigated Si FETs (see Insert in Fig. 1.)

We also observed the detection of sub-THz signal (120 GHz) in longer channel Si MOSFETs (i.e. 800 nm.)[7]

VOLTAGE-TUNABLE EMISSION OF TERAHERTZ RADIATION

A InGaAs/AlInAs HEMT on InP substrate (with the gate length of 60 nm, the source-drain spacing of 1.3 µm, and the gate width of 50 µm[8]) was placed into a cyclotron emission spectrometer[9] kept at 4.2 K and totally isolated from the 300 K background radiation.

The emission occurred when the drain current (drain voltage) reached a certain threshold ~4.5 mA (~200 mV), see Fig. 2. The peak emission frequency shifted from ~0.42 THz up to ~1 THz with increasing source-drain voltage. The observed radiation power was in the nW range as compared to the pW power of the bulk InSb emitters used for calibration.

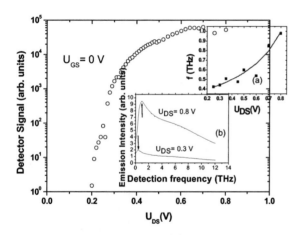

FIGURE 2. Threshold behavior of the THz emission signal as a function of U_{DS} at $U_{GS} = 0$ V and magnetic field $B = 1$ T. (a) Resonant frequency of the emission, f, from an InGaAs HEMT vs. U_{sd}. The continuous line is plotted using Eq.1.of Ref. 6. Spectra of emission from an InGaAs HEMT for different source–drain voltages, U_{sd}. The arrows mark emission maxima at 0.42 THz and 1.0 THz for U_{sd} equal to 0.3 V and 0.8 V, respectively.

The resonant frequency of the emission is determined by the sheet electron density under the gate and by the effective gate length L_{eff}. As the drain bias increases, the gate length modulation effect decreases the effective channel length, leading to an increased emission frequency. This model is in good agreement with measured data (see the top insert in Fig. 2.)

We also observe a long tail of the emission spectrum (up to 12 THz.) Possible explanations involve plasma wave instability in the ungated device channel and the plasma modes in the gate channel propagating under different angles with respect to the current flow.

We also performed the emission measurements in magnetic fields up to 6 T. The emission threshold voltage was close to the transistor saturation voltage that increased with magnetic field due to geometric magnetoresistance.

CONCLUSION

We demonstrated the THz detection by 30 nm Si MOSFETs. We also reported on a resonant THz emission from a 60 nm InGaAs HEMT and interpreted this emission as caused by the Dyakonov-Shur plasma wave instability.

REFERENCES

1. M. Dyakonov and M. S. Shur, Phys. Rev. Lett. **71**, 2465 (1993); M. Dyakonov and M. S. Shur, IEEE Trans. ED **43**, 380, (1996).
2. M. Dyakonov and M. S. Shur in "Terahertz Sources and Systems", Kluwer Academic Publishers-Netherlands ed. by R. E. Miles, p187-207 (2001).
3. W. Knap, Y. Deng, S. Rumyantsev, and M. S. Shur, Appl. Phys. Lett., **81**, 4637 (2002); W. Knap, Y. Deng, S. Rumyantsev , J.-Q. Lü, M. S. Shur, C. A. Saylor, L. C. Brunel, Appl. Phys. Lett. Vol. **80**, (2002)
4. W. Knap, V. Kachorovskii, Y. Deng, S. Rumyantsev, J.-Q. Lü, R. Gaska, M. S. Shur G. Simin, X. Hu and M. Asif Khan, C. A. Saylor, L. C. Brunel. J. Appl. Phys., **91**, 9346 (2002)
5. X.G. Peralta, S.J. Allen, M.C. Wanke, J.A. Simmons, M.P. Lilly, J.L. Reno, P.J. Burke, and J.P. Eisenstein. Appl. Phys. Lett. **81**, 1627 (2002)
6. W. Knap, J. Lusakowski, T. Parenty, S. Bollaert, A. Cappy, V.V. Popov, and M. S. Shur. Appl. Phys. Lett. **84**, 3523 (2004)
7. W. Knap, F. Teppe, Y.M. Meziani, N. Dyakonova, J. Lusakowski, F. Boeuf, T. Skotnicki, D. Maude, S. Rumyantsev and M.S. Shur, accepted for publication in APL
8. T. Parenty, S. Bollaert, J. Mateos, X. Wallart, A. Cappy, Proc. of Indium Phosphide and Related Material (IPRM) Conference, Nara Japan, May 2001, pp 626-629.
9. W. Knap, D. Dur, A. Raymond, C. Meny, J. Leotin, S. Huant, B. Etienne, Review of Scientific Instruments, **63**, 3293 (1992)

Big Light: Optical Coherence Over Very Large Areas in Photonic-Crystal Distributed-Feedback Lasers

W. W. Bewley, J. R. Lindle, C. S. Kim, I. Vurgaftman, C. L. Canedy, M. Kim, and J. R. Meyer

Code 5613, Naval Research Laboratory
Washington, DC 20375

Abstract. Recent progress in the development of edge-emitting and surface-emitting photonic-crystal distributed-feedback lasers is reviewed. For an edge-emitting device with a mid-IR "W" active region, the beam quality is no worse than 8 times the diffraction limit for stripes as wide as 600 μm.

INTRODUCTION

Maintaining optical coherence in a semiconductor laser with a large lateral area is extremely challenging, owing to the tendency of the optical mode to break up into numerous filaments that lase independently of one another. When a given region experiences a negative fluctuation in its carrier concentration, the local refractive index increases and the mode is incrementally attracted to that region. That in turn increases the stimulated emission rate which decreases the carrier density further, producing a feedback that ultimately collapses the mode into a narrow filament. In order to combat this effect, we have enforced optical coherence over large areas via diffraction by a relatively shallow two-dimensional grating, in the photonic-crystal distributed-feedback (PCDFB) laser. In the edge-emitting (EE) PCDFB laser, the symmetry axes of the rectangular lattice are tilted at an angle to the facet. Three distinct diffraction processes strongly couple the optical field across the similarly tilted gain stripe. In the surface-emitting (SE) PCDFB lasers, the grating is hexagonal, facets are not required, and the output beam has circular symmetry.

EDGE-EMITTING PCDFB LASERS

The simulated projection of substantially enhanced optical coherence has been confirmed experimentally. In the present experiment, electron-beam lithography was used to pattern a 2nd-order PCDFB grating onto an antimonide type-II "W" quantum well laser emitting at $\lambda \approx 3.7$ μm. We note that it is particularly challenging to produce a good-quality optical beam from a mid-IR "W" laser due to a relatively large product of the linewidth enhancement factor (LEF) and internal loss [1]. The tilt angle of the grating was 16°, and the measured dimensions (53% duty cycle) result in the following set of coupling coefficients: $|\kappa_1'| = 7$ cm^{-1}, $|\kappa_2'| = 100$ cm^{-1}, and $|\kappa_3'| = 7$ cm^{-1}.

Temperature tuning of the energy gap was sufficient to bring the peak of the gain spectrum into resonance with the grating wavelength. For pulsed optical pumping at 10 times the lasing threshold (I_{th}), the output beam is essentially diffraction-limited (DL) up to a stripe width of 150 μm (roughly 10 times the limit for an unpatterned Fabry-Perot device), and remains no worse than 8xDL for stripes as wide as 600 μm. In that limit the far-field profile is nearly Gaussian, with a divergence angle of only 1.2°. The results for the beam etendue of several PCDFB and α-DFB [2] devices are shown in Fig. 1. The excellent beam quality obtained for the present PCDFB device (filled circles) comes with only a reasonable decrease in the differential quantum efficiency. The efficiency is over 40% of that in an unpatterned EE device from the same wafer for stripes wider than 250 μm. Therefore, the present result constitutes a considerable improvement over all previous infrared semiconductor lasers with such broad stripes.

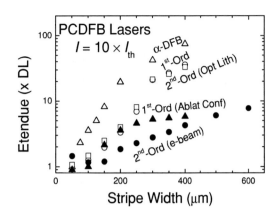

FIGURE 1. Etendue (in units of the diffraction limit) as a function of stripe width for several α-DFB and 1st-order and 2nd-order EE PCDFB lasers operating at 10 I_{th}.

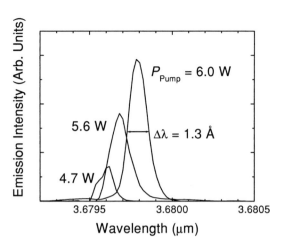

FIGURE 2. Emission spectrum of the SE PCDFB laser for several pump powers under cw optical pumping with a spot diameter of 600 μm.

SURFACE-EMITTING PCDFB LASERS

Whereas EE PCDFB lasers show very promising results, they produce strongly elliptical beams, since the growth-direction optical aperture is several orders of magnitude smaller than the emission stripe. By contrast, for optimized grating parameters the SE PCDFB laser is expected to produce a circular output beams without any requirement for facet definition. A moderately large product of the LEF and the internal loss is actually an advantage, since it helps to stabilize the in-phase mode [3]. Furthermore, numerical simulations indicate that relatively weak coupling into the surface-emitting direction is preferred.

A SE PCDFB laser with the lowest-order grating capable of surface emission has been fabricated using electron-beam lithography. In spite of the rather narrow emission line (1.3 Å full width at half maximum, FWHM) shown in Fig. 2, the beam quality was rather poor. Near-field experiments demonstrate that the lasing spreads out far beyond the pump spot, which is attributed to an insufficiently strong diffractive coupling in the plane of the device. However, for pump spot diameters of 600-800 μm the differential quantum efficiency was as high as 60% of that for an edge-emitting Fabry-Perot device. Another SE PCDFB laser, in which the efficiency was considerably lower, displayed a much narrower far-field emission pattern. Figure 3 shows that the FWHM of the divergence angle for this device was 0.9°, which corresponds to <8 times the diffraction limit. Further optimization is necessary in order to realize the full potential of these devices.

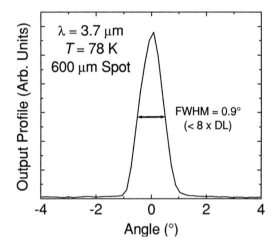

FIGURE 3. Far-field pattern of an earlier SE PCDFB laser optically pumped with a 600-μm-wide spot at a temperature of 78 K.

REFERENCES

1. Vurgaftman I., and Meyer, J. R., "Photonic-crystal distributed-feedback quantum cascade lasers", *IEEE J. Quantum Electron.* **38**, 592-602 (2002).
2. Bartolo R. E., Bewley W. W., Vurgaftman I., Felix C. L., Meyer J. R., and Yang M. J., "Mid-infrared angled-grating distributed-feedback laser", *Appl. Phys. Lett.* **76**, 3164-3166 (2000).
3. Vurgaftman I., and Meyer J. R., "Design optimization for high-brightness surface-emitting photonic-crystal distributed-feedback lasers", *IEEE J. Quantum Electron.* **39**, 689-700, (2003).

Novel Resonant-Tunnelling Quantum Dot Photon Detectors For Quantum Information Technology

J C Blakesley, P See, A J Shields, B E Kardynal, P Atkinson, I Farrer, D A Ritchie

Toshiba Research Europe Ltd, 260 Cambridge Science Park, Milton Road, Cambridge CB4 0WE. UK.
Cavendish Laboratory, University of Cambridge, Madingley Road, Cambridge CB3 0HE. UK.

Abstract. We have created single-photon detectors based on micron-size GaAs/AlGaAs resonant tunnelling diodes containing a layer of self-assembled InAs quantum dots. A quantum efficiency of 12.5% has been achieved with a dark count rate of about 10^{-3} per second in devices with 10nm AlGaAs barriers at 4K. With optimisation, we expect that the quantum efficiency of the devices can be significantly improved. This type of device could eventually find application in efficient quantum communication or computation.

INTRODUCTION

Most single-photon detectors used for quantum information applications use avalanche processes to amplify the charge generated by the photons. These detectors generate errors due to afterpulses caused by secondary avalanches triggered by trapped charges. In addition, no avalanche detector could be used for an application in which the spin-state of the photocarriers needs to be preserved.

Semiconductor quantum dots have potential for application to quantum information technology as they have very few confined electronic states, which can have long lifetimes. Self-assembled quantum dots can be easily incorporated into semiconductor heterostructures, making them very versatile. He we present results of studies on a new type of single-photon detector that incorporates quantum dots into resonant tunnelling diodes [1] (RTDs). The devices could overcome some of the problems inherent in avalanche detectors.

DEVICE STRUCTURE AND FABRICATION

Devices were made by incorporating a layer of InAs quantum dots into a GaAs/AlGaAs RTD. Heterostructures were grown on a semi-insulating GaAs substrate by molecular-beam epitaxy. A typical structure consisted of: 200nm GaAs buffer, 10nm AlAs etch-stop layer, 230nm GaAs emitter with graded n-type doping from 1×10^{18} to $1 \times 10^{16} cm^{-3}$, 20nm undoped GaAs spacer, 10nm $Al_{0.33}Ga_{0.67}As$ barrier, 10nm GaAs well, 10nm $Al_{0.33}Ga_{0.67}As$ barrier, 2nm GaAs spacer, InAs self-assembled quantum dot layer, 300nm undoped GaAs intrinsic region and 50nm n+ doped GaAs collector. The layer of quantum dots was grown in the Stranski-Krastnow mode [2] to give a density of approximately 100 dots per square micron.

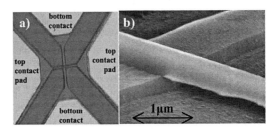

FIGURE 1. Cross-wire resonant tunnelling devices. **a)** optical image. **b)** scanning electron microscope image of active tunnelling area, intersection between bridge contact and line mesa.

Devices were fabricated using a cross wire design (see fig.1) similar to that used by Wang et al. [3]. This was achieved by using a combination of non-selective and selective wet-etches to create a suspended bridge contact to the top of the active tunnel area. Comparisons of tunnel currents with large-area diodes confirmed that active tunnel areas were about $1\mu m^2$. This design had the advantage that the active tunnel area was not obscured by opaque metal contacts.

EXPERIMENTAL RESULTS AND DISCUSSION

We observed the current-voltage characteristics of the RTDs in the presence of low-intensity light. The resonant peak in forward bias (applying a positive bias to the collector side of the structure) shifted to a lower voltage as the devices were illuminated by light with wavelength shorter than the GaAs bandgap (fig.2a).

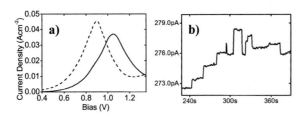

FIGURE 2. a) Resonant tunnelling peak of a large-area device with 10nm AlGaAs barriers at 4K. In dark (solid line) and under weak illumination (dashed line). **b)** Tunnel current versus time of a cross-wire device under illumination by very weak light.

We measured current through the devices as a function of time while applying a constant bias below the centre of the resonant peak. We saw a change in the device current under illumination, consistent with the shift in the peak bias. However, in cross-wire devices, we saw many discrete step-like changes in current instead of a constant smooth change (fig.2b). Each positive step was followed by a negative step of the same size after a delay of about 10 to 100 seconds. After we stopped illuminating the devices, the positive steps ceased but the negative steps continued to occur until the current had returned to its initial value.

We believe that the steps are caused by charge trapping in the layer of quantum dots. Initially, in the dark, electrons are trapped in the dots. Each of these creates an electric dipole with mirror charges on the emitter side of the barriers. This creates an electric field through the barriers that opposes the applied bias locally to the dot. At the correct bias the dipole brings the barriers out of resonance and reduces the tunnel current.

When the devices are illuminated, carriers are generated in the thick intrinsic region on the collector side of the barriers. The electric field due to the external bias separates the charges and drains the electrons to the collector contact. The photoholes, however, are attracted to the trapped charge in the quantum dots. The holes recombine with the trapped electrons, negating the dipole effect and bringing the barriers closer to resonance. This produces the positive steps that are observed in the tunnel current. Thus, each step corresponds to the detection of a single photon. The negative steps that follow the positive steps are due to the repopulation of the dots by the tunnelling current. The length of the recombination time is due to the low current density passing through the barriers.

By differentiating the signal with respect to time and applying an appropriate discriminator level, we were able to count the number of steps observed as a function of photon flux incident on the devices. The rate of positive steps had a linear dependence on the photon flux (fig. 3). With the discriminator level optimised for quantum efficiency, we achieved a maximum quantum efficiency of 12.5%. By increasing the discriminator level we could reach dark-count rates of about $10^{-4} s^{-1}$ while keeping quantum efficiency greater than 5%.

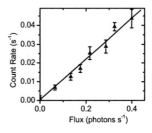

FIGURE 3. Count rate versus flux of 550nm photons incident on the active device area of a cross-wire device at 4K.

We have also observed single-photon detection in devices with 4nm barriers using a cryogenic amplifier. These devices have tunnel currents 10^4 times larger than those of the 10nm barrier devices, and correspondingly larger steps. This allows single shot read out of the detector using conventional photon counting electronics with approximately microsecond detection time resolution.

REFERENCES

1. H. Mizuta and T. Tanoue, *The Physics and Applications of Resonant Tunnelling Diodes*, Cambridge University Press, Cambridge, 1995.
2. D. Bimberg, M. Grundmann and N. N. Ledentsov, *Quantum Dot Heterostructures*, Wiley, New York, 1999.
3. Wang et al., *Appl. Phys. Lett.* **65**, 1124-1126 (1994).

The Dynamical Diffraction Effect in a Two-Dimensional Photonic Crystals

O.A. Usov, M.V. Maximov

Ioffe Physico-Technical Institute, RAS, Polytechnical str. 26, 194021 St. Petersburg, Russia

Abstract. The anomalous enhancement of absorption of some low-frequency molecular vibrations of liquid crystals inside a two-dimensional macroporous silicon photonic crystals is shown to be multiple scattering or dynamical diffraction of photonic mode with maximum of the electric field in the region of low index refraction. The enhancement of optical absorption is believed to increase sensitivity of devices used for conclusive identification and quantification of a small molecular concentration species in liquid and gas phase mixtures.

INTRODUCTION

Enhanced optical absorption of low-frequency molecular vibrations of liquid crystals (LCs) infiltrated into channels of two-dimensional (2D) macroporous silicon photonic crystals (PCs) observed with Fourier transform infrared (FTIR) spectroscopy [1-2] is shown to be an effect of dynamical diffraction (DD) or multiple scattering of photons of selected dispersion mode of PCs. This effect correlates with X-ray diffraction experiment in atomic crystals for Laue geometry well known as anomalous transmission (Borrmann effect) and enhanced absorption [3]. The DD theory predicts two resonance waveguiding modes inside the PCs: one has node and one has antinode at a lower refractive index plane. But only for PCs in the optical range it is possible to insert LC molecules to check the electric field distribution inside pores. Some experimental studies were oriented for turning photonic band gap (PBG) by means of temperature or external electric field dependence of index of refraction of LCs infiltrated into porous systems. The PBG dependence on temperature was found from transmission or reflection measurements using Bragg geometry as discussed by S.W. Leonard et al [4-5]. But to find anomalous molecular absorption, photonic dispersion relations or other multiple beam diffraction properties of photons in a dispersive media the Laue geometry has some advantages [3]. The anomalous IR light absorption of some low-frequency molecular vibrations of LCs [1-2] in 2D macroporous silicon PCs is explained by unusual electric field distribution with maximum in the center of pores with LCs specific for forward diffracted light, which propagates in the region of lower index refraction of 2D PCs. It is important to underline that the effect of enhancement of optical absorption also correlates with one of efficient light extraction scheme used to enhance quantum efficiency of light-emitting diodes. The extraction efficiency and absorption is strongly dependence on the angle between the propagation direction of emission light and the PC lattice. But for absorption case we can observe only maximum enhancement because it is difficult to separate generation and extraction regions to use the angle dependence as it was done in the case of light emission in 2D PCs [6].

EXPERIMENTAL

Macroporous silicon PCs experimentally studied by Perova et al [1-2] consists of a regular cylindrical pores of 3-4.5 μm diameter and 200-250 μm depth, arranged in triangular lattice of 12 μm period, and encapsulated with two types of LCs: ferroelectric (FLC) mixture SCE-8 and triphenylene based discotic (DLC) H5T-NO$_2$. Fourier transform infrared measurements were performed in the range of 450-

4000 cm^{-1}. A 20-fold absorption enhancement of the low-frequency vibrational bands was observed in the region 600-900 cm^{-1} for both types of LCs. The enhancement of absorption of molecular vibrations of LCs is supposed to be a result of specific electric field distribution in the pores and increase of light paths due to multiple scattering and interlayer tunneling of photons. To compare different geometry and influence of the PC lattice parameters the results of S.W. Leonard et al [4-5] were also analyzed. Their 2D PCs possess the same triangular lattice, but with less parameters: lattice constant equal to 2.3 μm, diameter of pores in the range 1.58-2.13 μm, which infiltrated by E7 type LC with refractive index 1.49-1.69. They measured transmission and PBG in the Γ-K and Γ-M directions; also found some LC absorption bands in the range 3.4-7.2 μm, but no enhancement of absorption obtained. It is perhaps very small due very thin samples (only 8 PC layers with thickness about 18.4 μm) and limited light paths inside the pores with LCs.

RESULTS AND DISCUSSION

The main question is how to find the frequency of PC mode and electric field distribution, which control the absorption enhancement effect. Using plane wave (PW) method we have numerically calculated photon dispersion of 2D triangular structure PCs for all symmetry points of Brillouin zone for different PC parameters. We have obtained that the frequency band about 500-800 cm^{-1} of enhanced molecular absorption positioned in the region of the lowest air (donor) photon bands. For PCs fabricated by S.W. Leonard et al [3-4] this effect should be also observed in the higher frequency range about 2500-4000 cm^{-1}, if change the experimental set up. For analysis of absorption enhancement effect we used Purcell enhancement factor F_p, corrected by Boroditsky et al [6], which account photon degeneracy of high symmetry band edges. In the weak coupling regime the modified Purcell factor is described as

$$F_p = \frac{3g}{4\pi^2}\left(\frac{\lambda_m}{n}\right)^3\left(\frac{Q_m}{V_m}\right) \quad (1)$$

where Q_m is the mode quality factor, g is mode degeneracy, λ is the wavelength, n is the refractive index, and V_m is the mode volume. The Purcell enhancement factor is shown to be anisotropic and its maximum value to characterize the enhancement of light absorption due to coupling of radiation modes to phase matched guided modes, which absorbed by molecules. The F_p(Γ-K) is obtained to be about 3 times more than F_p(Γ-M) and it determines the 20-fold enhancement of absorption in the region 600-900 cm^{-1} of the low-frequency vibration bands. The value of Purcell enhancement factor F_p obtained by Boroditsky et al [6] is supposed to be equal to F_p(Γ-M) and thus also qualitatively confirm the value of anomalous absorption enhancement.

Changing the structure of PCs (triangular to square) or parameters (r/a, a/λ) it is possible to change the frequency range of the effect or to design PC for observation it in the selected frequency range. The enhancement of optical absorption is believed to increase sensitivity of devices used for conclusive identification and quantification of a small molecular concentration species in liquid and gas phase mixtures.

ACKNOWLEDGMENTS

The work is partially supported by Scientific program "New Materials and Structures".

REFERENCES

1. Perova T.S., Astrova E.A., and Usov O.A., "Infrared spectra of liquid crystals incapsulated into channels of macroporous silicon", Microsys. Tekh. **11**, 13-15 (2002).

2. Perova T.S., Astrova E.A., Tsvetkov A.G., Tkachenko A.G., Vij J.K., and Kumar S., "Orientation of discotic and ferroelectric liquid crystals in macroporous silicon matrix", Phys.Sol.State, **44**, 1145-1150 (2002).

3. Pinsker Z. G., Dynamical diffraction scattering of X-rays in crystals, NY, Springer-Verlag, 1978.

4. Leonard S.V., Mondia J.P., van Driel H.M., Toader O., John S., Bush K., Birner A., Gosele O., and Lehman V., "Tunable two-dimensional photonic crystals using liquid-crystal infiltration", Phys. Rev. **B61**, R2389-2392 (2000).

5. Jamois C., Wehrspohn R.B., Andreani L.C., Hermann C., Hess O., and Gösele U., "Silicon-based two-dimensional photonic crystal waveguides", Photonics and Nanostructures **1**, 1-13 (2003).

6. Boroditsky, M., Vrijen R., Krauss T.F., Coccioli R., Bhat R., and Yablonovitch E., "Spontaneous Emission Extraction and Purcell Enhancement from Thin-Film two-dimensional photonic crystals", J. Lightwave Tech., 17, 2096-2113 (1999).

Piezoelectric effect on the lasing characteristics of (111)B InGaAs/AlGaAs laser diodes

G. Deligeorgis[1], G.E. Dialynas[1], N. Le Thomas[2], Z. Hatzopoulos[1], and N.T. Pelekanos[1,3]

[1] *Microelectronics Research Group, FORTH-IESL, P.O Box 1527, 71110 Heraklion, Greece*
[2] *Departement de Recherche Fondamentale sur la Matière Condensée, CEA/Grenoble, France*
[3] *Materials Science and Technology Department, University of Crete, P.O. Box 2208, 71003 Heraklion, Greece*

Abstract. In this work we investigate both experimentally and theoretically the effect of the piezoelectric (PZ) field on the lasing characteristics of InGaAs (111)B quantum wells. We show that, under certain conditions, the existence of the PZ field can be beneficial and lead to substantial threshold current reduction in (111)-grown InGaAs/AlGaAs laser diodes, compared to their (100) counterparts.

INTRODUCTION

Strained InGaAs quantum wells (QWs) grown on (111)B GaAs have received considerable attention in recent years, due to the existence of piezoelectric (PZ) fields [1], which modify strongly the QW electronic band structure and optical properties and make these QWs appealing for applications in electro-optics [2] and nonlinear optics [3,4]. In this work, we examine both experimentally and theoretically the effect of the built-in PZ field on the gain spectrum and threshold current density of (111)B laser diodes (LDs).

EXPERIMENTAL RESULTS

A series of identical QW laser structures was grown on (100) and (111)B GaAs substrates by molecular beam epitaxy. The structures that will be discussed here consist of a 10nm thick $In_{0.1}Ga_{0.9}As$ active QW, sandwiched between two undoped 125nm thick $Al_{0.15}Ga_{0.85}As$ barrier layers. Two 1.6μm thick $Al_{0.3}Ga_{0.7}As$ cladding layers, p and n doped, were grown on either side of the active region. LDs with different cavity lengths and mesa widths have been fabricated, and electroluminescence spectra have been recorded for temperatures ranging from 17 to 300K. The electrical pumping was performed using 500ns pulses at a repetition rate of 10KHz, to avoid heating effects, and current densities required to reach lasing threshold were determined.

The main finding of this work is that in a large number of identical (111)B and (100) LDs of different mesa widths and cavity lengths, we have systematically observed significantly lower threshold currents in the (111)B case. This is a rather unanticipated result, since gain reduction caused by the PZ field in the (111)B case is expected to lead to higher thresholds. In Fig.1, we plot as example the experimental threshold current densities as a function of temperature for two identical (111)B and (100) LDs, with the same active region, mesa width (40μm), and cavity length (1.5mm).

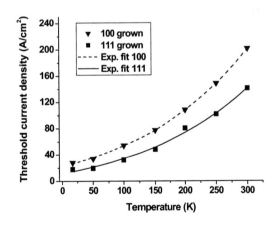

FIGURE 1. Measured threshold current densities versus temperature for identical (111)B and (100) laser diodes. The curves through the data points are exponential fits.

Clearly, the threshold current in the (111)B device remains for all temperatures much lower than its (100) counterpart. As explained in the theoretical part, this threshold reduction for the (111)B case can be attributed to the combination of the beneficial action of the PZ field and the lighter in-plane heavy-hole mass.

In addition, we have observed that the threshold reduction in the (111)B case declines gradually with decreasing cavity length. This can be attributed to the higher optical losses in the shorter cavities, requiring higher gain in order to achieve lasing, which as predicted by our model beneath, "dilutes" the beneficial action of the PZ field in the (111) oriented devices.

THEORETICAL MODEL

To account for the experimental results, we employed a model based on a self-consistent Schroedinger-Poisson solution, utilized to determine the modified QW potential profile, in the presence of the PZ field and the high carrier densities required for lasing. The model also incorporates strain effects on band structure and effective mass dependence on growth direction. Utilizing the transition matrix method, gain and spontaneous emission spectra were calculated, for several injection carrier densities, and the radiative component of threshold current was obtained [5]. In our analysis, non-radiative effects were neglected.

In Fig.2, we plot the calculated threshold current densities as a function of temperature, both for (111) and (100) oriented structures, assuming a gain coefficient value of $500cm^{-1}$. Figure 2 reproduces rather well the experimental results of Fig.1. Our model simulations show that there are two main reasons for the lower threshold current densities obtained in the (111) devices. First reason is the PZ effect in the (111) QWs, which causes not only the above mentioned gain reduction for a given carrier density, but also a significant increase of carrier lifetime in the QW. Under certain circumstances, the latter can overrides the gain reduction, leading to lower current densities, necessary to reach lasing in a (111)B LD. The second reason for the observed reduction is that (111)-grown InGaAs QWs exhibit reduced in-plane effective heavy hole mass, which also results to further decrease of the threshold current density.

Our modelling further shows that with increasing cavity losses the beneficial effect of the PZ field diminishes and the threshold current density difference between (111) and (100) oriented devices is predicted to shrink, in agreement with our experimental observations.

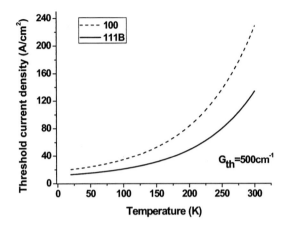

FIGURE 2. Calculated threshold current density J_{th} as a function of temperature for a 10nm $In_{0.1}Ga_{0.9}As/Al_{0.15}Ga_{0.85}As$ QW, grown on (111)B and (100) GaAs, assuming a threshold gain $G_{th}=500cm^{-1}$.

CONCLUSION

The effect of the built-in piezoelectric field on $In_{0.1}Ga_{0.9}As/Al_{0.15}Ga_{0.85}As$ QW laser performance has been investigated experimentally and theoretically. We have observed systematically lower threshold currents in (111)B laser diodes with piezoelectric QWs, compared to (100) devices without any internal field. Our model gain analysis based on a self-consistent treatment of Schroedinger-Poisson equations, accounted well for most of our experimental findings.

REFERENCES

1. E.A. Caridi, T.Y Chang, K.W. Goossen, L.F. Eastman, Appl. Phys. Lett. **56**, 659-661 (1990)
2. D.L. Smith and C. Mailhiot, Phys. Rev. Lett. **58**, 1264-1267 (1987)
3. C. Mailhiot and D.L. Smith, Phys. Rev. B **37**, 10415-10418 (1988)
4. A.S. Pabla, J. Woodhead, E.A. Khoo, R. Grey, J.P.R. David, G.J. Rees, Appl. Phys. Lett. **68**, 1595-1597 (1996)
5. W. W. Chow, H. Amano, T. Takeuchi, J. Han, Appl. Phys. Lett. **75**, 244-246 (1999)

Resonant-photon tunneling effect at 1.5 micron observed in GaAs/AlGaAs multi-layered structure containing InGaSb quantum dots

Naokatsu Yamamoto, Kouichi Akahane, Shin-ichiro Gozu, Naoki Ohtani

National Institute of Information and Communications Technology (NICT), Dept. of Basic and Advanced Research
4-2-1 Nukui-kitamachi, Koganei, Tokyo, Japan

Abstract. We demonstrated that a resonant-photon tunneling (RPT) effect for a 1.5 μm-waveband can be observed in a GaAs/AlGaAs multi-layered structure containing InGaSb quantum dots (QDs) and that the resonant-incidence angle can be controlled all-optically with a control light irradiation into the QDs active layer. These results clearly suggest the possibility of creating RPT devices for all-optical signal processing in the optical communication wavebands.

INTRODUCTION

The photon-tunneling effect has been investigated in multi-layered dielectric structures [1]. The authors previously demonstrated numerically that all-optical switches (AOSs) and all-optical memories (AOMs) operations can be achieved using the resonant photon-tunneling (RPT) effect in GaAs/AlGaAs multi-layered structures [2]. These all-optical devices use the coupling of evanescent waves as the near-field optics in multi-layered semiconductor structures. Recently, we also created InGaSb quantum dots (QDs) embedded in a GaAs matrix, which then showed long-wavelength laser emission around 1.3 μm [3, 4]. Therefore, we expected a GaAs/AlGaAs multi-layered structure containing InGaSb QDs as an active layer to enable AOS operation for optical-communication wavelengths. In this paper, we investigate the RPT effect in the 1.5 μm-waveband in a GaAs/AlGaAs multi-layered structure containing InGaSb QDs. We also demonstrate all-optical control of the resonant-incidence angle of the RPT.

EXPERIMENTAL

We fabricated multi-layered samples by molecular beam epitaxy. Figure 1 shows the schematic structure of the samples. An AlAs barrier layer and a GaAs drain layer were grown on a GaAs (001) substrate at 580°C. A 2ML $In_{0.5}Ga_{0.5}Sb$ QD layer was grown on the GaAs surface at 400°C. During this growth, Si atom irradiation was applied to form the high-density InGaSb QD layer [5]. To form the active layer, seven-periodical-stacked InGaSb QDs embedded in the GaAs matrix were fabricated. A top AlAs barrier layer and a GaAs guide layer were then grown at 540°C. Therefore, the InGaSb QD active layer and GaAs drain layer were sandwiched between the top and bottom

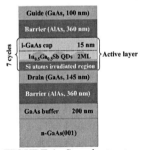

FIGURE 1. Sample structure

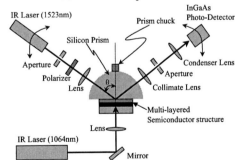

FIGURE 2. Infrared Total Reflection Measurements

AlAs barrier layers. We measured the RPT effect in the multi-layered samples with an infrared (IR) total reflection measurement system (Fig. 2). The sample surface was in direct contact with a hemispheric Si prism, and set at the center of the measurement system. Input light from a 1523-nm IR laser was irradiated onto the sample surface at an angle of θ through the Si prism. The intensity of the total reflected output light was measured by an InGaAs photodetector. In addition, we irradiated control light from a 1064-nm IR laser into the InGaSb QD active layer through the GaAs substrate to realize all-optical control of the resonant-incidence angle.

RESULTS AND DISCUSSION

Figure 3 shows the RPT effect in a multi-layered sample containing InGaSb QDs. The critical angle was 58.9° between the upper guide- and barrier-layers at 1523 nm. A sharp dip at an angle of 63.03° can be clearly observed in the total reflection region. Additionally, we calculated a traveling light-wave in the multi-layered sample by using a transfer-matrix method [2]. The theoretical reflection spectrum is also shown in Fig. 3. The resonant-incidence angle of the experimental result accorded substantially with the theoretical one. Therefore, the sharp dip was due to the RPT effect in the multi-layered sample. The evanescent optical waves were expected to penetrate into the double barrier layers at the resonant-incidence angle. However, the experimental resonant-dip became much more highly reflective compared with the theoretical result at the resonant-incidence angle. This is why the aperture widths affected the degree of input light parallelness.

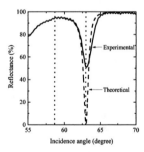

FIGURE 3. Resonant Photon Tunneling Dip

The shift of the resonant-incidence angle with the control light irradiation is shown in Fig. 4. The resonant-incidence angle was observed at 63.03° without the control light irradiation. A 0.08° shift was observed with the control light irradiation, and this experimental result agreed with the theoretical one when the refractive index change of the QD-active layer was assumed to be Δn=0.007 with the control

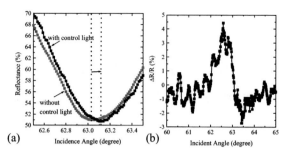

FIGURE 4. Shift of resonant angle (a), and all-optical reflectance change (b)

light irradiation (0.16 μW/μm^2). Figure 4(b) shows the reflectance-change (ΔR/R) with and without the control light irradiation near the resonant-incidence angle. A maximum ΔR/R of 4.5% was observed. From our findings, we believe that an all-optical switching device for optical communication wavelength operation can be realized by using RPT in a GaAs/AlAs multi-layered structure containing InGaSb QDs.

CONCLUSION

We observed the RPT effect in a GaAs/AlGaAs multi-layered structure containing InGaSb QDs for the 1.5 μm-waveband. We successfully demonstrated that the controlling of the resonant-incidence angle and the reflectance-change for the optical communication wavebands can be controlled all-optically by the control light irradiation into the QD active layer. These results indicate that the realizing the all-optical devices for the optical communication by using the RPT effect and the QDs active layer.

ACKNOWLEDGMENTS

We thank Mr. S. Miyashita for his technical assistance as well as Dr. T. Itabe and Dr. M. Izutsu at NICT for their encouragement throughout this work.

REFERENCES

1. S. Hayashi, H. Kurokawa, and H. Oga, Opt. Rev. 6, 1999, p.204.
2. N. Yamamoto, and N. Ohtani, Jpn. J. Appl. Phys. 43, 2004, pp.1393-1397.
3. N. Yamamoto, K. Akahane, S. Gozu, and N. Ohtani, Proc. of 16th IPRM, 2004, p564
4. N. Yamamoto, K. Akahane, S. Gozu, and N. Ohtani, Electron. Lett. 4, 18, 2004, p.1120
5. N. Yamamoto, K. Akahane, and N. Ohtani, Physica E21, 2004, p.204.

Amorphous Si/ Multicrystalline Si Heterojunctons for Photovoltaic Device Applications

Mahdi Farrokh Baroughi, Siva Sivoththaman

Department of E&CE, University of Waterloo, Waterloo, Ontario, N2L3G1, Canada

Abstract. Amorphous silicon/ multicrystalline silicon heterojunction diodes and photovoltaic cells are fabricated and characterized. Dark and illuminated current-voltage-temperature characteristics, high frequency capacitance-voltage and spectral response characteristics of the fabricated structures are employed for analysis of the a-Si/mc-Si heterojunctions. Interface quality is accessed by dark current-voltage-temperature characteristics and high frequency capacitance voltage measurement. Spectral response measurement on and off grain boundary is exploited to measure grain boundary degradation in heterojunction solar cells.

INTRODUCTION

mc-Si based solar cells supply more than 30% of world's solar cell market. Current technologies employ high temperature diffusion processes for realizing solar cells[1,2]. In most of those materials however, the presence of large number of grains, grain boundaries, and crystallographic defects necessitates defect passivation, normally hydrogenation passivation. However, this imposes a temperature (T) limit for any post-passivation processes such as pn junction diffusion at high-T. Implementation of amorphous Si (a-Si)/crystalline Si heterojunctions (HJ) in place of diffused HJs in defective Si can keep the process temperature low thereby preserving the defect passivation. Unlike a-Si/single c-Si HJs[3,4,5,6,7], little work have been reported on a-Si/mc-Si HJ solar cells[8]. This work presents low temperature realization and characterization of a-Si/mc-Si HJ solar cells. Dark and illuminated current-voltage-temperature (IVT), high frequency capacitance-voltage (CV), and External Quantum Efficiency (EQE) characteristics provide a unique work frame for qualifying HJs for photovoltaic applications.

EXPERIMENTAL

Devices with two different structures are fabricated for electrical and photovoltaic characterization of HJs. Figure 1(a) shows a HJ diode utilizing a p-type mc-Si wafer, an ultra thin intrinsic a-Si, and a thin (n⁺) a-Si with aluminum layers as back and front contacts. Figure 1(b) shows a PV cell structure with addition of a transparent conductive oxide (TCO) layer providing a low resistive path for carriers. Detailed fabrication information is brought elsewhere[9].

RESULTS ANS DISCUSSTIN

1. IVT Characterization

Dark IVT characteristic of a-Si/c-Si HJ reveals valuable information about carrier transport mechanisms in HJs. Like a-Si/c-Si HJ[9,10,11], drift-diffusion, recombination at the heterointerface and Multi Tunneling Capture Emission (MTCE) are the most important carrier transport mechanisms in a-Si/mc-Si HJs. Temperature dependence of the diode saturation current and ideality factor are employed to investigate dominant carrier transport regimes in different bias conditions. IVT characteristics of a 4mm² a-Si / mc-Si HJ diode are shown in figure 2. Measurement results reveal bias dependence of dominant transport mechanism: current is recombination-dominated at low forward bias (V_A < 0.25 V at 27°C), and drift-diffusion dominated at high bias regime (0.25V-0.5V). Relatively large ideality factor (η=1.35) implies that passivation of mc-Si surface at heterointerface is incomplete and there is noticeable carrier recombination at interface.

Dark IVT of HJ diodes, with different sizes, on and off grain boundary doesn't show any noticeable change because of grain boundary degradation. This is due to the fact that area current of the diode is much larger than grain boundary leakage current.

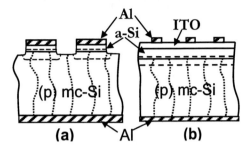

Figure 1: The structure of fabricated devices, (a) mesa etched heterojunction diode, (b) Heterojunction solar cell

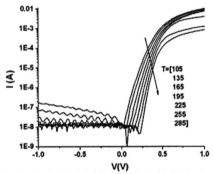

Figure 2: Dark IVT curves of the HJ diode

2. High Frequency CV

High Frequency CV (HFCV) characteristic is employed to qualify heterointerface quality by extracting HJ diode built in potential [9]. Figure 3 show the HFCV characteristic of the HJ diode with 2mm x 2mm area. This characterization results in 0.23V built in potential that is much lower than what is normally expected for a (n+) a-Si / (p) c-Si HJ diode (typical value: 0.65V). Low built in potential is another indication of low quality interface between a-Si and (p) mc-Si HJ. The interface quality is normally improved by inserting an optimized ultra thin intrinsic a-Si layer between (p) mc-Si and (n+) a-Si. Like IVT measurement, HFCV characteristic of HJ diodes on and off grain boundary doesn't show any noticeable change due to grain boundary degradation.

3. Illuminated IV and spectral response

Illuminated IV of the HJ solar cells showed open circuit voltages of 550mV. Spectral response was measured using a micrometer-scale slit at various parts on the device, on- and off- grain boundaries. Carriers that are photo-generated far away from junction (by large λ photons) are more likely to recombine in the neighborhood of grain-boundaries since they need to diffuse longer distances towards the junction. A well-passivated grain boundary should reduce this effect. Figure 4 shows EQE of device on and off grain boundary. The ratio of off grain boundary/on grain boundary EQE is shown in figure 4 as normalized EQE.

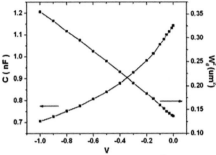

Figure 3. HFCV Characteristics of 4mm^2 HJ structure (W_d: depletion layer width)

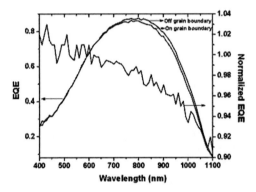

Figure 4: EQE on- and off- grain boundary and normalized EQE

CONCLUSION

a-Si/mc-Si HJ diode and solar cell structures have been fabricated and characterized. Interface quality and transport mechanisms are stuied by IVT measurements. A low built-in potential of 0.23V, obtained by HFCV, indicates a defective hetero-interface pointing to the need for a better interfacial passivation. Spectral response measurements show a lower response in the neighborhood of grain boundaries.

REFERENCES

1. J. Nijs et al., ITED, **46,** 10 (1999)
2. M. A. Green, Prog. in PV: Res.& App, **8,** 127 (2000)
3. M. Kolter et al., Proc. 13th Euro. PVSEC, 1995;1526-1529
4. B. Jagannathan et al., Sol. Egy. Mat. & solar cells, **46,** 289 (1997)
5. R. Hussein et al., Tech. Dig. 11th Int. PVSEC 1999; 351
6. D. Rosa et al., Pro. of 2nd WCPEC, 1998; 1583-1586
7. N. Jensen et al., Prog. in PV: Res. & App., **10,** 1 (2002)
8. D. Rosa et al., Proc. of 2nd WCPEC, 1998; 2440-2443
9. M. F. Baroughi et al., JVST, **22,** 1015 (2004)

High-index-contrast Optical Waveguides on Silicon

Antti Säynätjoki[1], Sanna Arpiainen[2], Jouni Ahopelto[2], Harri Lipsanen[1]

[1]*Optoelectronics laboratory, Helsinki University of Technology, P.O.Box 3500, 02015 HUT, Finland*
[2]*VTT Information Technology, P.O.Box 1208, 02044 VTT, Finland*

Abstract. Optical waveguides with a high refractive index contrast were fabricated on silicon-on-insulator (SOI) and polysilicon. The attenuation of 2dB/cm was achieved for SOI, and 11 dB/cm for hydrogenated polysilicon waveguides. Unlike in SOI waveguides, where the main contribution to attenuation is due to the surface scattering, a significant part of attenuation in polysilicon originates from absorption and scattering in the material.

INTRODUCTION

While developed for the needs of microelectronics, the silicon-on-insulator (SOI) wafers are excellent substrates for optical waveguides. The single-crystalline, atomically smooth SOI-layer works as the waveguide core with a high refractive index contrast to both silicon dioxide and air. For the same reason, SOI is also commonly used in two-dimensional photonic crystals that have prospects for dense optical integration in the near future.

Though SOI wafers are close to the perfect in material quality, they are rather expensive. In applications where higher optical attenuation can be tolerated, the SOI-layer might be replaced by polycrystalline silicon (poly-Si) that can be grown on almost any substrate by standard methods of microelectronics fabrication. An optical loss of 9 dB/cm has already been reached in poly-Si strip waveguides [1]. In this work we have fabricated nominally 220 nm thick poly-Si strip waveguides with different widths and radii of curvature. These are compared with 190 nm thick SOI waveguides for the transmission loss and its mechanisms.

EXPERIMENTAL

Waveguide Fabrication

Polysilicon waveguides were processed on <100> silicon substrates that were covered by a 2-micron thick, chemical-mechanically polished PECVD SiO_2. The non-conventional choice of oxide was made to provide a cladding layer with a sufficient thickness. To form polysilicon, a 220-nm layer of LPCVD amorphous silicon was deposited on the oxide and crystallized by two-step annealing in N_2 ambient. The first annealing step was at 600 °C for 16 hours and the second at 1100 °C for another 16 hours, following the procedure represented by Liao et al. [1]. The strip waveguides were defined by reactive-ion-etching (RIE) through the poly-Si (Fig.1). After patterning, one set of wafers was hydrogenated in N_2:H_2 3:1 atmosphere at 425 °C for 30 minutes.

SOI waveguides were processed by a single RIE step on SOI Unibond wafers manufactured by SOITEC with a 190 nm thick Si layer on a 3 μm thick buried oxide.

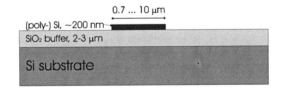

FIGURE 1. Schematic cross section of the waveguides.

Optical Transmission Measurements

Optical transmission properties were measured by coupling light into the end facets of the waveguides. Optical loss was verified by linear regression from four sets of waveguides with different length. The variation of poly-Si film parameters from one waveguide set to another may have caused some error into the loss values. Two light sources were used: a

10-mW tunable semiconductor laser operating at the wavelength of 1550 nm, and spontaneous emission from an erbium doped fiber amplifier at a wavelength range of 1530 to 1590 nm.

The single-crystalline SOI waveguides had an optical loss of about 2 dB/cm, whereas the average loss in poly-Si waveguides was around 14 dB/cm and reduced to 11 dB/cm after hydrogenation. In SOI, the lowest attenuation was measured from 7 µm wide waveguides, and the loss increased significantly at widths smaller than 5 µm. In poly-Si, a slight decrease in the attenuation was measured with decreasing the waveguide width down to 3 µm. The light was guided through curves with the curvature radius down to 40 µm with no measurable excess loss, after which a drastic decrease of transmission was observed.

DISCUSSION

The loss in the 7-µm wide or wider SOI waveguides was around 2 dB/cm. The RMS surface roughness of the SOI is specified to less than 0.2 nm, which, according to the estimate derived by Tien [2], results in a scattering loss of 1.2 dB/cm. Since bulk absorption in silicon is minimal at this wavelength, the rest of the loss is addressed to scattering in the waveguide sidewalls.

The loss in polysilicon waveguides is a combination of several components, all related to the grain morphology. The RMS roughness of the as deposited amorphous silicon was only 0.25 nm, but increased to around 0.7 nm due to the grain formation in crystallization. This surface roughness may alone result in a loss of about 5 dB/cm [2]. In polysilicon, significant attenuation is also resulted by absorption and scattering in the grain boundaries. The absorption can be reduced by passivating the silicon dangling bonds with hydrogen; annealing in a hydrogenating atmosphere reduced the absorption loss by 3 dB/cm. Grain boundary scattering may be reduced by adjusting the grain size and the width of the grain boundaries by tailoring the crystallization temperatures and times.

The increase in optical loss with reducing waveguide width is due to roughness in the etched sidewalls of the waveguide stripes. The scanning electron microscope image in Fig. 2 shows that the amplitude of roughness on the sidewall is about 10 nm.

The high refractive index contrast between silicon and air confines the light tightly into the waveguide, and this was seen as high transmission even in relatively tight bends.

While small waveguide dimensions and transmission through tight bends were to some extent realized using conventional index-guiding, even smaller dimensions may be achieved using photonic crystal waveguides. Polysilicon may be considered as lossless in the length scale of the multiple Bragg reflections giving rise to the photonic crystal phenomena. Photonic crystals also have a nature of reducing photon states in the structure, which may inhibit the somewhat inevitable scattering in the polycrystalline silicon.

FIGURE 2. Tilted scanning electron microscope image of the etched sidewall in polysilicon film.

CONCLUSIONS

Vertically single-mode poly-Si strip waveguides were fabricated. The 14 dB/cm attenuation measured from polysilicon waveguides was reduced to 11 dB/cm by hydrogenation. Though the attenuation is relatively high in an absolute scale, the small size of photonic crystal devices may enable the use of poly-Si in photonics with an acceptable total attenuation.

ACKNOWLEDGMENTS

This work has been partially funded by the Academy of Finland research project PhC-OPTICS and the EU-IST Project PHAT grant Nr 510162. The authors wish to thank M. Kapulainen, K. Solehmainen, O. Reentilä and Y. Cui for assistance in the measurements.

REFERENCES

1. L. Liao et al., *Journal of Electronic Materials* **29**, 1380 (2000).
2. P.K. Tien, *Appl. Opt.* **10**, 2408 (1971).

Origin of Efficient Light Emission from Si pn Diodes Prepared by Ion Implantation

T. Dekorsy, J. M. Sun, W. Skorupa, A. Mücklich, B. Schmidt, and M. Helm

Institute of Ion Beam Physics and Materials Research, Forschungszentrum Rossendorf, PO Box 510119, D-01314 Dresden, Germany

Abstract. Electroluminescence with power efficiencies larger than 0.1 % is observed from silicon pn diodes prepared by boron implantation. The implanted boron concentration is above the solubility limit for the post-implantation annealing temperature leading to the formation of boron clusters during annealing. The electroluminescence from electron-hole pairs exhibits an anomalous increase in the total intensity with increasing temperature. This behavior is explained by the thermal release of carriers trapped at local potential minima related to the boron clusters.

INTRODUCTION

Light emission from Si has attracted considerable attention in the past few years due to the future potential in on-chip and inter-chip optical interconnects [1]. Electroluminescence (EL) with attractive power efficiencies up to the range of 0.1 % to 1 % has been observed from high-purity Si diodes owing to a reduced non-radiative recombination rate [2] and in Si pn diodes prepared by ion implantation [3]. We focus on the latter approach and investigate the origin of the relatively high EL efficiency in Si pn diodes prepared by high-dose boron implantation, especially on the intriguing and anomalous increase of the EL for a temperature increase up to room temperature.

DIODE PREPARATION

The silicon pn diodes were prepared by high-dose boron implantation into (001) oriented Sb-doped n-type silicon substrates with a resistivity of 0.1 Ωcm. The implantation is performed at a tilt angle of 7° through a 50 nm thermally grown SiO_2 layer. Boron doses between 2×10^{13} and 3×10^{17} cm^{-2} were implanted at an energy of 25 keV. All samples were subsequently furnace annealed at 1050°C for 20 minutes and processed into 1 mm diameter diodes with aluminum ring contacts on top.

STRUCTURAL CHARACTERIZATION

Figure 1 shows an XTEM image of the sample created by boron implantation at 25 keV with a dose of 4×10^{15} cm^{-2}. The implantation induced defects extend down to 180 nm below the surface. Some individual dark spots with a diameter of 10 to 20 nm are observed at the end of the dislocation lines close to the pn junction after annealing. The origin of the contrast of these dark spots is attributed to high strain in an area of locally high density of Si-B interstitial clusters, Si interstitials or to strain fields around extended defects.

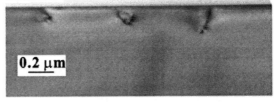

FIGURE 1. XTEM image of the annealed sample implanted at 25 keV with a boron dose of 4×10^{15} cm^{-2}.

The relative boron concentration at the end of the extended defects was analyzed by EDX using a focused electron beam of 70 nm in diameter. Although no quantitative analysis can be made, the EDX spectrum reveals a high boron concentration at the dark spot compared to the region free of defects. Such a localized high boron concentration is attributed to boron gettering during the nucleation of the extended defects, as observed in heavily boron doped silicon wafers pre-amorphized by Ge^+ and Si^+.[5,6]

ELECTROLUMINESCENCE

The EL spectra at 12 K from the silicon pn diodes prepared by boron implantation at different doses between 2×10^{13} cm^{-2} and 4×10^{15} cm^{-2} are shown in Fig. 2. At implantation doses $> 5\times10^{14}$ cm^{-2}, the spectra exhibit a peak from the transverse optical (TO) phonon-assisted free exciton recombination at 1.1 eV (FE^{TO}) and two broader asymmetric EL peaks close to 1.05 eV and 0.95 eV from TO phonon-assisted recombination of excitons bound to traps, P_I^{TO} and P_{II}^{TO}, respectively. At the lowest implantation doses of 2×10^{13} cm^{-2}, no luminescence from bound excitons is observed in the EL spectrum. Above an implantation dose of 5×10^{14} cm^{-2}, P_I^{TO} and P_{II}^{TO} increase rapidly with further increasing the boron doses. The peak energies of P_I^{TO} and P_{II}^{TO} also change with increasing the boron doses. These results indicate that both peaks are strongly correlated to the traps created by high-dose boron implantation and the subsequent annealing.

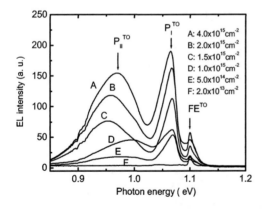

FIGURE 2. EL spectra from silicon pn diodes prepared by boron implantation at an energy of 25 keV and different doses as given in the figure. All samples are annealed at 1050 °C for 20 minutes. The spectra are recorded under forward bias at a current of 50 mA.

The peak height of the bound-exciton peaks P_I^{TO} and P_{II}^{TO} as well as the overall integrated EL intensity of the FE^{TO} peak and its phonon replicas is plotted as a function of temperature in Fig. 3. The P_I^{TO} peak decreases from 15 K and is completely thermally quenched at 80 K, while the P_{II}^{TO} peak starts to decrease at 80 K and is thermally quenched at a temperature of 260 K. Associated with the decrease of these two peaks is an increase of the FE^{TO} peak. This increase of the band-edge EL intensity is in contrast to the temperature dependence of the PL from the n-type substrate, which shows no trap-related luminescence. The EL intensity of the FE^{TO} peak shows a two-step increase with rising temperature in close correlation with the decrease of the two bound-exciton peaks. This correlation indicates that the increase of the band-edge free electron-hole recombination comes from the thermal dissociation of bound excitons with increasing temperature. Our results reflect a low recombination rate of the spatially indirect bound excitons in the pn diodes, which have a recombination rate over 100 times lower than the thermal emission rate at 220 K as calculated by a rate equation model considering the transition between bound excitons and free excitons/electron-hole pairs [7].

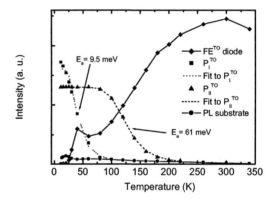

FIGURE 3. Dependence of the EL intensity from different bands versus the lattice temperature for sample A of Fig. 2. Measurements are performed at 50 mA. The dotted and dashed lines are theoretical fits with the activation energies given in the figure, solid lines are guides to the eye. The temperature dependence of the photoluminescence from the band-edge recombination of n-type silicon substrate is also shown.

REFERENCES

1. A. Polman, *Nature Materials* **1**, 10 (2002).
2. M. A. Green, et al., *Nature* **412**, 805 (2001).
3. W.L. Ng, et al., *Nature* **410**, 192 (2001).
4. J. M. Sun et al., *Appl. Phys. Lett.* **83**, 3885 (2003).
5. C. Bonafos et al., *Appl. Phys. Lett.* **71**, 365 (1997).
6. J. Xia, et al, *J. Appl. Phys.* **85**, 7597-7603 (1999).
7. J. M. Sun et al., submitted to *Phys. Rev. B*.

Temperature and polarization dependence of the optical gain and optically pumped lasing in GaInNAs/GaAs MQW structures

J. Kvietkova[*], M. Hetterich[*], A.Yu. Egorov[†], H. Riechert[†], G. Leibiger[**] and V. Gottschalch[**]

[*]*Institut für Angewandte Physik and Center for Functional Nanostructures (CFN), Universität Karlsruhe, D-76131 Karlsruhe, Germany*
[†]*Infineon Technologies, D-81730 München, Germany*
[**]*Fakultät für Chemie und Mineralogie, Universität Leipzig, D-04103 Leipzig, Germany*

Abstract. We present the experimental investigation of the optical gain in GaInNAs/GaAs multi-quantum well laser structures using the variable stripe length method. The low temperature gain was typically in the order of 100 cm^{-1}, while at room temperature lower gain values of $\sim$ 50 cm^{-1} were observed. Optically pumped lasing was detected from cleaved resonator samples and a T_0 value of approximately 90 K was obtained. The emitted light was predominantly TE-polarized as expected for compressively strained quantum wells.

INTRODUCTION

The GaInNAs material system has in recent years been successfully applied in light emitting devices operating in the near infrared region important for optical communications (see, e.g. [1, 2, 3]). In comparison to the InGaAsP/InP material system previously used for light sources in the near infrared spectral range, GaInNAs-based lasers have several advantages: the possibility of pseudomorphic growth on GaAs substrates, the use of GaAs/AlGaAs Bragg mirrors in vertical lasers, higher temperature stability, and as theoretically predicted, also higher material gain [4]. We report on the experimental investigation of the optical gain and its polarization and on the temperature dependence of the optically pumped lasing in GaInNAs/GaAs multi-quantum well (MQW) laser structures.

EXPERIMENTAL

The samples used in this work were grown on GaAs (001) substrates using either solid-source molecular beam epitaxy (MBE) with an RF-coupled plasma source for nitrogen, or metalorganic vapor-phase epitaxy (MOVPE). The MBE-sample consists of two $Ga_{0.65}In_{0.35}N_{0.017}As_{0.983}$ quantum wells (6.5 nm) and 20 nm thick $Ga_{0.94}In_{0.06}N_{0.018}As_{0.982}$ barriers positioned in the middle of a 300 nm thick GaAs layer. This undoped region is embedded between p- and n- type $Al_{0.34}Ga_{0.66}As$ cladding layers. The MOVPE-sample has two $Ga_{0.7}In_{0.3}N_{0.001}As_{0.999}$ quantum wells (6 nm) divided by a GaAs barrier (25 nm) and embedded in a 130 nm thick undoped GaAs layer. This active region is placed into an $Al_{0.35}Ga_{0.65}As$ waveguide. The N and In mole fraction in the GaInNAs layers were determined using both x-ray diffraction and precalibrated photoluminescence investigations of the corresponding InGaAs structures. In order to study optically pumped lasing the MBE-sample was cleaved into narrow stripes, the facets forming a resonator with a width of several hundred micrometers.

The spectra of the optical gain were measured using the variable stripe length method [5]. The sample was mounted in a cryostat and excited in a stripe-like geometry using a pulsed YAG:Nd laser (1064 nm). The length of the stripe was varied while the width was kept constant at about 100 μm. The emission from the edge of the sample was dispersed by a monochromator and detected by a liquid nitrogen cooled InGaAs diode array. The spectral dependence of the gain was obtained from the ratio of two emission spectra measured for the stripe lengths L and 2L using the following expression [6]:

$$g(\hbar\omega) = \frac{1}{L}\ln\left(\frac{I(\hbar\omega,2L)}{I(\hbar\omega,L)} - 1\right) \quad (1)$$

RESULTS AND DISCUSSION

Figure 1 shows the room temperature transverse electrical (TE) and transverse magnetic (TM) polarized gain spectra of the MOVPE-sample. For the TE-polarized gain a maximum value of approximately 70 cm^{-1} was obtained for an excitation density of 5.5 MW/cm^2, while the TM-polarized gain is practically zero. This is due to the compressive strain in the quantum wells leading to a heavy-hole–light-hole splitting. As a result the light holes are only weakly confined or even unconfined [7, 8]. The values of the optical gain we obtained are in agreement with those presented in [9] measured by the Hakki-Paoli method, as well as with the theoretical and experimental data shown in [10].

The peak of the gain spectra is positioned at a wavelength of 1140 nm which corresponds to the very low concentration of nitrogen in the QW layers.

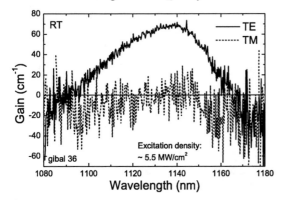

FIGURE 1. Room-temperature gain spectra of the MOVPE-sample measured for TE (solid line) and TM (dashed line) polarizations.

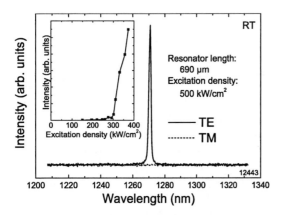

FIGURE 2. Room-temperature lasing spectra of the MBE-sample for TE (solid line) and TM (dashed line) polarization. The inset shows the dependence of the emission intensity on excitation density.

Optically pumped lasing from the cleaved samples was studied in the temperature range from 5 K up to room temperature. In Fig. 2 a room temperature spectrum is shown for an MBE-sample with a resonator length of 690 μm and for an excitation density of 500 kW/cm^2. For the TE polarization a sharp laser peak can be observed positioned at a wavelength of 1270 nm, while the TM-polarized emission is zero, as expected. The inset of Fig. 2 depicts the dependence of the peak emission on the excitation density, showing a lasing threshold of about 300 kW/cm^2. From the temperature-dependent measurement of the threshold excitation intensity we obtained the T_0 value of approximately 90 K. This is in agreement with other experimental values observed in electrically pumped lasers [11].

SUMMARY

We presented experimental optical gain spectra of a GaInNAs/GaAs MQW laser structure, with a room temperature peak value of 70 cm^{-1}. The temperature dependence of the optically pumped lasing yields a T_0 value of 90 K for a wavelength of 1270 nm. Both, the gain and lasing spectra show predominantly TE polarization, as expected for compressively strained QWs.

ACKNOWLEDGMENTS

This work was supported by the Deutsche Forschungsgemeinschaft (DFG) within the Emmy Noether program and the Center for Functional Nanostructures (CFN), project A2.2.

REFERENCES

1. Borchert, B., Egorov, A. Y., Illek, S., Komainda, M., and Riechert, H., *Electron. Lett.*, **35**, 2204–2206 (1999).
2. Fischer, M., Reinhardt, M., and Forchel, A., *IEEE Photonics Technol. Lett.*, **12**, 1313–1315 (2000).
3. Choquette, K. D., et al., *Electron. Lett.*, **36**, 1388–1390 (2000).
4. Hader, J., Koch, S. W., Moloney, J. V., and O'Reilly, E. P., *Appl. Phys. Lett.*, **77**, 630–632 (2000).
5. Shaklee, K. L., and Leheny, R. F., *Appl. Phys. Lett.*, **77**, 475–477 (1971).
6. Hvam, J. M., *J. Appl. Phys.*, **49**, 3124–3126 (1978).
7. Hetterich, M., Grau, A., Egorov, A. Y., and Riechert, H., *J. Appl. Phys.*, **94**, 1810–1813 (2003).
8. Chow, W. W., Jones, E. D., Modine, N. A., Allerman, A. A., and Kurtz, S. R., *Appl. Phys. Lett.*, **75**, 2891–2893 (1999).
9. Gerhardt, N., Hofmann, M. R., and Rühle, W. W., *IEE Proceedings - Optoelectron.*, **150**, 45–48 (2003).
10. Hofmann, M. R., et al., *IEEE J. Quantum Electron.*, **38**, 213–221 (2002).
11. Fehse, R., et al., *Electron. Lett.*, **37**, 1518–1520 (2001).

Enhanced Electroabsorption in MQW Structures Containing an *nipi* Delta Doping Superlattice

C. V-B. Tribuzy, M. C. L. Areiza, S. M. Landi, M. Borgström, M. P. Pires and P. L. Souza

LabSem – CETUC – Pontifícia Universidade Católica do Rio de Janeiro,
R. Marquês de São Vicente, 225, Rio de Janeiro, RJ 22453-900, Brazil

Abstract. It is experimentally observed that the introduction of an nipi delta doping superlattice in the multiple quantum well (MQW) region of an amplitude modulator improves the device performance in terms of contrast ratio.

INTRODUCTION

The demand for high speed in optical communication requires minimization of the voltage operation, *i.e* maximization of the change in absorption coefficient, $\Delta\alpha$, for a certain applied AC bias.

Batty and Allsopp [1] theoretically predicted that the introduction of an *nipi* delta doping superlattice into a MQW structure, where the *n*- and *p*-type delta layers are introduced in the center of the QWs and of the barriers, should double the Stark shift. This effect can only be observed if the *n* and *p* doping levels are well balanced [2].

We report the experimental observation of an enhancement of $\Delta\alpha$ with the introduction of an *nipi* delta doping superlattice in a GaAs/AlGaAs MQW structure (DMQW).

EXPERIMENTAL DETAILS

The investigated MOVPE-grown GaAs/AlGaAs DMQW structures have 20 periods with 100 Å and 50 Å thick QWs and barriers, respectively. A detailed description of the samples is published elsewhere [3].

The Stark-shift of the processed photodiodes was determined from photocurrent (PC) experiments as a function of the applied reverse voltage with light propagating along the growth axis. To quantify the values of $\Delta\alpha$, transmission measurements are performed. For that, a hole on the backside of the samples was etched to eliminate the absorption by the GaAs substrate and buffer layer.

RESULTS AND DISCUSSION

PC measurements of the photodiodes, which essentially reveal the absorption, were carried out. A pre-bias of 2 volts was used to sweep out free carriers from the MQW region and guarantee that it is depleted. A red-shift of the PC curves as the reverse applied voltage increases was clearly observed. However, it was less pronounced than that for the undoped structure, apparently, in contradiction with the predicted increase in the Stark shift with the introduction of the doping superlattice. The reason for the reduced Stark shift in the delta doped samples is the presence of an excess net fixed charge within the MQW region of the sample, despite the thorough doping calibration.

Nevertheless, the important parameter to be optimized is $\Delta\alpha$, which does not depend solely on the Stark shift, but also on the overlap of the electron and hole wavefunctions. In order to determine $\Delta\alpha$, one should indirectly extract the absolute value of α from the transmission measurements, since the spectra are modulated by multiple internal reflections.

To achieve that, first the PC spectra were fit to

determine the qualitative real and the imaginary parts of the refractive index, $n(\lambda)$ and $k(\lambda)$, which have both an excitonic and a 2D contribution from the QWs. The obtained functions $k(\lambda)$ and $n(\lambda)$ are used as input to, subsequently, determine the transmission spectrum.

A simple model was used to calculate the transmission spectra which is given by $T^*(\lambda,d).T(\lambda,d)$, where

$$T(\lambda,d) = t_{12}(\lambda).t_{23}(\lambda).\frac{\exp(i0.5.b(\lambda,d))}{1+r_{12}(\lambda).r_{23}(\lambda).\exp(ib(\lambda,d))} \quad (1)$$

t_{12}, t_{23}, r_{12} and r_{23} are the transmission and reflection Fresnel coefficients for normal incidence. The indexes 12 and 23 indicate, respectively, the light incident from medium 1 (air) to medium 2 (sample) and from medium 2 to medium 3 (air). d is the material thickness and b is the qualitative complex refractive index which, in turn, is given by: $b(\lambda,d)=4\pi d/\lambda$ $[n(\lambda)+ik(\lambda)]$.

An AlGaAs bulk contribution was also taken into account using published values [4]. 2D and bulk contributions were weighed up according to the GaAs and AlGaAs layer thicknesses to determine the effective $n(\lambda)$ and $k(\lambda)$ of the sample.

The absorption coefficient, $\alpha(\lambda)$, is then quantitatively obtained from: $\alpha(\lambda)=4\pi\, k(\lambda)\,/\lambda$ by adjusting the fitting parameters, namely, the thickness d of the sample and the absolute values for $n(\lambda)$ and $k(\lambda)$. Their final values are chosen to be such which best fit the experimental transmission data. In Fig. 1 the experimental and calculated transmission spectra for the DMQW (a) and the undoped reference (b) samples, are depicted, respectively.

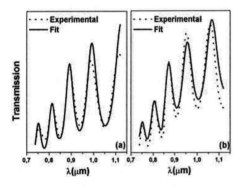

FIGURE 1. Transmission spectra for the (a) DMQW and (b) reference samples. The dashed line corresponds to the experimental results while the solid line is the result of the fitting procedure.

Once $\alpha(\lambda)$ at zero applied voltage was obtained, the PC spectra were calibrated in cm^{-1}, as described elsewhere [5]. $\Delta\alpha$ as a function of the applied reverse bias was obtained from the calibrated PC spectra by subtracting its value for higher reverse voltages from that for the lowest non zero reverse voltage (pre-bias). Figure 2 shows the dependence of $\Delta\alpha$ on the applied reverse voltage (ΔV) for the undoped reference sample and for the DMQW. A detuning of 30 meV was used. One clearly observes an enhancement of $\Delta\alpha$ for the DMQW structure although an enhancement of the Stark-shift has not yet been observed[3]. An improvement of a third is observed in $\Delta\alpha/\Delta V$ for a ΔV equal to –4V. Since the contrast ratio in dB (CR) varies linearly with $\Delta\alpha$, a better device performance is expected. The enhancement of $\Delta\alpha$ can be explained by the overlap integral of the wavefuctions corresponding to the first electron and heavy hole levels.

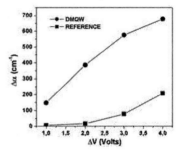

FIGURE 2. Dependence of $\Delta\alpha$ on the applied reverse voltage (ΔV) for the reference and for the DMQW samples.

CONCLUSIONS

We have experimentally demonstrated that the introduction of an *nipi* delta-doping superlattice in GaAs/AlGaAs MQW enhances $\Delta\alpha$ with applied reverse bias by a factor as large as 3, despite the fact that the Stark-shift for the investigated structures is smaller than that for an undoped sample. This enhancement is attributed to the fact that the reduction of the overlap between the electron and the hole wavefunctions with the applied voltage is less pronounced for the DMQW. Our results indicate that the DMQW structures should lead to an overall better performance of amplitude modulators.

REFERENCES

1. W. Batty *et al*, Electronic Lett. **29** 2066-2067 (1993).
2. C. V.-B. Tribuzy *et al*, *Physica E* **11**, 261-267(2001).
3. C.V-B.Tribuzy *et al*, *Appl. Phys. .Lett.* **84**, 3256-3258 (2004).
4. S. Adachi, J. Appl. Phys **66**, 6030-6040 (1989).
5. M. P. Pires *et al*, J. Lightw. Technol. **18**, 598–603 (2000).

Carrier recombination in InGaAs(P) Quantum Well Laser Structures: Band gap and Temperature Dependence

S.J. Sweeney, D. A. Lock and A.R. Adams

Advanced Technology Institute, University of Surrey, Guildford, Surrey, GU2 7XH, UK

Abstract. Using a combination of temperature and pressure dependence measurements, we investigate the relative importance of recombination processes in InGaAs-based QW lasers. We find that radiative and Auger recombination are important in high quality InGaAs material. At 1.5μm, Auger recombination accounts for 80% I_{th} at room temperature reducing to ~50% at 1.3μm and ~15% at 980nm. We also find that Auger recombination dominates the temperature dependence of I_{th} around room temperature over the entire operating wavelength range studied (980nm-1.5μm).

INTRODUCTION

InGaAs(P)-based quantum wells (QWs) are used in semiconductor lasers over a very wide wavelength range (~900-1700nm) by growing on either GaAs or InP substrates. It is of significant scientific and technological interest to understand carrier recombination in this material. One of the most important parameters associated with semiconductor lasers is the threshold current, I_{th}, above which stimulated emission dominates. Minimising I_{th} maximizes the electro-optic conversion efficiency. Hence it is vital to understand which carrier recombination processes dominate I_{th} and the extent to which they depend on parameters such as the operating wavelength (λ) and temperature (T). In this study, we performed measurements on 980nm InGaAs/GaAs, 1.3μm InGaAsP/InP and 1.5μm InGaAs/InP quantum well (QW) lasers. By measuring I_{th} and the spontaneous emission as a function of T, we determined the relative importance of the radiative and non-radiative recombination paths. Hydrostatic pressure measurements were used to investigate their dependence on band gap (E_g).

TEMPERATURE DEPENDENCE

Assuming equal electron and hole densities, the threshold current of semiconductor lasers can in general be written as; $I=eV(An+Bn^2+Cn^3)$ where n is the carrier density, e is the electronic charge and V is the active volume. The An term corresponds to recombination via defects and impurities, Bn^2 is due to spontaneous (radiative) emission and Cn^3 is due to

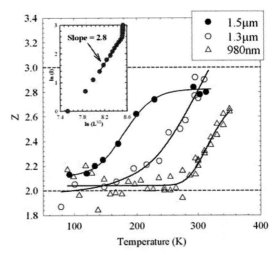

FIGURE 1. Power law dependence (Z) of the threshold current on carrier density for the 980nm, 1.3μm and 1.5μm lasers. The inset illustrates how Z is calculated.

Auger recombination whereby the energy of a recombining electron and hole excites a third electron or hole. Since the emitted spontaneous emission, L, is directly proportional to the radiative current, we may write that $L \propto n^2$ and hence, $n \propto L^{1/2}$. Thus, by plotting a graph of $\ln I$ versus $\ln L^{1/2}$, the slope (Z) provides a direct measure of the power dependence of I on n. L can be determined experimentally by measuring the un-amplified emission from a "window" milled into the substrate of a semiconductor laser [1]. The inset of Fig. 1 shows such a plot for the 1.5μm device at 300K where Z=2.8. This is consistent with a dominant Auger recombination process, $\propto n^3$. The results of repeating the Z measurement over a wide T range are shown in Fig. 1. From this plot we observe that at each wavelength at the lowest temperatures, Z~2 consistent with the current being dominated by radiative recombination. For the 1.5μm device, above ~130K, Z increases with T and stabilizes at a value of ~2.8 at room temperature. This is consistent with a transition from radiatively dominated behaviour at low T to Auger recombination dominating at and above room temperature, RT. Similar behaviour is apparent for the 1.3μm and 980nm lasers where we observe that Z increase above ~2 at a temperature of ~170K and ~280K respectively. The fact that this occurs at increasingly higher temperatures as λ decreases is consistent with the strong band gap (E_g) dependence of Auger recombination. Furthermore, given that over this T range, Z does not drop below a value of two for these devices we conclude that defect related recombination is negligible in these devices at the high carrier densities required for lasing threshold. This is in contrast to materials such as InGaAsN where defect-related recombination can account for 50% I_{th} at RT for N fractions ~2% [2].

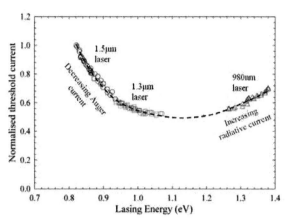

FIGURE 3. Band gap dependence of I_{th} in the nominally 980nm, 1.3μm and 1.5μm lasers. The line is a guide to the eye.

BAND GAP DEPENDENCE

Hydrostatic pressure is a useful tool to investigate the band gap, E_g, dependence of recombination processes in semiconductor lasers since E_g increases with increasing pressure. Because of the increase in E_g and electron effective mass with pressure I_{rad} increases with pressure. In contrast, due to the strong decrease in the Auger coefficient, C, with E_g, I_{Aug} decreases with pressure. In Fig. 3, we plot the normalised RT pressure dependence of I_{th} for the 980nm, 1.3μm and 1.5μm lasers, plotted versus lasing energy. From this graph it is clear that with increasing E_g, I_{th} decreases for the 1.3μm and 1.5μm devices due to the reduction in Auger recombination. In contrast, for the 980nm device where radiative recombination dominates, I_{th} increases with band gap, due to the increase in I_{rad}.

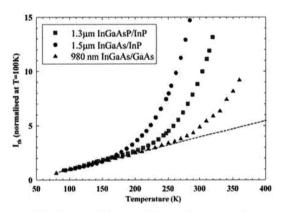

FIGURE 2. $I_{th}(T)$ for the 980nm, 1.3μm and 1.5μm lasers. The dashed line is the projected ideal $I_{rad}(T)$.

In Fig. 2 we plot the measured $I_{th}(T)$ for the three lasers (normalised at 100K). For all three devices, at low T we find that $I_{th} \propto T$ as expected for an ideal QW for which the radiative current, $I_{rad} \propto T$ [1]. Above a certain "break-point" temperature, T_B [1] I_{th} increases super-linearly. Furthermore, it can be seen that for all of the devices, at T_B, Z increases above 2 as observed in Fig. 1. Thus, at low T, both I_{th} and its temperature sensitivity are governed by radiative recombination. Above T_B, Auger recombination begins to contribute to I_{th} and gives rise to the super-linear increase in I_{th}. By extrapolating the low temperature linear variation of I_{th} we deduce that at RT, Auger recombination accounts for ~15%, ~50% and ~80% of I_{th} in the 980nm, 1.3μm and 1.5μm lasers, respectively. This is consistent with the strong E_g dependence of Auger recombination.

CONCLUSIONS

In summary, from a combination of temperature and pressure dependence measurements we have shown the increasingly important role that Auger recombination plays in InGaAs based lasers as the operating wavelength increases from 980nm to 1.5μm.

ACKNOWLEDGMENTS

The authors gratefully acknowledge EPSRC (UK) and the Nuffield foundation NUF-NAL scheme for financial support.

REFERENCES

1. S. J. Sweeney *et al*, IEEE Phot. Tech. Lett., **10**, p1076 (1998).
2. R. Fehse *et al*, Elec. Lett., **37**, p1518 (2001).

Growth and Characterization of 1.3μm Multi-Layer Quantum Dots Lasers Incorporating High Growth Temperature Spacer Layers

I. R. Sellers[#], H. Y. Liu[*], D. J. Mowbray[#], T. J. Badcock[#], K. M. Groom[*], M. Hopkinson[*], M. Gutiérrez[*], M. S. Skolnick[#], R. Beanland[+], D. T. Childs[+] and D. J. Robbins[+]

[#]*Department of Physics and Astronomy, University of Sheffield, Sheffield, S3 7RH, United Kingdom*
[*]*Department of Electronic and Electrical Engineering, University of Sheffield, Sheffield S1 3JD, United Kingdom*
[+]*Bookham Technology plc, Caswell, Towcester, Northamptonshire, NN12 8EQ, United Kingdom*

Abstract. The use of high growth temperature GaAs spacer layers is shown to significantly improve the performance of 1.3μm multi-layer quantum dot lasers by preventing the formation of threading dislocations. As-cleaved devices exhibit a cw room temperature (300K) threshold current density (J_{th}) of 39 Acm^{-2} and operate up to 105°C. High reflectivity coated facet devices operate at 300K with a cw J_{th} of 17 Acm^{-2}.

INTRODUCTION

Quantum dots (QDs) provide a viable route for the fabrication of 1.3μm GaAs based lasers [1]. However the low intrinsic QD gain typically requires the use of low-loss cavity designs or multi-layer devices. In the latter case the nature of the spacer layer is critical in achieving uniform layers and preventing defect formation in these highly strained systems. In this paper we demonstrate that the use of GaAs spacer layers grown at an elevated temperature significantly improves the device performance.

RESULTS

The devices were grown by solid-source molecular beam epitaxy and consisted of 3.0 InAs monolayers grown within 8nm $In_{0.15}Ga_{0.85}As$ quantum wells - dot-in-a-well (DWELL) structures. Devices contained 3 or 5 DWELLs separated by 50nm of GaAs. In one set of structures the GaAs spacer layers were grown at the same temperature (510°C) as the InAs and $In_{0.15}Ga_{0.85}As$. In a second set the growth temperature was increased to 580°C after the first 15nm, being reduced back to 510°C before the growth of the next DWELL. The use of this high growth temperature spacer layer (HGTSL) is found to significantly improve the device performance [2]. $Al_{0.4}Ga_{0.6}As$ cladding layers, grown at 620°C, completed the structures.

Figure 1 shows a cross-sectional transmission electron (TEM) micrograph of a device consisting of 5 DWELLs grown without HGTSLs. A threading dislocation is observed, originating in a pair of defective QDs in the second layer. As defective QDs in one layer generally result in the generation of even larger defective dots in subsequent layers, the areal density of the defects increases with DWELL number, ~10^7 and ~10^9 cm^{-2} for 3 and 5 DWELLs respectively. In contrast the introduction of the HGTSLs inhibits defect formation, an upper limit of 1x10^6 cm^{-2} being placed on their density in the HGTSL-containing structures.

Without HGTSLs the growth surface immediately prior to the growth of a second QD layer is found to exhibit a subnanometer surface roughness, as observed in AFM studies of uncapped structures. This roughness results from the low surface mobility of the Ga atoms which is insufficient to completely planarise the surface following the growth of the first QD layer [3]. It is believed that QD nucleation occurs preferentially in surface pits, resulting in the growth of a subset of large defective QDs. Pits are observed above defect QDs in TEM images and these pits further modify the nucleation of dots in subsequent layers. In contrast, enhanced Ga mobility at the higher growth temperature used for the final stage of the HGTSLs results in a better planarisation of the surface, and the elimination of the defective QDs.

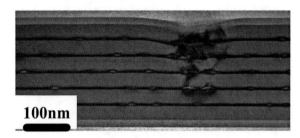

FIGURE 1. Cross-sectional TEM image of a 5 DWELL structure grown without HGTSLs.

The introduction of the HGTSLs increases the spontaneous emission intensity of the QDs but maintains the emission wavelength and linewidth. Hence the HGTSLs remove the small subset of dislocated QDs but do not significantly affect the optical properties of the remaining QDs.

Figure 2 compares the lasing characteristics of 5mm long ridge cavity devices with and without HGTSLs. For devices with 3 DWELLs the threshold current density (J_{th}) is always lower for the device with HGTSLs, a factor ~5 lower at room temperature (300K). Devices with 5 DWELLS and no HGTSLs contain too many defects to lase at 300K, only pulsed operation up to 190K is possible. In contrast a 5 DWELL device with HGTSLs exhibits excellent characteristics, with 300K cw J_{th}, λ_{las} and T_o of 39Acm^{-2}, 1.307µm and 111K respectively. A 300K pulsed J_{th} of 31Acm^{-2} is obtained, with cw operation up to 105°C (measurement system limited). A 3 DWELL 2mm long high reflectivity (HR) facet coated (R≈90%) device gives a 300K cw J_{th} of 17Acm^{-2} (32.5Acm^{-2} for an as-cleaved device). The characteristics of the present devices are competitive with the best reported literature values, e.g. 24Acm^{-2} (cw) for a 19.2mm cavity length λ_{las}=1.28µm DWELL [4] and 19Acm^{-2} (cw) 1.13mm HR facets λ_{las}=1.33µm, atomic layer epitaxy (ALE) grown [5]. Hence the present HGTSL-containing devices would appear to represent the lowest J_{th} for both as-grown and HR coated >1.3µm QD lasers. This enhanced performance is a result of the removal of the small fraction of defective QDs, which act as non-radiative recombination centres and possibly photon scattering centres which act to increase the cavity internal loss.

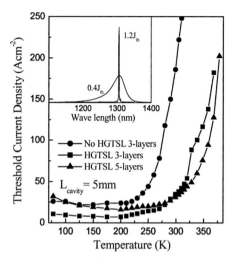

FIGURE 2. Comparison of the temperature dependence of J_{th} for devices with and without HGTSLs. The inset shows emission spectra of a 5 DWELL device with HGTSLs.

ACKNOWLEDGMENTS

This work is supported by the U.K. Engineering and Physical Sciences Research Council (EPSRC), grant GR/S49308/01, and the European Commission GROWTH programme NANOMAT project, contract G5RD-CT-2001-00545.

REFERENCES

1. V M Ustinov, A E Zhukov, A Yu Egorov and N A Maleev, *Quantum Dot Lasers*, Oxford University Press, Oxford, 2003.
2. H Y Liu, I R Sellers, T J Badcock, D J Mowbray, M S Skolnick, K M Groom, M Gutiérrez, M Hopkinson, J S Ng, J P R David and R Beanland, Appl. Phys. Lett. 26th July **85,** (2004).
3. E C Le Ru, A J Bennett, C Roberts and R Murray, J. Appl. Phys. **91,** 1365 (2002)
4. X D Huang, A Stintz, C P Hains, G T Liu, J Cheng and K J Malloy, Electron. Letts. **36,** 41 (2000)
5. G Park, O B Shchekin, D L Huffaker and D G Deppe, IEEE Photon. Technol. Letts. **13,** 230 (2000)

Self-consistent Calculation of Band Diagram and Carrier Distribution of Type-II Interband Cascade Lasers

Peng Peng, Yao-Ming Mu, and S.S.Pei

Texas Center for Superconductivity and Advanced Materials, and Department of Electrical & Computer Engineering, University of Houston, Houston, TX 77204-5004

Abstract. By solving the one-dimensional time-independent Schrödinger equation and Poisson equation simultaneously, we have numerically simulated the energy-band diagram and steady-state carrier distribution of the type-II Interband cascade lasers based on InAs/InGaSb/AlSb multi-quantum wells. The energy levels and wavefunctions of electron and hole in the active region of type-II interband cascade lasers were obtained self-consistently. Our results show that the effects of carrier distribution on band structure should be taken into account in the simulation and design of laser device.

INTRODUCTION

In the mid-infrared range, type-II interband quantum cascade lasers (ICLs) have shown their potential for the applications in chemical sensing and environmental monitoring, especially in the wavelength range of 3-5μm [1]. Early theoretical modeling suggested that type-II ICLs can achieve low threshold current density and high performance at room temperature [2,3]. At present, the performance of type-II ICLs has been improved significantly, but there still has much room for optimization. More accurate method is necessary to simulate and optimize the design of device.

In the previous theoretical investigations of ICLs, a uniform electrical field (linear band profile) was assumed [2,7,9]. In fact, the band will be bended due to the internal carrier distribution. To investigate the effect of carrier distribution on band structure and wavefunction, we have numerically simulated the energy-band diagram and the carrier distribution in ICLs by simultaneously solving the one-dimensional time-independent Schrödinger equation and Poisson equation in this paper. The self-consistent solution was obtained using the three-point finite-difference method.

SELF-CONSISTENT CALCULATION

Due to the complicacy of type-II ICL structures, a nonuniform mesh has to be adopted in the self-consistent calculation. [4]

Assuming that z_η is the position of interface, Schrödinger equation can be written as:

$$-\frac{1}{m^*d^2}\psi_j(z_{i-1}) + [\frac{2}{m^*d^2} + V(z_i)]\psi_j(z_i) - \frac{1}{m^*d^2}\psi_j(z_{i+1}) = E_j\psi_j(z_i) \quad (1)$$

for $z_i \neq z_\eta$ & $z_{\eta\pm1}$.

At the interface and its neighbors, we have

$$-\frac{1}{m_L^*d_L^2}\psi_j(z_{\eta-2}) + [\frac{1}{m_L^*d_L^2}(2 - \frac{1}{1+\alpha_{LR}}) + V(z_{\eta-1})]\psi_j(z_{\eta-1}) - \frac{1}{m_L^*d_L^2}\sqrt{\frac{d_L}{d_R}}\frac{\alpha_{LR}}{1+\alpha_{LR}}\psi_j(z_{\eta+1}) = E_j\psi_j(z_{\eta-1}) \quad (2)$$

$$-\frac{1}{m_L^*d_L^2}\sqrt{\frac{d_L}{d_R}}\frac{\alpha_{LR}}{1+\alpha_{LR}}\psi_j(z_{\eta-1}) + [\frac{1}{m_R^*d_R^2}(2 - \frac{1}{1+\alpha_{LR}}) + V(z_{\eta+1})]\psi_j(z_{\eta+1}) - \frac{1}{m_R^*d_R^2}\psi_j(z_{\eta+2}) = E_j\psi_j(z_{\eta+1}) \quad (3)$$

$$\psi_j(z_\eta) = [\psi_j(z_{\eta-1}) + \alpha_{LR} \cdot \psi_j(z_{\eta+1})]/(1+\alpha_{LR}) \quad (4)$$

where $\alpha_{LR} = m_L^*d_L / m_R^*d_R$, d_L and d_R are the mesh lengths of the left and right side of the interface, respectively.

Poisson equation can be expressed as

$$\kappa_i[V_H(z_{i-1}) - 2V_H(z_i) + V_H(z_{i+1})]/d_i^2 = 8\pi[N_D(z_i) - N_A(z_i) - n(z_i) + p(z_i)] \quad (5)$$

for $z_i \neq z_\eta$, where κ is the dielectric constant;

$$\frac{2\kappa_L}{d_L(d_R+d_L)}V_H(z_{\eta-1}) - [\frac{2\kappa_L}{d_L(d_R+d_L)} + \frac{2\kappa_R}{d_R(d_R+d_L)}]V_H(z_\eta) + \frac{2\kappa_R}{d_R(d_R+d_L)}V_H(z_{\eta+1}) = 8\pi[N_D(z_\eta) - N_A(z_\eta) - n(z_\eta) + p(z_\eta)] \quad (6)$$

for the interface, where V_H is the Hartree potential.

The self-consistent results were obtained by solving the above Schrödinger and Poisson equations simultaneously.

RESULTS AND DISCUSSION

We calculate three stages of a typical type-II ICL structure in Ref [5,6]. The self-consistent calculation was performed using single-band model at T=80K and strain effects were considered [7,9]. The self-consistent results including the band profile and the steady-state carrier distribution are shown in Fig. 1. Figure 2 shows energy levels and wavefunctions of the active region in detail. A bias of 0.39V per stage is applied to satisfy the threshold carrier (electron) density ($\sim 3.3 \times 10^{11}$ cm^{-2}) [7]. In our calculation, the sign of electron and hole density is assumed as + and -, respectively.

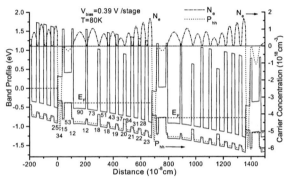

FIGURE 1. The band profile and steady-state carrier distribution of the type-II ICL under a 0.39V bias per stage at T=80K.

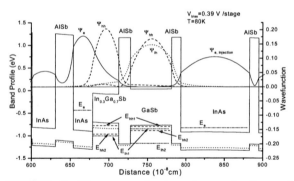

FIGURE 2. The energy levels and wavefunctions of electron, heavy hole, and light hole in the active region of the type-II ICL under a 0.39V bias per stage at T=80K.

Figure 1 shows that the carriers in injection regions can't be ignored. The built-in electric field generated by the holes in the InGaSb & GaSb wells would be compensated not only by the electrons in the active region InAs well, but also gradually by the excess electron concentrations near the left end of the n-doped injection region. The field in the GaSb well and the first InAs well of the injection region is opposite to the external bias. That results in the band bending upwards in the GaSb well and the first several layers of the injection region. The self-consistent results show that near the threshold, the average electric field is about -134.7 kV/cm and -79.2 kV/cm in the InAs well and Ga$_{0.7}$In$_{0.3}$Sb well of the active region, respectively. But it reverses to +10 kV/cm in the GaSb well. In the linear band profile approximation, a uniform electric field ($\sim$ -75kV/cm [7,9]) is assumed across the whole structure. This approximation would cause deviation, especially for the energy levels in the GaSb hole wells and the minibands in injection regions.

In previous literatures [8], electrical neutrality condition was considered only in the active regions. In our self-consistent calculation, it is satisfied for whole structure. The calculated heavy hole density in Ga$_{0.7}$In$_{0.3}$Sb well is up to -12.8$\times 10^{11}$cm^{-2}, which is 3 times of the electron density in the InAs well of the active region. This result will affect the theoretical estimation of the optical gain and threshold carrier density.

At T=80K, the lasing wavelength is about 3.58μm by the linear band profile approximation, but our self-consistent result is 3.65μm, which is more close to the experimental wavelength 3.85μm [6].

In the ICLs, energy levels have to be designed precisely. The self-consistent calculations show that carrier distribution influences the band profile of structure. Our self-consistent band profile can be adopted as the input of multiband k•p simulation to obtain more accurate band structures, wavefunctions, and optical properties of device.

ACKNOWLEDGMENTS

This work was supported by the AFOSR MURI Grand No. F49620-00-1-0331 and Texas Advanced Technology Program.

REFERENCES

1. Yang R.Q., Hill C.J., Yang B.H., Wong C.M., Muller R.E., and Echtermach P.M., *Appl. Phys. Letters* **84**, 3699-3701 (2004)
2. Meyer J.R., Vurgaftman I., Yang R.Q., and Ram-Mohan L.R., *Electron Lett.***32**, 45-46 (1996)
3. Vurgaftman I., Meyer J.R., and Ram-Mohan L.R., *IEEE Photonics Technol. Letters* **9**, 170-172 (1997).
4. Tan I.H., Snider G.L., Chang L.D., and Hu E.L., *J.Appl. Phys.* **68**, 4071-4076 (1990)
5. Yang B.H., Zhang D., Yang R.Q., Lin C-H., Murry S.J., and Pei S.S., *Appl. Phys. Letters* **72**, 2220-2222 (1998)
6. Yang R.Q., Yang B.H., Zhang D., Lin C.H., Murry J., Wu H. and Pei S.S., *Appl. Phys. Letters* **71**, 2409-2411 (1997)
7. Mu Y.M, and Yang R.Q., *J. Appl. Phys.* **84**, 5357-5359 (1998)
8. Liu Guobin, Chuang Shun-Lien, *Phys Rev. B* **65**, 165220-1 —165220-10 (2002)
9. Mu Y.M, and Yang R.Q., *Proc. SPIE* **3625**, 811-822 (1999).

Study On Low Bias Avalanche Multiplication In Modulation Doped Quantum-Dot Infrared Photodetectors

Yong Hoon Kang, Uk Hyun Lee, Joon Ho Oum, and Songcheol Hong

Department of Electrical Engineering and Computer Science, Korea Advanced Institute of Science and Technology, Taejon, 305-701, Korea

Abstract. An avalanche gain mechanism of InAs/GaAs quantum dot infrared photodetector is presented. We observed a low voltage avalanche process from the modulation doped n-i-n quantum-dot infrared photodetector. This is due to the remained electrons at modulation doped region.

INTRODUCTION

The Quantum Dot Infrared Photodetector (QDIP) have large responsivity (R), as large as 1 A/W, was reported in several papers, and it was believed that avalanche process gives a large responsivity [1, 2]. We also observed a large responsivity at the same QDIP structure [1] by an avalanche process. In addition, we observed a different avalanche process due to modulation doping.

CONTENTS

Both an epitaxial layer and self-assembled InAs QDs for QDIP were grown on a semi-insulating GaAs substrate along the direction of (100) using a RIBER 32P molecular beam epitaxy (MBE) system. The structure of QDIP is n-i-n to maximum electron transport. The absorption region of a QDIP consists of 5 periods of InAs/GaAs quantum dots and a 40 nm barrier. Two QDIPs have different doping structure to compare the response. One is δ-doped with 5×10^{17} cm^{-3} silicon lies 6 nm below the QDs, another is direct doped with 5×10^{17} cm^{-3} silicon at QDs itself. The mesa was fabricated by a standard photolithography technique with a wet etching technique. We used an AuGe/Ni/Au alloy to form an ohmic contact for electrodes.

The bias dependence of photoresponse is measured by Lock-in-amplifier system equipped with an optical chopper. We obtain the responsivity-bias curves by choosing peak points of photoresponse at the different bias voltage. Figure 1 shows the responsivity R with respect to the applied bias voltage V_a, The scatters are the experimental values measured at a sample temperature 20 K, the lines are the theoretical values. The photo-excited electrons are accelerated by electric field and interact with the ground state electrons via the Coulomb potential. After interaction, the incident electrons lose energy while the bound electrons are promoted. These excited electrons can then go out thereby creating gain and producing total avalanche gain along the QDIP structure. An Arrhenius plot is represented in Fig. 2 for a bias voltage of 0.2 V. In Fig. 2, the activation energy of modulation doped QDIP is very different from that of direct doped one. In the case of direct doped QDIP, the activation energy is very similar to the energy of photoresponse peak ~5 µm. But modulation doped QDIP shows the much smaller activation energy. We believe this difference comes from the remained electrons at modulated conduction band by the δ-doping. Because δ-doped region has a thickness of 3 nm, there are 1.5×10^{11} cm^{-2} electrons. But the QD density of each absorption layer is approximately 10^{10} cm^{-2}, many electrons (~1.3×10^{11} cm^{-2}) remain in δ-doped region. The electron transition from the δ-doped region to the QDs region eventually makes a conduction band modulation and there is a local potential.

For calculating the responsivity including an avalanche process, we have to calculate the ionization coefficient using equation (6) of reference 3. Even if we deal with a QD, we can apply the same approximation to the transition rate and obtain the same formula of an ionization coefficient. As a result, in the case of a direct doped QDIP, we used equation (1) for responsivity-bias curve. To fit that in the modulation doped QDIP we have to change the form of equation (1) as

$$R = \left(\frac{1}{h\nu}\right) \alpha f p \frac{1}{5} \left\{ \begin{array}{l} 4(1+\beta) + 3(1+\beta)(1+\beta') + \\ 2(1+\beta)(1+\beta')^2 + (1+\beta)(1+\beta')^3 \end{array} \right\} \quad (2)$$

where β' and β are a function of the potential of QDs and δ-doped regions, respectively [3].

In order to compare experiment with theory we first fit the data shown in Fig. 1. The agreement between theory and experiment is excellent for both the magnitude as well as the shape of the avalanche gain responsivity curve. In particular the gain saturation at high voltages is due to the decrease in ionization coefficient at high incident energies. Because the electron density of modulation doped region would be easily decreased by phonon scattering and dark current, the saturated responsivity of modulation doped QDIP is smaller than that of direct doped one. The fitted curves with equation (1), (2) agree with the experimental result very well. From this fit we determine $p=1$, $\gamma \cong 1$, $\sigma \cong 35$ meV, and $E_b \cong 0$ as a common fitting parameter. And we determined the activation energy of (2) as 63 meV. This value is similar to the value from Arrhenius plot in Fig. 2. From the comparison between (2), which includes the sufficient structural information, we can find that a modulation doped QDIP really has an avalanche process at δ-doped region and, eventually, low bias avalanche multiplication is possible. This paper shows the importance of the modulation doping of a QDIP, which allows an avalanche gain process at low voltages. Figure 1 shows the responsivity of (2) is much larger than that of (1) at bias voltage from 0.5 V to 1.3 V.

In summary, the avalanche gain in InAs/GaAs QDIP was studied with the responsivity-bias curve. Directly-doped QDIP shows a conventional avalanche process from each QD layers, while modulation-doped QDIP shows an avalanche process at low bias voltages. It is found that δ-doped region contribute to the avalanche multiplication at low voltages. The possibility of low voltage operation of QDIP is demonstrated.

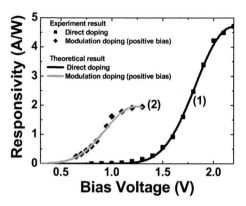

FIGURE 1. Plot of responsivity R against applied bias voltage V. The scatters are the experimental values measured at a sample temperature 20 K, the lines are the theoretical values.

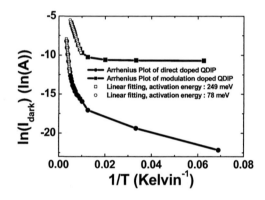

FIGURE 2. The comparison of activation energy between the directly doped QDIP and the modulation doped QDIP.

ACKNOWLEDGMENTS

This work was supported, in part, by KISTEP (under Nano-Structure Technology Projects) and the MOE BK21 program. The authors would like to thank J. K. Kim for his technical assistance.

REFERENCES

1. D. Pan, E. Towe, and S. Kennerly, Appl. Phys. Lett. **75**, 2719 (1999)
2. S. Y. Wang, S. D. Lin, H. W. Wu, and C. P. Lee, Appl. Phys. Lett. 78, 1023 (2001)
3. B. F. Levine, K. K. Choi, C. G. Bethea, J. Walker, and R. J. Malik, Appl. Phys. Lett. **51**, 934 (1987)

Type-II Interband Cascade Lasers: From Concept to Devices

Rui Q. Yang

Jet Propulsion Laboratory, California Institute of Technology, Pasadena, CA 91109, USA

Abstract. I briefly discuss the underlying physics of type-II interband cascade lasers and review their development from concept to high performance devices.

INTRODUCTION

It has been ten years since the proposal of type-II interband cascade (IC) lasers in 1994 [1]. Significant progress has been achieved toward developing efficient mid-IR type-II IC lasers [2-7] in meeting requirements in several practical applications such as gas sensing, IR countermeasures, environmental monitoring, communications, medical diagnostics and chemical warfare monitoring. Still, the high-device-performance projected by early theories [8-9] has yet to be realized, and extensive research and development need to be carried out to deepen our understanding and to further improve type-II IC lasers. Nevertheless, it may be instructive to discuss, from retrospective view, how type-II IC lasers have been developed from concept to devices, although it will be brief and incomplete.

CONCEPT DEVELOPMENT

Using interband tunneling facilitated by the broken band-gap alignment in type-II quantum wells (QWs) as shown in Fig. 1, IC lasers reuse injected electrons in cascade stages for photon generation with high quantum efficiency. Unlike intraband quantum cascade (QC) lasers [10], IC lasers use optical transitions between the conduction and valence bands with opposite dispersion curvatures without involving fast phonon scattering. Similar to intraband QC lasers, carrier concentration required for threshold is lower than conventional diode lasers, due to many serially connected stages with uniform carrier injection over every stage. Hence, losses are reduced in IC lasers, making it possible to significantly lower threshold current density. Type-II IC lasers can be constructed from the nearly lattice-matched InAs/GaSb/AlSb III-V material system. Due to a large conduction-band offset and type-II band-edge alignment in this Sb-based system, excellent carrier confinement can be achieved in IC lasers and their emission wavelengths can be tailored in a wide wavelength range, in principle from mid-IR (as short as 2.5 μm) to far-IR.

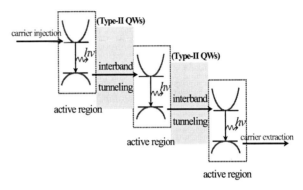

FIGURE 1. Illustration of the type-II IC laser concept.

Several aspects were considered in conceiving type-II IC lasers. These include specific semiconductor material system, carrier transport, internal absorption losses, nonradiative processes, optical transitions and their interplays. As a matter of fact, the IC laser concept was evolved from the author's pursuit of intersubband lasing in type-II QW system [11-12]. The concept formation process benefited from early investigations of interband tunneling in Sb-based type-II QWs, and from technical literature on relevant subjects such as conventional mid-IR diode lasers. Some details of historical background were reflected in Ref. 12. It should be noted that the proposal of type-II IC lasers was initially driven largely by the author's curiosity and interest accompanied with desires of excitement and success when facing challenges.

EARLY EXPERIMENTAL EFFORTS

Experimental explorations of mid-IR IC lasers started in early 1995 at the University of Houston (UH), where a Riber 32 molecular beam epitaxy (MBE) system was available for the growth of Sb-based QW structures. Collaborative efforts between several groups from 1995 to 1997 led to the initial success in proof-of-concept with encouraging results such as relatively high quantum efficiencies (>200%), peak power of ~0.5 W/facet, near room temperature (~286 K) pulsed operation, electroluminescence at a wavelength of 15 μm [2]. Also, some issues were encountered associated with long growth time (>15 hours), poorly controlled material quality, imperfect device design, lack of complete understanding and very preliminary device fabrication [2-3], which hampered new progress in 1998.

In early of 1999, efforts in developing type-II IC lasers began at Army Research Laboratory (ARL) where good experimental facilities including a Varian Gen-II MBE system were available. With experience gained in early investigations of mid-IR IC lasers and improved device design, as well as possibly better controlled material quality, significant progress was quickly achieved at ARL in terms of cw operation, higher quantum efficiency (>600%), peak power (~6 W/facet), power conversion efficiency (>14%) and reproducibility [3]. The progress was then slowed down to some extent for various reasons during the telecommunication bubble period.

RECENT ADVANCES

The progress in IC laser development was revived in the late of 2001 with the demonstration of a room temperature IC laser near 3.5 μm [3] and subsequent advancement in terms of higher cw operation temperature (214 K) and power efficiency (23%) [7]. With the setup of a new Gen-III MBE system in 2002 at Jet Propulsion Laboratory (JPL), fast progress has been made there [4-6]. Some outstanding performance features include low threshold current densities (e.g. ~9 A/cm^2 at 80 K), above room temperature pulsed operation (e.g. 325 K limited by cryostat setup), cw operation up to 217 K as shown in Fig.2, and extension of lasing wavelength in the 5.1-5.6 μm region with cw operation up to 165 K [4-6]. Cw operation of mid-IR single-mode DFB IC lasers have been demonstrated and exploited in the application for sensitive detection of some molecules such as CH_4 and H_2CO [4-6].

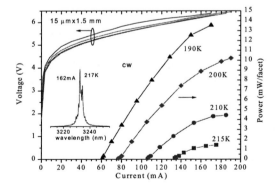

FIGURE 2. *I-V-L* characteristics & spectrum of an IC laser.

CONCLUDING REMARKS

The successful growth of IC laser samples in different MBE systems at three places (i.e. Riber 32 at UH, Varian Gen-II at ARL, Applied-EPI Gen-III at JPL) confirmed the general principle and concept of type-II IC lasers. Significant progress in developing high-performance IC lasers has been made following a nonlinear relationship with time due to mixed reasons from technical and non-technical aspects. The main challenge remains to achieve cw operation at room temperature for meeting requirements of many practical applications. However, when contrasted with other semiconductor laser technologies such as InP- and GaAs-based QC lasers, the amount of efforts expended in developing Sb-based IC lasers has been very limited. Still, there is significant room for improvement considering that many parameters in device design and MBE growth have not been optimized, as well as in device fabrication and packaging.

ACKNOWLEDGMENTS

I am grateful to many people for their comments, contributions, sharing knowledge and facilities, support and encouragement. This work was performed at JPL, California Institute of Technology, and was supported in part by NASA, AEMC Technology Development Program, and Enabling Concepts & Technologies Program.

REFERENCES

1. R. Q. Yang, at *7th Inter. Conf. on Superlattices, Microstructures and Microdevices*, Banff, Canada, Aug. 1994; *Superlattices and Microstruct.* **17**, 77-83 (1995).
2. R. Q. Yang, *Microelectronics J.* **30**, 1043-1056 (1999); and references therein.
3. R. Q. Yang, et al., *IEEE J. Quantum Electron.* **38**, 559-568 (2002); and reference therein.
4. R. Q. Yang, C. J. Hill, C. M. Wong, *Proc. SPIE* **5365**, 218-227 (2004); and references therein.
5. R. Q. Yang, et al, *Appl. Phys. Lett.* **84**, 3699 (2004).
6. R. Q. Yang, and C. J. Hill, at *The 6th International Conference on Mid-Infrared Optoelectronics Materials and Devices*, St Petersburg, Russia, June, 2004.
7. J. L. Bradshaw, et al., *Physica E* **20**, 479-583 (2004).
8. J. R. Meyer, I. Vurgaftman, R. Q. Yang, and L. R. Ram-Mohan, *Electronics Letters*, **32**, 45-46 (1996).
9. I. Vurgaftman, J. R. Meyer, and L. R. Ram-Mohan, *IEEE Photo. Tech. Lett.* **9**, 170-172 (1997).
10. J. Faist, et al., *Science* **264**, 553-556 (1994).
11. R. Q. Yang, J. M. Xu, *Appl. Phys. Lett.* **59**, 181 (1991).
12. R. Q. Yang, "Novel concepts and structures for infrared lasers", in *Long Wavelength Infrared Emitters Based on Quantum Wells and Superlattices*, edited by M. Helm, Gordon & Breach, Singapore, 2000, pp. 13-64.

Spectral hole burning by storage of electrons or holes

T. Warming*, W. Wieczorek*, M. Geller*, A. Zhukov[†], V.M. Ustinov[†] and D. Bimberg*

*Institut für Festkörperphysik, Technische Universität Berlin, Germany
[†]A.F.Ioffe Physico-Technical Institute RAS, St-Petersburg, Russia

Abstract. Spectral hole burning in spectra of self-organized InAs/GaAs quantum dots is demonstrated. Resonant charging with holes and for the first time with electrons is observed as saturation in the absorption spectra. This enables investigations on longer time scales than the lifetime of radiative recombination, in view of quantum dot memory devices [1] or spin storage. Varying temperatures and excitation densities reveal the character of carrier escape mechanisms.

InAs/GaAs QDs in an electric field show different tunnel escape rates for electrons and holes, due to the different effective masses. Excitons, generated by resonant excitation in a subensemble of QDs, decay due to tunneling of the lighter electron. The remaining hole is stored in the quantum dot [2]. The imbalance of the tunnel rates can be enhanced by an additional AlGaAs barrier on the p-doped side, or overcompensated by a barrier on the n-doped side, permitting wavelength selective storage of a single electron in a quantum dot.

The presence of a spectator hole or electron leads to a shift of the transition energy of the QD [3]. Consequently wavelength selective charging of a subensemble leads to a spectral hole in absorption. Here we use photocurrent spectroscopy [4] to resolve the spectrally selective saturation [5].

The investigated samples were grown by molecular beam epitaxy on GaAs(001) substrates. The QDs are formed by deposition of 2.5 ML InAs, yielding a QD area density of $\sim 5 \cdot 10^{10}\,\text{cm}^{-2}$. The InAs/GaAs QDs are embedded in the center of the 300 nm thick intrinsic region of a p-i-n structure. The growth temperature was lowered from 585 °C to 485 °C for the deposition of the InAs QD layer and the first 7 nm of the GaAs cap. In photoluminescence (PL), the QD ensemble shows a ground state transition energy of $\sim 1.13\,\text{eV}$ with a FWHM of 60 meV.

Two structures are studied. Sample H is designed for the storage of holes. A 30 nm thick $Al_{0.3}Ga_{0.7}As$ tunnel barrier is included on the p-side separated from the QDs by a 10 nm GaAs spacer. To suppress the tunnel emission of the electrons, sample E has a 75 nm thick $Al_{0.35}Ga_{0.65}As$ tunnel barrier with a 7 nm spacer on the n-doped side of the QDs. The schematic band structures are shown in the insets of Fig. 1. High resolution measurements of the spectral selective charging are shown in Fig. 1. We performed two-color experiments, using an attenuated cw-Ti:Sapphire laser as a tunable pump source and a tungsten lamp dispersed by a 1 m spectrometer as probe excitation (Fig. 1a). Sample E necessitates higher probe excitation densities so an attenuated Ti:Sapphire laser for probe excitation and a Nd:YAG Laser as pump

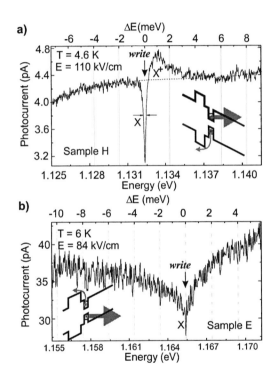

FIGURE 1. Resonant excitation accumulates holes a) or electrons b). This prevents further absorption and leads to spectral hole in absorption.

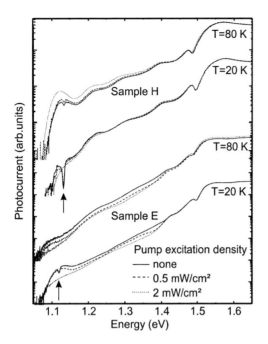

FIGURE 2. Spectral holes due to electron and hole storage at two temperature with varying write excitation densities.

source are used (Fig. 1b).

The better signal to noise ratio in Fig. 1a) reveals the structure of the spectral hole. In addition to the narrow hole induced absorption is visible on the high-energy side shifted by ~ 1 meV. As pointed out above, the saturation is not the result of a high exciton density, but optical charge accumulation. Here, the resonantly excited QDs are positively charged. The spectral hole is, thus, a consequence of the renormalization of the exciton transition by a spectator hole and the absorption is shifted by the trion binding energy rather than saturated. Note, that the renormalization prevents doubly charged QDs as long as the trion binding energy (~ 1 meV) exceeds the FWHM of the pump source ($\sim 100 \mu$eV). The photo-induced absorption can be attributed to the generation of the positive trion complex. Note, that the FWHM of the photo-induced absorption is much broader (~ 2 meV) than the spectral hole (190 μeV). The inhomogeneous broadening of the trion binding energy is attributed to QDs with different combinations of size, shape, and composition having the same ground state transition energy. The induced absorption due to the negative trion complex, expected on the low energy side of the spectral hole in Fig. 1b) is not yet resolvable.

The temperature and write density dependence of the photocurrent spectra reveal the participating carrier escape mechanisms. Fig. 2 shows the photocurrent spectra of both samples at 20 K and 80 K with varying write densities. Sample H shows a sharp spectral hole at low temperature; the hole depth increases with augmenting write density. The photocurrent of the ground state beside the hole rises with ascending write excitation densitie. At higher temperatures, the hole depth decreases, and the photocurrent beneath the hole decreases with rising excitation densities. At 100 K the spectral hole disappears.

Sample E also shows an increase of the photocurrent beneath the spectral hole with ascending excitation, even at higher temperatures. The spectral hole is well resolvable at temperatures up to 70 K.

The photocurrent depends on the absorption of the QDs, and in this regime, where we have long carrier retention times, on those, because the QDs become occupied and further absorption is suppressed until the carrier leave the QD. In Sample H the hole escape is dominated by optical intraband absorption at low temperatures and moderate fields (up to 120 kV/cm). The escape probability rises with increasing excitation and leads to increasing photocurrent. Additional excitation below the QD ground state absorption (at 0.98 eV) also leads to higher photocurrent.

With rising temperature, the probability for thermal escape of the stored carriers grows. Thermal escape differs from intraband absorption and tunnel escape, where the electric field extracts the carriers in the growth direction and prevents recapture. Thermal activated holes are retained by the AlGaAs Barrier and are recaptured by other QDs. This thermal redistribution leads to an off-resonant occupation of the QD ensemble. The absorption of the whole ensemble is reduced.

Due to the smaller effective mass of the electrons, the tunnel probability for electrons is much smaller than for holes. For thermally activated electrons tunneling through the top of the barrier is faster than recapturing, thus activated electrons leave the QDs in a combined process [6]; off-resonant occupation is not observable.

The experiments demonstrated the spectrally selective saturation of the absorption due to the storage of electrons or holes. At higher temperature thermal redistribution is observed for holes, but not for electrons, due to higher tunnel probabilities. The impact of optical intraband absorption is revealed in excitation density experiments.

This work was supported by the NANOMAT project of the European Commission's Growth Progammes, Contract Number G5RD-CT-2001-00545, and SFB 296 of DFG.

REFERENCES

1. Muto, S., *Jpn. J. Appl. Phys.*, **34**, L210–L212 (1995).
2. Kapteyn, C., et al., *Physica E*, **13**, 259 (2002).
3. Regelman, D. V., et al., *Phys. Rev. B*, **64**, 165301 (2001).
4. Fry, P. W., et al., *Appl. Phys. Lett.*, **77**, 4344 (2000).
5. Heitz, R., et al., *Physica E*, **21**, 215 (2004).
6. Kapteyn, C., et al., *Appl. Phys. Lett.*, **76**, 1573 (2000).

Buried Long-Wavelength Infrared HgCdTe P-on-n Heterojunctions

J. Rutkowski, P. Madejczyk, W. Gawron, L. Kubiak, A. Piotrowski*

Institute of Applied Physics, Military University of Technology, 2 Kaliskiego Str., 00-908 Warsaw, Poland
* Vigo System S.A., 3 Swietlikow Str., 01-389 Warsaw, Poland

Abstract. State-of-the-art long wavelength infrared (LWIR) photovoltaic detectors fabricated in HgCdTe grown by both liquid phase epitaxy (LPE) and metalorganic vapor phase epitaxy (MOVPE) have been described. To take the advantage of the strengths of the buried homojunction and the heterojunction epitaxy techniques, a new process has been developed for producing the double layer planar heterojunction (DLPH) HgCdTe photodiodes.

INTRODUCTION

Current and future needs for IR imaging applications require producible, high-performance photovoltaic HgCdTe focal plane arrays with small pixel dimensions [1]. The heterojunction mesa technology is rather straightforward, but it can be difficult to control the passivation process, especially for small-area devices. An important development in the evolution of HgCdTe photodiode technology is the capability to grow double layer heterostructures with formation of planar photodiodes by implantation [2]. The benefits of the DLPH architecture include a reduction in tunnelling current and surface generation-recombination currents. We have developed a new process that omits the ion-implantation step and allows formation of a controllable p-n junction by selective grow of the p-type cap layer through a CdTe mask.

EXPERIMENTAL

Figure 1 is a schematic illustration of the planar HgCdTe diode heterostructure. An important aspect of the DLPH approach is a planar P-doped/n-doped device geometry that includes a wide-bandgap cap layer over a narrow-bandgap base layer, which is the active device layer. The junction buried beneath the top wide gap layer overcome the technologically difficult passivation stage. The long wavelength $Hg_{1-x}Cd_xTe$ active layer has been LPE grown on the lattice-matched (111)B CdZnTe substrate from Te-rich solutions. The dipping LPE semi-closed system was used. A detail discussion of the LPE growth procedures can be found in Ref. [3]. Then, the active layer was covered with CdTe isolation layer grown by MOVPE technique. Next, the standard photolithographic techniques and etching in a 5% Br:ethylene glycol were used for define the active area on the base layer. After the chemical etching, the wide-gap cap p-type layer was grown on base layer in openings in CdTe layer. The HgCdTe layer was grown in horizontal MOVPE reactor by the interdiffused multilayer process (IMP) using diisopropyltellurium (DiPTe) and dimethylcadmium (DMCd) precursors, and elemental Hg. For acceptor doping with arsenic, the precursor AsH_3 has been used.

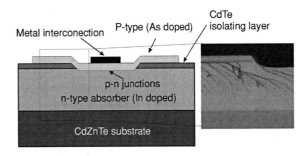

FIGURE 1. Schematic drawing of the planar DLPH HgCdTe diode configuration with cross section view of the structure.

At the end of the DLPH fabrication process the structure was annealed under Hg overpressure at 260°C for 24 h in a high-pressure system to annihilate Hg vacancies formed in the HgCdTe lattice during growth. The planar devices consist of a thick (10-15 µm) $Hg_{1-x}Cd_xTe$ (x=0.22) n-type layer capped with a thin (1 µm) wider band gap p-type layer (x=0.28). Electrical contacts were made with gold on top of the p-type capping layer and indium on the n-type layer. The test devices were the following areas: 100×100 µm², 200×200 µm², 300×300 µm².

Average composition was determined from an infrared absorption measurement in the central area of the layers. Chemical analysis and the Cd, Hg and Te profile compositions at different depths of the epitaxial layers were performed using secondary ion mass spectrometry. Transport properties were measured in temperature range of 77–300 K using the Van der Pauw arrangement. The n-type absorbing base region was deliberately doped with indium at a level of about 10^{15} cm^{-3}. The p–type layer is doped with arsenic at a level of about 5×10^{17} cm^{-3}.

DEVICE RESULTS

The performance of DLPH HgCdTe photodiodes at temperature 77 K has been analyzed. The spectral response was measured with a Fourier transform IR spectrometer. Back-illuminated quantum efficiency values of about 70 % were measured at 77 K for a device with 10 µm cutoff wavelength. This λ_{co} value is consistent with the composition value of the narrow-gap layer (x=0.228) calculated from the absorption edge of the room temperature IR transmission spectrum.

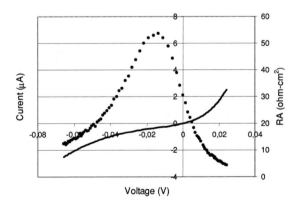

FIGURE 2. Typical current-voltage and dynamic resistance-voltage plots for a DLPH photodiode at 77 K.

The typical current-voltage and dynamic resistance curves are illustrated in Fig. 2. The maximum R_0A value obtained was 30 ohm cm² for 10 µm cutoff wavelength detectors at 77 K. Detailed analysis of the current characteristics of these diodes as a function of temperature show that they have high performance, and their dark current is diffusion-limited down to 77 K.

Finally, the device characteristics of DLPH structures were compared with "mesa" devices obtained with one fabrication process. Both device types show similar characteristics, as can be seen in Table 1, that confirm the feasibility of fabricating the DLPH configuration.

TABLE 1. The Comparison of Device Results at 77 K.

Parameter	DLPH	Mesa
Quantum Efficiency	71%	67%
R_0A Product (ohm cm²)	30	35
Detectivity (cmHz$^{1/2}$/W)	3×10^{10}	2×10^{10}

CONCLUSIONS

The new planar technology significantly simplifies the fabrication process of the photodiodes and connects the advantages of LPE and MOCVD processes. High performance LWIR detectors operated at 77 K have been obtained.

ACKNOWLEDGMENTS

The financial support for this research was provided by the Polish State Committee for Scientific Research under Contract No. 4 T08A 031 23.

REFERENCES

1. Rogalski, A., Adamiec, K., and Rutkowski, J., *Narrow-Gap Semiconductor Photodiodes*, Bellingham, SPIE Press, 2000, pp. 237-335.

2. Arias, M., Pasko, J.G., Zandian, M., Shin, S.H., Williams, G.M., Bubulac, L.O., DeWames, R.E., and Tennant, W.E., *Appl. Phys. Letters* **62**, 976-978 (1993).

3. Kubiak, L. , Wenus, J., Rutkowski, J. and Rogalski, A., „LPE growth of $Hg_{1-x}Cd_xTe$ heterostructures" in *Crystal Growth and Epitaxy*, edited by M.A. Herman, Vienna, 2003, pp. 141-153.

Improvement of Fabrication Method of Resonant Cavity Enhanced Photodiodes for Bi-directional Optical Interconnects

Il-Sug Chung[†,*], Yong Tak Lee[†], Jae-Eun Kim[**] and Hae Yong Park[**]

*ischung@gist.ac.kr

[†]Gwangju Institute of Science and Technology, Department of Information and Communications, Gwangju, 500-712, Republic of Korea

[**]Korea Advanced Institute of Science and Technology, Department of Physics, Daejeon, 305-701, Republic of Korea

Abstract. Reflectivity of a quarter wave mirror (QWM) can be controlled by choosing the outermost layer of the QWM appropriately. For a GaAs/Al$_{0.75}$Ga$_{0.25}$As QWM, depositing a TiO$_2$ layer as the incoming outermost layer decreases the reflectivity to the same degree as removing 18.5 layers. Therefore, an efficient method to fabricate resonant cavity enhanced photodiodes for bi-directional optical interconnects is proposed. This method needs no etching process for quantum efficiency tuning, while giving a quantum efficiency of approximately 90% using currently available deposition technique.

Resonant cavity enhanced photodiodes (RCE PD's) have attracted considerable attention since they provide both a high quantum efficiency and a high device speed, which are impossible in conventional p-i-n photodiodes. In addition, RCE PD's can be fabricated using vertical-cavity surface-emitting laser (VCSEL) epi wafer, which makes both VCSEL and RCE PD good candidates for bi-directional optical interconnects.

Owing to the absorption of EM waves in the absorption layer within RCE PD, the quantum efficiency η of a RCE PD is maximum when the following relation is satisfied at the peak wavelength, λ_0[1]:

$$R_{\text{top}}(\lambda_0) = e^{-\alpha d} R_{\text{bot}}(\lambda_0), \quad (1)$$

where R_{top} and R_{bot} are the reflectivity of top and bottom distributed Bragg reflectors (DBR's) of the RCE PD, respectively, and α and d are the absorption coefficient and the thickness of quantum well layers, respectively. It has been reported that for top emitting VCSEL and RCE PD pairs, some layers of the top DBR within RCE PD are etched to reduce R_{top} and satisfy the above relation[2](see FIGURE 1.(a)), whereas for bottom emitting VCSEL and RCE PD pairs, some layers of the bottom DBR are selectively oxidized[3]. In this study, we propose a novel method to fabricate top emitting VCSEL and RCE PD pairs which has several advantages over reported methods.

Recently, we showed that depositing a dielectric layer onto a top DBR breaks the periodicity in reflection phase

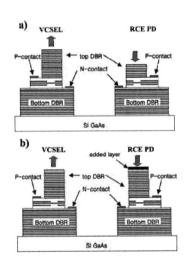

FIGURE 1. Schematic diagram of a bi-directional top-emitting VCSEL and RCE PD: (a) some layers of the top DBR of the RCE PD are etched, and (b) a dielectric layer is deposited onto the top DBR of RCE PD.

shift at the outermost layer of the top DBR, leading to the loss in reflectivity of the top DBR[4]. Using this property, RCE PD with high quantum efficiency can be fabricated without etching as shown in FIGURE 1.(b).

Here, the etching and the depositing methods are compared, assuming RCE PD's are fabricated using a 980nm VCSEL epi wafer. Our design of 980nm VCSEL epi

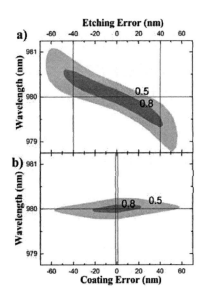

FIGURE 2. Quantum efficiency, η, as aa function of (a) etching error and (b) coating error.

wafer has two $In_{0.2}Ga_{0.8}As$ quantum wells, each with a thickness of 8nm, and the top DBR is $(HL)^{25}$, where H and L refer to a GaAs layer and an $Al_{0.75}Ga_{0.25}As$ layer, respectively. Following the reported etching method[2], $(HL)^9$ must be etched to satisfy the Eq.(1), whereas depositing a TiO_2 layer with quarter-wave thickness and a refractive index of 2.24 is equivalent to eliminating $(HL)^{8.1}$.

FIGURE 2.(a) shows that the peak wavelength of the RCE PD differs from the operating wavelength of the VCSEL, i.e., 980nm, as the top DBR is not etched to the optimal value. For example, when the top DBR is etched by more than 40nm, which is the common precision of III-V semiconductor etching facilities, the quantum efficiency at 980nm is lower than 50%. Since the theoretical value of the full width at half maximum (FWHM) for VCSEL is on the order of sub-nm, it is very important for the optical interconnects that the peak wavelength of RCE PD must be the same as that of VCSEL.

As for deposition method, the peak wavelength does not change, although the dielectric layer is deposited to a thickness that is slightly different from the optimal. In addition, the thickness of deposited dielectric layer can be controlled to sub-nm precision when ellisometry is used. As a result, a quantum efficiency at 980nm is guaranteed to be approximately 90% (FIGURE 2.(b)).

In order to break the periodicity in reflection phase shift between the air and the deposited dielectric layer, the deposited layer must have quarter-wave thickness. Its refractive index must be smaller than that of H, and be chosen to satisfy the Eq.(1) more closely. When the top DBR has a non-periodic layer like $0.5H$ in $0.5H(LH)^{25}$, the optimal thickness of the deposited layer is slightly different from quarter-wave thickness and must be determined by Yee's transfer matrix method[5]. For more details, see the reference [4].

This work has been partially supported by MOST through the Tera-level Nanodevices project.

REFERENCES

1. M. S. Ünlü and S. Strite, *J. Appl. Phys.*, **78**, 607 (1995).
2. T. Knödl, H. K. H. Choy, J. L. Pan, R. King, R. Jäger, G. Lullo, J. F. Ahadian, R. J. Ram, C. G. Fonstad, Jr., and K. J. Ebeling, *IEEE PHOTONIC. TECH. L.*, **11**, 1289 (1999).
3. O. Sjölund, D. A. Louderback, E. R. Hegblom, J. Ko, and L. A. Coldren, *Appl. Phys. Lett.*, **73**, 1 (1998).
4. I.-S. Chung, Y. Lee, J.-E. Kim, and H. Park, *J. Appl. Phys.*, **96**, 2423 (2004).
5. P. Yee, *Optical Waves in Layered Media*, (John Wiley & Sons, 1988), Chap. 5.

Real Time Read-Out of Single Photon Absorption by a Field Effect Transistor with a Layer of Quantum Dots

B. Kardynał[a], A. J. Shields[a], N. S. Beattie[a,b], I. Farrer[b], D. A. Ritchie[b]

[a]*Cambridge Research Laboratory, Toshiba Research Europe Ltd, Cambridge, CB4 3HS, UK*
[b]*Cavendish Laboratory, University of Cambridge, Cambridge, CB3 0HE, UK*

Abstract. We present results of photon counting with a single photon detector based on field effect transistor with a layer of quantum dots in close proximity to the channel. Detection of a photon is achieved when photo-hole is captured by negatively charged InAs dot leading to a step-like increase of the current in the channel of the transistor. We use a transimpedance amplifier with a cryogenic stage and ac coupling on the input to convert the current steps of 1-2 nA height into voltage peaks with microsecond time resolution. We show that single photon counting with 0.3% efficiency, dark count rate below 10^{-8} ns^{-1} is then achieved, while the jitter limited by the circuit can be as low as 8.5 ns.

INTRODUCTION

Operation of many devices, such as optical memories [1], lasers [2] or infra-red detectors [3], relies on the charge storage in InAs dots. It was also shown that AlGaAs/GaAs high electron mobility transistor with a layer of quantum dots in close proximity of the electron channel can be used to detect single photons [4]. In this paper we demonstrate photon counting with sub-microsecond response times with such quantum dot field effect transistor (QDFET) using a custom amplifier with a cryogenic input stage [5]. We show the QDFET/amplifier combination to have a low dark count rate and jitter limited by the capacitances of the circuit.

EXPERIMENTAL RESULTS

In a quantum dot QDFET the source-drain conduction in the quantum well occurs through hopping in a disordered potential created by highly charged dots in the vicinity of the channel. In devices in which the active region contains just a few hundred quantum dots and whose dimensions are comparable to the hopping length, the interactions between the electrons stored in the dots and electrons in the channel are strong enough to result in a small increase in the source-drain current when a photo-generated hole recombines with an electron trapped in a dot [4].

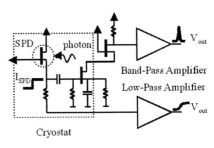

FIGURE 1. Schematic diagram of the amplifier used for photon counting.

In order to use QDFET for photon counting steps in the source-drain current need to be converted into voltage peaks. This has been achieved using the amplifier shown in Figure 1, which consists of a cascode pre-amplifier followed by a band-pass amplifier [5]. The input transistor of the cascode could be placed very near the QDFET, which is operated at 4.2 K, thereby reducing the effect of the lead capacitance between the cryogenic and room temperature parts of the circuit. This allows radio frequency operation of the detector, even though it has a channel resistance in excess of 100 kΩ. Signal from the detector is ac coupled to the input of the amplifier.

The signal on the output of the cascode is passed to a band-pass amplifier, which serves to shape the output pulse and to reduce the noise. Reset pulse necessary for refilling the dots with electrons after illumination [4] was sent based on the measurements of total current of the detector with a low-pass amplifier.

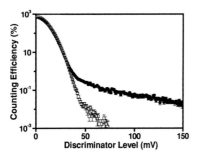

FIGURE 2. Counting efficiency of the single photon (●) and dark counts (□) as a function of the discriminator level.

The device studied here is modulation doped GaAs/AsGaAs quantum well heretostructure, which had a dot density of 1.4 10^{14} m^{-2}, separated from the 20 nm thick GaAs channel by a 10 nm thick AlGaAs barrier. The wafer was processed into a transistor with 1 μm wide mesa between a source and drain and gated with 1 μm long NiCr semitransparent gate. The active area of the device is defined by the overlap of the mesa and the gate. The QDFET was biased with 20 mV across the source-drain and -0.6 V between the gate and drain. Each detection cycle was initialized by a 1.4 V and 1 μs reset gate pulse. The detector was illuminated with pulses from a laser diode emitting light at 684 nm, with the laser clock of 400 kHz. Each pulse had on average of 0.1 photons per 1 μm^2 (area of the detector).

Figure 2 shows results of photon counting as a function of discriminator level. Each measurement was performed for 500 ns starting with the arrival of the laser pulse to count photons and simultaneously 1 μs later to measure the dark counts. For a given discriminator level a count is added whenever the rising edge of the peak from the output of the detector circuit crossed the value of the discriminator level.

Maximum efficiency of the device of 0.3% is obtained when the dark count rate is 0.015% at 50 mV discriminator level. Measured efficiency of the device is limited by reflection from the gate and low absorption in GaAs well and corresponds to 21% of photons absorbed in the well being detected.

The main source of the dark counts in the system at low discriminator levels (below 47 mV) is the amplifier noise. Second contribution is caused by random telegraph signal in the detector itself and it can be very small depending on the gate voltage. [5]. It drops abruptly with the discriminator level in Figure 2 and when it dropped below resolution (10^{-8} ns^{-1}, at 83 mV), the photon counting efficiency is still 0.09.

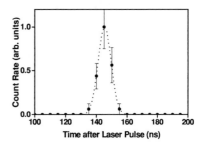

FIGURE 3. Time resolved single photon counting (■). Dotted line shows Gaussian fit to the data.

The jitter of the detection was estimated in time resolved counting (Figure 3), when counting was performed in 5 ns windows starting at different times after the laser pulse, at 150 mV discriminator level. When fastest amplifier was used all the counts occur within 20 ns. A Gaussian fit to the data estimates the full width at half height of the peak to 8.5 ns. This value is limited by the capacitance of the circuit.

SUMMARY

In this paper we present result of photon counting with quantum dot field effect transistor. Conversion and amplification of the single photon signal from the detector using an amplifier with cryogenic input stage allows counting with maximum efficiency of 0.3% and efficiency of 0.09% when dark count rate drops below 10^{-8} ns^{-1} and with jitter of 8.5 ns. Both the dark count rate and the jitter are limited by the amplifier.

REFERENCES

1. G. Yusa and H. Sakaki, Appl. Phys. Lett. **70,** 345 (1997).
2. D. Bimberg, M. Grundmann, N. N. Ledentsov in "Photonic Devices" in *Quantum Dot Heterostructures*, John Wiley and Sons, 1999, pp.279-302.
3. J. Phillips, K. Kamath and P. Bhattacharya, Appl. Phys. Lett. **72,** 2020 (1998).
4. A. J. Shields, M. P. O'Sullivan, I. Farrer, D. A. Ritchie, R. A. Hogg, M. L. Leadbeater, C. E. Norman, and M. Pepper, Appl. Phys. Lett. **76,** 3673 (2000).
5. B. E. Kardynał and A. J. Shields, N. S. Beattie, I. Farrer, K. Cooper and D. A. Ritchie, Appl. Phys. Lett. **84,** 419 (2004).

The Polarization-dependence of the Gain in Quantum Well Lasers

Fredrik Boxberg *, Roman Tereshonkov* and Jukka Tulkki*

Helsinki University of Technology, Espoo, Finland

Abstract. The governing role of the conduction band - heavy-hole band (C-HH) coupling on the polarization of gain in quantum well lasers has been predicted in the phenomenological model of Asada *et al*. In their model, based on bulk band structure arguments, the C-HH coupling makes the transition dipole moment orthogonal to the electron wave vector and thereby guaranties the conservation of angular momentum. We have made a quantitative study of the gain polarization using an 8-band $\mathbf{k} \cdot \mathbf{p}$ envelope wave function method. Our calculation shows that Asada's model is qualitatively correct while substantial quantitative differences are found.

INTRODUCTION

The polarization of gain in quantum well (QW) lasers has conventionally been analyzed within the model developed by Asada *et al*.[1] In their model the polarization of gain is governed by the symmetry change of the heavy hole band due to the electric dipole coupling of the conduction and valence bands. However, this result is exact within the 8-band model only for a bulk semiconductor and its use in the analysis of the polarization of light in QW lasers is based on phenomenological arguments. The model of Asada *et al*. gives still good agreement with experiments.[2]

The dominant role of conduction band - valence band interaction on the polarization of gain is obviously preserved also in semiconductor quantum structures. The influence of the conduction band - valence band interaction on the gain polarization has generally been overlooked in numerical multiband calculations. This has prompted us to carry out a complete 8-band calculation.[3] of the polarization of gain in QW lasers and analyze in detail the approximations involved in the model of Asada *et al*.

KANE'S MODEL FOR BULK SEMICONDUCTORS

Kane defined a linearised 8-band $\mathbf{k} \cdot \mathbf{p}$ Hamiltonian for bulk semiconductors,[4] *i.e.*, non-diagonal terms proportional to $\mathbf{k}^2$ were omitted). Within this model, the eigenstates and -energies can be solved analytically for arbitrary value of $\mathbf{k}$. The C band becomes a weakly mixed state of $|S\rangle$ and $|LH\rangle$ components. The HH eigenstate accounting for the C-HH coupling is obtained from the

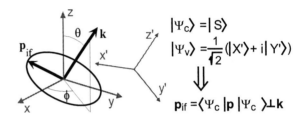

FIGURE 1. The coupling of the C and the HH band leads to orthogonality of the electron wave vector $\mathbf{k}$ and the optical dipole moment $\mathbf{p}_{if}$.

non-coupled HH-eigenstate by rotating it into a coordinate system where $k_x = k_y = 0$, *i.e.*, $z' \parallel \mathbf{k}$ as shown in Fig. 1. The rotation of the HH state makes the transition dipole moment orthogonal to $\mathbf{k}$ and conserves the total angular momentum in an electric dipole interband transition. The influence of the rotation in the linear regime on the *energy* of the HH-band is very small. However, the C-HH coupling breaks the symmetry of the HH-wave function and this leads to fundamental changes in the polarization of light emission. Figure 2 shows the band mixing coefficients for bulk GaAs using both Kane's model and the full 8-band model.

ASADA'S MODEL FOR QUANTUM WELLS

Asada *et al*. extended Kane's model to QWs,[1] *i.e.*, the leading terms in the gain polarization were derived from the from the coupling of the C ($|1/2,\pm1/2\rangle$) and HH ($|3/2,\pm3/2\rangle$) bands of bulk semiconductors. The result

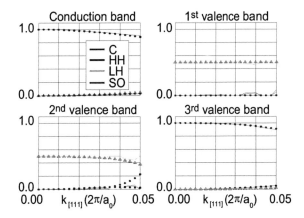

FIGURE 2. Comparison of the band mixing with Kane's linear in bf k coupling model (markers) and the full 8-band $\mathbf{k}\cdot\mathbf{p}$ model (solid lines) for bulk GaAs.

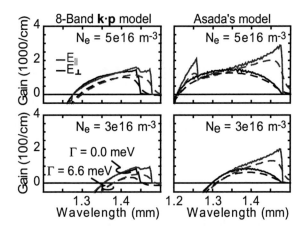

FIGURE 4. Gain in the unstrained AlInAs/GaInAs QW as a function of the photon wave length, using the full 8-band $\mathbf{k}\cdot\mathbf{p}$ model (left) and Asada's model (right). The red and black lines correspond to polarization parallel and orthogonal to the QW plane respectively.

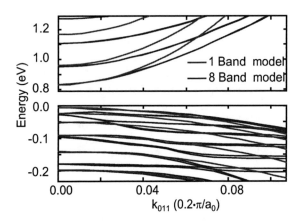

FIGURE 3. Energy bands of the unstrained AlInAs/GaInAs QW using Asada's 1-band model (red lines) and the full 8-band $\mathbf{k}\cdot\mathbf{p}$ model (blue lines).

8-BAND

The eigenstates $\Psi_i(\mathbf{k})$ and energy bands $E_i(\mathbf{k})$ of bulk material and a QW were solved using an 8-band $\mathbf{k}\cdot\mathbf{p}$ model.[3] The model includes both strain-effects and zinc-blend lattice related lack of inversion symmetry. For bulk material the band structure was solved for $\mathbf{k}$ along X and L crystal directions. The band structure of the QW was solved for a discrete set of values for $k = [k_x, k_y]$, where $k_x, k_y \in [0..k_{max}]$. The total gain was then derived by integrating the optical transition rates for all transitions in the relevant $\mathbf{k}$ space.

becomes approximative in any semiconductor quantum structure since the mixing of the valence bands cannot be omitted even at the band edge. The energies and eigenstates were solved in an effective-mass, 1-band model. They assumed that the conduction bands were rotated, pure $|S'\rangle$ bands and that the valence bands were $|HH'\rangle$ states, like in Kane's model. However, for a single QW, the wave vector is well defined only in the plane of the QW. The effect of the QW confinement on the rotation of the $|HH\rangle$ states had to be approximated by an effective $\mathbf{k}$ component perpendicular to the plane of the QW. The model contains also several other approximations like neglecting the LH bands from the gain calculation. Figure 3 shows the energy bands of an 10 nm wide, unstrained AlInAs/GaInAs QW using Asada's model and the full 8-band model.

CONCLUSIONS

A comparison of our 8-band $\mathbf{k}\cdot\mathbf{p}$ calculation with results based on Asada's model (see Fig. 4) shows a fair qualitative agreement especially when the natural life time effect is accounted for in the gain calculation. On quantitative level discrepancy is substantial and therefore any modeling that aims at predicting the ration of TM and TE gain polarization should be based on 8-band $\mathbf{k}\cdot\mathbf{p}$ theory.

REFERENCES

1. Asada, M., Kameyama, A., and Suematsu, Y., *IEEE J. OF Quant. Electron.*, **20**, 745–753 (1984).
2. Yamada, M., Ogita, S., Yamagishi, M., and Tabata, K., *IEEE J. OF Quant. Electron.*, **21**, 640–645 (1985).
3. Bahder, T., *Phys. Rev. B*, **41**, 11992–12001 (1990).
4. Kane, E. O., *J. Phys. Chem. Solids*, **1**, 249–261 (1957).

GaAs/Al$_{0.45}$Ga$_{0.55}$As Double Phonon Resonance Quantum Cascade Laser

D. Indjin[1], A. Mirčetić[2], P. Harrison[1], R. W. Kelsall[1], Z. Ikonić[1], V. D. Jovanović[1], V. Milanović[2], M. Giehler[3], R. Hey[3] and H.T. Grahn[3]

[1]*School of Electronic and Electrical Engineering, University of Leeds, Leeds LS2 9JT, UK*
[2]*Faculty of Electrical Engineering, P.O. Box 35-54, 11120 Belgrade, Serbia and Montenegro*
[3]*Paul-Drude-Institut für Festkörperelektronik, Hausvogteiplatz 5-7, 10117, Berlin, Germany*

Abstract. In this paper we show that the idea of a mid-infrared quantum-cascade laser with gain regions based on a double-phonon resonance can be implemented in the GaAs/AlGaAs system. In contrast to the usual GaAs/AlGaAs laser active region design, which involves a triple quantum well active region, here we identify an optimal heterostructure design by using a simulated annealing algorithm which is programmed to maximize the laser gain for a given wavelength and for subband spacings prescribed to satisfy the double-phonon resonance condition. The output characteristics are calculated using a full self-consistent rate equation model of the intersubband electron transport. Initial devices grown according to this design show laser action up to 240K in pulsed mode with good agreement between the calculated and measured characteristics.

INTRODUCTION

Mid-infrared GaAs/AlGaAs quantum cascade lasers (QCLs) have not yet achieved the same high temperature performance as InGaAs/AlInAs devices, although considerable research effort has resulted in room temperature pulsed mode operation of a 9μm GaAs/Al$_{0.45}$Ga$_{0.55}$As QCL [1]. A recent publication suggests that InP-based mid-infrared QCLs with gain regions based on a double-phonon (2LO) resonance have improved temperature characteristics over devices which use a single optical phonon resonance for depopulation of the lower laser subband [2]. In this paper we show that the same idea can be transferred to GaAs/AlGaAs mid-infrared QCLs.

MODEL AND DESIGNS

In contrast to the usual GaAs/AlGaAs QCL active region design, which involves a triple quantum well active region (TQW) [1], here we identify an optimal heterostructure design by using a simulated annealing algorithm which is programmed to maximize the laser gain for a given wavelength and for subband spacings prescribed to satisfy the double-phonon resonance condition. The active region of the QCL consists of four quantum wells as shown in Fig.1.

In order to extract the output characteristics, a detailed analysis of intersubband electron scattering transport in the QCL is applied using a full self-consistent rate equation model [3,4]. Our approach includes electron-LO phonon and electron-electron scattering between all injector/collector, active region and weakly localized continuum-like states, and has been shown to give excellent agreement with measured gain coefficients and threshold currents for the reference 11 μm TQW GaAs/AlGaAs QCL [5]. Using this model, we have calculated the laser gain and electric field-current density output characteristics for the new and reference structures. A noticeable improvement in performance is predicted for the double-phonon resonance QCL, compared to the TQW device. The modal gain is larger than in TQW structure for a range of current densities, which leads to a reduction in threshold current, see Fig. 2.

The structure shown in Fig. 1 has been grown (30 periods) with a standard double plasmon waveguide configuration. A very good agreement between the calculated and measured current-voltage characteristics of the device has been obtained (see Fig.3). The observed emission wavelength $\lambda \approx 10.4\mu m$ (see Fig.4) is in close agreement with the calculated $\lambda \approx 10.2\mu m$. The threshold current at 77K is 1.95A and the maximum operating temperature is 240K. The doping density in this initial sample was relatively low ($N_s \sim 3.8 \times 10^{11} cm^{-2}$), and it is anticipated that room-temperature operation can be achieved by a modest increase in this value [6].

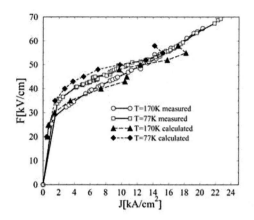

FIGURE 3. The calculated and measured current-voltage (current density - electric field) characteristics of the QCL structure shown in Fig. 1, at T=77K and T=170K.

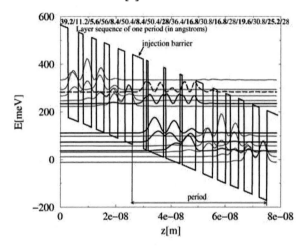

FIGURE 1. A schematic diagram of the quasi-bound energy levels and associated wave functions squared of a $\lambda \approx 10.2\mu m$ GaAs/Al$_{0.45}$Ga$_{0.55}$As 2LO QCL.

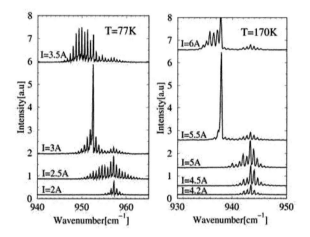

FIGURE 4. Measured emission spectra of the structure shown in Fig. 1 for various drive currents, at T=77K (left) and T=170K (right). The area of the laser stripe was $2235 \times 18 \mu m^2$.

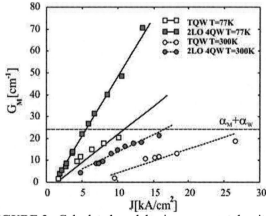

FIGURE 2. Calculated modal gain vs. current density dependence for the double-phonon and reference [5] structures.

REFERENCES

1. H. Page, C. Backer, A. Robertson, G. Glastre, V. Ortiz, and C. Sirtori, Appl. Phys. Lett. 78, 3529 (2001).
2. 2. J. Faist, D. Hofstetter, M. Back, T. Aellen, M. Rochat, and S. Blaser, IEEE J. Quantum Elect., 38, (2002).
3. D. Indjin, P. Harrison, R. W. Kelsall, and Z. Ikonic, Appl. Phys. Lett., 81, 400 (2002).
4. D. Indjin, S. Tomic, Z. Ikonic, P. Harrison, R. W. Kelsall, V. Milanovic, and S. Kocinac, Appl. Phys. Lett., 81, 2163 (2002).
5. P. Kruck, H. Page, C. Sirtori, S. Barbieri, M. Stellmacher, J. Nagle, Appl. Phys. Lett., 76, 3340 (2000).
6. M. Giehler, R. Hey, H. Kostial, S. Cronenberg, T. Ohtsuka, L. Schrottke, and H. T. Grahn, Appl. Phys. Lett., 82, 671 (2003).

Very High Temperature Operation of ~ 5.75 µm Quantum Cascade Lasers

A. Friedrich[1], G. Scarpa[2], G. Boehm[1] and M.-C. Amann[1]

[1]*Walter-Schottky Institut, Technische Universitaet Muenchen, Am Coulombwall, D-85748 Garching, Germany*
[2]*Lehrstuhl fuer Nanoelektronik, Technische Universitaet Muenchen, Arcisstr. 21, D-80333 Muenchen, Germany*

Abstract. We have fabricated GaInAs/AlInAs strain-compensated quantum cascade lasers with InP and GaInAs cladding layers using solid-source molecular-beam epitaxy (MBE). Low threshold current densities and high temperature operation of uncoated devices, with a record value of 490 K, have been achieved in pulsed mode.

INTRODUCTION

Quantum cascade (QC) lasers are among the most promising light sources in the mid- and far-infrared spectral region for applications based on absorption lines of gas molecules or atmospheric transmission windows [1-2]. Despite eminent improvements obtained in the last few years, high-temperature operation still represents a challenge for further optimization. A crucial point in the design of high-quality QC lasers is the reduction of the waveguide losses. We report on a QC laser with an improved waveguide design, making use of InP as cladding layer. The lasers are realised in the strain-compensated GaInAs/AlInAs material system with an active region based on two-phonon resonant emission [3]. Laser action up to 490 K with a decrease of the threshold current density, compared to previous devices, is achieved.

DEVICE DESIGN AND FABRICATION

By designing waveguides with only GaInAs, one can take advantage of the free-carrier plasma effect, but along with the reduction of the refractive index in highly doped layers goes a strong increase in the absorption. Therefore, to reduce the optical losses and simultaneously enhance the TM-mode confinement factor, a thick, low-doped InP-layer has been included in the waveguide. Fig. 1 shows the fundamental TM-mode and the refractive index profile. The calculated values of the absorption losses and the confinement factor are $\alpha_W = 3.1 cm^{-1}$ and $\Gamma_{AR} = 62\%$. For comparable samples with the same active region but waveguides consisting of GaInAs the values have been $\alpha_W = 6.5 cm^{-1}$ and $\Gamma_{AR} = 47\%$ [4]. Hence, by the significant improvement of the Γ_{AR}/α_W ratio, a distinct reduction of the threshold current density can be expected and laser performance can be improved.

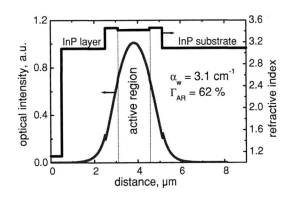

FIGURE 1. TM-mode profile and refractive index vs. growth direction, starting from the contact layer

The whole structure, including the 2 µm thick InP-cladding layer, was grown in a Varian solid-source MBE system on a low-doped InP substrate (n = $1 \times 10^{17} cm^{-3}$). After growth, ridge-waveguide lasers were fabricated by conventional photolithography and nonselective wet chemical etching using a HBr:H$_2$O$_2$:H$_2$O solution. A 250 nm thick SiO$_2$-layer was used as insulation. After evaporation of Ti/Pt/Au contacts the wafer was thinned to 120 µm and a Ge/Au/Ni/Au bottom-contact metallization was made. All devices were mounted ridge-side up on copper heat sinks, which were placed in a cold finger cryostat with an uncoated ZnSe window, typically operated between 77 K and 500 K.

DEVICE CHARACTERISTICS

A typical pulsed light output-current P(I) characteristic of a representative uncoated device with 26 µm ridge width and 1.5 mm cavity length is shown in Fig. 2. The applied pulse width was 250 ns and the repetition frequency 250 Hz. At room-temperature the emission wavelength was about 5.75 µm. The lasers work up to the record high heat sink temperature of 490 K. Note that the maximum operation temperature is set-up-limited. In fact, the tin-solder needed to stick the fixture for current injection on the heat sink melts at temperatures of about 490 K. Previous uncoated samples with GaInAs waveguides worked up to 450 K and coated devices up to 470 K [5].

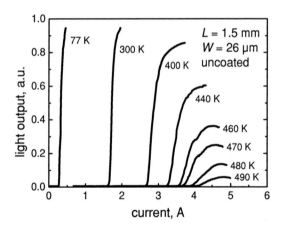

FIGURE 2. Pulsed light output of an uncoated quantum cascade laser vs. current at various heat sink temperatures

A plot of the threshold current density at various heat sink temperatures for an uncoated 3 mm long and 26 µm wide representative device is shown in Fig. 3. Low threshold current densities of 0.44 kA/cm^2 and 3.65 kA/cm^2 at 77 K and 300 K respectively, have been obtained. Devices of 2.5 mm cavity length and 30 µm width without InP-cladding showed best values like 1 kA/cm^2 at 77 K and 4.2 kA/cm^2 at 300 K [5]. The exponential fit curve for the determination of T$_0$ is also displayed.

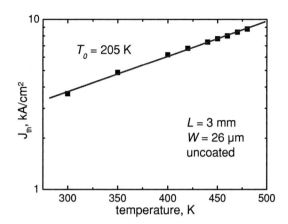

FIGURE 3. Threshold current density vs. heat sink temperature

Laser devices with high-reflection coating [5] were also tested. The maximum operation temperature in pulsed mode was 480 K, due to the melting of the solder. Low threshold current densities like 0.15 kA/cm^2 at 77 K and 2.47 kA/cm^2 at 300 K have been achieved for a 26 µm wide and 2 mm long device.

In conclusion, we have fabricated high-performance InP-based QC lasers using InP and GaInAs as cladding materials, with low threshold current densities and record high operation temperatures.

REFERENCES

1. Blaser, S., Hofstetter, D., Beck, M., Faist, J., "Free-space optical data link using Peltier-cooled quantum cascade laser" in *Electron. Lett.*, 2001, 37, pp. 778-780.

2. Knoll, B. and Keilmann, F.: "Near-field probing of vibrational absorption for chemical microscopy" in *Nature*, 1999, 399, pp.134-137.

3. Hofstetter, D., Beck, M., Aellen, T., Faist, J.: "High-temperature operation of distributed feedback quantum-cascade lasers at 5.3 µm" in *Appl. Phys. Lett.*, 2001, 78, pp. 396-398.

4. Scarpa, G., Ulbrich, N., Boehm, G., Abstreiter, G., Amann, M.-C.: "Low-loss GaInAs-based waveguides for high-performance 5.5 µm InP-based quantum cascade lasers" in *IEE Proc.-Optoelectron.*, 2003, 150, pp. 284-287.

5. Scarpa, G., Ulbrich, N., Rosskopf, J., Boehm, G., Abstreiter, G., Amann, M.-C.: "High-performance 5.5 µm quantum cascade lasers with high reflection coating" in *IEE Proc.-Optoelectron.*, 2002, 149, pp. 201-205.

Microscopic modeling of THz quantum cascade lasers and other optoelectronic quantum devices

Rita C. Iotti and Fausto Rossi

Istituto Nazionale per la Fisica della Materia and Dipartimento di Fisica Politecnico di Torino, Corso Duca degli Abruzzi 24, 10129 Torino, Italy

Abstract. The most advanced microscopic treatments of optoelectronic quantum devices are reviewed, with special emphasis on the design and simulation of novel quantum-cascade devices operating in the Terahertz spectral region. After recalling the fundamentals of hot-electron transport versus nonequilibrium energy-relaxation and carrier-redistribution in such devices, a few simulated experiments are presented and discussed.

FUNDAMENTALS

Recent developments in nanoscience/technology push device miniaturization toward regimes where the traditional semiclassical transport treatments can no longer be employed, and more rigorous quantum-transport approaches are imperative [1]. However, in spite of the potential quantum-mechanical nature of carrier dynamics in the core region of typical nanostructured devices, their overall behavior is often the result of a non-trivial interplay between phase coherence and energy relaxation/dephasing [2]. It follows that, in general, a proper treatment of such novel nanoscale devices requires a theoretical modelling able to properly account for both coherent and incoherent —i.e., phase-breaking— processes on the same footing within a many-body picture.

More precisely, the idealized behavior of a so-called "quantum device" [3] is usually described via the elementary physical picture of the square-well potential and/or in terms of a simple quantum-mechanical n-level system. For a quantitative investigation of state-of-the-art quantum optoelectronic devices, however, two features strongly influence and modify such simplified scenario: (i) the intrinsic many-body nature of the carrier system under investigation, and (ii) the potential coupling of the electronic subsystem of interest with a variety of interaction mechanisms, including the presence of spatial boundaries [4].

The wide family of so-called quantum devices can be divided into two main classes: (i) semiconductor devices characterized by a genuine quantum-mechanical behavior of their carrier subsystem, and (ii) low-dimensional nanostructures whose transport dynamics may be safely treated within the semiclassical picture.

Devices in the first class are natural candidates for the implementation of quantum information/computation processing, thanks to the weak coupling of the carrier subsystem with the host material. These include, in particular, semiconductor quantum-dot structures, for which all-optical implementations have been recently proposed [5]. In this case, the typical quantum-mechanical carrier dynamics is only weakly disturbed by decoherence processes; therefore, the latter are usually described in terms of extremely simplified models.

In contrast, the second class of quantum devices —in spite of their partially discrete energy spectrum due to spatial carrier confinement— exhibits a carrier dynamics which can be still described via a semiclassical scattering picture. Such optoelectronic quantum devices include multi-quantum-well (MQW) and superlattice structures, like quantum-cascade lasers (QCLs) [6]. In this case, we deal with a strong interplay between intra-subband energy relaxation/thermalization and inter-subband hot-carrier redistribution; it follows that for a quantitative description of such non-trivial non-equilibrium scenario the latter need to be treated via microscopic and fully three-dimensional (3D) models.

In this paper we shall primarily focus on this second class of quantum devices, providing a comprehensive microscopic theory of charge transport in semiconductor nanostructures.

THEORETICAL APPROACH AND SIMULATION STRATEGY

As anticipated, semiconductor-based quantum devices are usually modelled by analogy with $n-$level atomic systems. Their theoretical description is thus often grounded on purely macroscopic models: The various carrier concentrations within each subband are the only relevant quantities, and their time evolution is described by a set of rate equations, in terms of phenomenological injection/loss contributions and scattering rates. As pointed out in [7], this rate-equation scheme can only operate as an *a posteriori* fitting procedure, rather than a quantitative and predictive tool. Indeed, contrary to the behaviour of a multi-level system, the electron dynamics in these quantum devices occurs within a multi-subband structure, and the existence of transverse, *i.e.*, in-plane, degrees of freedom should be properly taken into account. A rigorous treatment of the 3D nature of the problem is therefore imperative.

A 3D description within a kinetic Boltzmann-like treatment has been originally proposed in [7], then generalized in [8] and further extended to include quantum effects in [9]. In this scheme, the fundamental ingredient is the carrier distribution function f_α, α being a global quantum number labelling single-particle states. To be more specific, we focus on a prototypical quantum-cascade laser design. Its core region is made up of a periodic repetition of several identical stages, each consisting of a MQW structure in which an active region and an electron-injecting and -collecting part may be identified. In this scheme, the label α comprises the stage index λ, the subband index ν within each stage, and the in-plane wavevector $\mathbf{k}$.

The time evolution of the carrier distribution function f_α is governed by the following set of kinetic equations:

$$\frac{d}{dt}f_\alpha = \sum_{\alpha'}\left[P_{\alpha,\alpha'}f_{\alpha'} - P_{\alpha',\alpha}f_\alpha\right] . \qquad (1)$$

Here, $P_{\alpha',\alpha}$ is the scattering probability for a process connecting the state with in-plane wavevector $\mathbf{k}$ in the νth subband of the λth stage, to the state $\mathbf{k}'\nu'$ of the λ'th one.

Although eq. (1) corresponds to the more complete description of the problem and allows to kinetically study electron dynamics within the full core region of the device, semi-phenomenological models have been also proposed [7]. They can still provide valuable information on the microscopic carrier dynamics, with the advantage of being much less computer-time consuming. This is valid, in particular, when the region of interest corresponds to a portion of the whole structure, like, *e.g.*, the bare active region of each stage of the device. Limiting the kinetic treatment to a subset $\bar\alpha$ of the whole multisubband structure, the kinetic equation in (1) can be rewritten as

$$\frac{d}{dt}f_{\bar\alpha} = [g_{\bar\alpha} - \Gamma_{\bar\alpha} f_{\bar\alpha}]_{i/l} + \sum_{\bar\alpha'}\left[P_{\bar\alpha,\bar\alpha'}f_{\bar\alpha'} - P_{\bar\alpha',\bar\alpha}f_{\bar\alpha}\right] . \qquad (2)$$

Here, the first two terms describe injection and loss of carriers with wavevector $\mathbf{k}$ in miniband ν, while the last one accounts for in- and out-scattering processes between states $\mathbf{k}\nu$ and $\mathbf{k}'\nu'$. Contrary to a fully macroscopic model, this formulation provides a description of intra- and inter-miniband scattering, in the $\bar\alpha$ subset, in terms of microscopic ingredients only. However, the injection/loss contributions in eq. (2), coupling the active region with the injector/collector, are treated on a partially phenomenological level: The g and Γ functions are defined within the same kinetic picture $\mathbf{k}\nu$ and are adjusted to reproduce the experimentally measured current density across the device.

The kinetic description proposed in eq. (2) came out to be quite useful to address the microscopic nature of the hot-carrier relaxation within a portion of the structure, like, *e.g.*, the device active region [7]. However, due to the presence of free-parameters coupling this region of interest to the rest of the device, it does not allow to address the nature of the physical mechanisms governing charge transport through injector/active-region/collector interfaces. To this end, the partially phenomenological model has to be replaced by the fully microscopic description of the whole MQW core structure in eq. (1).

The periodicity of the real structure allows one to assume $f_{\lambda'\nu\mathbf{k}} = f_{\lambda\nu\mathbf{k}}$, that is the carrier distribution function does not depend on the stage index. Under this assumption, eq. (1) can then be rewritten as:

$$\begin{aligned}\frac{d}{dt}f_{\lambda\nu\mathbf{k}} =\ & \sum_{\nu'\mathbf{k}'}\left[P_{\lambda\nu\mathbf{k},\lambda\nu'\mathbf{k}'}f_{\lambda\nu'\mathbf{k}'} - P_{\lambda\nu'\mathbf{k}',\lambda\nu\mathbf{k}}f_{\lambda\nu\mathbf{k}}\right] \\ & + \sum_{\nu'\mathbf{k}';\lambda'\neq\lambda} P_{\lambda\nu\mathbf{k},\lambda'\nu'\mathbf{k}'}f_{\lambda\nu'\mathbf{k}'} \\ & - \sum_{\nu'\mathbf{k}';\lambda'\neq\lambda} P_{\lambda'\nu'\mathbf{k}',\lambda\nu\mathbf{k}}f_{\lambda\nu\mathbf{k}} . \qquad (3)\end{aligned}$$

Here, the first term accounts for all intra-stage scattering dynamics, while the second and third one correspond, respectively, to injection and loss contributions, coupling this stage with the neighbouring ones. They are the microscopic equivalent of the g and Γ functions of the partially phenomenological model in eq. (2). As we can see, by imposing the periodic boundary conditions previously mentioned, we are able to "close the circuit" without resorting to phenomenological parameters. As for the $P_{\lambda'\nu'\mathbf{k}',\lambda\nu\mathbf{k}}$, it is usually sufficient to limit the interstage ($\lambda \neq \lambda'$) scattering to just nearest-neighbour coupling ($\lambda' = \lambda \pm 1$).

In view of their Boltzmann-like structures, Eqs. (1) and (3) as well as (2) can be "sampled" by means of a proper Monte Carlo (MC) simulation scheme [10]. Indeed, the latter has proven to be a very powerful technique allowing the inclusion, at a kinetic level, of a large variety of scattering mechanisms.

More specifically, benefitting from the translational symmetry previously discussed, we can simulate carrier transport over the central —$\lambda = 0$— stage only.

Every time a carrier in state $\nu,\mathbf{k}$ undergoes an interstage scattering process (i.e., $0,\nu,\mathbf{k} \to \pm 1, \nu', \mathbf{k}'$), it is properly re-injected into the central region ($0, \nu, \mathbf{k} \to 0, \nu', \mathbf{k}'$) and the corresponding electron charge $\pm e$ will contribute to the current through the device. This charge-conserving scheme allows for a purely microscopic —i.e, free of phenomenological input parameters— evaluation of the device performances such as the gain spectrum or the current-voltage characteristics. The current density j across the whole structure, for example, can be obtained as a pure output of the simulation, just by a proper sampling of the in- and out-stage scattering processes:

$$j \propto \sum_{\nu\nu'\mathbf{k}\mathbf{k}'} \left[\sum_{\lambda'>\lambda} P_{\lambda'\nu'\mathbf{k}',\lambda\nu\mathbf{k}} f_{\lambda\nu\mathbf{k}} - \sum_{\lambda'<\lambda} P_{\lambda'\nu'\mathbf{k}',\lambda\nu\mathbf{k}} f_{\lambda\nu\mathbf{k}} \right]. \quad (4)$$

In the following Section we shall present and discuss simulated experiments based on the semiclassical-transport approach here reviewed. In order to properly describe the complex interplay between carrier thermalization and energy-relaxation/dephasing in QCL structures, we have included in our simulation all relevant interaction mechanisms in a fully 3D fashion: intra- and intersubband carrier-phonon (c-p) as well as carrier-carrier (c-c) scattering processes.

TOWARDS THZ SOURCES

The parameter-free simulation scheme reviewed in the previous Section has been successfully applied to the modelling of electron dynamics in several state-of-the-art mid-infrared QCL structures, allowing for a detailed investigation of the main physical mechanisms responsible for their operation. For this reason, it appeared a reliable tool for the characterization and optimization of novel designs. One of the main issues in this respect regarded the quest for a QCL emitting in the THz region of the electromagnetic spectrum.

In principle, QCLs can be designed to emit at any wavelength over an extremely wide range, using the same combination of materials and only varying the heterostructure design. However, the translation of the cascade scheme into devices operating at photon energies below the longitudinal-optical-phonon threshold of the host material ($\hbar\omega_{LO} \simeq 36$ meV in GaAs) is definitely not straightforward. In this configuration, in fact, the main difficulty in achieving a population inversion regime in the active region, is that optical-phonon emission cannot be used to selectively depopulate the MQW excited states, since it acts with equal efficiency on both the upper and the lower laser subbands. On the contrary, the complex and non-intuitive interplay between various competing non-radiative relaxation channels has to

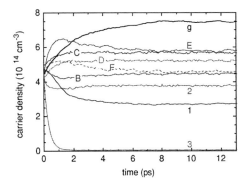

FIGURE 1. Time evolution of the simulated carrier densities in the various subbands of the MQW structure of Ref. [12], in the presence of both c-p and c-c scattering mechanisms.

be taken into account to properly evaluate the performances of new designs and proposals. For this reason, although electroluminescence in the THz region of the electromagnetic spectrum has been detected from a variety of QC structures in the last years, lasing has not been reported until very recently [11].

Starting from the above considerations, we have recently proposed a prototypical design for a QCL emitting at $h\nu \sim 18$ meV, corresponding to a frequency of ~ 4.4 THz [12]. It is based on a vertical-transition configuration, which is known to lead to larger dipole matrix elements and narrower linewidths, and employs a conventional chirped-superlattice design. In order to minimize the density of electrons in the lower laser subband, 1, we employ a dense miniband, with seven subbands (g,B,C,D,E,F,1, in order of increasing energy). The latter provides a large phase space where electrons scattered either from the upper laser subband, 2, or directly from the injector can spread.

Figure 1 shows the results of our simulated experiments as far as the population of each subband is concerned. In our "charge conserving" scheme, we start the simulation by assuming the total number of carriers to be equally distributed among the different subbands; the electron distribution functions then evolve according to Eq. (3) and a steady-state condition is eventually reached, leading to the desired $2 \to 1$ population inversion. Such effect is due to the non-trivial interplay between the c-p and c-c scattering channels, since a qualitatively different scenario is obtained when only c-p interaction is taken into account. In the latter case, electrons would accumulate in subband 1, thus preventing from a population-inversion regime to be established [12].

To gain a better insight into this synergic co-operation, it is useful to examine the distribution function of the

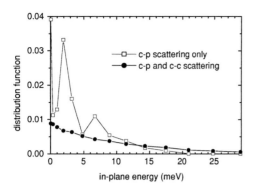
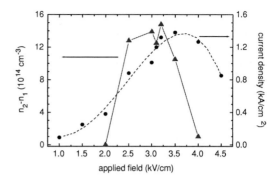

FIGURE 2. Electron distribution function vs in-plane energy, for carriers in subband 1 of the design proposed in Ref. [12].

FIGURE 3. Simulated current density (discs) and population inversion (triangles) vs applied field, for the design proposed in Ref. [12].

electrons in subband 1. This is reported in Fig. 2 as a function of the in-plane energy. When c-p scattering is the only active mechanisms, electrons relax down in the subband and are mainly localized within less than 15 meV from its bottom. Depletion of the lower laser states due to photon emission into the injector miniband, whose width amounts to ~18 meV, is therefore highly inhibited. Carrier-carrier interaction, on its side, acts as a thermalization mechanism within the subband. It spreads the distribution function over a wider energy range, therefore providing electrons with the necessary in-plane momentum to further relax, by phonon emission, out of the lower laser subband.

The miniband dispersion of this design is chosen as large as possible compatibly with the need of avoiding cross absorption. This suppresses thermal backfilling, and provides a large operating range of currents and voltages. This is evident from Fig. 3, where the simulated current-voltage characteristics is reported. The trend of the curve shows features which are typical of these devices: A high-resistivity nonlinear behaviour at low fields, due to level misalignment, an intermediate linear region, in which high mobilities are achieved, and a negative-differential resistance regime when the applied fields are so high to suppress resonant tunneling. Also shown in Fig. 3 is the significant population inversion that this structure is predicted to exhibit throughout the whole linear range of the current-voltage characteristics. These features made this prototypical design a promising candidate for the realization of the first THz QCL. Indeed, lasing action was demonstrated shortly after its proposal [11].

As final remark, we stress that the proposed theoretical approach can be easily extended to describe —in addition to the nonequilibrium carrier dynamics— the non trivial carrier-light coupling of the electron subsystem with the cavity modes of the lasing device. Indeed, within the standard particle-like picture of the MC technique one should be able to simulate the number of photons as well. Work in this direction is in progress.

ACKNOWLEDGMENTS

We are grateful to Rüdeger Köhler and Alessandro Tredicucci for their relevant contribution to some of the work presented in this review.

REFERENCES

1. See, e.g., *Hot Carriers in Semiconductor Nanostructures: Physics and Applications*, edited by J. Shah, Academic Press inc., Boston (1992); *Theory of Transport Properties of Semiconductor Nanostructures*, edited by E. Schöll, Chapman and Hall, London (1998).
2. See, e.g., F. Rossi and T. Kuhn, Rev. Mod. Phys. **74**, 895 (2002) and references therein.
3. See, e.g., *Physics of Quantum Electron Devices*, edited by F. Capasso, Springer, Berlin (1990) and references therein.
4. See, e.g., F. Rossi, A. Di Carlo, and P. Lugli, Phys. Rev. Lett. **80**, 3348 (1998) and references therein.
5. See, e.g., E. Biolatti, R.C. Iotti, P. Zanardi, and F. Rossi, Phys. Rev. Lett. **85**, 5647 (2000) and references therein.
6. See, e.g., C. Gmachl, F. Capasso, D.L. Sivco, and A.Y. Cho, Rep. Prog. Phys. **64**, 1533 (2001) and references therein.
7. R.C. Iotti and F. Rossi, Appl. Phys. Lett. **76**, 2265 (2000).
8. R.C. Iotti and F. Rossi, Appl. Phys. Lett. **78**, 2902 (2001).
9. R.C. Iotti and F. Rossi, Phys. Rev. Lett. **87**, 146603 (2001).
10. C. Jacoboni and P. Lugli, *The Monte Carlo method for semiconductor device simulation*, Springer, Wien (1989).
11. R. Köhler *et al.*, Nature **417**, 156 (2002).
12. R. Köhler *et al.*, Appl. Phys. Lett. **79**, 3920 (2001).

Terahertz quantum cascade laser emitting at 160 μm in strong magnetic field

Giacomo Scalari*, Stepháne Blaser*, Lorenzo Sirigu*, Marcel Graf*, Lassaad Ajili*, Jerôme Faist*, Harvey Beere[†], Edmund Linfield[†], David Ritchie[†] and Giles Davies[**]

*Institute of Physics, University of Neuchâtel, CH-2000 Neuchâtel, Switzerland
[†]Cavendish Laboratory, University of Cambridge, Madingley Road, Cambridge CB3 0HE, UK
[**]School of Electronic and Electrical Engineering, University of Leeds, Leeds LS2 9JT, UK

Abstract. Intersubband transitions and magnetic quantization are exploited to realize an extremely low threshold quantum cascade laser emitting at $\lambda \simeq 160$ μm. Non-radiative Auger processes are efficiently quenched when the system reaches the condition of strong in-plane confinement, allowing laser action with threshold current densities as low as 1 A/cm^2 at 4.2 K at 13 T and pulsed operating temperatures of 65 K. Characteristic features in light emission and transport are recognized as signatures of single and two-electron scattering processes between Landau levels.

INTRODUCTION

The proposal of modifying, by appropriate dimensional confinement, the density of states of the active medium of a semiconductor laser in order to get better radiative efficiency dates the early 80's [1]. In parallel, the possibility to tailor the electronic properties with the use of the band structure engineering [2] led to the development, from the early proposal of Kazarinov and Suris [3], of the intersubband unipolar quantum cascade (QC) laser [4]. In the QC laser, one of the main limiting factor and the principal cause of the high threshold current density is the longitudinal optical (LO) phonon scattering that is very efficient. An intersubband system, where the carriers do not need to relax to reach the gap to recombine emitting photon like an interband laser, seems to be an ideal candidate to benefit from 3D confinement. Such a confinement can be produced by applying a magnetic field perpendicularly to the layers of the heterostructure breaking the in-plane parabolic dispersion of the electrons in discrete Landau levels. The possibility to control, by means of a magnetic field, the LO phonon scattering in QC lasers has been recently demonstrated [5]. The extension of the QC lasers to the low frequency region below the LO phonon energy [6], has in fact made possibile the realization of three-dimensional strongly confined systems and recent studies of THz QC lasers in a perpendicular magnetic field [7, 8, 9] have shown significant reduction of the lasing threshold in function of the applied perpendicular B field. Following those results, a QC structure that benefits from a 3D carrier confinement arising from Landau quantization has been conceived [10] for the operation at very low photon energies where population inversion is difficult to achieve.

EXPERIMENTAL RESULTS

Here we present a laser structure emitting at 1.9 THz ($\lambda \simeq 160$ μm, $h\nu$=7.9 meV) where the condition of strong confinement in the three dimensions is reached for magnetic fields above 7.4 T. The design strategy follows the guidelines exposed in [10]; the active region is formed by a large (55 nm) single quantum well and the electrons are injected in the second excited state. Then they can radiatively decay on the first excited state ($\Delta E_{32} = 7.9$ meV) emitting a photon and relax in a five-quantum-well injector. The energy spacing E_{21} is tuned to the value of $\Delta E_{21} = 5$ meV to show a inter-Landau level resonance at 2.9 T that shortens the lower state lifetime and enhances population inversion, because for this value of the field the cyclotron energy equals the energy spacing E_{21}. The fan plot of the active region involves the three levels of the single quantum well(Fig.1 (a)), and shows that the first excited Landau level of the ground state $|1,1\rangle$ crosses the state $|3,0\rangle$ for a magnetic field value of 7.4 T ($E_{cycl}(7.4T) = 12.9$ meV $= \Delta E_{31}$); starting from this field value the structure enters a region where the energy of the in-plane confinement due to magnetic field is higher than the overall energy spacing in the active region $E_{31} = 12.9$ meV. The structure has been grown by MBE in the Al$_{0.15}$Ga$_{0.85}$As/GaAs material system on a semi-insulating substrate in order to exploit the

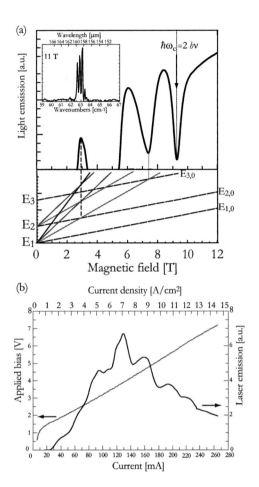

FIGURE 1. (a) Laser emission as a function of magnetic field at a 4 V bias. Note the resonances at 7.4 T ($\hbar\omega_c = E_{31}$) and 9.3 T ($\hbar\omega_c = 2h\nu$). Inset: typical spectrum of a 4.7 mm long, 420 μm device at 11 T. (b) Pulsed LI curve of a device at 13.2 T and 4.2 K temperature.

single plasmon waveguide concept [6], then processed in ridge stripes (100-500 μm wide, 0.5-4.7 mm long) and inserted in the bore of a superconducting cryostat [10]. Measurements of laser emission in function of magnetic field are reported in Fig.1: the laser turns on, as expected, at 2.9 T, then again after 5.5 T and performs a series of oscillations. It is easy to identify the light minimum at 7.4 T as due to the lifetime reduction of the upper lasing state [7, 8, 9] due to the resonance $|3,0\rangle \longrightarrow |1,1\rangle$. The other minimum at 9.3 T is a signature of a 2 electron scattering event that has been calculated theoretically [11] and observed only in magnetotransport at very high magnetic field values. Threshold current oscillates as light emission but with opposite phase, and the very low values reached at high fields ($J_{thresh} < 10$ A/cm^2 for $B > 6T$) prove the efficiency of the magnetic confinement in quenching non-radiative processes. After 9.3 T the laser threshold keeps decreasing with applied magnetic field and reaches values lower than 1 A/cm^2 above 13 T. Measurements of laser emission in function of the temperature have been performed, yielding a maximum pulsed operating temperature of 65 K at a magnetic field value of 11 T.

CONCLUSIONS

The broad current dynamic range (see Fig. 1 (b)) and the extremely reduced value of the threshold current of the device show the potentiality of the 3D confinement for the realization of an extremely efficient laser, that can be further improved by high quality factor optical resonators and double metal waveguides [12]. The extension of the same concept at longer wavelengths seems to be one viable way towards low frequency-low threshold terahertz lasers based on 3D carrier confinement, achievable also by nanofabrication technologies.

REFERENCES

1. Arakawa, Y., and Sakaki, H., *Appl. Phys. Lett.*, **40**, 939–941 (1982).
2. Capasso, F., *Science*, **235**, 172–176 (1987).
3. Kazarinov, R., and Suris, R., *Sov. Phys. Semicond.*, **5**, 707–709 (1971).
4. Faist, J., Capasso, F., Sivco, D., Sirtori, C., Hutchinson, A., and Cho, A., *Science*, **264**, 553–556 (1994).
5. Becker, C., Sirtori, C., Drachenko, O., Rylkov, V., Smirnov, D., and Leotin, J., *Appl. Phys. Lett.*, **81**, 2941–2943 (2002).
6. Köhler, R., Tredicucci, A., Beltram, F., Beere, H., Linfield, E., Davies, A., Ritchie, D., Iotti, R., and Rossi, F., *Nature*, **417**, 156–159 (2002).
7. Scalari, G., Blaser, S., Rochat, M., Ajili, L., Willenberg, H., Hofstetter, D., Faist, J., Beere, H., Davies, A., Linfield, E., and Ritchie, D., in *Proceedings of the 26th International Conference on the Physics of Semiconductors*, 2002, CD.
8. Alton, J., Barbieri, S., Fowler, J., Beere, H., Muscat, J., Linfield, E., Ritchie, D., Davies, G., Köhler, R., and Tredicucci, A., *Phys. Rev. B*, **68**, 081303-1–081303-4 (2003).
9. Tamosiunas, V., Zobl, R., Ulrich, J., Unterrainer, K., Colombelli, R., Gmachl, C., West, K., Pfeiffer, L., and Capasso, F., *Appl. Phys. Lett.*, **83**, 3873–3875 (2003).
10. Scalari, G., Blaser, S., Ajili, L., Faist, J., Beere, H., Linfield, E., Ritchie, D., and Davies, G., *Appl. Phys. Lett.*, **83**, 3453–3455 (2003).
11. Kempa, K., Zhou, Y., Engelbrecht, J., Bakshi, P., Ha, H., Moser, J., Naughton, M., Ulrich, J., Strasser, G., Gornik, E., and Unterrainer, K., *Phys. Rev. Lett.*, **88**, 226803-1–226803-4 (2002).
12. Unterrainer, K., Colombelli, R., Gmachl, C., Capasso, F., Hwang, H., Sergent, A., Sivco, D., and Cho, A., *Appl. Phys. Lett.*, **80**, 3060–3062 (2002).

Optical gain spectra due to a one-dimensional electron-hole plasma in quantum-wire lasers

Yuhei Hayamizu[*], Masahiro Yoshita[*], Hirotake Itoh[*], Hidefumi Akiyama[*], Loren N. Pfeiffer[†] and Ken W. West[†]

[*]Institute for Solid State Physics, University of Tokyo, CREST, JST, 5-1-5 Kashiwanoha, Kashiwa, Chiba, 277-8581, JAPAN
[†]Bell Laboratories, Lucent Technologies, 600 Mountain Avenue, Murray Hill, NJ 07974, USA

Abstract. We report the first measurement of evolutional optical gain spectra for various electron-hole densities in T-shaped quantum-wire lasers with high structural uniformity at 5 K. At the high excitation power, a broad absorption peak and a sharp gain peak are observed instead of discrete excitonic absorption peaks. It indicates that an electron–hole plasma yield the gain of quantum-wire lasers. Additionally, spectral shape of the gain peak is symmetric, and shows no similarity with one-dimensional (1D) density of states. It suggests influence of strong Coulomb correlations in a quantum wire.

Quantum-wire lasers are expected to show high performance due to sharply peaked gain spectra based on the one-dimensional (1D) density of states (DOS) of non-interacting electrons and holes. However, there has been no experimental evidence for such gain spectra. Furthermore, recent several theoretical and experimental reports [1] suggested that strong 1D Coulomb interactions among carriers influence gain spectra and lasing mechanisms. In this study, we present the first measurement of gain spectra in T-shaped quantum wire (T-wire) lasers with high structural uniformity at 5 K by means of Cassidy's method [2], and experimentally confirm the effect of strong Coulomb correlations in the quantum-wire by the obtained gain spectra.

T-wire lasers were fabricated by the cleaved edge overgrowth method with molecular beam epitaxy. We study two kinds of T-wire-laser structures, a 20-wire laser and a single-wire laser, where each quantum wire is formed at a T-intersection of a 6 nm (110) GaAs quantum well (arm well) and a 14 nm (001) $Al_{0.07}Ga_{0.93}As$ quantum well (stem well) [3]. The 20 wires or the single wire are embedded in the center of a T-shaped optical waveguide. Their cavity length was 500 μm and the laser cavity mirrors were left uncoated. As shown in the inset of Fig. 1 (b), excitation light from a cw Titanium Sapphire laser was focused into a filament shape by a cylindrical lens and an objective lens from the top (110) surface to pump the whole cavity uniformly. The excitation light with the energy of 1.631 eV is absorbed in the wire and the arm well, and the excited carriers in the arm well flow into the wire. Spontaneous emission emitted perpendicularly to the waveguide was collected by the same objective

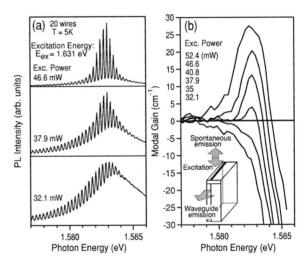

FIGURE 1. Waveguide-emission spectra (a) and modal gain spectra obtained by using the Cassidy's method (b) from a 20-wire laser for various excitation powers. The 1.631 eV excitation light mainly excites the arm well. The inset shows a schematic of the experimental arrangement.

lens and detected via a spectrometer. Waveguide emission which came out through one of the cavity mirrors and had polarization parallel to the arm well was detected via another objective lens, a polarizer, and a spectrometer with high spectral resolution of 0.01 meV.

Figure 1 (a) shows waveguide-emission spectra of the 20-wire laser for various excitation powers. Intensity oscillations in the waveguide-emission spectra depend on photon energy and excitation powers. On the basis

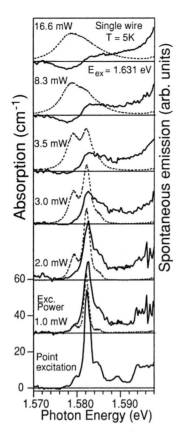

FIGURE 2. Absorption (solid) and spontaneous emission (dotted) spectra of a single quantum-wire laser for various excitation powers.

of Cassidy's method [2], we derived energy-dependent gain coefficients from analysis of peak-to-valley ratios in the intensity oscillations. Figure 1 (b) shows the gain spectra obtained from the waveguide-emission spectra in Fig. 1 (a). Below the excitation power of 37.9 mW, the wires show no gain, but strong absorption in all energies. At 37.9 mW, a gain peak appears at 1.5825 eV with line width of 0.5 meV. As the excitation power increases up to 52.4 mW, the gain peaks become larger and broader. The observed shapes of the gain peaks are symmetric or have a tail on the low energy side, and never show the shape of 1D DOS. This suggests that gain spectra of high-density electrons and holes are strongly modified from the 1D DOS of non-interacting electrons and holes due to strong Coulomb correlations in the wire.

To reveal spectral evolution of gain at high electron-hole densities from absorption at low density, we measured gain/absorption spectra of the single-wire laser. Solid curves in Fig. 2 show the gain/absorption spectra for various excitation powers, where positive and negative values mean absorption and gain, respectively. The spectrum denoted as "point excitation" was measured by focusing the excitation light into a 1 μm spot, and hence represents an absorption spectrum for zero excitation power. It agrees very well with photoluminescence (PL) excitation spectra measured and analyzed previously [4]. A large peak at 1.582 eV is due to the ground-state exciton, and a small peak at 1.589 eV is to the first-excited-state exciton. A step at 1.594 eV corresponds to an absorption onset by higher exciton states and 1D continuum states.

As the excitation power is increased, the exciton peaks and the step become broader in width and weaker in peak intensity without energy shifts. Instead of the peak suppression, absorption in the energy regions between the peaks and the step is increased. Note that, at 3.0-3.5 mW excitation powers, a continuous absorption band is formed, which has an onset with a shoulder evolved from the exciton peak at the same energy. At high excitation powers (8.3-16.6 mW), a gain peak appears at 1.578 eV, which is analogous to the gain spectra observed for the 20-wire laser shown in Fig. 1 (b). Continuous absorption is observed on the higher energy side of the gain region. We believe that the observed spectra demonstrate formation of an electron-hole plasma. The two regions of gain and absorption most likely reflect Fermi filling of electrons and holes in their band edge, and the energy that separates the gain and absorption regions indicates the Fermi edge. The position of Fermi edge seems to show no shift, but stay at the same energy with the original energy position of the exciton, though we do not know the reason. This might relate to strong excitonic correlations still remaining in the plasma.

The dotted curves in Fig. 2 shows spontaneous emission spectra measured perpendicularly to the waveguide. They agree very well to the results of our previous micro-PL study [5]. At low excitation powers, a single exciton peak is observed at 1.582 eV. As the excitation power is increased, a biexciton peak emerges 3 meV below the exciton peak. Then, the biexciton peak gradually changes to an electron-hole plasma peak, showing significant broadening without any peak shift, while the exciton peak is quenched. Comparison of the solid and dotted curves shows that the significant broadening in PL and the creation of gain occur simultaneously with the formation of an electron-hole plasma.

REFERENCES

1. D. W. Wang *et al.*, *Phys. Rev. B.* **64**, 195313 (2001), and references therein.
2. D. T. Cassidy, *J. Appl. Phys.* **56**, 3096 (1984).
3. Y. Hayamizu *et al.*, *Appl. Phys. Lett.* **81**, 4937 (2002).
   H. Akiyama *et al.*, *Phys. Rev. B* **67**, 041302 (2003).
4. H. Itoh *et al.*, *Appl. Phys. Lett.* **83**, 2043 (2003).
   H. Akiyama *et al.*, *Appl. Phys. Lett.* **82**, 379 (2003).
5. M. Yoshita *et al.*, cond-mat/0402526 (2004).

Chemical and Biological Sensing Based on the Surface Photovoltage Measurement of the Si Surface Potential Barrier.

K.Nauka, Zhiyong Li, T.I.Kamins

Hewlett Packard Laboratories
Palo Alto, CA 94304, USA

Abstract. Chemical and biological species deposited on the crystalline silicon surface created surface barrier changes that were detected using the non-contact Surface Photovoltage technique. The magnitude of the surface barrier modifications provided a signature allowing quantification of the sensed species. The simplicity and sensitivity of this technique offers an exciting opportunity for a new type of low cost sensing devices.

INTRODUCTION

Recently, there has been an increased need for chemical and biological sensors. Particularly, there is a need for devices satisfying the following requirements: low manufacturing cost, device design allowing easy integration into an electronic control platform, sensitivity, ability to detect multiple species, speed, and reusability. The silicon surface has been known as a promising binding platform for a variety of chemical and biological compounds [1,2]. Silicon surface barrier measurement has been previously proposed for sensing small quantities of selected gases [3,4]. This work expands the technique further to aqueous solutions of a variety of inorganic chemicals, selected organic compounds, and biological DNA species.

EXPERIMENT AND RESULTS

High quality 150 mm diameter, p-type silicon wafers with resistivities of $1 \div 20$ Ω cm were used in the experiments. Wafers were either (001) or (111) oriented; some wafers were intentionally misoriented by 4° from the (111) plane towards [011]. In order to provide uniform surface termination before exposure to the investigated chemical and biological species, wafer surface was reconstructed in H_2 at 1100°C and subsequently cooled in H_2 to a temperature below 400°C using a load locked, lamp heated reactor. Previous experiments demonstrated that this heat treatment provided uniform hydrogen coverage [5,6].

In order to investigate the effect of chemical species wafers were dipped at room temperature into aqueous solutions of various inorganic acids and bases. Efficacy of the measurement was demonstrated by varying chemicals concentration (pH = $1 \div 6$ for acids and pH = $8 \div 13$ for bases) and exposure time. Similarly, the effect of the ionic species concentration (same pH, various chemicals) was investigated. Silicon wafers were also exposed to a few, selected organic compounds (alkenes – pentane, hexane, and octane). Silicon surfaces in contact with hot alkenes underwent thermal hydrosilylation providing at least partial monolayer termination [7].

Immediately after chemical exposure the surface potential barrier was determined by non-contact Surface Photovoltage (SPV) measurements [5,6]. Occasionally, the surface was blown off with nitrogen after chemical exposure but before the measurement. The SPV signal was obtained by illuminating the silicon surface with monochromatic light having photon energy below the silicon bandgap, and it was measured capacitively using a transparent conducting electrode placed above the silicon surface.

SPV sensing was also used in genomic experiments. In this application single strand, 12-group long DNA probes was anchored to a functionalized silicon surface [8]. Then, the barrier height was compared with the corresponding result obtained for the surfaces where DNA strands were hybridized using matched and one-base mismatched DNA chains.

The measured SPV signal corresponds to the difference between the surface potential barriers at the dark back surface and the front illuminated surface of silicon. At high photon flux, the illuminated surface barrier becomes negligible, and the SPV signal reaches a maximum corresponding to the equilibrium surface barrier height. This potential barrier is determined by the charges residing on the silicon surface or in its vicinity. Most chemical and biological species are capable of modifying the surface barrier, either by charge transfer while forming chemical bonds with the silicon or silicon oxide surface (chemisorption), or by electrostatic interactions (physisorption) with the surface states.

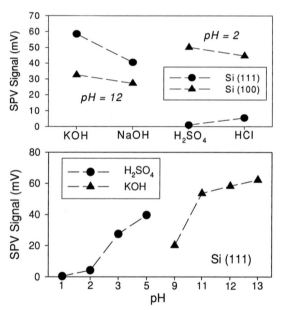

FIGURE 1. Effect of the type and concentration of inorganic species on the SPV signal.

The change of the silicon surface barrier is related to the type of chemical species and its concentration (Figure 1). It also depends on the silicon crystallographic orientation, the original surface roughness, and the cleaning that preceded the experiment. The surface barrier modification usually occurred within the first 10 – 20 seconds of chemical exposure, reaching steady-state after this period. Well-defined, reproducible modification of the SPV signal was observed for both acids and bases over the pH ranges tested. Similar results were obtained for the organic species. Figure 2 shows that the SPV signal can completely recover after sequential exposures to inorganic acids and bases.

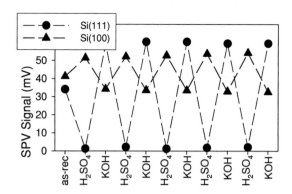

FIGURE 2. Change of the SPV signal after sequential acid and base treatments.

Figure 3 demonstrates application of SPV sensing in biological applications. Clearly distinguishable and reproducible changes of the surface potential barrier were observed when single oligonucleotide strands were attached to the surface, and then they were hybridized with either matched or single base mismatched DNA strands.

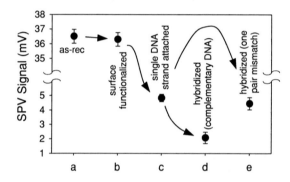

FIGURE 3. Surface barrier modification observed during the genomic experiment.

REFERENCES

1. Askenasy, G., Cahen, D., Cohen, R., Shanzer, A., Vilan, R., *Acc.Chem.Res.***35**, 121 (2002.
2. Buriak, J.M., *Chem.Rev.***102**, 1271 (2002).
3. Nicolas, D. Souteyrand, E., Martin, J.R., *Sens.Acc.***B44**, 507 (1997).
4. Adamowicz, B., Miczek, M., Brun, C., Gruzza, B., Hasegawa, H., *Thin Sol.Films* **436**, 101 (2003).
5. Nauka, K., Kamins, T.I., *J.Electrochem.Soc.***146**, 292 (1999).
6. Kamins, T.I., Nauka, K., Williams, R.S., *Appl.Phys.***A73**, 1 (2001).
7. Sung, M.M., Kluht, J., Yauw, O.W., *Langmuir* **13**, 6164 (1997).
8. Li, Z., Chen, Y., Li, X., Kamins, T.I., Nauka, K., Williams, R.S., *NanoLett.***4**, 245 (2004).

Towards a High Diffraction Efficiency of Photorefractive Multiple Quantum Wells

T. Z. Ward, P. Yu, S. Balasubramanian, M. Chandrasekhar, H.R. Chandrasekhar

Department of Physics and Astronomy, University of Missouri-Columbia, Columbia, Missouri 65211

Abstract. We propose a method to improve the diffraction efficiency of photorefractive multiple quantum well devices in the transverse-field geometry. Higher efficiencies have been achieved through systematic electrical modulation studies.

INTRODUCTION

Photorefractive quantum well (PRQW) devices have shown promise in laser based ultrasound detection and *in vivo* biomedical imaging [1-4]. In these applications, the device records dynamic holograms using adaptive interferometry. The device acts as a holographic beam combiner in two-wave and four-wave mixing configurations. The sensitivity of the system is determined by the minimum detectable intensity and diffraction efficiency [5]. A relatively low diffractive efficiency of the PRQW device reduces the system sensitivity. For example, in a coherent domain biomedical imaging system, a sensitivity of −80 dB has been measured for AlGaAs/GaAs PRQW devices. This gives a limited image depth in comparsion with conventional optical coherence tomography which shows a sensitivity of −110 dB, about 30 dB higher than a semiconductor based system. Although we can increase the reference intensity to gain more diffraction efficiency for the PRQW device in a wave-mixing setup, the main limitation in diffraction efficiency is the total intensity of the writing beams on the semiconductor devices. Because the devices operate under high applied electric fields and the laser intensities generate high photoconductivity, the devices experience Joule heating. The Joule heating must be kept below a safe value, else thermal run-away occurs, and the device may be destroyed.

When a laser and an electric field are applied to the PRQW device, the Joule heating is given by:

$$P_J = \frac{I_J}{h\nu} \alpha \mu \tau e E^2 H d w \quad (1)$$

where $h\nu$ is the incident photon energy, α is the absorption coefficient, $\mu\tau$ is the mobility-lifetime product of the dominant carrier, e is the electron charge, I_J is the incident intensity, and E is the applied electric field. The window dimensions included in the Joule heating are given as follows: H is the window height, w is the window width and d is the PRQW thickness. The critical heating depends on the electrical power per area. For PRQW devices fabricated with epoxy on glass, the maximum allowable electrical power density is $P_J = 100$ mW/cm^2.

Previously, higher Joule intensities could be achieved by working at a lower field or with shorter carrier lifetimes. Controlling the lifetime (and possibly the mobility) is a more practical means of increasing the Joule intensity. However, these procedures are limited by the conditions in device fabrication and only give limited improvement. To overcome the Joule heating problem and increase the diffraction efficiency, we used a low duty cycle pulsed electric field on the AlGaAs/GaAs devices.

EXPERIMENTS

We used standard techniques to measure Joule heating and differential transmission. A resistor of

1.03 MΩ was connected in series to the device. A pulsed electric field was applied across the device window. The device was illuminated by a He-Ne laser. A diode laser (HL832SE, Hitachi) was used in two-wave mixing experiments. The wavelength of the laser was tuned by temperature. The signal beam interfered with the reference beam at the PRQW with an intensity ratio of 1:3 and a fringe period of 30 μm. A Si detector (Det110, Thorlabs) with a 750nm long pass filter and a lock-in amplifier (SR510, Stanford Research) were used in two-wave mixing experiments.

RESULTS AND DISCUSSION

Figure 1 shows Joule heating versus duty cycle under different applied electric fields. The laser intensity was kept at 17 mW/cm^2. We also tested other intensities, and the results were comparable. Since the electric field was applied parallel to the quantum well layers (transverse or Franz-Keldysh geometry), the exciton broadened and changed the electroabsorption. Theoretically, if we only considered the lowest order of the electric field effect, the change in absorption should have been quadratic with the applied field. We observed a deviation from the quadratic function at an electric field of about 5.0kV/cm, which can be explained as a high order effect.

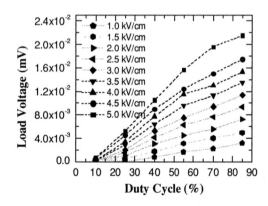

FIGURE 1. Load voltage as a function of duty cycle. Lines are for guidance.

In most wave-mixing applications of the PRQW devices, the relative intensity ratio between the signal and reference should be considered in addition to the total intensity; all should be optimized to obtain higher diffractive efficiencies. Figure 2 shows two-wave mixing as a function of duty cycle. The intensity ratio of 1:3 between the signal and reference was selected to ensure high differential transmission and two-beam coupling. As we expected, the average two-wave mixing signal decreased at lower duty cycles. However, the peak value remained nearly the same, even showing a slight increase at the lowest duty cycle. Also at higher duty cycles (about 60%), saturation of the two-wave mixing signal has been observed to correspond with Joule heating measurements. An estimate of a factor of 17 in diffraction efficiency improvement was observed by using this method when the same Joule heating value was maintained.

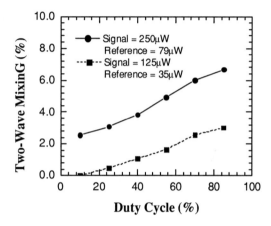

FIGURE 2. Two wave mixing as a function of duty cycle. Beam diameter 2mm each. Lines are for guidance.

In conclusion, we tested the concept of using a low duty cycle electric field to ease Joule heating constraints on PRQW devices. Increased diffraction efficiency has been observed, which allow using this method in wave-mixing applications.

REFERENCES

1. R. Jones, S. C. W. Hyde, M. J. Lynn, N .P. Barry, J. C. Dainty, P. M. W. French, K. M. Kwolek, D. D. Nolte, and M. R. Melloch, Appl. Phys. Lett. **69**, 1837-1839 (1996).
2. P. Yu, M. Mustata, P. M. W. French, J. J. Turek, M. R. Melloch, and D. D. Nolte, Appl. Phys. Lett. vol. 83, 575, (2003).
3. P. Yu, L. Peng, D. D. Nolte, and M. R. Melloch, Optics Letters, 28, 819, (2003).
4. P. Yu, L. Peng, M. Mustata, J. J. Turek, M. R. Melloch, and D. D. Nolte, Optics Letters, 29, 68, (2004).
5. D. D. Nolte, T. Cubel, L. J. Pyrak-Nolte, and M. R. Melloch, J. Opt. Soc. Am. B **18**, 195-205 (2001).

Functional Imaging Using InGaAs/GaAs Photorefractive Multiple Quantum Wells

S. Balasubramanian[1], S. Iwamoto[2], M. Chandrasekhar[1], H. R. Chandrasekhar[1], K. Kuroda[2], and P. Yu[1]

[1]*Department of Physics and Astronomy, University of Missouri-Columbia, Columbia MO 65211, USA*
[2]*Institute of Industrial Science, University of Tokyo, 4-6-1 Komaba, Meguro-ku, Tokyo, Japan*

Abstract. We propose the use of an InGaAs/GaAs photorefractive quantum well (PRQW) as an adaptive beam combiner for holographic optical coherence imaging applications. Holograms have been observed by using a diode laser and an interferometer. A weaker quantum confined exciton leads to the saturation of electroabsorption and hence diffraction, under a high external electric field, in the InGaAs PRQW. A careful choice of external electric field modulation seems to reduce this effect. We examine several characteristics that govern the use of an InGaAs PRQW in a functional imaging system.

INTRODUCTION

Holographic optical coherence imaging (OCI) can record full-frame depth-resolved without computed tomography, allowing real-time video display [1]. Holographic OCI based on AlGaAs/GaAs PRQWs has been demonstrated for imaging through tissue in a degenerate four-wave mixing configuration [1,2]. Although AlGaAs PRQWs can also operate in a non-degenerate FWM configuration [3], it is difficult to use them for functional imaging. The photon energy of the writing beams, being necessarily above the band-edge of AlGaAs, is in a range where light suffers strong scattering in biological tissue. InGaAs based multiple quantum wells have been shown to possess good photorefractive properties at 1064 nm [4]. It is therefore of interest to study the use of an InGaAs/GaAs PRQW in a coherence imaging system. This will enable one to write holograms at a longer wavelength than AlGaAs thereby reducing the scatter from tissue.

FUNCTIONAL IMAGING SETUP

Our functional imaging system consists of a low-coherence length light source (a broadband, ~50 nm, Light Emitting Diode) that writes the hologram using an imaging Michelson interferometer configuration. One of the two beams writing the hologram is from the sample and the other is the reference. The hologram is read by a laser diode at 1064 nm. Wavelength tunability is achieved by temperature tuning the laser diode enabling one to sweep the excitonic wavelength of the PRQW. The PRQW acts as the optical filter by passing the ballistic image-bearing light while rejecting the incoherent scattered background. A square aperture after the PRQW rejects the zero-orders. The image is then recorded from the diffracted first order using a CCD camera. Since the PRQW is just an optical filter, there is no reconstruction involved and one can acquire full-frame depth resolved images by changing the path length of the reference beam.

EXPERIMENTS AND DISCUSSION

The phenomenon of saturation in the FWM signal has been observed in InGaAs/GaAs PRQWs though the mechanism still remains unclear as the FWM saturates without the associated saturation of the electro-absorption [4]. Varying the duty cycle of the applied electric field could help in increasing the electric field cutoff at which the FWM saturates. This has important consequences to improving the sensitivity of our imaging system. To test this hypothesis the photocurrent was monitored across a 1.047 MΩ resistor. Unipolar square wave electric fields of 8 and 10 kV/cm were applied to an InGaAs/GaAs PRQW (similar to the one in ref. 2) at a modulation of 105 Hz. The laser diode was tuned to the excitonic absorption of the PRQW and an intensity of 2.23 mW/cm² was incident on the sample. Figure 1 shows the load voltage as a function of duty cycle. One can see a 20% increase in the photocurrent at 50% duty cycle when the electric field is increased from 8 to 10 kV/cm. The duty cycle was kept below 50% for the 10 kV/cm applied field to prevent possible surface damage to the sample.

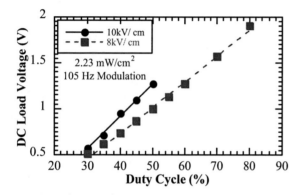

FIGURE 1. Load voltage as a function of duty cycle across a 1.047 MΩ resistor. The light intensity across the sample is 2.23 mW/cm².

Looking at Fig. 2, which shows the peak differential transmission (defined as the ratio of the change in transmission due to the electro-absorption under an applied electric field to the zero-field transmission) as a function of duty cycle, one sees that at 50% duty cycle there is nearly a 50% increase in the differential transmission when the field is increased from 8 to 10 kV/cm. Further the electro-absorption seems to start saturating at higher duty cycle values for 8 kV/cm and even decreases for duty cycles larger than 60%. Detailed time-resolved experiments will help us to understand the effect of hot-carriers on this saturation. Non-degenerate FWM experiments with a mirror in the place of the sample were done by writing gratings with a laser diode that was temperature tuned through 1064 nm at a fringe spacing of 71 μm. The FWM signal was read off the signal beam using a photodiode and lockin amplifier. An external unipolar square wave electric field of 10 kV/cm and frequency 1 kHz was applied in the plane of the PRQW. The total light intensity on the sample was 12.5 mW/cm² with a signal to reference ratio of 3.5. The FWM efficiency (defined as the ratio of the diffracted beam to the transmitted beam) is shown in Fig. 3 as a function of laser diode temperature. Laser diode noise is seen as

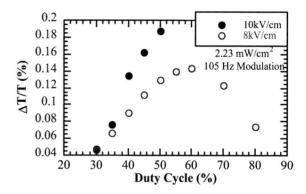

FIGURE 2. Differential transmission vs. duty cycle for applied fields of 8 and 10 kV/cm.

the dip near 29°C in the spectra. One can clearly see that there is an optimum duty cycle, ~ 40%, for a

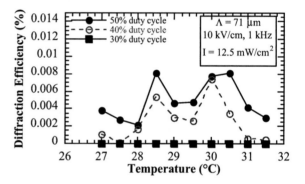

FIGURE 3. FWM efficiency as a function of laser diode temperature. The exciton is around 29 °C.

measurable FWM signal. Also correcting for laser fluctuations, the spectra for 40% and 50% duty cycles are on top of each other. This is evidence that by reducing the duty cycle, and thereby the joule heating, one can optimize the performance of the InGaAs PRQW without sacrificing on FWM efficiency. Further experiments are planned to improve beam shaping and stability to improve the FWM efficiency.

REFERENCES

1. P. Yu et al., *App. Phys. Lett,* **83**, 575, 2003.

2. P. Yu et al., *Opt. Lett,* **29**, 68, 2004.

3. R. Jones et al., *Opt. Lett,* **23**, 103, 1998.

4. S. Iwamoto et al., *Opt. Lett,* **24**, No. 5, 321, 1999.

Electric and Magnetic Manipulation of Biological Systems

H. Lee, T.P. Hunt, Y. Liu, D. Ham and R.M. Westervelt

Department of Physics and Division of Engineering and Applied Sciences
Harvard University, Cambridge, Massachusetts 02138

Abstract. New types of biological cell manipulation systems, a micropost matrix, a microelectromagnet matrix, and a microcoil array, were developed. The micropost matrix consists of post-shaped electrodes embedded in an insulating layer. With a separate ac voltage applied to each electrode, the micropost matrix generates dielectrophoretic force to trap and move individual biological cells. The microelectromagnet matrix consists of two arrays of straight wires aligned perpendicular to each other, that are covered with insulating layers. By independently controlling the current in each wire, the microelectromagnet matrix creates versatile magnetic fields to manipulate individual biological cells attached to magnetic beads. The microcoil array is a set of coils implemented in a foundry using a standard silicon fabrication technology. Current sources to the coils, and control circuits are integrated on a single chip, making the device self-contained. Versatile manipulation of biological cells was demonstrated using these devices by generating optimized electric or magnetic field patterns. A single yeast cell was trapped and positioned with microscopic resolution, and multiple yeast cells were trapped and independently moved along the separate paths for cell-sorting.

INTRODUCTION

The capability to manipulate individual biological cells in microfluidic systems is in ever-increasing demand for biological research and clinical practices. By combining microelectronics and microfluidics, we have developed programmable cell manipulation systems, a micropost matrix, a microelectromagnet matrix, and a microcoil array, that produce versatile electric or magnetic fields for cell manipulation.

ELECTRIC MANIPULATION

A micropost matrix generates local electric fields to controls the motion of individual biological cells by exerting dielectrophoretic force [1,2]. The micropost matrix consists of an array of post-shaped electrodes embedded in an insulator [Fig. 1(a)]. Each electrode is connected to an independent ac voltage source controlled by a computer. By adjusting the voltage in each post, the micropost matrix can create versatile electric field patterns to trap and move individual biological cells. Figure 1(b) shows the manipulation of a single yeast cell in a microfluidic chamber fabricated on top of the micropost matrix. Voltages to four posts were turned on and off sequentially, trapping and moving a cell in a square pattern.

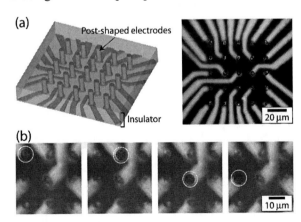

FIGURE 1. (a) Schematic and micrograph of a micropost matrix. The micropost matrix consists of post-shaped electrodes embedded in an insulator. (b) A yeast cell (circled, contrast enhanced) is moved clockwise in a square pattern above the micropost matrix.

MAGNETIC MANIPULATION

Biological cell manipulation using micron size magnetic beads is a common practice in biology and biomedicine [3]. Magnetic beads coated with antibodies can be attached to target biological cells with high selectivity. By applying an external magnetic field, the bead-bound cells can be manipulated in a noninvasive way.

Microelectromagnet Matrix

A microelectromagnet matrix generates local magnetic fields to control the motion of individual cells attached to magnetic beads [4,5]. The matrix consists of two arrays of straight wires separated and capped by insulating layers [Fig. 2(a)]. By independently controlling the current in each wire, the matrix produces magnetic fields that can be dynamically programmed. The sequence of images in Fig. 2(b) shows a cell-sorting operation using the matrix. A single peak in the magnetic field magnitude was created, trapping a viable yeast cell and two nonviable yeast cells. Subsequently, the single peak was split into two peaks: the viable cell was separated by moving one of the peaks, while the other peak held the nonviable cells.

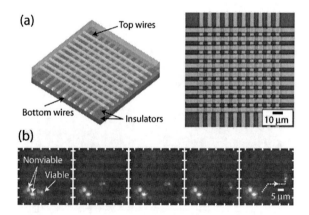

FIGURE 2. (a) Schematic and micrograph of a microelectromagnet matrix. The matrix consists of two arrays of wires covered with insulating layers. (b) By controlling the positions of two magnetic peaks, a viable yeast cell was separated from nonviable cells. White ticks indicate wire positions.

Microcoil Array

Using the advanced silicon fabrication technology, complex and self-contained manipulation systems can be implemented, that can individually manipulate a large number of biological cells inside a microfluidic channel. Figure 3 shows a schematic and micrographs of a microcoil array fabricated in foundry. To generate strong and local magnetic fields, three planar coils were vertically connected for each microcoil. Current sources to the array, and other auxiliary circuits were integrated in the same chip, considerably reducing the size and the complexity of an experiment setup.

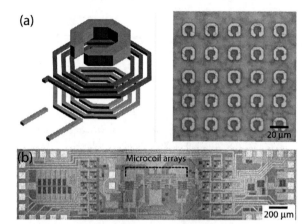

FIGURE 3. (a) Schematic and micrograph of a microcoil array. Three planar coils are vertically connected through vias to form a microcoil. (b) Micrograph of a whole chip with three coil arrays. Current sources and control circuits for the arrays are integrated in the chip.

ACKNOWLEDGMENTS

The authors thank DeVito and Feindt in Analog Device Inc., for the microcoil array fabrication. This work was supported by the Nanoscale Science and Engineering Center at Harvard under NSF Grant No. PHY-0117795.

REFERENCES

1. Pohl, H. A., *Dielectrophoresis*, Cambridge: Cambridge Press, 1978.

2. Hunt, T. P, Lee, H., and Westervelt, R. M., *Appl. Phys. Lett.,* submitted (2004).

3. Häfeli, U., *Scientific and Clinical Applications of Magnetic Carriers*, New York: Plenum, 1997.

4. Lee, C. S, Lee, H., and Westervelt, R. M., *Appl. Phys. Lett.* **79**, 3308-3310 (2001).

5. Lee, H., Purdon, A. M., and Westervelt, R. M., *Appl. Phys. Lett.* **85**, 1063-1065 (2004).

Ferroelectric Gates with Rewritable Domain Nanopatterns for Modulation of Transport Properties in GaN/AlGaN Heterostructures

I. Stolichnov, L. Malin, E. Colla, J. Baborowski, N. Setter, J-F. Carlin*

Laboratory of Ceramics, EPFL (Swiss Federal Institute of Technology), 1015 Lausanne, Switzerland
* Semiconductor Device Physics Group, EPFL, 1015 Lausanne, Switzerland

Abstract. The concept of field-effect transistor with ferroelectric gate in combination with the advanced technique of direct domain writing is applied for modulation of transport properties of 2D electron gas located close to the interface in a GaN/AlGaN heterostructure. The ferroelectric Pb(Zr,Ti)O$_3$ film grown on top of such heterostructure was poled in a controllable way by scanning probe microscope inducing depletion or accumulation effects in the 2D gas depending on the polarity orientation. The artificially created domain arrangements can be projected onto the 2D gas provoking local depletion underneath the poled area. This ferroelectric lithography is potentially interesting for a number of applications as a flexible and nondestructive way of making rewritable patterns on low-dimensional structures with nanoscale resolution. In this work we present the first results on the ferroelectric gate patterning and its impact on the charge concentration and mobility of the 2D gas in the GaN/AlGaN heterostructure.

INTRODUCTION

Ferroelectric materials integrated into the semiconductor media have been intensively studied over the last decade for numerous electronic applications. In particular, the general property of ferroelectrics to switch the spontaneous polarization by an external electric field is exploited in non-volatile memory devices [1]. One of the implementations of the ferroelectric memory device concept is the ferroelectric field effect transistor with the gate comprising a ferroelectric layer that can be poled positively or negatively provoking charge accumulation or depletion in the transistor channel. Such devices have been successfully realized on classical Si/SiO$_2$ system using the Pb(Zr,Ti)O$_3$ (PZT) ferroelectric layer as a gate material [2]. The major problem of commercialization of these devices is the integration due to high processing temperatures of ferroelectric materials, which are hardly compatible with the silicon technology. Apart from the nonvolatile memories, the ferroelectric gates may be interesting for nanostructure patterning through the polarization domain engineering. The polarization domain arrangements can be created artificially by local poling of the ferroelectric layer using a Scanning Probe Microscope (SPM). Earlier the SPM-assisted direct domain writing on the PZT film was successfully used for modification of conductive properties of metallic SrRuO$_3$ layer underneath [3].

In the present work we combine the ferroelectric gate with the semiconductor structure containing 2D electron gas close to the ferroelectric/semiconductor interface. For the first time we demonstrate the effect of 2D gas depletion induced by polarization domain writing on the ferroelectric gate, which opens new possibilities for nanostructure design through domain engineering by "ferroelectric lithography".

RESULTS AND DISCUSSION

For this experiment 400 nm thick polycrystalline PZT film was deposited on AlGaN/GaN heterostructure with a 2D electron gas located 20nm below the surface. Mesas with Hall bar geometry have been defined by electron cyclotron resonance reactive ion etching. The PZT film was poled using SPM technique with dc voltage of ±40V applied to the SPM conductive cantilever tip (Fig. 1).

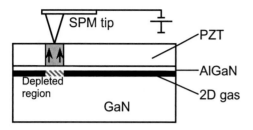

FIGURE 1. Local depletion of 2D electron gas, induced by artificially created polarization domains in PZT ferroelectric layer deposited on top of AlGaN/GaN heterostructure.

The created polarization pattern was controlled by piezoresponse force microscopy (PFM) [4] (Fig. 2).

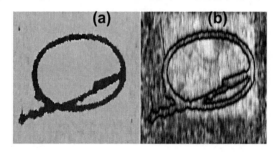

FIGURE 2. Direct writing of polarization domains on PZT film deposited on top of AlGaN/GaN heterostructure. The PFM images represent phase (a) and amplitude (b) of local piezoresponse. The scanned area is 7x7μm.

In order to evaluate the effect of the polarized ferroelectric film on the 2D gas carrier concentration the preferential polarization was induced in the area of 50x50 μm and then the concentration of electrons was measured by the Hall effect. First, measurements were done without poling in order to define the initial electron concentration. The concentration that was found to be 5×10^{12} cm^{-2} remained virtually unchanged within the temperature range from 77K to 298K. Then the studied area was poled with –40V (polarity corresponds to the 2D gas depletion) and the measurements were repeated. As the last step the polarity on the same area was inverted and then the third series of measurements was performed. Figure 3 shows that the electron concentration changes approximately by factor two as the sign of polarization in PZT switches. The electron mobility was virtually independent of the polarization and was measured to be 1200 cm^{2}V^{-1}s^{-1} at 298°K and 4200 cm^{2}V^{-1}s^{-1} at 77K. The observed temperature dependence of the depletion may be attributed to the depolarization effects and will be addressed in the upcoming papers.

To summarize, the experimental results suggest that the artificial domain pattern written on the ferroelectric gate can be projected directly onto the 2D gas. Hence, the arbitrary-shaped low-dimensional semiconductor structures can be defined with nanoscale resolution by domain engineering. The essential advantage of such "ferroelectric lithography" compared to the alternative techniques is that the created patterns are rewritable. The domain pattern can be modified or completely erased and rewritten without causing any damage to the sample, which opens new opportunities for experiments with semiconductor nanostructures as well as for device optimization. The application of the ferroelectric lithography to other systems with 2D gas with high mobility of carriers such as GaAs/AlGaAs system will be addressed in the further publications.

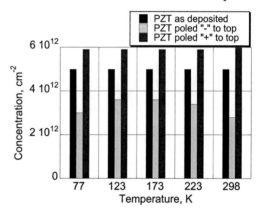

FIGURE 3. Effect of preferential top-to-bottom and bottom-to-top spontaneous polarization in the PZT film on the electron concentration in the 2D gas.

ACKNOWLEDGMENTS

This work was supported by Swiss National Sciences Foundation.

REFERENCES

1. O. Auciello, J. Scott and R. Ramesh, Physics Today **51**, 22-27 (1998).
2. N. Basit, H. Kim, and J. Blachere, Appl. Phys. Lett. **73**, 3941-3943 (1998).
3. C. Ahn, T. Tybell, L. Antognazza, K. Char, R. Hammond, M. Beasley, O. Fischer, and J.-M. Triscone, Science **276**, 1100-1103 (1997).
4. I. Stolichnov, E. Colla, A. Tagantsev, S. Bharadwaja, H. Seungbum, N. Setter, J. S. Cross, and M. Tsukada, Appl. Phys. Lett. **80**, 4804-4806 (2002).

Towards Tunneling Through a Single Dopant Atom

J. Caro[1], G.D.J. Smit[1], H. Sellier[1], R. Loo[2], M. Caymax[2], S. Rogge[1], and T.M. Klapwijk[1]

[1]*Kavli Institute of Nanoscience, Delft University of Technology, Lorentzweg 1, 2628 CJ Delft, The Netherlands*
[2]*IMEC, Kapeldreef 75, B-3001 Leuven, Belgium*

Abstract. Aiming for atom-based functionality, we study self-assembled $CoSi_2$/Si Schottky nanodiodes and CVD-grown Si δ-doped $p^+/p^-/p^+$ tunneling devices. Due to their smallness, the $CoSi_2$/Si diodes comprise only a limited number of dopant atoms in the Schottky barrier. Transport through the smaller diodes is dominated by randomly positioned individual dopant atoms, as reflected in device-to-device conductance fluctuations at 300 K and resonant tunneling peaks at 4.5 K. The layered structure of the δ-doped devices has the promise of better control over active atoms. Indeed, in large $p^+/p^-/p^+$ devices the boron atoms in the δ-layer induce resonant tunneling through the B^+ state of these atoms. The resonance position shifts to higher voltages in a magnetic field, which is interpreted as a diamagnetic shift.

INTRODUCTION

Atomic-scale electronics, based on the use of dopant atoms as bottom-up elements in a semiconductor, is very attractive because of the large Bohr orbit of the atoms. We aim to realize electronic functionality at the atomic scale by using individual dopant atoms in silicon and by manipulating their wave function. This question of atom based functionality is addressed by means of electrical transport experiments.

CONDUCTANCE FLUCTUATIONS IN SCHOTTKY NANODIODES

We have measured I-V curves of $CoSi_2$/Si Schottky diodes down to a diameter of 15 nm [1]. Self-assembled $CoSi_2$ islands are grown on a Si(111) substrate (n-type, $N_D \approx 2 \times 10^{18}$ cm^{-3}) by evaporating a submonolayer of Co onto the clean (7×7)-reconstructed surface and by subsequent annealing. The resulting islands, with diameters in the range 15-80 nm, define nano-scale epitaxial Schottky diodes. I-V curves of the diodes are obtained with a scanning tunneling microscope, by lowering the tip onto the islands. Both diode fabrication and electrical measurements with are made in ultra high vacuum.

Room temperature I-V curves, exemplified in the inset of Fig. 1, show weakly rectifying behavior. Large area diodes (area > 2000 nm^2) of equal size have the same zero-bias conductance per unit area (see Fig. 1). For smaller diodes the conductance shows pronounced device-to-device fluctuations, their magnitude increasing with decreasing area.

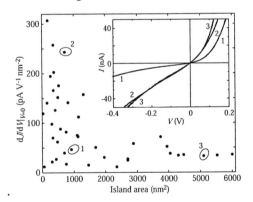

FIGURE 1. Size dependence of zero-bias conductance per unit area for Schottky nanodiodes. The inset shows some typical I-V curves, the numbering corresponding to that in the main graph.

These fluctuations cannot be due Schottky barrier inhomogeneities, interface defects or grain boudaries [1]. Rather, they arise from number and position fluctuations of the randomly distributed dopant atoms in the Schottky barrier. These atoms in the barrier locally cause a lower barrier lowering and thus a higher current. This interpretation is supported by a statistical model for the fluctuations based on random dopant positions [1]. It also agrees with the peaks we observe in the resonant tunneling spectra for these Schottky nanodiodes at 4.5 K [1].

RESONANT TUNNELING THROUGH THE B^+ LEVEL IN A δ-DOPED SILICON BARRIER

Inspired by the results on nanodiodes, we shifted to boron doped, CVD-grown silicon tunneling devices of the type $p^+/p^-/p^+$, the p^+ layers and p^- layer serving as electrodes and tunnel barrier, respectively. Centered in the barrier a boron δ-layer of density 1.7×10^{11} cm^{-2} was grown. For small enough diameters such devices enable vertical tunneling through a few and finally one dopant atom of the δ-layer. A local gate will enable manipulation of the wave function at the dopant atom, which will be reflected in the tunnel current.

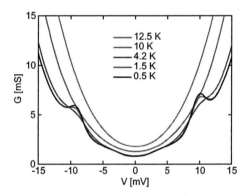

FIGURE 2. Conductance curves of a 400×400 μm^2 device, for the temperatures listed. With decreasing temperature the conductance decreases (most clear at higher biases) and a tunneling resonance develops at ±10 mV.

In large area devices, at 4.2 K and below, we observe a resonance at ± 10 mV, as demonstrated in Fig. 2. The resonance arises from tunneling through the B^+ state of the boron atoms in the δ-layer [2]. Such a state forms when a second hole weakly binds to a boron atom. This interpretation is based on the ionization energy E^+=6.7 meV of the B^+ state derived from the resonance position and the barrier height ϕ_B=11.7 meV obtained from measurements of activated transport over the barrier. This value of E^+ is consistent with [2] ionization energies measured for B^+ and D^- (the donor counterpart of B^+) at doping levels where the B^+ ions are interacting [3].

In the inset of Fig. 3 we plot the response of the tunneling spectra to a magnetic field oriented perpendicular to the layers. With increasing field the peak shifts to higher bias and becomes broader and weaker. The main panel shows the level shift of the B^+ state deduced from the resonance shift. Interpreting the shift as a diamagnetic shift, we derive proper values of the mean distance of the two holes to the B^- core [2].

Presently, we are miniaturizing these devices to the level of a few dopant atoms (diameter down to 100 nm). The first results indicate that below diameters of a few μm additional resonances show up in the spectra resulting from tunneling through surface states.

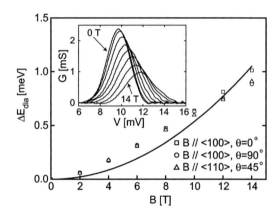

FIGURE 3. Magnetic field induced shift of the B^+ state, for the device of Fig. 2, for three in-plane field orientations θ (θ is the angle between the field and the current direction. The parabola is a fit of the expression for the diamagnetic shift of the B^+ state to the data points. The points are derived from curves as in the inset, which shows the field dependence of the resonance (field step is 2 T).

ACKNOWLEDGMENTS

This work is part of the research program of FOM, which is financially supported by NWO. One of us, S.R., wishes to acknowledge the Royal Netherlands Academy of Arts and Sciences for financial support.

REFERENCES

1. G.D.J. Smit *et al.*, Phys. Rev. B **69**, 035338 (2004).
2. J. Caro *et al.*, Phys. Rev. B **69**, 125324 (2004).
3. This is the case in the barrier of our devices, where the background doping outside the δ-layer is N_B=10^{17} cm^{-3}.

High-Speed and Non-Volatile Nano Electro-Mechanical Memory Incorporating Si Quantum Dots

Y. Tsuchiya[*], K. Takai[*], N. Momo[*], T. Nagami[*],
S. Yamaguchi[†], T. Shimada[†], H. Mizuta[¶], and S. Oda[*,¶]

[*]Quantum Nanoelectronics Research Center, Tokyo Institute of Technology, Meguro-ku, Tokyo 152-8552, Japan
[†]Central Research Laboratory, Hitachi Ltd., Kokubunji-shi, Tokyo 185-8601, Japan
[¶]Department of Physical Electronics, Tokyo Institute of Technology, Meguro-ku, Tokyo 152-8552, Japan

Abstract. Basic device characteristics were investigated for a new high-speed and non-volatile nano electro-mechanical systems (NEMS) memory device with nanocrystalline silicon (nc-Si) dots embedded in its movable floating gate beam. Over 1 GHz operation is possible due to the size reduction of the NEMS. From a simulation of mechanical properties of the movable floating gate beam, advantage of using the nc-Si dots array was shown for low power operation. The mechanical bistability of the fabricated SiO_2 beam was clearly observed in both experimental and simulation studies.

INTRODUCTION

Nano Electro-Mechanical Systems (NEMS) have a possibility of high-speed operation in the GHz regime since the characteristic frequencies are expected to increase with decreasing their dimensions [1]. We proposed a new non-volatile memory concept based on bistable operation of the NEMS structure combined with nanocrystalline-Si (nc-Si) dots [2]. In this paper, we study basic device characteristics which are essential for operating our NEMS memory device.

OPERATION PRINCIPLE

Schematics of the NEMS memory device are shown in Fig. 1. Our new memory features a mechanically bistable floating gate beam, which incorporates the nc-Si dots as single-electron charge storage. The beam is suspended in the cavity under the gate electrode and moves via electrostatic interactions between the gate electrode and the charge in the nc-Si dots. Positional displacement of the beam is sensed via a change in the drain current of the MOSFET underneath.

From a mechanical analysis assuming the maximum central displacement of 50 nm, the switching speed between two stable states was estimated to be ~ 0.5 ns for a SiO_2 beam with the dimension of $1.0 \times 1.0 \times 0.1$ μm^3. By optimizing both the beam structure and stored charge amount, we may build an extremely fast and non-volatile memory.

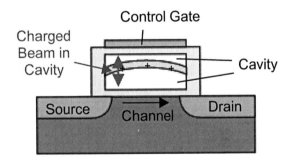

FIGURE 1. A schematic of a NEMS memory device

MECHANICAL PROPERTIES

Mechanical properties of the beam were investigated using a 3D finite element simulation [3]. We compared a nc-Si beam where a 2D nc-Si dot array was embedded inside a SiO_2 film and a simple poly-Si beam where a thin poly-Si sheet was placed between two SiO_2 layers. Calculated results under a constant uniaxial pressure are shown in Fig. 2. The large displacement was observed around at the center of the beams as shown in the rotated view. Note that the width of dark area around the center of the nc-Si beam is larger than that of the poly-Si beam. This indicates that a larger displacement is achievable with the nc-Si beam when an electric field is applied via the gate bias. Therefore, the nc-Si beam enables highly-controlled charge storage and is suitable for low power operation.

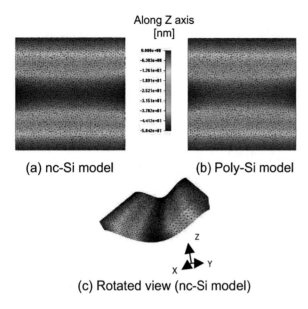

FIGURE 2. Beam deformation simulated for two different beam structures under uniaxial pressure; Top views of (a) nc-Si beam, and (b) poly-Si beam, and rotated view of the nc-Si beam (c)

BISTABILITY OF BEAM

Bistability of the beam is a key feature for a non-volatile memory application in this device. Figure 3(a) shows a single layer beam structure for testing its mechanical properties. After the undercut of the sacrificed Si layer underneath the beam, the most of the beams bent upward naturally as shown in Fig. 3(b), which is considered as a result of release of mechanical stress stored in Si/SiO$_2$ interface during thermal oxidation. Note that in the same test chip, several downward-bent beams were also observed as shown in Fig. 3(c). These observations strongly suggest that the beams fabricated by this method have bistable nature. This bistability was also demonstrated by using the numerical simulation which takes account of the nonlinear dependence of the beam deformation on the applied force.

SUMMARY

We have investigated basic device characteristics of the non-volatile NEMS memory device with the nc-Si dots embedded in its movable floating gate beam. We found that the beam with the embedded nc-Si dots array is suitable for the movable floating gate beam to achieve low power operation. The mechanical bistability was clearly observed for the fabricated SiO$_2$ beam.

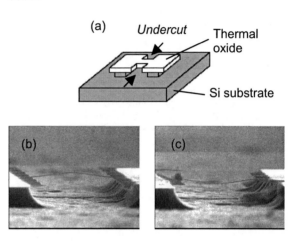

FIGURE 3. (a) A schematic illustration of the fabrication of a single layer test structure. (b) Upward-bent and (c) downward-bent SiO$_2$ beam after undercut process.

ACKNOWLEDGMENTS

The authors are grateful to Prof. Y. Higo, Prof. K. Takashima, and Dr. S. Koyama for discussion and helpful comments. This work has been supported by CREST program of the Japan Science and Technology Agency, and IRCP of the Japan Society for the Promotion of Science.

REFERENCES

1. Huang, X. M. H., Zorman, C. A., Mehregany, M., and Roukes, M. L., *Nature* **421**, 496 (2003).

2. Tsuchiya, Y., Takai, K., Momo, N., Yamaguchi, S., Shimada, T., Koyama, S., Takashima, K., Higo, Y., Mizuta, H., and Oda, S., in *2004 Silicon Nanoelectronics Workshop*, June 13-14, 2004, Hawaii, USA

2. http://www.j-insight.com/ & http://adventure.q.t.u-tokyo.ac.jp/

Physics of Deep Submicron CMOS VLSI

Dennis D. Buss

Texas Instruments Inc
Dallas, Texas

Abstract. The Integrated Circuit (IC) was invented in 1958, and modern CMOS was invented in 1980. The semiconductor physics that underlies the IC was discovered in the early part of the past century, and, by the early 60's, it was simplified and codified such that it could be used by engineers to design transistors of ever shrinking size and increasing performance. However, in recent years, the "Electrical Engineering Physics" of the 60's is becoming increasingly inadequate. Empirical corrections are being made to allow for quantum and non-equilibrium Boltzmann transport effects. Moreover, as features in CMOS transistors reach atomic dimensions, continuum physics is no longer adequate, and devices must be designed increasingly, at the atomic level. As transistors approach the end of scaling, the physics to design them will become increasingly complex, and Electrical Engineering Physics will no longer suffice.

INTRODUCTION TO CMOS VLSI TECHNOLOGY

In the past 34 years, since 1970, IC minimum feature size has shrunk by a factor of 70X from 6 um in 1970 to 90 nm today. (Fig 1) New generations of technology have been introduced every 2-3 years, and in every generation, the feature size has shrunk to approximately 70% of the feature size of the previous generation. A linear shrink of 70% results in an area shrink of 50% which results in 2X the number of die per wafer. Historically, the wafer cost has increased by 20% per generation so that the cost per die has decreased by 40%. This 40% reduction every 2-3 years has been the driving force that has brought about the proliferation of electronics worldwide to the point where billions of people around the globe can afford cell phones, PDAs and other sophisticated electronic products to access the internet. On average, over the past 34 years, feature size has shrunk to 70% of the previous generation every 2.9 years. However in recent years, this shrinking has accelerated, and the time between generations is down to two years. Table 1 shows how other aspects of the semiconductor industry have evolved as feature size has shrunk.

The device physics required to understand semiconductor devices and to develop new technology generations was codified in the early 60's. [1] The quantum nature of semiconductors was simplified to a

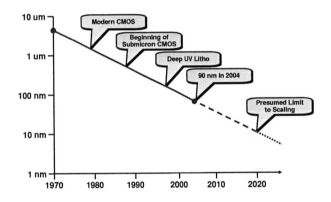

FIGURE 1. 34 years of scaling of minimum feature size.

	1970	Today	Change
Feature Size	6 um	90 nm	70x Reduction
Transistor Density			5000x Increase
Chip Size	~10 mm²	~400 mm²	40x Increase
Transistors/Chip	1000	200 M	200,000x Increase
Clock Frequency	100 kHz	> 1 GHz	>10,000x Increase
Power Dissipation	~100 mW	~100 W	~1000x Increase
Fab Cost	~$10 M	>$1 B	>100x Increase
WW IC Revenue	$700 M	$170 B	240x Increase
WW Electronics Revenue	$70 B	$1.1 T	16x Increase

TABLE 1. Changes in the semiconductor industry in 34 years.

bandgap, an effective mass for electrons and holes, and an effective density of states for conduction and valence bands. Transport was simplified to equilibrium Boltzmann transport and characterized by a mean time between scattering events τ. This simplified treatment, together with Fermi statistics, has been sufficient for R&D professionals in semiconductor technology development up to the present time. [2]

Today's state of the art VLSI is illustrated in Fig 2. This IC performs the central processing function in high performance servers. It has 256 million transistors, minimum gate length of 37 nm, Equivalent Oxide Thickness (EOT) of 12Å and 8 levels of Cu metal with Organo-Silicate Glass (OSG) low-k dielectric for insulation between metal lines.

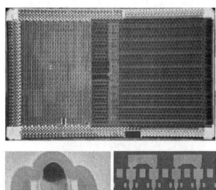

FIGURE 2. An example of today's state-of-the art IC

As indicated in Fig 1, scaling is expected to continue for another 6 generations beyond 90 nm to 10 nm, after which future scaling is dubious. The Semiconductor Industry Association (SIA) has charted its membership to develop an International Technology Roadmap for Semiconductors (ITRS) to the "end of the roadmap": ie, to about 10 nm. [3] The challenges to achieve the ITRS roadmap are formidable.

The remainder of the paper deals with the issues of physics that must be overcome in order to reach the end of the roadmap. It is these same set of issues that will eventually result in the end of scaling.

PHYSICS CHALLENGES TO CONTINUED VLSI SCALING

As pointed out in the now famous paper on device scaling [4], gate oxide must scale with gate length. In the late 60's, when gate length, L_g was 10 um, the gate oxide thickness was 120 nm (~1 % of L_g). Today, with gate length of 37 nm (see Fig 2), the gate insulator is a Plasma Nitrided Oxide (PNO) with Equivalent Oxide Thickness (EOT) of 1.2 nm (~3% of L_g). Gate insulator thickness has not scaled proportional to L_g, and looking forward, the lack of scalability of gate insulator is a major barrier to further scaling.

Scaling of gate insulator is related to voltage scaling. The advantages of voltage scaling are many. Logic delay is proportional to supply voltage V_{dd} provided the drive current I_{on} can be maintained as the voltage is dropped. In addition, there are tremendous power savings to be realized by decreasing voltage

$$\text{Power} \sim \tfrac{1}{2} C V_{dd}^2 F_c \qquad (1)$$

where F_c is the clock frequency. The key to achieving these advantages of voltage scaling is to maintain constant I_{on} as V_{dd} is reduced. I_{on} is proportional to the charge in the inversion layer Q_{inv} times the velocity of the charge. Furthermore, Q_{inv} is determined by the displacement in the gate insulator $D_{ins} = \varepsilon_{ins} E_{ins}$. Consequently, it is necessary to scale gate insulator thickness when voltage scales in order to achieve constant I_{on} as V_{dd} is reduced.

But as gate insulator shrinks, gate leakage current increases. As shown in Fig 3, as oxide shrinks from 30 Å to 10 Å, gate leakage increases by **8 orders of magnitude** to unacceptable levels. In an attempt to improve the trade off between leakage and EOT, Nitrogen has been added to the gate insulator. For Nitrogen concentrations that have been achieve to date (6-12 %), the leakage current has been reduced by 10X to 100X as shown in Fig 3. The need for further reduction in EOT has led to the search for so-called hi-k dielectrics which are suitable for use as gate insulators: materials such as HfO, AlO, HfSiO, HfSiON and others. Fig 3, shows that HfSiON reduces gate leakage by 100X to 10,000X compared to SiO_2.

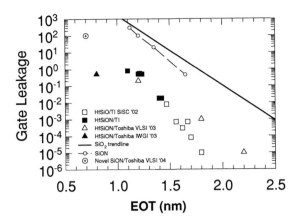

FIGURE 3. Gate tunneling current as a function of insulator thickness for a variety of insulator materials. [5]

However, a tremendous amount of knowledge exists relative to the Si-SiO₂ interface, and the move to hi-k gate insulator is fraught with difficulties.
- Reduction in mobility compared to SiO₂
- Shift in threshold of up to 800 mV for PMOS
- Uncertain reliability

Much additional research needs to be done to understand
- the atomic and electron orbital nature of the interface between hi-k insulators and Si
- the scattering mechanisms for carriers in the inversion layer and methods to achieve mobility on the "universal mobility" curve for hi-k insulators.
- The reliability of these films at thicknesses of a few monolayers and at electric fields ~60% of the breakdown field

The effects of Stark quantization in the inversion layer of Si were first studied in the early 60's [6]. However, Stark quantization increases the threshold for NMOS by only 75 mV, and this was not important until recently. Moreover, as device dimensions shrink, process induced stresses have an increasing effect on the mobility in the center of the channel. In modern IC processing, stress of up to 1 GPa is not uncommon. Some of these stresses increase mobility (and I_{on}), and other stresses decrease mobility. Stress engineering is becoming increasingly important in device development, and this requires an understanding of the energy bands and scattering matrix elements of Si in the presence of stress and in the presence of strong electric fields (~1 MV/cm) perpendicular to current flow. The case of electrons under stress is well understood [7]. In the absence of stress, the 001 electric field removes the degeneracy between electron minima at Δ, and the minimum energy for Δ_2 electrons is 75 meV lower than the minimum energy for Δ_4 electrons. However, because of the small density of states in the Δ_2 band, relative to the Δ_4 band, there is no substantial shift from Δ_2 to Δ_4, and the mobility does not increase substantially. However, if a biaxial tensile stress of 1.3 GPa is applied, the Δ_2 and Δ_4 minima split by an additional 135 meV. This situation is shown in Fig 4. It results in substantial shift of electrons into the Δ_2 bands, and results in 80% increase in mobility.

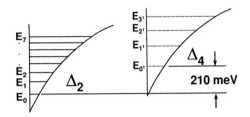

FIGURE 4. Energy bands for electrons in an MOS inversion layer in the presence of stress.

The behavior of holes is complicated by two factors. 1) The light hole and heavy hole valence bands are spherical and parabolic around Γ, but they become highly non-parabolic and non-isotropic at small distances in k-space away from Γ. 2) The valence bands, split by Stark splitting continue to interact strongly under stress. Notwithstanding this complexity, the full quantum treatment has been done [8, 9], and is required to optimize the stress for PMOS. The optimum stress for PMOS is uniaxial compressive stress parallel to the direction of current flow, and this effect has been shown to provide an increase of 35% in PMOS drive current when epitaxial SiGe replaces Si in the Drain Extension (DE) of the PMOS transistor. [10]

Perhaps the most serious limitation on scaling MOS transistors results from simple electrostatics. In the off state, transistor sub-threshold leakage is an exponential function of the barrier height. This results in a turn-off which is characterized by a sub-threshold slope. If EOT today was ~ 1% of gate length (as it was in the 70's), ideal sub-threshold slope of 60 mV (at room temperature) could be achieved. However, gate insulator thickness is not expected to scale as fast as L_g, and sub-threshold slope will continue to degrade. Consider a transistor with threshold of 0.3V and I_{on} = 1 mA/um. If the sub-threshold slope is 60 mV/decade, the transistor off current will be I_{off} = 10 nA/um. This is high but acceptable. If the sub-threshold slope is degraded to 100 mV/decade, I_{off} = 1uA/um, which is totally unacceptable.

The above discussion of electrostatics shows that the failure of gate insulator thickness to scale at the same rate as feature size is a major limitation to continued transistor scaling. This limitation can be mitigated somewhat using multi-gate transistor structures shown in Fig 5. In such structures, the gate has increased control over the channel, compared to source, drain or substrate, even if gate insulator thickness is not reduced. Such structures may be used to extend the scaling of CMOS beyond the planar limit, by 2X – 3X.

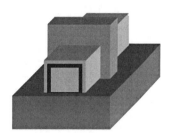

FIGURE 5. 3-D MOS structure proposed to increase gate coupling to the channel.

As transistor scaling approaches the end of the roadmap, there will be additional physics challenges.
- Metal gate electrodes will replace poly-Si, and the interface, which sets the effective work-function, is poorly understood today.
- Tunneling through the gate insulator and across the drain-to-substrate junction will increasingly dominate transistor behavior.
- The discrete positioning of dopant atoms in the transistor channel will increasingly degrade transistor performance.
- Transistors will become increasingly ballistic.
- And new materials will be introduced into the Source/Drain and channel.

CONCLUSION

For the past 35 years, transistors have been developed using "Electrical Engineering Physics", which was codified in the early 60's. However, as transistor scaling approaches the "End of the Roadmap", Electrical Engineering Physics is no longer sufficient. Transistor development increasingly requires
- sophisticated quantum physics
- non-equilibrium Boltzman transport
- material science at the atomic and electron orbital level

This has implications for physics education at the graduate level, and it provides career opportunities for condensed matter physicists in the semiconductor industry.

ACKNOWLEDGMENTS

The author is grateful for the many thoughtful contributions from colleagues at MIT and at Texas Instruments.

REFERENCES

1. Semiconductor Electronics Education Committee (SEEC) Books. Seven volumes published in 1964 by John Wiley & Sons Inc.

2. For example, Ben Streetman and Sanjay Bannerjee, Solid State Electronic Devices, Prentice Hall, 2000.

3. 2003 International Technology Roadmap for Semiconductors, Semiconductor Industry Association, 2003.

4. R. Denard et al. "Design of Ion-Implanted MOSFET's with Very Small Physical Dimensions", IEEE J. Solid-State Circuits, SC-9, pp 256-268.

5. Luigi Colombo. Private communication

6. F. Stern, "Quantum and Continuum Results for Inversion Layers", Proc of the Tenth International Conf on the Physics of Semiconductors, Aug 1970, pp 451-458.

7. S. Takagi, J.L. Hoyt, J. Welser, and J.F. Gibbons, "Comparative study of phonon-limited mobility of two-dimensional electrons in strained and unstrained Si metal-oxide-semiconductor field-effect transistors," J. Appl. Phys. 80, pp. 1567-1577, August 1996

8. M. V. Fischetti et al, "Six-band k·P calculation of the hole mobility in silicon inversion layers: Dependence of surface orientation, strain and silicon thickness", Jol of Applied Physics, 94, 15 July 2003, pp 1079-1095

9. M. Giles et al, "Understanding Stress Enhanced Performance in Intel 90 nm CMOS Technology", 2004 Symposium on VLSI Technology Digest of Technical Papers, pp 118-119.

10. P. R. Chidambaram et al, "35% Drive Current Improvement from Recessed-SiGe Drain Extensions on 37 nm Gate Length PMOS" 2004 Symposium on VLSI Technology Digest of Technical Papers, June 2004, to pp 48-49.

AUTHOR INDEX

A

Abboudy, S., 1483
Abdullin, K. A., 107
Abernathy, C. R., 1319, 1399
Abram, R. A., 373, 589, 1125, 1166
Abramof, E., 1397
Abramov, A., 1192
Abstreiter, G., 443, 892, 898, 900, 915, 923, 1501
Abulnasr, L., 1483
Achiba, Y., 1039
Acosta, D. R., 187
Adames, M., 403
Adams, A. R., 681, 985, 1545
Agarwal, K. C., 171
Ager III, J. W., 67, 73, 209, 611
Agrinskaja, N. V., 955
Aguado, R., 1411
Ahn, S., 1349
Ahopelto, J., 1537
Ahuja, R., 177
Aidala, K. E., 1461
Aihara, M., 729
Aikawa, H., 773
Ajili, L., 1573
Akabori, M., 1297, 1373
Akagi, K., 389
Akahane, K., 1533
Akai, Y., 1057
Akasaki, I., 263, 285
Akazaki, T., 397, 1162
Akera, H., 551, 553
Akimoto, K., 1059
Akimoto, R., 967, 971, 1137
Akimov, A. V., 1301
Akinaga, H., 1323
Akita, K., 1137
Akiyama, H., 885, 887, 1575
Akiyama, T., 389, 393
Akopian, N., 669
Aku-Leh, C., 1228
Al-Ahmadi, A. N., 761
Alawadhi, H., 73
Alberi, K., 223
Albert, B. E., 419
Alberto, H. V., 169, 193
Albrecht, F., 89
Albrecht, J. D., 359, 941, 1377
Albuquerque, E. L., 385
Alcalde, A. M., 637
Aleksiejunas, R., 241
Aleshkin, V. Y., 947, 949, 1214

Aleszkiewicz, M., 185
Alfaramawi, K., 1483
Allen, S. J., 900, 1143
Alshanskii, G. A., 1003
Alsina, F., 995
Alt, H. C., 237
Al-Thani, H. A., 161
Amand, T., 751, 1293, 1361
Amann, M.-C., 1567
Amano, H., 263, 285
Amo, A., 133, 1113, 1170
Amore Bonapasta, A., 199
Anda, E. V., 675
Ando, H., 1039
Ando, T., 537, 1029, 1059
André, R., 1113
Andrearczyk, T., 1271
Andreev, A. D., 681, 695
Andronikov, D., 973, 983
Aniel, F., 1123
Anwand, W., 99
Aoki, H., 649
Aoki, K., 129
Aoki, N., 475, 823, 913, 1033, 1035, 1053, 1083, 1091
Aoyagi, S., 325
Aoyagi, Y., 571
Apetrii, G., 923
Arakaki, H., 959
Arakawa, Y., 627, 635, 655, 695, 741, 743, 1401
Arapov, Y. G., 1003
Areiza, M. C. L., 1543
Arif, M., 1079
Arita, M., 1401
Arnaudov, B., 261, 285
Arndt, C., 823
Arpiainen, S., 1537
Artús, L., 293
As, D., 253
Asada, H., 349
Asahi, H., 623
Asahina, T., 1389
Asaka, K., 894
Asenov, A., 1497
Ashkenov, N., 165, 401
Ashkinadze, B. M., 415, 511
Ashley, T., 295
Ašmontas, S., 1204
Asmussen, Jr., J., 115
Atkinson, P., 665, 1527
Austing, D. G., 649
Averkiev, N. S., 955, 1405

Axt, V. M., 883, 907, 1242
Ayres de Campos, N., 169, 193

B

Babinski, A., 685
Baborowski, J., 1585
Bacher, G., 591, 927, 1427
Badcock, T. J., 1547
Bădescu, Ş. C., 763
Baer, N., 661, 725
Bagaev, V. S., 755
Bagrets, D. A., 1255
Baily, S. A., 1265
Bakarov, A. K., 457, 461, 579, 629, 693, 827
Baker, S. C., 1027
Balasubramanian, S., 1579, 1581
Ban, S. L., 233, 391
Bando, K., 894
Bang, K. I., 587
Bao, J. M., 1313, 1429
Barcz, A., 185
Bardyszewski, W., 1160
Bar-Joseph, I., 501
Barker, J. M., 227, 1263
Barker, J. R., 1493, 1497
Baron, F. A., 411
Baroughi, M. F., 1535
Barrick, T., 709, 973
Bartel, T., 701
Basnar, B., 1275
Bataineh, M., 115
Baumgartner, A., 509
Bausk, N., 585
Bayer, M., 661, 685, 1158, 1301
Bayir, M., 795
Beanland, R., 1547
Beattie, N. S., 1561
Beck, M., 1385
Becker, C. R., 1427
Becker, J. M., 715
Becker, P., 69
Beckman, S. P., 145
Beeman, J. W., 611
Beere, H. E., 1289, 1573
Beggs, D. M., 1166
Beling, C. D., 99
Bel'kov, V. V., 1309
Bell, A., 213, 217, 301, 863
Bellet, E., 365
Bellotti, E., 239
Belyaev, A. E., 439, 459
Belykh, V., 1192
Benedict, L. X., 1061
Benndorf, G., 183, 351

Bennett, A. J., 665
Benyoucef, M., 661, 725
Bergman, J. P., 991, 1319
Bernardot, F., 363, 1359
Berney, J. H., 957, 1149
Bernholc, J., 1331
Bernier, P., 1043
Bertolini, M., 1299, 1311
Bertoni, C. M., 375
Bertram, F., 301
Betbeder-Matibet, O., 1151, 1153
Bewley, W. W., 1525
Bezerra, G. A., 1095
Bhat, I., 937, 989
Bichler, M., 443, 892, 898, 900, 915
Biehne, G., 197
Bielejec, E., 453, 917, 921
Bimberg, D., 143, 147, 689, 701, 767, 769, 1555
Biquard, X., 365
Bird, J. P., 475, 801, 817, 905, 913, 1033, 1214
Birdwell, A. G., 109
Birner, S., 898
Birudavolu, S., 603
Bisi, O., 859
Blajnov, P. A., 1405
Blakesley, J. C., 1527
Blanchet, G. B., 1087
Blaser, S., 1573
Blattner, A. J., 347
Bleszynski, A. C., 779, 1461
Bloch, J., 707, 771
Blue, A., 207
Bobbert, P. A., 1101
Bobyl, A. V., 789
Bockowski, M., 251
Boebinger, G. S., 449
Boehm, G., 1567
Boeuf, F., 1523
Bogaart, E. W., 731
Bogusławski, P., 1331
Bollaert, S., 1523
Bonanni, A., 219
Bongiorno, A., 423
Borgström, M., 1543
Born, H., 633
Borseth, T. M., 181
Böttcher, R., 351
Bouanani-Rahbi, R., 363
Bougerol, C., 365
Boukari, H., 1311
Boulanger, D., 1147
Bourgognon, C., 1164
Bowdler, P. S., 101
Bowers, C. R., 519, 1455

Boxberg, F., 1563
Boxleitner, W., 471
Bracht, H., 97
Bragas, A. V., 1429
Brand, S., 373, 1125, 1166
Brasil, M. J. S. P., 483
Brauer, G., 99
Brendemühl, B., 1521
Brennan, K. F., 1049
Breunig, H. G., 1238, 1242
Brezna, W., 1275
Briddon, P. R., 91
Briggs, G. A. D., 695
Brown, E. R., 845
Brown, K. R., 1437
Bruchhausen, A., 1117
Brunkov, P. N., 789
Brunner, K., 657
Brunner, R., 801
Bruno, P., 1409
Bryce, B., 1437
Bryja, L., 1158
Bryksin, V. V., 1283
Buchanan, D. A., 495
Bucher, E., 1131
Buckle, L., 295
Buczkowski, A., 103
Bujkiewicz, S., 993
Bulutay, C., 231
Bundesmann, C., 165
Burdov, V. A., 857
Burke, T., 295
Burstein, E., 17
Buß, A., 559
Buss, D. D., 1591
Bussi, G., 1025, 1085
Butkute, R., 185
Buyanova, I. A., 259, 265, 271, 1319, 1387, 1399
Bykov, A. A., 457, 461, 579, 827
Byszewski, M., 1158

C

Cademartiri, L., 597
Caetano, E. W. S., 385, 1095, 1355
Cahill, J., 687
Cai, X. M., 319
Caldas, M. J., 1067, 1085
Caldwell, J. D., 519
Calleja, E., 221
Calleja, J. M., 221, 531, 715
Camacho B., A. S., 403
Campion, R. P., 333, 1371
Campman, K. L., 1143

Camps, I., 483
Canedy, C. L., 1525
Canonico, M., 1495
Cantarero, A., 617, 1017
Cantele, G., 859
Cao, H., 727
Capaz, R. B., 745, 1047, 1441
Capizzi, M., 265, 621
Cappy, A., 1523
Cardona, M., 0, 67, 1228
Carlin, J.-F., 1585
Caro, J., 1587
Carrier, P., 287
Cartoixà, X., 1421
Castner, T. G., 481
Cavada, B. S., 1095
Cavanna, A., 707, 771
Caymax, M., 1587
Celebi, Y. G., 255
Cerqueira, M. F., 857
Chakarvorty, R., 361, 1337
Chamarro, M., 1361
Chan, V., 1449
Chandrasekhar, H. R., 1579, 1581
Chandrasekhar, M., 1579, 1581
Chang, E., 1025
Chang, K. J., 95, 315, 735
Chang, R. K., 1517
Chang, S. K., 1489
Chang, Y.-C., 1421
Chao, K. A., 1192
Chaparro, S. A., 399
Charlton, M., 193
Charron, L., 1131
Cheah, K. W., 889
Chebotarev, A., 1184
Chebotareva, G., 1184
Chen, C., 157, 167, 297
Chen, C. Q., 217
Chen, C.-Y., 1027
Chen, G.-T., 1399
Chen, H., 1379
Chen, L., 297, 989, 1087
Chen, W. M., 259, 265, 271, 1319, 1387, 1399
Chen, X. D., 99
Chen, Y., 1351
Chen, Z. H., 1321
Cheng, E. S., 1192
Cheng, S.-J., 577
Cheng, W., 851
Chern, G. D., 1517
Chernyavsky, E., 1507
Chi, J., 331
Chicoine, M., 1491
Childs, D. T., 1547

Chinone, T., 299
Chiu, P. T., 347
Cho, H.-I., 445
Cho, M., 281
Cho, Y. H., 1515
Choi, M., 159
Choi, S. G., 125
Choi, S. W., 329
Chokomakoua, J. C., 533
Chong, T. C., 157, 167, 1381
Chovet, A., 1489
Chow, K. H., 169
Chow, W., 1204
Choy, W. C. H., 889
Christen, J., 301
Chrzan, D. C., 145, 611
Chung, I.-S., 1559
Chung, K. S., 1515
Chung, S. J., 223, 405, 533, 1379
Chyi, J.-I., 1399
Cibert, J., 365, 1160, 1168, 1299, 1311
Ciorga, M., 1303, 1413
Ciulin, V., 957
Claflin, B., 359
Clark, R. G., 1449
Clark, S. J., 373, 589, 1125
Clerjaud, B., 363
Cohen, E., 415, 511
Cohen, M. L., 3, 1047
Coleridge, P. T., 1009, 1347
Colla, E., 1585
Collins, C. J., 243
Combescot, M., 891, 1151, 1153
Cook, C. S., 65
Cooper, K., 665
Coquelin, M., 1015
Cornet, C., 787
Cota, E., 1411
Cottrell, S. P., 193
Cox, R. T., 1164
Cox, S. F. J., 169, 193, 255
Creffield, C., 805
Cremades, A., 1049
Cronenberger, S., 629
Crooker, S. A., 709, 973, 983, 1216
Cros, A., 617
Cucinotta, C., 1067
Cunningham, W., 207
Cuscó, R., 293
Czycholl, G., 1240

D

da Costa, J. A. P., 385, 1355
da Cunha Lima, I. C., 323

Dadgar, A., 143, 147
Dalakyan, A., 1192
Dalgarno, P. A., 645, 667
Dalmau, R. F., 211, 297
Dalpian, G. M., 225
Danckwerts, J., 1465
Daniel, B., 171
Danilchenko, B. A., 439
Danylyuk, S. V., 439, 459
Darton, R. C., 1027
DasGupta, K., 1289
da Silva, Jr., E. F., 177, 189, 837, 853, 1355, 1481, 1487
da Silva, Jr., J. B., 837
Das Sarma, S., 1441
Daudin, B., 617
Davies, G., 1573
Davies, J., 465
Davis, E. A., 193, 255
Davis, R. F., 297
Davydov, A., 109
Davydov, V. Y., 275
Däweritz, L., 235
D'Costa, V., 65, 217
Deacon, R. S., 1019, 1027
de Andrada e Silva, E. A., 1307
de Azevedo, W. M., 837
Deborde, J.-L., 437
Degani, M. H., 759
Degoli, E., 859
Dehaese, O., 987
Dekorsy, T., 235, 1539
Deligeorgis, G., 1531
de Lima, Jr., M. M., 1109
de Lima Filho, J. L., 1095
Del Sole, R., 859
Demikhovskii, V. Y., 961
Denker, U., 599
Denlinger, J., 303
de Oliveira, E. L., 1487
de Oliveira, M. C. F., 1095
DeSilva, A., 1097
Desjardins, P., 953
de Sousa, J. S., 853, 1487
de Sousa, R. L., 1069
de Souza, C. A., 959
Desrosiers, N., 1491
Destefani, C. F., 1325
Deubert, S., 681
de Vasconcelos, E. A., 837, 1487
Deveaud, B., 957, 1149
de Visser, A., 1003
Devreese, J. T., 673, 803
Deych, L. I., 929, 969
Dholakia, K., 1099
Diaconu, M., 351

Diakonov, A. M., 1212
Dialynas, G. E., 1531
Díaz-Arencibia, P., 933
Dietl, T., 56, 333, 1271, 1365, 1371
Dietz, N., 211
Di Gaspare, L., 1483
Dignam, M. M., 1176
Ding, D., 399
Ding, L., 765
Di Stefano, O., 975
Djurišić, A. B., 319, 879, 889
Dmitriev, A., 1005
Dobrowolska, M., 337, 361, 367, 703, 1337, 1357
Dobrowolski, W., 185
Doezema, R. E., 405
Döhler, G. H., 119, 149, 151, 959, 1220, 1385
Dolzhenko, D., 1202
Donchev, V., 705
dos Santos, B. E. C. A., 837
Drachenko, O., 1145
Dragosavac, M., 495
Dresselhaus, G., 7
Dresselhaus, M. S., 25
Drichko, I. L., 1212
Driscoll, D., 1220
Driscoll, D. C., 845
Driscoll, D. D., 775, 811, 1503
Drok, E., 439
Drouhin, H.-J., 1345
Du, M.-H., 625
Du, R. R., 917
Dubon, O. D., 223, 609, 1367
Dudiy, S. V., 265, 1121
Dujovne, I., 1087
Dultz, W., 1099
Dumchenko, D., 1131
Dupertuis, M.-A., 869, 881
Dvurechenskii, A. V., 583, 607, 1393
Dwir, B., 869
Dworzak, M., 633, 701
Dyakonova, N., 1523
Dynowska, E., 185

E

Eastman, L. F., 419
Eaves, L., 267, 293, 497, 993
Eble, B., 1293
Eckhause, T. A., 1190
Ediger, M., 645, 667, 739, 751
Edmonds, K. W., 333, 1371
Edwards, N. V., 111
Efanov, A. V., 137
Efros, A. L., 669

Egorov, A. Y., 963, 1541
Ehrenfreund, E., 669, 1174
Ekenberg, U., 1423
El Abidi, A., Ab., 1001
Elfving, A., 713
Elhassan, M., 801
El Kaaouachi, Ab., 1001
Ellenberger, C., 407
El-Wahidy, E. F., 1483
Elzerman, J. M., 44
Emin, D., 1265
Emiroglu, E., 1439
Emtsev, V. V., 275
Endicott, J., 267
Endo, A., 499, 1013
Endo, T., 339
Eng, K., 479
Enichlmair, H., 1275
Enriquez, H. B., 381
Ensslin, K., 407, 509, 775, 781, 811, 1503
Erenburg, S., 585
Erofeeva, I., 947, 949
Ertl, F., 1501
Esquinazi, P., 351, 1333
Eto, M., 821, 1447
Evaldsson, M., 1395, 1413
Evangelisti, F., 1483
Even, J., 787, 987
Ezawa, Z. F., 545, 563, 565

F

Fafard, S., 661
Fahy, S., 277
Fainstein, A., 1105, 1117
Faist, J., 1573
Fal'ko, V. I., 815
Fallahi, P., 779
Fanciulli, R., 1176
Fareed, Q., 241
Farias, G. A., 189, 759, 1355, 1481
Farias, G. D., 1095
Farrer, I., 1527, 1561
Fedorov, I. A., 1234
Fedorych, O. M., 335
Fehse, R., 985
Feil, T., 900
Felici, M., 265
Feng, Y., 1009
Fenn, J. G., 1182
Ferguson, A. J., 1449
Ferhat, M., 1315
Fernandez, J. R. L., 1095
Fernández, S., 221
Ferrand, D., 365, 1299, 1311

Ferreira da Silva,, 177
Ferry, D. K., 227, 801, 1263
Fikar, J., 753
Filatov, D. O., 619, 961
Filippone, F., 199
Filonovich, S. A., 675
Fischer, F., 443
Fischer, S. F., 437, 923
Fisher, P., 101
Fishman, G., 1123, 1345
Fletcher, R., 1009
Folliot, H., 787
Fomin, V. M., 673, 803
Fonoberov, V. A., 673
Forchel, A., 591, 661, 681, 725, 927, 1427
Foreman, B. A., 413
Förstner, J., 1465
Fortin, E., 1131
Fowler, B., 495
Fowler, D., 293, 497, 993
Foxon, C. T., 333, 1371
Franz, J., 0
Franz, R., 163
Frazier, R., 1319
Freeman, W., 1273
Freire, J. A. K., 759, 837, 1481
Freire, V. N., 189, 385, 759, 853, 1095, 1355, 1481, 1487
Friedland, K. J., 531
Friedrich, A., 1567
Fritsch, D., 201
Fritsche, R., 153
Fromhold, T. M., 993
Fronc, K., 677
Fu, D. J., 329
Fuerst, J., 137
Fuhrer, A., 781, 1503
Fujii, K., 127
Fujikawa, A., 967
Fujimoto, K., 349
Fujita, K., 173
Fujiwara, Y., 121, 131, 139
Fukuda, A., 563, 565
Fukuda, T., 655
Fukui, T., 877
Fukuma, Y., 349
Fung, K. K., 765
Fung, S., 99
Furdyna, J. K., 303, 337, 361, 367, 703, 1291, 1337, 1357, 1429
Furis, M., 709

G

Gabrielsen, S., 181
Gaj, J. A., 717, 1168, 1299, 1311
Galaktionov, E. A., 491
Galazka, R. R., 337
Galbraith, I., 749, 1236
Gallagher, B. L., 333, 1371
Gallart, M., 629
Galperin, Y. M., 1212, 1281
Galvão Gobato, Y., 483
Galzerani, J. C., 693
Gammon, D., 32
Gangilenka, V. R., 1097
Ganichev, S. D., 1309
Gao, X. P. A., 449, 1216
Garcia, B. A., 187
Garcia, J. C., 189
García, J. M., 621, 803
Garcia, R., 863
García-Cristóbal, A., 617, 1017
García-Rocha, M., 933
Garleff, J. K., 909
Garrett, G. A., 243
Garro, N., 617
Gartner, P., 661, 725
Gaska, R., 241
Gaskill, D. K., 419
Gatti, M., 859
Gaubas, E., 207
Gavrilenko, V., 947, 949, 1214
Gawron, W., 1557
Ge, W. K., 605, 765, 1011
Ge, Z., 367
Geelhaar, L., 497, 985
Geller, M., 1555
Geng, C., 387
George, T. F., 321
Gérard, J. M., 771, 1359
Gerardot, B., 669
Gerardot, B. D., 645, 667, 739, 751
Gerlovin, I. Y., 1417
Gershoni, D., 669, 1174
Gerthsen, D., 767
Geshi, M., 327
Gething, J. D., 557, 561
Ghali, M., 1383
Gherman, C., 1131
Ghosh, A., 1289
Ghosh, K., 1079
Gibson, M. C., 373, 1125
Giddings, A. D., 1371
Giebultowicz, T. M., 313, 337
Giedd, R. E., 1079
Giehler, M., 1565

Giglberger, S., 1309
Giguère, A., 1491
Gil, C., 1499, 1509
Gil, J. M., 169, 193
Gilinsky, A. M., 141, 629
Gilliot, P., 629
Giorgi, M., 1505
Giraud, R., 365
Girgis, A. M., 941
Girlanda, R., 671, 975
Gislason, H. P., 229, 433
Giustino, F., 423
Gladilin, V. N., 803
Glas, F., 583
Glavin, B. A., 999
Glembocki, O. J., 419
Glerup, M., 1043
Glosser, R., 109
Goel, N., 405, 533, 1379
Gogneau, N., 617
Goldman, R. S., 1190
Goldys, E. M., 261
Golka, S., 1275
Golnik, A., 1299
Goloshchapov, S. I., 81, 1186
Golub, L. E., 1309
Gomeniuk, Y. V., 237
Gomez-Iglesias, A., 1182
Gong, M., 99
Goodnick, S. M., 227
Gorelkinskii, Y. V., 107
Gornik, E., 471, 997, 1015
Goruganti, V., 331
Goss, J. P., 91
Gossard, A. C., 509, 775, 779, 811, 845, 1143, 1174, 1220, 1461, 1503
Goto, A., 1471
Gotthold, D. W., 419
Gottschalch, V., 455, 1541
Gou, W., 331
Govorov, A. O., 645, 683, 751, 813, 1379
Gozu, S.-i, 1533
Gradauskas, J., 1204
Graf, D., 811
Graf, M., 1573
Graf, P. A., 1127
Grahn, H., 1565
Granados, D., 621, 803
Grandjean, N., 719
Grau, A., 963
Grayson, M., 443, 915
Grbić, B., 407
Gregory, R. B., 111
Gribnikov, Z. S., 1200, 1473
Grillenberger, J., 89, 181
Groetzschel, R., 607

Groom, K. M., 1547
Groshaus, J. G., 501
Grossman, E. N., 1208
Grossner, U., 89, 181
Gruber, T., 991
Grundler, D., 431, 467
Grundmann, M., 165, 183, 197, 201, 351, 849, 875, 1333
Grynberg, M., 1129
Grzegory, I., 251
Gudimetta, S., 1499
Guezo, M., 987
Guffarth, F., 689, 769
Guha, S., 1079
Gui, Y. S., 829
Guillet, T., 891
Guimarães, F. E. G., 959
Guimarães, P. S. S., 1143
Guo, B. C., 605
Guo, S. P., 419
Gupta, J. A., 399, 1303
Gupta, R., 977, 1135
Gurudev Dutt, M. V., 32
Gurung, T., 657, 1327, 1357
Gusev, G. M., 461, 545, 789
Gutiérrez, M., 1547
Gutkin, A. A., 789
Gutowski, J., 1238, 1242, 1521
Guzman, E., 351
Gvozdić, D. M., 1423
Gwilliam, R., 249

H

Haboeck, U., 211
Haddad, G. I., 1200, 1473
Haegel, N. M., 1273
Hahn, E., 767
Hains, C. P., 603
Hajak, H., 898
Haller, E. E., 67, 73, 97, 209, 611
Halley, D., 365
Halm, S., 927
Ham, D., 1583
Hamada, I., 191
Hamhuis, G. J., 613
Hamill, S., 489
Hamilton, A. R., 1449
Hammerschmidt, T., 601
Han, I. K., 1489
Han, M. S., 203
Han, S.-H., 161
Han, Y., 841
Hanamoto, L. K., 1397
Hankiewicz, E. M., 1365

Hannewald, K., 1101
Hansel, S., 357, 795
Hansen, W., 451, 491, 807, 829
Hanson, M., 779, 1220
Hanson, M. P., 845
Hanson, R., 44
Hansson, G. V., 713
Haque, F., 657
Haque, I., 1246
Hara, M., 499
Hara, S., 877
Harada, S., 475, 1053
Harada, Y., 1353
Harayama, T., 475
Hardwick, D., 993
Harmand, J. C., 291
Harowitz, M., 697
Harris, J. J., 487
Harrison, P., 1204, 1565
Harus, G. I., 1003
Hasama, H., 1137
Hasanudin, 281
Hasegawa, A., 1459
Hasegawa, T., 855
Hashi, K., 1471
Hashii, S., 1083
Hashimoto, A., 1083
Hashimoto, K., 567, 1341
Hashimoto, Y., 1363
Hasko, D., 1439
Hasoon, F. S., 161
Hatano, T., 643, 785
Hatzopoulos, Z., 1531
Haug, R. J., 467, 503, 521, 559, 573, 715, 777, 809, 815
Hauschild, R., 896
Hautakangas, S., 261
Haverkort, J. E. M., 731
Hawrylak, P., 577, 685, 1303
Hayamizu, Y., 885, 887, 1575
Haywood, S. K., 977, 1135
He, H. T., 1011
He, R., 1087
Heben, M. J., 1031
Hein, G., 559, 1222, 1419
Heitmann, D., 467, 829, 1115, 1172
Heitsch, S., 183
Heller, E. J., 779, 1461
Helm, M., 235, 1539
Hemmi, M., 475
Henini, M., 561, 993
Henneberger, K., 163
Henriques, A. B., 979, 1019, 1397
Henriques Neto, J. M., 385
Heremans, J. J., 1379
Hermann, A. M., 161

Hernández-Calderón, I., 933
Herrle, T., 892
Herzinger, C. M., 201, 455
Hetterich, M., 171, 963, 1541
Hewaparakrama, K., 711
Hey, H., 995
Hey, R., 531, 1109, 1405, 1565
Heyn, C., 451, 467, 491, 807, 829, 1115
Hicks, J. L., 533
Higuchi, Y., 595
Hill, G., 687, 977
Hingerl, K., 219
Hiraka, K., 121
Hirakawa, K., 569, 635, 1071, 1224
Hirata, K., 1033
Hirayama, Y., 563, 565, 567, 825, 925, 1162, 1251, 1341
Hirjibehedin, C. F., 1087
Hirose, S., 743
Hirsch, A., 1222
Hjalmarson, H. P., 1269
Hoang, T. B., 677
Hoang, V. D., 1273
Hoch, M. J. R., 1317
Hochmut, H., 201
Hochmuth, H., 351
Hoffman, A., 211
Hoffmann, A., 633, 689, 701
Hofmann, T., 455
Högele, A., 751
Hohls, F., 503, 521, 559, 573, 777
Hoi, L. S., 843
Holland, M. C., 461, 489
Holtz, P. O., 705, 713
Holz, M., 431
Homma, Y., 931
Hommel, D., 1521
Honda, M., 783
Hönerlage, B., 629
Hong, S., 1551
Hong, Y. G., 259
Höntschel, J., 1501
Hopkinson, M., 267, 687, 977, 1547
Hori, H., 353, 355
Horio, N., 299
Horiuchi, K., 1083
Horsell, D. W., 1281
Hosea, T. J. C., 295
Hoshino, K., 627
Hosokawa, D., 129
Hours, J., 707, 771
Hovakimian, L. B., 1261
Hsieh, T.-P., 647
Hu, C.-M., 829
Hu, C. Y., 1453
Hu, X., 1441

Huang, C. F., 575
Huang, J., 779
Huang, S., 603
Huard, V., 1164
Huffaker, D. L., 603
Humlíček, J., 113, 753
Hummel, S., K., 103
Hunt, T. P., 1583
Hyomi, K., 1389

I

Iannaccone, G., 473
Ibáñez, J., 267, 293
Ichida, M., 1039
Ichikawa, T., 1075
Ignatev, I. V., 1417
Ihara, T., 887
Ihm, G., 735
Ihn, T., 407, 509, 775, 781, 811, 1503
Iijima, T., 1471
Ikegami, A., 1091
Ikezawa, M., 721, 729
Ikonić, Z., 1565
Ikonnikov, A., 947, 949, 1214
Ikushima, K., 505, 569
Imamura, H., 1409
Inada, M., 691, 855, 861
Inari, M., 877
Incze, A., 859
Indjin, D., 1565
Inoshita, T., 1435
Inoue, K., 1023, 1041
Inoue, M., 127, 817
Iotti, R. C., 671, 1469, 1569
Ishibashi, K., 801, 1021
Ishida, N., 1210
Ishida, S., 627, 635, 1287
Ishihara, S., 951
Ishihara, T., 1073
Ishi-Hayase, J., 1073
Ishii, H., 1285
Ishii, S., 1033
Ishiwata, Y., 1021
Ismail-Beigi, S., 1061
Itaya, S., 353, 355
Ito, T., 393
Itoh, H., 1575
Itoh, K. M., 38
Itskevich, I. E., 687, 965
Ivanchik, I., 1202
Ivanov, A. L., 757
Ivanov, S. V., 263, 991, 1387
Ivanov, Y. L., 955
Ivchenko, E. L., 1309

Iwamoto, S., 627, 1581
Iwase, Y., 823
Iwata, K., 565
Iwaya, S., 263
Iye, Y., 499, 773, 1013
Izadifard, M., 1399

J

Jackson, H. E., 657, 677, 711, 1327, 1357
Jacobs, P., 1313
Jaeckel, B., 153
Jaegermann, W., 153
Jahnke, F., 661, 725, 1240
Jakiela, R., 185
Jalil, M. B. A., 1381
Jancu, J.-M., 1307
Janik, E., 185, 1383
Jankó, B., 1291, 1337
Jantsch, W., 253, 1218
Jarjour, A., 695
Jaroszyński, J., 1271
Jasinski, J., 209
Jayasekera, T., 1279
Jeong, I. T., 1349
Jeong, T. S., 203
Jesson, D. E., 599
Ji, Z., 971
Jiang, H. X., 297
Jmerik, V. N., 263
Johnson, M. B., 533
Johnson, N. M., 1517
Johnson, S. R., 399, 1214
Jomard, F., 363
Jones, G. M., 1513
Jones, R., 91
Jones, W. B., 1121, 1127
Jorio, A., 25
Jouravlev, O. N., 1255
Jovanović, V., 1565
Jung, M., 635
Jungwirth, T., 1365
Junker, K., 111
Jusserand, B., 1105, 1117, 1295

K

Kacman, P., 313
Kagami, K., 343
Kageshima, H., 77, 389, 393
Kaidashev, E. M., 183, 197, 849, 875
Kakegawa, T., 1297
Kako, S., 627
Kalagin, A. K., 457

Kaliteevski, M. A., 1166
Kalt, H., 896, 935
Kalugin, N. G., 1222
Kambour, K., 1269
Kamins, T. I., 1577
Kaminska, E., 185
Kaminska, M., 251
Kamiyama, M., S., 263
Kampen, T. U., 1097
Kanamaru, S., 551
Kanbara, D., 1363, 1401
Kanda, Y., 79
Kane, B. E., 479, 1437
Kaneko, Y., 289
Kang, J., 315
Kang, T. W., 321, 329, 735, 793, 943, 1515
Kang, Y. H., 1551
Kanisawa, K., 825
Kapon, E., 869, 881, 919
Kappei, L., 1149
Karaiskaj, D., 67
Karczewski, G., 593, 677, 711, 973, 983, 1170, 1271, 1295, 1327, 1357, 1363
Kardynal, B. E., 1527, 1561
Karlsson, K. F., 705, 881, 919, 991
Karpierz, K., 1129
Karpovich, I. A., 961
Karrai, K., 645, 667, 739, 751
Kasic, A., 263
Kast, M., 471, 1015
Kasturiarachchi, T., 405
Katagiri, T., 523
Kataura, H., 1039
Katayama, S., 945
Katayama-Yoshida, H., 87, 105, 191, 315, 317
Katiyar, R. S., 623
Kato, K., 395
Kato, S., 1471
Kato, T., 1083
Katsumoto, S., 499, 773
Kauser, M. Z., 1049
Kavokin, A. V., 263, 1156, 1166
Kavokin, K. V., 557, 561
Kawaguchi, S., 867
Kawaguchi, Y., 635
Kawano, Y., 527
Kayanuma, I., 1389
Kayanuma, K., 1387
Kayanuma, Z. H., 1321
Kazukauskas, V., 207
Kearns, J., 103, 1267, 1457
Keeble, D. J., 193
Keeth, J. G., 1079
Kehl, T. W., 1079
Keller, D., 1301

Keller, M. W., 819
Kelly, M. J., 249
Kelsall, R. W., 1565
Kemmochi, K., 317
Kent, P. R. C., 1369
Kępa, H., 313
Kereselidze, T., 737
Khan, M. A., 217, 297
Khatami, S., 115
Kheng, K., 615, 629, 1164
Khlobystov, A. N., 1043
Khoi Le, V., 337
Khokhlov, D., 1202
Khomitskiy, D. V., 961
Kibis, O. V., 1045
Kičin, S., 781
Kida, E. S., 1033
Kida, M., 1035, 1053
Kida, N., 475
Kido, G., 1471
Kiesel, P., 149
Kieseling, F., 163
Kim, C. S., 1525
Kim, D., 205
Kim, D. H., 1349
Kim, D. Y., 1515
Kim, E. K., 639
Kim, G.-H., 575
Kim, H., 902
Kim, H.-M., 1515
Kim, J., 1399
Kim, J.-E., 1559
Kim, J. W., 321
Kim, K., 1121, 1127
Kim, K. W., 409, 411, 999, 1234
Kim, M., 1525
Kim, N., 321
Kim, S. S., 587
Kim, T. S., 1349
Kim, Y.-H., 625, 1031
Kim, Y.-S., 95
Kimura, A., 349
Kipp, T., 1115
Kirchheim, R., 151
Kirchner, C., 896, 991
Kirk, K. J., 465
Kirk, W. P., 175
Kirscht, F., 103, 1267
Kirste, A., 795
Kiselev, A. A., 409, 411
Kishimoto, T., 1459
Kita, T., 387, 1353
Kitaev, V., 597
Kitamura, T., 1391
Kitazawa, H., 1471
Klapwijk, T. M., 435, 1587

Klar, O., 1220
Klarer, D., 179
Klein, A., 153
Klein, N., 439, 459
Kleinert, P., 1283
Klimov, V. I., 709
Kling, R., 896, 991
Klingshirn, C., 171, 896, 935
Klix, W., 1501
Kloc, C., 1087
Klochikhin, A., 275, 935
Kłopotowski, Ł., 133, 1113
Klotsa, D. K., 1093
Knap, W., 1523
Kneip, M., 1158, 1301
Kneissl, M., 1517
Knorr, A., 1465
Kobayashi, K., 773
Kobayashi, N., 447
Kobayashi, T., 931
Kobori, H., 127, 867
Koch, F., 1188, 1196
Kochelap, V. A., 439, 999
Kocher, G., 253
Kochereshko, V., 973, 983
Koenraad, P. M., 803
Koguchi, N., 679
Köhler, K., 1204
Koiller, B., 745, 1441
Koizumi, A., 121, 131, 139
Komiyama, S., 505, 569, 635
Komori, Y., 555
Könemann, J., 809, 815
Konnikov, S. G., 789
Kono, J., 1143
Konofaos, N., 1463
Konttinen, J., 283
Kop'ev, P. S., 263
Korkusinski, M., 685
Koshida, N., 797
Koshino, M., 537
Kossacki, P., 1160, 1168, 1311
Kossacki, P. W., 1299
Kossut, J., 185, 593, 677, 711, 1129, 1170, 1327, 1357, 1363, 1383
Kosuge, T., 299
Kotera, N., 939
Kouvetakis, J., 65
Kouwenhoven, L. P., 44
Kovalev, A. E., 519
Kowalczyk, E., 185
Kowalik, K., 717
Koyama, M., 817, 855
Koyanagi, T., 349
Koyano, S., 325
Kozhanov, A., 1005, 1202

Kozlov, D., 947, 949
Kozub, V. I., 1281
Kozumi, S., 563
Krämer, S., 119
Křápek, V., 113, 753
Kratzer, P., 311, 601, 745
Krebs, O., 717, 1293
Krebs, R., 681
Kroemer, H., 123
Kröger, R., 1521
Krokhin, A. A., 993
Kronenwerth, O., 431
Kruglova, M. V., 619
Kruse, C., 1521
Kubiak, L., 1557
Kubisa, M., 1158
Kubo, K., 343
Kubo, Y., 1075
Kuchar, F., 801
Kudrawiec, R., 283, 291, 417
Kudryashev, V. M., 461
Kuhn, T., 155, 883, 907, 1242
Kuhns, P. L., 1317
Kuk, Y., 902
Kulatov, E., 365
Kulik, J., 111
Kulipanov, G., 585
Kulyuk, L., 1131
Kumada, N., 563, 565, 567
Kümmell, T., 591
Kunc, K., 1230
Kunze, U., 437, 923
Künzel, H., 1419
Kurakin, A. M., 459
Kuriyama, K., 289
Kuroda, K., 1581
Kuroda, N., 281
Kuroda, S., 345, 365, 1323, 1351
Kuroda, T., 299, 679, 1391
Kuroiwa, Y., 325
Kurtze, H., 661
Kusakabe, K., 327, 1359
Kushida, K., 289
Kutrowski, M., 361, 957, 1337
Kuznetsov, A. Y., 181, 789
Kuznetsov, O., 947, 949
Kvietkova, J., 1541
Kvon, Z. D., 445, 487, 491, 911
Kyrychenko, F. V., 1319

L

Labbé, C., 787, 987
Lacharmoise, P., 1105
Lai, K. T., 1135

Laikhtman, B., 981, 1198
Lamas, T. E., 545, 789, 979
Landi, S. M., 1543
Langbein, W., 1232
Lapilli, C. M., 543
Lapointe, J., 1303
Largeau, L., 363
La Rocca, G. C., 1307
Larsson, M., 713
Lassen, B., 871, 873
Lau, K. M., 765
Laurent, S., 1293
Lawler, H. M., 1194
Lax, B., 11
Lazić, S., 221
Leburton, J.-P., 44, 853
Lechner, C., 1339
LeCorre, A., 987
Lee, 902
Lee, B. C., 1349
Lee, B. W., 1049
Lee, C.P., 1192
Lee, H., 1583
Lee, J., 900
Lee, J. C., 329
Lee, J.-C., 647
Lee, J.-H., 445
Lee, J. I., 1489
Lee, M. B., 1489
Lee, S., 703
Lee, S. H., 943
Lee, S. J., 321, 329, 735, 793
Lee, S.-J., 639
Lee, U. H., 1551
Lee, W. C., 329
Lee, W. S., 203
Lee, Y. T., 1559
Leibiger, G., 455, 1541
Leifer, K., 919
Leite, J. R., 189, 323, 385, 461, 545, 789, 837, 1095, 1481
Leite Alves, H. W., 1069, 1095
Lemaître, A., 363, 583, 717, 1293
Lenzner, J., 849, 875
Leo, K., 1176
Léotin, J., 1145
LeRoy, B. J., 1461
Le Thomas, N., 1531
Leturcq, R., 811
Leung, Y. H., 319, 879, 889
Levi, D. H., 161
Levichev, V. V., 961
Levinson, Y., 501
Lew Yan Voon, L. C., 659, 871, 873
Leymarie, J., 263
Li, B. S., 1137

Li, C., 653, 937, 989
Li, D., 889
Li, H.-D., 843
Li, J., 1139
Li, L., 491
Li, L. H., 291
Li, L. J., 1043
Li, L.-J., 1027
Li, S. X., 209
Li, X., 32
Li, Y., 331
Li, Z., 1577
Liang, C.-T., 575
Liang, H., 765
Liang, W., 1263
Liang, X. X., 233
Liao, C. Y., 73, 611
Libal, A., 1291
Lichti, R. L., 169, 193, 255
Liddle, J. A., 609
Liew, T., 1381
Liliental-Weber, Z., 209, 611
Lilly, M. P., 453, 913, 917, 921
Lim, A. C. H., 977
Lim, K. S., 195, 587
Lim, W. L., 361, 367
Limpijumnong, S., 259
Lin, J.-F., 1033
Lin, J. Y., 297
Lin, R.-M., 647
Lin, W.-L., 337
Lin, Y., 549
Linder, E., 415, 511
Lindgren, T., 177
Lindle, J. R., 1525
Lindsay, A., 277, 455
Linfield, E. H., 491, 1255, 1573
Ling, C. C., 99
Linke, H., 823, 1257
Linnik, T. L., 999
Lino, A. T., 189
Lippold, G., 165
Lipsanen, H., 117, 1537
Lisauskas, A., 1204
Lischka, K., 253
Lisyansky, A. A., 929, 969
Littler, C. L., 109
Litton, C. W., 941
Liu, C., 1407
Liu, F., 841
Liu, H., 319
Liu, H. Y., 1547
Liu, M., 653
Liu, R., 215, 217, 1495
Liu, X., 303, 337, 361, 367, 1337
Liu, Y., 359, 1583

Liu, Z. T., 889
Llorens, J. M., 617
Loata, G., 1220
Lock, D. A., 1545
Löffler, T., 1220
Löfgren, A., 1257
Loginenko, O., 1188, 1196
Lohmeyer, H., 907, 1521
Löhneysen, H. v., 909
Löhr, S., 491, 829
Long, A. R., 465, 489
Loo, R., 1587
López-Richard, V., 637
Lord, J. S., 193
Lorenz, M., 183, 197, 201, 351, 401, 849, 875
Lorke, A., 651, 733
Loualiche, S., 787, 987
Louie, S. G., 1047, 1061
Lu, H., 209, 263, 285
Lu, J., 1317
Lu, J.-Q., 1495
Lu, X., 73, 843
Lu, Z. D., 765
Lucey, D. W., 847
Lundsgaard Hansen, J., 97
Luppi, E., 859
Łusakowski, J., 1523
Lüth, H., 439, 459
Lüttjohann, S., 733
Luukanen, A., 1208
Lyanda-Geller, Y., 763, 1427, 1431

M

Maan, J. C., 641
MacFadzean, S., 489
Machida, T., 505
MacKenzie, M., 489
Mackowski, S., 657, 677, 711, 1327, 1357
Macucci, M., 473
Madejczyk, P., 1557
Maeda, N., 447
Maeda, T., 551
Maehashi, K., 1023, 1041
Maemoto, T., 817
Maestre, D., 1049
Magnusson, B., 285
Magri, R., 123, 375, 859
Maia, Jr., F. F., 1355
Maingault, L., 615
Maire, N., 777
Majewski, J. A., 1403
Majid, A., 143, 147
Majkrzak, C. F., 313

Makarovsky, O., 497
Maki, H., 1021
Makino, T., 691, 861
Makler, S. S., 483, 675
Maksimov, A. A., 1301
Maksym, P., 649
Malikova, L., 109
Malin, L., 1585
Malissa, H., 1218
Malko, A., 869
Malzer, S., 119, 149, 151, 959, 1220, 1385
Mamaluy, D., 799
Mampazhy, A., 653
Manciu, F. S., 847
Mandal, K. C., 159
Mani, R. G., 517
Manning, R. J., 1182
Mano, T., 731
Mansurov, V. G., 719
Manzke, G., 163
Maranowski, K. D., 509, 1461
Marcet, S., 365
Marconcini, P., 473
Marderfeld, I., 1174
Mariani, E., 521
Marie, X., 751, 1293, 1361
Mariette, H., 365, 615, 629, 1353
Marko, I. P., 681
Marlow, C. A., 1257
Marques, G. E., 483, 637, 1325
Mars, D. E., 223
Marsal, L., 1353
Mart, V., 1267
Martín, M. D., 133, 1113, 1170
Martin, T. P., 823
Martinez, A. I., 187
Marui, H., 213, 279, 301, 421
Maruyama, W., 723
Masago, A., 87
Maślana, W., 1168, 1299
Masumoto, Y., 721, 723, 729, 894, 1351, 1417
Masut, R. A., 953
Mata, O. V., 839
Matagne, P., 44
Matsuda, K., 79
Matsuishi, K., 1075
Matsumoto, K., 855, 1023, 1041
Matsumoto, T., 1055
Matsushita, D., 395
Matsusue, T., 1363, 1401
Matsuzaka, S., 1453
Matthews, A. J., 557, 561
Mattila, M., 117
Maude, D. K., 815, 1523
Mauguin, O., 363

Maung, S. M., 945
Maximov, G. A., 619
Maximov, M. V., 1529
May, S. J., 347
Mazilu, A., M., 1182
Mazzucato, S., 621
Mbenkum, B. N., 401
McCombe, B. D., 847
McConville, D., 985
McFarland, R., 479
McGhee, E. J., 751
McMullen, T., 465, 489
Medeiros-Ribeiro, G., 1475
Mei, J., 215
Meier, C., 733
Meinhold, D., 1176
Meisels, R., 801
Melhuish, G. P., 461
Melnik, R., 871, 873, 1485
Mendach, S., 451
Mendez, E. E., 549
Menéndez, J., 65, 1228
Merc, U., 997
Merlin, R., 1190, 1228, 1313, 1429
Merrick, M., 295
Metzner, C., 1385
Meyer, J. R., 1525
Meyer, R. C., 405
Meyer, T. A., 67
Meziani, Y. M., 1523
Miao, Q., 1087
Michikita, T., 105
Michler, P., 661, 725
Mickevicius, J., 241
Mihara, T., 1035, 1053
Mikhov, M. K., 937
Milanović, V., 1565
Milekhin, A. G., 693
Miller, M., 1182
Miller, R. B., 1164
Mills, Jr., A. P., 449
Millunchick, J. M., 1315
Mil'shtein, S., 485, 1499, 1509, 1511
Minami, F., 679, 1459
Mino, H., 593, 967, 971, 1363, 1401
Minor, A., 223, 609
Miranda, R. P., 675
Mirčetić, A., 1565
Mirin, R. P., 819
Mishchenko, A. M., 457, 827
Mishra, U., 441
Misiewicz, J., 283, 291, 417, 1158
Miska, P., 787
Missous, M., 1135

Mitsumori, Y., 1459
Miyake, T., 1037
Miyamoto, K., 1035
Miyashita, N., 1517
Miyashita, S., 1251
Miyazawa, T., 643, 743
Mizrahi, U., 669
Mizuta, H., 797, 1589
Mochizuki, M., 1083, 1091
Moehl, S., 615
Molenkamp, L. W., 657, 1301, 1427
Molinari, E., 1025, 1067, 1085
Molnár, B., 1335
Momo, N., 1589
Monakhov, E. V., 789
Monemar, B., 261, 263, 285, 705, 991
Monroy, E., 617
Moon, C.-Y., 95
Moret, N., 869
Mori, N., 797
Morier-Genoud, F., 1149
Morimoto, T., 475, 913, 1053
Morino, M., 565
Morita, K., 1453
Morkoc, H., 1263
Morozov, S. V., 619
Morrison, M. A., 1279
Moskalenko, E. S., 705
Motohisa, J., 877
Moulton, W. G., 1317
Mourokh, L. G., 813, 905
Moutanabbir, O., 1491
Mowbray, D. J., 50, 1547
Moyer, H. P., 1208
Mu, Y.-M., 1549
Mücklich, A., 1539
Mudryi, A. V., 275
Muessig, H.-J., 75
Mukai, T., 215, 279, 421
Mukashev, B. N., 107
Mullen, K., 1279
Munekata, H., 307, 339
Muraki, K., 563, 565, 567, 1341
Muraoka, K., 395
Murayama, A., 1321, 1387, 1389
Murayama, M., 1391
Murdin, B. N., 295
Murphy, S. Q., 533
Murray, B., 823
Musikhin, Y. G., 789
Mussler, G., 235, 237
Myles, C. W., 1269
Myong, S. Y., 195

N

Na, J. H., 865
Nachtwei, G., 559, 1222, 1419
Nafidi, Ab., 1001
Nafidi, Ah., 1001
Nagahara, S., 1353
Nagai, M., 353
Nagami, T., 1589
Nagata, A., 931
Naito, R., 475
Nakada, K., 563
Nakagawa, T., 553
Nakahara, J.-i., 931
Nakai, M., 817
Nakamura, J., 371, 951
Nakamura, K., 131, 139
Nakaoka, T., 741
Nakarmi, M. L., 297
Nakasaki, Y., 395
Nakashima, A., 817
Nakata, Y., 643
Nakayama, M., 205
Nakayama, T., 173, 1089, 1285
Nakazato, K., 797
Nanao, T., 867
Naranjo, F. B., 221
Nardin, D., 621
Narukawa, Y., 215
Nathan, M. I., 359
Natori, A., 371, 951
Nauen, A., 777
Nauka, K., 1577
Nazarov, Y. V., 1255
Ndawana, M. L., 1259
Nee, T.-E., 647
Nekrutkina, O., 991
Nenashev, A. V., 1393
Neu, G., 719
Neumann, S., 119
Neverov, V. N., 1003
Newaz, A. K. M., 549
Nguyen, T., 677
Ni, W. X., 713
Nicholas, R. J., 1019, 1027, 1043
Nickolaenko, A. E., 141
Nieminen, R. M., 809
Nikiforov, A., 585
Nikiforov, A. I., 1212
Nikitenko, S., 585
Nikitin, A. Y., 719
Nikolaev, N. I., 757
Nikolaev, V. V., 1156
Nikolitchev, D. E., 619
Ning, C. Z., 1139
Ninno, D., 859
Nishi, Y., 649
Nishibayashi, K., 1321
Nishimoto, Y., 595
Nishioka, M., 655
Nishizawa, N., 345
Nitta, J., 549
Niu, Q., 539
Nobis, T., 849, 875
Noblitt, C., 159
Noborisaka, J., 877
Noh, S. K., 639, 793
Nomokonov, D. V., 457, 579, 827
Nomura, S., 397, 571, 1162
Nötzel, R., 613, 731
Noveski, V., 211
Novikov, P. L., 607
Novikov, S. V., 1371
Ntalaperas, D., 1463
Nuckolls, C., 1087
Nuntawong, N., 603
Nylandsted Larsen, A., 97

O

Oberli, D. Y., 869
Ochiai, Y., 475, 823, 913, 1033, 1035, 1053, 1083, 1091
Oda, M., 1089
Oda, S., 1589
Oe, K., 387
Off, J., 927
Offermans, P., 803
Oh, J. E., 943
Oheda, H., 85
Ohki, S., 1471
Ohno, H., 1453
Ohno, Y., 1023, 1041, 1453
Ohta, H., 121
Ohtani, N., 1533
Ohyama, T., 127
Oiwa, A., 339
Oka, Y., 1321, 1387, 1389
Okamoto, A., 1287
Okamoto, T., 469, 527, 555
Okuno, T., 1351
Oliveira, R. F., 979
Olshanetsky, E. B., 487, 519
Omiya, H., 215, 279, 421
Omling, P., 1257
Onari, S., 1075
Onida, G., 859
Onishchenko, E. E., 755
Ono, K., 925
Ooi, S., 1033
Oonishi, D., 823

O'Reilly, E. P., 277, 455, 695, 985
Orschel, B., 103, 1457
Ortner, G., 685
Osborn, K. D., 819
Ossau, W., 1301
Ossicini, S., 859
Ota, T., 643, 785
Oto, K., 523, 967
Otterburg, T., 869
Oulton, R., 661
Oum, J. H., 1551
Ozaki, N., 345
Ozawa, H., 951
Ozin, G. A., 597

P

Paarmann, A., 689
Pacher, C., 997, 1015
Pakuła, K., 257
Palacios, T., 441
Palmstrøm, C. J., 125
Pan, C.-C., 1399
Pan, J. L., 1178
Pang, Q., 605
Pang, Y., 1077
Parenty, T., 1523
Parfitt, D. G. W., 529, 1045
Park, C. M., 943
Park, H. Y., 1559
Park, Y. S., 203, 943
Parrot, R., 1147
Pascher, H., 137
Paskov, P. P., 261, 285
Paskova, T., 261, 285
Pasquarello, A., 423
Passow, T., 963
Paszkiewicz, B., 417
Paszkiewicz, R., 417
Patanè, A., 267, 293, 497, 993
Patel, A., 1455
Patriarche, G., 583
Patton, B., 1232
Paul, D. J., 495
Pavelescu, E. M., 283
Pearton, S. J., 1319, 1399
Peeters, F., 1335
Pei, S. S., 1549
Pelekanos, N. T., 1531
Pelucchi, E., 869
Peng, P., 1549
Pepe, I., 177
Pepper, M., 491, 495, 1289
Perales, O., 839
Perea, J. I., 699

Pereira, T. A. S., 1481
Perez, F., 1295
Permogorov, S., 935
Persson, C., 177
Pessa, M., 283
Peter, E., 707, 771
Peters, J. A., 1379
Petras, I., 1463
Petroff, P. M., 645, 667, 669, 705, 739, 751
Petrov, P. V., 955
Petruska, M., 709
Petter, K., 1115
Pfannkuche, D., 535
Pfeiffer, L. N., 415, 449, 511, 885, 887, 1313, 1575
Pickenhain, R., 197, 1333
Pieruccini, M., 671
Pierz, K., 715, 777
Piltz, J., 165
Pinczuk, A., 513, 1087
Pinheiro, J. A., 1095
Pinheiro, J. R., 1095
Pioda, A., 781
Pioro-Ladrière, M., 1303
Piotrowska, A., 185
Piotrowski, A., 1557
Piqueras, J., 1049
Pires, M. P., 1543
Piroto Duarte, J., 169, 193
Pistone, G., 975
Platero, G., 805, 1411
Platonov, A., 973
Platz, C., 787
Plaut, A. S., 461
Płochocka, P., 1168, 1299
Ploog, K. H., 235, 531, 1405
Plotnikov, A. E., 911
Pokatilov, E. P., 673
Polimeni, A., 265, 621
Pollak, F. H., 109
Ponce, F. A., 0, 213, 215, 217, 279, 301, 421, 863
Ponomarenko, L., 1003
Ponomarev, I. V., 929, 969
Poole, P. J., 1347
Popov, V., 1507, 1523
Popović, D., 1271
Pöppl, A., 351
Porowski, S., 251
Porras, D., 699
Portal, J. C., 445, 457, 487, 545, 579, 911
Porţeanu, H. E., 1188, 1196
Portella-Oberli, M. T., 957, 1149
Portnoi, M. E., 529, 557, 561, 1045, 1156
Potemski, M., 251, 685, 1158
Pötschke, K., 769

Poweleit, C., 1263
Pozder, S., 1495
Prado, S. J., 637
Prasad, P. N., 847
Prettl, W., 1309
Priller, H., 896
Prins, A. D., 931
Proietti Zaccaria, R., 1469
Proskuryakov, Y. Y., 491
Prost, W., 119
Przezdziecka, E., 185
Przybylińska, H., 253
Puhle, C., 357
Pulci, O., 859
Puller, V. I., 905
Pupysheva, O., 1005
Pusep, Y. A., 959, 1007
Puska, M. J., 809

Q

Qi, Y. D., 765
Qin, J., 1513
Qin, W., 111
Quinn, J. J., 525, 541
Quivy, A. A., 545, 789, 979

R

Radmilovic, V., 609
Radulovic, N., 1485
Radzewicz, C., 1168
Rahm, A., 849, 875
Rahman, M., 207
Raker, T., 155
Ramdas, A. K., 73
Ramierez, A. P., 449
Ram-Mohan, L. R., 941
Ramon, G., 1431
Rapoport, I., 103, 1267, 1457
Rappl, P. H. O., 1397
Rappoport, T. G., 1291
Räsänen, E., 809
Rastelli, A., 599
Rauh, R. D., 159
Reason, M., 1190
Rebelo, L. M., 1355
Redliński, P., 1291, 1337
Ree, D. D., 719
Reese, O., 659
Regelman, D. V., 669
Reinecke, T. L., 763, 1343, 1427, 1431, 1451
Reinwald, M., 781, 892, 898, 900, 1172
Reißmann, L., 767

Reithmaier, J. P., 591, 681
Reitmeyer, Z. J., 297
Ren, F., 1319, 1399
Ren, H.-W., 721, 723
Ren, S.-F., 851
Renard, V., 445, 487, 911
Renner, F. H., 1220
Reno, J. L., 453, 519, 917, 921
Reno, J. R., 913
Reusch, T. C. G., 151
Reuss, F., 896, 991
Reuter, D., 407, 437, 503, 641, 651, 733, 1158
Reznitsky, A., 935
Rheinländer, B., 201
Rhode, M., 467
Ribeiro, E., 1475
Ribeiro, M. B., 959
Rice, J. H., 695, 865
Richard, S., 1123, 1345
Richter, K., 1375
Riechert, H., 497, 963, 985, 1541
Riemann, H., 67, 73
Riffe, D. M., 1226
Righi, M. C., 375
Riposan, A., 1315
Ristein, J., 377
Ritchie, D. A., 491, 557, 575, 665, 1255, 1289, 1527, 1561, 1573
Robbe, V., 461
Robbins, D. J., 1547
Robbins, P. D., 709
Robinson, J. T., 609
Robinson, J. W., 695, 865
Rocher, N., 461
Röder, U., 179
Rodrigues, S. C. P., 323
Rodriguez, S., 73
Rodt, S., 767, 769
Rogach, A. L., 747
Rogge, S., 1587
Rolo, A. G., 747
Romanov, D. A., 1133
Romanov, K. S., 955
Römer, R. A., 1093, 1259
Rontani, M., 643
Roshko, S. H., 1255
Roskos, H., 1220
Roskos, H. G., 1204
Ross, G. G., 1491
Ross, Jr., J. H., 331
Rossi, F., 671, 1469, 1569
Rössler, U., 1339
Roth, S. F., 915
Rougemaille, N., 1345
Rowe, A. C. H., 135, 477

Rückmann, I., 1238, 1242
Ruden, P. P., 359, 1049, 1377
Rudin, S., 239, 243, 1451
Rudra, A., 919
Ruh, E., 775
Ruini, A., 1025, 1067, 1085
Rumyantsev, S., 1523
Rutkowski, J., 1557
Ryabchenko, S. M., 1329
Rybchenko, S. I., 687, 965
Ryczko, K., 1158
Ryzhkov, V., 1192

S

Saarinen, K., 261
Sabathil, M., 799
Sabbah, A. J., 1226
Sablikov, V. A., 911
Sachdeva, R., 91
Sachrajda, A. S., 1303, 1413
Sadki, K., E. S., 1033
Sadofyev, Y. G., 399, 1214
Sadowski, J., 335
Sadowski, M. L., 257
Saeki, A., 1041
Safonov, S. S., 491, 1255
Sahoo, Y., 847
Saiki, K., 281
Saito, R., 25
Saito, S., 1037, 1055, 1057
Saito, T., 741
Saitoh, T., 447
Sakai, S., 207
Saku, T., 563, 565, 567, 1162, 1341
Sakuma, H., 569
Sakuma, Y., 743
Salinero, V. T., 373
Saminadayar, K., 1164
Sampath, A. V., 243
Samson, G., 937, 989
Samuelson, L., 1257
Sanada, H., 1453
Sands, D., 965
Sandu, T., 175
Sanguinetti, S., 679
Sankowski, P., 313
Sano, A., 773
Santoprete, R., 745
Santos, M. B., 405, 533, 1379
Santos, P. V., 995, 1017, 1109
Sarachik, M. P., 435
Saraniti, M., 227
Sarkar, D., 531, 715
Sasa, S., 817
Sasaki, M., 1459
Sasaki, T., 475, 913, 1033, 1035, 1053, 1083
Sasaki, T. K., 1091
Satanin, A. M., 857
Sato, D., 397, 1162
Sato, H., 349
Sato, K., 317, 325
Sauer, R., 179
Sauerwald, A., 591
Savasta, S., 671, 975
Savchenko, A. K., 491, 1255, 1281
Sawabe, T., 894
Sawada, A., 563, 565
Sawaki, N., 595, 1154
Sawicki, M., 333, 1371
Säynätjoki, A., 1537
Scalari, G., 1573
Scalbert, D., 1383
Scarpa, G., 1567
Scarpulla, M. A., 223, 1367
Schaff, W. J., 209, 263, 285
Schäffler, F., 467
Schallenberg, T., 657
Scheffler, M., 311, 745
Scherbakov, A. V., 1301
Schiettekatte, F., 1491
Schleser, R., 775
Schlesser, R., 211, 297
Schliwa, A., 767, 769
Schmeißer, D., 75
Schmidegg, K., 219
Schmidt, B., 1539
Schmidt, H., 201, 351
Schmidt, O. G., 599
Schmidt-Grund, R., 201
Schmitzer, H., 1099
Schmult, S., 892
Schneider, P., 1309
Schoenfeld, W. V., 705
Scholz, F., 387, 927
Scholz, R., 1307
Schömig, H., 927
Schramm, A., 807
Schreiber, M., 1259
Schröter, P., 1172
Schubert, M., 165, 201, 401, 455
Schuh, D., 443, 892, 898, 900, 915, 923
Schulhauser, C., 751
Schüller, C., 1115, 1172
Schulman, J. N., 1208
Schulthess, T. C., 1369
Schulz, S., 807
Schulze, S., 629, 693
Schulze-Wischeler, F., 503, 521
Schumacher, S., 1240
Schuster, R., 898

Schwab, M., 661
Schweizer, H., 387
Schwertberger, R., 591
Scolfaro, L. M. R., 189, 323, 385, 1481
Seabra, A. C., 789
Seamons, J. A., 453, 917, 921
Sebald, K., 1521
See, P., 1527
Seemann, M., 163
Seghier, D., 229, 433
Seguin, R., 767
Seike, M., 317
Seki, S., 1041
Sekine, N., 1224
Sekizawa, K., 325
Seliuta, D., 1204
Sellers, I. R., 1547
Sellier, H., 1587
Semenov, Y. G., 1329
Senawiratne, J., 211
Senba, S., 349
Senellart, P., 707, 717, 771
Sénès, M., 1359
Seo, K., 1321
Setter, N., 1585
Setzer, A., 351
Sevik, C., 231
Shabaev, A., 669
Shailos, A., 905, 913
Sham, L. J., 32, 1431
Shamirzaev, T. S., 137, 141, 629
Shao, J., 835, 843
Sharp, I. D., 611
Sheard, F. W., 993
Shegai, O. A., 141
Shelushinina, N. G., 1003
Shen, H., 243
Shen, H.-T., 647
Shen, R., 593
Shen, S., 367
Sheng, C. X., 1081
Sheng, W., 577, 685
Shengurov, V. G., 619
Shi, J., 1063
Shibasaki, I., 1287
Shields, A. J., 665, 1527, 1561
Shields, P. A., 1019, 1027
Shimada, T., 1589
Shimizu, T., 1471
Shimomura, T., 205
Shirai, K., 87
Shiraishi, K., 389, 427, 1433
Shirley, E. L., 1194
Shoji, O., 325
Shon, Y., 321
Shorubalko, I., 1257

Shtinkov, N., 953
Shtrikman, H., 501
Shubina, T. V., 263, 991
Shumway, J., 697
Shur, M. S., 241, 947, 949, 1523
Shvartsman, L. D., 981, 1133, 1198
Sielemann, R., 89
Sigrist, M., 781, 1503
Silov, A. Y., 1405
Silvestri, H. H., 97
Simmons, J. A., 519, 913
Simmons, M. Y., 491
Simon, S. H., 461
Simserides, C., 341
Singh, J., 441
Singh, M., 441
Singh, M. R., 1246
Singh, S. P., 839
Sinning, S., 235
Sinova, J., 1365
Sipahi, G. M., 323
Sipatov, A. Y., 313
Sirigu, L., 1573
Širmulis, E., 1204
Sirtori, C., 1145
Sitar, Z., 211, 297
Sitter, H., 219
Sivoththaman, S., 1535
Skolnick, M. S., 50, 687, 1547
Skorupa, W., 99, 1539
Skotnicki, T., 1523
Skromme, B. J., 297, 937, 989
Skuras, E., 465
Smagina, Z. V., 607
Smirnov, A., 159, 813
Smirnov, D., 1145
Smirnov, I. Y., 1212
Smit, G. D. J., 1587
Smith, A., 757
Smith, D. L., 1377
Smith, J. D., 695
Smith, J. M., 645, 667
Smith, K., 207
Smith, L. M., 657, 677, 711, 1327, 1357
Smith, T., 1009
Smoliner, J., 1275
Soares, J. A. N. T., 837
Södervall, U., 261
Sogawa, T., 995
Sohn, J. Y., 1216
Sokolov, V. N., 999, 1234
Solin, S. A., 135, 477
Solnyshkov, D. D., 263
Solomon, G. S., 727
Somers, A., 591
Somisetty, S., 485

Song, H. Z., 643
Song, H.-Z., 741
Song, W., 549
Song, Y., 835
Song, Y. J., 902
Soni, R. K., 623
Sonkusare, A., 965
Sopanen, M., 117
Sotomayor, N. M., 461, 789
Soukiassian, P. G., 381
Souma, K., I., 1389
Souma, S., 735, 793
Sousa, J. S., 1481
Souza, P. L., 1543
Souza Dantas, N., 177
Souza de Almeida, J., 177
Spataru, C. D., 1047, 1061
Spemann, D., 183, 201, 351
Sque, S. J., 91
Srinivasan, S., 279, 421
Srivastava, S. K., 125
Stanley, C. R., 461, 489
Stanton, C. J., 1319
Stapleton, S. P., 993
Starr, C., 1099
Stavrinou, P. N., 977
Steel, D. G., 32
Steen, C., 149
Steenson, P., 1204
Steiner, T. D., 941
Stellmach, C., 559, 1222
Stenzel, R., 1501
Stępniewski, R., 1158
Stevens, M. R., 213, 301
Stifter, D., 219
Stoffel, M., 599
Stolichnov, I., 1585
Stolz, H., 163
Stopa, M., 643, 783, 785, 925
Stopford, P., 489
Stotz, J. A. H., 995
Strassburg, M., 211, 701
Strasser, G., 471, 1015
Stringer, E. A., 1271
Strittmatter, A., 701, 767
Studart, N., 1143, 1445
Studenikin, S., 1303
Studenikin, S. A., 1347
Su, B., 1172
Suemune, I., 293
Sugimura, A., 691, 855, 861, 867
Sukhodub, G., 573
Sun, J. M., 1539
Sun, L., 1437
Sun, W., 297
Sun, Z. Z., 1011

Sung, M. G., 1349
Sürgers, C., 909
Suto, F., 729
Sužiedėlis, A., 1204
Suzuki, K., 1373
Suzuki, M., 131, 139, 565, 1021, 1206
Suzuki, N., 327, 1359
Suzuki, Y., 1391
Suzuura, H., 551, 553
Svensson, B. G., 89, 181
Sweeney, S. J., 681, 985, 1545
Sweyllam, A., 1483
Syassen, K., 1230
Syperek, M., 417
Szczytko, J., 1149
Szot, M., 1129

T

Tackeuchi, A., 299, 1391
Tagawa, S., 1041
Taguchi, A., 77
Takagahara, T., 1391
Takagi, H., 1154
Takahashi, M., 343
Takahashi, Y., 325
Takai, K., 1589
Takano, F., 1323
Takano, Y., 325
Takase, K., 325
Takata, M., 691
Takatsu, M., 643, 743
Takayanagi, H., 397, 1162, 1433
Takechi, H., 339
Takeda, K., 1287
Takeda, Y., 121, 131, 139
Takemoto, K., 743
Takeyama, S., 593, 967, 971, 1363
Takita, K., 345, 1323, 1351
Talapin, D. V., 747
Tallman, R. E., 847
Tamošiūnas, V., 1204
Tamulaitis, G., 241
Tamura, H., 397, 1162, 1433
Tan, S. G., 1381
Tanaka, A., 349
Tanaka, H., 1154
Tanaka, K., 939
Tanaka, S., 213, 279, 301, 421
Tangney, P., 1047
Taniguchi, K., 299
Taniguchi, M., 349
Tartakovskii, A. I., 687
Tartakovskii, I. I., 1301
Tarucha, S., 643, 649, 783, 785, 925, 1435

Tatarenko, S., 1160, 1164, 1168, 1299, 1311
Tatebayashi, J., 655
Taylor, P., 103, 1267, 1457
Taylor, R. A., 695, 865
Taylor, R. P., 823, 1257
Tchelidze, T., 737
Teepe, M., 1517
Tegude, F. J., 119
Teisseire, M., 719
Tejedor, C., 699
Tenishev, L., 935
Tenne, D. A., 629, 693
Teo, K. L., 157, 167, 1381
Teppe, F., 1383, 1523
Terai, Y., 1351
Terasawa, D., 563
Tereshonkov, R., 1563
Terreault, B., 1491
Testelin, C., 1361
Thaler, G., 1319, 1399
Theodoropoulos, K., 1463
Thevenard, L., 363
Thewalt, M. L. W., 67
Theys, B., 363
Thinh, N. Q., 259
Thomas, A. C., 863
Thonke, K., 179, 831
Timm, C., 1365
Ting, D. Z.-Y., 1421
Tisnek, T. V., 81, 1186
Titov, A., 365
Titova, L. V., 361, 1337
Tiwald, T. E., 111
Tkachenko, V. A., 911
Tlaczala, M., 417
Tokizaki, T., 1323
Tokura, Y., 785, 925, 1251
Tomar, M. S., 839
Tomita, T., 1387, 1389
Tong, M. H., 1077
Tonouchi, M., 131, 139, 1206
Toropov, A. A., 991, 1387
Toropov, A. I., 133, 457, 461, 629, 693
Torres, M. O., 177
Trallero-Giner, C., 637, 1230
Tran-Anh, T., 795
Tranitz, H.-P., 892, 900, 1172, 1244
Tribollet, J., 1361
Tribuzy, C. V-B., 1543
Trigo, M., 1190
Tripathy, S., 1244
Tronc, P., 719
Tsakalidis, A., 1463
Tsaousidou, M., 1009
Tsen, K. T., 1263
Tsoi, S., 73

Tsubaki, K., 447
Tsubota, T., 523
Tsuchiya, Y., 1589
Tsui, Y., 435
Tsuji, K., 349
Tsuji, Y., 469
Tsukamoto, H., 327
Tsuneyuki, S., 389
Tsuya, D., 1021
Tu, C. W., 259, 265, 271
Tulkki, J., 1563
Tulupenko, V., 1192
Turner, M. S., 1093
Twardowski, A., 251

U

Uchida, K., 931
Uchiyama, C., 729
Ueda, A., 1447
Uematsu, M., 389
Uesugi, K., 293, 497
Ueta, A. Y., 1397
Uetake, A., 1389
Ulbrich, R. G., 151, 909
Ulloa, S. E., 683, 761, 781, 1325, 1445, 1465
Ullrich, C. A., 1141
Ulrich, S. M., 661, 725
Umansky, V., 501
Umeno, A., 1071
Umezu, I., 691, 855, 861, 867
Unitt, D. C., 665
Uno, S., 797
Urbaszek, B., 751, 1293, 1361
Uryu, S., 1029
Usher, A., 557, 561
Uskova, E. A., 1003
Usov, O. A., 1529
Ustinov, V. M., 789, 955, 1555
Usuki, T., 643, 741, 743
Utsumi, Y., 1409

V

Vaitkus, J., 207
Valcheva, E., 285
Valušis, G., 1204
van der Meulen, H. P., 531, 715
Vandersypen, L. M. K., 44
Van der Werf, D. P., 193
van Lippen, T., 613
Van Nostrand, J. E., 359
Vardeny, Z. V., 1063, 1077, 1081, 1407
Vasanelli, A., 1145

Vasilevskiy, M. I., 675, 747, 857
Vasilopoulos, P., 1335, 1415
Vasilyev, Y., 1222
Vasson, A., 263
Veinger, A. I., 81, 1186
Veksler, D., 947, 949
Verbin, S. Y., 935, 1417
Vercik, A., 483
Verma, A., 1049
Vermeire, B., 1507
Vertiatchikh, A. V., 419
Viale, Y., 629
Vickers, R. E. M., 101
Vieira, G. S., 1143
Vilan, S., 669
Vilão, R. C., 169, 193
Villas-Bôas, J. M., 683, 1143, 1445
Viña, L., 133, 1113, 1170
Vinokur, V. M., 1281
Viret, M., 363
Vitkalov, S. A., 435
Vitusevich, S. A., 439, 459
Vladimirova, M., 1383
Vogl, P., 799, 898, 1403
Vogt, F., 1419
Voisin, P., 717, 1293, 1307
von Borczyskowski, C., 629
von Freymann, G., 597
von Löhneysen, H., 907
von Middendorff, C., 455
von Ortenberg, M., 357, 795
von Wenckstern, H., 183, 197, 351, 1333
Vorona, I., 259
Voss, T., 1238, 1242
Vurgaftman, I., 1525
Výborný, K., 535

W

Waag, A., 896, 991, 1301
Wada, K., 77
Wada, N., 135
Wada, O., 387, 1353
Wagner, G., 875
Wagner, H. P., 1097, 1099, 1244
Wahlstrand, J. K., 1190, 1313
Wakida, H., 1039
Walls, J. D., 779
Walukiewicz, W., 209, 223, 303, 367, 1367
Wang, D. L., 765
Wang, D. M., 1313
Wang, F. J., 1063
Wang, J.-B., 399
Wang, J. N., 605, 765, 1011
Wang, K. L., 411
Wang, K. Y., 333, 1371
Wang, Q., 841
Wang, R., 937, 989
Wang, S. T., 391
Wang, X. F., 1415
Wang, X.-Q., 589
Wang, X. R., 539, 1011
Wang, Y., 239, 387
Wang, Y. J., 405
Wang, Y. Q., 1011
Warburton, R. J., 645, 667, 739, 751
Ward, T. Z., 1579
Warming, T., 689, 1555
Wasilewski, Z. R., 661, 685, 1009, 1303
Watanabe, J., 281
Watkins, G. D., 245
Watling, J. R., 1497
Watson, D., 1202
Wegscheider, W., 781, 892, 898, 900, 1172, 1309
Wei, F., 727
Wei, S.-H., 225, 287
Weichselbaum, A., 781, 1465
Weidinger, A., 169, 193
Weiner, A. M., 1176
Weinhold, S., 197, 1333
Weinstein, B. A., 847
Weiss, D., 1309
Welsch, H., 451
Welsch, M. K., 1427
Weman, H., 881, 919
Wenderoth, M., 151, 909
Wessels, B. W., 347
West, K. W., 449, 885, 887, 1313, 1575
Westervelt, R. M., 779, 1461, 1583
Westfahl, Jr., H., 1475
Wexler, C., 543
Whelan, S., 249
Wibbelhoff, O. S., 651
Wieck, A. D., 407, 437, 503, 641, 651, 733, 1158
Wieczorek, W., 1555
Wiersig, J., 661, 725
Wijewardane, H. O., 1141
Wilamowski, Z., 335, 1218
Wilbrandt, P.-J., 151
Wilde, M. A., 467
Wilkinson, P. B., 993
Willatzen, M., 659, 871, 873, 1485
Willems van Beveren, L. H., 44
Williams, D., 1439
Williams, D. P., 695
Wilson, A. J., 711
Wilson, J. A., 465, 489
Wiltshire, J. G., 1043
Wimmer, M., 1375

Winkelnkemper, M., 767
Winking, L., 151
Wirthmann, A., 829
Wischmeier, L., 1238
Witowski, A. M., 257
Witthuhn, W., 89
Woggon, U., 1232
Wójs, A., 525, 541
Wojtowicz, T., 303, 361, 367, 593, 957, 1271, 1291, 1337, 1363
Wolos, A., 251
Wolter, J. H., 613, 731, 803, 1405
Woo, J. C., 793, 1349
Woods, L. M., 1343
Woods, N. J., 487
Wraback, M., 239, 243
Wright, M. H., 1289
Wróbel, J., 677, 1271
Wu, D., 1063
Wu, H., 311
Wu, J., 209
Wu, S.-N., 835
Wu, Y., 32
Wu, Y.-F., 647
Wu, Y.-R., 441
Wühr, J., 883
Wysmolek, A., 251

X

Xie, M. H., 319, 879, 889
Xie, Z., 727
Xin, H. P., 265
Xin, Y. C., 603
Xiong, G., 539
Xiong, Z. H., 1063
Xu, Q., 611

Y

Yablonovitch, E., 409, 411
Yabushita, T., 299
Yakimov, A. I., 585, 1212
Yakovlev, D. R., 1301
Yakunin, M. V., 1003
Yamada, K., 643, 649, 785
Yamada, S., 1297, 1373
Yamagiwa, M., 679
Yamaguchi, H., 825, 1251
Yamaguchi, K., 1391
Yamaguchi, M., 397, 595, 1154, 1162
Yamaguchi, S., 1589
Yamaguchi, T., 785
Yamakawa, S., 227

Yamamoto, H., 971
Yamamoto, M., 783, 925
Yamamoto, N., 1533
Yamamoto, Y., 353, 355
Yamamuro, T., 1363
Yamashita, K., 387
Yamauchi, J., 83
Yan, J., 249
Yan, Y., 843
Yan, Z. W., 233
Yanase, A., 317
Yang, C., 1407
Yang, C.-H., 653, 1513
Yang, C. L., 605, 765
Yang, H., 319
Yang, J., 297
Yang, J. W., 217
Yang, L., 1497
Yang, M.-J., 653, 1513
Yang, R. Q., 1553
Yang, S. H., 605
Yasin, S., 695
Yee, K., 337, 361
Yee, K. J., 1337
Yeh, N.-T., 647
Yen, S. T., 1192
Yi, D. O., 611
Yilmaz, D. E., 231
Yoh, K., 1315
Yokoyama, N., 643, 743
Yoshida, M., 121
Yoshita, M., 885, 887, 1575
Youn, C. J., 203
Yu, G., 1347
Yu, K., 1495
Yu, K. M., 209, 223, 303, 367, 611, 1367
Yu, L. S., 765
Yu, P., 1579, 1581
Yu, P. Y., 851
Yusa, G., 1341

Z

Zabrodskii, A. G., 81, 1186
Zafar Iqbal, M., 143, 147
Zahn, D. R. T., 629, 693
Zakharov, D. N., 209, 611
Zander, T., 807
Zanelatto, G., 693
Zavada, J. M., 1234, 1399
Zawadzki, P., 1303, 1413
Zehnder, C., 829
Zeitler, U., 467, 503, 641
Zelensky, S. E., 439
Zeller, J., 896

Zhang, J., 487
Zhang, J. P., 241
Zhang, J.-Z., 749, 1236
Zhang, L.-X., 44
Zhang, S. B., 259, 625, 1031
Zhang, S.-L., 835, 843
Zhang, X. H., 405
Zhang, X. X., 319
Zhang, Y., 1214
Zhang, Y.-H., 399
Zhao, J., 1228
Zhao, Y., 625
Zhu, J. J., 319

Zhu, M., 557, 841
Zhukov, A. E., 789, 1555
Zhuravlev, K. S., 133, 137, 141, 629, 719
Zibold, T., 799
Ziese, M., 1333
Zimmer, M., 1095
Zimmer, P., 633
Zinovieva, A. F., 1393
Zinovyev, V. A., 607
Zollner, S., 65, 111, 1495
Zozoulenko, I. V., 1395, 1413
Zunger, A., 123, 265, 1121
Zvonkov, B. N., 1003